HANDBOOK OF OCCUPATIONAL SAFETY AND HEALTH

HANDBOOK OF OCCUPATIONAL SAFETY AND HEALTH

Edited by

LAWRENCE SLOTE, Eng.Sc.D., P.E.
New York University

A Wiley-Interscience Publication

JOHN WILEY & SONS

New York Chichester Brisbane Toronto Singapore

Library of Congress Cataloging-in-Publication Data:

Handbook of occupational safety and health.

"A Wiley-Interscience publication."
Includes index.
1. Industrial hygiene—Handbooks, manuals, etc.
2. Industrial safety—Handbooks, manuals, etc.
I. Slote, Lawrence. [DNLM: 1. Accidents, Occupational—prevention & control. 2. Occupational Health Services—organization & administration. WA 485 H2356]
RC967.H26 1987 613.6′2 86-29003
ISBN 0-471-81029-0

Printed in the United States of America

10 9 8 7 6 5 4 3 2 1

Contributors

MAHMOUD A. AYOUB, North Carolina State University, Raleigh, North Carolina
BRADFORD L. BARICK, Wausau Insurance Companies, Wausau, Wisconsin
LEWIS BASS, The Kairos Company, Mountain View, California
MICHAEL F. BIANCARDI, Wausau, Wisconsin
BARBARA P. BILLAUER, Stroock & Stroock & Lavan New York, New York
KENNETH M. BURGESS, Alcoa, Pittsburgh, Pennsylvania
JOHN R. CANADA, North Carolina State University, Raleigh, North Carolina
KENNETH S. COHEN, Consulting Health Services, El Cajon, California
SHIRLEY A. CONIBEAR, University of Illinois College of Medicine, Chicago, Illinois
BERNADETTE CONNOR, United States Steel Corp., Homestead, Pennsylvania
ALPHONSE A. CORONA, JR., Consultant, Yardley, Pennsylvania
CATHERINE D. GADDY, General Physics Corp., Columbia, Maryland
BRAD T. GARBER, University of New Haven, New Haven, Connecticut
ROGER C. JENSEN, National Institute of Occupational Safety and Health, Morgantown, West Virginia
ROBERT D. JOHNSON, Northern Telecom, Inc., Nashville, Tennessee
JOHN W. JONES, The St. Paul Insurance Companies, St. Paul, Minnesota
RICHARD A. LEMEN, National Institute of Occupational Safety and Health, Cincinnati, Ohio
A. JAMES MCKNIGHT, National Public Services Research Institute, Landover, Maryland
DAVID E. MILLER, Mobil Oil Corporation, Princeton, New Jersey
VICTORIA COOPER MUSSELMAN, Carnow, Conibear & Associates, Ltd., Chicago, Illinois
JERRY W. NEWMAN, National Institute of Occupational Safety and Health, Cincinnati, Ohio
JAMES C. NOLTER, Marsh & McLennan, Protection Consultants, Coral Gables, Florida
RICHARD J. PETRUCCO, Mobil Oil Corporation, Princeton, New Jersey
JACK B. RE VELLE, Hughes Aircraft Company, Fullerton, California
FOSTER C. RINEFORT, Eastern Illinois University, Charleston, Illinois

Sheridan J. Rodgers, Mine Safety Appliances, Co., Evans City, Pennsylvania
Laurie R. Sachtleben, Gulf Oil Corporation, Pittsburgh, Pennsylvania
Margaret C. Samways, NUS Corporation, Pittsburgh, Pennsylvania
Theodore F. Schoenborn, National Institute of Occupational Safety and Health, Cincinnati, Ohio
Lothar R. Schroeder, General Physics Corp., Columbia, Maryland
Arthur B. Shostak, Drexel University, Philadelphia, Pennsylvania
William E. Tarrants, N.H.T.S.A., U.S. Department of Transportation, Washington, D.C.
Donald M. Weekes, Jr., TRC Environmental Consultants, Inc., East Hartford, Connecticut

Preface

The purpose of this handbook is to provide a basic reference source that in part uses techniques and methods from various disciplines applicable to occupational safety and health. This handbook is not intended as an elementary text on occupational safety and health. Rather the presumption is that the reader is a mature, intelligent person who is not necessarily a full-time safety practitioner. It is also directed to the full-time safety practitioners with broad general knowledge of the field, who will use the handbook to review and update their understanding of specific topics as the need arises. The need for such a handbook is highlighted by the growing scope and diversity of the field.

It is well known that technological developments are progressing much faster than the knowledge in occupational safety and health required to combat the potential harm from these developments. No longer is it possible for one person to be well versed in the entire spectrum of methods applicable to occupational safety and health. On the contrary, specialists evolving in various branches of safety and health have barely enough time to keep up with developments in their own subject area and can have only a passing acquaintance with other disciplines' techniques and methods.

Serving as editor-in-chief for this undertaking has been a very satisfying experience for me. The preparation of a handbook of this sort entailed the generous cooperation of a number of individuals. All the participants are listed, of course, and the work of each contributing author is readily identifiable. On the other hand, the work of the members of the Editorial Advisory Board, which is less visible, has also been invaluable. This Advisory Board consisted of Robert J. Firenze, Frank Goldsmith, John V. Grimaldi, Earl D. Heath, Jacqueline Messite, Marvin Mills, Aldo Osti, and William E. Tarrants.

Last and most important, I wish to thank the contributors for the high quality of their material and for their understanding and patience during the manuscript editing. It is mostly because of their efforts that I believe the result will be an important contribution to our profession and hopefully will help compensate for the tremendous amount of work put forth by so many for so long.

LAWRENCE SLOTE

New York, New York
May 1987

Contents

HANDBOOK OF OCCUPATIONAL SAFETY AND HEALTH

PART I

OCCUPATIONAL HEALTH

CHAPTER 1

How to Deal with the Troubled Employee

KENNETH M. BURGESS
Aluminum Company of America
Pittsburgh, Pennsylvania

LAURIE R. SACHTLEBEN
Gulf Oil Corporation
Pittsburgh, Pennsylvania

BERNADETTE CONNOR
Consultant/United States Steel Corporation-Mon Valley Works
Pittsburgh, Pennsylvania

1.1 INTRODUCTION

If you're a manager or supervisor, chances are that sooner or later you will deal with a troubled employee. Simply defined, a troubled employee is any worker, from manual laborer to high-paid executive, who is affected by a personal or emotional problem that diminishes his or her effectiveness on the job.

The situations that trigger employee personal problems can vary widely. They can be job-related or have nothing whatsoever to do with work. Likewise, the type of person who is affected and the problematic behavior that he or she exhibits are variables. Consider, for example, these cases:

Bill is the original three-martini luncher. Several times a week he heads out to meet a client at 11:30 a.m. and returns around 2:00 p.m., smelling of alcohol. For the rest of the afternoon he stays in his office, reading, making phone calls, or chatting with coworkers. Last week Bill's boss even found him nodding off at his desk.

Renee hasn't been her usual self since she started having personal problems with her boyfriend. Her supervisor estimates that Renee spends up to an hour each day talking by phone to friends and relatives about her on-again, off-again relationship. Once an efficient and cheerful worker, Renee now habitually comes to work 10 to 15 minutes late, often looking as if she has been crying. She's becoming distracted and error prone.

Pete, a shipping clerk, has always been a reliable employee. His boss, however, thinks that Pete may be pilfering small items from company shipments. Although the boss doesn't want to fire Pete, he has heard rumors that the clerk is maintaining a costly cocaine habit.

It's obvious that Bill, Renee, and Pete have problems that are serious enough to make them less-than-productive employees. What should be equally obvious is that because these employees have problems, their employers have a problem, too.

1.1.1 Troubled Employees: How Widespread Is the Problem?

If troubled employees were a rare phenomenon, managers might be able to ignore them or "work around" them. Unfortunately, statistics show that employee personal problems are quite common:

In a random survey of 18,000 persons, the National Institute of Mental Health (NIMH) found that one-fifth of those questioned had suffered a mental health problem of some kind during the previous six months. (U.S. News and World Report, Oct. 15, 1984: 18).

As much as 10% of the work force in heavy industry includes alcoholics and drug abusers, according to the National Institute of Alcoholism and Alcohol Abuse (NIAAA). It is also estimated that up to 18% of all industrial workers have personal problems serious enough to interfere with job performance (Masi, 1981: 14).

The problems of alcoholic, emotionally disturbed or otherwise troubled employees are too prevalent to ignore. What's more, this situation is too costly to overlook. It has been estimated that a troubled employee can cost his or her company the equivalent of three months' wages per year in lateness, absenteeism, lost productivity, errors in judgment, accidents, compensation claims, and so on (Wrich, 1974: 34). Estimates of the full extent of this loss vary, but one recent study placed it in excess of $46 billion annually (Cork, 1984: 61). In addition:

Alcoholics are absent from the job sixteen times more often than nonalcoholics; file five times as many compensation claims; receive three times the sickness benefits; and, have an accident rate four times greater. Alcoholics also tend to be repeatedly involved in the grievance process (*Nation's Business,* May 1974).

Employee alcoholism results in 36 million days of disability each year (Follman, 1978: 81).

Furthermore, the costs associated with troubled employees often go beyond dollars and cents. The seriously disturbed person is a hazard. One source estimates that 80–90% of all industrial accidents have a mental base (Follman, 1978: 80) while roughly half of all traffic fatalities involve drivers under the influence of alcohol or other drugs (Poley et al., 1979: 30).

Even morale suffers in the presence of troubled employees. Common complaints about problem coworkers include "You just can't count on her anymore," "He's so unpredictable; I try to avoid him," and "I can't understand how our boss lets him get away with that kind of behavior."

1.1.2 What Is Your Role?

Few managers and supervisors, of course, intentionally allow their employees to "get away" with behavior that harms work performance. Yet many of us are threatened by the idea of intervening in an employee's personal life. In an effort to avoid confrontation, we may deny that problem employees really are that much of a problem. We accommodate their erratic behavior, sometimes to the point where it becomes "normal" to us. We may even cover up for the problem employee's shortcomings. Or we may wish that we could simply fire the source of trouble.

Nevertheless, there are compelling reasons why we as managers and supervisors should concern ourselves with—*and act on*—troubled employees. A problem worker may be a perfectly good employee who is experiencing an acute but temporary difficulty. Letting that employee continue to work in his or her impaired condition is unfair both to the worker and to the company that invests time and money in its work force.

Firing the troubled employee isn't always the solution, either. Considering the high percentage of workers who experience personal problems at some point in their careers, termination may not even be practical.

In addition, for those of us dealing with Bargaining Unit employees, recent rulings by arbitrators make it clear that companies are responsible for properly handling dismissal cases involving alcoholics and drug abusers. Many companies have been forced to reinstate addicted employees (*with* back wages) due to improper dismissal.

The Federal Rehabilitation Act of 1973, which explicitly prohibits discrimination based on current or former "handicap" or "disability," recognizes alcoholism and drug addiction as covered handicaps. Thus, the Act protects from discrimination persons who are in treatment for such problems and even (in some cases) alcoholics and drug addicts who have not sought treatment. Some states have followed this example as well.

As a final point, any supervisor or manager must realize that employee performance problems are a direct reflection on his or her own peformance. It is not your fault that one of your employees abuses alcohol or suffers from an emotional disturbance. Nevertheless, when that employee's problem affects your work area, you must deal with the situation, just as you would deal with a broken piece of machinery, a missed deadline, or any other management problem.

The challenge of dealing with troubled employees is often frustrating to managers and supervisors, and for good reason. After all, you're a business manager, not a psychologist or an arbitrator. You probably have neither the time nor the desire to grapple with the personal problems of individual employees. Your main concern is to keep your business function running smoothly.

Fortunately, your normal managerial role—that of keeping business running smoothly—is precisely the role you should assume in helping the troubled employee.

In fact, *the manager or supervisor's only goal in dealing with a troubled worker should be to solve the work performance problem created by that worker.*

Put another way, your role is to normalize your workplace. This means discouraging counterproductive behaviors and reinforcing positive, task-oriented ones. It means dealing quickly and directly with absenteeism, tardiness, poor quality work, disruptive outbursts, and other negative behaviors. And, of course, it means following your company's policies (or union/management agreement) on handling personnel problems.

At first glance, this philosophy may seem somewhat limiting. After all, the troubled employee is more than a work performance problem. He or she may also be a victim, a person in dire need of help, a valued worker, a friend. Conversely, he or she may be a troublemaker, a poor worker, someone you've never really liked.

It's difficult to remove your own biases and emotions when dealing with a problem worker. However, as the following case studies demonstrate, we can do more harm than good when we involve ourselves in workers' problems on any but a highly professional level.

1.1.3 What Can Go Wrong?

CASE 1

Steve, a supervisor, thought he knew Phil, his senior technician, well. They had come up the ranks together, bowled on the company team, even logged some hours together after work in a local tavern. Now, after 18 years with the company, Phil was beginning to come to work late and take a lot of sick time. He even missed a deadline, and the division manager had "read the riot act" to Steve because of a resulting complaint from a customer.

Steve got together with Phil for a "heart to heart," but Phil's behavior changed for only a few weeks. Steve began to lose faith in his one-time best worker and friend. Yet his documentation on this disciplinary problem was weak, and he knew why: he thought he could turn Phil around without getting him into trouble.

Phil had promised that he would complete a project by the end of the week, but called off Friday with the flu. Steve and another technician worked all day to meet Phil's commitment, and for the first time Steve realized how poor Phil's work had become.

On Monday morning, Phil had his wife call him off again. Steve began to realize that Phil was indeed becoming a problem employee, and that Steve's efforts to "carry" Phil were costing Steve the respect of others in his department.

CASE 2
Kathy, a laborer in her mid-30s, was suffering from a bona fide mental health problem, paranoid delusions (suspiciousness), which tended to make her somewhat withdrawn and distant with coworkers. Kathy generally was able to hide her conditions well, but she began questioning orders from Bill, her supervisor. Bill also noticed that it was taking Kathy far too long to finish tasks. Thinking that Kathy was simply a "crybaby," Bill began keeping a closer eye on her. He started giving her tougher jobs to break her of what he felt was prima donna behavior. Bill believed that his "solution" was effective, and he failed to document any disciplinary actions.

One day Bill ordered Kathy to climb a scaffold to grease a piece of heavy machinery. Kathy refused until Bill ordered her to do the job or go home. Kathy then obeyed, but only after informing Bill of her fear of heights. Near the top of the scaffolding, Kathy froze. Despite Bill's orders to come down, Kathy remained on the scaffolding until the plant's emergency technicians removed her. By that time, the assistant superintendent, safety manager, and plant security had arrived. Kathy was taken to the emergency room of a local hospital and remained out of work for a week. With the help of her union, she filed both a grievance and an EEO charge against Bill.

1.1.4 A Careful Balance

As you can see, the manager who deals with the troubled employee often must strike a careful balance among his or her own conflicting views of the right way to act. Playing amateur psychologist, launching a vendetta against a problem worker, or trying to "help" him or her by covering up are all tactics that can lead to emotional, organizational, and legal disaster.

Again, your only obligation in dealing with the troubled employee is to handle those problems that have a negative effect on the employee's job performance or that of his or her coworkers. This can be done, however, in a compassionate and constructive way, as the following section demonstrates.

1.2 THE HELP PROCESS

An effective supervisor knows how to get the best performance from his or her employees. Exacting good—or more accurately, improved—performance from the troubled employee is not always easy, but it is a distinct part of the supervisory process. It requires a combination of observation, communication, and action: recognizing and documenting the problem, planning for and engaging in a confrontation, and following through on expectations that the problem be resolved.

1.2.1 Step 1: Recognize and Document

We have already established that a troubled employee is any worker whose performance suffers because of personal or emotional problems. How can you recognize the types of poor performance and undesirable behavior that may point to such underlying problems?

The safest and wisest procedure is to evaluate employee performance problems according to documentable, *tangible violations of rules and policies governing safety or conduct.* Put another way, the performance and behavior "problems" that you observe should be problems as defined by your company's standards, not yours. Too many supervisors fall into the trap of making themselves "judge and jury" when it comes to dealing with troubled employees. It is imperative that you treat every performance problem consistently.

Table 1.1. lists many of the performance and behavior "red flags" encountered by supervisors and managers. Use the table as a checklist to observe counterproductive behaviors and to document them *as they occur.*

Figure 1.1. is a sample documentation form. It shows the importance of recording specific problems, times and dates (where appropriate) and your comments and actions.

Note that in both tables, performance problems are listed in order from the most tangible and documentable (i.e., absenteeism and tardiness) to the more subjective. This reflects the unarguable fact that many behavior and performance problems exist mainly in the eye of the beholder, making them difficult to document.

Obviously, objectivity is extremely important at this stage in the help process. If you find that your evaluation of a performance problem is skewed toward more subjective criteria, you would do well to stop and reconsider your position. By whose standards, for example, does an employe have poor relationships with coworkers? Is personality, or attitude, the only area in which this employee exhibits undesirable behavior? Remember, this is an evaluation not a witch hunt.

If you believe that an employee has demonstrated a pattern of the listed counterproductive behaviors, a work performance problem certainly exists. Whether this performance problem is based on some mental or emotional difficulty that your worker may have, however, has not yet been established. While some instances of poor work performance are the result of personal or emotional prob-

TABLE 1.1. "Red Flags": Job Performance Problems to Watch for and Document

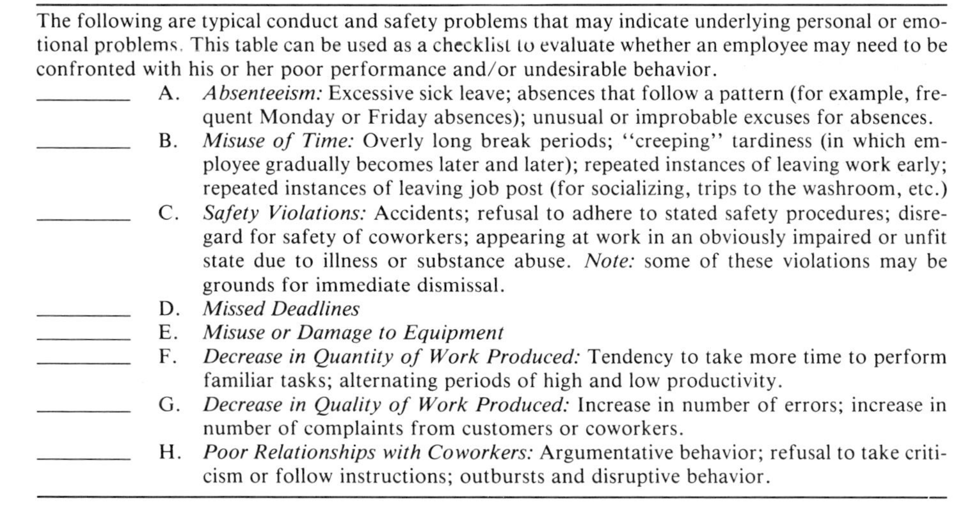

The following are typical conduct and safety problems that may indicate underlying personal or emotional problems. This table can be used as a checklist to evaluate whether an employee may need to be confronted with his or her poor performance and/or undesirable behavior.

________ A. *Absenteeism:* Excessive sick leave; absences that follow a pattern (for example, frequent Monday or Friday absences); unusual or improbable excuses for absences.

________ B. *Misuse of Time:* Overly long break periods; "creeping" tardiness (in which employee gradually becomes later and later); repeated instances of leaving work early; repeated instances of leaving job post (for socializing, trips to the washroom, etc.)

________ C. *Safety Violations:* Accidents; refusal to adhere to stated safety procedures; disregard for safety of coworkers; appearing at work in an obviously impaired or unfit state due to illness or substance abuse. *Note:* some of these violations may be grounds for immediate dismissal.

________ D. *Missed Deadlines*

________ E. *Misuse or Damage to Equipment*

________ F. *Decrease in Quantity of Work Produced:* Tendency to take more time to perform familiar tasks; alternating periods of high and low productivity.

________ G. *Decrease in Quality of Work Produced:* Increase in number of errors; increase in number of complaints from customers or coworkers.

________ H. *Poor Relationships with Coworkers:* Argumentative behavior; refusal to take criticism or follow instructions; outbursts and disruptive behavior.

lems, many are not. Everyone is entitled to an "off" day caused by minor physical illness, fatigue, boredom, or distraction over some non-work-related event.

Section 1.3, Common Problem Areas, explains how certain personal and emotional problems manifest themselves at work. This information may be useful to you in understanding some of the many sources of personal difficulties. Do not, however, make the mistake of recognizing a performance problem and then trying to "shop around" for a set of symptoms to explain its cause. Again, remember that although it is helpful to have some knowledge about addiction and common mental and emotional problems, your job is not to diagnose these problems, only to handle their job-related side effects.

Despite the rather narrow range of your involvement, however, also recognize that as a manager or supervisor you may in fact be quite closely tied to your employee's problem. While you are not the cause of the problem, you can become a complication in an already difficult situation—consider the example given previously of the mentally ill woman whose supervisor goaded her into climbing a scaffold despite her fear. Thus, as you evaluate the troubled employee's situation, consider his or her work history and recognize your role in that history. Examine your own biases and your interactions with the employee.

Finally, recognize that even if you are above reproach in your dealings with this employee, you are now responsible for intervening to solve his or her performance problem. Occasionally performance problems caused by emotional and personal difficulties "work themselves out" without outside assistance. More often, however, they do not. Often they grow worse. Just as you would not wait around for a piece of broken equipment to fix itself, you can't afford to wait until a troubled employee straightens out and becomes productive again.

1.2.2 Step 2: Plan and Confront

Once you have assessed that a particular employee has a work performance problem that may be caused by underlying personal or emotional difficulties, you must plan to formally confront the employee with your evaluation. Although the word "confront" sometimes carries a negative implication, a business confrontation is simply a meeting in which a problem is presented and potential solutions are discussed.

Confrontation, of course, is a built-in function of your role as manager or supervisor. If you are dealing with a troubled employee who has been exhibiting spotty or deteriorating job performance, you are already likely to have confronted him or her at least once by this stage. The key element in this particular confrontation, however, is to plan for and structure it so that (a) you are at all times in control, and (b) the confrontation results in positive, constructive action.

In planning for such a confrontation, equip yourself with information, not opinion or judgment. Use the existing documentation of the individual's performance and behavior problems. Refer again to Table 1.1. and Figure 1.1. to assist in further documentation. Record pertinent actions, discussions, dates, and times. You will be much better prepared for a confrontation if you can, for example, point

Summary of Job Performance Problems

Date: 1/10/84
Name: Charles B. Employee
Department: Shipping
Supervisor: James Supervisor
Period Covered: 10/1/83 through 12/21/83

Type of Problem	Record of Dates (where appropriate)	Comment (include disciplinary action taken, if any)
Attendance		
Unexplained absences of several days.	10/2 10/3 (10/15) 10/16 11/13 11/14	Claimed the flu. Subsequent events proved this to be questionable.
Absences of a day or half a day.	(11/19) (11/26) 11/28 (12/3)	
Absences on Mondays or Fridays.	Monday and Friday absences circled above.	
Leaving work early.		
Longer lunch periods.		
Unexplained disappearances from job in working hours.	12/20	Found sleeping behind packing cases. Oral reprimand. Hard to find on shipping dock sometimes.
Productivity		
Overall lowered productivity.	No specific dates.	Other men covering up?
More sporadic work pace.		
Increase in mistakes.	11/8 12/19	Shipments did not meet deadline because of mistakes. Note made in file.
Lower quality of work.		
Reliability		
Increasing inclination to put things off.		
Missing important meetings and deadlines.		
Neglect of details formerly attended to.	11/7 12/4 12/18	Invoices not properly processed on three occasions.
Attitude		
Tendency to blame other employees.	12/19	Blamed fellow employee for late shipment, although the shipment was not that employee's area of responsibility.
Avoidance of supervisor or associates.		
Quarrelsome and suspicious.	No specific dates.	Involved in fist fight on lunch hour -heard unofficially on "grapevine".

Signed: James Supervisor

Source: Gulf Oil Corporation

FIGURE 1.1. Sample form for documentation of performance problems. (From Gulf Oil Corporation.)

out to the employee that he or she has been late for work six times in the past two weeks than if you say, "You seem to be very uncooperative lately."

As you document your employee's problem, resist the temptation to *over*document. "Building a case" against the individual should not be necessary if your evaluation of the problem is legitimate. Furthermore, your overdocumentation may be seen by management or union as a personal vendetta against a disliked employee.

It may be helpful at this point to remind yourself that you are confronting a *problem,* not a person. Your ultimate goal is to use your managerial skills to help the employee help himself or herself in solving a *work performance* problem. You are under no obligation to take on whatever personal problem may be behind the situation. Nevertheless, the human element that comes into play during confrontations such as these can be rather intimidating. Don't be surprised if you feel somewhat uncomfortable both before and during the actual confrontation.

The following is a guideline for conducting the confrontation:

Suggested Confrontation Procedure

A. Review the history of the work performance problem with the employee. State as precisely and factually as you can what the employee has done that violates company policy or procedure. Have your documentation present and refer to it as necessary. Avoid the tendency, however, to constantly cite incidents from your file. This will automatically put the employee on the defensive. Remember that this is a business meeting, not a trial.

B. Invite the employee to discuss the work performance problem with you. Attempt to clarify and agree upon what is wrong. Ask "what" questions rather than "why" questions, for example, "What do you think is the problem here?" instead of "Why are you constantly late for work?"

The reason for avoiding "why" questions is twofold: first, most of us, when asking someone *why* they do something we disapprove of, can't help sounding at least slightly accusatory. Your "why" automatically puts the employee on the defensive. Furthermore, although indeed you do want to hear the employee's reasons for his or her poor performance, an open-ended "why" is like an invitation to the employee to discuss personal issues. Remember that in this confrontation, it is *not* your goal to "get to the bottom" of your employee's personal affairs.

C. Listen with empathy. Although you should not encourage the employee to "unload" his or her intimate problems on you, you must be willing to hear the worker's explanation of the problem as he or she sees it. Sometimes the explanation will be quite logical, and you'll quickly realize that the work performance problem at hand can be solved easily. At other times, however, it is this part of the conversation that can clue you in to the fact that you are indeed dealing with a troubled worker.

D. Acknowledge but *do not analayze* the employee's problem. If an employee tells you, for example, that she has been late for work recently because she and her boyfriend have been fighting, all you need to say is, "Well, it sounds like this is not an easy time for you." You can then continue by saying, "Still, we *do* expect you to be here on time."

E. State the consequences of continued problem behavior, and indicate that the situation must be changed. This is the point where you must frankly state your (and your company's) expectations of the worker and specifically refer to discipline that may be taken if the expectations aren't met.

F. Ask the employee for realistic recommendations on how he or she can work to solve the work performance problem. Encourage the worker to be specific; instead of a vague promise to "work harder" or to "try to be on time," solicit a commitment for particular actions that will result in desired improvements. This sets the stage for *you* to make your own recommendations, including referring the employee to an outside agency, Employee Assistance Program, or other source of help for personal and emotional problems.

G. Select a time period over which the employee will try to improve his or her performance and set a date to review progress. Then document the entire confrontation and the actions you have recommended.

H. Follow up with the agreed-upon review. If the employee's improvement has been satisfactory, document the improvement and close the case.

Table 1.2. provides additional guidance for this confrontation process, and Table 1.3. creates a sample "script" of an effective confrontation.

1.2.3 Step 3: Follow Through

The confrontation technique just described is a businesslike, results-oriented approach that does not attempt to delve into the employee's personal or emotional background. It keeps you within your area of expertise—confronting and solving work performance problems—so that you can stay in control of the situation. Finally, it creates a standard that you can apply consistently to a variety of work performance problems.

As you might expect, however, this approach does not always work instantly. Some employees are unwilling to improve their performance after a single warning. And some troubled employees are actually *unable* to change their behavior. The alcoholic who arrives at work every Monday morning with a hangover, for example, is unlikely to suddenly change that behavior on the basis of a single confronta-

TABLE 1.2. Confronting the Employee: Dos and Don'ts

DO	focus solely on declining job performance and the offer of help to improve the performance.
DO	have on hand written documentation of the declining job performance so you can let the record speak for itself.
DO	maintain a firm and formal, yet considerate, attitude. If the interview becomes a casual or intimate conversation, the impact of the message will be lessened.
DO	explain that help is available for whatever personal or emotional problem the employee may have.
DO	explain that the employee must decide on his or her own whether to seek assistance but that either way, job performance must improve.
DON'T	try to find out "what is wrong" with the employee, and don't allow yourself to get involved in the employee's personal life.
DON'T	make generalizations or insinuations about the employee's performance.
DON'T	moralize. Restrict your criticism to job performance.
DON'T	be misled by sympathy-evoking tactics. Stay focused on your right to expect appropriate behavior and satisfactory job performance.
DON'T	threaten discipline unless you're willing and able to carry out the threat.
DON'T	cover up for the employee. This type of "help" is misguided kindness that can lead to serious delay in real help reaching the person.

TABLE 1.3. Sample "Script" for an Effective Confrontation

Supervisor: I have been reviewing your work attendance record, Ann, and I see that you have missed a lot of days in the past two months. I also have noted from your time cards that you have been coming in late pretty consistently in the mornings. This is really a big change in your attendance. Can you help me understand what's going on, Ann? In the past you've been one of my most reliable workers. I could always depend on you to get the job done.

Ann: Well, you know how it is when you have kids—one thing after another. It started a couple of months ago. My oldest boy came down with the flu, and I just couldn't make good baby-sitting arrangements. So I took some time off then. No sooner did he get well then my other child became sick—same story. I'm sure you can understand that.

Supervisor: Yes, I can, Ann; but you've been able to handle both your home and job responsibilities much better in the past, which is part of the reason I wanted to discuss this with you. You haven't explained your lateness.

Ann: Oh yes, I also had a whole series of car problems. And anyway, you know how difficult getting out of the house in the morning with two kids can be.

Supervisor: I can see now that you've had some personal things interfering with your job. What I think we should do now is to come up with some ways to resolve whatever has been the complicating factors. What do you suggest, Ann?

Ann: The first thing I need to do is get my car in for a real overhaul. Then I need to line up some reliable child care for emergency situations. I think that should take care of all of this.

Supervisor: I think your ideas are good, and maybe there are some other things that you can do as well. You know how I value you as an employee. I don't want you to further jeopardize your job. I am also going to recommend that you make an appointment with a professional counselor. You see, Ann, it will give you an opportunity to look at this situation more closely and maybe come up with some additional recommendations.

The recent events at work have put us both in a difficult position. If this absenteeism and lateness are not remedied, then I will have no choice but to let you go. I am really hoping that we can iron out whatever has been the problem behind this. Maybe an open discussion with a professional will give you the opportunity to look more closely at what's been going on and to come up with some other ways to resolve it. How does this sound to you?

Ann: Well, I don't think the situation is that bad that I have to see a "shrink." Don't you think that's extreme? I think I can work it out on my own.

Supervisor: Well, Ann, I'm recommending the counseling. Now I'm going to make a notation on your personnel record about our discussion today and include the suggestions both you and I have made. However, I want you to know that if no real progress is made in one month, then I will have to follow company policy in this matter, Ann, and terminate your employment. I just want you to have every opportunity to get your work behavior back to standard. Think it over, Ann. Let's set up an appointment for one month from today—say at 8:00 a.m. sharp.

tion. Even if he or she *wants* to, the physical ability to change will be missing until further help is received.

Thus, the confrontation process may have to be repeated, even though the employee may in fact be working to cope with his or her personal problems. No matter what the employee's situation may be, however, your overall goal and strategy in dealing with problem employees should remain the same: to solve the work performance problem these difficulties create. Remember that as long as you follow and document company policy at every step, you *always* retain the right as a manager or supervisor to expect and demand that your employees maintain a satisfactory level of job performance.

Thus, your role in following up on your meeting with a troubled employee is the same role you're already comfortable with: that of supervising the worker's job performance. Monitor work done, document problems, and if necessary, reconfront the employee. Most important, follow this procedure with fairness and *consistency,* that is, without special favors and without discrimination.

1.2.4 Making Referrals for Further Help

If you have a company Employee Assistance Program (EAP) or a company counselor, a referral to this person is a natural progression of disciplinary action. The Referral Resources section to this chapter goes into considerable detail on EAPs and how they work. These programs have been dramatically effective in helping troubled employees, particularly those with alcohol and addiction problems. During their relatively short existence, EAPs have aided great numbers of workers in saving their jobs, their self-respect, even their lives.

If you have followed the complete help process just described but do not have an Employee Assistance Program or company counselor, you may find it helpful to refer the employee to your company medical department. Most such departments, even small ones, are able to recognize troubled employees and steer them toward sources of help for their underlying problems.

Inform your medical representative of the counterproductive behavior patterns you have observed and the disciplinary actions you have taken. Then, when you make the actual referral, inform the employee that you are asking for the medical department's help to give the employee the best possible opportunity to change his or her behavior. Emphasize that this referral (and all others) will be kept in confidence, as will any treatment the employee receives.

If you have no company resources available for employee referral, you would do well to familiarize yourself with helping resources in your community. This process will require very little of your time, and will enable you to give direction to the employee whose continuing poor work performance may reflect a serious personal problem.

Remember, as a supervisor or manager, it is your responsibility to address work performance problems. If the performance problems continue and you must reconfront the employee, inform the person that such continued behavior *will result in further disciplinary action, including termination.* This is also the most appropriate time to suggest that "the problem may be too difficult for the person to handle alone," and, "should he or she wish to seek help for any personal problem that may be affecting work performance, such help is available in the community."

Making this recommendation will not be easy for the manager/supervisor, especially the first time. It is most important that the discussion be handled in a very caring and professional way. Be certain to inform the employee that his or her performance is unacceptable, and that your recommendation is being made because he or she has been unable to rectify the problem through the regular supervisory channel. Also, inform the employee that his or her job is at stake, and that seeking counseling assistance at this time may be the only way to get his or her work, and possibly personal life more under control.

If the employee begins to talk to you about personal problems at this stage of confrontation, remain attentive, but inform the person that you are not a professional therapist. Suggest that he or she will benefit from more professional assistance. It is extremely important to assure the employee that you will respect confidentiality regarding your discussion, but that you will document in the record that you have discussed the referral recommendation.

If the employee becomes angered by your recommendation, and we have experienced such emotional responses, reiterate that it is your responsibility to correct employee work performance problems. Then, restate your position that the person has been unable to meet minimal performance standards, and that such behavior will no longer be tolerated. Also, as with the employee who more readily accepted your recommendation, inform the employee that it will be his or her decision, not yours, as to whether to seek help, and that you will document that you have made a referral recommendation.

Section 1.4., Referral Resources, will enable you to begin your search for suitable community referral groups and will also discuss the appropriate use and limitations of each type of referral.

1.3 COMMON PROBLEM AREAS

There are many reasons why an employee's work performance can deteriorate. Often (but not always) such deterioration is a byproduct of personal and/or emotional difficulties. Sources of such difficulty may include problems with alcoholism and addiction, emotional/psychiatric disturbances, marital/family discord (including financial and legal difficulty), and the often-overlooked problems related to stress, relocation, promotion/transfer, and job loss.

Whether these difficulties are long- or short-term, they can only detract from the ability to cope, leaving the employee less energetic and less motivated to perform his or her job. It is not our intention to ask you to identify or diagnose these conditions through an individual's work performance pattern; that would be impossible. Furthermore, resolution of these problems is best left to professionals (psychiatrists, psychologists, counselors, lawyers and financial consultants, to name a few).

It is, however, your job as a supervisor to monitor employee work performance; this is *your* area of expertise. The purpose of this section, then, is to increase your general understanding of common mental and emotional problems so that you can apply this knowledge throughout the process of helping the troubled employee to improve his or her job performance.

1.3.1 Alcoholism, Drug Abuse, and Other Addictive Behaviors

Addiction to alcohol and other drugs has been described medically as a style of living that implies "continuing and overwhelming involvement" with a substance (*Merck Manual,* 1982). This involvement usually leads to a significant risk or physical or mental harm that can only be prevented by stopping usage, something which the addict is unwilling or unable to do.

Addiction to alcohol or drugs is always an unhealthy, even dangerous condition. Unfortunately, many of us have a limited understanding of the addict, how addiction problems manifest themselves, and how these conditions can be treated.

Addiction problems can be viewed in two ways: first as symptoms of an underlying problem such as stress, emotional illness, physical illness or injury; and second as direct problems that in turn cause a whole range of other problems in the addicted person's marriage, job, finances, and social life.

Addictions may take many forms, the most prevalent being alcohol and drug addiction. In the latter case it makes no difference whether the drug of choice is legal or illegal. Many legalized substances such as prescribed tranquilizers can be as addictive as cocaine or heroin if misused.

Physical dependence does not occur with all drugs. When it does, however, the addict needs to use more of the abused drug as time goes on to achieve the drug's original effect. If the physically dependent addict attempts to stop using the drug, symptoms of withdrawal occur. Withdrawal from some drugs, especially alcohol, can be quite dangerous and usually requires medical attention.

Although a drug or alcohol abuser may or may not be addicted physically, it is safe to say that such a person is driven to use the drug repeatedly, either to produce pleasure or to avoid uncomfortable feelings.

We can become addicted to other things, too. Gambling, certain eating disorders, and smoking can all become addictive. However, these types of addiction will not be specifically addressed here because the addictive process is fairly similar for all types of addictions—as is the potential negative impact of addiction on the work setting.

Alcohol Abuse

Alcohol is probably the most abused of all drugs. Alcoholism is the number one work-related addictive problem, and it appears to be spreading. Recent investigations into the problem of alcoholism indicate that the rise in the number of alcoholics is proportionately greater than the increase in the American population (Connor, 1980).

Alcoholism is a treatable disease and not a moral problem. The American Medical Association recognized alcoholism as a disease as early as 1956. It has been a long slow process for the disease concept of alcoholism to be accepted. As a result, alcoholism goes untreated in many persons, ruining their mental and physical health, destroying family relationships, and of course, often causing a loss of job.

Progress continues to be made through public education efforts and the ever-increasing number of recovering alcoholics who are willing to "come out of the closet." While this trend is encouraging, the Office of Technology Assessment estimates that 85% of all persons who have serious problems related to the abuse of alcohol still are not receiving treatment (*Corporate Commentary,* June 1984).

The most serious and perhaps the least understood complications of alcoholism and heavy drinking are health problems. Initially, an alcoholic's health may be affected in subtle ways such as changes in eating and sleeping patterns, night sweating, forgetfulness, shortened attention span, or a tendency to become easily fatigued. If these signs are ignored and the drinking pattern remains unchanged, more serious complications may develop.

The National Council on Alcoholism has noted that long-term excessive alcohol use can adversely affect every major body organ. The following health and safety problems are the most common:

Liver damage, especially cirrhosis

Heart disease

High blood pressure

Brain damage, including confusion, memory loss, and psychosis

Increased rate of accidents (auto accidents, drownings, and falls)

Obviously, moderation is necessary to prevent such complications. But since drinking "in moderation" is a very subjective definition, a closer look at the warning signs of an impending problem with alcohol is needed.

A person does not have to be labeled "alcoholic" or fit any sort of stereotype to have a drinking problem. The Education Department of the United Steelworkers Union published a checklist of signs that one's drinking is becoming unhealthy. Some of these indicators are:

The inability to stop at one or two drinks

Drinking alone

Attempts to hide drinking

Tendency to gulp drinks

Lateness or absenteeism at work

Increased irritability

Problems at home related to drinking behavior

Hand tremors and increased nervousness

Drunken driving incidents/arrests

Repeated attempts at abstinence

Loss of memory of events while drinking

Remember that any one of these signs is not reason to believe that a problem exists. A combination of signs, however, should be taken as an important warning. The first step in dealing with any problem, including alcoholism, is recognizing that the problem exists.

Drug Abuse

Tia Schneider Denenberg and R. V. Denenberg wrote in their book, *Alcohol and Drugs: Issues in the Workplace*, that the cost of alcohol and drug addiction in industry is thought to be at least $100 billion annually. Just as alcoholism is believed to be on the increase, so is the incidence of other drug abuses. A recent estimate by the National Institute on Drug Abuse states that fully one-third of Americans over the age of 12 have used marijuana, hallucinogens, cocaine, heroin, or psychotherapeutic drugs (legal drugs used for nonmedical purposes). Contrast this information with statistics from the 1960s, which stated that less than 4% of the American public had used an illegal substance (Denenberg and Denenberg, 1983).

Tia Denenberg further points out that the individuals in these statistics are the same people who go to work every day in our factories and offices. Thus, unless treatment and/or rehabilitation occurs, the employer ends up with a troubled employee who costs money, lowers production, and may be a hazard. As a manager or supervisor, you are in a position to spot the troubled employee through job performance. Supervisors should not, however, attempt to diagnose a substance abuse problem or to recommend treatment.

For background information only, a brief review of the major categories of drugs, some of their effects on users, and some of the dangers involved follows. Keep in mind that it is fairly common for an addict to use several drugs, including alcohol, in combination. As a result, behavioral patterns of drug abusers are somewhat unpredictable and may vary according to the particular combination of drugs used.

Marijuana: Symptoms include altered perception, dilated pupils, lack of concentration, mood swings, and changes in appetite, especially cravings for sweets. Long-term use may create a psychological dependence, and an overdose may create serious psychological problems.

Hallucinogens (LSD, MDA, PCP, Peyote, Mescaline): These are mood- and perception- altering drugs, frequently producing a trancelike state, anxiety, confusion, and tremors. Users experience either euphoria or depression. Tolerance, or a need to increase the amount taken, develops quickly, Flashbacks can occur even after the user has ceased taking the drug. High risks of accidental overdoses or suicide exist among users.

Amphetamines (Bennies, Dexies, Speed): Users experience anxiety, excitability, rapid speech (often difficult to understand), restlessness, tremors, sweating, and running nose. These drugs carry high risk of physical dependence and medical problems include heart disease and infections. Psychological dependence often accompanies the abuse patterns.

Heroin: Symptoms include stupor, drowsiness, watery eyes, running nose, puffed eyelids, loss of appetite, itching, scratch marks, nausea, pinpoint pupils, and confusion. Physical and psychological dependence can develop. Long-term users may develop malnutrition, infection, and hepatitis.

Barbiturates: Users exhibit decreased alertness, drowsiness, confusion, slurred speech, lack of coordination, mildly dilated pupils, and a slowing down of reflex actions and muscle control. Physical and psychological dependence is likely to occur with long-term use. Barbiturate use carries a high risk of accidents due to slowed reaction and confusion. Accidental overdoses may lead to respiratory arrest, convulsions, or death.

Dealing with the Addicted Employee

As the preceding descriptions imply, abuse of nearly any drug, including alcohol, can lead to physical or emotional crisis. As a supervisor or manager, your job is to identify troubled employees—some of whom probably have addiction problems—through normal work evaluation procedures *before* a crisis develops in the workplace. A supervisory confrontation such as that outlined earlier in this chapter will then be necessary.

Remember that addictive and alcoholic employees have many ploys to counteract appropriate supervisory action. The addict's most important mechanism is denying that a problem exists. It is really very easy to understand why denial is so crucial. If denial stops, then the person must confront his or her problem. It is very frightening to recognize that you must give up the very thing that for a long time was your source of comfort (or at least your crutch).

The process of denial requires the addict to ward off every effort to be confronted with whatever he or she fears, often requiring a great deal of manipulation and deceit. If you have ever worked with an addict, you may already be familiar with their implausible explanations for inappropriate behavior, their cries of being unfairly attacked or criticized, their pleas for "just one more chance." The denying addict may try nearly any tactic to put you off just long enough to delay an actual confrontation and return to the "safety" of his or her addiction.

It should be noted here that addiction is, with very few exceptions, not self-treatable. Thus, a referral to a help program, be it your Employee Assistance Program, the company nurse, Alcoholics Anonymous, or any of a score of other sources, is a crucial part of proper procedure for dealing with addicts. Because addictions, particularly alcohol or drug abuse, are such complex and widespread problems, and because treatment for addiction tends to be long and difficult process, we strongly recommend that you contact the appropriate referral resource (see Section 1.4.) *immediately* if you believe that an employee has an addiction problem.

1.3.2 Family Problems and Related Legal and Financial Issues

Just as problems at work can have a definite effect on a worker's home life, problems at home can adversely affect job performance. One of the more obvious family-related problems is marital discord. Marriages have developmental histories not unlike those of growing children. Couples undergo normal and predictable stages in their relationships. Just as a child is likely to pass through the "terrible two" stage, a couple is likely to experience rough times in their relationship as well. Some couples weather these stages of growth and find them to be productive and beneficial. Many others, though, find their situation unbearable and decide to terminate the relationship through divorce.

Whether a couple is attempting to work through a difficult time or has opted for divorce, however, the strain of either will inevitably rob the involved employee of energy and concentration that would otherwise be directed toward his or her job. Some of the difficulties a couple may go through include:

Pregnancy and childbirth
Parenting problems (including children's school problems)
Family communications problems
Sexual dysfunctions
Dealing with aged parents
Serious illness and/or death
Money management problems

Keep in mind that these problems eventually confront nearly every family, and many families deal with them quite well. In other cases, however, too many problems converge at once, or the family members are not strong enough to deal with them in a healthy way.

In many cases of problem-plagued families, we often see that addiction (to drugs, alcohol or other substances or behaviors) is also involved. In such cases the original problems often are compounded by financial and legal entanglements because of the expense of maintaining an alcohol or drug habit.

As this stage, further complications often arise: impaired judgment and physical exhaustion. These often lead to on- and off-the-job accidents, which cause even more havoc with the family budget.

Legal and financial problems also occur in "normal,"healthy families; but in the troubled family we see a higher proportion of drug-related offenses and misdemeanor and felony charges (theft, assualt and battery, and so on). The troubled employee with work productivity problems usually harbors a cluster of difficulties that he or she obviously cannot cope with adequately.

As a supervisor, you probably hear about your workers' large and small home-related problems every day, either in person or "through the grapevine." Confronting an employee whose performance suffers because of such problems can be a particularly sensitive issue. On one hand, we all have families and home lives. We're easily tempted to sympathize a bit too emphatically with the worker who is going through a divorce or having financial difficulties. We may find ourselves continuously giving the employee just one more chance to improve his or her declining work productivity. This may seem the humane thing to do, yet it is certainly not fair to your other workers; and it may serve only to weaken your department's overall effectiveness.

On the other hand, we may tend to shy away from the troubled employee whose problems are rooted at home. Many supervisors, recognizing that it is unprofessional to become intimately involved with their employee's nonwork lives, intentionally avoid confronting or mentioning such problems. Unfortunately, these supervisors may be blinded to situations in which a very real, very serious problem at home carries over to an employee's work performance.

The key here is to remember that your job is to monitor and to set high standards of work performance. Thus, if an employee has a home-based problem that adversely affects him or her while at work, you must confront the worker. You will be doing him or her a genuine favor, particularly if, in the confrontation, you are able to refer the worker to a family counselor or other professional trained to help the worker.

1.3.3 Changes in Job Status: Promotions, Transfers, and Job Loss

All new situations create anxiety. Those that affect a person's job can be the most stressful of all. Think for a moment of the people you supervise and the roles their jobs play in their lives. For some people, a job is everything: a source of income, of pride, even of identity. Even for those who don't identify closely with their work, a job is still an important routine.

For each of your workers, a job change, whether it is a promotion, a transfer, or the loss of that job, will have profound impact. As a supervisor, you can witness firsthand this impact and its repercussions on job performance.

Promotions and Transfers

A promotion or a transfer to a new location (including arrival as a newly hired employee) often creates a conflict between expectations and reality. Consider this case:

Janet, age 30, has been a typist in the steno pool for eight years. She's an excellent typist and a pleasant, if rather shy, person. She has no prior supervisory experience.

When the former head of the steno pool retires, Janet is promoted to replace her. The personnel manager assumes that Janet, because of her experience in the pool, can take over the job with little or no training. Janet also is eager to take on the new job.

But three months after Janet has assumed duties as head stenographer, several crises have developed. Janet has been accused of favoring her old friends by giving them less work. She also has hesitated to reprimand certain workers for tardiness; as a result, some of the typists now habitually come to work late. Productivity in the steno pool has decreased. The personnel manager considers firing Janet, but Janet beats him to the punch. She says that if she can't have her old job back, she'll quit. "I know I'm no good at this," she says.

The expectation-v.-reality cycle in Janet's case is evident. Both she and management undertook a new venture with great optimism, despite the fact that Janet may not have been sufficiently qualified and certainly was not properly trained for her new job. When things didn't work out as expected, both parties "crashed," concluding hastily that the situation was irreparable.

Similarly, most promotions and transfers begin on a wave of good feeling, particularly on the part of the employee. He or she feels flattered, challenged, and hopeful that the new situation will be rewarding.

Almost inevitably, however, the employee will feel self-doubt, even if he or she is doing fairly well. A new job, after all, makes new demands. Here are just a few thoughts that the promoted employee may have:

> I'm just not right for this new job. Management may not know it yet, but soon they'll see that I don't have the needed skills.
>
> I miss my old coworkers. Now I have to prove myself to a whole new group.
>
> Things were easier on the old job. I wonder if I'll ever be comfortable here.

The transferred worker is even more vulnerable, because a transfer involves much more than a change in job conditions. Relocations is physically and emotionally grueling for the employee and each family member. Spouses and children of relocated employees usually must surrender their jobs, their friends, their comfortable lifestyle. These problems often assume crisis proportions, particularly for teenage children or any family member who feels he or she has been uprooted unfairly or too often.

If worry, fear of inadequacy and family problems become too much, the promoted or transferred employee may respond with a number of negative behaviors:

Withdrawal: Faced with a new situation that he or she sees as hostile, the employee "lies low," refusing to acknowledge or become a part of the new work environment. He or she may even withdraw through alcohol abuse or similar destructive patterns.
Overcompensation: Wanting badly to succeed, the employee becomes a workaholic and often demands the same of subordinates. This behavior often is an attempt to screen out the unpleasantness that the employee must face as part of a promotion or transfer.
Hostility: This behavior often combines elements of both withdrawal and overcompensation. It stems from resentment over the changes and problems that inevitably accompany a promotion or transfer.

Any of these behaviors eventually must affect job performance. If you supervise an employee who is having trouble handling a job change, you might do well to remember that these situations, even more so than employee problems caused by factors such as alcoholism, are a direct reflection on you. Although you may have had nothing to do with the employee's promotion or transfer, that worker is now a highly visible part of your work force. If there's a related work performance problem, you must solve it.

The right way to handle such problems is not significantly different from various management techniques you may already use:

1. *Lay some groundwork.* It shouldn't matter whether the new employee under your charge was brought in with or without your approval. What's important is that you're now in charge and you have a right to know the employee's abilities and shortcomings.

   If necessary, reinterview the employee as if he or she were a job candidate. Assess how you think he or she will fit into your department. Be frank; state *your* expectations. Then ask, "What elements of this job do you feel will take the most getting used to?"
2. *Monitor job performance.* This, of course, is the mainstay of your job as a manager. The promoted or transferred employee, however, comes to you "with strings" — that is, management has special expectations for him or her. Therefore you must pay particularly close and frequent attention to the employee's performance.

   Create a system that enables you to keep tabs on the worker's progress. Ask around on how the worker is doing (and don't forget to ask the employee himself or herself). Document performance so you'll be able to show the worker what he or she is doing right or wrong.
3. *Assist.* As already stated, promoted and transferred employees are vulnerable and visible. You can almost assume that, in addition to feeling pressure in his or her new job, your employee also is undergoing a certain amount of emotional tension. Don't stand back waiting for the employee to fail. Instead, apply your skills in training, supervising, and counseling the worker.

Job Loss

The employee who is retiring, about to be laid off, or leaving you for a new opportunity faces a potentially devastating situation. As already mentioned, many people identify extremely closely with their jobs; they *are* their jobs. Losing a job can be profoundly shocking and damaging to such a person's self-esteem.

And, as is the case when a person discovers he or she is dying, the employee who loses a job may go through these "stages of dying' as described by Elisabeth Kubler-Ross in *On Death and Dying:*

1. *Denial:* The employee insists, "No, not me." He or she resists the reality of the job loss.
2. *Rage and anger:* "Why me?" The employee resents the fact that others have jobs while he or she will not. This is also a stage where the employee may exhibit considerable bitterness toward the company and management.
3. *Bargaining:* "Yes, me, but. . . ." The employee may attempt to strike a deal to stay on for a few days, weeks, months, or even longer.
4. *Depression:* "Yes, me." First the employee exhibits regret over work not finished, wrongs committed, and so on. Then he or she enters a quiet stage of preparing to leave.
5. *Acceptance:* "My time is very close now and it's all right." The employee who accepts his or her job loss is probably neither happy nor unhappy. In her book, Dr. Ross describes this stage as not a resignation but a victory.

Naturally, you will not see each of these stages clearly exhibited with every employee who leaves your workplace. Some workers, especially those who do not identify strongly with their jobs, will seem quite blasé about their departures.

The irony of supervising an employee "on the way out," of course, is that no matter which of the preceding behaviors he or she exhibits, you won't have to deal with them for long. Nevertheless, you have every right to expect good performance from all of your employees, even the "lame ducks." Furthermore, the employee who is about to leave your group can profoundly affect the morale of those who stay behind.

Keep these steps in mind when dealing with an employee's loss of job:

1. *Balance professionalism with concern for the employee's future.* You are not responsible for solving the employee's problems or for finding him or her a new job. Nevertheless, an insensitive (or indifferent) attitude won't win you any support from either the worker who's leaving *or* the coworkers who are not. Keep in mind that the employee who is about to lose a job may be feeling guilty, fearful, and useless. If possible, refer the worker to outplacement agencies and other sources of help.
2. *Let the worker know what's expected in his or her last days.* Some "lame duck" employees are able to take large amounts of accumulated sick leave before they part. Others generally ease up on their work pace and performance. If situations such as these create problems in safety, scheduling, or productivity, it's your job to make the demands necessary to keep work running smoothly.
3. *Make a clean break.* Whether he or she was a valued worker or deadwood, the employee in question will no longer be a part of your work force. Try to make the parting amiable, but leave little question that it is a parting. Discourage any tendencies for the employee to cling to his or her old workplace. Naturally you can't discourage the ex-employee from socializing with former coworkers, but this should not take place on company time.

1.3.4 Stress

To some extent, stress has a natural place in a work environment. After all, where would we be if meeting deadlines, making important presentations, and exchanging ideas with our coworkers didn't create a certain amount of anxiety and tension? Stress is fundamentally a *survival* mechanism, our bodies' physical reaction to external pressure or change. As such, stress can be the motivator that keeps us alert, creative, and productive.

Nevertheless, in the past 30 years health experts have come to recognize that stress overload can have damaging mental and physical side effects. The American Academy of Family Physicians estimates that two-thirds of all office visits to family doctors are the results of stress-related symptoms: headaches, digestive problems, sleep disorders, and joint and muscular pains, to name a few.

Furthermore, according to a *Time* magazine cover story on stress (*Time*, June 6, 1983), stress-related absenteeism, medical expenses, and lost productivity cost U.S. businesses more than $750 per worker annually. *Time* also noted that the nation's three top-selling drugs are all stress medications: Tagamet for ulcers, Inderal for hyperventilation and Valium, a tranquilizer.

Stress is such a fact of modern life that it has been called an epidemic. Unfortunately, because stress is so pervasive, and because our lives are so full of stressors (situations that cause stress reactions), some people have turned stress into a catch-all excuse for nearly every imaginable physical or mental illness. You've probably heard, "I've been under a lot of stress lately" as an employee's explanation for everything from having a backache to running a red light on the way into work.

Luckily, we as supervisors are not obligated to diagnose our employees' stress-related problems. It is wise, however, to recognize certain basic truths about stress and its effects on the workplace.

1. *Stress is a distinct physical phenomenon.* As mentioned earlier, stress is our bodies' survival mechanism for dealing with change. When confronted with an unfamiliar situation, you first evaluate it to see whether it poses a threat. If you perceive that it does not, you react and adapt accordingly. Whether you realize it or not, this seemingly simple process is a form of stress.

The stress that concerns us, however, comes when you view a new situation as a threat. When this happens, you breathe faster, your heartbeat quickens, your muscles tense, and adrenalin pours into your bloodstream. These reactions prepare you either to fight or to run away from the threat you face.

The obvious problem, however, is that most work-related threats do not require us to fight or flee. Much as we might *wish* to take these options, we instead find ourselves compromising, tolerating, and finessing our way out of trouble. Despite these measures, we still cannot always evade what threatens us. Our stress becomes *distress.*

All of this takes its toll on our bodies. Our tension builds up, but is seldom released. The resulting symptoms, in addition to those already named, can include pounding heart, dizziness, fatigue, restlessness, trembling, and abnormal eating habits.

2. *Stressors come in every form.* We already know that stress on the job is plentiful. Stress at home is no less common. In fact, most of the employee problems discussed earlier in this chapter are in and of themselves stressful. The worker who tries to hide a drinking problem, for example, will most likely feel stress. So will the employee whose spouse or child is ill.

Even positive events can be stressful. A promotion, the birth of a child, even a reduction of your workload can all create undercurrents of tension. The key is to remember that most stress is a reaction to what you *perceive* as harmful. What is stressful to you may not be to your coworkers. Furthermore, what is stressful to you today may not be so tomorrow.

3. *Stress becomes a problem when it overloads our systems.* Stress is a normal reaction. The healthy body (and mind) can handle a certain amount of stress through a variety of coping mechanisms. In fact, it is neither desirable nor advisable to try to eliminate all stress from your life.

But, like a disease, stress can prey upon us when our emotional or physical resistance is low. A series of stressful events can accumulate and overwhelm. When that happens, our coping mechanisms may fail. Thus, the employee who usually can handle high-pressure situations may suddenly become very upset over a relatively minor problem.

Likewise, stress overload can cause certain behavioral tendencies to become uncontrollable. Let's say, for example, that one of your employees has a tendency to be highly critical of fellow workers in a particular minority group. Normally this employee recognizes this behavior as undesirable and attempts to control it. But under the cumulative effects of stress, the employee loses this control. The result may be that he or she takes out frustration on these minority employees in a way that he or she later regrets.

4. *Stress can be managed.* Even major, longstanding overloads of stress can be dealt with. "Stress management" is the term commonly used to describe the various methods of coping. Many stress management techniques are simple behavior modifications that any employee can learn: planning ahead to avoid particularly stressful situations, for example, or learning to take a breather in the middle of a hectic day.

Other stress management techniques require more involvement. Employees who exhibit physical symptoms of stress should see a physician, who may prescribe a new diet and exercise regimen as well as medication.

Talking over the causes of tension and stress is also important. Mental health agencies and employee assistance programs are particularly attuned to this function. Most, in fact, offer stress management seminars and workshops specifically designed to help working people.

Finally, managers themselves can do much to relieve their workers' job-related stress. Adequate standards and supervision, a healthy and pleasant work environment, and open channels of communication can go a long way toward easing the daily tensions of most jobs. The alert supervisor watches his or her work force carefully to see whether any individuals need assistance that goes beyond these basic requirements.

1.3.5 Emotional Problems

Several categories of emotional problems have primary impact on the work setting. These categories are based on symptoms that effect employee behavior. For the most part, the actual causes of these problems are unknown, but they can be resolved, or at least managed, with professional therapy and/or prescribed medication. Stress, already discussed in this chapter, is often the precipitating factor, making the employee's life unmanageable, and forcing the person to adopt an unhealthy coping style.

Emotional problems can generally be classified as either neurotic or psychotic disorders. Psychotic disorders involve a loss of contact with reality, are often accompanied by delusions or hallucinations, and require prompt treatment by a professional.

Neurotic disorders are emotional disturbances of all types, other than psychoses. Neurosis causes the individual to feel psychological pain or discomfort far beyond what is appropriate for the situation, experience, or object in question. While it is very difficult to draw a clear distinction between mentally ill and mentally healthy people, the overwhelming number of people who suffer from emotional disturbances suffer from some sort of neurotic disorder.

In an attempt at brevity, the primary categories to be discussed in this section include: (1) anxiety, (2) phobia (irrational fear), and (3) depression.

Anxiety

According to psychiatric literature, it is estimated that between 2 and 4% of the population suffers from some type of anxiety disorder. Anxiety problems may appear to have no visible cause, but they are very real for the person experiencing them. The anxiety may be related to the person's job or work situation, family, marriage, and so on, and tends to cause excessive tension. Anxiety will often be accompanied by some physical symptoms (e.g., sweating, heart palpitations, shaking). This combination of anxiety and physical symptoms may cause the person to experience a nagging sense of hopelessness, and he or she may feel helpless to do anything about the situation.

Phobia

Phobic disorders cause the person to experience a persistent, irrational fear of a specific activity, situation, experience, or object. Again, such fear may have no logical base, but it is very real to the person experiencing a phobia (fear of heights, closed spaces, crowds, etc.). This fear will usually push the person to isolate himself from the outside world. While it is normal for a person to experience occasional fear, when such fear begins to interfere with one's ability to carry out a normal life, the problem requires professional treatment. Short-term treatment by a professional therapist will help the neurotic individual to break this pattern and begin to experience a more rewarding life.

Depression

All of us will feel depressed now and then—it is a very natural reaction to life's tensions. Depression, in fact, appears to be a very common problem in our society. Such feelings of sadness (melancholia) go away very quickly for most of us, but for others (perhaps as many as 20% of us) depression lingers. The depressed person may not even recognize his or her own symptoms, or may be afraid to appear weak and/or vulnerable.

While most of us can handle our depression successfully, many people have much more serious periods of depression. Some of these people are still functional (able to meet day-to-day responsibilities), but their depression becomes much more severe over time. Others become so severely depressed that they cannot face the stress of daily life. Such people often fall into the pitfall of alcohol and drug abuse; others may even become suicidal.

Depression can also manifest itself in various forms. One fairly common form of severe depression, manic-depressive illness, will cause the person to experience mood swings, which will vary from lengthy periods of severe depression to periods of elation. During the elation period, the person may exhibit "speeded-up" behavior (i.e., exaggerated body movements, excessive talking, laughing, etc.). This form of depression, because of the mood swings, causes both the person and those around him or her to become confused, because such actions are inconsistent with depression in general. During either phase of this illness (depression/elation), the person may become out of touch with reality, and manifest signs of psychosis.

In short, when the symptoms of emotional difficulty are recognized in your employee, it is time to insist that the person seek professional help. As with any illness, treatment for an emotional problem is most successful when it is begun early. Playing amateur psychologist should be avoided, as these conditions are serious, and can only be treated by a physician, psychologist, or therapist (counselor). If you have a company counselor or Employee Assistance Program, referral should be made at the earliest possible time. Your discussion with the employee should be private, supportive, and firm. If you have a physician and/or facility nurse, these individuals are to be consulted. Referral can then be made to your medical representatives for employee evaluation and further referral for treatment.

In lieu of company medical resources, community mental health centers, mental health clinics, and general hospitals are appropriate locations for evaluation and treatment.

1.4 REFERRAL RESOURCES

When continuing poor work performance has been observed and documented, the manager or supervisor must confront the employee with the evidence that has been compiled. As discussed earlier in this chapter, the confrontation is an opportunity both to manage the problem and to offer the employee help for what may be a serious personal problem. The way you handle this confrontation is most important, and the actual discussion must take place in a private setting.

At the referral stage of confrontation, you should not attempt to argue with the employee or try to convince him or her to follow your recommendations. The discussion should focus on a review of the evidence of poor work performance, and leave the employee with a full understanding of the problem. Additionally, the employee needs to know that *further disciplinary action can result in discharge.* While you may be aware of, or sense, a personal problem, you must refrain from attempting to diagnose the employee's situation.

Whether or not disciplinary action can be held in abeyance depends on the nature of the infraction. It is important though, to inform the employee that while he or she may reject your suggestion, you will document that you recommended that the employee seek help. In short, while the decision to seek help is up to the employee, he or she must also know that appropriate disciplinary action will be taken should work performance fail to improve.

For those of you with an Employe Assistance program, company counselor, or Medical department, the referral process can be direct. Explain to the employee that you recommend that he or she see the counselor or meet with the medical representative to discuss any problem contributing to poor work performance. Assure the employee that you will not pry into his or her problem, but that you will contact the referral source to confirm that such appointment was kept. Also, explain to the person that you will inform the referral source as to the nature of your referral, and the potential seriousness of continuing poor work performance.

If the employee refuses to follow your recommendation, and work problems continue, your handling of the matter must follow your company's policies regarding such action (and/or Labor Contract Agreement). This same caution holds true for the employee who accepts referral for help, but whose work performance fails to improve.

At this point, if you find that you have no company resources at your disposal, recommend that the employee seek help through existing community treatment programs. These programs, which generally offer short-term outpatient care, will help the employee to better diagnose his or her situation, develop a treatment plan, and, if necessary, provide a short-term acute care program for serious mental/emotional and/or drug and alcohol problems. The following discussion of both outpatient and inpatient care providers is presented to help you to better assess your community services network.

1.4.1 Outpatient Care

Most serious personal problems can be managed by referring the employee to outpatient care. This type of counseling assistance does not involve hospitalization, and appointments can usually be made without interference with the employee's work schedule.

Outpatient counseling is provided by professional therapists (psychiatrists, psychologists, and other qualified human service professionals), and, depending on the problem encountered, typically helps the employee to improve his or her performance in very few sessions. Obviously, when a more serious emotional problem is involved, the employee may need to remain away from the job, and even extended treatment may be indicated.

Outpatient therapy involves a one-to-one relationship between a patient and therapist, and includes assessment (diagnosis), development of a treatment plan, the treatment process, and evaluation of outcome. The relationship between the patient and therapist is confidential in nature, and you should neither expect nor attempt to elicit confidential information from the treatment provider.

When suggesting to the problem employee that he or she seek help from an outpatient clinic or therapist, inform the person that you will expect to receive word that he or she made an initial appointment, and that you will inform the therapist or program representative of the nature of your referral request. By making this statement at the time of your referral, the employee with be aware that he or she should ask the therapist to inform you of his or her attendance, and expect you to communicate the extent of disciplinary involvement. This information will be very useful in helping the therapist to make an initial evaluation and in establishing both a treatment plan and schedule. Remember, though, that such information can only be exchanged with appropriate written permission from the employee.

Outpatient Mental Health Clinics are easy to locate, and can usually be found in your local telephone directory. The amount of time you will spend on developing a referral resource network is minimal, and the responsible supervisor or manager will complete this task prior to encountering a need for the service. When you contact a local resource, you can generally request that a representative from the agency visit your work location. During this visit, you will need to have a copy of your company's medical benefits package for the agency's review, should future referrals need to be made. You will also need to explain just how future referrals for evaluation and treatment will be made to the group; you will also need to know each other's expectations.

In the event that no such agency is available in your community, you may wish to find an individual practitioner to whom such referrals can be made. The process of getting to know each other is the same as with the community agency, and can be accomplished in a very short period of time.

Most employees who receive treatment for alcohol and drug abuse problems also become involved in Alcoholics Anonymous (AA). This self-help group has been extremely beneficial in the treatment of addiction problems, and employees who become involved in AA participate actively for the remainder of their lives. A recovering alcoholic who is active in AA can become an extremely important part of your referral network. While members of the AA community remain anonymous, a phone call to your local group will inform you of meetings in your area and will help you in referring an employee should the need arise. Additionally, gaining knowledge about this valuable organization may result in a voluntary commitment on the part of a recovering employee to serve as one of your referral resources.

1.4.2 Inpatient Care

Inpatient care is typically provided by hospitals and/or free-standing alcohol and drug rehabilitation programs. An inpatient referral is indicated when the employee's problem is severe enough that he or she cannot continue at work, and must be hospitalized for a chronic condition (alcoholism, drug addiction, or serious mental or emotional problems). The inpatient stay may last from a few days to a month or more, and such programs are generally staffed by physicians, psychologists, and professional therapists.

While it would be rare for a supervisor or manager to make an inpatient referral, it is important to know that when an employee is referred to inpatient, just as with any stay in a hospital, arrangements need to be made to insure the payment of medical and sick-leave benefits. In some cases, the employee may not be able to return to work after an inpatient stay, and continuing arrangements have to be made to maintain the employee's disability payments.

1.4.3 Employee Assistance Programs

Employee Assistance Programs (EAPs) provide an answer to the problems troubled employees cause to business and industry.

Appropriately implemented and staffed, an EAP can help to control the enormous losses businesses incur from employee absenteeism, decreased production, health care claims, and accidents and injuries on and off the job. Through supervisor identification of poor work performance, intervention, and referral for treatment and recovery, EAPs have shown very high rates of success in working with problem employees.

These programs were originally started to help employees who were affected by the disease of alcoholism. In the 1930s and early 1940s, when AA (Alcoholics Anonymous) became established as the most successful group to work with alcoholics, industrial alcoholism programs based on the principles of AA began to emerge. These programs typically used the help of recovering alcoholics to identify and rehabilitate alcoholic coworkers, and they quickly became recognized as an effective tool for dealing with employee alcoholism.

The EAP concept evolved as industry began to realize the cost effectiveness of company-based alcoholism programs. EAPs generally provide evaluation and referral services for employees suffering from a wide range of problems: alcoholism, drug abuse or addiction, mental health problems, emotional problems, family difficulties (including financial and legal problems), and other personal difficulties that adversely affect work performance.

Historically, EAPs have been located within the employer organization. More recently, though, some EAPs have been contracted through service providers located in the community. An approach often used by smaller employers has been the joining together of a number of companies to provide employee assistance services through a consortium arrangement.

Programs have been designed for all types of organizations: small businesses; major corporations; labor unions; city, county, state, and federal government agencies; occupational groups and professional organizations. In companies with bargaining unit employees, unions have been involved in the design and development of the EAP.

According to the Association of Labor/Management Administrators and Consultants on Alcoholism (ALMACA), a nationwide group that supports EAP development, more than 5000 U.S. companies currently have some type of employee assistance or counseling program. By and large, these programs are recognized as successful both from a human perspective and in cost effectiveness. EAPs, when they are available, are the most appropriate referral resource for supervisors using the disciplinary approach in dealing with problem employees.

The success of the EAP movement has also been attributed to the programs' ability to "bring together" companies and unions. The EAP typically fosters positive management/union relationships, promotes resolution of problems, maintains confidentiality and employee dignity, and provides a return on mutual investment of time and money.

How does an EAP work? In general, whether an in-house (company counselor) or external (contracted program) model is used, the process is quite similar. The EAP counselor is most often a professionally trained individual with an understanding of a wide range of human problems. Typically, he or she possesses an advanced degree in counseling, social work, or psychology, as well as an understanding of the business or industrial setting.

A major function of the EAP counselor is to provide consultation to managers and supervisors involved with problem employees. Generally, the counselor has already talked to the supervisor about the work performance problem before the supervisor refers the employee for counseling. If not, the counselor will ask the manager or supervisor for the employee's work and disciplinary history before the initial evaluation session.

In all cases, referral to the EAP is voluntary. Often, however, the supervisor reinforces the employee's decision to seek help by informing him or her that further work performance problems can result in discipline or termination. The EAP is presented to the employee as a resource to help resolve any problems that may be contributing to or causing continued poor performance. The referral is never made in lieu of further disciplinary action; such action will still be taken if performance does not improve. Thus, the EAP is designed to help any employee with personal problems, but *not* to provide an escape or cover for employees who are not willing to follow company policy regarding work performance.

At the time of referral, it becomes the counselor's responsibility to assess the employee and to make a suitable referral for treatment. The EAP counselor continues to monitor the employee while the treatment continues and to make certain that he or she is complying with general recommendations. While the clinical counselor will keep the supervisor informed about the employee's adherence to EAP recommendations, actual assessment and treatment information should remain strictly confidential. Employee records maintained by the EAP counselor must be kept separate from other company files, and in accord with federal regulations (42 CFR, Part 2), drug and alcohol abuse records are protected from any form of release without specific written consent.

For more information on EAPs, including information on how to start an EAP in your company, consult any of the EAP-related books or magazines listed in the *Reference Materials* section below.

1.4.4 Reference Materials

These publications may be useful to the manager or supervisor who wants to learn more about dealing with troubled workers.

General

Follmann, J. F., *Helping the Troubled Employee,* Performance Resource Press, Inc., Troy, MI, 1978. This book is designed to help the employee, the employee's family and his or her employer to recognize and deal with mental and emotional disorders. It examines forms of treatment and sources of help, including ways to finance care and treatment.

Occupational Health & Safety magazine, Medical Publications, Inc., Waco, TX. This monthly magazine takes the supervisory point of view in examining a variety of health and safety issues. Employee personal and emotional problems such as alcohol abuse and stress disorders are frequently featured. Articles are geared toward efficiency and cost effectiveness.

Alcoholism and Addiction Problems

Follman, J. F., *Alcoholics and Business,* Performance Resource Press, Inc., Troy, MI, 1976. This book addresses the attitudes of employers and organized labor toward alcoholic employees. Issues examined include insurance and health benefit plans as they relate to cost of the treatment.

Gold, M. S., M.D., *800-Cocaine,* Bantam Books, Inc., New York, 1984. The founder of a toll-free national hotline for cocaine abusers, Dr. Gold is very knowledgeable on the country's fastest-growing drug problem. *800-Cocaine* answers many questions you may have about cocaine, presents the growing problem of cocaine use in the workplace, and discusses the difficulty of finding good treatment programs for abusers.

Trice, H. M., and Roman, P. M., *Spirits and Demons at Work: Alcohol and Other Drugs on the Job,* Publications Division, New York State School of Industrial Labor Relations, Cornell University, Ithaca, NY, 1979. This book examines the relationship between substance abuse and work performance. Job-based risks and behaviors of problem drinkers and drug users, supervisory reactions, and strategies for constructive confrontation are presented.

Twerski, A. J., M.D., *Caution: "Kindness" Can Be Dangerous to the Alcoholic,* Prentice-Hall, Englewood Cliffs, NJ, 1981. Using case studies and simple, straightforward language, Twerski writes for those who regularly come in contact with problem drinkers. He specifically addresses the destructive behavior patterns alcoholics exhibit, and the ways in which well-intentioned friends can inadvertently allow and even encourage these patterns to continue.

Stress

Friedman, M., M.D., and Rosenman, R. H., M.D., *Type A Behavior and Your Heart,* Knopf, New York, 1974. Friedman and Rosenman, both cardiologists, contend that the majority of heart attack victims exhibit a combined personality, lifestyle, and attitude known as Type A behavior. This book provides insight into the ways we handle stress, and it enables each reader to assess his or her own personality type.

Jacobson, E., M.D., *You Must Relax,* McGraw-Hill, New York, 1976. Jacobson asserts that health disorders caused by tension and stress are more common than most of us realize. He also writes, however, that we can diminish the negative effects of stress in our lives through muscular relaxation techniques. The result of more than 60 years' research in the field, *You Must Relax* is interesting and instructive.

Counseling and Employee Assistance Programs (EAPs)

Dunkin, W. S., *The EAP Manual,* National Council on Alcoholism, New York, 1982. This small handbook is a practical, step-by-step guide to establishing an effective employee alcoholism/assistance program. Sample policies, procedures, communications and recordkeeping forms are included.

Employee Assistance Digest, Performance Resource Press, Troy, MI. This bimonthly magazine covers a wealth of issues concerning troubled employees and the formal programs companies create to help them. Many articles on topics such as stress, alcoholism, and burnout may be of general interest to supervisors and mangers.

Employee Assistance Program: A Model, U.S. Brewers Association, Inc., Washington, DC, 1980. This overview of a model EAP was written primarily for managers who are not familiar with EAPs, alcoholism, and mental health problems. It describes how to set up and expand a program, as well as the costs and benefits of EAPs.

Employee Assistance Program: Theory and Operation, The Aluminum Company of America, Pittsburgh, PA, 1983. (Available from ALMACA, 1800 N. Kent Street, Suite 907, Arlington, VA. 22209.) Developed as a public service document by ALCOA, this manual will prove valuable to those interested in developing an EAP at the work site. *Theory and Operation* provides an introduction to

EAP, the reasons behind the recent growth of these programs, and a step-by-step approach to program development.

Muldoon, J. A., and Berdie, M., *Effective Employee Assistance: A Comprehensive Guide for the Employer*, CompCare Publications, Minneapolis, 1980. This concise yet comprehensive guide explains the benefits of having an EAP, discusses some of the most commonly encountered employee problems (alcoholism, family problems, psychological/emotional problems, etc.) and outlines the steps involved in developing an EAP, including needs assessment, proposals, and support gathering.

Presnall, L. F., *Occupational Counseling and Referral Systems*, Utah Alcoholism Foundation, Salt Lake City, UT, 1981. This book provides excellent reading for those interested in industrial-based counseling programs. A history of the occupational counseling movement moves quickly to a discussion of the key elements of a good program.

REFERENCES

"*Alcoholism: Problem and Promise*," The United Steelworkers of America Pamphlet, #PR-172-R, Education Department, United Steelworkers of America, Pittsburgh, Pa.

"Business's Multibillion Dollar Hangover," *Nation's Business*, May 1974.

Connor, B., "A Course in Humanistic Education for Alcohol Specialists", *Journal of Alcohol and Drug Education*, Fall 1980.

Cook, P. J., The Economics of Alcohol Consumption and Abuse, in *Alcoholism and Related Problems: Issues for the American Public*, Prentice-Hall, Englewood Cliffs, NJ, 1984, p. 61.

Corporate Commentary, Q.T.A. Reports Alcoholism Treatment Valuable, But Effectiveness Does Not Always Rise With Cost **1**,1 June 1984.

Dennenberg, T. S. and R.V. Dennenberg, *Alcohol and Drugs: Issues in the Workplace*, Bureau of National Affairs, Inc., Washington, DC, 1983.

Follman, J. F., Jr., *The Economics of Industrial Health*, AMACON Publications, New York, 1978.

Kubler-Ross, E., On Death and Dying, in *Death: The Final Stage of Growth*, by E. Kubler-Ross, Prentice-Hall, Englewood Cliffs, NJ, 1975.

Masi, D., *Human Services in Industry*, Lexington Books, Lexington, MA, 1981, p. 14.

"Mental Disorders: 1 in 5 is a Victim," *U.S. News and World Report*, **97**, 16, October 15, 1984, p. 18.

Merck Manual of Diagnosis and Therapy, Merck, Sharp and Dohme Research Laboratories, Merck and Co., Inc., Rahway, NJ, 1982.

Perham, J., Battling Employee Alcoholism, *The ALMACAN*, Association of Labor-Management Administrators and Consultants on Alcoholism, Vol. 12/10, October 1982.

Poley, W. G. Lea, and G. Vibe, *Alcoholism: A Treatment Manual*, Gardner Press, New York, 1979, p. 30.

Stress: Can We Cope?, *Time*, June 6, 1983.

Wrich, J. T., *The Employee Assistance Program*, Hazelden Foundation, Center City, MINN, 1974, p. 34.

CHAPTER 2

How to Manage Blue-Collar Stress

ARTHUR B. SHOSTAK
Drexel University
Philadelphia, Pennsylvania

Blue-collar stress can be reduced rapidly, effectively, and economically if its root sources are understood clearly, its challenge to outmoded management styles recognized honestly, and its sensitive alleviation designated as one of the highest possible corporate priorities.

2.1 SOURCES OF NEGATIVE STRESS

Much of value is understood about stressful workplace conditions that produce a "psychological or physical reaction . . . usually unpleasant and sometimes productive of symptoms of emotional or physiological disability" (McLean, 1979). Eight leading negative stressors can usefully be divided into two groups, objective and subjective, both of which engender fear in blue-collar workers, undermine their self-esteem, and lead many to worry that they are losing control of valued aspects of their lives.

2.1.1 Objective Negative Stressors

Blue-collar workers focus the attention of stress researchers on compensation complaints, workplace risks, workplace shoddiness, and fear of job loss . . . all of which have haunted the scene for decades past (Hershey, 1932).

Income Inequities

Few blue-collarites escape corrosive doubts about the fairness of their net take-home pay, a topic that regularly heads ranked lists of workplace sources of job discontent (Wallick, 1972; Shostak, 1980). Throughout the postwar period, and especially in the 1970s, inflation undermined earnings so severely that in 1978 one scholar concluded that a rise in local taxation could "very well have meant no rise at all in disposable income for workers in many areas over the past ten years" (Gappert, 1978: 47).

More recently, 60% of those workers who regained jobs after being laid off between 1979 and 1983 reported the necessity of accepting paycuts of 20% or more (Harton, 1984: 10). Since the late 1960s, some 67% of all new jobs have been in industries that paid average annual wages and salaries of less than $13,600 (in 1980 dollars). In 1969, only 45% of all jobs were in that low category. By contrast, "smokestack" industries, which paid average wages of more than $20,250—nearly all of them in manufacturing—grew the least after 1970 (Harrison and Bluestone, 1984). Labor Department economists affirm that the 20 fastest growing jobs now pay annual wages that average fully $5000 less than the 20 occupations in steepest decline (Kuttner, 1984: 33). Along with bottom-line misgivings about income adequacy, many blue-collarites harbor considerable envy—and some deep-seated resentment—about the earnings and "perks" enjoyed by higher-ups. Anecdotes abound in factories or truck terminals, anecdotes fed and inflated by TV hyperbole about astronomical levels of compensation voted for themselves by managers and executives. Much ire is stirred by media revelations about executive stock compensation schemes, such glittering perks as corporate health spas and vacation sites, and other desirable rewards.

To make matters even worse, another negative stressor here concerns roll-backs agreed to during the 1980–1983 recession. Between those workers who gave up previously negotiated gains in wages or fringes, and those many others who accepted a freeze (one that seldom recognized cost-of-living increases), blue-collarites lost considerable ground, especially in their perception of power and leverage in compensation realities. Relieved to have jobs to go to while millions of others were losing theirs, most blue-collarites learned anew how little clout they have in wage and fringe negotiations when a short-fused economy behaves in erratic and volatile ways.

Health and Safety Hazards

Second only to compensation as a negative stressor, this topic is complicated by frequent scare stories about workplace chemicals newly suspected of being hazardous to handlers. Blue-collarites learn from the media's "breathless" coverage that over 14 million of them are exposed daily to materials on lists of possibly toxic character, and about 100,000 workers may die prematurely every year from industry-related diseases readily avoided. As well, another 300,000 workers may contract disabling industry-related diseases over the same 12-month period (O'Toole, 1974: 26).

Even without the "silent violence" of occupational diseases, blue-collar work remains a nerve-racking source of life-and-limb hazard: over 12,000 workers lose their lives annually in workplace accidents, over 100,000 workers are permanently disabled every year, and employers report more than 5 million occupational injuries annually (Engler, 1984). Between dangerous machinery and transport equipment, extremes of heat and cold, fumes and radiation exposure, pesticides and other toxic ingredients, the stress toll here is enormous.

Workplace Ambience

Unpleasant working conditions, the third highest-ranked objective stressor in most polls, stir blue-collar anger at physical discomfort, odors, noise, general neglect, and a double standard that protects high housekeeping standards for white-collarites, but allows lower standards for hard-hat employees. Stress sources are as plain as factory windows gritty with accumulated grime, a toolshed floor marred by traces of everything dropped on it over recent weeks, odors from industrial processes that poor venting barely reaches, and the unrelenting abuse of roaring, whirring, pounding, whistling, and staccato noises.

Less obvious complaints have workers distressed by poor lighting, incessant drafts, extreme temperature variations, poor maintenance of plant bathrooms and parking lots, and the slowness with which disabled pay phones or vending machines in the workplace are repaired. Overall, far too many blue-collarites suspect that little of this makes much difference to their supervisors, at least while accident, turnover, and productivity levels remain within acceptable limits.

Job Security Anxieties

Haunted by the expectation that the economy may soon find their labor unnecessary, millions of blue-collarites worry that they may be next to join the job-seeking unemployed masses. And, as many concede, "There's only one thing worse than the stress of holding down an unpleasant job. And that's the stress of holding down no job" (Smith, 1983: 126).

American blue-collarites recognize that jobs in manufacturing are being lost as this nation loses its competitive advantage in autos, apparel industries, cement, rubber, synthetic fibers, and others (Magaziner and Reich, 1983). Many know from media coverage that while the labor force grew by 24% between 1968 and 1978, manufacturing employment grew by less than 4%. Many fear that "a section of industrial workers, primarily male blue-collar workers in the older manufacturing centers, are becoming redundant as the U.S. moves toward an employment structure increasingly dominated by poorly paid jobs that are frequently sex-stereotyped as female" (Appelbaum, 1984). Many suspect that a large number of the 2.5 million manufacturing jobs lost since 1979 will never be recovered. And all are chilled by media-featured pronouncements by influential management consultants who insist "in the current competitive environment, companies will either reduce their employment by 25% or 100%" (Salisbury, 1984).

Largely because of the business slumps in 1979 and in 1980–1983, about 5.1 million workers with at least three years' tenure (and 6.4 million with less tenure) were dismissed. While 60% of the 11.5 million were reemployed by January, 1984, the burden of joblessness clearly fell disproportionately on the industrial "rust bowl" states, workers in traditional smoke-stack industries, and the blue-collar middle class (Hartson, 1984).

Those who look ahead "through a glass darkly" dwell on jobs threatened by robotics, numeric control systems, CAD-CAM applications, and other programmable automation technologies (Schwartz and Neikirk, 1983). In the auto industry alone, for example, experts estimate that as many as 25,000 robots may require 5000 new human "caretakers" by 1990, but at a loss of 45,000 other auto industry jobs (Appelbaum, 1984). Little wonder, then, that a specialist in worker stress believes "unemployment and the threat of job loss are exquisitely threatening to many; seriously disrupting to others" (McLean, 1979: 55).

2.1.2 Subjective Negative Stressors

Four aspects of work—status level, the quality of supervision, sociability quotient, and level of job satisfaction—explain much about blue-collar stress not otherwise accounted for by objective factors.

Work Status

Drawing on the five dimensions Americans commonly use in assessing the status of a job—money, power, prestige, nature of the work, and nature of job prerequisites (such as necessary schooling)—the public thinks little of manual work. Blue-collar jobs are ranked at the bottom of the occupational status ladder, and many workers know and resent this. In a society where prestige, power, and privilege are located high above the factory floor, many blue-collarites believe competition in status assignment has been stacked against them from the start.

Stress follows when workers take to heart the relatively low opinion American society has of their livelihood. As well, a major prop to self-esteem, worker pride in craft, seems steadily lessened by technological advances that dilute worker skills. Self-denigration therefore grows, threatening to raise the toll already taken in low morale and lowered productivity by "low-status" blues.

Supervision

While much about foremen, plant managers, and other direct superordinates exasperate subordinates, two items in particular vex blue-collarites: the enervating pettiness of various work rules, and the enervating nature of relentless pressure for more and more production.

While workplaces vary widely in some ways, many blue-collar settings are marred by similar outmoded and childish work rules which can resemble a Marine Corps boot camp approach to supervision. Rigid rules exist about how often workers can use the bathroom, whether they can or cannot talk with coworkers, when they can begin to wash up or line up to punch out at the day's end, and so forth and so on. Prohibitions against horseplay, smoking, and other adult prerogatives remind manual workers that no other segment of the work force has its workday so closely regulated and policed.

Negative stress also flows from an ill-managed expectation of greater and greater output. Blue-collarites resist much of this pressure because their disillusioning past has many convinced that higher turnout leads primarily to still greater demands, ad infinitum. Certain supervisors browbeat, intimidate, provoke, or in other heavy-handed ways push unreasonable production targets—and, not incidentally, stir considerable (debilitating) stress. Workers respond with covert sabotage and a costly guerilla war between "us" and "them."

Sociability

Next only to wages and fringes, workers value the opportunity they get from the job to socialize with buddies, pals, cronies, and their "old boy" or "old girl" circle of coworkers. The leeway to "shmooze" (that is, to do friendly things unrelated directly to assigned work) is regarded by many blue-collarites as perhaps their highest-order workplace need (Schrank, 1978). When these needs can be met, a worker has the security of coworker support, earthy camaraderie, and an invaluable source of tips on "where influence lies, where respect is deserved, and where support is forthcoming" (Miller and Fry, 1977: 472).

Distress enters when the common need to be part of a community at work is thwarted by sexism, racism, or ageism, and many workers succumb to prejudice-fed internecine warfare. Especially stressful are rumors that minorities are not only given preference in hiring, but that they later enjoy special treatment when promotions are being considered. Certain white male blue-collarites fear that gains by women, nonwhites, and older workers come primarily at their expense. The absence of a rigorous and sensitive effort by the employer to counter such suspicions leaves the work force seriously divided among bitterly antagonistic cliques.

JOB SATISFACTION

Subjective negative stress is revealed here by the fervid wish blue-collarites have to see their children avoid following in their paths, and their insistence that they would not live their work lives over in the same way if they could do it again. Researchers believe "the message is clear: The Blue-Collar Blues are predominantly associated with those working conditions that discourage good work performance, impede personal growth, and stifle autonomy and creativity. Having relatively poor fringe benefits is also important" (Seashore and Barnowe, 1974: 549).

To be sure, some blue-collar dissatisfaction here is commended by researchers who view it as evidence of a healthy refusal to accept the least worthy aspects of the scene. Workers are thought to "feel angry at being exploited, resentful at being mistreated, and frustrated at the limited control they do have. It is because they feel *human* (rather than like defeated robots or interchangeable mechanical parts) that they have these feelings." (Reiff, 1976: 48). Unfortunately, when these conditions are unremedied, such feelings often lead to ever-higher levels of absenteeism, restlessness, unacceptable work, grievance filing, and "retirement at work," a costly combination that explains much about low productivity and poor morale.

2.2 REFORM RESPONSES TO NEGATIVE STRESS

Given this litany of woes, this (incomplete) list of eight major blue-collar problem areas, how might concerned parties (management, labor, and rank and file alike) respond in a cost-conscious, creative, and effective fashion? While only a dozen promising answers will be highlighted below, they should demonstrate the remarkable range and feasibility of available, field-tested reform responses (Stoner and Fry, 1983).

2.2.1 Income Inequity Options

Three in particular—gainsharing plans, medical cost containment plans, and comparable worth plans—merit special attention.

GAINSHARING PLANS

Used by companies for over 50 years, more gainsharing plans have been implemented in the past 5 years than in the preceding half-century. Their success (whether in the Scanlon, Rucker, improshare, or other such variety) draws on their distinct ability to share both the responsibility *and* the (financial) rewards for making work process/work site improvements.

Gainsharing plans basically combine custom-tailored programs of employee involvement with attractive financial formulas for bright ideas that boost projects. Most of the plans solicit employee ideas to solve company problems, and share resulting gains through a periodic cash bonus. A rare study in 1984 of 33 such plans found that 80% had measurable improvements of 20–30% in some hard measure of productivity, cost savings, or quality. About 75% had achieved a decline in grievances, an enhanced work climate, and a pronounced drop in negative stress level: ". . . the results were often dramatic and across the board" (Bullock, 1984: 3).

Gainsharing plans seemed to work best if voted on by employees prior to implementation. A formal involvement structure, such as an employee task force, was vital, as was the option of paying no bonus in hard times. Outside consultants were used in about nine out of ten cases, and four out of five plans were in unionized plants. Increasing sophistication in computer-aided measurement tools has considerably strengthened bonus logic, details, and credibility. All in all, researchers believe it "safe to predict a dramatic increase in gainsharing activity in the future" (Bullock, 1984: 4).

HEALTH CARE COST CONTAINMENT

Typical of the overdue and innovative tactics being taken to relieve income stress are joint labor–management efforts to reduce health care costs. Intent on protecting workers' hard-won health benefits, many unions and companies invent, improve, and capitalize on mutually acceptable solutions to the cost containment challenge.

With the cost of medical services going up 15–20% a year, and labor contracts falling behind in their ability to cover escalating expenses, blue-collar pressure for pocketbook relief grows more and more incessant: conventional cost-cutting reforms, such as second opinion programs and prehospitalization systems, no longer suffice.

Instead, progressive labor and management representatives are forming joint committees to recommend new cost containment measures (including cost-sharing, if potentially viable), and to participate in local coalitions that spearhead such efforts as the development of alternative delivery systems. Health fairs and other health promotion and education activities try to win worker support for cost cuts, even as special state-wide conferences are sponsored to encourage expanded enrollment in HMOs (Klett, 1984).

Related attention commonly is paid to early detection of disease (hypertension screening, periodic physical exams, etc.), control of biological risk (hypertension treatment, cardiac rehabilitation, etc.), detection of high risk behavior (health risk profiles through screening instruments, etc.), and the development of a healthy corporate culture (noise control, day care facilities, smoking restrictions, flex-time, job sharing, etc.). Happily, research suggests that the workplace is a *very* effective prevention and treatment site, and, "although more difficult to ascertain, many of these health improvements have translated into dollar savings" (Rosen, 1984: 7).

Comparable Worth

On the books now in four states (Iowa, Minnesota, New Mexico, and Washington), and under active legislative review in 1984 in nearly 30 others, comparable worth legislation posits the novel idea that different jobs can be compared in such a way that "equal pay for equal work" can become "equal pay for comparable worth." Proponents want employers to be required by law to use job evaluation techniques to compare dissimilar jobs and adjust wage levels, especially in the case of underpaid female workers (the 80% of all women workers concentrated in only 20 of the Labor Department's 427 job categories).

Employers intent on getting their house in order *before* the emergence of a comparable worth fracas might want to consider adapting the 1983 AT&T/Communication Workers Union model.

Following years of bitter labor relations, the parties resolved in 1980 to begin a new era of cooperation and joint problem solving. One result was the formation of a joint national Occupational Job Evaluation Committee to research, develop, and make recommendations concerning the design and implementation of an improved job evaluation plan. The Committee agreed that no wages would be reduced as a result of its work, and all employees would have training opportunities to improve their skills. As well, the psychological and emotional effects of technological change would be considered in any job evaluation, and all employees would have the right to appeal the scoring, job description, and relative worth of their jobs. The key to the entire process, Committee members agreed, was holding interviews with workers, who are "the only real experts on the jobs being studied." While fully aware of the huge scope of their undertaking, the Committee persists in maintaining that with the joint commitment of CWA and the new Bell companies, "the task is not unsurmountable" (Straw and Foged, 1982).

Perhaps the last word in the matter can go to a conservative business newspaper, which editorialized in 1984 that "the appearance of legislation such as 'comparable worth' is often the direct result of failed private sector and free market efforts to correct inequities and discrimination. The best solution is better management and employee relations, and a rigorous reevaluation of pay scales. If business fails to heed the warnings issued by the appearance of this legislation, then the troublesome intervention created by such bills will be its justly deserved reward" (*Philadelphia Business Journal*, 1984).

2.2.2 Health and Safety Options

Especially promising are four variations on a theme, the first of which appeals to the pocketbook; the second, to the psyche; the third, to the desire to behave better; and the fourth, to the right to know more . . . all the better to reduce health stressors and enhance the workplace quality of life.

Wellness Pay

Interest is growing in the explicit use of a carrot, a cash reward for desirable behavior. Bonuses for workers who make an extra effort not to get sick enough to miss work are still rare, but are steadily gaining positive attention.

Quaker Oats employees, for example, could earn a bonus of as much as $300 in 1983, and, if they kept medical bills below a specified target, a dividend worth $100 or so. Designed to discourage unnecessary trips to the doctor, "the fender-benders of the health business," the wellness pay plans help reduce stress by putting the employer clearly on record in support of employee illness prevention, rather than only sickness-recovery efforts (Goodman, 1983). When incorporated into a far-reaching health services program, the company's wellness pay campaign could achieve handsome dividends in attendance gains for minor costs in earned bonuses.

Employee Assistance Plans (EAP)

First developed as early as 1917, the nation's more than 7,500 EAP programs are hailed today as sound, profit earning, health promoting investments.

Designed to help low productivity employees overcome work damaging personal problems, the EAP offers free and confidential counseling to employees and their dependents, as well as referrals to outside community sources and follow-up counseling sessions. Expertise over a broad span of areas is entailed: job, marital, family, financial, legal, consumerism, health, alcoholism, drug, preretirement, and career change questions. Naturally, as a leading EAP pioneer explains, "providing employees a relief valve to vent frustration and reduce stress is a basic concept of an employee assistance program" (Wagner, 1982).

EAPs are especially valuable for their in-house contribution to supervisory training. The best of them offer fully subscribed courses for first-level managers on basic counseling skills (active listening and nonverbal communication skills, interpersonal inventory, helping relationships, ice breaking, referral methods, etc.). Attendees learn how to identify disfunctional employee behavior, make sensitive early referrals, and represent to employees just what the EAP offers.

Thanks to the over half-century of experience American companies have gained with EAPs, numerous guidelines exist to buttress the effort; for example, it is advisable to create a local association of EAP programs to provide continuing education sessions, guest presenters, and counselors to meet the counseling needs of EAP staffers. Similarly, housing an EAP program in a company's training division can provide counselees with a certain amount of anonymity and counselors a valuable change of pace in the form of occasional training assignments. Certain companies, however, are better off not to conduct their own program, but should instead contract out to external specialists (such as Info-Now, a Philadelphia EAP that serves 19 different companies with counseling services for about $10 per year per employee) (Roessel, 1983).

Naturally, EAPs have shortcomings, not the least of which include blue-collar suspicion of company motives, blue-collar insistence on strict confidentiality, and blue-collar skepticism about overreliance by EAP staff on "psychobabble" and talk cure modalities. Concerned employers can respond to these criticisms with a frank, unsentimental expression of interest in both employees' well-being and productivity, or, as Busch explains: "An employee's personal problems are private *unless* they cause the employee's job performance to decline and deteriorate. When that happens, the personal problems become a matter of concern for the company" (Busch, 1981). Companies can also offer written pledges of absolute confidentiality, the better to earn the trust and assure the anonymity necessary if employees are to use the EAP. Finally, reliance can be placed on behavior-modification programs that minimize the verbal explorations blue-collarites distrust, and emphasize instead modifying observable behaviors, as through operant conditioning, aversion therapy, "contract" setting, and the like.

Much to the credit of EAP practitioners the program's track record indicates a bonanza can be mined at pennies on the dollar. To estimate the potential cost saving of an EAP, an organization can

> Simply divide the average total number of employees into your annual payroll to derive the average employee wage.
> Multiply the average annual wage by 17% of your total employees (statistical average of troubled employees) to obtain the payroll to troubled employees.
> Next, take 25% of the troubled employee payroll to obtain your present loss due to troubled employees.
> Finally, take 50% of this loss to identify the amount you could save through an EAP.
>
> Wagner, *Personnel Administration,* Nov. 1982, 59–64.

Cost-effective and field proven, EAPs can generate a return of $4 to $16 for every dollar spent on referral and counseling services (Shriver, 1984). For a fraction of recruiting and training costs, then, EAPs help companies refurbish troubled personnel, upgrade supervision, and significantly improve the stress climate at (and away) from work. Available at present, however, to only 12% of the labor force, EAPs remain a major untapped resource (Shriver, 1984).

Workshops on Stress Management

Professional consultants are making headway adapting their anti-stress message, techniques, exercises, and "homework" to the particular profile of blue-collarites, and more and more companies and unions are giving this option a fair try.

Skilled workshop leaders, backed by strong endorsements from recent clients, are helping hard-boiled and skeptical manual workers learn how to "get into your calm scene," "entertain your mind," practice deep breathing and other meditation techniques, and grasp the notion that "the fears we have are what keep us from relieving our stress." Deliberately crisp and jargon-free, their message is strong in pithy action formulas ("If you can't flee or fight, then float."). Many borrow a major AA guideline (learn what you can and cannot control), and urge exploration by the listener of his or her own "comfort zone," or range of stress comfortable to that person. Above all, the workshop leader explains that reducing negative stress is *both* an individual *and* an organizational responsibility. "Homework" consists of mental and physical exercises custom tailored to each participant *and* constructive programmatic reforms discussed by workshop members eager to recommend them to management.

Noteworthy about such workshops is the presence of blue-collarites, especially males, who would never seek mental health or psychological counseling. But they will give this option a try if it is offered on work time at the workplace, has the endorsement of the local union or popular workmates, has no mental illness stigma attached, is free and voluntary, avoids "Mickey Mouse" material and a patronizing tone, and combines earthy advice, humor, and optimism with the imprimatur of company sponsorship.

"Right-to-Know" Legislation

On the legislative front the major reform option is a controversial piece of legislation sometimes known as the "Worker and Community Right-to-Know Act." While details vary in the bills passed thus far by

20 states and various cities, a bill passed by the Illinois legislature in January 1984 is typical: Employers must alert workers to toxic substances, send the state a list of toxic substances they use, and accept state mediation services when appropriate. When companies are slow to send their lists, the state puts pressure on by contacting their insurance companies. Especially encouraging is the fact that in the law's first year of operation, 45 of 46 charges lodged by workers who did not get adequate information were settled by artful negotiation and conciliation, largely to the satisfaction and lowered stress of all (Trost, 1984).

Compromise and conciliation carry the parties far in this potentially acrimonious matter. Instructive is the example of leading chemical companies and their role in the 1984 drama of Pennsylvania action on a proposed bill. On learning details of a first draft prepared by the state AFL-CIO and Philaposh, an activist citizen-labor coalition, the chemical firms indicated a willingness to negotiate details and support passage of a compromise document, long before legislators had an opportunity to debate and vote on any version of the bill.

Confidential negotiations helped the chemical firms win the exemption of farm suppliers, the elimination of on-site testing, improved protection for trade secrets, and a lengthening time requirements for chemical labeling. Union negotiators, in turn, won the inclusion of potentially hazardous chemicals in the bill's coverage and the critical acquiesence of potential opponents (the chemical firms) in the bill's eventual passage. Both sides complimented one another for reducing stress by their give and take, and indicated their satisfaction with this way of achieving change.

2.2.3 Job Security Options

Stress relief here pivots around three exciting possibilities, the first of which changes the focus of the parties, the second changes the terms of the bargain, and the third changes the letter of the law.

Employment Security

Progress begins with the substitution of a new concept and perspective—employment security—for the outmoded and counterproductive idea of job security. The latter implies an implausible right to cling to a particular job, while the former helps blue-collarites realize they may have to change jobs—possibly even shift temporarily to lower-priority tasks—if they are to weather troubled times on any payroll at all.

Employment security, the praise-winning approach of IBM, Hewlett-Packard, and several dozen other major firms, relieves "job loss jitters" with its pledge of no layoffs despite business distress. The companies keep their work force lean and flexible, with all workers trained to fill a variety of jobs. If and when the companies face temporary declines in business, they defer layoffs by reassigning employees, calling in work previously contracted out, intensifying their market efforts, and, as a last resort, asking everyone, from the CEO down, to voluntarily reduce their paid working hours so as to preserve the job holding of all.

Research has found that, contrary to popular belief, this approach does not reduce corporate profits and productivity, but instead, "opens the door to rapid technological change and productivity improvement" (Zager, 1984: 3). Employees can stop fearing overnight job loss, and start supporting the sort of continual change companies need to remain competitive.

Negotiated Security

Thanks to appropriate media attention, millions of blue-collarites know about and applaud the pioneering contracts signed in 1984 in the auto industry, contracts that lead the way where employment security pledges are concerned.

In return for valuable union concessions on work rule changes and deliberately modest wage and fringe gains, the parties agreed to avoid plant shutdowns and reduce the exporting of jobs that could be done in the United States. They agreed to promote interplant transfers of employees displaced by automation, and invest heavily in worker retraining programs. For the first time ever, the companies agreed to create a fund for loans to displaced auto workers eager to open their own small businesses (Buss, 1984).

Typical of agenda innovations is the UAW-Buick joint retraining and reassessment program, an integral part of Buick's 5-year business plan to achieve the best union-management relationship in the entire industry. Buick's Employee Development Center, located on the grounds of the Buick plant, accepts employees for up to one year of retraining during which they draw their full salary and benefits. As many blue-collarites need to complete their basic schooling, the Center offers academic as well as technical courses. And as Buick has proprietary knowledge of the job skills it expects to need months and years ahead, the Center is able to retrain workers for specific posts not yet available, but scheduled to be filled in the very near future.

As these new props to job security help build trust between rank and filers and Buick, the union local has been able to cancel its earlier opposition to participation in quality work circles. Before the company opened the Center in 1982, the local felt that previous decades of bitter labor relations justified its knee-jerk rejection of any new management fad (like a quality work circle) that might result in layoffs. But with the Center's existence as concrete evidence of a genuine prosecurity commitment by top management, the UAW local has agreed to endorse participation in the company's 200 quality work circles which will better help the company improve its competitive position.

PLANT-CLOSING LEGISLATION

Under consideration in 1984 in 38 states, and on the books in Maine, Massachusetts, and Wisconsin, plant-closing laws exist throughout Western Europe where American-owned businesses have decades of unexceptional experience coexisting with them.

Massachusetts, a state where argument is commonplace and political debate is rarely on simmer, surprised observers and set a model for the nation when, in 1984, it passed a plant-closing law that seemed to have benefits for everyone. Credit was given to a behind-the-scenes compromise achieved inside the Governor's Commission on the Future of Mature Industries, a 36-person group of business, labor, government, and public interest citizens. This group kept debate focused on how to protect the state's job creation possibilities even while relieving the human toll of shutdowns.

After six drafts the Commission highlighted a four-page "social contract" that encourages, but does *not* require, companies to give employees at least 90 days notice of a factory or office closing. To encourage compliance, only companies that sign the social compact are supposed to have access to various state financing agencies, such as the one that issues industrial revenue bonds. As well, these companies will be eligible for public money that may help them find new products to produce. Employees in turn are guaranteed an extra three months of employer-sponsored health insurance coverage after a shutdown, and, if they receive less than 90 days notice of job loss, they will get extra severance compensation paid from a new state fund.

A spokesman for the state AFL-CIO later explained their preference for a one-year mandatory notice of closing, along with their judgment that the voluntary social contract approach was a "fair settlement. The workers are protected no matter what happens." Similarly, a business representative noted that "once labor came to understand that mandatory plant-closing notice was not acceptable, business started to focus on what we could do to be more socially responsible. We met somewhere in the middle." And a university administrator who cochaired the Commission observed: "It's not a neat, trim package, but it helps us move ahead on an issue where everyone gains something" (Carlson, 1984).

2.2.4 Supervision Reform Options

Two stress-relieving options merit particular attention here, as the first illustrates the interdependence of elements of a company, while the second highlights the strategic place supervisory personnel have in blue-collar realities.

SHARED AUTHORITY SYSTEMS

The present command and control relationship between management and workers—a deeply imbedded caste separation—is giving way in leading companies to dramatic experiments with thoroughly revised supervision approaches (Peters and Waterman, Jr., 1982). Some businesses are drawn to job redesign efforts, while others prefer the Scanlon Plan or similar productivity-bonus schemes. A very small number are trying autonomous work teams, while hundreds of firms are experimenting with quality work circles (Maccoby, 1984). Whatever the approach, all of these models of cooperative representation help worksites shift to a shared authority system, and this, in turn, helps expose and relieve much negative stress (Nowak, 1984).

Research on worker participation techniques warns against assuming conversion that looks simple and quick, when it is neither (Caplan et al., 1975). But the movement continues to draw strength from the stature of notable corporations that have been promoting these aids for years—companies as diverse as GM, Polaroid, Dana, Herman Miller, TRW, and Proctor & Gamble. They boast of handsome rewards in reduced absenteeism and higher product quality. "They also perceived benefits that cannot always be measured, such as the unknowable savings from an idea generated by a quality circle that prevented a product recall. And they have seen that employee participation has, over the long run, made entire organizations more effective" (The Editors of *Fortune,* 1984 ed.: 108).

Advice on how to use these varied techniques best is plentiful and field tested; for instance, avoid undue haste in measuring or quantifying immediate gains, lest the process be upset; expect implementation to take several years; and recognize that securing the acceptance of foreman and middle managers is *the* single greatest obstacle to achieving a new work culture. Union participation at the earliest possible stage is critical, as is visible commitment from top management and a climate of employment security throughout the enterprise (Truell, 1984).

When these tenets are honored, companies find rewards throughout the entire corporate culture: ". . . blue-collar workers are no longer just workers: they become the lowest level of management. Because the company needs fewer administrators to supervise them, it can fashion a leaner and more responsive organization, with clearer and faster communication up and down the chain" (The Editors of *Fortune,* 1984 ed.: 116). Little wonder, therefore, that stress-reduction proponents welcome worker participation techniques, and think them likely to "yield a reliable alternative managerial style, one necessary for the future of U.S. productivity" (Gorlin and Schein, 1984).

Supervision Redirection

An agenda for supervisory moves against negative stressors should include efforts to:

Make blue-collarites "visible" and help them attain a feeling of significance.

Practice consensus-building skills rather than rely on hierarchical posturing.

Evaluate foremen on their contribution to people development, along with low levels of worker grievances, low absenteeism, the development of work skills, and the improved quality of product and service.

Encourage teamwork by helping employees realize and value their interdependencies.

Commit to continuous educational programs to help foreman and workers alike acquire skills to aid interpersonal relationships.

Even supervisors in the forefront of workplace antistress efforts need help putting their convictions and hunches into practice. Only as their own negative stress level is lowered, and their efforts receive high-quality, top-level support, can their campaign to humanize the company's work culture possibly succeed (Pascarella, 1984: 194-196).

Happily, research suggests that more and more companies realize that they "can't depend on automatic growth in the market and the ability to push prices up to cover their costs. They're recognizing they have to use people differently than in the past" (Hoerr, 1984).

2.3 SUMMARY

Much can be done to mitigate blue-collar stress through reform campaigns realistic in goals, ample in resources, and patient in perspective. Above all, successful campaign leaders will understand that "there is no such thing as an occupational stress problem which can be solved for all time and forgotten about, like a mathematical problem of two plus equals four. There are only occupational stress issues—never fully delineated, never completely resolved, always changing, always in need of alert accommodation." (Siu, in Shostak, 1980: 7-8). And these issues are always challenging us to get on with the job, to make the progress readily available *now* to achieve blue-collar work that truly honors us all.

REFERENCES

Appelbaum, E., "High Tech" and the Structural Employment Problems of the Eighties, in Eileen Collins, Ed., *American Jobs and the Changing Industrial Base*, Ballinger, New York, 1984.

Bullock, R. J., Gainsharing—A Successful Track Record, *World of Work Report*, **9,** 8, August, 1984, pp. 3-4.

Busch, E. J., Developing an Employee Assistance Program, *Personnel Journal*, **60,** 9: 1981, pp. 708-711.

Buss, D. D., GM Labor Contract Contains Several Elements of Surprise, *Wall Street Journal*, September 25, 1985, p. 7.

Caplan, R. D. et al., *Job Demands and Worker Health: Main Effects and Occupational Differences*. NIOSH Research Report, Government Printing Office, Washington, D.C., 1975.

Carlson, E., Massachusetts Deftly Handles Volatile Issue of Plant Closings, *Wall Street Journal*, June 26, 1984, p. 37.

Engler, R., *A Job Safety and Health Bill of Rights*, Philadelphia Area Project on Occupational Safety and Health, Philadelphia, 1984.

Gappert, G., *Post-Affluent America: The Social Economy of the Future*. Franklin Watts, New York, 1978.

Goodman, E., Bad Idea: Bonuses for Staying Healthy, *The Philadelphia Inquirer*, February 4, 1983, p. 13A.

Gorlin H. and L. Schein, *Innovations in Managing Human Resources*, Conference Board, New York, 1984.

Harrison, B. and B. Bluestone, "More Jobs, Lower Wages, *New York Times*, June 19, 1984, p. A27.

Hartson, M., Recessions Took 5.1 Million Jobs Over Five Years, *Philadelphia Inquirer*, December 1, 1984, pp. 1, 10.

Hersey, R. B., *Worker's Emotions in Shop and Home: A Study of Individual Workers from the Psychological and Physiological Standpoint*, University of Pennsylvania Press, Philadelphia, 1932.

Hoerr, J., Why Recovery Isn't Getting Cooperation in the Workplace, *Business Week*, February 20, 1984, p. 32.

Klett, S. V., The Labor Factor, *Business and Health,* September 1984, pp. 31-35.

Kuttner, R., Jobs, *Dissent,* Winter, 1984, pp. 30-41.

Maccoby, M., Helping Labor and Management Set Up a Quality-of-Worklife Program, *Monthly Labor Review,* March, 1984, pp. 28-32.

Magaziner, I. C. and R. R. Reich, *Minding America's Business,* Vintage Books, New York, 1983.

McLean, A. A., *Work Stress,* Addison-Wesley, Reading, MA, 1979.

Miller, J. and L. F. Fry, Work-Reflect Consequences of Influence, Respect, and Solidarity in Two Law Enforcement Agencies," *Sociology of Work and Occupations,* November, 1977, pp. 470-478.

Nowak, M. F., Worker Participation and Its Potential Application in the United States, *Labor Law Journal,* March, 1984, pp. 148-166.

O'Toole, J. Ed., *Work in America,* MIT Press, Cambridge, MA, 1974.

Pascarella, P., *The New Achievers: Creating a Modern Work Ethic,* Free Press, New York, 1984.

Peters, T. J. and C. Waterman, Jr., *In Search of Excellence,* Harper and Row, New York, 1982.

Philadelphia Business Journal, Comparable Worth Bills are the Wrong Solution, (Editorial), September 3-9, 1984, p. 4.

Reiff, R., Alienation and Dehumanization, In Widick, B. J., Ed., *Auto Work and Its Discontents,* Johns Hopkins University Press, Baltimore, MD, 1976, pp. 45-51.

Roessel, C., Local Firms Get "Tough" with Alcoholics, *Philadelphia Business Journal,* January 31-February 6, 1983, p. 6.

Rosen, R. H., Worksite Health Promotion: Fact or Fantasy, *Corporate Commentary,* **1,** 1, June, 1984, pp. 1-8.

Salisbury, D. F., The Coming Industrial Revolution, *The Christian Science Monitor,* November 20, 1984, pp. 37, 44.

Schrank, R., *Ten Thousand Work Days,* MIT Press, Cambridge, MA, 1978.

Schwartz, G. G. and W. Neikirk, *The Work Revolution: The Future of Work in the Post-Industrial Society,* Rawson Associates, New York, 1983.

Seashore, S. and J. T. Barnowe, Demographic and Job Factors Associated with the "Blue Collar Blues," In Robert P. Guinn, and Linda J. Shepard, Eds., *The 1972-73 Quality of Employment Survey,* ISR, Ann Arbor, MI, 1974.

Shostak, A. B., *Blue-Collar Stress,* Addison-Wesley, Reading, MA, 1980.

Shriver, J., Employers are Helping Addicted Workers, *USA Today,* January 12, 1983, p. 5D.

Siu, R. as quoted in Arthur B. Shostak, *Blue-Collar Stress,* Addison-Wesley, Reading, MA, 1980.

Smith, R. E., *Workrights,* E. P. Dutton, New York, 1983.

Stoner, C. R. and Fred L. Fry, Developing a Corporate Policy for Managing Stress, *Personnel,* **60,** 3, May-June, 1983, pp. 66-76.

Straw, R. J. and Lorel E. Foged, Job Evaluation: One Union's Experience, *ILR Report,* Spring, 1982, pp. 24-26.

The Editors of *Fortune, Working Smarter,* Penguin Books, New York, 1984 ed.

Trost, C., Labor Letter, *Wall Street Journal,* November 20, 1984, p. 1.

Truell, G. F., *Building and Managing Productive Work Teams,* PAT Publications, Buffalo, 1984.

Wagner, W. G., Assisting Employees with Personnel Problems, *Personnel Administrator,* **27,** 11, November 1982, pp. 59-64.

Wallick, F., *The American Worker: An Endangered Species,* Ballantine, New York, 1972.

Zager, R., *Employment Security in a Free Economy,* Work in America Institute, Inc., Scarsdale, NY, 1984.

PART II

ERGONOMICS

CHAPTER 3

How to Apply Ergonomic Principles to Minimize Human Error and Maximize Human Efficiency

LOTHAR R. SCHROEDER
General Physics Corporation
Columbia, Maryland

CATHERINE D. GADDY
General Physics Corporation
Columbia, Maryland

3.1 INTRODUCTION TO ERGONOMICS

Workplace designs that do not take ergonomic principles into consideration are likely to lead to an increase in errors and accidents and a decrease in safety and efficiency. Error-prone designs (1) place demands on performance that exceed capabilities of the user, (2) violate the user's expectancies based on his or her past experience, and (3) make the task unnecessarily difficult, unpleasant, or dangerous.

Ergonomics is an approach to workplace design and evaluation from the perspective of the worker or operator, whose primary goals are to minimize human error and to maximize human efficiency. For the purpose of this chapter, the word "ergonomics" will include information from the fields of human factors, engineering psychology, and to some extent, industrial engineering.

The goal of this chapter is to provide practical ergonomic guidelines for the safety practitioner to use in evaluating workplace design against ergonomic criteria, and for improving the workplace accordingly. To accomplish this purpose, the chapter has two main objectives:

1. To provide ergonomic principles for workplace design
2. To describe ergonomic methods for use by the safety practitioner

The remainder of this chapter is divided into two parts: Ergonomic Principles and Ergonomic Methods. In the first of these sections the role of the human in the system is discussed as well as the human capabilities and limitations that affect work. Ergonomic principles derived from research on humans at work are then presented. Principles applicable to displays, controls, warnings, and workstation and workspace design are also noted. The section on methods includes data collection and analysis approaches available to the safety practitioner to use in identifying ergonomic deficiencies. Then a discussion of corrective actions and follow-up evaluation for these is provided.

Other improvements may be gained through human factors analyses of personnel selection, job design, organizational management, training, and procedures. This chapter does not include discussion of these issues. The reader is referred to Dunnette (1983) for these topics and to U.S. NRC (1982) for procedures. A complementary approach to ergonomic analysis is human reliability analysis, which focuses on the probability of worker error.

This chapter is written to assist the safety practitioner in evaluating the existing systems in the workplace and implementing changes. Therefore, this chapter assumes that the practitioner is in a "*backfit*" situation, that is, changes are to be made to an existing workplace. As a result of this assumption, practical methods for evaluating the existing workplace are provided. Comparatively minor changes or enhancements, as opposed to major redesign of the workplace, are suggested. Because these minor changes can have an appreciable effect on safety and productivity and are relatively inexpensive, it is well worth the effort to incorporate them as improvements.

3.2 ERGONOMIC PRINCIPLES

3.2.1 The Human As A System Element

The systems approach applied to ergonomics treated humans and machines as components interacting to bring about some desired objective. In this section, we will provide background information on the role of the individual as a system element with a focus on traditional treatments of the human-machine interface. The goal is to provide a framework for discussing the potential capabilities and limitations of humans operating under varied circumstances and environments.

The Human-Machine Interface

Most contemporary conceptualizations of the human-machine interface are extensions of a fairly simple model provided by Franklin Taylor (1957). At the workstation level, the model focuses on the areas where the machine component(s) of the human-machine system communicate with the human component(s)—the various displays, and areas where the human component(s) communicates with the machine component(s)—equipment controls.

From an information processing perspective, the individual in the human-machine system receives information from displays via sensory mechanisms. Humans process this information using perceptual and cognitive mechanisms, and may act upon the information (human motor mechanisms) by manipulating controls. Figure 3.1, a simple representation of the interface for the basic human-machine elements, provides a framework for discussing some of the more important human capabilities and limitations that impact on the workplace.

Also included in Figure 3.1 is a workspace-environment interface. This boundary represents the interaction of an individual operating at a workstation level with other individuals, other workstations, and overall environmental factors. Subsequent sections of this chapter will address issues related to workstation integration and the impact of workspace environment (e.g. noise level, temperatures) on individual performance.

Human Capabilities and Limitations

The operator receives information about the equipment he or she is using and the overall work environment through the senses. The practitioner interested in taking full advantage of human sensory capabilities should be aware of a large body of research that explains how individuals deal with inputs received from the environment.

Although we commonly think of five sense modalities, there are other senses, such as body movement and body position, that have been extensively studied and are quite important in certain specific circumstances.

In most industrial applications, however, vision is probably used most frequently as a display input channel, followed by hearing (audition). When there is an option to choose between one and the other, a

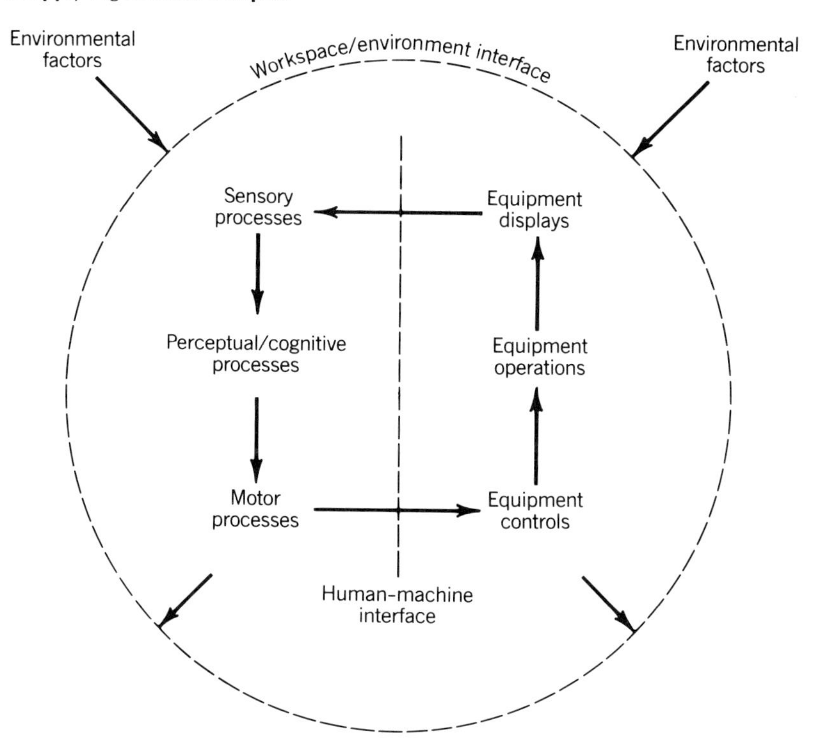

FIGURE 3.1. A diagram representing the elements of a human-machine interface.

consideration of the relative advantages of each can be beneficial. For example, auditory signals are most appropriate as warning signals, since reaction time is fastest for auditory signals and the receiver need not be stationary to receive the message. For lengthy or complex information, evidence exists that the visual mode of presentation is superior. See McCormick (1976) for a more complete discussion of these applications.

Several other key issues should be considered in attempting to understand how the human senses respond to input. First, ranges of sensation or the capacity of the individual to detect the input should be considered. For example, if an alarm is not loud enough compared to background noise the operator may not detect it. Similar limitations are apparent in each of our modalities.

Some of the earliest psychological investigation attempted to establish the exact relationship between sensory experience and direct physical measures. These researchers found that the quantities of light or sound that were physically measurable were not always seen or heard. To account for this, there developed the concept of a threshold, or minimum value, above which a stimulus had to rise in order to be perceived.

The pioneer investigations also uncovered the finding that for inputs above threshold, equal increments in stimulus magnitude do not produce equal increments in sensation over the entire range of stimuli. For example, it requires more than twice the physical sound level (approximately 10-fold) to generate a tone that appears to be twice as loud to an observer.

In most ergonomic applications, the detection by an individual of the presence of a signal is a necessary but not sufficient condition for making an appropriate response. Detection is usually accompanied by discrimination, or recognition. The demands for discrimination can be either on a relative or an absolute basis. For *relative* judgments, the task is simply to discriminate whether two inputs or signals differ in some way, for example, discriminating which of two auditory alarms is louder. On the other hand, *absolute* judgments are made without a reference stimulus available, such as identifying the loudness of the auditory alarm. Unfortunately, while people can make far more relative than absolute discriminations, they are more frequently called upon to perform the latter in human-machine environments. This finding has important implications for coding of displays.

Miller (1956) in a classic article, surveyed a number of studies that described a person's ability to make absolute judgments. He reported that approximately "seven, plus or minus two" signals varying along a single dimension such as intensity or frequency could be reliably identified. It should be noted, however, that when more than one stimulus dimension is used, the number of absolutely identifiable signals increases. Table 3.1 lists the number of reliable absolute judgments that can be made for visual and auditory modalities for single stimulus dimensions and different combinations of dimensions. Thus, for example, if colors are used to code high-pressure steam pipes where errors of identification could be very serious or costly, no more than nine different color hues (dominant wavelengths) should be applied. The coding scheme can be made to transmit more information if multidimensional stimuli are used, for example, size, brightness, and hue. Nevertheless, one should not increase the number of stimulus categories indefinitely in coding applications. The addition of new dimensions reaches a point of diminishing returns, resulting in confusion on the part of the user.

At the next stage in the diagram of the human-machine system presented in Figure 3.1, the person processes the information before making the appropriate response. It should be recognized, however, that these cognitive processes cannot always be neatly separated from the sensory inputs and the resultant outputs or actions. For example, the process of perception includes aspects of sensation and, as a cognitive function, the attachment of meaning to what is sensed.

The nature of the intervening cognitive functions varies with the situation, and may include storing the information in memory and making judgments and decisions based on it.

A basic understanding of the properties of human memory is one consideration important to the practitioner interested in optimizing the human-machine system. Although as many as four memory storage systems have been proposed (Ervin and Anders, 1970), for most practical applications it is convenient to think of two basic systems: short-term memory and long-term memory. The two systems are usually differentiated on the basis of capacity and duration. Short-term memory is a limited capacity storage system that loses information rapidly in the absence of sustained attention and rehearsal. The limited capacity of short-term memory is commonly called *memory span*, or the number of items that can be recalled immediately without error. Rehearsal and decay are determiners of the time in short-term memory, but so is interference from incoming information. Individuals who quickly store and retrieve transitory information on the job in order to control equipment or make operational decisions rely heavily on short-term memory.

Long-term memory has a large capacity and stored information is relatively permanent. Information can usually be retrieved from long-term memory by an active cognitive search process. Often the search is easy, as when we immediately recall an item of information. At other times it can be difficult, as when an item is "on the tip of the tongue." Speed and accuracy of recall can be vastly improved if the individual rehearses the information and if it is originally presented in meaningful "chunks."

Even though evidence indicates that information in short-term memory is lost very rapidly, short-term recall can be improved by proper task and equipment design.

Some suggestions include:

1. Improve the discriminability of key items on a display or panel that must be monitored sequentially. A long search time will result in more items being lost from short-term memory.
2. Reduce the number of inputs that must be recalled.
3. Reduce the interval between the presentation of the display or information and its required recall.
4. Increase the meaningfulness or familiarity of the information or display.

For a decision to be executed by the human in the system, some action must be taken. Psychologists have devoted substantial effort developing a clear conception of the structure of human abilities that

TABLE 3.1. Absolute Judgments of Various Stimulus Dimensions for Visual and Auditory Sense Modalities

Sense Modality	Stimulus Dimension(s)	No. of Levels That Can Be Discriminated on Absolute Basis
Vision (single dimension)	Luminance	5
	Dominant wavelength	9
	Line length	7-8
	Area	6
	Dot position in space	10
	Direction of line inclination	7-11
Vision (multiple dimensions)	Size, brightness, and hue (varied together)	18
Audition (single dimension)	Intensity	5
	Pitch	7
Audition (multiple dimensions)	Loudness and pitch	9

Source: Adapted from Van Cott and Kinkade (1972).

underlie skilled performance on different tasks, but with only limited success. Fleishman (1966) has identified as many as 11 factors that account for a major part of the variance in human motor performance: control precision, multilimb coordinations, response orientation, reaction time, speed of arm movement, rate of control, manual dexterity, finger dexterity, arm-hand steadiness, wrist/finger speed, and aiming. These factors have been found to be relatively independent of each other (Lundy and Trumbo, 1980); in other words, individuals possessing a high level of ability on one factor may not possess a similar level of ability on a second factor. It has also been discovered that as an individual practices and becomes more proficient at a given task, he or she is more likely to use skills and abilities that are highly task specific.

The emphasis on the identification of human performance abilities may be valuable in developing human-machine systems. If the demands that will be placed on the human can be determined by analysis of operator tasks, as described in Section 3.3.2, then it is possible to develop training programs to prepare workers to meet these demands. A second approach, more useful in the early stages of workplace design, is to identify the characteristics and abilities of the least skilled performers who would be likely to use the facilities or equipment (perhaps the 5th percentile on the behavior or characteristics of interest). Using this strategy, the workplace designer insures that the facilities and tasks can be effectively and safely operated by these individuals. One can apply the same general reasoning to other kinds of motor activities and performance criteria, such as strength, endurance, and task accuracy requirements. A discussion of design strategies that take into account individual variability in body sizes is provided later in this section.

Two job-related conditions that have a significant impact on human capabilities to receive, process, and act upon information are *fatigue* and *boredom*. The concept of fatigue stems from the study of muscular performance. If a muscle is exercised continuously for a long time, it gradually loses its ability to contract. After a period of rest, the strength of the muscle generally returns. Tremor, tension, and pain often accompany this state. It has also been suggested that an analogous condition of fatigue, mental fatigue, results from extended use of central nervous system functions. Perceptual changes, slowing of performance, irregularity of timing, disorganization of performance, and, paradoxically, temporary improvement of performance all have been attributed to mental fatigue (Welford, 1976). Modifying work rest periods, changing the nature of work activities and the working environment are potential means of preventing or reducing fatigue.

Boredom is a psychological state that is difficult to quantify. Although an employee may be physically active, he or she may still be bored, because of the repetitive nature of the work, the lack of challenge in the job, the lack of sensory stimulation, or a combination of these factors. Extreme conditions of boredom have been generated in the laboratory in studies of sensory deprivation and prolonged vigilance. Boredom, like fatigue, may degrade performance and increase its variability. Again, with attention to the design of the job and the workplace, conditions that typically lead to boredom may be eliminated or greatly reduced.

In evaluating a new item of equipment or in deciding whether to automate a human task, human limitations as well as the system requirements and environment should be considered. *Functional allocation* is the process of deciding the relative emphasis to place on the human vs. the machine in selecting components to fit particular task requirements. A number of authors beginning with Fitts (1951) have attempted to develop criteria to assist the analyst in specifying the allocation of functions. These criteria typically compare the capabilities of humans with those of machines and tend to be somewhat general. For example, on matters of power, speed, and consistency there is no match—the machine is far superior, and other things being equal, should be selected for functions that depend on these characteristics. The relative advantage of the machine is also apparent for routine, repetitive, or very precise operations, or for storing and retrieving large amounts of information in short periods. The human, on the other hand, is extremely adaptable and able to react to unexpected, low probability events that the machine may fail to detect. People are far superior to machines in perceiving patterns and making generalizations about them, although this may be changed through advances in computer systems.

If there is clear-cut functional superiority on the part of either the machine or human, then the selection between the two is clear, assuming factors such as cost, design feasibility, and development time are within acceptable limits. In actual practice, however, the use of comparison lists is seldom directly applied. The proviso "other things being equal" seems rarely realized. The design solution is almost always based on a resolution of other constraints given by the problem; tradeoffs occur between speed and safety, power and weight, operability and ease of maintenance, comfort and costs. A human may be more proficient at performing fine manipulations of a control, but if this operation must be performed in a hostile or dangerous environment, then the situation might very well dictate the use of a machine to perform the function.

ANTHROPOMETRY

The human imposes still another challenge for the workplace designer or safety practitioner, that is, the structural and functional variability among people. Reliable anthropometric data (data concerning the measurement of physical features and functions of the body) can help deal with this challenge.

No attempt will be made here to reproduce the voluminous anthropometric data that have been generated by numerous investigators. Instead, a sampling of data will be presented, along with general principles for its application. The data are also heavily weighted towards military populations, largely due to the completeness of this database and its accessibility to designers. The designer must keep in mind, however, that the groups measured should be representative of the equipment users. Additional sources of data can be found in human factors handbooks (Van Cott and Kinkade, 1972; Woodson, 1981) and anthropometric references (Damon, 1966).

Anthropometric data for design and evaluation are generally expressed in percentiles. As a minimum, the equipment should attempt to accommodate from the 5th to the 95th percentile of the user population; it is usually impossible to accommodate the extremes. Special clothing needs must also be taken into account; a number of anthropometric tables are based upon body dimensions without clothing.

Structural body dimensions have been tabulated for subjects in fixed, standardized positions. Tables 3.2 and 3.3 represent body dimensions of standing and seated military personnel respectively (see the corresponding Figures 3.2 and 3.3). Data are in centimeters and inches for both male and female populations. The specific reference should be consulted for explanations of methods of data collection and more detailed descriptions of populations measured.

Attempts have been made to provide practitioners with body dimensions taken from dynamic body positions. Table 3.4 and Figure 3.4 are sample anthropometrics applicable to specially restricted workspaces. The 5th and 95th percentile values for military personnel are given in both centimeters and inches. Although these values are for fully clothed subjects, suitable allowances should be made for heavy clothing or special protective clothing.

The reader should note that additional data are available on muscle strength and muscle power, weight-lifting capacity, etc. (Grandjean, 1981; Olishifaki, 1979, Chap. 13; and Woodson, 1981). These data are highly situation-specific, that is, it varies even for the same person according to body position.

In applying anthropometric data, McCormick (1976) offers three general strategies: provide for an adjustable range, design for extreme individuals, or design for the average. The first principle suggests that equipment in the workplace be made adjustable to account for the body sizes of a large segment of the user population (usually 90%). At first glance, this approach is the most compelling. If the worker is aware that the equipment is adjustable and is given the opportunity to make the adjustment, then this strategy will best serve the individual's needs. The adjustment of chair height or illumination level on CRT screens are examples of the application of this principle. Unfortunately, it is not always technically or economically feasible to build this feature into workplace design. It is, however, an option more designers are considering.

The second principle, design for extreme individuals, is applied in situations where some limiting factor necessitates accommodating individuals at one extreme or the other of some anthropometric characteristic. For example, the maximum height for controls on panels is usually dictated by the arm reach of the 5th percentile female. In this case 95% of the population (including the largest male and female users) would be able to reach the control without a step stool or other aids. On the other hand, the minimum dimension for a piece of equipment should usually accommodate the user at the upper

TABLE 3.2. Standing Body Dimensions

	5th Percentile				95th Percentile			
	Ground Troops		Females		Ground Troops		Females	
Standing Body Dimensions	Cm	In.	Cm	In.	Cm	In.	Cm	In.
1. Stature	162.8	64.1	152.4	60.0	185.6	73.1	174.1	68.5
2. Eye height (standing)	151.1	59.5	140.9	55.5	173.3	68.2	162.2	63.9
3. Shoulder (acromale) height	133.8	52.6	123.0	48.4	154.2	60.7	143.7	56.6
4. Chest (nipple) height[a]	117.9	46.4	109.3	43.0	136.5	53.7	127.8	50.3
5. Elbow (radial) height	101.0	39.8	94.9	37.4	117.8	46.4	110.7	43.6
6. Waist height	96.6	38.0	93.1	36.6	115.2	45.3	110.3	43.4
7. Crotch height	76.3	30.0	68.1	26.8	91.8	36.1	83.9	33.0
8. Gluteal furrow height	73.3	28.8	66.4	26.2	87.7	34.5	81.0	31.9
9. Kneecap height	47.5	18.7	43.8	17.2	58.6	23.1	52.5	20.7
10. Calf height	31.1	12.2	29.0	11.4	40.6	16.0	36.8	14.4
11. Functional reach	72.6	28.6	64.0	25.2	90.9	35.8	80.4	31.7
12. Functional reach extended	84.2	33.2	73.5	28.9	101.2	39.8	92.7	36.5

Source: Adapted from DOD, 1981.
[a]Bustpoint Height for Women

TABLE 3.3. Seated Body Dimensions

	5th Percentile				95th Percentile			
	Ground Troops		Females		Ground Troops		Females	
Seated Body Dimensions	Cm	In.	Cm	In.	Cm	In.	Cm	In.
13. Vertical arm reach sitting	128.6	50.6	117.4	46.2	147.8	58.2	139.4	54.9
14. Sitting height erect	83.5	32.9	79.0	31.1	96.9	38.2	90.9	35.8
15. Sitting height relaxed	81.5	32.1	77.5	30.5	94.8	37.3	89.7	35.3
16. Eye height sitting erect	72.0	28.3	67.7	26.6	84.6	33.3	79.1	31.2
17. Eye height sitting relaxed	70.0	27.6	96.2	26.1	82.5	32.5	77.9	30.7
18. Midshoulder height	56.6	22.3	53.7	21.2	67.7	26.7	62.5	24.6
19. Shoulder height sitting	54.2	21.3	48.9	19.6	65.4	25.7	60.3	23.7
20. Shoulder-elbow length	33.3	13.1	30.6	12.1	40.2	15.8	36.6	14.4
21. Elbow-grip length	31.7	12.5	29.6	11.6	36.3	15.1	35.4	14.0
22. Elbow-fingertip length	43.8	17.3	40.0	15.7	52.0	20.5	47.5	18.7
23. Elbow rest height	17.5	6.9	16.1	6.4	26.0	11.0	26.9	10.6
24. Thigh clearance height	–	–	10.4	4.1	–	–	17.5	6.9
25. Knee height sitting	49.7	19.6	46.9	18.5	60.2	23.7	55.5	21.8
26. Popliteal height	39.7	15.6	36.0	15.0	50.0	19.7	45.7	18.0
27. Buttock-knee length	54.9	21.6	53.1	20.9	65.8	25.9	63.2	24.9
28. Buttock-popliteal length	45.5	17.9	43.4	17.1	54.5	21.5	52.6	20.7
29. Functional leg length	110.6	43.5	96.6	38.2	127.7	50.3	118.6	46.7

Source: Adapted from DOD, 1981.

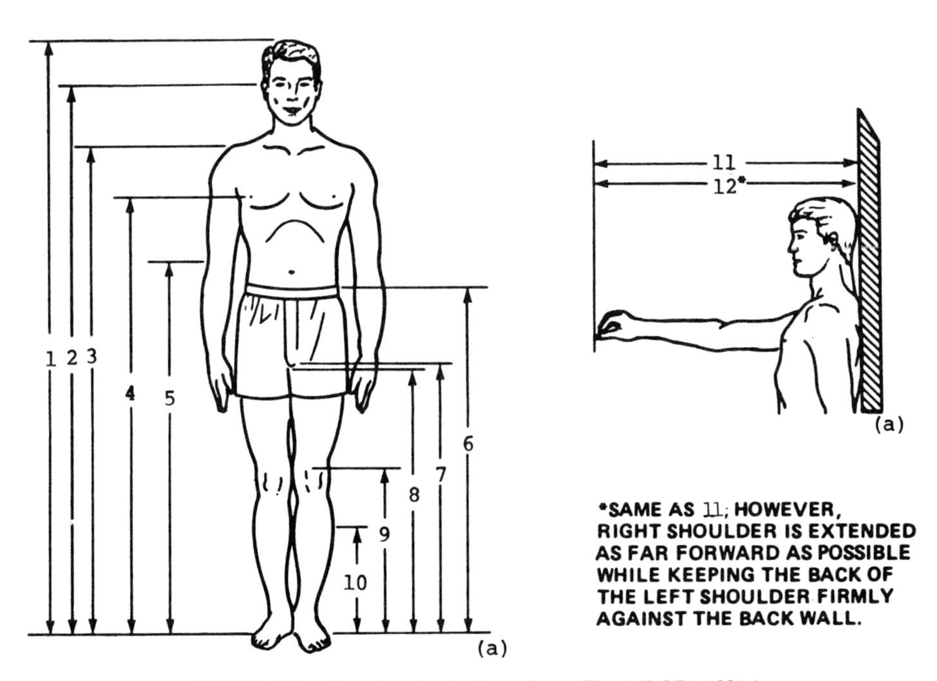

FIGURE 3.2. Standing body dimensions. (From DOD, 1981.)

95th percentile on the dimension of interest. Clearance on an aisle or door is often designed with this in mind.

The last approach, design for the average user, is applied when it is not feasible to design using the first or second principles. In this case, the appropriate dimension is the measure of that characteristic defined by the 50th percentile user. One application of this strategy has been in defining work bench or desk heights.

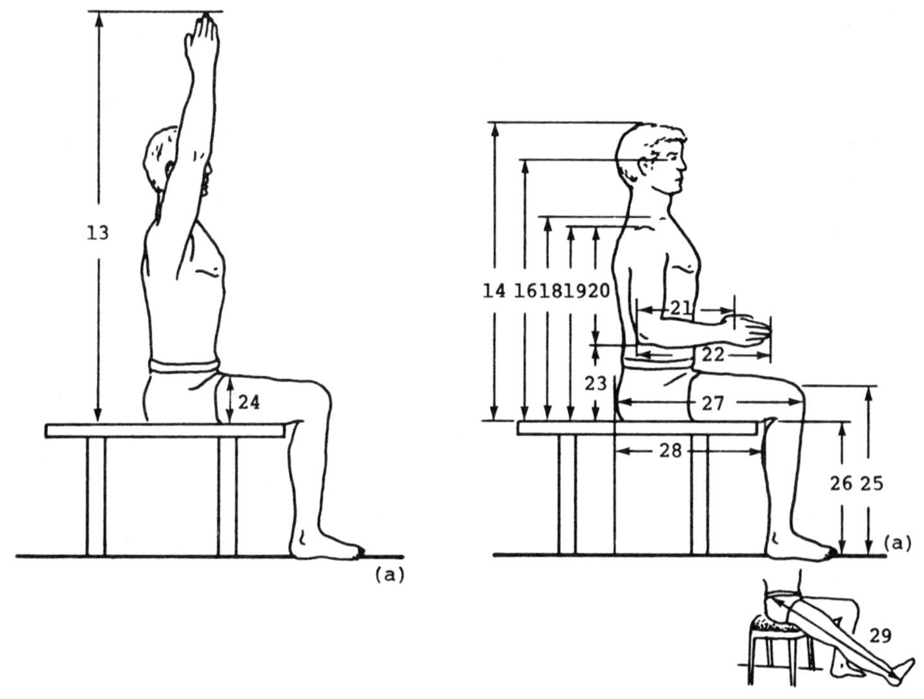

FIGURE 3.3. Seated body dimensions. (From DOD, 1981.)

TABLE 3.4. Anthropometric Data for Working Positions[a]

Position	5th Percentile Men Cm	5th Percentile Men In.	5th Percentile Women Cm	5th Percentile Women In.	95th Percentile Men Cm	95th Percentile Men In.	95th Percentile Women Cm	95th Percentile Women In.
1. Bent torso breadth	40.9	16.1	36.8	14.5	48.3	19.0	43.5	17.1
2. Bent torso height	125.6	49.4	112.7	44.4	149.8	59.0	138.6	54.6
3. Kneeling leg length	63.9	25.2	50.2	23.3	75.5	29.7	70.5	27.8
4. Kneeling leg height	121.9	48.0	114.5	45.1	136.9	53.9	130.3	51.3
5. Horizontal length, knees bent	150.8	59.4	140.3	55.2	173.0	66.1	163.8	64.5
6. Bent knee height, supine	44.7	17.6	41.3	16.3	53.5	21.1	49.6	19.5

[a]See Figure 3.4 for illustration.

THE ROLE OF PAST EXPERIENCE

The fact that humans bring a number of unique experiences and expectancies into the workplace has important implications for establishing a safe and productive work environment. In most cases our past experiences are beneficial to job performance. The practitioner can insure that this is more likely to occur if population stereotypes and transfer effects are considered in facility designs or workplace enhancements.

A population stereotype is a learned tendency among members of a given population to respond in a predictable fashion to a class of stimuli. This response tendency usually develops over an individual's lifetime. Flipping a light switch up instead of down to turn a light on is a simple illustration. Table 3.5 lists population stereotypes for motion and color. A work environment that forces an employee to respond in a manner that contradicts this learned tendency is likely to lead to performance problems. When stereotypes are violated, reaction time is slower, the action may be perceived as more difficult and frustrating, and there is a greater tendency for error. Even if a different response is learned on the job, there is a tendency among employees to revert back to stereotypical behavior patterns at times of stress.

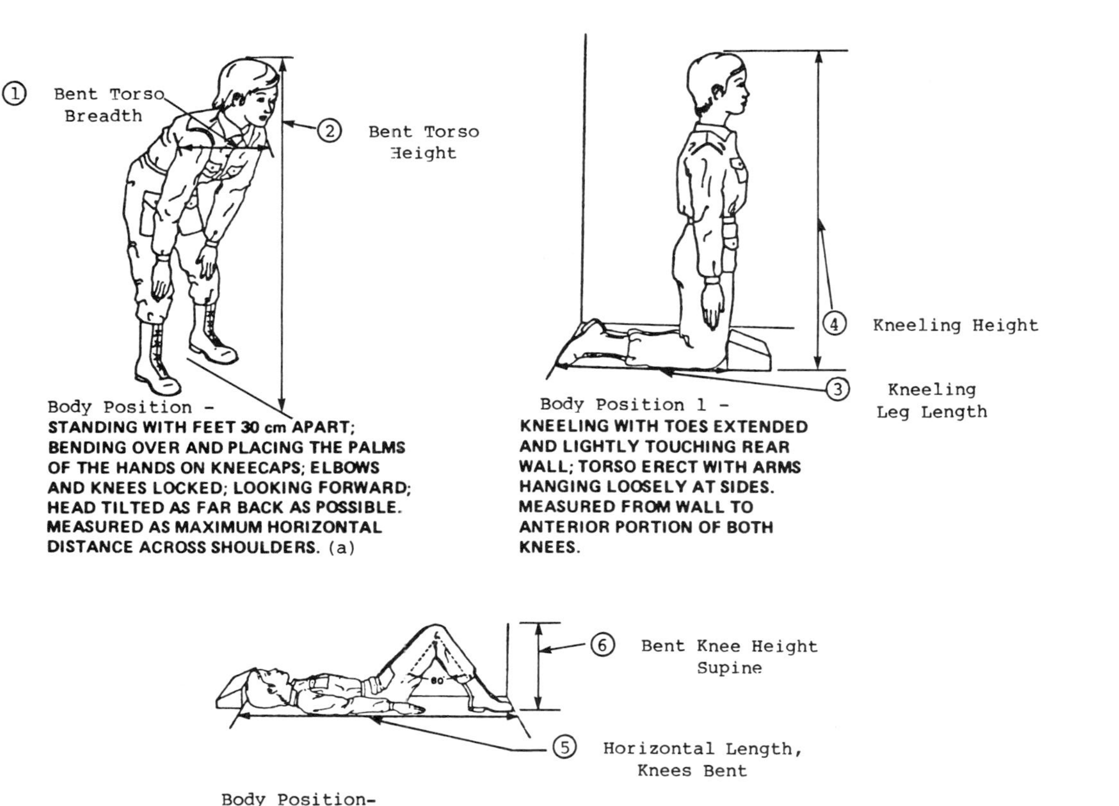

FIGURE 3.4. Anthropometric data for working positions. (From DOD, 1981.)

TABLE 3.5. Population Stereotypes for Motion and Color

Class	Example	Stereotype
Motion[a]	On, start, run, Open	Up, right forward, clockwise pull
	Off, stop, close	Down, left backward, counterclockwise
	Right	Clockwise, right
	Left	Counterclockwise, left
	Raise	Up
	Lower	Down
	Increase	Forward, up, right, clockwise
	Decrease	Backward, down, left, counterclockwise
Color[b]	Red	Stop, hot, danger
	Yellow	Caution, near, safe
	Green	Go, safe, on
	Blue	Cold, off, far

Source:
[a]From U.S. NRC, 1981.
[b]Adapted from Bergum, 1981.

The redesign of displays and controls that do not violate strong populational stereotypes should result in substantial improvements in human reliability (Swain and Guttmann, 1980).

An additional factor to consider is negative transfer. *Negative transfer* refers to the detrimental impact of previous knowledge and experience on learning new information.

Much of the evidence from psychological research, collected in association with verbal and motor skills performance, concludes that negative transfer results when new responses are coupled to old stimuli. For example, moving old equipment to new locations may cause difficulty when operators attempt new movement patterns using this equipment. The strongest negative transfer occurs when a particular response is reversed. For example, if the automatic transmission in your first car were arranged so that the shift lever must be pulled all the way *down* for the car to be placed in reverse, and your new car has just the opposite arrangement (requiring you to push the lever *up* to go into reverse) you have the greatest potential for negative transfer. As with well-learned stereotypical behavior, negative transfer is most likely to occur under conditions of stress. One obvious solution to the problem of negative transfer is, wherever possible, to standardize designs, with consideration also being given to population stereotype behavior.

The literature also suggests that the degree of similarity between the old and new equipment configurations will influence the extent of negative transfer. Interestingly, where new responses may potentially lead to negative transfer, the designer may be more successful if he or she also radically changes the stimulus conditions (Ellis, 1972). Unfortunately, there is very little evidence regarding the influence of negative transfer in real-world industrial environments. In all likelihood, well-planned changes in equipment controls and displays will minimize such effects and performance will gradually improve with retraining (Sawyer et al., 1982).

3.2.2 Principles of Workspace Design

This section will provide some basic ergonomic principles for enhancing the design and safe operation of work facilities and equipment. The material is organized first at the component level, then the workstation level, and finally at the workspace level.

Visual Displays

Prior to evaluating a particular display, the information requirements of the task(s) to be performed must be known. (Task analysis is discussed in Section 3.3.2.) Questions to be asked about information requirements usually include the following:

1. What parameter or information must be displayed?
2. Is the information changing?
3. How rapidly and accurately must this information be received and processed by the human?
4. What is the range of the parameter or extent of the information to be diplayed?
5. Must historical or trend information be provided?

For existing human-machine systems, this data is generally known.

Displays can be categorized as either dynamic or static. The emphasis of this section will be on dynamic displays that provide information about the changing state of some equipment or system.

Table 3.6 is a summary of the types of dynamic displays that are indicated by various information requirements. As can be seen, where precise quantitative information is required, a counter is usually appropriate. Meters are often used to provide quantitative information, but precision may be sacrificed and they become less applicable for extremely large ranges. Although mechanical counters are available, the trend is towards greater use of LEDs (light-emitting diodes) to present numerical information. As a matter of fact, the designer should be aware of the potential to overuse LEDs in situations where other display formats would be more appropriate.

Meters are categorized by shape and whether they possess a moving or a fixed scale. Unless an extremely large range must be presented, moving-pointer fixed-scale meters are to be preferred over fixed-pointer moving-scale indicators. Meter designs are typically round, horizontal, or vertical or some variant of these shapes (e.g. semicircular or open window). In addition to providing an estimate of the value of the variable, meters can also serve as the basis for qualitative reading in certain applications. For example, they may be used for determining the status of the parameter in terms of a limited number of predetermined ranges, or for observing trends or rates of change. Qualitative reading can be improved by slicing the entire continuum of values into a limited number of ranges and coding them according to some general scheme. Figure 3.5 provides an example of the use of this meter banding approach.

Status lights are the preferred display where only information about changes in discrete states of the variable without quantitative information is needed. A "power on" indicator light and a warning indicator light (usually also flashing) are examples. Equipment or system status should be inferred by illuminated indicators, *never* by the absence of illumination. The color coding of indicator lights should also conform to population stereotypes.

Representational displays are useful when operators or users need a working model of the process or system. They provide a good overall assessment of system functioning and how components interact. Maps and mimic displays are examples. These displays, most suitable for large, complex control systems, enable the user to locate information quickly and to identify more readily problems in system functioning. One of their drawbacks, however is that system revisions are difficult to backfit into these displays.

The last visual display category considered is computer generated cathode ray tube (CRT) displays. It is apparent that the CRT is rapidly replacing many of the conventional displays in human-machine systems. It is very flexible, and, if properly programmed, a computer can display to a CRT or printer a vast amount of information in short time using a variety of formats.

Once a display mode has been selected, the designer will want to optimize the manner in which the information is presented. Some major concerns to address in this regard include display usability, readability, and meaningfulness.

Usability of displays is enhanced when scale units are consistent with the degree of precision needed by the user and the range of the parameter spans the expected operational range of values. Preferred linear scales are those with 1, 2, or 5 unit progressions of scale values or decimal multiples of these values (McCormick, 1976). Extraneous information should be avoided in a display and information should be

TABLE 3.6. Application of Various Types of Dynamic Visual Displays

Information Requirement	Type	Examples
Need for precise readings over a wide dynamic range	Quantitative	Counter, Meter
Need to establish the approximate value, direction, or rate of change of the parameter	Qualitative	Meter, Recorder
Need to represent discrete independent categories	Status	Indicator lights
Need to provide a realistic representation of the entire system and how components interact	Representational	Mimic diagram, Map
Need to present large quantities of information in a variety of formats; need to track targets	Alphanumeric or Symbolic	CRT displays

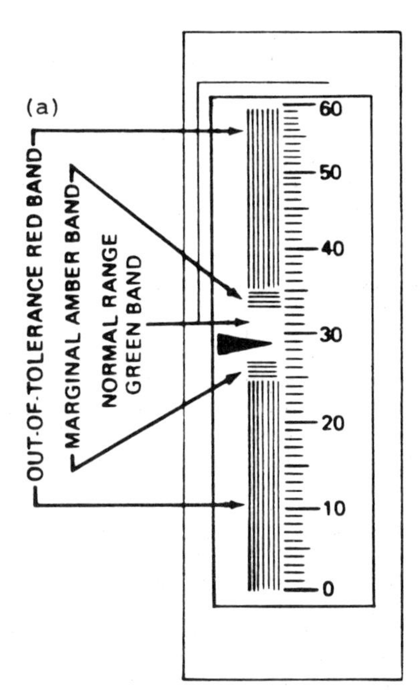

FIGURE 3.5. Example of meter banding. (From U.S. NRC, 1981.)

presented so that the need to apply conversion factors or perform mental calculations is eliminated. The direction and placement of scale values should follow population stereotypes; for example, values on a horizontal scale should increase from left to right.

The second factor is *readibility*. Readability can be improved by focusing on the size, style, and contrast of characters. The recommended character height is dependent upon a number of factors including illumination, contrast, distance from observer, exposure time, and importance. The character style should be simple and consistently applied. Except for lengthy prose, all characters should be uppercase letters. The recommended width-to-height ratio for capital letters falls between 1 : 1 and 3 : 5, while the standard recommendation for numerals is about 3 : 5 (McCormick, 1976). Finally, under normal levels of illumination, black letters on a white background afford the best contrast and legibility.

It is apparent that the *meaningfulness* of the messages displayed can also have a significant effect upon user performance. The symbols and terms displayed should be familiar to the user and applied consistently.

Controls

The second human interface with equipment in a human–machine environment is the control system. Various types of controls enable the operator to govern a process or to select information. The selection of the type of control (e.g., lever or toggle switch) must also be appropriate for the task. This selection undoubtedly will vary as a function of the degree of force required, accuracy, whether continuous adjustments need to be made, the range required, and other considerations. Table 3.7 summarizes the applicable common control types given specific task requirements.

Detailed discussions of the sizes recommended and the forces required to operate specific types of controls are available in numerous references (e.g. DOD, 1981; and Woodson, 1981).

However, a number of principles are applicable to most types of controls. These are listed below:

1. Controls should be of the type normally anticipated by the operator. In addition, required control movements should conform to the population stereotypes discussed previously.
2. The force, speed, accuracy, and range requirements of controls should fall within the limits of the least capable operator. Control sensitivity should not exceed the need.
3. Critical controls should be designed so as to prevent accidental activation. For example, make use of shields and guards, and optimize placement.
4. Activating controls should result in a positive indication (click, detent, etc.). Operators should also receive a positive indication that there has been a system response.

TABLE 3.7. Application of Common Control Types

Control Requirement	Recommended Control
1. Momentary activation	Pushbutton/Legend pushbutton
2. Selection of 2 discrete positions where space limitations are severe	Toggle switch, slide switches or rocker switches
3. On-off control that requires more force than can be applied by a finger-operated control	Push-pull control
4. Selection of 2 or more discrete functions	Rotary selector switch
5. Prevention of unauthorized operation	Key-operated switch
6. Secondary control, when hands are likely to be occupied or where control operation requires great force	Foot pedal or foot switch
7. Compact digital control input device and readout of these inputs	Thumbwheel
8. Precise continuous adjustments	Knob
9. Continuous adjustments where accuracy and speed requirements are not great and space is a premium	Concentrically mounted knob
10. Application of large amounts of force/displacement or multidimensional movements required	Lever
11. Provision for control of position	Ball control
12. Numerous rotations of a control	Crank
13. Provision of high breakout or rotational force	Handwheels
14. Entering of alphanumeric data	Keyboard

5. Controls should be resistant to slippage (e.g. serrated or knurled edges).
6. Controls should be compatible with any emergency gear or clothing that is necessary and should also be able to withstand occasional abuse.
7. Controls should be readily identifiable.

The last point will be considered in greater depth, since in some operational environments correct and rapid identification is a matter of life and death.

Basically, the identification of controls is a coding problem. The main coding methods available are location, shape, size, mode of operation, labeling, and color. The use of a coding mode for a particular application is governed by the relative advantages and disadvantages of each type of code. Table 3.8 summarizes these factors. Note that where coding is used to differentiate among controls, the application must be applied consistently throughout the human-machine environment. Some examples of the use of controls coding include:

1. To identify controls belonging to the same functional class (e.g., all pump speed controls).
2. To identify controls belonging to the same subsystems (e.g., all controls in the electrical system).
3. To identify frequently used controls.
4. To identify controls used in emergency situations or controls with particular functional importance.
5. To highlight controls that are used together (e.g., reset switch and corresponding control).
6. To associate controls with corresponding displays.

Warnings

In industrial environments, warnings are necessary to alert individuals to impending danger, to alert an operator to a critical change in system or equipment status, and to remind individuals that a critical action must be taken. Warnings can be transmitted visually or auditorily, or some combination of the two.

As discussed earlier, auditory alarms have a distinct advantage because they do not require the intended hearer to be facing the signal source. Common audio alarms include the horn, whistle, siren, buzzer, chime, and oscillator. The specific signal device chosen depends on the environment to which it will be applied (see Woodson, 1981). Whichever signal is selected, it should serve two functions—attention getting and problem identification. Recommended characteristics of audio warning signals are listed in Table 3.9. Although verbal warning signals also can be applied, these will not be treated here.

TABLE 3.8. Advantages and Disadvantages of Various Types of Coding

	Location	Shape	Size	Mode of Operation	Labeling	Color
Advantages						
Improves visual identification	**X**	**X**	**X**		X	X
Improves nonvisual identification (tactual and kinesthetic)	X	X	X	X		
Helps standardization	X	X	X	X	X	X
Aids identification under low levels of illumination and colored lighting	X	X	X	X	(When transilluminated)	
May aid in identifying control position (settings)		X		X	X	
Requires little (if any) training: is not subject to forgetting					X	
Disadvantages						
May require extra space	X	X	X	X	X	
Affects manipulation of the control (ease of use)	X	X	X	X		
Limited in number of available coding categories	X	X	X	X		X
May be less effective if operator wears gloves		X	X	X		X
Controls must be viewed (i.e., must be within visual areas and with adequate illumination present)					X	X

Source: DOD, 1981.

TABLE 3.9. Characteristics of Audio Warning Signals

Characteristics	Recommendation
Frequency	Use frequencies between 200 and 5000 Hz and, if possible, between 500 and 3000 Hz. When signals must travel over 300 m, use frequencies below 1000 Hz. Frequencies below 500 Hz should be used when signals must bend around obstacles or pass through partitions (DOD, 1981).
Intensity	A signal-to-noise ratio of at least 20 db should be provided in at least one octave band between 200 and 5000 Hz at the operating position of the intended receiver. Signals should not be so intense as to cause discomfort or ringing in the ears as an after effect (DOD, 1981).
Discriminability	Use signals with frequencies different from those that dominate background noise. A modulated signal (1 to 8 beeps per second or warbling sounds varying from 1 to 3 times per second) can be used (McCormick, 1976).

In applications where color is used to code emergency conditions or states that approach emergencies, the colors red, yellow or amber, and green have the following connotations:

1. *Red:* for unsafe, danger, immediate action required, or an indication that a critical parameter is out of tolerance.
2. *Amber* (Yellow): for hazardous condition (potentially unsafe), caution, attention required, or an indication that a marginal value or parameter exists.
3. *Green:* for safe condition, no operator action required, or an indication that a parameter is within tolerance.

A flashing red light should be reserved for situations of extreme danger. In this case, the recommended flash rate is three to five flashes per second with approximately equal on and off times (Woodson, 1981).

Warning labels may also be used in hazardous environments. They should be conspicuously mounted adjacent to any equipment that presents a hazard to personnel, for example, high voltage, toxic vapors, or heat. Their design should follow effective application of ergonomic labeling criteria (see Table 3.12 in Section 3.3.1).

Considerations in Workstation Design

Depending on task requirements, workstation designs fall into 3 major categories, sitdown, standup, or sit-stand configurations. Guidelines based on anthropometric data are available for designing cabinets, consoles, desks, and other workstations (see Woodson, 1981).

The layout of workstations involves tradeoffs among a number of factors. Decisions must be made concerning placement of the component controls and displays to provide for safe and productive performance. To the extent possible, the following functional principles should be considered.

1. *Grouping by task sequence:* Controls and displays should be assigned to workstations so as to minimize operator movement.
2. *Grouping by system function:* Within the constraints of grouping by task sequence, controls and displays should be assigned in functionally organized groups.
3. *Grouping by importance and frequency of use:* Within the constraints of grouping by task sequence and system function, controls and displays should be assigned to workstation position depending on their importance and frequency of use.

Workstation layout strategies should be applied consistently and should make the most effective use of human sensory and manipulation capabilities.

The layout of equipment should also be based upon user expectations. Expectations will generally be met when components have a left-to-right or top-to-bottom arrangement. Such equipment should be identified in alphabetic or numeric sequences (e.g. components A, B, C, D, or 1, 2, 3, 4).

Finally, the designer needs to consider the proper integration of related controls and displays. Controls and displays that are used together should be close together. The preferred arrangement is to have a display located above its corresponding control, a location that will not obscure the display information when the control is being manipulated. If control and display equipment must be separated, then the relative position and ordering of display components should be equivalent to their corresponding control components.

Workspace Design Principles

The overall workspace can be improved by proper attention to ergonomic principles. Areas of concern at this level include how best to integrate several work areas and how to insure that ambient environmental conditions fall within acceptable ranges. Refer to subsequent sections in this chapter for methods to survey and analyze workspace conditions.

The optimal location of equipment in the workspace is best evaluated in dynamic settings in which workers are observed interacting with each other and with equipment. However, a number of ergonomic principles relating to workspace design generalize across a range of industrial applications. General principles for workplace area designs should:

1. Allow for sufficient maneuvering space between work areas.
2. Eliminate the accidental activation of equipment.
3. Provide adequate access space and visual clearance to important display.
4. Provide adequate space for maintenance activities.

Thermal comfort, noise, and lighting are among the most important environmental factors to assess in occupational settings.

Assuming radiant temperature on the job is not a factor, then thermal comfort is influenced primarily by air temperature, air humidity, and the rate of air movement. Within facilities where light work is being performed, provisions should be made to maintain an effective temperature of approximately 65°F (18°C) and ranging between 60-68°F (15.6-20°C). Active jobs or those requiring exceptionally heavy labor may necessitate the use of a temperature range of 55-60°F (12.8-15.6°C) while relatively sedentary workers usually prefer temperature in the range of 67-73°F (19.4-22.8°C) (Shackel, 1984).

Relative humidity has little effect on thermal comfort at normal temperatures. It should, however, decrease with rising temperatures, and should remain above 15% to prevent drying and irritation of body tissues.

If possible, air should move past individuals at approximately 65 feet (20 meters) per minute, but not more than 100 feet (30 meters) per minute. Ventilation or other protective measures should keep vapors, dust, fumes, and gasses within specified limits.

Control of noise is a critical factor in industrial environments, because repeated exposure to noise or very high noise levels can damage hearing. Noise can also have an effect on job performance. In attempting to establish acceptable limits on noise, the question of criteria for acceptability becomes important. The Occupational Safety and Health Administration (OSHA) has established a set of permissible noise exposure limits for individuals working in industry that depend on the duration of exposure (Federal Register, 1971). Additional standards such as those based on interference with communications are also available and should be consulted in specific applications.

In specifying lighting requirements, most guidelines are expressed in terms of illumination rather than luminance. *Illumination* is a measure of the intensity of light falling on a surface, whereas *luminance* is a measure of the brightness of a surface and is a function of the amount of reflected light.

For good performance and visual comfort, work environments should provide a suitable level of illumination, little glare, a fairly uniform distribution of light, and freedom from flickering light. Suitable illumination depends on the requirements of the task. Table 3.10 provides illumination requirements for a number of common tasks and work areas. It should be noted that CRT workstations require special attention with regard to lighting. See Stammerjohn et al. (1981) for a discussion of these issues.

3.3 ERGONOMIC METHODS

Formal research methods were used in laboratory and field settings to gather and analyze the data that formed the basis for the principles discussed earlier in this chapter. Less formal methods of data collection and analysis are presented in this section, because the safety practitioner may have limited background in experimental design and statistics or limited resources on the job for conducting research. The practitioner interested in a more formal discussion of behavioral science research is referred to Kerlinger (1973).

This section describes four phases used to identify and correct ergonomic deficiencies:

1. Data collection
2. Ergonomic analysis
3. Corrective and preventive actions
4. Evaluation of changes and development of conventions

These phases are involved in the application of the ergonomic principles discussed earlier. As an example, if noise is considered a potential ergonomic problem in the workplace, data collection is taking measurements with a sound level meter; ergonomic analysis is comparison of readings to appropriate standards; corrective and preventive actions might be application of noise-reduction techniques such as installation of sound-dampening ceiling panels; and evaluation of changes could be done by taking new measurements.

The example above assumes that a problem, noise, had already been identified. In many instances, a safety practitioner may not know the causes of a problem. In these cases, the four phrases may also be applied, although more detailed investigations will be required during data collection and more effort will be required during analysis, in the attempt to locate or isolate ergonomic deficiencies. Multiple deficiencies may have to be prioritized so that the practitioner can focus his or her attention on the most serious deficiencies first.

Activities in each of the four phases are described in the sections that follow. Emphasis throughout the sections is on relatively straightforward, inexpensive, and minimally time-consuming approaches for identification and correction of deficiencies.

3.3.1 Data Collection

Data collection may be necessary for two primary reasons. First, the nature of ergonomic deficiencies may not be readily apparent. Second, at a more specific level of detail, deficiencies may be apparent but their severity, location, or conditions of occurrence are not known. In this second case, diagnostic data collection should be planned. Whether for investigative or diagnostic purposes, several methods of data collection are suitable within the practical limitations of the typical workplace. The practitioner can collect data through surveys of:

1. Historical data
2. Workplace
3. Workers
4. Worker tasks

Depending on the type of problem suspected, the nature of the workplace, or the availability of resources, one of these data collection surveys may be more appropriate than the others.

TABLE 3.10. Sample Illumination Requirements

Task or Work Areas	Recommended lux (fc)	Minimum lux (fc)
Assembly, general		
Coarse	540 (50)	325 (30)
Medium	810 (75)	540 (50)
Fine	1075 (100)	810 (75)
Precise	3230 (300)	2155 (200)
Bench work		
Rough	540 (50)	325 (30)
Medium	810 (75)	540 (50)
Fine	1615 (150)	1075 (100)
Extra fine	3230 (300)	2155 (200)
Inspection tasks, general		
Rough	540 (50)	325 (30)
Medium	1075 (100)	540 (50)
Fine	2155 (200)	1075 (100)
Extra fine	3230 (300)	2155 (200)
Machine operation, automatic	540 (50)	325 (30)
Electrical equipment testing	540 (50)	325 (30)
Repair work		
General	540 (50)	325 (30)
Instrument	2155 (200)	1075 (100)
Testing		
Rough	540 (50)	325 (30)
Fine	1075 (100)	540 (50)
Extra fine	2155 (200)	1075 (100)
Office work, general	755 (70)	540 (50)
Reading		
Large print	325 (30)	110 (10)
Newspaper	540 (50)	325 (30)
Handwritten reports, pencil	755 (70)	540 (50)
Small type	755 (70)	540 (50)
Prolonged reading	755 (70)	540 (50)
Transcribing and tabulating	1075 (100)	540 (50)
Business machine operation	1075 (100)	540 (50)
Switchboards	540 (50)	325 (30)
Panels		
Front	540 (50)	325 (30)
Rear	325 (30)	110 (10)
Console surface	540 (50)	325 (30)
Service areas, general	215 (20)	110 (10)
Storage (general warehouse)	110 (10)	55 (5)
Stairways and hallways	215 (20)	110 (10)

Source: Adapted from DOD, 1981.

Survey of Historical Data

Accident report forms are one source of data. Since report forms may not contain the type or level of detail of information necessary for analysis, additional sources of historical data may be needed.

As follow-up to an accident reporting form, the safety practitioner can review the event to determine whether factors that have the potential to contribute to worker performance problems were involved. Interviews with witnesses, supervisors, or workers may be necessary to gather additional information about potential ergonomic deficiencies. A form such as that shown in Table 3.11 may be useful to the practitioner in gathering background information. The entries in Table 3.11 are self-explanatory, except for "Error Categorization." The type of error may be categorized as (1) omission (omits a step) or (2) commission (performs a step incorrectly); more specific types of commission error are (a) extraneous, (b) sequential, or (c) time. However, for the purposes of the sample form, Table 3.11, error categorization can serve as one additional data point in trying to determine what went wrong. Familiarity with the ergonomic principles and guidelines described in this chapter will assist the safety practitioner in identifying ergonomic deficiencies that may have contributed to the accident.

Interviewing may be necessary as a follow-up to the accident report form for identifying potential ergonomic deficiencies. Interview guidelines are presented in the section on Survey of Workers.

An additional approach, known as the *critical incident technique*, focuses on the survey of "near accidents." Using this approach, users are questioned as to whether they have ever made an error or mistake using the equipment. Responses then can be analyzed to establish what conditions might have led to an accident. This technique can easily be integrated with an overall survey of the workers.

Survey of Workplace

A second source of data is obtained through a survey of the workplace. Ergonomic principles relevant to the specific workplace and/or the problem can be listed in a checklist format, which serves as a reminder to the safety practitioner of the variables to consider when he or she inspects the workplace. It also provides documentation for later reference. A sample checklist (adapted from U.S. NRC, 1981) is provided in Table 3.12. In actual application, space should be provided next to each checklist item for an indication of compliance ("Yes" or "No") and comments. Considerations in developing and applying a checklist are discussed in this section.

The source of checklist items can be the principles described and referenced earlier in this chapter, or another ergonomic reference if additional information is needed (e.g., DOD, 1981; Hammer, 1976; U.S. NRC, 1981; Van Cott and Kinkade, 1972).

Items in the workplace checklist may cover:

1. Displays
2. Controls
3. Warnings and labels
4. Workstation design
5. Workspace design, layout, communications, environmental characteristics

Items for an ergonomics checklist can be drawn from these general areas as applicable to the specific industrial setting. Items selected should be objective and based on observable or measurable attributes of the workplace.

Some checklist items can be completed by an observer who is minimally knowledgeable of the workplace. No special qualifications are necessary in order to count the increments between scale markings on a display. However, many items will require a thorough knowledge of the workplace and its operation. For example, to determine whether a scale range on a display is adequate, an observer must know the range of operating conditions represented by the scaled variable. For this reason, it is usually recom-

TABLE 3.11. Sample Ergonomic Deficiency Summary Report for Use In Historical Data Survey

Report Number: Report Date: Occurrence Date:
Error Categorization: Workstation:
Equipment or Instruments Involved:
Major Systems Involved:
Identification of Occurrence:
Summarize Events Preceding Occurrence:
Summarize Events During Occurrence:
Summarize Consequences of Occurrence:
Identification of Probable Cause:
Corrective Action Taken/Proposed:
Additional Recommendations:

TABLE 3.12. Sample Ergonomic Checklist Items

(1) Displays

1.1 The display has:
   A. The capability to distinguish among significant levels of the information.
   B. The required level of precision, without excess precision.
   C. Feedback for any deliberate movement of a control.

1.2 The display:
   A. Conforms to direction-of-movement conventions (Table 3.5, Section 3.2.2).
   B. Indicates the rate and limits of movement.

1.3 The display is easy to:
   A. Locate.
   B. Read without parallax from the operating position.

1.4 The display provides indication of display failure (e.g., off-scale indication).

1.5 Display response time lag:
   A. Displays reflect in real time the time lag, if any, between actuation of the control and the change in system conditions.
   B. There is no time lag between a system condition change and the display indication.
   C. If there is a lag time between control actuation and ultimate system state, there is an immediate indication of the process and direction of the parameter change.

1.6 Display color coding conforms to the ergonomic conventions, including number of colors, type of colors, conformance to population stereotypes, and contrast considerations.

1.7 Information printed on the display face:
   A. Provides only a concise identification of the parameter displayed, the units shown, and any transformation required (i.e., no extraneous or unnecessary information).
   B. Conforms to a standard abbreviations list.

1.8 Scale readings:
   A. Provide the degree of precision and accuracy needed.
   B. Avoid conversions.
   C. Use percentage indication only when the parameter is meaningfully reflected by percentage.
   D. Provide for the display of scale values that span the expected range of operational parameters.
   E. When multiplied or divided by a power of 10, are clearly marked on the display.
   F. Minimize the display of normal random variations in display performance.
   G. Use compatible scales for displays of the same parameter.
   H. Avoid multiscale indicators.

1.9 Scale markings:
   A. No more than 9 scale graduations separate numerals.
   B. Major and minor graduations are used if there are up to 4 graduations between numerals.
   C. Major, intermediate, and minor graduations are used if there are 5 or more graduations between numerals.
   D. Successive values are indicated by graduations in units (e.g., 1, 2, 3, 4, 5) or those values multiplied by some power of 10.
   E. Scale markings on linear and circular scales are oriented with numerals presented vertically.
   F. Where pointer movement is more than 360°, the zero point is located at the 12 o'clock position.
   G. Where positive and negative values are displayed around a zero or null position, the zero or null point is located at the 12 o'clock position.
   H. Where the scale covers less than a full rotation of the pointer, scale endpoints are indicated by a break in the scale at least one numbered interval in length, oriented at the 6 o'clock position.

1.10 Zones, such as "Operating Range" and "Danger Range" are marked and color coded.

1.11 Pointers:
   A. Use simple pointer tips.
   B. Minimize concealment of scale graduation marks or numerals.
   C. Extend to within about 1/16 in. of, but do not overlap, the smallest graduation marks on the scale.
   D. Are mounted to avoid parallax errors.
   E. Pointer/background contrast and pointer size are adequate to permit rapid recognition of pointer position.
   F. Moving-scale, fixed-pointer meters are avoided.

1.12 Bulb-test capability is provided for light indicators with safety implications.

TABLE 3.12. Sample Ergonomic Checklist Items (cont.)

1.13 Bulb replacement is rapid and convenient with power on and without hazard to personnel or equipment.
1.14 Reflections from light sources do not cause light indicators to appear to be glowing when they are off, or vice versa.
1.15 System/equipment status is inferred from illuminated indicators, not from the absence of illumination.
1.16 Legend light indicators:
  A. Light indicator intensity is at least 10% greater than surrounding panel, as measured by a spot photometer.
  B. Legends on light indicators are legible under ambient illumination with indicator light off.
  C. Symbolic legends are unambiguous.
  D. Legend text is limited to no more than 3 lines.
  E. Legends are worded to tell the status indicated by the glowing light.
  F. Legend light indicators are distinguished from legend pushbuttons.
1.17 Graphic recorders:
  A. Graphic recorders are used to record trend information and material that may be needed for later reference.
  B. Pen, ink, and paper are of a quality to provide a clear, distinct, and reliable marking.
1.18 Counters are used when there is a need for quick, precise reading of a quantitative value and trend information is not needed.
  A. Numerals are read horizontally.
  B. Rate of change on electronic counters is less than 2 per second. Numbers on drum counters change by snap action.
  C. Character-to-background contrast ratio is 15:1 minimum and 20:1 preferred, as measured by a spot photometer.
1.19 Computer displays meet ergonomic guidelines for hardware and software format, content, and command language. (e.g., IBM, 1984; U.S. NRC, 1981).

(2) Controls

2.1 Controls provide:
  A. Ease of adjustment.
  B. Sufficient range of control.
  C. Required level of precision, without excess precision.
  D. Operability in sufficient time, under expected dynamic conditions, and within the limits of manual dexterity, coordination, and reaction time.
  E. Sufficient durability to retain their appearance, "feel," and functional characteristics during their service life.
  F. Compatibility with emergency gear.
2.2 Accidental activation of a control is prevented by:
  A. Control is located where the operator will not strike or move it accidentally.
  B. Control is contained within a recess or barrier.
  C. Movable cover or guard does not interfere with the operation of the control.
  D. Control is interlocking.
  E. Control is resistant to movement.
  F. When a strict sequential activation is necessary, locks are provided to prevent the control from passing through a position.
  G. Rotary action controls are used when linear or pushbutton controls would be subject to inadvertent activation and fixed protective structures cannot be used.
  H. Safety or lock wires are avoided.
2.3 Controls are separated to:
  A. Allow access to adjacent controls.
  B. Prevent inadvertent actuation of adjacent controls.
  C. Allow simultaneous actuation where necessary.
2.4 Visibility:
  A. Control is easy to locate.
  B. Control setting is easy to read, without parallax, from the operating position.
2.5 Indication is provided of activation for a control (e.g., snap feel; audible click; integral light; pointer, if rotary control).
2.6 Direction of movement of a control follows movement conventions in Section 3.2.1 and the rate and limits of movement of the control are indicated.
2.7 Emergency controls are distinctive.
2.8 Color coding and shape coding guidelines (Section 3.2.2) are followed.

TABLE 3.12. Sample Ergonomic Checklist Items (cont.)

2.9 Specifications for specific controls should be consulted if applicable, e.g., for pushbuttons, legend pushbuttons, key-operated controls, continuous adjustment rotary controls, rotary selector controls, thumbwheels, slide switches, toggle switches, rocker switches, or other (e.g., U.S. NRC, 1981).

(3) Warnings and labels

3.1 A hierarchical labeling scheme is used to reduce confusion, operator search time, and redundancy (e.g., major labels are used to identify major systems or operator workstations, and component labels are used to identify each discrete panel or console element).

3.2 Associated controls and displays used in the same sequence of operations are labeled with the same alphabetic or numeric sequence.

3.3 Label control positions:
A. All discrete functional control positions are identified.
B. Direction of motion (increase, decrease) for continuous motion rotary controls is identified.

3.4 Access openings are labeled to identify the function of items accessible through it.

3.5 All danger, warning, and safety instruction labels are in accordance with appropriate safety standards.

3.6 Labels are positioned consistently, close to the components they describe, and located so they can be seen during operation.

3.7 Spatial orientation:
A. Labels are oriented horizontally so that they may be read quickly and easily from left to right.
B. Vertical orientation is used only where space is limited.
C. Curved patterns of labeling are avoided.

3.8 Labels provide the function of equipment items and engineering characteristics or nomenclature, if needed for clarity.

3.9 Wording on labels expresses exactly what action is intended.

3.10 A list of standard names, acronyms, and part/system numbers is in place, administratively controlled, and used consistently in procedures and within and across pieces of equipment.

3.11 Abstract symbols are used only if they have a commonly accepted meaning for all intended users (e.g., %).

3.12 Wording on labels is concise without compromising meaning.

3.13 When labels containing similar words, abbreviations, or acronyms are located close to each other, coding or selection of different words is used to reduce confusion.

3.14 A temporary label is used when necessary to identify out-of-service equipment, to accommodate unique, one-time plant activities, or to improve operator understanding and efficiency.
A. Temporary labels conform to good ergonomic principles and are administratively controlled.
B. Tag-outs, if required, do not obscure the label associated with the nonoperable device or any adjacent devices or labels.

3.15 Mimics are used to integrate system components into functionally oriented diagrams that reflect component relationships.

3.16 Mimics meet ergonomic guidelines for color coding, symbols, readibility, and consistency.

3.17 Label lettering character height is 0.004″ × viewing distance.

3.18 Character dimensions:
A. Letter width-to-height ratio is between 1:1 and 3:5.
B. Numeral width-to-height ratio is 3:5, except for the numeral, "4", which is one stroke width wider, and the numeral "1", which is one stroke in width.
C. Stroke width-to-character height ratio is between 1:6 and 1:8.

3.19 Type style is upper-case only.

3.20 Spacing is at least:
A. One stroke width between characters.
B. One character width between words.
C. One-half character height between lines.
D. One-half character height for top and bottom borders.
E. One-half character width for side borders.

3.21 Under normal illumination levels, black lettering on a white background is used to provide maximum readability.

(4) Workstation Design

4.1 Controls and displays are grouped according to:
A. Task sequence.
B. System function.
C. Importance and frequency of use.

TABLE 3.12. Sample Ergonomic Checklist Items (cont.)

4.2 The following methods are used to assist in recognition and identification of controls and displays:
   A. Spacing between groups of components.
   B. Demarcation of groups of controls and displays with a contrasting line.
   C. Color shading consistent with the color coding conventions.

4.3 Controls and displays are arranged:
   A. In a logical order.
   B. In sequence, such as left-to-right or top-to-bottom, if appropriate.

4.4 A consistent layout is used for identical or similar control or display sets at all locations; mirror imaging is avoided.

4.5 Strings of similar components:
   A. Components are grouped in horizontal rows no more than 20 in. long.
   B. Strings of more than 5 components are broken up into smaller strings by spacing, demarcation, or labeling.

4.6 Placement on vertical panels:
   A. Controls are placed between 34 in. and 70 in. above the floor.
   B. Controls requiring precise or frequent operation and emergency controls are placed between 34 in. and 53 in. above the floor.
   C. Displays are placed between 41 in. and 70 in. above the floor.
   D. Displays that must be read frequently or precisely are placed between 50 in. and 65 in. above the floor.

4.7 The association of controls and displays is established by:
   A. Location.
   B. Labeling.
   C. Coding.
   D. Demarcation.
   E. Consistency with operator expectations.

4.8 A symmetrical layout is used for associated controls and displays which are used in the same sequence.

4.9 Tools are ergonomically designed (e.g., Olishifski, 1979).

(5) Workspace Design

5.1 Furniture and equipment layout is appropriate for circulation patterns.

5.2 Workstation design, clearances, and envelopes meet anthropometric guidelines as described in Section 3.2.2 (see also Van Cott and Kinkade, 1972).
   A. The distance from the back of any desk to an opposing surface, is 36 in. or greater.
   B. Lateral space for a seated worker at a desk is at least 30 in.
   C. The distance from a row of equipment or panel to an opposing surface or wall is at least 50 in.
   D. At least 8 ft is provided between opposing rows of equipment where more than one person must work at the same time.
   E. Maximum lateral spread of controls and displays used by a single, standing worker should be 72 in. or less.

5.3 Operator protective equipment meets ergonomic design guidelines, is checked periodically, is available in sufficient quantity, and is accessible.

5.4 Temperature and humidity are maintained within an acceptable comfort zone.

5.5 Temperature at head and floor level do not differ by more than 10°F.

5.6 Ventilation quantity and velocity are maintained within an acceptable level (see Section 3.2.2).

5.7 Illumination levels are appropriate for the task, e.g., 75 footcandles for a seated operator workstation (see Section 3.2.2).

5.8 Luminance ratios are appropriate, e.g., task area vs. adjacent darker surroundings provides a contrast ratio of 3:1.

5.9 Glare is minimized.

5.10 Emergency lighting, when needed, provides at least 10 footcandles at primary workstations.

5.11 Background noise level is kept to a minimum (see Van Cott & Kinkade, 1972).

5.12 Communication equipment is provided to support tasks and meets ergonomic guidelines.

5.13 Emergency messages are given priority transmission.

5.14 Communication is possible with emergency gear.

mended that a user of the workstation serve as a resource person for the individual completing the checklist. The worker should have experience in operation of the system or equipment in the workplace.

A number of workers should be questioned about the workplace to provide cross-validation of information among users. Usually, the more individuals that provide information for a particular item, the more likely the safety practitioner is to collect valid data.

Some checklist items can be completed by talking to the worker. Other items may be best addressed by observation in the workplace, so the worker can point out what he or she is referring to. For items that require information about operating conditions or conditions of use, some level of simulation of actual events may be necessary.

Simulation involves walking through actual events with some degree of realism. A simulation can range from an enactment of system operation to a computer-driven replica of the workplace that mimics actual operation.

Sometimes it is not practical to evaluate every aspect of the workplace and its operation or use with the checklist. Therefore some strategy must be used for choosing representative items for evaluation. For example, one might limit simulation discussions to a sample of emergency events or tasks that have led to accidents, or one might measure noise levels at sample times throughout the shift. The logic behind the sampling strategy should be based on the purpose and priorities of the evaluation. This is a matter of judgment that must be properly handled by the practitioner.

Some checklist items can be completed on a workplace-wide basis, such as layout items. Other items will require evaluation on a workstation-by-workstation basis and perhaps later for consistency across workstations. Still others, such as characteristics of specific types of equipment, can be evaluated once. If the equipment is duplicated it need not be evaluated again. For example, the dimensions of a switch can be taken once and when the same switch is found at another workstation, its dimensions need not be taken again. Of course, such aspects as location of switches in proximity to each other may change from workstation to workstation and these unique aspects may have to be evaluated on a case-by-case basis.

Evaluation of some checklist items may require special equipment. A sound level meter or recorder is necessary for measuring noise, and a light meter to measure illuminance. A spot photometer is used to measure luminance, or glare, while a graphic recorder may be used to measure temperature and humidity. Such equipment may be purchased, or rented if necessary. Tape measure, ruler, and protractor are also needed for certain workstation measurements.

Once a checklist is developed or compiled and such methods as talk-throughs or walk-throughs are chosen for completing it, the checklist is ready to be used. Observers should be careful to document discrepancies from checklist items. Since time spent in the workplace talking with users may be limited, observers must be careful to document as much information as possible without unnecessarily taking up users' time.

Once the checklist is completed, the observer should be encouraged to write up his or her findings as soon as possible. Often notes taken hurriedly in the workplace make more sense shortly thereafter than a few weeks later. Findings should be written up for further analysis.

The write-up should include a thorough description of the ergonomic deficiency: the ergonomic principle or guideline involved, the amount by which the finding does not meet the guideline, a description of the type and number of pieces of equipment or components involved and their location, and any information on past accidents or the potential for error and implications for safety attributed to the deficiency. The write-up may be needed for justification to management of the need for improvement.

Survey of Workers

A third source of information about workplace ergonomics is the workers themselves. Users or operators of a workstation can provide valuable experiences and insights about the workplace.

Since opinions from workers are likely to be varied, a representative sample should be used, that is, workers with a range of ability or experience. They should be aware that while it may not be possible to address or implement some of their specific complaints or suggestions, their opinions are important and will be considered during analysis of ergonomic deficiencies.

Of course it must be realized that workers do not always know what is best, either for themselves or for the use or operation of a workstation. For example, one worker may think that flexibility in a computer input program would make the job easier but may not realize that this flexibility might increase the potential for input errors.

With these cautions in mind, two types of worker survey methods will be described, the questionnaire and the interview. These methods share some considerations in common, such as the development of questions, and also have some unique aspects, such as the format of presentation.

Questionnaires are often chosen to gather worker opinions because they can be relatively inexpensive to administer, workers can complete them at their convenience, worker responses can be kept anonymous relatively easily and convincingly, and documentation of opinions is provided. Interviews are often more time consuming to administer, or at least less flexible in administration, and workers may not feel comfortable criticizing the workplace.

Conversely, questionnaires may not yield thorough information, since the worker may not want to write out details and may not ask for clarification about questions. Also, the administrator of the questionnaire may not be readily available to follow up on a response from the worker to ask for elaboration or examples. In an interview, the interviewer can clarify questions or ask the individual for more detail.

As a rule of thumb, the more structured the responses required, the more suitable a questionnaire is. When questions are exploratory and expected responses are open-ended, an interview is probably more suitable.

Guidance on writing questions for either questionnaires or interviews is provided below.

1. Write questions that will provide useful answers. For example, if a yes or no answer is not sufficient, be sure to ask for details.
2. Choose the appropriate type of question, structured (response choices provided) or open-ended.
3. Pretest the questions on a sample of workers to make certain their content is clear and unambiguous.
4. Present one question at a time to avoid complexity and ambiguity.
5. Avoid questions that are leading, threatening, or can be influenced by social desirability. Leading questions prompt a particular response. Threatening questions put the respondent on guard. Socially desirable questions yield an acceptable answer which may or may not be true.
6. Make certain potential respondents have the necessary background to understand the questions.

Attention to these considerations in developing questions will increase the likelihood of the generation of reliable and valid information. Questions yield reliable data when they evoke similar responses from the same individual at different times, or when the same question elicits similar types of answers from different individuals. The actual responses may change to reflect the reality of a changing workplace, but the meaning of the question stays the same.

Questions yield valid data when the answers are appropriate to the problem being investigated. Validity can be established by comparing information from different sources. For example, information from workers and their supervisors can be compared to determine similarities and differences in their perspectives. Then an objective evaluation of the information can be made.

Survey of Worker Tasks

A fourth source of information is a survey of tasks that are performed in the workplace. Three approaches can be taken to gather this type of information: participation, expert judgment, and observation. Many of the considerations involved in surveys of the workplace and the worker are also applicable to surveys of worker tasks. The unique approach in the survey of worker tasks is the orientation toward what is done, rather than what the workplace is like or what the workers think about the workplace.

For some tasks, participation by the safety practitioner may be possible. Knowledge from firsthand experience can provide valuable information about ergonomic aspects of the workplace. When the safety practitioner can assume the role of the user, he or she can get to know the realities of the workplace and may gain insights into options for change. Since many tasks require specialized training, this approach may not be available.

"Expert" judgment, that is, judgment of users or operators of a workstation, can also be used to collect information about tasks. When experts are consulted, a structured and systematic format for collecting data should be used, and biases should be avoided. A more formal approach to gathering expert judgment than that presented here can be found in Comer et al. (1984).

A sample format for gathering task descriptions from experts is provided in Table 3.13. The column headings should be chosen to reflect needs for task information specific to the workplace. For example, if the workstation under evaluation is a computer CRT display, the heading "Information and Control Resources" would be appropriate. If the workplace is a shop floor, "Equipment and Tools Needed" would be appropriate. Likewise, the level of detail of the task description should reflect the level of the evaluation. Sometimes overall tasks can be described. Other times, tasks must be broken down into step-by-step actions (see Meister, 1971 for a more detailed practical discussion). The purpose of developing a format is to provide consistent, useful information about tasks from experts.

Procedures, training materials, or other documentation of tasks often are helpful in filling out a task description. However, care must be taken not to carry biased or incorrect information from a source into the analysis. When possible, arrange for an objective review of the task description to make certain essential tasks are presented appropriately. The subject matter expert can perform this review.

TABLE 3.13. Sample Task Description Format Column Headings

Reference to Procedure	Action To Be Taken	Needed Resources	Available Resources	Location	Potential Ergonomic Deficiencies

The third potential source of data for the task survey is observation of worker performance in the workplace. A task description format such as the one described previously (shown in Table 3.13) can also be used to structure observation. The observer must be aware that his or her presence may affect performance. The less obtrusive the observer the less likely the observer's presence is to affect task performance.

3.3.2 Ergonomic Analysis

Ergonomic analysis is relatively straightforward for data collected through surveys of historical data, the workplace, and the worker. The survey of worker tasks requires a task analysis to identify ergonomic deficiencies. Deficiencies identified based on the data collection surveys and subsequent analysis can be prioritized to provide an order in which they should be addressed.

Task Analysis

Data collected from surveys of historical data, the workplace, or workers will yield results that are in a form suitable for prioritization (as described below). The data must be compiled, but relatively little additional analytical work will be required. In a survey of worker tasks, however, task-descriptive data will need to be analyzed before useful results are gained.

Task analysis involves an evaluation of task-descriptive data to analyze whether task requirements are met. If the purpose of the task analysis is to identify ergonomic shortcomings in the workplace, the analyst will review the task-descriptive data to answer such questions as those in Table 3.14. The process will lead to identification of potential ergonomic problems.

More formal analyses are also available to assist in addressing some of these questions. For instance, link analysis (Chapanis, 1959) focuses on worker movement patterns and is valuable in analyzing workspace layout (question 2, Table 3.14).

Prioritization of Findings

Unless an ergonomic evaluation is focused on only one problem or a simple set of problems, the results will need to be prioritized.

Findings can be prioritized on the basis of two factors (DOD, 1977):

1. Criticality (e.g., degree of danger)
2. Probability of occurrence (e.g., of an accident)

Judgments of criticality and probability of occurrence should be made based on the best available data for the system or workplace (e.g. historical data). Then ergonomic deficiencies with the potential for high criticality events and high probability of occurrence can be addressed first. Deficiencies judged able to produce high criticality low probability events might be the next priority, followed by low criticality high probability deficiencies. Deficiencies judged to be low both in criticality and probability can be analyzed on a case-by-case basis to determine if they interact with other deficiencies, and thereby warrant attention, or if they are of such minimal consequence that action may not be justified.

Once priorities have been established, alternatives for corrective actions can be developed. If some lower priority findings are relatively easily corrected, these may be addressed earlier.

Caution must be exercised in the development of corrective actions. Changes to the workplace may

TABLE 3.14. Sample Questions To Be Addressed During Task Analysis

1. Are equipment and other resources available to support task performance?
2. Does workstation arrangement minimize interference among workers?
3. Are controls or displays that are frequently or precisely operated or read located conveniently?
4. Are workers given adequate warning and time to respond to warnings before a serious problem develops?
5. Does each control provide sufficient but not overly precise range of control? Does each display provide a sufficient but not overly precise indication?
6. Do workers have all the information (displayed, written, or verbally communicated) they need to perform the task successfully?
7. Are danger, warning, and safety instruction labels provided?
8. Does the computer system respond promptly to the user or provide a response delay message?
9. Are equipment and workstation resources organized by task sequence, function, importance, or frequency of use?
10. Is the time lag between worker action and workstation or system response minimal?
11. Are tasks omitted or committed that are not covered in training or procedures?
12. Do workers in a group interact successfully to accomplish the task?

create additional problems. For example, changing the color coding on a workstation to make it more consistent across workstations may result in negative transfer. Thus, the benefit of a change must be compared to the negative effects it might create.

3.3.3 Corrective and Preventive Actions

Enhancements to the workplace can provide short-term or long-term correction for ergonomic problems. Enhancements include:

1. Providing warning labels that are ergonomically designed.
2. Color-coding or shape-coding controls.
3. Adding operating band coding to displays (e.g., to show normal or dangerous ranges on a scale).
4. Adding demarcation lines around groups of associated controls and displays.
5. Installing protective covers over controls or moving mechanical parts.
6. Replacing or rearranging components.
7. Adding alarms.
8. Standardizing coding or content of labels.
9. Rearranging workstations or furniture.
10. Controlling noise level.
11. Adjusting illumination level.
12. Providing protective equipment.
13. Adding glare screens to computer displays.
14. Ensuring proper electrical grounding.

More significant changes to the workplace, such as design modifications or purchase of new equipment, may be necessary to backfit ergonomic considerations into a workplace. However, enhancements that are relatively inexpensive can produce significant error reduction.

Mock-ups of enhancements or other changes (drawings of the planned change or a model of the change) are recommended to evaluate the potential effectiveness of the improvement and to examine the practical aspects of the change.

One potential problem in making any change to the workplace is worker resistance. Discussing the reasons for change with the users or operators and, if possible, involving the workers in decision making tends to alleviate this problem.

Management resistance may also be encountered. Careful documentation and presentation of the priority of ergonomic deficiencies—their potential for accidents in terms of criticality and probability of occurrence—can be used to convince management of the need for attention.

3.3.4 Evaluation of Changes

Follow-up data collection using the same choice of methods discussed in Section 3.3.1 will provide an indication of the effectiveness of the change. Sufficient time should be allowed after implementation for workers to get accustomed to the change.

Documentation of ergonomic guidelines and conventions established for a specific workplace can be useful as a future reference. If a later change is planned, these guidelines can be consulted to ensure that the change meets ergonomic criteria and is consistent with approved ergonomic conventions established for the particular workplace.

ACKNOWLEDGMENTS

The authors wish to express appreciation to their employer, General Physics Corporation, and especially to Mr. John Beakes for his personal and corporate support and to Ms. Debbie Ruskus for her typing support. Dr. Pete Doyle provided helpful review comments on the chapter. Ms. Becky Pardoe helped condense the ergonomics checklist. The chapter represents the views of its authors.

REFERENCES

Bergum, B. O., Population Stereotypes: An Attempt to Measure and Define, *Proceedings of the Human Factors Society-25th Annual Meeting,* October, 1981, pp. 662-665.

Chapanis, A., *Research Techniques in Human Engineering,* John Hopkins Press, Baltimore, 1959.

Comer, M. K., D. A. Seaver, W. G. Stillwell, and C. D. Gaddy, *Generating Human Reliability Estimates Using Expert Judgment,* NUREG/CR-3688, U.S. Nuclear Regulatory Commission, Washington, D.C., 1984.

Damon, A., *The Human Body in Equipment,* Harvard University Press, Cambridge, MA, 1966.

DOD, *Military Standard: Human Engineering Design Criteria for Military Systems, Equipment and Facilities,* MIL-STD-1472C, Department of Defense, Washington, D.C., 1981.

DOD, *Military Standards: System Safety Program Requirements,* MIL-STD-882A, Department of Defense, Washington, D.C., 1977.

Dunnette, M. D., Ed., *Handbook of Industrial and Organizational Psychology,* John Wiley, New York, 1983.

Ellis, H. C., *Fundamentals of Human Learning and Cognitions,* W. C. Brown Co., Dubuque, IO, 1972.

Ervin, F. R. and T. R. Anders, Normal and Pathological Memory: Data and a Conceptual Scheme, in F. S. Schmidt, Ed., *The Neurosciences—Second Study Program,* Rockerfeller University Press, New York, 1970.

FEDERAL REGISTER, **36,**105, May 29, 1971.

Fetts, P. M., Ed., *Human Engineering for an Effective Air-Navigation and Traffic-Control System,* National Research Council, Washington, D.C., 1951.

Fleishman, E. A., *Human Abilities and the Acquisition of Skill,* in E. A. Bilodean, Ed., *Acquisition of Skill,* Academic Press, New York, 1966.

Hammer, W., *Occupational Safety Management and Engineering,* Prentice-Hall, Englewood Cliffs, NJ, 1976.

Hilgendorf, L., Information Input and Response Time, *Ergonomics,* **9,** 1, 1966.

IBM, *Human Factors of Workstations with Visual Displays,* International Business Machines Corporation, San Jose, CA, 1984.

Kerlinger, N., *Foundations of Behavioral Research,* Holt, Rinehart & Winston, New York, 1973.

Landy, F. J. and D. A. Trumbo, *The Psychology of Work Behavior,* Dorsey Press, Homewood, IL, 1980.

McCormick, E. J., *Human Factors in Engineering and Design,* McGraw-Hill, New York, 1976.

Meister, D., *Human Factors: Theory and Practice,* John Wiley, New York, 1971.

Miller, G. A., The Magic Number Seven, Plus or Minus Two. Some Limits on Our Capacity for Processing Information, *Psychological Review,* **63,** 1, 1956.

Olishifski, J. B., Ed., *Fundamentals of Industrial Hygiene,* National Safety Council, Chicago, 1979.

Sawyer, C. R., R. F. Pain, H. P. Van Cott, Control Room Modification and Transfer of Training, *Transaction of the 1982 Winter ANS Meeting,* **43,** 1982.

Shackel, B., Ed., *Applied Ergonomics Handbook,* Butterworth Scientific, London, 1984.

Stammerjohn, L. W., M. J. Smith, and B. G. F. Cohen, Evaluation of Work Station Design Factors in VDT Operations, *Human Factors,* **23,** 4, 1981.

Swain, A. D. and H. E. Guttmann, *Handbook of Human Reliability Analysis with Emphasis on Nuclear Power Plant Applications,* NUREG/CR-1278, U.S. Nuclear Regulatory Commission, Washington, D.C., 1983.

Taylor, F. V., Psychology and the Design of Machines, *American Psychologist,* **21,** 2, 1957.

U.S. NRC (Nuclear Regulatory Commission), *Guidelines for Control Room Design Reviews,* NUREG-0700, U.S. Nuclear Regulatory Commission, Washington, D.C., 1981.

U.S. NRC (Nuclear Regulatory Commission), *Guidelines for the Preparation of Emergency Operating Procedures,* NUREG-0899, U.S. Nuclear Regulatory Commission, Washington, D.C., 1982.

Van Cott, H. P. and R. G. Kinkade, Eds., *Human Engineering Guide to Equipment Design,* American Institutes for Research, Washington, D.C., 1972.

Welford, A. T., *Skilled Performance: Perceptual and Motor Skills,* Scott-Foresman and Co., Glenview, IL, 1976.

Woodson, W. E., *Human Factors Design Handbook,* McGraw-Hill, New York, 1981.

PART III

INSURANCE

CHAPTER 4

How to Provide Liability Insurance Coverage

DONALD M. WEEKES, JR.
TRC Environmental Consultants, Inc.
East Hartford, Connecticut

For the safety practitioner, insurance can be the great unknown. The concepts and theories that govern insurance are often so alien to the safety and health professional that even the jargon and terminology common to insurance are incomprehensible. Yet this strange system of protection of one's assets has a profound effect upon the job of any individual involved with safety and health.

This chapter will examine liability insurance and will explain as thoroughly as possible the relationship between this form of insurance and the responsibilities of the safety/health professional. This will be done by exploring three areas: exposures, coverages, and mechanics.

Each area will be divided into categories, which will be examined in light of the effect they may have on the job of the safety practitioner.

4.1 EXPOSURES

The safety practitioner will be familiar with *exposures.* It is the area in which the practitioner can have the most notable effect by reducing and controlling the amount of exposures that may result in a claim. In safety, examining exposures is comparable to safety surveying, that is, looking for potential accident hazards such as tripping or failing to use two-hand controls for potentially dangerous operations. In health, exposures are similiar to air sampling for contaminants around a potentially hazardous chemical operation.

By delineating the exposures, an underwriter is attempting to find all the possible ways that a loss could occur under the policy to be placed for the potential insured. The underwriter is using previous experience and knowledge of the potential insured's operations to determine if the risk is worth the premium to be generated.

Thus the first step in any successful underwriter's job is recognition of the exposures. How the safety practitioner can assist in the recognition process will be shown in each of the exposure categories.

Exposures can be divided into three major categories: employees, customers, and third parties.

4.1.1 Employees

The *employees* category is the one most familiar to safety and health professionals. In fact, many of the current safety and health efforts undertaken by industry, insurance companies, and various associations or groups were begun shortly after the introduction of Workers' Compensation insurance in the early 1900s. Many state legislatures required such insurance before a firm could operate within the state. It was in response to the costs of insurance that many companies instituted safety departments. Due to the high number of claims from Workers' Compensation, many insurance companies started safety departments to assist their clients in the proper safety and health techniques.

So insurance and safety and health have worked in concert for many years. However, each has developed its own method of dealing with the problem of employee injuries. For example, safety and health professionals examine losses in order to take measures to prevent future accidents. While the underwriters are in favor of that practice, they look at the losses from the viewpoint of a premium/loss ratio. The method of underwriting (use of deductibles, exclusions, etc.) needs to be adjusted to reduce the losses anticipated next year.

Both are looking at the same data but are seeing different problems, due to their differing functions. However, the work of the safety practitioner can be very useful to the underwriter in determining future trends in losses. Let us look at some employee exposure categories and illustrate this.

Bodily Injury by Accident

Data concerning the accidental losses at an insured's facility are often kept by the safety practitioner, who is responsible for keeping company records so that accidents can be reported to the federal or state occupational safety and health department. In addition, accident investigations are often conducted by the safety practitioner to determine the cause of the accident and to take appropriate action to reduce or eliminate the cause. Accident trends and statistics are also the bailiwick of the professional.

For the underwriter, these statistics and trends form the basis for an opinion of the risk. By reviewing these factors, the underwriter will be able to determine whether the account is an insurable risk, at what limits, and at what deductible. Finally, the accident records can have a large effect on premium.

For example, take the case of a 67-year-old facility with 5 floors. The safety director notes that 25% of the injuries sustained by employees occur on the staircases, which are located at either end of the building. Upon review of the accident reports, it is observed that 67% of the injuries are a result of trips and falls. The safety director investigates the staircases and discovers that the stairs are hard metal with treads that have worn out over the years. The surface is now very slippery and the edges worn so that a slip directs the foot outward and downward.

The underwriter for the Workers' Compensation Insurance of this firm notes that 25% of the claims and 30% of the money paid in claims last year were as a result of staircase falls. It is also noted that the facility is 67 years old and has 10 staircases. The potential for loss next year is still high.

However, if the safety director convinces the company that the staircases must be covered with a slip resistant surface, this could have a vast effect on the employee exposure. The key is to get information to the underwriter when the evaluation for next year's premium and policy is being made.

Occupational Disease

The industrial hygienist and occupational health practitioner (nurse or doctor) are mainly involved with this employee exposure. Unlike accidents, which tend to injure the employee immediately, these injuries may not develop until many years after the initial exposure. Both the occupational health practitioner and the underwriter must think about the past exposures at the facility that may result in present occupational disease. In addition, the occupational health practitioner must deal with the current exposures in order to prevent future disease.

Statistics on this exposure tend to be nebulous at best, due to the lack of verifiable connection between the disease and the cause. Occupational disease potential is evaluated much as a detective looks for clues to a murder.

The occupational health practitioner can examine the evidence biologically, in the results of tests on the employees, and atmospherically, by the results of air sampling throughout the plant. Measurable changes in these two critical areas may indicate that problems that need attention exist.

The underwriter is often not well versed on the technical aspects of occupational disease. Often, the underwriter will react to the toxicity of the materials at a facility and not necessarily the exposure of the employees. It is the responsibility of the occupational health specialist to interpret the sampling data for the underwriter and put it into understandable terms.

Vehicular Accidents

As it is a specialty among safety practitioners (i.e., traffic safety manager), so it is in insurance (commercial automobile liability underwriter). Statistically, this is a well-documented risk; careful record keeping is a must for all who are involved with the evaluation of vehicular accidents.

The basic units are similar to the bodily-injury-by-accident statistics, which is the number of accidents per million man-hours worked. For vehicular accidents, statistics are kept as the number of accidents per million man-hours on the road. Investigations of accidents tend to be the same also, in that the investigator is trying to determine, in both cases, the "how, who, when, where, and why" of the accident. The main difference is among vehicular accidents there are consistencies and trends that may not exist among employee work-related accidents.

The commercial auto underwriter looks at the statistics and the accident reports from the viewpoint of the potential for additional accidents. Often the actions taken by the company to correct problem areas and help prevent recurrences are most important. The safety practitioner can be most helpful by reporting what was done.

For example, one company's delivery trucks were involved in 24 vehicular accidents last year. Of these, 10 were found to be caused by failure of the brake system in short stopping circumstances. The traffic safety manager documented this and presented it to the vice-president of transportation, who recommended that the maintenance of the brakes be increased from once every 4 months to once every 3 months. The company does its own maintenance, and the maintenance crew was told to look particularly for worn brakeshoes, found in 8 of the 10 brake failures.

In addition, the drivers were asked to test their brakes every day and prior to every long trip. They were required to keep a maintenance sheet on their vehicle and submit it weekly. As a result of the increased maintenance, brake-failure accidents were reduced to 2 in the following year.

While the traffic safety manager did the important task of recognition, evaluation, and recommendation for this problem, it is also necessary that the job be documented. A well-written report on this maintenance change may result in a vastly different premium.

Off-Site Accidents

Off-site accidents are placed statistically into other categories, such as Workers' Compensation or Automobile Liability. However, often the cause of an off-site accident is quite different from the cause of one occurring on-site or involving a vehicle. For example, sales personnel make visits to facilities at which they may not be given appropriate safety equipment. All the safety precautions at the plant and extra care taken while driving will not help at the nonowned facility.

Also, many firms employ maintenance and service personnel who must also work at nonowned sites. This is particularly true with construction companies employing craft workers. Often, the individual worker may be completely alone while performing difficult and potentially dangerous tasks.

The underwriter looks at these types of companies as having a high potential for loss, so in general, such risks are avoided. However, the safety/health practitioner can mitigate the situation by providing a well-organized, well-written safety plan for these employees who work away from the owned facility. This may include recommendations for safety apparatus, personal protective equipment, procedures for emergencies, and so on.

In general, inclusion of such a written plan would help convince an underwriter that the company is taking action to reduce employee accidents off-site. If implementing the plan results in reduced incidents, the firm has been successful in two ways, (1) improved employee safety, and (2) reduced insurance costs.

Medical

The safety/health professional has an indirect effect upon the employee medical benefits plan. Employee illness, and some family illness as well, may be related directly back to exposures on the job. This is true not only for chemicals, physical agents, and work-related diseases, but also for biological contamination as the result of employee-to-employee contact.

The accident and health underwriter is looking for more than a good record of employee health. The underwriter is also looking for prevention plans that reduce the employee's potential for illness or accident, either on the job or off. A thorough and complete medical program should include CPR Training, employee medical testing, and "lifestyle modification" programs, which include dietary and exercise programs. The safety aspect of home life should also be emphasized by educational programs for fire safety in the home and first aid.

In general, the safety and health practicioner should assist the medical program by providing the expertise in loss prevention that is needed to reduce medical costs both at work and at home.

4.1.2 Customers

In general, interaction between the customers of a firm and the company's safety and health personnel is minimal. This is due primarily to the traditional role of the safety and health professionals, which was to evaluate the workplace and make recommendations to control hazards.

Today, however, many of the activities of the safety and health practitioner directly or indirectly affect the customer. This is due largely to the reaction of underwriters to large losses from customers. The underwriter requires stricter controls and better engineering to help prevent future losses of the same nature.

In quality control for products, in safety measures to protect passers-by at construction sites, and in protective devices for factory visitors, you can see the work of the professional affecting the customer. The goals of the underwriter and the safety and health professional are the same: reduced exposure for the customer and reduced liability for the company and its insurance carrier.

Product Liability

Product hazards mean bodily injury or property damage arising from the insured's products or reliance upon representation or warranty made at any time with respect to the product, provided that the loss occurs away from the insured's premises and after physical possession of the products has been relinquished to others. The insured's products means goods or products manufactured, sold, handled, or distributed by the insured or by others trading under the insured's name, including its container.

In recent years, product liability has received a great deal of publicity due to court rulings that have favored the plaintiffs. For the general liability underwriter, this has resulted in a very cautious attitude towards any risk with the possibility of loss. In general, the insurance industry has been attempting to get all firms with product exposure to undertake quality control measures to reduce the number of opportunities for loss. In addition, the Claims departments have been requesting that companies have preplanned responses to emergency situations such as product tampering or sabotage. Product recall systems that address the public information problem are also requested by insurance representatives of their larger clients with potential product-loss situations.

In these areas, the safety and health practitioner can be very helpful. Many can apply the principles of their fields to the product liability question, by recognizing the risk, evaluating the products and their manufacturing process, and recommending quality-control measures to reduce the risk. In plants, the occupational safety personnel often are also responsible for product safety. In both cases, the objective is the same: the reduction of probable loss by means of control methods, in particular, engineering controls.

The key item from an underwriting viewpoint is that the implementation and evaluation of the controls are documented in writing. This allows the underwriter to feel comfortable with the risk, because the written documentation can be utilized in a court case to indicate the quality control measures taken by the insured. This may not exonerate the company totally if there is a defective product, but it may help to reduce the amount of monetary damages, particularly punitive damages to the plaintiff. The written material may show that the insured took all reasonable precautions to insure that a defective product did not get on the market.

The safety and health professional may also be helpful in the planning and implementing product safety material and instructions for product use to be included in the packaging. Thorough and easy to understand safety guidelines may help the user avoid accidents and, again, the instructions may be helpful in a law suit. Failure to follow instructions may indicate negligence on the part of the user.

Finally, the safety and health professional can be of assistance in the design of an effective product recall program, particularly when the product may cause bodily harm if handled in a certain way. The technical terms and safety instructions that need to be disseminated should be written or, as a minimum, reviewed by the safety and health personnel to insure that they are technically correct.

Completed Operations

Completed operations are that portion of the operations performed by or on behalf of an insured under a contract at the site of the operations that have been completed. Additionally, it can be defined as that portion of the work that has been put to its intended use by any person or organization other than another contractor or subcontractor engaged in performing operations for a principal as part of the same project. Finally, operations that may require further service or maintenance work, repair, or replacement because of any defect or deficiency, but which are otherwise complete, shall also be deemed completed.

Although it has not received as much publicity as product liability, this also is a growing area of litigation, due primarily to the growth of the service industry, which has now overtaken the manufacturing industry as the primary employer in the United States. More and more employees are involved in providing a service (e.g., repair and maintenance of equipment) than in manufacture.

This type of operation presents problems both to the underwriter and the company's safety and health professionals due to the lack of supervision over the employees. Many of the losses as the result of completed operations occur because the employees do not exercise the care necessary to insure that the job is safe. For example, a contractor agrees to repair the ceiling and roof of an office building. The contract specifications require that a ventilation system be constructed through the roof in order to vent the office below. The new insulation effectively seals in most of the conditioned air as well as air including smoking contaminants, solvent fumes, carbon dioxide, and carbon monoxide. The vents were designed to open periodically to allow air change in the room at a determined rate, based upon the expected needs of the occupants. During the construction, the electrician and the heat/ventilation/air conditioning (HVAC) engineer, subcontractors of the insured, each thought that the other connected the wires from the thermostat to a particular vent. The general contractor did not check to make sure that it was connected. Since the construction took place in the spring, the vent was not required to perform to its fullest extent.

However, the first day that the temperature outside was over 90°C, the failure of the vent to perform caused considerable discomfort among the occupants of the room directly under the vent. The temperature inside, due to the body heat and lack of air circulation, was hotter than outside. Several people passed out and the rest suffered from headaches, fevers, sweating, and heat exhaustion. The employer sent them all home. It was three days before the HVAC engineer could return to the site and it was two weeks more before the missing wires were discovered.

The employer sued the general contractor to collect damages to his business and for the medical payments for his employees. The employer could probably collect, because the occurrence was due to the negligence of the subcontractor working for the general contractor and the failure of the general contractor to inspect the wiring before finishing the job.

The safety and health professional, in a traditional arrangement, would not be responsible for such failures. However, many professionals are now part of loss prevention, or risk management, departments. This risk and the potential for loss associated with it fall under the jurisdiction of such departments.

For the underwriter, the risk management of the completed operations hazards is important as an indication that the company is attempting to control losses. The underwriter looks for a written completed-operations-loss-prevention manual of all contractors accepted for coverage. In addition, the underwriter requires that all subcontractors submit certificates of insurance, prior to the job, showing that they have coverage if there is a problem.

In general, completed operations represents a new area for safety and health professionals, which has not been fully addressed yet. To go beyond the Workers' Compensation implications requires that the professional think as a risk manager. In that regard, the exposures to the customer at a construction site or during a service visit can be equally important as the exposures to the worker and, therefore, should be evaluated. The evaluation should be followed by careful and thorough recommendations to control the potential for damage to the customers' property and any bodily injury that may result.

On-Site Visitor Accidents

In many ways, these accidents are similar to employee accidents that occur on the premises. Many of the same hazards that exist for the company's employees will cause problems for the visitor. Unfortunately, personal protective equipment for visitors is often inadequate or lacking, and visitors have not been trained to handle the potentially hazardous conditions within the facility. This can result in serious bodily injury losses that can and should be easily prevented.

The key for the safety and health practitioner is involvement. The professional should be involved in setting up safety procedures for all visitors. For the underwriter, a written procedure is essential to an understanding and evaluation of the facility. Both are looking for written steps that will help to insure the safety of visitors while they are in the plant.

Although a visitor may not be exposed to the hazards of the operations to the same extent as the workers, it is always best to provide the visitor with the same protection as the workers. Although this may be overprotection, it does provide the visitor with a margin of safety to compensate for a possible lack of knowledge of the dangers associated with the operation. At the same time, it will tell workers that all personnel, not just themselves, are required to wear the personal protective equipment, helping to reinforce the need for the employees to wear the protection.

Educational and training materials on the potential for harm should always be given to visitors at the beginning of the visit. The greater the danger, the more training should be done. Training should also be given on the correct use of all protective equipment.

The safety and health professional should be responsible for the technical matters associated with personal protective equipment and training materials, and should insure that the correct equipment is used and maintained. Also, educational materials should be reviewed for technical correctness. All instructions, procedures, and manuals should be carefully written and their use documented.

The underwriter will review any losses that may have occurred, but the documented loss prevention techniques will also be evaluated. A well-organized and well-written program may help to sway an underwriter concerning the commitment of the company to visitor safety.

4.1.3 Third Parties

The third major class of risk exposure is the largest group in terms of potential claimants and the one that receives the least amount of attention by the safety and health professional. That is the *third party* situation. A third party, by definition, is anyone who is not an employee (first party) or a customer (second party.) This large nebulous group of individuals does not have direct contact with the company but may be affected indirectly by the actions of the company.

In general, the claims resulting from third parties are the result of long-term conditions or policies of the insured. For example, a class action concerning racial discrimination at a plant will occur only after a number of people, over a period of time, have been discriminated against. Two exceptions occur frequently: sudden, accidental pollution and third party involvement in incidents with first or second parties. An example of the second case is a vehicular accident where an employee drives into a parked car. This could result in a third party suit.

The underwriters for general liability, auto liability, and environmental impairment liability policies are interested in third party liability. In each case, the underwriter is looking for the methods used by the insured to limit its liability by preventive methods and effective loss management techniques. It is in these areas that the safety and health professional, acting as a risk manager, can provide expertise that may make a difference.

For a risk manager, it is important to recognize each of the following areas as potential loss situations:

Environmental impairment
Third party involvement (in employee or customer incidents)
Class action suits
Stockholders' suits
Professional liability

Environmental Impairment

Environmental matters and their relationship to insurance is a relatively new topic. Until 1973, the insurance industry did not address this in its policies and coverage was provided in effect, although it was not the subject of many claims. However, since 1973, the number of environmentally related claims and the problems associated with them have been growing very quickly.

In 1973, the Insurance Service Office (ISO) promulgated a general liability policy that for the first time addressed the pollution question. That policy, which is used as the standard throughout the industry, excluded all pollution on and into the land, air or water, except for those incidents that were "sudden and accidental." Those words were undefined in the policy.

Until that time, the general liability policy was silent on pollution and it was considered likely that any claims for bodily injury and property damage would be handled by this policy. With the 1973 revision, that was supposed to be the case no longer, but the courts saw it differently.

At the same time, certain insurance companies began to fill the gap in coverage that the exclusion created by offering a pollution liability policy, which was designed to provide gradual pollution coverage. By these two developments, the insurance industry had split environmental pollution incidents into two, sudden and accidental situations and gradual problems. The question became, where does one (sudden and accidental) end and the other (gradual) begin?

This question remains today, and for the general liability underwriter it has become very important. The courts have ruled in numerous cases that "sudden and accidental" can be interpreted to cover a variety of gradual pollution situations that ISO never intended to cover. The general liability policy has been paying for these situations over the past few years, and the GL underwriter has become very aware of the problems associated with certain risks. Very careful underwriting with regard to pollution has been the rule in the industry for the past few years.

In addition, the pollution liability underwriter has also become more aware of the difficulties associated with certain classes of business and pollution. Many of these policies have resulted in large claims, which have greatly affected the profitability of this type of insurance. The underwriter has become very cautious about the type of account that is acceptable.

The basis for the underwriting of a risk with potential pollution problems is the *risk assessment report*. The report is written either by an outside consultant or by an internal insurance representative. However, in either case, the company's safety and health professional staff can have a major impact on the type of information in the report and how it is presented.

The key question for the risk assessor is, how does the company control its wastes and discharges to reduce exposure to the outside? The safety and health professional should have input into the company's action plan for wastes and discharge. The toxicological data on the effects of chemicals can be gathered and presented by the professional. The emergency response plan in case of an incident would also benefit from the work of the safety and health staff. They should work hand in hand with the environmental staff of the company, if there is one, to control the exposures to the public. If there is no environmental personnel, it will often be the responsibility of the safety and health practitioners to handle these matters.

At any rate, it is the responsibility of the risk manager to insure that the full and complete data are presented to the risk assessor for the report. In this way, the underwriter can have the information necessary to make the correct decision about premiums, exclusions, and deductibles for the account.

Third Party Involvement

The key item here for both the underwriter and the safety and health professional is correct loss and claims management. Beyond what was described earlier regarding controls over employer or customer exposures, further control of third party exposures that result due to actions of the employee or customer is difficult. When an incident occurs despite the best efforts to control the exposure, it is important that the third party involved, the innocent bystander, be handled initially by the employee or customer, and by the Claims Department of the insured or the insurance company when the claim is made.

By initiating and implementing steps that result in better handling, the safety and health professional may help to reduce the third party's pain and anguish over the loss. For example, a concerned and courteous truck driver involved in an accident with a motor vehicle may help to dispel the anger of the car driver. In addition, if someone is hurt, a first-aid-trained driver may be able to help with the injury until professional help arrives.

In catastrophic occurrences such as an explosion, a tank car overturning and spilling a toxic chemical, or contamination of a product by one of the company's chemicals, the safety and health specialist can be helpful in two ways: providing professional advice on the plan or action to reduce and control the exposure, and serving as a technical spokesperson on the incident to the public. First, the professional can consult on ways to clean up and remove the substance to minimize the exposure to the public. Second, the professional can explain in reasoned and dispassionate fashion what is taking place at the site and what is being done to alleviate the damage.

The underwriter looks for printed material indicating that the company has plans for effective loss management. The underwriter wants to see that each employee or customer has been given adequate information in order to handle any situation that comes up involving a third party. For employees, this may involve a training session that discusses what to do in an emergency situation. The preparedness of the company and its employees for potential disasters is important.

For customers, this may require quick and easy access to necessary information from the company when an accident occurs. Toll-free customer information numbers or hot lines have been effective.

In addition, the company should prepare a method of handling the public information detail. Knowledgeable, confident spokespersons who can speak intelligently to the technical issues should be ready to step in when a catastrophe occurs. In many cases, the safety and health professional fits the bill.

Class Action Suits

Initially, it may seem farfetched that the safety and health professional can affect the losses involved with the class action type of third party suit. However, a review of past class actions indicates that the specialist may be helpful. For example, two of the largest class action suits ever initiated involved environmental matters, the asbestos litigation against the asbestos manufacturers, and the Vietnam veterans' suit against the manufacturers of Agent Orange. In both cases, epidemiological and toxicological data gathered by safety and health experts indicated that potential problems were developing. In each case, the information was not utilized fully by the management of those companies.

A class action suit can be filed by groups on behalf of people, things, or places that cannot file suit themselves, for example, the environmental class action suits against polluters on behalf of rivers being contaminated with PCBs, on behalf of wild birds whose eggs were softened due to the parents' exposure to DDT, or on behalf of whole cities that are being polluted with various airborne contaminants being emitted by a class of business prevalent in that area. In many of these cases, the filing of the suit required that the company think about its potential liability. This, in turn, resulted in a review of the policies that caused the situation to develop. Eventually, the safety and health personnel may be asked for their opinion of the situation and advice on what the company could do now to alleviate the situation.

For the specialist, the ideal method of handling these cases is not to let them develop in the first place. This requires that the professional have input into the managerial decision-making process. Whenever a decision is to be made that may affect safety and health, whether that of employees, customers, or third parties, the safety and health department should be asked for an opinion on the matter. Thus, decisions concerning policies can be made with all the necessary input placed before the managers.

Unforeseen situations that result in class action suits may still develop. However, it is likely that many others will be sidestepped with the foreknowledge provided by the specialists.

In addition, when class action suits involving safety and health issues are filed, the staff should be involved as soon as possible. A full professional evaluation of the issues involved may help the company decide what action to take. The safety and health professionals can also serve as public spokespersons to answer technical questions about the suit or can provide technical data to those who will be answering the questions.

For the underwriter, the key items are the company's previous suits, its reaction to them, and the company's plan to avoid future suits by involving safety and health personnel in decisions. Lawsuits that may affect the profitability of the company by a minimum of 10% of its gross income must be listed in a Securities and Exchange Commission (SEC) 10-K report and in the company's annual report. The information should be complete enough so that when the underwriter reads it, additional questions are kept to a minimum. However, if additional explanation is needed to explore further the technical aspects of the problem, safety and health personnel should be involved in writing the report.

All input of the safety and health professionals should be documented and available to the underwriter for review, if requested. Companies with a good system for communication on safety and health matters are more likely to be treated as good accounts.

Stockholders' Suits

Directors and officers of companies often make decisions that will affect the way the firm conducts business, what kind of business it conducts, and with whom it conducts business. In some cases, these decisions may affect the price of the company stock adversely. For example, the sale of a large facility may hinge on the decision of the board of directors. If the decision to sell at a particular price causes the stock to drop 10 points, it is possible that there may be a stockholders' suit.

Certain decisions of the directors and officers may result in environmental problems. When these problems are substantial enough to cause a drop in the stock price, a lawsuit may result. For example, the purchase of a facility with an ongoing health exposure problem might cause such a price drop. Consider the purchase of an older bottling plant by a new soft-drink manufacturer. The old plant has had a serious noise problem since it opened. Thirty percent of the employees decide to file a class action suit against the company for hearing loss. A stockholders' suit may charge that the directors and officers did not investigate the plant fully before purchasing it and therefore, the stock of the company went down 20% when the hearing loss suit was filed.

A careful review by the safety and health staff of the potential for losses at a facility prior to purchase can help the directors and officers in their decision-making process. It should be noted that almost all Directors and Officers Liability policies exclude coverage for any and all pollution incidents. When they make decisions that result in adverse environmental effects, the resultant lawsuits may prove to be personally costly to the directors and officers.

When a lawsuit involving the directors and officers is filed for an environmental reason, the previous company actions that are the basis for the suit should be examined by the safety and health department. The actions may have been well thought out, planned, and justified under the circumstances. The defense of the claim may be based upon the professional staff's evaluation of these actions. A frank and honest report on the situation from the staff will be helpful to the directors and officers in deciding what action should be taken.

For the director and officers liability underwriter, the key items reviewed are the company's annual report and its SEC 10-K report. Lawsuits against the directors and officers should be carefully explained in these reports. In suits involving safety and health matters, the professional staff should have input into the explanation of those suits. In particular, the wording should be reviewed for technical correctness. The D & O underwriter will also want to know what has been done to prevent possible stockholders' lawsuits in terms of the type of communication between the board of directors and the professional safety and health staff.

Professional Liability

Lawsuits against the professional staff (e.g. architects, engineers, doctors, etc.) can result from safety and health situations for which the staff is responsible. Such a suit can be filed against the safety and health professional, as well.

Many such suits are related to injuries or illnesses allegedly caused by the professional due to negligence, errors, or omissions that occurred during a project, for example, the collapse of an overhanging walkway in a public place, causing bodily injury to patrons under the walkway. It may be found that the collapse was due to incorrect calculations by the architect on the weight tolerances of the concrete supports.

For a safety and health practitioner, the suit may allege that the professional acted in a negligent manner by failing to recommend correction of a situation that resulted in injury. Such suits are usually filed in conjunction with another claim against the company, and have been increasing in numbers in recent years. One of the purposes of the suit is to warn other safety and health personnel that they also could be sued if they are negligent.

The safety and health personnel must provide techniques that will require the professionals on the company staff to take into account the possibility of injury while they perform their jobs. These procedures should direct the professionals to see the safety and health implications of their actions, and should become a part of the work habits of all the professionals, so that these implications are always considered. It is suspected that in this way many errors and omissions and, therefore, negligence could be avoided.

These techniques may include review of the professional's work by the safety and health staff, who will be able to determine if additional exposures exist that the professional has overlooked. Also, the specialist can get a better "handle" on the professional's job and may have recommendations as to what the professional should be doing in addition as part of the procedures to reduce safety and health errors and omissions.

To reduce their own errors and omissions, the safety and health staff should be audited internally periodically. The same good work habits that incorporate safety and health into all actions must be instilled into the safety and health professionals as well.

For the professional liability underwriter, the key items are the previous losses under professional liability policies, the reaction of the company to them, and the methods the company has initiated to prevent further losses. In general, the underwriter is looking to see what potential loss situations exist at the insured and what the company has done to reduce this potential.

4.2 COVERAGES

The exposures listed in the first section indicate the variety of problems that may affect the company from a safety and health perspective. This section will address the available insurance that provides liability coverage for each exposure.

First, a review will be made of the key sections in all insurance policies, including those areas that are common in all policies and the aspects of each. Next, each major coverage provided today by the insurance industry will be reviewed. The key differences between policies and the basis for rating each type of policy will be examined.

For the safety and health practitioner, knowledge about different policies is critical when discussing exposures and their relationship to the profitability of a company. The comptroller or insurance buyer of the firm will want to know if changes in the engineering controls will affect the insurability of the company and the size of the premium. Each should work closely with the other in order to provide support for each one's positions with regard to such matters as boiler safety, workers' compensation insurance costs, and claims settling for third party involvement in traffic accidents.

4.2.1 General Requirements

When purchasing a policy, one of the key questions is whether it is an "ISO Standard" policy or a "Manuscript" form. The Insurance Services Office, or ISO, disseminates sample policies produced by their committees, which consist of representatives of a variety of insurance companies. Many companies will place their own logo on the policy and issue it as their own. The advantage of doing this is that ISO has received approval from the state insurance boards for the policy to be issued within the state. In addition, the policy is acceptable to a majority of the insurance field, thereby reducing friction concerning exclusions. Third, a track record with regard to claims and the courts' interpretation of the policy will be developed relatively quickly due to the large number of insureds who have received that policy.

A "Manuscript" policy is one that has been written (1) by the insurance company for its own clients, or (2) jointly with the insured to handle the uniqueness of the insurance client's liability situation. The insurance company issues its own policy when there is no ISO Standard policy for that type of coverage, or when the insurance company wants to change the ISO form to make the coverage broader or more restrictive. For example, a number of policies for environmental impairment liability were issued in the mid-1970s before ISO came out with a standardized version in 1978. Still later, policies came out that were broader than the ISO form.

Sometimes the Manuscript form addresses the uniqueness of the liability associated with the business. For example, a crop-dusting outfit may need specific coverage to address errant sprays that may affect the neighbors' crops instead of the designated site. A modification in the contractual liability section may be necessary.

In general, all policies, whether they are ISO Standard or Manuscripted, have the following five features:

1. Insuring agreements
2. Limits and deductibles
3. Exclusions
4. Definitions
5. Claims procedures

Insuring Agreements

As defined in the Fire, Casualty and Surety Bulletins issued by the National Underwriter Company, Insuring Agreements are defined as:

> (The Insuring Agreement) expresses the insurer's promise to pay on behalf of the insured for damages and the insurer's duty to settle or defend claims against the insured alleging bodily

injury or property damage covered under the policy, even if the claims are groundless, false, or fraudulent.

This is the basis for the coverage offered under the policy. Any and all coverage should be laid out in this section so that it is clear to all who read the policy what circumstances would allow a claim against the policy. This is the heart of the policy. If it is not covered here, it is not covered.

This is why the insuring agreement should be reviewed carefully *before* the policy is bought. For example, if a company buys a general liability policy, the insuring agreement should be checked to make sure coverage is provided for the full limit for all damages that the insured may be legally obligated to pay that result from bodily injury or property damage occurring during the policy period. The language should be straightforward, with no unexplained qualifications or clauses that limit the coverage significantly.

Additionally, the insuring agreement should be evaluated to determine if it is an "occurrence" form or a "claims-made" form. This is a critical difference, which may become more important due to the emergence of more and more claims-made forms for all lines of coverage.

The occurrence form provides coverage (e.g. bodily injury (BI) and property damage (PD)) for occurrences during the policy period. This seems fine until you realize that no coverage is provided if the BI and PD occur before the policy period, regardless of whether the resulting claim is made during the policy period. In effect, that means that not all claims made now are covered by the present policy. This becomes particularly important if the previous policy provided inadequate limits to cover the resultant claim.

At the same time, remember that coverage is provided for a claim made after the policy period ends, so long as the occurrence took place during the policy period. For example, many of the long-term occupational disease cases have been filed against all the policies dating back to the employees' first date of employment. The limits available on each of the policies may be applicable to these cases all the way through the last date of employment.

Claims-made policies pay for claims brought against the insured during the policy period, regardless of when the BI or PD occurs. But once the policy expires, no coverage is available for any claim made after the expiration date, regardless of when the occurrence took place. The claims-made form "cuts the tail" of claims, which means that the insurance company no longer has to worry about a policy years after it was issued. Once it expires, the right to file a claim expires.

As you see, each type of policy has its advantages and disadvantages. Thus it is important to know what type of form your company has and important that your company is aware of the restrictions and advantages of the form.

Another item to be aware of in the insuring agreements is the "retroactive date," which is important for the claims-made form. This is the date before which no coverage is afforded. For the insurance company, this again cuts the other end of the claims by limiting the retroactive coverage available under this policy. This date should be differentiated from the "inception date," which is the date the policy begins coverage. The retroactive date is usually a date set chronologically before the inception date and coverage is provided for the time period between those dates.

For a claims-made policy, this time period should be sufficient to provide your company with coverage for most occurrences that might generate claims during the policy period. For example, a gradual pollution incident that can take place over ten years should be covered for a minimum of ten years. This length of time may not always be available, but it is important to recognize the limitations on the policy if it is not. In all cases, it is best to ask about availability.

The clauses outlining the defense-costs coverage should also be reviewed carefully. Some policies offer to defend all claims and include the cost in the policy limits. In effect, the limits of liability available to pay a claim are being depleted by the costs to defend the claim. Defense costs (or claims expenses) may be offered as a sublimit of the policy, which means that only the amount noted in the sublimit is available for these expenses. Others offer to pay defense costs in addition to the limit of liability, which means the full limits to pay a claim remain intact. The applicability of the deductible to the defense costs should also be checked. In general, the best option, if available and cost effective, is defense costs in addition to the limits of liability with no deductible applied to it. The insuring agreements should make that clear.

Limits and Deductibles

This section will delineate how much money is available to pay for the damages that are found to be covered by the insuring agreements. It also shows, by a deductible provision, how much the insured is responsible for before the insurance company will begin to pay. In general, the insurance company will pay the deductible amount, if it is anticipated that a substantial claim amount will be paid. The insured will then reimburse the insurance company for that payment.

The limits of liability can be divided into two categories, (1) per "occurrence," "incident," or "claim" limits, or (2) "aggregate" limits.

Occurrence limits are used for occurrence policies, which mean that only the sum listed at this limit is available to pay for all damages that result from any one occurrence. The intent is to limit the amount to be paid by the insurance company for any one situation to that amount. This works well when there is one occurrence and one claimant, because the policy cannot pay to that one claimant any more than is on the policy limit line. However, this is complicated when there are multiple claimants for one occurrence, or multiple occurrences involving one claimant, or multiple occurrences involving multiple claimants.

In some cases, the courts have ruled that each occurrence must be treated separately, and therefore, the full limit is applicable for each occurrence. In other cases, judges have ruled that each claimant must be treated separately, which in effect means that the full limit is available for each and every claim. Obviously, this can be quite expensive for the insurance company, particularly if the insurance company has insured a company for a period of time. In that case, each limit for each policy may be available for each claimant or for each incident, thereby increasing the insurance company's liability substantially.

The insurance industry is addressing this problem in two ways, (1) by the "claims made" policy, where there is a per incident limit, or (2) by placing a "per claim" limit on the occurrence policy.

The *per incident* limit theoretically reduces the insurance company's liability to one set limit for each situation regardless of the number of claimants involved. Each claimant in a catastrophic situation would not receive the full limits available for the damages suffered during the incident. For example, a policy with a $1 million per incident limit would pay only $1 million total when any incident took place.

Again, the courts have interpreted the clause differently than expected. Some courts have decided that each and every property damage or bodily injury is considered a separate "incident," thus making the limit applicable in full to each claimant. Currently, the "claims made" aspect of the policy disallows pyramiding of the limits, which can occur with a long-term insured on an occurrence policy.

The *per claim* limit of liability is the latest attempt by the insurance industry to decrease the potential for loss. All claims for damages sustained by any person or organization as the result of any one incident will be deemed to have been made when the first of those claims is made. In other words, all claims resulting from one occurrence will be treated as one claim and the maximum possible loss will be the per claim limit.

The second limit listed on a policy is the aggregate limit. The *aggregate limit* is the most that the insurer will pay during the policy year regardless of the number of occurrences.

There are two types of aggregate limits, (1) "per site" aggregate, and (2) "per policy" aggregate.

The *per site aggregate*, important when the insured has multiple locations, indicates the total money available to pay damages claims for incidents that occurred at that one site. For example, if you purchase a policy with a "per site" aggregate of $1 million and your company owns four sites, that means that each site is protected under that policy for up to $1 million. The disadvantage of this type of policy is that the premiums can be quite high.

The *per policy aggregate* limits the insurance protection to the aggregate limit, regardless of the number of sites. Using the same conditions listed above, only $1 million dollars in protection would be available for any losses at all four sites. This policy is less expensive but does not offer the protection of the per site aggregate.

Regardless of the policy, it is important to clarify what types of limits are listed. Since type can have a major effect on premiums, claims services desired, the extent of loss control available from the insurance company, and a variety of other factors, the safety and health practitioner should always be aware of the types of limits listed on the company policies.

Exclusions

Just as the insuring agreement indicates what is covered, the exclusions section states what is not covered. This area of the policy is often twice as long as the insuring agreement section. Insurance companies have learned, through litigation, that when the policy is silent on an area, the courts will find that coverage exists. Therefore, the policy will list every possible area that the insurance company will not cover.

Many different types of exclusions are applicable to the circumstances of the insured and the type of policy to be written. However, there are some general characteristics of exclusions and some common exclusions that appear in many policies, including:

1. Liabilities that are covered by other policies.
2. Liabilities that took place in the past or at past locations.
3. Contractual liability.
4. Liabilities that take place outside the coverage territory.
5. Liabilities resulting from damages that are intentional or expected by the insured.
6. Liabilities that result from war.

In general, insurance policies should not provide overlapping coverages for the same liability. From the insurance company viewpoint, there is the possibility that the insured or the claimant could collect twice for the same occurrence. This is particularly a problem when one insurance company provides a number of different policies for the same insured.

From the insured's viewpoint, the duplication may be costly in terms of additional premiums for each policy; however, this is rarely a problem because of the exclusions in each policy that make the policy mutually exclusive of others. For example, a policy providing general liability coverage would exclude coverage for employees' liability (covered under workers' compensation), vehicular liability (commercial auto), injury to products or work performed (separate products liability policy), pollution except sudden and accidental (pollution liability), and liquor liability (separate liquor liability cover). If the exposure exists at the insured, the exclusion on one policy should be covered by another policy.

However, some exposures that are excluded cannot be covered elsewhere, because the insurance industry as a whole will not provide insurance for that exposure. For example, war is considered a unique situation where considerable damage and bodily injury, completely out of the control of the insured, can take place. Because of the lack of control and the total destruction that is possible, all insurance policies exclude this exposure.

Most policies contain specific time periods of coverage, usually running from the inception date or retroactive date to the expiration date. Nothing outside this time period is covered and most policies state that in an exclusion. Many policies include a list of designated locations that are covered under the terms of the policy, and all other sites, whether owned by the insured or not, are not covered. This would also apply to sites that are no longer the property of the insured, whether they have been sold, abandoned, or given away. Even with policies that do not designate sites, an "alienated premises" exclusion is usually included in the policy.

For the majority of insureds, a coverage territory of the United States, its territories and possessions, and Canada is sufficient. Most policies issued by U.S. insurance companies will restrict coverage to these areas. This includes any legal suits that may be filed in other jurisdictions. The defense costs and all liabilities associated with these suits are excluded under the policy.

One of the most important exclusions is the *contractual liability exclusion.* John W. Hinley has this to say about contracts:

> The contractual agreement is an indemnifying clause or clauses forming a part of a complete contract or the agreement may be a separate contract in itself. . . Such agreements vary widely according to their terms and in many instances are quite complex and give rise to conflicting opinions as to the extent of the assumed liability which in many cases may be quite beyond reason. . . .
>
> *Proceedings of the Casualty Actuarial Society,* Volume XV, Number 51, p. 151

The comprehensive general liability policy for example, does not include coverage for contracts other than incidental contracts. The insured under a contract can become liable in three ways: (1) negligence, where the insured bears responsibility for the incident; (2) statute, such as workers' compensation laws; and (3) by assuming the liability of others. A separate contractual liability policy must be written to provide coverage when the insured has assumed liability under other than an incidental contract.

The definitions of an incidental contract includes any of the following contracts, if in writing:

1. A lease of premises.
2. An easement agreement, except in connection with construction or demolition operations on or adjacent to a railroad.
3. An agreement to indemnify a municipality, if required by ordinance.
4. A sidetrack agreement.
5. An elevator maintenance agreement.

Definitions

This section allows the insurance company to define what the terms in the other sections of the policy mean. It is particularly important to review this, because the expected coverage may be restricted by what is defined. Conversely, this is also the area where additional coverage may be hidden among the clauses used to define the terms of the policy. It is also important to note what is not defined, because when a policy is "silent" on something, often more coverage than expected is in the policy.

For example, the definition of insured site may include any and all sites currently occupied by the insured. This would be considered "blanket" coverage and would be part of a very broad policy. In some policies, any new sites acquired during the policy year would be covered automatically.

In other policies, only the specific locations as designated in the policy are covered. This restricts coverage and is called a "site-specific" policy. It is imperative that the accompanying list of sites be reviewed carefully, to insure that the necessary locations are covered. In addition, any new locations acquired during the term of the policy must be added by endorsement.

The definition of "property damage" may include a clause, "including consequential loss or use thereof." This means that part of the losses paid can include compensation for the time that the insured is unable to use the property while it is being cleaned up. This may be one area forgotten when filing a claim and additional monies for damages could be collected.

One subject often not addressed in policies is the issue of punitive damages. Many policies do not mention or define them. Many courts have ruled that "damages" covered in the policy include punitive damages. This is not true if the term, "compensatory damages" is used, as this term is interpreted to exclude punitive damages. However, it should be noted that many states do not allow the exclusion of punitive damages.

An example of a missing definition is the phrase, "sudden and accidental," which is undefined in the ISO General Liability policy applying to pollution. All pollution was excluded except for "sudden and accidental" pollution, which was never defined. Court interpretations have indicated that "gradual" pollution may be covered under the "accidental" portion of the clause. This was not the insurance industry's intention when it decided upon those words.

At any rate, it is worthwhile to review this section because these situations may prove to be important when a claim arises.

Claims Procedures

In order to file a claim under the terms of the policy, it is necessary to know the correct procedure of notification. In addition, we will note what the insured must do to facilitate the claim, and also what the insured cannot do.

In general, written notice of the claim must be sent to the insurance company, outlining the circumstances of the claim. This includes time, place, and nature of the incident, injury, or damage sustained by any persons or organizations (including their names and addresses) and names and addresses of available witnesses. All summons, notices, demands, or other processes must also be forwarded.

The insured must also cooperate fully with the insurance company's representatives with regard to settlements, conducting suits or proceedings, or enforcing the right of contribution or indemnity against anyone else who may be liable for the damages. The insured must attend hearings and trials and assist in securing and giving evidence and obtaining the attendance of witnesses.

The one thing that the insured *cannot* do is make any payment or incur any expense unless it is approved by the insurance company. This should be made clear to all that it may affect, so that no disagreement will arise later as to the extent of the liability that the insurance company will assume under the policy.

All in all, this is straightforward, but it is important to be sure that the correct procedures are completed and no adverse actions are taken.

4.2.2 Uniqueness of Each Coverage

This section will discuss the various available insurance coverages that can be purchased from most property and casualty (P&C) insurance companies. Keep in mind that the previous section discussed the general format of all policies and what they all have in common. What will be discussed here is the way each type of insurance differs from the others.

Each coverage requires different data for the underwriter to evaluate the risk. That data will be outlined under each coverage type. This is the area in which the occupational safety and health professional can be of most help. Much of the data relate directly to the exposures of the risks, and the practitioner will be the employee most aware of the situation of that exposure. The various coverages to be examined are:

1. General liability
2. Workers' compensation
3. Commercial auto
4. Umbrella
5. Special casualty
6. Accident and health
7. Environmental impairment

General Liability

General Liability is the most universal of all policies and also, in many ways, the most comprehensive. In fact, it is often called Comprehensive General Liability (CGL). The first standard form for this coverage was made available in 1940 by two precursors of the Insurance Service Office (ISO). Currently, the rules and rates of the insurance are written in the General Liability Division of the *Commercial Lines Manual.*

It was named "comprehensive" for two reasons, (1) it contained one comprehensive insuring agreement covering all hazards within the scope of the insuring agreement that were not otherwise excluded, and (2) it provided automatic coverage for new locations and business activities of the named insured that arose after the policy inception. This was in contrast to the "schedule" liability policies that were available up to then, which required separate agreements for each specific hazard.

In general, the CGL provides coverage for bodily injury and property damage of third parties that are the responsibility of the insured. It covers the hazards of premises and operations, independent contractors, and products and completed operations.

The "premises and operations" hazard encompasses liability for accidental bodily injury or property damage that results either from a condition on the insured's premises or from the insured's operations in progress, whether on or away from the insured's premises. Many of the exclusions in the CGL are applicable primarily to premises and operations coverage, including contractual liability, automobiles, liquor liability, pollution, employee injuries, damage to property in the insured's care, and others.

The "independent contractors" hazard exists whenever the insured hires an independent contractor to do work on behalf of the insured. The insured has potential liability for injury to others as a result of the work performed by the contractor.

The "products liability" hazard exists for any insured that manufactures, sells, handles, or distributes goods or products. Any bodily injury or property damage that arises from the goods or products is the insured's liability. It should be noted that the CGL form restricts by exclusion the extent of coverage available under this policy.

The "completed operations" hazard is the insured's potential liability for bodily injury or property damage arising out of the insured's completed work, for example, the bodily injury and property damage that may result from the collapse of a building after its completion. Again, like the "products liability" hazard, the coverage is restricted by exclusion.

A CGL policy consists of three documents, (1) a Declarations Page, (2) the policy jacket, which contains the definitions and provisions common to various standard coverage parts, and (3) the CGL coverage part, which contains the insuring agreement, exclusions, and other provisions specific to CGL coverage.

The Declarations Page outlines the data on the risk itself, such as name, address, inception, expiration, and retroactive dates, policy limits and deductibles, broker's name and address, commissions, and other such information. Such a page is attached to almost every type of insurance policy. This is a legal document and a record that the insured does have insurance, so it is imperative that all data is correct.

The policy jacket is the standard terms and conditions of the policy. Section 4.2.1 covered these items in depth. Suffice it to say that all sections of the policy jacket should be reviewed carefully to determine what is and what is not covered by the policy.

The coverage section is where the coverage becomes specific for the insured in question. The broker, representing the insured, and the underwriter negotiate the actual scope of coverage. In the CGL policy, this is where the other coverage parts (contractual liability, broad form property damage, etc.) are added. In contrast, various exclusions that restrict coverage can also be added.

These additional coverages and exclusions must be studied carefully to determine what is actually covered by the policy. For example, although it has been said that the CGL form covers the products and completed operations hazard, the insurance forms for the two major business groups that employ safety and health practitioners do not contain this coverage. The form used for Manufacturers and Contractors (M & C) and that used for Owners, Landlords, and Tenants (OL & T) must carry a separate section in order to provide this coverage. Similarly, OL & T and M & C require special arrangements for full coverage of the independent contractors hazard.

Though this all sounds complicated, which it is, one rule must be kept in mind when reviewing a policy: "if it ain't excluded, it's covered!" Once you look at the policy language with that in mind, the exclusions can be read one at a time and their effect on the basic coverage noted. In the same way, all coverage parts can be read for content with regard to the additional coverage extended by the endorsement. It may be helpful to keep a score card on the additions and subtractions so that the effectiveness of overall policy can be evaluated.

At any rate, it is helpful for the safety and health professional to have a good working knowledge of what a CGL policy does and does not cover. There are two reasons for this, (1) the professional then can determine if the exposures that exist at the company are adequately covered by insurance under the terms and conditions of the policy, and (2) the professional can assist in gathering the information needed to present clearly and concisely to the underwriter what is taking place at the account.

The information given to the underwriter is known as the "submission." For comprehensive general liability coverage, the following are necessary as part of the submission:

1. Application
2. Financial data on the firm
3. Loss data and analysis
4. Requests for unique coverage features
5. The firm's loss control program information

Depending upon the company, the application can be quite extensive or very short. However, in most cases, the pertinent information on the firm's activities are still necessary. For example, the company's annual gross sales figure would always be requested, as well as the main business activity, number of employees, size of facility(ies), number of vehicles, location of facility(ies), and types and numbers of products. Additionally, the application may request data on previous business activity(ies), previous business location(s), plans for expansion, growth figures in sales, personnel, and products, and other business related matters.

The applications are usually designed to elicit basic facts about the company, but not necessarily the in-depth data needed to determine ultimately the acceptability of a risk or what premium should be requested. Sometimes the facts asked for relate to the exposures at the facility and how they are controlled. For example, there may be a question concerning premises exposure regarding the protection of passers-by from bodily injury that may result due to the activities of employees at a site. The safety and health professional may be the individual best equipped to discuss the exposure and the controls that have been instituted to reduce the exposure.

The application may request two items where the professional can be helpful, (1) loss data and (2) loss control program information.

Loss data is any and all information concerning situations that either resulted or could have resulted in a claim against the policy. For CGL losses, this may involve insurance company records or internal records indicating all the particulars of the loss. This would include the cause of the loss, actual dollar amounts paid, the anticipated future payments (called *reserves*), the anticipated future claims related to incidents that took place during the policy period (called *incurred but not reported* (IBNR) losses), and any other pertinent data. As will be seen later, this information is the basis for setting premiums for the risk.

Remember that the current CGL carrier (insurance company) does not automatically provide this data to any new carrier who has been approached to replace it. The company interested in getting premium "quotations" from a new carrier must supply the loss data to them. The "old" carrier, therefore, has an advantage over the others in that they have the loss experience data, perhaps for a number of years. Of course, if that record is poor, it would be anticipated that, at renewal, the data would be used against the insured. The insurance company often has computer readouts on current losses including all the pertinent details. It would behoove all insureds to request these runs on a regular basis for evaluation purposes.

The occupational safety and health professionals on the insured's staff should be part of the evaluation process. The practitioners can evaluate accident trends and note in a report what has been done to alleviate the exposure. In addition, for large losses, a full inspection report and analysis can be completed by the professional staff. This report should be part of the submission made to the underwriter, who may be swayed by the actions listed in the report to control the number and severity of the accidents. Substantial differences in premium may be involved.

Also, a section of the submission should be devoted to enumerating and describing the elements of the loss control program of the insured. The focus should be on those parts of the program that relate to the coverage being requested. For example, a general liability submission should show how the exposures to third parties are addressed by the Loss Control department. A systematic review of the procedures for quality control of products would be important if products liability is included in the request for insurance.

The safety and health personnel should have direct input to the submission in these areas. The practitioner should be able to complete the information in a technical fashion and add to the completeness and thoroughness of the submission.

The financial data for the firm is considered the jursidiction of the comptroller or financial officer of the insured. This usually involves the annual report of the company and certain SEC reports, such as the 10-K. The main function of the safety and health professional with regard to this data is to make sure the numbers for the Safety and Health department's budget are correct. Also, under the Legal Proceedings section of the 10-K, the company may list various claims that may affect the financial capabilities of the firm. The professionals should make sure that any loss information is correct and that the company has listed all relevant losses.

The Loss Control department should also be part of decision making about unique coverage features. Situations that have a large potential for loss may not be covered under the provisions of the standard CGL policy. For example, a large part of a construction company's business is conducted by contract and therefore, coverage for contractual liability is important. Such a situation would be apparent to the broker and the insurance buyer for the insured. However, other circumstances may be more visible to the professional that others may not see.

For example, the practitioner may be aware of preliminary research data that indicate the possibility of a higher toxicity than previously known for an ingredient in one of the company's products. The ingredient may be removed immediately, but the possibility of future liability associated with the past practices of using that ingredient still exists. The professional would be able to outline the problems and evaluate the potential for loss. The insurance buyer may request "past practices" products coverage, using the professional's report as the basis for the submission.

Once the underwriter has all the information in the submission and any additional data that are requested, a decision is made as to the acceptability of the risk. In general, an underwriter is given guidelines by the home office as to the types of business that the insurance company considers to be good risks and poor risks. These rules are usually contained in booklets that are periodically updated to reflect changes in the judicial and public environment with regard to the exposures associated with a particular type of account.

Often there is a grading system, which indicates what current thinking is on the potential for loss at a class of business. For example, a facility with a large storage tank filled with flammable liquid would be rated as a high hazard, and a neighborhood grocery store would not. These classifications come into play early in the submission process. Once the class of business is known, the underwriter will review the underwriting guidelines to determine whether or not the risk is in an acceptable class.

Another source of classification of risk is *Best's Underwriting Manual.* This book, also updated yearly, classifies many different types of businesses with regard to the standard lines of coverage (GL, Workers' Compensation, Commercial Auto, etc.), and assigns a rating from 1 (excellent) to 10 (poor) to each type of exposure for each business class. This rating is often used to establish parameters for premium rates.

Also keep in mind that the insurance company often establishes rates different from the standard rates sent to the State Insurance Board for approval by ISO or other insurance groups. This change in rate, called a "deviation," may be lower or higher than the established rates depending upon the current market for this product line. The underwriter is thus allowed certain leeway with regard to setting premiums, which often depends on the quality of the data included in the submission. Also the underwriter will often compare the risk to others previously evaluated to determine where this one fits in the scale. For example, if this risk controls the exposure through engineering devices better than another, that may be reflected in the premium to be charged.

The rating device that underlies CGL coverage is the insurer's right to conduct an audit at policy expiration and assess additional premium if necessary. Ordinarily, at the beginning of each policy period the CGL insurer surveys the insured's existing or anticipated exposures and charges an appropriate premium called the "deposit premium." At the end of the policy period, the insurer conducts an audit to determine actual exposures and using this information, either charges an additional premium or refunds part of the deposit premium.

An additional premium may be charged, for instance, if, during the audit, an additional location or operation not anticipated in the deposit premium is discovered. Similarly, if a location closes down during the policy period, a refund may be in order. The Safety and Health department of the insured should be prepared to provide any additional data concerning new operations and their engineering controls requested by the insurer. This will enable the underwriters to get a clear picture of what is taking place and should assist the underwriter in setting a fair and equitable additional premium.

All in all, the safety and health practitioner can be very helpful in providing data to the Corporate Insurance department on the existing exposures and evaluating the potential for loss. Specific tasks involving loss data analysis and large loss reports can be performed best by the professional staff. This information will help the underwriter in fully evaluating the risk. The safety and health professional should be aware of this and should press the Insurance department about the quality of the information going into the submission. In the long run, that input can result in better data going to the insurance company, which may have a salutary effect on the premium to be charged.

Workers' Compensation

This coverage is the one most familiar to the safety and health field. Because of the cost of this insurance, Safety and Health departments of manufacturing companies were created, as were the Engineering departments in insurance companies. However, in many ways the coverage still remains somewhat mysterious to the practitioner, and the principles that govern its use and pricing are foreign. This section will attempt to dispel some of that strangeness so that the professional will have a good working knowledge of the product.

Although the first workers' compensation (WC) policies were issued around 1911, the current standard form was only issued in April 1984. Prior to that, the standard policy had not been revised since 1954. Officially called the Workers' Compensation and Employers Liability Policy, the new form is the result of a compromise between employers and employees on this critical issue. The employees give up the right to sue their employers for employment-related injuries and the employers agree to pay for the damages through insurance, under a statute mechanism providing specifically scheduled benefits. Insurance has been, and most likely will continue to be, the most effective method for an employer to compensate employees and their families for work-related injuries or diseases. The Employee Liability portion of the policy protects employers against suits filed to collect damages that are not compensable under the WC coverage.

The revision of the policy, which also utilized "easy read," or simplified language, was done by the National Council on Compensation Insurance (NCCI). It should be noted that all states and territories have WC statutes. But six states—Nevada, North Dakota, Ohio, Washington, West Virginia, and Wyoming—administer their own programs and are called "monopolistic" states. In effect, all employees in these states are covered under the state insurance plan, not by private company insurance.

The policy contains three distinct areas of coverage, (1) workers' compensation, (2) employers liability, and (3) other states' insurance.

Under Part 1, the workers' compensation (WC) section, the insurer agrees to assume liability imposed upon the insured by the workers' compensation laws of the states listed on the declarations page. The language is simple and straightforward: "We will pay promptly when due the benefits required of you by the Workers' Compensation Law." *Workers' Compensation* is defined as the laws of each state listed on the declarations page. Coverage applies to bodily injury by accident and bodily injury by disease (including resulting death).

There are two restrictions: (1) bodily injury by accident must occur during the policy period; and (2) bodily injury by disease must be caused or aggravated by conditions of employment, with the employee's last day of exposure falling during the policy period. This section also covers reasonable expenses incurred at the insurer's request, premiums for appeal bonds, litigation costs taxed against the insured, and interest on a judgment as required by law. The insurer also has a right and duty to defend the insured in any proceeding for benefits payable under the policy.

The employers liability portion of the policy, Part 2, is similar to other liability insuring agreements. This section covers the insured for liabilities arising out of the injury of an employee in the course of employment not covered by workers' compensation. The insured damages include: (1) "Third-Party-Over" suits, which involve the insured's liability for damages claimed against a third party by one of the insured's employees; (2) damages assessed for care and loss of services; and (3) consequential bodily injury to a spouse or other close relative of the injured employee. In addition, this part provides coverage for "dual capacity" actions, which are damages assessed because of bodily injury to an employee that arises out of and in the course of employment, claimed against the insured in a capacity other than an employee.

Part 3 of the policy provides other states' coverage. This is necessary because Part 1 applies only to obligations imposed on the insured in the states listed on the declarations page. If an insured is confronted with a claim from a state not listed, there would be no coverage.

It has been suggested that a statement as follows be inserted in the declarations: "All states except Nevada, North Dakota, Ohio, Washington, West Virginia, Wyoming (private insurance is not permitted in these monopolistic states), and states designated in Item 3.A of the Information Declarations Page." This should avoid a loss uninsured due to an oversight. Note that insurers operating in only a limited number of states will not be able to include this broad statement.

Part 4 is "Your (The Insured's) duties if injury occurs." Part 5 is "Premium," and Part 6 is "Conditions." Both of the parts are subject to exclusions listed under Section F, "Payments You Must Make." These limitations include any payments in excess of the benefits regularly required by Workers' Compensation statutes due to: (1) serious and willful misconduct by the insured; (2) the knowing employment of a person in violation of the law; (3) failure to comply with health and safety laws or regulations; or (4) the discharge, coercement, or other discrimination against employees in violation of the WC law.

There is no coverage for liability of another assumed under a contract; however, this does not apply to a warranty that the insured's work will be done in a workmanlike manner. Punitive damages are excluded, as is coverage for bodily injury to an employee while employed in violation of the law with the insured's "actual knowledge." For the Employee's Liability section, there are additional exclusions relating to no coverage for damages imposed by a Workers' Compensation, Occupational Disease, Unemployment Compensation, or Disability Benefits law. In addition, there is no coverage for bodily injury intentionally caused or aggravated by the insured, or for damages arising out of the discharge of, coercion of, or discrimination against any employee in violation of the liability.

Although most companies consider WC to be the exclusive remedy for work-related disabilities, there are several reasons why Employees' Liability coverage is desirable. Some states do not make WC insurance compulsory or do not require coverage unless there are three or more employees. In some cases, an on-the-job injury or disease is not considered to be work related and therefore is not compensable. The employee may still sue if there is reason to believe that the employer should be accountable.

Certain state laws have been interpreted to permit suits against employers by spouses and dependents of disabled workers to collect damages for themselves beyond what the employee has gotten. The basis of such suits may be loss of companionship, comfort, and affection.

Finally, employers are being confronted with claims and suits when an injured employee sues a negligent third party for compensation beyond WC coverage. The third party may in turn sue the employer for contributory negligence.

From this review of the WC policy, it should be clear how the occupational safety and health professional can be of help regarding the coverage. All the covered items should fall within the jurisdiction of the practitioner. Perhaps the most important clause in the policy from the viewpoint of the professional is the exclusion related to the failure to comply with health and safety laws or regulations. Underwriters are becoming increasingly knowledgeable about these laws and denial of a claim due to a violation of a regulation may cause an uproar with the insured. However, under the terms of this policy, such a possibility does exist. The practitioner should utilize this possibility to press for the necessary controls to eliminate such violations.

In addition, the safety and health staff can be helpful in the submission process. Many of the same features required of the general liability submission are required for WC coverage, for instance, an application, loss data and analysis, and needed coverage features. The safety and health personnel can help to prepare the needed data and review it to insure its high technical quality.

This would be particularly true with the loss data and analysis, an area in which the staff is probably already spending much of its time. The same information used to justify expenditure for new engineering control equipment can also be geared towards any underwriting submission. Again, insurance company computer loss runs can be helpful as data sources. The interpretation of the information to indicate trends, to report on large losses, and to show the engineering changes that have been made to address the problems should be done by the health and safety staff.

In addition to this data, the submission should also include a written safety and health program and a safety training program. These would be good indications of the insured's attitude towards safety and health and would demonstrate that the insured has addressed the major potential loss areas. For the underwriter, any deviations from the "book" premium will be based upon the evaluation of the additional material by the underwriter. Often the underwriter will be influenced favorably by a well-written and thorough program that provides good guidelines to the insured's employees about safety and health matters.

Conversely, a poorly written and disorganized program can affect the underwriter's opinion on the risk adversely. In either case, the underwriter will request an insurance loss control representative to visit the facility to investigate what is actually taking place. This usually occurs after the underwriter has indicated an interest in the risk and the broker has been given an approximate premium amount, called a "preliminary premium indication." This is before the risk is "bound," which means that the coverage is in effect.

The visit of the loss control representative is the most important responsibility of the Safety and Health Department regarding securing Workers' Compensation Insurance. Other responsibilities, including loss data analysis and application review, are important, but the report filed by the loss control rep can overwhelm all the previous work, whether it is favorable or not. The safety and health staff must make sure that the loss control rep understands the prospective insured's programs and can see the future plans to address any problems that may arise.

First and foremost, the safety and health practitioner should accompany the insurance company representative on the inspection and answer all questions as completely as possible. Although the rep may not be an expert on the particular operations at your facility, remember that the rep is a professional, who will see hundreds of facilities in a year. The questions asked often reflect this experience as well as the information being requested by the underwriter.

Second, the professional should make sure that the supervisors of the operations examined are informed in advance of the visit and asked to cooperate with the inspector. The failure to cooperate is often cited by the loss control rep as an example of poor management practices and will adversely effect the opinion of risk. Selected managers of main operations should be available for interviews by the representative, if requested, in order to provide a description of the operation.

Lastly, a wrap-up session should be conducted with the rep. If the loss control rep has recommendations for the operations, these should be noted and, if necessary, discussed with the rep. Any misconceptions about the operations of the loss control programs should be cleared up at this time. The session will enable the safety and health practitioner to report back to management about what they may expect to find in the loss control rep's report.

Once the underwriter has the loss control report, the finalization of the premium begins. It should be noted that the rating basis for all WC premiums is the total remuneration earned during the policy period by employees and officers of the insured, excepting those specifically excluded from coverage under the policy.

The National Council on Compensation Insurance is a statistical agency that accumulates and prepares data for the various state boards for consideration in setting "manual" rates. To accumulate the necessary data, the insurance companies must file a Unit Statistical Plan, which lists, for each class of business, the payroll and incurred losses. The National Council takes the latest available data for a two-year period as a basis for computing future manual rates. These rates are promulgated on a regular basis and are the rates used initially by all insurance companies.

Within a given business class, there will be a wide variety of companies who may have had different loss experience. To accommodate this possibility, the underwriter will use "experience rating." An *experience rate* is a manual rate modified on the basis of the deviation of the loss experience of an individual risk from that of the average established by the classification. If the accident history shows a loss cost below normal for the class, a credit is allowed, whereas unfavorable experience results in a debit. This credit or debit is applied to the manual rate to find the adjusted, or experience, rate.

The larger the risk, the more likely it is that it will have credible experience data, whereas small risks with few employees and fewer hours worked per year may not have good data. A good rule of thumb, therefore, is that the larger the risk, the larger will be its credit for good experience and the larger its debit for poor experience.

Experience rating thus enhances the importance of a good safety and health program aimed at reducing the frequency and severity of accidents. Such a program can have a direct effect on the premium. On the other hand, laxity in enforcement of a program will eventually be reflected in the loss experience and, therefore, in the premium.

One additional rating system may be affected favorably by the safety and health professional by the loss control. *Retrospective rating* is a plan or method that permits adjustment of the final premium on the basis of the employer's own loss experience, subject to maximum and minimum limits. Such plans can be valuable to insureds with modest premium who feel that their experience will be better than average and would like to benefit from the expected lower loss ratios. For larger risks, the chief advantage will be providing legal liability to satisfy government regulation while still gearing the premium to the actual losses of the particular insured.

Whatever method is used to calculate the premium, one general principle stands out: the lower the losses, the lower the premium. For the safety and health practitioner, this should be the hammer to use during budget crunch time. There is a "bottom-line" effect that the professional can point to when negotiating for additional personnel, equipment, or a larger say in the engineering controls that are used. Safety and health can then be seen as a direct money-saving device that is absolutely vital in reducing costs, particularly insurance costs. Of course, the key is to make sure that the occupational safety and health program is capable of reducing the number of accidents and controlling the severity of those losses that do occur. But once that program is in place, the case can be made about the relationship between a strong program and good loss experience.

Business Auto

For many companies, the transportation of raw materials, products, and employees is an important component of the firm's business. For others, that moving of materials and people *is* the business. In either case, the accidents that can result on the road are multitudinous and the Business Auto policy is the means used to protect the company against the resultant losses.

The major components of this type of policy are:

1. Physical damage
2. Automobile liability

Automobile physical damage insurance for commercial and public vehicles involves four basic coverages, comprehensive, specified perils, collision, and towing. The insurer agrees in the insuring agreement to pay for loss to a covered auto or its equipment under any one of the first three coverages listed above.

Comprehensive coverage means that the insurer will pay for loss from any cause except collision or overturn. No list of perils that specifies, one way or another, what is or is not covered, is attached.

On the other hand, *specific perils* coverage does list what is covered. These perils are: (1) fire or explosion; (2) theft; (3) windstorm, hail, or earthquake; (4) flood; (5) mischief or vandalism; (6) sinking, burning, collision, or derailment of any conveyance transporting the covered auto. These are offered as a package because experience shows that most insureds who do not buy comprehensive coverage select all of the above.

Collision coverage involves the damage caused by a covered auto's collision with another object or by its overturn.

The key item for a business auto policy is the selection of the covered autos. There are five physical damage insurance categories:

1. Owned autos only.
2. Owned private passenger autos only.
3. Owned autos other than private passenger autos only.
4. Specifically described autos (basic auto).
5. Hired autos only.

By picking one of these categories, the insured narrows the coverage that has been purchased. It is further limited by the "Persons Insured" section. This defines the three categories of insured individuals as follows:

1. The persons or organization as named on the policy for any covered auto.
2. Those other than the named insured who have been given permission to use the covered auto, such as employees, executive officers, those who service or park the covered auto, and those who move property to or from the covered auto.
3. Anyone who is liable for the conduct of an insured described above is an insured but only to the extent of that liability.

These two clauses make it very clear as to what is or is not covered.

The exclusions for physical damage are as follows:

1. Wear and tear, freezing, mechanical or electrical breakdown, unless these are caused by other loss covered by the policy.

2. Blowouts, punctures, and other road damage to tires, unless the damage is caused by other loss covered by the policy.
3. Loss caused by declared or undeclared war.
4. Loss caused by nuclear weapon.
5. Nonpermanent installed tape decks, loss of tapes, or loss of CBs, mobile radios, or telephone equipment.

Automobile liability provides coverage for bodily injury and property damage to third parties caused by the insured's covered autos and the persons insured. In many ways, this is similar to commercial general liability in the scope of coverage. However, all coverage for this policy is restricted to the covered autos and the persons insured.

Nine automobile liability classifications from which the insured must choose:

1. Any auto (comprehensive auto liability coverage).
2. Owned autos only.
3. Owned private passenger autos only.
4. Owned autos other than private passenger autos.
5. Owned autos subjected to no fault.
6. Owned autos subject to compulsory uninsured motorist law.
7. Specifically described autos (basic auto).
8. Hired autos only.
9. Nonowned autos only.

The persons-insured section of the physical damage section is applicable to automobile liability. For purposes of contrast, your personal auto liability policy would most likely be a "Basic Auto" policy, whereas your company, if it has a fleet, most likely has a "Comprehensive Auto" policy.

The exclusions are familiar, including exclusions of contractor liability and pollution, and a broad form nuclear energy liability exclusion. Also, no coverage is afforded for injuries covered under Workers' Compensation, Disability Benefits, or similar laws. No bodily injuries to employees or to fellow employees is included for coverage. In addition, there is an exclusion for damage to property owned or transported by the insured.

As part of the overall loss control visit to a facility, the insurance company representative may review the auto losses, review the safety and health procedures, and inspect such fleet areas as garages, depots, and repair facilities. The purpose of this will be the same as the visits to the manufacturing areas for workers' compensation, that is, to clarify for the underwriter in a report what is happening at the facility. The company's traffic safety manager or other individual responsible for the safety and health of the drivers should meet with the loss control representative during that visit to insure that correct and complete data are being gathered.

The traffic safety manager should be prepared to show the rep a written safety and health program that outlines the major potential exposures and the procedures used by the company to control those exposures. For example, the daily inspection of the vehicles should be described and any forms used as verification of the review should be included in the written program material. Also, the procedures to follow in an accident should be carefully delineated and the training methods used to present the procedures to the drivers should be shown.

The loss control rep's report will form a major part of the material that the underwriter will want in order to evaluate and rate a business auto risk fairly. To review, the data needed for a submission should include:

1. Application (includes complete data on all vehicles).
2. Loss data and analysis.
3. Motor vehicle registration for all covered autos.
4. Drivers' licenses and records for all person insured.
5. Safety and health program.
6. Training program.

With these materials and the loss control representative's written report, the underwriter will have a good picture of what is going on at the facility and can then decide whether or not it is an acceptable risk. If it is considered acceptable, the account would be rated to determine the premiums.

The rating of these coverages is based upon the rates set forth in ISO's *Commercial Lines Manual.* The classification tables in this book list all industries and their class code for each line of business, including auto. For Business Auto, the main factors that are used to rate a risk are those for trucks, and those for autos. For risks with trucks, the factors are:

1. Size of the trucks (by weight)
2. Business class (in classification tables)
3. Radius class (distances traveled)
4. Zones (geographic location of main depots)
5. Collision vs. noncollision coverage

For risks with private passenger autos, the factors are:

1. Original cost
2. Age group (up to 6 years)
3. Deductibles per type of coverage
4. Comprehensive vs. collision only vs. combined coverages
5. Geographic location

Each and every truck and private passenger auto are rated separately and included in the cost of the premium. In addition, the underwriter will request Motor Vehicle Registrations (MVRs) on all vehicles to be sure that the autos and trucks are properly registered and, in many states, inspected. Data on all drivers will also be requested, including previous driving records of violations, revocations, and the like. Copies of current drivers' licenses often provide this type of data.

The Safety and Health department plays an important part in obtaining insurance for commercial auto exposures. The data accumulated by that department on auto losses are vital in determining the acceptability of the risk, and then in deciding the preliminary premium to be offered. Interaction between the insurance loss control representative and the department focuses the rep's report to underwriting on the aspects of the loss control program that help prevent and reduce the frequency and severity of auto accidents. Often the extent to which the loss control program is seen as being effective sways an underwriter about an account.

Umbrella

The name of this line of coverage is very descriptive. An *umbrella* policy provides coverage over other applicable policies for an insured against liabilities that result from damages to third parties. In other words, this is excess insurance, which only comes into play if the money paid out on a loss exceeds the primary policy's liability limits.

For example, a CGL policy may provide coverage only for the first $1 million in losses. Any losses in a policy year exceeding that amount would not be covered by this policy. A "first umbrella" policy may pick up the next $4 million in limits to bring the total available limits for CGL losses to $5 million. Various umbrellas may be placed above the first umbrella in a manner called *layering*. A particular layer may often have a number of insurance companies participating in it. For example, the second $5 million in excess (or "x of" in insurance jargon) of the first $5 million may be split on a 50/50 basis between two insurance companies. They will share everything—the limit, the premium, and the losses, on a 50-50 basis.

Depending upon the insured, umbrellas over the initial set of casualty policies (CGL, Basic Auto, etc.) may extend upwards to $200 million in limits, or even more. For many companies, limits that high are not necessary and are too expensive. In general, limits of $20 million to $50 million are satisfactory and are not likely to be exceeded in even the worst catastrophe.

For the underwriter, the key data that are needed to evaluate the risk are:

1. Underlying policy coverages.
2. Price for each layer of coverage.
3. Loss data on past events that entered the layer to be covered.
4. Any unique endorsements of the underlying policies.

The rating is generally based upon the price of the layers below the one being quoted. A good rule of thumb is that the next $5 million of coverage will cost one half the price being asked per million for the previous $5 million. For example, if a layer of $5M x of $20M ($5 million in excess of $20 million of coverage) is priced at $30,000, the next $5 million ($5M x of $25M) will most likely be priced at $15,000.

In insurance jargon, there is a "working layer," which is the layer where losses would normally be expected, and there is a "catastrophe layer," which only comes into play if a huge disaster takes place. The working layer, as would be expected, is priced much higher than the catastrophe layer, so certain insurers may be more interested in the working layer. These insurers tend to be very selective on the risks they choose. Other insurers are only interested in "providing capacity," which means that their limits are available only at very high layers, well above the working layer. These insurers are making money on writing large numbers of risks.

The occupational health and safety practitioner can have the same impact on the umbrella carriers, or insurance companies, as on the "primary," or first policy, insurers. A good loss control program backed by statistical evidence of the results will still affect the pricing of the layers. It may also affect the limits at which the working layer ends, because a well-controlled risk should result in lower frequency and less severe losses.

The professional should be aware of the company's umbrella program and the limits that are currently available in case of a loss. If a situation exists in which a loss would exhaust the limits of liability, the safety and health practitioner should so indicate to the risk manager. If the potential for loss can be reduced, it should be; but if it cannot, the risk manager may wish to increase the limits by purchasing more insurance.

In any event, umbrella coverage is important to the safety and health staff because knowledge of it may dictate some of the actions taken to control catastrophic occurrences. A thorough safety and health professional should be aware of the concepts involved in the writing of this type of coverage.

Special Casualty

Three major types of coverages exist in this area:

1. Directors and officers liability policy (D & O).
2. Errors and omissions liability policy (E & O), also known as professional liability.
3. Excess workers' compensation.

The first, D & O, provides coverage for the insureds' directors and officers against stockholders' law suits. The intent of this policy is to allow the executives of the insured to do what is necessary to protect the firm without fear of the stockholders, particularly minority stockholders who are against the current company management. A good example is a take-over attempt that the directors and officers of the company are attempting to ward off. If the executives are successful yet the stock loses value due to their actions, it is possible that the directors and officers may be sued.

In general, this is a financial situation that does not involve the Occupational Safety and Health department, who are usually far removed from such dealings. However, on occasion the directors may make decisions that are based on data assembled by the safety and health staff. For example, a decision by a large asbestos firm to file for bankruptcy may be based in part on the health analysis of the workers who might have had contact with asbestos and will be diagnosed as having mesothelioma, cancer of the pleura. Decisions of this magnitude may not happen at your firm, but other decisions of importance may be based on the data you have gathered. A plant may be closed in part because the accident rate is high due to the age of the equipment.

Your information must be correct because of the impact it may have in the hands of the directors or officers. Very careful checking should be done on all statistics because the use to which they may be put is unknown. You can be certain that if the data are incorrect and the wrong decision is made, the directors will make sure that you are informed.

The underwriter will look at the annual report and the 10-K financial report when evaluating the risk. As noted previously, these documents pinpoint environmental problems that may affect the financial condition of the company. The safety and health staff should make sure that the information is complete and technically correct. The underwriter will also be looking at new acquisitions to determine what effect on the bottom line these may have. Again, a full analysis by the engineering staff, including safety and health, may assuage the underwriters' trepidation with regard to the potential for loss.

In short, while this is not a major area of concern for the safety and health professional, it should be addressed and the correct actions taken to insure that the professional does not add to the directors' and officers' loss potential.

The Errors and Omissions (E & O) policy, also called Professional Liability, will be of interest to the safety and health practitioner because of the coverage provided in case of the professional's own errors and omissions. This policy pays for the damages (bodily injury and property damage) that the insured become legally obligated to pay as the result of any negligent act, error, or omission of the insured solely while performing the professional services described in the policy. There are two conditions outlined in the insuring agreement, (1) time of act or omission, and (2) territorial scope.

There is one exception to the condition that the error or omission occurs during the policy period, that is, that it occurred prior to the policy period but after the retroactive date specified in the policy. In addition, the insured must have had no knowledge of it at policy inception and no other valid and collectible insurance applies to this act or omission.

The negligent act, error, or omission, as well as the resulting damage, must occur within the continental United States, Hawaii, or Canada. Also, the suit for damages must be brought in a court within the same territory.

The E & O policy contains the usual exclusions: intentional acts of the insured; obligations under Workers' Compensation or other similar laws; aircraft and motor vehicle exposures; pollution exposures; and contractual liability. But some exclusions are unique to this coverage.

No previous claims are covered, nor any circumstances of which the insured is aware and which may result in any claim. Damages against the insured arising from libel, slander, invasion of privacy, and assault and battery are excluded. Failure to complete work on time is excluded. Dishonest, fraudulent, criminal, or wrongful acts or omissions committed intentionally by or at the direction of the insured are excluded.

Specifically, there are three related *exclusions* concerning professional acts:

1. Claims caused by acts not arising out the professional services to which the insurance applies are excluded.
2. Loss emanating from any insured's conduct of business owned by the insured is not covered except those professional acts insured in the policy.
3. Liability arising out of the conduct of any individual, partnership, corporation, or joint ventures of which the insured is a member is not covered unless such operations are specifically insured in this policy.

It should also be noted that most of these are "claims-made" policies, which means that the claim must be made during the policy period in order for the coverage to apply. Professional liability policies are divided into various professional classes, such as architects and engineer, lawyers, nurses, and so on. Only the class of individuals specified on the policy is covered by the provisions.

The submissions should include an application and current loss data. The application for this coverage is very important because all the statements are considered the insured's agreements, representations, and warranties. It is therefore considered part of the policy. As such, any statements on the application that are found to be erroneous will void the policy.

All safety and health personnel should be covered by a professional liability policy for their errors and omissions. A copy of the policy should be examined by the professional to make sure that coverage is adequate and no gaps in coverage exist. In addition, the professional should take steps to reduce the potential for loss from any professional within the company covered under this policy. For example, the engineers on staff should have a quality control system that checks their reports for errors before they are sent. Significant recommendations concerning operations by safety engineers should have managerial approval and peer review before being placed in reports.

Although these may seem like trivial matters in the overall scheme of things, note that large losses involving substantial bodily injury and property damage can result if the professional does not perform the job correctly. For example, an error in the weight distribution characteristics of a platform can result in collapse of the platform when a heavier than normal weight is placed on it. Tolerances and safety factors are very important, and errors there can have severe results.

The last casualty policy to be discussed is Excess Workers' Compensation. As can be judged by its name, it covers the same losses as the Workers' Compensation policy but at a higher limit of liability. These limits may be necessary when the state-imposed limits for an injury are found to be inadequate by a jury, or the potential for injuries at the insured exceeds the state limit for aggregate.

In the first situation, an *excess incident* limit policy may be necessary. This would provide a higher cover of insurance for any one injury that an insured may sustain. The *excess aggregate* limit policy would provide coverage for all damages for injuries sustained that exceed the previous policy's aggregate. The purpose of this is to provide coverage in case of a catastrophe that may injure many employees.

Excess workers' compensation sells to large companies in the monopolistic workers' compensation states because of the possibility of a particular company exceeding the state's limit of liability. In particular, the larger firms buy excess aggregate policies because of their number of employees.

The underwriter requires that the submission include the underlying Workers' Compensation coverages and pricing and the loss data on incidents that have affected the layer in the past. The safety and health professional can provide the same assistance as that required in the Workers' Compensation submission. For example, a thorough and technically correct report on the losses will help show the underwriter what steps have been taken to reduce the potential for occurrence of similar incidents in the future.

In general, however, it should be remembered that all three special casualty policies are "special occasion" policies, which do not have a significant effect on the day-to-day existence of a company. For the safety and health professional, the issues associated with these policies should be a low priority, after the majority of the work has been completed on the basic exposures: auto, employees, and the public. But the exposures should be examined and the policies read so that the professional can be sure that all significant exposures are uncovered and evaluated.

Accident and Health

Many companies are offering to their employees group health plans that help reduce the cost of health insurance by spreading it out over a large group. The purpose of the policy is to provide coverage for the medical expenses incurred by the company's employees and their dependents. In addition, many policies offer options for accidental death, including flight insurance and other forms of travel insurance.

Also, there is a newer coverage, offering unique features that may appeal to a select group, called *kidnap and ransom* insurance. This provides funds for expenses incurred by the insured in the case of a kidnapped employee. Such a policy may be particularly appealing to companies with executives overseas.

In the health area, interrelationship exists between the health of the employee on the job and off. Company sponsored exercise groups and preventive medicine clinics help to support the loss control program at work. Physical exams and medical tests are part of both the on- and off-the-job health plan, the purpose of which is to maintain the employee's well-being and thus have a more productive, happier worker.

The safety and health practitioner should be instrumental in setting up the off-the-job health plan and should be able to correlate it successfully with the on-the-job loss control program. If the employee works with solvents at work and in addition is an artist using oil-based paints, it's likely that the occupational exposure is being compounded by the nonoccupational one. Additional blood exams for this worker may be indicated to determine what effect, if any, the combined exposure may have on the white blood count.

The underwriter for this policy is looking for data on the preventive measures undertaken by the company to control employee accidents and illnesses off the job. For example, a home safety program emphasizing measures to prevent such accidents as trips and falls, electrical shocks, fire, and so on, should be a good addition to a submission for an accident and health group plan. In addition, the following should be part of the submission:

1. Application, as a group.
2. Loss data for the group over a period of time.
3. Any unique coverage features desired by the company for the group.

Like many other coverages, each submission will look pretty much like the next if the information provided includes only the three items listed above. It is the added data on preventive measures, provided by the safety and health professional, that makes the submission stand out to the underwriter. A good program that emphasizes the employer's actions to help the employees maintain their health both on the job and off is likely to impress the underwriter. This may allow the underwriter leeway with regard to the conditions and premiums of the program and may save money for the insured and the insured's employees.

Environmental Impairment Liability (Pollution Liability)

This policy, one of the newest insurance products, has already had a pronounced effect upon the marketplace. In ten years, a $40 million market has developed, although significant changes are taking place that portend a change for the worse in the future viability of the product line. At any rate, the purpose of the policy is to provide coverage for the insured for obligations incurred due to property damage, bodily injury, and cleanup as the result of gradual pollution incidents. It is expected that in time sudden and accidental pollution incidents will also be covered, which means that, in effect, all pollution would be covered by one policy. Currently, sudden and accidental pollution is covered under the CGL policy but it is expected that an absolute pollution exclusion soon will be used on CGL policies, meaning that only pollution liability policies will provide pollution coverage.

This is a third party policy, which means that the insured itself is not protected from its own damage. It is also a claims-made policy, often with a retroactive date. The claim must be filed during the policy period and only incidents taking place after the retroactive date are provided coverage. There are many exclusions, including obligations under Workers' Compensation or other similar laws; intentional acts of the insured; aircraft, watercraft, and motor vehicle exposures; contractual liability; products liability; exposures arising from premises no longer owned by the insured; oil- or water-well exposures; nuclear exposures; and exposures arising due to the insured's noncompliance with federal or state laws. There is a territorial restriction to the United States, Canada and Mexico.

Despite these restrictions, the policy is considered quite broad, or liberal, in that all pollution on the land, in the air, or in the water is covered. In some cases, the hazardous wastes sent to a landfill are also covered for any liabilities they may incur. Because of this breadth, the underwriter must be very careful about the types of accounts written for this policy. The submission material required is quite extensive and must include:

1. An application.
2. Financial data (10-K and annual report).
3. Loss data on both sudden and accidental and gradual pollution incidents.
4. A risk assessment report or loss control reports on pollution exposures and controls.

The key item is the risk assessment report. In most cases, this must be completed by a consultant who is hired either by the potential insured or the underwriter. Since the policy is relatively new, the basis for premiums is not firmly established in manuals, but it is very dependent upon the underwriter's evaluation of the risk. Because of that, the risk assessment report is the starting point for the underwriter.

For the safety and health practitioner, the risk assessor's visit must be treated much like the visit of the insurance company's loss control representative to review the facility for Workers' Compensation. The necessary data must be available for review and the effected departments should have representatives available to explain the processes. The professional should take the time to be with the risk assessor as much as possible in order to answer any questions concerning the safety and health aspects of the plant. At the end, a wrap-up session should be held so that the risk assessor's recommendations can be discussed and plans can be made to begin implementing them as soon as possible.

All in all, the safety and health practitioner can be a valuable counterpoint during the risk assessor's visit. Garbled or misinterpreted data can be corrected and a good impression can be made of the emphasis the company is placing on the environmental area.

The practitioner can also help in filling out the application. Much of the data requested in the forms relate directly or indirectly to the responsibilities of the Safety and Health department. For example, the application may request water quality sample results from the water treatment facility. The professional can acquire the latest results and make sure that the data are clearly presented and technically correct. This expertise could also be utilized when assembling the loss data. The professional can update the situation since the loss, and indicate what the company has done to prevent future losses.

It should be noted again that the future of this cover is clouded, mainly due to the substantial losses that have been sustained on both pollution liability and general liability with regard to "sudden and accidental" pollution incidents. Since the inception of the pollution exclusion in 1973, which allowed coverage for sudden and accidental pollution on the CGL form but supposedly no gradual pollution coverage, many courts have ruled that the CGL language also allowed coverage for gradual pollution under the "accidental" part of the clause. In other words, the words "sudden" and "accidental" were separated, instead of being read as one clause, and coverage was afforded under both words. Payments for damages under the "accidental" section for gradual pollution have been enormous. Certainly, it can be seen that the potential for losses due to pollution is indeed great.

This section has provided a picture of various policies as they exist now. Be aware that this is strictly an overview and the actual underwriting and policy wording are likely to be far more complicated. It is suggested that those interested in more details on a particular policy should contact the ISO for information.

Now that we have reviewed the exposures that may exist where we work, and have taken a look at the various policies that will provide insurance coverage for these exposures, what remains is the mechanics of the actual purchase of the policies. The next section addresses this question.

4.3 MECHANICS

For the safety and health professional with risk management responsibilities, the purchasing of insurance can be a harrowing experience. As can be seen from the previous section, insurance has its own jargon, which is often difficult for an outsider to understand. In addition, many of the provisions of the policies are quite legalistic and subject to interpretation not only by nonlegal personnel but by lawyers as well.

Although insurance can be a tangled web, it is extremely helpful when losses arise. Therefore, the necessity of having the correct policies with the adequate limits of liability far outweighs the hassle it may be to deal with any problems that may arise. The main rule when dealing with insurance is, use common sense. If the terms and conditions do not make sense to you, have your broker or the underwriter explain them. If the premium seems too high, find out the basis for the premium. If an exclusion seems particularly tricky and endlessly confusing to you, have it clarified by a lawyer. Insurance is a common sense proposition, and if it does not make sense to you, imagine what a jury might do with the provisions of your policy.

Your main contact with the insurance community will be your insurance agent. However, when purchasing a policy, there are three main components that must be taken into account:

1. The broker/agent.
2. The insurance company.
3. The specific underwriter assigned to your account.

This section will examine the relationship that you should establish with each of these, what information you should have about them, and what you can expect them to do for you. It is always important to keep in mind that your company pays the premium by which each of the three live.

4.3.1 The Broker/Agent

To begin, there is a difference between a broker and an agent. A broker usually works for a nationwide firm with offices everywhere and can work with all the major insurance companies. These large firms are called "alphabet houses" because they often are called by their initials, such as M & M for Marsh & McLennan and A & A for Alexander & Alexander. An agent is usually a much smaller firm that may work with a number of insurance companies or with only one.

The chief advantage of a broker is the size of the firm, which allows it to do many things that a smaller agent cannot do for you, such as loss control, claim services, and computer loss runs. The chief disadvantage may also be size, because your company's insurance program may not receive as much attention as a larger company's insurance program. This reflects the fact that the basis for profit for the broker is commission, and the larger the insurance program, the larger the commission.

For an agent, on the other hand, your insurance program may be the largest piece of business the firm handles. Therefore, the amount of time and attention spent on it may be much greater. However, the disadvantage of using an agent is getting access to the insurance companies and products you may want.

An intermediate solution is what is known as the "second line" agencies, which are large regional agencies. These firms have access to the majority of insurance companies, can often match the services provided by the brokerage houses, and are still small enough to spend the time on a moderate-sized account.

Whatever route you go, there are three items that you should check on before choosing a broker/agent:

1. The services that can be provided.
2. The experience the firm has.
3. The market access of the firm.

Many agents/brokers have additional services they can offer you besides placing your business with an insurance company. Many firms have loss control personnel on staff or as a separate entity, which can be called upon by the client to provide inspections and evaluations. The accessibility of these representatives and their costs should be discussed and, if possible, determined in advance.

Claim services can be extremely valuable for lines of insurance in which you can expect losses. Workers' Compensation losses at a 1000 employee factory, for example, may require a full-time staff member to handle the paperwork and oversee the program. Many broker/agents are equipped to deal with such a situation. In addition, some brokers/agents have claims specialists who can investigate a loss and make recommendations concerning settlements. This can help to resolve many of the disputes between parties before they become lawsuits.

Brokers and agents may also have access through computer terminals to loss data generated by the insurance company in terms of loss reserves (money set aside to pay a claim at a future date when settled), losses actually paid, information on the incidents that generated the losses, and what is called Incurred But Not Reported (IBNR) losses. The last figure is a statistically derived estimate of the amount of losses that have occurred during the policy period but will not be reported until the policy period is over. Such information is often available only on computer terminals to an agent who has a very close relationship with a particular insurance company, although all agents/brokers can get this information run on one of their insureds if they request it. It often takes time to arrive, whereas the computer terminal is usually immediate.

The risk manager should ask about all services even if there is no immediate intention of utilizing any particular service. It is unknown what future services may be required. Those services that the risk manager wants upon policy inception should be delineated and priced in advance by the broker/agent. No surprise costs should arise later.

The risk manager should have a good picture of the coverages needed before choosing the broker/agent. In that way, the risk manager can inquire about the experience of the broker/agent with placing such coverages. This is particularly true if the risk manager is planning to purchase a policy that is out of the ordinary, such as professional liability or umbrellas at high limits of liability. Also, the broker/agent should be familiar with your company's type of exposure and your company's financial status, product lines, and history. This information should be presented to the broker/agent and the risk manager should attempt to gauge the broker/agent's ability to understand what is taking place at the company and what exposures need to be insured.

A professional broker/agent can assist the risk manager by evaluating the company's risk and determining the best method of safeguarding the company against loss. A thorough broker/agent will ask many questions of the risk manager and others to discover any hidden exposures and to clarify those exposures that may not be readily apparent. The best agent/brokers are thinking "exposure first, commission second."

Once the risk manager has been satisfied about the services and the experience of the agent/broker, the next item is to find out what markets the agent/broker has access to. It may not be necessary to have a broker/agent who is capable of working with every insurance company if the agent/broker can work with all the markets with the coverages you need. This can easily be discovered by asking the agent/broker which companies it represents. If you do not hear the name of an insurance company you are particularly interested in, ask the broker/agent about that company.

Access is important because in some lines of coverage, there are only a few markets available that will write that line of business. If that coverage is important to you and that particular agent/broker cannot get it, that may not be the broker/agent for you. Of course, you always have the option of splitting the lines of coverage between two brokers. The disadvantage to this is that certain companies will not offer the special lines of coverages unless you also buy the standard (CGL, Auto, WC) lines from the company. It is therefore best to make sure your agent/broker can get quotes on all the lines you need.

It should be noted that certain brokers/agents have an arrangement with an insurance company to market certain lines of coverage with special features including pricing. Although the term "exclusive" is often used with regard to these arrangements, this is often not the case in the sense that any other agent with a "broker of record" letter that authorizes representation can approach the insurance company for a price for that product. The insurance company cannot legally offer differing pricing to different agents for the same risk. However, certain coverage features may be different and that may add up to a premium differential.

At any rate, a "broker of record" letter from you will allow the broker/agent to begin sending submissions to the available markets. You then must direct the broker/agent as to which insurance companies to approach.

4.3.2 Insurance Companies

The key decision on insurance companies is whether to approach one company for all the policies you need or to approach different carriers for different policies. The advantage of putting all policies at one insurance company is the possibility of discounted pricing for the whole package based upon the sheer volume of premium. Insurance companies are aware that many incidents affect a variety of policies because of their interlocking nature. It is expected that there is a better chance to get a handle on a loss if the group of policies is all in one company.

In addition, a large premium will enable the insurance carrier to have a pool of cash available if an incident that affects any one line takes place. That money can be earning revenue until such time as a loss is paid, thus enabling the insurance company to have another source of income.

From the viewpoint of the insured, the advantage of having one carrier is the ease of claims handling. All losses would be reported in one fashion to one company and there would be no need to determine the applicability of coverage since all policies are in one place.

Conversely, the advantage of placing different lines of coverage with different carriers is that each company may provide you with the exact coverage you need, particularly with regard to "exotic" coverages that not everyone needs. A *manuscripted*, or designed, policy may provide coverage that an ISO manual form does not. Depending on the needs of your company, this flexibility in coverage may be important.

Also, with each line of coverage, competition between carriers may be more intense than between carriers who can offer package deals. The premium one carrier may offer on WC may be substantially lower than the price offered on WC as part of a package. That can vary considerably from year to year, but if your company is price sensitive with regard to insurance, this consideration may be important.

Another important factor in choosing insurance companies should be the financial stability of the carrier. Each year, Best's Insurance Guides, Inc., rates each insurance company on the basis of its financial conditions. A+ is the highest rating. Best's takes into account the gross writing premium, the net profit, the loss reserves set aside, investment income, and other fiscal considerations in making the rating. As a rule of thumb, it would be best to work only with companies that have Best's ratings of B+ or better. These ratings can be discovered from a variety of sources, including your broker; make sure that the rating is for the most recent year.

In addition, annual reports of the insurance companies will give you a better picture of the company as a whole or as part of a larger entity. Many insurance companies today are part of a large financial services company or are a subsidiary of a conglomerate. A review of the annual reports of the larger companies will indicate what priority the insurance company has within the larger firm.

Also, the annual report may indicate the direction of the insurance company regarding certain lines of coverage. For example, the company may announce an increased interest in General Liability Insurance for small to medium-sized firms in the low-hazard group areas. If your facility fits that description, this may be a good carrier to approach about your insurance needs. Also, new products are often discussed in the report, and these may also fulfill a need of your company. A fuller spreadsheet on the financial state of the insurance company is also included in the report.

Insurance companies have many services that are often available to the insureds for little or no charge. Brochures and booklets concerning these services should be obtained and reviewed. Some are duplicates of the ones that the broker/agent can provide (loss control, claims, loss runs) and may be used in lieu of the broker's services. Also, carriers can provide training and educational materials to supplement what you have and enhance the insurance-related program. These include defensive driver training, safety and health off-the-job, first aid and cardiopulmonary resuscitation (CPR) and other courses.

One of the primary services, loss control, can be quite extensive in terms of the range of specialties. For example, a typical Loss Control Department will have an industrial hygienist, a fire prevention specialist, a boiler and machinery specialist, a fleet specialist, an environmental specialist, and others. Each of these can be called upon by the insured to investigate certain potential problem areas and make recommendations to correct the situation. Often, the cost of their services can be negotiated as part of the premium paid up front. A service visit program can be established, and the various specialists' visit can be scheduled in advance. To do this, it will be necessary to know about the capabilities of the loss control staff in advance. Your broker should have up-to-date information on each insurance company's services, which you should request for review. These services may be important enough to your plant to justify the expenditure of extra premium dollars in order to obtain them.

In today's market, it is often just as important to know the reinsurers that back each product line you will be buying as it is to know the insurance carrier. Reinsurance is the insurance company's insurance company, in that the reinsurer will pay the losses on a certain portion of the limits of liability in return for a piece of the premium. For example, on a $1 million limit, the insurance company may pay the first $250,000 of losses and the reinsurer may pay the next $750,000 of losses. As you can see, on a large loss, the reinsurer would be even more at risk than the insurance company.

Therefore, the type of reinsurer used is important to the insured because that reinsurer is at a risk for any loss that occurs. The reinsurer must also be financially stable and capable of absorbing the loss if it occurs. Best's also rates reinsurers and the same rule of thumb—B+ or better—applies to them. Additionally, it is preferable that the reinsurer be domestic as opposed to foreign, for ease of claims handling. However, it must be admitted that many of the most stable reinsurers (including those on the London market and in Bermuda) are foreign. The key point is that the risk manager should know what reinsurers back each line of coverage that has been purchased, so that any news concerning a reinsurer can be viewed in light of that knowledge. Reinsurers are generally smaller firms than insurance companies and their rate of bankruptcy is higher. The financial soundness of the companies backing your policies is of the utmost importance.

All the data must be weighed before any final decisions are made on which insurance companies and their policies should be the ones you as the risk manager should choose. Fortunately, most of the information is easily obtainable for review. If the insurance terminology becomes too tangled, your insurance agent should be able to straighten it out for you.

4.3.3 Underwriters

Ultimately, the individual who will have the most impact upon the premium and the details of your insurance program is the underwriter. The underwriter will evaluate the material sent in as part of the submission and will decide whether your company is a viable risk for the insurance company to insure against certain types of damages. It is therefore essential that you are on good terms with the underwriters who have been assigned to your account. Their need for information and documentation must be met in order that you can put together the best insurance package possible for your company.

In most insurance transactions, the insured does not meet with the underwriter. The correspondence is handled through the broker/agent and, in most cases, this system works well. However, it should be noted that if you are having a problem understanding what is needed by the underwriter, you should meet directly with the underwriter to discuss that problem. For example, the underwriter may be asking about operations that are only a minor part of your overall plant and you are unable to understand what is wanted. The underwriter may be concerned unnecessarily about this procedure, but may be pressing for this data because it was mentioned in the submission but not elaborated upon. A quick meeting on the matter may suffice to clear the air.

Because of the sheer volume of paperwork that an underwriter will see in a typical year, the initial submission must make a good impression so that the underwriter can distinguish your submission from others. To accomplish this, the underwriter will measure the data on three points:

1. Adequacy of the information for the coverage desired.
2. Accuracy of the information.
3. No loose ends left in the data.

The adequacy of the information relates, first and foremost, to the items essential to a submission for each line of coverage reviewed in the previous section. Without this initial set of data, the underwriter will not be able to proceed with the submission. This is where the two rules of underwriting come into play, particularly the second, "underwrite to sell" and "underwrite to cover your assets." The first rule indicates that pricing has to be within the current marketplace pricing, otherwise there will be no sale. The second rule state that the file's data must justify the pricing that the underwriter has determined.

Like many departments in other companies, underwriters are audited by their managers and supervisors frequently, by top management periodically, and by the state insurance board on a scheduled basis. The file must show how the premium was derived and all scheduled credits and adjustments must be documented.

That is one reason why the underwriter may request additional follow-up data after the initial submission. The way to limit that request is to make sure that the facts and figures most likely to be wanted by the underwriter are in the submission in the first place. Also, the underwriter should be able to find the information easily without a massive search. Do not attempt to overwhelm the underwriter with tons of reports that may not be relevant to the type of insurance being requested. That will put off an underwriter as much as a lack of data. A well-organized and thorough submission may include an index, clear, concise text, and uncomplicated, precise figures and charts.

The accuracy of the data should be documented with collaborating outside sources, if possible. A report on Workers' Compensation slip-and-fall accidents done by a consultant would reinforce an in-house statement on the status of control measures to reduce such incidents. Actual wastewater sample test results in a chart will back up an indication that the company is in compliance with state and federal waste discharge standards. When outside collaboration is not possible, statements concerning an item should indicate the basis for the remarks. There should be no inaccuracies in the submission, if at all possible.

The bane of the underwriter's existence is loose ends. The underwriter must question anything for which the data is inadequate or inaccurate so that the evaluation of the risk and the judgments made by the underwriter cannot be questioned later. Any unclear statement will be picked out, and if the evidence for such a statement is not presented, the underwriter will request such evidence. When inquiries come from the underwriter, the reply should also provide complete and accurate answers to all questions. In addition, all answers should be documented so that no more doubts will remain in the underwriter's mind about this matter.

When meetings with the underwriter are required in the initial stages of contact, a good first impression is important. You should be prepared for the meeting with all the necessary data available for review. It is always best to include the broker/agent in all meetings, so that the broker/agent, as a third party, can break an impasse and steer the conversations to a productive conclusion. Also, if a specific area is in question, the person within your company with the best information on that situation should be present at the meeting.

The underwriter's product is the terms and conditions of the quotation on your account, including the premium. Once you have this, it is up to you to react to the quotation. Generally, you are given a period of time, usually 30 days, to get back to the underwriter with your answer. If you are dissatisfied with the quotation, you shoud ask the basis for the pricing and why certain terms (particularly exclusions and deductibles) were included. The answers to these inquiries may give you some leeway with regard to negotiations. For example, the underwriter may be justifying a high premium on the basis that the company is not taking steps to combat an exposure that is resulting in frequent accidents. You may be able to show the measures that have been taken or will be taken in the future to reduce this exposure.

The risk manager should be suspicious of a very low price as well. It is possible that the coverage from the company for this line is very restrictive. Alternatively, the underwriter may be inexperienced with such coverage and an audit down the line may force cancellation of the policy. The search would begin all over again. A third reason may be that the insurance company is underpricing the rest of the market in order to write a large volume in a short period of time, and then withdraw from the market. This again would leave you high and dry, with the possibility of no coverage.

A reasonable price and good terms and conditions with a solid basis in the facts and figures of the account should be the goal of the risk manager. Once that is accomplished to everyone's satisfaction, the account can be "bound," or insurance purchased. A binder should be signed by the insurance company outlining all the terms, conditions, and prices in writing. The risk manager should make sure that the binder is complete and accurate before accepting it. Generally, a binder indicates coverage for a period of time running from 30 to 90 days. Within that time, a policy as described earlier will be assembled by the underwriter and other members of the insurance company. Again, the risk manager should verify that all statements made in the policy and accompanying material are complete and accurate. A careful reading of all endorsements and exclusions, the declarations page, and certificates of insurance should be done by the risk manager or the staff. Any discrepancies or inaccuracies should be immediately reported to the broker/agent and the underwriter for correction.

4.4 CONCLUSION

The purpose of this chapter was to provide the safety and health practitioner with practical knowledge about liability insurance and to enable the practitioner to use that knowledge to adequately protect the assets of the company through liability insurance policies. Three major aspects of insurance—exposures, coverages, and mechanics—were explored in depth. Examples were given to illustrate many of the points to be made about these subjects.

By illuminating the subject in this fashion, it is hoped that safety and health professionals will have a better understanding of insurance in their future dealings. If insurance is, as they say, a "necessary evil," it is best to know the evil as thoroughly as possible in order to conquer it.

One final point: real-life situations involving insurance are almost never exactly as they are written in a book. It is only by experiencing the actual placement of a book of insurance business that an individual can say that he or she knows what insurance is about. The material in this chapter can serve only as a beginning when delving into this subject. But even a beginning is better than attempting to dive into insurance without any knowledge at all.

BIBLIOGRAPHY

1. *Analysis of Workers' Compensation Laws,* Washington, D.C.: U.S. Chamber of Commerce, 1983.
2. *Best's Insurance Report—Property-Casualty,* Oldwick, NJ: A. M. Best, 1983.
3. *Best's Loss Control Engineering Manual,* Oldwick, NJ: A. M. Best, 1984.
4. Castle, G., R. F. Cushman, and P. R. Kensicki, *The Business Insurance Handbook,* Homewood, IL: Dow Jones-Irwin, 1981.
5. *Fire, Casualty and Surety Bulletins,* 3rd Printing, Cincinnati: The National Underwriter Company, 1980.
6. Hinley, J. W., "Problems in Relation to Contractual Liability Insurance," in *Proceedings of the Casualty Actuarial Society,* Volume XXV, Number 51, published by the Society, 1935.
7. Insurance Service Office, *Commercial Lines Manual,* 3rd ed., New York: Insurance Service Office, 1984.
8. *Rimco's Manual of Rules, Classifications and Interpretations for Workers' Compensation Insurance,* Dallas, TX: International Risk Management Institute, 1984.
9. Williams, C. A., Jr., G. L. Head, R. C. Horn, and G. W. Glendenning, *Principles of Risk Management and Insurance,* Vol. II, 2nd ed., Malvern, PA: American Institute for Property and Liability Underwriters, 1981.

PART IV

OCCUPATIONAL SAFETY AND HEALTH MANAGEMENT

CHAPTER 5

How to Conduct an Accident Investigation

JAMES C. NOLTER
M&M Protection Consultants
Coral Cables, Florida

ROBERT D. JOHNSON
Northern Telecom Inc.
Nashville, Tennessee

5.1 INTRODUCTION

Let there be no doubt as to the need for effective accident investigation. Although our national trends show a decline in the total number of accidental deaths and workplace accidents, associated dollar costs, both direct and indirect, continue to rise. According to the National Safety Council, the latest dollar loss estimate in which work-related deaths and work-related disabling injuries occurred is $31.4 billion in 1982. The Council also states the cost of all accidents (work and nonwork related) is spiraling close to $90 billion per year (National Safety Council, 1983). Figures 5.1 and 5.2 and 5.3 represent the Council's best estimate as to the cost of the accident phenomenon to our nation. Note the indirect loss estimates from work accidents represented on Figure 5.2.

Included in these statistics are thousands of individual firms across the United States employing

Costs of accidents in 1982

Accidents in which deaths or disabling injuries occurred, together with noninjury motor-vehicle accidents and fires, cost the nation in 1982, at least

$88.4 billion (billion)

Motor-vehicle accidents **$41.6**

This cost figure includes wage loss, medical expense, insurance administration cost, and property damage from moving motor-vehicle accidents. Not included are the cost of public agencies such as police and fire departments, courts, indirect losses to employers of off-the-job accidents to employees, the value of cargo losses in commercial vehicles, and damages awarded in excess of direct losses. Fire damage to parked motor-vehicles is not included here but is distributed to the other classes.

Work accidents **$31.4**

This cost figure includes wage loss, medical expense, insurance administration cost, fire loss, and an estimate of indirect costs arising out of work accidents. Not included is the value of property damage other than fire loss, and indirect loss from fires.

Home accidents . . . . **$ 9.9**

This cost figure includes wage loss, medical expense, health insurance administration cost, and fire loss. Not included are the costs of property damage other than fire loss, and the indirect cost to employers of off-the-job accidents to employees.

Public accidents . . . . **$ 7.0**

This cost figure includes wage loss, medical expense, health insurance administration cost, and fire loss. Not included are the costs of property damage other than fire loss, and the indirect cost to employers of off-the-job accidents to employees.

Certain Costs[a] of Accidents by Class, 1982 ($ billions)

Cost	TOTAL[b]	Motor-Vehicle	Work	Home	Public Nonmotor-Vehicle
Total	**$88.4**	**$41.6**	**$31.4**	**$9.9**	**$7.0**
Wage loss	23.7	12.3	5.2	3.7	3.7
Medical expense	12.1	3.9	3.6	2.8	2.1
Insurance administration	15.3	9.2	5.9	0.1	0.1
Fire loss	6.4	(c)	2.0	3.3	1.1
Motor-veh. prop. damage	16.2	16.2	(c)	(c)	(c)
Indirect work loss	14.7	(c)	14.7	(c)	(c)

Source: National Safety Council estimates (rounded) based on information from the National Center for Health Statistics, state industrial commissions, state traffic authorities, state departments of health, insurance companies and associations, industrial establishments, and other sources.

[a]Cost estimates are not comparable with those of previous years. As additional or more precise data become available they are used from that year forward, but previously estimated figures are not revised.

[b]Duplications between work and motor-vehicle and home and motor-vehicle are eliminated in the totals.

[c]Not included, see comments by class of accident above.

FIGURE 5.1. The cost of the accident phenomenon to the United States (1982). (Reprinted with the permission of The National Safety Council.)

1982 accident cost components

TOTAL—ALL ACCIDENTS[b] .... **$88.4 billion**

These costs include: (billion)

Wage loss **$23.7**

Since, theoretically, a worker's contribution to the wealth of the nation is measured in terms of wages, then the total of wages lost due to accidents provides a measure of this lost productivity. For nonfatal injuries, actual wage losses are used; for fatalities and permanent disabilities, the figure used is the present value of all future earnings lost.

Medical expense **$12.1**

Doctor fees, hospital charges, the cost of medicines, and all other medical expenses incurred as the result of accidental injuries are included.

Insurance administration cost **$15.3**

This is the difference between premiums paid to insurance companies and claims paid out by them; it is their cost of doing business and is a part of the accident cost total. Claims paid by insurance companies are not identified separately, as every claim is compensation for losses such as wages, medical expenses, property damage, etc., which are included in other categories above and below.

Property damage in motor-vehicle accidents **$16.2**

Includes the value of property damage to vehicles from moving motor-vehicle accidents. The damage is valued at the cost to repair the vehicle or the market value of the vehicle when damage exceeds its market value. The cost of minor damage (such as scratches or dents incurred while parking) is considered part of the normal wear and tear to vehicles and is not included.

Fire loss .... **$ 6.4**

Includes losses from building fires of $5.7 billion and from nonbuilding fires, of $0.7 billion. By class of accident these totals break down as follows. Building: work $1.7 billion, home $3.2 billion, public $0.8 billion. Nonbuilding: work $0.3 billion, home $0.1 billion, public $0.3 billion.

Indirect loss from work accidents **$14.7**

This is the money value of time lost by noninjured workers. Includes time spent filling out accident reports, giving first aid to injured workers, and time lost due to production slowdowns. This loss is conservatively estimated as equal to the sum of lost wages, medical expenses, and insurance administration cost of work accidents.

See source and footnotes on page 4.

FIGURE 5.2. Cost components of the accident phenomenon to the United States (1982). (Reprinted with the permission of The National Safety Council.)

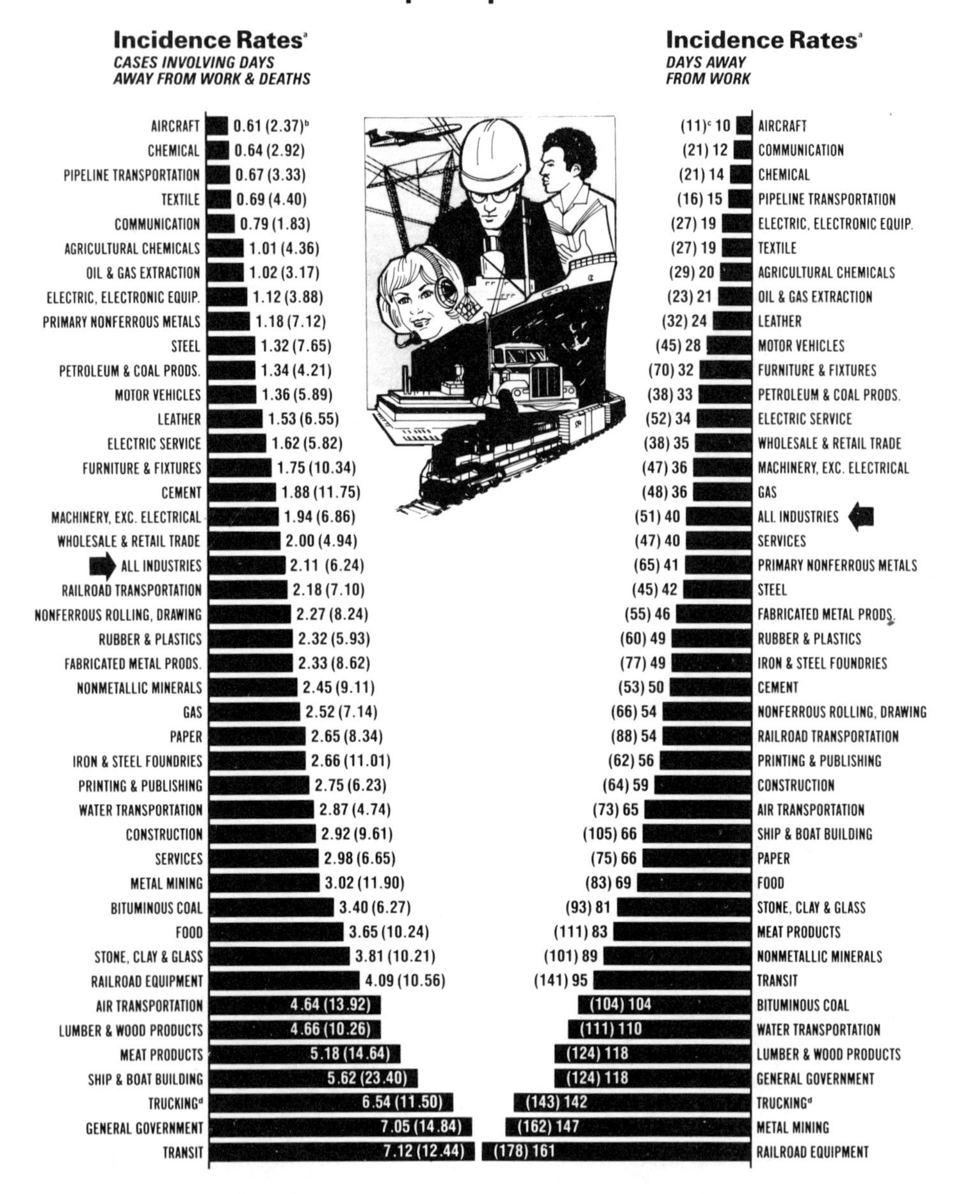

Source: Based on reports of National Safety Council members participating in the *Award Plan for Recognizing Good Occupational Safety Records.* These rates should not be interpreted as representative of the industries listed or of Council member companies. Data complied in accordance with OSHA recordkeeping definitions. See page 30.

[a]Incidence rates per 100 full-time employees, using 200,000 employee hours as the equivalent.

[b]Rates in parentheses are Total Recordable Cases.

[c]Rates in parentheses are Total Lost Workdays.

[d]Excludes SIC code 4215.

FIGURE 5.3. Industry incidence rate comparisons (1982). (Reprinted with the permission of The National Safety Council.)

millions of men and women. Many of these firms have had and continue to have employee accidents or have recognized their exposure to potential adverse experience, dictating the implementation of sound procedures to guard against such occurrences. Withholding moral and humanitarian arguments, the cost of these accidents is an unproductive drain on our national, local, and plant-level economies. Rapid changes in materials, machines, and processes have multiplied the areas of risk or uncertainty for all concerned.

The recent tragedy in Bhopal, India, which resulted in over 2000 lost lives and approximately 150,000 injuries, reinforces our thought that, given today's technology, how could the worst industrial accident in history have occurred? Hopefully, a proper accident investigation will yield the relevant causal factors so that corrective action can be taken to prevent similar occurrences.

Most organizations have accident investigation procedures. However, these procedures are often based upon antiquated accident causation theories and do not inspire the investigator to find and correct the defects in the management system that allow hazards to exist and accidents to occur.

The purpose of this chapter is to explore the process of effective accident investigation and analysis techniques and the procedures that will guide the investigator in gathering facts, analyzing causes, and suggesting effective corrective actions.

Several industries have adopted excellent accident investigation and analysis techniques that have helped lead those industries to a point where accidents have been minimized. Such industries as nuclear, aerospace, and energy use systems safety techniques. These techniques, though somewhat simplified, will be discussed and proposed for use in general industry.

5.1.1 Definition of an Accident

Before we begin to investigate accidents, it would seem appropriate that the investigator should be familiar with the definition of *accident*. A most basic dictionary definition of accident is "an unexpected, unwanted event" or perhaps "an event occurring by chance or from unknown causes." For our purposes, we see immediately that part of this definition is not valid as we cannot accept unknown causes. The investigator's job is to uncover accident cause(s). All accidents do have causes.

As we will see, "accident" definitions vary in order to accommodate the sometimes differing philosophies of various authors. However, the safety profession would most likely describe its current philosophies as evolutions or refinements to safety knowledge over a period of time. Neither safety in general nor accident investigation specifically can be considered an exact science.

DeReamer defines accident as the end product of a sequence of acts or events that result in some consequence judged to be "undesirable," such as minor injury, major injury, property damage, interruption, production delay, or undue wear and tear (DeReamer, 1980) W. G. Johnson's definition refers to a "mishap" as an unwanted transfer of energy that produces injury to persons, damage to property, degradation of ongoing process, or other unwanted losses (Johnson, 1980). Grimaldi and Simonds have stated:

> Damage need not accompany an occurrence for it to be given the label (accident). A semantic problem therefore emerges. While many people view every case of unexpected damage as accidental, others will not. An auto mechanic, who washes engine parts in gasoline and smokes while doing it, will be said by some to have had an accident when burned in the inevitable fire this practice can cause. Others will say that the event could not be termed an "accident," since it was predictable therefore not unexpected.
>
> *Grimaldi and Simons, 1975.*

5.1.2 Causes of an Accident

In a manner similar to the change in accident definition, methodologies for the recognition of accident causes have also changed over the years. Recently, F. A. Manuele concluded that accident causation is somewhat complex and requires broad consideration of personnel, equipment, materials, and the environment and, particularly, management decision-making that impacts on design, installation and maintenance, accepted operational practices, performance accountability, training, and so on (Manuele, 1982). Mr. Manuele's writing is a condensed collection and perhaps a culmination of numerous writings by safety theorists over the last 50 years or so. In the beginning, safety practitioners recognized that accidents were caused by physical conditions. As a result, they began to identify and correct physical conditions judged to be hazardous. In the early 1930s however, H. W. Heinrich recognized that *people* cause accidents, not just conditions. He stated:

> The occurrence of an injury invariably results from a completed sequence of factors, the last one of these being the injury itself. The accident which caused the injury is in turn invariably caused or permitted directly by the unsafe act of a person and/or a mechanical or physical hazard.
>
> *Heinrich, 1950.*

Heinrich's accident-factors sequence was represented by five dominos:

1. Ancestry/social environment
2. Fault of a person
3. Unsafe act/condition
4. Accident
5. Injury

The objective, of course, was to remove one of the dominos, which in turn would stop the "falling" sequence leading to injury. Like electricity following the path of least resistance, it was quickly recognized that the easiest factor to remove was "unsafe act or condition." Hence, a practical approach to accident causation was born and, by the nature of its simplicity, is still being applied today. Heinrich went on to say, "in accident prevention, the bull's-eye of the target is in the middle of the sequence—an unsafe act of a person or a mechanical or physical hazard," but subsequently added, with less emphasis, that there were "underlying accident causes." What were these causes? A common accident example given by Dan Peterson is perhaps most practical. An investigation of a fall from a ladder using the Heinrich approach reveals that the unsafe act was climbing a defective ladder. The unsafe condition is a defective ladder. The correction is getting rid of the defective ladder. Looking at the same accident and searching for multiple causes (underlying causes) we ask:

1. Why was the defective ladder not found during normal inspections?
2. Why did the supervisor allow its use?
3. Didn't the injured employee know it should not be used?
4. Was the employee properly trained?
5. Was the employee reminded not to use the ladder?
6. Did the superior examine the job first?

The answers to these and other questions would lead to the following kinds of corrections (Peterson, 1978):

1. An improved inspection procedure
2. Improved training
3. A better definition of responsibilities
4. Prejob planning by supervision

Perhaps the reader can add additional questions and thereby add further to the list of corrections. The point to be remembered is that every accident has multiple contributing factors.

5.1.3 What is an Accident Investigation?

An accident investigation is basically the supervisor's analysis and account of an accident based on the factual information gathered by a thorough and conscientious examination of all factors involved. It is not a mere reptition of the worker's explanation of the accident. True accident investigation includes the objective evaluation of all facts, opinions, statements, and related information, as well as definite action steps to be taken to prevent a recurrence.

The time for accident investigation is always as soon as possible. The less time between accident and investigation, the better the information that can be obtained. Facts are clearer, more details remembered, and the conditions are nearest those at the time of the accident. The only situations that should be permitted to delay the investigation are those in which medical treatment is needed or the worker is emotionally upset.

5.1.4 Management Weaknesses

Bird and Loftus (1976) established "four major elements of subsystems in the total business operation"—people, equipment, material, environment—which "individually or in combination provide the source of causes that contribute to a downgrading incident." Are these not the same elements management must control in order to be effective? The same authors put forth a domino sequence of their own:

1. Lack of management control
2. Basic cause(s)—Origin(s)
3. Immediate cause(s)—symptoms

4. Incidents—contact
5. People-property-loss

It is important to note that the first domino of the accident causation sequence now takes into consideration weakness(es) of the management system. Unless we consider the management system as a fundamental root cause of accidents, we may find we are uncovering accident symptoms rather than causes.

For some time, it has been recognized that accidents are a nonnecessary business expense. The system safety analysis technique, Management Oversight and Risk Tree (MORT), presents certain concepts that lead to a new definition of accident. The elements of the definition that evolves, according to Johnson, are:

1. An unwanted transfer of energy
2. Because of a lack of barrier and/or controls
3. Producing injury to persons, property, or process
4. Preceded by sequences of planning and operational errors, which:
   a. Failed to adjust to changes in physical or human factors
   b. Produced unsafe conditions and/or unsafe acts
5. Arising out of the risk in an activity
6. Interrupting or downgrading the activity

This definition is of interest in that it describes the causes of unsafe conditions and unsafe acts as sequences of planning and operational errors. These errors, most assuredly, have been allowed to exist due to defects in management's system of control.

The points to be made by reviewing these two definitions are:

1. Accidents involve not only injuries but property damage, interruption, production delays, and other unwanted losses.
2. The basic causes of accidents are defects in the management system.

If an accident investigation program or procedure is to be effective, the procedure must investigate accidents that result in losses other than personal injury, and the investigator/analyst must be trained and motivated to identify defects in the management system as the basic causes of accidents. Providing the investigator with the appropriate training is easily accomplished. However, motivation is more difficult. This is examined in Section 5.4 of this chapter.

5.1.5 Why Investigate?

The investigator/analyst should have the purpose of the procedure clearly in mind before the investigation begins. In occupational injury accidents, the ultimate goal of the investigation should be to determine the basic causes—the management system defects—and how these causes can be eliminated to prevent a recurrence.

Investigation of accidents is the responsibility of all levels of management and the concern of every employee, but the line supervisor's unique position gives him or her special priority and responsibility in this function. First-line supervisors in the area where an accident occurs have certain qualifications and advantages other members of management do not have.

1. They know the most about the situation, have daily contact and familiarity with the people, machines, and materials and environment involved. They know the standard practices and circumstances in the area as well as the hazards.
2. They have a personal interest in identifying accident causes. To good supervisors, accidents are not simply figures and statistics, they are their people, their machines, their material, their environment. Accident investigations focus a welcome light on the conditions and hazards that could endanger the lives of their workers or damage their equipment and material.
3. They can take the most immediate action to prevent an accident from recurring. Being in direct control of the people, procedures, and property in the area gives supervisors the advantage of taking immediate corrective action and the greatest opportunity for effective follow-up.
4. They can communicate more effectively with their workers. Though a worker may be "employed by" the company, he or she "works for" the line supervisor, and knows that the supervisor is interested in his safety. To the workers, the front-line supervisors are the "management" they know best. They speak their language and even more important, understand it. In accident reporting and investigation the worker can "tell it like it is" to a front-line supervisor.

According to Ferry and others, the reasons for having the supervisor make the investigation are also the same reasons why the supervisor should *not* be making the investigation. The supervisor's reputation is on the line. Some of the causes uncovered are bound to reflect in some way on his or her method of operation. The supervisor's closeness to the situation may preclude an open and unbiased approach to the supervisor-caused elements that exist. The more thorough the investigation, the more likely the supervisor is to be implicated and found to be contributing to the event. It is hard to be objective when you are blowing the whistle on yourself. While it should never be the objective of an investigation to "hang" someone, there cannot help but be some stigma when one's own operation is involved. It requires all the fortitude and objectivity a supervisor can muster to carry out an investigation of his or her own department (Ferry, 1981). In organizations where the complexity of accident causation is understood, finding fault is minimized. The fault or defect is considered to be an element of the system, which may have allowed or possibly encouraged an individual to err.

The front-line supervisor has been taught since day 1 that unsafe acts and unsafe conditions cause accidents (remember Heinrich's domino sequence previously discussed). So, following instructions, the supervisor typically determines that the accident was caused by an unsafe act of the employee and that the appropriate corrective action would be to retrain the employee or tell the employee to be more careful. Typically, the supervisor will not address the basic cause of why the employee performed the unsafe act or why the unsafe condition was allowed to exist.

In order to arrive at the ultimate goal of identifying and correcting management system defects, a clearly identified procedure for reporting incidents, gathering facts, analyzing information, identifying causes, suggesting corrective actions, and evaluating corrective actions should be established. This procedure should be well planned, set forth in writing, and clearly understood by everyone involved.

No matter how conscientious supervisors may be, they cannot investigate accidents until they are aware of them. Certainly, major property damage and serious personal injury are seldom a problem from the standpoint of reporting. But what about the minor injury or property damage, the apparently unimportant little accident? It is in this "unspectacular" area that failure to report usually occurs.

There is no such thing as an unimportant accident. The immediate results, or effect, of the accident may be classified as minor, serious, or major, but this in no way means that the accident itself is unimportant. The cable that suddenly breaks probably started with an unsafe wire that frayed. The infection that possibly requires amputation may well have started with a small cut or scratch. But beyond these obvious dangers are many others. The unreported accident cannot be investigated or its cause(s) corrected. It becomes a time bomb, waiting to be triggered again. The fact that the results of the accident were minor may be pure blind luck.

5.2 THE ACCIDENT INVESTIGATION

5.2.1 Planning the Investigation

The process of accident investigation should be planned well in advance. This planning involves items that must be prepared before an accident occurs. Such items are:

1. Accident reporting policy
2. Investigator/Analyst assignment
3. Investigator/Analyst training
4. Assembly of tools for gathering facts
5. Development of report forms
6. Distribution of reports and recommendations
7. Evaluation of report conclusions and recommendations
8. Follow-up on recommendations

Accident Reporting Policy

The accident reporting policy should be developed and effectively communicated so that workers and members of management understand that all accidents should be reported to supervision immediately. This is extremely important, due to the need for preservation of evidence and accurate recounts of the incident from the injured and/or witnesses.

Communicating this policy should be accomplished through:

1. Orientation and training procedures
2. Rules and regulations
3. Periodic safety meetings
4. Employee publications
5. Informational signs on bulletin boards and in first aid areas.

Late reporting of accidents severly hampers the fact-gathering steps of the investigation process. However, if through the accident investigation process a trend of late reporting is identified, a weakness in the management system for developing, communicating, and enforcing policy will be evident.

Investigator/Analyst Assignment

The preplan must clearly designate who is responsible for investigating and analyzing the accident. First-level supervision is most often assigned this responsibility, which seems appropriate, since this person is the management representative closest to the day-to-day activities that resulted in the mishap. Supervisors are generally more familiar with their employees, the work areas, and the job procedures then any other member of management. If the first-line supervisor has responsibilities for production, quality control, and employee relations in the management system, this person should also have safety responsibilities, one of which is accident investigation. This is not meant to imply that the first-line supervisor should have total responsibility. In all instances, other members of management should be responsible to guide and assist the supervisor. Others involved should include:

1. *Safety coordinator* to assist the investigator in fact finding, causal analysis, and development of conclusions and recommendations.
2. *Personnel* to supply such pertinent facts concerning the individual as preexisting conditions, previous job assignments, mental capabilities, physical capabilities, and so on, which may have had a role in accident causation.
3. *Engineering* to supply detailed process information as necessary and to assist in developing effective engineering solutions to any problems identified.
4. *Purchasing* to supply information concerning purchasing procedures that may have contributed to the incident, such as safety equipment specifications, hazardous materials requisitioned, or procedures, materials packaging specifications, and so on.
5. *Maintenance* to assist in determining if the lack of preventive maintenance or any maintenance procedure contributed to the incident.

This list is not all-inclusive and is dependent upon the particular organization. However, the need for involvement of other managers and staff in the accident investigation and analysis process is evident.

Investigator/Analyst Training

An effective program for accident investigation and analysis cannot be implemented unless the investigator has been trained to recognize the complexity of accident causation. Any training program should address:

1. Reporting policies
2. Responsibilities
3. Accident causation theory
4. Report preparation
5. Interviewing techniques
6. General methods of hazard control
7. Use of tools (cameras, measurement devices, video tapes, recorders, etc.)
8. Fact analysis
9. Recommendation development
10. Follow-up procedures

Fact-Gathering Tools

The accident investigation preplan should address the availability of tools for use in gathering pertinent facts for the investigation.

Ted Ferry suggests having an accident investigation kit available for immediate use. This kit could include:

1. Camera and film
2. Clipboard, paper, and pencil
3. Copy of regulations and standard operation procedure
4. Magnifying glass
5. Reporting form
6. Sturdy gloves
7. High visibility tapes

8. First aid kit
9. Cassette recorder with spare cassettes
10. Graph paper
11. Ruler and tapemeasure
12. Identification tags for parts
13. Scotch tape
14. Specimen containers
15. Compass

In some instances it would be appropriate to include such environmental measuring devices as a noise level meter, a velometer, a pump with detector tubes, a light meter, and so on. The selection of this equipment is dependent upon the industry and the hazards likely to be encountered. Appropriate personal protective equipment (hard hats, safety glasses or goggles, respirators, coveralls) may also be needed.

Report Forms

Investigation report forms and checklists are necessary but often are found to be inadequate for investigation purposes. This is especially true for sections that address causal analysis and corrective actions. A more detailed discussion of the problems and suggestions for an accident investigation report form have been included in Section 5.2.6.

The preplan should address the development and availability of accident investigation report forms.

Distribution of Reports and Recommendations

When the accident investigation report is complete, it should be distributed for top-management review and action. The distribution procedure should be well thought out and developed in advance to assure that all involved parties will be included.

Depending upon the type and sensitivity of the report, certain parties should be excluded from the distribution. For example, it would probably not be desirable to distribute an internal report suggesting management system weaknesses to all employees or anywhere outside the organization.

Generally the number of copies of the report should be kept to a minimum. The report can be circulated within the management organization for the review process. Extra copies or copies of certain portions of the report can be made as necessary. Due to the possibility of litigation, which exists with any accident, the duplication and circulation of an investigation report should be decided in advance and based upon a need-to-know. Otherwise, the exposure to possible court actions may compromise the validity of facts included in the report.

Specific procedures for handling such serious accidents as multiple injuries, fatalities, environmental incidents, serious fires, and so on, should be established. The procedure for an unusual accident should address the following:

1. Reporting to corporate and/or division headquarters, emergency and regulatory authorities.
2. Contact with the news media.
3. Contact with affected families and employees.
4. Contact with the surrounding community residents.

The specific procedure should designate individuals responsible for these contacts and specify how the contacts are to be made, the type of information that should be made available, and to whom the individual responsible for the contact can turn for advice.

Very often the company's legal counsel and public relations department will need to be involved in any contacts. Inaccurate statements released during or shortly after an emergency can have a serious detrimental effect upon public relations.

Evaluation of Report Conclusions and Recommendations

The preplan for accident investigation should stress how the report conclusions and recommendations are to be evaluated and how the recommendations are to be completed. Whose responsibility is it to determine if the report conclusions and recommendations are appropriate? Usually, this would be left up to the safety coordinator or a safety committee. Additional conclusions and recommendations may be necessary. Many times the ultimate responsibility for the completion of recommendations will be left to the plant manager or top management authority, who should be responsible for assigning the responsibility for carrying out the suggested corrective action to a person or department.

Follow-up of Recommendations

After the recommendations have been submitted and action has been taken to complete the recommendations, it is essential that someone evaluate the system to determine if the corrective action has been completed as recommended and whether this corrective action is appropriate. Usually this will be left to the safety coordinator or safety committee, but the process should also involve the accident investigator/analyst who suggested the corrective action and also the immediate supervisor in the area.

A document on planning for accident investigation was prepared for the United States Energy Resource and Development Administration by the System Safety Development Center, which describes a systems approach to preparation for accident investigation (Nertney and Fielding, 1976). This document suggests many activities that have been discussed here, as well as a logic tree for clarifying the procedures. The planning tree is divided into eight areas:

1. Reporting and classifying events
2. Preservation of evidence
3. Structuring the board charter
4. Establishing the board
5. Conducting the investigation
6. Training personnel
7. Establishing reporting procedures and controls
8. Conducting Postinvestigation activities (related to the board)

5.2.2 Gathering Facts

At the Scene

The way in which the facts at the scene of an accident are gathered can have a profound influence on the final outcome of the accident investigation. The identification of the proper causative factors and appropriate corrective actions will be possible only if the at the scene accident investigation is well done.

It is important that the investigator take enough time to assess the situation at the scene to insure that another accident does not result from the hazards that might be present. However, the investigator should be aware that conditions tend to change very rapidly and that the evidence must be preserved as quickly as practical. The investigator should not attempt to gather evidence before emergency response personnel have completed their tasks and the scene is safe. Then, the first priority of the investigator should be to preserve the evidence.

Preservation of Evidence

After the accident has occurred, the accident scene may be crowded with spectators and others trying to clear the scene in order to get back to normal operations. The investigator should have the authority and be prepared to clear the scene of spectators and delay cleanup efforts until this evidence has been properly preserved.

Evidence may be physical, for example a broken handrail or missing machine guards. It may be environmental: poor weather conditions, poor lighting, oxygen deficiency. Evidence may involve personnel who witnessed the accident or witnessed the occurrences just before or just after the accident. Securing statements from those persons is an important part of preservation of evidence, as is obtaining any required help from an expert. Listed below are several methods for collecting and preserving evidence:

1. Making illustrations
2. Photographing
3. Collecting specimens
4. Recording weather conditions, if applicable
5. Securing important documents
6. Securing data tapes from computer operated equipment
7. Measuring environmental conditions

These are a few methods for gathering physical evidence at the scene of an accident. Other methods may be applicable, depending upon the type of accident and the industry in which the accident is involved.

Interviewing Witnesses

Witnesses must be interviewed as soon as possible after an accident has occurred, since their stories tend to be more truthful and less biased if they are interviewed shortly after the incident. The investigator must realize that the witness may be reluctant to discuss all the facts, out of fear of reprimand or discipline for himself or for a coworker. This fear can be eased if the investigator recognizes the possibility of its existence and uses the appropriate interviewing techniques.

Under no circumstances should the investigator attempt to interview an injured person until that person has received appropriate first-aid treatment and is in a stable condition. It might be better also to delay interviewing of witnesses who may be distraught. In these cases, a bit of judgment is necessary on the investigator's part. The basic rules for interviewing witnesses and the injured personnel are listed below:

1. *Identify the purpose of the investigation.* The injured person or witness can be placed at ease if they realize that the purpose of the investigation is to identify facts for the prevention of a recurrence. Fault finding and placing the blame are not the purpose of an investigation. This should be made clear to the witness.
2. *Get the witness's own version of the accident.* The investigator must allow the witness to tell his/her story. Interruptions and interjections from the investigator are not helpful in obtaining the witness's point of view.
3. Ask open-ended questions. Questions that cannot be answered with a simple yes or no should be asked. The witnesses should be asked to expound upon what they know.
4. *The investigator should repeat his or her understanding of what the witness has relayed.* This technique will allow the investigator and the witness to agree upon the story described by the witness.
5. *Allow the witness or injured person to feel as though he or she has had a part in the investigation.* The witness should be asked to identify the cause of the accident, the conclusions, and appropriate corrective actions. If the investigator asks for the witness's assistance in completing the investigation, the witness will be more likely to cooperate freely.

Investigators who are front-line supervisors have probably already developed some skills in interviewing and discussing problems with their personnel. They learn quickly to be good judges of character and can evaluate a witness's statements. They may be able to confirm the witness's statements with physical evidence or other witnesses' statements. This is good practice and should be encouraged. Discrepancies in the witnesses' stories should be handled with tact. It would be very detrimental to the accident investigation process if the witnesses were accused of being untruthful.

Photography

Photography is probably one of the most frequently used methods of preserving evidence in accident investigation procedures. If well done, photography can be a very useful source for evaluating the evidence at a later date. However, poorly done photographs can be worthless and a waste of time.

There are several good references on the subject of accident photography. Wood has done an excellent job in describing the technique for the use of photography in accident investigation in a recent article which should be reviewed (Wood, 1984).

5.2.3 Analyzing Causal Factors

Analyzing the causal factors involved in an accident is the most difficult yet the most important task that must be undertaken by the investigator. Often the investigator tries to identify *one* casual factor that contributed to the accident. If multiple causes are identified, the investigator may try to identify the main contributing cause. This is an error in judgment and planning.

If the goal of accident investigation is to prevent a recurrence of similar incidents and learn how to correct deficiencies in the management system, *all* contributing factors become important in the accident prevention process. What may be an insignificant contributing factor in today's accident may be the major contributing factor in a future accident. Therefore, it is prudent to attempt to identify all causal factors and develop control measure for each.

From the investigator's point of view the "all-cause" approach emphasizes the need to seek all causal factors and not to concentrate on one area to the exclusion of others. Ferry states that this approach is estimated to have raised the average number of correctable causal factors found in a mishap investigation from four to eight, a doubling of efficiency.

Many accident investigation procedures used in industry today require only that the investigator obtain the information necessary to complete the OSHA 200 Log or the insurance carrier's first report of injury. These procedures are dreadfully inadequate in identifying all causal factors, relating those facts to defects in the management system, and developing the appropriate actions to be taken.

Recent developments in accident analysis techniques have been based upon the *systems safety approach.* Techniques such as TOR (Technique of Operations Review) and MORT (Management Oversight and Risk Tree) have been developed to address the basic causes of accidents—the defects in the management system. Although these techniques are complex, most management level personnel can understand and use them to improve the accident analysis procedure.

Technique of Operations Review

Technique of operation review (TOR) was developed to identify management involvement in degrading situations, whether those situations be production problems, quality problems, or accident situations.

This technique does not focus on operator error or machine malfunctions, but rather places the focus upon the management system where those malfunctions took place. The procedure involves four steps:

1. State
2. Trace
3. Eliminate
4. Seek

State. In this step, the analyst is asked briefly to describe what happened. Questions are asked until every analyst is satisfied with the understanding of the incident.

Trace. The analysts are referred to a TOR Analysis Sheet, a listing of operational errors that focus on management systems. The analysts then are asked to identify which of the operational errors listed was the main contributing cause of the accident. When the main cause has been agreed upon, the other contributing factors, listed upon the analysis sheet, are traced. The analysts make a decision whether those contributing factors are actual factors in this particular accident or not.

The TOR Analysis Sheet is divided into eight sections, as illustrated in Figure 5.4. Each section lists a group of factors that may or may not contribute to the accident being investigated. The general categories of factors listed for each section are:

1. Coaching
2. Responsibility
3. Authority
4. Supervision
5. Disorder
6. Operational
7. Personal Traits
8. Management

The analysts are asked to begin the *trace* step by selecting the most appropriate contributing factor. TOR will lead the analysts into other areas, where more contributing factors will be uncovered. By the end of the process, the analysts will decide that a defect in the management system has contributed to the accident.

After *trace* is completed, the next step, *eliminate,* can begin.

Eliminate. In this step the analysts study the factors elicited during the trace step in more depth, and eliminate any factors that clearly do not apply in this case. The factors that are eliminated may still be problems, although they should be treated separately for this analysis purpose.

Seek. The final step in the TOR analysis is to seek to develop solutions to problems identified. The TOR analysis procedure offers no clearcut solutions, although by identifying problems, solutions may become more evident.

Peterson's example of TOR Analysis:

> To illustrate TOR analysis, picture an incident as a manager of Excel Devices Company first hears of it. A customer has received a palette load of devices as ordered but not one of the devices contained its essential electric motor. A man was dispatched with a box of motors to install them on the spot. Here's an irate customer, phone calls, travel expense, overtime, waste, and inefficiency. Instead of blaming someone, he sits down with his supervisors in a TOR session.
>
> The incident is described and brief question-answer discussion reveals the facts of the case. Analysis of anything requires facts. So the first five minutes or so is devoted to facts, not blame.
>
> Next, the manager directs his supervisors to the TOR form and directs them to select one number which they consider to be the direct proximate cause of the incident. He asks them to STATE IMMEDIATE CAUSE, the first step of TOR analysis. Discussion and disagreement ensue. The form must be learned, words and meanings hashed, and insights gained. This first step may absorb five or ten minutes (the manager should press his supervisors for consensus agreement at this point, and not let discussion degenerate into an aimless bull session). By pressing his demand for a decision, the group finally agreed on item 43 as the IMMEDIATE CAUSE. This number, 43, is jotted at the top of a sheet of paper and the TRACE STEP begins as illustrated below. Do not discuss each item in depth. The TRACE STEP begins as a hasty overview. The manager should press for a prompt decision, a minute or two on each item. Cause the group to

TOR TECHNIC OF OPERATIONS REVIEW

1 COACHING

Code	Cause	Related codes
10	Unusual situation, failure to coach (new man, tool, equipment, process, material, etc.)	44, 24, 62
11	No instruction. No instruction available for particular situation	44, 22, 24, 80
12	Training not formulated or need not foreseen	24, 34, 86
13	Correction. Failure to correct or failure to see need to correct	42, 20, 30
14	Instruction inadequate. Instruction was attempted but result shows it didn't take	15, 16, 42
15	Supervisor failed to tell why	44, 24, 83
16	Supervisor failed to listen	11, 81
17		
18		
19		

3 AUTHORITY (Power to decide)

Code	Cause	Related codes
30	Bypassing, conflicting orders, too many bosses	44, 13
31	Decision too far above the problem	36, 83, 85
32	Authority inadequate to cope with the situation	81, 83
33	Decision exceeded authority	20, 26, 14
34	Decision evaded, problem dumped on the boss	36, 14, 85
35	Orders failed to produce desired result. Not clear, not understood, or not followed	40, 46, 13, 15
36	Subordinates fail to exercise their power to decide	26, 12, 83, 85
37		
38		
39		

2 RESPONSIBILITY

Code	Cause	Related codes
20	Duties and tasks not clear	44, 34, 14, 53
21	Conflicting goals	80,
22	Responsibility, not clear or failure to accept	26, 14, 54, 82
23	Dual responsibility	47, 34, 13
24	Pressure of immediate tasks obscures full scope of responsibilities	36, 12, 51
25	Buck passing, responsibility not tied down	44, 26, 55, 60
26	Job descriptions inadequate	80, 86
27		
28		
29		

4 SUPERVISION

Code	Cause	Related codes
40	Morale. Tension, insecurity, lack of faith in the supervisor and the future of the job	15, 56, 64, 80
41	Conduct. Supervisor sets poor example	13, 84
42	Unsafe Acts. Failure to observe and correct	24, 11, 52
43	Rules. Failure to make necessary rules, or to publicize them. Inadequate follow-up and enforcement. Unfair enforcement or weak discipline	25, 36, 12, 52
44	Initiative. Failure to see problems and exert an influence on them	22, 34, 30
45	Honest error. Failure to act, or action turned out to be wrong	10, 12, 15, 81
46	Team spirit. Men are not pulling with the supervisor	40, 21, 56
47	Co-operation. Poor co-operation. Failure to plan for co-ordination	23, 25, 15, 66
48		
49		

FIGURE 5.4. (Technique of operations review (TOR) cause codes. (From Peterson, 1978. Reprinted by permission of McGraw-Hill Book Company.)

5 DISORDER

51 Work Flow. Inefficient or hazardous layout, scheduling, arrangement, stacking, piling, routing, storing, etc. 41, 24, 31, 80

52 Conditions. Inefficient or unsafe due to faulty inspection, supervisory action, or maintenance 21, 32, 14, 86

53 Property loss. Accidental breakage or damage due to faulty procedure, inspection, supervision, or maintenance 43, 20, 80

54 Clutter. Anything unnecessary in the work area. (Excess materials, defective tools and equipment, excess due to faulty work flow, etc.) 44, 36, 80

55 Lack. Absence of anything needed. (Proper tools, protective equipment, guards, fire equipment, bins, scrap barrels, janitorial service, etc.) 44, 36, 80

56 Voluntary compliance. Work group sees no advantage to themselves 40, 15, 41

57

58

59

6 OPERATIONAL

60 Job procedure. Awkward, unsafe, inefficient, poorly planned 44, 32

61 Work load. Pace too fast, too slow, or erratic 44, 51, 63

62 New procedure. New or unusual tasks or hazards not yet understood 43, 44

63 Short handed. High turnover or absenteeism 80, 40, 61

64 Unattractive jobs. Job conditions or rewards are not competitive 81, 46

65 Job placement. Hasty or improper job selection and placement 80, 86

66 Co-ordination. Departments inadvertently create problems for each other (production, maintenance, purchasing, personnel, sales, etc.) 45, 35, 13

67

68

69

7 PERSONAL TRAITS

(When accident occurs)

70 Physical condition — strength, agility, poor reaction, clumsy, etc. 44, 26, 65

71 Health — sick, tired, taking medicine 44, 24, 65

72 Impairment — amputee, vision, hearing, heart, diabetic, epileptic, hernia, etc. 44, 24, 65

73 Alcohol — (If definite facts are known) 80

74 Personality — excitable, lazy, goof-off, unhappy, easily distracted, impulsive, anxious, irritable, complacent, etc. 44, 13

75 Adjustment — aggressive, show off, stubborn, insolent, scorns advice and instruction, defies authority, antisocial, argues, timid, etc. 44, 13

76 Work habits — sloppy. Confusion and disorder in work area. Careless of tools, equipment and procedure 44, 13

77 Work assignment — unsuited for this particular individual 42, 65

78

79

8 MANAGEMENT

80 Policy. Failure to assert a management will prior to the situation at hand 24, 81, 83

81 Goals. Not clear, or not projected as an "action image" 83, 86

82 Accountability. Failure to measure or appraise results 36

83 Span of attention. Too many irons in the fire. Inadequate delegation. Inadequate development of subordinates 12, 86

84 Performance appraisals. Inadequate or dwell excessively on short range performance 20, 65

85 Mistakes. Failure to support and encourage subordinates to exercise their power to decide 36

86 Staffing. Assign full or part-time responsibility for related functions 66

87

88

89

FIGURE 5.4. *(Continued)*

decide briskly whether each item did or did not contribute to the incident, whether it is "in" or "out." Keep insisting on that decision to control aimless bull sessioning. Trace numbers are listed vertically as they develop. In this illustration, explanatory notes indicate the discussion and thinking that took place.

43 Rules
25 Buck-passing
36
~~12~~
~~52~~
44
~~26~~
~~55~~
60

Each box should have received a sticker indicating incomplete assembly. Instead, one sticker had been attached to the whole palette load. [For trace numbers follow item 43 on the TOR form, Figure 5.4.] These are listed vertically at left.

ITEM 25. Buck-passing contributed because it was not clear who should observe the sticker: the forklift operator, the loading foreman, or the assembly foreman. (Since item 25 is "in," four more trace numbers result.)

ITEM 36 Authority was ill used, the supervisors decided, since this rule violation is widely overlooked. Add numbers 26, 12, 83,85 to the list.

~~26~~
~~12~~
83
85
22
34
30
44
32
36

ITEM 12 "out," not a factor; hence crossed out. Note: Discussion of just four points has generated a long list looming ahead. Have faith; the TRACE STEP does end.

Press on, urging "in" or "out" decisions; you want insight and hasty overview, not in-depth discussion.

ITEM 52. "out." Conditions not at fault.

ITEM 44. "in." Initiative at fault. Add numbers 22, 34, 30 to the list.

ITEM 26 "out." Job descriptions were adequate.

ITEM 55. "out."

ITEM 60. "in." Assembly line frequently failed to receive needed parts thus producing incomplete devices at the end of the line. Add 44, 32 to the list.

ITEMS 26 and 12. "out" as previously decided.

ITEM 85. "in." Supervisors decide that mistakes were blamed, but inaction was overlooked. Add 36 to the list as shown on the TOR form.

Tracing is not yet finished, but it is more exciting to observe this process in a group than it is to read it, so let us stop. In practice, the TRACE STEP continues until it runs out in either of two ways. Usually, the "outs" overtake the "ins" and you come to the end. Or, sometimes, the final number on the list repeats the number at the top (in this case, item 43), and you have come full circle back to the beginning.

So far, we have illustrated the STATE step and have demonstrated the TRACE step, even though we leave it unfinished in the illustration. When the TRACE step is completed, two steps remain. To illustrate the next step, ELIMINATION, let me complete tracing without explanation. In brief, item 30 was accepted as contributing to the incident, and this in turn generated item 13 as "in." Thus from all the items assessed, we end up with the following list of operating errors which contributed to the incident, the following list of "in" items:

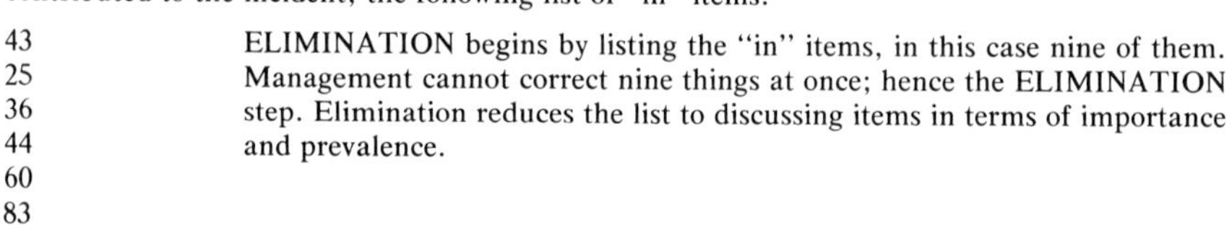

43
25
36
44
60
83
85
30
13

ELIMINATION begins by listing the "in" items, in this case nine of them. Management cannot correct nine things at once; hence the ELIMINATION step. Elimination reduces the list to discussing items in terms of importance and prevalence.

In this case, for example, items 13 and 30 were regarded as exceptions, not prevalent supervisory performance. Items 44, 36, and 25 were regarded for the moment as selfcorrective as a result of the insights gained in the TOR session. The elimination meant discussion and insights in depth, which can be appreciated by noting the meanings attached to the numbers. The elimination slid into the final step. SEEK FEASIBLE CORRECTIVE ACTION.

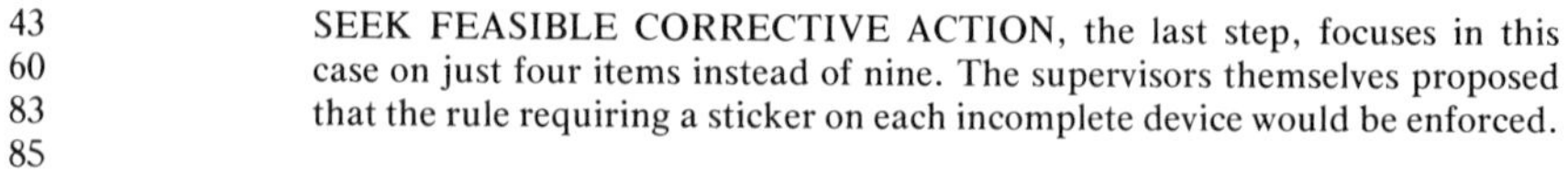

43
60
83
85

SEEK FEASIBLE CORRECTIVE ACTION, the last step, focuses in this case on just four items instead of nine. The supervisors themselves proposed that the rule requiring a sticker on each incomplete device would be enforced.

They further proposed to put the stickers on the side from which the forklift driver approached. Thus he would be sure to see them.

The manager began an investigation to determine why parts frequently failed to appear at the assembly line in proper coordination. He was also left pondering items 83 and 85, which had stayed "in," revealing that his supervisors felt that mistakes were noted but failure to act was largely overlooked.

I believe that TOR helps to identify a problem by tracing symptoms back to underlying root causes. A problem correctly defined is usually half solved. TOR helps in identifying and in defining problems by searching for root causes, causes that lead eventually to accidents and numerous other kinds of management losses.

Peterson, *Techniques of Safety Management.*

Management Oversight and Risk Tree

Management oversight and risk tree (MORT) is a more detailed procedure then TOR in that MORT attempts to describe all aspects of a perfect management system for safety. A risk tree is used to illustrate that accidents are a product of oversights and ommissions or assumed risks. *Assumed risks* are those hazards that management is aware of but has decided to accept the risk; either the corrective action is impossible, or it is not cost effective.

Figure 5.5 illustrates the basic logic of the technique. Oversights and omissions are caused by specific control factors and management system factors being less than adequate (LTA). The tree then continues with specific control factors and specific management system factors that could be at fault.

The MORT technique is accompanied by a list of questions that refer back to the logic chart. By answering these questions, the analysts will identify weaknesses in the management system and can propose improvements that will eliminate the causes of the accident.

The MORT analytical technique is difficult for the average supervisor and can be rather time consuming. It is certainly an approach that should be used for major accidents; however a simplified version would be appropriate for the routine accident that the first-line supervisor normally is required to investigate. A simplified procedure for supervisors has been described in the literature and may work well (Johnson, 1984).

In this simplified MORT technique, the analyst is asked to consider if defects in the elements of the safety management system could have contributed to the accident in question. When asked specific rather leading questions about the safety management system, the first-line supervisor can identify contributing causes beyond the unsafe act/unsafe condition. When these management-system defects can be identified as routine, appropriate corrective actions can be prescribed by the first line-supervisor.

Figure 5.6 is a logic chart for this simplified MORT procedure. This chart, along with the questions in Figure 5.7, can be used to direct an analyst's thought processes during the analysis.

The method described in the simplified MORT procedures referenced rarely take more then one hour to complete. On average, 30 minutes should be adequate. Surely the first-line supervisor can spare 30-60 minutes to analyze a defect in the management system and describe methods of control.

Several other methods of accident analysis are currently available. Some are being developed in conjunction with computer programs so that the paperwork is kept to a minimum. Any technique that focues on management-system defects and multiple causes of accidents may prove to be effective. The investigator/analyst must understand that unsafe acts and unsafe conditions are the proximate causes of accidents, but that these causes result from the basic causes, which are management-system defects.

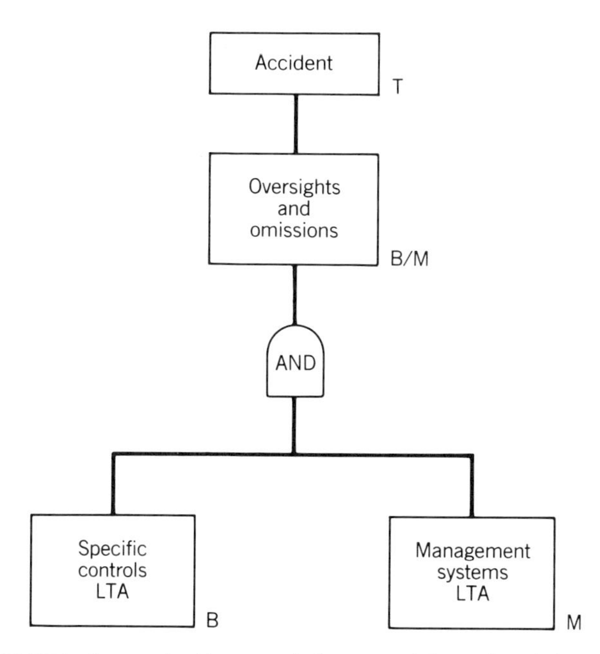

FIGURE 5.5. MORT logic tree. Accidents result from oversights and omissions which result from specific control factors and management system factors being less than adequate (LTA).

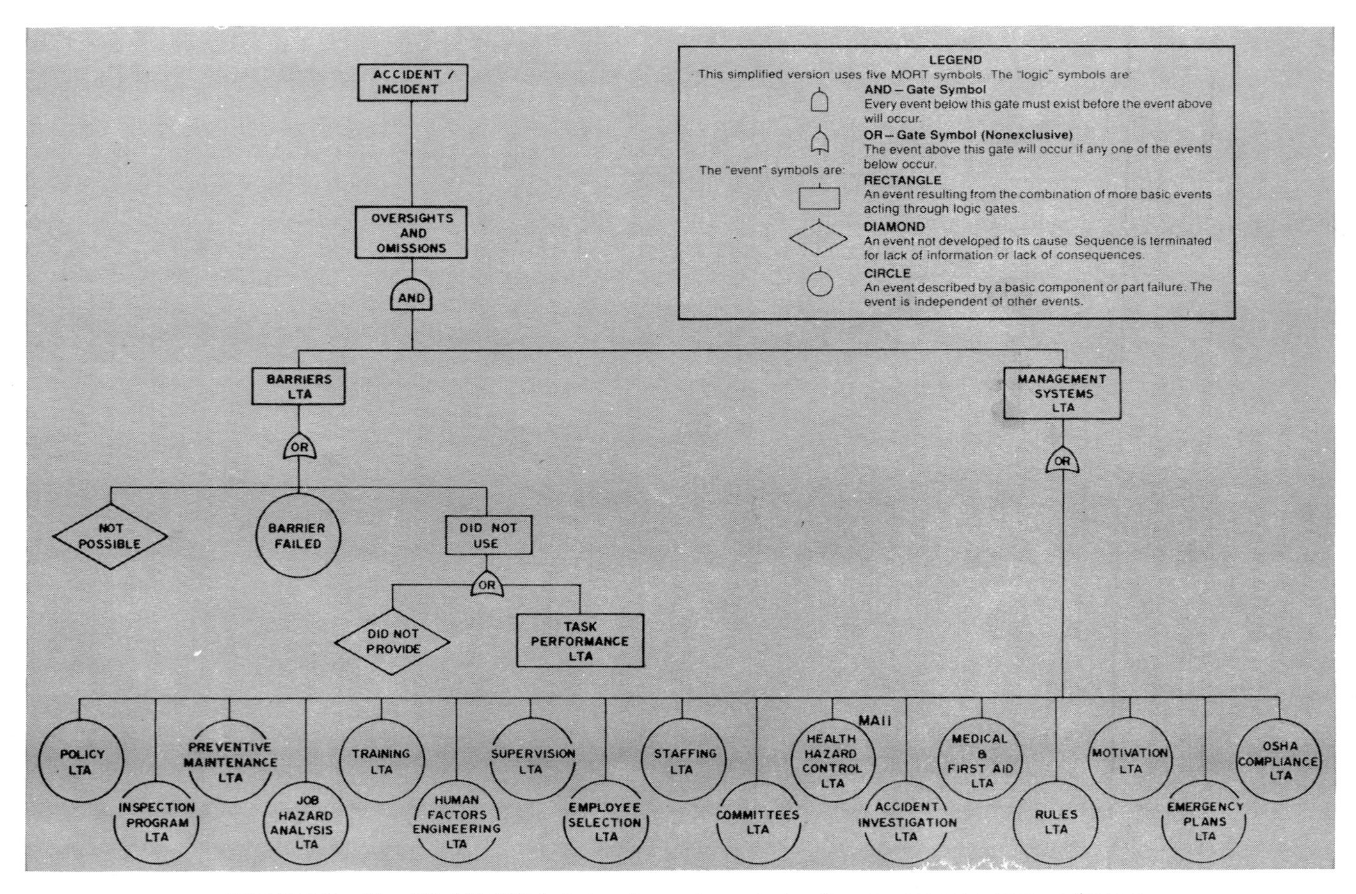

FIGURE 5.6. Simplified MORT logic tree for use in analyzing the management system weaknesses that contribute to the accident sequence. (From Johnson, 1984.)

ACCIDENT ANALYSIS QUESTIONS

Fundamental Questions

- What happened?
- Why?
- What were the losses?

Oversights and Omissions

- The fundamental causes of accidents result from oversights and omissions. Oversights and omissions result from barriers being less than adequate (LTA) and management systems being LTA.

Barriers LTA

- Were there adequate barriers on the unwanted energy?
- What were the specific barriers?

None Possible

- Top management must assume the risk for design in which no barriers to unwanted energy flow were possible.

Barrier Failed

- Did the barrier prevent the transfer of energy as designed?

Did Not Use

- Were barriers used?
  a) *Did not provide:*
  Were barriers provided where possible? (Top management must assume risk for failure to provide barriers.)
  b) *Task Performance Error*
  Were the provided barriers against the transfer of unwanted energy used properly?

Management Systems LTA

- Were all the factors of the management system necessary, sufficient, and organized in such a manner as to assure that the overall program would be effective?

Policy LTA

- Was there a written, up-to-date policy with a broad enough scope to address major problems likely to be encountered?
- Could it be implemented without conflict?

Inspection LTA

- Was there adequate inspection of equipment, processes, utilities, operations, etc.?
- Was the plan LTA? Was execution of the plan LTA?
- Are inspectors trained in hazard recognition?

Preventive Maintenance

- Was there adequate PM of equipment, processes, utilities, operations, etc.?

Job Hazard Analysis

- Had the job been analyzed?
- Did the analysis identify the hazard and appropriate safe work procedure?

Training

- Was the employee adequately trained in safe work procedures and hazard identification?
- Was the supervisor adequately trained?
- Were inspectors properly trained?

Human Factors Engineering

- Was consideration given in design, plan, and procedures to human characteristics as they interface with machine and environmental characteristics?

Supervision

- Was worksite supervision adequate?
- Were necessary supportive services adequate?

Employee Selection

- Were methods of personnel selection and placement adequate?
- Were the safety-related job requirements adequately defined so as to select and place the individual with adequate qualifications?

Staffing

- Did the safety staff have the proper organizational status and are they educated, experienced, and promotable?

Committees

- Were committees used to improve safety understanding and attitudes?
- Did these groups have a positive action orientation toward real-life problems?

Health Hazard Controls

- Were people and objects free from physical stresses caused by health hazards?

Accident Investigation

- Was the Accident Investigation program LTA?
- Have similar accidents occurred which were not properly analyzed for cause and appropriate corrective action?

Medical and First Aid

- Was the medical program adequate to detect any physical impairment of the employee?
- Was first aid rendered in a timely fashion to control the severity of the injury?

Rules

- Were safety rules published and enforced concerning performance related to the accident?

Employee Motivation

- Were the employee motivation, participation, and acceptance adequate?

Emergency Plans

- Was there adequate emergency action to prevent a second accident from occurring?
- Was the plan executed adequately?
- Was the plan designed adequately?

OSHA Compliance

- Were there written procedures to assure compliance with applicable OSHA requirements?

FIGURE 5.7. List of analysis questions for use with simplified MORT logic tree. (From Johnson, 1984.)

5.2.4 Conclusions

The appropriate analysis of an accident should lead to significant conclusions. The conclusions are based upon the clear evidence identified during the analysis, as well as other factors that may not have contributed clearly, but have been evaluated. Conclusions may be positive. For example, due to the good training of the crane operator and riggers, the load travel path was cleared of pedestrians before the load was lifted and no injuries resulted from the dropped load.

Of course, conclusions will identify weaknesses in the system more often then strengths. After all, an accident investigation is the result of an unwanted event or, in progressive organizations, the identification of the possibility of an unwanted event.

As described earlier, there are multiple contributing factors involved in all accidents. Separate conclusions should be developed to address each contributing factor identified. Some contributing factors may appear to be more important in the particular incidents being investigated. However, since the investigation is an evaluation of the system, all factors are important and should be stressed equally.

5.2.5 Recommendations

It has been suggested that the investigator may not be the best person to recommend corrective actions (Wood, 1979). In the occupational setting the first-line supervisor and/or safety coordinator are often saddled with the tasks of investigation and development of recommendations. There may be conflict in instances where the conclusions point out the investigator has not fulfilled assigned management duties. Upper management review is necessary, therefore, to insure that the investigator neither escapes corrective action nor places too much blame upon him or her.

Recommendations that address each conclusion drawn from the analysis should be developed. These sections should be practical, cost-effective solutions to the problem.

It is suggested that the recommendations be listed one at a time and that one person be assigned to complete the steps necessary to comply with a single recommendation. In many cases, the front-line supervisor will not have the authority to make this assignment—another good reason for upper management review and participation.

The recommendations submitted should lead to the ultimate purpose of the investigation procedure, suggestions for corrective action that must address the multiple contributing factors and result in their control or elimination. No investigation can be considered complete until recommendations are submitted and corrective action has been taken.

5.2.6 Reporting

If the accident investigator is a professional who spends a great deal of time investigating accidents, a narrative report is preferable. A narrative report should be divided into sections:

1. Summary
2. Detailed description of accident
3. Analysis of evidence
4. Conclusions
5. Recommendations

Most accident investigators/analysts are not professional investigators. They are first-line supervisors who have many duties and time restrictions. These investigators need the guidance offered by accident investigation forms and checklists.

The report forms should be divided into sections:

1. *General information.* Employee identification, dates, time, department, shift, and so on should be recorded. Also, information for OSHA and insurance carrier reports should be included in this section.
2. *Description of accident.* Allows the investigator to record a brief but thorough description of the event.
3. *Analysis of causal factors.* This section should guide the investigator to basic accident causes—management system defects.
4. *Conclusions.* The investigator should list the conclusions drawn from the analysis of causal factors.
5. *Recommendations.* Individual suggestions that address each causal factor should be listed.
6. *Approvals.* Signatures of the investigator, involved staff members, and upper management should indicate their review and approval.

7. *Follow-up.* The report should include a section in which investigator records the action taken as a result of the investigation. The form could be developed so that this section can be placed in a follow-up file for later recall.

The specifics of an investigation form will differ in almost every organization. Therefore, a particular form is not suggested. However, these rules should be used when developing a form:

1. *Do not* limit the report form to one page. Although one page is convenient, enough information seldom can be recorded to convey adequately the steps in the process.
2. *Do* guide the investigator to identify multiple accident causes. Many current forms motivate the investigator to check one from Column A, one from Column B, and one from Column C and perceive that the investigation is complete.
3. *List* areas of the management system to be considered. As in MORT, ask questions concerning inspection, maintenance, design, personnel selection and training, etc.
4. *Do not* omit upper management review and approval of the report.
5. *Insure* that a system for follow-up is part of the reporting mechanism.

The report is often the only part of the investigation that is available after a short period of time. Reports must be accurate, legible, and carefully thought out to be effective in preventing accidents. Otherwise, the investigation process may be a useless exercise in paper work, which should be avoided.

5.2.7 Accident Investigation Program Management

As in any element of the safety system, the accident investigation program must be properly managed. Management must take four basic steps to insure that the program is properly managed. These steps are:

1. *Define policy.* Policy can be defined in a simple statement that is distributed, reviewed, and enforced. The policy statement should clearly indicate that all accidents will be reported and investigated immediately and that accidents result from management system defects that are to be identified and corrected during the investigation process.
2. *Assign responsibility.* Upper management must decide on who will investigate the accidents, who will approve the reports, and who will be responsible for carrying out the corrective actions. This should be clearly stated in job descriptions or separate lists of responsibilities.
3. *Train those responsible.* Everyone who must investigate accidents should be trained in the procedures for gathering facts, analyzing causes, developing corrective actions, and reporting results. In particular, the training should stress causal analysis. Without good analysis, conclusions and recommendations may be worthless.
4. *Hold those responsible accountable for performance.* As in any assigned responsibility, performance of the associated duties must be evaluated periodically. If performance of assigned responsibilities is not measured, the duties become unimportant and will cease to be completed. An excellent method for evaluating and measuring an investigator's work is available (Bird, 1984.) The results of this evaluation must be tied into the motivation promotions of those responsible. Motivational factors such as promotions continued growth, and continued assignment of responsibility are affected by most performance appraisal systems.

5.3 THE REALITIES OF ACCIDENT INVESTIGATION

Most of this chapter thus far has stressed meaningful assessments of all mishaps, regardless of whether such mishaps result in a near-miss, property damage, production delay, or injury. We understand that this assertion no doubt has created conflicts and confusion in the minds of many readers. The ideal approach is to examine and investigate all undesired events, but the real world places limitations on time, staffing, and other resources. So the reader may very well ask. What really is to be investigated and how far should the investigation be carried? The answers vary accordingly to the beliefs of management within any one organization.

5.3.1 Management Philosophy

Though the rapid growth of technology includes new accident investigation methods as well as production materials, machines, and processes, it appears that the development of these production factors is rapidly exceeding our abilities to conduct effective accident investigations. Ferry states:

> The advancement of inflation and the advent of new investigative tools has made good investigation possible and necessary, but perhaps a luxury. For the investigator who might explain to a cost-conscious management the higher cost of a thorough investigation as well as the need for such investigation calls for more then the mere skills of investigating mishaps.
>
> Tight money reduces the availability of investigative funds at the very time at which there is increased need for a good prevention effort, based at least in part on thorough investigation. The message to investigators is for more thorough investigations using the best available techniques. It calls for getting the maximum amount of usable information for a minimum cost without jeopardizing the investigation results. Justification for good investigation must be worked out in terms of payoff, fewer mishaps, and increased effectiveness of operations. This situation can be compounded by an excellent mishap experience in some organizations, meaning that it becomes increasingly difficult to make reductions in mishap rates. An increased effort just to stay even or make modest reductions will not be cost-effective to management unless they are extremely far-sighted. Few will say it, but there is a tendency to accept some level of mishap loss as a cost of doing business. Management may be sympathetic to an intensified investigation effort, but the effort may just not seem worthwhile.
>
> One solution to this myriad of problems is better investigation with roughly the same resources. This is done by knowing precisely what to look for, how to go about it in an efficient manner, and what techniques to use for the resources invested.
>
> Ferry, *Modern Accident Investigation and Analysis*

There are undoubtedly two extreme management philosphies concerning accident investigation. First, there are those who do not accept the need for effective accident investigation but still succumb to economic and legal pressures for conducting such investigations. This "going through the motions" is perhaps one of the worst management defects of all. This is the same management that believes accidents are a cost of doing business and ask why more time and money should be spent on investigating something that has cost too much already. On the positive side is management that recognizes that investigating accidents does have value but that unfortunately the investigating process is reactionary in nature. This same management seeks not only to benefit from the investigating process but also to identify hazards and management-system defects before accidents result. Job Safety Analysis (JSA) is a proven technique utilized by progressive managements in industry today to identify hazards and develop associated control measures.

5.3.2 Job Safety Analysis

Job Safety Analysis (JSA) or Job Hazard Analysis (JHA), a systematic study of work procedures on a job, is utilized by those firms who desire to identify and control hazards before such hazards result in injury. JSA or JHA today is often referred to simply as Total Job Analysis (TJA) because the benefits of TJA are intended to exceed those of JSA or JHA by encompassing safety as one of the many inseparable parts of management's and supervisor's jobs.

In the concept of the supervisor's job, safety is not considered a separate and distinct responsibility, but rather an integral part of many other responsibilities such as maintaining productivity, quality, and morale. A job performed without regard to the prevention of property damage or personal injury could hardly be termed efficient. Efficiency demands maximum use and control of personnel, equipment, materials, and the environment in any job. A proper JSA will determine the hazards associated with each of the preceding factors in addition to quality and production factors.

There are four basic steps in a JSA:

1. Select the job to be analyzed.
2. Break down the job into a series of steps.
3. Determine factors that relate to production, quality, safety.
4. Eliminate or otherwise control factors conducive to error or potential losses.

Job Selection. The job to be selected can be based on any one of the following factors:

1. *Accident frequency.* A job that has produced a large number of accidents is a prime candidate. The higher the number of accidents-the higher the priority.
2. *Severity of accidents.* Jobs that have experienced a serious or costly accident should be analyzed.
3. *Severe injury potential.* Some jobs may have no history of accidents but exhibit the potential for a serious accident.
4. *New jobs.* JSA will reveal hazards and prevent accidents on new jobs before poor accident experience develops.

5. *Seasonal or infrequently done jobs.* These jobs and their associated hazards are often overlooked until they result in a serious accident. Analysis can identify acceptable hazards.
6. *Manual material-handling jobs.* Jobs that have a high degree of manual handling should be analyzed. Emphasis should be placed on job design to eliminate as much of the manual handling as possible.

Breaking Down The Job Into Steps. Any job can be examined critically when it is broken down into basic steps. One should keep in mind that each job must be looked at systematically, that is, the job should be broken down into steps that describe what is done, in what order. The action should be described but details should be excluded. Only basic steps need to be described—neither too detailed nor too broad.

The key to job breakdown is to select the appropriate worker to observe. This person should be experienced, capable, and cooperative. Brief the worker on the purpose of the JSA and ask him or her to participate in the development of the JSA. Common errors when preparing a JSA are that the job steps are broken down too fine or not fine enough.

Identified Production/Quality/Safety Factors

Can anyone be struck by something (chain, hook, etc.) or be contacted by anything (electricity, hot metal, acid, etc.)?

Can anyone strike against or make contact with anything?

Can the employee be caught between, caught in, or caught on, anything?

Can the employee slip, trip, or fall in any way?

Can the employee strain or overexert?

Can the employee be exposed to such health hazards as radiation, toxic fumes, chemicals, gas, heat smoke, and so on.

Can the employee injure any other employees?

Eliminate and/or Control. The final step in the JSA is to eliminate or control the hazard by one or more of the following methods:

Eliminate
Substitute
Isolate
Guard
Administration
Personal Protective Equipment
Training and Education

The initial benefits of a JSA will be noticed while the analysis is still being developed. Supervisors will find that they are learning more about the jobs they supervise. Workers will take more interest and pride in a job when they are asked to help in its analysis. With these comes an increase in ideas and suggestions for ways of increasing productivity, improving quality, eliminating waste, and improving safety. As a consequence, costs are usually reduced. A JSA can be used for job instruction, as it enables anyone to learn how to do the job in the best way and in the shortest time. It also standardizes the job procedure so that everyone learns to do the job the same way.

5.4 SUMMARY

The accident investigation process does not have to be totally reactive. If well managed, the process can prove effective in preventing future accidents, and can identify weakness in the management system that affect production, quality, and employee relations, as well as safety.

> While it does not seem practical to expect line-supervisors to become professionals in accident investigation, a better understanding of the breadth of accident causation, resulting from the training programs conducted by safety professionals, would promote more relevant questions and a greater effectiveness.
>
> Although a much better job can be done in training line-supervisors and other management personnel to more effectively investigate accidents, there is a practical limit to realistic expectations. Surely, the line-supervisor and those to whom supervisors report have a responsibility for

accident investigation. For those accidents which have a potential for, or have achieved, serious adverse results, the counsel of a safety professional is needed. It's assumed that, to fulfill that responsibility, the safety professional will have mastered a management systems approach to accident investigation.

Too many safety professionals have been obsessed with unsafe acts and unsafe conditions as accident causes. Ted Ferry wrote that, "While the traditionalist will seek unsafe acts or conditions, the systems person will look at what went wrong with the system, perceiving something wrong with the system's operation or organization that allowed the mishap to take place."

If your accident cause determinations and your subsequent analyses are principally directed toward unsafe acts and unsafe conditions, and don't go further into management systems, your investigation and analyses procedures are obsolete.

Manuele, *Professional Safety Magazine,* October 1982

Mr. Manuele concludes with certainty that the quality of supervision and management training with regard to accident causation should be improved. Though we agree wholeheartedly, moving back one step, it appears the availability and/or frequency of training needs some improvement as well. Safety training, in our opinion, is sadly overlooked in many businesses and industries, even to the extreme of no training at all. We need hear only once such a management response as "we just cannot take our supervisors off the job for this kind of training" to know that we have yet an additional problem.

Before any king of safety training, whether for management, supervisors or employees is recommended, objectives of such training should be stated specifically and understood. The National Safety Council's *Accident Prevention Manual for Industrial Operations* (1981), lists the general objectives of safety training as follows:

1. To involve supervisors in the company's accident prevention program
2. To establish the supervisor as being the key person in preventing accidents.
3. To get supervisors to understand the nature of their safety responsibilities.
4. To provide supervisors with information on causes of accidents and occupational health hazards and methods of prevention.
5. To give supervisors an opportunity to consider current problems of accident prevention and to develop solutions based upon their own and other's experience.
6. To help supervisors gain skill in accident prevention activities.
7. To help supervisors do the safety job in their own departments.

To this, we would like to add an eighth objective, *accident investigation training should emphasize the management system rather then the employee.*

It should be understood that an evaluation of management systems with regard to accident investigation will require training. As well, the investigation process systems as proposed in this chapter will require additional time, certainly more time than if no investigations have been conducted in the past. The results, however, will be a much improved and more efficient management system.

REFERENCES

Bird, F. E. Jr., *Management Guide to Loss Control,* Institute Press, Atlanta, 1984.

Bird, F. E. Jr. and Loftus, R. A., *Loss Control Management,* Institute Press, Loganville, GA, 1976

DeReamer, R., *Modern Safety and Health Technology,* Wiley, New York, 1980.

Ferry, T., S., *Modern Accident Investigation and Analysis,* Wiley, New York, 1981.

Grimaldi and Simonds, *Safety Management,* 3rd ed., Richard D. Irwin, Homewood, IL, 1975.

Heinrich, H. W., *Industrial Accident Prevention,* 3rd ed., McGraw-Hill, New York, 1950.

Johnson, R. D., Simplifying MORT for Supervisors, *National Safety News,* **130,** No. 3, September 1984.

Johnson, W. G., *MORT—Safety Assurance Systems,* Marcel Dekker, New York, 1980.

J. W. G., *Mort—Safety Assurance Systems,* Marcel Dekker, New York, 1980.

Manuele, F. A., Accident Investigation and Analysis, an Evaluative Review, *Professional Safety Magazine,* American Society of Safety Engineers, October 1982.

National Safety Council, *Accident Prevention Manual for Industrial Operations,* 8th ed., Chicago, 1981.

National Safety Council, Chicago, *Accident Facts,* 1983.

Nertney, R. J. and J. R. Fielding, *A Contractor Guide to Advance Preparation for Accident Investigations;* U.S. Energy Research and Development Administration, 1976.

Peterson, D., *Techniques of Safety Management,* 2nd ed., McGraw-Hill, New York, 1978.

Wood, R. H., How Does the Investigator Develop Recommendations?, *Annual Seminar of the Society of Air Safety Investigators Proceedings,* Montreal, ISASI, September 1979.

Wood, R. H., Accident Photography, in Ted S. Ferry, Ed., *Readings in Accident Investigation,* Charles C. Thomas, Springfield, IL, 1984.

CHAPTER 6

How to Establish Industrial Loss Prevention and Fire Protection

David E. Miller
Mobil Oil Corporation
Princeton, New Jersey

Richard J. Petrucco
Mobil Oil Corporation
Princeton, New Jersey

Alphonse A. Corona, Jr.
Consultant
Yardley, Pennsylvania

The information provided in this chapter was current and reliable at the time of publication. Regulatory and technical requirements are subject to change, hence, the information detailed herein is only in compliance with current regulations in effect at time of publication. Consult OSHA regulations or seek legal counsel for accurate information on current requirements. The writers disclaim any liability resulting from reliance on obsolescent or other inaccurate information presented in this chapter.

6.1 INTRODUCTION

6.1.1 Basic Process Safety

The integration of safety into any plant, especially on processing toxic or flammable liquids or gases, involves several disciplines. The initial requirement is to determine the toxicological and flammability properties of the products. The next is to determine whether the products can be processed safely. These determinations may require review by a toxicologist, industrial hygienist, process engineer, equipment design engineers, an instrument engineer, and a safety engineer, among others.

6.1.2 Equipment Safety

After the process has been approved as safe, then safety must be built into the plant hardware. Pressure vessels, pumps, fired heaters, and other equipment must be designed to withstand the operating conditions, which include abnormal upsets such as exothermic reactions, failure of instrument controls, and human failures. In anticipation of such failures, safety alarms and shutdown controls must be designed into the process. Wherever possible, automatic safety shutdowns should fail in the safe position. Where practical, backup manual systems should be provided

6.1.3 Operator Safety

Next, operators must be trained in the safe operation of the unit, in how to detect incipient equipment problems, and finally, in what action to take when an emergency arises. Procedures must be written for normal plant operation and also for any anticipated emergency. These procedures should be available to all affected personnel and periodically reviewed and updated to reflect changing conditions. Management should publicize to plant personnel that they support these procedures and take the necessary steps to assure that the procedures are being practiced.

6.1.4 Safety Audits and Training Programs

In addition to day-to-day safe plant operations, periodic safety audits should be made and firefighters be kept trained to respond to any emergency. An emergency response plan should be developed utilizing plant and municipal medical staff, plant security staff, and mutual aid provided by other local industries. The emergency response plan should be tested periodically to insure effectiveness.

Safety cannot be initially designed into a plant and then forgotten. Maintaining safety is an on-going daily struggle. Even when it is maintained to the highest degree, there is no guarantee that accidents will not occur. However, the converse is true: *If safety is not observed, accidents are inevitable.* Plants for the most part are still run by people. Even if plants become completely automated with robots and controlled by computers, people will still be required to assure safety.

6.2 TYPES OF EMERGENCIES

Emergencies may vary from minor incidents, such as a small flammable liquid spill or flange leak, to a major disaster such as a massive release of toxic material or a large vapor-cloud ignition. Some emergencies can be prevented or at least minimized, while others, which involve parties outside the plant or natural disasters, often cannot be avoided. However, if the type of emergency cannot be controlled, at a minimum plans should be developed and people trained for action if the incident occurs. Some types of emergencies are:

- Flammable and combustible liquid leaks and spills, with or without ignition.
- Release of toxic materials.
- Flammable vapor-cloud formation, with or without ignition. Ignition may result in deflagration or detonation.
- Equipment failures resulting in unit shutdown, fire, or deflagration.
- Accidents resulting in injury or loss of life.
- Spills causing damage to the environment.
- Loss of plant electrical or steam power, cooling water, or instrument air, resulting in unplanned sudden shutdown of one or more process units and possibly a fire.
- Natural disasters such as floods, hurricanes, earthquakes, blizzards.

6.3 RISK

6.3.1 Definition

A risk is the possibility of an undesirable occurrence. The only justification for taking a particular risk is that the reward clearly exceeds the penalty if the associated accident takes place. In order to decide whether to take a particular risk, it must be quantified carefully.

6.3.2 Quantification

To quantify a risk, it is necessary first to determine what the risk is and the extent of damage in a worst-case scenario. Next it is necessary to determine the probability of the accident occurring. In evaluating the probability of an event, there are *two current popular hypotheses:* (1) A disaster that never happened, is about to happen. (2) A disaster that has never happened, won't happen. Both of these hypotheses are false. Three Mile Island proved that a nuclear reactor meltdown was possible and Chernobyl made it a reality. A likely risk should not be accepted unless the penalty of the incident is also accepted. If the loss of property and production is acceptable, then the risk is acceptable. Of course, this also assumes that all reasonable precautions have been taken to eliminate or mitigate the penalty. Pure chance-taking is never good business practice, and when the risk involves the possible loss of life, then it should not be accepted.

6.3.3 Methods for Risk Quantification

In order to quantify a risk, some of the following tools can be used:

Past experience:

What has caused an incident, how often it can be expected to occur, and what the expected extent of damage is. To make this determination, a database should be constructed from experience within the plant or company and throughout identical and similar industries. This is a costly, time-consuming procedure, and where very few incidents have occurred, the validity of the information may be questionable. Some data on the reliability of instruments appears in Anyakora, Engel, and Lees, (1981). Probabilities for equipment failure can be found in Browning, (1969).

Logic models.

These are concerned primarily with the "on/off" connection between variables and incidents and are useful for analyzing complex systems.

Two common techniques used in logic models are Failure Modes and Effects Analysis (FMEA) and Fault Tree Analysis (FTA).

FMEA breaks down a system into its simplest components and asks "what if" questions regarding all possible failure modes for each component. The probability of a failure occurring in the system is the sum of the failure probabilities for each component.

The advantages of FMEA are that it highlights components with the highest probability of failure and permits improving these components. FMEA results in a systematic piece-by-piece review of each component and permits establishing inspection and test programs to monitor the performance of high-failure components.

Fault tree analysis, which goes further than FMEA in that it identifies failures in subsystems, is often used in conjunction with FMEA. It determines the consequence of two or more component failures occurring simultaneously.

6.4 HAZARD IMPACT

6.4.1 Identifying Potential Hazards

As stated previously, the risk-taking decision should be based not only on the probability of an incident occurring, but also on the potential impact or consequences of the incident. *Consequence analysis* can be the most difficult phase of overall hazards analysis. Some of the most serious hazards involve the formation of toxic and flammable vapors. It is often extremely difficult to determine the maximum quantity of flammable vapor that may be involved in an explosion or the maximum quantity of a toxic release that could be expected. As an example, the percentage of a total gas release that can be in the flammable range at any given time has been quoted by various sources as between 3 and 10%. Thus the assumed amount of explosives could be incorrect by a factor of 3. One method of calculating peak overpressure of an explosion is to determine TNT equivalent based on the heat of combustion per pound of the chemical, related to TNT. This is a rough correlation, but is used because more is known about the pressure waves and destruction caused by the detonation of TNT than that of any other explosive.

6.4.2 Consequence Analysis

From the calculated extent and intensity of potential damage, the impact on both plant/personnel and the surrounding community can be determined. Using this data, plans can be formulated to cope with the results.

The determination of the impact on plant and personnel should consider the following factors:

- The number of people exposed in the plant onsite areas and in administration buildings, shops, warehouses, laboratories, etc., both during the day and at night.
- The extent of property damage within the plant.
- In-plant and outside first aid response.
- The time required for cleanup and repair or replacement of damaged equipment and the lost production cost.
- The effect on employees who may lose work if a portion of the plant is shut down for an extended period.
- Backlash legislation, that is, the passing of local, state, or federal laws that would increase the cost of doing business. An example is mandating additional safety features that are not necessary, forcing the plant to relocate or shut down.

In determining the impact of an incident on the surrounding community, the following factors should be considered:

- The maximum number of people outside the plant exposed to a hazard day and night.
- The type of people exposed and the probability of panic, i.e., adults, children, senior citizens, handicapped.
- Medical response: the time required to evacuate injured as casualties increase.
- Extent of property damage outside plant, including lost production and lost work time.
- Damage to the company's overall public image.
- Passage of laws that could cause considerable added cost without appreciable added safety.
- Permanent loss of some markets from boycotts and lost production.
- Personal injury and damage suits.

6.5 LOSS PREVENTION

6.5.1 Loss Prevention Objectives

The objectives of loss prevention are:

a. To minimize the risk of fires and explosion in industrial plants where flammable and combustible materials are processes, stored, and shipped. Examples are petroleum refineries, petrochemical and chemical plants liquid fuel storage terminals. Flammable and combustible liquids are defined in NFPA Standard 321 (1982).

b. Prevent gas and liquid releases that could result in the formation of large flammable and toxic vapor clouds. Such clouds could expand past the plant boundaries and expose to serious hazards the public and property outside the plant.

6.5.2 Developing a Loss Prevention Program

The development of a loss prevention program that will meet the objectives stated in Section 6.5.1 should start with the philosophy of the plant design and operation. If the plant is designed to operate with numerous automatic controls and a minimum of personnel, then emphasis may be made on automatic safety and emergency shutdown systems. If the philosophy is to automate safety systems, then a sophisticated inspection and maintenance program should be established to assure that such automatic systems will function during an emergency and that frequent false alarms will not occur. On the other hand, if sufficient personnel are available both in the control room and in the process areas to detect and cope with emergencies, then manual shutdown and isolation systems may be advisable. Manual systems are usually more reliable than automatic, require less maintenance, and give trained operating personnel the option of activating these emergency protection systems. One disadvantage to manual systems is that, in some cases, operators may not have sufficient time to determine the cause of the hazard and take the proper emergency corrective measures. It must be emphasized that reliance on manual systems demands extensive training and auditing to ensure employee familiarization with systems operation.

A loss prevention program is implemented by:

a. Identifying all possible plant hazards.
b. Evaluating the probability and risks of these hazards.
c. Eliminating or controlling these hazards.

6.5.3 Hazard Identification and Probability

Identification of hazards and their probability is facilitated by:

EXPERIENCE WITH SIMILAR UNITS
Some chemical processes have an inherent hazard such as high pressure or temperature, an exothermic, or "runaway," type of reaction, involve the formation of a toxic material, or use hazardous reactants such as corrosive acids.

PLANT ACCIDENT REPORTS
Most companies maintain plant accident reports, which may provide a good reference for anticipating hazards and estimating probability. Many companies are computerizing these accident reports to facilitate access and compile probability indices.

STEP-BY-STEP REVIEW OF PROCESS
A detailed review of each step of the process should be made by following through process flow and piping and instrument drawings. This review can be more effective if made with the process engineer familiar with the process and the operator familiar with operations.

SEMINARS AND MEETINGS
Seminars and meetings periodically conducted by trade organizations and professional societies such as the American Petroleum Institute, The National Petroleum Refiners Association, the American Society of Mechanical Engineers, American Institute of Chemical Engineers, and the Society of Fire Protection Engineers. These organizations provide access to a cross-section of experience and practices throughout similar industries.

PUBLISHED PAPERS
Papers published in journals often identify problems at the inception stage. This permits correction of problems without having to repeat the "learning experience."

INSURERS
The insurer of a particular plant usually has compiled a history of losses with documentation of events causing these losses of all of the sites insured by the company. Thus, the insurer's experience may point out similar problems that occurred in other plants.

6.5.4 Hazard Elimination and Control

After a hazard has been identified, the potential results should be assessed to determine which risks are acceptable and which are not. Risks involving human life and health are not acceptable. Some risks involving property only may be acceptable to a limited degree. Providing no fire protection for a plant and letting it burn down on the premise that it is fully insured and can be rebuilt is *not* an acceptable risk. This is true even when economics are against fire protection.

Some methods used to eliminate or control hazards are:

a. *Changing the process conditions where possible.* For instance, reducing the pressure or temperature, or substituting a less hazardous material for one or more of the reactants may be possible.

b. *Changing the design of process equipment.* Using pumps with double seals or can-type pumps for hazardous materials; using storage tanks with no bottom piping connections for refrigerated hydrocarbon storage, such as LNG; eliminating piping sliding and bellows-type expansion joints and proprietary couplings having soft elastomeric ring gaskets; using self-reinforcing forged nozzles rather than nozzle reinforcing pads for high pressure vessels and exchangers; upgrading equipment materials, such as using higher alloys for high temperature equipment and equipment exposed to highly corrosive liquids and gases.

c. *Installing remote operated safety shutoff valves or interlocks that automatically operate such valves when an emergency occurs.* A few examples are remote operated valves on bottom outlets of vessels and storage tanks containing large volumes of flammable or toxic liquids; check valves in large diameter piping to prevent backflow if a break occurs (only when flow is always in the same direction); excess flow valves that automatically close at high flow velocity resulting from a pipe rupture; automatic valves interlocked with a pressure sensor to shut-in a line when the pressure drops because of a piping failure; air-operated or spring-loaded valves equipped with a fusible plug or link to close when exposed to fire.

d. *Installing fixed automatic fire-fighting equipment to control or extinguish a fire and minimize losses.* Some protective systems are: automatic fixed halon extinguishing systems in computer rooms or inside gas turbine generator housings; sprinkler systems in buildings; high-density fixed water sprays designed for fire intensity control; low-density water sprays to cool shells of pressure vessels when exposed to fire to prevent rupture due to overheating the metal.

6.5.5 Safe Location of Onsite Plant Facilities

Siting the Process Unit

Onsite plant facilities are considered to be those located within the limits of the process units, commonly call the battery limits.

When locating a new process unit, a number of factors should be considered. First, prevailing winds and terrain should influence the location.

Where there is a distinct prevailing wind, process units should be located up-wind of tankage area, because flammable vapors resulting from a storage tank pumpover could be ignited by hot or fired process equipment. Where there is a pronounced slope to the terrain, process units should be located at a higher elevation than the tankage, to prevent a flammable or combustible liquid spill from running into the process unit.

Another factor to be considered is the proximity of process units to the surrounding properties both within and outside the plant. Where possible, process units should be located at least 200 ft (60 m) from the property line, plant administration building and other offices, maintenance shops, and warehouse.

The rectangular block system of process unit arrangement with access roadways on all sides is the preferred layout to facilitate maintenance and fire fighting. A good spacing between nonintegrated process units handling flammable liquids is about 100 ft (30 m) for safety requirements: integrated units designed to start up and shut down simultaneously may be 50 ft (15 m) apart. Greater spacing for nonintegrated units is necessary to perform welding and other hot maintenance work while the adjacent unit is on-stream.

Control Buildings

Large modern plants commonly utilize a single control building from which several process units are operated. This method assures that more than one operator will always be available in an emergency. However, if the control building is exposed to a fire or explosion from an adjacent process unit, the control of several process units could be lost simultaneously. Therefore, it becomes more urgent to provide better protection for a central control building.

This can be accomplished by providing greater spacing between the building and process units and designing the building to be blast resistant without windows. A spacing of 100 ft (30 m) is usually considered adequate. When locating the control building, remember that all process units should be easily accessible to the operators; otherwise they will not be able to make corrective actions inside a process unit during an emergency.

Process Unit Layout

The usual layout of a process unit is a rectangular plot with the process-unit pipe rack running through the center of the unit plot, in the long direction. The plant main pipe rack is always off the process unit plot, to protect it from damage in a process-unit fire. Process equipment is laid out on both sides of the process-unit pipe rack. This design keeps pipe runs to a minimum and improves access to equipment from under the rack. No process equipment should be located under the rack, so that access will not be impeded, and piping will not be exposed to a fire resulting from equipment failure.

Process Equipment Layout

Table 6.1 shows a suggested minimum safe spacing between process equipment. This information must be used with discretion, however, and the particular application should be carefully considered. The size of the equipment, severity of operating conditions, and materials being processed have an important bearing on safe spacing. Equipment spacing less than that shown in Table 6.1 does not in itself mean

TABLE 6.1. Suggested Minimum Spacing Between Process Equipment

Equipment	From	Spacing Ft	Spacing m
Fired heaters and	Pumps, exchangers, gas compressor, pressure vessels, air coolers, main pipeways	50	15
	Fired heaters	25[b]	7.5[b]
	Pumps[a]	10	3
Gas compressors and	Pumps, air coolers	25	7.5
	Pressure vessels	30	9
	Exchangers, main pipeways	30	9
	Fired heaters	50	15
	Gas compressors	6	2
Pressure vessels and	Fired heaters	50	15
	Main pipeways, gas compressors	30	9
	Exchangers,[a] pumps[a] pressure vessels,	15	4.5
	Air coolers Pumps, exchangers	10	3
Pumps and (under auto-ignition temperature)	Fired heaters	50	15
	Gas compressors	15	4.5
	Exchangers,[a]	15	4.5
	main pipeway	15	4.5
Gas compressors and	Air coolers Exchangers,[a] pressure vessels, air coolers	10	3
	Pumps[a]	5	1.5
	Pumps	3	1
Pumps[a] and	Main pipeway	30	9
	Gas compressors	25	7.5
	Exchangers, pressure vessels, air coolers	15	4.5
	Fired heaters	10	3
	Pumps	5	1.5

[a]Equipment handling flammable liquids over auto-ignition temperature (A.I.T.).

[b]Spacing only when a heater may be shut down for entry and maintenance while the adjacent heater is in operation. When heaters shut down simultaneously for maintenance, the spacing may be reduced to 5 ft (1.5 m).

that an unsafe condition exists, but it is an indication that a safety review may be recommended. When the spacing is less than that which is normally considered safe, compensation often can be made by remote or automatic emergency shutdown (ESD), automatic fire protection or detection, or some other means short of relocation.

Process Heaters

A fired process heater is usually the most hazardous equipment on the process-unit site, since it is an ever-present ignition source for flammable vapors released from other equipment. Also the process coil is subject to severe temperatures and, frequently, very high pressure. The preferred location for heaters is at the periphery of the plot, at one end, where it is easier to maintain a free buffer zone between heaters and other equipment and also provide greater safety for fire fighters.

Pumps

Pumps handling flammable liquids are also potential problems, because fires have occurred frequently from seal leaks. The preferred location of pumps is at the edge of the process unit pipe rack, just outside the rack, with the driver end towards the rack. Although this location exposes the pipe rack to a pump fire, it has the advantages of minimizing piping and providing easy access under the rack in an emergency.

Pumps should not be located beneath other equipment such as pressure vessels and air coolers.

Gas Compressors

Fires from seal leaks or flange leaks are less common than those from pumps and fired heaters, but it is good practice to locate compressors at the periphery of the plot, somewhat isolated from other equipment containing flammable liquids and gases. This is basically defensive, to protect a very costly piece of equipment, which cannot be replaced or repaired readily, from fires originating at other equipment.

Pressure Vessels

Pressure vessels, as a source of fires, have a low hazard incidence. However, many vessels, such as crude oil desalters and accumulators, may contain a large volume of flammable liquid. If such vessels are exposed to fire they can rupture, resulting in a BLEVE (boiling liquid expanding vapor explosion). This type of failure usually results in extensive damage to the process area and sometimes loss of life to operators and fire fighters. For this reason it is good practice to keep such vessels well away from fired heaters, but usually they may be as close as 10–15 ft from other equipment. Providing extra fire protection, such as fixed water monitors or fixed water sprays, is often done to protect those higher risk vessels from fire exposure.

Air Coolers

Fin-fan air coolers have a relatively good safety record. The most common hazard is from a leak between a tube and a tube sheet due to a poorly rolled tube. These leaks are usually small and easily corrected without damage.

Air coolers may be safely located 10 to 15 ft from other equipment, with the exception of fired heaters. Most air coolers are located above the top level of the process-unit pipe rack. In this location space can be saved, permitting more generous spacing of other equipment inside the process unit. Preferably pumps should not be located beneath coolers. When this cannot be avoided, fixed water sprays designed for high fire intensity control should be installed over the pump.

Heat Exchangers

Shell and tube heat exchangers also have a good safety record. Most incidents involve a shell flange leak, particularly in high temperature hydrogen service. Exchangers are usually located 15–20 ft from most equipment except fired heaters. They are often stacked in two or three levels to save space. Other equipment should not be placed above exchangers located at grade.

6.5.6 Process Equipment and Facilities Design

Designing Safety into Process Equipment Operation

The safe design of most process equipment is usually assured by recognized codes such as the ASME Boiler and Pressure Vessel Code, ASTM and ANSI standards, and Tubular Exchanger Manufacturers Association (TEMA). The finished product is hydrostatically tested, inspected by a registered inspector, and stamped. However, after the equipment is installed, it is the responsibility of the user to assure safe operation. This section discusses design features to protect equipment from poor operation, faulty fabrication, inadequate maintenance, and circumstances beyond the control of the user.

Fired Heaters

Most heater firebox explosions occur during light-off, usually due to the failure of the operator to follow the recommended light-off procedure. After an unscheduled shutdown, the temptation is great to take a "short cut" to make up lost production. Therefore, it is mandatory that the company establish a clear operating procedure and see to its enforcement. A "prover" type system to assure that all pilot and main burner fuel valves are completely closed may be installed. This involves a one-inch valved bypass around the pilot control valve. When the bypass valve is opened, the pilot control valve cannot be opened unless the pressure in the bypass reaches operating pressure. After about 30 seconds, a timer automatically closes the fuel bypass valve. Such systems can improve safety on light-off, but are no substitute for a good procedure, enforced.

Other safety features normally recommended are:

Low Fuel Pressure Shutdown. A valve on both the pilot and main burner fuel lines can be interlocked to sequentially shut down on main burner fuel low pressure and low pilot fuel. These valves, commonly called "chopper valves," require resetting in the field so that an operator must go to the site and determine what caused the valve to trip.

Coil Backflow Prevention. The most effective way to extinguish a heater fire resulting from a process coil rupture, is to shut off flow in the coil. This can be done by the use of a check valve on the coil outlet and a remote operated valve on the inlet to the coil. In fouling service, where deposits such as coke are laid down inside the coil, it is necessary to install a remote operated valve on the coil outlet, because a check valve may fail to close. Ability to close valves remotely, that is, from the control room or in the field (at least 75 ft, or 23 m, away from the heater) may be provided.

Snuffing Steam. A provision for injecting steam for purging flammable vapors in the fire box prior to lightoff should be present. This steam can be used also for snuffing out a fire resulting from a ruptured process tube. Valves for injecting steam may be manual and are usually located a minimum of 50 ft (15 m) from the heater to permit access during a fire. On large heaters it may be advisable to install remote operated valves for quick steam injection.

Remote Damper Control. Remote control of the heater stack damper should be considered, at least for large duty heaters (over 20 million BTU/h) and especially for heaters sharing a common stack. Besides helping to control the fire and limit damage inside ducting, improved operating economy will result.

Stack Thermocouple. A thermocouple with high temperature alarm may be used to detect a fire resulting from a ruptured coil. A thermocouple becomes more important with a windowless control room and little or no manpower in the field, because it provides quick fire detection.

Pressure Vessels

Some pressure vessels contain large quantities of flammable or toxic liquids that would create a serious hazard if released to the atmosphere. Although pressure vessels are inherently safe, precautions must be taken to prevent leaks and spills. Some protective systems are:

Shut-off Valves. A shut-off valve is usually provided on all bottom connections so that the vessel contents can be controlled in the case where piping has holed-through due to corrosion or damaged by an accident. A shut-off valve should seriously be considered when the vessel contents of flammable or combustible liquid exceeds 50 barrels (8 m^3). The valve is usually operated manually, but in the case of large diameter piping a remote operated valve should be considered.

Depressuring Valves. A depressuring valve is often used to reduce pressure inside an uninsulated vessel when it is exposed to fire. The vessel relief valve, set to open at the vessel maximum allowable working pressure, will not protect the vessel during a fire of long duration. If the vessel metal becomes overheated (above 1000°F, 540°C) the shell may rupture at the normal operating pressure. The valve is usually remotely operated and discharges to the flare header. When the valve is located close to the vessel, the valve operator may require fireproofing to assure that it functions properly when exposed to fire. Guidelines for designing depressuring systems are provided in API Recommended Practice 521, 1982.

Unless the normal operating pressure of the vessel is over 100 psig (690 kPa), it becomes difficult to reduce pressure effectively before the metal temperature rises to a high level; otherwise, the valve and depressuring line would have to be of very large diameter.

Pressure Relieving Systems. To prevent pressure vessels, exchangers, pumps, and other pressure-containing equipment from being overpressured and damaged, relief valves are installed. These valves are generally spring loaded devices that open when the pressure of the system reaches the maximum allowable working pressure. Relief valves should be located as specified in the ASME *Boiler and Pressure*

Vessel Code, Section VII, Division 1, API *Recommended Practice* 521 Section 2, and API *Recommended Practice* 520, Part 1, Sections 4, 5, and Appendix B.

The relieving load for each valve should be calculated based on upset conditions such as electrical or steam power failure, cooling water failure, blocked outlet, or fire. The valve is sized for a single maximum load. The total release from valves in the process unit is collected in the unit flare header, which then goes offsite to the flare knockout drum and is burned at the flare.

It is desirable to release only clean, nontoxic flammable vapors into the atmosphere. This can be done safely only if the vapors are:

- Lighter than air
- Heavier than air but molecular weight not exceeding 80 and with a discharge velocity of 500 ft/s (150 m/s) to disperse vapors

Water Draws. Water accumulates in the bottom of some pressure vessels and periodically must be drawn off to an open sewer. This operation may be performed by manually operating a valve or by a control valve set to open automatically on high water level and close on low level. Both methods can be hazardous. The manual system requires that an operator remain at the valve whenever water is being drawn, to prevent the escape of flammable liquids or gases when the water level falls too low. If the operator is not alert, a combustible vapor cloud can form, when the vessel contains a light vapor pressure material, and expose both the operator and equipment. This hazard can be reduced by keeping the water draw line small, preferably one inch; installing a low-level alarm to warn the operator; and keeping the water draw valve as far as practical from the point of emergence (usually an open drain hub going to the sewer). When the product is a high vapor pressure material that auto-refrigerates, there should be two water draw valves at least two feet apart. The downstream valve is opened first and then the upstream valve is opened. If the down stream valve freezes in the open position, the upstream valve can be used to shut off flow.

While the precautions mentioned above enhance safety during manual water drawing, the responsibility for safety lies mainly with management. The following items may be utilized to assure the required protection:

- Training personnel.
- Warning signs.
- Written procedures.
- Fresh-air breathing equipment.
- Color coding piping to identify hazardous materials.
- Plugging or capping drains.
- Toxic or combustible gas analyzers with visual and audible alarms.

When water is drawn automatically from vessels to an open sewer, the control valve may fail in the open position, permitting combustible vapors to escape. When automatic water draws are used, the vessel usually should have low water level guage to alarm in both the field and in the control room.

Whenever a vessel contains a toxic liquid material, the water draw should discharge to a closed system, but never to the flare header, which is generally designed for vapor releases only. Even with other less hazardous materials, it is preferable to discharge to a closed system, when feasible.

Compressors

Compressors generally require protection from fire exposure from more hazardous equipment. However, some inherent hazards should be guarded against. The most serious is a seal leak that could cause the formation of a vapor cloud, resulting in a fire or deflagration. The severity of the event can be minimized by installing a remote shut-off valve on the compressor inlet and a check valve on the outlet to stop the flow of process gas.

If possible, compressors should not be installed inside closed buildings. When a shelter is advisable, the sides should be open to facilitate ventilation. Sometimes it becomes necessary to install compressors indoors in very cold climates or to install enclosures around them for noise supression. In these cases, the building or enclosure should be ventilated, sometimes using fans, to prevent the accumulation of combustible vapors. The installation of combustible gas analyzers should be considered, to detect any accumulation in the early stages and permit shutting down the compressor. An inerting and fire extinguishing system using halon or CO_2 may be advisable inside noise-suppression enclosures.

Pumps

Most pump fires result from seal leaks, which usually are a result of lack of proper maintenance. An experienced operator or maintenance person can detect a bad seal or bearing and shut down the pump before a failure occurs.

Pumps should be located outdoors where possible, rather than indoors, where combustible vapors can accumulate.

All pumps handling flammable liquids should have steel casings. Cast iron can crack readily when exposed to fire, releasing additional fuel.

Air Coolers

Fin fan air coolers are prone to damage from any nearby fire, because the fan will induce the hot combustion gases, exposing the tubes. A clearly identified fan shut-off switch located at grade, about 50 ft (15 m) from the cooler, will permit shutting down in an emergency.

Exchangers

Shell and tube heat exchangers sometimes require cleaning while the process unit remains on stream. If this is anticipated, provisions should be made to isolate the exchanger using blinds to protect personnel. A valve alone should never be relied upon for tight shut-off.

Piping

Piping should be adequately supported, guided, and anchored. The practice of hanging piping from rods should be avoided, since such supports are vulnerable to failure in a fire. The use of sliding and bellows expansion joints and proprietary couplings using rubber rings should also be avoided, because they are prone to failure. Preferably, expansion loops should be used. Small size piping can be damaged easily by vibration and impact and should be adequately supported.

Emergency Shutdown Valves

Emergency shutdown (ESD) valves may be manual, solenoid, pneumatic, or motor operated. Solenoid and pneumatic operated valves should be designed to fail in the safe position. The safe position is not always obvious. Should a valve for depressuring a pressure vessel fail in the open or closed position? The safety engineer prefers the valve to fail *open* in case it becomes inoperable in a fire, but the operator wants the valve to fail *closed*, to prevent all depressuring valves from opening upon loss of instrument air. Both conditions create a hazard. This situation may be resolved by installing back-up air cylinders or installing depressuring valves in non-fire-exposed areas.

Where these valves cannot be made to fail safe, they should be fireproofed to assure operation during a fire. Fireproofing should include electrical power and valve operator, but not the steel valve body. When stainless steel instrument air tubing is used, it usually requires no fireproofing.

Electric motor operated valves fail in the last position and must be fireproofed to assure operation.

ESD's are used to isolate systems where a hazard exists, such as at fired heaters, compressors, vessels containing large quantities of flammable materials or toxic materials, or long pipelines where a break could occur, and so forth.

Battery Limit Valves

Battery limit valves are used to isolate a process unit from the main pipe rack and other process units to permit safe shutdown. In some applications they are used to shut in the entire plant. They can also be used in an emergency to shut off feed to a process unit during a fire. The valves usually are located on top of the unit pipe rack before joining the main pipe rack. Since they are used infrequently, they are generally manually operated and should be located in a fire-safe area to be tenable in an emergency.

Drainage

Grading within the process unit is important, not only for the removal of storm water, but also for the removal of flammable and combustible liquid spills. The preferred design of the concrete slab is to have the high point of paving beneath the centerline of the process unit pipe rack with catch basins located on both sides of (but not under) the rack. This design will cause any piping spill to run from beneath the rack and also prevent a spill from some other source from running beneath the rack. No catch basin should be located beneath equipment, especially fired heaters. Catch basins should be fire stopped by a turned down elbow or over-under weir, both having a six-inch liquid seal. These liquid seals will prevent a fire from going through the entire sewer and also stop an explosion at the first seal. The seal or fire stop is usually located at the inlet to the catch basin, with a straight-through outlet, so that flammable and combustible liquids will not be trapped in the catch basin.

Storm water and oily water may be combined in a single sewer system, but it is becoming common to have separate storm and oily water sewers. With secondary and tertiary water treatment systems being installed, economics dictate segregating the two types of water to reduce the quantity of water being treated.

The storm water system should be designed to handle both the maximum rate of rainfall expected and the fire water-rate anticipated, but not simultaneously. The fire water-drainage requirement should take into account the fixed monitors, fixed water sprays, and hose lines that are likely to be utilized. This does not include the total fire water that could be used by activating all fire-water systems simultaneously, but the equipment expected to be required to fight a fire. A total of 4000 gal/min (908 m^3/h) is often used to fight a fire in one process unit. Catch basins are often designed to handle 500 gal/min and the maximum drainage area for one catch basin usually does not exceed 3000 ft^2 (279 m^2).

Always assume that a sewer is in the explosive range. Whenever hot work is performed in the area, adjacent catch basins should be covered with a tarpaulin over which sand is spread to maintain the seal.

Electrical Classification of Areas

The classification of hazardous locations for electrical equipment is defined in NFPA 30—1984, Flammable and Combustible Liquids Code and NFPA 70—1987 National Electric Code, Article 500, and API 500A, Classification of Locations for Electrical Installations in Petroleum Refineries, 1982. Once an area has been classified, the National Electric Code specifies the type of electrical equipment that can be installed safely. However, the problem often is in determining the extent of the classified area. Classification is by Class, Group, and Division as shown in Table 6.2.

Most areas inside plants handling flammable liquids will be classified as Division II. Division I areas are those where heavier-than-air gases are present in below grade areas such as sumps, open trenches, and sewers; enclosed areas containing equipment handling flammables such as compressor enclosures and pump houses; and immediately around vents. With lighter-than-air gases, Division I areas might be found in the roof portion of an inadequately ventilated compressor house.

Electrical switches and other sparking devices located in Division I and II areas should be explosion proof or intrinsically safe. Motors located in Division II areas should be nonsparking type rated for Division II service; motors in Division I areas should be explosion proof.

"Explosion proof" does not mean that an explosion cannot occur. If flammable vapors penetrate into the electrical housing an explosion will occur, but it will be contained within the housing and not ignite vapors outside. However, the electrical equipment inside the housing may be destroyed.

Control Buildings

Since control buildings contain numerous electrical sparking devices, it is important to exclude flammable gases. This is accomplished by keeping the inside of the building at a slightly higher pressure (usually 1/4–1/2 in. of water). The pressurization air must be taken from a gas-free area, and the intake is usually at least 25 ft (7.6 m) above grade. It is preferable to use two air blowers, one of which is a standby, interlocked to start the standby blower if the operating one fails. An alarm should warn the operator when a blower has failed so that it may be repaired. A vestibule at each doorway is preferred to maintain pressure when a door is opened.

Pressurization is preferable to installing explosion-proof electrical equipment in the control room. Not only is explosion-proof equipment expensive, but the explosion-proof rating of such equipment is difficult to maintain over a period of time.

Many central control buildings constructed today are designed to be blast resistant. The objective of

TABLE 6.2 Classification of Areas for Electric Installations

Class	Presence of
I	Flammable gases or vapors
II	Combustible dust
III	Ignitable fibers or flyings
Group	**Atmospheres containing**
A	Acetylene
B	Butadiene, ethylene oxide, propylene oxide, acrolein, or hydrogen
C	Cyclopropane, ethyl ether, or ethylene
D	Acetone, alcohol, ammonia, benzene, benzol, butane, gasoline, hexane, lacquer solvent vapors, naphtha, natural gas, propane
E	Combustible metal dusts having resistivity of less than 10^5 ohm-cm
G	Combustible dusts having resistivity of 10^5 ohm-cm or greater
Division	
1	Flammable atmosphere under normal conditions
2	Flammable atmosphere only under abnormal conditions
Unclassified	Locations that need not be classified Division 1 or 2

blast-resistant construction is to protect personnel in the building from an explosion in a process unit and also to permit an orderly shutdown of all other units. Such buildings are designed for the usual maximum anticipated wind and snow loads and also a short-time peak blast over-pressure. Determining the design peak over-pressure may be difficult. The pressure anticipated will depend upon the distance of the control building from the process units and a realistic determination of what flammable material and how much could be involved. A hazards analysis of all involved units may be advisable in order to make a realistic determination of the possible blast severity. When the critical material processed has a tendency to form a high-velocity (above the speed of sound) detonation and the control building is close to the source (50 to 100 ft), then a blast-resistant design may not be practical. The only solution in this case is to locate the control building further from the blast source (200 ft or more).

Blast-resistant construction usually requires the elimination of windows. Although some operators object to running a plant they can't see, it has been proven that an operator can "see" almost anything that goes wrong in the process unit from instrument data. With the use of computerized data processing, instantaneous information on what is happening in the unit has been greatly expanded.

6.5.7 Safe Design of Offsite Facilities

Definition

Offsite facilities are those that contribute to the operation and sustenance of the plant processing units. They are sited apart from process units to protect them from exposure to fire and explosions and to assure their continued performance during emergencies. Offsite facilities include the utility area providing plant steam, electricity, cooling water, and air; bulk storage; flares; shipping facilities such as marine wharf, truck, and tank car loading areas; electrical substations; fire water; cooling towers; water treatment; and administration buildings, warehouses, store houses, maintenance shops, and fire house.

Boilers

Boilers providing plant steam for the operation of steam turbines, process steam, snuffing steam for extinguishing fires, and purge steam to remove flammable vapors in pressure vessels and other equipment must be kept operating, especially during an emergency. Snuffing steam must be available when a process coil in a fired heater ruptures, and standby pumps with steam turbine drivers must be available if a total electrical power failure occurs.

To help maintain continued operation, boilers should be located in a safe area, usually a minimum of 100 ft (30 m) from process units and 200 ft (60 m) from storage tanks and truck and tank car loading racks. Since fire box explosions are prone to occur in boilers, other off-site facilities must be protected. The administration building, warehouse, shops, cooling towers, process control buildings, and process unit battery limits are usually located a minimum of 100 ft from boilers.

To avoid boiler explosions from an accumulation of fuel in the firebox, safety interlocks are usually installed to automatically shut down at the following conditions:

- Low fuel pressure.
- Low water level.
- Loss of induced and forced draft combustion air fans.
- Flameout as indicated by ultraviolet flame detectors.
- High furnace pressure.

These safety interlocks and safe lightoff procedures are described in detail in NFPA 85A-1982, NFPA 85B-1984, and NFPA 85D-1984. In addition to shutting down the boiler, visible and audible alarms should indicate the nature of the malfuncition. Other less critical upset conditions should be designed to alarm only without automatic shut down.

Plant Electrical Supply

The plant electrical supply must be made reliable. Cooling water pumps, process unit pumps, air cooler fans, cooling tower fans, boiler combustion air fans, and motor driven compressors depend upon a reliable electrical power supply. A total loss of electricity could have serious results in many plants. To protect against this, it is customary to have two electrical supply lines into a plant in case one line is severed. In addition, it may be advisable to have two separate sources of electrical power supply, such as purchased power and generated power. Sometimes purchased power is integrated into an electrical grid system supplied by different utility companies. Such a system is extremely reliable.

The main power lines should be routed through the plant in a safe area. They should not be near process units, nor should they run through storage tank diked areas. They should also be protected from drainage ditch fires by culverts or concrete slabs over ditches, because large spills of flammable liquids may go into these ditches.

Electrical Substations

Electrical substations contain switchgear for process units, and are usually critical for control of units. They may be large central substations for several process units, smaller buildings housing switchgear for only one or two process units, or small outdoor racks. The outdoor racks usually contain explosion-proof electrical equipment when they are located in areas where flammable vapors could accumulate. Indoor electrical equipment is usually unclassified. Therefore the building should be located in a safe area and is usually pressurized to exclude flammable vapors. Smoke detectors are often installed inside substations, especially central stations, to detect an incipient fire quickly.

A safe spacing for a central control substation from process units is about 200 ft (60 m). A unit substation is usually located close to the process unit it serves.

Plant Cooling Water

The cooling water supply is also important to the safety of a plant. Loss of cooling water will result in overheating such process equipment as heat exchangers, and, therefore, could cause a fire. It can also result in high run-down temperature to storage tanks, causing flammable vapors to excape to the atmosphere. The use of standby diesel cooling water pumps can avoid or mitigate this situation.

Cooling Towers

Cooling towers are generally subjected to two types of exposures:

- Class A fires when the tower is shut down and wooden members are dry.
- Class B fires when hydrocarbon liquids get into the basin due to an exchanger tube leak.

Of these two exposures, the first is more common. Where the cooling tower is of prime importance for plant operation, or is very large and involves a significant investment, it is advisable to consider a fixed sprinkler system. NFPA 214-1983, Standard on Water Cooling Towers, covers the design of sprinklers for cooling towers.

Instrument Air

The supply of instrument air must be reliable; without it, it may not be possible to operate emergency shut-down systems. One method of providing a sure source of air is with backup by air cylinders.

A number of explosions have occurred over the years inside lubricated instrument air compressors, especially reciprocal compressors. A detailed description of this phenonemon can be found in Report of Joint NGPA-API Engine Manufacturers' Task Force on Engine Starting Air Line Explosions (1963) and the paper, "Fire-Resistant Lubes Reduce Danger of Refinery Explosions (McCoy and Handly, 1985). Although the causes of air compressor explosions were determined over twenty years ago, they still, unfortunately, occur. These explosions have resulted when the compressor lubricating oil reached autoignition temperature (AIT). Several conditions may cause this to happen such as:

- Using a lubricating oil having a low AIT, such as mineral oil.
- Fouling of intercoolers on the cooling surface, permitting the air temperature rises above design.
- Rust deposits inside piping downstream of compressor discharge. Rust deposits reduce the AIT of lubricating oil.

High AIT lubricating oils are available, but consult the compressor vendor before using them, to assure compatibility with compressor components such as seals, gasket, etc.

Many users specify unlubricated compressors for supplying instrument air.

Storage Tank Farms

Bulk storage of liquid feed and finished products is usually in vertical atmospheric and low-pressure storage tanks grouped in areas called tank farms. It is good practice to group similar materials for safe storage; low-flash-point flammable materials should be segregated from higher flash-point combustible liquids. Since low-flash-point liquids are the most hazardous, a fire occurring in such a tank is not likely to involve the high-flash liquids as well. Minimum spacing between storage tanks is specified in NFPA 30-1984, Flammable and Combustible Liquids. It is preferable to locate grouped storage tanks at the periphery of the plant site, downwind from process units and at a lower grade then process units, so that a major uncontrolled spill will not flow toward the process units. A minimum spacing of 200 ft (60 m) from process units, administration building, and shops is suggested.

Flammable liquids, having a closed cup flash point below 100°F (37.7°C), are usually stored in open top floating roof tanks, because these tanks provide safer storage than cone roof tanks. Combustible liquids ae usually stored in cone roof tanks, because these materials are less hazardous and the tank vapor space is usually not in the flammable range. High vapor pressure liquids, such as propane and butane, are stored in horizontal pressure vessels or in spherical vessels under pressure.

Atmospheric and low-pressure storage tanks should be surrounded by dikes or bunds to impound any liquid spills. When sufficient land is available, a diversion diking system may be used. Instead of impounding a spill, the liquid is allowed to flow by gravity to a safe area. However, a spill should not be diverted around other tankage. The volume of an impounding dike should be great enough to contain a complete spill of the tank within the dike when the tank is full. When more than one tank is located in a dike, then the dike should be designed to contain the complete spill of only one tank, that having the largest volume. The more hazardous tanks, such as crude storage, heated, and slop tanks are usually individually diked. Dikes may be of earthen, concrete, or steel construction. One advantage of earthen dikes with sloped walls is that it is easier for personnel working inside the dike to escape in an emergency. However, if not stabilized and maintained, earthen dikes erode from the weather, and sometimes are breached for maintenance and then not repaired. Reinforced concrete and steel dikes are used, usually where land is limited. Such dikes require little maintenance, but impede escape of personnel and are more likely to trap flammable vapors inside the dike.

Drainage is an important consideration in designing a tank farm. The diked area should be graded so that a spill will flow away from tanks. A fire-stopped drain, such as a turned-down elbow with a six-inch liquid seal, should be provided to drain water from inside the dike. The drain should have a valve, which is normally kept closed to impound a spill, and is located outside the dike. Storm water drainage in tank farms is often through open ditches located outside of dikes. When drainage ditches are used, they should be periodically fire stopped to avoid an uncontrolled fire through the ditch if a large flammable liquid release occurs.

Some precautions to be taken in tank farms are:

- Locate pumps outside of dikes. A pump seal leak can initiate a fire which may involve the tank.
- Piping inside dikes should be permitted only for tanks located within the dike.
- Avoid all threaded and proprietary couplings using elastomeric ring gaskets. These joints will leak or pull apart in a fire.
- To protect against ignition from external sources, use PV (pressure-vacuum) vents on top of cone roof tanks rather than flame arrestors. Flame arresters often freeze in cold climates. They also can plug or collect dirt in all environments.
- Open top floating roof tanks should have the floating roof electrically bonded to the shell. Where bonding is not inherent, such as by a metallic weather shield, it should be installed as shown in NFPA-1983, Standard for Low Expansion Foam and Combined Agent Systems.
- Tank pumpovers have occurred, especially where receipt is from a marine tanker or pipeline over which the plant has no direct control. In these circumstances the use of high-level alarms on such tanks should be considered.
- On heated tanks the liquid temperature should be maintained at least 25°F (14°C) below or above 212°F (100°C). Storage around 212°F can result in an explosion from the rapid formation of steam if water accumulates within the tank.
- When initially filling a cone roof tank, the inlet velocity should be kept below 3 ft/s (91 cm/s) to avoid a buildup of static electricity, until there is assurance that the inlet is covered with at least 3 ft (0.9 m) of product.
- Do not gauge the tank for at least 15 minutes after filling in order to permit a static charge to drain off.

Flare Systems

The purpose of the flare system is to dispose of large releases of flammable, combustible, and toxic vapors resulting from an upset operating condition within a processing unit. Normally the flare is not in use, but it is necessary as the only safe way of disposing of large quantities of gases suddenly released voluntarily or involuntarily. API RP 521 (1982) Guide For Pressure-Relieving and Depressuring Systems, provides guidelines for the design of flare systems. The flare system consists of:

- A closed pressure-relief system for collecting gas discharges from process unit relief valves, depressuring valves, pressure control vents, and onsite liquid knockout drums.
- Onsite liquid knockout drums where liquid is discharged from a process unit.
- Flare header from process units to convey the released vapor to the flare knockout drum.
- Flare knockout drum to remove any liquid carried over from process units or condensed in the flare header.
- Flare header from the flare knockout drum to the flare.
- Seals or purge gas or both to prevent a flashback.
- Flash stack with flashback protection such as a molecular or kinetic seal.
- Pilot, igniter, and steam for smokeless burning.

In designing the flare system, the maximum relief load should be determined based on process unit:

- Loss of cooling water.
- Loss of electrical power.
- Loss of steam supply.
- Process unit fire.
- A blocked valve.

In determining relief loads, only the single largest of the upset conditions listed above should be applied; cases resulting from double jeopardy should not be considered unless they are expected to occur simultaneously.

Factors that affect emergency relief loading are:

- In considering layout of cooling water lines and supply, can complete loss of cooling water occur?
- Is the electrical supply reliable with more than one source of supply, so a total electrical outage will not occur?
- Do critical pumps with electric motor drives have steam turbine driven spares?
- Will air coolers still provide sufficient cooling due to natural convection if the fan stops due to loss of electrical power?
- Will process unit valves remain in the safe position (open or closed) if the above utility failures occur?

Large plants usually have more than one elevated flare, so that emergency repairs can be made without shutting down the plant. The location of elevated flares is important, not only because liquid sometimes gets carried out the stack, but also because of the high heat radiation that results when emergency reliefs occur. Recommended design flare radiation levels are shown in API RP-521 (1982), Table III.

Marine Terminals

At large plants located on waterways, crude or other liquid charge for process units arrives by marine tanker and finished products are shipped out by tanker or barge.

Because of the large volumes of flammable liquids handled by marine tankers (100,000 tons and larger) safety is extremely important. Some safety considerations at a marine wharf are:

- *Location:* At least 200 ft (60 m) from process units and other plant facilities.
- *Control room:* In an area with good escape routes, pressurized to prevent gas infiltration.
- *Safety shutdowns:* Capability to remotely shutdown all valves on wharf with remote operated back-up valves on shore in case valves on wharf are involved in a fire.
- *Communications:* Provide good communications with ship and terminal.
- *Loading/unloading facilities:* For large liquid quantities, use all-steel loading arms rather than cargo hoses. When cargo hoses are used they should be the approved type certified by vendor. For very hazardous materials, consider quick disconnect and shutoff on end of loading arms.
- *Area classification:* All electrical equipment should meet the area classification. Refer to API RP-500A (1982).
- *Wharf piping:* Use all steel piping and minimize flanged joints; avoid threaded connections. Provide for piping flexibility taking into account the motion of the wharf.
- *Drainage:* Paving should be graded to carry a spill from beneath manifolds and piped into a closed tank where the spill can be pumped to shore.
- *Static electricity and stray currents:* Provide bonding, grounding, and insulating flanges as indicated in API RP 2003 (1982).
- *Provide fire protection commensurate with importance of wharf and exposure hazards:* Consider fire water requirements and the need for fixed water and foam monitors (manual or remote operated); hand, wheeled, and skid-mounted dry chemical extinguishers; under-pier foam/water sprinkler system.

Tank Car and Truck Loading Racks

At chemical plants some raw materials for processing arrive by rail tank car. Finished products often are shipped out by rail and tank trucks. Some precautions and considerations to be taken at rail and truck terminals are:

- Locate in safe area, about 200 ft or more from other plant facilities.
- Observe safe loading and unloading procedures. Refer to API RP 2003 (1982) for grounding and bonding requirements.

- All electrical equipment should meet area classification (API 500A, 1982).
- Provide remote emergency shutdown of pumps and closing of all valves.
- For truck loading facilities, establish safe truck route to avoid accidents inside plant. If possible, a separate plant gate should be used for truck terminals.
- Provide fire protection commensurate with hazard. Consider need for fixed water monitors, hand and wheeled dry chemical fire extinguishers, fixed foam/water sprays at loading points.
- Pavement should be graded to carry spills away from trucks and into a plant sewer system, properly fire stopped.

6.6 FIRE PROTECTION

The overall plant fire protection system should be designed to meet the requirements to contain and extinguish the most serious fires that could occur. Where there is adequate manpower and a well-trained plant fire department is maintained, fire fighting systems may be manual. Where the manpower is limited and high hazards exist, automatic and remote systems are prudent. The basic elements of fire fighting systems are as follows:

Fire Water Supply

The first step is to provide a reliable fire water supply. For small low-hazard chemical plants the municipal water supply may be adequate. The installation of a fire water storage tank is usually advisable. The quantity of water stored should be determined by the design flow rate to the fire main and the length of time expected to fight a major fire. This time varies from 4–8 h, depending upon the size of the plant and the existing hazards. About 3000 gal/min (680 m^3/h) normally would be adequate for a small chemical plant and a minimum of 4000–5000 gal/min (909–1135 m^3h) would be required for a petrochemical plant or refinery. The design flow should be based on the assumption that only one process unit at a time will be on fire. Although there is a possibility of a fire spreading from one unit to another, the probability is low if units are adequately spaced. It is also unlikely that adequate manpower and fire fighting equipment would be available to fight two process unit fires of major magnitude. In determining the flow rate, the total of fixed monitors, water sprays, and hose streams should be determined for each process unit. However, this flow should be based on the equipment expected to be required for any single type fire, and not the total firefighting equipment installed onsite.

Where possible, more than one fire water source should be available, such as a fire water storage tank and a river, sea, or treated water pond or basin.

Fire Pumps

Fire water pumps should be selected on the basis of reliability and meet the requirements of NFPA 20, *Standard for the Installation of Centrifugal Fire Pumps,* (1983). Electric-motor-driven pumps require little maintenance and are easily started, but should not be relied upon where power failures can occur. Diesel-driven pumps are very reliable, but cost more than electric driven and require more maintenance. Steam-turbine-driven pumps should be used only where there is a reliable source of steam, especially during a fire when steam requirements for spare pumps may be high.

It is good practice to install more than one fire pump as backup. Often one electric pump and one or more diesel pumps are installed. Typical pumps sizes vary between 1500 and 3000 gal/min (340–680 m^3/h).

When a fire water system is required to be pressured constantly for immediate use, then a jockey pump should be installed. When the fire main pressure cannot be maintained by the jockey pump, then the main fire pumps should start automatically and sequentially upon demand. A constantly running jockey pump is preferred to an intermediate running pump with an air pressure tank. Jockey pumps should be electric with a minimum capacity of 100–200 gal/min (23–45 m^3h) to avoid frequent startup of fire pumps. Fire pumps should be installed in a fire safe area at least 200 ft from any fire hazard.

Fire Mains and Hydrants

The fire main system is usually laid out in a grid pattern. With this design, water supply is from two or more directions, which permits smaller diameter pipe. Reliability is also improved by installing diversion valves, which can be used to isolate a main break at any location and continue fighting a fire.

Steel pipe is the preferred material for resistance to mechanical damage. Pipe sections should be welded together and flanges used at valves, hydrants, and other fittings. Friction type flexible couplings with elastomeric ring gaskets should be avoided where possible. Underground mains should have exterior surfaces coated and wrapped for corrosion protection. In addition, cathodic protection is usually provided. When salt water is used, the pipe interior should be cement or epoxy lined. Plastic pipe should be used with caution. Use a good quality plastic pipe and make sure by careful inspection that all joints are sound and do not leak. Fire mains around process units preferably should be underground to protect them against fires, explosions, and mechanical damage. In less hazardous offsite areas, fire mains are often above grade. Mains generally are 8-in. minimum size, except that in low hazard offsite areas 6-in. mains may be used when the required flow rate is well known.

Fire hydrants usually have two 2½-in. hose connections and a 4½ or 5-in. fire truck pumper connection. They should be located on fire mains looped around process units and spaced 150-200 ft (45-60 m) apart. They should also be located on mains looped around tank storage areas on about 200-300 ft (60-90 m) spacing. Hydrants around buildings and other offsite areas should be located as the hazard demands. They should be located near plant roads to be accessible to fire trucks. Hose threads should be compatible with the municipal fire department and mutual aid services. When this is not feasible, hose coupling adapters should be available. Wet barrel hydrants, without underground shut-off valves, should have a valve on each hose connection.

Fire Monitors

Fixed water monitors should be installed around process units to protect specific pieces of equipment such as heates, exchanger banks, vertical and horizontal pressure vessels, and compressors. They are also used in offsite areas to protect marine tankers and equipment on marine wharfs, at truck terminals to protect the loading facilities and trucks, tank cars at tank car loading racks, in tank storage farms to protect LPG tanks, at gasoline treating plants, around TEL (tetraethyl lead) facilities and cooling towers.

Around process units it is preferable to locate monitors outside the battery limits, where firefighters are in a less exposed location. Where the process area is congested, it may be necessary to install some monitors inside the unit and also in elevated locations. Elevated monitors usually are operable from grade so that firefighters can readily escape.

The commonly used monitor size is 500 gal/min (113 m^3/h). This size is adequate for cooling most equipment and does not drain the water supply from other firefighting equipment. Larger monitors, 1000 gal/min (227 m^3/h), are sometimes used to protect such high hazards as marine wharfs and tankers.

Two-wheeled portable foam/water monitors are useful for towing to a fire as needed. The 500 and 1000 gal/min sizes are common, and are usually kept in the plant firehouse when not needed. The monitor generally is supplied with the foam/water solution properly proportioned by a fire truck.

Fixed Water Sprays

Fixed water sprays have the following uses:

- Cooling metal to prevent distortion, rupture, or buckling when fire exposed.
- Controlling fire intensity.
- Dispersing combustible and toxic vapor clouds.
- Maintaining the integrity of electrical cable exposed to fire.
- Cooling Class A materials below their ignition temperature to extinguish a fire.

Water sprays should be designed in accordance with NFPA 15-1985, Standard for Water Spray Fixed Systems for Fire Protection. Water densities for various applications are shown in Table 6.3.

Water sprays for cooling metal or applied to:

- Uninsulated pressure vessels containing flammable and toxic liquids.
- Uninsulated pressure storage vessels containing light ends such as propane and butane.
- Compressors.

Water sprays to control fire intensity are applied to:

- Pumps,
- Hot piping manifolds, particularly when the liquid is above autoignition temperature.

Vapor cloud dispersal applications are:

- Toxic liquids or gases that could escape from a vessel, pump, compressor, manifold, or other equipment.
- Flammable liquids under very high pressure, which could leak out of flanges and pump seals to form a large vapor or mist cloud.

TABLE 6.3. Recommended Water Spray Densities for Cooling, Fire Intensity Control, and Fire Extinguishment[a]

		Water Density	
		gal/min/ft²	L/min/m²
Cooling			
Pressure vessels, compressors		0.25	10.2
Horizontal steel structures		0.10	4.1
Vertical steel structures		0.25	10.2
Single level pipe racks		0.25[b]	10.2[b]
Two level pipe racks	Lower level	0.20[b]	8.2[b]
	Upper level	0.15[b]	6.1[b]
Fire Intensity Control			
Pumps, hot piping manifolds, etc.		0.5	20.4
Vapor Cloud Dispersal			
Leaks from flanges, seals, etc. (see NFPA 15, par. A-4-4.4)		0.6 min	24.4 min
Electrical Cable and Transformers			
Cable (unshielded)		0.3	12.2
Cable (flame shield beneath)		0.15	6.1
Transformers (on all surface)		0.25	10.2
Transformers (on ground surface)		0.15	6.1
Extinguishment			
Conveyor belts (top of top belt and bottom of bottom belt)		0.25	10.2

[a]Refer to NFPA 15 for extent of water coverage and locations of spray nozzels.
[b]Applied to underside of piping.

Electrical cable and transformer protection are used to:

- Limit the loss of grouped electrical cable runs in tunnels or other areas difficult to reach for fire-fighting.
- Maintain the operability of grouped electrical cables during a fire.
- Maintain integrity of transformers.

Fire extinguishment applications are:

- Extinguish fires on conveyor belts carrying combustible materials such as coal.
- Protect long conveyor belts from being destroyed by fire caused by hot roller bearings.
- Extinguish fires in wooden cooling towers.

Activation of water sprays may be manual or automatic. Manual activation is usually by a quarter-turn ball valve located in a fire safe area. Automatic activation may be by heat or smoke detectors.

Advantage of manual operation:

- Less expensive than automatic.
- More reliable.
- Less maintenance required.
- Permits operator to exercise judgment and prevents false alarms.

Advantage of automatic operation:

- Limit losses by quick activation.
- Frees operator to shut down the unit.
- Protects areas normally unattended or during off hours.

When considering water spray protection, keep in mind that sufficient water must be available so that the fire water supply will not be depleted in an emergency. Also, drainage of spray water must be considered; otherwise a hydrocarbon fire could be spread throughout the area.

Live Hose Reels

Live hose reels hold 50–100 ft (15–30 m) of noncollapsible hard rubber hose, usually 1½ in. diameter. The water line is pressured up to the hose inlet, and to change the hose all that is required is to open the valve at the hose inlet. With hard rubber hose, not all of the hose needs to be removed from the reel in order to use it. Live hose reels have a flow rate of only about 100 gal/min (23 m^3/h), and are not suitable for fighting large fires. Their primary purpose is in fighting small incipient fires to provide holding action unitl the fire department arrives.

Some applications for live hose reels are:

- At process units, where they are usually installed under the unit pipe rack for easy access and can be deployed on either side of the rack.
- On marine wharfs, to protect loading areas and piping manifolds.
- Inside buildings, such as warehouses and storehouses, where immediate fire hose response is desired.
- At truck loading racks, where manpower is limited and quick response is desired.
- At areas where small gas containers, such as propane, are filled.

In cold climates winterizing is required, usually by electric heat tracing the above-ground portion of the water supply piping.

Portable Fire Extinguishers

Portable fire extinguishers for industrial use range from the small 5 lb (2.3 kg) hand-held unit to the 350 lb (160 kg) wheeled dry chemical unit. NFPA 10-1984, Standard for Portable Fire Extinguishers, covers selection, installation, inspections, maintenance, and testing. The frequently used types of extinguishing materials are:

- Potassium and sodium bicarbonate dry chemical, used for Class B fires such as most flammable liquids and gases.
- Multipurpose dry chemical, used for Class A, B, and C fires.
- Halon 1211(BCF), used for electrical equipment.
- CO_2, used for electrical equipment.
- Pressurized water, used for Class A fires, usually in office areas containing paper and other combustible materials.
- AFFF (aqueous film forming foam), used where securement of a small flammable liquid fire is desired.

UL tests and classifies fire extinguishers regarding the types of fires and fire area that can be extinguished. A UL classification, such as 120-B:C, indicates that a type B or C fire, 120 ft in area, can be put out by a trained operator using the extinguisher. For a more complete explanation of the UL rating, see NFPA 10, Paragraph A-1-4.2.

Some plant applications for extinguishers are:

- *Office buildings:* Five pound (2.3 kg) multipurpose dry chemical extinguishers located a maximum of about 75 ft (23 m) from any hazard. Pressurized water extinguishers, 2½ gallons (10 L), also may be used, but are not as effective as dry chemical.
- *Warehouses and maintenance buildings:* Thirty pound (9 kg) multipurpose dry chemical extinguishers. Where Class B materials are stored and handled, use 30 lb potassium bicarbonate. Locate extinguishers not more than 50 ft (15 m) from any hazard.
- *Laboratories:* At least 9 lb BCF and one 30 lb potassium dry chemical extinguisher or more, depending upon size of laboratory.
- *Process areas:* Locate 30 lb potassium bicarbonate extinguishers so that at least one extinguisher is not more than 50 ft (15 m) from all hazards. Also provide one or two 350 lb potassium bicarbonate wheeled extinguishers in each process unit. Extinguishers mounted on pipe rack columns are accessible and easily located. Also locate extinguishers at each fired heater, compressor, in pump areas and at transformers. At least one 22 lb BCF extinguisher for electrical fires should be located in control rooms and outside substations.
- *Offsite areas:* One 30 lb multipurpose extinguisher at top deck stairway landing of cooling towers. Locate one or more 30 lb potassium bicarbonate extinguishers at boilers and at offsite pump areas.
- *Marine terminals:* One or two 30 lb potassium bicarbonate extinguishers at each cargo loading area, plus one 30 lb unit in the control room and near piping manifolds. Also locate one 350 lb

potassium bicarbonate wheeled extinguisher near cargo holes and loading arms. Large capacity 2000 lb (909 kg) skid-mounted dry chemical units are frequently used on wharfs, particularly where access for mobile fire fighting equipment is poor.

The use of dry chemical extinguishers in areas where computers and other electrical equipment is located should be avoided, because the cleanup required may be more expensive than fire damage.

Extinguishers must be properly mounted and the location clearly marked for quick identification. They should be periodically inspected and tagged. After using, they should be immediately recharged and replaced.

Fire Hose

Soft fire hose is used for hand hose lines, to supply water from hydrants or a fire truck to portable fire-fighting equipment such as wheeled monitors, and to supply water from hydrants to fire trucks. The basic hose sizes are 1½, 2½, and 3 in. Three-inch hose is used primarily to supply water from hydrants to fire trucks and is equipped with 2½ in. couplings. Fire hose should meet the requirements of NFPA 1961, Standard for Fire Hose. Hose couplings should be in accordance with NFPA 1963, Standard for Screw Threads and Gaskets for Fire Hose Connections.

Single jacket hose consisting of all synthetic fiber with rubber tube liner is commonly used. Some hose is impregnated with polyvinyl chloride, Hypalon, or other plastic coating to reduce wear.

Hose is often stored in hose houses or carts in the plant area where it is expected to be used. The disadvantage of this practice is that a considerable inventory of hose may be stored throughout the plant, and often is not maintained, due to lack of inspection. In large plants with a well-trained fire department and one or more fire trucks, all hose is carried on fire trucks. This is the preferred practice, because the hose is tested and maintained in good condition and is available wherever it is needed.

Fire Foam

Fire foams are used to extinguish and secure hydrocarbon pool fires such as spills and atmospheric storage tank fires. Types of foam concentrates are:

- *Regular protein foam:* Used for spill fires and storage tank fires for top application only. Protein foam is less used since the development of synthetic foams with superior properties.
- *Fluoroprotein foam:* This foam serves the same uses as protein foam, but is generally more effective and can be used for subsurface application for cone roof storage tank fires.
- *AFFF (Aqueous Film Forming Foam):* This foam is a combination of fluorocarbon surfactants and synthetic foaming agents that quickly spread across the surface of a hydrocarbon pool fire. It has quicker control and extinguishment than fluoroprotein foam, but does not have as good sealing properties against hot metal, such as the shell of a storage tank on fire.
- *Alcohol type foam:* Used for fires involving alcohol and other polar solvent liquids that break down other foams. It can also be used against hydrocarbon fires.
- *High expansion foam:* This foam is used at expansion rates of 100–800 to 1. It is intended primarily for use on Class A materials in confined areas such as warehouses and other buildings. Because of its low density, it can be blown away easily when applied outdoors.

Foam concentrate is supplied in two concentrations, 3% and 6%. The quality and cost of both foams is equivalent, but 3% foam has the advantage of making double the quantity of foam when stored in equal size containers. This becomes important on mobile fire fighting equipment where allowable weight is limited. NFPA 11, Standard for Low Expansion Foam and Combined Agent Systems, covers the characteristics of foam producing agents and the requirements for design, installation and operation of foam extinguishing systems.

Foam solution can be proportioned and generated by:

- *Fixed proportioning systems:* One such fixed system, balanced pressure proportioning, consists of a foam concentrate storage tank, a foam concentrate pump, and a control valve to proportion the foam. Another fixed system, pressure proportioning, consists of a foam concentrate storage tank with an internal membrane or bladder and a venturi for proportioning foam. Water pressure on the bladder is used to force foam out of the liquid storage tank.
- *Mobile proportioning systems:* These are installed on fire trucks and have the advantage of being able to proportion foam solution wherever there is a water supply such as from hydrants or a pond. Balanced pressure proportioning is usually used on fire trucks.
- *Fixed and portable line proportioners:* This is an inexpensive proportioning method, designed for a predetermined water discharge at a predetermined water pressure. Water flowing through a venturi creates a vacuum that picks up foam concentrate through a connection in the side of the venturi. In a fixed installation, a foam concentrate tank supplies one or more venturis. In the

portable system, a tube connected to the venturi is inserted into a drum of foam concentrate. The portable pickup tube is useful for converting a hose water line to a foam line by using a foam-making nozzle on the end of the hose. This arrangement is useful for small hydrocarbon spill fires.

Some applications for foam are:

- *Cone roof and covered floating roof storage tanks:* Foam chambers installed at the top of the shell expand foam solution to extinguish tank fires. The system must be designed to cover the entire liquid surface at the prescribed rate. A fixed proportioning system or a fire truck may be used to supply foam solution to the proportioners.
- *Subsurface foam:* A high back-pressure foam maker may be used to generate foam and inject it into the bottom of a cone roof storage tank. Fluoroprotein foam is preferred for this application. Subsurface application has the advantage of being simpler to install, because a product line to the tank may be used without having to weld a new connection onto the tank. Foam may be proportioned by a fixed system or a fire truck.
- *Catenary and over-the-top foam systems:* These systems are used to extinguish rim fires in open top floating roof tanks. In the catenary system, foam is deposited at the floating roof seal by rigid pipe and flexible metal hose to adjust for movement of the floating roof. The over-the-top system deposits foam at the top of the shell, on the inside, and the foam runs down the shell and around the roof seal. Foam may be supplied by fixed system or a fire truck.
- *Foam/water sprinkler systems:* These systems are used mainly beneath marine wharfs and at truck loading racks. They are usually supplied by a fixed foam proportioning system.
- *Fire trucks:* Trucks are used to proportion foam and apply it to fires in process units or in storage tanks, marine terminals, and other plant offsite areas. Foam is generated and applied by hose lines with air aspirating foam nozzles or from turret monitors mounted on the truck. In this capacity, the foam fire truck becomes a very effective fire-fighting apparatus.

Halon and Carbon Dioxide Extinguishing Systems

Halon and carbon dioxide total flooding systems are used where valuable equipment is located, in unattended indoor locations, and where a clean agent is required for extinguishment.

The two basic extinguishing agents used are Halon 1301 and carbon dioxide. Halon 1211 is also used sometimes, but it has a greater toxicity and should be avoided whenever personnel could be present. Halon 1301 is nontoxic and is preferred to CO_2, but is considerably more expensive.

NFPA 12-A, *Standard for Halon 1301 Fire Extinguishing Systems* (1985), contains minimum requirements for halogenated agent extinguishing systems.

NFPA 12, *Standard on Carbon Dioxide Extinguishing Systems* (1985), contains requirements for CO_2 extinguishing systems.

Halon and carbon dioxide flooding systems are used for:

- Computer rooms, where a clean agent is required.
- Gas turbine enclosures, which would not be safe to enter during a fire.
- Storage vaults.

Fire Detectors

Fire detectors are used to give warning that a fire has occurred in unattended places and also to activate automatic fire-fighting equipment. They are usually located in remote high-risk areas or in high-investment areas. Typical locations are computer rooms, grouped shipping pumps, tank truck loading terminals at automatic water spray installations, warehouses and other storage areas, and gas compressor enclosures. Detectors operate on one of three principles: reaction to heat, flame radiation, or products of combustion. Types of detectors and characteristics are as follows:

Heat Detectors

- *Fusible link or plug:* Very reliable and needs no power source, but is slow to respond.
- *Fixed temperature:* Reliable and low cost but slow and affected by wind.
- *Rate of rise:* Will detect slow temperature rises and rapid rises faster than fixed temperature detectors. Affected by wind and should be confined to indoor use.

Smoke Detectors

- *Ionization:* Early warning, will detect smoldering fires, but subject to false alarms where internal combustion equipment is used or personnel are smoking; limited to indoor use.

- *Photoelectric:* Early warning, will detect smoldering fires, subject to false alarms from dusts; limited to indoors.
- *Ultraviolet (UV):* Fast response, high sensitivity, self-testing, but response retarded by thick smoke and false alarms from welding, lightning, etc.
- *Infrared (IR):* Fast response, but very prone to false alarms from hot surfaces such as furnace firebrick and hot engine manifolds.
- *Combined UV/IR:* High-speed response and sensitivity, low false alarms, but thick smoke obscures range, is expensive.

Plant Fire Alarms

A plant fire alarm system is necessary to notify all personnel that a fire has occurred and to report the location. Methods used to report a fire are:

- *Telephone:* The preferred system utilizes a dedicated number to report a fire. The call is usually received in the control building and sometimes also the gatehouse. The coded number dialed also activates audible fire horns or sirens in the plant.
- *Radio:* Walkie-talkie radios may be used to report a fire to the control room and also activate the plant fire alarm.
- *Manual fire alarm stations:* Pull box or push-button alarms located throughout the plant can be used to report a fire and sound the audible alarm. In large plants the location of the activated alarm station often appears on a dedicated computer CRT.

Fire Trucks

Medium-sized and large plants often have one or more fire trucks and a well-trained fire fighting crew. The designated fire fighters usually perform other duties in the plant and respond to the firehouse when the plant fire alarm system is sounded. The design of plant fire trucks varies with the types of fires expected. Some types of trucks used are:

- *Foam truck:* Used primarily to fight hydrocarbon spill fires and atmospheric storage tank fires, it carries a large volume, up to 1000 gal (3800 L), of foam concentrate. Water is supplied to the truck from hydrants and the pressure is boosted by a fire water pump (usually 1000 gal/min) driven by the truck engine. Foam solution is proportioned by a foam concentrate pump and proportioning system.
- *Dry chemical truck:* Used primarily for rapid fire knockdown and extinguishment of process unit fires. A large capacity of potassium bicarbonate dry chemical is carried on the truck, up to 2000 lb (900 kg). The truck also usually contains a fire water pump to supply cooling water streams.
- *Combination foam/dry chemical truck:* Combines the advantages of both the foam and dry chemical trucks. It contains 200 lb of dry chemical and 1000 gal of foam concentrate and is effective for fighting both process unit and tank fires.
- *Triple agent truck:* Contains a large quantity of dry chemical, foam concentrate, and premixed AFFF solution. This truck is very effective in fighting various types of fires, but because of weight limitations, some compromises have to be made. The dry chemical capacity is 1800–2000 lb, the foam concentrate tank holds about 500 gal (1900 L) and the premix tank holds about 200 gal (760 L) of AFFF/water solution, which is pressurized for immediate use by nitrogen cylinders when activated. This truck can quickly extinguish a spill fire followed by securing with premixed AFFF. It can also be used to fight a tank fire when an outside backup source of foam concentrate is provided by a foam trailer or nurse truck.

Fireproofing

Fireproofing consists of a passive insulating coating over steel support elements and pressure-containing components to protect them from high temperature exposure by retarding heat transfer to the protected member.

The objectives of fireproofing are:

- To prevent the collapse of structures and equipment and limit the spread of fire.
- To prevent the release of flammable or toxic liquids from failed equipment.
- To secure escape routes for personnel.

Fireproofing is more critical in fire exposed areas where flammable liquids are processed, stored, or shipped and where a fire could occur from a leak or spill. In determining the need for fireproofing, it is necessary to consider the characteristics of the materials being handled, the severity of operating conditions, the quantity of fuel contained in equipment and piping, and the replacement value of plant facilities and business interruption.

Some types of fireproofing materials available are:

- *Dense concrete:* This consists of Portland cement, sand, and aggregate having a dried density of 140-150 lb/ft^3 (2240-2400 kg/m^3). Concrete may be formed, troweled or gun applied.
  Concrete requires no unusual skill to apply and has high resistance to impact and other abuse. Its high density makes it an inefficient insulator and requires more structural support than lighter weight materials.
- *Lightweight concrete:* Also called cementitious materials, they consist of a lightweight aggregate such as perlite or vermiculite and a cement binder. The density dry is from 25-80 lb/ft^3 (400-1280 kg/m^3). Since they are lightweight, they are very good insulators. However, the low density also results in reduced strength, less resistance to impact than dense concrete, and high porosity, which leads to moisture penetration. These materials usually require a weather protective top coat to prevent corrosion of substrate and spalling in freezing climates. In selecting the density of the mix, the properties of fire protection vs. strength must be considered.
- *Intumescent and subliming mastics:* These mastics provide heat barriers through one or more of the following mechanisms:
  Intumescence: materials expand to several times their volume when exposed to heat and form a protective insulating char at the barrier facing the fire.
  Subliming: materials that absorb large amounts of heat while transferring directly from a solid to a gaseous state.
  Mastics are sprayed on the substrate to form a thin coat, usually 3/16 in.-1/2 in. (5-13 mm). The main advantages are light weight, thin coats, and speed of application. Since proper bonding to the substrate and careful control of coat thickness are very important, only vendor-approved and experienced applicators should be used. Some mastics contain a flammable solvent and should not be applied near such ignition sources as fired heaters and boilers, during operation.
- *Preformed inorganic panels:* Precast or compressed fire resistant panels composed of a lightweight aggregate and a cement binder or a compressed inorganic insulating material such as calcium silicate.
  Panels are attached to the substrate by mechanical fasteners designed to withstand fire exposure without appreciable loss of strength. When used outdoors, an external weather coating system may be required. Also, all joints should be caulked with a mastic. These materials have the advantage of requiring no time-consuming curing or drying cycle. However, they are labor intensive when applied to existing units having instruments or other equipment supported on structural steel columns.

The thickness of fireproofing required for protection of a substrate is expressed in hours of protection, that is, the time until the steel reaches a temperature of 1000°F (538°C). Until recently, the fire test used to determine the hours of protection was ASTM E119 (American Society of Testing and Materials, 1983). Several years ago tests conducted by some oil companies determined that the E119 time-temperature curve rate of rise was too slow to predict accurately the hours of protection in a high temperature-rate-rise hydrocarbon fire (Warren and Corona, 1975). Underwriters Laboratories has recently developed a high rise fire test, UL 1709, which gives a more realistic evaluation of fireproofing materials in a hydrocarbon type fire (Underwriters Laboratories 1709, 1984). Fire ratings of all materials tested by UL are listed in the *Fire Resistance Directory* (Underwriters Laboratories, 1984).

In determining fireproofing requirements, the following factors should be considered:

- Volume of flammable or combustible liquid likely to be involved.
- Characteristics of material spilled.
- Ability of drainage system to remove spill.
- Congestion of equipment.
- Severity of operating conditions.
- Importance of the facility.
- Ability to isolate the leak with safety shutoff valves.

Some applications of fireproofing in fire exposed areas are:

- *Multilevel equipment structures:* Structures supporting equipment containing hazardous materials should be fireproofed, usually up to the equipment support level. Structures supporting non-hazardous equipment are usually fireproofed from grade to a height of 30-40 ft (9-12 m).

- *Pipe racks:* Process unit pipe racks generally should have columns and a least the first level of beams supporting pipe fireproofed. When a quantity of large diameter pipe (10 in. and larger) is carried on a second or third level, then consideration should be given to extending fireproofing to upper levels.
- *Equipment supports:* Pressure vessel skirts and legs, air cooler legs, legs of fired heaters, and high saddle supports over 3 ft high (1 m) are often fireproofed.
- *Grouped instrument cable:* Preferably, this should run underground. When located on the pipe rack, fireproofing should be considered.
- *Emergency valves:* The preferred design is to fail in the safe position. When this cannot be done, the valve operator and power supply should be fireproofed. Steel valve bodies do not require fireproofing. In designing fireproofing for electrical cable, it should be kept in mind that the insulation deteriorates at about 320°F (160°C).

6.7 AUDITS

Periodic audits are recommended to assure that engineering controls are being maintained and also to determine if any upgrading of safety controls is advisable. Preferably an operator familiar with unit operation, and sometimes a process design engineer, should accompany the safety or loss prevention engineer during the audit. The safety audit should include:

- Piping and instrument drawings to verify that safety shutdown systems have been installed and maintained in accordance with the latest drawings.
- Process unit plot plans and equipment lists to identify equipment.
- Check list of emergency safety shutdown systems, so that an inspection of these facilities can determine if they are maintained in good operating condition.
- Review of plant safety procedures to assure that they are kept current with any changes in plant operations.
- Visual inspection and testing of plant fire protection systems, such as fixed water sprays, water monitors, fire pumps, fire mains, etc.
- Review of plant accident and fire reports to determine what types of accidents occur most frequently.

6.8 TRAINING

Continuous training of plant personnel is very important to assure that proper actions are taken in any emergency. A good training program will teach personnel to take the correct action instinctively in any emergency, because all emergencies that can arise have been anticipated and carefully thought out prior to their occurrence. The following types of preplanning and training are necessary.

- Training of process unit operators regarding corrective measures to take in all unit upset conditions and in emergency shutdown of the unit or plant. The unit operating manual should contain detailed operational procedures for all emergency conditions.
- Training of the plant fire-fighting crew in the operation of all plant fire-fighting equipment. A fire training ground should be established where various types of fires can be started and extinguished by the plant fire-fighting crew. It is usually beneficial to invite the local municipal fire department for training exercises, in order to familiarize them with the types of hazards in the plant and the use of plant-fire fighting equipment.
  Other plant personnel should be trained in the use of first aid fire-fighting equipment, such as fire extinguishers.
- Training of the plant security force regarding what action to take in emergencies, such as making the accident area accessible to municipal and mutual aid fire companies, and keeping the area free of people who have no useful task to perform in the emergency.
- Training personnel in first aid so that they can administer the proper minor treatment until trained medical help arrives. This training serves a useful purpose not only in the plant, but also in the home and elsewhere outside the plant.
- Instructing the medical department (nurse, doctor and ambulance crew) in actions to be taken in emergencies and accidents. They should be familiar with the effects of all of toxic and hazardous materials used in the plant and the recommended treatment.
- The affected plant personnel should be trained in the proper use of all plant safety equipment such as air masks, safety showers and fire blankets.

6.9 EMERGENCY RESPONSE PLAN

6.9.1 Objectives

The basic goals of an effective emergency response are twofold:

1. Minimize injury to personnel.
2. Minimize damage to facility and therefore return to normal operation as soon as possible.

Accomplishing these objectives requires an extensive planning effort prior to the emergency. It does little good to attempt to develop an emergency plan when a disaster is imminent, or worse, after one has occurred. It is axiomatic that while hazards in facilities can be reduced, risk is an element of everyday existence and, therefore, cannot be totally eliminated. To keep the risk to an absolute minimum, advance planning is essential. The development of such a plan should address all possible emergencies (fire, explosion, flood, personnel injuries, etc.) that could occur and the development must include:

- Assessment of risk.
- Facility organization for dealing with emergencies.
- Assessment of resources.
- Training.
- Emergency headquarters.
- Security.
- Public relations.

6.9.2 Risk Assessment

This assessment must enumerate all the potential problems that may be encountered at a facility, including those that could occur in the surrounding community and affect the plant. The magnitude of the potential problem must be established as well as the probability of the event occurring, enabling management to prioritize its effort in plan development.

6.9.3 Facility Organization

A management team must be established to cope with emergencies. Ideally, it should include staff members from production, utilities, fire and safety, medical, and maintenance. In small facilities some organization members may have multiple roles. In any event, roles and authority must be clearly defined. In addition, the organization listing must include contact phone numbers when staff members are not on site. More importantly, each emergency organization member must have a staff alternate in the event that the primary team member cannot be contacted or is otherwise unaccessible.

6.9.4 Assessment of Resources

A factual current knowledge of plant emergency response equipment is essential to any emergency preparedness program. Not only must the capability and utility be known, but plant personnel must know how to use it. In addition, to the extent possible, plant personnel must access the capability, availability, and utility of the emergency organization that may be called upon to assist in an emergency. Following this, a meeting involving all potential respondents to an emergency should be held. The objectives of the meeting are to familiarize them with the facility, resources at the facility, resources in the community, and to define the role of the various groups.

6.9.5 Training

Plant personnel must be proficient in several areas, including knowledge of the emergency alerting system, familiarity with respect to the emergency organization, shut-down procedures, fire fighting, first aid, etc. Each of these must be outlined in concise clear terms. Plant personnel must be thoroughly trained in the use of emergency equipment and the carrying out of the emergency response procedures. Community response personnel should also be informed of these plant operations that may have an impact on their role in the response effort. A clear chain of command is essential.

6.9.6 Emergency Headquarters

In large facilities there should be at least two widely separated areas that would serve as group headquarters for dealing with emergencies. The locations should be equipped with all the necessary equipment to enable the staff to be in contact with plant personnel as well as mutual aid organizations. These access

lines should be dedicated, so that essential communication will be readily available to the response group.
Once the above elements are in place, it is essential that joint exercises be held to evaluate the effectiveness of the plan and to correct any shortcomings that may be present.

6.9.7 Security Considerations

Provisions should be made to have a security component to the emergency response team. Typically, security deals with such hazards as sabotage, terrorism, and so on; however, more typically the primary function in most emergencies is to limit access to those people and that equipment that will assist in coping with and resolving the emergency. To accomplish this requires:

- Identifying key plant and community personnel who can be activated instantaneously.
- Establishing a group having security responsibility or other plant personnel who assume security roles in an emergency.

6.9.8 Public relations

An important element of any emergency plan is appointing an individual who is responsible for informing both plant and community personnel about aspects of the emergency. A person must be designated to deal with the public/media. To effectively accomplish this, he or she must have access to the highest levels of management dealing with the emergency.

REFERENCES

Anyakora, S. N., Engel, G. F. M., and Lees, F. P., Some Data on the Reliability of Instruments in the Chemical Plant Environment, *Chemical Engineer,* November 1981.

API (American Petroleum Institute), 500A, *Classification of Locations for Electrical Installations in Petroleum Refineries,* 1982.

API RP 520, *Recommended Practice for the Design and Installation of Pressure-Relieving Systems in Refineries,* 1976.

API RP 521, *Guide for Pressure Relieving and Depressuring Systems,* 1982.

API RP 2003, *Protection Against Ignitions Arising Out of Static, Lightning, and Stray Currents,* 1982.

ASME (American Society of Mechanical Engineers), *Boiler and Pressure Vessel Code,* Section VII, Pressure Vessels, Division 1 and 2.

ASTM (American Society of Testing and Materials), E119, *Fire Tests* or *Building Construction and Materials,* 1983.

Browning, R., Reactive Probabilities of Loss Incidents, *Chemical Engineering,* p. 134, December 15, 1969.

McCoy, C. S., and Hanly, F. J., Fire-Resistant Lubes Reduce Danger of Refinery Explosions, *Oil and Gas Journal,* November 10, 1985, pp. 191–197.

NFPA (National Fire Protection Association); 10, Standard for Portable Fire Extinguisher, 1984.

NFPA 11, *Standard for Low Expansion Foam and Combined Agent Systems,* 1983.

NFPA 12, *Standard on Carbon Dioxide Extinguishing Systems,* 1980.

NFPA 12-A, *Standard for Halon 1301 Fire Extinguishing Systems,* 1985.

NFPA 15, *Standard for Water Spray Fixed Systems for Fire Protection,* 1985.

NFPA 20, *Standard for the Installation of Centrifugal Fire Pumps,* 1983.

NFPA 30, *Flammable and Combustible Liquids Code,* 1984.

NFPA 70, *National Electric Code,* 1987.

NFPA 85A, *Standard for Prevention of Furnace Explosions in Fuel Oil and Natural Gas-Fired Single Burner Boiler-Furnaces,* 1982.

NFPA 85B, *Standard for Prevention of Furnace Explosions in Natural Gas-Fired Multiple Burner Boiler-Furnaces,* 1984.

NFPA 85D, *Standard for Prevention of Furnace Explosions in Fuel Oil-Fired Multiple Burner Boiler-Furnaces,* 1984.

NFPA 214, *Standard on Water Cooling Towers,* 1983.

NFPA (National Fire Protection Association), 321, *Standard on Basic Classification of Flammable and Combustible Liquids,* 1982.

NFPA, *Standard for Fire Hoses,* 1961–1979.

NFPA, *Standard for Screw Threads and Gaskets for Fire Hose Connections,* 1963–1979.

Report of Joint NGPA-API Engine Manufacturers' Task Force on Engine Starting Air Line Explosions, presented at 42nd Annual Convention, Natural Gas Association, March 20–22, 19683, Houston and at Midyear Meeting, The American Petroleum Institute, Safety Committee, May 6–9, 1963, Dallas.

Underwriters Laboratories, Publications Stock, *Fire Resistance Directory,* 1984.

Underwriters Laboratories UL 1709, *Structural Steel Protected for Resistance to Rapid Temperature Rise Fires,* 1984.

Warren, J. H., and Corona, A. A., This Method Test Fire Protective Coatings, *Hydrocarbon Processing,* January 1975.

CHAPTER 7

How to Tap Occupational Safety and Health Resources

KENNETH S. COHEN
Consulting Health Services
El Cajon, California

7.1 DEFINING THE PROBLEM

The safety and health professions are not narrowly defined sciences with closely drawn lines of separation between classic academic disciplines. Those who come into the fields are generally drawn from a variety of science or engineering disciplines. This focus, narrowed by the dictates of academic structure, rapidly broadens in the demanding arena of the workplace. In the absence of direct personal resource, the conscientious professional will (or should) turn to "outside" help in filling the informational gaps needed to do the job.

The "help" resources available to the safety and health practitioner spans a wide range, from printed and electronic media, to consultants and agencies whose services vary in cost from free to very expensive. Selecting the "right" resource is the challenge. Once the selection is made, the responsibility still lies with the professional to insure the continuing quality of the resource. The safety and health professional has a duty to those entrusted to their watchful care. The exercise of that duty does not allow for excuses as to why the duty was not met, but rather implies that the failure to uphold that duty is negligence. The true professional knows when to say I don't know, and then proceeds to look for outside help.

This chapter will deal with the various needs and available resources that address the many and varied disciplines of the safety and health community. Although it would be impossible to give a complete list of existing sources of help, those which are either representative of a class or unique in what they provide will be showcased for your review and further investigation. *I will not attempt to give you my wisdom, but rather lead you to the threshold of your own!*

7.1.1 The Safety Discipline

The specialty we know as Safety is typically made up of two major areas, technical and management. Although many additional titles and responsibilities impinge on Safety, the need for resources typically falls into these two areas.

Technical Specialties

The technical aspects of safety can be basically broken up into three main categories: engineering, fire science, and program implementation. All of these areas offer a myriad of resources ranging from text and journal articles to specialty consultants who deal with broad and narrow issues of most problem areas.

Engineering Specialties. The domain of the engineer—and many safety professionals are either licensed or registered as Professional Engineer (PE)—extends over the physical environment of the workplace. The many aspects of the worksite, summarized in Table 7.1, are typically designed, implemented, and controlled by one of the three main engineering divisions. These divisions begin as mechanical, civil, and electrical, but evolved as the roots of a great tree into subspecialties such as industrial engineers (mechanical) and further into the sub-subspecialty of packaging engineer and still further focused to a microspecialty as in "blister pack/vacuum forming." Perhaps your problem is a heated element on a piece of packaging equipment, which needs guarding while still allowing the machine to operate. Alterations to the original machine design pose major liability-sharing issues, yet the workers are getting burned. Where do you go for help?

A major resource with heavy engineering backup is the manufacturer, who often will provide a great deal of consultive assistance without charge when a safety issue is presented. It must be remembered that the manufacturer has a great deal to lose from product liability litigation when one of their products damages your workers.

Local Engineering Society members can also prove to be valuable resources, both in defining the problem and identifying a consultant who may be able to address specific issues.

Colleges and universities with schools of engineering can also serve as sources of talent from both the faculty and student pools. In addition to consultant talent, most campuses contain extensive library facilities. Sometimes you may need only a standard or reference to solve a problem and the access to a good library can be invaluable.

TABLE 7.1. The Technical Areas Impacting on the Safety Discipline

Housekeeping Material Handling Material Storage
Walkways Machinery and Equipment Electrical Equipment
Welding Equipment Hand Tools Ladders Exits
Floors Platforms Railings Lighting Ventilation
Valves Protective Clothing and Equipment
Dust Fumes Gases Vapors
Explosion Hazards First Aid Hand and Power Trucks
Fire Fighting Equipment Guards and Safety Devices

Professional societies, such as the American Society of Safety Engineers (ASSE), will often have specialized divisions or sections devoted to engineering practice. To aid in the selection of a consultant, ASSE publishes a periodic listing of the members of each of its divisions.

Fire Science. No one person is totally expert in fire science. Specialization ranges from on and off the job fire prevention to postfire arson investigation. The academic preparation for these areas is difficult to come by, since few schools offer curricula in the specific fire sciences, and most "experts" have graduated from the School of Hard Knocks or are ex-firefighters. This in no way diminishes their skill or knowledge, but rather speaks to your inability to find some person with a vast string of initials trailing after his or her name.

One look at the sixteen-volume National Fire Code will tell you about the diversity of interest and basic science that goes into the field of "fire." The volumes of code themselves are an excellent guide and reference. Additional volumes from the National Fire Protection Association (NFPA), for instance the *Industrial Fire Protection Handbook,* can often define and solve the problem before you have to seek outside help.

An important pool of available talent and consultants is the Society of Fire Prevention Engineers. The unique makeup of their membership includes firefighters, equipment suppliers, academicians, and consulting engineers, who meet regularly in most urban centers to share information and fire scene conditions.

Safety Program Implementation. We can all read a textbook, an article, or listen to a consultant, but "our" program may just not want to work the way it's supposed to!

Where do you turn for help? Such a problem deserves a special type of solution. In my experience, you can have all the right ingredients and still not have a working safety program. You stare at each part and know it should work, but all together they don't fit. Now is the time to get a consultant who says, "I've done it before, and I can show you how to do it so it'll work in your location." Many top corporate Safety Managers retire into an active consulting practice. Don't ignore the combined years of experience these individuals represent. They've probably seen everything, including every problem in safety, and that degree of experience demands respect. Sometimes these retired safety persons join other retired specialists and donate their services to keep busy in their retirement.

Another way to get such unique talent is "cross-pollenation" with other safety and health professionals in your town, industrial park, or neighborhood. Inviting your counterpart to inspect your plant, as a courtesy that you will exchange in kind, can reveal the classic situation where, from routine contact, you are overlooking a hazard that a "stranger" will spot immediately. There are very few secrets in the safety field. It takes one of us to see the problem, and another to understand it. This can also be elaborated upon in the development of a relationship of "uncertainty sharing." Often we need only to share a problem with an understanding and knowledgeable someone, consultant or colleague, and the solution suddenly becomes clear.

Safety Management

Everyone wants to be a manager, but only a few really know how to be one. Management seminars, springing up around the country, profess to transfer instantly "managerial skills" to all attendees, but that's just not possible, no matter how inspiring the presentation. Individual skill must be developed in a number of areas demanded of the safety perspective, and hearing or sitting through a one- to two-day presentation may or may not fill the gap in experience. These seminars can be excellent points of departure, which can lead you to any number of other resources mentioned by the lecturer, referenced in the handout materials, or gleaned from fellow attendees. Four major items of safety management will be dealt with from a resource perspective.

Safety Training. Possibly the greatest challenge to management in the field of Industrial Relations is fighting the desire to terminate a worker when that worker makes a mistake. The question posed by moral and ethical standards is, "Was the employee ever trained how to do it RIGHT?" That may seem naive to the uninitiated, but in most accident investigations, lack of training looms high on the list of precipitating factors. Supervisors routinely suggest on accident investigation forms that the employee involved needs further training. Top management asks why the employee wasn't trained properly in the first place.

Training resources are only now becoming prevalent in the world of safety. New audiovisual materials are being generated as fast as emerging regulations demand them. Each new Occupational Safety and Health Administration (OSHA) standard seems to carry with it a requirement for training the worker in the particular hazard.

Not all safety and health professionals are able to stand before a group and be the teacher, as is often required. One on one they do very well, but in front of a group they fall apart. This person needs help. A recent article (Cohen, 1982) described the current "revolution" going on in safety training through the use of videotape. Today's worker is "socialized" to the network productions they see in their homes and will not tolerate films or slide programs that look like 8 mm home movies. Surveys indicate that just appearing on a television monitor reflects the aura of credibility to the viewing worker.

High quality in-house productions are on the rise in many large and small companies, with the aid of an internal audio-visual department or by contract with an professional production company. The quality of the output varies on a gradient of cost that runs from $100 per "finished minute" to over $3000 per minute. The final product reflects that age-old adage, You get what you pay for.

Less costly programs, such as slide and sync-tape, can be effective but need gentle care and handling to insure their acceptance. Some of the professional slide production and computer generated graphics companies produce "starter kit" safety programs, which need only some slides taken around your workspaces. A simple script follows the computer slides with a short description of what type of photo of your plant would go best in a particular position. A few days after you shoot some slides, insert them between the "stock" slides, and read the script into a microphone. You have a program. Not necessarily a network quality production, but it's yours, it's quick, it's inexpensive, and it features recognizable workers from your location, which heightens viewer interest.

This newly developed slide-narration program can be polished still further by making a video transfer. Again we are faced with an option situation—inexpensive or moderate cost—done either in-house or by an outside consultant. The major problem to overcome is the slide-change "blackout," which is exaggerated by the video transfer. The least costly conversion is to place a video camera in front of a rear-projection screen and a microphone at the speaker of the tapeplayer. You have a canned program, but with lots of blackout. This can be eliminated by the use of a "dissolve" system and video chain, which is usually the purview of the postproduction video house. By using two projectors, one slide fades into another onto the screen, producing a professional image without the bothersome blackouts.

Motivational Programs. *No matter what we do, we can't get our employees interested and involved in the safety program!*

How many times have we heard that old line from all phases of industry? The techniques used to motivate employees to participate in the safety effort range from the issuance of trading stamps to expensive paid holidays or cash equivalents.

Many companies who sell "incentive ware" will act as consultants to your program and advise on the range of gift products that might be well suited to the employees at your site. A gift or reward is usually given in exchange for some safety achievement at a department level on up to the entire corporation. This same concept can also be tied into a general "we care" promotion, which is used by all functions of management to correct problems, such as an adverse safety record, or to heighten morale among employees for increased quality and productivity.

The nature of the gift is generally paramount in employee acceptance of the program. If the gift is attractive to both the employees and their spouses, additional reinforcement supports the effort. This is most often the case with the use of trading stamps. These merchant incentives, still popular in many parts of the country, can be collected and used for valuable and varied gifts. Numbers of stamps are issued periodically, based upon some safety milestone such as Time-Loss Injuries. This system is interestingly presented in a NIOSH videotape (NIOSH, 1982) depicting several successful programs using this form of motivation.

Safety incentive Gift Catalogs are numerous and can be referenced in most of the major Safety and Health publications. These catalogs are generally for the officials of the organization, but very slick, full-color glossy brochures are available for employee perusal after a program is in place.

For those who believe employees are motivated best by their stomachs, one company tried a unique program through their in-house lunchroom. A "Bingo game" was established throughout the company with risk to each department being the motivating factor for cash prizes. All employees had Bingo cards, and numbers were called only on successive days of no time-loss injuries. When an injury occurred, that "game" was cancelled and a new one started. During the course of a current game winners were awarded various amounts of cash based upon the difficulty of the Bingo pattern. During the course of a "game" month, each department that remained accident-free would receive a free "special" lunch for each employee. The program *worked* and both accidents and time-loss injuries decreased.

Risk Management. The concept of risk management has recently enlarged to mean total protection of workforce, the workplace, and the company. It goes far beyond the limits of the Safety/Security domain of years past and now encompasses everything from insurance programs to public liability. The flood of publications available to interested safety and health professionals is significant. Although most risk management consultants began in the insurance field, the tide is shifting to diversified consultants coming from other loss-control areas such as in-plant and Corporate Safety positions.

The time to seek help in risk management is not immediately after experiencing a major risk crisis, but *before!* The best of the risk management consultants are themselves only "consultant coordinators." The diversity of risk in any one company spans so many areas that the quality of a risk consultant must be judged by those other experts available to your overall needs. No one person is an expert in everything, and that goes for risk management as well. If your consultant covers everything from safety to legal liability, and all points between, I would have questions about the skill level he or she might have in any one area.

A valuable source of consultants might be your insurance carrier. Many companies offer both loss control and risk management advice as a self-serving adjunct to their coverage. Other organizations, such as Risk Insurance Management Society (RIMS) and the National Safety Management Society (NSMS), can also be sources of consultant referral, seminars, and helpful publications. This area of consultancy typically deals with the pure liability of an organization and often lacks the specific focus needed in everyday problems.

Safety Program Design. Typically the design phase of a quality program in safety and health begins with a management commitment to success. With this assumption under our belts, we can move forward to getting the job done. We will need input from the various organizational components of our company and our first level of "consultant staff" will come from within. The players on our team will generally be made up from Industrial Relations, Production, Facilities, Safety, Health and Medical, Top Management, Quality Control, Material Handling, Marketing, and any other appropriate departments in your organization. Every team needs a captain, and it's not always the best idea to have the safety department lead the discussions, but often getting the "boss" involved as the discussion leader helps to get the program going.

Often bringing in an outside consultant in the design phase of a program is not a good idea. Unless the consultant is very familiar with your type of organization, much time may be wasted. Again, this is a good time for the cross-pollenation theory of consultancy mentioned previously. If you can find someone from your industry or a similar one to implement or aid in the initial design of your program, many hours can be saved. A quality safety and health program is like a fine suit. In order for it to fit well and look good it must be patterned and tailored to the body it protects; to do anything less will produce an ill-fitting item that brings pride to no one.

7.1.2 Health Disciplines

The industrial health field is divided into two major areas: Industrial Hygiene and Allied Medical specialties. Although each has its unique areas of skills application, the interaction and "team" approach is invaluable to the maintenance of workers' health. It should be stressed that though today's industrial health department may be most evident in regulating and medicating the postinjury worker, the essence of occupational health is and should be *preventive* medicine! We must always strive to look forward in our view of safety and health rather than looking back.

Industrial Hygiene

Over the past fifteen years since OSHA, the practice of Industrial Hygiene has undergone a series of major evolutionary modifications. A true dichotomous population exists within the ranks of industrial hygiene practitioners: (1) a senior group of hygiene specialists, those who were able to affect change and employee health prior to the existence of "instant-readout-digital-monitoring instrumentation with flashing red, amber, and green lights; and (2) a junior group, fresh out of an academic program consisting of didactics and little practical experience, who are enamoured of the mountains of analytical data they can collect and reduce rather than their workplace implications. Due to the needs of regulatory compliance, industry grabs onto the latter group as an alternative to the high-cost advice of the seasoned professional. All too often, the new hygienists find themselves alone in a world that "doesn't look like what they taught us in school!" Most hi-tech journeymen and professionals hone their skills on the rough stone of apprenticeship. In the sellers' market of Industrial Hygiene, the buyer accepts a degree or certification as sufficient qualification for workplace problems that carry with them huge potential liability.

Where does a prudent company look for competent evaluation of its industrial health problems? This is a question with few easy answers. The insurance carrier provides limited services, directly proportional to the size of the tendered premium. The consultative branch of a regulatory agency will hire those persons who are attracted to the "policeman" policy or are in need of training, gained at the client's expense. Consulting firms are generally expensive, and are often biased, since the worse the situation is made to appear, the more work can be contracted. A dilemma, at best, but let's look at some of the options and quality-control steps that might be taken to avert some of the pitfalls.

Recognition, Evaluation, and Control. The tried and true precepts of Industrial Hygiene which approach any occupational health exposure in terms of Recognition, Evaluation, and Control are hopefully used by all those who are in active practice. Emphasis should be directed to the Recognition phase, as primary over the others, due to the fact that the problem doesn't exist until it can be compared or referenced to prior experience. This is where the value of a Senior Hygiene specialist becomes of value. Selecting a consultant for a problem within the workplace should be done on a hierachical basis as illustrated in Figure 7.1. A major source of highly qualified consultants is the Council of Industrial

FIGURE 7.1. The heirarchy of selecting an industrial hygiene consultant.

Hygiene Consultants (CIHC) (see References) who publish an annual roster of its members and their peer-reviewed specialties.

Much greater latitude exists in the evaluation phase, better known as the "monitoring" phase, than the other phases in the selection of consultant resources. Once the problems agent, or etiology, has been identified, any number of options can be exercised to determine the extent of the problem. Numerous testing and analytical laboratories are available to take on outside work. Laboratory use demands continuing responsibility for maintaining quality control, by the user, in spite of professed accuracy of proficiency testing. Many laboratories also have certified and non-certified Industrial Hygienists who rent out on an hourly rate for specific tests or evaluations.

Control technology and systems engineering have a far more complex basis of consultation than do the two areas discussed previously. The design of an acoustic enclosure or a ventilation system must be done, if it is to be done properly—and in some states "legally"—by registered or professional engineers (PEs). Many industrial hygienists do have a dual certification (i.e., CIH and PE) which allows them to cover several areas of consultation practice. Check out the legal requirements in your area, and pick an appropriate consultant before you have to redo the work.

Additional text resources can also be of great assistance in dealing with control projects. Of the many texts available on ventilation, the single most valuable is the ACGIH Ventilation Manual, which is currently in its 18th edition.

Industrial Hygiene Program Design. The degree and extent of industrial hygiene involvement within a company is often related to the chemical exposure, though not necessarily. Many nonchemical, primarily office, locations have significant health problems such as "closed building syndrome." Obviously the degree of emphasis and prioritization of time and funds should be related to the highest risk situations first, which are more obvious with direct chemical exposure. The section above, Safety Program Design, will be similarly applicable.

Legal Investigation. Litigation of the occupational health exposure is complex and time consuming for the industrial hygienist. Most feared by practitioners is the prospect of a case going to court. It takes a special personality to hold up under the onslaught of cross-examination by a crafty opposing counsel who appears to know as much about industrial hygiene as you do. When selecting a consultant, it is important to know how he or she may stand up in court if the investigation should lead there. Ask about court experience, familiarity with trial and deposition, and experience with "chain of evidence" documentation.

A key issue in the selection of a consultant for a potential legal matter is the consultant's ability to deal with "demonstrative evidence." Photographic and video skills are key qualifications for a prospective consultant, as the additional cost of photographic talent will push up the cost of the consulting.

Another key issue in selecting an industrial hygiene consultant for a potential legal case is the type and number of cases worked in the past. Has the consultant been "court qualified" as an expert in the field you need? Has the expert worked more for the defendant or the plaintiff or *both* equally? Another important consideration is the potential for conflict of interest between your organization and any other of the expert's clients. These questions should be settled long before the actual consultation begins.

Medical Consultation

The whole area of medical involvement in the workplace can be a voyage through uncharted waters for those who don't have a skilled navigator to help on the trip. Many of the occupational medicine aspects of safety and health go virtually ignored by most professionals until the condition either goes away or becomes aggravated to the point of litigation. My advice is, get *competent* help, a trained and board-qualified physician to set your course early in the voyage.

Medical Surveillance. Numerous standards call for biological and medical surveillance of the worker for contamination and effect of the toxic materials in the workplace. These blood tests and biological parameters are not only the legal responsibility of a physician, but are also governed by patient privacy statutes. What tests are right, how often they should be done, and what they mean, are not decisions to be made by a physician without industrial background. Occupationally trained physicians are very rare, even in major metropolitan areas. A primary source of referrals is the local chapter of the American Occupational Medical Association (AOMA). A few of the major medical schools are beginning to develop Occupational Medicine clinics, which may supply the needed consultive personnel. Board certification in occupational medicine is a subdivision of preventive internal medicine, and is recognized throughout the World. Beware of the general practitioner who claims a vast industrial background. This level of experience may be solely of the "lump and bump" variety, obtained by patching up industrial accident victims at a local clinic. The depth of specialized medical knowledge required for medical surveillance goes far beyond acute symptomology; the physician must be able to project the potential toxicology of an exposure.

Employee Assistance Programs. The concept of an organized Employee Assistance Program is truly a reflection of our times. The need for distraction, diversion, or escape provided by abuse substances is steadily becoming a greater problem in the industrial community. The response of management has changed from, ignore the problem, to the realization that the employee is an asset more readily "salvaged" on the job than off. Employee Assistance Program clinics are springing up in-house as parts of an industrial relations department and as free-standing clinics that contract with industry to make psychiatric and psychological consulting services available to all employees.

In this important area of service, credentials are an obvious first step toward validating a specific program. Identifying references who have been satisfied with the offered service is critical, demanding more than a casual phone conversation. Get a letter from the reference or reject the program as one not worth involvement.

A critical point, as with all medical services, is the program's ability to assure patient privacy over a long period of time. If the life of program records is shorter than you believe is appropriate—generally ten years past the expected date of an employee's termination—then it should be understood that custodianship of all records will revert to your company for archival purposes.

A caution with Employee Assistance Programs is made in light of several pending cases involving persons "rejected" by insurance carriers for certain types of personal coverage (e.g., disability) on the basis of having been in counseling, analysis, or psychotherapy. Legal sources consulted in this regard believe that an employee thus rejected will have recourse back to the employer to the extent of the desired coverage. It's worth checking out with local legal counsel.

Acute Treatment Services. Often we do not have a choice of which doctor or medical facility will treat our workers when they are involved in a workplace accident, injury, or exposure. Police and fire personnel may direct victims to the nearest trauma center, if one exists in your area, or to the "closest" hospital with an emergency ward. It generally pays, in time to treatment as well as knowledge of treatment, to inform the medical staff of your local hospital of any peculiar problems or unique chemical situations that may be encountered.

A good way to familiarize medical personnel with your company or plant is to invite them to tour the location. This will clarify the risks to which employees are subject, making treatment quicker in case of an accident.

Occasionally free-standing medical facilities will treat minor "lump and bump" injuries. These clinics are generally located within industrial areas and are seldom more than several minutes away from your plant. The scope of service they offer should be determined *before an accident,* to maximize the level of medical care.

Preplacement Examinations. An ounce of prevention can significantly outweigh thousands of dollars of cure. This is reflected in the reduced compensation costs when medical screening is applied to high-risk operations. It is important to underline at this point, that the purpose of the exam is *not* to exclude an employee from being hired, but to prevent putting an already hired employee into an area of jeopardy.

Many conditions that would become evident at the time of an examination are medical reasons for an employee not to work with a particular chemical or do a uniquely strenuous job.

Preplacement exams can be performed by a wide variety of medical specialists including your in-house doctor, the free-standing clinic, or consultant specialists. Within the category of consultant specialist are the "traveling road show" examination vans, which work throughout the country. Make sure the physician directly in charge of the examinations is qualified for your needs. All doctors can do examinations for lead exposure, but board certified occupational physicians can generally do it better.

Legal Investigations. The need often arises for medical testimony or investigation of a worker's compensation case, a personal injury action, or third-party action against your organization due to medical damages. When this happens, you need someone who will not only survive but win in the litigation arena.

Few physicians survive well under direct examination, let alone cross-examination. The witness box is an unnerving place where the art of medicine looks less and less like science with each succeeding question. The responsive witness answers yes or no. Seldom, in the experience of most physicians, is anything medical a yes or no, black or white answer and as a result, a doctor is often a poor witness.

Finding an occupational specialist who also has done some court testifying is not difficult. If you scan the ads in plaintiff's publications you will find just about any type of specialist you desire. The best selection, made in concert with your local or company counsel, is often the doctor who also runs a clinical practice. Juries typically look with disfavor on the "professional witness," both from a credibility standpoint as well as monetary considerations.

Legal Resources

The major problem associated with finding or using legal resources is that we all too often wait till the last moment, and then it may be too late.

The time to establish a relationship with your legal counsel is at the onset of a problem, not two thirds of the way through it. The evidence is cold, the witnesses are forgetful, and opposing side is already armed and ready to do battle. A retainer often makes early onset involvement easier. The prepaid retainer is like a line of credit, which can be drawn against for a variety of small to major services. It's worth considering for small as well as large organizations.

General Liability Counsel. Most companies or organizations have either a house counsel or an attorney (or firm of attorneys) who is consulted regarding any number of issues that require legal advice. The areas of specialization vary slightly, but most often is a general practitioner. Occasionally corporate counsel will have such special skills as Security and Exchange Commission (SEC) or tax law experience. Seldom will counsel have a particular interest or experience in the area of either general or specific liability. If your counselor is "good" and honest, he or she will suggest some form of outside consultation in areas of deficiency. No one can be all things to all persons, and professionals are the last to admit they can't do it all!

Workers' Compensation. Workers' compensation attorneys are specialists who keep up with a major body of law that is in a constant state of flux. Companies with multistate locations must realize that the law differs from state to state and may require a number of attorneys to deal with the complex web of cases a large company can accumulate.

The law of workers' compensation is written with specific bias toward the worker. This requires an innovative approach to evidence collection and presentation on the part of the attorney handling the case. The value of the damages also impacts the level of effort used in a case. This option shrinks in light of a trend in many jurisdictions to use the "outcome" of a workers' compensation case as the foundation for subsequent third-party actions. To this end, many compensation cases are being fought with a vigor unprecedented in the past.

Demonstrative evidence is becoming a much larger part of the workers' compensation attorney's armor before today's Administrative Law Judge (ALJ) who hears the case. The applicant's counsel may well plead the workplace as a ". . . dank and dirty, sweatshop with a character conducive to injury and illness!" This can easily be reversed by a videotape showing the existing work area, after laying a foundation for assuming that the current scenes are as they were during the period in question.

Workers' compensation attorneys are not in the habit of bringing in outside "experts," due to the high cost to their client. This cost, however, when looked at in the broader perspective of the full litigation picture, pales in comparison to the potential costs of a third-party action. It may well save the company thousands of dollars in potential judgments by bringing in an outside plaintiff's attorney to "inspect" the vulnerability potential as they would if a case were actually being sued.

Product Liability. The various entities within a company rarely communicate regarding safety within the plant and without. Accidents that are a result of workers handling a product may well foreshadow what may happen in the hands of the normal consumer. When the consumer is injured, the product may be judged "defective." The review of a product by the company attorney can sometimes eliminate the chance of "defect."

Should the in-house counsel feel unprepared for the task of product liability evaluation, a good place to turn would be a plaintiff's attorney, who sues on this theory on a daily basis. That type of consultant usually has a cadre of already qualified products experts waiting in the wings. What an advantage to have the product "picked apart" on your behalf rather than in the courtroom.

Professional Liability. In our litigious world, numerous kinds of suits are invented to circumvent the protection afforded by existing law. Workers' compensation affords the employer a shield of immunity from suit in most states. It becomes possible to penetrate the "shield" under certain circumstances such as *fraudulent misrepresentation* and willful, wanton *negligence.* It appears that the *easiest way to show negligence* is through the acts of agent of management required to act in a "professional" capacity such as a doctor or a safety engineer. Legal expertise is of particular importance in evaluating the relative vulnerability of the professional within the organization. Once the vulnerable areas are identified, standards of conduct to comply with the existing "standard of care" in the community can be set. Often, after the fact, we find well-intentioned professional employees behaving in a way they believed was the company policy. Legal expertise is needed before, not after the fact of an incident that be interpreted as negligence.

7.2 FINDING THE "RIGHT" RESOURCE

Everyone you talk to before you make your selection, believes that *their* resource is the best, and can tell you what *your* mistakes were if the selection "turns sour." Only *you* know what's best for your application. With a little advance planning and data collection, a quality resource/consultant can be obtained and used to the best advantage of you and your company. The following guidelines can offer a "yellow brick road" to finding the "right" resource.

7.2.1 Internal Resources

Sometimes we cannot see the forest for the trees! A company succeeds because of the expertise it applies in producing its service or product. Look to your own resources before paying a high price for the talent you require.

Local Talent

We need expertise, engineers, lawyers, doctors, safety engineers, industrial hygienists, and the like. Don't forget to look within your company or at surrounding companies, with which you may have established some form of reciprocal relationship.

Unfortunately, no catalog of employee expertise exists to aid our search. Perhaps the chemist we talk to over lunch each day is also an expert in violent reactions related to an explosion experienced in the plant. We just didn't think to ask. A variety of talent often goes unnoticed when we assume that our problems relate only to our scope of interest.

Corporate Talent

"We can do it ourselves, we don't need Big Brother to help us!" is the plaintiff's cry from the divisional officers who actually need help but do not want to admit to their corporate superiors that they can't find the appropriate resource. It might even seem better to lose the war than to admit that in one small battle they are going down to defeat. The very nature of the corporate structure is that of maximal utilization of resources and it doesn't pay to have to replicate talent in every division when temporary loan of a single person can benefit all concerned.

It never hurts to ask if a specific talent or resource is available within the corporate structure. A major corporation had reports of serious compensation problems related to backstrain in one of its divisions. Finding an appropriate consultant only increased the cost and the number of variables to be dealt with, but provided no cure for the problem. While attending a marketing conference within the corporation, the general manager of this division discussed his problem with other managers. This uncertainty sharing lead to the discovery of a product-design engineer within the corporation who had a great deal of interest in material handling devices and was eager to get involved with solving the problem. A three-week loan of one employee after a bit of conversation and the problem disappeared. It's not always that easy, but worth a try.

The Insurance Industry

We task an insurance carrier to indemnify us in time of loss, for which we pay reasonably high dollar premiums. The carrier, in turn, has a duty to aid us in our loss control effort, since they paint themselves as the experts in understanding "risk."

Most carriers have large staffs of loss control specialists, safety engineers, hygienists, nurses, and the like to aid in reducing the insured's potential for loss. Generally assistance is directly proportional to the premium dollar and the potential liability, but small policies also can get a bit of attention if they scream loud enough.

7.2.2 Governmental Resources

As taxpayers, businesses have a right to take full advantage of the agency talent provided by government to answer the problems facing its constituency. The bureaucracy may not always have the answer we need, but it's worth exploring their talent pool first, as the price is right.

OSHA Consulting Service

Either by Federal agency contract or state agreement program there is a consulting service providing "free" help in dealing with the safety and health problems related to the workplace. Upon request to the local Occupational Safety and Health Administration office, either a safety engineer or industrial hygiene specialist will be made available to assist a business in solving the problems identified or suspected.

The consultive branch of OSHA differs from the regulatory arm in that it is precluded from either issuing citations or informing its sister agency unless it sees an extreme hazard or imminent danger. Many organizations admit to fearing the regulatory reprisals without knowing about the division that exists between the branches.

Local Public Health Service

Many public health agencies are gearing up to deal with environmental and occupational diseases. Staff additions in the area of epidemiology and industrial hygiene are expanding the "classic" role of the health department from the area of infectious disease and sanitation to hazardous waste spills and community right-to-know chemical regulation. A major movement is underway to cross-train sanitarians into a dual role of food inspection and chemical hazards monitoring. In many jurisdictions, the local Health Department is being asked to deal with the inventory of hazardous chemicals and processes in order to aid emergency services personnel with the catastrophic potential in our hi-tech industrial environment.

Fire Jurisdiction

Fire prevention inspections have long been an asset and resource to the industrial safety department in preventing massive property losses that can occur in a fire. Property loss and *preservation of life safety* is a complex field, which extends beyond the adherence to local code and regulation. Industry today deals with large quantities of both flammable liquid and uniquely toxic combustible building materials. Standards evolve daily as the reality of the fire scene changes our perception of fire loading and retardation.

The expertise of the local fire department can often be utilized to review the risk due to fire or arson in the industrial setting. The local fire authority can also be an excellent source of training for personnel and internal "brigade" activity.

7.2.3 Private Consultants

Wherever you look in today's business or industrial environment, you'll find consultants, experts claiming the ability to "get you on the right track," and all it takes is money!

What you're buying, in essence, is the expertise of someone who has gone through the process (successfully) and can now avoid all of the pitfalls. For this service, you will pay at a rate arrived at by the consultant consistent with his or her success rate. The concept of "what the traffic will bear" continues to dominate the pricing market. A high-priced expert is either worth what he or she asks, or will starve. The busy consultant—and be sure to verify that the business is real and not just a smokescreen—is usually getting a high price from satisfied clients.

Local Societies or Associations

This gathering of specialists is often for the purpose of professional development, but can be merely for social purposes. Seeking the "right" expert from the pack is a job requiring skill. Professionals seldom give "bad" recommendations about their fellow members, so *how* the recommendation is offered is often more significant than what is actually said. Most members of a professional community will be very brief when asked about a colleague they cannot truly, in their hearts, say is a quality expert. Again, you must evaluate the input and its validity.

The Yellow Pages

Finding the right consultant or service in the phone book is asking for trouble and highly unlikely. The "good" ones don't advertise, but use a word of mouth to enlarge their business.

The only reasonable way of getting a list of consultants is from a professional directory, as described earlier. Then choose several persons on the list who are closest to your geographic area. Beware of the temptation to "save money" by a selection of the closest consultant in order to avoid transportation charges. The old adage still applies: You get what you pay for!

NETWORKING WITH COLLEAGUES

The very best means of consultant validation is to ask someone you trust. Get together with your colleagues and ask whether anyone has used the tentative selection. Was he or she everything that was expected? Were there problems? Were time requirements met? Was the fee appropriate for the quality of the work? These and many more questions specifically related to your application are critical to the proper selection of a consultant who will represent both the company and you. If your friends won't tell you the straight facts, then no one will.

7.3 QUALIFYING THE RESOURCE

Whatever the nature of the resource, consultant, textbook, on-line database, or so forth, the resource reflects your choice and can make or break you in the eyes of the company management. You bring in the "outsider" and expose other management personnel. Will it bring you the credit you deserve? It will, only if you've done your homework on validating the resources qualifications.

Establishing a *due diligence* file is critical to the qualification process and your survival in the event the resource fails to perform as advertised. This file should be a secure location for keeping *all* the justification information on the resource selected. Document any and all conversations, comments, or other information gathered on the resource, both for evaluation purposes as well as reevaluation, should the resource prove less than adequate to the task.

7.3.1 Establishing The "CYA" File

Another name for the file mentioned above is a *CYA file.* CYA is short for "cover your anatomy," and refers to guarding your own flank. Hiring consultants can be a tricky business. It is not at all unusual for a member of management to be fired along with a consultant who fails in the task for which he or she was hired. Someone has to share the blame, and it doesn't need to be you if you cover yourself with good documentation on the consultant selection process. The mere breadth of your CYA file will indicate your diligence in the process of resource selection. It's also hard to wade through a good deal of material and not glean from it what the true nature of the resource is. Don't be shallow in your searching; the information you gather is critical to a reasonable selection process and the success of the mission.

7.3.2 Good and Bad References

Since the collection of references on the resource is very important, whether they are formally written or notes from conversations, it's worth taking a few moments to discuss the nature of a "reference."

When evaluating a reference it is important to note where the emphasis is placed. If you have indicated the nature of your need to the person supplying the reference, then you should have the information about the resource in those specific terms. If the information in a reference is not given in these terms, consider this a point against the resource in question. When giving a reference for someone felt to be less than deserving of a "glowing" reply, the typical response is to talk around the issues and give a variety of nonrelated comments. Read between the lines to get at the meat of the reference.

I am also suspicious of the overly exuberant reference, which goes far beyond the task defined. It could truly represent a consultant of exceptional quality, but might equally mean a potential conflict of interest between the candidate and the reference. Weigh the good comments against the bad and make your own decision.

7.3.3 Resumés

We all want to show our good face first for all to see! Look for this tendency when evaluating an autobiographical resumé. Resumé consultants, and there are such creatures, advise that you list the best first and omit anything that might look bad. An evaluator should read a resumé in the same manner. Does the consultant list his education toward the end of the resumé in order to hide it? Has the consultant published in peer review journals where his colleagues can openly judge the skills he or she professes? What employment is listed, all jobs or just those that might give good referrals? All of these factors and more should be weighed in the reading of a resumé.

The *curriculum vitae* (CV) is nothing more than a fancy version of the resumé written in academic style. Typically, professionals consider that the resumé is from someone out of work and looking, while the CV represents someone who has a job, but is always looking.

7.4 EVALUATING THE END PRODUCT

The consultant looks like a good choice, the project must be done, what could go wrong? Almost anything. Establish from the start what you want as the "end product" of the job. If you're paying for the work, get your money's worth. If the consultation is a *pro bono* service, you deserve what you paid for!

7.4.1 Get and Follow a Contract

Regardless of the resource type, it pays to get a formal agreement or contract defining the scope of work and when it will be completed. Such details as written reports should be defined in advance. These conditions obviously are not possible with consultation from an agency such as OSHA/Public Health/University consultancies. Most "commercial" consultants have contracts that spell out in detail what you should get, and how long it will take to get it.

7.4.2 Performance Standards

Two major categories exist in the standard of work delivery for a consultant, *specification* and *performance*. The former spells out what is intended in the way of work activity, and the latter suggests what must be the outcome of the consultation. It may be necessary to get both levels defined in some consultive relationships in order to monitor progress as well as the final outcome.

7.4.3 Demand and Keep Good Records

Documentation can make or break a consulting project. Good record keeping is a sign of a quality consultant. When every job is documented as if it would end up in court, there is seldom any need to redo a forgotten detail. Good records keep a good consultant on-track by providing an audit trail, which can be followed by any and all interested parties to the work.

Notes taken at the job site must be maintained by the consultant or the laboratory for an agreed-upon period of time or transferred to the company as custodian of record.

7.5 ESTIMATING AND DEALING WITH RESOURCE COST

The cost of a consulting service is judged inexpensive while the problem is visible, but considered prohibitively expensive after the problem is solved and the consultant is waiting for payment. A true theory of relativity could be generated from this premise and is the reason that many consultants require payment in advance!

There is no such thing as a free lunch or a worthwhile "free" consultation. Getting what you pay for seems to apply only when a "second" consultant must take over the job and try to pick up the pieces. Many professionals are dedicated and provide pro bono consulting to persons with no visible means of support, and that service is usually high quality beyond the limits placed by a paying job. Additional "free" services can be offered when the consultant is in a student/teacher relationship. Beyond that which has already been discussed you should expect to pay the "right price for the right quality job."

7.5.1 Retainer Agreements

Pre-payment of consulting fees can often effect a significant reduction (discount) on the total cost of using an expert. This payment, in advance, constitutes a retainer of services which can be used over a period of time. An example of this is illustrated in Figure 7.2. Many additional services are associated with a retainer, such as telephone consultation, calendar priority, and the like. The retainer usually runs for a period of one year and it is the company's responsibility to use the days during that period.

7.5.2 Legal Costs

Many consultants will require indemnification or defense from legal actions associated with the work performed. This is also augmented with individual Professional Liability coverage carried by the consultant.

TEN CONSULTING DAYS @ $850	$8500.00
TEN DAY RETAINER (Prepaid)	7500.00
SAVING PER YEAR	$1000.00

FIGURE 7.2. Consultant retainer costs over a one-year period.

It is good practice to have consulting agreements reviewed by your legal department to prevent any misunderstanding during or after the work has begun.

7.5.3 Laboratory Fees

Much of the safety and health consulting done in the modern hi-tech workplace requires some form of laboratory confirmation or identification.

Laboratory services can be contracted for in many ways, but we will limit our discussion to the two most common. The consultant generally has a laboratory that has given good service in the past and will be used in this effort. It can, and should be arranged in advance, for either the consultant or the company to be billed for the service.

Typically, when the consultant contracts for the service, the cost is augmented with an "interpretation" fee, unless the lab belongs to the consultant. When the company pays for the lab work, the cost is billed direct and the results are copied both to the company and the consultant.

The preferred, and less likely to be compromised, method of payment is the latter.

7.5.4 Computer Time Share Fees

Computer costs are often incurred when a consultant must do specialized research about your specific problem. This work is typically on a time-share basis with a major database such as the National Library of Medicine or the Lockheed Information Services. Charges range from $22/h up to $85/h, depending on the information required. Some lawyers will provide legal computer services, which may be as much as $135/h.

The fee is generally proportional to the amount of data recovered on your topic. You may also be asked to pay the professional's time on top of time-share costs, but this should be determined in advance of the work.

7.6 QUALITY CONTROL OF RESOURCES

This is the final stage in developing a sound working relationship with consultants and other resources. Letting everyone know in advance that close and careful scrutiny will be exercised throughout the project will keep everyone on their toes.

7.6.1 Staying in Control

The consultant may appear to be in control of the project, but the client must always have the ultimate say as to the direction of the project. This can generate serious ethical problems between the client and the consultant when opinions differ over professional needs. This should be clearly defined at the beginning of a relationship. The consultant is regulated not only by professional liability obligations but also by ethical codes of the profession.

The only option open to the consultant, when these conditions are breached, is to withdraw from the job. This clearly states to the world that a major problem existed, which was beyond negotiation. In a thirteen-year consulting practice, only twice have I felt the need to divorce myself from a client who, in my opinion, was placing employees at risk and would not alter the situation.

7.6.2 Uncertainty Sharing

As discussed previously, the concept of uncertainty sharing can be the most powerful tool in completing a project with everyone's best interests at heart. If, at the onset of a job, both parties agree to discuss difficulties as they arise, the fewest number of egos will get bruised in the process. Set the stage early on and the benefits will be forthcoming.

7.6.3 Summary

The quality of any safety and health resource will be judged by its users. A word-of-mouth reputation will carry a good consultant for years without ever needing a bit of advertising. Less-than-good consultants last a short while and then are never heard from again.

Pick your resources carefully, and stick with the tried and true rather than "looking for that larger bone imaged in the water" that can be had a bit less expensively. It may turn out to be only an image, after all.

REFERENCES

ACGIH, *Industrial Ventilation: A Manual of Recommended Practice,* 18th ed., 1984.

CIHC, Membership Roster, Council of Industrial Hygiene Consultants, P.O. Box 1625, El Cajon, CA 92022.

Cohen, Kenneth S., Videotape Revolution in Safety and Health Training, *Occupational Health and Safety,* June 1982.

NIOSH, *Loss Control Behavior Management,* 3/4" Videotape, 1982.

CHAPTER 8

How to Evaluate Your Occupational Safety and Health Program

WILLIAM E. TARRANTS
National Highway Traffic Safety Administration
U.S. Department of Transportation
Washington, D.C.

8.1 INTRODUCTION

The objective of this chapter is to provide safety professionals, managers, and others with practical tools for planning, organizing, and conducting evaluations of safety and health programs. More specifically, this chapter is concerned with describing in nontechnical terms the concepts, methods, and techniques of evaluation, demonstrating, step by step, various procedures for designing and conducting evaluations of safety and health programs, and suggesting ways of planning and organizing the evaluation function. Issues relating to what evaluation can and cannot accomplish, how to recognize good evaluations and avoid bad ones, and how to integrate evaluation results with planning and management systems will also be addressed.

8.1.1 A Definition of Evaluation

The term "evaluation" is commonly used to describe a variety of judgmental activities, usually involving an assessment of some degree of quality in relation to a desired goal. Closely associated with this term is the concept of "deciding," that is, making a judgment about whether a specific accident countermeasure is effective, whether a certain safety program is "paying off," or whether to continue, modify, or eliminate a certain safety-related activity. A more pragmatic view, perhaps, is that evaluation precedes decision making and that a more appropriate definition of evaluation is "comparison." Typically, in carrying out the evaluation process programs or activities are compared with each other, with a selected "criterion," or standard of effectiveness, or with a goal that determines success. Following these comparisons, a decision is made concerning which program or activity is superior, or whether a specific one is producing appropriate results in terms of the established objectives. Ultimately a decision is made whether to adopt a new program or countermeasure, modify or expand an existing program, or eliminate or reduce the level of effort involved in a currently operating program. If evaluation precedes these decisions, the choice is more likely to be correct than if no evaluation was conducted. Thus we can conclude that evaluation is vital to good program management, requiring competent decision making.

In order to plan and conduct useful evaluations, three things are needed: (1) an item (activity, project, countermeasure, or program) to compare, (2) a standard or criterion with which to compare it, and (3) an established objective for use in determining when or whether success is achieved. Then the safety professional can answer such questions as, "Is this program equal to, poorer than, or better than the acceptable or desired standards or criteria? Is the program achieving its objective? Should I modify or eliminate this program? Which, among two or more programs, is most cost beneficial?" The process of answering one or more of these questions will produce an evaluation. In summary, we can define evaluation as the process of comparing various projects, activities, countermeasures, or programs with criteria and objectives that have been agreed upon as being acceptable for purposes of measuring achievement.

8.1.2 Purposes of Evaluation

According to Carol Weiss (1972) the purpose of evaluation is to measure the effects of a program or project against the objectives that it was designed to achieve and to make decisions about its continuation and improvement. According to Weiss, there are four key features in this definition. (1) "To measure" refers to the methods used in evaluation. (2) "The effects" refers to program *outcomes* that can be measured objectively, thus avoiding subjective value judgments by program personnel. (3) "The comparison of effects with objectives" stresses use of explicit *criteria* for judging the program's success. (4) "The purpose of evaluation" is served by its contribution to decision making that will improve future programs and, subsequently, future operations of the organization.

Use of systematic formal evaluation procedures is the best way to learn from past experience. In the field of occupational safety and health, we can identify several specific reasons for conducting evaluations. We can evaluate:

1. Proposed new projects or programs to determine their effectiveness in comparison with existing projects or programs.
2. Old or existing projects or programs for improvement or expansion to increase their effectiveness in achieving their objectives.
3. Existing projects to assess or measure their cause-and-effect ("bottom line") impact.
4. Existing projects to determine in quantitative terms what has been accomplished and at what cost.

5. Various projects or activities within a program to help select which alternative(s) best achieve program objectives.
6. The entire program to determine its value to the organization.

Evaluation may be viewed as a prerequisite to planning, thereby making it an essential part of the management process. Because evaluation requires both a statement of objectives and systematic collection of data relating to the achievement of objectives, it enables managers to maintain the project's and/or program's direction as well as assess its short- and long-term usefulness. Thus evaluation serves as a guide for making changes in project or program effort or emphasis. It is a way of increasing effectiveness by learning and profiting from past experiences. Evaluation is a necessary and integral part of overall program and project management, and program or project planning is vital in the overall management process. Thus we can conclude that program and project management and evaluation are inseparable if managers are to be successful.

8.1.3 Program vs. Project Evaluation

The title of this section includes the word "program." We have used the term "project" as well as "program" in our discussion of evaluation. These two terms are very closely related. A program normally consists of a number of projects, each making some contribution to the program's objectives. It is also possible that the various projects within a program will interact among themselves to produce a synergistic effect, with the end result being more than would be expected by arithmetically summing the separate effects of each project. Often evaluation takes place at the individual project level, since that is where changes are often initiated within a program. We rarely throw out an entire safety and health program and start over with a completely new one. More typically, we identify a new countermeasure, decide that an existing countermeasure or activity might not be producing as we would like, or we wish to expand the level of effort at the project level while other projects (countermeasures, activities, etc.) within the overall program continue. In these instances, evaluation at the project level may be most appropriate.

Evaluation of a program consisting of two or more (multiple) projects may also be desired. Program evaluation involves examining multiple program components (projects, countermeasures, etc.) concurrently in order to assess their combined impact on program objectives. While the methodology may be somewhat more complex, the anticipated results are the same, namely, to achieve some measure of the impact of the total program on the objectives of that program.

8.2 THE ORGANIZATION AND MANAGEMENT OF EVALUATION

Evaluation is part of the management process; therefore, evaluation decisions are essentially management decisions. The nature of such decisions, particularly those involving the scope and complexity of an evaluation effort, requires that a manager go through the same process that he or she would follow in making any major management decision. Consideration must be given to such issues as personnel requirements, financial resources needed, the organization structure required for evaluation, and how the evaluation program will be managed. In addition, consideration must be given to anticipated evaluation problems and gaining management acceptance of evaluation as a desirable activity within the organization. As with all administrative functions, evaluation requires time, money, facilities, and personnel. Unlike many administrative functions, however, evaluation is often not recognized as a top priority by managers. As a result, an administrator must often use considerable ingenuity to evaluate a program or project, given built-in system constraints, particularly on money and personnel. Unfortunately, the need for conducting an evaluation is often not recognized until a program or project is in trouble. At that time, bringing about the necessary corrections is often very difficult.

8.2.1 Personnel Requirements

Since evaluation is a relatively sophisticated task, a knowledgeable manager will make certain that the staff members are qualified to plan and conduct the evaluations and perform the analyses needed to produce valid results. If basic responsibility for evaluation is placed with the Director of Occupational Safety and Health, then higher-level management must assume that this staff person has the necessary qualifications or has available qualified staff specialists in evaluation to assist in planning, organizing, and implementing the evaluation program. It may be necessary to hire a person with mathematical and systems analysis skills on an as-needed basis, perhaps as a consultant. The point here is that the qualifications and experience of available personnel resources should be carefully analyzed to determine whether additional personnel or additional training of the present staff may be in order.

The basic evaluation tasks to be performed suggest certain personnel requirements for the evaluation function. In general, two basic skill areas can be defined. The first requires an understanding of and proficiency in experimental design. This requirement presumes a knowledge of basic statistical principles and quantitative methods necessary for evaluation design. The second skill area is data collection and processing. Evaluation requires the collection and analysis of data based on a predetermined study design. In larger organizations, it is likely that persons with these skill qualifications will be available to actually conduct the evaluations or to assist and/or train other existing staff members to perform this function. In smaller organizations, it may be necessary to hire an outside experienced evaluator.

8.2.2 Financing Evaluation

Just as evaluation is an integral part of management and operations, the cost of performing an evaluation must be considered an integral part of program and project costs. Project planning should include provisions for adequate funding of the evaluation component. For funding purposes, we might identify three broad categories of projects based on the level of evaluation required:

1. Projects requiring minimal evaluation. This type of project may only require a monitoring operation.
2. Projects that provide for a definite evaluation plan requiring some data collection, but not a sophisticated experimental design. This type of project may only require some comparative analyses.
3. Projects that require a relatively detailed, sophisticated evaluation design involving experimental techniques. Projects of this type may require funds to hire an experienced evaluator.

The actual evaluation costs will vary considerably, depending on the nature of the project or the complexity of the program. Costs will be minimal for projects in the first category and may be substantial for those in the third category. Ideally the costs of evaluation will be built into the budget at the planning stage of the project or program, including the funds necessary to provide evaluation training for existing staff members.

8.2.3 Anticipating and Solving Evaluation Management Problems

Organization, personnel resources, and funding may be the most easily solved problems. One of the most difficult management problems is that people are reluctant to participate in the evaluation process. Creators of projects and programs tend to expect too much success from their efforts, thereby producing an initial atmosphere that makes it almost impossible for a manager to admit that his or her undertaking is not a glowing success. This expectation in turn creates a situation in which an outside evaluator is unwelcome. It may also create a tendency for the manager who evaluates his or her own program to consciously or unconsciously build a favorable bias into the evaluation process and/or the data output in order to indicate that the project or program was successful.

There are no simple solutions to this problem. The best approach might be to work toward creating an environment in which emphasis is placed on problem identification, problem solving, and the view that evaluation is a means to an end and not an end in itself. A successful evaluation produces information that is useful for planning and managing occupational safety and health projects and programs. If the evaluation produces useful information, then the evaluation is a success even if the project in its present form did not achieve all of its objectives. Emphasis should be placed on the identification and solution of problems and not on the establishment of personal blame for not achieving complete project or program success. The concept of program evaluation is being addressed more frequently in the occupational safety and health literature. One thought often expressed is that the safety performance of today's large organizations needs to be improved and that evaluation is required to give all management levels the same kinds of information about safety, or the lack of it, as they are receiving concerning the quality, quantity, and effectiveness of other aspects of the company's operations. The days of justifying safety programming and budgeting decisions based on the personal judgment and experience of the Safety Director are rapidly drawing to a close. What is needed is a clear definition of the objectives a program is expected to accomplish and then a courageous, hard-nosed examination of actual results against initial expectations. Evaluation should occur at every step of program planning, organization, and development, including goal setting, program testing, and program operation. Evaluation, if it is conducted at all, is usually started at the point at which a program is in full operation and delivering services or, as was mentioned previously, when the program appears to be in trouble.

From these observations, it should be clear that there are three important challenges for managers in their efforts toward making evaluation work:

1. There is a need to develop systematic information and decision systems for management that build evaluation into every step of program planning, development, and implementation.
2. There is difficulty in interpreting evaluation results and in summarizing these results in a form

suitable for presentation to the decision-makers at each level of management within an organization.

3. Program managers at various levels feel threatened by the process and results of evaluation as they relate to their area of responsibility.

The term "evaluation" carries with it a rather negative connotation, particularly among managers with program design and implementation responsibilities. They often regard evaluation as a third-party criticism of their efforts, with the objective of seeking reasons to criticize, frustrate, or terminate their programs.

Management information specialists working in the field of evaluation can do much to counteract the fears associated with evaluation. Program evaluations will tell managers if they are achieving the results they desire.

The current period of expanding knowledge and tightening budgets in the occupational safety and health field is fertile for the advancement and expansion of evaluation efforts. The conflicts will probably become worse rather than better in the immediate future. Therefore, if new knowledge derived from evaluation is to find its way into active safety and health programs and wasteful programs are to be eliminated, ongoing activities must be appraised to determine what reallocation of resources can take place. Similarly, new programs must be carefully pretested and evaluated at the pilot stage before they are put into practice.

8.3 TYPES AND LEVELS OF EVALUATION

Evaluation has been defined as the process of comparing the results of a project or program with its originally established objectives. While objectives may be expressed in different ways, it is generally accepted that for adequate project or program planning and evaluation, objectives should be stated in terms of the expected effects. It is important to distinguish between the activities and the effects of a project or program in order to understand the distinctions between different types of evaluations.

A project usually involves the completion of certain tasks and activities that are specifically defined and measurable, for example, the installation of guards on certain pieces of equipment or the training of a given number of first-line supervisors in procedures for meeting their safety responsibilities. In these two examples, project activities would be considered complete when the guards are installed and working properly or the first-line supervisors are trained. These activities are designed to contribute to the ultimate objective: a reduction in occupational injuries and work injury losses involving the newly guarded machines and an overall reduction in accidents and accident losses by workers in the units under the responsibility of the supervisors trained. Loss reduction is an ultimate objective because it is indirectly associated with the installation of guards and the training of supervisors.

Projects may also have immediate or intermediate objectives. The immediate objective of the supervisor training program is an increase in knowledge and skills of the supervisors. An intermediate objective is the implementation or use of those skills in their daily work. Meeting the immediate and intermediate objectives should lead to achievement of the ultimate objective, worker accident loss reduction.

8.3.1 Stages in Intelligent Problem Solving

Evaluation represents an attempt to utilize the scientific method for the purpose of assessing the usefulness of a project or program in achieving its objective. Researchers generally recognize five stages that recur in intelligent problem solving (Tarrants, 1979):

1. *Planning* (including problem identification and analysis): We actually define the problem and establish its parameters in the planning stage. It is essential that planning include establishing procedures for evaluation. Planning also includes the selection of at least one criterion of effectiveness and the prescription of appropriate methods of measurement.
2. *Observation:* The evaluator observes what happens as the independent or task variable (the new safety program or accident countermeasure) is introduced to the experimental group and the dependent variable (criterion) changes or remains stable over time.
3. *Hypotheses:* In order to explain the facts observed, conjectures are formulated into a hypothesis, or tentative proposition, expressing the relationship believed to have been detected in the data. In formulating hypotheses, prime consideration must be given to the criterion or measurement standard to be used in conducting the evaluation.
4. *Prediction:* Based on the hypothesis or theory, we make deductions concerning the consequences of the hypothesis we have formulated. Here is where experience, knowledge, and perspicuity are important.

5. *Verification:* The evaluator collects new data and tests the predictions made from the hypothesis. The essence of verification is to test the relationship between the variables identified by the hypothesis.

This process of scientific problem solving applied to accident control problems enables the safety professional to gain knowledge through experiments, as opposed to intuitive speculation and the application of trial-and-error loss prevention methods.

8.3.2 Types of Evaluation

Evaluation is the process of systematically examining project or program activities and measuring their effects. Two principal types of evaluations are applicable to the evaluation of occupational safety and health projects and programs:

1. *Administrative Evaluation:* A judgement of value or worth based on comparisons of actual task accomplishments to established performance goals.
2. *Effectiveness (impact) Evaluation:* A determination of the extent to which task operations have contributed to the achievement of project goals or program objectives.

Such evaluations are best thought of as a continuum in which the elements of one type form the basis for the next. Thus an effectiveness evaluation would include the steps necessary for an administrative evaluation.

8.3.3 Administrative Evaluation

Administrative evaluation generally focuses on two aspects of project or program activity:

1. *Performance or Output:* Specific outputs of the project or program measured in quantifiable terms, for example, the number of supervisors completing a safety training course, the number of 5-minute safety talks presented, etc.
2. *Schedule:* The time necessary for the completion of each task as examined and compared with the project plan.

Although administrative evaluations are important to the management of occupational safety and health programs, the application of effectiveness (impact) evaluation techniques is essential to the determination of how well a program or project is functioning.

8.3.4 Effectiveness Evaluation

Effectiveness evaluations are used to determine which countermeasures actually reduce accident losses (ultimate objective) or achieve some immediate or intermediate objective. This knowledge aids in the selection, continuation, expansion, or reduction of a specific safety project or program.

An effectiveness evaluation focuses on the effect(s) of performing certain activities in terms of their impact on project or program goals and objectives. For example, a safety training program is initiated for workers in a machine shop for the purpose of improving safety performance. We first conduct an administrative evaluation to determine the number of workers satisfactorily completing the training. Project effectiveness would then be measured by:

1. Assessing the knowledge, skills, and perhaps attitude changes brought about by completing the training course.
2. Evaluating improvements in unsafe behavior engaged in by the workers completing the course.
3. Determining the degree of achievement of the ultimate objective of reduced accident losses (reduced frequency and severity of disabling and/or minor injuries).

Administrative evaluation documents the actual occurrence of the results of project activity. Effectiveness evaluation documents the degree to which the objectives of the project or program were actually achieved.

8.4 MEASUREMENT IN EVALUATION

Measurement is simply defined as the process of assigning numbers to objects or events according to rules (Stevens, 1951). Measurement is basic to all evaluation activities, since without it, we cannot quantify our observations. Measurement is an absolute prerequisite for control and prediction. The degree to which we are able to control real-world events and predict their occurrence is determined by our ability to measure them. Measurement is primarily a descriptive process. It allows us to qualify, order, and quantify certain events and ultimately use the results as a basis for control and prediction of actual performance.

Generally, measures are needed to identify project or program performance levels. Is the accident prevention countermeasure paying off? A second purpose of safety performance measurement is to provide information concerning changes in the safety state within an organization or operation. A valid and reliable measure of these changes permits evaluation of the effectiveness of accident prevention efforts over time. A measure of the total safety state would, of course, include accidents that have the potential for producing loss as well as those that result in actual loss during the evaluation period. At any given moment, accidents that have a potential for loss may produce a loss. Which exposure results in loss and the degree of severity of future losses are influenced by chance factors. Therefore, measurement techniques that are more sensitive to the fundamental behavioral and condition malfunctions that may in time contribute to an accident loss are much more useful as indices of safety performance.

8.4.1 Characteristics of Good Measuring Technique

It is possible to postulate a number of characteristics of measurement techniques that can serve as standards or criteria with which to judge the appropriateness of any given instrument. The following are postulated characteristics of a good measurement technique without regard for their relative importance (Tarrants, 1980).

1. *The measurement instrument should have administrative feasibility.* This means that one must be able to construct and use it.
2. *The measurement instrument should be adapted to the range of the characteristics to be evaluated.* The instrument should include within its range all of the values important to the intended application.
3. *The unit of measure should be constant throughout the range to be evaluated.* It should yield readings that are expressed in the same units.
4. *The measurement criterion must be quantifiable.* A quantitative evaluation is preferable to a qualitative evaluation in that it has greater precision.
5. *The measurement technique must be sensitive enough to detect differences and to serve as a criterion for evaluation.* The ideal measure of safety performance must be sensitive enough to detect meaningful changes in environmental and behavioral conditions.
6. *The measurement technique must be capable of duplication with the same results obtained from the same items measured.* In other words, it should be consistent or reliable over time and thus provide minimum variability when measuring the same condition.
7. *The measure must be valid or measuring what is purported to be measured.* What should be measured in the safety field? What is safety success? The criteria of safety success, and thus the determination of what is to be measured, should include the minimization of all unsafe behavior and unsafe conditions and their consequences, as well as minimization of accident losses resulting from disabling injuries, recordable incidents, first-aid cases, and fatalities.
8. *The measurement instrument should yield results that are free from error.* A concentrated effort should be made to reduce both biased and compensating errors.
9. *The measurement instrument should be efficient.* This requires that the cost of obtaining and using the instrument is consistent with the benefits to be gained.
10. *The measurement instrument must be understandable.* This requires that both the measurement criterion and the evaluation techniques can be comprehended by persons charged with the responsibility of using the instrument or approving it for use.

These characteristics of a good criterion or measure are rarely achieved by any one instrument of evaluation. Often only a combination of measures can provide a reasonable compromise.

8.4.2 Standard Safety Measurement Techniques

Several indices of safety performance are now available for use in evaluations: number of disabling injuries, incidence rate, severity rate, injury frequency rate, accident cost, number of deaths, number of first-aid cases, the ratio of severity to frequency, and the "serious injury index," which includes information about accidents resulting in major injuries regardless of the degree of disability involved.

The calculation most often used to measure safety performance under the standards developed by the Occupational Safety and Health Administration is the Incidence Rate, IR. The IR is defined as:

$$\text{IR} = \frac{\text{No. recordable illness or injuries} \times 200{,}000}{\text{No. man-hours worked}} \tag{8.1}$$

The 200,000 in the numerator is equivalent to 100 full-time employees working 40 hours a week for 50 weeks. The denominator automatically adjusts for differences in hours of exposure.

A measure of the severity of injury and occupational illness losses can be obtained by use of the formula

$$\text{SR} = \frac{\text{No. man-days lost} \times 200{,}000}{\text{No. man-hours worked}} \tag{8.2}$$

Data used for these measures are normally collected on a monthly, quarterly, or annual basis for use in quantifying the recordable injuries and providing comparative incidence data during successive time periods. Normally the data used in making these calculations are obtained from OSHA Forms 101 and 102, which most employers are required to complete.

8.4.3 Noninjurious Measures of Safety Performance

An accident-causal-factor identification technique is needed that will identify noninjurious accidents as well as those involving disabling injuries. This technique should also be capable of identifying unsafe conditions or defects in the environment that have an accident-producing potential. The general justification for the collection of noninjurious accident data is that the severity of accident results appears to be largely fortuitous. Accidents from the same causes can recur with high frequency without a resulting injury. Which accident does produce an injury appears to be determined largely by chance, in most situations. This suggests that the real importance of any accident is that it identifies a situation that could potentially result in an injury or property damage loss. Whether a given accident does or does not result in a loss from each occurrence in less significant. What is needed is an evaluation technique that identifies causal factors associated with noninjurious accidents. Then this information can be used as a basis for an accident prevention program designed to remove these causes before more severe, loss-type accidents occur. The more frequent opportunity to detect causes at the "no injury" stage should lead to the desirable objective, total loss prevention, with most accident problems being identified for correction before any losses actually occur. If the same techniques could be used to identify causal factors involved in injurious accidents of varying severity, then the entire range of the accident severity spectrum could be represented by a single method of accident-causal-factor evaluation. Two noninjurious measures of safety performance have been developed and tested in actual industrial plant environments and have been found to meet these requirements: the critical incident technique and behavior sampling.

The Critical Incident Technique

The critical incident technique is a method of identifying errors and/or unsafe conditions that contribute to accidents within a given population by means of a stratified random sample of participant-observers selected from within this population. These participant-observers are selected from the major plant departments so that a representative sample of operations existing within different hazard categories can be obtained.

In applying the technique, an interviewer questions a number of persons who have performed particular activities or jobs within certain environments and asks them to recall and describe unsafe errors they have made or observed, or unsafe conditions that have come to their attention. The incidents described are classified into hazard categories from which potential accident problems are defined. The Critical Incident Technique is described in detail in Section 8.6.3.

Behavior Sampling

Another measurement technique used in the evaluation of safety projects and programs is called *behavior sampling*, or *activity sampling*. Similar to the critical incident technique, the application of this evaluation method does not depend on the occurrence of accident losses.

Safety professionals generally accept the premise that accidents are causally related to behavior. Even the existence of so-called "unsafe conditions" is a testimony to the fact that behavioral acts of omission or commission occurred that produced this result or allowed it to happen. It is the behavior itself that serves as the best criterion for evaluating safety performance, not a relatively rare disabling injury. When a recordable work injury *does* occur, the acts that contributed to it at the moment of or directly preceding the accident are still of major importance. A method of evaluation is needed that allows the direct assessment of unsafe behaviors. If we accept that the degree of change in behavior in a positive (safe) direction is a measure of the effectiveness of an accident prevention program, then ways of *detecting* this change should serve as a guide for the assessment of the safety program itself. Such a method has been derived by using statistical control techniques in combination with randomly selected observations of worker behavior. Depending on the observation method chosen, the behavior data collected can be used to compute either the percent of workers involved in unsafe behavior or the percent of time that the observed workers were behaving unsafely. New countermeasures, projects, or programs aimed at positive behavior change are introduced, new observations collected, and the results plotted on statistical control charts to determine whether worker behavior has improved significantly. Behavior sampling is described in detail in Section 8.6.4.

8.5 STEPS IN THE EVALUATION PROCESS

The more closely evaluations approach the scientific model, that is, the more complete and comprehensive the data, the better the data analysis, the more objective the criteria, and the better the controls exercised to eliminate the influence of variables extraneous to the one (or more) variables purposely introduced in the population studied, the more useful the evaluation will be. In general, an evaluation must be adapted to the needs of those who will use the results, the amount of time available, a reasonable expenditure of money and personnel resources, the characteristics of the specific program or countermeasure to be evaluated, the characteristics of the population studied, and the general needs and priorities of the company. In spite of certain adaptations that must be made when a program or countermeasure is evaluated at a given location, it is possible to postulate a general set of steps or a step-by-step guide for use in planning, designing and conducting an evaluation of a safety program or countermeasure. These steps provide, in effect, a "do-it-yourself" procedure for systematically setting up and carrying out the evaluation process. Where highly technical matters are involved, as in an intricate experimental design or a sophisticated data analysis technique, general guidelines have been included, along with a few easy-to-locate references for readers interested in pursuing the topic themselves. In these instances, use of a specialist in statistical methods and experimental design is strongly suggested.

The following eight steps should enable the safety professional and manager to understand the evaluation process and make the appropriate decisions for the conduct of useful evaluations. Although the steps are arranged in numerical order, some will be performed somewhat concurrently. In following steps 1 and 2, for example, it would be difficult to select a program or countermeasure for evaluation without defining the purposes and limitations of the evaluation at the same time. It is also usually necessary to refer back and forth from one step to another during the course of planning and conducting the evaluation to make certain that all of the elements necessary for an effective evaluation are included.

8.5.1 Step 1: Select the Countermeasures, Projects, or Programs to be Evaluated

Obviously, because of limitations of time, personnel, and other resources it is not possible to evaluate all countermeasures, projects, and programs at once. Also, the benefits of evaluation are usually greater for some activities than for others. Evaluations, especially effectiveness evaluations, can be expensive to conduct and thus must be considered from a cost/benefit point of view. Because evaluation resources are limited, activities must be chosen for evaluation according to a priority based on their worthiness and the decision-making information needs of management.

Currently Operating Vs. Proposed New Projects

In deciding what should be evaluated, consideration should be given to currently operating, continuing projects rather than proposed new projects. It may be useful during the early stages of a new evaluation activity to consider allocating some resources to evaluating proposed new programs as well as to evaluating those programs that have been in operation for awhile. Ideally, evaluation procedures should be built in at the planning stage of a new project or program. More often a countermeasure is selected for implementation based on an immediate need identified from an analysis of accident causal factors. The countermeasure is usually selected for application based on the experience and knowledge of the safety professional or from suggestions for problem solutions contained in accident prevention manuals, standards, or other references. Frequently the implemented countermeasure becomes "institutionalized" and is continued indefinitely, even though it may have ceased to achieve its intended goal, or an alternative countermeasure might come to be more appropriate. A preferred approach would be to conduct a pilot test of a proposed new program or project, evaluate its impact in terms of accomplishing the desired objectives, and then consider its more universal application based on the outcome of the pilot test. Even though the results of a pilot program evaluation are highly positive, a plan for including evaluation as an essential component of general program implementation should be established so that a measure of continuing success, or lack of it, can be obtained over time. By doing so, programs that cease to be useful can be identified early and can be replaced by more effective alternative programs. It is important to know how effective a new program is before it is implemented on a large scale. It is also important to know when an established program ceases to be useful.

Suggested Project Selection Criteria

As stated previously, it is usually beneficial to pilot test and evaluate a proposed new program or countermeasure before it is implemented on a broad scale throughout the organization. This will allow adjustments to be made in program content, or perhaps lead to its rejection, before incurring the expense of large-scale implementation. For continuing fully operating programs, it is useful to consider some specific criteria for selecting those that will be evaluated on a priority basis. Three suggested criteria for use in deciding which ongoing projects to evaluate are (1) project cost, (2) importance of the project, and (3) the magnitude of the project as defined by the scale of its application.

A costly project needs to be evaluated to determine if it is, in fact, cost-beneficial in accomplishing its objectives in a timely, efficient manner. If it is not achieving its intended results, then the money required to operate it should perhaps be shifted to another project or program where it can be used more productively. The results of a good evaluation will help the manager make this decision. Also, a part of the cost criterion in selecting a project to evaluate is the cost of the evaluation itself. Typically, evaluation costs should not exceed 10–20% of the operating costs of a continuing project. Evaluation costs associated with a pilot project may exceed this amount when the project is expected to be repeated several times or implemented on a large scale after being proved effective.

The importance of the project in terms of its estimated potential significance in reducing accident losses (deaths, injuries, property damage, production stoppage, etc.) is a major selection criterion for evaluation. Sometimes a project or countermeasure is selected for implementation because it is flashy and draws considerable attention to the safety program and perhaps to the Safety Director. Other countermeasures are implemented because managers *expect* them to be in place as part of an ongoing company safety effort. A more appropriate index of importance is the success of the program in reducing accidents and the various losses they produce.

The third selection criterion to consider is the magnitude of a project's application within the company. This criterion is somewhat related to both cost and importance. A broadly applied countermeasure may be both costly and highly important in terms of the number of employees it reaches. If a project is implemented throughout a large, multiplant corporation, then it may rank fairly high in priority in contrast to a project that is applied only on a limited scale.

Procedural Guidelines for Use in Applying the Selection Process

The following steps are suggested for use in selecting ongoing, continuing projects for evaluation:

1. Prepare descriptive statements for each project under consideration for evaluation. This statement should include a summary of the nature, scope, and functions of the project, a statement of the project objectives, a description of the major activities included in the project, and the length of the project's operation.
2. Establish an estimate of the cost of each project.
3. Estimate the current and potential success of each project in reducing accidents and their resulting losses.
4. Estimate the cost of conducting an evaluation of each project.
5. Estimate the magnitude of each project's application within the company.
6. Based on an assessment of the information collected from steps 1–5 for each project, establish a priority ranking for evaluation among the projects assessed.

Summary of Step 1

1. In the early stages of a new evaluation effort, consider currently operating vs. proposed new projects as candidates for priority evaluations. Ideally all new projects should have evaluation built in at the planning stage. New projects should also be pilot tested and evaluated before being implemented on a large-scale basis. Currently operating ongoing programs or projects should be identified for priority evaluation based on certain predetermined selection criteria.
2. Suggested criteria for selecting ongoing projects for priority evaluation include:
   a. Project cost.
   b. Importance of the project in terms of its actual or potential impact on reducing accidents and their resulting losses.
   c. The magnitude of the project's application within the company.
3. Procedural guidelines for applying the selection process:
   a. Prepare descriptive statements for each project candidate for evaluation.
   b. Estimate the cost of each candidate project.
   c. Estimate the current and potential success of each project in reducing accidents and their losses.
   d. Estimate the evaluation costs for each project.

e. Estimate the magnitude of each project's application.
f. Based on an analysis of the previous steps, establish a priority ranking for project evaluations.

8.5.2 Step 2: Define the Purposes and Limitations of the Evaluation

Once a project has been selected for evaluation, whether it is a pilot project to test a new countermeasure before its large-scale implementation or an ongoing project selected for priority evaluation, it is necessary to establish the most appropriate type and level of evaluation to be conducted. In general, consideration should be given to such factors as (1) the kind of information required to make management decisions, (2) the magnitude of the evaluation effort, (3) the urgency of obtaining evaluation results, and (4) the constraints of resources available for conducting the evaluation.

These factors determine the nature, quality, and usefulness of the evaluation and will have an influence on each subsequent step in the evaluation process. A clear definition of the purposes and limitations of the evaluation to be conducted will assist in designing a good evaluation and will help derive the most benefit from the evaluation itself.

Determine What Kinds of Information are Needed

Administrative evaluations should be conducted to measure and assess the outputs of a project (as opposed to project effectiveness or impact on predetermined criteria). The progress of a project in terms of the achievement of certain production milestones can also be assessed by conducting an administrative evaluation. Administrative evaluation involves measures of performance, scheduling, and costs associated with a project. Measures of performance might include such factors as the number of persons affected by the project, number of machines or other equipment used, and/or the number of tasks to be performed. Scheduling measures might include such factors as the number of days (weeks, months, etc.) required to conduct the project from start to finish, number of days (weeks, months, etc.) required to complete each milestone or task, and/or the number of days (weeks, months, etc.) required to achieve (complete) each project objective. Administrative evaluations might also include the costs of such items as personnel, equipment, materials; the overall cost of achieving each milestone or project objective; and the total cost of the project.

Projects involving the application of a countermeasure to achieve an impact on reducing accidents and the resulting losses should be subject to effectiveness evaluation. Information produced by effectiveness (impact) evaluations should include the degree of progress toward the achievement of immediate, intermediate, and ultimate project objectives, the impact (effects) of the project on such evaluation criteria as unsafe behavior, unsafe conditions, human errors, number of accidents, number of "near misses" or noninjurious accidents, lost time resulting from accidents, accident costs (both direct and indirect), and the impact of the project on sustaining and/or enhancing the mission of the organization.

Decide on the Desired Level of Evaluation Effort

Evaluations cost money. If no evaluation specialist is available within the organization to assist in evaluation planning, study design, and data collection and analysis, then one should be hired on a consulting basis. A minimum level of evaluation effort is usually essential to achieve meaningful, timely evaluation results. In addition to selecting an evaluation plan and study design that is most appropriate to the project, consideration must be given to the evaluation criteria, units of measurement to be used, method of data collection and recording, use of a control group if necessary, methods of analyzing the data, and the interpretations and conclusions drawn from the evaluation results. An assessment of these factors will help to determine the appropriate level of evaluation effort in terms of personnel, time, and money.

The Urgency of Obtaining Evaluation Results

While an appropriate effectiveness (impact) evaluation built in at the planning stage of a new project or incorporated into the regular operation of an ongoing project will be expected to produce data (measured results) on a regular, periodic basis, it may be useful to time the production of evaluation results to coincide with the need for information during the planning–budgeting cycle within the organization. This will assist the managers in making decisions about project continuations, new project planning, and operating budgets associated with future time periods. Obtaining the results of administrative evaluations should not be a problem since the development, operation, and progress of a project in terms of its outputs can easily be monitored on a weekly, monthly, quarterly, or any other appropriate time basis. Effectiveness evaluations that focus on the effects (impacts) of results rather than activities or outputs may present a problem in the timing of the production of information, since evaluations are tied into the life or operating cycle of the project. Careful attention to project scheduling might help with this problem.

The Constraints of Resources Available for Conducting the Evaluation

Some limitations may be placed on evaluations because of the constraints of time, money and access to information. Many projects require considerable time in order to produce valid and reliable results. This is particularly true when evaluative criteria such as disabling work injuries are used. Short time periods combined with small worker populations will not provide sufficient exposure to produce meaningful effectiveness results. Time constraints should be estimated on the basis of past experience with similar projects applied to a similar worker population size.

The level of evaluation activity will always be affected by budgetary considerations. Required budget levels for conducting effectiveness evaluations can be estimated by studying past evaluations and, allowing for inflation, applying their cost analyses to the proposed current-project plan.

Another constraint or limitation on evaluation that should be considered at this stage is the possible sources of data required to conduct the evaluation. What evaluation criteria will be used? What units of measure will be required? Will additional outside help be needed? How will the data be collected? These questions about the availability of data and information accessibility should be answered, at least partially, at this early stage of project planning since they are critical to the nature, quality, and usefulness of the evaluations. Management at all levels as well as the safety professionals concerned with or involved in the evaluation must understand the factors to be evaluated, how they are to be evaluated, and the time schedule for producing the evaluation results. Unless these factors are known and understood at the beginning, the eventual evaluation results and the final evaluation report may vary so widely in focus and quality from expectations that they will have a negative effect on the company decision-makers. The person or persons who will utilize the results of the evaluation should participate in decisions concerning the type and level of evaluation that will best suit the project.

Summary of Step 2

In establishing the most appropriate type and level of evaluation to be conducted, consideration should be given to such factors as:

1. The kind of information required.
2. The magnitude of the evaluation effort.
3. The urgency of obtaining evaluation results.
4. The constraints of resources available for conducting the evaluation.

These factors, which determine the nature, quality, and usefulness of the evaluation, will influence each subsequent step in the evaluation process. Decisions on the purpose and limitations of the evaluation should be made immediately following the step 1 decisions. The person or persons who will utilize the results of the evaluation should participate in decisions concerning the type and level of evaluation that will best suit the project.

8.5.3 Step 3: Define Project Objectives and Determine Evaluation Criteria

Occupational safety and health programs and projects have the ultimate objective of preventing accident losses (e.g. fatalities, disabling injuries, work time losses, equipment damage, product losses, indirect and direct accident costs, etc.) Pain and suffering associated with accident injuries and deaths are also important prevention objectives, although their units of measure are somewhat less precise. Normally accidents and their losses result from a multitude of causes and contributing factors. It is thus unlikely that any single countermeasure project will solve the problem by itself. One issue of concern relates to the fact that the ultimate objective of most countermeasure projects is to create situations in which accident losses (injuries, deaths, property damage, etc.) do not occur. It is impossible to measure a nonevent. The problem is further complicated by the fact that the causal factors associated with accidents do not always result in losses each time they appear. For example, an operator may disengage a machine guard in order to have easier access to the point of operation *without* an injury occurring for a substantial period of time. Yet the *potential* for loss is constantly present as long as this unsafe condition exists. While it is important to measure loss-type events, it is much more important to determine the existence of behaviors and conditions that have the potential for producing *future* losses so that actions can be taken to correct potential accident loss problems at the no-loss stage. Only by following this procedure can the ultimate objective of accident loss control ever hope to be achieved. In defining countermeasure project objectives, attention must be given to identifying and utilizing evaluation criteria that are not totally dependent on the occurrence of losses to be measured. Thus, while the ultimate objective of accident countermeasure projects is to reduce or eliminate losses resulting from injuries and deaths, the immediate objective is to reduce or eliminate the causal factors (unsafe conditions and human errors) that have the potential for producing future losses.

Defining Project Objectives

The development of a model for determining project objectives can be divided into the following phases:

1. Define the problem.
2. Categorize or classify the major causal factors associated with that problem.

3. Select one or more causal factors to be removed or controlled as the potential objective(s) of the countermeasure project.
4. Select one or more countermeasures judged to have the potential for removing or controlling these causal factors.
5. Define evaluation criteria related to the achievement of the countermeasure project objective(s) in quantifiable terms.

It is important to describe the objectives in terms that are meaningful to persons involved in conducting, evaluating, and approving the countermeasure project. These are the persons most closely involved with the project's ultimate success and continued operation and the ones that have the most influence on its success.

Once the objectives are defined, they should be clearly expressed in writing and circulated among those concerned with the various aspects of the project. The statement of objectives should include the following:

1. A description of the project's immediate and ultimate objectives, in quantitative terms whenever possible. The advantage of expressing objectives in quantitative or numerical terms is that it allows all interested persons to understand and share the same expectations concerning the project's outcome.
2. A definition of the target population expected to be affected by the project. Some projects are limited in the scope of their operation (e.g., a pilot test of a countermeasure project on a limited experimental and control group). Others may have a broader scope of application (e.g., all operators of engine lathes in the plant).
3. A statement of the period of time within which the objectives are to be achieved. Sufficient operating time should be provided to permit the acquisition of data based on the selected evaluation criteria. For example, if disabling work injuries resulting from accidents are selected as the criterion, a substantial time period may be required for the project, since disabling injuries are relatively infrequent events and it takes a long time for enough injurious accidents to accumulate so that results can be measured. In contrast, if unsafe behavior is selected as the evaluation criterion, the influence of the countermeasure project can be determined (evaluated) within a relatively short time following its implementation. Within the overall time scheduled for project evaluation, it is possible to establish milestone points for achieving immediate and ultimate objectives. If the objectives are related to the milestones in a realistic fashion, project progress can be monitored with a greater degree of precision. For example, the percent reduction in workers involved in unsafe behaviors within a given work environment can be measured at preselected milestone points to determine the influence of the countermeasure project on this criterion, a point when the percent reduction in unsafe behavior is stabilized can be identified, and the longevity of this stability (i.e. the point when the percent of unsafe behavior increases) can also be determined.

Determining Evaluation Criteria

Evaluation criteria must be defined based on the achievement of the immediate and ultimate objectives, the evaluation criteria must be stated in quantifiable terms, and the methods of measurement must be established relative to the selected criteria. The first criterion should relate to the causal factors associated with the immediate objective of the countermeasure project. If reduction in unsafe behavior of workers is established as the immediate objective, then percent of unsafe behavior before (baseline data) and after implementation of the countermeasure project should be selected as the evaluation criterion. The reduction of accident losses to a certain preestablished level might be selected as the ultimate criterion of effectiveness. In this instance, baseline data should be collected to establish the level of accident loss prior to implementation of the project. Then, during the implementation process, similar data should be collected at periodic intervals to assess the impact of the countermeasure on this criterion.

Identifying Appropriate Measurement Techniques

Project effectiveness can be measured in terms of the degree to which it achieves its immediate objective(s) as well as the influence the project exerts on the ultimate objective(s). One or more criteria should be established as an indicator for determining when an objective has actually been accomplished. Measurement techniques are then applied based on these criteria. Behavior sampling techniques can be applied to determine the influence of the countermeasure project on the immediate criterion of percent of unsafe behavior (see Section 8.6.4). Similarly, the long-term influence of the project on accident loss reduction can be measured by assessing the change in recordable work injuries over time, or the change in recorded first-aid cases before vs. after the project is implemented. Similarly, a safety training program for workers can be measured in terms of the immediate knowledge and skill objectives of the course. Attitude change can also be measured. The impact of the course on unsafe behavior, and ultimately on the reduction in accident losses, can be measured in the process of evaluating the achievement of these longer-range objectives.

When a large target population and/or a substantial time period are involved, it may be useful and appropriate to measure loss-type events as an index of safety performance. Work injury frequency rates can be calculated and the percent reduction in relative frequency rates over time can be calculated. For additional techniques for measuring safety performance, see Tarrants (1980).

Summary of Step 3

1. Define project objectives by:
   a. Defining the problem.
   b. Determining the major causal factors associated with that problem.
   c. Selecting one or more causal factors to be removed or controlled as the objective of the project.
   d. Selecting one or more countermeasures appropriate for removing or controlling the causal factors.
2. The statement of objectives should include:
   a. A description of the project's immediate and ultimate objectives, in quantitative terms.
   b. A definition of the target population covered by the project.
   c. A statement of the period of time within which the objectives are to be achieved.
3. Identify the evaluation criteria:
   a. Definition should be based on the achievement of the immediate and ultimate objectives.
   b. The evaluation criteria must be stated in quantifiable terms.
   c. Establish the units of measurement relative to the selected criteria.
4. Identify the appropriate measurement techniques.

8.5.4 Step 4: Design the Evaluation Study

The evaluation study process should permit a determination of whether the results of the project are attributable to the project activities or to some other factors. A reduction in the frequency of work injuries, for example, might be attributable to a countermeasure project or may well have been produced by a statement of interest by a plant manager, a change in accident reporting, or some other factor totally unrelated to the countermeasure project itself. In order to reduce confusion and error, projects selected for effectiveness evaluation should preferably be designed as experiments and should follow the scientific rules of experimental design. Where the conditions for experimental evaluation cannot be met, consideration may be given to other study designs.

Types of Study Designs

The manner in which a project evaluation study is to be conducted is called the design. The study design specifies how variables affecting the outcome of the project will be controlled. According to Isaac and Michael (1971) there are three basic types of study designs: experimental, causal-comparative, and quasi-experimental. These authors define these three types of study design as follows:

1. *True experimental design* investigates possible cause-effect relationships by exposing one or more experimental groups to one or more "treatment conditions" (project activities designed to influence or change behavior) and by comparing the results to one or more control groups not receiving the treatment, thus controlling for other variables that may have a possible influence on the outcome.
2. *Causal-comparative design* seeks to establish cause-and-effect relationships by observing an effect and searching back through data for possible (and plausible) causal factors.
3. *Quasi-experimental design* approximates the conditions of the true experiment in a setting that does not allow control of all relevant variables.

The Experimental Design. An experiment is a knowledge process that involves proposing one or more hypotheses and accepting or rejecting those hypotheses based on external criteria and whether or not the observed differences are probably due to the independent or task variable introduced (the countermeasure activity) or probably can be accounted for by chance factors. In the experiment one or more variables (elements introduced as part of the experiment) are arranged so that their effects on the criterion variable can be observed and measured. The observation and measurement of the impact of the variables on the evaluation criteria provide the basis for accepting or rejecting the hypothesis or assumption about the causal relationship between the task variable and the criterion variable. An experiment may involve the observation and measurement of a number of elements, all of which are "variable" in that they can be changed, or substituted, according to the kind of experiment to be performed and the measurements (data) desired from the experiment. It should be understood that experimental results cannot be considered as absolute proof that an assumption or hypothesis can be accepted or rejected. The results of an experiment are usually stated in terms of some probability of error.

The Causal-Comparative Design. The causal-comparative design (also known as the ex-post-facto or after-the-fact design) does not make use of control groups. An example of this design is the measurement of Incidence Rate in a department after the completion of a countermeasure project, such as the introduction of a poster program. If the safety director or evaluator concluded that the poster program caused a reduction in Incidence Rate, and therefore concluded that the project was effective, he or she may be open to criticism on the basis that other explanations of this result (hypotheses) are possible. For example, a chance remark by the plant manager concerning certain hoped-for behavior changes may have influenced the safety behavior of workers in the department; a small department may not experience a recordable work injury by chance, especially when a relatively short period of time is involved, even though the basic unsafe conditions or behaviors remain unchanged; or the immediate supervisor may have corrected the problem quite independent of the poster campaign's influence.

The Quasi-Experimental Design. This study design approximates the conditions of a true experimental design in situations that do not allow for the control and/or manipulation of all relevant variables. The classic source of information concerning this design is a book by Donald T. Campbell and Julian Stanley, *Experimental and Quasi-Experimental Designs for Research* (1966). Random assignment to experimental and control groups is the essential feature of true experiments because it provides the best assurance that the two groups are alike in the unknown variables so that the observed differences following introduction of the countermeasure can be attributed to the program with some confidence and a degree of precision. However, there are numerous situations in the real world where random assignment is very difficult to achieve. In these situations, the quasi-experimental may be the preferred evaluation design. In a quasi-experiment, assignment to treatment occurs in some nonrandom fashion, usually because certain individuals or groups are assigned to a treatment (or countermeasure) by professionals who believe they should receive it. A control group may be established by finding a "matched" group, or the experimental group may serve as its own control group. The main problem with quasi-experiments is that the nonrandom assignment makes it difficult to draw causal inferences. However, some design features can help evaluators rule out certain alternative interpretations of the obtained relationships between the independent (task) variable and the dependent (criterion) variable. They therefore help us to draw less ambiguous causal inferences (see Campbell and Stanley, 1966).

Design the Actual Evaluation Study

A good evaluation design is the key to successful effectiveness evaluation. The main function of the design is to eliminate possible explanations of the evaluation results other than the one associated with the task variable (countermeasure) applied. The steps to follow in preparing the evaluation design are as follows:

1. *Conduct preliminary planning and procedures development studies.* Appropriate initial issues to consider include an analysis of the problem based on problem identification studies, accident reports, inspection reports, critical incident studies and other sources of information relevant to existing accident and potential accident problems. Based on these analyses, the safety professional/evaluator will:
   a. Define the specific occupational safety and health problem.
   b. Identify the objectives relating to the problem solution.
   c. Identify all potential countermeasures or activities that appear to be operationally feasible.

The product of this step will be a clear definition of the problem to be solved, a documentation of possible objectives relating to the solution, and a description of the most feasible potential countermeasures available for use in achieving a solution to the problem.

2. *Select and fully define the most appropriate countermeasure.* Based on the problem analysis, a specific target group of workers is identified as the recipient of the countermeasure, and a single countermeasure program is selected from among the alternative countermeasures defined in the previous step. A preliminary design for the project should be developed, including refinements of the countermeasure description and cost estimates.
3. *Prepare a detailed project design.* A detailed statement of the project design is now prepared, including the formulation of a project evaluation plan. The evaluation plan should specify the project objectives, identify the specific questions to be answered, define the performance targets and evaluation criteria, list the sequence of activities (milestones) on a projected time scale, describe the research design, and estimate evaluation costs and resources required. The level of project effort is determined in terms of sample size and activity duration required to achieve project impact.

4. *Prepare the final countermeasure program package for approval.* Once the detailed project design is completed, the final step before conducting the evaluation project is obtaining approval from the various levels of management who are able to commit funds and other resources necessary for its implementation.

Define Data Sources and Determine the Types of Data to be Collected

After designing the project for study, the next step (perhaps taken concurrently with project design) is to define the sources of data to be used. Data are needed that will assist in the identification of possible solutions to the problem; therefore consideration must be given to the use to which the information will be put. Factors that determine data requirements include:

1. Objectives of the project.
2. Criteria to be used for evaluation.
3. Units of measure to be used.
4. Quantity of information required.
5. Quality (degree of detail) of information required.
6. Design of the evaluation procedures.

In light of these factors, the evaluator should consider what data are available, what additional data are required for the project, and how to bridge the gap between these two levels of data. Generally, data required for a project are based on a consideration of:

Resources employed during the project.
Methods used in the evaluation.
A measure of the accomplishments of the project objectives.
The impact of the project on the immediate and ultimate project objectives.

Data concerning project resources should include a description of the resource and an estimate of the costs of personnel, equipment, space, and facilities used in the project. Data on the project's methods should include a detailed description of all activities performed (including data on performance, selection of control groups, and randomization procedures if appropriate), and a statement of operating costs. Data on project accomplishments should relate to the expected accomplishments stated in the project objectives and identified in the evaluation criteria. These data are vital to the assessment of project effectiveness. Both treatment- and control-group information should be included, as appropriate. Impact data include measures of the effects of the project on the immediate and ultimate objectives identified in the statement of objectives. When analyzed, data on project accomplishments, together with impact data, produce an effectiveness evaluation indicating whether or not the actual accomplishments of the project produced the desired results.

Summary of Step 4

1. Conduct preliminary planning and procedures development studies.
   a. Define the specific occupational safety and health problem.
   b. Identify the objectives relating to the problem solution.
   c. Identify potential countermeasures that appear to be operationally feasible for use in achieving a problem solution.
2. Select and fully define the most appropriate countermeasure. Based on problem analysis:
   a. Define a specific target group of workers.
   b. Select a countermeasure program for implementation within the target group.
3. Prepare a detailed project design, including an evaluation plan.
   a. Specify project objectives.
   b. Identify questions to be answered.
   c. Define performance targets and evaluation criteria.
   d. Define units of measure to be used.
   e. List activities sequence, including project milestones, on a time scale.
   f. Describe the research design (using control groups where possible).
   g. Estimate evaluation costs and resources required based on such variables as sample size and length of time required to achieve project impact.
   h. Prepare the final program package for approval.

4. Define data sources and determine the types of data to be collected.
   a. Criteria that determine data requirements include:
      i. Project objectives.
      ii. Evaluation criteria.
      iii. Units of measure.
      iv. Quantity of information required.
      v. Quality (degree of detail) of information required.
      vi. Design of evaluation procedures
   b. Data sources should be considered in light of:
      i. Available data.
      ii. Required data.
      iii. Additional data needed to bridge the gap.
   c. Types of data to be collected are based on considerations of:
      i. Resources employed during the project.
      ii. Methods used in the evaluation.
      iii. Measures of accomplishment of project objectives.
      iv. The impact of the project on immediate and ultimate objectives.

8.5.5 Step 5: Design the Data Analysis System

The data analysis methods and procedures used in an evaluation depend on two factors: (1) the objectives of the project and (2) the project data required and available. Different types of data are amenable to different types of analysis. Therefore decisions concerning data analysis methods must be made along with the decisions made in steps 3 and 4. How the data are analyzed will depend on the data collection methods used, the nature of the control and experimental groups, and on what questions the evaluation is expected to answer. Decisions should be made on these issues at the project planning stage. The collected data must be analyzed to determine whether or not the designated criteria for achievement have been met. In step 7, some of the statistical methods that can be used in data analysis will be discussed.

Establish Data Analysis Procedures

The primary purpose of data analysis is to determine with some degree of confidence whether or not a given countermeasure or program had a significant effect on a problem. The basic design of the data analysis techniques and procedures, that is, the way in which collected data will demonstrate whether the criteria for project achievement have been met, should be established before project activities begin. The evaluation of most occupational safety and health projects preferably should involve at least a quasi-experimental design. An experimental procedure that analyzes data produced from such a design will determine (1) the difference or relationship, if any, that exists between variables in the project, and (2) the effect of that difference or relationship upon the problem selected.

Define the Level of Measurement

It is also important to identify the measurement techniques to be applied to the evaluation criteria. This includes the definition of the level of measurement of the data to be collected, that is, are the data to be collected measured on a nominal, ordinal, interval or ratio scale? Data collected on a measurement scale that does not assume an underlying normal distribution will limit the choices of statistical tests. For nonnormal distributions, a set of nonparametric statistical tests are available for data analysis. The design of the data analysis system will, of course, affect the methods of data collection identified in step 6 as well as the statistical procedures applied in the data analysis phase (step 7).

As a general rule, a qualified statistician should be consulted when defining the data analysis techniques and procedures. Input should also be provided by available computer personnel, the project director, and other concerned management personnel. Involvement of these interested individuals at the planning stage will help insure that the best procedures have been applied, that the data will be collected properly, and that the data analysis techniques and procedures are most appropriate for the selected study design.

Summary of Step 5

1. Translate general evaluation questions into specific, quantifiable questions to be answered.
2. Establish data analysis procedures that will determine the difference or relationship, if any, that exists between variables in the project and the effect of that difference on the problem selected.

3. Define the level of measurement to be used (i.e. are the data to be measured on a nominal, ordinal, interval, or ratio scale?)
4. Document the proposed analysis techniques for use in step 6.

8.5.6 Step 6: Collect Data

In many instances the design of the evaluation project and the data analysis system will determine the most appropriate data collection method. Techniques of data collection will vary according to the evaluation criterion selected, which, in turn, is based on the impact objectives of the project. In some situations data need not be collected for every single case under potential consideration. It is possible to *sample* data from a large number of cases and produce essentially the same results as would be obtained from examining all available data. Sampling is a procedure in which a limited number of representative cases are drawn for study from a large "population" of cases that are available for observation.

Sampling is conducted because the population of interest is too large to be feasibly studied in its entirety. By examining a smaller, representative portion of the population, information acquired from the sample can be used to draw conclusions about the entire population.

Some of the conditions that determine the most appropriate sample size include:

1. Frequency of occurrence.
2. Size of the population under study.
3. Variability of the units within the population with respect to the measurement criterion.
4. Desired accuracy of the results.

Random Sampling

One method of obtaining a representative measure of the population is to select the sample randomly. In a pure random sample, all the cases or subjects selected for study have an equal and independent chance of being selected. A random sample can be selected by one of several procedures, the most common being to assign numbers to the units to be studied within the population, enter a table of random numbers at a random point, and select every x^{th} number (e.g., every fifth number) as the identified sample unit.

As an example, suppose a behavior sampling study is conducted within a punch press department. The population under study consists of the punch press operators. The measurement criterion is the percent of time the workers are behaving unsafely. This time can be determined by selecting the observation times at random from among all the times available during that portion of the shift when work is actually being conducted (eliminating rest periods, lunch hours, etc.). It would be difficult to observe the operators all the time during a shift. It is more appropriate to make the observations at random points in time and then draw an inference concerning the proportion of time the operators are behaving unsafely. It is also possible to determine in advance how large the sample size must be in order to achieve the desired degree of accuracy.

Stratified Random Sampling

When two or more ways of classifying data exist within the population studied (e.g. operators working on different machines, male vs. female operators, etc.) it may be desirable to represent each category proportionately in the sample. In this situation the population is divided into different classes, or strata, and a proportionate number of cases are drawn at random from each stratum. Percentages can be established for each stratum so that the percentage or proportion of subjects within each sample stratification is the same as the percentage or proportion of the individuals within the entire population. This procedure is known as *stratified random sampling.* Its use generally permits a smaller total sample size to be selected while attaining the same degree of accuracy.

If every person or unit within the population were exactly alike, it would be necessary only to examine one person or unit to make a determination or inference about the whole population. In actual situations, the individual members of a population are rarely, if ever, exactly alike on the variable(s) studied. As a result, by chance alone, whenever a *small* portion of a population is sampled, that portion will differ somewhat from the makeup of the total population. This is called *sampling error.* Sampling error is distinguished from nonchance variations due to variables introduced or injected into the population, such as a new safety education program, a poster campaign, or biased sampling techniques. The use of large enough samples is one way of reducing sampling error. Larger samples are more likely to be representative of the total population and therefore provide better capability of generalizing the results.

When Data Should be Collected

Data should be collected before, during, and after program activities begin. It is important that some data are collected before project activities begin, even when control groups are used in the evaluation study design. This procedure is important for two reasons: (1) it provides prestudy data for use in comparing the before and after stages of the project, and assessing the magnitude of the project's impact;

and (2) even when control groups are used, the "before" data are useful in verifying the assumption that initially both the treatment group and control group are similar in the important variables to be studied.

Data should be collected during the project so that the rate and magnitude of change in the target audience can be assessed. Finally, data should be collected after the project to measure the effectiveness, or impact, of the program introduced.

Other Data Sources and Collection Methods

Use of evaluation criteria other than observed behavior sampling requires various data collection methods, depending on the units of measure and the goals of the project. When a very large population is to be studied (for example, all production line employees of General Motors or all off-the-job motor vehicle operations within the Department of the Army) indexes of safety performance that are insensitive and relatively rare events in a small population become useful as program impact criteria. For example, disabling work injuries or recordable work injuries as defined by the Occupational Safety and Health Act of 1970 may occur frequently enough to produce sufficient units of measure for evaluation purposes. Similarly, off-the-job disabling injuries occurring to all enlisted U.S. Army personnel stationed in the continental United States may provide an adequate, sensitive index for measuring the impact of a program or countermeasure designed to reduce these losses. However, as the population to be studied decreases in size, the exposure of persons to potential loss-producing sources becomes insufficient to produce a sufficient number of loss-type accidents within a reasonable period of time. In these cases, other measurement criteria should be selected, such as minor or first-aid cases, critical incidents, or direct observations of unsafe behavior and/or unsafe conditions. When disabling injuries or first-aid cases are used as evaluation criteria, it is important to make certain that any reductions occurring during and after the countermeasure activities are introduced are not produced by failure to report the accident or failure to report to the dispensary for first-aid treatment. Reporting accuracy (reliability) must be held constant before, during, and after the introduction of the program or countermeasure.

Summary of Step 6

1. Determine whether 100% data collection or data sampling is the most appropriate method of data collection.
2. If sampling is used, determine whether pure random sampling or stratified random sampling procedures will be followed.
3. Select an appropriate sample size.
4. Collect data before (baseline data), during, and after program or countermeasure activities are introduced.
5. Consider and utilize other data sources and collection methods, as appropriate (e.g. disabling or lost-time injuries, first-aid cases, critical incidents, and direct observations of unsafe behavior and/or unsafe conditions).

8.5.7 Step 7: Present and Analyze Data

The purpose of data presentation and analysis is to draw conclusions from the data collected. The amount and type of data analysis required to produce a useful evaluation depends on the objectives and achievements of the project, the design of the evaluation study, and the conclusions to be drawn when the project is completed. This step is concerned with how to present the data (i.e., what should the data say, how should the data be summarized, in tables, graphs, charts, etc.), how to analyze and interpret the data, and how to look for possible sources of error or misapplication of techniques in processing and presenting the data and in drawing conclusions from the results.

Data analysis procedures must be considered at the study planning stage. The types of analysis that can be conducted are determined by such study design variables as the evaluation criteria, units of measure used, types of data collected, size of the sample, stratification of the sample, how the data were collected, and so on. Most analytical techniques require that data be collected in a certain form for the technique to be used. Thus it is important at the study design stage to determine the type of analysis to be conducted so that the data collection format, sample size, and so on used in the study will permit the desired type of data analysis.

Determine the Methods of Data Summary and Presentation

After all data have been collected, the evaluator must decide how to display the raw data so that it can be understood, analyzed, and interpreted. As a general rule, data should be displayed in such a way that the answers to questions addressed in the evaluation design are readily communicated to the intended audience. In showing trends over time, for example, it is more appropriate to present the data in the form of a graph than in a table. However, to show differences in terms of frequencies or percentages, a comparison table might be most appropriate.

When counting and comparing various units of measure, several data forms are available for selection. Some of the more common forms include frequency, percentages, proportions, probabilities, and ratios. *Frequency* is the number of items in a given category, for example, the number of recordable work injuries in a given department, the number of persons completing safety training in the plant, the number of first-aid cases reported during the month, or the number of off-the-job traffic accidents sustained by employees in a given company. Frequencies are well understood and useful in making comparisons; however, the population base, or units of exposure from which the frequencies are drawn, must be the same. When comparing work injury frequencies among several companies, those with the larger number of workers are more likely to experience a greater frequency of work injuries than smaller companies because of their larger exposure.

A *percent* is the number of units falling in a particular category per 100 units of exposure. Percent means "per hundred," for example the number of recordable injuries per 100 workers in a particular plant. A *proportion* is a part or fraction of the whole expressed as a decimal between 0 and 1. A proportion is 1/100 of a percent and a percent is 100 times a proportion. Proportions are less familiar to the general reader than percent. Percents are also useful when referring to overall gain or loss, while proportions are always expressed as parts of something and can never exceed the total, which is 1.0. An advantage of proportions is their relation to probabilities. Every probability can be expressed in the form of a proportion. A *probability* is the expected or theoretical likelihood that a given event will occur. In tossing a coin, the probability that a head will occur is $1/2$, or one chance in two. Thus the probability of a head (or a tail) occurring in a single toss is $1/2$, or 0.5. A *ratio* is a fraction that expresses the relationship of one part of the whole to another part. Ratios are similar to proportions, except that a proportion is restricted to the ratio of a part to the whole (1.0), whereas a ratio is not so restricted. Ratios describe basic rates and relationships. The OSHA Incidence Rate is an index number of recordable work injury experiences expressed as the ratio of number of recordable work injuries per 200,000 man hours of work exposure. For example, 5 recordable work injuries per 200,000 man-hours divided by 2,000,000 total man-hours worked produces an Incidence Rate of 0.5. This is shown as follows:

$$\text{IR} = \frac{\text{N} \times 200{,}000}{MH}$$

N = Number of recordable work injuries or illnesses
MH = Man-hours = Total hours worked by all employees during the reference time period (e.g., one year)
200,000 = Base for 100 full-time equivalent workers working 40 hours per week, 50 weeks per year.
In this case,

$$\text{IR} = \frac{5 \times 200{,}000}{2{,}000{,}000} = \frac{1{,}000{,}000}{2{,}000{,}000}$$
$$= 1/2 = 0.5$$

Summarize the Data

Data may be summarized in the form of tables, graphs, charts, etc. One type of table permits the data to be presented in their complete, original form. Thus the reader can check the raw data or perform operations other than the ones presented in the report. A second type of table presents a summary of the original data, including such descriptive statistics as means, medians, ranges, and standard deviations, with the data grouped in one or more meaningful ways. Headings for a summary table should be clear and concise and provide an understandable label for the information presented. A third type of table is an aggregate table that presents the findings included in several previous tables. This final summary table serves as a basis for presenting the major conclusions of the evaluation.

When organizing a table for presentation, consideration should be given to the major points to be emphasized. There is also the practical consideration of keeping the table dimensions consistent with the page dimensions while maintaining readable type size. Common figures used in presenting data include bar diagrams, pie diagrams, trend charts, and frequency distributions.

A figure commonly used for displaying frequencies or percentages is the bar chart, or histogram. When percentages are displayed in bar form, the number of units represented by each bar (N) is often included at the top or end of each bar. Pie diagrams are often used to show proportions of the total, with the entire 360° "pie" representing 100%. Each segment of the circle represents the number or percent of units in that category. Pie diagrams are restricted to displaying the proportions of a total or whole, while the bar chart also presents proportions of parts to the whole. Trend charts show changes in frequencies, percentages, or proportions over time. Changing conditions, as well as a continuous time arrangement of data, are shown on this type of chart. Movement of the line up or down (or level) over time provides an obvious pictorial indication of trend. A systematic arrangement of data that exhibits the division of the values of a variable into classes and indicates the number of cases in each class is called a *frequency distribution.* A class interval, that is, the number of units within one class of data that separates it from another, is determined for each frequency distribution according to the manner in which the data are collected and certain established customs. For example, preferred ranges, by custom, are 1, 2, 3, 5, 10, and 20 units.

Compute Appropriate Statistics to Describe and Interpret the Data

Displayed data are useful in providing the reader with a picture of the results of an evaluation. In order to interpret or make inferences concerning the data, however, more must be done with the numbers collected. Two general types of statistical analysis can be applied to the data: *descriptive* and *inferential.* Descriptive statistics, as the name implies, provides us with more detailed descriptions of the data. These include measures of average and measures of spread, or dispersion. There are three measures of average: the mean, median, and mode. The *mode* is simply the value or values that occur most frequently. If the distribution has two concentrations of recurring values, the distribution is bimodal; if more than two, it is multimodal. There need not be any mode. The *median* of a distribution is the middle value, or the point at which one half of the values fall above and one half fall below. The *arithmetic mean* (often called simply the mean) is the average most typically considered. It is determined by calculating the sum of all the values or scores and dividing by the number of values or scores in the distribution. In symbolic form:

$$\text{Arithmetic Mean} = \bar{X} = \frac{\Sigma X}{N}$$

The mean is the only one of the three averages that includes every value in the entire distribution in its calculation. Since any change in the value of any item in the data collection will be reflected in the sum of those values, and hence in the mean, this average has a considerable advantage over the mode and median as a statistical descriptor.

In addition to measures of average, measures of the spread, or dispersion, that is, the variation or scattering of a distribution of frequencies about their average, are also important as descriptive statistics. Two common indicators of spread or dispersion are the *range* and the *standard deviation.* The range is the easiest measure of dispersion to establish. It is simply the interval between the upper and lower limits of the distribution, or the spread between the highest and lowest values. Although it is easy to calculate, the range also is often unreliable as a descriptive measure, since it is strongly influenced by one or more abnormally high or low values. A more reliable measure of variability is the standard deviation.

The standard deviation is the average of all the deviations from the mean in a distribution. It is defined mathematically as the square root of the arithmetic mean of the squared deviations of measurements about the mean. In symbolic form:

$$\text{S.D.} = \sqrt{\frac{\Sigma X^2 - \dfrac{(\Sigma X)^2}{N}}{N}}$$

where

X = the value or raw score and
N = the number of cases in the distribution.

The square of the standard deviation is referred to as the *variance.* Stated another way, the variance is the mean of the squared deviations. The variance has many important characteristics that are utilized in the testing of hypotheses in inferential statistics.

Knowing the standard deviation makes it possible to describe the way the values are spread within a distribution. If the standard deviation is large, the values in the distribution are widely dispersed. If it is small, the values are less variable. Many characteristics of human behavior are distributed in a large population according to a bell-shaped curve often referred to as the "normal curve" (see Figure 8.1).

In a true normal distribution, 34% of the cases will be on each side of the mean within one standard deviation. Approximately 47.5% of the cases fall between the mean score and plus two S.D. units and another 47.5% fall within minus two S.D. units of the mean. Accepting the assumption that the distribution of data approximates a normal distribution, opens the door to the use of a whole class of evaluation tools used in inferential statistics to determine the significance of differences.

Statistical significance is a term used to describe the probability of an event occurring either by chance or as the result of given conditions. If a number resulting from an appropriate test is found to be statistically nonsignificant, the difference that it quantifies can reasonably be attributed to chance. However, if a number is statistically significant, the difference that it quantifies probably occurred as the result of certain conditions. Statistical significance is expressed in terms of probability, usually given

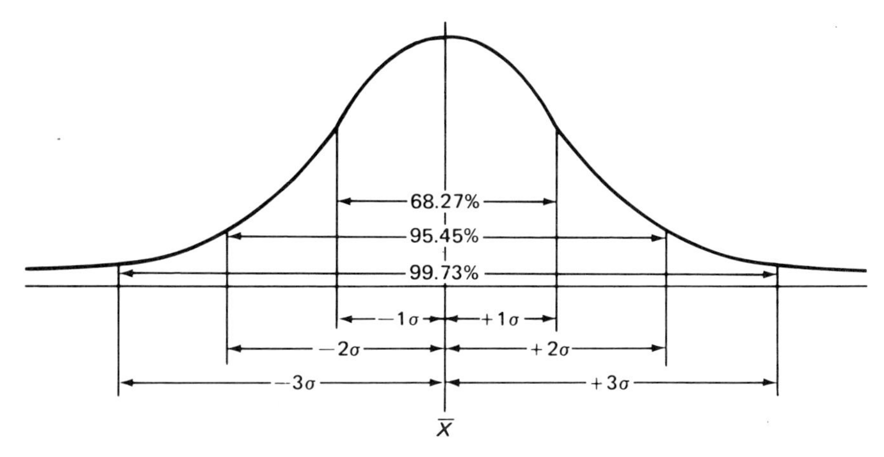

FIGURE 8.1. The normal curve.

by a decimal between 0.00 and 1.00. For example, if the differences in units of measure between an experimental and a control group are statistically significant at the 0.05 level, this means that the probability that this difference is due to chance and not the conditions introduced is 5 out of 100. In other words, we can state that the introduced conditions are effective with an assurance that this expectation will be correct 95 times out of 100. The probability level selected for statistical significance is arbitrary; however, by tradition the commonly accepted levels are 0.05 (5 times out of 100), 0.01 (1 time out of 100), and sometimes 0.10 (10 times out of 100). The desired level of statistical significance should be established before conducting the statistical test used in analyzing the data. The value of using a test of statistical significance is that it permits the evaluator to determine the likelihood that a change resulted from the introduction of a countermeasure rather than from some chance fluctuation in the evaluation criterion (e.g. number of disabling injuries, number of first-aid cases, number of observed unsafe behaviors, etc.).

When testing the significance of differences for statistical purposes, evaluators generally begin with a statement called the null hypothesis. The null hypothesis states that there is *no difference* between the group receiving the countermeasure or other treatment and the group not receiving this change. The test of statistical significance is used to estimate the probability of error if the null hypothesis is accepted (i.e., that there are no differences between the groups) or rejected (i.e., that there are differences between the groups). Thus, if our countermeasure really works, we would expect to reject the null hypothesis, with a certain probability that we are wrong.

Many techniques are available for testing the statistical significance of differences between groups. The appropriate technique in a given evaluation depends on the nature of the data collected, the statistical design of the project, and the form of the data distribution. Numerous textbooks have been written on the various statistical techniques used in evaluations. Obviously, all of the appropriate statistical tools cannot be covered in this chapter. In order to provide a model for use in effectiveness (impact) evaluations of accident countermeasures, two of the more common analytical techniques—the *t*-test and chi-square—will be presented in brief form. For a more detailed discussion of statistical testing, the reader is referred to the textbooks referenced at the end of this chapter.

The *t*-test is used in effectiveness (impact) evaluations to determine whether or not the difference between a treatment group and a control group is due to chance. More specifically, it is a statistical technique for testing the null hypothesis that the mean values from two groups do not differ in a statistically significant way. The *t* is a value that represents the ratio of the difference between the means and the standard error of the means for both groups. The computational formula for the *t*-test is:

$$t\text{-test} = \frac{\bar{X}_a - \bar{X}_b}{\sqrt{\left[\dfrac{\Sigma X_a^2 - \dfrac{(\Sigma X_a)^2}{N_a} + \Sigma X_b^2 - \dfrac{(\Sigma X_b)^2}{N_b}}{N_a + N_b - 2}\right] \cdot \left[\dfrac{1}{N_a} + \dfrac{1}{N_b}\right]}} \tag{8.3}$$

Where

$\bar{X}_a$ = the mean of group *a*
$\bar{X}_b$ = the mean of group *b*
N_a = the number of values in group *a*

N_b = the number of values in group b
ΣX_a^2 = the sum of the squared values of group a
ΣX_b^2 = the sum of the squared values of group b
$(\Sigma X_a)^2$ = the square of the sum of the scores in group a
$(\Sigma X_b)^2$ = the square of the sum of the scores in group b

To determine the significance of the difference in the means of the two groups, a table of t values is used. The significance of the value obtained from the calculated t-test is determined by referring to a t-table, which gives the size of the t-test calculated value needed for significance at each level of confidence for each level of df, or "degree of freedom." Degrees of freedom (df) is the number of observations that are free to be varied within a set of data after all constraints have been established. For t-tests, df = the number of observations or values in both groups minus 2, or symbolically, $df = N_a + N_b - 2$. While there are different ways of calculating t-test values, depending upon the design of the evaluation study, t simply indicates the probability level at which a given t as large as the one observed could have occurred by chance. Thus in the "results" section of the evaluation report, a statement such as the following should be included: "The results of the t-test indicate a significant difference between the countermeasure group and the control group ($t = 4.16$, $df = 14$, $p < .01$)." This means that there is less than one chance in 100 that the actual difference between the two means might be due to chance and not to the influence of the countermeasure. Therefore, by rejecting the null hypothesis and concluding that the two groups did differ in terms of the measurement criterion, the chance of being in error would be less than 1 out of 100. An example of an evaluation using the t-test is included at the end of this chapter.

Chi-square, symbolized by the Greek letter χ^2, is a test for statistical difference between observed (O) and expected (E) frequencies based on the formula:

$$\chi^2 = \Sigma\left[\frac{(O - E)^2}{E}\right] \tag{8.4}$$

This test compares the observed number of cases falling in each category with the expected number of cases under the null hypothesis to determine whether the actual distribution differs significantly from the expected. The χ^2 test may also be used to determine whether two observed frequency distributions are significantly related to each other. Chi-square is most commonly used with frequency data, that is, data that are classified according to the number of times a variable is observed (e.g., frequency of first-aid cases, frequency of recordable injuries, etc.) as compared with the number of times it could be expected to be observed by chance. The frequency data are usually displayed in the form of a matrix such as the following:

	f_o		
	Number of Workers With First-aid Injuries	Number of Workers Without First-aid Injuries	Total
Experimental	2	13	15
Control	9	6	15
Total	11	19	

Any number of rows and columns can be used (the example above contains two rows and two columns and thus is called a 2×2 χ^2). For each cell in the matrix an expected value is determined by the formula:

$$f_e = \frac{(\Sigma f_r)\ (\Sigma f_k)}{N_{\text{total}}} \tag{8.5}$$

where

Σf_r = the sum of the row frequencies
Σf_k = the sum of the column frequencies
N_{total} = the total number of frequencies in the matrix

The column and row totals in the f_e matrix are the same as those in the f_o matrix. For this 2 × 2 chi-square data, the expected frequencies are:

f_e

	Number of Workers With First-aid Injuries	Number of Workers Without First-aid Injuries	Total
Experimental	5.5	9.5	15
Control	5.5	9.5	15
Total	11	19	

From the observed and expected values, a value for χ^2 reflects the magnitude of the difference between the observed value and the expected value. The significance of the χ^2 value is determined by referring to a table of χ^2 values, which gives the size of the χ^2 needed for significance at each level of confidence for each level of degrees of freedom (*df*). The value of *df* for the chi-square test is found by multiplying the number of rows minus 1 by the number of columns minus 1, $df = (r-1)(c-1)$. It should be noted that whenever we have a 2 × 2 matrix in which at least one cell contains a frequency of 10 or less, the Yates Correction Formula should be used. The Yates Correction Formula consists of adding half a case to the smallest frequency of the 2 × 2 table and adjusting all other frequencies so that the row and column totals remain the same. This correction is only applicable to the 2 × 2 matrix. With Yates Correction, the χ^2 formula becomes:

$$\chi^2 = \Sigma\left[\frac{(|f_o - f_e| - 0.5)^2}{f_e}\right] \tag{8.6}$$

An example of an evaluation using the chi-square test is included at the end of this chapter.

The *t*-test and chi-square test are two of the most commonly used tests for the significance of differences between groups. Other techniques include the analysis of variance *F*-test, analysis of covariance, sign test, and the Mann-Whitney *U* test. The same basic procedures are followed in applying each of these tests:

1. A test value is computed.
2. This value is translated by means of established tables into a probability value (p).
3. This value is then used to accept or reject the null hypothesis and to draw conclusions about the meaning of the observed differences between the groups.

Quasi-experimental study designs, such as interrupted time series analysis, can also be used to determine the significance of change in the criterion variable over time. For information concerning the use of this evaluation research method, the reader is referred to Campbell and Stanley, 1963, 1966; Cook and Campbell, 1976; Glass, Willson, and Gottman, 1975; and McCleary and Hay, 1980.

Identify Possible Threats to Validity

Throughout the course of conducting a countermeasure evaluation project, any number of problems may arise to threaten the validity of the evaluation. While some of these problems are common to all experimental research, some are particularly prevalent in the real-world settings of occupational safety and health programs. The evaluator in this field should pay special attention to detecting these potential threats to validity at the project site and make necessary improvements in the project design to counter these difficulties. Special care should be taken to follow the study design carefully during the operational phase of the project.

Webster's Dictionary defines the term *valid* as being "sound, effective, well-grounded on principals or evidence, and able to withstand criticism or objections." In the field of statistics, validity means the extent to which an evaluation measure actually measures what it is presumed to measure. The goal of the evaluator is to produce an evaluation that is as valid as possible within the constraints of the operational environment. A threat to validity is any factor or event that introduces an error or bias to the evaluation. These biases can make interpretation of the evaluation results more difficult, in that the bias itself may produce an alternative explanation for any observed effects. A good evaluation design should eliminate alternative explanations of evaluation results, and thus control or minimize the threats to validity. Factors threatening validity can be categorized as internal and external. A threat to internal validity is any nonproject factor that might have caused the results or make it difficult to identify the contributions of the project or countermeasure to the results. Threats to external validity are those factors that make it difficult to generalize the study results outside the limits of the current evaluation project. Internal validity is probably most important to attain, since it is crucial to determine whether or not the project or countermeasure had an effect.

Some of the specific problems that can threaten evaluation validity include:

1. The evaluation measure isn't collectable, valid, or reliable.
2. The sample size is too small to permit detection of countermeasure effects.
3. The selected groups for the study are not representative of the target population to which application of project results is desired.
4. When time comparisons are involved, the pre- and post-test time periods are not comparable. An extremely high or low evaluation measure in the preproject or baseline time period may be vulnerable to regression-to-the-mean problems, i.e., the statistical tendency of extreme values to return to the mean value upon subsequent measurement.
5. When treatment and control groups are involved in the evaluation, the two groups selected are not truly comparable. If the groups are not comparable, it is difficult to conclude that subsequent changes in the experimental group were due to the project's activities.
6. Confounding factors or events may have an impact on the project. Nonproject related extraneous events that affect the evaluation measures may occur. One example of a confounding factor is the classic "Hawthorne effect," in which selection for participation in the study in itself influenced worker performance and thus biased the study results.
7. Selection bias is introduced to the two groups, including differential dropout rates.
8. The record keeping system breaks down, producing inaccurate or missing data.
9. There may be pressure to bias the results in favor of the project expectations.
10. Statistical techniques are inappropriately applied.

Summarize the Results, Discuss (Interpret) the Findings and Conclusions, and Discuss Implications for the Program

The process of data analysis consists of two major phases: (1) the selection and application of statistical tests to data and (2) the interpretation of evaluation results as a basis for formulation of meaningful conclusions regarding countermeasure (project) effectiveness. The evaluator must use the most appropriate statistical techniques to insure that the countermeasure effects will be identified. Results must be interpreted in ways that produce the most accurate assessment of the project. Inferences must be made with reasonable confidence concerning the causal factors involved and the relationships between events. Development of conclusions regarding countermeasure effects requires weighing various relevant factors, some of which are not clearly evident in the results of the statistical tests. It is understandable that, after having invested a substantial amount of effort in an activity, there is considerable pressure to achieve its objectives and disappointment when project impact is not demonstrated. However, it should be understood that information about project failures can be just as valuable as information about successes. Clear evidence of failure will permit the program manager to avoid continuing the use of ineffective countermeasures.

Finally, the evaluator should review all findings with program managers, operations personnel, and others concerned about the results of countermeasure activities. These persons may raise questions that will provide further insight into the causes or meaning of a particular finding.

Summary of Step 7

1. Determine methods for presenting and analyzing data at the planning stage of the project.
2. Summarize the data in the form of tables, graphs, charts, and so on.
3. Compute appropriate statistics to describe and interpret the data.
4. Identify possible threats to validity.
5. Summarize the results, discuss (interpret) the findings and conclusions, and discuss the implications of the findings for the program.

8.5.8 Step 8: Produce the Evaluation Report and Integrate the Evaluation Results Into the Management System

Preparation of the evaluation report should begin when the evaluation project itself is initiated. By keeping in close touch with the project as it progresses, taking notes, and summarizing observations at the end of each step, the evaluator will be assured that all relevant information is systematically recorded and the procedures and results obtained in each step are properly noted. The evaluation report itself should be produced immediately upon completion of the data analysis step. This will assure timely feedback of evaluation results to managers, planners, and project personnel. It may also be useful to produce brief interim progress reports as each step is completed. Evaluation reports should be as concise as possible, readable by their intended audiences, communicate failures as well as successes, and prepared and distributed in a timely manner.

Suggested Outline for the Evaluation Report

The following is a suggested topical outline for a project evaluation report:

1. Introduction.
2. Abstract and/or a more detailed executive summary.
3. Definition and discussion of the problem, including background information as appropriate.
4. Presentation of the administrative evaluation of project performance, (including the quantification of activities initiated and completed, personnel data, time requirements for achieving each objective, cost of achieving each objective, problems encountered, etc.).
5. Presentation of the effectiveness (impact) evaluation of project performance (including summary of evaluation methods used, evaluation design, evaluation criteria used, units of measurement, data collection and analysis, summary of results, objectives achieved, problems encountered, etc.).
6. Conclusions and recommendations, including a summary of project results and impact, a summary of successes and failures, unexpected side effects of the project, recommendations for improving future evaluation projects, and quantified support and justification for the conclusions.

Integrating Evaluation Results Into the Management System

A major purpose of evaluation is to provide managers with sufficient information about the results of various activities, projects, countermeasures, etc. in order to make sound, knowledgeable decisions concerning future similar efforts. Information obtained through evaluation is important to the planning process and provides important information for future evaluations. Perhaps the simplest way to integrate administrative evaluation results into management systems is for the evaluator to prepare and present to managers regular periodic reports on the actual task accomplishments compared to established performance goals, including unit cost and operational efficiency. Effectiveness (impact) evaluation results can be integrated into the management planning and decision-making process soon after the completion of data analysis and the analysis of evaluation results.

Integration of evaluation results into planning and management systems involves six steps, as follows:

1. Define the planning process or establish one if none exists.
2. Establish the relationship between the planning process and the evaluation process.
3. Integrate evaluation results into the planning cycle.
4. Time the evaluation process to allow results to be available at the appropriate time during the planning cycle.
5. Utilize evaluation results in management decision making.
6. Identify obstacles to management acceptance of evaluation and develop plans and procedures for overcoming them.

Summary of Step 8

1. Begin preparation of the evaluation report when the evaluation project is initiated.
2. Produce the completed evaluation report immediately following completion of the data analysis step.
3. Suggested outline for the evaluation report:
   a. Introduction.
   b. Abstract and/or Executive Summary.
   c. Definition and discussion of the problem.
   d. Presentation of the administrative evaluation of project performance.
   e. Presentation of the effectiveness (impact) evaluation of project performance.
   f. Conclusions and recommendations.
4. Integrate the evaluation results into management systems.
   a. Define the planning process.
   b. Establish the relationship between the planning and evaluation processes.
   c. Integrate evaluation results into the planning cycle.
   d. Time the evaluation process to allow results to be available at the appropriate time during the planning cycle.

e. Utilize evaluation results in management decision making.
f. Identify and develop procedures for overcoming obstacles to management acceptance of evaluation.

8.6 EVALUATION TECHNIQUES AND EXAMPLES

8.6.1 Using the *t*-test

PROBLEM
The safety director has developed a new worker safety training course and he would like to evaluate its impact on a limited target audience before conducting the course on a larger scale. Two similar departments are selected for the evaluation study. Department *B* is given the training program (1 hour a week for 12 weeks) and Department *A* is given no safety training. Other relevant factors that may influence the criterion variable (number of first-aid cases) are held constant for both groups.

Assumptions

1. Type of work and amount of exposure in each department is equivalent.
2. Worker past accident experience is equivalent for each department.
3. Work experience is equivalent in each group.
4. The only safety activity to which the workers in both departments are exposed is the safety training course in Department *B*.
5. Independence between samples.
6. The population studied approximates a normal distribution.

Measure
Number of first-aid cases reported in each department. The measures are taken during the 12 consecutive weeks of training.

Data Collection
The data collected are shown in the following table:

NUMBER OF FIRST-AID CASES REPORTED EACH WEEK

Week	Department A	Department B
1	65	50
2	25	55
3	42	42
4	64	71
5	72	76
6	95	24
7	100	14
8	25	58
9	56	62
10	50	45
11	25	12
12	36	15
	$\Sigma X_A = 655$	$\Sigma X_B = 524$

Hypothesis
The difference between the number of reported first-aid cases between Department *A* and Department *B* is due to chance.

Test Statistic

$$t\text{-test} = \frac{\bar{X}_A - \bar{X}_B}{\sqrt{\left[\dfrac{\Sigma X_A^2 - \dfrac{(\Sigma X_A)^2}{N_A} + \Sigma X_B^2 - \dfrac{(\Sigma X_B)^2}{N_B}}{N_A + N_B - 2}\right] \cdot \left[\dfrac{1}{N_A} + \dfrac{1}{N_B}\right]}}$$

Hypothesis Test

1. The null hypothesis (H_0) states that there is no significant difference between the number of first-aid cases experienced by Department A workers and the number experienced by Department B workers during the 12-week period studied. The alternative hypothesis (H_1) states that the number of first-aid cases in the department receiving the training (Department B) is significantly lower than the number experienced by the department not receiving the training (Department A). In symbolic form:

$$H_0: P(B) = P(A) = 1/2$$

$$H_1: P(B) < P(A)$$

2. The test statistic chosen is the t-test where:

$\bar{X}_A$ = the mean of Department A
$\bar{X}_B$ = the mean of Department B
N_A = the number of weeks of observation in Department A
N_B = the number of weeks of observation in Department B
ΣX_A^2 = the sum of squared values of first-aid cases in Department A
ΣX_B^2 = the sum of squared values of first-aid cases in Department B
$(\Sigma X_A)^2$ = the square of the sum of the first-aid cases in Department A
$(\Sigma X_B{}^2)$ = the square of the sum of the first-aid cases in Department B
df = degrees of freedom = $N_A + N_B - 2$

3. A significance level of 0.05 is chosen. The sample size (N) in this example is 12 weekly collections of first-aid cases.

4. The sampling distribution under the null hypothesis is given by the t-distribution with $df = N_A + N_B - 2$. Tables of the critical values of t from the sampling distribution, together with their associated probabilities of occurrence under H_0, are presented in the appendix of most statistics texts.

5. Since the alternative hypothesis (H_1) indicates the predicted direction of the difference between the two departments, a one-tailed test is indicated. In a one-tailed test the region of rejection is entirely at one end (or tail) of the sampling distribution. The size of the region of rejection is expressed by α, the level of significance, in this case, 0.05. If $\alpha = 0.05$, then the size of the region of rejection is 5% of the entire space included under the curve in the distribution. This is illustrated pictorially as follows:

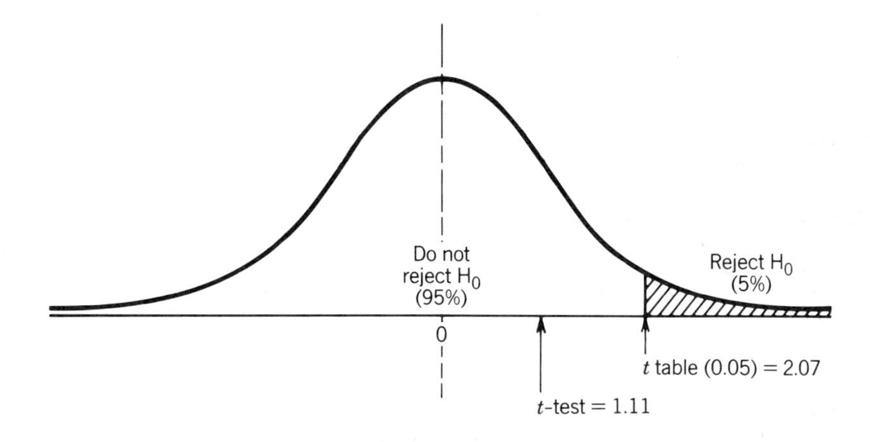

6. Using the collected data from the two departments, we compute the value of the statistical test as follows:

$$t = \frac{\bar{X}_A - \bar{X}_B}{\sqrt{\left[\dfrac{\Sigma X_A^2 - \dfrac{(\Sigma X_A)^2}{N_A} + \Sigma X_B^2 - \dfrac{(\Sigma X_B)^2}{N_B}}{N_A + N_B - 2}\right] \cdot \left[\dfrac{1}{N_A} + \dfrac{1}{N_B}\right]}}$$

where

$$\bar{X}_A = 54.6$$
$$\bar{X}_B = 43.6$$
$$N_A = 12$$
$$N_B = 12$$
$$\Sigma X_A^2 = 43{,}101$$
$$\Sigma X_B^2 = 28{,}480$$
$$(\Sigma X_A)^2 = 429{,}025$$
$$(\Sigma X_B)^2 = 274{,}576$$
$$df = N_A + N_B - 2 = 12 + 12 - 2 = 22$$

$$t = \frac{54.6 - 43.6}{\sqrt{\left[\dfrac{43{,}101 - \dfrac{429{,}025}{12} + 28{,}480 - \dfrac{274{,}576}{12}}{12 + 12 - 2}\right] \cdot \left[\dfrac{1}{12} + \dfrac{1}{12}\right]}}$$

$$t = \frac{11}{\sqrt{\left[\dfrac{43{,}101 - 35{,}752 + 28{,}480 - 22{,}881}{22}\right] \cdot [0.167]}}$$

$$t = \frac{11}{\sqrt{\left[\dfrac{7{,}349 + 5{,}599}{22}\right] \cdot [0.167]}}$$

$$t = \frac{11}{\sqrt{\left[\dfrac{12{,}948}{22}\right] \cdot [0.167]}}$$

$$t = \frac{11}{\sqrt{[588.54][0.167]}}$$

$$t = \frac{11}{\sqrt{98.29}}$$

$$t = \frac{11}{9.9}$$

$$t_{\text{test}} = 1.11$$

7. The computed t-value is compared with the table t-value for the sampling distribution of the test statistic at the 0.05 level of significance (see figure). At 0.05 significance level for $df = 22$, t-table $= 2.07$.
8. *Decision:* Since the value of the test is smaller than that required for significance at the 0.05 level as indicated by the tabled value, the test falls in the region of the curve in which the null hypothesis is not rejected. Thus these findings indicate that there is no significant difference between the number of first-aid cases in the departments with trained and untrained workers. In symbolic form:

$$t\text{-test} = 1.11 < t\text{-table}\ (.05) = 2.07$$

Thus we do not reject H_0. This means that, all other factors held constant, and within the limits and validity of the assumptions, with two samples containing 12 observations of first-aid cases each, safety training apparently had no significant (nonchance) influence on the first-aid cases in Department B.

8.6.2 Using Chi-Square

PROBLEM
First-aid cases for a given month were distributed by shifts among four departments as follows:

Dept.	First Shift	Second Shift
A	10	5
B	10	20
C	5	25
D	5	10

Is it likely that these differences in accidents could have arisen by chance?

f_o (observed)

Dept.	First Shift		Second Shift		Total
A	10	1	5	2	15
B	10	3	20	4	30
C	5	5	25	6	30
D	5	7	10	8	15
Total	30		60		90

fe (expected)

Dept.	First Shift		Second Shift		Total
A	5	1	10	2	15
B	10	3	20	4	30
C	10	5	20	6	30
D	5	7	10	8	15
Total	30		60		90

$$f_e = \frac{(\Sigma f_r)(\Sigma f_k)}{N \text{ total}}$$

$$\chi^2 = \Sigma\left[\frac{(f_o - f_e)^2}{f_e}\right]$$

Cell	f_o	f_e	$f_o - f_e$	$(f_o - f_e)^2$	$\frac{(f_o - f_e)^2}{f_e}$
1	10	5	5	25	5.0
2	5	10	−5	25	2.5
3	10	10	0	0	0
4	20	20	0	0	0
5	5	10	−5	25	2.5
6	25	20	5	25	1.25
7	5	5	0	0	0
8	10	10	0	0	0

$$\chi^2 = \Sigma\left[\frac{(f_o - f_e)^2}{f_e}\right] = 11.25$$

$$df = (r - 1)(c - 1)$$
$$= (4 - 1)(2 - 1) = 3 \cdot 1 = 3$$

for $df = 3$, χ^2 table = 7.82 at 0.05 level
χ^2 table = 11.34 at 0.01 level

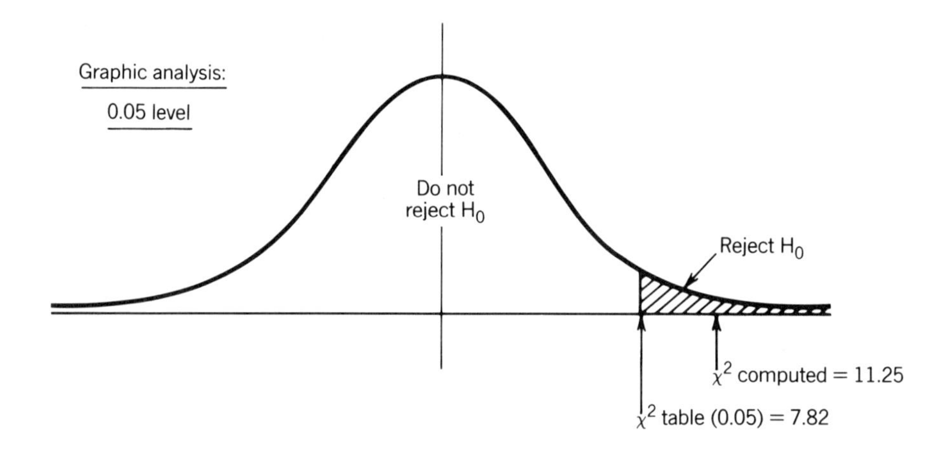

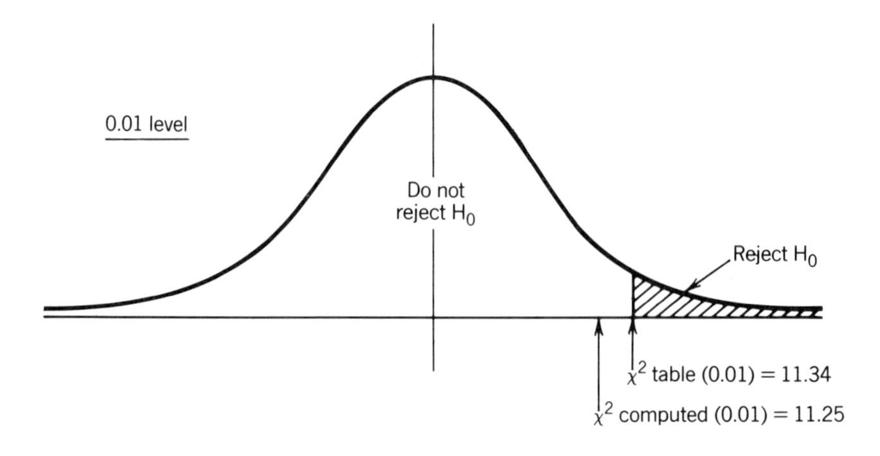

Decision
Since χ^2 computed > χ^2 table at 0.05 and < χ^2 table at 0.01: Reject null hypothesis (H_o) at 0.05 level. Do not reject null hypothesis (H_0) at 0.01 level.

8.6.3 Using the Critical Incident Technique

An evaluation procedure known as the Critical Incident Technique (CIT) is a method of identifying errors and/or unsafe conditions that contribute to accidents within a given population by means of a stratified random sample of participant-observers selected from within this population. These participant-observers are selected from the major plant departments so that a representative sample of operations existing within different hazard categories can be obtained.

In applying the technique, an interviewer questions a number of persons who have performed particular activities or jobs within certain environments, and asks them to recall and describe unsafe errors they have made or observed or unsafe conditions that have come to their attention in connection with various operations in the plant. A stratified random sample procedure is used to select the participant-observers in order that a representative sample of workers can be interviewed. The participant-observers are encouraged to describe as many specific "critical incidents" as he or she can recall, regardless of whether or not they resulted in injury or property damage.

It is quite appropriate to stimulate the thought processes of the participant-observer by providing a list of unsafe acts and unsafe conditions that, by definition, have produced death, injury, or property damage within similar types of environments or operations at any location. These can be obtained from accident prevention handbooks, safety textbooks, or from the accident records of any large organization. The key to a successful evaluation is to emphasize to the participant-observer that incidents that have *actually been observed* should be described in sufficient detail to provide the same or higher levels of causal information than would be provided by an accident investigation report involving a disabling or fatal injury. The interviewer should be prepared to ask probing questions so that complete information describing all relevant causal factors can be obtained.

The incidents described by a number of participant-observers are transcribed on tape and classified into hazard categories from which potential accident problem areas are defined. One or more accident countermeasures can then be introduced as a program to solve these identified problems. The number of participant-observers to be interviewed can be determined by plotting a cumulative frequency distribution of new incidents introduced by each successive person interviewed. A decision can be made concerning the adequacy of the participant-observer sample size based on the absence of new incidents. A reapplication of the CIT using a new stratified random sample may be conducted periodically in order to detect new accident problem areas, or for use as a measure of the effectiveness of a prevention program in overcoming the previously identified problems.

The careful selection of a representative sample of participant-observers each time, based on valid stratification criteria, should enable the evaluator to make an inference concerning the state of the entire population of the organization from which the sample was selected at that particular time. The frequency of reapplication of the CIT is a function of the degree of variance in the types of critical incidents over time. A reapplication should perhaps follow any major alteration in the size, function, or mission of the organization studied. The conduct of a new study would also be appropriate following the introduction of a group of countermeasures based on the analysis of causal factors associated with the previously revealed critical incidents. The CIT would then function as an evaluation tool to determine the impact of the new countermeasures on the potential or actual loss-producing incidents within the departments or activities studied. The objective of the CIT is to discover critical causal factors, that is, those that have contributed to an actual or potential loss-producing accident or have the potential for contributing to future loss-producing accidents.

The results of several research studies utilizing the CIT have shown that:

1. CIT dependably reveals causal factors in terms of errors and/or unsafe conditions which lead to occupational accidents.
2. CIT is able to identify causal factors associated with both injurious and noninjurious accidents.
3. Use of the CIT reveals a greater amount of information about accident causes than presently used methods of accident study and provides a more sensitive method for evaluating total accident and/or *potential* accident performance.
4. The causes of noninjurious accidents as identified by the CIT can be used to identify sources of potentially injurious accidents.
5. Use of the CIT to identify accident causes is both feasible and efficient.

The discovery that CIT is sensitive to the identification of accident causal factors that have a *potential* for accident loss but have not yet produced a loss provides a significant advance in our ability to evaluate the safety performance of an organization. The existence of an injury or property damage loss, or perhaps a death, is no longer a necessary condition for measuring and appraising safety performance, or for identifying the presence of unsafe behaviors and conditions. By using CIT, the safety professional is able to identify and examine accident problems before the fact instead of after the fact in terms of their injury-producing or property-damaging consequences. By concentrating on the identification of problems associated with *loss potential* or "*near-miss*" accidents, the evaluator is removed from having to rely on evaluation techniques based on the probabilistic, fortuitous, rare-event accident manifestations of disabling injury, fatal, and property damaging events. True loss control becomes possible, since there is no need to wait for a loss to occur before safety performance can be evaluated and accident problems can be identified.

8.6.4 Using Behavior Sampling

Behavior sampling consists of making a number of observations of behavior at random points in time. Instantaneous decisions classifying the observed behavior as "safe" or "unsafe" are made in its given environment. In the random method of selecting times for making the observations, every point in time during which work is performed has an equal chance of being selected for observation. This procedure allows an inference to be made concerning the state of the entire time period or population from which the sample was chosen. One of two basic measurement techniques may be chosen: (1) at random points in time, walk through the department or unit to be evaluated, observe every working person, and classify that person's behavior as either safe or unsafe; (2) at random points in time, observe the behavior of a group of the same workers each time and classify the behavior of each worker as safe or unsafe. Thus it is possible, in the first method, to compute the percent of workers involved in unsafe behavior, and in the second method to compute the percent of time that observed workers were behaving unsafely.

Three basic procedures must be followed in order to use behavior sampling as an effectiveness or impact evaluation method:

1. Accumulate data on unsafe behavior for a sufficient number of successive periods to discriminate between differences in proportions of unsafe behaviors.
2. Apply a reasonable random fluctuation as a yardstick to determine when an observed proportion of unsafe behaviors differs from other observed proportions sufficiently to indicate a real change in the behavioral system.
3. Determine periods of stability in behavioral acts in order to distinguish the time when a new level of stability supplants the old level.

Behavior control charts are constructed based on the mean ($\bar{p}$) percent of time each worker is involved in unsafe acts, or the mean percent unsafe behavior for the entire group as recorded during the observation period. Using a 95% level of confidence (5% control limits), the upper control limit (UCL) Eq. 8.7 and lower control limit (LCL) Eq. 8.8 are calculated as follows:

$$\mathrm{UCL} = \bar{p} + 1.96\sqrt{\frac{\bar{p}(1-\bar{p})}{n}} \tag{8.7}$$

$$\mathrm{LCL} = p - 1.96\sqrt{\frac{\bar{p}(1-\bar{p})}{n}} \tag{8.8}$$

A determination of stability is made in terms of the percent of unsafe behavior for the test period, that is, the percent of unsafe behavior is plotted to determine if all points fall within the upper and lower control limits. If the situation is not stable, minor changes in the accident prevention program are introduced to the population studied and the sampling study is repeated until stability in percent of unsafe behavior is reached.

Having achieved stability, a major countermeasure is introduced, with the behavior sampling study repeated on a weekly or other appropriate time basis during and after introduction of the new program. The percent of unsafe behavior is plotted on the control chart for each period following initiation of the program and a determination is made concerning whether or not there has been a significant improvement in unsafe behavior. Finally, the behavior sampling study is repeated periodically, with data plotted on the control chart to assure that the percent of unsafe behavior remains stable at the minimum level.

In applying behavior sampling procedures, a trial study should be conducted when stability is reached to determine the number of observations or readings required in order to meet the predetermined accuracy and confidence level requirements. For example, the number of readings required for *a* ± 10% accuracy at the 95% level of confidence is computed as follows:

$$N = \frac{4\,(1-p)}{S^2 p} \tag{8.9}$$

where

N = Number of readings
S = Desired accuracy = 0.10
p = Percent unsafe acts observed during the trial study

If the trial study indicates that the workers were behaving unsafely 40% of the time (mean), then the number of readings required can be determined by substituting 0.40 for p in the equation above:

$$N = \frac{4\,(1 - 0.40)}{(0.10)^2(0.40)}$$

$$= \frac{2.4}{0.004}$$

$$= 600 \text{ readings required}$$

If 20 operators were observed in the study, the 600 observations required for $\pm$ 10% accuracy at the 95% confidence level can be obtained by observing the 20 operators during each of 10 trips per day for three days (10 trips $\times$ 20 operators = 200 observations/day $\times$ 3 days = 600 readings).

In behavior sampling studies, instability or a significant (nonchance) increase in level of unsafe behavior signals trouble, and a search for the contributing causal factors concentrates on the time period when the change occurred. If the causal factors are properly identified and corrected, the mean proportion of unsafe behavior would be expected to drop to its previous level of stability. If a new countermeasure remains effective over time in reducing unsafe behavior, a new mean may be established at a lower level of stability. The objective is not to maintain a certain level of stability but, instead, to induce positive change in worker behavior. If a new countermeasure (safety program) intended to improve unsafe behavior is introduced, hopefully a significant (nonchance) downward shift in the mean proportion of unsafe behavior will be achieved.

REFERENCES

Campbell, D. T. and J. Stanley, *Experimental and Quasi-Experimental Designs for Research,* Rand-McNally, Chicago, 1966.

Cook, T. D. and D. T. Campbell, *Quasi-Experimentation: Design and Analysis Issues for Field Settings,* Rand-McNally, Chicago, 1979.

Glass, G. V., V. L. Willson, and J. M. Gottman, *Design and Analysis of Time Series Experiments,* Colorado Associated University Press, Boulder, CO, 1975.

Isaac, S. and W. B. Michael, *Handbook in Research and Evaluation,* Knapp Publishing, San Diego, CA, 1971.

McCleary, R. and R. A. Hay, Jr., *Applied Time Series Analysis for the Social Sciences,* Sage Publications, Beverly Hills, CA, 1980.

Stevens, S. S. (Ed.), *Handbook of Experimental Psychology,* Wiley, New York, 1951.

Tarrants, W. E., *An Evaluation of the Critical Incident Technique as a Method for Identifying Industrial Accident Causal Factors,* Unpublished Doctoral Dissertation, New York University, 1963 (Also available from University Microfilms, Inc., Ann Arbor, Michigan).

Tarrants, W. E., An Introduction to Applied Safety Research, in *Professional Safety,* July 1979, pp. 38–43.

Tarrants, W. E., Applying Measurement Concepts to the Appraisal of Safety Performance, *Journal of the American Society of Safety Engineers,* May 1965, pp. 15–22.

Tarrants, W. E., The Evaluation of Safety Program Effectiveness, in Peterson, D. and Goodale, J., *Readings in Industrial Accident Prevention,* McGraw-Hill, New York, 1980.

Tarrants, W. E., *The Measurement of Safety Performance,* Garland STPM Press, New York, 1980.

Weiss, C. H., *Evaluation Research,* Prentice-Hall, Englewood Cliffs, NJ, 1972.

BIBLIOGRAPHY

Austin, M. J., G. Cox, N. Gottlieb, J. D. Hawkins, J. M. Kruzich, and R. Rauch, *Evaluating Your Agency's Programs,* Sage Publications, Beverly Hills, CA, 1981.

Ciarlo, James A. (Ed.), *Utilizing Evaluation: Concepts and Measurement Techniques,* Sage Publications, Beverly Hills, CA, 1981.

Edwards, A. L., *Statistical Analysis, 4th ed.,* Holt, Rinehart & Winston, New York, 1974.

Fink, A. and J. Kosecoff, *An Evaluation Primer,* Sage Publications, Beverly Hills, CA, 1980.

Gay, L. R., *Educational Evaluation and Measurement: Competencies for Analysis and Application,* Merrill, Columbus, OH, 1980.

Grimaldi, J. V. and R. H. Simonds, *Safety Management*, 4th ed., Richard D. Irwin, Homewood, IL, 1984.

Hays, W. L., *Statistics for the Social Sciences*, 2nd ed., Holt, Rinehart & Winston, New York, 1973.

Kerlinger, F., *Foundations of Behavioral Research*, 2nd ed., Holt, Rinehart & Winston, New York, 1973.

Moore, D. S., *Statistics: Concepts and Controversies*, W. H. Freeman, San Francisco, 1979.

Morris, L., C. Fitz-Gibbon and M. Henerson, *Program Evaluation Kit*, 8 volumes, Sage Publications, Beverly Hills, CA, 1978.

Pitz, G. F. and J. McKillip, *Decision Analysis for Program Evaluators*, Sage Publications, Beverly Hills, CA, 1984.

Rossi, P. H., H. E. Freeman, and S. R. Wright, *Evaluation: A Systematic Approach*, Sage Publications, Beverly Hills, CA, 1979.

Rutman, L., *Evaluation Research Methods: A Basic Guide*, 2nd ed., Sage Publications, Beverly Hills, CA, 1984.

Rutman, L., *Planning Useful Evaluations*, Sage Publications, Beverly Hills, CA, 1980.

Smith, N. L., *New Techniques for Evaluation*, Sage Publications, Beverly Hills, CA, 1981.

Struening, E. L. and M. Guttentag (Eds.), *Handbook of Evaluation Research, Volumes I and II*, Sage Publications, Beverly Hills, CA, 1975.

Suchman, E. A., *Evaluative Research*, Russell Sage Foundation, New York, 1967.

Thorndike, R. M., *Data Collection and Analysis: Basic Statistics*, Gardner Press, New York, 1982.

Weiss, C. H. (Ed.), *Evaluating Action Programs: Readings in Social Action and Education*, Allyn & Bacon, Reading, MA, 1972.

Winer, B. J., *Statistical Principles in Experimental Design*, 2nd ed., McGraw-Hill, New York, 1971.

CHAPTER 9

How to Ensure and Maintain Quality in a Medical Surveillance Program

SHIRLEY A. CONIBEAR
University of Illinois College of Medicine
Chicago, Illinois

9.1 REASONS FOR DOING MEDICAL SURVEILLANCE

The terms *medical surveillance* and *biological monitoring* can be used interchangeably to refer to the systematic medical testing of workers exposed to potentially toxic materials used in their workplace, to determine if increased absorption has occurred or if signs of disease are developing. The crudest kind of medical surveillance is to wait for frank signs of disease to develop, such as complaints of vomiting, or physical signs such as dermatitis. Historically, this constituted medical surveillance. Today, the essential purpose of medical surveillance is to detect early signs of movement along a continuum from health to disease. It is now recognized that before frank and sometimes irreversible disease develops, certain biochemical or functional changes can be observed in the exposed person. For instance, elevated serum cholesterol levels do not indicate current illness but rather that the individual is at increased risk of developing heart disease in the future. Another example is that of zinc protoporphyrin (ZPP) values in lead exposure. An elevated ZPP level is not a disease state in itself but indicates increased lead absorption. Initially, its elevation precedes such serious signs of lead poisoning as peripheral neuropathy.

A major portion of occupational health practice involves looking for these early and ever more subtle signs of change in the body in order to interrupt exposure and provide wider margins of safety. As soon as early signs of disease or evidence of increased risk of disease are detected, efforts to remove or limit exposure can be undertaken. The affected individual may then be advised to change his life style (quit smoking, stop drinking alcohol, etc.), to change jobs, or to begin drug therapy.

Medical surveillance yields two distinct types of information. The first involves biochemical or functional changes resulting from exposure, as discussed above. The second type of information gathered during medical surveillance is quantitation of the amount of material absorbed into the body. This includes measuring the original material in body fluids or tissues such as blood levels or urinary arsenic levels, or measuring metabolic byproducts of a material such as with solvents like benzene and toluene. In both cases, small fluctuations in measured quantities may result in decisions to remove a worker from the job, or to install elaborate engineering control systems. The accuracy of the information generated thus becomes critical, and results are likely to be subject to careful scrutiny.

Medical surveillance provides the basis for decision making that may have a major affect on the lives of individual workers and may cause the employer to spend time and money modifying the workplace. Legal rights and obligations inevitably become involved in decisions of such consequence. The test results that these decisions are based on must be accurate and reliable. The purpose of this chapter is to describe how to achieve an acceptable level of quality in data generated by medical surveillance testing and how to maintain this quality over time.

Quality in a medical surveillance program means doing the right procedure(s) at the right time(s) in the right way, recording results accurately, interpreting them correctly, and telling everyone who needs or has the right to know about the results. Unfortunately, there is no universal agreement among professionals about what constitutes many of the "rights" in this definition. To a certain extent, the goals and objectives of the medical surveillance program and the reason it has been initiated will determine what is appropriate. The parameters by which quality is judged may vary from situation to situation. It is, therefore, important to define these goals and objectives.

For exposure to certain materials, medical surveillance is required by government regulations. The Occupational Safety and Health Administration (OSHA) requires medical monitoring for exposure to asbestos, noise, lead, cotton dust and coal dust, for instance. Medical monitoring is also a prerequisite for respirator wear. The regulations concerning medical surveillance for respirator wear are nonspecific and leave test selection and evaluation up to the physician. In the case of monitoring lead-exposed workers, the law is very specific about which tests must be done (blood lead), which laboratories are approved to make the measurements, the frequency of testing, and the action to be taken at various blood lead levels. The noise standard specifies acceptable test equipment, test facilities and level of training of the tester (OSHA 2206, *General Industry Standards*).

Medical surveillance is frequently done to document the state of a person's health at a particular point in time concerning the existence of a specific medical condition. The preplacement exam and exit exam done before the worker leaves employment or upon retiring have this objective.

Medical surveillance may be used to detect unexpected or new diseases in workers working with a new and relatively untested compound. The startup of a new production process calls for this type of medical surveillance in which a variety of organ systems are surveyed. Frequent testing is called for in the face of great uncertainty as to dose-response relationships.

Some persons are clearly more susceptible to certain exposures than others. For instance, some workers are more likely to develop an allergic response. Medical monitoring may try to screen workers to detect these hypersusceptibles. Persons with alpha-1-antitrypsin deficiencies or those with a previous history of allergy may be at high risk for pulmonary irritants, for instance. If unable to detect high-risk individuals, ongoing monitoring tries to identify persons who show signs of rapidly deteriorating health parameters. For example, at a given noise exposure, some individuals will experience more rapid permanent threshold shift than others, although these people cannot be identified prior to exposure. In this case, reliability of the test becomes absolutely critical, since small decrements over time are significant.

Medical monitoring is sometimes done strictly as a benefit, or "perk," for supervisory personnel or executives. The goal of such a program may be to monitor for diseases common in the population surveyed. Such a screening exercise calls for tests with high sensitivity for the diseases of interest.

Another major function of medical surveillance is as a check on environmental monitoring. By quantitating the amount of material or metabolite present in body tissues and comparing these to expected levels estimated from air concentrations, one can assess the relative importance of skin and gastrointestinal absorption, as well as the effectiveness of personal and industrial hygiene measures. Since environmental monitoring is usually done intermittently, it is not necessarily representative of a worker's total exposure. For some materials, say metals, a body burden accumulates and is a good indication of long-term exposure of the worker. Thus medical surveillance in combination with environmental monitoring can play a critical role in assessing actual exposure and absorption.

Medical surveillance frequently serves a legal purpose as well as a medical one. A series of normal pulmonary function tests or blood lead levels below 40 μg (micrograms) provides a good defense against claims of workplace related illness. Medical surveillance data used as evidence in litigation will undergo great scrutiny. Meticulous written documentation of all quality control measures as well as test results is needed if this use is contemplated.

Finally, medical surveillance is carried out to determine fitness to work at a particular job. These examinations are likely to include such things as pulmonary function testing to determine fitness to wear a respirator, measurement of strength or agility to determine fitness to carry out heavy physical labor, or vision screening to detect impairment. Since biological monitoring is used to determine whether or not a person may begin or continue to work at a job, it is likely to be challenged, either legally or by a trade union, whenever an unfavorable determination is made. This places a premium upon accurate and reliable test results.

These different purposes of medical surveillance influence the criteria for what constitutes quality by shifting the emphasis. The next section will discuss each element in the definition of quality in a medical surveillance program in terms of these various situations.

9.2 THE ELEMENTS THAT DETERMINE QUALITY IN A MEDICAL SURVEILLANCE PROGRAM

In the last fifty years, tens of thousands of new chemicals have been introduced into the workplace. Very little emphasis has been placed on testing them for long-term, chronic effects in animal models. At any rate, extrapolation from animals to humans is tricky. The problems of long latency between exposure and the development of disease, as in the case of cancer, and the similarity of human response to toxic agents with other common diseases makes the diagnosis of occupational disease a difficult one. Medical surveillance programs must choose tests in spite of uncertainty about exactly what the expected disease or risk factor is, as well as what the best way to measure it may be. Medical testing continues to evolve at a rapid rate and new testing technology becomes available almost daily. One frequently encounters old, outdated test procedures, current accepted procedures, and new developmental procedures being used and reported on simultaneously. In the face of this rapid turnover of technology, one must choose and carry out testing.

In order to select appropriate tests and types of information to gather, one must know what the potential exposure is. This means knowing, for each raw material and final and waste product in the facility, the quantities, physical characteristics, possible contaminants and their concentrations, and intermediate or byproducts that may be formed. Additionally, the likely exposure level of the work force should be characterized. This means looking at the frequency of use of the chemicals on a daily, monthly, or annual basis, the extent of exposure workers have to the material, whether protective equipment is worn, and whether gastrointestinal as well as skin absorption is possible. After the workplace has been characterized in this manner, one must look at the toxicological literature to determine what organ systems are likely to be at risk. If the material is new, there may be no human experience and one will have to extrapolate the likely target organs from animal studies. A toxicologist or occupational medicine physician can assist in determining the likely effects. Sometimes this is done by looking at compounds with similar molecular structure or properties and extrapolating from their known effects. When looking at the literature, one must remember that old, current, and new test systems are likely to be reported on. In some cases, one must select old, outdated technology because it is required by OSHA or other regulations. The selection of new technology should be made with caution, and consideration should be given to running accepted, standard tests simultaneously until the new technology is better accepted. Some new tests, which may be very attractive, lack valid population-based reference ranges, and so must be selected with caution.

Other considerations in test selection include the acceptability of the test to the worker. Procedures that are time consuming, uncomfortable, or painful are not likely to be well received and may have a high refusal rate. Nerve conduction velocity measurement is a good way to monitor for early signs of peripheral neuropathy, but it is unpleasant, even painful, and not likely to be accepted on a repeated basis by the workers. Tests that have risks must be weighed in terms of their potential benefit to the worker. For instance, tests that deliver a dose of radiation, such as CAT scans, should be chosen only if a clear benefit to the worker can be described. Biopsies, sometimes used to quantitate levels of material in the body such as fat biopsies for PCBs or lung biopsies to detect asbestos or beryllium, should be carried out only under special circumstances.

Tests must be simple enough to be carried out accurately time after time. For instance, the 24 hour collection of urine for measurement of heavy metals or solvent metabolites is unlikely to be successful because of the problem of proper collection. Carrying a bottle for urine collection may be embarrassing or a nuisance to the worker and an incomplete specimen is quite likely. Sperm analysis is another test that is difficult to do because of the specimen collection procedure. The worker must be asked to abstain from sexual intercourse for at least 48 hours prior to specimen collection in order to standardize the result. Many workers may find this objectionable and they may fail to comply, thus introducing an unquantifiable error into the measurements. Further, semen specimens must be analyzed within minutes after collection for motility measurements to be accurate, necessitating the worker producing the specimen at the test facility. This may be objectionable or impossible for some persons. Tests such as these may be too intrusive into personal life to be practical.

A further consideration about test performance is the level of training of the tester. Neurobehavioral tests are one of the only ways of assessing central nervous system dysfunction. However, they must be administered by persons trained in psychology and can be quite time consuming. To test a work force of 100 might require three weeks of a psychologist's time. If the plant is in a remote area and if such an individual is not readily available, this type of testing may not be practical to select in a medical surveillance program.

Before selecting a test or procedure, one must know not only what the test measures and how accurately it does so, but also the entire sequence of events necessary to collect an adequate specimen as well as perform the test accurately.

Two standard measures of a test's ability to detect the presence of a certain disease are sensitivity and specificity. Ideally, a test gives a positive result in all persons who have the disease and a negative result in all persons who do not have the disease. However, because of the inherent variability in a population as well as the existence of other disease states in which the test may be positive, one is inevitably faced with the problem of people having a positive test result who do not have the disease in question (false positives), as well as people having a negative test result who do have the disease (false negatives). These possible outcomes are shown in Figure 9.1. Sensitivity and specificity are used to measure a test's ability to separate persons with a certain disease from those without. *Sensitivity* is defined as the true positives divided by the true positives plus the false negatives, and is a measure of how good the test is at detecting persons who have the disease. In general, a good screening test should have a high sensitivity so that it doesn't miss anyone who has the disease, although it may falsely identify a certain number of people who do not.

Specificity is defined as the true negatives divided by the true negatives plus the false positives. It is a measure of the test's ability to identify persons who do not have the disease in question. A highly specific test has few false positive results. Sensitivity and specificity are partially reciprocal and can be manipulated by changing the cutoff level for normal. (A full discussion of this is beyond the scope of this chapter and the reader is referred to Galen and Gambino, 1975 and Henry, 1979.) Sensitivity and specificity are specific to the test, the test technique, and the disease. Thus, a blood lead level done by atomic absorption has a different sensitivity and specificity than that done by the Delves cup method. Sensitivity and specificity rates are published for various tests and should be sought out and considered when selecting tests.

Test frequency depends in part upon the pattern of workplace exposure. Intermittent exposures may require test frequency to be tailored to the production process. Testing may be triggered by an unusual or emergency situation such as a spill or loss of integrity of a protective garment. A knowledge of the half-life, storage and excretion pattern of the material in question is a good guide to frequency of testing. Pre- and postshift measurements are frequently carried out where the half-life of the material in the body is less than eight hours. In general, the higher the exposure levels, the more frequent the monitoring should be, since disease can develop or body burden can accumulate rapidly. Also, the more uncertain the situation, the more frequent the monitoring should be to detect the first and earliest signs of problems.

Another consideration is the rapidity of response of the organ system likely to be affected. Materials that cause acute or immediate response, such as irritating or asphyxiant gases, require measurement

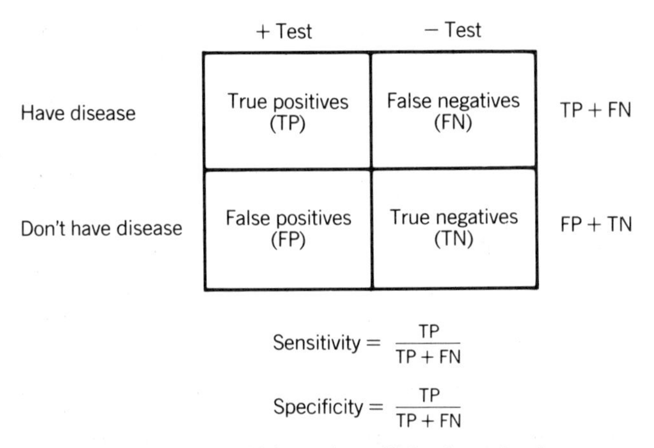

FIGURE 9.1. Sensitivity and specificity determination.

immediately after exposure. Testing may be scheduled so as to measure short-term reversible changes such as signs of bronchospasm pre- and postshift, as measured by pulmonary function.

Practical questions of equipment availability may also dictate frequency of testing. Many companies carry out annual audiometric testing using a mobile van. If the plant is not large enough to warrant having its own audiometric booth, this may be the only practical way to test individuals. The frequency of testing will be dictated by the number of times a year that it is economic to have the van come around.

A major determinant of quality is the proper performance of the test. Specific problems for common tests will be discussed in detail in Section 9.4. The two qualities sought in test performance are accuracy and reliability. *Accuracy* is the ability to arrive at the true or correct value. To a certain extent, accuracy is dependent upon the test methodology and cannot be improved beyond certain limits. One way to ensure accuracy is to compare to an outside standard. This may mean calibrating equipment, as in the case of pulmonary function testing, or measuring spiked specimens, as in the case of laboratory samples. The skill of the operator and the cooperation of the subject can also affect accuracy. Using personnel trained or certified as in the case of "A" or "B" readers for chest x-ray readings increases accuracy.

The second measure of test performance, *reliability*, measures the ability to arrive at the same measurement time after time. It is especially important in medical surveillance. Data that are reliably *in*accurate may still be of use in trending over time. For instance, if a pulmonary function test consistently underestimates forced vital capacity (FVC) by 175 ml (milliliters), time trends can still be analyzed accurately. Reliability depends upon using standardized procedures rigidly adhered to over time. Reliability can be ensured by using the same equipment, the same laboratory, and the same technique. It is important to have written procedures for performing each test and to educate staff members in the importance of adhering to them. Reliability can be assessed by doing repeat testing, submitting uniquely labelled split specimens, observing procedures in the field, and analyzing data for inconsistent results over time.

Complete and accurate documentation of test procedures as well as of test results is essential. Record the date and time of the test, the name of the tester, and the name and at least one other unique identifier of the individual being tested. Many plants employ brothers, father and son, or others with the same name.

Confidentiality of test results and medical findings must be addressed in the process of record keeping. All employees have the legal right to know results of any tests carried out on them. They should be informed of the nature, purpose, and any risk of testing. This process should be documented in the medical record using consent to be examined and release of information forms. Ideally, the record should show evidence of every instance in which testing has been carried out, the employee has been informed of results, or records have been released. Records must be kept in a secure manner with access limited to medical personnel or those who require access for insurance purposes. Medical information should be released on a need to know basis only. Sometimes medical surveillance uncovers diseases unrelated to the workplace, such as an elevated glucose. The program must notify the individual so that he or she may take necessary follow-up action. If the finding is not work related and does not affect the individual's ability to perform the job, the employer is not informed. Employers are notified of all work-related conditions or conditions that may affect workers' ability to perform their jobs safely. Employees have the right to view their own medical records and also have the right to have a copy sent to their personal physician.

Accurate record keeping is facilitated by well-planned and well-designed forms that prompt the tester to record appropriate information consistently and legibly. Many methods for automating test results have been developed, and in many instances results can be recorded directly on a computer disk at the time of testing. The advantage of this is that information is accurately recorded in a complete manner with no chance of transcription errors. However, a contemporaneous written record should also be generated to guard against the possible loss of an electronic file, either through erasure or problems in the initial recording. In addition, there is some question as to the legality of electronic files, since they can be changed without leaving a trace.

Handling large data sets with multiple measurements over time necessitates the use of a computer file, and computer files necessitate a uniform recording system rather than free-form data entry. This should be considered in designing forms. It is best to provide a list of choices or questions that can be answered yes or no. Categories must be mutually exclusive. Questions should ask for specific, quantitative information where possible. Ask "How many drinks per day?", not "Are you a heavy drinker?"

Information must be interpreted on two levels, individual and group. A physician must make a decision about the importance, relevance, and meaning of each test result for the individual and inform him or her of the appropriate action to take, in a letter or an interview. This closes the preventive loop of testing, risk identification, risk reduction, and retesting, by allowing people to take necessary action to reduce or eliminate identified risk factors. In no circumstances should a person simply be given a copy of the results without any interpretation.

Analysis of the entire group may be as simple as developing descriptive statistics showing the frequency of abnormalities among various groups at the plant, or may involve more sophisticated statistical analysis comparing before and after exposure or exposed with nonexposed workers. Looking at trends over a period of years is also an important type of analysis. In generating such reports, individual workers should not be identified by name or in any way that allows their results to become known to everyone.

9.3 ORGANIZING A MEDICAL SURVEILLANCE PROGRAM

A critical aspect of a medical surveillance program is establishing a reporting structure with clearly defined lines of authority and responsibilities. While this may seem too obvious to mention, unfortunately it is frequently not the case. Responsibility for matters having to do with occupational safety and health has frequently been delegated to persons with no training or expertise in the field. Programs have had a tendency to evolve by small increments over time as problems arise. This ad hoc kind of growth frequently results in erratic patterns of testing resulting in situations where testing goes on long after exposure has ceased or necessary testing is not begun when new situations arise.

A variety of organizational structures is possible in occupational health and safety, but some basic rules are thought to be applicable by most practitioners in the field. One of these is that health and safety personnel should not report directly to a person with production responsibilities such as a plant manager. Health and safety considerations have a tendency to result in slower production and the expenditure of funds for engineering or personal protective controls. This is frequently antithetical to the plant manager's goal of producing more product at less cost. It is desirable to interpose an organizational buffer between health and safety and production. Usually, this means that health and safety personnel report to the personnel director at the plant or through the corporate structure to other health and safety personnel. Larger companies sometimes have a separate chain of command for health and safety, so that a plant medical director reports to a regional medical director, who reports to a corporate medical director. Either reporting structure seems to work well and is acceptable.

The principal organizational guideline is that there should be a single individual responsible for the medical surveillance program corporate wide, and that authority and responsibility pass through this individual to the program at the local level. As in all organizations, responsibility must be accompanied by authority, otherwise chaos will result.

One of the functions of the program director is to see that procedures are standardized from facility to facility and from year to year. This is necessary to analyze trends over time and compare health effects from site to site. It also ensures a certain basic level of quality from site to site and makes cost-effective use of funds easier. Subsequent sections of this chapter will discuss specific criteria for commonly used tests, which can be used as a starting point for establishing required protocols.

The second responsibility of the program director is to review tests and procedures at least annually in order to add, eliminate, or upgrade testing modalities as exposures change, laws regarding medical surveillance are added, or medical tests themselves improve. If the program director is not a medical professional, some outside assistance will probably be required to accomplish this. Feedback from the professionals throughout the organization should be solicited to take advantage of their expertise as well as their "hands on" experience. Consultants should also be evaluated annually, since turnover in their personnel can greatly affect the quality of service one gets. Equipment should be evaluated on the basis of its performance in the field. One must also assess the skills of in-house personnel and provide the means and the incentive for them to keep their knowledge base current. If funds permit, one trip to a local or national professional meeting is desirable, since it allows professionals to interact with others in the field, exchange new ideas, and stay current with the developing technology. In special circumstances, in-house educational programs may be arranged to provide group cohesiveness as well as to standardize test procedures. The third task of a program director is to continuously monitor and evaluate the data generated by medical surveillance. (Specific audit procedures are suggested on a test-by-test basis in subsequent sections.) The program director should have access to all information generated by every part of the medical surveillance program. In some cases, to ensure personal confidentiality, this data will have to be edited so that no individual can be identified. This is easily done through scrambling of identifiers, such as social security numbers, by a computer software program. It is essential that all data be available for quality control checks and that the selection of a particular file for audit is made on a random basis without any foreknowledge. The audit procedure should also include a site visit, preferably annually, or at least once every two years. A formal report should be generated as a result of the visit or the audit procedures, with results and corrective actions recorded for the files.

The director of a medical surveillance program must ensure that each unit of the program has the necessary personnel, equipment, and facilities to carry out the specified program. Usually one or more consultants will be necessary to provide such services as clinical laboratory testing or x-rays. The particulars of how to do this have been frequently discussed and written about in the literature. See Howe, 1975, for an in-depth description of existing practices and recommendations.

9.4 ENSURING AND MAINTAINING QUALITY IN VARIOUS TEST SYSTEMS

9.4.1 History and Physical Exam

The history of the patient's past and current medical problems and symptoms is the single most important piece of information in making a diagnosis. Since the purpose of medical surveillance is not primarily to diagnose disease, the patient's history may be somewhat abbreviated. Part of its purpose is to discover confounding, preexisting medical conditions that may simulate or obscure workplace-related disease. The history can also pinpoint acute exposure by picking up complaints of irritation of the mucous membranes or the skin. In certain types of disease, clinical tests are too crude to measure early changes and the first sign of developing occupational illness may, in fact, be the onset of symptoms. Occupational asthma may first evidence itself by intermittent chest tightness or mild shortness of breath before decrements in pulmonary function can be measured. Neurologic problems, particularly of the central nervous system, may first be noticed by the patient. Complaints such as dizziness, personality change, loss of balance, or incoordination are difficult to quantitate through medical diagnostic testing, and one must rely on the history as one of the earliest indicators of these problems.

The history is usually taken by a physician, but sometimes a nurse can do so. The ability to elicit a complete and accurate history is an essential part of every clinically trained physician's expertise. The physician learns a pattern of questioning that surveys all organ systems followed by more detailed questioning in any area that elicits a positive response. Some nurses are more skilled at history taking than others. Many nurses have received special training in the area of physical diagnosis, which includes history taking.

In a medical surveillance program, it is best to formalize this pattern of questions by writing them down. Focus on questions pertinent to the organ systems most likely to be affected by the exposure in question and abbreviate followup questions in areas of less interest. By writing the questions, one ensures that the same questions will be asked year after year and that none will be omitted.

One must decide between a self-administered questionnaire or an interviewer-administered one. Generally, the quality of the interviewer-administered questionnaire is better, although it is more expensive because it consumes professional time. The interviewer-administered questionnaire is more accurate and reproducible, especially if there are problems of literacy or non-English speakers in the workforce. It also allows the professional an opportunity to explore positive responses in more depth and to record this as anecdotal information. A preprinted questionnaire allows one to standardize the history from facility to facility and also allows standardization should professional personnel change. A preprinted form facilitates computerization of the data and also allows comparison of responses elicited by several interviewers. If interviewer bias is suspected, the patterns of responses across interviewers can be compared to detect this kind of problem.

The only facility needed to take a proper medical history is a quiet private room where two persons can sit comfortably. Workers may not admit to certain symptoms, such as impotence, if they think others may overhear.

The questionnaire should be piloted before it is finalized to determine if the language is suitable for the ethnic and cultural background of the workers. For example, a British respiratory questionnaire is frequently used to quantitate bronchitis. It contains a question about "catarrh," a term that is meaningless to most American workers. One should consider whether or not language barriers may exist in the population and provide translations and bilingual interviewers where needed. The questionnaire should provide unambiguous response categories without any overlap. Ask enough information to allow the calculation of pack-years smoked and ounces of alcohol consumed per day. A well-laid-out form requiring a minimum of writing will reduce the possibility of misinterpretation or loss of information due to illegible handwriting.

In order to ensure accuracy in data entry, a blind reentry for verification by the data entry person is advised. It is unlikely that the same typographic error would occur twice on exactly the same key stroke. Wherever discrepancies between the first and second entries occur, the original record can be checked.

The physical examination is generally much less sensitive than the history and provides information about more advanced problems. The physical exam requires a clean, quiet, private area with an exam table comfortable for the patient and the examiner. Diagnostic equipment for the examination is standard and it is best left to the examiner to make a personal choice. Physical examinations are usually performed either by a licensed physician trained in clinical medicine or by a nurse who has received special training in physical diagnosis. Nurses who have not received special training should not routinely be allowed to perform physical examinations. In some states, physician's assistants are allowed to carry out physical examinations. This may be a viable option in certain facilities.

The physical examination is, like the history, a series of measurements and examinations learned by every clinically trained physician. However, the depth of the examination normally depends upon patients' complaints or whether abnormal findings are present. In order to focus the examination on the organ systems of interest and ensure complete recording of results or measurements, a preprinted form is recommended. This allows standardization from facility to facility and also from professional to professional, and can provide a uniform grading system or quantification of measurements of interest. For instance, in workers potentially exposed to agents causing testicular atrophy, like DBCP, it is important to measure and record the size of each testicle to enable year-to-year comparison. Liver size and the briskness of reflex response are also important quantifiable entities. A score or measurement provides more information than simply reporting that liver size is normal or that there is no sign of testicular atrophy. More precise recording allows for better monitoring. It also enables one to detect intraexaminer variability. If physician A examines a mix of exposed and unexposed persons and yet consistently measures liver sizes larger than those measured by physician B who examines a similar population, one must investigate the techniques of the two physicians and arrive at some standard form of measurement. A preprinted form also allows for ease of computerization, which has the same benefits as a computerized history.

It is critical to select the right questions and the right examination procedures for the preprinted forms, so knowledgeable professionals to assist in the design should be sought. Pilot test the forms to determine whether or not they are complete, unambiguous, and well organized. Get feedback from the people who are using them regarding layout and completeness. Once in use, a computer program can be used to check for intraobserver differences, missing or inconsistent information, and extremely abnormal results. Accuracy can be evaluated by comparing permanent features such as scars or deformities in the physical examination or birth dates and past medical history items in the history with previous exams.

9.4.2 X-rays

X-ray equipment is expensive even for units that do limited examinations. Unless you have an extremely large facility with a lot of occupational injuries, it is unlikely that you will have an x-ray unit in your medical department. Therefore, you will have to contract out for this service, generally to a local hospital, clinic, or a radiologist's private office. Currently, the chest x-ray is the only radiologic examination used on a regular basis in medical surveillance. This discussion will focus on evaluating a facility for its ability to take and interpret appropriate chest x-rays. Many improvements have developed in x-rays within the last 15 to 20 years but because of the expense, some very old equipment is still in use. Asking the age of existing equipment will give you some indication about whether this is a problem. Make sure that a standard 14″ × 17″ film is being taken. Some older units take smaller films, which are not acceptable. Older equipment may deliver a larger dose of radiation to the patient and use older types of film with poorer resolution. Both are undesirable.

Many states have training or licensing requirements for persons who take x-rays. Check the laws in the state in which you are operating. Taking an x-ray is similar to taking a photograph and is subject to similar types of problems. The quality of the picture is dependent upon the skill of the operator, as well as upon the quality of the film and the equipment. If a film is overexposed, lung markings will be "burnt out" and early signs of pneumoconiosis will be obliterated. A film which is underexposed will show too many markings, which may be interpreted as pneumoconiosis that in fact does not exist. It is important that similar exposures be used consistently from year to year in the same person to allow better comparison of films. Additionally, the individual must be positioned properly in front of the film so that the costophrenic angles, or lower edges of the lung, are not cut off the picture. This can be difficult with a large individual. A good operator is aware of and corrects for these types of problems. Good practice also requires shielding the patient's reproductive organs during the procedure to cut down on the amount of radiation.

Chest x-rays should be interpreted by a physician with training either in radiology or pulmonary medicine. The National Institute for Occupational Safety and Health (NIOSH) sponsors a three-day training seminar for physicians on how to read chest x-rays. This program instructs physicians in the use of a standardized format called the ILO-Cincinnati Classification of Radiographs of Pneumoconiosis (OSHA Series, 22, 1980). It provides a standardized scoring system for quantitating all types of lung diseases that may result from exposure to such fibrogenic dusts as asbestos, coal, and silica. It has been in wide use throughout the United States and other countries for a number of years, and is used by the U.S. Department of Labor in determining if a coal miner is eligible for disability payments. An ILO reading is required when applying for benefits under this program. While it is not specifically required by any other OSHA standards at this time, it is certainly standard practice in occupational health. It is advisable to use this system for a number of reasons. It allows for quantitative comparison of series of chest x-rays. It is a well-known, standardized system that can be used by a variety of radiologists or certified readers and result in uniform interpretations. It lends itself to statistical analysis by reducing chest x-ray findings to a series of scores. Certified readers are characterized as either "A" or "B" readers. An "A" reader is one who has taken the course sponsored by NIOSH and the American College of Radiology on how to use the system. A "B" reader is one who has subsequently passed an examination using the scoring system. The NIOSH facility in Morgantown, West Virginia publishes a list of "B" readers located throughout the United States.

It is important to maintain the original films of all old chest x-rays for each worker. Even though the ILO reading is standardized, there is no substitute for viewing the actual film. The employer should have the original x-rays in its possession or under its control. Some OSHA regulations, notably that for asbestos, require films and records to be kept for at least 20 years after exposure ceases. Many medical facilities have a policy of destroying x-rays after a designated period, which may be as little as five years. X-rays are bulky to store and contain valuable amounts of silver, so they are frequently sold in bulk for silver reclamation. If films are stored by a radiologist or clinic, it is best to have a written agreement not to destroy any films without prior notification. In general, copies of x-rays are less satisfactory than originals and are not acceptable substitutes.

Workers must be informed of any abnormalities described on their chest x-ray. This includes even early and clinically insignificant signs of developing pneumoconiosis. It is the examining physician's or requesting physician's responsibility to properly inform these individuals. A system of notification should be agreed upon ahead of time with either the examining or the requesting physician about how this will be done. This constitutes good ethical practice as well as meeting legal requirements.

9.4.3 Pulmonary Function Testing

The most commonly used test for measuring lung function in a medical surveillance program is that of spirometry. Spirometry measures lung volume as the forced vital capacity (FVC), as well as lung compliance as the forced expiratory volume at one second (FEV_1) divided by forced vital capacity (FEV_1/FVC). FVC decreases with the loss of lung volume that occurs in restrictive lung diseases resulting from exposure to various fibrogenic dusts such as silica and asbestos. FEV_1/FVC decreases in obstructive lung disease, an asthma-like condition developing from exposure to plant and animal products, cigarettes, and certain chemicals such as toluene-di-isocyanate, phtallic anhydride, and others. This condition may have an irreversible component and turn into emphysema eventually. A variety of other measurements of lung compliance such as forced expiratory flow between 25% and 75% (FEF 25–75) and peak expiratory flow rate (PEFR) are also routinely measured during spirometry and are considered by some to be more sensitive or earlier measures of obstructive lung disease. Various reference ranges have been developed by a variety of investigators to define normal lung function. These are referred to as *nomograms*, and are known by the name of the person who developed them. The nomograms of Morris, Knudson, and Cory are the most commonly used. Nomograms predict lung function using a formula based on the height, age, sex, and sometimes race, of the tested person. This formula is derived by testing a large group of supposedly normal persons and then expressing their lung function as a function of age, sex, and height. Using different nomograms will result in different predicted lung volumes for the same individual (Kanner and Morris, 1975). Therefore, it is important to know what nomogram has been used in the past and to use the same nomogram from year to year and facility to facility. It is not disastrous if different nomograms have been used previously, as long as the actual lung volumes have been recorded. The predicted value can always be recalculated using the formula for the selected nomogram, but it is essential that the actual lung volumes be recorded for each test.

The pulmonary function test is a noninvasive, painless procedure, which carries almost no risk to the individual tested. The results are dependent upon the degree of effort of the individual being tested and therefore, dependent upon the understanding and willingness of the person. Part of the training of the tester is geared toward eliciting a maximum effort from the subject and determining whether or not a best and consistent effort has been given.

No special preparation of the subject is necessary prior to testing, although it is best to request that individuals not smoke for several hours prior to having the test done. Sometimes before- and after-shift tests are made to measure change in lung compliance or evidence of obstructive changes developing as a result of exposure in the workplace. In these cases, it is especially important to record the time of the test and any exposure the worker may have had prior to testing. Also, individuals may have some decrease in their lung compliance if they are suffering from respiratory infections at the time of testing. This should also be noted in the record. Persons known to be asthmatic and/or being treated with certain drugs will have a variable response depending upon the state of their disease and the dose and timing of their medication. These facts should also be recorded.

No special facility is necessary for pulmonary function testing. It can be done in any clean area free of dust and fumes where a desk top is available. A wide variety of spirometers is available to carry out pulmonary function testing. The prices range from $500–$50,000. Basically, two types of spirometers are in use, classified according to the way in which they measure volume. One type of device measures volume directly and derives measurements of flow. These instruments use displacement of a bellows or piston to measure exhaled volume. Generally, they are simple to operate, inexpensive, and hold calibration well. Because of their accuracy and reliability, they are generally the best option for a medical surveillance program. The second type measures air flow and derives volume using a variety of devices. In general, this type of spirometer measures midtracing flow more accurately but requires frequent calibration, is more expensive, and can be disrupted easily. NIOSH tests pulmonary function equipment for reliability and accuracy and periodically publishes error rates for each spirometer. In general, an error of less than 3% is acceptable. "Smart" pulmonary function equipment with built-in microprocessors are more costly but provide error diagnostics, test procedure prompting, electronic recording of results, and calculation of various lung volumes.

Most pulmonary function equipment generates a graph or tracing that shows time on the x-axis and volume on the y-axis. It is essential that a tracing be printed or displayed for each trial and that a written record of the best tracing be generated and saved. The tester needs to see each curve as it is generated in the testing process to determine whether the worker has understood the instructions and is responding maximally. When selecting pulmonary function equipment, these requirements should be considered.

Calculations of volume are made by locating specified time-volume coordinates on the tracing. This can be done manually, or electronically by the spirometer. When doing large numbers of tests, electronic calculation saves time and is at least as accurate as manual calculation. A spirometer with a microprocessor frequently stores the selected nomogram and prints a comparison of the measured volumes with the predicted volumes calculated on the basis of the person's age, height, and sex. A nomogram is selected when the equipment is installed. Manual calculation requires standard tables or a formula.

It is generally felt to be necessary to make corrections for body temperature and ambient air pressure (BTPS). These corrections can be done manually using a formula or by the spirometer's microprocessor.

All pulmonary function test results should contain the nomogram used to generate expected volumes, a printout of the actual volume for each parameter, the height, age, race, and sex of the tested individual, the air temperature and pressure, and whether or not correction for BTPS has been made. The tracing of the selected trial should be printed and saved. If the data is to be computerized, it is not necessary to enter predicted values for forced vital capacity or the ratio of FEV_1 to FVC. Nomogram values can be computer generated using the desired formula, and ratios of interest can be calculated. This saves entry time and focuses attention on measured volumes, the critical numbers.

Since measurement of lung function is effort dependent, the training of the tester is particularly important. The OSHA cotton dust standard requires that nonphysicians who perform pulmonary function tests take a NIOSH-approved training course in proper procedures. While this certification is not required for other exposures, it has become an informal standard and should be required for all personnel responsible for pulmonary function testing. NIOSH-approved courses are offered three or four times a year by various organizations and cost about $300. They provide didactic and practical training in proper procedures for performing spirometry using a variety of equipment. Course completion ensures standardized test procedures and knowledgeable personnel. No legal restrictions exist on what profession is allowed to perform pulmonary function testing. It may be done by a registered nurse, a licensed practical nurse, a respiratory care technician, a medical technologist, or even someone who is not trained in the health sciences. For this reason, when selecting a facility or setting up an in-house program to perform pulmonary function testing, inquire about the training and certification of the individual who will be performing the test.

Spirometry consists of maximum inhalation followed by rapid, maximum exhalation into a mouthpiece connected to the spirometer. The tester coaches the person being tested and watches the resulting curve to determine whether a maximum effort has been given. Some possible problems include terminating the test before maximum exhalation has occurred, improper positioning of the mouthpiece, or inconsistent effort. Three efforts are made and test results must be within $\pm 5\%$ of each other or more test curves must be obtained until consistent results are achieved. Various criteria exist for determining which effort is selected as the "best" or most representative effort. These criteria must be chosen by qualified personnel familiar with the goals and objectives of the program.

Equipment should be standardized at least daily using a calibrated syringe that delivers 3 L of air to the machine. The chart speed can be checked with a stopwatch. These procedures should be carried out each time the machine is used if less than daily, each time it is moved, or daily. The results should be recorded after each calibration. In this manner, leaks or improper chart speed, the main equipment problems, can be detected.

It is essential that the worker's height be measured with shoes off prior to each test session. Measuring with shoes on will introduce an error into the calculation by predicting a larger volume. Likewise, asking the worker is not adequate and is likely to lead to an overestimate of height by as much as one to two inches. It is not uncommon to look back through a series of pulmonary function tests done over the years and find the recorded height of the subject varying by one to two inches. This will cause predicted values to be in error and thus all observed-to-predicted percentages will also be incorrect. However, all is not lost as long as the person's correct height is known and the actual lung volumes have been recorded. Predicted values can be recalculated using the correct height and compared to actual lung volumes, allowing the correct interpretation.

A quick quality check for consistency and proper technique can be made by assembling a series of pulmonary function tests done over time on a single individual. Record the height, age, and measured forced vital capacity for each year in sequence. The heights should all be within half an inch of each other. The ages should be correct at each test session. The forced vital capacity should show a downward decline of 30–40 mL per year. This decline is expected with age and is taken into consideration by the nomogram. A steeper decline may indicate developing restrictive lung disease, but an increase in forced vital capacity or an up and down variation indicates either a problem with test procedure or possibly some type of medical problem at the time of testing, such as a severe asthma attack, respiratory infection, or neurologic or musculoskeletal abnormality. Any increase in forced vital capacity should be investigated to see if any medical cause can be determined. Check several other worker's records in the same way. If they show a similar pattern, this strongly suggests a problem with the test procedure. Do not try to use forced expiratory volume at one second (FEV_1) in a similar way as a check on test procedure. It can vary widely from day to day or hour to hour depending on exposure to materials that cause bronchospasm.

In general, changing equipment, tester, or test facility should not result in consistent differences in measurements. A consistent difference between facilities, testers, or equipment must also be investigated. Make sure that you know who performed each test and what equipment they used so that these comparisons can be made.

9.4.4 Electrocardiogram

The electrocardiogram (EKG) is a recording of the polarization of the heart muscle as it contracts and relaxes. It provides information on the size, shape, orientation, rhythm, and conduction system of the heart. It is relatively simple to perform and requires only a quiet, clean, private area where the worker can lie down comfortably with chest and limbs exposed. No hazard or discomfort is attached to having a cardiogram done. The test equipment consists of an electrocardiogram. The technology for measuring the electrical events is basically the same in all equipment. The price varies according to the amount of electronic manipulation of the signal, labeling, and interpretation subsequently given. The least automated EKG equipment leaves the most room for operator error. The more sophisticated equipment tends to ensure proper labeling of the leads, complete and accurate patient data, and a better quality tracing.

There are no standard training programs for persons performing EKGs. Since the procedure is noninvasive and there is no risk of harm to the patient, technicians or ancillary health personnel are frequently allowed to perform the test after some instruction.

Some common errors or problems include failure to standardize the vertical axis at the beginning of each tracing, selecting the wrong chart speed, improper or absent labeling of various leads, improper positioning of the leads on the patient, wandering baseline on the tracing, and 60-cycle electrical interference in the signal. These problems can be eliminated or controlled through proper training and supervision of the operator. Most problems are easily detected by looking at a few tracings. A random sample of tracings should be reviewed periodically for quality control. The physician who interprets the cardiogram can also be asked to comment on their quality.

The interpretation of EKGs can be done by any clinically trained physician, but a cardiologist is desirable. Most communities now have a cardiologist available or tracings can be sent away. EKG interpretations are somewhat standardized in that certain measurements of rates, intervals, and axis are always recorded, but otherwise the interpretation is a free-form, narrative type. This makes computerization and comparison of tracings between years or groups difficult. In a medical surveillance program, it is desirable to codify EKG interpretations either by developing a form for use by the physician-reader or by extracting information from the free-form interpretation. Standard recording systems exist, most notably the Minnesota system by Blackman, but none are completely satisfactory. *The International Classification of Diseases* (ICD) provides a good basis for building a coding system and has the advantage of being widely used by hospitals and insurance companies.

Recently, microprocessors have been developed that can provide a computerized reading of the EKG. These interpretations always require a second reading by a physician. Most programs are prone to "over read" cardiograms and, on occasion, may totally misinterpret certain problems. The EKG is especially prone to yielding false positives in young individuals and computerized interpretations can be especially inaccurate in this group.

The medical record should always include the original EKG tracing as well as the interpretation. Subtle changes may develop over a period of time and old tracings are particularly valuable for side-by-side comparison by the physician in these cases.

9.4.5 Audiometric Testing

The OSHA noise standard requires preplacement baseline and annual audiometric testing for persons exposed to more than 90 dB averaged over eight hours. This testing must consist of air conduction threshold measurements made at 500, 1000, 2000, 3000, 4000, and 6000 Hz (OSHA 2206, *General Industry Standards*). To carry out this type of testing, one must have a noise attenuating booth in which the worker sits and an audiometer that presents the proper tones with varying degrees of loudness. Manually operated audiometers can be purchased for under $1000. New instruments are virtually all standardized to the 1969 ANSI standard. This standardization is built into the machine and corresponds to the nomogram used in spirometry.

The noise attenuating booth must meet strenuous OSHA regulations. It is usually installed by the seller and sound levels inside the booth should be tested on final installation. The seller should provide written documentation of the noise levels in the booth at a variety of frequencies. If the booth must be moved for any reason, this process must be repeated. If outside consultants do audiometric testing, ask to see results of sound levels measured in their booth. If a mobile testing van is used, be careful not to park it in an area where low frequency sound may be transmitted to the booth. In general, low frequencies are the hardest to attenuate and may require repositioning of the booth. Under no circumstances should testing be done without a booth. Some industrial clinics may test in a so-called "quiet room" or use an in-the-ear otoscope-audiometer. These do not satisfy OSHA requirements, cannot be standardized, and are not acceptable alternatives.

The audiometer can be calibrated in two ways: (1) by using electronic equipment such as a sound meter to measure the loudness of tones generated by the audiometer, or (2) use a biologic control. This simply means measuring the hearing threshold of a non-noise-exposed, nonhearing-impaired employee who is likely to be available for retesting. Whenever one wishes to standardize the equipment, this employee is retested and his or her audiogram compared to the previous one. A written record should be kept of all standardization procedures, which should be done at least weekly.

When large numbers of audiograms must be performed, automated or computer-assisted audiometers may be used. These may be "self-testing," meaning that the person is put through the test guided only by instruction from the electronic equipment's software. A trained operator must be in the vicinity and available to review the output and to retest manually if the person is having difficulty following instructions.

OSHA regulations specify that only trained audiologists or those who have completed an audiologic training course are acceptable testers. The course lasts for three days and is open to everyone regardless of medical background. It is conducted by a variety of organizations throughout the United States as demand warrants. Any nurse, medical technologist, or other medical technical person who will be doing audiometric testing definitely should take this course, since the techniques of audiologic testing are not routinely taught to most health professionals. If an outside vendor is selected to do the testing, inquire as to whether the individual doing the audiograms has been certified or is an audiologist.

Exposure to loud noise can cause a temporary hearing loss, referred to as *temporary threshold shift,* which is indistinguishable from permanent threshold shift on an audiogram. The severity of temporary threshold shift is directly proportional to the loudness and the duration of noise exposure. Time to recover is proportional to the severity of the loss. In most cases of workplace exposure, the temporary loss will be gone 12–16 hours after exposure. This is the reason that the OSHA regulations require 16 hours since last noise exposure or use of hearing protection prior to testing. Noise exposure outside of work, failure to wear hearing protection, or inadequate protection at work can all result in falsely high threshold measurements on the audiogram. To avoid this problem, ask each person prior to testing about noise exposure on and off the job in the last 12–16 hours. Don't test persons with significant noise exposure in this time period. Give them hearing protection, instruct them on its use, ask them to avoid noise exposure prior to testing, and reschedule. Record history of noise exposure in the last 12–16 hours for all tested persons. Check for past problems with temporary threshold shift by scanning each person's audiograms in sequence. If hearing improves more than 5 dB at any threshold from test to test, you may have a technical problem. The frequency most sensitive to noise is 4000 Hz and is likely to show a problem first.

The tester should look in the subject's ear canals with a speculum prior to testing. If the canal is blocked by wax, do not test. This can be a common problem where in-the-ear hearing protection is used, since it tends to pack wax into the ear. Suggest ear drops and reschedule testing.

9.4.6 Laboratory Testing

With the exception of some of the inhaled dusts, almost every industrial material can be absorbed into the body to a greater or lesser extent. Toxic materials may enter the body by inhalation, after which they pass directly into the bloodstream, through skin contact in which they penetrate intact skin and go into the blood stream, or by ingestion of contaminated food or by swallowing material coughed out of the lung, in which case they are absorbed into the bloodstream through the stomach or intestine. The science of toxicology studies and describes the natural history of each chemical, following it from absorption through distribution throughout the body, metabolism or breakdown, storage in various parts of the body, and finally, excretion. Many laboratory tests used in medical surveillance are based on quantitating the toxic material during this cycle of absorption, distribution, storage, breakdown and excretion. Several problems exist with such tests. For example, there can be wide individual variability in absorption. The state of a person's nutrition affects absorption. Persons who are calcium deficient tend to absorb lead from the stomach in greater quantities. Or there may be significant age differences in absorption. These differences, which are greatest in infancy and childhood, tend to even out as persons get older. For instance, the pH in the stomach of infants is much higher, that is, less acidic, than in adults and this may accelerate absorption of certain metals. Some genetic differences exist between individuals, particularly in the speed with which their enzymes are able to break down certain organic compounds. Some individuals have inducible enzymes, which can be speeded up to metabolize foreign compounds at 10 to 100 times the rate of other people. Some persons have lesser capacity to excrete toxic materials or their metabolites because of preexistent kidney disease. Since most metals are excreted via the kidney, a decrease in kidney function can increase body burden, the total amount stored, in these persons. In the case of persons who have accumulated a large body burden, blood and urine levels may remain high years after exposure stops as the body rids itself of what it has stored up. This can occur with metals and such fat-soluble materials DDT, PCBs, dieldrin, and others.

Other laboratory tests are chosen to demonstrate the effect of the toxic material on the body. Genetic differences may greatly influence these responses. For instance, persons with a fairly common genetic disorder known as G-6-PD deficiency cannot reverse the effects of aniline or nitrite compounds on their red blood cells. These persons will show a more severe response to the same dose of the same material as compared to a normal person.

Many of the reactions of the body to toxic chemicals are not specific to a particular chemical and may result from other naturally occurring diseases. For instance, such chlorinated solvents as carbon tetrachloride can cause a hepatitis-like condition. Lead can cause anemia. Toluene diisocynate can cause an asthma-like response. Every medical surveillance program that involves laboratory testing will inevitably turn up a number of individuals who have abnormal test results. Some of these results will be false positives, that is, the person will have an abnormal test result but will not have any disease, and others will have diseases that are not related to the workplace. Separating these from work-related conditions is the challenge to and responsibility of the physician and other professionals in charge of the program.

Because of the variety of laboratory tests performed in a medical surveillance program, an outside clinical laboratory is inevitably used. It is the responsibility of the physician, with help from others in the medical surveillance program, to select appropriate tests, collect specimens in an appropriate manner, and arrange for and coordinate their shipment to the laboratory so that they arrive in good condition. A good clinical laboratory will provide the proper specimen containers and any special instructions on collection and preservation techniques.

Selecting a good laboratory is key to obtaining reliable data. Before selecting a laboratory, it is a good idea to meet with the service representative and someone with technical expertise from the laboratory. They should be able and willing to provide reference ranges relevant to the population you are going to be testing. These references ranges define the upper and lower limits of what is considered to be normal, or disease free. Always ask the size of the population used to develop reference ranges. In general, the smaller the population used, the less certain you can be of a reference range. The reference range should be based on a population that the laboratory has actually sampled and measured. Some laboratories try to cut corners by using reference ranges published by other laboratories. This is not as reliable and is generally not an acceptable alternative. Techniques may vary slightly from laboratory to laboratory, resulting in small shifts in the distribution of test results. These small differences may not be important to the clinician treating individual patients, but in carrying out epidemiologic analysis of group data, these differences will become troublesome and are thus unacceptable.

The laboratory should have daily quality control checks that allow them to standardize procedures from day to day. Should your testing program find elevated BUNs in every person who was tested on July 24, you want to be able to call up the laboratory and have them verify that the BUN analysis was not running falsely high on every sample tested on that day. The laboratory should be willing to allow you to send a certain number of split specimens in each batch to verify the reliability of the test. In some cases, you may also want to send a few control specimens from persons known to be normal to verify the shipment and preservation procedure. These additional tests should be run at no cost or at a greatly reduced rate.

The laboratory personnel should be available and able to answer all of your questions about specimen collection, preservation, and test procedure. A good laboratory does not resent such questions. They should be able to answer questions about the accuracy of the technique they use and be willing to advise you about the existence of other techniques. The laboratory should have a policy of automatically rerunning each abnormal test result for verification before it is ever sent to you. They should also be willing to save unused portions of specimens for a short time, in case additional testing is desirable. The laboratory should be willing to notify you by phone if values are greatly abnormal, represent an emergency situation, or exceed some prearranged value.

Quality control procedures within the laboratory are usually extensive and their proper design and execution cannot be discussed in this chapter. See OSHA, 22, 1980 and Kenner and Mirros, 1975, for laboratory methods and procedures.

Clinical laboratory testing is a very competitive business and laboratories are frequently willing to negotiate prices if a large number of specimens is guaranteed. Prices for a test may vary by a factor of two or three. Unfortunately, the lower prices may not include the same level of quality control or analytic precision, but these differences may not be evident to the unwary until it's too late. For example, screening urine for signs of recreational drug use has become common in certain industries and some labs offer to test for a battery of substances for a very small fee. Screening results are then used as the basis for hiring or keeping on an employee. Inaccurate results may result in lawsuits later on. It can be disastrous to discover that the laboratory mixed up specimens or that the precision of the technique is such that prescription drugs may have been mistaken for recreational drugs. The best policy is to pay a fair price for the product that you require. If the price seems unbelievably low of if you get quotes that vary widely, be careful. You must investigate whether you are getting the quality of testing you require. You cannot assess quality by looking at a number on a laboratory report.

Technically, some specimens do not require a trained medical professional for collection. These include specimens like urine, breath, hair, and semen. Blood drawing requires a trained phlebotomist, a nurse with phlebotomy skills, or a physician. However, it is advisable to use a trained health professional to collect and preserve all specimens whenever possible. Where this is impractical, such as in remote locations where frequent sampling is required, one or two individuals should be instructed in proper collection and preservation procedures before being allowed to take charge of the project. A trained individual understands what is essential in the process vs. what is important vs. what is trivial and can be safely ignored. For instance, if a 24-hour urine is being collected and the worker has omitted one collection, it might be possible to estimate total volume and send the specimen anyway. On the other hand, if the analysis is for porphyrins and the urine was exposed to light, the specimen is useless and should be discarded.

Many materials such as heavy metals or metabolites of organics such as trichlorethylene are excreted in the urine. The concentration of urine changes throughout the day as people drink water, eat, sweat, or otherwise change their state of hydration. More concentrated urine will contain a higher concentration of the toxic material of interest than will dilute urine. In order to eliminate this variability, all urinary concentrations of toxics should be expressed in terms of their concentration in urine with a specific gravity of 1.024. In order to standardize to this specific gravity, it is necessary to measure the specific gravity in the test urine and calculate a conversion factor for the concentration of the toxic material.

In some cases, it is desirable to express concentration of the toxic in terms of weight per milliliter of urine excreted, which tells you the total amount eliminated each day. This is frequently used for metal excretion. To make this calculation, the volume of urine excreted in 24 hours must be measured. To do this, the individual must carry a container and collect all urine. Since this is cumbersome and socially unacceptable, there is a tendency to not collect every specimen. Collections that are under 1000 mL for a 24-hour period for an adult are suspected of being incomplete.

Another problem with using urine to measure excretion of toxic materials is that of contamination. With the heavy metals, one is measuring microgram quantities. This means that a small amount of dust falling into the specimen can raise the concentration two- or threefold. Contamination is a particular problem if the specimens are collected at work. Workers may have metal dust on their hands or clothing that can contaminate the specimen. Contamination may also occur if the wrong container is used. The heavy metals are ubiquitous in the environment and one must use specially washed and packaged containers to be sure that no extraneous material is present. Plastic may absorb the toxic or may emit undesirable material into the specimen. Some of the preservative used can either bind with the material to be measured or may actually contain the material itself, so proper container selection is critical. Finally, proper handling and preservation of the specimen is necessary to insure that no changes occur in the specimen on the way to the laboratory. In the case of urine, bacterial contamination can be a problem if the specimen is not refrigerated and shipped promptly.

Breath is sometimes used to measure concentrations of metabolites or toxic materials. It is especially useful for organic compounds like the solvents. In most cases, the breath is analyzed on the spot as the employee exhales into the analytic equipment. This has the least opportunity for error. Otherwise, the exhaled air must be sealed in a container that doesn't leak or already contain atmospheric contaminants. These problems can be formidable, so exhaled air is usually not sent away for analysis.

Such heavy metals as mercury, arsenic, and lead accumulate in hair because of their biochemical affinity for sulfhydryl bonds, which are numerous in hair. These materials are actually incorporated into the hair shaft as it grows. Reference ranges in nonexposed populations have been developed for some of these materials. Because they are easy to obtain and require no special preservation, hair samples have been used to monitor exposure in populations located in remote areas. The main problem with hair specimens is that of external contamination. Employees should be instructed not to go into the work area, to wash their hair, and not to use any hair care products before taking a specimen. Some hair care products contain metals. For instance, one of the self-applied hair colorants for men actually contains lead oxide as the active ingredient. Shampoo may contain selenium, which can interfere with analysis. It is a good idea to clip the hair specimen at the scalp and label the scalp end. If levels are high, the laboratory may be able to section the hair shaft and determine whether concentrations are evenly distributed along it or whether exposure is recent (near scalp) or long past (away from scalp).

Most laboratory testing requires a sample of venous blood. Some procedures require only several drops of blood from the finger or ear. For instance, zinc protoporphyrin measurements require only a few drops of blood and have become popular in place of blood lead levels, partly for this reason. Many enzymes and waste products measured in blood change from hour to hour or day to day. Studies have shown that such seemingly inconsequential differences as whether the patient is lying down or standing, has recently exercised heavily, or recently slept can affect some test results by as much as 20%. The order in which tubes of blood are drawn and the time for which the tourniquet is left on the arm can also affect results. It is well known that the time of the last meal can affect certain tests, such as those for glucose and cholesterol. Persons are required to fast for 12 hours prior to having their blood drawn in order to standardize these results. Modest alcohol ingestion within the previous several days can result in elevated liver enzymes, even in persons who are not alcoholic. Many medications can alter results as well.

In a medical surveillance program, the goal should be to eliminate or control as many of these variables as possible. The best procedure is always to draw morning fasting specimens before work. This provides standardization for time of day, exercise, and time since last meal. The laboratory technician should develop a protocol for drawing blood that is used invariably. Usually this means having the person sitting comfortably in a chair while drawing tubes in a set order.

The second area where error can be introduced is in preparation and preservation of specimens. A variety of preservatives are used, depending upon the test. One may require whole blood, either clotted or anticoagulated, or serum, or plasma. For some tests, the specimen must be frozen. Many of the hormones, for instance, decompose rapidly if the specimen is not frozen. Other specimens should be kept cool but not allowed to freeze, particularly if one wants to look at the cells. Freezing breaks down the cells and makes the specimen useless. The laboratory technician, in conjunction with the clinical laboratory, will set up a protocol for each test procedure and should be able to deal with all these potential problems.

Contamination of blood specimens can also be a problem. The person drawing the blood should be certain to clean the arm thoroughly to avoid getting a small amount of dust on the needle on the way into the vein. Test tubes should remain stoppered while around potentially contaminating materials.

One common problem with laboratory specimens occurs in facilities that do not centrifuge blood before sending specimens to the laboratory. Serum is used for many analyses and requires that the clotted cells be removed from the specimen. This is usually done by centrifuging the test tube and decanting off the serum, a straw-colored liquid which remains on top. This procedure is best done immediately after the blood is drawn, but can wait up to six hours. To spare both the expense of owning a centrifuge and the time spent processing the blood, some medical facilities send the blood to the lab in its unseparated form. The longer the specimen is allowed to sit, the more cells burst and spill their contents into the serum. This results in falsely elevated values for serum potassium and some of the enzymes such as LDH. Elevated potassium is relatively rare in the general population and is a good way to check on the quality of specimens reaching the laboratory. Scan a series of laboratory results for potassium levels. If many of them are high, check to see whether or not the laboratory has noted that the blood was hemolyzed, that is, contained many broken red cells when it arrived. This represents a significant problem with specimen preservation. Recently, so-called separator tubes have been developed in which a gel forms an impermeable barrier between the cells and the serum in the test tube as the tube is centrifuged. The whole tube is sent to the laboratory without actually decanting off the serum. Sometimes the seal between the serum and cells may break or be incomplete, again resulting in hemolysis. This can be detected in the same way. It is a good idea to require that the technician decant the serum into another container to avoid this problem.

For some tests, such as carboxyhemoglobin or blood gases, the time between specimen collection and analysis is critical and is very short. For this reason, arterial blood gases are usually done in a hospital, which has the analytical equipment on hand. Carboxyhemoglobin provides a measurement of carbon monoxide exposure. In order to do it properly, the specimen must be shipped to the laboratory promptly. It may be desirable to send a control specimen along. A nonexposed nonsmoker might be sent as a negative control to assure that the result is yielding the expected values. This is especially useful in situations where the tendency is for an old or improperly preserved specimen to become more and more abnormal. By sending a known normal, one can gauge the contribution of the preservation process to the abnormal findings.

9.5 SELECTING A FACILITY OR VENDOR

If the plant is too small to warrant hiring in-house medical personnel or if the budget will not support it, an outside facility or consultant will have to be selected. There are two basic alternatives: (1) to contract with a local clinic or physician and (2) to contract with a mobile testing van. To a certain extent, the choice will be predetermined by the location of the plant and the availability of appropriately trained medical professionals. The advantages of using a local facility are that the physicians are usually available to take care of injuries and illnesses, preplacement examinations can be carried out providing a preexposure baseline, and there is more flexibility in scheduling examinations so that no one is missed because of vacation or illness at the time of examination. Exam frequency can be increased at any time to accommodate special situations.

Some of the advantages of using a mobile testing van are the opportunity for standardized testing throughout the company, no loss of work time because of travel to and from the exam site, the ability to perform a large volume of tests in a short period of time, and the convenience of dealing with only one contractor for multiple sites. While some of these factors do not involve quality per se, they can affect the overall quality of a surveillance program by influencing participation rates. One must consider the turnover rate in the workforce, the nature and extent of the hazard, and the type and extent of testing that is contemplated. Review these in terms of the overall goals and the objectives of the medical surveillance program. One option may clearly be superior to the other at this point. Use the questionnaire (Table 9.1) to evaluate the quality of testing done by either a local facility or a mobile van. In some cases, weaknesses

TABLE 9.1. Quality Evaluation Questionnaire

I. X-Ray

1. Do you either take or evaluate X-rays?
   - Yes 1
   - No 2
2. Who takes your X-rays?

   ____
   (Name) (Title)
3. Is the person who takes your X-rays:
   - Formally trained X-ray technician 1
   - Other . . . (describe) 2

   ____
4. What size of chest film do you take?

   ____
5. Who interprets the X-rays?

   a. Name ______________________________

   b. Board certified in roentgenology?
      - Yes 1
      - No 2

   c. "A" reader?
      - Yes 1
      - No 2

   d. "B" reader?
      - Yes 1
      - No 2
6. How long do you keep old films? ______________________________
7. Do you notify anyone before they are destroyed?

   ____

II. Pulmonary Function Testing

1. Do you or your staff conduct pulmonary function testing in your facility?
   - Yes 1
   - No 2

   (If yes, please send a representative sample of the tracing and report you submit.)
2. Please describe the type of pulmonary function equipment you use by completing the blanks below:

   a. Equipment manufacturer ______________________________

   b. Model number ______________________________

   c. Year of purchase ______________________________
3. The nomogram used to calculate results is:
   - Morris 1
   - Knudsun 2
   - Corey 3
   - Other . . . (Specify) 4

   ____
4. The number of pulmonary function trials routinely used in an individual examination is:

   ____
5. What criteria do you use for acceptable variability among trial results?

   ____

6. The calculations for pulmonary function test results are:
   - Automated 1
   - Manual 2

   (If manual, calculations for pulmonary function test results are done by:

   Name ____________________

   Title or Profession ____________________
7. When performing pulmonary function testing, do you:
   - Measure the patient's height 1
   - Ask the patient for height 2
8. When performing pulmonary function testing, do you use nose clips?
   - Yes 1
   - No 2
9. When performing pulmonary function testing, do you have the patient
   - Sitting 1
   - Standing 2
10. Do you use bronchodilator
    - Routinely 1
    - If the PFT is abnormal 2
    - Never 3
11. Are results converted to BTPS?
    - Yes 1
    - No 2
12. Does your pulmonary function machine automatically do the conversion?
    - Yes 1
    - No 2
13. List all members of your staff who perform pulmonary function testing:

    1. ____________________
       (Name)

       ____
       (Professional or educational level)

       ____
       (Nature of special training for pulmonary function testing)
14. Has this person successfully completed the NIOSH Pulmonary Function Technician Training Course?
    - Yes 1
      (If yes, indicate date: ____________________)
    - No 2
15. Describe the procedure for calibration.

    ____
16. How often do you calibrate your pulmonary function testing equipment?

    ____

III. PHYSICIAN EXAMINATIONS

1. Do you examine employees at your facility?
   - Yes 1
   - No 2
2. If you are in solo practice, complete only for yourself. If you are in a group or clinic please complete for *all physicians* who might examine employees.

   Person 1. ____________________
   (Name) (Degree and year)

   ____
   (Specialty board certification) (Year)

   ____
   (Specialty board certification) (Year)

TABLE 9.1. *(continued)*

3. Do any nonphysicians conduct examinations?
   No 1
   Yes 2
   (If yes, please give name and profession:)

   ____

   ____

4. Do you use a form to record your results for the physical?
   No 1
   Yes 2
   (Please provide a copy)
5. Do you use a form to record the patient's history?
   (please provide a copy)
   No 1
   Yes 2
6. Who has access to employees' medical records?

   ____

   ____

   ____

7. Where are these charts and medical records stored?

   ____

   ____

8. Do you routinely inform the patient of abnormal findings or test results?
   Yes 1
   If yes, please describe how: ______________________________

   ____

   ____

   No 2
9. Do you routinely inform anyone else of these abnormal findings?
   Yes 1
   If yes, who: ______________________________

   How: ______________________________

   ____

   No 2
10. How long do you keep patient charts and medical records?

    ____

    ____

11. When files are purged, what do you do with patient charts and medical records?

    ____

    ____

12. Do you have a policy of notifying anyone before records are purged?
    Yes 1
    No 2

IV. LABORATORY TESTING

1. Do you perform or collect specimens for laboratory analysis?
   Yes 1
   No 2

TABLE 9.1. *(continued)*

2. What laboratory performs the analysis?

   Firm ____
   (Name) (City and State)

   Is this laboratory certified by the Center for Disease Control?
   Yes 1
   No 2

   Is the laboratory licensed by the State Health Department in the state where it is located?
   Yes 1
   No 2

   List all other certifications held by the laboratory:

   American College of Pathologists? ______________________

   Joint Commission on Hospital Accreditation? ______________________

   Other: ______________________

3. Do you do any laboratory analysis in your facility?
   Yes 1
   (If yes, list which tests and who does the analysis.)

   ____

   ____

   No 2

4. Do you routinely centrifuge blood when serum or plasma is needed before sending the specimen to the laboratory?
   Yes 1
   No 2

5. Do you make slides for differential counting?
   Yes 1
   No 2

6. What is the average amount of time elapsed between specimen collection and analysis?

V. AUDIOMETRIC TESTING

1. Do you or your staff conduct audiometric testing at your facility?
   Yes 1
   (If yes, please send a representative sample of the report you submit.)
   No 2

2. Describe the type of audiometer used in your facility:

   a. Equipment manufacturer ______________________

   b. Year purchased ______________________

   c. ANSI 1969 standardized? ______________________

3. Is you audiometric equipment automated?
   Yes 1
   No 2

4. Is the test self-recording?
   Yes 1
   No 2

5. List the frequencies at which you perform audiometric testing.

   ____

   ____

TABLE 9.1. *(continued)*

6. Where is your audiometric testing conducted?
   Quiet room 1
   Booth 2
   Other . . . (Specify) 3

   ____

7. If you answered "2" for the above question, has your audiometric booth been tested for sound attenuation?
   Yes 1
   No 2
8. Does you audiometric booth meet OSHA noise standard requirements?
   Yes 1
   No 2
9. List all members of your staff who perform audiometric testing.

   Person 1. ______________________________
   (Name)

   ____
   (Professional or educational level)
10. Have they successfully completed the audiometric technician training course?
    No 1
    Yes 2

    Give date ______________________________
11. Do members of your staff calibrate audiometric testing equipment?
    Yes . . (Specify how often) 1

    ____
    No 2
12. Describe your office procedure for calibrating your audiometric testing equipment.

    ____

    ____

    ____
13. Does someone do a speculum examination of the ears before testing?
    Yes 1

    (If yes, who? ______________________________
    No 2

can be readily corrected or procedures changed, for instance using a different nomogram or centrifuging blood before sending it to the laboratory. Some deficiencies, such as lack of a noise attenuating booth or untrained personnel, are expensive to correct and probably cannot be changed to accommodate your protocol.

Sometimes, it may be necessary to use several local sources to perform the required testing. For instance, x-rays may have to be done at a local hospital, audiograms at an audiologist's office, and the rest of the testing at the physician's office. This is likely to result in scheduling difficulties and a lot of employee time away from work. If a large number of persons require testing frequently, such an arrangement is probably untenable. The more data sources there are, the more likely information is to get lost or mixed up, and the more likely people are to miss their appointment or not get scheduled for necessary testing. Such problems do not exist with a van, but if the company does not or cannot satisfactorily perform all the testing you require, it is best to investigate other companies.

Another important issue is that of responsibility for follow-up. It is vital to the success of any medical surveillance program that workers be informed of all abnormal results, that they be given copies of the results to give to their primary care physician, and that they understand the relative importance or urgency attached to each abnormality. Strictly speaking, it is the physician ordering the test who has these responsibilities. However, when using a mobile testing van, follow-up is usually accomplished by a form letter with no personal attention given to the employee. Since many employees do not have a personal physician, they may fail to take action or to fully understand the implications of abnormal findings. Without effective feedback to the employee, much of the potential preventive benefits of the program will be lost. An interested local physician can carry out face-to-face notification as well as perform the testing. This can be a significant factor in the success of a medical surveillance program.

Before selecting either a local facility or mobile van, it is a good idea to visit the facility. You should tour the examination areas and meet the personnel who will be carrying out any part of the medical surveillance program. Make it a point to see all the equipment that will be used in the testing process. Be sure to look at the waiting area to see if it is clean and pleasant. Listen to the receptionist talking on the phone or to patients. Look at the medical records area to see whether it appears well organized and of adequate size. Try to elicit all the information listed in the questionnaire (Table 9.1). Remember that this facility will be representing your company. Ask yourself if employees will be comfortable going there and if the image projected is that of a high-quality professional organization. Employees' impression of the care they receive will affect participation rates and overall attitude toward the program. Since much medical testing is dependent upon the personnel who carry it out, and since personnel can turn over rapidly, every facility should be reevaluated on an annual basis. The questionnaire (Table 9.1) can be mailed out, unless sampling of test results indicate a developing problem.

9.6 HOW TO BE COST-EFFECTIVE

Carrying out an effective, high-quality medical surveillance program can be costly. Mobile testing vans, usually the cheapest option, charge $90–$120 per employee for a basic exam. Alternatively, investment in equipment, personnel, and space are necessary to do the work in-house. Laboratory tests and outside consulting add to the cost. Since testing must be performed at regular intervals, there is little cost saving over time. These costs are not essential to the operation of a company, so they must be justified in some way. Usually this means comparing them with estimates of potential cost of disease in terms of lost work time, increased insurance premiums, fines, and lawsuits. This is the cost-benefit approach. It answers the question, "How does the cost of the program compare with costs occurring as a consequence of not having it?" Such a comparison is not generally acceptable when the cost can include loss of life or health in addition to financial loss. Thus, medical surveillance programs are usually evaluated on the basis of whether or not they are cost-effective. This approach asks, "Are we accomplishing our goals for the least cost?" To determine this, one first needs to quantitate the desired effect.

Primary prevention refers to the process of finding risk factors or early disease while it is still in a reversible state, eliminating the risk factor or cause of the disease, and retesting to determine how much improvement has occurred. In an occupational health program good industrial hygiene, engineering controls, and personal hygiene practices will result in low risk and little or no disease. Medical surveillance provides verification of what the health and safety professional knows should be happening on the basis of environmental monitoring. Thus, in a truly effective health and safety program, there should be very little increased risk or occupationally related disease for the program to detect. You know that your health and safety program is effective when medical surveillance finds no problems. Quantitating just how much future disease you have prevented is difficult. Sometimes historical data will be available to demonstrate problems that predated the program. Where this is available, it certainly should be used to quantitate reduced illness. One of the pitfalls of this analysis is that a new program may detect a lot of chronic disease caused by earlier exposures, causing diseases rates to rise sharply compared to preexamination levels.

Medical testing in itself does not constitute a prevention program. The other components of a health and safety program must be in place to close the loop of detection, intervention, and retesting. Whether the cost of these programs is added into the cost-effectiveness equation depends more on the administrative structure and how budgets are organized than on fixed rules.

In its narrowest sense, cost-effectiveness in a medical surveillance program can be defined as provision of required testing for the least cost. This is contrary to some of the guidelines used in general preventive medicine programs to decide about cost-effectiveness of screening in the general population. For instance, a screening test to detect a disease that is not treatable is not recommended in the general population, nor where the number of people with the disease is very small, or where the disease does not have serious consequences. None of these rules are applicable in deciding whether medical surveillance tests are cost-effective, because medical surveillance is required to document a negative as much as to find a positive. It may quantitate solvent exposure or periodically document normal hearing. If it does this as accurately and economically as possible, it is cost-effective.

One of the ways to ensure cost effectiveness is to purchase exactly what you need. Do not buy more or fewer services or greater or lesser quality of service than required. For example, you may calculate the cost of an audiogram as $8.00 a test if done in-house vs. $5.00 a test if done as part of an examination on a mobile van. However, if the testing booth on the van does not meet OSHA specifications, all the tests will be invalid as far as meeting OSHA regulations. The cheaper examination is no bargain since its results are essentially worthless and cost-effectiveness is not achieved.

Further, a number of employees may be absent or unable to be tested because of noise exposure on the one or two days that the van is on site. The price for retesting these individuals at a local facility plus the lost work time traveling to and from the facility may bring the cost of the mobile van exam up to a level comparable to the in-house test. In addition, repeat audiograms at more frequent intervals throughout the year may detect temporary threshold shifts and identify workers with inadequate hearing protection or detect specially susceptible persons sooner. The value of this to a comprehensive hearing conversation program can be significant.

Alternatively, consider an example of buying more service than is needed. A local hospital may charge $100 for a chest x-ray compared to $20 on a mobile van. The hospital cost includes expensive overhead like 24-hour service, the latest equipment, access to emergency services, etc. These services are not needed and do not increase the quality of the end product, a chest x-ray. The mobile van is a more cost-effective choice.

Be sure to compare like with like when comparing prices. Adding a microprocessor to pulmonary function equipment may add $5000 to the cost of a $1000 machine, but will save more than this in the staff time required to calculate volumes and predicted values with the manual equipment. A more subtle comparison is at the level of personnel training. A nurse who has completed the audiometric technician training course and the pulmonary function training course may ask for a larger salary than a nurse with no diagnostic skills. Time lost in training the inexperienced nurse or problems with accuracy may more than make up for the difference in salary. Remember that consultants have similar costs. A difference in hourly rates or test prices frequently reflects the training and experience of the staff.

Prices for medical or diagnostic testing often vary widely by geographic location. The cost of an electrocardiogram in a rural area may differ by a factor of two or three from that in an urban area. It is a good idea to find out what the usual and the customary fees are in the area before deciding whether or not a price is reasonable.

Try to take advantage of economy of scale whenever possible. If your organization has 5000 employees in 10 different locations around the United States, it may be possible to negotiate a better price for the purchase of equipment or laboratory services on a corporate-wide basis than to allow each facility to make its own arrangement. Economy of scale is a critical factor in deciding whether to set up an in-house program with your own personnel and equipment or to hire a consultant or outside vendor. A rule of thumb in occupational health is that a plant with 500 workers requires a full-time nurse on site and a plant with 1000 or more workers requires a full-time physician, depending in part upon the exposures. Number of employees is not the only indicator of need. A facility that has 500 workers but handles highly toxic materials will require a more intense program than a facility with 2000 white-collar workers.

The goals and objective of the medical surveillance program as discussed in Section 9.2 will help to define what constitutes an effective program. A program that generates inadequate, unreliable data is of no use to anyone and only provides a false sense of security to all concerned. It is not cost-effective, whatever its price.

REFERENCES

Galen, R. S., and Gambino, R. S., *Beyond Normality: The Predictive Value and Efficiency of Medical Diagnoses,* John Wiley, New York, 1975.

Henry, J. B., Todd, Sanford, Davidsohn; *Clinical Diagnosis and Management by Laboratory Methods,* Vol. I, 16th ed., W. B. Saunders, Philadelphia, 1979.

Howe, H. F., Organization and Operation of an Occupational Health Program, Part I, *Journal of Occupational Medicine,* pp. 2–64, Occupational Health Institute, June 1975.

Occupational Safety and Health Series, No. 22 (Rev.), *Guidelines for The Use of ILO International Classification of Radiographs of Pneumoconioses,* 1980.

OSHA Safety and Health Standards (29 CFR 1910), OSHA 2206, *General Industry Standards.*

Kanner, R. and H. Morris, Eds., *Clinical Pulmonary Function Testing, A Manual of Uniform Laboratory Procedures for the Intermountain Area,* Salt Lake City, 1975.

National Center for Health Statistics, *The International Classification of Diseases, Clinical Modifications,* ICO-9-CM. 250 9th Rev., Commission on Professional and Hospital Activities, Ann Arbor, MI, 1970.

CHAPTER 10

How to Select and Use Personal Protective Garments

SHERIDAN J. RODGERS
MSA Research Corporation
Evans City, Pennsylvania

10.1 INTRODUCTION

10.1.1 History of Chemical Protective Clothing

Skin is one of the major organs of the human body, with a surface area of approximately 1.8 m^2. The skin serves to protect the muscular structure and internal organs from corrosive, toxic, and other potentially harmful hazardous chemicals as well as physical threats. The Bureau of Labor Statistics reports that over 40% of all compensation claims are due to dermatological problems and that approximately 60% of all claims paid are a result of skin-related problems. These facts show that the need to protect workers' skin from exposure to hazardous chemical liquids and vapors is not only humanitarian, but also economic.

The use of personal protective equipment dates back hundreds of centuries. Workers in historic times used cloth masks to prevent inhalation and ingestion of dust particles, and gloves made of sheepskin and goat skin were commonly used to protect the hands, knees, and so on. Leather aprons were employed by workers exposed to acids, alkalis, and other corrosive chemicals. Today the worker can be completely protected with garments ranging from basic chemical resistant work clothing to fully encapsulating ensembles.

The advent of various natural and synthetic rubbers and polymers has resulted in a higher degree of protection for workers who may be exposed to hazardous chemicals. But even today, the generalized recommendations "wear protective clothing" or "wear rubber gloves" are commonly used. In many cases, these recommendations have led to a false sense of security on the part of the worker, and the use of inappropriate protective clothing has resulted in the worker being needlessly exposed to hazardous or toxic chemicals. It does appear that the days of "generic protection" are rapidly becoming history, and it is the intent of this chapter to provide guidelines on the selection of the proper garment and garment material that a worker should use when exposed to a hazardous chemical threat.

In the past two decades, there has been a significant increase in the testing of protective garments and garment materials against a broad spectrum of hazardous chemicals. As recently as 1981, ASTM published a standard for permeation testing of protective garment materials (ASTM F739-81). Various government agencies such as the U.S. Coast Guard, EPA, NIOSH, and the U.S. Army have funded studies to determine the protective capabilities of various materials against toxic chemicals. Some manufacturers and vendors of protective garments have initiated in-house programs to evaluate protection factors provided by complete protective ensembles and to assist customers in the selection of the most appropriate protective garment for a specific use.

10.1.2 Engineering Controls Versus Protective Clothing

It must be stressed that use of engineering controls is the preferred method of protecting a worker from hazardous chemical exposures. Exhaust hoods and localized ventilation systems can direct potentially hazardous fumes and vapors away from the work area. Splash shields and drip pans will minimize exposure of the worker to contact with liquids. Remote manipulation, long-handled tools and implements, and robotics can reduce worker exposure. Yet there are some instances where engineering controls are impractical or impossible to implement, and in these cases protective garments are the only recourse to provide a safe and healthful environment for the worker.

Examples of instances in which engineering controls are not practical recently have been given nationwide media coverage. The Love Canal incident, train accidents where corrosive, toxic, and flammable chemicals are released, cleanup of hazardous waste dumps, containment and recovery of oil and other hazardous chemical spills in waterways are but a few examples of instances where engineering controls cannot be implemented. In the everyday work place, employees are required to enter areas where exposure to both liquid splash and vapor is possible. Production workers and maintenance personnel are especially susceptible to this type of exposure. Such workers must be provided with garments that will ensure protection against a host of hazardous chemicals.

10.1.3 How to Put This Information Into Practice

Thousands of hazardous chemicals are being manufactured, shipped, and used throughout the world. The Registry of Toxic Effects of Chemical Substances lists toxicity information on nearly 60,000 chemicals. (2) HAZARDLINE has developed a computerized data base on 3000 chemicals as of May 1984, and is adding 2500 new entries each week (Tathen and Lewis, 1983). With the number of hazardous chemicals used throughout industry and with the wide variety of protective garment designs and materials that are commercially available, it is little wonder that the health and safety professional is faced with difficult, and in many cases ill-defined, decisions.

The "How To" of selecting a protective garment involves an analysis of the type of protection required, and an evaluation of the commercially available protective equipment. Manufacturers can provide information on available types of garments, prices of ensembles, and in many cases, recommendations on type of garment material for a specific hazardous chemical exposure. However, the task goes beyond merely accumulating information from manufacturers; it involves interaction between the user and the supplier. Only the user can provide detailed information on the chemical being used, the chemical class, the physical form, the usage and exposure profile, and the degree of protection required. The intent of this chapter is to present guidelines and procedures for both users and manufacturers in the selection of protective garments.

10.2 CHEMICAL PROTECTIVE CLOTHING

10.2.1 Introduction

Personal protective equipment (PPE) is a generalized category that covers all such protective devices as respirators, self-contained breathing apparatus, goggles, face shields, gloves, hoods, splash suits, fully encapsulating ensembles, and so on. Discussions in this chapter relate to chemical protective clothing (CPC), and so are limited to hoods, splash suits, encapsulating ensembles, and other CPC items; accessory components such as cooling systems and visor materials are also addressed.

A broad range of CPC is available from various manufacturers. Protective clothing is tailored to meet specific requirements based on the degree of protection required for the user; therefore both the design and material of the clothing must be considered. Other features of CPC to be addressed include techniques of seam fabrication, type of closures, accessory items such as boots and gloves, and other CPC related factors.

10.2.2 Types of CPC Available

Various types of CPC are available, and the user must decide on which type to purchase, based on the degree of protection required. These decisions will be influenced by the hazardous material being used, the exposure time of the individual user, the usage profile, and other factors. Procedures for selecting specific items of CPC are addressed later in this chapter.

CPC has been developed to protect against a variety of materials, including radioactive and irritating dusts, toxic chemical vapors, liquid splashes and deluge with hazardous chemical liquids, and other threats to the health and safety of the worker. The types of CPC range from basic work clothes to fully encapsulating ensembles, with a wide variety of designs between these two extremes.

Figure 10.1 shows some of the basic items that are to provide splash protection for specific areas of the body, including chest, legs, and arms. These are available in such materials as PVC, natural rubber,

FIGURE 10.1. Basic protective items for individual areas of the body.

and neoprene. Splash resistance is provided against acids, caustics, oils, greases, gasoline, and solvents, depending upon the material of construction.

Figure 10.2 shows such typical items of protective work clothing as shirt and trousers, lab coats and coveralls. The material is fabricated with a fiber that is resistant to attack by a number of chemicals, but the woven fabric is not resistant to penetration by liquids and vapors.

Chemical/acid resistant hoods are shown in Figure 10.3. Both hoods provide protection to the worker's head, eyes, neck, shoulders, back, and chest. Typical materials of construction include PVC, neoprene, and natural rubber.

The components of an anticontamination coverall are shown in Figure 10.4. The ensemble consists of a coverall, boots, and hood. A filter-type respirator provides protection against inhalation of potentially hazardous dusts and particulates.

Reusable/disposable protective clothing made of Tyvek, cellulosic materials, and coated Tyvek is available (Figure 10.5). These are used routinely in cleanrooms, laboratories and electronic component manufacturing plants, and to a limited degree where splash protection is required. Reusable/disposable items include shirts, pants, smocks, aprons, hoods, coveralls, sleeve protectors, head covers, snoods, and so on.

As the need for protection increases, suit designs become more sophisticated (Figure 10.6). The Sunray suit is made of vinyl-coated nylon and is resistant to oils, acids, solvents, and other chemicals. The rubber utility suit is neoprene on cotton, and the seams are vulcanized throughout. The rubberized protective suit is a two-piece neoprene-on-cotton ensemble and provides head-to-toe protection for maintenance and repair assignments. Respiratory protection, in the latter case, is provided by either filter-type gas masks or air supplied through an umbilical.

When maximum protection is required, fully encapsulating suits are the choice (Figure 10.7). These garments, depending upon the construction material, are resistant to acids, alkalis, solvents, fuels, oxidizers, carcinogens, and the like. Since the ensembles are totally encapsulating, breathing air must be supplied either with a self-contained breathing apparatus or a supplied air system. In addition, body cooling is required to minimize the effects of heat stress imposed by the lack of evaporative cooling.

The examples presented here are representative of but a few CPC designs available throughout the protective clothing industry. But they do exemplify the variety of choices available to the health and safety professional, and demonstrate the dilemma which these personnel may face in selecting the proper ensemble for a specific application.

10.3 PROTECTIVE GARMENT MATERIALS

10.3.1 Introduction

The health and safety professional is faced not only with a variety of garment designs, but also numbers of garment materials. Natural and synthetic rubbers and elastomers have been developed, and are evolv-

Shirt and Trousers

Coveralls

Lab Coat

FIGURE 10.2. Basic protective work clothing.

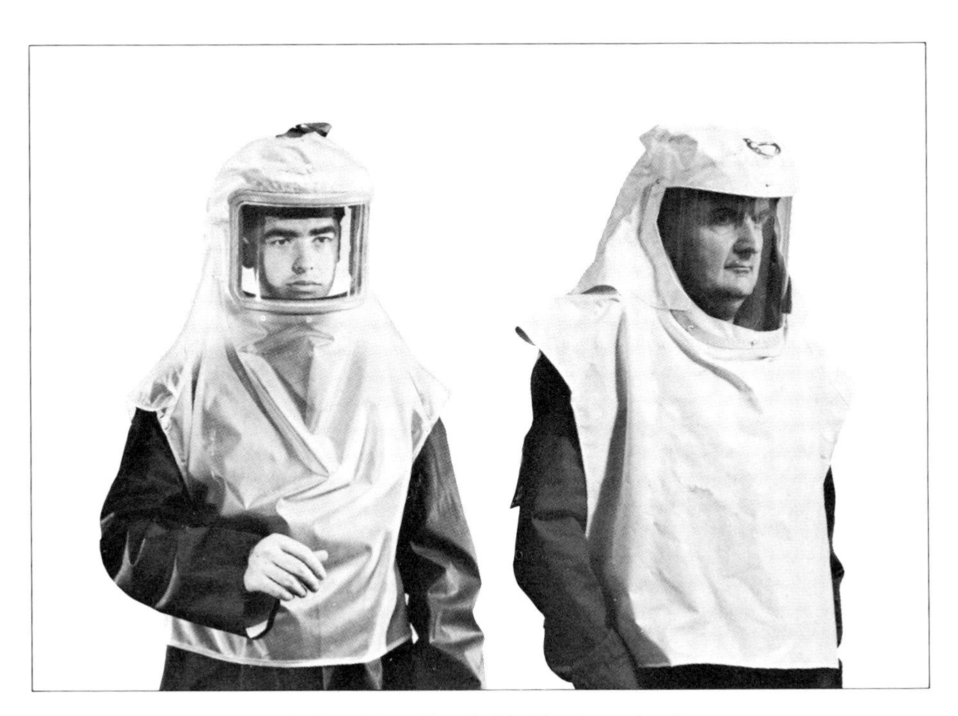

FIGURE 10.3. Chemical/acid resistant hoods.

FIGURE 10.4. Anticontamination coverall.

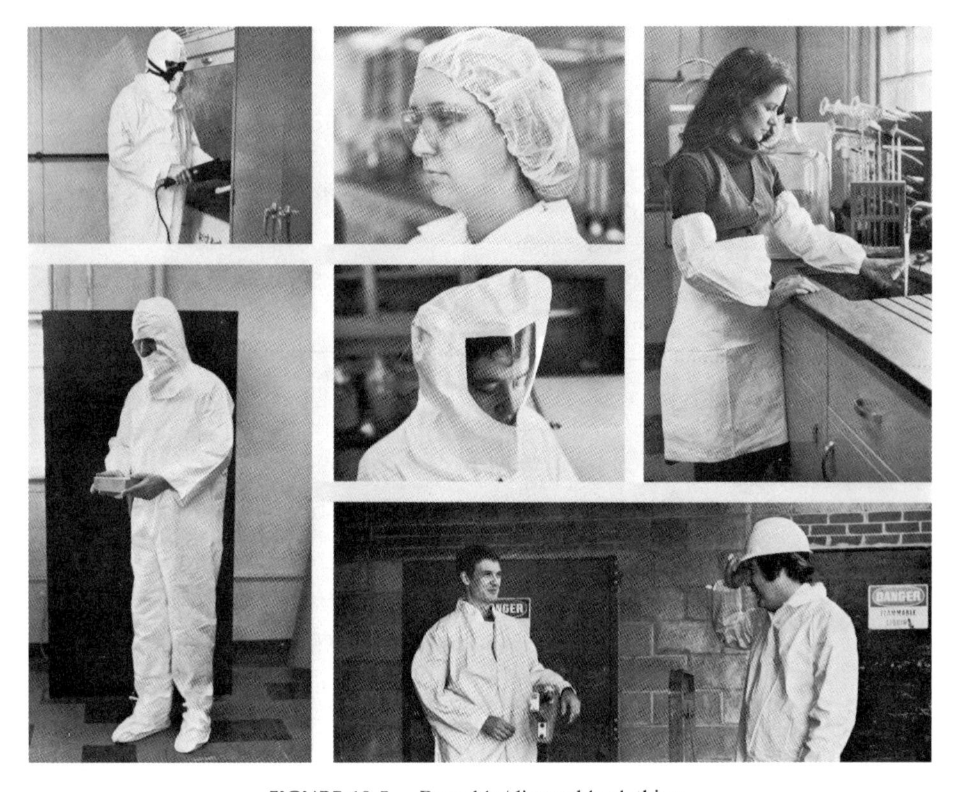

FIGURE 10.5. Reusable/disposable clothing.

ing continuously. These materials may be supported or unsupported, and in some cases combinations of materials are used to provide a maximum degree of protection.

In spite of the evolving technology in protective garment materials, the following facts must be considered:

1. All plastic and rubber materials are permeable to some degree.
2. No single material is resistant to all chemicals.
3. For some chemicals, no acceptable garment material is currently available to provide adequate protection for the user.

How, then, does a health and safety professional choose a garment material? A review of the history of garment materials and a summary of the state of the art for garment materials development and testing can provide a means for rational selection of a specific ensemble.

10.3.2 Materials Testing and Classification

Materials used for protective garments include, but are not limited to:

- Butyl
- Chlorinated polyethylene
- Natural rubber
- Neoprene
- Nitrile
- Polyethylene
- Polyurethane
- Polyvinyl alcohol
- PVC

- Saran
- Treated woven fabrics
- Composites of the above

These materials may be unsupported, or supported on such substrates as cotton, Nomex, polyester, nylon, and so on. Supported materials, in general, are more puncture- and tear-resistant than unsupported materials.

CPC manufacturers generally provide a rating for garments, including descriptors such as Excellent, Good, Fair, Poor, Not Recommended. Unfortunately, there is no standard means of classification within the industry. Some manufacturers use only three classifications, while others use all of the classifications. One manufacturer may designate an Excellent rating for a material that is resistant for two hours, while another manufacturer may require an eight-hour resistance for the same rating.

To further compound the classification problem, different procedures have been used to evaluate garment material resistance. In the past, degradation testing was used as a means of rating materials. This involved immersing a test swatch of the material in a chemical for a predetermined time, and measuring changes such as obvious disintegration, swelling, weight change, elongation, tensile strength, and so on. This type of test did indicate gross degradation, but did not reveal the permeation resistance of the material. *Permeation* is the movement of liquids or vapors through a garment material, and occurs on a molecular level by sorption into the surface of the material, diffusion through the material, and desorption on the opposite side of the material. *Penetration* is the passage of vapors or liquids through zippers, seams, pinholes, or other imperfections in the protective clothing material. Permeation resistance and penetration resistance are extremely important factors when dealing with fully encapsulating suits in working with highly toxic or corrosive chemicals.

In 1981, ASTM published a standard for permeation testing of protective garment materials (ASTM, F739-81). The procedure consists of clamping the material between two test cells and challenging one side of the material with a chemical liquid or vapor. The opposite side of the material is moni-

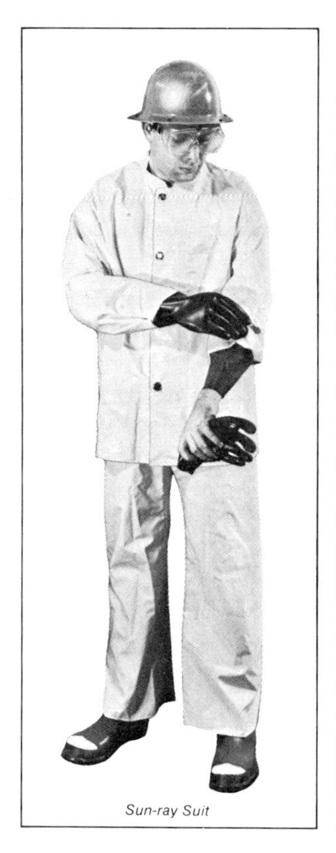

FIGURE 10.6. Chemical protective garments.

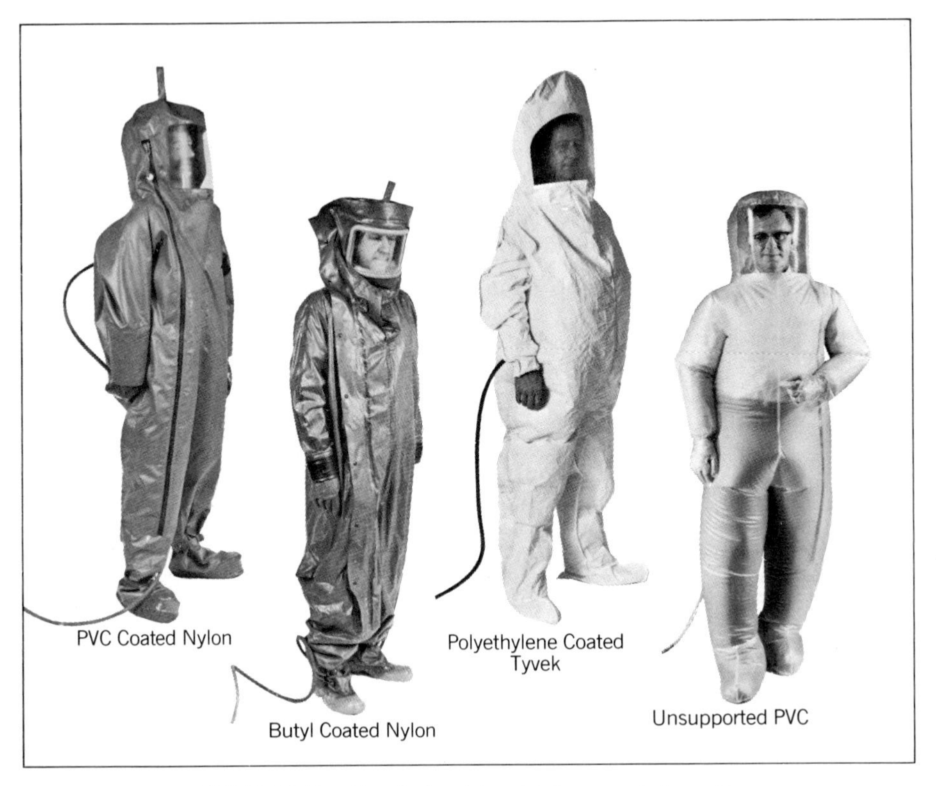

FIGURE 10.7. Chemical resistant total-encapsulating suits.

tored for the breakthrough of the chemical using such analytical techniques as gas chromatography, total hydrocarbon analysis, UV and IR spectroscopy, and others.

Breakthrough times as well as steady-state permeation rates are determined with this procedure. Both factors are important when selecting a protective garment material. Breakthrough time is an indication of how long a material can be exposed to a chemical before that chemical permeates to the inside of the material. Steady-state permeation is a measure of the rate at which a chemical passes through the material after breakthrough has occurred. One material can possess a long breakthrough time but once breakthrough has occurred, it can exhibit a very high permeation rate. This can be illustrated as follows:

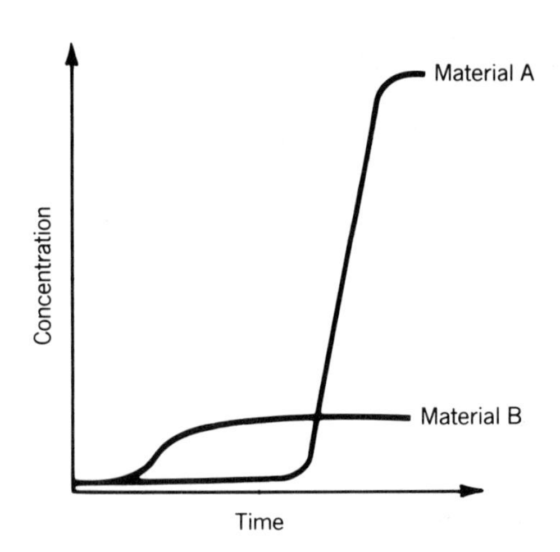

In this example, Material A resisted breakthrough for a long period of time, but once breakthrough occurred, the permeation rate was high. Material B exhibited just the opposite behavior, short breakthrough time and low permeation rate.

The ideal situation is to have a garment material with a very long breakthrough time and a very low permeation rate after breakthrough occurs. Unfortunately, this is rarely the case, and the user is faced with a decision based on a number of factors including the usage profile, the nature of the challenge material, the severity of the chemical threat, and so on.

The American Conference of Governmental Industrial Hygienists (ACGIH) publishes the *Threshold Limit Values for Chemical Substances and Physical Agents in the Work Environment.* These values are useful in assessing the risk of a worker exposed to a hazardous chemical vapor; concentrations in the workplace can often be maintained below these levels with proper engineering controls. The risk associated with exposure to chemical splashes, spills, and cleanup operations is more difficult to define, because: (1) the area of the garment in contact with the liquid chemical varies from incident to incident, and (2) no authoritative guidelines have evolved on an acceptable permeation rate or flux through garments.

At a recent ASTM symposium on protective clothing, Forsberg (1984) suggested a classification based on permeation rate:

Class	Level of Permeation Rate	Permeation Rate ($mg/m^2/min$)
1	Very low	<10
2	Low	11-100
3	Moderate	101-1000
4	High	1001-10,000
5	Very high	>10,000

Schlatter (1982) has suggested a similar classification:

Classification	Permeation Rate ($mg/m^2/min$)
Excellent	<9
Very good	<90
Good	<900
Fair	<9000
Poor	<90,000
Not recommended	>90,000

These are qualitative ratings, and do not address the relative toxicity of each individual hazardous chemical. However, they are initial steps in the recognition of the need to classify garment materials according to permeation rate, and it is projected that a standardized classification system will evolve. However, until manufacturers and suppliers test permeation of garment materials in the same fashion and report classifications in a uniform, standardized manner, the health and safety professional will be faced with assessing the protective characteristics of different garment materials and making subjective decisions on the best information available.

10.4 GARMENT SELECTION PROCESS

10.4.1 Introduction

The CPC selection process includes evaluation of the chemical threat, a description of the usage and exposure profile, and an estimate of the degree of protection required. Once these items have been addressed, the type of CPC required and the preferred garment material can be specified. Other factors such as body cooling, air supply, and communications also must be considered.

10.4.2 Evaluation of the Chemical Threat

The first step in selecting a protective garment is to determine the degree of threat imposed by the hazardous chemical. The threat is a function of a number of factors, including:

1. Chemical or chemical family.
2. Physical form: solid, liquid, vapor.
3. Physiological action: irritant, corrosive, toxic, carcinogen.

4. Extent of hazard: Threshold Limit Value (TLV), Permissible Exposure Level (PEL), Short Term Exposure Limit (STEL).

This information, along with the exposure profile of the worker, is used to assess the degree of protection required.

Chemicals of the same family tend to affect the body in the same fashion, for example, mineral acids cause irritations or burns; aromatic hydrocarbons attack the central nervous systems, liver, and kidneys; and organophosphates act as cholinesterase inhibitors. It should be noted, however, that members of the same family may present different degrees of hazard. The TLVs for the aromatic hydrocarbons benezene, toluene, and xylene are 10, 100, and 100 ppm, respectively. The mineral acids have different TLVs depending on how they attack the skin. Pesticides such as parathion and malathion are of the same chemical family as some of the chemical warfare agents, yet the agents are extremely more toxic than the pesticides.

How does a safety engineer or an industrial hygienist establish the degree of chemical threat? Manufacturers and routine users of chemicals should have little problem with this, since they are aware of the hazards associated with the chemical being produced and used. Other safety personnel such as those employed by EPA, OSHA, USCG, and local fire departments are faced with a greater dilemma: they may not be familiar with the hazards associated with the great variety of chemicals that may be encountered during spill cleanup and disposal and/or during rescue operations.

Sources are available that reveal that types of threats presented by various chemicals:

- ACGIH *Threshold Limit Values*
- Manufacturers' Material Safety Data Sheets (Available from chemical manufacturers)
- Chemical Hazardous Response Information System (CHRIS)
- OSHA Guide to Chemical Hazards
- RTECS (Tathen and Lewis, 1983)
- HAZARDLINE (Bransford, 1984)

Information from these sources will assist in identifying the chemical and/or chemical class, determining the mode of action, estimating the toxicity of the chemical, and establishing the degree of threat.

10.4.3 Usage Profile

In selecting a protective garment, the usage profile must be considered. Typical profiles, in order of increasing severity, would include but not be limited to:

1. Normal production
2. Splash/escape
3. Entry/maintenance
4. Cleanup/disposal

Exposures during routine production activities will be minimal, and are likely to be limited to vapor concentrations at or below the PEL. Other assigned work activities could result in intermittent exposures when the worker is splashed with a hazardous chemical. The worker would normally exit the area, use a safety shower, and take off the protective clothing. Maintenance activities require entry into an area that might be contaminated with a hazardous chemical and might result in frequent exposures; the possibility of additional exposure to the chemical could result from leaking plumbing, a valve failure, and so on. Workers involved in cleanup and disposal operations must expect long-term, direct, and continuous exposure of the protective ensemble to hazardous chemicals. Firemen may be exposed to both liquid and high-vapor concentrations of unidentified chemicals. Given that the chemical threat and the usage profile have been defined, how does one determine the degree of protection?

10.4.4 Degree of Protection

Under a U.S. Coast Guard sponsored program, a rationale was established for defining the degree of protection for the Response Team members (MSA Report CG-D-58-80). References were reviewed, and ultimately the Toxic Hazards Rating according to Sax, 1975, was used as the primary basis for recommending the degree of protection required. Table 10.1 lists the ratings developed by Sax, and Table 10.2 presents guidelines for degree of protection.

Note that the recommended degrees of protection were based on the assumption that the worker would be exposed to the liquid or gaseous hazardous chemical for periods of up to two hours; this would be a typical usage profile for members of a cleanup Response Team. Lesser degrees of protection may be acceptable for short-term exposure, but this is a subjective decision required of safety personnel.

TABLE 10.1. Toxic Hazard Ratings According to Sax

Rating	Effect	Comments
0	None	(a) No harm under any conditions; (b) Harmful only under unusual conditions or overwhelming dosage.
1	Slight	Causes readily reversible changes which disappear after end of exposure.
2	Moderate	May involve both irreversible and reversible changes not severe enough to cause death or permanent injury.
3	High	May cause death or permanent injury after very short exposure to small quantities.
4	Unknown	No information on humans considered valid by authors.

TABLE 10.2. Guidelines for Degree or Protection Required

Toxicity Rating	Effect of Exposure	Protective Response[a]
0	Skin absorption or irritation	No requirement
1	Skin absorption or irritation	Nonsealed
2	Skin absorption or irritation	Nonsealed
3	Skin absorption or irritation	Fully encapsulating suit[b]

[a]Carcinogens or suspected carcinogens require a fully encapsulating suit.

[b]Fully encapsulating sealed suit indicates need for air supply.

The study established that of the 985 hazardous chemicals on the CHRIS list, 403 would require fully encapsulating suits. Based on a pass-fail criterion of three hours with no breakthrough, butyl rubber exhibited the following performance characteristics:

- Acceptable: 155
- Nonacceptable: 187
- Unidentified: 61

This indicates that alternative garment materials would be required to provide adequate protection against the complete list of chemicals; fortunately, other garment materials are available.

10.4.5 Garment Material Selection

Numbers of studies on garment materials have been and are being conducted by government agencies as well as private industry. In the past, material ratings were based on exposure/degradation tests that did not reveal the permeation resistance of a material against a chemical, and in some cases resulted in erroneous decisions. Studies have shown that rapid permeation may occur with no visible evidence of material degradation, and conversely, the material may resist permeation in spite of being severely damaged.

Recently permeation testing has become the standard for rating garment materials. An ACGIH publication, *Guidelines for the Selection of Chemical Protective Clothing,* contains a comprehensive listing of protective garment materials chemically resistant to permeation. However, one should not assume that butyl is butyl is butyl. The thickness of the garment materials, whether the material is supported or unsupported, and even individual formulations control both breaktime and permeation rate. Thus, in selecting a garment material, the manufacturer should be queried regarding the resistance of *his* specific material against the chemical of interest.

Table 10.3 presents one manufacturer's chemical resistance chart for a fully encapsulating ensemble of either Vautex (Viton on nylon on neoprene) or Betex (butyl on nylon on neoprene) (MSA Data Sheet 13-02-07). Breakthrough times in minutes are listed and a five-level rating system, ranging from Excellent to Not Recommended, is used. Since these are fully encapsulating ensembles, and the wearer would be limited to an in-suit time of less than three hours, any material/chemical combination showing a breakthrough time of greater than 180 minutes is given an Excellent rating.

TABLE 10.3. Chemical Resistance Chart

Chemical Resistance Chart	23 mil 19.1 oz/yd^2 Vautex		19 mil 15.9 oz/yd^2 Betex	
	Breakthrough time in minutes	Rating	Breakthrough time in minutes	Rating
Strong mineral acid:				
HCl (concentrated) hydrochloric acid	480+	E	480+	E
H_2SO_4 (concentrated) sulfuric acid	480+	E	480+	E
HF (concentrated) hydrofluoric acid	480+	E	480+	E
H_2SO_4, fuming, 30% SO_3 dissolved	480+	E	480+	E
Strong oxidizing acid:				
HNO_3	480+	E	480+	E
Chlorine:				
Chlorine	480+	E	480+	E
Aromatic hydrocarbon:				
Benzene	210	E	8	N/R
Halogenated hydrocarbon:				
Ethylene dibromide	480+	E	45	P
Aliphatic amine:				
Isobutyl amine	33	P	75	F
Cyclic ester:				
Tetrahydrofuran	7	N/R	10	N/R
Aldehyde:				
Isodecaldehyde	480+	E	480+	E
Substituted allyl:				
Allyl Alcohol	480+	E	480+	E
Aromatic amine:				
Aniline	480+	E	480+	E
Nitrile:				
Acetonitrile	45	N/R	480+	E
Acrylate:				
Ethyl acrylate	12	N/R	60	F
Ketone:				
Methyl ethyl ketone	4	N/R	5	N/R
Alcohol:				
Methyl alcohol	480+	E	480+	E
Phenol:				
Phenol	480+	E	480+	E
Ammonia:				
Ammonia gas	390	E	480+	E
Aliphatic hydrocarbon:				
Gasoline	480+	E	20	N/R

Key to rating:
E, Excellent—No breakthrough for minimum of 3 hours (180 min)
G, Good—No breakthrough for minimum of 2 hours (120 min)
F, Fair—No breakthrough for minimum of 1 hour (60 min)
P, Poor—No breakthrough for minimum of 1/2 hour (30 min)
NR, N/R—Means challenge substance may break through fabric and expose wearer to skin contact within a few minutes after exposure to the challenge substance.

Source: MSA Data Sheet 13-02-07.

10.4.6 Ensemble Design

Many garmet designs are available, with various options including types of closures, techniques of seam fabrication, provisions for a self-contained breathing apparatus (SCBA), and so on. With nonencapsulating suits, such features are of minimal importance. On the other hand, they are very important with totally encapsulating suits where the maximum degree of protection is required.

Where the degree of protection required is low and the use of a nonencapsulating suit, for example, a splash suit, is deemed to be acceptable, some penetration of vapor or aerosol can be tolerated. In such cases, stitched seams and button or snap closures are acceptable. Where a high degree of protection is required and fully encapsulating suits are to be used, penetration of a hazardous chemical cannot be tolerated. Impervious seams are required and may include the following fabrication techniques:

1. Thermal bonding
2. Impact bonding
3. RF bonding
4. Stitched and glued
5. Stitched and taped

A high degree of protection also establishes the need for such penetration-resistant closures as:

1. Double offset zippers
2. Zippers with an overlay
3. Zipper/Zip-lok® configurations

Consideration is being given to methods of determining the penetration resistance afforded by various suit designs. Corn oil aerosol has been used to measure the protection factors of suit designs to liquid aerosol penetration. Sulfur hexafluoride (SF_6) and Freon have been used to determine the resistance of seams and closures to penetration by chemical vapors. A kit for pressurizing the suit and monitoring pressure loss to determine garment integrity is available. ASTM is currently developing a test procedure using either a soap bubble method or ammonia gas to determine leakage of encapsulating garments.

10.4.7 Accessory Equipment

Once the threat has been established, the degree of protection defined, the garment material selected and the suit design specified, accessory equipment must be considered. This includes such items as:

1. Overgloves and boots
2. Communication system
3. Respiratory support (supplied or self-contained)
4. Cooling
5. Visor

Cooling is of particular importance with fully encapsulating suits, since heat stress will limit the functional time within the garment. When a supplied-air umbilical is available, forced air or a vortex cooling tube can be used (Figure 10.8). When an umbilical cannot be used, ice vests or a cooled liquid circulating system can supply the necessary cooling.

Although the visor is an integral part of an encapsulating suit, options do exist with respect to the material:

- Polycarbonate
- PVC
- Acetate
- Surlyn
- Methyl methacrylate

Selection of the proper visor material requires the same consideration of permeation resistance as does the suit material. In addition, visor material may become fogged or crazed, so a replaceable visor is important.

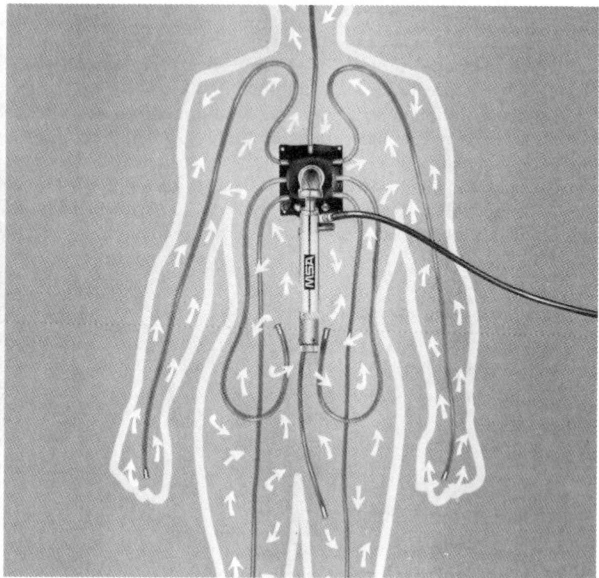

FIGURE 10.8. Vortex tube cooling system.

10.4.8 Selection Process

The information needed to recommend a specific protective ensemble has been presented. Now I will give examples of how to utilize this information, using a questionnaire developed by MSA (Figures 10.9 and 10.10). This was an actual exercise in garment selection with only the name of the purchaser being deleted. In the first instance, it was established that workers would be exposed to liquid and vapor forms of aniline on a daily, routine basis. Aniline is an irritant and corrosive to the skin; the information sources cited suggested that total body protection would be required for the intended usage of the chemical. All sources of information indicated that butyl rubber was permeation resistant to aniline. The review of the factors impacting on suit selection led to the recommendation of a specific garment: a fully encapsulating butyl ensemble.

In the second case, the worker was exposed intermittently to the possibility of being splashed with acetic acid. In the event of a splash, the worker would immediately exit the work area and activate a nearby deluge shower. Thus, although the chemical is a strong irritant, the usage profile suggested that only splash protection would be required, and therefore a splash suit (pants and jacket) was recommended. A variety of materials resist permeation by acetic acid. Since PVC is the least expensive of the candidate protective materials, a Sun-ray suit (PVC coated) was recommended as the most cost-effective means of providing splash protection to the user. No visor material was specified since the suit is nonencapsulating, but a face shield was recommended.

In these two examples, sufficient information was available on the chemical threat, the usage profile, and the garment material's resistance to the chemical. Unfortunately, this is not always the case. In many instances, there will be insufficient information on the health effects of the chemical and perhaps no information on the type of material suitable for use with the chemical. In such cases, it would be prudent to provide maximum protection, that is, full body encapsulation, as an initial procedure; subsequent experience could indicate that a lesser degree of protection would be acceptable. In addition, swatches of candidate materials should be tested against the chemical to provide guidance in selection of the material to be used for the ensemble.

10.5 USE AND CARE OF PROTECTIVE CLOTHING

10.5.1 Use of Protective Clothing

As with any item of personal protective equipment, proper training is required to ensure that the equipment is used properly. Manufacturer's recommendations should be followed closely, and routine training sessions should be instituted. Proper training in the use of CPC includes:

1. Procedures for donning the equipment.
2. Methods of sealing (snaps, buttons, zippers).
3. Use of accessory items (communications, SCBA).
4. Procedures for doffing the equipment.

With nonencapsulating suits, donning, for example, the pants and coat is an exercise similar to normal dressing. Donning a fully encapsulating ensemble is a more difficult procedure, and frequently requires the use of a "buddy system." Accessory equipment such as overgloves, overboots, body cooling, breathing apparatus, and communications equipment further complicates the donning procedure. Assistance from a coworker will ensure that all the accessory equipment is installed and that the ensemble has been correctly donned and sealed. The buddy system will also minimize the time required for suiting up.

Prior to donning an item of CPC, the user should inspect the garment and accessory equipment. Any imperfections such as tears, split seams, punctures, cracked or crazed lenses, and so on, dictate that the ensemble should not be used. Blistered or tacky surfaces would indicate that the suit is not acceptable for use.

When wearing a suit in a contaminated environment, the user should avoid snagging or tearing the garment. If the integrity of the suit is compromised, the user should exit the area, remove the suit, and wash any contaminant from the skin.

Once the mission is completed, the user will exit the area and prepare to doff the suit. Again the buddy system is recommended to assist the user in decontamination and ensemble removal. The garment should be washed with, at least, water and/or a water-detergent mixture to remove gross surface

PROTECTIVE GARMENT QUESTIONNAIRE

I. COMPANY INFORMATION:

NAME ______________________

ADDRESS ______________________

____

CONTACT ______________________ TITLE ______________________

II. CHEMICAL INFORMATION:

CHEMICAL NAME Aniline

TRADE OR GENERIC NAME ______________________

CHEMICAL CLASS Aromatic amine

PHYSICAL FORM:

SOLID ______

LIQUID X

VAPOR X

COMBINATION ______

TLV or PEL 10 mg/m3 STEL 20 mg/m3

2 ppm 5 ppm

THREAT:

NUISANCE ______

IRRITANT X

CORROSIVE X

PERCUTANEOUS X

TOXIC X

FIGURE 10.9. Protective garment questionnaire (aniline).

Protective Garment Questionnaire (continued)

III. EXPOSURE PROFILE:

EXPOSURE: VAPOR ____X____
SPLASH ____X____
DELUGE ________

FREQUENCY: LIMITED ________
INTERMITTENT ________
CONTINUOUS __X__

USAGE: STANDARD ________
ESCAPE ________
MAINTENANCE/REPAIR ________
ENTRY/CLEANUP ________
OTHER Open pans containing 2 or more gallons

IV. DEGREE OF PROTECTION:

SAXS RATING 3 - Strong irritant
OSHA GUIDELINE Impervious clothing
USCG (CG-D-58-80) Sealed suit
MATERIAL SAFETY DATA SHEET Impervious clothing
TYPE OF PROTECTION REQUIRED Fully encapsulating suit

FIGURE 10.9. *(Continued)*

contamination. Additional neutralization with baking soda following an acid exposure or vinegar following a base exposure is recommended. After gross surface contamination has been removed, the ensemble can be removed with the assistance of a fellow worker. These procedures will minimize the possibility of exposure by both the user and assistant during doffing of the ensemble.

10.5.2 Decontamination and Storage

The process of surface decontamination, which has been discussed, involves removal of superficial contamination from the outer portion of the garment. In some cases, the chemical may have partially penetrated into the garment material, and in such instances surface decontamination is ineffective in removing the sorbed chemical. The chemical is free to permeate in either direction within the material, and can ultimately migrate to the inner surface. Future donning of the ensemble could result in contamination of the user's skin.

Although some studies have been done on decontamination, particularly in the military, little information, on decontamination procedures for reusable chemical protective clothing is available. Laundering with water and detergent is ineffective in removing absorbed chemical from elastometer material. Freon cleaning has been used with mixed results, and it haws been reported that Freon may degrade some garment materials. Warm air will expedite the permeation of the chemical through the material, and remove contamination that has migrated to the surface. Guidelines for decontamination procedures are limited at this time, although an ASTM Committee on Protective Clothing has formed a task force to develop procedures for decontamination.

Protective Garment Questionnaire (continued)

V. GARMENT RECOMMENDATION:

BEST AVAILABLE MATERIAL BASED ON:

MSA IN-HOUSE TESTING :	CHEMICAL	Aniline
	MATERIAL	B
	BREAK TIME	8+ hours
ACGIH	CHEMICAL	Aniline
	MATERIAL	B,NR, Neo, Nit
	RATING	RR
HAZARDLINE	CHEMICAL	Aniline
	MATERIAL	B, NR, Neo
	RATING	Excellent/good
MATERIAL RECOMMENDATION		B
MSA TRADENAME		Chempruf I or Chempruf II
VISOR		Polycarbonate

VI. ACCESSORIES:

COMMUNICATION	Yes
SCBA	
AIR SUPPLY	Yes
COOLING	Yes - Vortex

Key:
B = Butyl
CPE = Chlorinated polyethylene
Neo = Neoprene
Nit = Nitrile
NR = Natural rubber
PU = Polyurethane
PVC = Polyvinyl chloride
Sar = Saranex
V = Viton
RR = Break time greater than 1 hr

FIGURE 10.9. *(Continued)*

After protective garments have been doffed and decontaminated, they must be stored properly to ensure that suit integrity is maintained. Elastomeric materials are subject to degradation by heat and UV light, and should be stored in a cool, dark area, in a hanging position to avoid creating stress at creases.

10.6 SUMMARY

The science of selecting the appropriate chemical protective clothing for a given application is in its infancy. However, manufacturers, suppliers, and users are becoming increasingly aware of the capabilities as well as the limitations of protective equipment. It is now recognized that no single garment material will provide protection against all hazardous chemicals, and that some types of material provide better protection against a specific chemical than other types. Manufacturers and suppliers are providing increasingly more detailed information on the performance of garment materials. Users are endeavoring to select the proper type of ensemble to provide the worker with an adequate degree of protection.

Existing technology can help in rational selection of the proper protective clothing design and material. Basically, the steps required to achieve the user's goal of selecting a garment include:

- Define the chemical threat.
  - Identify the chemical or chemical class.
  - Specify the physical form of the chemical, that is, solid, liquid, or vapor.
  - Acquire information on the mode of action of the chemical, that is, irritant, corrosive, or toxic.
  - Search the literature for supporting information such as TLV or STEL.

PROTECTIVE GARMENT QUESTIONNAIRE

I. COMPANY INFORMATION:

NAME ____________________

ADDRESS ____________________

____

CONTACT ____________________ TITLE ____________________

II. CHEMICAL INFORMATION:

CHEMICAL NAME Glacial acetic acid

TRADE OR GENERIC NAME Vinegar acid

CHEMICAL CLASS Carboxylic acid

PHYSICAL FORM:

SOLID ____________________

LIQUID X

VAPOR X

COMBINATION ____________________

TLV or PEL 25 mg/m^3 STEL 37 mg/m^3

10 ppm 15 ppm

THREAT:

NUISANCE ____________________

IRRITANT X

CORROSIVE X

PERCUTANEOUS ____________________

TOXIC ____________________

III. EXPOSURE PROFILE:

EXPOSURE: VAPOR ____________________

SPLASH X

DELUGE ____________________

FREQUENCY: LIMITED X

INTERMITTENT ____________________

CONTINUOUS ____________________

USAGE: STANDARD X (bulk liquid transfer)

ESCAPE ____________________

MAINTENANCE/REPAIR ____________________

ENTRY/CLEANUP ____________________

OTHER ____________________

FIGURE 10.10. Protective garment questionnaire (glacial acetic acid).

Protective Garment Questionnaire (continued)

IV. DEGREE OF PROTECTION:

SAXS RATING 3 - Strong irritant

OSHA GUIDELINE Impervious clothing

USCG (CG-D-58-80) Sealed suit

MATERIAL SAFETY DATA SHEET Rubber gloves, goggles

TYPE OF PROTECTION REQUIRED Splash suit

V. GARMENT RECOMMENDATION:

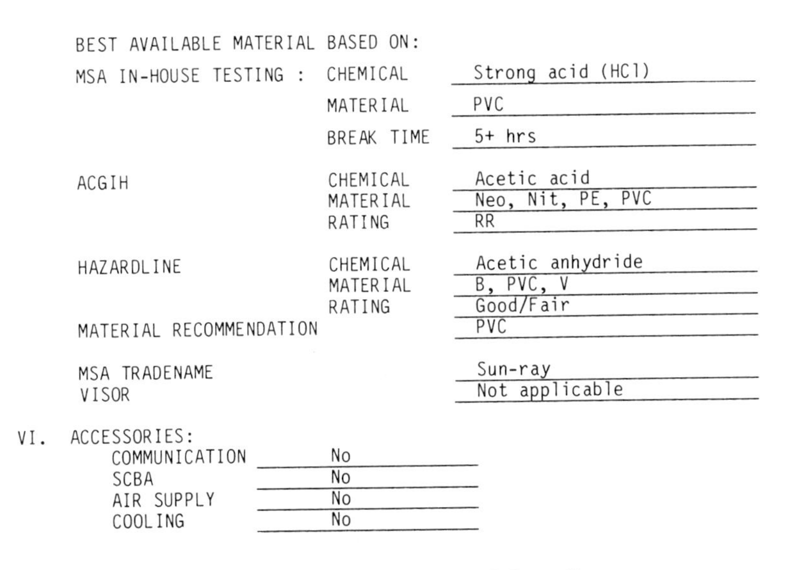

BEST AVAILABLE MATERIAL BASED ON:

MSA IN-HOUSE TESTING :	CHEMICAL	Strong acid (HCl)
	MATERIAL	PVC
	BREAK TIME	5+ hrs
ACGIH	CHEMICAL	Acetic acid
	MATERIAL	Neo, Nit, PE, PVC
	RATING	RR
HAZARDLINE	CHEMICAL	Acetic anhydride
	MATERIAL	B, PVC, V
	RATING	Good/Fair
MATERIAL RECOMMENDATION		PVC
MSA TRADENAME		Sun-ray
VISOR		Not applicable

VI. ACCESSORIES:

COMMUNICATION	No
SCBA	No
AIR SUPPLY	No
COOLING	No

Key:
B = Butyl
CPE = Chlorinated polyethylene
Neo = Neoprene
Nit = Nitrile
NR = Natural rubber
PU = Polyurethane
PVC = Polyvinyl chloride
Sar = Saranex
V = Viton
RR = Break time greater than 1 hr

FIGURE 10.10. *(Continued)*

- Define the exposure profile.
  - Determine whether the user will be exposed to vapor, splash, or deluge of the chemical.
  - Specify the frequency of exposure, that is, limited, intermittent, or continuous.
  - Specify the type of exposure, for instance, escape, maintenance/repair, or entry/clean-up.
- Determine the degree of protection required.
  - OSHA guidelines.
  - Material safety data sheet.
  - User experience.
  - Other.
- Select garment design and material.
  - Vendor's data sheet.
  - ACGIH *Guildelines for the Selection of Chemical Protective Clothing.*
  - HAZARDLINE.

TABLE 10.4. List of Vendors

Allied Glove Corp., 325 E. Chicago St., Milwaukee, WI 53202
American Optical Corp., Safety Products Division, 14 Mechanic Street, Southbridge, MA 01550
American Scientific Products, Div. of American Hospital Supply, 1430 Waukegan Road, McGaw Park, IL 60085
Ansell Corporation, Drawer 3221, Newport, Wilmington, DE 19804
*Arbill Industries, 2207 W. Glenwood Avenue, Philadelphia, PA 19132
B. F. Goodrich, Engineered Systems Division, 500 S. Main St., Department 0716, Akron, OH 44318
Barry Manufacturing Co., Ltd. 920 Lakeshore Road East, Mississauga, Ontario, Canada
Body-Guard, Division of Lion Uniform Co., 3323 W. Warner Avenue., Santa Ana, CA 92702
Boss Manufacturing Company, 221 W. First Street, Kewanee, IL 61443
Cesco Safety Products, 100 E. 16th Street, P.O. Box 1237, Kansas City, MO 64141
Charkate Glove and Specialty Company, 2615 123rd Street, Flushing, NY 11354
*Chem-Pro of East Wind, Inc., 33 Bassett Street, Clayton, DE 19938
Clean Room Products, 56 Penataquit Ave., Bay Shore, NY 11706
Cofish International, Ind., P.O. Box 13, East Haddam, CT 06423
Comasec, Drawer 10, Enfield, CN 06082
Dayton Flexible Products, 2210 Arbor Boulevard, Dayton, OH 45439
Defense Apparel, 286 Murphy Road, Hartford, CT 06114
Direct Safety Company, 6005 Martway, P.O. Box 2994, Shawnee Mission, KS 66202
*Disposables, Inc., 14 Locust St., Manhasset, NY 11030
Dorsey Safe-T-Shoe Company, 1222 Market Street, Chattanooga, TN 37401
Dow Chemical Company, 2020 Dow Center, Midland, MI 48640
*Durafab, Inc., Box 658, Cleburne, TX 76031
E. I. duPont deNemours & Co., Inc. Elastomers Division, Wilmington, DE 19898
Eastco Industrial Safety Corp., 26-15A 124th Street, Flushing, NY 11354
*Encon Manufacturing Co., Box 3826, Houston, TX 77253
Edmont Div. Beckton, Dickinson & Co., 1300 Walnut Street, Coshocton, OH 43812
Fairway Products, 303 Arch Street, Hillsdale, MI 49242
Falcon Industries, Inc., 401 Isom Road, San Antonio, TX 78216
*Fire-Chem Mfg., Inc., Jasper and York Streets, Philadelphia, PA 19125
Fisher Scientific Company, 711 Forbes Avenue, Pittsburgh, PA 15219
Frommelt Industries, Inc., Safety Products Division, P.O. Box 658, 2455 Kerper Boulevard, Dubuque, IA 52001
General Scientific Equipment Company, Lilmkiln Pike and Williams Avenue, Philadelphia, PA 19150
Glover Latex, Inc., Box 167, 514 S. Rose Street, Anaheim, CA 92805
Goodyear Rubber Products Corp., 329 McCarter Highway, Newark, NJ 07114
Granet Div., ESB, Inc., 25 Loring Drive, P.O. Box 588, Framingham, MA 01701
Greene Rubber Co., 160 Second Street, Cambridge, MA 02138
Hodgman, 207 E. Wolf St., Yorkville, IL 60560
Holcomb Safety Garment Co., 328 S. Jefferson St., Chicago, IL 60606
Hub Safety Equipment, 121 Liberty St., Box 454, South Quincy, MA 02269
*ILC Cover, P.O. Box 266, Frederica, DE 19946
Industrial Products Co., 21 Cabot Blvd., Langhorne, PA 19047
Industrial Safety and Equipment, 1376 Newbrecht Street, Lima, OH 45801
Intek Industries, 333 Sylvan Avenue, Englewood Cliffs, NJ 07632
Interex Corp., 3 Strathmore Rd., Natick, MA 01760
International Playtex Corp., Glove Division, 700 Fairfield Avenue, Stamford, CT 06902
Jomac Products, Inc., 863 Easton Rd., Warrington, PA 18976
Jordan David Safety Products, P.O. Box 400, Warrington, PA 18976
Kappler Disposables, Inc., P.O. Box 218, Guntersville, AL 35976
Kimberly-Clark Corp., Industrial Garments Fabrics, 1400 Holbomb Bridge Rd., Roswell, GA 30076
La Cross Rubber Mills Co., Box 1328, La Cross, WI 54601

TABLE 10.4. ***(continued)***

Magid Glove and Safety Mfg. Co., Inc., 2060 N. Kolmar Ave., Chicago, IL 60639
*Major Safety Service, Inc., 4500 Patent Rd., Norfolk, VA 23502
*Melco, Inc., 6603 Governor Printz Blvd., Wilmington, DE 19809
Miller Products Co., 209 Warren St., New York, NY 10007
*Mine Safety Appliances Co., 600 Penn Center Blvd., Pittsburgh, PA 15235
Monte Glove Company, Drawer 409, Maben, MS 39750
*National Draeger, Inc., Box 120, Pittsburgh, PA 15230
Neese Industries, Inc., P.O. Box 628, Gonzales, LA 70737
Norton Co., Safety Products Division, 2000 Plainfield Pike, Cranston, RI 02920
Oak Medical Supply Co., The Oak Rubber Company, 219 S. Sycamore St., Ravenna, OH 44266
OKI Supply Co., 7584 Reihold Dr., Cincinnati, OH 45237
Pendergast Safety Equipment Co., 6913 Tulip St., Philadelphia, PA 19135
Pioneer Industrial Products, Division of Brunswick Corp., 512 East Tiffin St., Willard, OH 44890
Plastex Protective Products, Inc., 9 Grand St., Garfield, NJ 02706
Plastimayd Corp., 2216 S.E. Seventh Ave., Portland, OR 97214
Protexall Company, Box 307, Green Lake, WI 54941
Pulmosan Safety Equipment Corp., Box 869, Flushing, NY 11354
PPG Industries, Inc., Chemicals Group, One Gateway Center, Pittsburgh, PA 15222
Rainfair, Inc., P.O. Box 1647, Racine, WS 53401
Ranger Rubber Co., 1100 E. Main St., Endicott, NY 13760
Record Industrial Co., 1020 Eighth Avenue, Box 407, King of Prussia, PA 19406
Renco Corp., 2060 Fairfax Avenue, Cherry Hill, NJ 08003
Robar Protective Products, 2213 W. Glenwood Avenue, Philadelphia, PA 19132
*Safety Clothing and Equipment Co., 4900 Campbell Road, Willoughby, OH 44094
W. H. Salisbury & Co., Box 1060, 7420 N. Long Avenue, Skokie, IL 60077
Singer Safety Co., 3800 N. Milwaukee Ave., Chicago, IL 60641
Standard Glove & Safety Equipment Corp., 34301 Lakeland Drive, Eastlake, OH 44094
*Standard Safety Equipment Co., Box 188, 431 N. Quentin Rd., Palatine, IL 60078
Steel Grip Safety Apparel Co., Inc., 700 Garfield St., Box 833, Danville, IL 61832
Surety Rubber Co., Box 97-T-3, Carrollton, OH 44615
SGL Homalite Industries, 11 Brookside Drive, Wilmington, DE 19804
Texier Glove Company, Inc., Box 166, 839 Highway 22, North Plainfield, NJ 07061
Tingley Rubber Corp., P.O. Box 100, South Plainfield, NJ 07080
Tracies Co., 100 Cabot St., Holyoke, MA 01040
United States Safety Service Co., 1535 Walnut St., P.O. Box 1237, Kansas City, MO 64108
Unitex Products, Inc., Division of Angelica Inc., Box 5310, Philadelphia, PA 19142
Vidaro Corp., 333 Martinel Dr., P.O. Box 535, Kent, OH 44240
*Wheeler Protective Apparel, Inc., 4330 W. Belmont Ave., Chicago, IL 60641
Worklon, Division Superior Surgical Mfg. Co., Seminole Blvd., 100th Terrace, Seminole, FL 33542

*Fully encapsulating suit.

- Specify accessory equipment
  - Communication.
  - SCBA.
  - Air supply.
  - Cooling.

Selecting the proper protective garment for a specific application can be accomplished best through user-vendor interaction. The user is best qualified to specify the requirements for the suit, and the vendor can make recommendations on the type of garment material, as well as the specific item of protective clothing which will best meet the user's needs. A list of vendors of CPC is presented in Table 10.4; suppliers of fully encapsulating ensembles are indicated with an asterisk.

REFERENCES

ACGIH, *Guidelines for the Selection of Chemical Protective Clothing,* Vol. I: Field Guide; Vol. II: Technical and Reference Manual, 1983.

ACGIH, *Threshold Limit Values for Chemical Substances in Work Air,* 1982.

ASTM, Resistance of Protective Clothing Materials to Permeation by Hazardous Liquid Chemicals, ASTM Designation: F739-81.

Bransford, J. B., Jr., Database Provides Information on Hazardous Chemicals, *Industrial Hygiene News,* May 1984.

CHRIS Hazardous Chemical Data, Department of Transportation, M16465.12, U.S. Government Printing Office, October 1978.

Dangerous Properties of Industrial Materials, 4th ed., Sax, 1975.

DHHS (NIOSH) Publication No. 81-123, *Occupational Health Guidelines for Chemical Hazards,* U.S. Government Printing Office.

Forsberg, K., The Development of Criteria for Selection of Chemical Protective Clothing Based on Permeation Data and Risk Assessment, Presented at the International ASTM Symposium on the Performance of Protective Clothing, Raleigh, NC, July 1984.

MSA Research Corp., Material Development Study for a Hazardous Chemical Protective Clothing Outfit, Report No. CG-D-58-80, August 1980.

Schlatter, C. N., Preliminary Guide to the Permeation Resistance of Edmont Gloves and Protective Clothing, May 1982.

Tathen, R. L. and R. J. Lewis, Sr., Eds., *Registry of Toxic Effects of Chemical Substances,* Vol. 1, U.S. Dept. of health and Human Services, June 1983.

PART V

OCCUPATIONAL SAFETY AND HEALTH INFORMATION

CHAPTER 11

How to Locate Information Sources for Occupational Safety and Health

THEODORE F. SCHOENBORN

JERRY NEWMAN

RICHARD A. LEMEN

National Institute for Occupational Safety and Health

Cincinnati, Ohio

11.14.23 Chemical Engineering Abstracts
11.14.24 Chemical Exposure
11.14.25 Chemical Regulations and Guidelines System
11.14.26 Chemicals in Human Tissues and Fluids
11.14.27 CHEMLINE
11.14.28 CHEMNAME
11.14.29 CHEMSEARCH
11.14.30 CHEMSIS
11.14.31 CHEMZERO
11.14.32 CIS
11.14.33 CISDOC
11.14.34 CLAIMS COMPOUNDS REGISTRY
11.14.35 CLAIMS/UNITERM
11.14.36 COLD
11.14.37 COMPENDEX
11.14.38 Conference Papers Index
11.14.39 CROSS
11.14.40 CTCP
11.14.41 Current Biotechnology Abstracts
11.14.42 DBI
11.14.43 DIALINDEX
11.14.44 DIRLINE
11.14.45 Dissertation Abstracts On Line
11.14.46 DOE Energy
11.14.47 EDB
11.14.48 EEIS
11.14.49 EI Engineering Meetings
11.14.50 EIX (Compendex)
11.14.51 EMI
11.14.52 ENERGYLINE
11.14.53 ENVIROLINE
11.14.54 Environmental Bibliography
11.14.55 ERIC
11.14.56 ETI
11.14.57 EUCAS
11.14.58 EXCERPTA MEDICA
11.14.59 FEDREG
11.14.60 FOODS ADLIBRA
11.14.61 FRSS
11.14.62 FSTA
11.14.63 HAZARDLINE
11.14.64 Health Audio-Visual Online Catalog
11.14.65 HSELINE
11.14.66 INSPEC
11.14.67 International Pharmaceutical Abstracts
11.14.68 IRCS
11.14.69 ISMEC
11.14.70 Kirk-Othmer Encyclopedia of Chemical Technology
11.14.71 Laboratory Hazards
11.14.72 LABORDOC
11.14.73 Life Science Collection
11.14.74 MEDLINE
11.14.75 Mental Health Abstracts
11.14.76 MESH
11.14.77 METADEX
11.14.78 NASA
11.14.79 NASA Tech Briefs
11.14.80 NCMH
11.14.81 NIOSHTIC
11.14.82 NRC
11.14.83 NSIC
11.14.84 NTIS
11.14.85 Nursing and Allied Health
11.14.86 OHMTADS
11.14.87 PAPERCHEM

11.1 INTRODUCTION

All professions increasingly suffer from an information problem because of the vast amount of scientific information available. Occupational safety and health, because it crosses many disciplines, suffers more than most. The problem is compounded by the fact that the disciplines are not closely related conceptually. The analysis of a specific occupational safety and health problem may require drawing on information from such widely disparate subject areas as medicine, industrial hygiene, behavioral science, and engineering. The regulatory environment that dominates the field contributes another significant barrier to the selection and use of information.

Thus, interdisciplinary barriers significantly hinder the transfer of needed information relevant to the practice of occupational safety and health. At the same time, the legalities involved increase the need for exhaustive information regarding hazards, corrective actions, and worker protection.

This chapter is intended to serve as a guide for individuals who must keep up with the information explosion in occupational safety and health. The materials are organized on the premise that two distinctly different kinds of information are of interest. First, there is information on standards, regulations, industrial and labor policies, workers' compensation, and related matters. Second, there is the specific technical information used to support standards development and hazard control programs. These two types of information are quite different and are identified using different approaches and information sources. Each, however, contributes to an overall understanding of the field.

The section describing technical information (11.3) presents the most useful sources of identifying the bulk of the world's literature in specific subject areas. Here the relevant computerized data bases are listed first. By pursuing the leads identified in the data base descriptions one can be reasonably assured of obtaining more than three-fourths of the world literature pertinent to the inquiry. The data bases cited are listed according to subject area and serve to narrow a search to those citations most likely to contribute significant information. Only those data bases comprised primarily of English language articles were included in this summary because English has become the language of science, especially in occupational safety and health. However, since not every information problem in occupational safety and health requires exhaustive information, key reference texts and key journals are listed by subject area to direct the user to the most appropriate source of general reference information.

The information sources used most frequently by the National Institute for Occupational Safety and Health (NIOSH) information specialists in responding to information queries are identified as the core resources. Many excellent reference texts exist to serve the field, such as the International Labour Organization's *Encyclopedia of Occupational Safety and Health* and *Patty's Industrial Hygiene and Toxicology*. If, for example, the information problem is a reference question about the effects of a hazardous agent or the constituents of an industrial product, there may be little need for access to the on-line information systems when the core reference resources are available.

The needs of the user with respect to detail and coverage dictate the appropriate methods and tools needed to obtain the desired information. If the user wants to minimize the probability of missing relevant information, then exhaustive searches of many computerized data bases are appropriate. To maximize the probability of retrieving useful information as it is published, Systematic Dissemination of Information (SDI) services are appropriate. SDI services provide relevant abstracts as they are added to a data base. They can be automatically selected on the basis of a previously established search profile. If timely information is essential, comprehensive newsletters and reporting services covering the regulatory activities are often the most appropriate. Several of these newsletters and reporting services are included in the following discussions. Certain users want to minimize the information-gathering task or require assistance in summarizing or analyzing the reference material. For this group, the standard reference texts or a specialized technical information service such as the Toxicity Data Center at Oak Ridge National Laboratories are the most appropriate methods.

This chapter is divided into two main sections to reflect the different types of information. The first (11.2) describes the general information sources that apply to the field of occupational safety and health so generally that they cannot be separated by individual subject area. Included in this section are general interest magazines and newsletters, regulatory agencies, voluntary standard-setting groups, relevant organizations, on-line data bases, sources of information about U.S. regulatory activities, and information about litigation and judicial matters. Information relevant to occupational safety and health programs, economic considerations, legal considerations, organizational considerations, policy considerations, and workers' compensation are also provided here.

The second section of the chapter (11.3) identifies the scientific and technical information sources in the field of occupational safety and health. For the purpose of organization, this section is subdivided into eleven separate categories, the first of which describes general technical reference sources relevant to the entire field. The remaining ten categories describe the information sources most relevant to the individual subject areas: behavioral sciences (11.4), chemistry/physics (11.5), engineering controls (11.6), epidemiology/surveillance (11.7), ergonomics (11.8), industrial hygiene (11.9), medicine/nursing (11.10), safety (11.11), toxicology (11.12), and training (11.13). Each category includes sources of technical information from computerized data bases, reference texts, related organizations, and technical journals.

11.2 SECTION ONE: GENERAL INFORMATION

In addition to the formal technical sources (texts, journals, and data bases), news and general feature publications about occupational safety and health can be found in the trade press and in the many newsletters that serve the field. The regulatory, economic, and scientific issues are all routinely presented in these publications. It should be remembered, however, that the best sources of information concerning the regulations are the regulatory agencies. They provide authoritative guidance concerning compliance with the standards and distribute exact copies of the regulations. The following six regulatory agencies are briefly described below: the Occupational Safety and Health Administration (OSHA); the Mine Safety and Health Administration (MSHA); the Environmental Protection Agency (EPA); the Nuclear Regulatory Commission (NRC); the U.S. Coast Guard (USCG); and the National Institute for Occupational Safety and Health (NIOSH). Strictly speaking, NIOSH is not a regulatory agency, but it is included because it makes recommendations for standards and contributes its expertise to the regulatory process.

In addition to these regulatory agencies, there are voluntary (consensus) standard-setting organizations that routinely develop recommendations in occupational health and safety. These consensus groups include the American Conference of Governmental Industrial Hygienists (ACGIH), which develops influential recommendations for workplace safety and health exposure limits; the American National Standards Institute (ANSI); and the National Fire Protection Association (NFPA). Many of the consensus standards developed by these groups were used as the basis for the occupational safety and health standards now enforced by OSHA. The Chemical Manufacturers Association, for example, operates the Chemical Transportation Emergency Center (CHEMTREC), which provides emergency advice to those who must contend with chemical spills. Although technically not a standard-setting organization, it is included in this group because of its influence and function in this area.

Supplementing the standards and guidelines issued by the official agencies are the many reporting services in occupational safety and health. These reporting services have advantages over the *Federal Register*, the official source of information on U.S. regulatory activity, in that they feature frequent updates in convenient loose-leaf binders. They also have extensive indexes to help track the development of standards over time and to provide current information about judicial interpretations of the standards. The reporting services also supplement the information about litigation and judicial interpretations of Occupational Safety and Health Review Commission (OSHRC) decisions.

Finally, the first section describes useful computer data bases and includes a bibliography of standards and regulations, general reference texts, and monographs that relate to the administrative and legal issues of occupational safety and health standards, workers' compensation benefits, and safety and health program administration. While this set of data bases and texts is of considerable interest to many occupational safety and health managers, it may be of only passing interest to those involved in the day-to-day identification, control, and prevention of occupational illness and injury.

Occupational safety and health issues are also widely reported in newspapers and newsmagazines, but the coverage is not at all comprehensive. There are, however, several nontechnical occupational safety and health trade publications that provide useful coverage of the field, especially the administrative issues, compliance issues, and, through advertisements, equipment availability. These trade publications are focused directly on news and features concerning the regulations and policies of the field.

11.2.1 Trade Publications

Chemical and Engineering News, American Chemical Society, 1155 16th Street, N.W., Washington, DC 20036. This is a weekly magazine that reports on current topics in the field of chemical engineering. Subjects include economic trends and regulatory activities that are of interest to the industry. Occupational safety and health issues and topics are routinely featured.

Health and Safety at Work, Madaren Publishers Ltd., David House, 69-77 High Street, Craydon C9 1QH, England. This British trade publication focuses on industrial health and safety concerns, industry news, and current topics.

Industrial Engineering, Institute of Industrial Engineers, 25 Technology Park, Atlanta, GA 30092. This trade publication focuses on concerns of the production process, process automation, and new production methods.

Issues in Science and Technology, National Academy of Sciences, National Academy of Engineering, Institute of Medicine, 2101 Constitution Avenue, Washington, DC 20418. This quarterly helps to inform the interested public and raise the quality of private and public decision making by providing a forum for discussion and debate. The discussions reflect the views of the authors, not a consensus of opinion or the opinion of any institute.

National Safety and Health News, National Safety Council, 444 N. Michigan Avenue, Chicago, IL 60611. This publication considers all aspects of safety and health issues of concern to those responsible for or interested in workplace safety and health.

Occupational Hazards, Penton IPC, 1111 Chester Avenue, Cleveland, OH 44114. This is a trade publication reporting on the practical aspects of industrial safety, health and hygiene, and plant protection. Product advertisements and short articles on occupational safety and health issues are presented.

Occupational Health, BT Journals, 1 St. Anne's Road, Eastbourne, E. Sussex, BN 21A3UN, England. This British publication provides news briefs, short discussions, and editorials related to current topics.

Occupational Health and Safety, Stevens Publishing Corp., 3700 W. Waco Drive, Box 7573, Waco, TX 76710. This monthly publication reports the news and presents current issues in occupational health. Referenced articles provide a general overview of the subject.

Occupational Safety and Health, Royal Society for the Prevention of Accidents, Occupational Safety Division, Cannon House, The Priory Queensway, Birmingham B4 A685, England. This publication covers news and occupational safety and health issues in the United Kingdom.

Ohio Monitor, Industrial Commission, Division of Safety and Hygiene, 246 N. High Street, Columbus, OH 43215. The *Ohio Monitor* discusses selected topics of interest to industry and health and safety professionals.

11.2.2 Newsletters

Newsletters are the best source of up-to-date information about developments in the field. They routinely cover new developments in standards and report on current research efforts. Many newsletters also provide guidance about program implementation and good work practices. The following newsletters are valuable sources of information about regulatory activities and individuals in the news.

Chemical Regulation Reporter, Bureau of National Affairs, 1231 25th Street, N.W. Washington, DC 20037. This weekly report provides a review of the current regulatory activities affecting manufacturers and users of chemical products.

Chemical Substances Control, Bureau of National Affairs, 1231 25th Street, N.W., Washington, DC 20037. Concerns related to the handling of chemicals, from premanufacture and use through transport and disposal, are reported in this biweekly. Sources of information, emergency procedures, record-keeping, and reporting requirements are presented.

Computers in Safety and Health, Kenneth Cohen, ed., Consulting Health Media, P.O. Box 1625, El Cajon, CA 92022. This newsletter reports on new applications of microcomputers in occupational safety and health. It concentrates on industrial hygiene applications.

Environmental Health Letter, Gershon W. Fishbein, ed./publisher, Environews, 1331 Pennsylvania Avenue, N.W., Suite 509, Washington, DC 20004. This newsletter was among the original publications in the field. It provides excellent coverage of news and issues concerning the general environment, including occupational safety and health.

The *Epidemiology Monitor*, Rodger Bernier, ed./publisher, 2861 Templar Knight Drive, Tucker, GA 30084. This publication provides current reports on activities in the field of epidemiology.

Genetic Engineering Letter, Gershon W. Fishbein, ed./publisher, Environews, 1331 Pennsylvania Avenue, N.W., Suite 509, Washington, DC 20004. This is a bimonthly publication recapping scientific issues, business concerns, and product announcements.

Genetic Technology News, Technical Insights, Kenneth A. Kovaly, publisher, 32 North Dean Street, Englewood, NJ 07631. This publication provides current news on genetic technology and a forum for discussion of related business activity.

Hazardous Waste News, Leonard A. Eiserer, publisher, Business Publishers, 951 Pershing Drive, Silver Spring, MD 20910. A weekly publication covering current events and reports related to hazardous waste, it includes summaries of legislative, governmental, and corporate activities.

Hazardous Waste Report, Aspen Systems Corporation, 1600 Research Boulevard, Rockville, MD 20805. This current topics report provides information on a range of hazardous waste issues including legislative, legal, and government activities.

Hazards Review, David Owen, ed., Elsevier International Bulletins, Elsevier Science Publishers, B.V. Amsterdam, 52 Vanderbilt Avenue, New York, NY 10017. This international bulletin reports on all matters pertaining to hazards associated with the production, processing, and use of rubber and plastic material.

Industrial Hygiene News Report, D. L. Flournoy, ed./publisher, Flournoy Publishers, Flournoy and Associates, 1845 W. Morse Avenue, Chicago, IL 60626. This publication reports on current methodology and research in the recognition, evaluation, and control of workplace hazards.

Job Safety and Health, Bureau of National Affairs, 1231 25th Street, N.W., Washington, DC 20037. This biweekly newsletter serves as an advisory bulletin on health and safety policies, procedures, and practices. Very helpful in understanding activities under the Occupational Safety and Health Act, it describes legal cases, program activities, and important developments in the field.

Job Safety & Health Report, Leonard A. Eiserer, publisher, Business Publishers, 951 Pershing Drive, Silver Spring, MD 20910. This biweekly report summarizes current occupational safety and health-related topics. Coverage includes legislation, economics, medicine, and technology.

Loss Prevention and Control, Bureau of National Affairs, 1231 25th Street, N.W., Washington, DC 20037. This biweekly information bulletin provides guidance on loss prevention procedures, management issues, and court decisions. It also covers risk identification, prevention, disaster management, and product liability.

Occupational Health and Safety Letter, Gershon W. Fishbein, ed./publisher, Environews, Inc., 1331 Pennsylvania Avenue, N.W., Suite 509, Washington DC 20004. Current activities related to occupational health and safety are highlighted in this newsletter, which provides one of the best summaries of news and issues in the field.

OSHA Compliance Letter, Bureau of Business Practice, 24 Rope Ferry Road, Waterford, CT 06385. This newsletter reports the interpretations of OSHA standards based on OSHA compliance officers' enforcement. The issues surrounding various interpretations of standards are well covered.

Pesticide and Toxic Chemical News, Food Chemical News, 1101 Pennsylvania Avenue, S.E., Washington, DC 20003. This is a weekly report on hazardous wastes, pesticides, toxic substances, and general issues of regulation and legislation.

State Regulation Report, Leonard A. Eiserer, publisher, Business Publishers, 951 Pershing Drive, Silver Spring, MD 20910. This is a biweekly report on state regulatory and disposal activities related to toxic substances and hazardous waste.

Toxic Material News, Leonard A. Eiserer, publisher, Business Publishers, 951 Pershing Drive, Silver Spring, MD 20910. Legislative and governmental regulatory activities related to toxic substances are covered in this weekly business newsletter.

TSCA Chemicals-in-Progress Bulletin, Environmental Protection Agency, Industrial Assistance Office, Washington, DC 20460. Recent developments and near-term plans concerning the Toxic Substances Control Act are discussed in this bimonthly news bulletin.

11.2.3 Regulatory Agencies

The regulatory agencies are an invaluable source of official materials and technical guidance. The principal U.S. Regulatory agency for occupational safety and health is the **Occupational Safety Health Ad-**

ministration (OSHA), an agency in the Department of Labor. It was created, as was the National Institute for Occupational Safety and Health (NIOSH), by the Occupational Safety and Health Act of 1970 (Public Law 91-596). The coverage of OSHA is very broad. It excludes only those workplaces covered by other agencies that had regulatory authority at the time of the passage of the OSHA Act. For example, the Mine Safety and Health Administration, the Nuclear Regulatory Commission, and the U.S. Coast Guard continue to provide the regulatory environment for protecting the health and safety of mine and quarry workers, nuclear plant workers, and merchant seamen. OSHA makes available to the public extensive information about compliance and guidance for good work practices through its Technical Data Center and its regional and area offices. OSHA is also responsible for training employers and employees in good work practices. A description of the OSHA Training Institute is provided in the section on training (11.13). OSHA, 200 Constitution Avenue, N.W., Washington, DC 20210.

The **Mine Safety and Health Administration** (MSHA) is responsible for the safety and health of underground miners, surface miners, and quarry workers. The MSHA, created by the Mine Safety and Health Act of 1977 (PL 95-164), is also located in the Department of Labor. Generally, the occupational safety and health standards of MSHA are similar to those enforced by OSHA except that requirements for limiting sources of ignition are considerably more stringent. Information about compliance and technical assistance is available from the MSHA regional and area offices. A description of the MSHA Training Institute is provided in the section on training (11.13). MSHA, Room 1237-A, 4015 Wilson Boulevard, Arlington, VA 22203.

The **Nuclear Regulatory Commission** (NRC) is specifically charged with promoting safety in the use, handling, and transportation of radioactive materials under the authority of the Atomic Energy Act of 1954. This includes certain work practices and the control of potential exposures in the workplace. Information about safety and health concerns associated with nuclear radiation sources can be obtained from the NRC. NRC, 1717 H Street, N.W., Washington, DC 20555.

The **Environmental Protection Agency** (EPA) does not have specific responsibility for occupational safety and health. However, it is responsible for air and water pollution, hazardous waste disposal, noise, radiation, toxic substances, and the licensing of pesticides, fungicides, and rodenticides, so it has published a wealth of information on the toxicity of chemicals and the hazards of physical agents. The EPA is an important information resource. The standard-setting activity of EPA often complements activities in the other regulatory agencies. EPA, 401 M Street, N.W., Washington, DC 20460.

The **Department of Transportation** (DOT), operating principally under the Hazardous Material Transportation Act, exercises regulatory authority to protect the public against risks associated with the transportation of hazardous materials. It regulates compliance with hazardous materials transport regulations and conducts training and education programs to support the Department's regulatory function and training for shipper and container manufacturers regarding inspection and compliance. Materials Transportation Bureau, Department of Transportation, 400 Seventh Street, S.W., Washington, DC 20590.

The **U.S. Coast Guard** (USCG), a part of the Department of Transportation, regulates the transportation of hazardous materials on navigable waters. It maintains a Chemical Hazards Response Information System (CHRIS) that provides information in the form of safety data sheets for those who work with chemical spills. This system is one of the best sources of information regarding chemical emergencies. U.S. Coast Guard, Washington, DC 20593.

The **Occupational Safety and Health Review Commission** (OSHRC), established by the Occupational Safety and Health Act of 1970, is the appellate body for compliance actions. The *Decisions of the Commission,* published annually, are an important resource because they provide elaboration of the existing standards. It is through the decisions of the Occupational Safety and Health Review Commission that the case law for OSHA compliance is established. Occupational Safety and Health Review Commission, 1825 K Street, N.W., Washington, DC 20006.

The **Federal Mine Safety and Health Review Commission** (FMSHRC), established by the Federal Mine Safety and Health Act, is the appellate body for compliance actions, issued by the Mine Safety and Health Administration. Federal Mine Safety and Health Review Commission, 6th Floor, 1730 K Street, N.W., Washington, DC 20006.

11.2.4 Research Agencies

The **National Institute for Occupational Safety and Health** (NIOSH), a federal agency established by the Occupational Safety and Health Act of 1970, holds a unique position because it is neither a regulatory agency nor a voluntary standard-setting organization. While NIOSH has mandated research functions, its principal role is to make recommendations to OSHA and MSHA concerning needed workplace standards. These recommendations, usually in the form of a report series entitled *Criteria for a Recommended Standard for Occupational Exposure to . . .* provide a basis for establishing new or revised standards. These criteria documents often recommend lower exposure limits than the enforceable limits established by OSHA. The recommendations also include guidance for good work practice and personal protection. To date, over 100 of these documents have been produced and transmitted to the Department of Labor. National Institute for Occupational Safety and Health, 1600 Clifton Avenue, Atlanta, GA 30333.

The **Bureau of Mines** (BOM), in the Department of Interior, was established to promote mine safety and health. The Bureau conducts research in mine safety both in its own facilities and through research grants. Included among these programs are projects by the states to enlarge or intensify safety education programs and to implement programs for the advancement of health and safety in mines. Bureau of Mines, 2401 E Street, N.W., Washington, DC 20241.

11.2.5 Voluntary Standards Organizations

The voluntary standards organizations occupy an important place in the occupational safety and health arena. Their consensus standards were the basis for most of the regulations now enforced by OSHA. Under the provisions of the Occupational Safety and Health Act of 1970, the national consensus standards in force in 1968 were adopted as the OSHA standards.

The **American National Standards Institute** (ANSI) is the widest-ranging national consensus standard-setting organization. It serves as a clearinghouse for nationally coordinated voluntary safety, engineering, and industrial standards and identifies as national standards those standards developed by appropriate groups. These standards include definitions, terminology, symbols and abbreviations, performance criteria, and methodology. ANSI efforts include standards related to construction, chemical manufacturing, and exposure limits for industrial chemicals. These standards represent the attainable levels of protection agreed upon by a broad cross section of industrial, academic, and health experts. The standards produced by ANSI are established using a consensus methodology. ANSI also represents United States interests in international standardization work and publishes the *Reporter, Standards Action,* and *Catalog of Standards.* ANSI, 1430 Broadway, New York, NY 10018.

The **American Conference of Governmental Industrial Hygienists** (ACGIH) produces the well-known *List of Threshold Limit Values* (TLV's®). The TLV's® reflect the professional judgment of ACGIH membership and are adopted throughout the world as enforceable limits. The ACGIH also produces many technical publications, which are described in the technical information section. Information about the TLV® list and other publications can be obtained by writing ACGIH. ACGIH, P.O. Box 1937, Cincinnati, OH 45201.

The **National Fire Protection Association** (NFPA) produces the National Fire Code using the national consensus method. This influential code was adopted by OSHA as part of its enforceable standards. The code provides significant information about safe practices in dealing with flammables and explosives. NFPA membership is drawn from the fire service, business, industry, and interested individuals in the fields of insurance, government, architecture, and engineering. The Association develops and publishes standards prepared by technical committees and conducts fire safety education programs. It also provides information on fire protection; compiles annual statistics on causes and occurrences of fires, fire deaths and fire fighter casualties; runs a field service on technical issues; and sponsors seminars. Publications and educational materials include *Fire Command, Fire News, Fire Journal,* and *Fire Technology* as well as technical committee reports, catalogs of publications, and visual aids. NFPA, Batterymarch Park, Quincy, MA 02269.

11.2.6 Official Regulatory Agency Announcements and Notices

In addition to the individual agency information sources, the newsletters, and the trade publications, the official information sources warrant attention. These are discussed in order of their specificity and rank in terms of their official standings. They include the *Federal Register* and the *Code of Federal Regulations.* The information in these sources is routinely reported by the professional and specialized reporting services and the data bases focusing on specific regulatory and guidance information.

The *Federal Register* is the official source of information on U.S. regulatory activity. This invaluable source of information contains notices of hearings, notices of proposed rule making, and final rules (regulations) covering the entire range of federal activity.

The *Code of Federal Regulations* (CFR) is the annual codification of the final rules published in the *Federal Register.* A running index of the CFR parts affected is published daily in the *Federal Register.* The chapters of the CFR that apply most to occupational safety and health are Titles 29 and 30. The *Federal Register* and the *Code of Federal Regulations* can be obtained from the Superintendent of Documents, U.S. Government Printing Office (GPO), Washington, DC 20402.

11.2.7 Reporting Services

Because the *Federal Register* covers the entire range of federal rule making, from tariffs to rules of the Federal Power Commission, various occupational safety and health reporting services provide summaries of the federal regulatory activities pertaining to this field. These reporting services not only compile the regulations as they are promulgated based on the law, but also annotate the regulations and legislation based upon the case laws established as the result of decisions of the review commissions and the federal courts. These reporting services are the easiest way to stay abreast of the regulatory developments in the field. The most important are listed below.

Chemical Regulation Reporter, Bureau of National Affairs, 1231 25th Street, N.W., Washington, DC 20037. This weekly report contains information on legislative, regulatory, and private sector activities relating to chemical and pesticide controls in the environment and in industry. Information is provided on testing and manufacturing, records, transportation, and disposal. Documentation includes the text of regulations and laws.

Employment Safety and Health Guide, Commerce Clearing House, 4025 W. Peterson Avenue, Chicago, IL 60646. This service describes the legal, regulatory, and administrative action arising from the implementation of the Occupational Safety and Health Act, the Mine Safety and Health Act, and other legislation governing safety and health in employment. It follows the decisions of the review commissions and reports on related regulatory activity. Current legislation, regulatory agency activities, congressional actions, and decisions of the Occupational Safety and Health Review Commission are covered in this weekly report.

Environment Reporter, Bureau of National Affairs, 1231 25th Street, N.W., Washington, DC 20037. Information on air and water pollution, solid and hazardous waste disposal, land use, control of radiation hazards, surface mining, and natural resource utilization is presented in the *Environment Reporter.* Technical developments, legislative activity, judicial decisions, and regulatory agency actions are also covered in this weekly.

Hazardous Materials Transportation, Bureau of National Affairs, 1231 25th Street, N.W., Washington, DC 20037. Documentation of rules and reports on current developments in the transportation of hazardous materials are provided in this monthly bulletin. Specific information is provided on the requirements for the various modes of transport including rail, air, ship, highway, and pipeline. Also provided is a cross-reference table for rules covering thousands of materials.

Index to Government Regulation, Bureau of National Affairs, 1231 25th Street, N.W., Washington, DC 20037. Decisions of the Department of Energy, Consumer Product Safety Commission, the EPA, OSHA, DOT, and Food and Drug Administration concerning the handling and use of chemicals are reported in this index. Information on chemicals is indexed by generic names, industrial names, and the Chemical Abstract Service Registry Number.

International Environment Reporter, Bureau of National Affairs, 1231 25th Street, N.W., Washington, DC 20037. This monthly service provides information on developments in environmental protection and pollution control. The multivolume file documents the text of laws and regulations and includes other official materials.

International Hazards Materials Transport Manual, Bureau of National Affairs, 1231 25th Street, N.W., Washington, DC 20037. This service publishes monthly bulletins on the requirements for transporting hazardous substances worldwide. The manual includes a list of 3000 commonly carried hazardous materials cross-referenced for easy use by hazard class, U.N. identification number, subsidiary risk, etc.

Mine Safety and Health Reporter, Bureau of National Affairs, 1231 25th Street, N.W., Washington, DC 20037. Safety and health regulations in mining and MSHA activities are reported biweekly. Copies of relevant legislation as well as federal Mine Safety and Health Review Commission decisions are provided.

Noise Regulation Reporter, Bureau of National Affairs, 1231 25th Street, N.W., Washington, DC 20037. This service updates information on noise control laws and regulations biweekly. Industrial noise as well as motor carrier, highway, airport, aircraft, and railroad noise are covered. Also included is information on audiometric testing, noise control technology, and record-keeping requirements.

Occupational Disease Digest, National Council on Compensation Insurance (NCCI), One Penn Plaza, New York, NY 10119. This report summarizes current judicial decisions, legislative and committee activities, and trends with respect to occupational diseases.

Occupational Safety and Health Reporter, Bureau of National Affairs, 1231 25th Street, N.W., Washington, DC 20037. This is a comprehensive source of information on compliance, prevention, and changing requirements. Each weekly issue includes reports on standards, variances, research, policies, and labor relations. In addition, a two-volume reference set is updated regularly.

OSHA Compliance Reporter, Commerce Clearing House, Chicago, IL 60646. This report provides comprehensive information on OSHA compliance activities and the decisions of the Occupational Safety and Health Review Commission (OSHRC) and the courts.

Safety and the Work Force, National Council on Compensation Insurance (NCCI), One Penn Plaza, New York, NY 10119. This report provides an analysis of compensation benefits and benefit administration as they relate to employee and employer claims activity.

U.S. Occupational Safety and Health Review Commission: Administrative Law Judge and Commission Decisions, OSHRC Subscription Address: Superintendent of Documents, U.S. Government Printing Office, Washington, DC 20402. This is the official compilation of the decisions of the Occupational Safety and Health Review Commission.

Workers' Compensation Cost Containment Guide, National Council on Compensation Insurance

(NCCI), One Penn Plaza, New York, NY 10119. This guide provides articles and discussions on cost containment issues including defense council selection, returning the employee to work, and other techniques to reduce the cost of workers' compensation.

Workers' Compensation Law Reporter, Commerce Clearing House, Chicago, IL 60646. This report provides ready access to information on workers' compensation decisions including administrative agencies, executive agencies, and the courts. These decisions involve both occupational disease and employers' liability. Workers' compensation statute summaries and occupational disease statutes are also provided.

11.2.8 Computer Data Bases

In addition to the reporting services, several on-line computer data bases provide ready access to the standards and regulations covering occupational safety and health. Those that provide U.S. and international standards or information about federal rule making are listed here. Data bases that provide technical information primarily about the sciences supporting occupational safety and health are included in section two of this chapter.

The explosion of scientific information has made it virtually impossible for the practitioner to obtain access to all of the information published. Not only is the volume of information too great to effectively review, the cost of the journals is prohibitive. Only large university libraries can be expected to have a significant portion of the literature. As a result, electronic information retrieval systems have become a necessity for those seeking information on all but the most common topics.

Until recently, five data base vendors based in the United States have provided virtually all of the files of interest to the occupational safety and health professional. These are Dialog Information Services (3460 Hillview Avenue, Palo Alto, CA 94304); the National Library of Medicine (NLM) (8600 Rockville Pike, Bethesda, MD 20209); DOE RECON (Department of Energy, Technical Information Center, P.O. Box 62, Oak Ridge, TN 37830); Bibliographic Retrieval Services (BRS) (BRS Information Technologies, 1200 Route 7, Latham, NY 12110); and System Development Corporation (SDC) Search Services (System Development Corporation, 2500 Colorado Avenue, Santa Monica, CA 90406). New vendors that were originally available only in Europe now make their information systems available in the United States. These include Pergamon Infoline (Pergamon International Information Corporation, 1340 Old Chain Bridge Road, McLean, VA 22101); Questel (1625 Eye Street, N.W., Suite 719, Wash-

TABLE 11.1. Regulatory Information Data Bases

Name	Producer	No. of Records	Coverage
CHEMICAL REGULATIONS AND GUIDELINES SYSTEM (DIALOG)	CRC Systems, Inc.	12,000/yr	81-present
CIS (Questel)	International Labour Organization/CIS	2,100/yr	74-present
CISDOC (ESA)	International Labour Organization/CIS	2,100/yr	74-present
FEDREG (SDC)	Capitol Services International	35,000/yr	77-present
FRSS (CIS, ICIS)	Federal Register Search System	160,000 References to proposed or final rules	77-present
HAZARDLINE (BRS)	Occupational Health Services, Inc.	3,000 chemical substances	Not available (N.A.)
HSELINE (Pergamon, ESA)	Health and Safety Executive	47,000	77-present
OHMTADS (CIS, ICIS)	EPA, Superfund	1,334 substances	N.A.
RTECS (NLM, CIS, ICIS)	National Institute for Occupational Safety and Health (NIOSH)	70,029	N.A.

Source: Data in this table were obtained from catalogs and manuals of the identified vendors. Specific citations for each data base and vendor are given in the alphabetical descriptions provided in section 11.14.

ington, DC 20006); and the Information Retrieval Service of the European Space Agency Information Retrieval Service, ESA-IRS, (Esrin, P.O. Box 64, 1-00044, Frascati, Italy). Hundreds of different data bases on subjects ranging from art history to zoology are available from these vendors.

Two other vendors, Information Consultants (ICI) (1133 15th Street, N.W., Suite 300, Washington, DC 20005) and Chemical Information Systems (CIS) (7215 York Road, Baltimore, MD 21212), offer similar highly specialized information systems (ICI and ICIS) containing data about regulations and the physical characteristics and toxicity of various chemicals. These systems differ from the ones mentioned above in that they provide data with references rather than abstracts and bibliographic information.

Each data base vendor has a different mix of data bases. Even when the same data base is available from different vendors, the periods covered and the implementation can be quite different. There is a trend for the vendors to include larger portions of the data bases in single files as the costs of computer storage decline. The following table lists the data bases that contain regulatory information important to occupational safety and health personnel. The availability by vendor and the periods of coverage are included for reference.

11.2.9 Occupational Safety and Health Programs

Many books discuss issues surrounding the delivery of occupational safety and health services. Information about policy and economic considerations, which in large measure provide the incentive for committing the personnel and financial resources to control workplace illnesses and injuries, is not often found in the newsletters, reporting services, or sources of technical information. Individual data bases and journals are also limited in their coverage of these issues. Reference texts are the most useful source of this information. This topic, therefore, is included under general information sources rather than as part of the technical information section.

The reference texts presented here mainly address the issues and concerns of administrators and organizers of worker occupational health and safety functions. The bibliography that follows covers the traditional management concerns as well as the relationships among occupational safety and health functions and other employee benefits, especially the health-related benefits. Much of what is presented originated in an effort to reduce health care costs, to increase productivity, and to improve relations between employers and employees.

The references have been grouped according to economic, legal, organizational, policy, and workers' compensation considerations.

Reference Texts: Economic Considerations

This section lists references on the economic impact of occupational safety and health standards, including rising health care costs.

Baram, M. S., "The Use of Cost-Benefit Analysis in Regulatory Decision-Making is Proving Harmful to Public Health," in *Management of Assessed Risk for Carcinogens,* U.S. Nicholson, ed., The New York Academy of Sciences, New York, NY, 1981.

Bolle, M. J., *OSHA Reform: An Economic Analysis of the Occupational Safety and Health Act,* Report No. 81-162E, U.S. Congress, Library of Congress, Congressional Research Service, Washington, DC, August 6, 1981.

Charles River Associates, *Economic and Environmental Analysis of the Current OSHA Lead Standard,* CRA Project No. 5356.60, contract report prepared for the Occupational Safety and Health Administration, U.S. Department of Labor, September 1982.

Egdahl, R. H., and D. C. Walsh, eds., *Containing Health Benefit Costs: The Self-Insurance Option,* Springer Series on Industry and Health Care, Vol. 6, Springer-Verlag, New York, NY, July 1979.

Egdahl, R. H., and D. C. Walsh, eds., *Industry and Health Care, Vol. 2: Health Cost Management and Medical Practice Patterns,* Ballinger Publications, Cambridge, MA, May 1985.

Egdahl, R. H., *Payer, Provider, Consumer: Industry Confronts Health Care Costs,* Springer Series in Industry and Health Care, No. 1, Springer-Verlag, New York, NY, December 1977.

Freedman, A., *Industry Response to Health Risk,* the Conference Board, New York, NY, 1981.

Green, M., and N. Waitzman, *Business War on the Law: An Analysis of the Benefits of Federal Health/Safety Enforcement,* 2nd ed., Corporate Accountability Research Group, Washington, DC, 1981.

Hattis, D., C. Mitchell, J. McCleary-Jones, et al., *Control of Occupational Exposures to Formaldehyde: A Case Study of Methodology for Assessing the Health and Economic Impacts of OSHA Health Standards,* report prepared for the U.S. Department of Labor, Massachusetts Institute of Technology, Center for Policy Alternatives, Cambridge, MA, April 1981.

Hickey, J. L. S., M. D. Wright, J. L. Warren, et al., Research Triangle Institute, *Abrasive Blasting, Technological Feasibility Assessment and Economic Impact Analysis of the Proposed Federal Standard,* Contract No. J-9-F-6-0225, report prepared for the Occupational Safety and Health Administration, U.S. Department of Labor, Washington, DC, September 1978.

Interdepartmental Workers' Compensation Task Force, *Research Report of the Interdepartmental Workers' Compensation*, U.S. Government Printing Office, Washington, DC, June 1979.

Lanzillotti, R. F., ed., *Economic Effects of Government-Mandated Costs*, University Presses of Florida, Gainesville, FL, 1977.

MacLean, D., and M. Sagoff, "A Critique of Cost-Benefit Analysis as a Technique for Determining Health Standards under the Occupational Safety and Health Act," (Working Paper #7) in *Preventing Illness and Injury in the Workplace*, contract report prepared for the Office of Technology Assessment, U.S. Congress, Washington, DC, April 1983.

Mishan, E. J., *Cost-Benefit Analysis*, Praeger Publisher, New York, NY, 1971.

Ruttenberg, R., "Compliance with the OSHA Cotton Dust Rule: The Role of Productivity Improving Technology," (Working Paper #18) in *Preventing Illness and Injury in the Workplace*, contract report prepared for the Office of Technology Assessment, U.S. Congress, Washington, DC, March 1983.

Settle, R. F., *Benefits and Costs of the Federal Asbestos Standard*, Contract No. B-9-M-5-0540, report prepared for the Office of the Assistant Secretary for Policy, Evaluation, and Research, U.S. Department of Labor, Washington, DC, February 1975.

Swartzman, D., R. Liroff, and K. Croke, *Cost-Benefit Analysis and Environmental Regulations*, Conservation Foundation, Washington, DC, 1982.

Warner, K. E., and B. R. Luce, *Cost-Benefit and Cost Effectiveness Analysis in Health Care*, Health Administration Press, Ann Arbor, MI, 1982.

Wiedenbaum, M. L., *The Costs of Government Regulation of Business*, study prepared for the Subcommittee on Economic Growth and Stabilization, Joint Economic Committee, U.S. Congress, U.S. Government Printing Office, Washington, DC, 1978.

REFERENCE TEXTS: LEGAL CONSIDERATIONS

The legal considerations references cover product liability and disease litigation as well as the legal aspects of compliance. They also provide general guidance to employers with respect to occupational safety and health law.

Baram, M. S., and J. R. Miyares, *The Legal Framework for Determining Unreasonable Risk from Carcinogenic Chemicals*, report prepared for the Office of Technology Assessment, U.S. Congress, Washington, DC, May 16, 1980.

Cohen, H., U.S. Congress, Library of Congress, Congressional Research Service, "Products Liability: A Legal Overview," Issue Brief No. IB77021, Washington, DC, October 7, 1982.

Connerton, M., and M. MacCarthy, *Cost-Benefit Analysis and Regulation: Expressway to Reform or Blind Alley?* National Policy Exchange, Washington, DC, 1982.

Harter, P., "Non-Regulatory Legal Incentives for the Adoption of Occupational Safety and Health Control Technologies," (Working Paper #8) in *Preventing Illness and Injury in the Workplace*, contract report prepared for the Office of Technology Assessment, U.S. Congress, Washington, DC, April 1983.

Hemenway, D., "Monitoring and Compliance: The Political Economy of Inspection," in Breit, W., ed., *Political Economy and Public Policy*, Vol. 4, JAI Press, April 1985.

Kelman, S., *Regulating America, Regulating Sweden*, MIT Press, Cambridge, MA, 1981.

Mintz, B. W., *OSHA: History, Law, and Policy*, Bureau of National Affairs, Washington, DC, 1984.

Nothstein, G., *The Law of Occupational Safety and Health*, Macmillan, The Free Press, New York, NY, 1981.

Rothstein, M. A., *Legal Issues Raised by Biochemical and Cytogenetic Testing in the Workplace*, contract report prepared for the Office of Technology Assessment, U.S. Congress, Washington, DC, 1982.

Rothstein, M. A., *Occupational Safety and Health Law*, 2nd ed., West Publishing Co., St. Paul, MN, 1983.

Smith, R. S., *The Occupational Safety and Health Act: Its Goals and its Achievements*, American Enterprise Institute, Washington, DC, 1976.

U.S. Congress, General Accounting Office, *Informal Settlement of OSHA Citations: Comments on the Legal Basis and other Selected Issues*, HRD-85-11, Washington, DC, October 26, 1984.

REFERENCE TEXTS: ORGANIZATIONAL CONSIDERATIONS

Organizational considerations include concerns and issues of occupational safety and health administration that relate to effectively protecting the health and safety of workers on the job.

Asfahl, R., ed., *Industrial Safety and Health Management*, rev. ed., January 1984.

Bacow, L. S., *Bargaining for Job Safety and Health*, MIT Press, Cambridge, MA, 1980.

Bird, F., *Management Guide to Loss Control*, Institute Press, Atlanta, GA, 1974.

Chelius, J. R., and R. S. Smith, "Experience Rating and Injury Prevention," in J. D. Worrall, ed., *Safety and the Work Force*, ILR Press, Ithaca, NY, 1983.

Coe, C. K., *Cutting Costs with a Safety Program*, University of Georgia Institute of Government, Athens, GA, 1977.

Egdahl, R. H., ed., *Background Papers on Industry's Changing Role in Health Care Delivery*, Springer Series on Industry and Health Care, No. 3, Springer-Verlag, New York, NY, November 1977.

Egdahl, R. H., and D. C. Walsh, eds., *Corporate Medical Departments: A Changing Agenda*, Industry and Health Care Series, Vol. 1, Ballinger Publishers, Cambridge, MA, April 1983.

Egdahl, R. H., and D. C. Walsh, eds., *Industry and HMOs: A Natural Alliance*, Springer Series on Industry and Health Care, Vol. 5, Springer-Verlag, New York, NY, April 1978.

Fessenden-Raden, J., and B. Gert, eds., *A Philosophical Approach to the Management of Occupational Health Hazards*, Transaction Books, New Brunswick, NJ, 1984.

Gardner, A. W., ed., *Current Approaches to Occupational Health*, Vol. 2, PSB Publishing Company, 1982.

Goble, R., D. Hattis, M. Ballew, and D. Thurston, "Implementation of the Occupational Lead Exposure Standard" (Working Paper #16) in *Preventing Illness and Injury in the Workplace*, contract report prepared for the Office of Technology Assessment, U.S. Congress, Washington, DC, November 1983.

Kent, M. B., "A History of Occupational Safety and Health in the United States," (Working Paper #4) in *Preventing Illness and Injury in the Workplace*, contract report prepared for the Office of Technology Assessment, U.S. Congress, Washington, DC, April 1983.

Kochan, T., L. Dyer, and D. Lipsky, *The Effectiveness of Union-Management Safety and Health Committees*, W. E. Upjohn Institute for Employment Research, Kalamazoo, MI, 1977.

Mendeloff, J., "An Analysis of OSHA Health Inspection Data" (Working Paper #2) in *Preventing Illness and Injury in the Workplace*, contract report prepared for the Office of Technology Assessment, U.S. Congress, Washington, DC, April 1983.

Policy Research Incorporated, *Protecting Workers' Health: Federal, State and Local Compliance-Assistance Programs*, Contract No. HHS-100-79-0172, report submitted to the Office of the Assistant Secretary for Planning and Evaluation, Department of Health and Human Services, January 1981.

Priest, W. C., "Computer Teleconferencing as a Mechanism to Improve Information Transfer for Workplace Safety and Health," (Working Paper #12) in *Preventing Illness and Injury in the Workplace*, contract report prepared for the Office of Technology Assessment, U.S. Congress, Washington, DC, May 1983.

U.S. Congress, General Accounting Office, *Labor Needs to Manage Its Workplace Consultation Program Better*, HRD-78-155, U.S. Government Printing Office, Washington, DC, December 18, 1978.

U.S. Congress, General Accounting Office, *How Effective Are OSHA's Complaint Procedures?*, HRD-79-48, U.S. Government Printing Office, Washington, DC, April 9, 1979.

U.S. Congress, General Accounting Office, *Workplace Inspection Program Weak in Detecting and Correcting Serious Hazards*, HRD-78-34, U.S. Government Printing Office, Washington, DC, May 19, 1978.

U.S. Congress, House Committee on Interstate and Foreign Commerce, Subcommittee on Oversight and Investigations, *Cost-Benefit Analysis: Wonder Tool or Mirage?*, Committee Print No. 96-IFC 62, U.S. Government Printing Office, Washington, DC, December 1980.

U.S. Congress, Office of Technology Assessment, *Technologies and Management Strategies for Hazardous Waste Control*, U.S. Government Printing Office, Washington, DC, March 1983.

U.S. Department of Labor, Occupational Safety and Health Administration, "Employer Abatement Assistance," Instruction No. CPL 2.53, U.S. Department of Labor, Washington, DC, July 15, 1982.

U.S. Department of Labor, Occupational Safety and Health Administration, *OSHA: Meeting its Mandate*, U.S. Department of Labor, Washington, DC, 1983.

U.S. Department of Labor, Occupational Safety and Health Administration, "Health Standards; Methods of Compliance," *Federal Register*, 48:7473–7476, U.S. Government Printing Office, Washington, DC, February 14, 1983.

Walsh, D. C., and R. H. Egdahl, eds., *Women, Work and Health: Challenge to Corporate Policy*, Springer Series on Industry and Health, Vol. 8, Springer-Verlag, October 1980.

REFERENCE TEXTS: POLICY CONSIDERATIONS

Policy considerations are public trends and attitudes that affect the implementation or operation of the safety and health function in industry.

Ashford, N. A., *Crisis in the Workplace: Occupational Disease and Injury,* MIT Press, Cambridge, MA, 1976.

Ashford, N., C. Hill, W. Mendez, et al., *Benefits of Environmental, Health, and Safety Regulation,* study prepared by the Center for Policy Alternatives, Massachusetts Institute of Technology, for the U.S. Senate, Committee on Governmental Affairs, Committee Print, Washington, DC, March 25, 1980.

Ashford, N. A., D. Hattis, G. Heaton, et al., *Evaluating Chemical Regulations: Trade-Off Analysis and Impact Assessment for Environmental Decision-Making,* contract report to the Council on Environmental Quality and the Environmental Protection Agency by the Center for Policy Alternatives, Massachusetts Institute of Technology, Cambridge, MA, July 1980.

Badaracco, J. L., Jr., *A Study of Adversarial and Cooperative Relationships Between Business and Government in Four Countries,* prepared for the Office of Technology, Strategy, and Evaluation, U.S. Department of Commerce, Washington, DC, 1981.

Bacow, L. S., *Bargaining for Job Safety and Health,* MIT Press, Cambridge, MA, 1980.

Berman, D. M., *Death on the Job: Occupational Health and Safety Struggles in the United States,* Monthly Review Press, New York, NY, 1978.

Bluestone, B., and B. Harrison, *The Deindustrialization of America,* Basic Books, New York, NY, 1982.

Bolle, M. J., "OSHA Reform: Dismantlement or Overhaul?" Issue Brief No. IB81068, U.S. Congress, Library of Congress, Congressional Research Service, Washington, DC, September 30, 1982.

Brown, M. S., "Setting Occupational Health Standards: The Vinyl Chloride Case," in *Controversy: Politics of Technical Decisions,* 2nd ed., D. Nelkin, ed., Sage Publications, Beverly Hills, CA, 1984.

Bureau of National Affairs, *Right to Know: A Regulatory Update on Providing Chemical Hazard Information,* Bureau of National Affairs, Washington, DC, 1985.

Bureau of National Affairs, *Safety Policies and the Impact of OSHA,* Bureau of National Affairs, Washington, DC, 1977.

Cornell, N., R. Noll, and B. Weingast, "Safety Regulations," in *Setting National Priorities: The Next Ten Years,* H. Owen and C. L. Schultze, eds., Brookings Institution, Washington, DC, 1976.

Doniger, D. D., *The Law and Policy of Toxic Substances Control,* Resources for the Future, Washington, DC, 1978.

Egdahl, W. H., and D. C. Walsh, eds., *Health Services and Health Hazards: The Employee's Need to Know,* Springer Series on Industry and Health Care, Vol. 4, Springer-Verlag, New York, NY, 1978.

Goldbeck, W. B., *A Business Perspective on Industry & Health Care,* Springer Series on Industry and Health Care, Vol. 2, Springer-Verlag, 1978.

Goldbeck, W. B., *Industry's Voice in Health Policy,* Springer Series on Industry and Health Care, Vol. 7, Springer-Verlag, 1979.

Green, M., and N. Waitzman, *Business War on the Law: An Analysis of the Benefits of Federal Health/Safety Enforcement,* 2nd ed., Corporate Accountability Research Group, Washington, DC, 1981.

Kneese, A. V., and C. L. Schultze, *Pollution, Prices, and Public Policy,* Brookings Institution, Washington, DC, 1975.

Lave, J. R., and L. B. Lave, "Decision Frameworks to Enhance Occupational Health and Safety Regulations," (Working Paper #6) in *Preventing Illness and Injury in the Workplace,* contract report prepared for the Office of Technology Assessment, U.S. Congress, Washington, DC, February 1983.

Lave, L. B., *The Strategy of Social Regulation,* Brookings Institution, Washington, DC, 1981.

MacAvoy, P., ed., *OSHA Safety Regulations: Report of the Presidential Task Force,* American Enterprise Institute, Washington, DC, 1977.

MacLaury, J., ed., *Protecting People at Work: A Reader in Occupational Safety and Health,* U.S. Department of Labor, Washington, DC, 1980.

MacLean, D., "A History of the Consideration of Economic Impacts in the Determination of Feasibility of Standards Under the OSH Act," (Working Paper #5) in *Preventing Illness and Injury in the Workplace,* contract report prepared for the Office of Technology Assessment, U.S. Congress, Washington, DC, April 1983.

McCaffrey, D. P., *OSHA and the Politics of Health Regulation,* Plenum Press, New York, NY, 1982.

Mendeloff, J., *Regulating Safety: An Economic and Political Analysis of Occupational Safety and Health Policy,* MIT Press, Cambridge, MA, 1979.

Miller, J. C., III, and B. Yandle, *Benefit-Cost Analyses of Social Regulation*, American Enterprise Institute, Washington, DC, 1979.

Page, J., and M. O'Brien, *Bitter Wages: Ralph Nader's Study Group Report on Disease and Injury on the Job*, Grossman, New York, NY, 1973.

Schultze, C. L., *The Public Use of Private Interest*, Brookings Institution, Washington, DC, 1977.

Schwartz, J. B., and S. S. Epstein, "Problems in Assessing Risk from Occupational and Environmental Exposure to Carcinogens," in *Banbury Report 9: Quantification of Occupational Cancer*, R. Peto and M. Schneiderman, eds., Cold Spring Harbor Laboratory, New York, NY, 1981.

Simon, P. J., *Reagan in the Workplace: Unraveling the Health and Safety Net*, Center for Study of Responsive Law, Washington, DC, 1983.

Smith, R. S., "Protecting Workers' Health and Safety," in *Instead of Regulation*, R. W. Poole, Jr., ed., Lexington Books, D.C. Heath & Co., Lexington, MA, 1982.

Tolchin, S. J., and M. Tolchin, *Dismantling America: The Rush to Deregulate*, Houghton Mifflin, Boston, MA, 1983.

U.S. Congress, General Accounting Office, *Slow Progress Likely in the Development of Standards for Toxic Substances and Harmful Physical Agents Found in Workplaces*, B-163375, U.S. Government Printing Office, Washington, DC, September 28, 1973.

U.S. Congress, General Accounting Office, *States' Protection of Workers Needs Improvement*, HRD-76-161, U.S. Government Printing Office, Washington, DC, September 9, 1976.

U.S. Congress, House Committee on Interstate and Foreign Commerce, Subcommittee on Oversight and Investigations and Subcommittee on Consumer Protection and Finance, *Use of Cost-Benefit Analysis by Regulatory Agencies*, Joint Hearings, July 30, October 10 and 24, 1979, U.S. Government Printing Office, Washington, DC, 1980.

Wallick, F., *The American Worker: An Endangered Species*, Ballantine Books, New York, NY, 1972, republished with additional material under the title *Don't Let Your Job Kill You*, Progressive Press, Washington, DC, 1984.

Reference Texts: Workers' Compensation

The workers' compensation references presented here address both the economic incentives that influence implementation of workplace safety and health programs and the various private and public attempts to assess the effect of workers' compensation on the health and welfare of employees.

Barth, P. S., with H. A. Hunt, *Worker's Compensation and Work-Related Illnesses and Diseases*, MIT Press, Cambridge, MA, 1980.

Burton, J. F., "Compensation for Permanent Partial Disabilities," in J. D. Worral, ed., *Safety and the Work Force*, ILR Press, Ithaca, NY, 1983.

Castrovinci, J. L., "Prelude to Welfare Capitalism: The Role of Business in the Enactment of Workmen's Compensation Legislation in Illinois, 1905–12," in F. R. Borel and S. J. Diner, eds., *Compassion and Responsibility*, University of Chicago Press, Chicago, IL, 1980.

Chelius, J. R., "An Empirical Analysis of Safety Regulation," in *Supplemental Studies*, National Commission on State Workmen's Compensation Laws, U.S. Government Printing Office, Washington, DC, 1973.

Chelius, J. R., *Workplace Safety and Health: The Role of Workers' Compensation*, American Enterprise Institute, Washington, DC, 1977.

Corn, J. K., "Historical Perspective: The Evaluation of the Definition of Certain Work-Related Illnesses," (Working Paper #3) in *Preventing Illness and Injury in the Workplace*, contract report prepared for the Office of Technology Assessment, U.S. Congress, Washington, DC, July 1983.

Darling-Hammond, L., and T. J. Kniesner, *The Law and Economics of Workers' Compensation*, Rand Corporation, The Institute for Civil Justice, Santa Monica, CA, 1980.

Dorsey, S., "Employment Hazards and Fringe Benefits: Further Tests for Compensating Differentials," in J. D. Worall, ed., *Safety and the Work Force*, ILR Press, Ithaca, NY, 1983.

Hughes, J. T., "Byssinosis Compensation in North Carolina," in J. S. Lee and W. N. Rom, eds., *Legal and Ethical Dilemmas in Occupational Health*, Ann Arbor Science, Ann Arbor, MI, 1982.

Interdepartmental Workers' Compensation Task Force, *Research Report of the Interdepartmental Workers' Compensation*, U.S. Goverment Printing Office, Washington, DC, June 1979.

National Commission on State Workmen's Compensation Laws, *Compendium on Workmen's Compensation*, U.S. Government Printing Office, Washington, DC, 1973.

National Commission on State Workmen's Compensation Laws, *Report*, U.S. Government Printing Office, Washington, DC, 1973.

National Commission on State Workmen's Compensation Laws, *Supplemental Studies,* U.S. Government Printing Office, Washington, DC, 1973.

Russell, L. B., "Pricing Industrial Accidents," in *Supplemental Studies,* National Commission on State Workmen's Compensation Laws, U.S. Government Printing Office, Washington, DC, 1973.

Selikoff, I. J., et al., *Disability Compensation for Asbestos-Associated Disease in the United States,* Contract No. J-9-M-8—0165, report prepared for the Office of the Assistant Secretary for Policy, Evaluation, and Research, U.S. Department of Labor, Washington, DC, June 1981.

Strand, S. H., and W. G. Johnson, *An Analysis of the Safety Incentive Provided by Experience Rating under the Workers' Compensation Program,* report prepared for the Assistant Secretary for Policy, Evaluation and Research, U.S. Department of Labor, Washington, DC, March 1980.

U.S. Chamber of Commerce, *Analysis of Workers' Compensation Laws: 1983,* U.S. Chamber of Commerce, Washington, DC, 1983.

Victor, R. B., *Workers' Compensation and Workplace Safety: The Nature of Employer Financial Incentives,* Rand, Santa Monica, CA, 1982.

Victor R. B., L. R. Cohen, and C. E. Phelps, *Workers' Compensation and Workplace Safety: Some Lessons From Economic Theory,* Rand, Santa Monica, CA, 1982.

Worrall, J. D., ed., *Safety and the Work Force: Incentives and Disincentives in Workers' Compensation,* ILR Press, Ithaca, NY, 1983.

11.3 SECTION TWO: TECHNICAL INFORMATION

The occupational safety and health reference materials and the technical information resources in this chapter are organized according to the following scheme. First, the core references, those sources that are multidisciplinary in coverage, are presented under the heading "General References." These core references are subdivided into the following four subheadings: "Data Bases," "References," "Organizations," and "Journals."

Second, the references and information sources specific to the subject areas closely related to occupational safety and health — behavioral sciences (11.4), chemistry/physics (11.5), engineering control (11.6), epidemiology (11.7), ergonomics (11.8), industrial hygiene (11.9), medicine/nursing (11.10), safety (11.11), toxicology (11.12), and training (11.13) — are presented under separate headings to provide quick access and eliminate the need to scan the entire reference list or chapter. This also provides easy access to materials for those working outside their professional specialty.

11.3.1 Core Data Bases

Since occupational safety and health is a field that covers many disciplines, important information can be located in almost all on-line data bases containing scientific information. The data bases described in this subsection are those that are multidisciplinary and contain the most significant collections of occupational safety and health information. Where appropriate, these data bases are repeated in the subsequent subsections pertaining to specialized subject areas. The data bases listed in later sections also contain many new entries that supplement the core data bases. These additional listings by subject area provide in-depth coverage of specific topics. The data bases that focus on the collection of regulatory information are listed here as a matter of convenience.

The explosion of scientific information has made it virtually impossible for the practitioner to obtain access to all of the information published. Not only is the volume of information too great to effectively review, the cost of the journals is prohibitive. Only large university libraries can be expected to have a significant portion of the literature. As a result, electronic information retrieval systems have become a necessity for those seeking information on all but the most common topics.

Until recently, five data base vendors based in the United States have provided virtually all of the files of interest to the occupational safety and health professional. These are Dialog Information Services (3460 Hillview Avenue, Palo Alto, CA 94304); the National Library of Medicine (NLM) (8600 Rockville Pike, Bethesda, MD 20209); DOE RECON (Department of Energy, Technical Information Center, P.O. Box 62, Oak Ridge, TN 37830); Bibliographic Retrieval Services (BRS) (BRS Information Technologies, 1200 Route 7, Latham, NY 12110); and System Development Corporation (SDC) Search Services (System Development Corporation, 2500 Colorado Avenue, Santa Monica, CA 90406). New vendors that were originally available only in Europe now make their information systems available in the United States. These include Pergamon Infoline (Pergamon International Information Corporation, 1340 Old Chain Bridge Road, McLean, VA 22101); Questel (1625 Eye Street, N.W., Suite 719, Washington, DC 20006); and the Information Retrieval Service of the European Space Agency Information Retrieval Service, ESA-IRS, (Esrin, P.O. Box 64, 1-00044, Frascati, Italy). Hundreds of different data bases on subjects ranging from art history to zoology are available from these vendors.

Two other vendors, Information Consultants (ICI) (1133 15th Street, N.W., Suite 300, Washington,

DC 20005) and Chemical Information Systems (CIS) (7215 York Road, Baltimore, MD 21212), offer similar highly specialized information systems (ICI and ICIS) containing data about regulations and the physical characteristics and toxicity of various chemicals. These systems differ from the ones mentioned above in that they provide data with references rather than abstracts and bibliographic information.

Each data base vendor has a different mix of data bases. Even when the same data base is available from different vendors, the periods covered and the implementation can be quite different. There is a trend for the vendors to include larger portions of the data bases in single files as the costs of computer storage decline. The following table lists the core data bases that contain information relevant to occupational safety and health. The availability by vendor and the periods of coverage are included for reference.

11.3.2 Core Reference Texts

The general reference texts included in this section are those that comprise the basic bookshelf for an occupational safety and health reference room. They cover multiple subject areas and are also included in the specific subject areas. The books and monographs published by NIOSH are generally available from the Superintendent of Documents, U.S. Government Printing Office (GPO), Washington, DC 20402, or the National Technical Information Service (NTIS), Springfield, VA 22161. The GPO and/or NTIS stock number is included for all NIOSH publications. For reference texts and monographs of private publishers, the necessary bibliographic information is provided.

American Conference of Governmental Industrial Hygienists, *Industrial Ventilation, a Manual of Recommended Practice*, Committee on Industrial Ventilation, Lansing, MI, new editions biennially (18th edition, 1984). This standard reference on industrial ventilation focuses on the design, maintenance, and evaluation of ventilation systems. Each new edition includes the current research data in the field. The manual is written to serve as a complete reference to industrial ventilation and eliminate the need for other references. Topics included are general principles of ventilation, dilution ventilation, ventilation for heat control, hood design, design of hoods for recirculated air, construction specifications, and information on the field testing of ventilation systems, fans, and air cleaning devices.

American Conference of Governmental Industrial Hygienists, *The Documentation of the Threshold Limit Values*, 4th ed., American Conference of Governmental Industrial Hygienists, Cincinnati, OH, 1980. This publication is in the form of a loose-leaf binder to accommodate the annual supplements. The documentation provides the basis for the influential ACGIH recommendations. The presentation is concise, summarizing the animal or human toxicity studies that serve as the basis for the exposure limit recommendations. Permissible exposure limits for both the United States and other countries are listed.

American Conference of Governmental Industrial Hygienists, *TLVs®: Threshold Limit Values for Chemical Substances and Physical Agents in the Workroom Environment with Intended Changes for 19*—(new editions annually), American Conference of Governmental Industrial Hygienists, Cincinnati, Ohio. This publication is one of the most influential in the field. The 1985-86 edition contains these sections: "Threshold Limit Values for Chemical Substances in the Work Environment Adopted by ACGIH with Intended Changes for 1985-86," "Biological Exposure Indices Proposed by ACGIH for 1985-86," and "Threshold Limit Values for Physical Agents in the Work Environment Adopted by ACGIH with intended Changes for 1985-86."

AIHA, *Hygienic Guide Series*, Hygienic Guides Committee, American Industrial Hygiene Association, Akron, OH, 1958, 2 vols. This series includes hygienic guides on more than 180 substances. It is widely used by industrial hygienists and includes chemical and physical properties, sampling methods, emergency treatment, and medical practice. The series, which is updated periodically, is the basis for the published AIHA exposure limit recommendations.

Atherly, G.R.C., *Occupational Health and Safety Concepts*, Applied Science Publishers, London, UK, 1978. This basic text, developed from an introductory course in occupational safety and health, presents the field from the viewpoint of the practitioner in the United Kingdom. The text covers the entire field of occupational safety and health, including body defense mechanisms, occupational cancer, and strategies for environmental control.

Burgess, W. A., *Recognition of Health Hazards in Industry: A Review of Materials and Processes*, Wiley-Interscience, New York, NY, 1981. Over thirty industrial operations are reviewed in this book, including materials used, engineering controls, possible worker hazards, proper work practices, and protective equipment. Technical appendices and a subject index are included.

Deutsche Forschungsgemeinschaft, *Maximum Concentration at the Workplace and Biological Tolerance Values for Working Materials 1984*, VCH Publishers, Deerfield Beach, FL, 1984. This publication lists the maximum allowable concentrations (MAK values) of the Federal Republic of Germany for chemicals and gases, vapors, or particulate matter in the air of a working area. MAK values are determined and revised annually by a commission of experts from all relevant scientific disciplines. This commission is appointed by Deutsche Forschungsgemeinschaft (the German Science Foundation).

Epstein, S., *Politics of Cancer*, Sierra Book Club Books, San Francisco, CA, 1978. This publication stresses the complexity of the human decision-making process with respect to the issues of cancer and

TABLE 11.2. Core Data Bases

Name	Producer	No. of records	Coverage
ACADEMIC AMERICAN ENCYCLOPEDIA (BRS, DIALOG)	Grolier Electronic Publishing, Inc.	Full text of encyclopedia	Not available (N.A.)
BIOSIS (BRS)	Biosciences Information Services	N.A. N.A.	70–77 78–present
BIOSIS (ESA)	Biosciences Information Services	200,000/yr	73–present
BIOSIS (SDC)	Biosciences Information Services	200,000/yr	69–73 74–79 80–present
BIOSIS PREVIEWS (DIALOG)	Biosciences Information Services	2 million 1.4 million 328,000/yr	69–76 77–81 82–present
CAS (SDC)	Chemical Abstracts	1.3 million 1.7 million 2.2 million 340,000/yr	67–71 72–76 77–81 82–present
CA SEARCH (BRS)	Chemical Abstracts	N.A. N.A.	70–76 77–present
CA SEARCH (DIALOG)	Chemical Abstracts	1.3 million 1.7 million 1.2 million .9 million 400,000/yr	67–71 72–76 77–79 80–81 82–present
CA SEARCH (PERGAMON)	Chemical Abstracts	1.7 million 400,000/yr	72–76 77–present
CASSI	Chemical Abstracts	50,000	1900–present
CHEMABS (ESA)	Chemical Abstracts	6 million	67–present
CHEMICAL REGULATIONS AND GUIDELINES SYSTEM (DIALOG)	CRC Systems, Inc.	12,000/yr	81–present
CHEMLINE	Chemical Abstracts	614,058	N.A.
CHEMNAME[a] (DIALOG)	Chemical Abstracts	100,000/yr	67–present
CHEMSEARCH (DIALOG)	Chemical Abstracts	45,000 substances	6 weeks
CHEMSIS (DIALOG)	Chemical Abstracts (Singly Indexed Substances)	800,000 1.2 million 1.4 million Approximately 200,000	67–71 72–76 77–81 82–present
CHEMZERO (DIALOG)	Chemical Abstracts	1.2 million substances	67–present
CIS (Questel)	International Labour Organization/CIS	2,100/yr	74–present
CISDOC (ESA)	International Labour Organization/CIS	2,100/yr	74–present

TABLE 11.2. *(continued)*

Name	Producer	No. of records	Coverage
CONFERENCE PAPERS INDEX (ESA, DIALOG)	Cambridge Scientific Abstracts	120,000/yr	73-present
CROSS[a]	Bibliographic Retrieval Service	Data base index	N.A.
CTCP (CIS, ICIS)	Clinical Toxicology of Commercial Products	18,000 products in commerce	N.A.
DBI[a] (SDC)	Systems Development Corporation	Data base index	N.A.
DIALINDEX[a] (DIALOG)	Dialog Information Services	Data base index	N.A.
DIRLINE (NLM)	National Library of Medicine	14,000	N.A.
DISSERTATION ABSTRACTS (DIALOG, BRS)	University Microfilms International	42,000/yr	1861-present
ENVIRONMENTAL BIBLIOGRAPHY (DIALOG)	Environmental Studies Institute	24,000/yr	73-present
EUCAS (QUESTEL)	Chemical Abstracts	1.3 million 1.7 million 2.2 million 340,000/yr	67-71 72-76 77-81 82-present
ERIC (SDC, BRS, DIALOG)	Educational Resources Information Center	30,000/yr	66-present
EXCERPTA MEDICA (DIALOG)	Excerpta Medica	1.2 million 240,000/yr	74-79 80-present
EXCERPTA MEDICA (BRS)	Excerpta Medica	240,000/yr	80-present
FEDREG (SDC)	Capitol Services International	35,000/yr	77-present
FOODS ADLIBRA (DIALOG)	K & M Publications, Inc.	3,600/yr	74-present
FRSS (CIS, ICIS)	Federal Register Search System	160,000 References to proposed or final rules	77-present
HAZARDLINE (BRS)	Occupational Health Services, Inc.	3,000 chemical substances	N.A.
HSELINE (PERGAMON, ESA)	Health and Safety Executive	47,000	77-present

TABLE 11.2. *(continued)*

Name	Producer	No. of records	Coverage
KIRK-OTHMER ENCYCLOPEDIA OF CHEMICAL TECHNOLOGY (BRS)	John Wiley & Sons, Inc.	Full text of encyclopedia	3rd ed. 76
LABORDOC (ESA, QUESTEL, SDC)	International Labour Organization	4,800/yr	65–present
LIFE SCIENCES COLLECTION (DIALOG)	Cambridge Scientific Abstracts	680,000	78–present
MEDLINE	National Library of Medicine	140,000/yr	83–present
MED80		804,874	80–82
MED77		775,167	77–79
MED75		642,710	75–76
MED71		883,906	71–74
MED66 (NLM)		954,650	66–70
MEDLINE (DIALOG)	National Library of Medicine	1.5 million	66–72
		1.7 million	73–79
		140,000/yr	80–present
MEDLINE (BRS)	National Library of Medicine	954,650	66–70
		883,906	71–74
		N.A.	75–78
		140,000/yr	79–present
MESH[a]	National Library of Medicine	controlled vocabulary	N.A.
METADEX (DIALOG)	American Society for Metals	577,000	74–present
NIOSHTIC (DIALOG, Pergamon)	National Institute for Occupational Safety and Health (NIOSH)	12,000/yr	1900–present
NRC (DOE)	National Referral Center, Library of Congress	2,000	current
NTIS (DIALOG)	National Technical Information Service	60,000/yr	64–present
NTIS (ESA)	National Technical Information Service	60,000/yr	69–present
NTIS (SDC, BRS)	National Technical Information Service	60,000/yr	70–present
PASCAL (QUESTEL)	Centre de Documentation Scientifique et Technique	1.7 million	73–76
		2.6 million	77–present
POLLUTION ABSTRACTS (ESA, DIALOG, BRS)	Cambridge Scientific Abstracts	8,400/yr	70–present

TABLE 11.2. ***(continued)***

Name	Producer	No. of records	Coverage
RTECS (NLM, CIS, ICIS)	National Institute for Occupational Safety and Health (NIOSH)	70,029	N.A.
SCIENCE CITATION INDEX (DIALOG)	Institute for Scientific Information	1.8 million 482,000/yr	74-77 78-present
SANSS[a] (CIS, ICIS)	Chemical Information System, Inc.	1 million chemicals in system	N.A.
SUPERINDEX (BRS)	Superindex, Inc.	2,000 books	N.A.
TOXICOLOGY DATA BANK (NLM)	National Library of Medicine	4,112 chemicals	N.A.
TOXLINE TOXBACK76 TOXBACK65 (NLM)	National Library of Medicine	140,000/yr 555,808 598,424	81-present 76-80 65-75
VOLUNTARY STANDARDS INFORMATION NETWORK (BRS)	Information Handling Services	over 500	N.A.

Source: Data in this table were obtained from catalogs and manuals of the identified vendors. Specific citations for each data base and vendor are given in the alphabetical descriptions provided in section 11.14.

[a]These data bases are primarily search aids for the technical information specialist using more than one data base on the system.

carcinogenesis. Decisions about social, political, and economic concerns are treated in light of the need to regulate consumer products, workplace exposures, and pollution in the general environment.

Finkle, A. J., ed., *Hamilton and Hardy's Industrial Toxicology*, 4th rev. ed., Wright-PSG Publishing, Littleton, MA, 1982. In this revised edition of the classic reference text on occupational safety and health, clinical aspects of occupationally induced diseases are described and updated. This text focuses on occupational diseases with particular emphasis on thirty metals. Discussions of chemical compounds, radiation, dusts, and biological hazards are presented. An extensive bibliography and a glossary of occupational medicine terms are also included.

Gosselin, R. E., H. C. Hodge, R. P. Smith, and M. N. Gleason, *Clinical Toxicology of Commercial Products*, 5th ed., Williams & Wilkins, Baltimore, MD, 1984. This standard reference work in clinical toxicology is designed to help the physician deal quickly and effectively with acute chemical poisonings by commercial products. The sections of the book are color coded for easy access. The sections include "First Aid and General Emergency Treatment," "Ingredients Index," "Therapeutics Index," "Supportive Treatment," "Trade Name Index," "General Formulations" (a guide to the probable composition of products not specifically listed by trade name), and "Manufacturers Index." This work is kept up to date by the monthly "Bulletin of Supplementary Material."

Hunt, V. R., *Work and the Health of Women*, CRC Press, Boca Raton, FL, 1980. This work provides insight into the industries in which female workers are numerous and discusses the work-related problems they experience. The hazards of these industries are noted, and a section of the book is devoted to the reproductive concerns of both males and females.

Hunter, D., *Diseases of Occupations*, 6th ed., Hodder and Stoughton, London, UK, 1976. A key reference for both occupational medicine and industrial hygiene, this text is one of the most complete references on the signs and symptoms of work-related diseases. It also provides historical descriptions for occupational medicine.

Key, M. M., A. F. Henschel, J. Butler, R. N. Ligo, and I. R. Tabershaw, eds., *Occupational Diseases: A Guide to Their Recognition*, 2nd ed., National Institute for Occupational Safety and Health, Cincinnati, OH, 1977. Available from NTIS PB 83-129-528. This book is an easy-to-use reference about the causes of occupational diseases. It was written as an aid to physicians, nurses, and other health professionals to help identify and control occupational diseases. Sections include discussions of the routes of entry of chemical agents, chemical hazards, biological hazards, chemical carcinogens, pesticides, physical agents, and occupational dermatosis.

Kirk, H., and R. Othmer, eds., *Encyclopedia of Chemical Technology*, 3rd ed., Wiley-Interscience, New York, NY, 1978–83, 23 vols. This encyclopedia is the single most comprehensive reference for chemical engineering in existence. The 23 volumes cover the entire spectrum of chemical engineering with extensive discussions and clear illustrations. This is the best single reference for information about manufacturing processes for plastics and chemicals.

Kusnetz, S., and M. K. Hutchinson, eds., *A Guide to the Work-Relatedness of Disease*, rev. ed., National Institute for Occupational Safety and Health, Cincinnati, OH, 1979. Available from GPO 017-033-00177-4, NTIS PB 298561. This guide was prepared as an aid to state agencies, physicians, and others concerned with workers' compensation for occupational diseases. Methods for collecting, organizing, and using evidence to determine the work-relatedness of a given disease are described. Known work-related exposures to chemical agents such as asbestos, benzene, crystalline silica, inorganic lead, and others are used as examples of the methods. DHEW (NIOSH) Publication No. 79–116.

LeDou, J., ed., *Occupational Health Law: A Guide for Industry*, Marcel Dekker, New York, NY, 1981. This guide is for occupational safety and health professionals not trained in the law. This guide to legislation and regulations affecting all aspects of the health of workers includes chapters on workers' compensation laws, the functions of the Occupational Safety and Health (OSH) Act and regulatory agencies, legal aspects of health maintenance organizations, employee health records, malpractice, and labor arbitration, among others.

Mackison, F. W., R. S. Stricoff, and L. J. Partridge, Jr., eds., *NIOSH/OSHA Pocket Guide to Chemical Hazards*, National Institute for Occupational Safety and Health, Cincinnati, OH, 1978. Available from NTIS PB 83-105-338, and *NIOSH/OSHA Occupational Health Guidelines for Chemical Hazards*, NIOSH, Cincinnati, OH, 1981. Available from NTIS PB 83-154-609. These two publications are the most widely distributed works ever developed by NIOSH. The *Pocket Guide* presents the information in a summary tabular format and the *Guidelines* present similar information in an expanded narrative format. The information presented includes the OSHA permissible exposure limit the NIOSH recommended exposure limit (if available), and the ACGIH threshold limit value for 1978. Also presented are chemical and physical properties, health hazard information (including target organs and routes of exposure), respiratory protection recommendations, and personal protection and sanitation practices. Reference is also made to the established NIOSH analytical method for each substance. The *Pocket Guide* and the *Guidelines* cover the most common industrial chemicals and the majority of the chemicals regulated by OSHA. It is a significant, easy-to-use reference source for this specific information. The *Pocket Guide* has been revised to include additional chemical substances and to reflect new information developed subsequent to the original publication.

The Merck Index: An Encyclopedia of Chemicals and Drugs, 9th ed., Merck & Co., Rahway, NJ, 1976. This volume is a valuable reference for information about the toxicity and physical properties of chemicals. Since it includes drugs as well as industrial chemicals, it contains information not often found in the other reference sources listed in this general discussion.

Nothstein, G. Z., *The Law of Occupational Safety and Health*, Free Press, New York, NY, 1981. The 1970 Occupational Safety and Health Act (OSH Act) is the subject of this publication. Coverage includes legislative background, enforcement plans, standards of the Act, variances, requirements for record-keeping and reporting, NIOSH responsibilities, civil liability, and other elements of the OSH Act.

Office of Technology Assessment, *Preventing Illness and Injury in the Workplace*, OTA-H-256, Congress of the United States, Office of Technology Assessment, Washington, DC, 1985. This is an excellent monograph on the status of efforts to define, assess, and remedy occupational injury and illness problems in the workplace. It contributes a broader perspective to the field of occupational safety and health by providing a rationale for preventing hazards before they occur. This publication summarizes the current program approaches as well as current data and knowledge about the magnitude of the occupational safety and health problem. It also provides a useful review of the relevant legislative history.

Parmeggiani, L., ed., *Encyclopedia of Occupational Safety and Health*, 2 vols., 3rd rev. ed., International Labour Organization, Washington, DC, 1983. This is the latest version of a comprehensive two-volume reference work with concise, easy-to-understand entries.

Patty's Industrial Hygiene and Toxicology, 3rd rev. ed., John Wiley and Sons, New York, NY, 1978–82, 3 vols. A comprehensive reference on industrial hygiene and industrial toxicology, this work is both an important reference for those in the field and a significant text on the principles and practice of industrial hygiene. Volume 1, *General Principles*, G. E. Clayton and F. E. Clayton, eds., provides the knowledge base necessary for the practice of industrial hygiene, including fundamentals of epidemiology, mode of entry and action of toxic materials, effects of noise, ionizing radiation, heat stress, and other stressors. Evaluation and control measures, industrial hygiene survey procedures, air sampling and analysis, engineering controls, respiratory protective equipment, and air pollution control are also thoroughly discussed. Volume 2, *Toxicology*, G. E. Clayton and F. E. Clayton, eds., is an extensive reference on industrial toxicology. Three separate books comprise Volume 2; each chapter discusses the toxicology of compounds of similar chemical structure. Extensive subject and chemical indexes are included in Volume 2. Volume 3, *Theory and Rationale of Industrial Hygiene Practice*, L. J. Cralley and L. V. Cralley, eds., provides insight into quantitative matters such as emission inventories, analytical measurements, evaluation of exposures to chemical agents, ionizing radiation, noise and vibration, pressure, biologic agents, surveillance programs, and data extrapolation.

Proctor, N. H., and J. P. Hughes, *Chemical Hazards of the Workplace*, Lippincott, Philadelphia, PA, 1978. Based upon the material developed by the NIOSH/OSHA Standards Completion Program, this book describes medical surveillance strategies for dealing with occupational exposures. Although somewhat out of date, it is still an excellent reference on the subject of medical surveillance.

Sax, N. I., *Dangerous Properties of Industrial Materials*, 6th ed., Van Nostrand Reinhold, New York, NY, 1983. This is a valuable quick reference work that contains information on the chemical and physical hazards of the most common industrial chemicals. Over 16,000 chemicals are listed in the general chemical section with information on each including acute toxicity, flammability, synonyms, descriptions, and physical constants. The text also includes short sections on the control of radiation hazards and air contaminants as well as sections on fire protection, labeling, and identification of hazardous materials.

Schilling, R. S. F., ed., *Occupational Health Practice*, 2nd ed., Butterworths, London, UK, 1981. This updated version includes chapters on occupational health services, effects of chemical exposures on organ systems, procedures for employee health screening, and methods and applications of epidemiology in occupational settings. Also included are chapters on industrial hygiene, ergonomics, and personal protective equipment.

Tatken, R. L., and R. J. Lewis, Sr., eds., *NIOSH Registry of Toxic Effects of Chemical Substances (RTECS)* National Institute for Occupational Safety and Health, Cincinnati, OH, 1981–82 (annual supplements). Available in paper only from NTIS PB 85-21871. A quarterly microfiche version is available only by annual subscription. Order the microfiche by title only. The title is *Microfiche Edition, Registry of Toxic Effects of Chemical Substances.* RTECS in the largest single source of toxicity information in print. It includes acute and chronic toxicity information coded from the literature as well as tumorogenic data for more than 70,000 substances and 210,000 synonyms. In addition, environmental limits from OSHA, MSHA, NIOSH, and ACGIH are included, as are references to the International Agency for Research on Cancer (IARC) monographs and the substantial risk information provided pursuant of section 8(e) of the Toxic Substances Control Act (TSCA). The information in this publication is available in a variety of formats. Both paper copy and microfiche are cited here, but RTECS is also available on-line from various sources as described in the data base portions of this chapter.

Ullmann's Encyclopedia of Industrial Chemistry, 5th ed., VCH Publishers, Deerfield Beach, FL, 1984. This is an international encyclopedia. The 5th edition, for which only the first volume is available, will consist of 28 alphabetically ordered volumes covering all branches of chemical and industrial products. In addition, eight basic knowledge volumes will be organized by subject, including chemical engineering fundamentals, unit operations, chemical reaction engineering and materials science, analytical methods, and environmental protection and plant safety. New volumes will be published at the rate of three or four per year. A cumulative index will be published annually.

U.S. Department of Health and Human Services, Public Health Service, Centers for Disease Control, National Institute for Occupational Safety and Health, *The Industrial Environment: Its Evaluation and Control*, National Institute for Occupational Safety and Health, Cincinnati, OH, 1973. Available from GPO 017-001-00396-4. Published as a basic text for graduate study in the field of occupational safety and health, this book is currently being revised but remains one of the most comprehensive treatments of the field in textbook format. The first section of the book provides a review of mathematics, chemistry, biochemistry, physiology, and toxicology as they relate to the practice of occupational safety and health, specifically industrial hygiene. The remaining chapters concentrate on the methodology of industrial hygiene practice, especially the evaluation of the harmful effects of physical agents and air contaminants.

Zenz, C., ed., *Occupational Medicine: Principles and Practical Applications*, Year Book Medical Publishers, Chicago, IL, 1975. This book is an introductory text of occupational medicine with the following sections: administrative considerations, employee health services for small business, occupational safety, the occupational health nurse, and clinical occupational medicine. Also included are physical assessment of workers, occupational lung diseases, the physical and chemical environment, and psychosocial considerations.

11.3.3 Core Organizations

This section describes several of the principal organizations that focus on the field of occupational safety and health. Included are professional societies and labor unions that have substantial activity related to the field as well as interest groups that advocate occupational safety and health principles and programs. Also included are organizations that participate materially in the field but may not have occupational safety and health as their primary focus. The organizations listed here often produce publications and sponsor workshops, meetings, and symposia.

Academy of Hazard Control Management (AHCM), 5010A Nicholson Lane, Rockville, MD 20852. The Academy was founded in 1981 to certify hazard control managers in all aspects of safety and health. It also promotes professional development and the exchange of information within the field.

American Academy of Occupational Medicine (AAOM), 2340 S. Arlington Heights Road, Suite 400, Arlington Heights, IL 60005. The AAOM, founded in 1946, has a current membership of 700 physicians whose primary interest is occupational medicine. The academy promotes maintenance and improvement of the health of industrial workers. Publications include the *Journal of Occupational Medicine* and a membership directory.

American Association of Occupational Health Nurses (AAOHN), 3500 Piedmont Road, N.E., Atlanta, GA 30305. The AAOHN, founded in 1942, has about 12,000 members and a staff of 11. This membership includes registered professional nurses employed by business and industrial firms, nurse educators, and others interested in occupational health nursing. They maintain a library on occupational health and general nursing and produce *Occupational Health Nursing* and a newsletter.

American Conference of Governmental Industrial Hygienists (ACGIH). Bldg, D-5, 6500 Glenway Avenue, Cincinnati, OH 45211. This is a professional society of governmental industrial hygienists, educators, and others conducting research in industrial hygiene. The conference functions mainly as a medium for the exchange of ideas and the promotion of standards and techniques in industrial health. Publications include transactions of the annual meeting as well as manuals, guides, and special studies. ACGIH publishes the TLV® list.

American Federation of Labor and Congress of Industrial Organizations (AFL-CIO), Department of Health and Safety and Social Security, 815 16th Street, N.W., Washington, DC 20006. The Department of Health and Safety is the center for AFL-CIO coordination of occupational safety and health issues. This division provides the staff for the Occupational Safety and Health Committee.

American Industrial Health Council (AIHC), 1075 Central Park Avenue, Scarsdale, NY 10583. This group of industrial firms is united to help the government develop scientifically valid methods of identifying potential industrial carcinogens and adopting effective, practical regulatory controls for such substances.

American Industrial Hygiene Association (AIHA), 475 Wolf Ledges Parkway, Akron, OH 44311. This professional society of 6000 industrial hygienists promotes the study and control of environmental factors affecting the health and well-being of industrial workers. Publications include the *American Industrial Hygiene Association Journal,* a newsletter, and guides and manuals on problems related to industrial hygiene.

American Insurance Association (AIA), 85 John Street, New York, NY 10038. This association represents insurance companies providing property and liability insurance and loss control. It publishes information bulletins on industrial safety rules, fire prevention, codes and ordinances, special hazards, workers' compensation, and state financial responsibility laws.

American Occupational Medical Association (AOMA), 2340 S. Arlington Heights Road, Arlington Heights, IL 60005. The AOMA, founded in 1915, has 4000 members and a staff of 14. The membership includes physicians in the field of occupational medicine. Its goal is to ensure high ethical standards by practitioners in the field and to promote widespread understanding of the value of quality medical care for workers. The Association conducts scientific sessions and postgraduate seminars. Publications include the *AOMA Report, the Journal of Occupational Medicine,* a membership directory, booklets, monographs, and reports.

American Public Health Association (APHA), 1015 15th Street, N.W., Washington, DC 20005. The APHA, founded in 1872, is a professional organization of about 30,000 physicians, nurses, educators, sanitary engineers, environmentalists, social workers, industrial hygienists, and other community health specialists, as well as interested consumers. The organization seeks to protect and promote personal and environmental health. Services include establishment of uniform practices and procedures, development of the etiology of communicable diseases, research in many areas of public health, and exploration of medical care programs and their relationships to public health. Publications include the *American Journal of Public Health, The Nation's Health, Washington Newsletter,* directories, books, manuals, and pamphlets.

Center for Occupational Hazards, Inc., Five Beekman Street, New York, NY 10038. The Center was established in 1977 to gather and disseminate information about health hazards encountered by artists, craftsmen, teachers, children, and others working with art materials. It provides on-site assessments of the health and safety features of facilities used by artists, craftsmen, and students. The Center maintains a reference library, contributes articles to newspapers and art publications, and presents workshops, courses, and lectures. Publications include *Arts Hazards Newsletter,* books, pamphlets, articles, and data sheets on specific hazards.

Chemical Manufacturers Association (CMA), 2501 M Street, N.W., Washington, DC 20037. The Association, founded in 1872, administers research of importance to chemical manufacturing, including air and water pollution control. The association also promotes in-plant safety and operates the *Chemical Transportation Emergency Center (CHEMTREC).*

Industrial Health Foundation (IHF), 34 Penn Circle W., Pittsburgh, PA 15206. This is a research and service organization of industrial companies organized for the advancement of health in industry. The Foundation maintains a research laboratory for member companies where the foundation's staff conducts studies to prevent industrial diseases and improve working conditions. It offers continuing education programs for nurses, industrial hygiene engineers, and physicians and also provides extensive information on such subjects as analytical methods, control procedures, health hazards, and toxicity. The foundation also provides engineering and medical services, compiles statistics on epidemiology studies, and maintains a library and several committees. Publications include the *Industrial Hygiene Digest*, bulletins, technical papers, and symposia proceedings.

International Labour Organization (ILO), c/o Washington Br. Office, 1750 New York Avenue, N.W., Washington, DC 20006. The ILO, a specialized agency associated with the United Nations, was founded in 1919 and presently consists of 150 member nations and a staff of almost 3000. The organization promotes voluntary cooperation among nations for the purpose of improving labor conditions and raising living standards. Emphasis is placed on vocational training, research and investigation, voluntary international labor codes, occupational safety and health, and vocational rehabilitation. The ILO has established the International Occupational Safety and Health Information Centre (CIS), the International Institute for Labor Studies, and the International Centre for Advanced Technical and Vocational Training. The organization maintains an extensive library with a large collection of government documents, pamphlets, and computerized information services.

Life Extension Foundation (LEF), 1185 Avenue of the Americas, New York, NY 10036. The Foundation was established to conduct research and provide education in preventive medicine and to study health examination records to correlate the health quotient to such factors as age, sex, locale, and occupation. It compiles statistics on various medical studies and provides articles on educational health issues for general publication and for house organs.

National Environmental Health Association (NEHA), 1200 Lincoln Street, Suite 704, Denver, CO 80203. NEHA consists of 4000 environmental health professionals who work for public health agencies providing public health education or environmental inspection services. The organization promotes educational and professional curricula for colleges and universities and provides self-paced learning modules for field professionals. Sections include air, land, and water quality; environmental management; food; general environmental health; injury prevention; occupational health; and institutional environmental health. Publications include the *Journal of Environmental Health* and a newsletter.

Occupational Health Institute (OHI), 2340 S. Arlington Heights Road, Arlington Heights, IL 60005. The Institute was founded in 1945 to provide professional guidance to industry in the development and maintenance of occupational health programs.

Occupational Medical Administrators' Association (OMAA), c/o Robert F. Gaglione, Medical Administration, New York Telephone, 1095 Avenue of the Americas, Room 2557, New York, NY 10036. Founded in 1959, this association includes medical administrators in industry or government united to stimulate the discussion of issues related to occupational medical administration and to promote sound principles and standards in the administration of occupational medical programs and facilities. The Association also encourages research and educational activities related to these topics.

Society for Occupational and Environmental Health (SOEH), 2021 K Street, N.W., Suite 305, Washington, DC 20006. This society has about 500 members, including scientists, academicians, and industry and labor representatives, who seek to improve the quality of both working and living places. The Society operates as a neutral forum for conferences involving all aspects of occupational health. It focuses public attention on scientific, social, and regulatory problems as well as supporting studies on specific categories of hazards, on methodologies for assessment of health effects, and on diseases associated with particular jobs. The Society also sponsors conferences and workshops and publishes a newsletter and conference proceedings.

Synthetic Organic Chemical Manufacturers Association (SOCMA), 1075 Central Park Avenue, Scarsdale, NY 10583. This group includes manufacturers of synthetic organic chemicals made from coal, natural gas, crude petroleum, and certain natural substances and their derivatives. Committees deal with emerging issues, environmental quality, occupational safety and health, and toxic substances.

Women's Occupational Health Resource Center (WOHRC), School of Public Health, 21 Audubon Avenue, 3rd Floor, Columbia University, New York, NY 10032. This organization has about 1000 members and serves as a clearinghouse for women's occupational health and safety issues. Its principal aim is to increase awareness of health and safety hazards women face in the workplace. Members advise manufacturers on design standards of safety equipment, offer technical assistance in setting up programs to develop occupational health awareness, and serve on a speaker's bureau. The group sponsors workshops, seminars, and panels on such topics as personal protective equipment, reproductive hazards, industrial safety, and the hazards of household work. It maintains a library with a computerized bibliographic information system and publishes a newsletter and fact sheets.

11.3.4 Core Journals

The most important journals in the field of occupational safety and health fall more logically within the respective subject area sections than in the general references section. Therefore, only the few journals that cover the sciences generally or clearly address several professional disciplines are included in this overview section. Many of the journals selected for inclusion are among the most frequently cited journals in NIOSH's on-line bibliographic data base, NIOSHTIC, which is now publicly available through several commercial vendors. NIOSHTIC collects the relevant English language literature that supports research in occupational safety and health.

The *American Industrial Hygiene Association Journal* focuses on all aspects of the field of industrial hygiene practice, policy, and procedures as well as reports of specific investigations and field activities. This publication is included among the most frequently cited journals in NIOSHTIC.

The *American Journal of Public Health* is the only professional journal included among the general reference materials. It is listed here because of the many disciplines that are covered by the field of public health. The journal includes important articles from such fields as epidemiology, surveillance, and engineering controls. This publication is included among the most frequently cited journals in NIOSHTIC.

Analytical Chemistry covers all branches of analytical chemistry. Articles may be entirely theoretical or be reports of laboratory experiments that support, argue, refute, or extend established theory. Correspondence, aids for analytical chemists, and articles on the use of computers are included. This publication is included among the most frequently cited journals in NIOSHTIC.

The *Annals of the New York Academy of Sciences* provides an open forum for discussions of selected occupational safety and health topics. This is an important monograph series because of the depth of coverage of the selected topics. This publication is included among the most frequently cited journals in NIOSHTIC.

Archives of Environmental Health presents articles concerned with the effects of environmental agents on human health and includes epidemiological, clinical, and experimental studies of man. This publication is included among the most frequently cited journals in NIOSHTIC.

The *British Journal of Industrial Medicine* provides original contributions relevant to occupational and environmental medicine, including toxicological studies of industrial and agricultural chemicals as well as articles on specific medical experiences and ethical considerations. This publication is included among the most frequently cited journals in NIOSHTIC.

Cancer Research presents the results of original experimental, clinical, and statistical studies of interest to cancer researchers. It also includes review articles. Announcements and reports of meetings of interest are included, and the availability of fellowships and scholarships is noted. Proceedings of meetings and symposia are published as external supplements to the journal. It is among the most frequently abstracted journals in both NIOSHTIC and TOXLINE.

Health Physics is the official journal of the Health Physics Society. Published monthly, this publication focuses on the theoretical and applied fields of health physics. This publication is included among the most frequently cited journals in NIOSHTIC.

The *International Archives of Occupational and Environmental Health* publishes review articles, original papers, short communications, documentation of international meetings, and activities related to occupational or ambient environmental problems. Its publication priorities include chemical and epidemiological studies related to morbidity or mortality, and also the estimation of health risk. This publication is included among the most frequently cited journals in NIOSHTIC.

The *Journal of Occupational Medicine* is the official publication of the American Occupational Medical Association and of the American Academy of Occupational Medicine. It presents articles on all phases of occupational medicine and occupational health practice. This publication is included among the most frequently cited journals in NIOSHTIC.

Lancet is an international journal of science and medicine that covers current research findings not published elsewhere. The journal covers a wide range of topics. It is included among the most frequently cited journals in NIOSHTIC.

Nature contains original articles on a wide range of topics. It is among the most frequently cited publications in NIOSHTIC. Articles may be either reports of major research developments or broader discussions of progress. "Letters" consist of short reports of research of particular interest, while "Matters Arising" is a feature permitting occasional discussion of papers previously appearing in the journal.

The *Scandinavian Journal of Work, Environment and Health* is published by the Finnish Institute of Occupational Health. It includes original articles in the field of occupational safety and health from throughout the Western world and encourages full-length discussions of important issues and research topics. This publication is included among the most frequently cited journals in NIOSHTIC.

Science publishes original material in five categories: articles, reports, letters, technical comments, and book reviews. Articles are written in journalistic format with the finding first and supporting details and arguments second. The concise reports incorporate information of news value to the scientific community or of unusual interest to the specialist. Letters provide a forum for discussion of matters of general interest to scientists, while the technical comments section contains letters concerning papers previously published in the journal. This publication is included among the most frequently cited journals in NIOSHTIC.

Toxicology and Applied Pharmacology, an official journal of the Society of Toxicology, focuses on original scientific research on changes in tissue structure or function resulting from the administration of chemicals, drugs, or natural products. This publication is included among the most frequently cited journals in NIOSHTIC.

11.4 BEHAVIORAL SCIENCE

This section deals with the worker's perception of and adaptation to the work environment. It includes the development and evaluation of stress control efforts and the development of psychometric and physiologic tests used to rate the cognitive demands of work in terms of its stress significance.

Most of the literature in this area is recent; the bulk of the scientific and professional information is found in the periodical literature. Therefore, the mainstay for information in these areas is the professional journal, which is most conveniently identified through appropriate use of the on-line data bases.

Health promotion, disease prevention, and risk reduction are fields in which behavioral science is a major component. These fields support the provision of facilities and programs that urge individuals to participate in behavior and activities that encourage the maintenance and improvement of personal health.

11.4.1 Behavioral Science Data Bases

The explosion of scientific information has made it virtually impossible for the practitioner to obtain access to all of the information published. Not only is the volume of information too great to effectively review, the cost of the journals is prohibitive. Only large university libraries can be expected to have a significant portion of the literature. As a result, electronic information retrieval systems have become a necessity for those seeking information on all but the most common topics.

Until recently, five data base vendors based in the United States have provided virtually all of the files of interest to the occupational safety and health professional. These are Dialog Information Services (3460 Hillview Avenue, Palo Alto, CA 94304), the National Library of Medicine (NLM) (8600 Rockville Pike, Bethesda, MD 20209), DOE RECON (Department of Energy, Technical Information Center, P.O. Box 62, Oak Ridge, TN 37830), Bibliographic Retrieval Services (BRS) (BRS Information Technologies, 1200 Route 7, Latham, NY 12110), and System Development Corporation (SDC) Search Services (System Development Corporation, 2500 Colorado Avenue, Santa Monica, CA 90406). New vendors that were originally available only in Europe now make their information systems available in the United States. These include Pergamon Infoline, (Pergamon International Information Corporation, 1340 Old Chain Bridge Road, Mclean, VA 22101), Questel (1625 Eye Street, N.W., Suite 719, Washington, DC 20006), and Information Retrieval Service of the European Space Agency Information Retrieval Service, ESA-IRS, (Esrin, P.O. Box 64, 1-00044, Frascati, Italy). Hundreds of different data bases on subjects ranging from art history to zoology are available from these vendors.

Two other vendors, Information Consultants (ICI) (1133 15th Street, N.W., Suite 300, Washington, DC 20005) and Chemical Information Systems (CIS) (7215 York Road, Baltimore, MD 21212), offer similar highly specialized information systems (ICI and ICIS) containing data about regulations and the physical characteristics and toxicity of various chemicals. These systems differ from the ones mentioned above in that they provide data with references rather than abstracts and bibliographic information.

Each data base vendor has a different mix of data bases. Even when the same data base is available from different vendors, the periods covered and the implementation can be quite different. There is a trend for the vendors to include larger portions of the data bases in single files as the costs of computer storage decline. The following table lists the data bases that contain behavorial science information relevant to occupational safety and health. The availability by vendor and the periods of coverage are included for reference.

11.4.2 Behavioral Science Reference Texts

Cooper, C. L., et al., eds., *A Behavioral Approach to the Management of Stress: A Practical Guide to Techniques,* John Wiley and Sons, New York, NY, 1982. This book is a practical guide to the recognition, assessment, and management of stress.

Cooper, C. L., and R. Payne, *Current Concerns in Occupational Stress,* John Wiley and Sons, New York, NY, 1980. This book is one of a series of studies of occupational stress. The series is intended to provide a forum for international researchers to discuss their work on occupational stress and health. The series is targeted at a range of disciplines, including occupational psychology and sociology, occupational medicine, management, and personnel, among others.

Cooper, C. L., and R. Payne, eds., *Stress at Work,* John Wiley and Sons, New York, NY, 1980. This book is the first in a series of publications on studies of occupational stress. It includes contributions

TABLE 11.3 Behavioral Science Data Bases

Name	Producer	No. of Records	Coverage
BIOSIS (SDC)	Biosciences Information Services	200,000/yr	69-73 74-79 80-present
BIOSIS (ESA)	Biosciences Information Services	200,000/yr	73-present
BIOSIS (BRS)	Biosciences Information Services	Not Available (N.A.) N.A.	70-77 78-present
BIOSIS PREVIEWS (DIALOG)	Biosciences Information Services	2 million 1.4 million 328,000/yr	69-76 77-81 82-present
CIS (Questel)	International Labour Organization/CIS	2,100/yr	74-present
CISDOC (ESA)	International Labour Organization/CIS	2,100/yr	74-present
ERIC (SDC, BRS, DIALOG)	Educational Resources Information Center	30,000/yr	66-present
HSELINE (PERGAMON, ESA)	Health and Safety Executive	47,000	77-present
MENTAL HEALTH ABSTRACTS (DIALOG)	National Clearinghouse for Mental Health Information (to 82) Plenum Data Co. to present	463,000	69-present
NCMH (BRS)	National Clearinghouse for Mental Health	N.A.	69-82
NIOSHTIC (DIALOG, Pergamon)	National Institute for Occupational Safety and Health (NIOSH)	12,000/yr	1900-present
PSYCINFO (SDC, DIALOG, BRS)	American Psychological Association	24,000/yr	67-present
SOCIOLOGICAL ABSTRACTS (DIALOG, BRS)	Sociological Abstracts, Inc.	8,500/yr	63-present
SOCIAL SCISEARCH (DIALOG, BRS)	Institute for Scientific Information	120,000/yr	72-present

Source: Data in this table were obtained from catalogs and manuals of the identified vendors. Specific citations for each data base and vendor are given in the alphabetical descriptions provided in Section 11.14.

from an international group of authors and covers the following subject areas: stress at work in perspective, stress factors in the person's environment, stress factors in the person dealing with stressors and strains, and issues in research on stress at work.

French, J. R. P., Jr., *The Mechanisms of Job Stress and Strain,* Wiley Interscience, New York, NY, 1982. This book presents the results of an analysis of survey data from 23 companies including a total of more than 2000 people representing both white- and blue-collar workers. Using multivariate analysis, the authors describe paths from stressful events to adverse health outcomes. Also discussed are the methods used to identify these adverse health effects.

Gardell, B., and G. Johansson, eds., *Working Life: A Social Science Contribution to Work Reform,* Wiley Interscience, New York, NY, 1981. This is an international collection of papers providing an understanding of conditions found at work by people in Western societies. Also described are the consequences of the work experience in terms of the individual's health and social adjustment.

Krinsky, L. W., Ph.D., ed., *Stress and Productivity,* Human Sciences Press, New York, NY, 1984. This publication is the result of a conference entitled "Stress and Productivity" held in April, 1981. The conference addressed stress as it affects the employee, the employer, and the workplace. The effects of stress on efficient and dedicated employees and the employer's appropriate response were the principal foci.

Mackay, C., and T. Cox, eds., *Response to Stress: Occupational Aspects,* IPC Science and Technical press, Guildford Surrey, UK, 1979. This publication presents a collection of papers from the Ergonomics Society's conference, entitled "Psychophysiological Response to Occupational Stress," held at the University of Nottingham, September, 1978.

Marrow, A. J., ed., *The Failure of Success,* AMACOM, New York, NY, 1972. This book provides an overview of the reasons for, approach to, and benefits of implementing modern management techniques designed to deal effectively with issues that interfere with maximizing the productivity of a work force. The main sections are entitled "Organizations and the Quality of Life," "Harnessing the Skills of Behavioral Science," "Releasing Human Potential," "Two Basic Techniques," and "Organizational Development."

Marshall, J., and C. L. Cooper, eds., *Coping with Stress at Work: Case Studies from Industry,* Renouf U.S.A., Brookfield, VT, 1981. This book deals with industrial psychology and stress, both psychological and physiological. Sections of the book include: "Training," "Promoting Good Physical and Mental Health," "Remedial Action," and "Changing the Environment."

Pelletier, K. R., Ph.D., *Healthy People in Unhealthy Places,* Delacorte Press, New York, NY, 1984. This book presents information on the impact of the work experience on stress-related health conditions. Section titles include "Making the Workplace Healthier," "Work and Stress," "Health Promotion Programs," and "Healthy People in Healthy Places."

U.S. Department of Health and Human Services, Public Health Service, Centers for Disease Control, National Institute for Occupational Safety and Health, *Health Consequences of Shift Work,* DHEW (NIOSH) Publication No. 78-154, National Institute for Occupational Safety and Health, Cincinnati, OH, 1978. Available from NTIS PB 80-176-563/A07. This publication consists of a NIOSH contract report as received from the contractor. It is based on a 30-month study that investigated the effects of working unconventional hours on the psychological and physiological well-being of workers. Data on 1200 nurses and a similar number of food processors were collected by reviewing health and accident files and administering a lengthy questionnaire. The inquiry covered incidence and prevalence of physical complaints, illness histories, eating patterns, sleep patterns, medication usage, life-style and domestic patterns, and psychological profiles.

U.S. Department of Health and Human Services, Public Health Service, Centers for Disease Control, National Institute for Occupational Safety and Health, *Proceedings: Reducing Occupational Stress,* DHEW (NIOSH) Publication No. 78-140, National Institute for Occupational Safety and Health, Cincinnati, OH, 1978. Available from GPO 017-033-00303. This is the proceedings of a conference held in May of 1977. Some of the presentations are based on the personal experience of the author; others describe actual models or theories both with and without verification.

U.S. Department of Health and Human Services, Public Health Service, Centers for Disease Control, National Institute for Occupational Safety and Health, *Twenty-four Hour Workday: Proceedings of a Symposium on Variations in Work-sleep Schedules,* DHEW (NIOSH) Publication No. 81-127, National Institute for Occupational Safety and Health, Cincinnati, OH, 1981. Available from NTIS PB 83-104-794/A99. This publication is the result of a symposium on variations in work-sleep schedules held in 1979. Three areas of concern with respect to work-sleep schedules addressed at the symposium are shift work, sleep, and biological rhythms.

OTHER BEHAVORIAL SCIENCE REFERENCE TEXTS

Cooper, C. L., and M. J. Smith, eds., *Job Stress and Blue Collar Work,* John Wiley and Sons, New York, NY, 1985.

Erfurt, J. C., and A. Foote, eds., *Blood Pressure Control Programs in Industrial Settings,* University of Michigan Institute of Labor, Ann Arbor, MI, 1979.

Everly, G. S., and H. L. Robert, *Occupational Health Promotion: Health Behavior in the Workplace*, Feldman & Associates; S. M. Weiss—Frwd. John Wiley and Sons, 1985.

Interagency Task Force on Workplace Safety and Health, *Making Prevention Pay*, Washington, DC, 1978.

Ricci, P. F., ed., *Principles of Health Risk Assessment*, Prentice-Hall, Englewood Cliffs, NJ, 1985.

Shephard, R. J., *Fitness & Health in Industry*, Medicine and Sport Science Series, Vol. 21, S. Karger, New York, NY, 1985.

U.S. Department of Health and Human Services, *Promoting Health/Preventing Disease Objectives for the Nation*, U.S. Government Printing Office, Washington, DC, 1980.

Weis, W. L., and B. W. Miller, eds., *The Smoke-Free Workplace*, Prometheus Books, Buffalo, NY, 1985.

Wolfe, S. M., and L. Abrams, *1983 Survey of Fourteen Union Occupational Safety and Health Programs*, Public Citizen, Health Research Group, Washington, DC, January 1984.

11.4.3 Behavioral Science Organizations

American Center for Quality of Work Life (ACQWL), 1411 K Street, N.W., Suite 930, Washington, DC 20005. The Center was established in 1973 to stimulate an understanding of proven ways for unions and management to utilize workers in solving work problems. It promotes increased operational effectiveness and a committed and involved work force. The Center helps labor and management improve the quality of work environments. Major areas of concern are occupational safety and health services and training.

American Productivity Center (APC), 123 North Post Oak Lane, Houston, TX 77024. The Center was established in 1977 with sponsorship from 400 major corporations, foundations, and individuals to improve productivity and the quality of work life in the United States. The Center works with businesses, unions, and governmental agencies to find ways to improve productivity. It conducts research, gathers information through fieldwork, compiles statistics, and disseminates information. Center publications include case studies, productivity briefs and a productivity digest, newsletters, bibliographies, and materials on trends. The areas of concentration are productivity management, productivity measurement, white-collar productivity, labor/management, national policy, and employee involvement.

International Organization for the Study of Human Development (IOSHD), c/o Professor Frank Falkner, School of Public Health, University of California, Berkeley, CA 94720. This organization, founded in 1969, currently has a membership of 100. Members include individuals with a Ph.D. or M.D. in the biological, behavioral, or other social sciences. They are organized to promote and facilitate the interdisciplinary study of human development over the entire life cycle. The organization publishes a newsletter.

NTL Institute (NTLI), P.O. Box 9155, Rosslyn Station, Arlington, VA 22209. The Institute was established in 1947 as a nonprofit corporation offering training, consultation, research, and publication services. The Institute provides an adjunct staff who conduct more than 150 programs annually. These participants are individuals and teams from occupational groups and corporations, communities, and government at all levels. Its publications include the *Journal of Applied Behavioral Science*.

World Association for Social Psychiatry (WASP), 737 Las Alturas Del Sol, Santa Barbara, CA 93103. The Association, founded in 1972, has 3000 members interested in or contributing to anthropology, social work, nursing, occupational therapy, or social psychiatry. The group's objective is to study the nature of humanity and human cultures and to prevent and treat internal changes and behavior disorders. It also promotes worldwide collaboration among allied professionals and societies, distributes theoretical and practical information, and attempts to advance the physical, social, and philosophic well-being of humanity. The Association sponsors workshops, offers demonstrations of local treatment facilities, and publishes the *American Journal of Social Psychiatry*, the *French Journal of Social Psychiatry*, and the *International Journal of Social Psychiatry*.

11.4.4 Behavorial Science Journals

Human Relations: A Journal of Studies Toward the Integration of Social Sciences is international in scope, interdisciplinary, and provides articles reflecting a broad spectrum of concerns and approaches that serve to integrate the social sciences. Included are theoretical, methodological, and review articles.

The *Journal of Applied Behavior Analysis* focuses on original experimental research and the analysis of behavior as it relates to problems of social importance. It also includes other technical articles relevant to this area of research.

The *Journal of Applied Behavioral Science* provides a forum for the presentation of current activities in the field. A significant portion of this material is related to the working environment.

The *Journal of Behavior Medicine* is an interdisciplinary publication that presents information that improves the understanding of the relationships between physical health and illness utilizing the knowledge of the behavioral sciences. Specific treatment is given to work related to health maintenance behavior, self-regulation therapies, biofeedback, and sociocultural influences on health and illness. Articles cover both theoretical and experimental subjects; there are also review, technical, and methodological articles and case studies.

The *Journal of Occupational Behavior* provides a forum for the presentation and discussion of reports on and reviews of the growing research related to problems associated with the psychosocial aspects of the work experience. An effort is made to relate this research to the broader aspects of social change, personal growth, and the quality of the individual's working life. Social and industrial changes relate to the individual prospects in terms of its impact on growth and development, management styles, and personal self-management of the working experience. This publication is included among the most frequently cited journals in NIOSHTIC.

The *Journal of Occupational Medicine* is the official publication of the American Occupational Medical Association and of the American Academy of Occupational Medicine. It presents articles on all phases of occupational medicine and occupational health practice. This publication is included among the most frequently cited journals in NIOSHTIC.

The *Journal of Occupational Psychology* contains papers on empirical as well as conceptual works. It covers industrial, organizational, engineering, vocational, and personal psychology. Occupational counseling and the behavioral aspects of industrial relations, ergonomics, human factors, and industrial sociology are also given primary treatment. This publication is included among the most frequently cited journals in NIOSHTIC.

Personal Psychology: A Journal of Applied Research presents research methods, research results, and their application to the solution of personnel problems faced by business, industry, and government. Critical reviews are also included. Areas of interest include training, worker analysis, employee relations, and morale.

Social Science and Medicine was established to disseminate research and theoretical works in which sociology is directly related to problems of mental and physical health.

Work and Occupations: An International Sociological Journal focuses on sociological issues related to work and occupations. It includes both practical and theoretical discussions.

11.5 CHEMISTRY/PHYSICS

Knowledge of chemical processes, the chemical and physical nature of substances, and specific analytical techniques in necessary to conduct occupational safety and health activities. To select appropriate controls, the practitioner must be able to assess the safety hazards associated with human exposures during chemical production, storage, and disposal and know the analytical methods used to determine the nature of specific exposures in the industrial environment. For years, adherence to established exposure limits has been an important strategy in preventing occupational illness and injury. Methods identifying specific exposures in the occupational environment are vital to this prevention strategy. Documentation of these exposures justifies the establishment of control systems used to protect the worker. In fact, methods of identifying specific exposures are an integral part of standards and help to establish when environmental exposures to toxic substances have occurred or are likely to occur.

11.5.1 Chemistry/Physics Data Bases

The explosion of scientific information has made it virtually impossible for the practitioner to obtain access to all of the information published. Not only is the volume of information too great to effectively review, the cost of the journals is prohibitive. Only large university libraries can be expected to have a significant portion of the literature. As a result, electronic information retrieval systems have become a necessity for those seeking information on all but the most common topics.

Until recently, five data base vendors based in the United States have provided virtually all of the files of interest to the occupational safety and health professional. These are Dialog Information Services, (3460 Hillview Avenue, Palo Alto, CA 94304); the National Library of Medicine (NLM) (8600 Rockville Pike, Bethesda, MD 20209); DOE RECON (Department of Energy, Technical Information Center, P.O. Box 62, Oak Ridge, TN 37830); Bibliographic Retrieval Services (BRS) (BRS Information Technologies, 1200 Route 7, Latham, NY 12110); and System Development Corporation (SDC) Search Services (System Development Corporation, 2500 Colorado Avenue, Santa Monica, CA 90406). New vendors originally available only in Europe now make their information systems available in the United States. These include Pergamon Infoline, (Pergamon International Information Corporation, 1340 Old Chain Bridge Road, Mclean, VA 22101); Questel (1625 Eye Street, N.W., Suite 719, Washington, DC 20006); and Information Retrieval Service of the European Space Agency Information Retrieval Service, ESA-IRS, (Esrin, P.O. Box 64, 1-00044, Frascati, Italy). Hundreds of different data bases on subjects ranging from art history to zoology are available from these vendors.

Two other vendors, Information Consultants (ICI) (1133 15th Street, N.W., Suite 300, Washington, DC 20005) and Chemical Information Systems (CIS) (7215 York Road, Baltimore, MD 21212), offer

TABLE 11.4 Chemistry/Physics Data Bases

Name	Producer	No. of records	Coverage
AMERICAN CHEMICAL SOCIETY PRIMARY JOURNAL DATA BASE (BRS)	Bibliographic Retrieval Services	30,000 articles in full text	Not Available (N.A.)
APLIT (SDC)	American Petroleum Institute	360,000	64-present
AGRICOLA (CAIN) (DIALOG, BRS)	USDA, Science and Education Administration	1.1 million 120,000/yr	70-78 79-present
CAB (ESA)	Commonwealth Agricultural Bureaux	144,000/yr	73-present
CAB ABSTRACTS (DIALOG)	Commonwealth Agricultural Bureaux	144,000/yr	72-present
CAS (SDC)	Chemical Abstracts	1.3 million 1.7 million 2.2 million 340,000/yr	67-71 72-76 77-81 82-present
CA SEARCH (BRS)	Chemical Abstracts	N.A. N.A.	70-76 77-present
CA SEARCH (DIALOG)	Chemical Abstracts	1.3 million 1.7 million 1.2 million .9 million 400,000/yr	67-71 72-76 77-79 80-81 82-present
CA SEARCH (PERGAMON)	Chemical Abstracts	1.7 million 400,000/yr	72-76 77-present
CEH-ONLINE (SDC)	Stanford Research Institute	Full text of handbook	N.A.
CHEMABS (ESA)	Chemical Abstracts	6 million	67-present
CHEMDEX (SDC)	Chemical Abstracts	3 million compounds	72-present
CHEMICAL ENGINEERING ABSTRACTS (PERGAMON)	The Royal Society of Chemistry	63,000	70-present
CHEMICAL EXPOSURE (DIALOG)	Chemical Effects Information Center	14,370	74-present
CHEMLINE (NLM)	Chemical Abstracts	614,058	N.A.
CHEMNAME (DIALOG)	Chemical Abstracts	100,000/yr	67-present
CHEMSEARCH (DIALOG)	Chemical Abstracts	45,000 substances	6 weeks
CHEMSIS (DIALOG)	Chemical Abstracts (Singly Indexed Substances)	800,000 1.2 million 1.4 million approximately 200,000	67-71 72-76 77-81 82-present

TABLE 11.4 (*continued*)

Name	Producer	No. of records	Coverage
CHEMZERO (DIALOG)	Chemical Abstracts	1.2 million substances	67-present
CIS (Questel)	International Labour Organization/CIS	2,100/yr	74-present
CISDOC (ESA)	International Labour Organization/CIS	2,100/yr	74-present
CLAIMS COMPOUND REGISTRY (DIALOG)	IFI/Plenum Data Co.	13,963 compounds	N.A.
CLAIMS/ UNITERM (DIALOG)	IFI/Plenum Data Co.	1.5 million	50-present
CURRENT BIOTECHNOLGY ABSTRACTS (PERGAMON)	The Royal Society of Chemistry	7,000	83-present
DOE ENERGY (DIALOG)	TIC, ORNL, DOE (Energy Data Base)	144,000/yr	74-present
EDB (DOE)	TIC, ORNL, DOE (Energy Data Base)	192,000/yr	1900-present
EDB (SDC)	TIC, ORNL, DOE (Energy Data Base)	800,000	30s-present
EI ENGINEERING Meetings (SDC, ESA, DIALOG)	Engineering Information, Inc.	108,000/yr	82-present
EUCAS (QUESTEL)	Chemical Abstracts	1.3 million 1.7 million 2.2 million 340,000/yr	67-71 72-76 77-81 82-present
KIRK-OTHMER ENCYCLOPEDIA OF CHEMICAL TECHNOLOGY (BRS)	John Wiley and Sons, Inc.	Full text of encyclopedia	3rd ed. 76
LABORATORY HAZARDS BULLETIN (PERGAMON)	The Royal Society of Chemistry	2,000	81-present
MEDLINE MED80 MED77 MED75 MED71 MED66 (NLM)	National Library of Medicine	140,000/yr 804,874 775,167 642,710 883,906 954,650	83-present 80-82 77-79 75-76 71-74 66-70
MEDLINE (DIALOG)	National Library of Medicine	140,000/yr 1.7 million 1.5 million	80-present 73-79 66-72
MEDLINE (BRS)	National Library of Medicine	140,000/yr N.A. 883,906 954,650	79-present 75-78 71-74 66-70
METADEX (DIALOG)	American Society for Metals	577,000	74-present

TABLE 11.4 (*continued*)

Name	Producer	No. of records	Coverage
NASA (ESA)	NASA, Scientific and Technical Branch	N.A.	62-present
NIOSHTIC (DIALOG, Pergamon)	National Institute for Occupational Safety and Health (NIOSH)	12,000/yr	1900-present
PASCAL (QUESTEL)	Centre de Documentation Scientifique et Technique	1.7 million 2.6 million	73-76 77-present
PESTDOC/ PESTDOC II (SDC)	Derwent Publications, Ltd.	8,500/yr	68-present
RTECS (NLM, CIS, ICIS)	National Institute for Occupational Safety and Health (NIOSH)	70,029	N.A.
SANSS (CIS, ICIS)	Chemical Information System, Inc.	1 million chemicals in system	N.A.
SPHERE DERMAL AQUIRE ENVIROFATE (CIS, ICIS)	Office of Toxic Substances, U.S. EPA	 655 chemicals 1,900 chemicals 450 chemicals	 N.A. N.A. N.A.
TOXICOLOGY DATA BANK (NLM)	National Library of Medicine	4,112 chemicals	N.A.
TOXLINE TOXBACK76 TOXBACK65 (NLM)	National Library of Medicine	140,000/yr 555,808 598,424	81-present 76-80 65-75
TSCA INITIAL INVENTORY (DIALOG)	Office of Toxic Substances, U.S. EPA	43,278 substances in production	1979
TSCA PLUS (SDC)	Office of Toxic Substances, U.S. EPA	61,000 substances in production	75-present
TSCAPP (CIS, ICIS)	Office of Toxic Substances, U.S. EPA	53,000 substances in production	75-77

Source: Data in this table were obtained from catalogs and manuals of the identified vendors. Specific citations for each data base and vendor are given in the alphabetical descriptions provided in Section 11.14.

similar highly specialized information systems (ICI and ICIS) containing data about regulations and the physical characteristics and toxicity of various chemicals. These systems differ from the ones mentioned above in that they provide data with references rather than abstracts and bibliographic information.

Each data base vendor has a different mix of data bases. Even when the same data base is available from different vendors, the periods covered and the implementation can be quite different. There is a trend for the vendors to include larger portions of the data bases in single files as the costs of computer storage decline. The following table lists the data bases that contain chemistry/physics information relevant to occupational safety and health. The availability by vendor and the periods of coverage are included for reference.

11.5.2 Chemistry/Physics Reference Texts

Bennett, H., ed., *Chemical Formulary.* Chemical Publications, 23 vols. (current), New York, NY, 1933-81. A compendium of chemical formulas of many commercial industrial products, these volumes are useful for identifying the constituents of commercial and industrial products.

Eller, P. M., ed., *NIOSH Manual of Analytical Methods,* 3rd Edition, National Institute for Occupational Safety and Health, Cincinnati, OH, 1984. The third edition of the *NIOSH Manual of Analytical Methods* includes 102 revised methods for over 200 toxic substances. It also contains new chapters on quality assurance, air sampling, development and evaluation of methods, and biological samples. This edition is issued in loose-leaf to accommodate periodic additions and revised methods. The first page of each method section contains summary of sampling and measurement parameters for easy reference. Following this summary are lists of reagents and equipment, concise step-by-step instructions for sampling, sample preparation, measurements, calibration, quality control, and calculations.

Kent, J. A., *Handbook of Industrial Chemistry,* 7th ed., Van Nostrand Reinhold, New York, NY, 1974. Information on many common industrial processes can be found in this well-illustrated handbook. Industrial water supply, waste water technology, and air pollution control are also covered.

Lynch, A. L., *Biological Monitoring for Industrial Chemical Exposure Control,* CRC Press, Boca Raton, FL, 1974. This book presents the rationale and methodology for biological monitoring of humans exposed to toxic chemicals. Topics include collection and analysis of urine, blood, breath, and tissue specimens; physiological monitoring of the respiratory, circulatory, nervous, and genitourinary systems; biologic threshold limits; and quality control for sampling and laboratory analysis.

Sliney, D., and M. Wolbarst, *Safety with Lasers and Other Optical Sources: A Comprehensive Handbook,* Plenum Publications, New York, NY, 1980. This is a handbook on the effects of laser and other light sources on biological organisms. Chapters cover the fundamentals of optical physics, anatomy and physiology of the eye and skin, protective equipment, exposure limits, laser hazard classifications, special applications of lasers (medical, construction, industrial, consumer), and safety training.

Ullmann's Encyclopedia of Industrial Chemistry, 5th ed., VCH Publishers, Deerfield Beach, FL, 1984. This is an international encyclopedia. The 5th edition, for which only the first volume is available, will consist of 28 alphabetically ordered volumes covering all branches of chemical and industrial products. In addition, eight basic knowledge volumes, organized by subject, will cover topics including chemical engineering fundamentals, unit operations, chemical reaction engineering and materials science, analytical methods, and environmental protection and plant safety. New volumes will be published at the rate of three or four per year. A cumulative index will be published annually.

U.S. Department of Health and Human Services, Public Health Service, Centers for Disease Control, National Institute for Occupational Safety and Health, *NIOSH Manual of Sampling Data Sheets,* DHEW (NIOSH) Publication No. 77-159, National Institute for Occupational Safety and Health, Cincinnati, OH, 1977. Available from GPO 017-033-00226-6. Available from NTIS PB 274835. This manual contains 255 NIOSH Sampling Data Sheets, which serve as guides for evaluating compliance with standards that regulate occupational exposure to air contaminants.

U.S. Department of Health and Human Services, Public Health Service, Centers for Disease Control, National Institute for Occupational Safety and Health, *NIOSH Manual of Sampling Data Sheets,* Supplement to 1977 Edition, DHEW (NIOSH) Publication No. 78-189, National Institute for Occupational Safety and Health, Cincinnati, OH, 1977. Available from GPO 017-033-00319-0. Available from NTIS PB 297783. Forty-six Sampling Data Sheets that were not included in the original NIOSH Manual of Sampling Data Sheets are contained in this supplement. These data sheets recommend sampling methodology for measuring occupational exposure to air contaminants.

Weast, R. C., ed., *Handbook of Chemistry and Physics,* 64th ed., CRC Press, Boca Raton, FL, 1983. An important source of essential data on chemistry and physics, this handbook is updated annually.

Other Chemistry/Physics References

Association of Official Analytical Chemists, *Official Methods of Analysis,* 10th ed. and supplements. Association of Official Analytical Chemists, 1965. New editions through 13th ed., Washington, DC, 1980.

Bretherick, L., *Handbook of Reactive Chemical Hazards,* 2nd ed., Butterworths, Woburn, MA, 1979.

Hawley, G. E., ed., *The Condensed Chemical Dictionary,* 9th ed., Van Nostrand Reinhold Co., New York, NY, 1977.

Hutzinger, O., ed., *Reactions & Processes,* Vol. 2, Pt. C., the Handbook of Environmental Chemistry Series, Springer-Verlag, New York, NY, December 1985.

11.5.3 Chemistry/Physics Organizations

American Chemical Society, 1155 16th St., N.W., Washington, DC 20036. The Society, founded in 1876, includes 127,000 chemists, chemical engineers, and others interested in the field. It conducts studies and surveys, offers career counseling, and produces over 15 publications, including the Chemical Abstracts Service and the *Journal of the American Chemical Society*.

American Oil Chemists' Society (AOCS), 508 S. Sixth Street, Champaign, IL 61820. The Society's membership includes chemists, biochemists, chemical engineers, research directors, plant personnel, and others in laboratories and chemical processing industries concerned with animal, marine, and vegetable oils and fats and their extraction, refining, safety, packaging, quality control, and use in consumer and industrial products, including foods, drugs, paints, waxes, lubricants, soaps, and cosmetics. The Society sponsors international conferences and short courses, certifies chemists, distributes cooperative check samples, and sells official reagents. Publications include a journal, *Lipids;* a laboratory manual on methods of analysis; significant symposia; and short-course lectures.

Association of Official Analytical Chemists (AOAC), 1111 N. 19th Street, Suite 210, Arlington, VA 22209. The AOAC membership includes government, academic, and industrial analytical scientists who develop, test, and collaboratively study methods for analyzing fertilizers, foods, feeds, pesticides, drugs, cosmetics, and other products related to agriculture and public health. Annual spring training workshops are conducted to train chemists, microbiologists, and toxicologists in analytical methodology. Association publications include a newsletter; a journal, *Official Methods of Analysis of the AOAC;* and other reference works.

Chemical Manufacturers Association (CMA), 2501 M Street, N.W., Washington, DC 20037. This association administers research in areas of broad import to chemical manufacturing, such as air and water pollution control. It conducts committee studies, workshops, and technical symposia; promotes in-plant safety; and operates the Chemical Transportation Emergency Center (CHEMTREC) that supplies guidance to emergency personnel on handling transportation accidents involving chemicals.

Hazardous Waste Treatment Council (HWTC), 1919 Pennsylvania Avenue, N.W., Suite 300, Washington, DC 20006. The HWTC includes firms dedicated to the use of high technology in managing hazardous wastes and to the restricted use of land disposal facilities in the interest of protecting human health and the environment. The Council advocates reducing the volume of hazardous wastes and the use of alternative technologies in their treatment, including chemical and biological treatments, fixation, neutralization, reclamation, recycling, and thermal treatments such as incineration. It encourages land disposal prohibitions and the development and enforcement of regulations. It also promotes reductions in the volume of hazardous waste generated annually and expansion of the EPA hazardous waste list. The HWTC advocates use of treatment technology as a more cost-effective approach to Superfund site cleanups and works with state, national, and international officials and firms to help develop programs that utilize treatment and minimize land disposal. It also provides technical assistance to members, sponsors special studies, participates in federal litigation and regulatory development, and maintains a library of materials on new technologies.

Intersociety Committee on Methods for Air Sampling and Analysis (ICMASA), Box 612, 425 E. 25th Street, New York, NY 10010. This society includes environmental engineers, chemists, biologists, and physicists involved in the preparation, publication, and use of manuals of recommended methods of air sampling and analysis. They have published *Methods of Air Sampling and Analysis*, 2nd edition, 1977. The Society is affiliated with the Air Pollution Control Association, the American Chemical Society, the American Institute of Chemical Engineers, the American Public Works Association, the American Society of Civil Engineers, the American Society of Mechanical Engineers, the Association of Official Analytical Chemists, the Health Physics Society, and the Instrument Society of America.

Synthetic Organic Chemical Manufacturers Association (SOCMA), 1075 Central Park Avenue, Scarsdale, NY, 10583. The Association, formed in 1921, includes manufacturers of synthetic organic chemicals manufactured from coal, natural gas, crude petroleum, and certain natural substances such as vegetable oils, fats, proteins, carbohydrates, rosin, grains, and their derivatives. Committees deal with emerging issues, environmental quality, occupational safety and health, and toxic substances.

11.5.4 Chemistry/Physics Journals

The *Analyst* covers all phases of the theory and practice of analytical chemistry, including chemical, physical, and biological methods. This publication is included among the most frequently cited journals in NIOSHTIC.

Analytica Chemica Acta is an international journal devoted to all branches of analytical chemistry. Articles include original papers, short communications, and reviews dealing with every aspect of modern chemical analysis, both fundamental and applied. This publication is included among the most frequently cited journals in NIOSHTIC.

Analytical Chemistry covers all branches of analytical chemistry. Articles may be entirely theoretical or report on laboratory experiments that support, argue, refute, or extend established theory. Correspondence, aids for analytical chemists, and articles on the use of computers are included. This publication is included among the most frequently cited journals in NIOSHTIC.

The *Journal of the American Chemical Society* includes papers in all fields of chemistry. Priority is given to articles on "pure" chemistry as distinguished from "applied" chemistry. This publication is included among the most frequently cited journals in NIOSHTIC.

The *Journal of Biological Chemistry* publishes papers on a broad range of topics of general interest to biochemists. The papers are original contributions, but if additional information warrants complete publication of the study, it may be permissible for them to have been published as preliminary communications or presented at a poster session. This publication is included among the most frequently cited journals in NIOSHTIC.

The *Journal of Chromatographic Science* is published monthly by Preston Publications and concerns topics related to chromatography and its applications. This publication is included among the most frequently cited journals in NIOSHTIC.

The *Journal of Colloid and Interface Science* presents original papers, letters to the editors, and book reviews.

The *Journal of Laboratory and Clinical Medicine* is the official publication of the Central Society for Clinical Research. The journal publishes original manuscripts describing investigations in laboratory and clinical medicine. Laboratory methods are described in some articles.

The *Journal of Occupational Medicine* is the official publication of the American Occupational Medical Association and of the American Academy of Occupational Medicine. It presents articles on all phases of occupational medicine and occupational health practice. This publication is included among the most frequently cited journals in NIOSHTIC.

11.6 ENGINEERING CONTROL

This section covers engineering concepts and applications specifically designed to prevent workplace illness and injury by controlling exposure to physical and chemical hazards. It includes the evaluation of work practices and the design of controls that reduce or eliminate exposures to toxic substances and harmful physical agents.

The general approaches for eliminating or reducing worker exposure are substitution, isolation, ventilation, administrative controls, and personal protective equipment. Engineering focuses mainly on the development of techniques for controlling the work environment. The information sources presented in this subsection deal with these topics.

Specifications for items of personal protective equipment may be ordered from the American National Standards Institute (ANSI) (American National Standards Institute, 1430 Broadway, New York, NY 10018) and the American Society for Testing and Materials (ASTM) (American Society for Testing and Materials, 1916 Race Street, Philadelphia, PA 19103).

11.6.1 Engineering Control Data Bases

The explosion of scientific information has made it virtually impossible for the practitioner to obtain access to all of the information published. Not only is the volume of information too great to effectively review, the cost of the journals is prohibitive. Only large university libraries can be expected to have a significant portion of the literature. As a result, electronic information retrieval systems have become a necessity for those seeking information on all but the most common topics.

Until recently, five data base vendors based in the United States provided virtually all of the files of interest to the occupational safety and health professional. These are Dialog Information Services (3460 Hillview Avenue, Palo Alto, CA 94304); the National Library of Medicine (NLM) (8600 Rockville Pike, Bethesda, MD 20209); DOE RECON (Department of Energy, Technical Information Center, P.O. Box 62, Oak Ridge, TN 37830); Bibliographic Retrieval Services (BRS) (BRS Information Technologies, 1200 Route 7, Latham, NY 12110); and System Development Corporation (SDC) Search Services (System Development Corporation, 2500 Colorado Avenue, Santa Monica, CA 90406). New vendors originally available only in Europe now make their information systems available in the United States. They include Pergamon Infoline (Pergamon International Information Corporation, 1340 Old Chain Bridge Road, Mclean, VA 22101), Questel (1625 Eye Street, N.W., Suite 719, Washington, DC 20006), and Information Retrieval Service of the European Space Agency Information Retrieval Service, ESA-IRS, (Esrin, P.O. Box 64, 1-00044, Frascati, Italy). Hundreds of different data bases on subjects ranging from art history to zoology are available from these vendors.

Two other vendors, Information Consultants (ICI) (1133 15th St., N.W., Suite 300, Washington, DC 20005) and Chemical Information Systems (CIS) (7215 York Road, Baltimore, MD 21212) offer similar highly specialized information systems (ICI and ICIS) containing data about the physical characteristics, regulations, and toxicity of various chemicals. These systems differ from the ones mentioned above in that they provide data with references rather than abstracts and bibliographic information.

Each data base vendor has a different mix of data bases. Even when the same data base is available from different vendors, the periods covered and the implementation can be quite different. There is a trend for the vendors to include larger portions of the data bases in single files as the costs of computer storage decline. The following table lists the data bases that contain engineering control information relevant to occupational safety and health. The availability by vendor and the periods of coverage are included for reference.

TABLE 11.5. Engineering Control Data Bases

Name	Producer	No. of Records	Coverage
AQUALINE (DIALOG)	Water Research Centre	75,500(/85)	60-present
BIOTECHNOLOGY (SDC)	Derwent Publications, Ltd.	15,000/yr	82-present
CAS (SDC)	Chemical Abstracts	1.3 million	67-71
		1.7 million	72-76
		2.2 million	77-81
		340,000/yr	82-present
CA SEARCH (DIALOG)	Chemical Abstracts	1.3 million	67-71
		1.7 million	72-76
		1.2 million	77-79
		.9 million	80-81
		400,000/yr	82-present
CA SEARCH (PERGAMON)	Chemical Abstracts	1.7 million	72-76
		400,000/yr	77-present
CA SEARCH (BRS)	Chemical Abstracts	Not available (N.A.)	70-76
		N.A.	77-present
CHEMICAL ENGINEERING ABSTRACTS (PERGAMON)	The Royal Society of Chemistry	63,000	70-present
CIS (Questel)	International Labour Organization/CIS	2,100/yr	74-present
CISDOC (ESA)	International Labour Organization/CIS	2,100/yr	74-present
COMPENDEX (ESA)	Engineering Information, Inc.	100,000/yr	69-present
COMPENDEX (DIALOG, SDC)	Engineering Information, Inc.	100,000/yr	70-present
COMPENDEX (PERGAMON)	Engineering Information, Inc.	100,000/yr	73-present
CURRENT BIOTECHNOLOGY ABSTRACTS (PERGAMON)	The Royal Society of Chemistry	7,000	83-present
DIRLINE (NLM)	National Library of Medicine	14,000	N.A.
DISSERTATION ABSTRACTS (DIALOG, BRS)	University Microfilms International	42,000/yr	1861-present
DOE ENERGY (DIALOG)	TIC, ORNL, DOE (Energy Data Base)	144,000/yr	74-present
EDB (DOE)	TIC, ORNL, DOE (Energy Data Base)	192,000/yr	1900-present
EDB (SDC)	TIC, ORNL, DOE (Energy Data Base)	800,000	30s-present
EI ENGINEERING Meetings (SDC, ESA, DIALOG)	Engineering Information, Inc.	108,000/yr	82-present
EIX (COMPENDEX) (DOE)	Engineering Information, Inc.	100,000/yr	74-present
ENERGYLINE (SDC, ESA, DIALOG)	EIC/ Intelligence	5,250/yr	71-present

TABLE 11.5. *(continued)*

Name	Producer	No. of Records	Coverage
ENVIROLINE (SDC, ESA, DOE, DIALOG)	EIC/ Intelligence	110,000	71-present
ENVIRONMENTAL BIBLIOGRAPHY (DIALOG)	Environmental Studies Institute	24,000/yr	73-present
EUCAS (QUESTEL)	Chemical Abstracts	1.3 million 1.7 million 2.2 million 340,000/yr	67-71 72-76 77-81 82-present
FOODS ADLIBRA (DIALOG)	K & M Publications, Inc.	3,600/yr	74-present
FSTA (DIALOG, SDC)	International Food Information Services	20,000/yr	69-present
HSELINE (PERGAMON, ESA)	Health and Safety Executive	47,000	77-present
INSPEC (BRS)	IEEE	N.A. N.A.	70-76 77-present
INSPEC (DIALOG)	IEEE	1.1 million 180,000/yr	69-77 78-present
INSPEC (ESA)	IEEE	150,000/yr	71-present
INSPEC (SDC)	IEEE	150,000/yr	69-present
ISMEC (DIALOG)	Information Service in Mechanical Engineering	15,000/yr	73-present
KIRK-OTHMER ENCYCLOPEDIA OF CHEMICAL TECHNOLOGY (BRS)	John Wiley and Sons, Inc.	Full text of encyclopedia	3rd ed. 76
LABORDOC (ESA, QUESTEL, SDC)	International Labour Organization	4,800/yr	65-present
METADEX (DIALOG)	American Society for Metals	577,000	74-present
NASA (ESA)	NASA, Scientific and Technical Information Branch	N.A.	62-present
NASA TECH BRIEFS (DOE)	NASA Scientific and Technical Information Branch	N.A.	N.A.
NIOSHTIC (DIALOG, PERGAMON)	National Institute for Occupational Safety and Health (NIOSH)	12,000/yr	1900-present
NTIS (DIALOG)	National Technical Information Service	60,000/yr	64-present
NTIS (ESA)	National Technical Information Service	60,000/yr	69-present
NTIS (SDC, BRS)	National Technical Information Service	60,000/yr	70-present
PAPERCHEM (SDC, DIALOG)	The Institute of Paper Chemistry	12,000/yr	68-present
PASCAL (QUESTEL)	Centre de Documentation Scientifique et Technique	1.7 million 2.6 million	73-76 77-present

TABLE 11.5. *(continued)*

Name	Producer	No. of Records	Coverage
PHP (DOE)	American Petroleum Institute	7,947	77-80
PIRA (PERGAMON)	The Paper Industries Research Association	10,000/yr	75-present
POLLUTION ABSTRACTS (ESA, DIALOG, BRS)	Cambridge Scientific Abstracts	8,400/yr	70-present
RAPRA ABSTRACTS (PERGAMON)	Rubber and Plastics Research Association of Great Britain	23,000/yr	72-present
RBOT (BRS)	Cincinnati Millicron	N.A.	80-present
SAE (SDC)	Society of Automotive Engineers	800/yr	65-present
TEXTILE TECHNOLOGY DIGEST (DIALOG)	Institute of Textile Technology	97,500	78-present
TSCA INITIAL INVENTORY (DIALOG)	Office of Toxic Substances, U.S. EPA	43,278 substances in production	1979
TSCA PLUS (SDC)	Office of Toxic Substances, U.S. EPA	61,000 substances in production	75-present
TSCAPP (CIS, ICIS)	Office of Toxic Substances, U.S. EPA	53,000 substances in production	75-77
VOLUNTARY STANDARDS INFORMATION NETWORK (BRS)	Information Handling Services	over 500	N.A.
WATER RESOURCES ABSTRACTS (DIALOG, DOE)	U.S. Department of Interior	172,500	68-present
WATERLIT (SDC)	South African Water Information Centre	12,000/yr	76-present
WATERNET (DIALOG)	American Water Works Association	3,000/yr	71-present
WELDASEARCH (DIALOG)	The Welding Institute	4,500/yr	67-present
WORLD ALUMINUM ABSTRACTS (DIALOG)	American Society for Metals	102,000	68-present
WORLD SURFACE COATINGS ABSTRACTS (PERGAMON)	Paint Research Association	8,400/yr	76-present
WORLD TEXTILE ABSTRACTS (PERGAMON, DIALOG)	Shirley Institute	9,000/yr	70-present
ZINC, LEAD AND CADMIUM ABSTRACTS (PERGAMON)	Zinc Development Association	3,000/yr	75-present

Source: Data in this table were obtained from catalogs and manuals of the identified vendors. Specific citations for each data base and vendor are given in the alphabetical descriptions provided in section 11.14.

11.6.2 Engineering Control Reference Texts

Alden, J. L., and J. M. Kane, *Design of Industrial Ventilation Systems,* 5th ed., Industrial Press, New York, NY, 1982. This seven-volume set provides information on the design, maintenance, and evaluation of industrial ventilation systems. Chapters cover air; water; inorganic chemicals and nucleonics; nonmetallic ores; silicate industries and solid mineral fuels; metals and ores; petroleum and organic chemicals; natural organic materials and related synthetic products; wood, paper, and textiles; plastics and photographic materials; vegetable food products and luxuries; and edible oils, fats, and animal food products. The fifth edition is an extensive update of the classic guide. The manual contains comprehensive information on the requirements for designing, purchasing, and operating exhaust systems. This revised edition also covers general exhaust ventilation and makeup air supply. A new chapter addressing energy conservation has been added in this edition. This book is a good central source of information about the design of industrial ventilation systems.

Burgess, W. A., *Recognition of Health Hazards in Industry: A Review of Materials and Processes,* Wiley-Interscience, New York, NY, 1981. Over thirty industrial operations are reviewed in this book, including materials used, engineering controls, possible worker hazards, proper work practices, and protective equipment. Technical appendices and a subject index are included.

Cralley, L. J., and L. V. Cralley, eds., *Industrial Hygiene Aspects of Plant Operations,* Macmillan, New York, NY, 3 vols. Volume I, *Process Flows,* 1983; Volume II, *Unit Operations and Product Fabrication,* 1983; Volume III, *Equipment Selection, Layout, and Building Design,* 1984. This three-volume work describes basic industrial hygiene applications and includes photographs and diagrams pertaining to a wide variety of industrial processes.

Fawcett, H. H., and W. S. Wood, eds., *Safety and Accident Prevention in Chemical Operations,* 2nd ed., John Wiley and Sons, New York, NY, 1982. This book provides excellent source material for chemical laboratory safety officers, industrial toxicologists, and industrial hygienists. It is a thorough review of accident prevention strategies and contains reference material on toxicity and the TSCA law, hazardous chemical waste disposal, and developments in fire extinguishment.

Kirk, H., and R. Othmer, eds., *Encyclopedia of Chemical Technology,* 3rd ed., Wiley-Interscience, New York, NY, 1978–83, 23 vols. This encyclopedia is the single most comprehensive reference for chemical engineering in existence. The 23 volumes cover the entire spectrum of chemical engineering with extensive discussions and clear illustrations. This is the best single reference for information about manufacturing processes for plastics and chemicals.

McDermott, H. J., *Handbook of Ventilation for Contaminant Control,* Ann Arbor Science Publishers, Ann Arbor, MI, 1976. This text focuses on the design of ventilation systems for controlling worker exposures to industrial contaminants. It includes a survey of the theoretical basis for design requirements and methods for evaluating ventilation systems.

Parmeggiani, L., ed., *Encyclopedia of Occupational Safety and Health,* 3rd rev. ed., 2 vols., International Labour Organization, Washington, DC, 1983. This third edition (1983) is the latest version of a comprehensive two-volume reference work with concise, easy-to-understand entries.

Perry, R. H., and C. H. Chilton, eds., *Chemical Engineer's Handbook,* 5th ed., McGraw-Hill, New York, NY, 1973. A ready reference for engineers, this book includes comprehensive data on all aspects of chemical engineering.

Ullmann's Encyclopedia of Industrial Chemistry, 5th ed., VCH Publishers, Deerfield Beach, FL, 1984. This is an international encyclopedia. The fifth edition, of which only the first volume is available, will consist of 28 alphabetically ordered volumes covering all branches of chemical and industrial products. In addition, eight basic knowledge volumes cover such topics as chemical engineering fundamentals, unit operations, chemical reaction engineering and materials science, analytical methods, and environmental protection and plant safety. New volumes will be published at the rate of three or four per year. A cumulative index will be published annually.

U.S. Department of Health and Human Services, Public Health Service, Centers for Disease Control, National Institute for Occupational Safety and Health, *The Industrial Environment: Its Evaluation and Control,* National Institute for Occupational Safety and Health, Cincinnati, OH, 1973. Available from GPO 017-001-00396-4. Published as a basic text for graduate study in the field of occupational safety and health, this book is currently being revised but remains one of the most comprehensive treatments of the field in textbook format. The first section of the book provides a review of mathematics, chemistry, biochemistry, physiology, and toxicology as they relate to the practice of occupational safety and health, and specifically to industrial hygiene. The remaining chapters concentrate on the methodology of industrial hygiene practice, especially the evaluation of the harmful effects of physical agents and air contaminants.

Other Engineering Control References Texts

American National Standards Institute, *American National Standard Practices for Respiratory Protection,* Z88.2-1969, August 1969.

American National Standards Institute, *American National Standard Practices for Respiratory Protection,* Z88.2-1980, May 1980.

American Standards Association, *American Standard Safety Code for the Protection of Heads, Eyes, and Respiratory Organs,* Z2-1938, December 1938.

American Standards Association, *American Standard Safety Code for Head, Eye, and Respiratory Protection,* Z2.1-1959, November 1959.

Blackwell, D. S., and G. S. Rajhans, eds., *Practical Guide to Respiratory Usage in Industry,* Butterworth, Woburn, MA, July 1985.

Blair, A., *Abrasive Blasting Respiratory Protective Practices Survey,* Report No. NIOSH-TR-048-73, National Institute for Occupational Safety and Health, Cincinnati, OH, August 1973.

Brandt, A. D., *Industrial Health Engineering,* John Wiley and Sons, New York, NY, 1947.

Caplan, K., "Philosophy and Management of Engineering Controls," in L. J. Cralley and L. V. Cralley, eds., *Patty's Industrial Hygiene and Toxicology,* Vol. III, John Wiley and Sons, New York, NY, 1979.

Daly, B. B., *Woods Practical Guide to Fan Engineering,* International Publications Service, New York, NY, 1978.

Fawcett, H. H., "Respiratory Hazards and Protection," in A. Hack, O. D. Bradley, and A. Trujillo, *Respirator Studies for the Nuclear Regulatory Commission, Oct. 1, 1976–Sept. 30, 1977, Protection Factors for Supplied-Air Respirators,* Los Alamos Scientific Laboratory, Los Alamos, NM, January 1978.

Held, B. J., "Personal Protection," in L. J. Cralley and L. V. Cralley, eds., *Patty's Industrial Hygiene and Toxicology,* Vol. III, John Wiley and Sons, New York, NY, 1979.

Hickey, J., L. S. Rice, and C. H. Boehlecke, "Technologies for Controlling Worker Exposure to Silica" (Working Paper #17) in *Preventing Illness and Injury in the Workplace,* Contract No. J-9-F-6-0225, report prepared for the Occupational Safety and Health Administration, U.S. Department of Labor, Washington, DC, September 1978.

Krause, F. E., "The Prevention of PVC Reactor Fouling," in *Symposium Proceedings — Control Technology in the Plastics and Resins Industry,* DHHS (NIOSH) Publication No. 81-107, NIOSH, Cincinnati, OH, January 1981.

McQuiston, F. C., and J. D. Parker, *Heating, Ventilating, and Air Conditioning: Analysis and Design,* John Wiley and Sons, New York, NY, 1977.

Miller, M., ed., *Toxic Control, Volume IV: Toxic Control in the Eighties,* Gov. Insts., 1980.

National Research Council, *Prudent Practices for Handling Hazardous Chemicals in Laboratories,* National Academy Press, Washington, DC, 1981.

Peterson, J. E., "Principles for Controlling the Occupational Environment," in *The Industrial Environment: Its Evaluations and Control,* U.S. Department of Health, Education, and Welfare, National Institute for Occupational Safety and Health, U.S. Government Printing Office, Washington, DC, 1973.

Pritchard, J. A., *A Guide to Industrial Respiratory Protection,* DHEW (NIOSH) Publication No. 76-189, National Institute for Occupational Safety and Health, Cincinnati, OH, June 1976.

Purswell, J. L., and R. Stephens, "Health and Safety Control Technologies in the Workplace: Accident Causation and Injury Control," (Working Paper #11) in *Preventing Illness and Injury in the Workplace,* contract report prepared for the Office of Technology Assessment, U.S. Congress, Washington, DC, July 1983.

Rahjans, G. S., and G. M. Bragg, *Engineering Aspects of Asbestos Dust Control,* Ann Arbor Science Publishers, Ann Arbor, MI, 1978.

Soule, R. D., "Industrial Hygiene Engineering Controls," in L. J. Cralley and L. V. Cralley, eds., *Patty's Industrial Hygiene and Toxicology,* Vol. III, John Wiley and Sons, New York, NY, 1979.

Springborn Management Services, "Report on Workplace Protective Equipment and Clothing," (Working Paper #10) in *Preventing Illness and Injury in the Workplace,* contract report prepared for the Office of Technology Assessment, U.S. Congress, Washington, DC, August 1982.

Stamper, E., and R. L. Koral, eds., *Handbook of Air Conditioning, Heating, and Ventilating,* Industrial Press, New York, NY, 1979.

U.S. Department of Health and Human Services, Public Health Service, Centers for Disease Control, National Institute for Occupation Safety and Health, *An Evaluation of Occupational Health Hazard Control Technology for the Foundry Industry,* HEW (NIOSH) Publication No. 79-144, National Institute for Occupational Safety and Health, Cincinnati, OH, 1978. Available from NTIS PB 80-164-908/A08.

U.S. Department of Health and Human Services, Public Health Service, Centers for Disease Control, National Institute for Occupational Safety and Health, *Engineering Control and Work Practices Manual,* HEW (NIOSH) Publication No. 76-179, National Institute for Occupational Safety and Health, Cincinnati, OH, 1976. Available from NTIS PB 273-296/A08.

U.S. Department of Health and Human Services, Public Health Service, Centers for Disease Control, National Institute for Occupational Safety and Health, *Engineering Control of Occupational Safety and Health Hazards: Recommendations for Improving Engineering Practice, Education, and Research,* Summary Report of the Engineering Control Technology Workshop Technical Panel, DHHS (NIOSH) Publication No. 84-102, National Institute for Occupational Safety and Health, Cincinnati, OH, July 1983. Available from NTIS 85-125-615 1A02/125-615/A0Z.

U.S. Department of Health and Human Services, Public Health Service, Centers for Disease Control, National Institute for Occupational Safety and Health, *Recommended Industrial Ventilation Guidelines,* HEW (NIOSH) Publication No. 76-172, National Institute for Occupational Safety and Health, Cincinnati, OH, 1976. Available from NTIS PB 271-811/A07.

U.S. Department of Health and Human Services, Public Health Service, Centers for Disease Control, National Institute for Occupational Safety and Health, *Proceedings of the Symposium on Occupational Health Hazards Control Technology in the Foundry and Secondary Non-ferrous Smelting Industries,* DHHS (NIOSH) Publication No. 81-114, National Institute for Occupational Safety and Health, Cincinnati, OH, August 1981. Available from NTIS PB 81-167-710/A20.

U.S. Department of Health and Human Services, Public Health Service, Centers for Disease Control, National Institute for Occupational Safety and Health, *Symposium Proceedings: Control Technology in the Plastics and Resins Industry,* DHHS (NIOSH) Publication No. 81-107, National Institute for Occupational Safety and Health, Cincinnati, OH, 1981, January 1981. Available from NTIS PB 82-235-599/A15.

User Guide to Dust and Fume Control, State Mutual Book and Periodical Service, New York, NY, 1981.

Van Cott, H. P., and R. G. Kinkade, eds., *Human Engineering Guide to Equipment Design,* U.S. Government Printing Office, Washington, DC, 1972.

11.6.3 Engineering Control Organizations

American Society of Heating, Refrigeration and Air Conditioning Engineers, 1791 Tullie Circle, Atlanta, GA 30329. The Society, founded in 1894, consists of about 45,000 engineers in the field. It sponsors research and committees in specialized areas. Publications include a monthly journal, a handbook, and a membership roster.

American Society of Mechanical Engineers, 345 E. 47th Street, New York, NY 10017. The Society, founded in 1880, currently has a membership of over 100,000. It is a technical society that sponsors research, develops codes, sponsors the American National Standards Institute, and maintains a large library.

American Society of Safety Engineers, 850 Busse Highway, Park Ridge, IL 60068. The Society, founded in 1911, has over 18,000 members interested in safety engineering and safety program management. Publications include *Professional Safety,* monographs, and textbooks.

Center for Chemical Plant Safety, Institute of Chemical Engineers, 345 E. 47th Street, New York, NY 10017. The Center is operated by the Institute to focus attention on chemical industry safety. The Center was formed as a response to the concern for safety after the Bhopal incident.

Robotics International of SME (RI/SME), P.O. Box 930, One SME Drive, Dearborn, MI 48128. This group, established in 1980, presently has a membership of 10,000 engineers, managers, educators, and government officials interested in industrial robotics. Robotics International considers the following specific topics: human factors, standards, and industrial robotics safety. They conduct symposiums, seminars, and clinics on most aspects of industrial robots and disseminate technical information. The organization maintains a library of 2000 volumes. Major divisions include assembly, castings and foundries, coatings, human factors, machine loading and unloading, materials handling, research and development, and welding. The group's publications include *Robotics Today,* books, technical papers, and reports on industrial robots.

11.6.4 Engineering Control Journals

The *American Industrial Hygiene Association Journal* focuses on all aspects of industrial hygiene practice, policy, and procedures and reports on specific investigations and field activities. This publication is included among the most frequently cited journals in NIOSHTIC.

The *Journal of Applied Physics,* published by the American Institute of Physics, focuses on general physics and its applications to other sciences, engineering, and industry. Priority is given to new experimental results in applied physics or applications of physics to other disciplines.

The *Journal of Occupational Medicine* is the official publication of the American Occupational Medical Association and of the American Academy of Occupational Medicine. It presents articles on all phases of occupational medicine and occupational health practice. This publication is included among the most frequently cited journals in NIOSHTIC.

Noise Control Engineering is published by the Institute of Noise Control Engineering, Purdue University. This publication is included among the most frequently cited journals in NIOSHTIC.

Professional Safety, the official publication of the American Society of Safety Engineers, presents a forum for the discussion of safety engineering and safety program administration.

11.7 EPIDEMIOLOGY/SURVEILLANCE

The United States does not have a standardized data collection system for the surveillance of disease associated with the work environment. The epidemiology and surveillance of occupational illness and injury consists of the continuous observation and the collection, analysis, and dissemination of data. To successfully characterize the illness and injury experience of workers, there must be a well-planned and systematic activity. Observation over time provides the data necessary for the application of epidemiologic methods to analyze the data and report the type and extent of disease and injury occurring among workers and occupational groups.

The utility of surveillance programs and epidemiologic analysis in describing the health impact of the employment experience is clear. They assist in pinpointing the causative factors in specific types of disease and help identify the working populations at greatest risk. Surveillance programs and epidemiology programs also help identify emerging problems that indicate that prevention efforts have failed.

The references provided in this section on epidemiology and surveillance cover both concept and practice. Much of the relevant data and many pertinent discussions of methodology are relatively new; most of the references are found in the periodic literature. For a detailed study of these subjects it is often necessary to consult this current literature directly. The most efficient way to identify these materials is the use of structured queries of the appropriate on-line bibliographic data bases described below. In addition, considerable useful information is integrated with materials from other professional disciplines. For example, reference materials on toxicology often include examples and discussions of epidemiologic methods and surveillance programs.

11.7.1 Epidemiology/Surveillance Data Bases

The explosion of scientific information has made it virtually impossible for the practitioner to obtain access to all of the information published. Not only is the volume of information too great to effectively review, the cost of the journals is prohibitive. Only large university libraries can be expected to have a significant portion of the literature. As a result, electronic information retrieval systems have become a necessity for those seeking information on all but the most common topics.

Until recently, five data base vendors based in the United States have provided virtually all of the files of interest to the occupational safety and health professional. These are Dialog Information Services (3460 Hillview Avenue, Palo Alto, CA 94304); the National Library of Medicine (NLM) (8600 Rockville Pike, Bethesda, MD 20209); DOE RECON (Department of Energy, Technical Information Center, P.O. Box 62, Oak Ridge, TN 37830); Bibliographic Retrieval Services (BRS) (BRS Information Technologies, 1200 Route 7, Latham, NY 12110); and System Development Corporation (SDC) Search Services (System Development Corporation, 2500 Colorado Avenue, Santa Monica, CA 90406). New vendors that were originally available only in Europe now make their information systems available in the United States. These include Pergamon Infoline (Pergamon International Information Corporation, 1340 Old Chain Bridge Road, Mclean, VA 22101), Questel (1625 Eye Street, N.W., Suite 719, Washington, DC 20006), and Information Retrieval Service of the European Space Agency Information Retrieval Service, ESA-IRS, (Esrin, P.O. Box 64, 1-00044, Frascati, Italy). Hundreds of different data bases on subjects ranging from art history to zoology are available from these vendors.

Two other vendors, Information Consultants (ICI) (1133 15th St. N.W., Suite 300, Washington, DC 20005) and Chemical Information Systems (CIS) (7215 York Road, Baltimore, MD 21212), offer similar highly specialized information systems (ICI and ICIS) containing data about regulations and the physical characteristics and toxicity of various chemicals. These systems differ from the ones mentioned above in that they provide data with references rather than abstracts and bibliographic information.

Each data base vendor has a different mix of data bases. Even when the same data base is available from different vendors, the periods covered and the implementation can be quite different. There is a trend for the vendors to include larger portions of the data bases in single files as the costs of computer storage decline. The following table lists the data bases that contain epidemiology/surveillance information relevant to occupational safety and health. The availability by vendor and the periods of coverage are included for reference.

Frequently data bases dealing with the specific professional disciplines in the biological and medical sciences include some references valuable to the study and practice of occupational disease and injury surveillance and epidemiology. The following data bases are considered useful in identifying relevant materials on one or more aspects of either surveillance or epidemiology.

TABLE 11.6. Epidemiology/Surveillance Data Bases

Name	Producer	No. of Records	Coverage
BIOSIS (BRS)	Biosciences Information Services	Not available (N.A.) N.A.	70–77 78–present
BIOSIS (ESA)	Biosciences Information Services	200,000/yr	73–present
BIOSIS (SDC)	Biosciences Information Services	200,000/yr	69–73 74–79 80–present
BIOSIS PREVIEWS (DIALOG)	Biosciences Information Services	2 million 1.4 million 328,000/yr	69–76 77–81 82–present
CANCERLIT (NLM)	International Cancer Research Data Bank	432,518	63–present
CIS (Questel)	International Labour Organization/CIS	2,100/yr	74–present
CISDOC (ESA)	International Labour Organization/CIS	2,100/yr	74–present
EEIS (DOE)	TIRC, ORNL, DOE	5,141	60–present
EUCAS (QUESTEL)	Chemical Abstracts	1.3 million 1.7 million 2.2 million 340,000/yr	67–71 72–76 77–81 82–present
HSELINE (PERGAMON, ESA)	Health and Safety Executive	47,000	77–present
MEDLINE MED80 MED77 MED75 MED71 MED66 (NLM)	National Library of Medicine	140,000/yr 804,874 775,167 642,710 883,906 954,650	83–present 80–82 77–79 75–76 71–74 66–70
MEDLINE (DIALOG)	National Library of Medicine	140,000/yr 1.7 million 1.5 million	80–present 73–79 66–72
MEDLINE (BRS)	National Library of Medicine	140,000/yr N.A. 883,906 954,650	79–present 75–78 71–74 66–70
NIOSHTIC (DIALOG, PERGAMON)	National Institute for Occupational Safety and Health (NIOSH)	12,000/yr	1900–present
PSYCINFO (SDC, DIALOG, BRS)	American Psychological Association	24,000/yr	67–present
SOCIOLOGICAL ABSTRACTS (DIALOG, BRS)	Sociological Abstracts, Inc.	8,500/yr	63–present
SOCIAL SCISEARCH (DIALOG, BRS)	Institute for Scientific Information	120,000/yr	72–present

Source: Data in this table were obtained from catalogs and manuals of the identified vendors. Specific citations for each data base and vendor are given in the alphabetical descriptions provided in section 11.14.

11.7.2 Epidemiology/Surveillance Reference Texts

Austin, D. F., and S. Benson Werner, *Epidemiology for the Health Sciences,* Charles C. Thomas Publisher, Springfield, IL, 1973. This text was designed for an introductory course in epidemiology.

Bridbord, K., and J. French, eds., *Toxicological and Carcinogenic Health Hazards in the Workplace,* Chem-Orbital, Park Forest South, IL, 1979. These are the proceedings of the first annual NIOSH Scientific Symposium. Held in 1978, the symposium covered a wide variety of topics related to illnesses resulting from working with chemical and physical agents. Epidemiologic studies of illness in workers exposed to specific agents are also included.

Chiazze, L., Jr., F. E. Lundin, and D. Watkins, *Methods and Issues in Occupational and Environmental Epidemiology,* Butterworth Publishers, Woburn, MA, 1983. This book deals with the application of epidemiology to occupational and environmental health policy. Major areas covered include data sources, design and analysis of epidemiological studies, various sources of bias, multiple etiological factors, and impediments of epidemiology.

Coulston, F., and P. Shubik, eds., *Human Epidemiology and Animal Laboratory Correlations in Chemical Carcinogenesis,* Ablex Publications Corp., Norwood, NJ, 1980. This is a collection of papers presented at a conference on the application of data from laboratory studies to the study of human populations.

Fishbein, L., *Potential Industrial Carcinogens and Mutagens,* Elsevier, Amsterdam, 1979. This is a volume in the Studies in Environmental Science series. The purpose of this volume was to present information regarding carcinogenic and/or mutagenic organic chemical compounds used by industry. Epidemiology, risk-assessment, and combined effects are also treated.

Fox, J. P., C. E. Hall, and L. R. Elveback, *Epidemiology: Man and Disease,* Macmillan, New York, NY, 1970. This text provides an introduction to epidemiology covering study design, sampling, and data analysis.

Interagency Regulatory Liaison Group, *Epidemiology Research Projects Directory,* Vol. 1, Interagency Regulatory Liaison Group, Washington, DC, 1980. Available from NTIS PB 80-111305. The directory provides a listing of publicly and privately funded studies in all areas of epidemiology. Each entry description provides title, investigator, affiliation, address, grant/control number, description, and funding.

Key, M. M., A. F. Henschel, J. Butler, R. N. Ligo, and I. R. Tabershaw, eds., *Occupational Diseases: A Guide to Their Recognition,* 2nd ed., National Institute for Occupational Safety and Health, Cincinnati, OH, 1977. Available from NTIS PB 83-129-528. This book is an easy-to-use reference about the causes of occupational diseases. It was written as an aid to physicians, nurses, and other members of the health professions to help identify and control occupational diseases. Sections cover routes of entry of chemical agents, chemical hazards, biological hazards, chemical carcinogens, pesticides, physical agents, and occupational dermatosis.

Mausner, J. S., and A. K. Bahn, *Epidemiology: An Introductory Text,* W. B. Saunders Company, Philadelphia, PA, 1974. This text was designed for an introductory course in epidemiology and community medicine.

Monson, R. R., *Occupational Epidemiology,* CRC Press, Boca Raton, FL, 1982. This book presents the fundamentals of epidemiology as they relate to occupational health. It explores the historical basis of epidemiology, discusses the nature of epidemiologic data and standard methodology, and describes epidemiological methods as they apply to the study of workers.

Peto, R., and M. Schneiderman, eds., *Quantification of Occupational Cancer: Banbury Report 9,* Cold Spring Harbor Laboratory, Cold Spring Harbor, NY, 1981. This report is a comprehensive discussion of the methodological problems of estimating the occupational contribution to the incidence of cancer.

U.S. Department of Health and Human Services, Public Health Service, Centers for Disease Control, National Institute for Occupational Safety and Health, *A Conceptual Framework for Occupational Health Surveillance,* DHEW (NIOSH) Publication No. 78-135, National Institute for Occupational Safety and Health, Cincinnati, OH, 1978. Available from NTIS PB 83-175-414/A03 (National Technical Information Service, Springfield, VA 22161). This paper explores existing practice in occupational health surveillance. A flow diagram is presented that shows how various sources of data can be combined to facilitate a better understanding of occupational diseases.

U.S. Department of Health and Human Services, Public Health Service, Centers for Disease Control, National Institute for Occupational Safety and Health, *Reliability and Utilization of Occupational Disease Data,* DHEW (NIOSH) Publication No. 77-189, National Institute for Occupational Safety and Health, Cincinnati, OH, 1978. Available from NTIS PB 80-175-698/A08. This publication discusses existing occupational disease reporting and data collection and the trend toward improvement.

OTHER EPIDEMIOLOGY/SURVEILLANCE REFERENCE TEXTS

California Department of Industrial Relations, Division of Labor Statistics and Research, *California Logging Industry: Analysis of Work Injuries and Illnesses,* Research Bulletin #2, November 1978.

California Department of Industrial Relations, Division of Labor Statistics and Research, *California*

Sawmills and Planing Mills Industry: Analysis of Work Injuries and Illnesses, Research Bulletin #3, December 1978.

California Department of Industrial Relations, Division of Labor Statistics and Research, *California Fabricated Structural Metal Products Industry: Analysis of Work Injuries and Illnesses,* Research Bulletin #4, December 1978.

California Department of Industrial Relations, Division of Labor Statistics and Research, *California Meat Products Industry: Analysis of Work Injuries and Illnesses,* Research Bulletin #5, December 1979.

California Department of Industrial Relations, Division of Labor Statistics and Research, *California Roofing and Sheet Metal Work: Analysis of Work Injuries and Illnesses,* Research Bulletin #6, December 1982.

Gordon, J., A. Akman, and M. Brooks, *Industrial Safety Statistics: A Re-examination,* Praeger, New York, NY, 1971.

Haddon, W., Jr., E. A. Suchman, and D. Klein, eds., "The Epidemiology of Accidents," in *Accident Research Methods and Approaches,* Harper and Row, New York, NY, 1964.

Health Resources Administration, *Statistics Needed for Determining the Effects of the Environment on Health: Report of the Technical Consultant Panel to the United States National Committee on Vital and Health Statistics,* DHEW (HRA) Publication No. 77-1457, Series 4-No. 20, Health Resources Administration, Rockville, MD, 1977.

ICF, *Development of a Methodology for Production of Industry Specific Research Planning Information,* report prepared for the U.S. Environmental Protection Agency and the National Institute for Occupational Safety and Health, Washington, DC, April 1982.

Kronebusch, K., "Occupational Injury Data: Are We Collecting What We Need for Identification, Prevention, and Evaluation?" in *Priorities in Health Statistics,* pp. 124–129, Proceedings of the 19th National Meeting to the Public Health Conference on Records and Statistics, August 1983, DHHS (PHS) Publication No. 81-1214, National Center for Health Statistics, Hyattsville, MD, December 1983.

MacMahon, B., and T. F. Pugh, *Epidemiology: Principles and Methods,* Little, Brown and Co., Boston, MA, 1970.

McFarland, R. A., *The Epidemiology of Industrial Accidents,* Harvard School of Public Health, Cambridge, MA, 1965.

Mendeloff, J., "An Analysis of OSHA Health Inspection Data," (Working Paper #2) of *Preventing Illness and Injury in the Workplace,* contract report prepared for U.S. Congress, Office of Technology Assessment, Washington, DC, April 1983.

National Safety Council, *Accident Facts,* National Safety Council, Chicago, IL, 1985.

Ries, P. W., "Episodes of Persons Injured: United States, 1975" in *Advance Data,* Number 18, DHEW (PHS) Publication No. 78-1250, National Center for Health Statistics, Hyattsville, MD, March 7, 1978.

Rockette, H., *Mortality Among Coal Miners Covered by the UMWA Health and Retirement Funds,* DHEW (NIOSH) Publication No. 77-155, National Institute for Occupational Safety and Health, Cincinnati, OH, 1977. Available from NTIS PB 262-472/A07.

Safety Sciences, *Feasibility of Securing Research-Defining Accident Statistics,* DHHS (NIOSH) Publication No. 78-180, National Institute for Occupational Safety and Health, Cincinnati, OH, September 1978. Available from NTIS PB 297-814/A14.

U.S. Congress, General Accounting Office, *Better Data on Severity and Causes of Worker Safety and Health Problems Should be Obtained From Workplaces, HRD-76-118,* U.S. Government Printing Office, Washington, DC, August 12, 1976.

U.S. Congress, House Committee on Government Operations, *Occupational Illness Data Collection: Fragmented, Unreliable, and Seventy Years Behind Communicable Disease Surveillance,* House Report No. 98-1144, U.S. Government Printing Office, Washington, DC, 1984.

U.S. Congress, Office of Technology Assessment, *Assessment of Technologies for Determining Cancer Risks From the Environment,* OTA-H-138, U.S. Government Printing Office, Washington, DC, June 1981.

U.S. Department of Commerce, Bureau of the Census, *County Business Patterns, 1981: United States,* U.S. Government Printing Office, Washington, DC, 1982.

U.S. Department of Health and Human Services, Public Health Service, Centers for Disease Control, National Institute for Occupational Safety and Health, *A Model for the Identification of High Risk Occupational Groups Using RTECS and NOHS Data,* DHHS (NIOSH) Publication No. 83-117, National Institute for Occupational Safety and Health, Cincinnati, OH, 1983. Available from NTIS PB 84-190-685/A09.

U.S. Department of Health and Human Services, Public Health Service, Centers for Disease Control, National Institute for Occupational Safety and Health, *National Occupational Hazard Survey, 1972–1974, Vol. I, Survey Manual,* DHEW (NIOSH) Publication No. 74-127, National Institute for Occupational Safety and Health, Cincinnati, OH, 1974. Available from NTIS PB 274-241/A10.

U.S. Department of Health and Human Services, Public Health Service, Centers for Disease Control, National Institute for Occupational Safety and Health, *National Occupational Hazard Survey, 1972–1974, Vol. II, Data Editing and Data Base Development,* DHEW (NIOSH) Publication No. 77-213, National Institute for Occupational Safety and Health, Cincinnati, OH, 1977. Available from NTIS PB 274-819/A08.

U.S. Department of Health and Human Services, Public Health Service, Centers for Disease Control, National Institute for Occupational Safety and Health, *National Occupational Hazard Survey, 1972–1974, Vol. III, Survey Analysis and Supplemental Tables,* DHEW (NIOSH) Publication No. 78-114, National Institute for Occupational Safety and Health, Cincinnati, OH, 1977. Available from NTIS PB 82-299-881/A99.

U.S. Department of Health and Human Services, Public Health Service, Centers for Disease Control, National Institute for Occupational Safety and Health, "Leading Work-Related Diseases and Injuries" in the *Morbidity and Mortality Weekly Report,* 32:24–26, 32, January 21, 1983.

U.S. Department of Health and Human Services, Public Health Service, National Center for Health Statistics, *Current Estimates From the National Health Interview Survey,* National Center for Health Statistics, Hyattsville, MD, various years.

U.S. Department of Health, Education, and Welfare, Public Health Service, National Center for Health Statistics, *Persons Injured and Disability Days by Detailed Type and Class of Accident, United States, 1971–72,* National Center for Health Statistics, Hyattsville, MD, January 1976.

U.S. Department of Health, Education, and Welfare, Public Health Service, National Center for Health Statistics, *Vital Statistics of the U.S.,* National Center for Health Statistics, Hyattsville, MD, various years.

U.S. Department of Labor, Bureau of Labor Statistics, *Accidents Involving Eye Injuries,* Report No. 597, U.S. Government Printing Office, Washington, DC, 1980.

U.S. Department of Labor, Bureau of Labor Statistics, *Accidents Involving Face Injuries,* Report No. 604, U.S. Government Printing Office, Washington, DC, 1980.

U.S. Department of Labor, Bureau of Labor Statistics, *Accidents Involving Head Injuries,* Report No. 605, U.S. Government Printing Office, Washington, DC, 1980.

U.S. Department of Labor, Bureau of Labor Statistics, *Accidents Involving Foot Injuries,* Report No. 626, U.S. Government Printing Office, Washington, DC, January 1981.

U.S. Department of Labor, Bureau of Labor Statistics, *Handbook of Labor Statistics,* 1980.

U.S. Department of Labor, Bureau of Labor Statistics, *Work-Related Hand Injuries and Upper Extremity Amputations,* Bulletin No. 2160, U.S. Government Printing Office, Washington, DC, 1982.

U.S. Department of Labor, Bureau of Labor Statistics, *Back Injuries Associated with Lifting,* Bulletin No. 2144, U.S. Government Printing Office, Washington, DC, April 1982.

U.S. Department of Labor, Bureau of Labor Statistics, *Occupational Injuries and Illnesses in the United States by Industry, 1980,* Bulletin No. 2130, U.S. Government Printing Office, Washington, DC, April 1982.

U.S. Department of Labor, Bureau of Labor Statistics, *Occupational Injuries and Illnesses in the United States by Industry, 1981,* Bulletin No. 2164, U.S. Government Printing Office, Washington, DC, January 1983.

U.S. Department of Labor, Bureau of Labor Statistics, *Occupational Injuries and Illnesses in the United States by Industry, 1982,* Bulletin No. 2196, U.S. Government Printing Office, Washington, DC, April 1984.

U.S. Department of Labor, Bureau of Labor Statistics, "Recordkeeping Requirements under the Occupational Safety and Health Act of 1970," rev. ed., Bureau of Labor Statistics, 1983.

U.S. Department of Labor, Occupational Safety and Health Administration, *Occupational Safety and Health Statistics of the Federal Government, FY 1979,* (OSHA) Publication No. 2066, Occupational Safety and Health Administration, Washington, DC, January 1981.

Young, J. L., C. L., Percy, A. J., Asire, et al., *Cancer Incidence and Mortality in the United States, 1973–1977,* National Cancer Institute Monograph No. 57, National Cancer Institute, Bethesda, MD, 1981.

11.7.3 Epidemiology/Surveillance Organizations

American Epidemiological Society (AES), c/o Dr. Philip S. Brachman, Centers for Disease Control, 1600 Clifton Road, N.E., Atlanta, GA 30333. This Society, founded in 1927, has about 300 members. These individuals are from diverse disciplines and have a common interest in the study of diseases in human populations.

Conference of State and Territorial Epidemiologists (CSTE), c/o Dr. William E. Parkin, New Jersey Department of Health, CN 360, Trenton, NJ 08265. The Conference, founded in 1951, has a membership consisting of state epidemiologists. Their purpose is to establish closer working relationships among members, to consult with and advise appropriate disciplines in other health agencies, and to provide technical advice and assistance to the Association of State and Territorial Health Officials.

Industrial Health Foundation (IHF), 34 Penn Circle W., Pittsburgh, PA 15206. This is a research and service organization of industrial companies organized for the advancement of health in industry. The Foundation maintains a research laboratory for member companies where the foundation's staff conducts studies to prevent industrial diseases and improve working conditions. It offers continuing education programs for nurses, industrial hygiene engineers, and physicians and also provides extensive information on such subjects as analytical methods, control procedures, health hazards, and toxicity. The Foundation also provides engineering and medical services, compiles statistics on epidemiology studies, and maintains a library. Publications include the *Industrial Hygiene Digest*, bulletins, technical papers, and symposia proceedings.

International Epidemiological Association (IEA), c/o Paul D. Stolley, M.D., Room 229 L NEB/S2, School of Medicine, University of Pennsylvania, Philadelphia, PA 19104. The Association was founded in 1950 and has 1400 members. The group is interested in epidemiology and study methods and applications of disease control. The Association conducts regional research seminars and workshops in which epidemiologic progress is explored in depth.

The National Council on Compensation Insurance (NCCI), One Penn Plaza, New York, NY 10119. The Council is a statistical research and ratemaking organization sponsored by the insurance industry that functions as an educational resource on a variety of compensation topics. It produces manuals of procedures and rating plans, provides a reporting service, and maintains a data base on lost-time claims. The data base, "Detailed Claim Information" (DCI), includes data on a statistical sample of claims filed, in certain states, since April of 1979. Fifty-four data items are collected on each of the 375,000 claims selected to date. The data base is not available to the public but arrangement can be made with NCCI for customized reports. There is a fee for this service. NCCI also publishes "Workers' Compensation Claim Characteristics" that provides details on workers' compensation costs and cost containment.

Society for Epidemiologic Research (SER), c/o American Journal of Epidemiology, 550 N. Broadway, Suite 201, Baltimore, MD 21205. The Society was founded in 1967 and has a membership that includes epidemiologists, researchers, public health administrators, educators, mathematicians, and statisticians interested in epidemiologic research. The Society's purpose is to stimulate scientific interest in and promote the exchange of information about epidemiologic research.

11.7.4 Epidemiology/Surveillance Journals

The *American Journal of Epidemiology* publishes original field and laboratory studies on the occurrence and distribution of endemic and epidemic diseases. Articles on infectious and noninfectious diseases and statistical methodology are included. The "Reviews and Commentary" section contains reviews and editorials on various aspects of epidemiologic research. This publication is included among the most frequently cited journals in NIOSHTIC.

The *British Journal of Cancer* publishes papers containing new information that constitute a distinct contribution to knowledge and are relevant to the clinical, epidemiological, pathological, or molecular aspects of oncology. It also publishes short communications and invited book reviews. Letters arising from material of a scientific nature already published in the journal are also presented. This publication is included among the most frequently cited journals in NIOSHTIC.

Cancer Research presents the results of original experimental, clinical, and statistical studies and review articles of interest to cancer researchers. Announcements and reports of meetings of interest are included, and the availability of fellowships and scholarships is noted. Proceedings of meetings and symposia are published as external supplements to the journal in which the full expense is borne by the sponsoring agency. It is among the most frequently abstracted journals in both NIOSHTIC and TOXLINE.

The *International Journal of Epidemiology*, an official publication of the International Epidemiological Association, presents original work, reviews, and letters to the editor in the fields of research and teaching in epidemiology. This publication is included among the most frequently cited journals in NIOSHTIC.

The *Journal of Epidemiology and Community Health* is published by the British Medical Association and focuses on public health problems and their solutions. This focus includes the distribution and behavior of disease in human populations as well as the effect of social and environmental conditions on disease evolution. This publication is included among the most frequently cited journals in NIOSHTIC.

The *Journal of Hygiene* reports on research about subjects related to infectious diseases. Emphasis is given to clinical and social aspects of the epidemiology, prevention, and control of diseases.

The *Journal of Occupational Medicine* is the official publication of the American Occupational Medical Association and of the American Academy of Occupational Medicine. It presents articles on all phases of occupational medicine and occupational health practice. This publication is included among the most frequently cited journals in NIOSHTIC.

The *Morbidity and Mortality Weekly Report,* a weekly series with periodic supplements, provides provisional data from state health departments. The supplements are reports on interesting cases, environmental hazards, or public health problems. Examples of these supplements include the July 1985 summaries of NIOSH Current Intelligence Bulletins and the July 1985 listing of NIOSH Recommendations for Occupational Safety and Health Standards.

The *New England Journal of Medicine* publishes original materials, special articles, and brief reports on medical intelligence. These are accepted with the understanding that the articles have not been published or submitted for publication elsewhere. This publication is included among the most frequently cited journals in NIOSHTIC.

11.8 Ergonomics

Ergonomics is sometimes called human engineering, biomechanics, or human factors. The field draws on knowledge from anatomy, physiology, and the behavioral sciences. The man-machine interface is especially important to occupational safety and health because of the large number of workers' compensation claims that arise from manual materials handling, repetitive motion, and vibration. The information sources in this section have been identified as significant to understanding and preventing ergonomic stress. The strategies for reducing ergonomic impact range from improving equipment design to redesigning tasks to educating workers. Sources of information for understanding and preventing lower back injuries and other injuries related to unaided lifting and lowering are included in this section. Sources of information on cumulative trauma caused or aggravated by repeated twisting or awkward postures are also included. Vibration-associated injuries to workers exposed to whole body vibration and segmental vibration, such as those who use chain saws and jackhammers, is the other principal category covered.

The elimination of biomechanical stress is the primary objective of ergonomics. In some cases this can be accomplished through substituting machines for workers in manual materials handling. In other cases, improved equipment design is useful—it can eliminate vibration in hand tools, for example, and isolation systems can be used in vehicles to reduce whole body vibration. Biomechanical stress may also be minimized through the redesign of manual tasks or through the rotation of workers into jobs with different physical demands. The various strategies for control are based on knowledge from many academic disciplines. Only those information sources that contribute directly to the field of ergonomics are included here. Obviously, the fields of behavioral science (11.4), engineering (11.6), and medicine (11.10) also include important sources for ergonomics. The reader would do well to review the listings in these sections for further related sources of information.

11.8.1 Ergonomics Data Bases

The explosion of scientific information has made it virtually impossible for the practitioner to obtain access to all of the information published. Not only is the volume of information too great to effectively review, the cost of the journals is prohibitive. Only large university libraries can be expected to have a significant portion of the literature. As a result, electronic information retrieval systems have become a necessity for those seeking information on all but the most common topics.

Until recently, five data base vendors based in the United States have provided virtually all of the files of interest to the occupational safety and health professional. These are Dialog Information Services (3460 Hillview Avenue, Palo Alto, CA 94304); the National Library of Medicine (NLM) (8600 Rockville Pike, Bethesda, MD 20209); DOE RECON (Department of Energy, Technical Information Center, P.O. Box 62, Oak Ridge, TN 37830); Bibliographic Retrieval Services (BRS) (BRS Information Technologies, 1200 Route 7, Latham, NY 12110); and System Development Corporation (SDC) Search Services (System Development Corporation, 2500 Colorado Avenue, Santa Monica, CA 90406). New vendors that were originally available only in Europe now make their information systems available in the United States. These include Pergamon Infoline (Pergamon International Information Corporation, 1340 Old Chain Bridge Road, Mclean, VA 22101); Questel (1625 Eye Street, N.W., Suite 719, Washington, DC 20006); and Information Retrieval Service of the European Space Agency Information Retrieval Service, ESA-IRS, (Esrin, P.O. Box 64, 1-00044, Frascati, Italy). Hundreds of different data bases on subjects ranging from art history to zoology are available from these vendors.

TABLE 11.7. Ergonomics Data Bases

Name	Producer	No. of Records	Coverage
ACADEMIC AMERICAN ENCYCLOPEDIA (BRS, DIALOG)	Grolier Electronic Publishing Inc.	Full text of encyclopedia	Not available (N.A.)
AGRICOLA (CAIN) (DIALOG, BRS)	USDA, Science and Education Administration	1.1 million 120,000/yr	70–78 79–present
AVLINE (NLM)	National Library of Medicine	10,000	75–present
A-V ONLINE (DIALOG)	National Information Center for Educational Media	403,000	N.A.
CIS (Questel)	International Labour Organization/CIS	2,100/yr	74–present
CISDOC (ESA)	International Labour Organization/CIS	2,100/yr	74–present
COLD (SDC)	Cold Regions Research Safety Engineering Laboratory	6,000/yr	62–present
COMPENDEX (ESA)	Engineering Information, Inc.	100,000/yr	69–present
COMPENDEX (DIALOG, SDC)	Engineering Information, Inc.	100,000/yr	70–present
COMPENDEX (PERGAMON)	Engineering Information, Inc.	100,000/yr	73–present
CONFERENCE PAPERS INDEX (ESA, DIALOG)	Cambridge Scientific Abstracts	120,000/yr	73–present
DISSERTATION ABSTRACTS (DIALOG, BRS)	University Microfilms International	42,000/yr	1861–present
EI ENGINEERING MEETINGS (SDC, ESA, DIALOG)	Engineering Information, Inc.	108,000/yr	82–present
EIX (COMPENDEX) (DOE)	Engineering Information, Inc.	100,000/yr	74–present
ENVIRONMENTAL BIBLIOGRAPHY (DIALOG)	Environmental Studies Institute	24,000/yr	73–present
ERIC (SDC, BRS, DIALOG)	Educational Resources Information Center	30,000/yr	66–present
EXCERPTA MEDICA (DIALOG)	Excerpta Medica	1.2 million 240,000/yr	74–79 80–present

TABLE 11.7. *(continued)*

Name	Producer	No. of Records	Coverage
EXCERPTA MEDICA (BRS)	Excerpta Medica	240,000/yr	80–present
FOODS ADLIBRA (DIALOG)	K & M Publications, Inc.	3,600/yr	74–present
HEALTH AUDIOVISUAL ONLINE CATALOG (BRS)	Northeastern Ohio Universities College of Medicine	5,000	60–present
HSELINE (PERGAMON, ESA)	Health and Safety Executive	47,000	77–present
INSPEC (BRS)	IEEE	N.A. N.A.	70–76 77–present
INSPEC (DIALOG)	IEEE	1.1 million 180,000/yr	69–77 78–present
INSPEC (ESA)	IEEE	150,000/yr	71–present
INSPEC (SDC)	IEEE	150,000/yr	69–present
ISMEC (DIALOG)	Information Service in Mechanical Engineering	15,000/yr	73–present
LABORDOC (ESA, QUESTEL, SDC)	International Labour Organization	4,800/yr	65–present
LIFE SCIENCES COLLECTION (DIALOG)	Cambridge Scientific Abstracts	680,000	78–present
MEDLINE MED80 MED77 MED75 MED71 MED66 (NLM)	National Library of Medicine	140,000/yr 804,874 775,167 642,710 883,906 954,650	83–present 80–82 77–79 75–76 71–74 66–70
MEDLINE (DIALOG)	National Library of Medicine	140,000/yr 1.7 million 1.5 million	80–present 73–79 66–72
MEDLINE (BRS)	National Library of Medicine	140,000/yr N.A. 883,906 954,650	79–present 75–78 71–74 66–70
MENTAL HEALTH ABSTRACTS (DIALOG)	National Clearinghouse for Mental Health Information (to 82) Plenum Data Co. to present	463,000	69–present
NASA (ESA)	NASA Scientific and Technical Information Branch	N.A.	62–present

TABLE 11.7. *(continued)*

Name	Producer	No. of Records	Coverage
NASA TECH BRIEFS (DOE)	NASA Scientific and Technical Information Branch	N.A.	N.A.
NCMH (BRS)	National Clearinghouse for Mental Health	N.A.	69-82
NIOSHTIC (DIALOG, PERGAMON)	National Institute for Occupational Safety and Health (NIOSH)	12,000/yr	1900-present
NSIC (DOE)	Nuclear Science Information Center ORNL, DOE	12,000/yr	63-present
NTIS (DIALOG)	National Technical Information Service	60,000/yr	64-present
NTIS (ESA)	National Technical Information Service	60,000/yr	69-present
NTIS (SDC, BRS)	National Technical Information Service	60,000/yr	70-present
NURSING AND ALLIED HEALTH (DIALOG)	Cumulative Index to Nursing and Allied Health Literature	20,000	83-present
PASCAL (QUESTEL)	Centre de Documentation Scientifique et Technique	1.7 million 2.6 million	73-76 77-present
PIRA (PERGAMON)	The Paper Industries Research Association	10,000/yr	75-present
PSYCINFO (SDC, DIALOG, BRS)	American Psychological Association	24,000/yr	67-present
RBOT (BRS)	Cincinnati Millicron	N.A.	80-present
SAE (SDC)	Society of Automotive Engineers	800/yr	65-present
SCISEARCH (DIALOG)	Institute for Scientific Information	1.8 million 482,000/yr	74-77 78-present
SUPERINDEX (BRS)	Superindex, Inc.	2,000 book indexes	N.A.
TEXTILE TECHNOLOGY DIGEST (DIALOG)	Institute of Textile Technology	97,500	78-present

TABLE 11.7. *(continued)*

Name	Producer	No. of Records	Coverage
TRIS (DIALOG)	United States Department of Transportation	18,000/yr	77–present
WORLD TEXTILE ABSTRACTS (PERGAMON, DIALOG)	Shirley Institute	9,000/yr	70–present

Source: Data in this table were obtained from catalogs and manuals of the identified vendors. Specific citations for each data base and vendor are given in the alphabetical descriptions provided in section 11.14.

Two other vendors, Information Consultants (ICI) (1133 15th St., N.W., Suite 300, Washington, DC 20005) and Chemical Information Systems (CIS) (7215 York Road, Baltimore, MD 21212) offer similar highly specialized information systems (ICI and ICIS) containing data about regulations and the physical characteristics, and toxicity of various chemicals. These systems differ from the ones mentioned above in that they provide data with references rather than abstracts and bibliographic information.

Each data base vendor has a different mix of data bases. Even when the same data base is available from different vendors, the periods covered and the implementation can be quite different. There is a trend for the vendors to include larger portions of the data bases in single files as the costs of computer storage decline. The following table lists the data bases that contain ergonomics information relevant to occupational safety and health. The availability by vendor and the periods of coverage are included for reference.

11.8.2 Ergonomics Reference Texts

Astrand, P. O., and K. Rodahl, *Textbook of Work Physiology: Physiological Bases of Exercise,* 2nd ed., McGraw-Hill Book Co., New York, NY, 1977. This text includes chapters that describe the processes of energy liberation and transfer in the human body; muscular function; the functions of the circulatory, respiratory, and skeletal systems; the basis for assessment of physical work capacity; the relation of body dimensions to muscular work; the body's temperature regulation mechanisms; and factors affecting physical performance such as physical training, nutrition, altitude and pressure, and tobacco smoking.

Baetjer, A. M., *Women in Industry: Their Health and Efficiency,* W. B. Saunders, Philadelphia, PA, 1946. Reprint Edition, Arno Press, New York, NY, 1977. This classic text on the work physiology of women includes data on the physical characteristics and strength of women workers as well as information on absenteeism, injuries common among women workers, occupational diseases, and more. Although some of the information on medical care is dated, the majority of the content is still relevant.

Chaffin, D. B., and G. Andersson, *Occupational Biomechanics,* John Wiley and Sons, New York, NY, 1984. This text provides the basis for the evaluation and design of manual work. It reviews the biomechanics literature and work analysis systems.

Edwards, E., and F. P. Lees, eds., *The Human Operator in Process Control,* State Mutual Book & Periodical Service, New York, NY, 1974. This volume is a collection of papers on the physiological aspects of the design and evaluation of automated industrial systems. Included among the topics are measurement of visual attention of operators, man-machine interaction, analysis of work operations, and decision-making processes for operators.

Eastman Kodak Company, *Ergonomic Design for People at Work,* Vol. 1, Human Factors Section, Lifetime Learning Publications, Belmont, CA, 1983. This book, the first volume of a series, provides practical guidelines for design of work areas and equipment and tools that reduce fatigue and discomfort.

French, J. R. P., Jr., *The Mechanisms of Job Stress and Strain,* 1982. This book presents the results of an analysis of survey data on more than 2000 people from 23 companies, including both white- and blue-collar workers. Using multivariate analysis, the authors describe paths from stressful events to adverse health outcomes. Also discussed are the methods used to identify these adverse health effects.

Grandjean, E., *Fitting the Task to the Man,* 3rd ed., International Publications Service, New York, NY, 1980. This classic text was translated into English in 1969. It provides practical guidance derived from ergonomic research.

Kerslake, D. M., *The Stress of Hot Environments,* Cambridge University Press, New York, NY, 1982. The physiological effects of heat on the human body are the focus of this monograph. Chapters cover heat exchange in the environment, convection and evaporation of sweat, respiration and water loss, and physiological responses to heat and heat stress.

McCormick, E. J., and M. S. Sanders, *Human Factors in Engineering and Design,* 5th ed., McGraw-Hill, New York, NY, 1982. This is a widely used text in the field of applied ergonomics that gives detailed coverage of the functions of humans in manmade environments and the thoughtful design of the human environment. Chapters deal with information input and output in man-machine systems, considerations for design of instrumentation, work space design and arrangement, and the influence of environmental variations on human performance. The new edition has been thoroughly updated.

Parker, J. F., and V. R. West, eds., *Bioastronautics Data Book,* NASA SP-3006, National Aeronautics and Space Administration, Washington, DC, 1973. This reference is a compendium of anthropometry data resulting from NASA's manned space flight program. Included is data on the limitations of human physiological performance, response to external stimuli, and performance under environmental extremes. The coverage of this text makes it a significant reference work for ergonomics.

Roebuck, J. A., K. H. E. Kroemer, and W. G. Thomson, *Engineering Anthropometry Methods,* John Wiley and Sons, New York, NY, 1975. This work is a compilation of information on techniques and tools for measurements, generalized procedures, and application of fundamental engineering principles to the design of materials and equipment for the human environment.

Sell, R. C., J. E. Crawley, G. W. Crockford, and J. G. Fox, *Human Factors in Work Design and Protection,* Taylor & Francis, London, 1977. This text discusses four different applications of human factors knowledge: the design of controls and structure of cabs for cranes, steel production job design, design of personal protective clothing, and human factors applications to quality control procedures.

U.S. Department of Defense, Naval Sea Systems Command, *U.S. Diving Manual,* U.S. Superintendent of Documents, Washington, DC, 1975. This two-part manual of diving practice is frequently updated with current technical information. Volume I covers the history of diving, underwater physics, underwater physiology, operations planning, scuba diving, surface-supplied diving, diving emergencies, and air decompression. Volume 2 covers mixed gas diving theory, underwater breathing apparatus, mixed gas surface-supplied diving, deep diving systems, oxygen diving operations, surface-supplied decompression, and helium-oxygen saturation diving.

U.S. Department of Health and Human Services, Public Health Service, Centers for Disease Control, National Institute for Occupational Safety and Health, *A Work Practices Guide for Manual Lifting,* DHHS (NIOSH) Publications No. 81-122, National Institute for Occupational Safety and Health, Cincinnati, OH, 1981, (reprinted by the American Industrial Hygiene Association, Akron, OH, 1983). This is perhaps the most complete discussion of work practices for manual lifting. It includes a discussion of the epidemiology of musculoskeletal injury, biomechanical concepts, physiological considerations, and psychophysical lifting limits.

Winter, D. A., *Biomechanics of Human Movement,* John Wiley and Sons, New York, NY, 1979. This is a specialized reference on measurement and assessment of human physical activity. Topics included are techniques for measurement of mechanical work, energy and power, muscle mechanics, and electromyography. It includes an appendix of anthropometric data.

Woodson, W. E., *Human Factors Design Handbook: Information & Guidelines for the Design of Systems, Facilities, Equipment, & Products for Human Use,* McGraw-Hill, New York, NY, 1981. This handbook is a compilation of ergonomic information designed to be used as a reference guide. Five chapters are included: "Systems Conceptualization," "Subsystems Design," "Component and Product Design," "Human Factors Data," and "Human Engineering Methods."

Zenz, C., ed., *Occupational Medicine: Principles and Practical Applications,* Year Book Medical Publishers, Chicago, IL, 1975. This book is an introductory text of occupational medicine that covers administrative considerations, employee health services for small business, occupational safety, the occupational health nurse, and clinical occupational medicine. Also included are physical assessment of workers, occupational lung diseases, the physical and chemical environment, and psychosocial considerations.

Other Ergonomics References

Armstrong, T. J., and G. D., Langolf, "Ergonomics and Occupational Safety and Health," in W. N. Rom, A. D. Renzetti, and J. S. Lett, eds., *Environmental and Occupational Medicine,* Little, Brown and Co., Boston, MA, 1983.

Bailey, D. W., and Human Performance Associates, *Human Performance Engineering: A Guide for System Designers,* Prentice-Hall, Englewood Cliffs, NJ, 1982.

Bianchi, G., et al., *Man under Vibration: Suffering and Protection,* Elsevier, New York, NY, 1982.

Burgess, J. H., *Human Factors—Ergonomics for Building and Construction,* John Wiley and Sons, New York, NY, 1981.

Eastman Kodak Company, Health, Safety and Human Factors Laboratory, Human Factors Section, *Ergonomics Design for People at Work,* Lifetime Learning Publications, Belmont, CA, 1983.

Frankel, V. H., and M. Nordin, *Basic Biomechanics of the Skeletal System,* Lea and Febiger, Philadelphia, PA, 1980.

Fung, Y. C., *Biomechanics*, Springer-Verlag, New York, NY, 1981.

International Labour Organization, *Ergonomics Principles in the Design of Hand Tools*, International Labour Unipub, New York, NY, 1981.

Kantowitz, B., and R. Sorkin, *Human Factors: Understanding People-System Relationships*, John Wiley and Sons, New York, NY, 1983.

Keyserling, W. M., "Occupational Safety and Ergonomics," in B. S. Levy and D. H. Wegman, eds., *Occupational Health*, Little, Brown and Co., Boston, MA, 1983.

Lenihan, J., *Human Engineering: The Body Re-Examined*, Braziller, New York, NY, 1975.

Leveau, V., *Biomechanics of Human Motion*, 2nd ed., Saunders, Philadelphia, PA, 1977.

McCormick, E. J., and M. Sanders, *Human Factors in Engineering and Design*, McGraw-Hill Book Company, New York, NY, 1982.

Meister, D., *Human Factors: Theory & Practice*, John Wiley and Sons, New York, NY, 1971.

Miller, D. I., and R. C. Nelson, *Biomechanics of Sport*, Lea and Febiger, Philadelphia, PA, 1976.

National Research Council, *Video Displays, Work and Vision*, National Academy Press, Washington, DC, 1983.

Osborne, D. J., *Ergonomics at Work*, John Wiley and Sons, New York, NY, 1982.

Roebuck, J. A., K. H. E. Kroemer, and W. G. Thomson, *Engineering Anthropometry Methods*, Willey-Interscience, New York, NY, 1975.

Shephard, R. J., *Human Physiological Work Capacity*, Cambridge University Press, New York, NY, 1978.

The Metals Society, eds., *The Human Factor in Metal Plane Operations & Design*, State Mutual Book and Periodicals Service, New York, NY, 1977.

Tichauer, E. R., *The Biomechanical Basis of Ergonomics*, John Wiley and Sons, New York, NY, 1978.

Van Cott, H. P., and R. G. Kinkade, eds., *Human Engineering Guide to Equipment Design*, U.S. Government Printing Office, Washington, DC, 1972.

Woodson, W. E., and D. W. Conover, eds., *Human Engineering Guide for Equipment Designers*, 2nd rev. ed., University of California Press, Berkeley, CA, 1965.

11.8.3 Ergonomics Organizations

The **Human Factors Society,** P.O. Box 1369, Santa Monica, CA 90406. This society was founded in 1957 and consists of 3300 members. Membership includes psychologists, engineers, physiologists, and behavioral scientists who are interested in the use of human factors in the development of systems and devices.

International Ergonomics Association (IEA), c/o Harry L. Davis, Eastman Kodak Co., Kodak Park Div., Bldg. 69, Rochester, NY, 14650. The Association was founded in 1959 with 16 federated societies interested in the study of human work and the work environment. The association encourages interaction among specialists and cooperates with employers' associations, trade unions, professionals, management, and government groups to encourage the practical application of ergonomic sciences in industries. Publications include *Ergonomics*.

11.8.4 Ergonomics Journals

The *American Industrial Hygiene Association Journal* focuses on all aspects of the field of industrial hygiene practice, policy, and procedures as well as reports of specific investigations and field activities. This publication is included among the most frequently cited journals in NIOSHTIC.

The *American Journal of Applied Physiology* contains manuscripts on respiration, exercise, and environmental physiology. Letters to the editor are reviewed and, if accepted, are sent to the original author for rebuttal if appropriate.

The *American Journal of Physiology* accepts only original material that has not been published in other than an abstract. Manuscripts that describe new methods, new apparatus, techniques with physiological applicability, teaching methods, and critiques of teaching methods and techniques are considered special communications. This publication is included among the most frequently cited journals in NIOSHTIC.

Ergonomics is the official publication of the Ergonomics Society and the International Ergonomics Association. Its primary focus is the scientific study of human factors in relation to working environments and equipment design. Articles considered are reports of both pure and applied work related to all aspects of people at work and current news about ergonomics activities worldwide. This publication is included among the most frequently cited journals in NIOSHTIC.

The *European Journal of Applied Physiology and Occupational Physiology* focuses on original work in the field of applied human physiology. The current emphasis is on work and environmental physiology, including exercise as a physiological stimulus for adaptation and related articles on muscle biochemistry, and endocrine responses following exercise. Articles reporting on animal experiments that are related to the human condition are also published. This publication is included among the most frequently cited journals in NIOSHTIC.

Human Factors, the journal of the Human Factors Society, is a bimonthly publication of original articles about people in relation to machines and environments. Articles cover methodology, quantitative and qualitative approaches to theory, and empirical subjects. Evaluative reviews are also included. The format feature includes articles, brief papers, and special topical issues related to human factors. This publication is included among the most frequently cited journals in NIOSHTIC.

The *Journal of Applied Physiology,* a publication of the American Physiological Society, presents original papers that relate to normal or abnormal function in the following areas: respiratory physiology, nonrespiratory functions of the lungs, environmental physiology, temperature regulation, exercise physiology, and interdependence. Priority within each category is given to articles dealing with integrated and adaptive mechanisms and to articles dealing with human research. This publication is included among the most frequently cited journals in NIOSHTIC.

The *Journal of Occupational Medicine* is the official publication of the American Occupational Medical Association and of the American Academy of Occupational Medicine. It presents articles on all phases of occupational medicine and occupational health practice. This publication is included among the most frequently cited journals in NIOSHTIC.

The *Journal of Sound and Vibration* is known for the prompt publication of original papers on experimental or theoretical work related to sound or vibration. Papers on vibration and sound problems from the engineering, physical, physiological, and psychological disciplines are published.

Medicine and Science in Sports and Exercise is the official journal of the American College of Sports Medicine. It presents original research in a multidisciplinary discussion about the role of physical activity in human health and function. Included are brief reviews, symposia, and letters to the editor.

11.9 INDUSTRIAL HYGIENE

The information in this section applies to a narrow definition of industrial hygiene. The sources provided reflect knowledge related to the protection of workers' health and safety through the evaluation of the work environment. This section covers the recognition of workplace hazards and the evaluation of the workplace environment to reduce occupational illness and injury.

The reference sources include a variety of technical and scientific materials important to evaluating the work environment. The material is organized with important data bases listed first followed by reference texts, scientific organizations, and finally technical journals that support and contribute to the field.

The industrial hygiene reference sources include a variety of technical and scientific material. This wide range of subject matter results from the nature of industrial hygiene practice and from the rapid growth of the specialized literature in this field. For example, issues in chemistry, physics, engineering, biochemistry, and medicine individually or collectively may need to be addressed in the evaluation or control of potential occupational health and safety hazards associated with the modern industrial environment. While specific information related to the practice of industrial hygiene is presented here, the principal references for each of the enumerated disciplines are presented in subject area sections of this chapter.

11.9.1 Industrial Hygiene Data Bases

The explosion of scientific information has made it virtually impossible for the practitioner to obtain access to all of the information published. Not only is the volume of information too great to effectively review, the cost of the journals is prohibitive. Only large university libraries can be expected to have a significant portion of the literature. As a result, electronic information retrieval systems have become a necessity for those seeking information on all but the most common topics.

Until recently, five data base vendors based in the United States have provided virtually all of the files of interest to the occupational safety and health professional. These are Dialog Information Services (3460 Hillview Avenue, Palo Alto, CA 94304); the National Library of Medicine (NLM) (8600 Rockville Pike, Bethesda, MD 20209); DOE RECON (Department of Energy, Technical Information Center, P.O. Box 62, Oak Ridge, TN 37830); Bibliographic Retrieval Services (BRS) (BRS Information Technologies, 1200 Route 7, Latham, NY 12110); and System Development Corporation (SDC) Search Services (System Development Corporation, 2500 Colorado Avenue, Santa Monica, CA 90406). New vendors that were originally available only in Europe now make their information systems available in the United States. These include Pergamon Infoline (Pergamon International Information Corporation, 1340 Old Chain Bridge Road, Mclean, VA 22101); Questel (1625 Eye Street, N.W., Suite 719, Washington, DC 20006); and the Information Retrieval Service of the European Space Agency Information Retrieval Service, ESA-IRS, (Esrin, P.O. Box 64, 1-00044, Frascati, Italy). Hundreds of different data bases on subjects ranging from art history to zoology are available from these vendors.

TABLE 11.8. Industrial Hygiene Data Bases

Name	Producer	No. of Records	Coverage
APLIT (SDC)	American Petroleum Institute	360,000	64–present
APTIC (DIALOG)	U.S. EPA	87,000	66–76
BIOSIS (BRS)	Biosciences Information Services	Not available (N.A.)	70–77 78–present
BIOSIS (ESA)	Biosciences Information Services	200,000/yr	73–present
BIOSIS (SDC)	Biosciences Information Services	200,000/yr	69–73 74–79 80–present
BIOSIS PREVIEWS (DIALOG)	Biosciences Information Services	2 million 1.4 million 328,000/yr	69–76 77–81 82–present
CAB (ESA)	Commonwealth Agricultural Bureaux	144,000/yr	73–present
CAB ABSTRACTS (DIALOG)	Commonwealth Agricultural Bureaux	144,000/yr	72–present
CAS (SDC)	Chemical Abstracts	1.3 million 1.7 million 2.2 million 340,000/yr	67–71 72–76 77–81 82–present
CA SEARCH (BRS)	Chemical Abstracts	N.A. N.A.	70–76 77–present
CA SEARCH (DIALOG)	Chemical Abstracts	1.3 million 1.7 million 1.2 million .9 million 400,000/yr	67–71 72–76 77–79 80–81 82–present
CA SEARCH (PERGAMON)	Chemical Abstracts	1.7 million 400,000/yr	72–76 77–present
CEH ONLINE (SDC)	Stanford Research Institute	Full text of handbook	N.A.
CESARS (CIS, ICIS)	Michigan Department of Natural Resources	180 chemicals "evaluated"	N.A.
CHEMABS (ESA)	Chemical Abstracts	6 million	67–present
CHEMICAL ENGINEERING ABSTRACTS (PERGAMON)	The Royal Society of Chemistry	63,000	70–present
CHEMICAL EXPOSURE (DIALOG)	Chemical Effects Information Center	14,370	74–present
CIS (QUESTEL)	International Labour Organization/CIS	2,100/yr	74–present
CISDOC (ESA)	International Labour Organization/CIS	2,100/yr	74–present

TABLE 11.8. *(continued)*

Name	Producer	No. of Records	Coverage
CTCP (CIS, ICIS)	Clinical Toxicology of Commercial Products	18,000 products in commerce	N.A.
DOE ENERGY (DIALOG)	TIC, ORNL, DOE (Energy Data Base)	144,000/yr	74-present
EDB (DOE)	TIC, ORNL, DOE (Energy Data Base)	192,000/yr	1900-present
EDB (SDC)	TIC, ORNL, DOE (Energy Data Base)	800,000	30s-present
EEIS (DOE)	TIRC, ORNL, DOE	5,141	60-present
ENVIROLINE (SDC, ESA, DOE, DIALOG)	EIC/ Intelligence	110,000	71-present
ENVIRONMENTAL BIBLIOGRAPHY (DIALOG)	Environmental Studies Institute	24,000/yr	73-present
EUCAS (QUESTEL)	Chemical Abstracts	1.3 million 1.7 million 2.2 million 340,000/yr	67-71 72-76 77-81 82-present
FSTA (DIALOG, SDC)	International Food Information Services	20,000/yr	69-present
Hazardline (BRS)	Occupational Health Services, Inc.	3,000 chemical substances	N.A.
HSELINE (PERGAMON, ESA)	Health and Safety Executive	47,000	77-present
ISMEC (DIALOG)	Information Service in Mechanical Engineering	15,000/yr	73-present
KIRK-OTHMER ENCYCLOPEDIA OF CHEMICAL TECHNOLOGY (BRS)	John Wiley and Sons, Inc.	Full text of encyclopedia	3rd ed. 76
LABORDOC (ESA, QUESTEL, SDC)	International Labour Organization	4,800/yr	65-present
NIOSHTIC (DIALOG, PERGAMON)	National Institute for Occupational Safety and Health (NIOSH)	12,000/yr	1900-present
NSIC (DOE)	Nuclear Science Information Center, ORNL, DOE	12,000/yr	63-present

TABLE 11.8. ***(continued)***

Name	Producer	No. of Records	Coverage
NTIS (DIALOG)	National Technical Information Service	60,000/yr	64–present
NTIS (ESA)	National Technical Information Service	60,000/yr	69–present
NTIS (SDC, BRS)	National Technical Information Service	60,000/yr	70–present
OHMTADS (CIS, ICIS)	U.S. EPA, Superfund	1,334 substances	N.A.
PAPERCHEM (SDC, DIALOG)	The Institute of Paper Chemistry	12,000/yr	68–present
PHP (DOE)	American Petroleum Institute	7,947	77–80
POLLUTION ABSTRACTS (ESA, DIALOG, BRS)	Cambridge Scientific Abstracts	8,400/yr	70–present
RTECS (NLM, CIS, ICIS)	National Institute for Occupational Safety and Health (NIOSH)	70,029	N.A.
SAE (SDC)	Society of Automotive Engineers	800/yr	65–present
SPHERE	Office of Toxic Substances, U.S. EPA		
DERMAL		655 chemicals	N.A.
AQUIRE		1,900 chemicals	N.A.
ENVIROFATE (CIS, ICIS)		450 chemicals	N.A.
TELEGEN (ESA, DIALOG)	Environment Information Center	3,600/yr	76–present
TEXTILE TECHNOLOGY DIGEST (DIALOG)	Institute of Textile Technology	97,500	78–present
TOXICOLOGY DATA BANK (NLM)	National Library of Medicine	4,112 chemicals	N.A.
TOXLINE	National Library of Medicine	140,000/yr	81–present
TOXBACK76		555,808	76–80
TOXBACK65 (NLM)		598,424	65–75
TSCA INITIAL INVENTORY (DIALOG)	Office of Toxic Substances, U.S. EPA	43,278 substances in production	1979
TSCA PLUS (SDC)	Office of Toxic Substances, U.S. EPA	61,000 substances in production	75–present

TABLE 11.8. *(continued)*

Name	Producer	No. of Records	Coverage
TSCAPP (CIS, ICIS)	Office of Toxic Substances, U.S. EPA	53,000 substances in production	75–77
VOLUNTARY STANDARDS INFORMATION NETWORK (BRS)	Information Handling Services	over 500	N.A.
WATER RESOURCES ABSTRACTS (DIALOG, DOE)	United States Department of Interior	172,500	68–present
WATERLIT (SDC)	South African Water Information Centre	12,000/yr	76–present
WATERNET (DIALOG)	American Water Works Association	3,000/yr	71–present
ZINC, LEAD AND CADMIUM ABSTRACTS (PERGAMON)	Zinc Development Association	3,000/yr	75–present

Source: Data in this table were obtained from catalogs and manuals of the identified vendors. Specific citations for each data base and vendor are given in the alphabetical descriptions provided in section 11.14.

Two other vendors, Information Consultants (ICI) (1133 15th St. N.W., Suite 300, Washington, DC 20005) and Chemical Information Systems (CIS) (7215 York Road, Baltimore, MD 21212), offer similar highly specialized information systems (ICI and ICIS) containing data about regulations and the physical characteristics and toxicity of various chemicals. These systems differ from the ones mentioned above in that they provide data with references rather than abstracts and bibliographic information.

Each data base vendor has a different mix of data bases. Even when the same data base is available from different vendors, the periods covered and the implementation can be quite different. There is a trend for the vendors to include larger portions of the data bases in single files as the costs of computer storage decline. The following table lists the data bases that contain industrial hygiene information relevant to occupational safety and health. The availability by vendor and the periods of coverage are included for reference.

11.9.2 Industrial Hygiene Reference Texts

American Conference of Governmental Industrial Hygienists, *Air Sampling Instruments Handbook*, 6th ed., American Conference of Governmental Industrial Hygienists, Cincinnati, OH, 1983. This is the most up-to-date, comprehensive reference text available on instruments for sampling and evaluating industrial contaminants. Contents include sampling methodology and the operation and performance of equipment.

American Conference of Governmental Industrial Hygienists, *The Documentation of the Threshold Limit Values*, 4th ed., American Conference of Governmental Industrial Hygienists, Cincinnati, OH, 1980. This publication is in the form of a loose-leaf binder to accommodate the annual supplements. The documentation provides the basis for the influential ACGIH recommendations. The presentation is concise, summarizing the animal and human toxicity studies that serve as the basis for the exposure limit recommendations. Permissible exposure limits for both the U.S. and other countries are listed.

American Conference of Governmental Industrial Hygienists, *TLVs®: Threshold Limit Values for Chemical Substances and Physical Agents in the Workroom Environment with Intended Changes for 19*—(new editions annually), American Conference of Governmental Industrial Hygienists, Cincinnati, OH. This publication is one of the most influential in the field. The 1985–86 edition contains these sections: "Threshold Limit Values for Chemical Substances in the Work Environment Adopted by ACGIH with Intended Changes for 1985–86," "Biological Exposure Indices Proposed by ACGIH for

1985–86," and "Threshold Limit Values for Physical Agents in the Work Environment Adopted by ACGIH with Intended Changes for 1985–86."

American Industrial Hygiene Association, *Heating and Cooling for Man in Industry,* 2nd ed., American Industrial Hygiene Association, Akron, OH, 1975. This publication provides information on the design of working environments. Chapters cover human tolerance of heat and cold; heat and moisture control; types and operation of ventilation systems; and other subjects.

American Industrial Hygiene Association, *Industrial Noise Control Manual,* 3rd ed., American Industrial Hygiene Association, Akron, OH, 1975. This manual provides the basic information necessary to prevent and control noise exposure in the work environment. Chapters cover the physics of sound, instruments and techniques for sound measurement, noise surveys, measurement and control of vibration, anatomy and physiology of the ear, effects of noise on humans, hearing measurement, use of personal protective equipment, engineering controls, and the legal aspects of industrial noise control.

American Public Health Association, *Methods of Air Sampling and Analysis,* 2nd ed., American Public Health Association, Washington, DC, 1977. This standard reference for the industrial hygienist and air pollution control engineer is the result of the collaboration of several professional societies, including the APHA, AIHA, ACGIH, APCA, and others. Included are methods for ambient air sampling and analysis, methods for sampling chemicals in workplace air and in biological samples, and analysis of air pollutants. Many of the methods listed in Part III are adopted from the NIOSH Manual of Analytical Methods.

Beaulieu, H. J., and R. M. Buchan, *Quantitative Industrial Hygiene,* Garland Publishing, New York, NY, 1981. Calibration and use of monitoring equipment for evaluation of air contamination, ventilation system efficiency, heat, noise, illumination, and non-ionizing radiation in the industrial environment are the focus of this laboratory text. Respiratory protection for workers and industrial hygiene survey practices are also covered.

Burgess, W. A., *Recognition of Health Hazards in Industry: A Review of Materials and Processes,* Wiley-Interscience, New York, NY, 1981. Over thirty industrial operations are reviewed in this book, including materials used, engineering controls, possible hazards, proper work practices, and protective equipment. Technical appendices and a subject index are included.

Cheremisinoff, P. N., ed., *Air Particulate Instrumentation and Analysis,* Ann Arbor Science Publishers, Ann Arbor, MI, 1981. The methods and equipment available for particulate analysis are discussed in this publication.

Cralley, L. J., and L. V. Cralley, eds., *Industrial Hygiene Aspects of Plant Operations,* Macmillan, New York, NY. 3 vols. Volume I. *Process Flows,* 1983; Volume II, *Unit Operations and Product Fabrication,* 1983; Volume III, *Equipment Selection, Layout, and Building Design,* 1984. This three-volume work describes basic industrial hygiene applications and includes photographs and diagrams on a wide variety of industrial processes.

Deutsche Forschungsgemeinschaft, *Maximum Concentration at the Workplace and Biological Tolerance Values for Working Materials 1984,* VCH Publishers, Deerfield Beach, FL, 1984. This publication lists the Federal Republic of Germany's maximum allowable concentrations (MAK values) of chemicals and gases, vapors, or particulate matter in the air of a working area. MAK values are determined and revised annually by a commission of experts from all relevant scientific disciplines. This commission is appointed by Deutsche Forschungsgemeinschaft (the German Science Foundation).

Fawcett, H. H., and W. S. Wood, eds., *Safety and Accident Prevention in Chemical Operations,* 2nd ed., John Wiley and Sons, New York, NY, 1982. This book provides excellent source material for chemical laboratory safety officers, industrial toxicologists, and industrial hygienists. It is a thorough review of accident prevention strategies and contains reference material on toxicity and the TSCA law, hazardous chemical waste disposal, and developments in fire extinguishment.

Gomez, M., R. Duffy, and V. Trivelli, *At Work in Copper: Occupational Health and Safety in Copper Smelting,* 3 vols., Inform, New York, NY, 1979. This three-volume set presents a comprehensive study of the health and safety conditions in the copper-smelting industry. *Volume 1: Industry Performance* provides a summary of findings, evaluation criteria, hazards and controls, and the findings. *Volume 2: Smelter Performance* and *Volume 3: Smelter Performance* present information on the sixteen domestic smelter locations studied. This work describes the industry, political and economic conditions, health and safety hazards, and the industries' programs for controlling the hazards and protecting the workers.

Hunter, D., *Diseases of Occupations,* 6th ed., Hodder and Stoughton, London, UK, 1976. A key reference for both occupational medicine and industrial hygiene, this text is one of the most complete references on the signs and symptoms of work-related diseases.

Kerslake, D. M., *The Stress of Hot Environments,* Cambridge University Press, New York, NY, 1982. The physiological effects of heat on the human body are the focus of this monograph. Chapters cover heat exchange in the environment, convection, evaporation of sweat, respiration and water loss, physiological responses to heat, and heat stress.

Lawrence Berkeley Laboratory, *Instrumentation for Environmental Monitoring,* Lawrence Berkeley Laboratory, Berkeley, CA, 1979. Information on instrumentation and procedures for monitoring air and water quality in the environment are the subjects of this four-volume work. Volume 1, part 1, covers sulfur dioxide, nitrogen oxides, photochemical oxidants, carbon monoxide, and hydrogen. Volume 1, part 2, covers particulates, noise, and hazardous air pollutants including mercury, asbestos, beryllium, and lead. Volume 2 covers instrumentation and procedures for monitoring water quality. Volume 3 covers instrumentation for monitoring ionizing and non-ionizing radiation including accelerators, X-radiation, reactors, fuel reprocessing, alpha, beta, gamma, neutrons, dosimeters, radionuclides, microwaves, lasers, and ultraviolet. Volume 4 covers biomedical monitoring including gaseous pollutants, particulates, mercury, cadmium, lead, pesticides, liquid pollutants, and radiation. The work is updated with supplements.

Levy, B. S., and D. H. Wegman, eds., *Occupational Health: Recognizing and Preventing Work-Related Disease*, Little, Brown and Co., Boston, MA, 1983. This text provides an introduction to issues currently facing occupational health professionals. Major sections cover work and health, occupational health services in the United States, recognizing and preventing occupational disease, and hazardous workplace exposures. There is also a section on special issues such as labor's perspective on worker health, women and minority workers, disability evaluation, and workers' compensation.

Maclean, D., and D. Enslie-Smith, *Accidental Hypothermia*, Blackwell Scientific Publications, Oxford, UK, 1977. The regulation of body temperature, thermal comfort, the abnormal physiology of hypothermia, hypothermia from exposure, and other subjects are covered in the text on the prevention, recognition, and treatment of cold-related injuries.

McCrone, W. C., *The Asbestos Particle Atlas*, Ann Arbor Science Publishers, Ann Arbor, MI, 1980. The identification of asbestos and asbestos-containing materials is detailed in over 400 photomicrographs in this atlas. Also discussed are analytical methods and mineralogy.

Mercer, T., *Aerosol Technology in Hazard Evaluation*, Academic Press, New York, NY, 1973. Included in this publication are chapters on equipment for measuring respirable aerosols and on problems unique to the sampling and evaluation of aerosols. From the Monograph Series on Industrial Hygiene, this publication covers the physical properties, measurement, and sampling of aerosols.

Olishifski, J., ed., *Fundamentals of Industrial Hygiene*, 2nd ed., National Safety Council, Chicago, IL, 1979. The recognition, evaluation, and control of occupational health hazards are the foci of this text, which includes chapters on anatomy and physiology and the design of industrial hygiene programs.

Olishifski, J., and E. Hartford, eds., *Industrial Noise and Hearing Conservation*, National Safety Council, Chicago, IL, 1975. This comprehensive reference on industrial noise, vibration, and hearing conservation programs in industry includes sections on noise control, audiometry, measurement of sound, effects of noise on man, and industrial hearing conservation programs. Also included are appendices on federal occupational and environmental noise regulations, guidelines for noise control, and information on accreditation of occupational hearing conservation programs.

Patty's Industrial Hygiene and Toxicology, 3rd rev. ed., John Wiley and Sons, New York, NY, 1978–1982, 3 vols. A comprehensive reference on industrial hygiene and industrial toxicology, this work is both an important reference for those in the field and a significant text on the principles and practice of industrial hygiene. *Volume 1, General Principles*, G. E. Clayton and F. E. Clayton, eds., provides the knowledge base necessary for the practice of industrial hygiene. Included are the fundamentals of epidemiology; mode of entry and action of toxic materials; effects of noise; ionizing radiation, heat stress, and other stressors. Evaluation and control measures, industrial hygiene survey procedures, air sampling and analysis, engineering controls, respiratory protective equipment, and air pollution control are also thoroughly discussed. *Volume 2, Toxicology*, G. E. Clayton and F. E. Clayton, eds., is an extensive reference on industrial toxicology. Three separate books comprise Volume 2; each chapter discusses the toxicology of compounds of similar chemical structure. Extensive subject and chemical indexes are included in Volume 2. *Volume 3, Theory and Rationale of Industrial Hygiene Practice*, L. J. Cralley and L. V. Cralley, eds., provides insight into quantitative matters such as emission inventories, analytical measurements, evaluation of exposures to chemical agents, ionizing radiation, noise and vibration, pressure, biologic agents, surveillance programs, and data extrapolation.

Parmeggiani, L., ed., *Encyclopedia of Occupational Safety and Health*, 3rd rev. ed. 2 vols., International Labour Organization, Washington, DC, 1983. The third edition (1983) is the latest version of a comprehensive two-volume reference work with concise, easy-to-understand entries.

Peterson, J. E., *Industrial Health*, Prentice-Hall, Englewood Cliffs, NJ, 1977. This book covers all the salient principles but requires only a minimal scientific background. It provides a solid introduction to industrial health. In addition to chemicals, subjects such as abnormal pressure, noise, biothermal stress, and radiation are discussed.

Pfafflin, J. R., and E. N. Ziegler, eds., *Encyclopedia of Environmental Science and Engineering*, Gordon and Breach Science Publishers, New York, NY, 1976. 2 vols. A valuable reference for the environmental engineer, these two volumes provide an overview of environmental science and technology and also give in-depth treatment to specific topics.

Proctor, N. H., and J. P. Hughes, *Chemical Hazards of the Workplace*, Lippincott, Philadelphia, PA, 1978. Based on the material developed by the NIOSH/OSHA Standards Completion Program, this book describes medical surveillance strategies for dealing with occupational exposures. Although somewhat out of date, it remains an excellent reference on the subject of medical surveillance.

Silverman, L., C. E. Billings, and M. First, *Particle Size Analysis in Industrial Hygiene,* Academic Press, New York, NY, 1971. The analysis of particulate matter, with emphasis on aerosols, is the subject of this book. Techniques, methodology, and equipment and their application to industrial hygiene, health physics, and air pollution control are also discussed.

Sitting, M., *Hazardous and Toxic Effects of Industrial Chemicals,* Noyes Data Corp., Park Ridge, NJ, 1979. Data for over 250 industrial chemical agents, including physical description, synonyms, potential occupational exposures, permissible exposure limits, routes of entry, and harmful effects for each are described in this book.

Tatken, R. L., and R. J. Lewis, Sr., eds., *NIOSH Registry of Toxic Effects of Chemical Substances (RTECS),* Cincinnati, OH, NIOSH, 1981-82 (annual supplements). Available in paper only from NTIS PB 85-21871. A quarterly microfiche version is available only by annual subscription. Order the microfiche by title only. The title is *Microfiche Edition, Registry of Toxic Effects of Chemical Substances.* RTECS is the largest single source of toxicity information in print. It includes acute and chronic toxicity information coded from the literature as well as tumorogenic data for more then 70,000 substances and 210,000 synonyms. In addition, environmental limits from OSHA, MSHA, NIOSH, and ACGIH are included, as are references to the International Agency for Research and Cancer (IARC) monographs and the substantial risk information provided pursuant to section 8(e) of the Toxic Substances Control Act (TSCA). The information in this publication is available in a variety of formats. Both paper copy and microfiche are cited here, but RTECS is also available on line from various sources as described in the data base portions of this chapter.

Ullmann's Encyclopedia of Industrial Chemistry, 5th ed., VCH Publishers, Deerfield Beach, FL, 1984. This is an international encyclopedia. The fifth edition, for which only the first volume is available, will consist of 28 alphabetically ordered volumes covering all branches of chemical and industrial products. In addition, there are eight basic knowledge volumes covering such subjects as chemical engineering fundamentals, unit operations, chemical reaction engineering and materials science, analytical methods, and environmental protection and plant safety. New volumes will be published at the rate of three or four per year. A cumulative index will be published annually.

U.S. Coast Guard, Department of Transportation. *CHRIS—Hazardous Chemical Data, Manual II,* U.S. Government Printing Office, Washington, DC, October 1985. CHRIS provides information for those cleaning up chemical spills. In the form of safety data sheets, it is one of the best available sources of information about chemical emergencies.

U.S. Department of Health and Human Services, Public Health Service, Centers for Disease Control, National Institute for Occupational Safety and Health, *The Industrial Environment: Its Evaluation and Control,* Cincinnati, OH, NIOSH, 1973. Available from GPO 017-001-00396-4. Published as a basic text for graduate study in the field of occupational safety and health, this book is currently being revised but remains one of the most comprehensive treatments of the field in textbook format. The first section of the book provides a review of mathematics, chemistry, biochemistry, physiology, and toxicology as they relate to the practice of occupational safety and health and specifically to industrial hygiene. The remaining chapters concentrate on the methodology of industrial hygiene practice, especially the evaluation of the harmful effects of physical agents and air contaminants.

Verschueren, K., *Handbook of Environmental Data on Organic Chemicals,* Van Nostrand-Reinhold, New York, NY, 1977. Environmental data on over 1000 organic chemicals, including their physical and chemical properties, air and water pollution data, and information on biological effects on microorganisms, plants, animals, and humans are presented in this handbook. Tables of data on threshold odor concentrations and a section on quantification of odors are of interest to the industrial hygienist.

Waldron, H. A., and J. M. Harrington, eds., *Occupational Hygiene—An Introductory Text,* Blackwell Scientific Publications, Boston, MA, 1980. The principles of occupational hygiene are the focus of this primer. Included are chapters on the physics of gases and aerosols, air sampling, industrial ventilation, vibration and noise, industrial lighting, heat, ionizing and non-ionizing radiation, personal protective equipment, basic principles of epidemiology, and statistical record-keeping for the industrial hygienist.

Zenz, C., ed., *Occupational Medicine: Principles and Practical Applications,* Year Book Medical Publishers, Chicago, IL, 1975. This introductory text on occupational medicine has sections on administrative considerations, employee health services for small businesses, occupational safety, the occupational health nurse, and clinical occupational medicine. Also included are physical assessment of workers, occupational lung diseases, the physical and chemical environment, and psychosocial considerations.

Other Industrial Hygiene Reference Texts

American Welding Society, *Radiological Surveillance of Airborne Contaminants in the Working Environment,* Miami, FL, 1979.

Andonyev, S., and O. Filipyev, *Dust and Fume Generation in the Iron and Steel Industry,* Imported Publications, Chicago, IL, 1977.

Arndt, R., and L. Chapman, "Potential Office Hazards and Controls," (Working Paper #14) in *Preventing Illness and Injury in the Workplace,* contract report prepared for the Office of Technology Assessment, U.S. Congress, Washington, DC, September 1984.
Brandt, A. D., *Industrial Health Engineering,* John Wiley and Sons, New York, NY, 1947.
Bartone, John C., II, *Health & Medical Aspects of Chemical Industries: Subject Analysis & Research Guide,* ABBE Publications Association, Washington, DC, October 1984.
Berlin, A., ed., *Assessment of Toxic Agents at the Workplace,* Kluwer Academic, Boston, MA, February 1984.
Bhatt, H. G., R. M. Sykes, and T. L. Sweeny, eds., *Management of Toxics and Hazardous Wastes,* Lewis Publications, New York, NY, June 1985.
Buchan, H. J., and R. M. Garland, *Quantitative Industrial Hygiene,* Garland Publishing, New York, NY, 1981.
Cadre, K. D., *The Measurement of Airborne Particles,* Krieger Publishing, Melbourne, FL, 1983 (reprint of 1975 edition).
Caplan, K., "Philosophy and Management of Engineering Controls," in L. J. Cralley and L. V. Cralley, eds., *Patty's Industrial Hygiene and Toxicology,* Vol. III, John Wiley and Sons, New York, NY, 1979.
Cheremisinoff, P. N., and P. P. Cheremisinoff, eds., *Industrial Noise Control Handbook,* Ann Arbor Science Publishers, Ann Arbor, MI, 1977.
Choudhary, G., ed., *Chemical Hazards in the Workplace: Measurements and Control,* American Chemical Society, Washington, DC, 1981.
Clifton, R., "Asbestos," *Bureau of Mines Minerals Handbook,* U.S. Department of the Interior, Washington, DC, 1977 and 1982.
Craft, B. F., "Occupational and Environmental Health Standards," in B. S. Levy and D. H. Wegman, eds., *Occupational Health,* Little, Brown and Co., Boston, MA, 1983.
Feldman, A. S., and C. L. Grimes, eds., *Hearing Conservation in Industry,* Williams & Wilkins, Baltimore, MD, July 1985.
Fischbein, A., "Environmental and Occupational Lead Exposure," in W. N. Rom, A. D. Renzetti, J. S. Lee, et al., eds., *Environmental and Occupational Medicine,* Little, Brown and Co., Boston, MA, 1983.
Goldsmith, F., and L. Kerr, *Occupational Safety and Health: The Prevention and Control of Work-related Hazards,* Human Sciences Press, New York, NY, 1982.
Hamilton, A., *Exploring the Dangerous Trades,* Little, Brown and Co., Boston, MA, 1943.
Keller, J. J., and Associates, *Occupational Exposure Guide,* Occupational Exposure Guide Series, J. J. Keller, Neenah, WI, 1985.
LaGrega, M. D., and D. A. Long, eds., *Toxic & Hazardous Wastes: Proceedings of the Sixteenth Mid-Atlantic Industrial Waste Conference,* Technomic, Westport, CN, 1984.
Moreton, J., and N. A. Falla, *Analysis of Airborne Pollutants in Working Atmospheres,* State Mutual Book and Periodical Service, New York, NY, 1980. (For the Royal Society for Chemistry of England).
National Research Council, Committee on Indoor Pollutants, *Indoor Pollutants,* National Academy Press, Washington, DC, 1983.
National Research Council, *Video Displays, Work, and Vision,* National Academy Press, Washington, DC, 1983.
Rahjans, G. S., and J. Sullivan, *Asbestos Sampling and Analysis,* Ann Arbor Science Publishers, Ann Arbor, MI, 1981.
Ruttenberg, R., "Compliance with the OSHA Cotton Dust Rule: The Role of Productivity Improving Technology," (Working Paper #18) in *Preventing Illness and Injury in the Workplace,* contract report prepared for the Office of Technology Assessment, U.S. Congress, Washington, DC, March 1983.
Smith, T. J., "Industrial Hygiene," in B. S. Levy, and D. H. Wegman, eds., *Occupational Health: Recognizing and Preventing Work-Related Disease,* Little, Brown and Co., Boston, MA, 1983.
Viscusi, W. K., *Employment Hazards,* Harvard University Press, Cambridge, MA, 1980.
Walters, D. B., ed., *Safe Handling of Chemical Carcinogens, Mutagens, Teratogens, and Highly Toxic Substances,* 2 vols., Ann Arbor Science Publishers, Ann Arbor, MI, 1980.
Weiss, G., ed., *Hazardous Chemicals Data Book,* Noyes Data Corp., Park Ridge, NJ, 1981.

11.9.3 Industrial Hygiene Organizations

American Academy of Environmental Engineers (AAEE), P.O. Box 269, Annapolis, MD 21404. The Academy, established in 1955, presently has 2200 members. It is a group of registered professional engineers organized to improve the standards of environmental engineering, certify those with special knowledge of environmental engineering, and furnish lists of those certified to the public. Recognized areas of specialization include air pollution control, general environment, industrial hygiene, radiation protection, solid waste management, water supply, and waste water. Publications include *The Diplomate*.

American Conference of Governmental Industrial Hygienists (ACGIH) Bldg. D-5, 6500 Glenway Ave., Cincinnati, OH 45211. The conference, founded in 1938, is a professional society of educators, persons employed by official governmental units responsible for full-time programs of industrial hygiene, and others conducting research in industrial hygiene. It is devoted to the development of administrative and technical aspects of worker health protection, functioning mainly as a medium for the exchange of ideas and the promotion of standards and techniques in industrial health.

American Industrial Hygiene Association (AIHA), 475 Wolf Ledges Parkway, Akron, OH 44311. This is the principal U.S. professional society for industrial hygiene. Its wide range of interests includes industrial hygiene, air pollution, aerosols, analytical chemistry, biochemical assays, contaminant control, noise, radiation, and toxicology. It publishes the *American Industrial Hygiene Association Journal*.

Center for Occupational Hazards (COH), Five Beekman St., New York, NY 10038. The Center was established in 1977 to gather and disseminate information about health hazards encountered by artists, craftsmen, teachers, children, and others working with art materials. It provides on-site assessments of the health and safety features of facilities used by artists, craftsmen, and students. The Center maintains a reference library, contributes articles to newspapers and art publications, and presents workshops, courses, and lectures. Publications include *Arts Hazards Newsletter*, books, pamphlets, articles, and data sheets on specific hazards.

Intersociety Committee on Methods for Air Sampling and Analysis (ICMASA), Box 612, 425 E. 25th Street, New York, NY 10010. This society includes environmental engineers, chemists, biologists, and physicists involved in the preparation, publication, and use of manuals of recommended methods of air sampling and analysis. They have published *Methods of Air Sampling and Analysis*, 2nd ed., 1977. Committees focus on carbon, general techniques, halogens, metals, oxidants and nitrogen, particulate matter, radioactive standardization coordination, stationary sources, and sulphur. The Society is affiliated with the Air Pollution Control Association, the American Chemical Society, the American Institute of Chemical Engineers, the American Public Works Association, the American Society of Civil Engineers, the American Society of Mechanical Engineers, the Association of Official Analytical Chemists, the Health Physics Society, and the Instrument Society of America.

The **Radiation Health Information Project** (RHIP), 218 D St., S.E., Washington, DC 20003. RHIP was founded in 1978 as a project of the Environmental Policy Institute, which encourages reductions in ionizing radiation exposures and develops policy options for stronger federal and state occupational and public health radiation protection standards. The project publishes *Radiation Public Health and Safety* (proceedings) and reports.

11.9.4 Industrial Hygiene Journals

The *Acoustical Society of America Journal* is published by the American Institute of Physics. This publication is included among the most frequently cited journals in NIOSHTIC.

The *American Industrial Hygiene Association Journal* focuses on all aspects of the field of industrial hygiene practice, policy, and procedures as well as presenting reports of specific investigations and field activities. This publication is included among the most frequently cited journals in NIOSHTIC.

The *Annals of Occupational Hygiene* is an international journal published for the British Occupational Hygiene Society. It focuses on a broad spectrum of industrial hygiene issues, evaluations, and interventions. This publication is included among the most frequently cited journals in NIOSHTIC.

Archives of Environmental Health presents articles concerned with the effects of environmental agents on human health and includes epidemiological, clinical, and experimental studies of man. This publication is included among the most frequently cited journals in NIOSHTIC.

Atmospheric Environment includes selected research papers, review articles, and preliminary communications covering all aspects of the interactions between man and the atmosphere. Subjects covered include administration and economic and political considerations. This publication is included among the most frequently cited journals in NIOSHTIC.

Environmental Health Perspectives is published by the National Institute of Environmental Health Sciences. Each issue is devoted to a special topic pertaining to the effects of the environment on the health of man. This journal publishes conference and workshop proceedings, perspective statements, toxicological summaries, and reviews. This publication is included among the most frequently cited journals in NIOSHTIC.

Environmental Research is an international journal of environmental medicine and the environmental sciences. It presents original research in any discipline concerned with man-environment relationships. The emphasis is on articles pertaining to the health effects of environmental alterations. This publication is included among the most frequently cited journals in NIOSHTIC.

Environmental Science and Technology is published by the American Chemical Society. It provides editorial comment and articles on topics of current interest to industry and the public. The publication has a broad focus including research, directions in environmental science, developing technology, environmental processes, and the social, political, and economic aspects of environmental issues. This publication is included among the most frequently cited journals in NIOSHTIC.

Health Physics is the official journal of the Health Physics Society. Published monthly, it focuses on theoretical and applied work of health physics. This publication is included among the most frequently cited journals in NIOSHTIC.

The *Industrial Hygiene Digest*, made available by the Industrial Health Foundation, presents current literature abstracts on environmental and occupational health and safety.

The *International Archives of Occupational and Environmental Health* is published by Springer-Verlag. This publication is included among the most frequently cited journals in NIOSHTIC.

The *Journal of Environmental Science and Health* is published by Environmental Health Sciences, Marcel Dekker. This publication is included among the most frequently cited journals in NIOSHTIC.

The *Journal of Hazardous Materials* is published by Elsevier Scientific Publishing Co. This publication is included among the most frequently cited journals in NIOSHTIC.

The *Journal of Occupational Medicine* is the official publication of the American Occupational Medical Association and of the American Academy of Occupational Medicine. It presents articles on all phases of occupational medicine and occupational health practice. This publication is included among the most frequently cited journals in NIOSHTIC.

The *Journal of Sound and Vibration* is published by Academic Press. This publication is included among the most frequently cited journals in NIOSHTIC.

Medicina del Lavoro presents original work and reviews in the field of occupational health and industrial hygiene. This publication is included among the most frequently cited journals in NIOSHTIC.

Noise and Vibration Bulletin is published by Multi-Science Publishing Co. This publication is included among the most frequently cited journals in NIOSHTIC.

Sound and Vibration is published by Acoustical Publications. This publication is included among the most frequently cited journals in NIOSHTIC.

11.10 MEDICINE/NURSING

The occupational medical and nursing subject area is concerned with the direct provision of medical care services within the industrial/occupational environment. Specific activities include medical monitoring, preemployment evaluations, placement assessments, periodic health examinations, the management of medical records, and the comparison of disease syndromes with employment experience. Information about the nonmedical aspects of coordinating occupational safety and health services is contained in section 11.2.9.

Monitoring workers who have known exposures to toxic substances is of great importance in the management of worker health. New methods for monitoring contaminant levels in body tissues and fluids are constantly being developed. The journal literature is an excellent source of this new information.

11.10.1 Medicine/Nursing Data Bases

The explosion of scientific information has made it virtually impossible for the practitioner to obtain access to all of the information published. Not only is the volume of information too great to effectively review, the cost of the journals is prohibitive. Only large university libraries can be expected to have a significant portion of the literature. As a result, electronic information retrieval systems have become a necessity for those seeking information on all but the most common topics.

Until recently, five data base vendors based in the United States have provided virtually all of the files of interest to the occupational safety and health professional. These are Dialog Information Services (3460 Hillview Avenue, Palo Alto, CA 94304); the National Library of Medicine (NLM) (8600 Rockville Pike, Bethesda, MD 20209); DOE RECON (Department of Energy, Technical Information Center, P.O. Box 62, Oak Ridge, TN 37830); Bibliographic Retrieval Services (BRS) (BRS Information Technologies, 1200 Route 7, Latham, NY 12110); and System Development Corporation (SDC) Search Services (System Development Corporation, 2500 Colorado Avenue, Santa Monica, CA 90406). New vendors that were originally available only in Europe now make their information systems available in the United States. These include Pergamon Infoline (Pergamon International Information Corporation, 1340 Old Chain Bridge Road, Mclean, VA 22101); Questel (1625 Eye Street, N.W., Suite 719, Washington, DC 20006); and Information Retrieval Service of the European Space Agency Information Retrieval Service, ESA-IRS, (Esrin, P.O. Box 64, 1-00044, Frascati, Italy). Hundreds of different data bases on subjects ranging from art history to zoology are available from these vendors.

Two other vendors, Information Consultants (ICI) (1133 15th St., N.W., Suite 300, Washington, DC 20005) and Chemical Information System (CIS) (7215 York Road, Baltimore, MD 21212), offer similar highly specialized information systems (ICI and ICIS) containing data about regulations and the

TABLE 11.9. Medical/Nursing Data Bases

Name	Producer	No. of Records	Coverage
BIOSIS (BRS)	Biosciences Information Services	Not available N.A.	70–77 78–present
BIOSIS (ESA)	Biosciences Information Services	200,000/yr	73–present
BIOSIS (SDC)	Biosciences Information Services	200,000/yr	69–73 74–79 80–present
BIOSIS PREVIEWS (DIALOG)	Biosciences Information Services	2 million 1.4 million 328,000/yr	69–76 77–81 82–present
BIOTECHNOLOGY (SDC)	Derwent Publications, Ltd.	15,000/yr	82–present
CAB (ESA)	Commonwealth Agricultural Bureaux	144,000/yr	73–present
CAB ABSTRACTS (DIALOG)	Commonwealth Agricultural Bureaux	144,000/yr	72–present
CANCERLIT (NLM)	International Cancer Research Data Bank	432,518	63–present
CAS (SDC)	Chemical Abstracts	1.3 million 1.7 million 2.2 million 340,000/yr	67–71 72–76 77–81 82–present
CA SEARCH (BRS)	Chemical Abstracts	N.A. N.A.	70–76 77–present
CA SEARCH (DIALOG)	Chemical Abstracts	1.3 million 1.7 million 1.2 million .9 million 400,000/yr	67–71 72–76 77–79 80–81 82–present
CA SEARCH (PERGAMON)	Chemical Abstracts	1.7 million 400,000/yr	72–76 77–present
CHEMABS (ESA)	Chemical Abstracts	6 million	67–present
CIS (Questel)	International Labour Organization/CIS	2,100/yr	74–present
CISDOC (ESA)	International Labour Organization/CIS	2,100/yr	74–present
CURRENT BIOTECHNOLOGY ABSTRACTS (PERGAMON)	The Royal Society of Chemistry	7,000	83–present
DOE ENERGY (DIALOG)	TIC, ORNL, DOE (Energy Data Base)	144,000/yr	74–present
EDB (DOE)	TIC,ORNL, DOE (Energy Data Base)	192,000/yr	1900–present
EDB (SDC)	TIC,ORNL, DOE (Energy Data Base)	800,000	30s–present

TABLE 11.9. *(continued)*

Name	Producer	No. of Records	Coverage
EEIS (DOE)	TIRC, ORNL, DOE	5,141	60–present
EMI (DOE, also NLM in TOXLINE)	Environmental Mutagen Information Center, ORNL, DOE	43,720	50–present
EUCAS (QUESTEL)	Chemical Abstracts	1.3 million 1.7 million 2.2 million 340,000/yr	67–71 72–76 77–81 82–present
ETI (DOE, also NLM TOXLINE)	Environmental Teratology Information Center, ORNL	31,583	50–present
EXCERPTA MEDICA (DIALOG)	Excerpta Medica	1.2 million 240,000/yr	74–79 80–present
EXCERPTA MEDICA (BRS)	Excerpta Medica	240,000/yr	80–present
HSELINE (PERGAMON, ESA)	Health and Safety Executive	47,000	77–present
IRCS (BRS)	IRCS Medical Science	Full text of publications	81–present
LIFE SCIENCES COLLECTION (DIALOG)	Cambridge Scientific Abstracts	680,000	78–present
MEDLINE MED80 MED77 MED75 MED71 MED66 (NLM)	National Library of Medicine	140,000/yr 804,874 775,167 642,710 883,906 954,650	83–present 80–82 77–79 75–76 71–74 66–70
MEDLINE (DIALOG)	National Library of Medicine	140,000/yr 1.7 million 1.5 million	80–present 73–79 66–72
MEDLINE (BRS)	National Library of Medicine	140,000/yr N.A. 883,906 954,650	79–present 75–78 71–74 66–70
METADEX (DIALOG)	American Society for Metals	577,000	74–present
NASA (ESA)	NASA Scientific and Technical Information Branch	N.A.	62–present
NIOSHTIC (DIALOG, PERGAMON)	National Institute for Occupational Safety and Health (NIOSH)	12,000/yr	1900–present
NURSING AND ALLIED HEALTH (DIALOG)	Cumulative Index to Nursing and Allied Health Literature	20,000	83–present

TABLE 11.9. *(continued)*

Name	Producer	No. of Records	Coverage
SPHERE	Office		
DERMAL	of Toxic	655 chemicals	N.A.
AQUIRE	Substances,	1,900 chemicals	N.A.
ENVIROFATE	U.S. EPA	450 chemicals	N.A.
(CIS, ICIS)			
SUPERINDEX (BRS)	Superindex, Inc.	2,000 book indexes	N.A.
TELEGEN (ESA, DIALOG)	Environment Information Center	3,600/yr	76–present
TOXICOLOGY DATA BANK (NLM)	National Library of Medicine	4,112 chemicals	N.A.
TOXLINE	National	140,000/yr	81–present
TOXBACK76	Library of	555,808	76–80
TOXBACK65	Medicine	598,424	65–75
(NLM)			

Source: Data in this table were obtained from catalogs and manuals of the identified vendors. Specific citations for each data base and vendor are given in the alphabetical descriptions provided in Section 11.14.

physical characteristics and toxicity of various chemicals. These systems differ from the ones mentioned above in that they provide data with references rather then abstracts and bibliographic information.

Each data base vendor has a different mix of data bases. Even when the same data base is available from different vendors, the periods covered and the implementation can be quite different. There is a trend for the vendors to include larger portions of the data bases in single files as the costs of computer storage decline. The following table lists the data bases that contain medicine/nursing information relevant to occupational safety and health. The availability by vendor and the periods of coverage are included for reference.

11.10.2 Medicine/Nursing Reference Texts

Adams, R. M., *Occupational Skin Diseases,* Grune and Stratton, New York, NY, 1983. This text discusses occupational dermatitis, clinical aspects of occupational skin disease, and protocols for plant surveys. Additional chapters cover industrial substances that produce skin disorders.

Aitio, A., V. Riihimaki, and H. Vainio, eds., *Biological Monitoring and Surveillance of Workers Exposed to Chemicals,* Hemisphere Publishing, Washington, DC, 1984. This is the proceedings of an International Course on Biological Monitoring held in August 1980. It contains papers of interest to those responsible for biological monitoring programs. The papers describe theoretical considerations, methodology, and the monitoring of metals, metalloids, and industrial solvents. Monitoring for combined exposures is also discussed.

Baselt, R. C., *Biological Monitoring Methods for Industrial Chemicals,* Biomedical Publications, Davis, CA, 1980. This is a compendium of analytical methods for the biological monitoring of approximately 70 common industrial substances. Methods are presented for analysis of the substances in blood, plasma, serum, and urine.

Benenson, A. S., ed., *Control of Communicable Diseases in Man,* 13th ed., American Public Health Association, Washington, DC, 1981. This text covers known human communicable diseases and serves as a ready reference work on preventive measures, control of disease spread among individuals and during epidemics, disaster implications, and international measures useful in controlling communicable diseases.

Bennett, P. B., and D. H. Elliott, eds., *The Physiology and Medicine of Diving and Compressed Air Work,* 2nd ed., Balliere Tindall, London, UK, 1975. This is a text on the medical aspects of diving and the management of diving operations. Topics include the diving environment, underwater breathing apparatus, life support systems, effects of pressure, oxygen toxicity, effects of carbon dioxide, gas narcosis, decompression of underwater workers, decompression sickness, and effects of exposure to cold.

Berlin, A., A. H. Wolff, and Y. Hasegawa, eds., *The Use of Biological Specimens for the Assessment of Human Exposure to Environmental Pollutants,* Kluwer Publications, Boston, MA, 1979. This is the proceedings of the joint conference of the Commission of the European Communities, U.S. Environmental Protection Agency, and the World Health Organization, held in April, 1977. The presentations dealt with techniques for analyzing biological specimens for toxic substances such as pesticides, asbestos, mercury, organic carcinogens, organochlorine compounds, lead, and radioactivity. In addition, techniques for analysis of body tissues are detailed and current practice in analytical methods is reviewed.

Burgess, W. A., *Recognition of Health Hazards in Industry: A Review of Materials and Processes,* Wiley-Interscience, New York, NY, 1981. Over thirty industrial operations are reviewed in this book including materials used, engineering controls, possible worker hazards, proper work practices, and protective equipment. Technical appendices and a subject index are included.

Cain, C. D., ed., *Flint's Emergency Treatment and Management,* 6th ed., W. B. Saunders, Philadelphia, PA, 1980. This is an emergency medical reference discussing general medical principles and emergency treatment. There is a special section on treatment of acute poisoning.

Dosman, J. A., and D. J. Cotton, eds., *Occupational Pulmonary Disease: Focus on Grain Dust and Health,* Academic Press, New York, NY, 1980. This work is the proceedings of the International Symposium on Grain Dust and Health, held in Saskatoon, Saskatchewan, in November 1977. Included are papers on epidemiological and physiological methods in the diagnosis of pulmonary disease, pulmonary disease in grain workers, methods of environmental assessment, and health surveillance.

Fosereau, J., C. Benezra, and H. Mailbach, *Occupational Contact Dermatitis: Clinical and Chemical Aspects,* W. B. Saunders, Philadelphia, PA, 1982. This is a useful reference text on occupational dermatological disorders. The first section provides general background on clinical aspects of evaluation of skin disease, with emphasis on guidelines for the occupational physician. Later sections describe dermatological problems associated with 16 specific occupations.

Gee, J. B. L., W. K. C. Morgan, and S. M. Brooks, eds., *Occupational Lung Disease,* Raven Press, New York, NY, 1982. This text resulted from proceedings of the International Congress on Occupational Lung Diseases, held in April 1982. Topics covered were lung cancer, airway diseases, byssinosis, asbestosis, metals, epidemiology, and preventive medicine.

Hunter, D., *Diseases of Occupations,* 6th ed., Hodder and Stoughton, London, UK, 1976. A key reference for both occupational medicine and industrial hygiene, this text is one of the most complete references on the signs and symptoms of work-related diseases. It also provides historical descriptions of occupational medicine.

Last, J. M., et al. eds., *Maxcy-Rosenau Public Health and Preventive Medicine,* 11th ed., Appleton-Century-Crofts, New York, NY, 1980. This is a text on public health practice with sections on public health methods, communicable diseases, and environmental disease such as occupational respiratory disease, diseases associated with chemical substances, metals, and the physical environment. Other topics include regulation and control of occupational health, aerospace medicine, behavioral factors, and health care planning.

Levy, B. S., and D. H. Wegman, eds., *Occupational Health: Recognizing and Preventing Work-Related Disease,* Little, Brown and Co., Boston, MA, 1983. This text provides an introduction to issues currently facing occupational health professionals. Major sections cover work and health, occupational health services in the United States, recognizing and preventing occupational disease, and hazardous workplace exposures. There is also a section on special issues such as women and minority workers, disability evaluation, workers' compensation, and labor's perspective on worker health.

Lynch, A. L., *Biological Monitoring for Industrial Chemical Exposure Control,* CRC Press, Boca Raton, FL, 1974. This book presents the rationale and methodology for biological monitoring of humans exposed to toxic chemicals. Topics include collection and analysis of urine, blood, breath, and tissue specimens (including skin and hair); physiological monitoring of the respiratory, circulatory, nervous, and genitourinary systems; biologic threshold limits; and quality control for sampling and laboratory analysis.

Maibach, H., and G. A. Gellin, eds., *Occupational Dermatology,* Year Book Medical Publishers, Chicago, IL, 1982. A text on occupational skin diseases, including chapters covering dermatotoxicology and problems in specific industries.

Parkes, W. R., *Occupational Lung Disorders,* 2nd ed., Butterworths, London, UK, 1982. This book covers occupational lung diseases, providing reference information on their causes, consequences, and treatment; the nature and fate of inhaled particles; chest radiographs and diagnosis; disease pathogenesis and pathology; and characteristics of inert dusts. There are also sections on diseases caused by inert dusts; silicates; beryllium; organic agents; and metallic, chemical, and physical agents. Other chapters cover occupational asthma and lung cancer.

Parmeggiani, L., ed., *Encyclopedia of Occupational Safety and Health,* 3rd rev. ed. 2 vols., International Labour Organization, Washington, DC, 1983. The third edition (1983) is the latest version of a comprehensive two-volume reference work with concise, easy-to-understand entries.

Peterson, J. E., *Industrial Health,* Prentice-Hall, Englewood Cliffs, NJ, 1977. This book covers all the salient principles but requires only a minimal scientific background. It provides a solid introduction to industrial health. In addition to chemicals, subjects such as abnormal pressure, noise, biothermal stress, and radiation are discussed.

Rom, W. N., A. D. Renzetti, J. S. Lee, et al., eds., *Environmental and Occupational Medicine,* Little, Brown and Co., Boston, MA, 1983. The three major sections of this comprehensive text cover environmental and occupational disease, environmental and occupational exposures, and control of environmental and occupational exposures.

Rumack, B. H., ed., *Poisindex*, Microfiche Edition, Micromedex, Denver, CO. This microfiche edition is produced in association with the National Center for Poison Information, with updates to present. This is the basic reference for local poison control centers.

Schilling, R. S. F., ed., *Occupational Health Practice*, 2nd ed., Butterworths, London, UK, 1981. This updated version includes chapters on occupational health services, effects of chemical exposures on organ systems, procedures for employee health screening, and methods and applications of epidemiology in occupational settings. Also included are chapters on industrial hygiene, ergonomics, and personal protective equipment.

U.S. Department of Health and Human Services, Public Health Service, Centers for Disease Control, National Institute for Occupational Safety and Health, "Leading Work-Related Diseases and Injuries," in *Morbidity and Mortality Weekly Report*, 32:24-26, 32, January 21, 1983. This publication is included with the reference texts because it provides the basis for the NIOSH strategy to prevent work-related illnesses and injuries.

U.S. Congress, Office of Technology Assessment, *The Role of Genetic Testing in the Prevention of Occupational Disease*, GPO Stock No. OTA-8A-194, U.S. Government Printing Office, Washington, DC, April 1983. This is an introduction to genetic testing of workers to detect changes resulting from chemical exposure or to screen and identify those with inheritable disease potentially exacerbated by exposure. Current issues, objectives, and methodology are discussed.

U.S. Department of Defense, Naval Sea Systems Command, *U.S. Diving Manual*, U.S. Superintendent of Documents, Washington, DC, 1975. This two-part manual of diving practice is frequently updated with current technical information. Volume I covers the history of diving, underwater physics, underwater physiology, operations planning, scuba diving, surface-supplied diving, emergencies, and air decompression. Volume 2 covers mixed gas diving theory, underwater breathing apparatus, mixed gas surface-supplied diving, deep diving systems, oxygen diving operations, surface-supplied decompression, and helium-oxygen saturation diving.

Weill, H., and M. Turner, eds., *Occupational Lung Diseases: Research Approaches and Methods*, Marcel Dekker, Warwick, NY, 1981. New techniques in the diagnosis of lung diseases are the focus of this monograph. Immunologic techniques and inhalation and exercise testing as well as current research information are discussed. Other chapters of interest to research workers and public health policymakers include tissue mineral identification, epidemiologic methods, worker surveys, and statistical analysis.

Wyngaarden, J. B., and L. H. Smith, *Cecil Textbook of Medicine*, 16th ed., W. B. Saunders, Philadelphia, PA, 1982. 2 vols. This medical textbook is encyclopedic in its treatment of disease conditions. Articles are one to two pages long and are signed and referenced.

Zenz, C., ed., *Occupational Medicine: Principles and Practical Applications*, Year Book Medical Publishers, Chicago, IL, 1975. This book is an introductory text of occupational medicine with sections on administrative considerations, employee health services for small businesses, occupational safety, the occupational health nurse, and clinical occupational medicine. Also included are physical assessment of workers, occupational lung diseases, the physical and chemical environment, and psychosocial considerations.

Zenz, C., ed., *Developments in Occupational Medicine*, Year Book Medical Publishers, Chicago, IL, August 1980. This volume provides updated information on new developments in occupational medicine. Chapters are devoted to problems affecting the health of women workers, occupational hazards to reproductive health, the effects of toxic substances on organ systems, and other current topics.

Other Medicine/Nursing Reference Texts

Bartone, J. C., ed., *Occupational Medicine: International Survey with Medical Subject Directory & Bibliography*, American Health Research Institute, Ltd., 1985.

Frazier, C. A., *Occupational Asthma*, Van Nostrand-Reinhold, New York, NY, May 1980.

Gardner, A. W., *Current Approaches to Occupational Medicine*, Year Book Medical Publishers, Chicago, IL, August 1980.

Lecture Notes on Occupational Medicine, H. A. Waldron Series: Blackwell Scientific Publications, 2nd ed., Mosby Blackwell, St. Louis, MO, 1979.

Morgan, D. P., *Recognition and Management of Pesticide Poisonings*, 2nd ed., EPA 540/9-76-011, U.S. Government Printing Office, Washington, DC, August 1976.

Turk, M. H., ed., *Occupational Medicine: Surveillance, Diagnosis & Treatment*, F & S Press Book, Ballinger Publications, New York, NY, 1982.

Tyrer, L., *Synopsis of Occupational Medicine*, Wright-PSG Publishing Co., Littleton, MA, 1979.

11.10.3 Medicine/Nursing Organizations

American Academy of Occupational Medicine, 2340 S. Arlington Heights Rd., Suite 400, Arlington Heights, IL 60005. The Academy, established in 1946, has 700 members including physicians whose primary interest is in some phase of occupational medicine. It promotes maintenance and improvement of the health of industrial workers. Publications include the *Journal of Occupational Medicine.*

American Association of Occupational Health Nurses, 3500 Piedmont Rd., N.E., Atlanta, GA 30305. This Association, founded in 1942, has 11,500 members. The membership includes nurse educators, registered professional nurses employed by business and industrial firms, and others interested in occupational health nursing. Publications include a journal, *Occupational Health Nursing,* and a newsletter.

American Board of Preventive Medicine, c/o Stanley R. Mohler, M.D., Dept. of Community Medicine, P.O. Box 927, Wright State University, Dayton, OH 45401. The American Board of Preventive Medicine was founded in 1948 and has about 4200 diplomates. The Board determines elgibility requirements, administers examinations, and certifies physicians in the specialized fields of public health, aerospace medicine, occupational medicine, and general preventive medicine.

American College of Preventive Medicine, 1015 15th St., N.W., Suite 403, Washington, DC 20005. This is a professional society, founded in 1954, of 2000 medical doctors specializing in preventive medicine, public health, aerospace medicine, and occupational medicine.

American Industrial Health Council, 1075 Central Park Ave., Scarsdale, NY 10583. The Council, founded in 1977, has a membership of 170 industrial firms united to assist the government in developing scientifically valid methods of identifying potential industrial carcinogens and adopting effective, practical regulatory controls for such cacinogens.

American Lung Association, 1740 Broadway, New York, NY 10019. This federation of state and local associations has a membership of 7500 physicians, nurses, and laymen interested in the prevention and control of lung disease. The Association, founded in 1904, works with other organizations in planning and conducting programs in community services; public, professional, and patient education; and research. It maintains the American Thoracic Society as its medical section and makes policy recommendations regarding medical care of respiratory disease, occupational health, hazards of smoking, and air conservation. The Association also has an occupational health committee. Publications include the *American Review of Respiratory Diseases.*

American Society of Clinical Pathologists, 2100 W. Harrison, Chicago, IL 60612. This is a professional society of 35,000 clinical pathologists, clinical scientists, chemists, microbiologists, and medical technologists. The Society, formed in 1922, is organized to promote a wider application of pathology and laboratory medicine in the diagnosis and treatment of disease and to conduct educational programs and publish educational materials in the field of clinical and anatomical pathology. Publications include the *American Journal of Clinical Pathology, Laboratory Medicine,* and a newsletter.

American Thoracic Society, 1740 Broadway, New York, NY 10019. This is a medical section of the American Lung Association (see separate entry). Founded in 1905, the 8500 members are specialists in pulmonary diseases, public health physicians concerned with tuberculosis control, thoracic surgeons, and research workers in lung diseases. Committees focus on allergy and clinical immunology, clinical problems, environmental and occupational health, federal lung programs, learning resources, microbiology (infection and immunity), pediatrics, pulmonary nomenclature, respiratory care, respiratory structure and function, and tuberculosis. The Society's publications include the *American Review of Respiratory Diseases.*

College of American Pathologists, 7400 Skokie Blvd., Skokie, IL 60076. The College of American Pathologists, founded in 1947, includes physicians practicing the specialty of pathology. There are 9500 members. The group fosters improvement of education, research, and medical laboratory service to physicians, hospitals, and the public. Publications include *Pathologist* and a newsletter.

11.10.4 Medicine/Nursing Journals

Acta Dermato-Venereologica provides a collection of case reports and treatment regimens as well as experimental results related to the range of dermatologic disorders. It covers all aspects of dermatitis and neurology. Papers are published from authors throughout the world; short reports of specific interest are published without delay.

Acta Physiologica Scandinavica is published by the Scandinavian Society for Physiology and concentrates on all phases of physiology research.

The *American Journal of Medicine* focuses on reports of original research, instructive case reports, and comment or criticism regarding published work or editorials.

The *American Journal of Applied Physiology* publishes original material on respiration, exercise, and environmental physiology.

The *American Journal of Clinical Pathology* publishes research and case reports of interest to those in the clinical practice of pathology.

The *American Journal of Industrial Medicine* publishes reports of original research, instructive case reports, and comments or criticisms in response to published work or editorials. This publication is included among the most frequently cited journals in NIOSHTIC.

The *American Journal of Pathology,* published by the Association of Pathologists, covers experimental pathology including cell injury and cell death, inflammatory reactions, disturbances in circulation, host-parasite interactions, and neoplastic growth. Research results related to human disease are a priority.

The *American Review of Respiratory Disease* is published by the American Lung Association. It is the official journal of the American Thoracic Society. Included in this review are original articles on tuberculosis and respiratory disease, including laboratory and clinical research and clinical observations related to respiratory biology and disease. This publication is included among the most frequently cited journals in NIOSHTIC.

Analytical Biochemistry publishes original contributions of methods of interest to the biochemical sciences and all fields that impinge on biochemical investigation. Methods for sequencing and synthesizing DNA and RNA are among the topics covered. Invited review articles on timely topics in methodology are also published. Book reviews of titles directly related to analytical and preparative methods in biochemistry and such impinging fields as immunology and tissue culture are published as well.

The *Annals of Allergy* publishes original contributions of general interest to allergists. Letters to the editor contain general-interest information or present constructive criticisms on controversial issues. Book reviews are also published. This publication is included among the most frequently cited journals in NIOSHTIC.

The *Annals of Internal Medicine* contains papers on clinical laboratory, socioeconomic, cultural, and historical topics pertinent to internal medicine and related fields.

The *Archives of Dermatology* is published by the American Medical Association. It contains original research reports, short reports, and letters. This publication is included among the most frequently cited journals in NIOSHTIC.

The *Archives of Internal Medicine* is published by the American Medical Association. It includes original research and case reports applicable to the practice of internal medicine.

The *Archives of Pathology and Laboratory Medicine* is an official publication of the American Medical Association that covers a wide range of topics related to studies in pathology and laboratory medicine.

The *British Journal of Cancer* publishes papers containing new information that constitutes a distinct contribution to the clinical, epidemiological, pathological, or molecular aspects of oncology. It also publishes short communications and invited book reviews. This publication is included among the most frequently cited journals in NIOSHTIC.

The *British Journal of Dermatology* publishes original articles on all aspects of normal biology and the pathology of the skin. It is a vehicle for the publication of both experimental and clinical research and serves equally the laboratory worker and clinician. This publication is included among the most frequently cited journals in NIOSHTIC.

The *British Journal of Experimental Pathology* publishes original contributions describing the techniques and results of experimental research into the causation, diagnosis, and cure of disease in man. Papers come from the fields of bacteriology, biochemistry, pharmacology, physiology, and serology. This publication is included among the most frequently cited journals in NIOSHTIC.

The *British Journal of Industrial Medicine* provides original contributions relevant to occupational and environmental medicine, including toxicological studies of industrial and agricultural chemicals and reports pertaining to medical experiences and ethical considerations. This publication is included among the most frequently cited journals in NIOSHTIC.

The *British Medical Journal* publishes original research relevant to clinical medicine. The condensed reports section contains shorter articles with data of marginal interest deleted. Short reports are typically clinical case histories and brief or negative research findings. The medical practice section contains articles about the organization and assessment of medical work in hospitals, general practice, and the sociological aspects of medicine. This publication is included among the most frequently cited journals in NIOSHTIC.

Cancer publishes original articles on all aspects of cancer. Abstracts of the contributions to this journal are translated into Interlingua for *Biological Abstracts* and for the abstracting service of the *Journal of the American Medical Association.* In addition, abstracts from the International Cancer Research Data Bank may be included after modification.

Cancer Letters includes contributions of interest to a multidisciplinary audience. All areas of cancer research, including chemical carcinogenesis, environmental and occupational cancer, geographic pathology and cancer epidemology, cancer control, and cancer biology, are covered. This publication is included among the most frequently cited journals in NIOSHTIC.

Cancer Research contains the results of well-documented original experimental, clinical, and statistical studies. It also includes concise review articles on subjects of interest to cancer researches. Letters to the editor deal with subjects and issues of interest to the cancer researcher. Experimental data is kept to a minimum in letters and authors are given opportunity to reply in simultaneous publication. This publication is included among the most frequently cited journals in NIOSHTIC.

Carcinogenesis is a multidisciplinary publication covering all aspects of research relating to the prevention of cancer in man. It includes full papers and short communications on viral, physical, and chemical carcinogenesis and mutagenesis. This publication is included among the most frequently cited journals in NIOSHTIC.

Chest is published by the American College of Chest Physicians. It contains original research of interest to those concerned with research on diseases of the chest. This publication is included among the most frequently cited journals in NIOSHTIC.

Clinical Chemistry publishes contributions on the application of chemistry to the understanding of human health and disease. Articles contain original information on experimental or theoretical advances in clinical chemistry. Subjects include analytical techniques, instrumentation, data processing, statistical analyses of data, clinical investigations in which chemistry has played a major role, and experimental investigations of chemically oriented problems of human disease. Reviews, case reports, scientific notes, and letters to the editor are also published.

Clinical and Experimental Dermatology features the full text of orations by speakers of international renown, symposia on selected topics, invited reviews, book reviews, and correspondence.

Contact Dermatitis is published primarily for clinicians interested in the various aspects of environmental dermatitis. Subjects include both allergic and irritant (toxic) types of contact dermatitis, occupational (industrial) dermatitis, and consumer's dermatitis from cosmetics and other products. The journal states that its purpose is to promote and maintain communication among dermatologists, industrial physicians, allergists, and clinical immunologists as well as chemists and research workers involved in industry and the production of consumer goods. This publication is included among the most frequently cited journals in NIOSHTIC.

Experientia publishes short reports of general scientific interest. The contributions are brief original reports. This publication is included among the most frequently cited journals in NIOSHTIC.

Federation Proceedings is the official publication of the Federation of the American Society for Experimental Biology. It contains papers presented at the Society-sponsored symposium held at the annual meeting and at other meetings of the constituent societies. It also includes papers presented at conferences, seminars, and symposia sponsored by other groups within the areas of interest of the constituent societies.

GANN: Japanese Journal of Cancer Research includes original papers, notes, and communications written by members of the Japanese Cancer Association. This publication is included among the most frequently cited journals in NIOSHTIC.

The *International Journal of Cancer* is concerned with laboratory and clinical research on cancer. The journal publishes original full-length research papers and findings published elsewhere as abstracts or preliminary communications. Letters about papers that have appeared in the journal or expressing constructive criticism and suggestions concerning cancer research in general are also published. This publication is included among the most frequently cited journals in NIOSHTIC.

The *Journal of Allergy and Clinical Immunology* publishes original contributions in the field of allergy and clinical immunology. In addition, the journal publishes allergy rounds, which present diseases related to allergic or immunologic mechanisms. This publication is included among the most frequently cited journals in NIOSHTIC.

The *Journal of the American Medical Association* is published four times per month as a forum for the discussion of progress in clinical medicine, medical research, and other fields of interest to physicians. This publication is included among the most frequently cited journals in NIOSHTIC.

The *Journal of Applied Physiology,* a publication of the American Physiological Society, presents original papers related to normal or abnormal function in the following areas: respiratory physiology, nonrespiratory functions of the lungs, environmental physiology, temperature regulation, exercise physiology, and interdependence. Priority in each category is given to articles concerning integrated and adaptive mechanisms and to articles on human research. This publication is included among the most frequently cited journals in NIOSHTIC.

The *Journal of Clinical Investigation* publishes original papers on research pertinent to human biology and disease. Articles on animal research or *in vitro* techniques are published if they are relevant to human normal or abnormal biology.

The *Journal of Investigative Dermatology* publishes original papers and reviews pertinent to the normal and abnormal function of the skin. This publication is included among the most frequently cited journals in NIOSHTIC.

The *Journal of Laboratory and Clinical Medicine* is the official publication of the Central Society for Clinical Research. The journal publishes original manuscripts describing investigations in the broad field of laboratory and clinical medicine. Also included are articles on laboratory methods.

The *Journal of the National Cancer Institute* contains results of published research; since it is a U.S. government publication, all material published is in the public domain. This publication is included among the most frequently cited journals in NIOSHTIC.

The *Journal of Occupational Medicine* is the official publication of the American Occupational Medical Association and the American Academy of Occupational Medicine. It presents articles on all phases of occupational medicine and occupational health practice. This publication is included among the most frequently cited journals in NIOSHTIC.

The *Journal of the Society of Occupational Medicine* presents the British perspective and British research results on occupational medicine. This publication is included among the most frequently cited journals in NIOSHTIC.

Laboratory Investigations is published by the United States-Canadian Division of the International Academy of Pathology. It provides prompt publication of significant advances in pathology research. The journal includes reports of original research, brief communications, and special review articles.

Medicina del Lavoro presents original work and reviews in the field of occupational health and industrial hygiene. This publication is included among the most frequently cited journals in NIOSHTIC.

The *Medical Journal of Australia* is published by the Australian Medical Association. It publishes original contributions in the general field of medicine.

The *New England Journal of Medicine* publishes original articles, special articles, and brief reports in medical intelligence. This publication is included among the most frequently cited journals in NIOSHTIC.

The *Proceedings of the Society for Experimental Biology and Medicine* publishes original papers covering biochemistry, immunology, nutrition, physiology, pathology, pharmacology, and radiobiology among other related fields.

Public Health Reports is the journal of the U.S. Public Health Service and is published by the Department of Health and Human Services. Articles of value to public health, disease prevention, health promotion, medical care, and community medicine are published.

Thorax concentrates on original work on diseases of the thorax and relevant anatomical and physiological studies. This publication is included among the most frequently cited journals in NIOSHTIC.

11.11 SAFETY

The term "safety," as used in this section, refers to traumatic physical events and short-term chemical exposures that result in injury or acute illness. This section includes references on accidental releases of energy, equipment failure, or human error that lead to short-term chemical exposure or mechanical action that produces bodily harm. Chemical exposure could result from unplanned events such as transportation accidents, spills, equipment failure, chemical waste cleanup, or other accidents and short-term releases of chemical substances. Traumatic physical events include fires, explosions, exposure to electricity, falls, and contact with moving objects. The references provided below reflect this breadth of subject matter. References on personal protective equipment are included under engineering controls (section 11.6.2), and references about the administration of safety programs are included in the occupational safety and health programs section (section 11.2.9).

11.11.1 Safety Data Bases

The explosion of scientific information has made it virtually impossible for the practitioner to obtain access to all of the information published. Not only is the volume of information too great to effectively review, the cost of the journals is prohibitive. Only large university libraries can be expected to have a significant portion of the literature. As a result, electronic information retrieval systems have become a necessity for those seeking information on all but the most common topics.

Until recently, five data base vendors based in the United States have provided virtually all of the files of interest to the occupational safety and health professional. These are Dialog Information Services (3460 Hillview Avenue, Palo Alto, CA 94304); the National Library of Medicine (NLM) (8600 Rockville Pike, Bethesda, MD 20209); DOE RECON (Department of Energy, Technical Information Center, P.O. Box 62, Oak Ridge, TN 37830); Bibliographic Retrieval Services (BRS) (BRS Information Technologies, 1200 Route 7, Latham, NY 12110); and System Development Corporation (SDC) Search Services (System Development Corporation, 2500 Colorado Avenue, Santa Monica, CA 90406). New vendors that were originally available only in Europe now make their information systems available in the United States. These include Pergamon Infoline (Pergamon International Information Corporation, 1340 Old Chain Bridge Road, Mclean, VA 22101); Questel (1625 Eye Street, N.W., Suite 719, Washington, DC 20006); and Information Retrieval Service of the European Space Agency Information Retrieval Service, ESA-IRS, (Esrin, P.O. Box 64, 1-00044, Frascati, Italy). Hundreds of different data bases on subjects ranging from art history to zoology are available from these vendors.

Two other vendors, Information Consultants (ICI) (1133 15th St., N.W., Suite 300, Washington, DC 20005) and Chemical Information Systems (CIS) (7215 York Road, Baltimore, MD 21212), offer similar highly specialized information systems (ICI and ICIS) containing data about regulations and the

TABLE 11.10 Safety Data Bases

Name	Producer	No. of Records	Coverage
CIS (Questel)	International Labour Organization/CIS	2,100/yr	74-present
CISDOC (ESA)	International Labour Organization/CIS	2,100/yr	74-present
COLD (SDC)	Cold Regions Research Safety Engineering Laboratory	6,000/yr	62-present
CURRENT BIOTECHNOLOGY ABSTRACTS (PERGAMON)	The Royal Society of Chemistry	7,000	83-present
DOE ENERGY (DIALOG)	TIC, ORNL, DOE (Energy Data Base)	144,000/yr	74-present
EDB (DOE)	TIC, ORNL, DOE (Energy Data Base)	192,000/yr	1900-present
EDB (SDC)	TIC, ORNL, DOE (Energy Data Base)	800,000	30s-present
ENERGYLINE (SDC, ESA, DIALOG)	EIC/ Intelligence	5,250/yr	71-present
HAZARDLINE (BRS)	Occupational Health Services, Inc.	3,000 chemical substances	Not available (N.A.)
HEALTH AUDIOVISUAL ONLINE CATALOG (BRS)	Northeastern Ohio Universities College of Medicine	5,000 citations	60-present
HSELINE (PERGAMON, ESA)	Health and Safety Executive	47,000	77-present
KIRK-OTHMER ENCYCLOPEDIA OF CHEMICAL TECHNOLOGY (BRS)	John Wiley and Sons, Inc.	Full text of encyclopedia	3rd ed. 76
LABORDOC (ESA, QUESTEL, SDC)	International Labour Organization	4,800/yr	65-present
LABORATORY HAZARDS BULLETIN (PERGAMON)	The Royal Society of Chemistry	2,000(4/84)	81-present
NIOSHTIC (DIALOG, PERGAMON)	National Institute for Occupational Safety and Health (NIOSH)	12,000/yr	1900-present
NSIC (DOE)	Nuclear Science Information Center, ORNL, DOE	12,000/yr	63-present

TABLE 11.10 *(continued)*

Name	Producer	No. of Records	Coverage
NTIS (DIALOG)	National Technical Information Service	60,000/yr	64-present
NTIS (ESA)	National Technical Information Service	60,000/yr	69-present
NTIS (SDC, BRS)	National Technical Information Service	60,000/yr	70-present
PHP (DOE)	American Petroleum Institute	7,947	77-80
RBOT (BRS)	Cincinnati Millicron	N.A.	80-present
SAE (SDC)	Society of Automotive Engineers	800/yr	65-present
SAFETY (PERGAMON)	Cambridge Scientific Abstracts	34,500	81-present
TRIS (DIALOG)	United States Department of Transportation	18,000/yr	77-present
VOLUNTARY STANDARDS INFORMATION NETWORK (BRS)	Information Handling Services	over 500	N.A.
WATER RESOURCES ABSTRACTS (DIALOG, DOE)	United States Department of Interior	172,500	68-present

Source: Data in this table were obtained from catalogs and manuals of the identified vendors. Specific citations for each data base and vendor are given in the alphabetical descriptions provided in section 11.14.

physical characteristics and toxicity of various chemicals. These systems differ from the ones mentioned above in that they provide data with references rather than abstracts and bibliographic information.

Each database vendor has a different mix of data bases. Even when the same database is available from different vendors, the periods covered and the implementation can be quite different. There is a trend for the vendors to include larger portions of the databases in single files as the costs of computer storage decline. The following table lists the databases that contain safety information relevant to occupational safety and health. The availability by vendor and the periods of coverage are included for reference.

11.11.2 Safety Reference Texts

Alliance of American Insurers, *Handbook of Hazardous Materials, Fire, Safety, Health,* 2nd ed., Schaumberg, IL, 1983. This is a handbook of information on all types of hazardous substances. Threshold Limit Values, NFPA ratings for health hazards, flammability and reactivity, flash point, boiling point, vapor pressure, specific gravity, and odor characteristics are covered, as are the symptoms of acute and chronic exposure.

Alliance of American Insurers, *Handbook of Organic Industrial Solvents,* 5th ed., Schaumberg, IL, 1980. This is an information guide for some 250 organic solvents. Included are chemical properties, fire and explosion information, health hazards, procedures for evaluation and control of exposure, and waste disposal.

American Society of Testing and Materials, *Toxic and Hazardous Industrial Chemicals Safety Manual,* International Technical Information Institute, Tokyo, 1979. Information on over 700 industrial chemicals was compiled for this manual. Each entry contains the following information: chemical name, synonyms, formula, uses, properties, hazard criteria, toxicity criteria, handling and storage, emergency treatments and measures, management of spills and leakage, and disposal and waste treatment.

Bretherick, L., ed., *Hazards in the Chemical Laboratory,* 3rd ed., Royal Society of Chemistry, London, UK, 1981. This guide covers the major health and safety hazards encountered in the chemical laboratory. It includes chapters on planning and management of safety policy, fire protection, reactive chemical hazards, toxicology, first aid medical management, and ionizing radiation. A major part of this book is an extensive list for individual chemicals of toxic effects, reaction hazards, fire hazards, first aid, disposal, and emergency spill containment procedures.

Fawcett, H. H., and W. S. Wood, eds., *Safety and Accident Prevention in Chemical Operations,* 2nd ed., John Wiley and Sons, New York, NY, 1982. This book provides excellent source material for chemical laboratory safety officers, industrial toxicologists, and industrial hygenists. It is a thorough review of accident prevention strategies and contains reference material on toxicity and the TSCA law, hazardous chemical waste disposal, and developments in fire extinguishment.

Kirk, H., and R. Othmer, eds., *Encyclopedia of Chemical Technology,* 3rd ed., Wiley-Interscience, New York, NY, 1978-83, 23 vols. This encyclopedia is the single most comprehensive reference for chemical engineering in existence. The 23 volumes cover the entire spectrum of chemical engineering with extensive discussions and clear illustrations. This is the best single reference for information about manufacturing processes for plastics and chemicals.

McElroy, F. E., ed., *Accident Prevention Manual for Industrial Operations,* 8th ed., National Safety Council, Chicago, IL, 1981. This manual is a comprehensive reference on accident prevention. It uses the term "accident" in the sense of an unplanned event that interrupts the completion of an activity. According to this definition, injury or damage is not necessary. Management techniques for preventing accidents and the technical aspects of accident prevention are included.

National Fire Protection Association, *Fire Protection Guide on Hazardous Materials,* 7th ed., Quincy, MA, 1979. This handbook provides fast access to information on fire and explosion hazards for several thousand chemicals of industrial importance. It contains a trade name index as well as listings of generic names. Major sections are "Flashpoint Index of Trade Name Liquids" (with data on over 8800 products); "Fire Hazard Properties of Flammable Liquids, Gases and Volatile Solids"; "Hazardous Chemicals Data" (fire, toxicity, explosion, storage, and fire fighting data on over 388 chemicals); "Manual of Hazardous Chemical Reactions" (2350 mixtures of two or more chemicals); and "Recommended System for the Identification of the Fire Hazard of Materials."

National Research Council, Committee on Hazardous Substances in the Laboratory, *Prudent Practices for Disposal of Chemicals from Laboratories,* National Academy Press, Washington, DC, 1983. This guide to environmentally safe methods for chemical and biological waste disposal in laboratories includes procedures for recovering and recycling wastes; legalities; and the current DOT requirements for labeling hazardous substances.

National Research Council, Committee on Hazardous Substances in the Laboratory. *Prudent Practices for Handling Hazardous Chemicals in Laboratories,* National Academy Press, Washington, DC, 1981. This book contains useful information for handling hazardous chemical substances. Topics covered include current U.S. regulations governing worker exposure to and disposal of hazardous chemicals, management of laboratory safety programs, principles of safe laboratory practice, special procedures for working with toxic substances, procedures for working with compressed gases, emergency procedures and first aid, personal protective equipment, and laboratory ventilation. Procedures for procurement, storage, distribution, and disposal of laboratory chemicals are also covered.

Simonds, R. H., and J. V. Grimaldi, *Safety Management: Accident Cost and Control,* rev. ed., Richard D. Irwin, Homewood, IL, 1963. This is a revised and updated edition of the classic text. It was written as a textbook for courses in accident prevention and safety administration. The emphasis is on integrating safety principles with sound management practices.

Sliney, D., and M. Wolbarst, *Safety with Lasers and Other Optical Sources: A Comprehensive Handbook,* Plenum Publications, New York, NY, 1980. This is a handbook on the effects of laser and other light sources on biological organisms. Chapters cover the fundamentals of optical physics, anatomy and physiology of the eye and skin, protective equipment, exposure limits, laser hazard classifications, special applications of lasers (medical, construction, industrial, consumer), and safety training.

U.S. Coast Guard, Department of Transportation, *CHRIS—Hazardous Chemical Data, Manual II,* U.S. Government Printing Office, Washington, DC, October 1978. CHRIS provides information on cleaning up chemical spills. In the form of safety data sheets, it is one of the best available sources of information about chemical emergencies.

U.S. Department of Transportation, *A Guide to the Federal Hazardous Materials Transportation Regulatory Program,* January 1983. DOT-1-83-12. This publication, intended to help the layman understand the Federal Hazardous Materials Transportation Regulatory Program, provides definitions of regulated material, a historical overview of the legislation, and descriptions of the responsible federal agencies. Most of the document is devoted to explaining the requirements for shippers and carriers of hazardous materials.

OTHER SAFETY REFERENCES

American Chemical Society, *Toxic Chemicals in Explosives Facilities: Safety and Engineering Design,* Symposium Series, No. 96, American Chemical Society, Washington, DC, 1979.

Business Roundtable, *Improving Construction Safety Performance,* Report A-3 of the Construction Industry Cost Effectiveness Project, Business Roundtable, New York, NY, January 1982.

Ferry, T. S., and C. C. Thomas, *Safety Program Administration for Engineers & Managers: A Resource Guide for Establishing & Evaluating Safety Programs,* 1984.

Fuscaldo, A., B. J. Erlick, and B. Hindman, eds., *Laboratory Safety—Theory and Practice,* Academic Press, New York, NY, 1980.

Haddon, W., Jr., E. A. Sushman, and D. Klein, eds., *Accident Research Methods and Approaches,* Harper and Row, New York, NY, 1964.

Haddon, W., Jr., and S. P. Baker, "Injury Control," in D. W. Clark and B. MacMahon, eds., *Preventive and Community Medicine,* Little, Brown and Co., Boston, MA, 1981.

Heinrich, H. W., *Industrial Accident Prevention: A Scientific Approach,* 4th ed., McGraw-Hill Book Co., New York, NY, 1959.

Johnson, W. G., *Management Oversight and Risk Tree,* U.S. Government Printing Office, Washington, DC, 1973.

King, R. W., and J. Magid, *Industrial Hazard & Safety Handbook,* Butterworth, Woburn, MA, 1979.

Malasky, S. ed., *Systems Safety: Technology & Application,* Garland Publication, New York, NY, 1982.

McFarland, R. A., *The Epidemiology of Industrial Accidents,* Harvard School of Public Health, Cambridge, MA, 1965.

McElroy, F. E., ed., *Accident Prevention Manual for Industrial Operations: Engineering and Technology,* 8th ed., National Safety Council, Chicago, IL, 1980.

Occupational Safety & Health in Road Transport, Inland Transport Committee: Eleventh Session, Geneva, 1985: Report III, Series: Programme of Industrial Activities, Unipub, New York, NY, 1985.

Peterson, D., ed., *Human Error Reduction & Safety Management,* Garland Publications, New York, NY, 1981.

Petersen, D., *Safety Management—A Human Approach,* Doran, Englewood, NJ, 1975.

Quirk, T. J., ed., *Hazardous Materials Guide: Shipping, Materials Handling and Transportation,* rev. ed., J. J. Keller, Neenah, WI, 1979.

Ridley, J. R., *Safety at Work,* Butterworth, Woburn, MA, 1983.

Singleton, W. T., "Future Trends in Accident Research in European Countries," in *Proceedings of International Seminar on Occupational Accident Research,* Swedish Board of Occupational Safety and Health, Stockholm, Sweden, 1983.

Tarrants, W. E., The *Measurement of Safety Performance,* Garland Publications, New York, NY, 1980.

U.S. Congress, General Accounting Office, *How Can Workplace Injuries Be Prevented? The Answers May be in OSHA Files,* HRD-79-43, Government Printing Office, Washington, DC, May 3, 1979.

Viscusi, W. K., *Employment Hazards,* Harvard University Press, Cambridge, MA, 1980.

Zabetakis, M., *Safety Manual No. 4: Accident Prevention,* Mine Safety and Health Administration, Washington, DC, 1975.

11.11.3 Safety Organizations

Academy of Hazard Control Management (AHCM), 5010A Nicholson Ln., Rockville, MD 20852. The Academy was established in 1981 for certified hazard control managers in safety and health, including occupational and environmental safety and health, to promote professional development and the exchange of information.

American Society of Safety Engineers, 850 Busse Hyw., Park Ridge, IL 60068. The Society, founded in 1911, has over 18,000 members interested in safety engineering and safety program management. Publications include *Professional Safety,* monographs, and textbooks.

Board of Certified Safety Professionals (BCSP), 208 Burwash Ave., Savoy, IL 61874. This Board, established in 1969, offers the designation of Certified Safety Professional (CSP) to safety engineers, industrial hygienists, safety managers, and others who have passed examinations and met other established criteria. Over 6000 CSPs have been issued certificates. The Board conducts research in the evaluation of competency for professional practice in the safety professions and also compiles statistics.

Center for Chemical Plant Safety, Institute of Chemical Engineers, 345 E. 47th St., New York, NY 10017. The Center is operated by the Institute to focus attention on chemical industry safety. The Center was formed as a response to the concern for safety after the Bhopal incident.

National Fire Protection Association (NFPA), Batterymarch Park, Quincy, MA 02269. The Association was formed in 1896 and draws its membership from the fire service, business and industry, health care, and educational and other institutions. It also includes individuals in the fields of insurance, government, architecture, and engineering. The Association develops, publishes, and disseminates standards. It compiles annual statistics on causes and occupancies of fires, fire deaths, and fire fighter casualties. It provides field service by specialists. Publications of the Association include *Fire Command, Fire News, Fire Journal, Fire Technology,* technical committee reports, catalogs of publications and visual aids, a fire protection reference directory, national fire codes, and a yearbook. The Association also publishes committee lists, a fire protection handbook, text and reference books, standards, educational booklets, reports, posters, and the "Learn Not to Burn" curriculum. It also produces films, videotapes, and slide/tape packages.

National Institute for Farm Safety (NIFS), 460 Henry Mall, Madison, WI 53706. The Institute was founded in 1962 for professional farm safety specialists dedicated to improving the farm accident record through education, engineering, and research. It contributes to the Cooperative Standards Program, helps fund accident studies, and encourages industry support and research. The Institute conducts training for farm safety workers and other interested individuals and offers consulting services to those interested in farm safety.

National Safety Council (NSC), 444 N. Michigan Ave., Chicago, IL 60611. The Council, founded in 1913, has a membership of over 15,000. It focuses on efforts to reduce the number and severity of accidents and occupational illnesses by producing and disseminating information. Its publications include *National Safety News, Traffic Safety, Family Safety, Accident Facts,* newsletters, and various technical publications.

National Safety Management Society (NSMS), 6060 Duke St., Alexandria, VA 22304. The Society was founded in 1978 for individuals with managerial responsibilities related to safety loss control management, including professionals in the fields of education, medicine, computer technology, security, labor relations, and industrial hygiene. They promote new concepts of accident prevention and loss control and the role of safety in management. They advocate favorable cost/benefit returns where possible. Professional growth of the membership is a major objective of the society.

Safety Equipment Institute (SEI), 1901 N. Moore St., Arlington, VA 22209. This Institute, founded in 1981, works to foster and advance public interest in safety and protective equipment and to assist safety equipment industries and government agencies in their mutual goal of providing workers with the best possible protective equipment. They recognize products that are certified to meet appropriate standards and manage a third-party certification program.

11.11.4 Safety Journals

Accident Analysis and Prevention is an international safety journal. This publication is included among the most frequently cited journals in NIOSHTIC.

The *American Industrial Hygiene Association Journal* focuses on all aspects of the field of industrial hygiene practice, policy, and procedures. It publishes reports of specific investigations and field activities. This publication is included among the most frequently cited journals in NIOSHTIC.

Journal of Combustion Toxicology presents information on the characteristics of fire, the flammability of various materials, and the toxicology of combustion products. It also covers testing methods and apparatus and theories and techniques of fire prevention, retardation, and extinguishment. This publication is included among the most frequently cited journals in NIOSHTIC.

The Journal of Occupational Medicine is the official publication of the American Occupational Medical Association and of the American Academy of Occupational Medicine. It presents articles on all phases of occupational medicine and occupational health practice. This publication is included among the most frequently cited journals in NIOSHTIC.

Professional Safety, the official publication of the American Society of Safety Engineers, presents a forum for the discussion of safety engineering and safety program administration.

11.12 TOXICOLOGY

In response to the increased industrial use of chemicals, occupational safety and health practitioners have focused considerable attention on the toxicity of materials found in the work environment. This attention has been influenced by the need to understand the relative level of health risk associated with the use of such materials, which may produce problems ranging from short-term acute effects to cancer, cardiovascular diseases, genetic damage, reproductive disorders, and teratogenicity. The response of the human body to exposures to these chemicals is extremely difficult to assess and often requires combined knowledge of medical practitioners, chemists, and industrial hygienists.

The references included under toxicology were selected to reflect the range of information needed to understand industrial exposures. The contribution of the toxicology program of the National Library of Medicine must be acknowledged. Its efforts to catalog toxicology information have contributed substantially to the coverage of this section.

11.12.1 Toxicology Data Bases

The explosion of scientific information has made it virtually impossible for the practitioner to obtain access to all of the information published. Not only is the volume of information too great to effectively review, the cost of the journals is prohibitive. Only large university libraries can be expected to have a significant portion of the literature. As a result, electronic information retrieval systems have become a necessity for those seeking information on all but the most common topics.

Until recently, five data base vendors based in the United States have provided virtually all of the files of interest to the occupational safety and health professional. These are Dialog Information Services (3460 Hillview Avenue, Palo Alto, CA 94304); the National Library of Medicine (NLM) (8600 Rockville Pike, Bethesda, MD 20209); DOE RECON (Department of Energy, Technical Information Center, P.O. Box 62, Oak Ridge, TN 37830); Bibliographic Retrieval Services (BRS) (BRS Information Technologies, 1200 Route 7, Latham, NY 12110); and System Development Corporation (SDC) Search Services (System Development Corporation, 2500 Colorado Avenue, Santa Monica, CA 90406). New vendors originally available only in Europe now make their information systems available in the United States. These include Pergamon Infoline (Pergamon International Information Corporation, 1340 Old Chain Bridge Road, Mclean, VA 22101); Questel (1625 Eye Street, N.W., Suite 719, Washington, DC 20006); and Information Retrieval Service of the European Space Agency Information Retrieval Services, ESA-IRS (Esrin, P.O. Box 64, 1-00044, Frascati, Italy). Hundreds of different data bases on subjects ranging from art history to zoology are available from these vendors.

Two other vendors, Information Consultants (ICI) (1133 15th St., N.W., Suite 300, Washington, DC 20005) and Chemical Information Systems (CIS) (7215 York Road, Baltimore, MD 21212) offer similar highly specialized information systems (ICI and ICIS) containing data about regulations and the physical characteristics and toxicity of various chemicals. These systems differ from the ones mentioned above in that they provide data with references rather than abstracts and bibliographic information.

Each data base vendor has a different mix of data bases. Even when the same data base is available from different vendors, the periods covered and the implementation can be quite different. There is a trend for the vendors to include larger portions of the data bases in single files as the costs of computer storage decline. The following table lists the data bases that contain toxicology information relevant to occupational safety and health. The availability by vendor and the periods of coverage are included for reference.

11.12.2 Toxicology Reference Texts

American Conference of Governmental Industrial Hygienists, *The Documentation of the Threshold Limit Values*, 4th ed., American Conference of Governmental Industrial Hygienists, Cincinnati, OH, 1980. This publication is in the form of a loose-leaf binder to accommodate the annual supplements. The documentation provides the basis for the influential ACGIH recommendations. The presentation is concise, summarizing the animal or human toxicity studies that serve as the basis for the exposure limit recommendations. Permissible exposure limits for both the United States and other countries are listed.

American Conference of Governmental Industrial Hygienists, *TLVs®: Threshold Limit Values for Chemical Substances and Physical Agents in the Workroom Environment with Intended Changes for 19*—(New editions annually), American Conference of Governmental Industrial Hygienists, Cincinnati, OH. This publication is one of the most influential in the field. The 1985–86 edition contains these sections: "Threshold Limit Values for Chemical Substances in the Work Environment Adopted by ACGIH with Intended Changes for 1985–86," "Biological Exposure Indices Proposed by ACGIH for 1985–86," and "Threshold Limit Values for Physical Agents in the Work Environment Adopted by ACGIH with Intended Changes for 1985–86."

Anderson, K. E., and R. M. Scott, *Fundamentals of Industrial Toxicology*, Ann Arbor Science Publishers, Ann Arbor, MI, 1981. This text was written for persons with no background in the field. The work includes target organ systems, dose-response, classification and types of exposures, and contaminant identification.

Arena, J. M., *Poisoning: Toxicology, Symptoms, Treatments*, 4th ed., Charles C. Thomas, Springfield, IL, 1979. This publication presents valuable information on general considerations pertaining to poisoning. It includes information on pesticides, industrial and occupational hazards, drugs, cosmetics, and even poisonous plants and animals. This is a practical book that covers product descriptions as well as diagnosis and treatment. In addition to covering most of the frequently involved substances, such unique topics as crayons, golf balls, and toxic Christmas decorations are discussed.

TABLE 11.11 Toxicology Data Bases

Name	Producer	No. of Records	Coverage
AGRICOLA (CAIN) (DIALOG, BRS)	USDA, Science and Education Administration	1.1 million 120,000/yr	70–78 79–present
AMERICAN CHEMICAL SOCIETY PRIMARY JOURNAL DATA BASE (BRS)	Bibliographic Retrieval Services	30,000 articles in full text	Not available (N.A.)
APLIT (SDC)	American Petroleum Institute	360,000	64–present
APTIC (DIALOG)	U.S. EPA	87,000	66–76
AQUALINE (DIALOG)	Water Research Centre	75,500	60–present
BIOSIS (BRS)	Biosciences Information Services	N.A. N.A.	70–77 78–present
BIOSIS (ESA)	Biosciences Information Services	200,000/yr	73–present
BIOSIS (SDC)	Biosciences Information Services	200,000/yr	69–73 74–79 80–present
BIOSIS PREVIEWS (DIALOG)	Biosciences Information Services	2 million 1.4 million 328,000/yr	69–76 77–81 82–present
BIOTECHNOLOGY (SDC)	Derwent Publications, Ltd.	15,000/yr	82–present
CAB (ESA)	Commonwealth Agricultural Bureaux	144,000/yr	73–present
CAB ABSTRACTS (DIALOG)	Commonwealth Agricultural Bureaux	144,000/yr	72–present
CANCERLIT (NLM)	International Cancer Research Data Bank	432,518	63–present
CAS (SDC)	Chemical Abstracts	1.3 million 1.7 million 2.2 million 340,000/yr	67–71 72–76 77–81 82–present
CA SEARCH (BRS)	Chemical Abstracts	N.A. N.A.	70–76 77–present
CA SEARCH (DIALOG)	Chemical Abstracts	1.3 million 1.7 million 1.2 million .9 million 400,000/yr	67–71 72–76 77–79 80–81 82–present
CA SEARCH (PERGAMON)	Chemical Abstracts	1.7 million 400,000/yr	72–76 77–present
CCRIS (CIS, ICIS)	National Cancer Institute	882 assay results	N.A.
CESARS (CIS, ICIS)	Michigan Department of Natural Resources	180 chemicals "evaluated"	N.A.

TABLE 11.11 (*continued*)

Name	Producer	No. of Records	Coverage
CHEMABS (ESA)	Chemical Abstracts	6 million	67–present
CHEMICAL EXPOSURE (DIALOG)	Chemical Effects Information Center	14,370	74–present
CIS (Questel)	International Labour Organization/CIS	2,100/yr	74–present
CISDOC (ESA)	International Labour Organization/CIS	2,100/yr	74–present
CTCP (CIS, ICIS)	Clinical Toxicology of Commercial Products	18,000 products in commerce	N.A.
CURRENT BIOTECHNOLOGY ABSTRACTS (PERGAMON)	The Royal Society of Chemistry	7,000	83–present
EEIS (DOE)	TIRC, ORNL, DOE	5,141	60–present
EMI (DOE, also NLM In TOXLINE)	Environmental Mutagen Information Center, ORNL, DOE	43,720	50–present
ENVIROLINE (SDC, ESA, DOE, DIALOG)	EIC/Intelligence	110,000	71–present
ENVIRONMENTAL BIBLIOGRAPHY (DIALOG)	Environmental Studies Institute	24,000/yr	73–present
ETI (DOE, also NLM in TOXLINE)	Environmental Teratology Information Center, ORNL, DOE	31,583	50–present
EXCERPTA MEDICA (DIALOG)	Excerpta Medica	1.2 million 240,000/yr	74–79 80–present
EXCERPTA MEDICA (BRS)	Excerpta Medica	240,000/yr	80–present
FOODS ADLIBRA (DIALOG)	K & M Publications, Inc.	3,600/yr	74–present
FSTA (DIALOG, SDC)	International Food Information Services	20,000/yr	69–present
HAZARDLINE (BRS)	Occupational Health Services, Inc.	3,000 chemical substances	N.A.
HSELINE (PERGAMON, ESA)	Health and Safety Executive	47,000	77–present

TABLE 11.11 (*continued*)

Name	Producer	No. of Records	Coverage
INTERNATIONAL PHARMACEUTICAL ABSTRACTS (DIALOG, BRS)	American Society of Hospital Pharmacists	6,000/yr	70-present
KIRK—OTHMER ENCYCLOPEDIA OF CHEMICAL TECHNOLOGY (BRS)	John Wiley and Sons, Inc.	Full text of encyclopedia	3rd ed. 76
LIFE SCIENCES COLLECTION (DIALOG)	Cambridge Scientific Abstracts	680,000	78-present
MEDLINE	National Library of Medicine	140,000/yr	83-present
MED80		804,874	80-82
MED77		775,167	77-79
MED75		642,710	75-76
MED71		883,906	71-74
MED66		954,650	66-70
(NLM)			
MEDLINE (DIALOG)	National Library of Medicine	140,000/yr	80-present
		1.7 million	73-79
		1.5 million	66-72
MEDLINE (BRS)	National Library of Medicine	140,000/yr	79-present
		N.A.	75-78
		883,906	71-74
		954,650	66-70
NASA (ESA)	NASA Scientific and Technical Information Branch	N.A.	62-present
NIOSHTIC (DIALOG, PERGAMON)	National Institute for Occupational Safety and Health (NIOSH)	12,000/yr	1900-present
OHMTADS (CIS, ICIS)	U.S. EPA, Superfund	1,334 substances	N.A.
PESTDOC/ PESTDOC II (SDC)	Derwent Publications, Ltd.	8,500/yr	68-present
POLLUTION ABSTRACTS (ESA, DIALOG, BRS)	Cambridge Scientific Abstracts	8,400/yr	70-present
RTECS (NLM, CIS, ICIS)	National Institute for Occupational Safety and Health (NIOSH)	70,029	N.A.
SPHERE	Office of Toxic Substances, U.S. EPA		
DERMAL		655 chemicals	N.A.
AQUIRE		1,900 chemicals	N.A.
ENVIROFATE		450 chemicals	N.A.
(CIS, ICIS)			
TELEGEN (ESA, DIALOG)	Environment Information Center	3,600/yr	76-present
TOXICOLOGY DATA BANK (NLM)	National Library of Medicine	4,112 chemicals	N.A.

TABLE 11.11 Core Data Bases *(continued)*

Name	Producer	No. of records	Coverage
TOXLINE TOXBACK76 TOXBACK65 (NLM)	National Library of Medicine	140,000/yr 555,808 598,424	81-present 76-80 65-75
WATERLIT (SDC)	South African Water Information Centre	12,000/yr	76-present
WATERNET (DIALOG)	American Water Works Association	3,000/yr	71-present
ZINC, LEAD AND CADMIUM ABSTRACTS (PERGAMON)	Zinc Development Association	3,000/yr	75-present

Source: Data in this table were obtained from catalogs and manuals of the identified vendors. Specific citations for each data base and vendor are given in the alphabetical descriptions provided in section 11.14.

Ariens, E. J., and A. M. Simonis, *Introduction to General Toxicology,* rev. prntg., Academic, New York, NY, 1976. A general introduction to toxicology, this text provides a useful overview. The revised edition presents a new chapter on environmental toxicology.

Baselt, R. C., *Biological Monitoring Methods for Industrial Chemicals,* Biomedical Publications, Davis, CA, 1980. This is a compendium of analytical methods for the biological monitoring of approximately 70 common industrial substances. Methods are presented for analysis of the substances in blood, plasma, serum, and urine.

Berman, E., *Toxic Metals and Their Analysis,* Heyden & Son, London, UK, 1980. This monograph presents methods of human tissue analysis with chapters on 31 different metals. The chapters cover the metals, biochemical roles, toxicology, and distribution in the body. Methods discussed include colorimetry, fluorimetry, polarography, and spectroscopy. Many tables and references are included.

Bridbord, K., and J. French, eds., *Toxicological and Carcinogenic Health Hazards in the Workplace,* Chem-Orbital, Park Forest South, IL, 1979. The proceedings of the first annual NIOSH Scientific Symposium, held in 1978, this publication covers a wide variety of topics related to illnesses resulting from working with chemical and physical agents. Epidemiologic studies of illness in workers exposed to specific agents are also included.

Casarett, L. J., and J. Doull, *Toxicology: The Basic Science of Poisons,* Macmillan, New York, NY, 1975. This is a widely used textbook on toxicology of interest to researchers and others as a basic reference.

De Bruin, A., *Biochemical Toxicology of Environmental Agents,* Elsevier, Amsterdam, 1976. This treatment of biochemistry has a subject index and 13,000 references. Radiation effects are discussed in each chapter.

Doull, J., C. D. Klaassen, and M. O. Amdur, eds., *Casarett and Doull's Toxicology,* 2nd ed., Macmillan, New York, NY, 1980. This toxicology text provides a comprehensive treatment of the subject. The chapters are well referenced. The text provides a good survey of toxicology, covering organ physiology, dose-response, and industrial toxicology.

Dreisbach, R. H., *Handbook of Poisonings: Diagnosis and Treatment,* 9th ed., Lange, Los Altos, CA, 1977. This is a summary of information on the diagnosis and treatment of clinically important poisons. The work covers general considerations, agricultural poisons, industrial hazards, household hazards, medicinal poisons, and animal and plant hazards. Over 800 references are contained in this compact handbook.

Environmental Protection Agency, *Toxic Substances Control Act: Initial Inventory,* Environmental Protection Agency, Washington, DC, 1979. This is an EPA inventory of chemical substances in the U.S. not regulated by other agencies. The initial inventory contained 44,000 chemicals. June 1, 1979 was the deadline for manufacturers to include their chemicals in the inventory or be required to go through premanufacturing notification.

Fawcett, H. H., and W. S. Wood, eds., *Safety and Accident Prevention in Chemical Operations,* 2nd ed., John Wiley and Sons, New York, NY, 1982. This book provides excellent source material for chemical laboratory safety officers, industrial toxicologists, and industrial hygienists. It is a thorough review of accident prevention strategies and contains reference material on toxicity and the TSCA law, hazardous chemical waste disposal, and developments in fire extinguishment.

Finkle, A. J., ed., *Hamilton and Hardy's Industrial Toxicology,* 4th rev. ed., Wright-PSG Publishing, Littleton, MA, 1982. In this revised edition of the classic reference text on occupational safety and health, clinical aspects of occupationally induced diseases are described and updated. This text focuses on occupational diseases, placing particular emphasis on thirty metals. Discussions of chemical compounds, radiation, dusts, and biological hazards are presented. An extensive bibliography and glossary of occupational medicine terms are also included.

Fishbein, L., *Potential Industrial Carcinogens and Mutagens,* Elsevier, Amsterdam, 1979. This volume presents information regarding carcinogenic and/or mutagenic organic chemical compounds used by industry. It is arranged by structural categories. Epidemiology, risk assessment, and combined effects are also treated.

Florkin, M., and E. H. Stotz, eds., *Comprehensive Biochemistry,* Elsevier-North Holland, New York, NY, 1962. 34 vols. (current). This is a standard reference in biochemistry. Most volumes are divided into sections on physicochemical and organic aspects of biochemistry, chemistry of biological compounds, biochemical reaction mechanics, metabolism, and chemical biology.

Gosselin, R. E., H. C. Hodge, R. P. Smith, and M. N. Gleason, *Clinical Toxicology of Commercial Products,* 5th ed., Williams and Wilkins, Baltimore, MD, 1984. This standard reference work in clinical toxicology is designed to help the physician deal quickly and effectively with acute chemical poisonings by commerical products. The sections of the book are color coded for easy access. The sections include "First Aid and General Emergency Treatment," "Ingredients Index," "Therapeutics Index," "Supportive Treatment," "Trade Name Index," "General Formulations" (a guide to the probable composition of products not specifically listed by trade name), and "Manufacturers Index." This work is kept up to date by the monthy *Bulletin of Supplementary Material.*

Hayes, W. J., Jr., *Toxicology of Pesticides,* 3rd rev. ed., Williams and Wilkins, Baltimore, MD, 1975. This reference on pesticide toxicology addresses the principles and the general conditions of exposure and relates these to observed human health effects, the problems of diagnosis and treatment, and the means of preventing injury. It also provides brief outlines of the impact of the pesticides on domestic animals and wildlife. Other topics include diagnosis, treatment, and prevention, there is also a discussion of pesticide effects on wildlife. CAS registry numbers and chemical names are given for all compounds mentioned in the text.

Loomis, T. A., *Essentials of Toxicology,* 3rd ed., Lea & Febiger, Philadelphia, PA, 1978. Clear and easy to follow, this introductory text on general toxicology presents the concepts of forensic, environmental, and economic toxicology. In addition, basic statistics, dose-response relationships, and other basic principles are presented.

Nordberg, G., L. Friberg, and V. Vouk, eds., *Handbook on the Toxicology of Metals,* Elsevier-North Holland, New York, NY, 1979. The mechanism of metal toxicity is the focus of this text. Introductory chapters discuss the general chemistry of metals; sampling of metals in air, water, food, and body fluids; and dose-response relationships. Later chapters are devoted to individual metals of toxicologic importance.

Parmeggiani, L., ed., *Encyclopedia of Occupational Safety and Health,* 3rd rev. ed., International Labour Organization, Washington, DC, 2 vols. The third edition (1983) is the latest version of a comprehensive two-volume reference work with concise, easy-to-understand entries.

Patty's Industrial Hygiene and Toxicology, 3rd rev. ed., John Wiley and Sons, New York, NY, 1978–1982, 3 vols. A comprehensive reference on industrial hygiene and industrial toxicology, this work is both an important reference for those in the field and a significant text on the principles and practice of industrial hygiene. *Volume 1, General Principles,* G. E. Clayton and F. E. Clayton, eds., covers the knowledge base necessary for the practice of industrial hygiene. Topics covered include: fundamentals of epidemiology; mode of entry and action of toxic materials; and the effects of noise, ionizing radiation, heat stress, and other stressors. Evaluation and control measures, industrial hygiene survey procedures, air sampling and analysis, engineering controls, respiratory protective equipment, and air pollution control are also thoroughly discussed. *Volume 2, Toxicology,* G. E. Clayton and F. E. Clayton, eds., is an extensive reference on industrial toxicology. Three separate books comprise Volume 2; each chapter discusses the toxicology of compounds of similar chemical structure. Extensive subject and chemical indexes are included in Volume 2. *Volume 3, Theory and Rationale of Industrial Hygiene Practice,* L. J. Cralley and L. V. Cralley, eds., provides insight into quantitative matters such as emission inventories; analytical measurements; evaluation of exposures to chemical agents; ionizing radiation; noise and vibration; pressure; biologic agents; surveillance programs; and data extrapolation.

Punkett, E. R., *Handbook of Industrial Toxicology,* 2nd ed., Chemical Publishing Co., New York, NY, 1976. This is a chemical handbook. Descriptive data, occupational exposures, TLVs®, toxicities, and preventive measures are provided for each chemical listed.

Ross, S., and M. Pronin, eds., *Toxic Substances Sourcebook,* Environmental Information Center, New York, NY, 1978. Much of this book is devoted to abstracts of current literature. This source book also provides lists of periodicals, books, films, and data bases. Regulations, with excerpts from the *Federal Register* with an emphasis on TSCA regulations, are discussed.

Rumack, B. H., ed., *Poisindex,* Microfiche Edition, Micromedex, Denver, CO (periodic updates). Written in association with the National Center for Poison Information, this is the basic reference for local poison control centers. Updates keep the material current.

Sax, N. I., ed., *Cancer Causing Chemicals,* Van Nostrand-Reinhold, New York, NY, 1981. Similar to the *NIOSH Registry of Toxic Effects of Chemical Substances,* this listing of known or potential carcinogens includes brief toxicological data. The materials listed are cross-referenced by synonym and Chemical Abstracts Service (CAS) Registry Numbers. Controlling occupational exposures and regulations affecting use of carcinogens are discussed in some detail.

Sax, N. I., *Dangerous Properties of Industrial Materials,* 6th ed., Van Nostrand-Reinhold, New York, NY, 1983. This is a valuable quick reference work that contains information on the chemical and physical hazards of the most common industrial chemicals. Over 16,000 chemicals are listed under the general chemical section; the information on each includes acute toxicity, flammability, synonyms, descriptions, and physical constants. The text also includes short sections on the control of radiation hazards and air contaminants and sections on fire protection, labeling, and identification of hazardous materials.

Searle, C. E., ed., *Chemical Carcinogens,* ACS Monograph 173, American Chemical Society, Washington, DC, 1976. Topics presented in this survey of chemical carcinogens include bioassays for chemical carcinogens and carcinogens in plants, microorganisms, foods, and industrial compounds. This work is extensively referenced.

Sontag, J. L., ed., *Carcinogens in Industry and the Environment,* Marcel Dekker, New York, NY, 1981. The book presents readings on the identification, detection, control, and sociolegal concepts of carcinogens. Papers include statistical analysis of data, *in vitro* carcinogenesis tests, safety, and control implications of industrial carcinogens. Carcinogens in foods and cosmetics and naturally occurring carcinogens are also discussed.

Sunshine, I., ed., *Handbook of Analytical Toxicology,* CRC Press, Cleveland, OH, 1969. Physical and chemical properties of drugs and chemical hazards are listed. Methods for analysis of biological specimens are also discussed.

Sunshine, I., ed., *Methodology for Analytical Toxicology,* CRC Press, Cleveland, OH, 1975. This book describes the methods of analysis for drugs in biological fluids to determine if poisoning has occurred. It does not treat industrial toxicology and occupational hazards.

Tatken, R. L., and R. J. Lewis, Sr., eds., *NIOSH Registry of Toxic Effects of Chemical Substances* (RTECS), National Institute for Occupational Safety and Health, Cincinnati, OH, 1981–82 (annual supplements). Available in paper only from NTIS PB 85-21871. A quarterly microfiche version is available only by annual subscription. Order the microfiche by title only. The title is *Microfiche Edition, Registry of Toxic Effects of Chemical Substances.* RTECS is the largest single source of toxicity information in print. It includes acute and chronic toxicity information coded from the literature as well as tumorogenic data for more than 70,000 substances and 210,000 synonyms. Environmental limits from OSHA, MSHA, NIOSH, and ACGIH are included, as are references to the International Agency for Research on Cancer (IARC) monographs and the substantial risk information provided pursuant to Section 8(e) of the Toxic Substances Control Act (TSCA). The information in this publication is available in a variety of formats. Both paper copy and microfiche are cited here, but RTECS is also available on-line from various sources as described in the data base portions of this chapter.

Other Toxicology Reference Texts

Barlow, S. M., and F. M. Sullivan, eds., *Reproductive Hazards of Industrial Chemicals,* Academy Press, Santa Clara, CA, October 1982.

Browning, E., *Toxicity of Industrial Metals,* 2nd ed., Appleton-Century-Crofts, New York, NY, 1969.

Browning, E., *Toxicity and Metabolism of Industrial Solvents,* American Elsevier, New York, NY, 1965.

Deichmann, W. B., ed., *Toxicology & Occupational Medicine,* Developments in Toxicology and Environmental Science Series, Vol. 4, Elsevier, New York, NY, 1979.

Fairchild, E., *Agricultural Chemicals and Pesticides: A Handbook of the Toxic Effects,* J. K. Burgess, Englewood, NJ, 1978.

Friberg, L., G. R. Nordberg, and V. N. Vouk, *Handbook on the Toxicity of Metals,* Elsevier North Holland, New York, NY, 1979.

Goodwin, B. L., *Handbook of Intermediary Metabolism of Aromatic Compounds,* John Wiley and Sons, New York, NY, 1976.

Hemminki, K., M. Sorsa, and H. Vainio, eds., *Occupational Hazards and Reproduction,* Hemisphere Publishing, Washington, DC, 1985.

Le Serve, A., *Chemicals, Work & Cancer,* Van Nostrand-Reinhold, New York, NY, 1980.

Lefaux, R., *Practical Toxicology of Plastics,* CRC Press, Cleveland, OH, 1968.

Martin, H., and C. R. Worthing, eds., *Pesticide Manual,* 5th ed., British Crop Protection Council, Worchester, UK, 1977.

Oehme, L., *Toxicity of Heavy Metals,* Marcel Dekker, New York, NY; Part I, 1978; Part II, 1979.

Sanders, C., ed., *Pulmonary Toxicology of Respirable Particles: Proceedings,* U.S. Department of Energy, Washington, DC, 1980.

Shepard, T. H., *Catalog of Teratogenic Agents,* 3rd ed., Johns Hopkins University Press, Baltimore, MD, 1980.

Sutton, H. E., *Mutagenic Effects of Environmental Contaminants,* Academic Press, New York, NY, 1972.

U.S. Congress, Office of Technology Assessment, *The Information Content of Premanufacture Notices, Background Paper,* OTA-BP-H-17, U.S. Government Printing Office, Washington, DC, April 1983.

U.S. Department of Health and Human Services, Public Health Service, National Toxicology Program, *National Toxicology Program: Fiscal Year 1984 Annual Plan,* NTP-84-023, National Toxicology Program, Research Triangle Park, NC, February 1984.

Venugopal, B., and T. D. Luckey, *Metal Toxicity in Mammals, 2,* Plenum Press, New York, NY, 1978.

Worthing, C. R., ed., *Pesticide Manual,* 6th ed., British Crop Protection Council, Worchester, UK, 1979.

Wedeen, R. P., *Poison in the Pot: The Legacy of Lead,* Southern Illinois University Press, Carbondale, IL, November 1984.

Woodhead, A. D., C. J. Shellabarger, and V. P. Bond, eds., *Assessment of Risk from Low-Level Exposure to Radiation & Chemicals: A Critical Overview,* Basic Life Sciences Series, Vol. 33, Plenum Publishing, New York, NY, June 1985.

11.12.3 Toxicology Organizations

American Association of Poison Control Centers, c/o William O. Robertson, MD, Children's Orthopedic Hospital, P.O. Box C-5371, Seattle, WA 98105. This association establishes standards for poison control centers and poison information centers. It procures information on the ingredients and potential acute toxicity of substances that may be involved in accidental poisonings.

American Industrial Health Council (AIHC), 1075 Central Park Ave., Scarsdale, NY 10583. The Council, established in 1977, has 170 member industrial firms united to assist the government in developing scientifically valid methods to identify potential industrial carcinogens and adopt effective, practical regulatory controls for such carcinogens.

Chemical Hazard Response Information System (CHRIS), U.S. Coast Guard, 400 Seventh St., S.W., Washington, DC 20590. CHRIS provides information essential for informed decision-making by Coast Guard personnel and others during emergencies involving the transport of hazardous chemicals. Also included is nonemergency information necessary to achieve improved levels of safety in the bulk shipment of hazardous chemicals.

Chemical Industry Institute of Toxicology (CIIT), P.O. Box 12137, Research Triangle Park, NC 27709. The Institute, organized in 1974, has 33 member chemical companies. The Institute is dedicated to the scientific, objective study of toxicological issues associated with the manufacture, use, and disposal of selected commodity chemicals and the dissemination of the information developed. Specific goals include updating existing toxicological testing, developing improved test methodologies, training professional toxicologists, and serving the health and environmental needs of the public through toxicology research. It maintains scientific advisory panels and sponsors conferences and workshops.

Federation of American Societies of Experimental Biology (FASEB), 9650 Rockville Pike, Bethesda, MD 20814. The Federation was organized to bring together investigators in the biological and medical sciences to share the results of biological research through publications and scientific meetings. It is composed of the following seven professional societies: the American Physiological Society, the American Society of Biological Chemists, the American Society for Pharmacology and Experimental Therapeutics, the American Association of Pathologists, the American Institute of Nutrition, the American Association of Immunologists, and the American Society for Cell Biology.

Forum for the Advancement of Toxicology, Center for the Health Sciences, College of Pharmacy, University of Tennessee, Memphis, TN 38163. The Forum, founded in 1968, has a membership of 2500 toxicologists, faculty members, administrators, health professionals, and other scientists interested in promoting education and research in toxicology.

Genetic Toxicology Association (GTA), c/o R. Bruce Dickson, Cleary, Gottlieb, Steen and Hamilton, 1725 N. St., N.W., Washington, DC 20036. The Association, founded in 1971, has a membership of 325 academic, industrial, and government genetic toxicologists.

International Agency for Research on Cancer (IARC), World Health Organization, Distribution and Sales Service, Geneva 27, Switzerland. The IARC is an agency of the World Health Organization. It brings together experts on the carcinogenicity of various chemicals or classes of chemicals to evaluate the open scientific literature relating to the carcinogenic hazard of chemicals to man. The Agency publishes *IARC Monographs on the Evaluation of the Carcinogenic Risk of Chemicals to Man.*

International Society on Toxicology (IST), c/o Dr. Philip Rosenberg, Dept. of Pharmacology and Toxicology, University of Connecticut, Storrs, CT 06268. The Society was founded in 1961. Its membership includes over 500 biochemists, pharmacologists, immunologists, physicians, and others studying animal, plant, and microbial toxins. It seeks to advance knowledge of the properties of toxins and antitoxins derived from plant and animal tissues. Publications include *Toxicon* and a newsletter.

National Pesticide Information Clearinghouse (NPIC), P.O. Box 2031, San Benito, TX 78586. This is the information clearinghouse on pesticides, accident prevention, and preventive medicine. Research is conducted on the effects of pesticides on human and environmental health, with special emphasis on farm workers and occupational safety and health. It assists the medical community with information on pesticides and poisoning and sponsors programs for training clinical personnel. It provides a toll-free telephone number: 800-531-7790; in Texas 800-292-7664.

Society of Toxicology (SOT), 475 Wolf Ledges Pkwy., Akron, OH 44311. This society, formed in 1961, has 1850 members. Membership is restricted to those who have conducted and published original investigations in some phase of toxicology and have a continuing professional interest in the field. Publications include *Toxicology and Applied Pharmacology, Fundamental and Applied Toxicology,* and a newsletter.

Toxicology Forum (TF), 1575 Eye St., N.W., Suite 800, Washington, DC 20005. The Toxicology Forum, founded in 1975, has 123 members representing industry, government, universities, and laboratories throughout the world. Its purpose is to facilitate communication among scientific decision makers and aid in developing safety assessments and establishing regulations concerning toxicology. Areas of study and research have included epidemiology, biotechnology, genetics, carcinogens, saccharin, and caffeine. The Forum publishes transcripts of proceedings and maintains archives.

11.12.4 Toxicology Journals

Acta Pharmacologica et Toxicologica is published by the Scandinavian Pharmacological Societies. Articles are based on original research in pharmacology and toxicology.

The *American Industrial Hygiene Association Journal* focuses on all aspects of the field of industrial hygiene practice, policy, and procedures and reports on specific investigations and field activities. This publication is included among the most frequently cited journals in NIOSHTIC.

The *Archives of Environmental Contamination and Toxicology* focuses on matters related to air, water, and food contamination. This publication is included among the most frequently cited journals in NIOSHTIC.

The *Archives of Environmental Health* includes current research reports on environmental health. This publication is included among the most frequently cited journals in NIOSHTIC.

Archives of Toxicology is the official publication of the European Society of Toxicology. Studies include the mechanisms of toxicology, defined effects in man, new methods of analysis, new methods of treatment, and experimental studies. This publication is included among the most frequently cited journals in NIOSHTIC.

The *Biochemical Journal* publishes articles and papers in all fields of biochemistry, including new experimental results, descriptions of new experimental methods, and new interpretations of results. Priority is given to the development of biochemical concepts rather than to simply recording facts.

Biochemical Pharmacology publishes original research in the form of full-length papers, short communications, and invited comments. Included is research with animals, cells, subcellular components, enzymes, and model systems if they define the mechanism of drug action. Antimicrobial and antiviral studies are published, as are descriptive mathematical models. Research on the use of drugs to elucidate physiological and behavioral mechanisms is also considered. This publication is included among the most frequently cited journals in NIOSHTIC.

The *Bulletin of Environmental Contamination and Toxicology* seeks to provide early reports of discoveries in the fields of air, soil, water, and food contamination. Included are articles on methodology and other matters related to the effects of toxic materials in the total environment. This publication is included among the most frequently cited journals in NIOSHTIC.

The *British Medical Journal* publishes original research relevant to clinical medicine. The condensed reports section contains shorter/articles with data of marginal interest deleted. Reports in this section are typically clinical case histories and brief or negative research findings. The medical practice section contains articles about general practice, the organization and assessment of medical work in hospitals, and the sociological aspects of medicine. This publication is included among the most frequently cited journals in NIOSHTIC.

Canadian Journal of Physiology and Pharmacology contains definitive papers, notes, rapid communications, and reviews in the fields of physiology, pharmacology, nutrition, and toxicology.

Cancer Research presents review articles and the results of original experimental, clinical, and statistical studies of interest to cancer researchers. Announcements and reports of meetings of interest are included, and the availability of fellowships and scholarships is noted. Proceedings of meetings and symposia are published as external supplements to the journal. It is among the most frequently abstracted journals in both NIOSHTIC and TOXLINE.

Carcinogenesis publishes research reports on all areas of carcinogenicity. It is included among the most frequently cited journals in NIOSHTIC.

Clinical Toxicology concentrates on the pragmatic aspects of poisoning. It includes many case reports and articles on diagnosis and treatment. Issues often include news items and announcements.

CRC Critical Reviews in Toxicology contains lengthy critical evaluations. It is not uncommon to find several hundred references per review. The president of the Chemical Industry Institute of Toxicology (CIIT) serves as the editor. The referee for each article is identified.

Environmental Health Perspectives is published by the National Institute of Environmental Health Sciences. Its objectives are to communicate research findings of environmntal health significance and to inform the scientific community of potential health hazards. Conference and workshop proceedings, perspective statements, toxicological summaries, and reviews are also published. This publication is included among the most frequently cited journals in NIOSHTIC.

Experientia publishes brief original reports of general scientific interest. This publication is included among the most frequently cited journals in NIOSHTIC.

Experimental and Molecular Pathology presents articles on the disease process with emphasis on structural and biochemical alterations in mammalian tissues and fluids. Newer techniques of analytical chemistry, histochemistry, pharmacology, toxicology, electron microscopy, and pathology in man and animals are a priority. Articles on virology, edocrinology, immunology, nutrition, chemotherapy, and geographic pathology are also published.

Food and Cosmetics Toxicology publishes original, unpublished research in the field of food and cosmetics toxicology. Reviews are also published following peer review. This publication is included among the most frequently cited journals in NIOSHTIC.

Human Toxicology is a British publication by Macmillan Press, Ltd., containing editorials, papers, short communications, conference reports, and letters to the editor.

Journal of Applied Toxicology is an international forum devoted to research and methods. Direct clinical, industrial, and environmental applications are emphasized. This journal includes articles on toxicology, reproduction, mutagenesis, carcinogenesis, the environment, and biochemical mechanisms.

The *Journal of Agricultural and Food Chemistry*, a publication of the American Chemical Society, publishes articles in multidisciplinary areas concerned with chemistry in all aspects of agriculture and food including pesticides, toxicology, fertilizers, and wastes. This publication is included among the most frequently cited journals in NIOSHTIC.

Journal of Biological Chemistry publishes papers on a broad range of topics of general interest to biochemists. The papers are original contributions; if additional information warrants complete publication of a study, it may be permissible for a paper to have been published as a preliminary communication or presented at a poster session. This publication is included among the most frequently cited journals in NIOSHTIC.

The *Journal of Chromatographic Science* is published monthly by Preston Publications and concerns topics related to chromatography and its applications. This publication is included among the most frequently cited journals in NIOSHTIC.

Journal of Combustion Toxicology presents information on the characteristics of fire, the flammability of various materials, and the toxicology of combustion products. It also covers pertinent testing methods and apparatus, theories, and techniques of fire prevention, retardation, and extinguishment. This publication is included among the most frequently cited journals in NIOSHTIC.

The *Journal of Environmental Pathology and Toxicology and Oncology* is the official publication of the American College of Toxicology. Each of the following sections of the publication has a separate editor: environmental mutagenesis and genetic toxicology, nutritional toxicology and pharmacology, inhalation toxicology and environmental health, and dermal toxicology. It also includes book reviews and announcements. This publication is included among the most frequently cited journals in NIOSHTIC.

The *Journal of Nutrition* is the official publication of the American Institute of Nutrition, which presents original research on all aspects of nutrition.

The *Journal of Occupational Medicine* is a publication of the American Occupational Medical Association and the American Academy of Occupational Medicine. It presents articles on all phases of occupational medicine and occupational health practice. This publication is included among the most frequently cited journals in NIOSHTIC.

Journal of Pharmacology and Experimental Therapeutics publishes original papers on all aspects of chemical effects on biological systems. The terms "chemicals" and "biological systems" are defined as broadly as possible with this journal. This publication is included among the most frequently cited journals in NIOSHTIC.

Journal of Toxicology and Environmental Health focuses on the toxicological effect of natural and manmade environmental pollutants. Carcinogenesis, mutagenesis, teratology, neurotoxicity, environmental factors affecting health, and other toxicological phenomena are of special interest. Emphasis is also given to human health in the workplace. This publication is included among the most frequently cited journals in NIOSHTIC.

Medicina del Lavoro presents original work and reviews in the field of occupational health and industrial hygiene. This publication is included among the most frequently cited journals in NIOSHTIC.

Mutation Research is an international journal on mutagenesis, chromosome breakage, and related subjects. Papers on original fundamental research, review articles, and short communications are featured. It is among the most frequently abstracted journals in both NIOSHTIC and TOXLINE.

Teratology — The International Journal of Abnormal Development is the official journal of the Teratology Society. Published bimonthly, it contains original reports of studies in all areas of abnormal development and related fields. "Teratogen Update" is a feature dealing with current knowledge of known human teratogens. This publication is among the most frequently abstracted journals in both NIOSHTIC and TOXLINE.

Toxicology, an international journal, focuses on the effects of chemicals on living systems. Published are original papers on the biological effects of chemical compounds including food additives, pesticides, drugs, chemical contaminants, consumer products, and industrial chemicals. This publication is included among the most frequently cited journals in NIOSHTIC.

Toxicology and Applied Pharmacology, an official journal of the Society of Toxicology, focuses on original scientific research on changes in tissue structure or function resulting from the administration of chemicals, drugs, or natural products. This publication is included among the most frequently cited journals in NIOSHTIC.

Toxicology Letters, an international journal, publishes short reports on biochemical mechanisms of mammalian toxicity, focusing on research of sufficient importance and interest to warrant rapid publication. This publication is included among the most frequently cited journals in NIOSHTIC.

Toxicon is the official journal of the International Society on Toxicology. Its purpose is the exchange of knowledge on the poisons derived from animals, plants, and microorganisms.

11.13 OCCUPATIONAL SAFETY AND HEALTH TRAINING

Occupational safety and health training includes worker training and education as well as the training and education of safety and health professionals. The training needs of these groups are different, but the content and overall objective of training are the same. The occupational health and safety objectives of each revolve around the recognition, assessment, and control of workplace hazards to protect workers.

Meeting the training needs of workers and health professionals has been approached in many ways. The worker may receive employer-sponsored training and in some cases training from the unions. OSHA has primary responsibility for employee training. Although NIOSH has primary responsibility for professional training, many of its publications are used in worker training activities. Student needs have been addressed through programs using the training materials developed by NIOSH for high school science teachers and the joint NIOSH-OSHA program for training vocational education teachers.

The training of health and safety professionals is a principal responsibility of NIOSH; it is met primarily through the establishment of the NIOSH Educational Resource Centers. NIOSH also develops training materials for short courses addressing the needs of an interdisciplinary group including physicians, nurses, safety professionals, and industrial hygienists. In addition, NIOSH has attempted to integrate occupational safety and health information into the required business and engineering courses in professional schools.

Short-term training courses often receive publicity in newsletters or trade and professional journals, or they can be identified by contacting the appropriate professional organization. The notices of training, however, are usually specific to the interests of the publication or organization. Questions concerning the availability of specialized programs of instruction and the treatment of specific subjects can be referred to the institutions described in section 11.13.3.

11.13.1 Occupational Safety and Health Training Data Bases

The majority of the training opportunities related to occupational safety and health are short-term courses that are announced in trade journals and newsletters. These courses are not often the subject of articles that would be abstracted by the various data base vendors. Commercial on-line data bases, therefore, are not useful for identifying the types of training available. They will, however, identify sources of training material and, to a lesser extent, course curricula.

Until recently, five data base vendors based in the United States have provided virtually all of the files of interest to the occupational safety and health professional. These are Dialog Information Services (3460 Hillview Avenue, Palo Alto, CA 94304); the National Library of Medicine (NLM) (8600 Rockville Pike, Bethesda, MD 20209); DOE RECON (Department of Energy, Technical Information Center, P.O. Box 62, Oak Ridge, TN 37830); Bibliographic Retrieval Services (BRS) (BRS Information Technologies, 1200 Route 7, Latham, NY 12110); and System Development Corporation (SDC) Search Services (System Development Corporation, 2500 Colorado Avenue, Santa Monica, CA 90406). New vendors that were originally available only in Europe now make their information systems available in the United States. These include Pergamon Infoline (Pergamon International Information Corporation, 1340 Old Chain Bridge Road, Mclean, VA 22101); Questel (1625 Eye Street, N.W., Suite 719, Washington, DC 20006); and Information Retrieval Service of the European Space Agency Information Retrieval Service, ESA-IRS, (Esrin, P.O. Box 64, 1-00044, Frascati, Italy). Hundreds of different data bases on subjects ranging from art history to zoology are available from these vendors.

Each data base vendor has a different mix of data bases. Even when the same data base is available from different vendors, the periods covered and the implementation can be quite different. There is a trend for the vendors to include larger portions of the data bases in single files as the costs of computer storage decline. The following table lists the data bases that contain training information relevant to occupational safety and health. The availability by vendor and the periods of coverage are included for reference.

TABLE 11.12. Training Data Bases

Name	Producer	No. of Records	Coverage
AVLINE (NLM)	National Library of Medicine	10,000	75–present
A-V ONLINE (DIALOG)	National Information Center for Educational Media	403,000	64–present
CIS (Questel)	International Labour Organization/CIS	2,100/yr	74–present
CISDOC (ESA)	International Labour Organization/CIS	2,100/yr	74–present
ERIC (SDC, BRS, DIALOG)	Educational Resources Information Center	30,000/yr	66–present
HEALTH AUDIO-VISUAL ONLINE CATALOG (BRS)	Northeastern Ohio Universities College of Medicine	5,000 citations	60–present
HSELINE (PERGAMON, ESA)	Health and Safety Executive	47,000	77–present
NIOSHTIC (DIALOG, PERGAMON)	National Institute for Occupational Safety and Health (NIOSH)	12,000/yr	1900–present
NTIS (DIALOG)	National Technical Information Service	60,000/yr	64–present
NTIS (ESA)	National Technical Information Service	60,000/yr	69–present
NTIS (SDC, BRS)	National Technical Information Service	60,000/yr	70–present

Source: Data in this table were obtained from catalogs and manuals of the identified vendors. Specific citations for each data base and vendor are given in the alphabetical descriptions provided in section 11.14.

11.13.2 Occupational Safety and Health Training Reference Texts

Brief, R. S., *Basic Industrial Hygiene: A Training Manual*, Exxon Corporation, New York, NY, 1975. Distributed by the American Industrial Hygiene Association. This is a basic text for students in introductory industrial hygiene courses. It surveys the major areas of interest in this field and, through lab exercises, guides students in the use of monitoring equipment.

U.S. Department of Health and Human Services, Public Health Service, Centers for Disease Control, National Institute for Occupational Safety and Health, *Audiovisual Resources in Occupational Safety and Health: An Evaluative Guide*, DHHS (NIOSH) Publication No. 82-102, National Institute for Occupational Safety and Health, Cincinnati, OH. Reviews of occupational safety and health audiovisual training materials and information on the availability of these materials can be found in this guide. Available from NTIS PB 83-105-395.

U.S. Department of Health and Human Services, Public Health Service, Centers for Disease Control, National Institute for Occupational Safety and Health, *The Industrial Environment: Its Evaluation and Control*, National Institute for Occupational Safety and Health, Cincinnati, OH, 1973. Available from GPO 017-001-00396-4. Published as a basic text for graduate study in the field of occupational safety and health, this book is currently being revised but remains one of the most comprehensive treatments of the field in textbook format. The first section of the book provides a review of mathematics, chemistry, biochemistry, physiology, and toxicology as they relate to the practice of occupational safety and health and specifically to industrial hygiene. The remaining chapters concentrate on the methodology of industrial hygiene practice, especially the evaluation of the harmful effects of physical agents and air contaminants.

Other Training References

The training materials listed below were developed by NIOSH as part of its mandate to support the practice of occupational safety and health. They provide the basis for many of the training courses offered throughout the United States. Although attending training courses in person yields greater benefit, the materials listed can at least provide an overview of the course content. Materials listed below are all available from the National Technical Information Service, 5285 Port Royal Road, Springfield, VA 22161. The order number is provided.

U.S. Department of Health and Human Services, Public Health Service, Centers for Disease Control, National Institute for Occupational Safety and Health, *Analytical Methods Applications in Safety Engineering: A Training Monograph*, National Institute for Occupational Safety and Health, Cincinnati, OH. Available from NTIS PB 83-195-669.

U.S. Department of Health and Human Services, Public Health Service, Centers for Disease Control, National Institute for Occupational Safety and Health, *Applied Industrial Hygiene*, National Institute for Occupational Safety and Health, Cincinnati, OH. Available from NTIS PB 85-238-996.

U.S. Department of Health and Human Services, Public Health Service, Centers for Disease Control, National Institute for Occupational Safety and Health, *Basic Course: Microwaves and Lasers*, National Institute for Occupational Safety and Health, Cincinnati, OH. Available from NTIS PB 84-154-475.

U.S. Department of Health and Human Services, Public Health Service, Centers for Disease Control, National Institute for Occupational Safety and Health, *Basic Ionizing Radiation Measurements Course*, National Institute for Occupational Safety and Health, Cincinnati, OH. Available from NTIS PB 84-190-677.

U.S. Department of Health and Human Services, Public Health Service, Centers for Disease Control, National Institute for Ocupational Safety and Health, *Behavior Management for Occupational Safety and Health*, National Institute for Occupational Safety and Health, Cincinnati, OH. Available from NTIS PB 85-234-805.

U.S. Department of Health and Human Services, Public Health Service, Centers for Disease Control, National Institute for Occupational Safety and Health, *Control of Occupational Health Hazards in the Dry Cleaning Industry: Instructor's Guide*, National Institute for Occupational Safety and Health, Cincinnati, OH. Availalbe from NTIS PB 84-243-658.

U.S. Department of Health and Human Services, Public Health Service, Centers for Disease Control, National Institute for Occupational Safety and Health, *Control of the Occupational Environment: Instructor's Manual*, National Institute for Occupational Safety and Health, Cincinnati, OH. Available from NTIS PB 85-241-545.

U.S. Department of Health and Human Services, Public Health Service, Centers for Disease Control, National Institute for Occupational Safety and Health, *Course Administrator's Guide: Evaluation and Control of Accident Potential in the Workplace (513)*, National Institute for Occupational Safety and Health, Cincinnati, OH. Available from NTIS PB 83-151-779.

U.S. Department of Health and Human Services, Public Health Service, Centers for Disease Control, National Institute for Occupational Safety and Health, *Engineering Control of Occupational Safety and Health Hazards: Recommendations for Improving Engineering Practice, Education, and Research,* Summary Report of the Engineering Control Technology Workshop Technical Panel, DHHS (NIOSH) Publication No. 84-102, National Institute for Occupational Safety and Health, Cincinnati, OH, July 1983. Available from NTIS PB 85-125-615.

U.S. Department of Health and Human Services, Public Health Service, Centers for Disease Control, National Institute for Occupational Safety and Health, *Evaluation and Control of Occupational Accident Potential: Student Manual (513),* 2 vols., National Institute for Occupational Safety and Health, Cincinnati, OH. Available from NTIS (Vol. I, PB 85-239-002; Vol. II, PB 83-140-061).

U.S. Department of Health and Human Services, Public Health Service, Centers for Disease Control, National Institute for Occupational Safety and Health, *Evaluation of Occupational Hazards: Instructor's Manual,* National Institute for Occupational Safety and Health, Cincinnati, OH. Available from NTIS PB 83-141-358.

U.S. Department of Health and Human Services, Public Health Service, Centers for Disease Control, National Institute for Occupational Safety and Health, *Fundamentals of Occupational Health Nursing Practice,* National Institute for Occupational Safety and Health, Cincinnati, OH. Available from NTIS PB 83-209-957.

U.S. Department of Health and Human Services, Public Health Service, Centers for Disease Control, National Institute for Occupational Safety and Health, *Gas, Vapor, and Particulate Sampling (592),* National Institute for Occupational Safety and Health, Cincinnati, OH. Available from NTIS PB 85-246-031.

U.S. Department of Health and Human Services, Public Health Service, Centers for Disease Control, National Institute for Occupational Safety and Health, *A Guide for Developing a Training Program for Anhydrous Ammonia Workers,* DHHS (NIOSH) Publication No. 79-119, National Institute for Occupational Safety and Health, Cincinnati, OH. Available from NTIS PB 80-189-475.

U.S. Department of Health and Human Services, Public Health Service, Centers for Disease Control, National Institute for Occupational Safety and Health, *Heat Measurements Course,* National Institute for Occupational Safety and Health, Cincinnati, OH. Available from NTIS PB 84-155-035.

U.S. Department of Health and Human Services, Public Health Service, Centers for Disease Control, National Institute for Occupational Safety and Health, *How to Write a Laboratory Quality Control Manual (597),* National Institute for Occupational Safety and Health, Cincinnati, OH. Available from NTIS PB 85-178-283.

U.S. Department of Health and Human Services, Public Health Service Centers for Disease Control, National Institute for Occupational Safety and Health, *Human Factors and Systems Principles for Occupational Safety and Health,* National Institute for Occupational Safety and Health, Cincinnati, OH. Available from NTIS PB 85-235-117.

U.S. Department of Health and Human Services, Public Health Service, Centers for Disease Control, National Institute for Occupational Safety and Health, *Illumination Measurements Course,* National Institute for Occupational Safety and Health, Cincinnati, OH. Available from NTIS PB 84-155-027.

U.S. Department of Health and Human Services, Public Health Service, Centers for Disease Control, National Institute for Occupational Safety and Health, *Industrial Hygiene Chemistry (590),* National Institute for Occupational Safety and Health, Cincinnati, OH, August 1980. Available from NTIS (Laboratory Manual, PB 85-242-428; Lecture Manual, PB 85-239-200).

U.S. Department of Health and Human Services, Public Health Service, Centers for Disease Control, National Institute for Occupational Safety and Health, *Industrial Hygiene Engineering and Control (552): Student Manuals,* National Institute for Occupational Safety and Health, Cincinnati, OH. Available from NTIS (Introduction, PB 85-178-200; Ergomomics, PB 85-178-234; Industrial Illumination, PB 85-178-986; Industrial Ventilation, PB 85-178-218; Radiation, PB 85-177-616; Sound, PB 85-178-226); and Other Topics, PB 85-178-242).

U.S. Department of Health and Human Services, Public Health Service, Centers for Disease Control, National Institute for Occupational Safety and Health, *Industrial Hygiene Laboratory Quality Control (587),* National Institute for Occupational Safety and Health, Cincinnati, OH. Available from NTIS PB 85-178-846.

U.S. Department of Health and Human Services, Public Health Service, Centers for Disease Control, National Institute for Occupational Safety and Health, *Industrial Hygiene Review Manual: September 1982,* National Institute for Occupational Safety and Health, Cincinnati, OH. Available from NTIS PB 85-103-836.

U.S. Department of Health and Human Services, Public Health Service, Centers for Disease Control, National Institute for Occupational Safety and Health, *Industrial Hygiene Sampling, Decision-*

Making, Monitoring, and Recordkeeping: Sampling Methods (554), National Institute for Occupational Safety and Health, Cincinnati, OH. Available from NTIS PB 85-178-259.

U.S. Department of Health and Human Services, Public Health Service, Centers for Disease Control, National Institute for Occupational Safety and Health, *Industrial Hygiene Sampling, Decision-Making, Monitoring, and Recordkeeping: Sampling Strategies (553),* National Institute for Occupational Safety and Health, Cincinnati, OH. Available from NTIS PB 85-178-853.

U.S. Department of Health and Human Services, Public Health Service, Centers for Disease Control, National Institute for Occupational Safety and Health, *Industrial Hygiene Surveying Techniques: Lesson Plan and Instructor Notes, Instructor's Manual,* National Institute for Occupational Safety and Health, Cincinnati, OH. Available from NTIS PB 82-238-908.

U.S. Department of Health and Human Services, Public Health Service, Centers for Disease Control, National Institute for Occupational Safety and Health, *Industrial Noise (581),* National Institute for Occupational Safety and Health, Cincinnati, OH. Available from NTIS PB 85-165-520.

U.S. Department of Health and Human Services, Public Health Service, Centers for Disease Control, National Institute for Occupational Safety and Health, *Industrial Hygiene Chemistry Course: Instructor Manuals,* 18 lessons, National Institute for Occupational Safety and Health, Cincinnati, OH. Available from NTIS (Lesson #1, PB 82-147-554), (#2, PB 82-147-562), (#3, PB 82-147-570), (#4, PB 82-147-588), (#5, PB 82-147-596), (#6, PB 82-147-604), (#7, PB 82-147-612), (#8, PB 82-147-620), (#9, PB 82-147-638), (#10, PB 82-147-646), (#11, PB 82-147-653), (#12, PB 82-147-661), (#13, PB 82-147-679), (#14, PB 82-147-687), (#15, PB 82-147-695), (#16, PB 82-147-703), (#17, PB 82-147-711), (#18, PB 82-147-729).

U.S. Department of Health and Human Services, Public Health Service, Centers for Disease Control, National Institute for Occupational Safety and Health, *Industrial Hygiene Chemistry Course,* National Institute for Occupational Safety and Health, Cincinnati, OH. Available from NTIS (Course Manual, PB 82-143-512; Laboratory Manual, PB 82-143-520).

U.S. Department of Health and Human Services, Public Health Service, Centers for Disease Control, National Institute for Occupational Safety and Health, *Introduction to Occupational Health (509),* National Institute for Occupational Safety and Health, Cincinnati, OH. Available from NTIS PB 85-237-998.

U.S. Department of Health and Human Services, Public Health Service, Centers for Disease Control, National Institute for Occupational Safety and Health, *Ionizing Radiation (584),* National Institute for Occupational Safety and Health, Cincinnati, OH. Available from NTIS (Laboratory Manual, PB 83-132-019; Student Manual, PB 83-156-588).

U.S. Department of Health and Human Services, Public Health Service, Centers for Disease Control, National Institute for Occupational Safety and Health, *Laboratory Exercises for Courses in Occupational Safety and Health I and Occupational Safety and Health II,* National Institute for Occupational Safety and Health, Cincinnati, OH. Available from NTIS PB 83-135-921.

U.S. Department of Health and Human Services, Public Health Service, Centers for Disease Control, National Institute for Occupational Safety and Health, *Lectures on Sampling for Gases and Vapors,* National Institute for Occupational Safety and Health, Cininnati, OH. Available from NTIS PB 84-154-046.

U.S. Department of Health and Human Services, Public Health Service, Centers for Disease Control, National Institute for Occupational Safety and Health, *Maintaining and Donning Self-Contained Breathing Appartus,* National Institute for Occupational Safety and Health, Cincinnati, OH. Available from NTIS PB 85-164-754.

U.S. Department of Health and Human Services, Public Health Service, Centers for Disease Control, National Institute for Occupational Safety and Health, *Maintaining Facilities and Operations: Instructor's Resource Guide,* National Institute for Occupational Safety and Health, Cincinnati, OH. Available from NTIS PB 85-238-459.

U.S. Department of Health and Human Services, Public Health Service, Centers for Disease Control, National Institute for Occupational Safety and Health, *Manual of Safety and Health Hazards in the School Science Laboratory,* National Institute for Occupational Safety and Health, Cincinnati, OH. Available from NTIS PB 83-187-435.

U.S. Department of Health and Human Services, Public Health Service, Centers for Disease Control, National Institute for Occupational Safety and Health, *NIOSH Special Course: Polychlorinated Biphenyls in the Workplace,* National Institute for Occupational Safety and Health, Cincinnati, OH. Available from NTIS PB 84-187-681.

U.S. Department of Health and Human Services, Public Health Service, Centers for Disease Control, National Institute for Occupational Safety and Health, *NIOSH Manual of Spirometry in Occupational Medicine,* National Institute for Occupational Safety and Health, Cincinnati, OH. Available from NTIS (Manual, PB 83-197-137; Instructor's Guide, PB 83-197-152; Workbook, PB 83-188-052; Student Guide, PB 83-196-105).

U.S. Department of Health and Human Services, Public Health Service, Centers for Disease Control, National Institute for Occupational Safety and Health, *Nonionizing Radiation (583)*, National Institute for Occupational Safety and Health, Cincinnati, OH. Available from NTIS (Laboratory Manual, PB 83-127-506; Lesson Plan, PB 83-155-747).

U.S. Department of Health and Human Services, Public Health Service, Centers for Disease Control, National Institute for Occupational Safety and Health, *Occupational Carcinogens: Instructor's Resource Guide*, National Institute for Occupational Safety and Health, Cincinnati, OH. Available from NTIS PB 85-235-372.

U.S. Department of Health and Human Services, Public Health Service, Centers for Disease Control, National Institute for Occupational Safety and Health, *Occupational Health and Safety of Municipal Workers*, National Institute for Occupational Safety and Health, Cincinnati, OH. Available from NTIS PB 82-213-026.

U.S. Department of Health and Human Services, Public Health Service, Centers for Disease Control, National Institute for Occupational Safety and Health, *Occupational Health Nursing for Small Industry*, National Institute for Occupational Safety and Health, Cincinnati, OH. Available from NTIS PB 83-106-062.

U.S. Department of Health and Human Services, Public Health Service, Centers for Disease Control, National Institute for Occupational Safety and Health, *Occupational Health Nursing: Basic Theory and Update (351)*, National Institute for Occupational Safety and Health, Cincinnati, OH. Available from NTIS PB 83-196-824.

U.S. Department of Health and Human Services, Public Health Service, Centers for Disease Control, National Institute for Occupational Safety and Health, *Occupational Safety and Health: An Overview*, National Institute for Occupational Safety and Health, Cincinnati, OH. Available from NTIS PB 85-238-665.

U.S. Department of Health and Human Services, Public Health Service, Centers for Disease Control, National Institute for Occupational Safety and Health, *Particulate Contaminants: Basic Course*, National Institute for Occupational Safety and Health, Cincinnati, OH. Available from NTIS PB 84-153-493.

U.S. Department of Health and Human Services, Public Health Service, Centers for Disease Control, National Institute for Occupational Safety and Health, *Principles of Accident Potential Recognition*, National Institute for Occupational Safety and Health, Cincinnati, OH. Available from NTIS PB 83-102-244.

U.S. Department of Health and Human Services, Public Health Service, Centers for Disease Control, National Institute for Occupational Safety and Health, *Principles of Occupational Safety and Health Engineering: Instructor's Guide*, National Institute for Occupational Safety and Health, Cincinnati, OH. Available from NTIS PB 85-107-654.

U.S. Department of Health and Human Services, Public Health Service, Centers for Disease Control, National Institute for Occupational Safety and Health, *Principles of Ventilation*, National Institute for Occupational Safety and Health, Cincinnati, OH. Available from NTIS (Instructor's Manual, PB 84-156-272; Lecture Outline, PB 84-154-137).

U.S. Department of Health and Human Services, Public Health Service, Centers for Disease Control, National Institute for Occupational Safety and Health, *Recognition of Accident Potential in the Workplace due to Human Factors (512)*, National Institute for Occupational Safety and Health, Cincinnati, OH. Available from NTIS (Instructor Manual, PB 82-231-002; Student Manual, PB 85-219-137; Course Director Guideline, PB 82-227-158).

U.S. Department of Health and Human Services, Public Health Service, Centers for Disease Control, National Institute for Occupational Safety and Health, *Recognition of Occupational Health Hazards: Course Manual (510)*, 2 vols., National Institute for Occupational Safety and Health, Cincinnati, OH. Available from NTIS (Vol. I, PB 85-107-662; Vol. II, PB 84-238-401).

U.S. Department of Health and Human Services, Public Health Service, Centers for Disease Control, National Institute for Occupational Safety and Health, *A Resource Guide for Health Science Students: Occupational and Environmental Health*, DHHS (NIOSH) Publication No. 80-118, National Institute for Occupational Safety and Health, Cincinnati, OH. Available from NTIS PB 81-170-961.

U.S. Department of Health and Human Services, Public Health Service, Centers for Disease Control, National Institute for Occupational Safety and Health, *Safety and Health for Industrial/Vocational Education for Supervisors and Instructors*, National Institute for Occupational Safety and Health, Cincinnati, OH. Available from NTIS PB 83-188-128.

U.S. Department of Health and Human Services, Public Health Service, Centers for Disease Control, National Institute for Occupational Safety and Health, *Safety and Health in Confined Workspaces for the Construction Industry: Training Resource Manual,* National Institute for Occupational Safety and Health, Cincinnati, OH. Available from NTIS PB 85-235-240.

U.S. Department of Health and Human Services, Public Health Service, Centers for Disease Control, National Institute for Occupational Safety and Health, *Safety in the Laboratory (580),* National Institute for Occupational Safety and Health, Cincinnati, OH, May 1979. Available from NTIS PB 85-223-105.

U.S. Department of Health and Human Services, Public Health Service, Centers for Disease Control, National Institute for Occupational Safety and Health, *Safety Program Design and Management, Module I—People, Motivation, Training,* National Institute for Occupational Safety and Health, Cincinnati, OH. Available from NTIS (Course Manual, PB 84-234-797; Instructor's Manual, PB 85-242-030).

U.S. Department of Health and Human Services, Public Health Service, Centers for Disease Control, National Institute for Occupational Safety and Health, *Safety Program Design and Management, Module II: Man, Machine, Environment,* National Institute for Occupational Safety and Health, Cincinnati, OH. Available from NTIS (Course Manual, PB 85-238-988; Instructor's Manual, PB 85-234-789).

U.S. Department of Health and Human Services, Public Health Service, Centers for Disease Control, National Institute for Occupational Safety and Health, *Sampling and Evaluating Airborne Asbestos Dust (582),* National Institute for Occupational Safety and Health, Cincinnati, OH. Available from NTIS PB 85-246-023.

U.S. Department of Health and Human Services, Public Health Service, Centers for Disease Control, National Institute for Occupational Safety and Health, *Suggested Curricula for Industrial Hygienists and Safety Professionals,* National Institute for Occupational Safety and Health, Cincinnati, OH. Available from NTIS PB 82-240-748.

11.13.3 Occupational Safety and Health Training Organizations

American Center for Quality of Work Life (ACQWL), 1411 K St., N.W., Suite 930, Washington, DC 20005. The Center was established in 1973 to stimulate an understanding of proven ways for unions and management to utilize workers in solving work problems. It promotes increased operational effectiveness and a committed and involved work force. The Center helps labor and management improve the quality of work environments. Major areas of concern are occupational safety and health services, training, and educational services.

Center for Environmental Intern Programs (CEIP), 25 West St., Boston, MA 02111. The Center, established in 1972, has a membership of upper-level undergraduate, graduate, and doctoral students and recent graduates seeking professional experience in environmental management. The Center fosters the professional development of its members by offering full-time assignments with private industry and government nonprofit organizations for terms ranging from 12 to 52 weeks. Participants are employed by CEIP and are considered actual staff members of the sponsoring organization. Project areas include conservation services, public policy, community development, and technical services. Individual assignments include administrative/legal analysis, communications, community development, environmental assessments, hazardous substances, pollution control, and public/occupational health. The Center sponsors regional seminars and annual summer conferences.

Institute for Labor Education and Research, 853 Broadway, Room 2007, New York, NY 10003. The Institute, a nonprofit organization funded by foundation and government grants, is comprised of labor educators dedicated to bringing innovative education from the workers' point of view to working people.

Mine Safety and Health Administration Training Academy, Beckley, WV 25801. This Academy provides short-term training for mine safety and health inspectors. In 1979 the academy was transferred from the Department of the Interior to the Department of Labor. The Academy designs, develops, and conducts training programs that help government, labor, and industry reduce illness and injury in the mining industry. It is responsible for the training of mine health and safety inspectors, MSHA technical support staff, and other mining personnel. The Academy's Department of Continuing Education publishes a series of safety manuals, training modules, and audiovisuals covering a range of topics of interest to those involved in mining health and safety.

National Institute for Occupational Safety and Health, Division of Training and Manpower Development, 4676 Columbia Parkway, Cincinnati, OH 45226. This Division is responsible for training professional personnel in occupational safety and health. It conducts short-term training courses and develops course curricula that are provided to others to use in training professionals. In addition, course director's manuals, instructor's manuals, student manuals, and audiovisuals covering a range of topics of interest to those involved in health and safety have been developed. This Division is responsible for the administration of the Educational Resource Centers as part of its long-term strategy to increase the number of well-trained individuals in the field.

NIOSH-Sponsored Educational Resource Centers. These centers are supported by grants from the National Institute for Occupational Safety and Health to train professionals in the field of occupational safety and health. Graduate degree programs are offered in industrial hygiene, occupational health nursing, and safety engineering through each center. In addition, residencies in occupational medicine are offered. The educational resource centers also provide short-term training in many areas of occupational safety and health using courses developed by NIOSH and by the faculty of the participating university. The addresses of these institutions follow:

Alabama Educational Resource Center, Director, Continuing Education, University of Alabama in Birmingham, School of Public Health, Birmingham, AL 35294, 205-934-7209, ATTN: R. Kent Oestenstad.

Arizona Educational Resource Center, Coodinator for Continuing Education, Arizona Center for Occupational Safety and Health, Tucson, AZ 85724, 602-626-6835, ATTN: Hershella L. Horton, R.N.

California Educational Resource Center—Northern, Northern California Occupational Health Center, University of California, 2521 Channing Way, Berkeley, CA 94720, 415-642-5507, ATTN: Lela D. Morris, R.N., M.P.H.

California Educational Resource Center—Southern, University of Southern California, Institute of Safety and Systems Management, Extension and In-service Programs, University Park, MC 0021, Los Angeles, CA 90089-0021, 213-743-6523 or 213-743-6383, ATTN: Richard K. Brown, Ph.D.

Cincinnati Educational Resource Center, University of Cincinnati, ML 182, 231 Bethesda Avenue, Cincinnati, OH 45267-0182, 513-872-5733, ATTN: Kay M. Hayes, M.P.H.

Harvard Educational Resource Center, Office of Continuing Education, Harvard School of Public Health, 677 Huntington Avenue, Boston, MA 02115-9957, 617-732-1171, ATTN: Dr. Dale Moeller.

Illinois Educational Resource Center, University of Illinois at Chicago, Great Lakes Center for Occupational Safety and Health, P.O. Box 6998, Chicago, IL 60680, 312-996-0807, ATTN: Susanne J. Klein.

Johns Hopkins Educational Resource Center, Department of Environmental Health Sciences, Johns Hopkins School of Hygiene and Public Health, the Johns Hopkins University, 615 North Wolfe Street, Baltimore, MD 21205, 301-955-2609, ATTN: Dr. Jacqueline Corn.

Michigan Educational Resource Center, University of Michigan, Center for Occupational Health and Safety Engineering, 1205 Beal—IOE Building, Ann Arbor, MI 48109, 313-763-0567, ATTN: Randy Rabourn.

Minnesota Educational Resource Center, Midwest Center for Occupational Health and Safety, 640 Jackson Street, St. Paul, MN 55101, 612-221-3992, ATTN: Ruth K. McIntyre.

New York/New Jersey Educational Resource Center, Office of Consumer Health Education, University of Medicine and Dentistry of New Jersey, Rutgers Medical School, Piscataway, NJ 08854, 201-463-4500, ATTN: Dr. Audrey Gotsch.

North Carolina Educational Resource Center, Occupational Safety and Health Educational Resource Center, UNC—CH, 109 Conner Drive, Suite 1101, Chapel Hill, NC 27513, 919-962-2101, ATTN: Ted M. Williams

Texas Educational Resource Center, University of Texas, School of Public Health, Houston, TX 77225, 713-792-7450, ATTN: Geov L. Parrish.

Utah Educational Resource Center, Rocky Mountain Center for Occupational and Environmental Health, Building 512, University of Utah, Salt Lake City, UT 84112, 801-581-5710, ATTN: Connie Crandall.

Washington Educational Resource Center, Northwest Center for Occupational Health and Safety, Department of Environmental Health, University of Washington, SC-34, Seattle, WA 98195, 261-543-1069, ATTN: Sharon Morris.

Occupational Safety and Health Training Center, Des Plaines, IL 50015. The Center provides basic and advanced training in safety and health. Its courses are designed to develop an effective work force and to aid in professional development. Courses are provided for federal and state compliance officers, state consultants, and others including private employers, employees, and employee representatives. The subject areas covered include all aspects of safety and health inspection and compliance work.

11.13.4 Occupational Safety and Health Training Journals

The various professional journals, occupational safety and health trade publications, and newsletters often contain a section on training opportunities. However, the best sources of information about training, both long- and short-term, are the cited professional and trade organizations.

11.14 DATA BASE DESCRIPTIONS

11.14.1 Academic American Encyclopedia Data Base

The on-line version of the Academic American Encyclopedia, this data base provides the full text on line of over 30,000 entries. Every entry includes the tables, bibliographies, and cross-reference listings in the printed version. This data base corresponds to the latest print edition and is updated twice a year. (BRS cat., 1983; DIALOG cat., 1/85)

11.14.2 AGRICOLA

The AGRICOLA data base, formerly known as CAIN, is useful to the field of occupational safety and health when pesticide uses or engineering controls are of interest. It contains citations of journal articles, government reports, serials, monographs, pamphlets, and other material acquired by the National Agricultural Library (NAL). AGRICOLA offers worldwide coverage of agricultural literature including agricultural economics, rural sociology, agricultural products, animal industry, agricultural engineering, entomology, nutrition, forestry, pesticides, plant science, soils, and fertilizers. (BRS cat., 1983; DIALOG cat., 1/85)

11.14.3 American Chemical Society Primary Journal Data Base

Approximately 30,000 articles, appearing in eighteen primary ACS chemistry journals, are available in full text in this data base. Immediate access to the most current information in chemistry research is possible because the articles are available on line at the same time the journals are printed. In addition to the complete text of the article, each record contains abstracts, complete reference and footnote listings, captions, and extensive CAS Registry Number references. (BRS cat., 1983)

11.14.4 APLIT

This data base covers worldwide refining literature including petroleum refining, petrochemicals, air and water conservation, transportation and storage, and petroleum substitutes. Source documents include trade journals, technical journals, meeting papers, and government reports. (SDC cat., 5/83)

11.14.5 APTIC

APTIC is a data base on all aspects of air pollution and its effects, prevention, and control. APTIC comprehensively covered all sources of literature from 1966 to early 1976. From late 1976 through 1978, the 7000 records entered were drawn only from selected sources not covered by major discipline-oriented data bases. There are no updates after 1978. The data-base includes all abstracts that appeared in *Air Pollution Abstracts* (no longer published). Those wanting to do a comprehensive search should use APTIC for retrospective and "fugitive" information and use other on-line files for more recent conventional citations. (DIALOG cat., 1/85)

11.14.6 AQUALINE

This data base, produced in England, provides information on drinking water quality, water treatment, sewage systems, quality monitoring, and environmental protection. Included are citations from 600 periodicals, research reports, legislation, conference proceedings, and preprints, as well as books, monographs, dissertations, and standards and miscellaneous publications from water-related institutions worldwide. (DIALOG cat., 1/85)

11.14.7 AVLINE

This data base is the National Library of Medicine's catalog of nonprinted materials in the health sciences. The materials in the data base have been reviewed by content experts from the health sciences profession under the auspices of the American Medical Colleges. The data base also contains lecture-type materials with full bibliograpic information that have been technically evaluated but have not been appraised by content experts. (NLM Manual, 1/78)

11.14.8 A-V ONLINE

This data base was formerly called the National Information Center for Educational Media (NICEM). It covers the entire spectrum of educational material from preschool to graduate and professional school levels. (DIALOG cat., 1/85)

11.14.9 BIOSIS PREVIEWS

This data base contains citations from the major publications of Biosciences Information Service (BIOSIS), including Biological Abstracts (BA), Biological Abstracts/Reports, Reviews, Meetings (BA/RRM), and Bioresearch Index (BIOL). It is the major English language service providing comprehensive worldwide coverage of research in the life sciences. BIOSIS includes journal articles, research reports, reviews, conference papers, symposia, books, and other sources in biology, medicine, and interdisciplinary life sciences. Three types of information are included: original research reports in biological and biomedical fields; reviews of original research in biology, history, and philosophy of the biological and biomedical sciences; or documentation and retrieval of biological and biomedical information. (DIALOG cat., 1/85; SDC cat., 5/83; BRS cat., 1983)

11.14.10 BIOTECHNOLOGY

BIOTECHNOLOGY covers the technical aspects of the field and is based on the international patent literature, journals, and scientific reports. The data base includes information on biochemical engineering, industrial use of microorganisms, and fermentation. The file covers the period 1982 to the present with 15,000 items added each month. (SDC cat., 5/83)

11.14.11 CAB

CAB ABSTRACTS contains all records published in the 22 journals of the Commonwealth Agricultural Bureaux (CAB). CAB is a leading scientific information service in agriculture and certain fields of applied biology. Over 8500 journals in 37 languages are scanned, as well as books, reports, and other publications. Significant papers are abstracted, while less important works are reported with bibliographic details only. (DIALOG cat., 1/85)

11.14.12 CAB ABSTRACTS

CAB ABSTRACTS contains all records published in the 22 journals of the Commonwealth Agricultural Bureaux (CAB). CAB is a leading scientific information service in agriculture and certain fields of applied biology. Over 8500 journals in 37 languages are scanned, as well as books, reports, and other publications. Significant papers are abstracted, while less important works are reported with bibliographic details only. (DIALOG cat., 1/85)

11.14.13 CANCERLIT

The CANCERLIT data base (formerly called CANCERLINE) is sponsored by the International Cancer Research Data Bank (ICRDB) Program of the National Cancer Institute. It contains citations and abstracts of published literature discussing all aspects of cancer research that had appeared in *Cancer Therapy Abstracts* from 1967–73 and *Carcinogen Abstracts* from 1963–73. Since 1973 the scope of the data base has been enlarged to include all cancer-related articles, proceedings of meetings, symposia, reports, selected monographs, books, and theses. (NLM Technical Manual, 1/78)

11.14.14 CANCERPROJ

The CANCERPROJ data base is produced by the International Cancer Research Data Bank (ICRDB) program of the National Cancer Institute. It contains summaries of ongoing cancer research projects funded during the three most recent years. Both federally and privately funded projects are listed. (NLM cat., 7/82)

11.14.15 CAS

The Chemical Abstracts data base (CAS, CA SEARCH, CHEMABS, and EUCAS) provides comprehensive international coverage of literature published in all fields of chemistry. It covers chemical sciences literature from over 12,000 journals, patents from 26 countries, new books, conference proceedings, and government research reports. Coverage corresponds to the printed Chemical Abstracts Indexes. The contents make the Chemical Abstracts Data Base valuable for those interested in technology and life sciences as well as for researchers and chemists. (BRS cat., 1983; SDC cat., 5/83; Pergamon cat., 1983; DIALOG cat., 1/85)

11.14.16 CA SEARCH

The Chemical Abstracts data base (CAS, CA SEARCH, CHEMABS, and EUCAS) provides comprehensive international coverage of literature published in all fields of chemistry. It covers chemical sciences literature from over 12,000 journals, patents from 26 countries, new books, conference proceedings, and government research reports. Coverage corresponds to the printed Chemical Abstracts Indexes. The contents make the Chemical Abstracts Data Base valuable for those interested in technology and life sciences as well as for researchers and chemists. (BRS cat., 1983; SDC cat., 5/83; Pergamon cat., 1983; DIALOG cat., 1/85)

11.14.17 CASSI

The Chemical Abstracts Source Index contains the bibliographic and library holdings information for scientific and technical primary literature relevant to the chemical sciences. The titles listed in CASSI derive from publication and bibliographic information included in over 70 years of Chemical Abstracts, about 700 important biological journal titles monitored by the Bio-Sciences Information Service, and several hundred titles covering the pure and theoretical chemical literature from 1830 through 1940 from Chemisches Zentralblatt. (SDC cat., 5/83)

11.14.18 CCRIS

The Chemical Carcinogenesis Research Information System (CCRIS) contains assay results and test conditions for over 850 chemicals. The data selected are from accepted scientific protocols with bibliographic information. Tests include mutagenicity, carcinogenicity, tumor promotion, and cocarcinogenicity. (ICIS cat., 12/84)

11.14.19 CEH-ONLINE

The Chemical Economics Handbook Online provides comprehensive data and studies of the history, status, and outlook for more than 1300 major commodity and specialty chemicals. The majority of data contained in the data base is tabular and covers chemicals, pharmaceuticals, petroleum, natural gas, minerals and metals, engineering and construction, forest products, building materials, consumer products, automotive products, rubber, paint, plastics, food and agriculture, and transportation. (SDC cat., 5/83)

11.14.20 CESARS

The Chemical Evaluation Search and Retrieval System (CESARS) provides access to information on about 150 chemicals of importance used in the Great Lakes Basin. The information is detailed including physical and chemical data as well as toxicity and environmental fate. (ICIS cat., 12/84)

11.14.21 CHEMABS

The Chemical Abstracts data base (CAS, CA SEARCH, CHEMABS, and EUCAS) provides comprehensive international coverage of literature published in all fields of chemistry. It covers chemical sciences literature from over 12,000 journals, patents from 26 countries, new books, conference proceedings, and government research reports. Coverage corresponds to the printed Chemical Abstracts Indexes. The contents make the Chemical Abstracts Data Base valuable for those interested in technology and life sciences as well as for researchers and chemists. (BRS cat., 1983; SDC cat., 5/83; Pergamon cat., 1983; DIALOG cat., 1/85)

11.14.22 CHEMDEX

This data base is a chemical dictionary file that is a companion to the Chemical Abstracts data bases. All compounds cited in the literature from 1972 to date are included. Each record contains a Chemical Abstracts Service (CAS) Registry Number, the molecular formula, nomenclature for specific compounds in accordance with Chemical Abstracts' rigorous rules; and many common synonyms recognized by the Chemical Abstracts Service. The Registry Numbers retrieved can be used to search in the Chemical Abstracts files. (SDC cat., 5/83)

11.14.23 Chemical Engineering Abstracts

This data base covers the entire field of chemical engineering. In addition, articles covering the chemically related aspects of mechanical, civil, electrical, and instrumentation engineering are included. Over 100 worldwide chemical and process engineering journals are reviewed to compile this data base. (Pergamon cat., 1983)

11.14.24 Chemical Exposure

This is a comprehensive data base of chemicals that have been identified in both human tissues and body fluids and in food and domestic animals. This data base includes the range of body burden information. Information includes chemical properties, synonyms, Chemical Abstracts Service (CAS) Registry Numbers, formulas, tissue measured, and analytical method used. It is a useful tool for examining the toxicity of various substances and the effects of contaminants on both animals and humans. (DIALOG cat., 1/85)

11.14.25 Chemical Regulations and Guidelines System

The Chemical Regulations and Guidelines System (CRGS) is an authoritative index to U.S. federal regulatory material concerning the control of chemical substances. It covers federal statutes, promulgated regulations and available federal guidelines, standards, and support documents. This data base follows the regulatory cycle and includes an up-to-date reference to each document, including main documents and revisions published in the *Federal Register.* Each chemical cited in a regulatory document is indexed by name, CAS Registry Number, and a chemical role tag. This letter gives information on the context in which the substances appear in the document. CRGS also provides links between the statutes, the regulations promulgated under these statutes, and the support documents generated prior to the promulgation of a regulation. Each document is described in terms of publication data, title, abstract, index terms, and chemical identifiers. (DIALOG cat., 1/85)

11.14.26 Chemicals in Human Tissues and Fluids

This file contains information on chemicals identified in human biological media as well as reported body burdens of drugs, metals, pesticides, and other substances. Specific information includes half-life; use and source of the chemical; tissue concentrations (range and mean); comments pertaining to the concentrations, number, and mix of cases; and demographic data; as well as information on health, pathology, and morphology. Chemical Abstracts Service (CAS) Registry Numbers are also included. (DOE data base description, 2/83)

11.14.27 CHEMLINE

CHEMLINE is an on-line, interactive chemical dictionary file maintained by the National Library of Medicine under contract with the Chemical Abstracts Service (CAS). It provides a mechanism whereby information on over one million chemical substance names and corresponding CAS Registry Numbers representing over 525,000 unique substances can be searched and retrieved on line. It primarily contains records for chemicals that are identified by CAS Registry Numbers in the TOXLINE, RTECS, TDB, and MEDLINE data bases; it also contains the EPA Toxic Substances Control Act (TSCA) Inventory of Chemical Substances. The CHEMLINE file contains CAS Registry Numbers, molecular formulas, *Chemical Abstracts* (CA) index names, synonyms, various nomenclature and structural fragments, and locator information on that particular chemical substance. CHEMLINE also contains ring information, including the number of rings within a ring system, ring sizes, ring elemental compositions, and component line formulas. (NLM Fact Sheet, 7/83)

11.14.28 CHEMNAME

CHEMNAME is a chemical name dictionary. It contains a listing of all chemical substances cited more than once in Chemical Abstracts since 1967. It contains Chemical Abstracts Service (CAS) Registry Number, molecular formula, CA Substance Index Name, available synonyms, ring data, and other chemical substance information. The primary purpose of this file is to support specific substance searching and various forms of substructure searching. (DIALOG cat., 1/85)

11.14.29 CHEMSEARCH

CHEMSEARCH is a dictionary listing of the most recently cited chemical substances in Chemical Abstracts. It is a companion file to Chemname. For each substance listed, the following information is provided: Chemical Abstracts Service (CAS) Registry Number, molecular formula, and CA Substance Index Names. The primary purpose of this file is to provide access to chemical substance nomenclature using CAS Registry Numbers for new compounds not yet recorded in other sources. (DIALOG cat., 1/85)

11.14.30 CHEMSIS

The CHEMSIS files are chemical substance dictionaries of Singly Indexed Substances cited during the 8th, 9th, 10th, and 11th Collective Index Periods (corresponding to the equivalent Chemical Abstracts files for those periods). This data base is a companion to Chemname, which contains substances cited more than once since 1967, and to CHEMZERO, which contains substances for which there are no citations in Chemical Abstracts. Information provided for each substance listed includes CAS Registry Number, molecular formula, CA Substance Index Names, available synonyms, and ring data. The primary purpose of the files is to support specific substance searching and substructure searching. (DIALOG cat., 1/85)

11.14.31 CHEMZERO

CHEMZERO is a nonbibliographic file containing those chemical substances for which there are no citations in the Chemical Abstracts Files. This enables the searcher to identify Chemical Abstracts Service (CAS) Registry Number, synonyms, and additional information on substances registered by Chemical Abstracts but not cited in the computer-searchable Chemical Abstracts files. Complete ring data are included for each substance in this file. (DIALOG cat., 1/85)

11.14.32 CIS

The CIS (CISDOC) data base contains worldwide literature on occupational safety and health. It is developed by the International Labour Organization with the cooperation of 40 national centers. Topics covered include occupational health and safety standards, pathology and toxicity of chemical exposures, education, ergonomics, engineering controls, and work practices. (ESA cat., 1/84)

11.14.33 CISDOC

The CIS (CISDOC) data base contains worldwide literature on occupational safety and health. It is developed by the International Labour Organization with the cooperation of 40 national centers. Topics covered include occupational health and safety standards, pathology and toxicity of chemical exposures, education, ergonomics, engineering controls, and work practices. (ESA cat., 1/84)

11.14.34 Claims Compounds Registry

The Claims Compounds Registry is a dictionary of specific chemical compounds. Each record contains the main compound name, the molecular formula, synonyms, and the IFI compound term number. The file is designed to locate terms of interest for searching the Claims/Uniterm patent files. (DIALOG cat., 1/85)

11.14.35 CLAIMS/UNITERM

The CLAIMS/UNITERM files provide comprehensive coverage of all U.S. chemical and chemically related patents issued from 1950 to the present. Each patent has been assigned 20 or more descriptors (uniterms) from the full text of the patent documents. Uniterm may be searched in conjunction with information from the title, patent number, classification codes, and publication date. Chemical Abstracts references and foreign equivalents from Belgium, France, Great Britain, West Germany, and the Netherlands are included. CLAIMS/UNITERM is produced by the IFI/Plenum Data company. (DIALOG cat., 1/85)

11.14.36 COLD

The Cold Regions data base covers all disciplines dealing with Antarctica, the Antarctic Ocean and subantarctic islands, civil engineering in cold regions, and behavior and operation of materials and equipment in cold temperatures. Source material includes monographs, technical reports, journal articles, conference papers, patents, and maps. (SDC cat., 5/83)

11.14.37 COMPENDEX

The COMPENDEX data base is the *Engineering Index* in machine-readable form. It provides worldwide abstracted information covering all aspects of technology and all disciplines of engineering and applied science from the world's significant engineering and technological literature. Literature from approximately 3500 sources is covered, including journals, monographs, technical reports, and standards including publications of engineering societies and organizations, papers from the proceedings of conferences, and selected government reports and books. (DIALOG cat., 1/86; SDC cat., 5/83)

11.14.38 Conference Papers Index

Conference Papers Index covers about 100,000 papers of approximately 1000 scientific and technical meetings held worldwide each year. It provides a centralized source for information on current research results presented at conferences and meetings and provides titles of papers as well as names and addresses (when available) of the paper's authors. The corresponding printed publication is the monthly Conference Papers Index. (DIALOG cat., 1/85)

11.14.39 CROSS

The CROSS data base serves as a subject and term index to the citations in the individual BRS data bases and permits the identification of those data bases most appropriate for the specific subject of the inquiry. (BRS cat., 1983)

11.14.40 CTCP

The Clinical Toxicology of Commercial Products (CTCP) is based on the book by the same title and contains information on some 20,000 chemical products, chemical manufacturers, chemical uses, and composition. (ICIS cat., 12/84)

11.14.41 Current Biotechnology Abstracts

Current Biotechnology Abstracts provides scientific, technical, and commercial information relevant to the field. The data base is organized by industrial interest including pharmaceuticals, chemicals, food, and agriculture. It is updated monthly and currently has over 7000 records. (Pergamon cat., 6/84)

11.14.42 DBI

The Data Base Index (DBI) is the master index to all ORBIT Search Service data bases. It is used as a selection tool to obtain a list of appropriate data bases on a given subject. Data bases are printed in ranked order by number of occurrences for a chosen term. (SDC cat., 5/83)

11.14.43 DIALINDEX

DIALINDEX is the on-line information directory to Dialog data bases. It is a collection of the file indexes of all Dialog data bases. The data base covers all subject areas included in the scope of the Dialog system. It provides the number of postings for each search statement in each of the data bases specified. It therefore indicates which files would be the most productive for searching and provides information used to determine how broadly or narrowly to define a search strategy. (DIALOG cat., 1/85)

11.14.44 DIRLINE

This data base is the National Library of Medicine's on-line interactive directory of organizations providing information in specific subject areas. The records in DIRLINE contain the sources listed in the National Referral Center of the Library of Congress and the National Health Information Clearinghouse data base. Information centers referenced include public and private organizations. (NLM Fact Sheet, 11/84)

11.14.45 Dissertation Abstracts On Line

Dissertation Abstracts On Line is a definitive subject, title, and author guide to virtually every American dissertation accepted at an accredited institution since 1861, when academic doctoral degrees were first granted in the United States. Approximately 99 percent of all American dissertations are cited in this file. Masters theses have been selectively indexed since 1962. In addition, the data base provides access to citations for thousands of Canadian dissertations and an increasing number of papers accepted in institutions abroad. Abstracts are included for doctoral dissertation records from July 1980 (Dissertation Abstracts International, Volume 41, Number 1) to the present. (DIALOG cat., 1/85; BRS cat., 1983)

11.14.46 DOE Energy

The Energy Data Base (EDB), sometimes called DOE Energy, provides comprehensive coverage of literature, patents, monographs, and technical reports concerning all aspects of energy production, utilization, and conservation. This file contains information about all types of fuels, including a large amount of literature on alternate fuels; transportation and storage of fuels; political, environmental, statistical, and management aspects of energy; and information on World War II German synthetic fuels documents. The Energy Data Base corresponds in part to several abstract journals such as Atomindex, Energy Research Abstracts, Energy Abstracts for Policy Analysis, Fusion Energy Update, and Radioactive Waste Update. Approximately 50 percent of the records in the data base are not announced in these secondary source publications. Abstracts are included for records added from 1976 to the present. (DIALOG cat., 1/85; DOE/TIC-4586/R1, 6/83; SDC cat., 5/83)

11.14.47 EDB

The Energy Data Base (EDB), sometimes called DOE Energy, provides comprehensive coverage of literature, patents, monographs, and technical reports concerning all aspects of energy production, utilization, and conservation. This file contains information about all types of fuels, including a large amount of literature on alternate fuels; transportation and storage of fuels; political, environmental, statistical, and management aspects of energy; and information on World War II German synthetic fuels documents. The Energy Data Base corresponds in part to several abstract journals such as Atomindex, Energy Research Abstracts, Energy Abstracts for Policy Analysis, Fusion Energy Update, and Radioactive Waste Update. Approximately 50 percent of the records in the data base are not announced in these secondary source publications. Abstracts are included for records added from 1976 to the present. (DIALOG cat., 1/85; DOE/TIC-4586/R1, 6/83; SDC cat., 5/83)

11.14.48 EEIS

The Epidemiology Information System provides information on the distribution and health effects of food contaminants, especially unavoidable contaminants, and on natural toxicants in foods. Data is indexed from published literature and from unpublished documents in the files of the Epidemiology Unit of the Food and Drug Administration's Bureau of Foods. (DOE data base description, 7/82)

11.14.49 EI Engineering Meetings

EI Engineering Meetings covers significant publisher proceedings of engineering and technical conferences, symposia, meetings, and colloquia. Each meeting included is indexed in a main conference record. In addition, all papers from the meeting are individually indexed. From July 1982, every main conference record included in this data base is also referenced in a Compendex record that includes an abstract, the number of presented papers, and the EI conference number. Records in EI Engineering Meetings do not contain abstracts. The file is produced by Engineering Information. (DIALOG cat., 1/85; SDC cat., 5/83)

11.14.50 EIX (Compendex)

The Compendex data base is the Engineering Index in machine-readable form. It provides worldwide abstracted information covering all aspects of technology and all disciplines of engineering and applied science from the world's significant engineering and technological literature. Literature is covered from approximately 3500 sources including journals, monographs, technical reports, and standards. Publications of engineering societies and organizations, papers from the proceedings of conferences, selected government reports, and books are also included. (DIALOG cat., 1/86; SDC cat., 5/83)

11.14.51 EMI

The Environmental Mutagen Information (EMI) Center data base consists primarily of references from the open literature that report the testing of chemicals, biological agents, and some physical agents for mutagenicity. It also includes general references and methods papers on test systems and organisms. The data base includes some studies dealing exclusively with the mutagenicity of ionizing or ultraviolet radiation, but no attempt has been made to obtain and index all such work. The entire data base is a subset of the National Library of Medicine's toxicology data base, TOXLINE. (DOE/TIC-4586/R1, 2/83)

11.14.52 ENERGYLINE

ENERGYLINE provides broad coverage of scientific, engineering, political, and socioeconomic aspects of energy policy, resources, research and development, environmental impact, fuel transport, and industrial transportation. Its interdisciplinary approach offers access to over 5000 international primary and secondary sources. ENERGYLINE corresponds to energy materials from Environment Abstracts from 1971 through 1975, and all materials from Energy Information Abstracts, beginning in January, 1976. (DIALOG cat., 1/85; SDC cat., 5/83)

11.14.53 ENVIROLINE

ENVIROLINE contains abstract citations from Environment Abstracts, a publication of Environment Information Center. This data base provides interdisciplinary scientific, technical, and socioeconomic coverage of the major English language and the more important non-English language environmental and resource literature. ENVIROLINE includes approximately 10,000 unique abstracts per year drawn from over 5,000 periodicals, symposia, and government and institutional reports, including the *Federal Register* and the *Official Gazette.* (DIALOG cat., 1/85; SDC cat., 5/83)

11.14.54 Environmental Bibliography

This data base covers the fields of general human ecology, atmospheric studies, energy, land resources, nutrition, and health. It is a general reference data base in the field of environmental science, indexing approximately 300 journals. (DIALOG cat., 1/85)

11.14.55 ERIC

Educational Resources Information Center (ERIC) covers periodic literature, project and technical reports, speeches, unpublished manuscripts, and books in education and education-related areas. It contains over 500,000 citations. Approximately 650 locations throughout the United States have complete collections of ERIC microfiche. (DIALOG cat., 1/85; BRS cat., 1983)

11.14.56 ETI

The Environmental Teratology Information data base covers the science dealing with the causes, mechanisms, and manifestations of structural or functional alterations in animal development—that is, birth defects. The Environmental Teratology data base contains information gathered from the open literature reporting on the testing and evaluation for teratogenic activity resulting from chemical, biological, and physical agents or from dietary deficiencies in warm-blooded animals. It also contains information on the examination of offspring at or near birth for structural or functional anomalies. The entire data base is a subset of the National Library of Medicine's toxicology data base, TOXLINE. (DOE/TIC-4586/R1, 2/83)

11.14.57 EUCAS

The Chemical Abstracts data base (CAS, CA SEARCH, CHEMABS, and EUCAS) provides comprehensive international coverage of literature published in all fields of chemistry. It covers chemical sciences literature from over 12,000 journals, patents from 26 countries, new books, conference proceedings, and government research reports. Coverage corresponds to the printed Chemical Abstracts Indexes. The contents make the Chemical Abstracts Data Base valuable for those interested in technology and life sciences as well as for researchers and chemists. (BRS cat., 1983; SDC cat., 5/83; Pergamon cat., 1983; DIALOG cat., 1/85)

11.14.58 EXCERPTA MEDICA

EXCERPTA MEDICA offers broad coverage of biomedicine and related biological sciences, as well as health economics and administration, environmental and pollution control, forensic science, drug literature, and toxicology. About 230,000 records are added annually, 60 percent of which contain abstracts. Occupational safety and health professionals can effectively use the data base's worldwide coverage and inclusive range of biomedical subjects. (DIALOG cat., 1/85; BRS cat., 1983)

11.14.59 FEDREG

Federal Register Abstracts provides comprehensive coverage of federal regulatory actions published in the *Federal Register*. Included are rules, proposed rules, hearing notices, and meeting notices as well as presidential proclamations and executive orders. (DIALOG cat., 1/85)

11.14.60 FOODS ADLIBRA

The FOODS ADLIBRA data base contains information on food technology and packaging. All new food products introduced since early 1974 are included, along with nutritional and toxicological information. Information about every level of the food industry, including agribusiness, meat packing, millers, dairies, bakers, retailers, food service operators, processors, brokers, and equipment suppliers, is in this data base. Major significant research and technological advances, process methods, packaging, and patents are also included. FOODS ADLIBRA is also available in printed form. (DIALOG cat., 1/85)

11.14.61 FRSS

The Federal Register Search System (FRSS) provides a chemical index for the *Federal Register.* It includes 160,000 references to proposed or final regulations relating to chemicals since 1/1/77. (ICIS cat., 12/84)

11.14.62 FSTA

Food Science and Technology Abstracts (FSTA) provides access to research and new development literature in the areas related to food science and technology. Basic allied disciplines such as chemistry, physics, biochemistry, and agriculture are covered as well. Documents are from international sources and include patents, books, standards, conference proceedings, reports, and reviews. (DIALOG cat., 1/85; SDC cat., 5/83)

11.14.63 HAZARDLINE

HAZARDLINE provides information on over 3200 substances; about 1300 are added per year. The data base provides access to data from the publications of NIOSH, OSHA, EPA, and TSCA with emphasis on health effects and physical or chemical information that provides the basis for protection of those potentially exposed. Updates on new regulations and toxicology reports are included. (BRS cat., 1985)

11.14.64 Health Audiovisual Online Catalog

The Health Audiovisual Online Catalog contains the media package holdings of eight Ohio academic health sciences libraries. This data base provides access to information on more than 5000 audiovisual packages in medicine, nursing, psychology, and allied health fields. (BRS cat., 1983)

11.14.65 HSELINE

HSELINE covers the health and safety aspects of manufacturing, agriculture, production, industrial hygiene, engineering, mining, nuclear technology, and industrial air pollution. The data base is produced by the U.K.'s Health and Safety Executive. It covers technical literature such as conference proceedings and reports as well as standards, codes of practice guides, and legislation. (Pergamon Green Sheet, 4/84)

11.14.66 INSPEC

INSPEC (Information Services in Physics, Electrotechnology, Computers and Control) corresponds to the Science Abstract publications—*Electrical and Electronics Abstracts, Computer and Control Abstracts, IT Focus,* and *Physics Abstracts.* It provides access to international journal articles, conference reports, dissertations, and technical literature on physics, electrical/electronics engineering, and computers and control. Source documents are primarily journal articles; conference proceedings, reports, patents, and books are also included. (DIALOG cat., 1/85; BRS cat., 1983; SDC cat., 5/83)

11.14.67 International Pharmaceutical Abstracts

International Pharmaceutical Abstracts indexes and abstracts worldwide literature on all phases of the development and use of drugs from pharmaceutical, medical, and related journals. This data base provides information on drug therapy, toxicity, and institutional pharmacy practice. Legislation, regulation, technology, information processing, education, economics, and ethics as related to pharmaceutical science and practice are also covered. (BRS cat., 1983; DIALOG cat., 1/85)

11.14.68 IRCS

The IRCS Medical Science data base contains the full text of all articles published in the IRCS series from 1981. On-line availability occurs simultaneously with print publication, providing immediate access to current medical and biomedical research findings and methodologies. In addition to the complete text of the IRCS article, every IRCS citation includes all tables, figure legends, reference listings, and publication information. (BRS cat., 1983)

11.14.69 ISMEC

The ISMEC (Information Service in Mechanical Engineering) data base corresponds to the biweekly printed ISMEC Bulletin. ISMEC indexes significant articles on all aspects of mechanical engineering, production engineering, and engineering management from approximately 250 journals published throughout the world. Books, reports, and conference proceedings are also indexed. These sources are further supplemented by relevant material from more than 2000 periodicals in physics and engineering that are received by INSPEC. The primary emphasis is on comprehensive coverage of leading international journals and conferences on mechanical engineering subjects. (DIALOG cat., 1/85)

11.14.70 Kirk-Othmer Encyclopedia of Chemical Technology

KIRK is the full-text on-line version of the 25-volume *Kirk-Othmer Encyclopedia of Chemical Technology* (3rd ed.), which is recognized as the most authoritative and comprehensive reference work in its field. Full-text features allow precise retrieval and display of extensive tabular data, index terms, plus cited and general references. The entire 9 million word text is completely searchable. (BRS cat., 1983)

11.14.71 Laboratory Hazards Bulletin

The Laboratory Hazards Bulletin is a bibliographic data base containing information from the literature on hazards likely to be encountered by the chemical and biochemical laboratory research worker. Included are new safety precautions, biological hazards, hazardous chemical reactions, and new legislation. It is produced by the Royal Society of Chemistry. (Pergamon Green Sheet, 4/84)

11.14.72 LABORDOC

LABORDOC is produced by the International Labour Organization. It covers worldwide journal and monographic literature on labor and labor-related areas, including industrial relations, economic and social development, law, environment, earth sciences, and technological change. It also covers national legislation and European Economic Community directives. (Questel data base description, 4/83; SDC cat., 5/83)

11.14.73 Life Sciences Collection

The Life Sciences Collection contains the abstracts from 17 abstracting journals covering the fields of biology, medicine, biochemistry, ecology, toxicology, microbiology, and some aspects of agriculture and veterinary science. These abstracts cover journal articles, books, conference proceedings, and report literature. (DIALOG cat., 1/85)

11.14.74 MEDLINE

MEDLINE (MEDLARS on line) is produced by the U.S. National Library of Medicine (NLM). The MEDLINE data base is the most comprehensive on-line resource for national and international medical literature. MEDLINE covers all aspects of biomedicine, including the allied health fields as well as the biological and physical sciences, humanities, and information sciences as they relate to medicine and health care. MEDLINE is indexed using NLM's controlled vocabulary, MESH (Medical Subject Headings). Over 40 percent of the records added beginning in 1975 contain author abstracts taken directly from the published articles. (DIALOG cat., 1/85; BRS cat., 1983)

11.14.75 MENTAL HEALTH ABSTRACTS

MENTAL HEALTH ABSTRACTS provides references and abstracts on all aspects of mental health and mental illness. International in scope, the data base indexes and abstracts articles from over 1000 periodicals as well as books, research reports, and program data. Each record includes the bibliographic citation and abstract as well as descriptors and identifiers. (DIALOG cat., 1/85)

11.14.76 MESH

The Medical Subject Headings file, commonly known as MESH, is the National Library of Medicine's controlled vocabulary containing all descriptors and qualifiers used to index and catalog material for computer retrieval and for Index Medicus, Current Catalog, Audiovisuals Catalog, and other more specialized bibliographies. Approximately half of the MEDLARS data bases may be searched using MESH. (NLM Manual, 1/78)

11.14.77 METADEX

The METADEX (Metal Abstracts/Alloys Index) data base, produced by the American Society for Metals (ASM) and the Metals Society (London), provides comprehensive coverage of international metallurgy literature. Six basic categories of metallurgy are covered: material classes, processes, properties, specific alloy designations, intermetallic compounds, and metallurgical systems. Information from the USSR and Eastern European countries is included. (DIALOG cat., 1/85; SDC cat., 5/83)

11.14.78 NASA

The National Atmospheric and Space Administration (NASA) data base covers the space sciences, chemistry, engineering, and the life sciences. The materials are gathered from over 1300 journals and a range of scientific reports from 1962 to the present. (ESA-IRS cat., 1/84)

11.14.79 NASA Tech Briefs

This file contains information taken from the publication of the same name. The file contains new technologies developed as a result of the space program and provides brief reports on new and potential products, newly developed industrial processes, advances in basic and applied research, and improvements in shop and laboratory techniques. (DOE data base summaries, 2/82)

11.14.80 NCMH

The National Clearinghouse for Mental Health (NCMH) data base offers coverage of the entire mental health field. It includes audiovisuals, dissertations, government documents, journals, monographs, and reports. It has not been updated since 1982. (BRS cat., 1983)

11.14.81 NIOSHTIC

NIOSHTIC is a major information resource in the field of occupational safety and health. This data base includes comprehensive information about the field of occupational safety and health research. Information from the early 1900s is presented, including classic epidemiology and toxicology studies. All references cited in Criteria Documents, developed by the National Institute for Occupational Safety and Health (NIOSH), are included in the data base, as are special collections such as the Von Ottegen files, the references cited in Patty, Lee's occupational health nursing files, and Lloyd's occupational epidemiology files. All NIOSH documents are added to the data base and over 400 journal titles and 70,000 monographs and technical reports are included. (DIALOG cat., 1/85)

11.14.82 NRC

The National Referral Center (NRC) data base contains the self-approved descriptions of organizations qualified and willing to answer questions or otherwise provide information on virtually any subject. Comprehensive coverage is mainly of the United States, but the file also has some data covering foreign organizations. For each organization the file lists the name, address, phone numbers, areas of interest, holdings (data bases, literature and report collections, specimens, etc.), representative publications, types of information services offered, fees (if any), and restrictions (if any). The Center sees itself as an information broker and this data base as one of its major tools. (DOE data base description, 5/82)

11.14.83 NSIC

The Nuclear Safety Information Center (NSIC) data base contains information about the operating experience of nuclear power plants especially as it relates to operational safety. Information is taken from docket material, especially event-specific and safety experience information. Generic information on operational safety and analytical and research-oriented information on operational safety is also included. (DOE data base description, 5/82)

11.14.84 NTIS

The National Technical Information Service (NTIS) is the major resource for locating U.S.-government-sponsored research reports and studies in the physical sciences, technology, engineering, biological sciences, medicine, health sciences, agriculture, and social sciences. The data base is produced by the National Technical Information Service (NTIS) of the U.S. Department of Commerce. It is the principal on-line source for the public sale and dissemination of U.S.-government-sponsored research. The data base contains government-sponsored research and development and engineering reports as well as other analyses prepared by over 200 government agencies and their contractors or grantees. Included in the data base are technical reports, some reprints, federally sponsored translations, and foreign-language reports in areas of major technical interest. (DIALOG cat., 1/85; BRS cat., 1983; SDC cat., 5/83)

11.14.85 Nursing and Allied Health

The Nursing and Allied Health data base is designed to meet the needs of nursing and the allied health professions. It abstracts more than 300 English language nursing journals and primary journals in more than a dozen allied health disciplines. This data base is the on-line version of the Cumulative Index to Nursing and Allied Health Literature. (DIALOG cat., 1/85)

11.14.86 OHMTADS

The Oil and Hazardous Materials Technical Assistance Data System (OHMTADS) was developed to provide emergency information to workers responding to chemical spills. The information includes details on material identification, containment, and disposal. The file currently has over 1200 chemical records selected on the basis of spill history, production, exposure data, and toxicity. (ICIS cat., 12/84)

11.14.87 PAPERCHEM

PAPERCHEM is a comprehensive data base covering worldwide literature on pulp and paper technology. It includes literature dealing with raw materials, engineering and process control, process technology, and products of the pulp and paper industry. Nearly 1000 periodicals in more than 20 languages are screened, as are the patent gazettes of six major countries. The data base corresponds to the printed Abstract Bulletin of the Institute of Paper Chemistry. (DIALOG cat., 1/85; SDC cat., 5/83)

11.14.88 PASCAL

PASCAL is a multidisciplinary data base produced by the Centre de Documentation Scientifique et Technique. Its coverage includes the fields of agricultural science, physics, life science, chemistry, and engineering. The data base contains records with titles and descriptors in both French and English. (Questel Data Base Fact Sheet, 7/82)

11.14.89 PESTDOC/PESTDOC-II

The PESTDOC data base provides international coverage of the literature on pesticides, herbicides, and plant protection. The file includes information for agriculture chemical manufacturers (excluding fertilizers) extracted from journals, conference proceedings, and scientific reports. The data base has restricted access and is available only through subscription. (SDC cat., 5/83)

11.14.90 PHP

The Petroleum Handling and Processing (PHP) data base contains information on the transport, storage, and catalysis of petroleum as well as information on the environmental impacts of petroleum handling and processing. (DOE data base description, 9/82)

11.14.91 PIRA

The Paper Information Research Association (PIRA) data base contains abstracts of scientific and technical literature on all aspects of paper-making, board-making, packaging, and printing. It includes such subjects as raw materials and additives, processes, paper and board mills, printing works, effluent treatment, recycling, waste disposal, and materials handling. (Pergamon Green Sheet, 11/83)

11.14.92 Pollution Abstracts

Pollution Abstracts covers international technical literature on environmental science and technology. The data base is the machine-readable version of the printed Pollution Abstracts. It is a leading resource for references to abstracted environmental-related technical literature on pollution, its sources, and its control. Social and legal aspects of environmental pollution are also covered. (DIALOG cat., 1/85; BRS cat., 1983)

11.14.93 PSYCINFO

Psychological Abstracts (PSYCHINFO) covers the world's literature in psychology and behavioral science. Over 900 periodicals and 1500 books, technical reports, and monographs are reviewed each year to provide coverage of original research, reviews, discussions, theories, conference reports, panel discussions, case studies, and descriptions of apparatus. (DIALOG cat., 1/85; BRS cat., 1983; SDC cat., 5/83)

11.14.94 RAPRA Abstracts

The Rubber and Plastics Research Association (RAPRA) Abstracts covers all commercial and technical aspects of the rubber and plastics industries including materials processing and products. It includes such topics as industrial hazards and toxicology, environmental effects, processing technology, and polymer synthesis. It is produced by the Rubber and Plastics Research Association of Great Britain. (Pergamon Green Sheet, 7/84)

11.14.95 RBOT

The Robotics Information (RBOT) data base provides access to current literature on all aspects of robotics. The subject areas include sensor systems, technical innovations, and machine intelligence. The file covers the period from 1980 to the present and is updated monthly. (BRS cat., 1983)

11.14.96 RTECS

The Registry of Toxic Effects of Chemical Substances (RTECS) is the on-line interactive version of the National Institute for Occupational Safety and Health's printed work. It is structured around individual chemical records and includes both acute and chronic toxicity data extracted from the literature, as well as synonyms, chemical identifiers, exposure standards, and the status of the compounds under various federal regulations and programs. The hard copy version of RTECS is updated annually; quarterly microfiche editions are available. (NLM Pocket Card, 3/83)

11.14.97 SAE

This data base provides access to technical papers presented at Society of Automotive Engineers (SAE) meetings and conferences. Papers from the International Federation of Automobile Engineering Societies (FISITA) are also covered. Subject coverage includes the technology of automobiles and other self-propelled vehicles such as spacecraft, missiles, military equipment, trucks, tractors, chain saws, and mechanical equipment. Topics include vehicle safety, materials and structures, testing, and instrumentation. (SDC cat., 5/83)

11.14.98 SAFETY

Safety Science Abstracts provides coverage of the broad field of safety, including identifying, evaluating, and eliminating or controlling hazards. Included within the data base are industrial and occupational safety, transportation safety, environmental safety, and medical safety. Within these areas the data base covers liability information and topics such as pollution, waste disposal, fire, radiation, pesticides, genetics, toxicology, injuries, and diseases. (Pergamon Draft Green Sheet, 1/86)

11.14.99 SANSS

The Structure and Nomenclature Search System (SANSS) is a query system developed to provide access to over 100 separate sources of information and data on 350,000 unique substances. This system provides cross-references to the *Merck Index* and the *Aldrich Catalog of Fine Chemicals*. (ICIS cat., 12/84)

11.14.100 SCISEARCH

The on-line version of the *Science Citation Index* (SCISEARCH) is a multidisciplinary index to the literature of science and technology prepared by the Institute for Scientific Information (ISI). It contains all the records published in *Science Citation Index* and additional records from *Current Contents* that are not included in the printed version. This data base is distinguished by two unusual characteristics: first, journals indexed are carefully selected on the basis of several criteria including citation analysis. Second, in addition to more conventional retrieval methods, this data base permits searching by cited references. (DIALOG cat., 1/85)

11.14.101 SOCIAL SCISEARCH

The Social Science Citation Index (SOCIAL SCISEARCH) data base is an international, multidisciplinary index to the literature of the social and behavioral sciences. The data base is prepared by the Institute for Scientific Information (ISI) and corresponds to the printed *Social Science Citation Index*. The data base is unique in that it enables the user to retrieve all papers that have been cited. This gives users the capability of tracing research results both forward and backward in time. (DIALOG cat., 1/85; BRS cat., 1983)

11.14.102 SOCIOLOGICAL ABSTRACTS

Sociological Abstracts corresponds to the printed index of the same name. It covers the world's literature on sociology and related disciplines in the social and behavioral sciences. Over 1200 journals and other serial publications are scanned each year to provide coverage of original research, reviews, discussions, monographic publications, theory, conference reports, panel discussions, and case studies. SOCA records include extensive abstracts. (DIALOG cat., 1/85; BRS cat., 1983)

11.14.103 SPHERE

The Scientific Parameters for Health and the Environment, Retrieval and Estimation (SPHERE) database is an integrated file on toxicology developed by the EPA to support the risk assessment of chemicals. The file contains detailed tests, descriptions, and results data from the international literature published between 1970 and 1981. It contains several components, including Dermal Absorption (DERMAL), Aquatic Information Retrieval (AQUIRE), and Environmental Fate (ENVIROFATE). (ICIS cat., 12/84)

11.14.104 SUPERINDEX

SUPERINDEX is an interdisciplinary data base consisting of back-of-the-book indexes from almost 2000 professional reference books in science, engineering, and medicine. The several million index entries in the database provide rapid, page-specific access to important reference books from over 20 major publishers. (BRS cat., 1983)

11.14.105 TELEGEN

TELEGEN contains information related to the fields of genetic engineering and biotechnology. Scientific, technical, and socioeconomic information is derived from 5000 worldwide sources ranging from scientific and nontechnical journals to conferences, corporate and government reports, and independent laboratory reports since 1973. The data base corresponds to the annual *Telegen Reporter Review* and is updated monthly from the *Telegen Reporter* published by the Environment Information Center (EIC). (DIALOG cat., 1/85)

11.14.106 TERM

TERM is an integrated, controlled vocabulary data base and search aid that provides users with controlled vocabulary descriptors, codes, and natural language synonyms for concepts in the behavioral and social sciences. The data base incorporates the subject codes and headings from six major print vocabulary thesauri. For each indexed concept, the equivalent subject headings from every print thesaurus are listed, as are descriptive terms synonymous to the main concept. (BRS cat., 1983)

11.14.107 Textile Technology Digest

The Textile Technology Digest provides international coverage of the literature of textiles and related subjects. Selected articles from current technical sources summarize the technological advancements and applications of science and research to the textile industry. Coverage includes the various aspects of textile production and processing as well as the automation and management systems of these operations. Textile Technology Digest corresponds to the printed publication of the same name. (DIALOG cat., 1/85)

11.14.108 Toxicology Data Bank

The Toxicology Data Bank (TDB) is an on-line, interactive file extracted from secondary and tertiary sources of chemical, pharmacological, and toxicological data, environmental and occupational information, manufacturing and use data, and chemical and physical properties. These sources may be supplemented with information from the current literature. It is composed of over 4000 comprehensive chemical records reviewed by the TDB Peer Review Committee comprised of scientists presently or formerly members of the National Institute of Health's Division of Research Grants Toxicology Study Section. The information in TDB is taken verbatim from the source, and references to the sources are given. (NLM Fact Sheet, 7/83)

11.14.109 TOXLINE

The TOXLINE (Toxicology On Line) database is the National Library of Medicine's on-line, interactive collection of toxicological information containing references to published studies on human and animal toxicity, the effects of environmental chemicals and pollutants, adverse drug reactions, and analytical methodology. The information in TOXLINE is taken from eleven secondary sources, which formulate the following subfiles: Toxicity Bibliography, Chemical-Biological Activities, Pesticides Abstracts, International Pharmaceutical Abstracts, Abstracts on Health Effects of Environmental Pollutants, Environmental Mutagen Information Center File, Environmental Teratology Information Center File, Toxicology/Epidemiology Research Projects, Toxicology Document Data Depository, Toxic Materials Information Center File, and the Hayes File on Pesticides. (NLM Fact Sheet, 7/83)

11.14.110 TRIS

The Transportation Research Information Service (TRIS) provides research information on air, highway, rail, and maritime transport. Included are regulations and standards; energy, environmental, and safety concerns; and construction and maintenance technology. The records can be either abstracts or summaries of research projects. (DIALOG cat., 1/85)

11.14.111 TSCA Initial Inventory

The TSCA Initial Inventory file is a nonbibliographic dictionary listing of chemical substances derived from the initial inventory of the Toxic Substances Control Act (TSCA) Chemical Substance Inventory. Substances listed in this file were in commercial use as of June 1, 1979. For each substance listed the following information is provided: Chemical Abstracts Service (CAS) Registry Number, preferred name, synonyms, and molecular formula. (DIALOG cat., 1/85)

11.14.112 TSCA Plus

The TSCA Plus data base is a nonbibliographic dictionary listing of chemical substances derived from the Initial Inventory of the Toxic Substances Control Act (TSCA) Chemical Substance Inventory. Substances listed in this file include additions as of 1981. Plant and production data are also included in this file. For each substance listed, the Chemical Abstracts Service (CAS) Registry Number, preferred name, synonyms, and molecular formula are provided. (SDC cat., 1/85)

11.14.113 TSCAPP

The Toxic Substances Control Act Plant and Production (TSCAPP) data base contains information supplied by manufacturers of certain chemicals manufactured between 1975 and 1977, including drugs and pesticides. The portion of the information considered confidential is not available on this system. The file contains information from 127,000 production citations on 53,000 unique substances and over 6000 processors and manufacturers. (ICIS cat., 12/84)

11.14.114 Voluntary Standards Information Network

The Voluntary Standards Information Network data base contains information about standards under development by the American National Standards Institute. It will also contain those voluntary standards under development by the signatories to the GATT Agreement on Technical Barriers to Trade. All aspects of engineering and related services standards are covered. (BRS Aid Page, 10/83)

11.14.115 Water Resources Abstracts

This data base is prepared from materials collected by over 50 water research centers and institutes in the United States. Research topics covered include the water-related aspects of nuclear radiation and safety. Water Resources Abstracts is particularly strong in literature on water planning, water cycle, and water quality. (DIALOG cat., 1/85; DOE data base description, 7/82)

11.14.116 WATERLIT

This data base covers worldwide water resources literature collected by the South African Water Information Centre. Topics include the chemistry, biology, and physics of water; hydrologic cycle; engineering and construction; desalination; utilization; waste water and pollution; recycling; quality control; limnology; health; ecology; planning; management; and standards. (SDC cat., 5/83)

11.14.117 WATERNET

WATERNET provides a comprehensive index of the publications of the American Water Works Association (AWWA) and AWWA Research Foundation. Included are books and proceedings, journal articles, standards, manuals, handbooks, and water quality standard test methods. Emphasis is on technical reports and studies from water utilities, regulatory agencies, and research groups in the United States, its territories, and Canada. Mexican, Latin American, European, and Asian data are also reported. The data base, produced by AWWA, is the on-line counterpart of the hard copy index covering journals, AWWA articles from 1971 to the present, and all AWWA and AWWARF publications from 1973 to the present. Non-AWWA-published articles are indexed on a selected basis. Each record contains a brief abstract. (DIALOG cat., 1/85)

11.14.118 WELDASEARCH

WELDASEARCH is the data base of the Welding Institute. It is a comprehensive data base covering all aspects of the joining of metals and plastics. Welded design, welding metallurgy, fatigue, and fracture mechanics are included as well as related areas such as metals spraying and thermal cutting. The data base provides international coverage of journals, books, patents, theses, and conferences. (DIALOG cat., 1/85)

11.14.119 World Aluminum Abstracts

World Aluminum Abstracts provides coverage of the world's technical literature on aluminum ranging from processing (exclusive of mining) through end uses. The data base includes information abstracted from journals, patents, government reports, conference proceedings, dissertations, books, and other publications. (DIALOG cat., 1/85)

11.14.120 World Surface Coatings Abstracts

World Surface Coatings Abstracts is derived from the publication, *World Surface Coatings Abstracts* (WSCA), founded in 1928. It provides worldwide coverage of the significant literature on all aspects of coatings applied to materials. Subjects included in the data base are industrial hazards, pigments, inks, polymers, analytical methods, varnishes and lacquers, and storage and transport. Surface Coatings Abstracts is produced by the Paint Research Association. (Pergamon Green Sheet, 11/83)

11.14.121 World Textile Abstracts

World Textile Abstracts corresponds to the publication of the same name. It covers the science, technology, and end use of textiles and related materials. Emphasis is given to pollution, safety and health hazards, synthesis, and chemistry of fiber-forming polymers and chemical materials used in the textile industry, manufacturing processes in textile and related industries, and chemical and mechanical treatment of textile materials. The data base is produced by the Shirley Institute and collaborating research associations. (DIALOG cat., 1/85; Pergamon Green Sheet, 11/83)

11.14.122 Zinc, Lead, and Cadmium Abstracts

Zinc, Lead, and Cadmium Abstracts covers all aspects of the production, properties, and uses of these metals and their alloys and compounds. Specifically covered are health and hygiene, pollution control, environment, biochemistry, electroplating, hot dip coating with rubber and plastics, and others. It corresponds to the publication of the same name but contains additional records that do not appear in the hard copy. Information is obtained from journals, conference proceedings, and technical reports. (Pergamon Green Sheet, 11/83)

BIBLIOGRAPHY

Akey, Denise S., ed. *Encyclopedia of Associations*, 18th ed., 2 vols., Gale Research Company, Madison Heights, MI, 1984.

Anon., *Books in Print 1982–1983*, R. R. Bowker Company, New York, NY, 1982.

Bibliographic Retrieval Service, *Directory and Database Catalog*, 1983, Bibliographic Retrieval Service, Latham, NY 12110.

Chemical Abstracts, *Chemical Abstracts Service Source Index®: 1907–1984*, Chemical Abstracts Service, Columbus, OH, 1984.

DIALOG Information Services, Inc., *Database Catalog*, 1985, DIALOG Information Services, Inc., Palo Alto, CA 94304.

European Space Agency, *Information Retrieval Service Catalog*, 1984, Esrin, P.O. Box 64, 1-00044, Frascati, Italy.

Information Consultants, Inc., *Chemical Information System User's Guide*, 1984, Information Consultants, Inc., Washington, DC 20005.

Interagency Education Program Liaison Group, Task Force on Environmental Cancer and Heart and Lung Disease, *Environmental Health-Related Information: A Bibliographic Guide to Federal Sources for Health Professionals*, 1984. Available from NTIS, Springfield, VA 22616.

International Molders and Allied Workers Union (AFL-CIO-CLC), *Health and Safety Resource Directory: 1980*, International Molders and Allied Workers Union (AFL-CIO-CLC), Cincinnati, OH 45206.

Peck, T. P., *Occupational Safety and Health: A Guide to Information Sources*, Gale Research Company, Booktower, Detroit, MI, 1974.

Pergamon International Information Corporation, *Online Information Services Catalog*, Pergamon International Information Corporation, Mclean, VA 22101.

System Development Corporation, *Orbit® Search Service Databases Catalog*, 1983, System Development Corporation, Santa Monica, CA 90406.

Tucker, M. E., *Industrial Hygiene: A Guide to Technical Information Sources*, American Industrial Hygiene Association, Akron, OH, 1984.

Turabian, Kate L., *Manual for Writers of Term Papers, Theses, and Dissertations*, 4th ed., University of Chicago Press, Chicago, IL, 1973.

United Nations Educational, Scientific and Cultural Organization (UNESCO), Handbook for Information Systems and Services, United Nations Educational, Scientific and Cultural Organization, New York, NY, 1977.

U.S. Congress, Office of Technology Assessment, *Preventing Illness and Injury in the Workplace*, OTA-H-256, U.S. Government Printing Office, Washington, DC 20402, April, 1985.

U.S. Department of Health and Human Services, Public Health Service, National Institutes of Health, National Cancer Institute, *A Compilation of Journal Instructions for Authors*, 1980, NIH Publication No. 80-1991, National Cancer Institute, Bethesda, MD 20205.

U.S. Department of Health and Human Services, Public Health Service, National Institutes of Health, National Library of Medicine, *Chemical Toxicology Files Sampler*, 1984, National Library of Medicine, Bethesda, MD 20205.

U.S. Department of Health Education and Welfare, Public Health Service, Office of Disease Prevention and Health Promotion, *Health Information Resources in the Federal Government 1984*, National Health Information Clearinghouse, Washington, DC 20013.

U.S. Department of Health Education and Welfare, Public Health Service, Centers for Disease Control, *MMWR: Morbidity and Mortality Weekly Report*, Atlanta, GA 30333.

U.S. Department of Health and Human Services, U.S. Public Health Service, Centers for Disease Control, National Institute for Occupational Safety and Health, *NIOSH Publications Catalog*, 6th ed., DHHS (NIOSH) Publication No. 84-118, National Institute for Occupational Safety and Health, Cincinnati, OH, 1984.

U.S. Department of Health and Human Services, Public Health Service, National Library of Medicine, *Proceedings of the Symposium on Information Transfer in Toxicology*, 1982, PB 82-220-922, National Library of Medicine, Bethesda, MD 20205.

U.S. Department of Health Education and Welfare, Public Health Service, Office of Disease Prevention and Health Promotion, *Worksite Health Promotion: A Bibliography of Selected Books and Resources*, 1983, Office of Disease Prevention and Health Promotion, Washington, DC 20201.

Whittenburg, J. A., "Guidelines for Planning a Task-Oriented Information System," in C. E. Nelson and D. K. Pollock, eds., *Communication Among Scientists and Engineers*, D. C. Heath and Company, Lexington, MA, 1970.

CHAPTER 12

How to Use Workers' Compensation Data to Identify High-Risk Groups

ROGER C. JENSEN
Division of Safety Research
National Institute for Occupational Safety and Health
Morgantown, West Virginia

12.1 INTRODUCTION

Information about injuries is a fundamental need of safety programs. Its value is recognized in such diverse applications as highway design, motor vehicle traffic regulation, pedestrian walkway design, pilot training, power plant hazard control, consumer products design, sports equipment design, the making of rules for sports, building design, home safety programs, fire prevention, and workplace safety programs. Professionals working in these fields have in common a need for factual background information about past events to guide programmatic and technical decisions affecting future risk.

The informational needs of these applications led to the development of specialized systems for reporting, saving, and retrieving information about the undesired events of concern—for example, personal injuries, property damage cases, and critical incidents. These injury/incident information systems typically include their own unique (1) criteria for identifying reportable events, (2) reporting forms, (3) coding schemes, and (4) computer storage, retrieval, and access mechanisms.

The uniqueness of each system is attributable, in part, to the need to strike a balance between the burden of obtaining the information and the benefits derived from the information system. Although the trade-offs resulting from this balancing may not be explicit, the resulting systems clearly reflect that element of human nature that makes people unwilling to spend much time carefully filling out a report form unless the benefits are clear. Thus, it is not surprising that an information system based on reports of workplace injuries provided by shop foremen will differ substantially from one based on reports of critical incidents from commercial airline pilots.

12.1.1 Purpose of Injury Data Analysis

The uniqueness of each injury/incident information system is also partially attributable to the emphasis placed on various purposes. For example, nearly all motor vehicle incident information systems are based on data recorded by police officers on official "accident" report forms. The primary purpose of these reports is to provide documentation to support subsequent legal actions (civil suits and/or citations for violating traffic laws). The information o'tained from these forms can be and is used for other purposes, such as identifying needed changes in traffic laws and setting enforcement priorities, but in terms of emphasis these uses are secondary to the primary purpose. In contrast, reports of aircraft accident investigations by the National Transportation Safety Board are primarily for identifying hazardous situations that are (it is hoped) amenable to correction. One of the secondary purposes of these reports is their documentary value for use in civil litigation arising from the crash.

Occupational injury information systems clearly reflect an appreciation of the trade-offs between the burden of reporting and the benefits of having the information recorded. The systems currently in use reflect a preference for report forms that only request those items of information regarded as essential to satisfy the primary purpose of the system. Little if any attempt is made to obtain nonessential information that might prove valuable for secondary purposes such as research concerning the influence of various work-environment factors on risk of injury.

This background should make it easier to appreciate workers' compensation data as a source of information about injuries. The primary purpose of traditional workers' compensation reporting systems is to provide the facts required to determine eligibility for and amount of compensation. Consequently, the forms used for reporting injury information are designed primarily to obtain the information required to evaluate the injury in relation to the applicable criteria for compensation.

There are several secondary uses of workers' compensation claim reports. A well-established secondary use is employer-specific frequency tabulations (i.e., counts), which are used to estimate appropriate compensation insurance rates for individual employers or establishments. Individual employers also use injury frequency tabulations to identify their most common injury patterns and their major sources of injury-related costs. Another secondary use involves the pooling of data from multiple employers to characterize the major injury problems associated with particular types of operations in order to help employers focus injury prevention efforts. An example of this kind of information is shown below. Provided by one workers' compensation insurance company[1] and based on prior claim experience involving machine shops (SIC 3632), it is used to help their insured machine shop operators focus their safety resources on the major problems:

1. Handling materials was responsible for 23 percent of the injury cost. Over 55 percent of this cost was from back injuries.
2. Getting caught in machines resulted in 11 percent of the cost. About 30 percent of this cost involved presses, including punch presses.
3. Being struck by falling objects accounted for 11 percent of the injury cost.
4. The cost of injuries from striking against machines, pipe, iron or steel stock or parts containers, and other objects was 7 percent of the total cost.

Such information can be useful for directing safety resources toward the most significant problems. Additional and more specific information can be even more helpful. This chapter describes some practical techniques for getting useful information from workers' compensation claim records.

At the outset, however, it should be understood that an individual employment establishment will have a limited number of workers' compensation claims to use in any analysis. Smaller employers find it especially difficult to identify patterns from their compensation records. Consequently, workers' compensation claim data are often more useful when the claim experiences of multiple employers are pooled for analysis. To illustrate this point, consider a machine tool operation employing 90 workers. During the past eight years, one machine operator lost a finger in a mechanical power press. An analysis of this employer's injury data might suggest that efforts to prevent injuries associated with other hazards ought to have a higher priority than the hazards of their power press operations. However, by looking at injury data from multiple machine tool operations, it becomes clear that finger amputation involving manually operated mechanical power presses is a major problem that clearly deserves a high priority for injury prevention efforts.

Compensation claim data is best regarded as a source of information that supplements other information sources (e.g., injuries reported on the OSHA Log, clinic visits, first report of injury forms).

Typically, the compensation records can be used to identify a certain department as the major contributor of those work-related injuries that are moderately serious or worse. This is followed up by an analysis of the most complete injury reports maintained for that department to zero in on more specific problems that might be addressed by the safety program.

This chapter presents concepts for analyzing injury data generally and also specific techniques for analyzing records of workers' compensation claims in order to distinguish among different groups of workers on the basis of their respective risk of incurring compensable injuries.

12.1.2 Estimating Risk

The simplest way to analyze workers' compensation data is to categorize and count the raw number of cases. Lists or tables showing the number of cases in each category are called "frequency tables." An employer, for example, may categorize its compensation claims from the past year into a few categories and compare this information with similar records from prior years. Such an approach is often regarded as useful for identifying important changes in claim experience; for example, back injury claims may have increased while hand injury claims decreased. The principal value of this simple frequency approach is the identification of the more common injury patterns. The numerous shortcomings of the approach stem from the lack of meaningful information content—especially the lack of information about the comparative likelihood of an individual or group of individuals incurring a compensable injury during a specified time period.

This introduces the concept of risk. Generally, the word "risk" is used in different ways in different contexts.[2-5] Rather than attempt to review the many different definitions of the term, a simple definition built upon the basic mathematics of probability is offered. In the context of occupational injury analysis, the term "risk" refers to the *probability* that an individual will be harmed while working for a specified period of time or while performing a certain task. Examples of tasks are lifting a box, changing an anode in an aluminum reduction pot, or walking from point A to point B.

The word "safety" is also used in different ways in different contexts. To maintain consistency with the preceding definition of risk, safety can be defined as the complement of risk.[5] That is, for any specified task or person-time interval, an individual has a certain probability of being unharmed $P(UH)$ and a certain probability of being harmed $P(H)$ such that

$$P(UH) + P(H) = 1 \tag{12.1}$$

These definitions of safety and risk presuppose unambiguous criteria for assigning the outcome of each task or person-time interval to either the harmed or unharmed category.[4]

The direct measurement of occupational injury risk, $P(H)$, is relatively difficult. Currently, the most practical techniques utilize data from existing record systems to calculate various risk quantities[6] that are thought to be positive correlates of true risk, $P(H)$. Conceptually, the model for this relationship is

$$P(H) = a + b(RQ) + \text{error} \tag{12.2}$$

where RQ refers to an appropriate risk quantity and a and b are constants. Numerical values for these risk quantities are generally obtained from past injury and employment records. Estimates of an individual's risk in the future are often based on values of risk quantities derived from the past experience of groups of people with similar personal characteristics or exposures. In the occupational safety field, the most useful risk quantities are those based on a ratio with a numerator reflecting the incidence of harmful events and a denominator reflecting the opportunities for the harmful event.

The basic risk quantities used for analyzing workers' compensation data are ratios obtained by dividing the number of new cases among a group of workers during a time period by a value representing the extent of exposure. The specified time period is often one year, but it may be shorter or longer. The exposure value for the denominator is based on employees who would have been counted in the numerator if they had incurred a compensable injury during the applicable time period. For comparing the risks of different groups of workers using workers' compensation data, the usual units for the denominator are either (1) total persons, or (2) total person-hours, person-shifts, or person-years of work. If total persons is the denominator, the ratio is called an incidence ratio (*IR*). If an expression of person-time is the denominator, the ratio is called an incidence density (*ID*).[7] For example, the XYZ company might obtain an annual *IR* of compensation claims for hand injuries among its machinists by dividing the number of hand injury claims that arose during the year from the company's machinists by the total number of machinists employed by XYZ on a certain date during the year. Or, if adequate records are available, it is preferable to determine an *ID* by dividing the number of new cases from XYZ's machinists that occurred during the year by the total number of person-hours of work by the company's machinists during the year. If any of these machinists had more than one claim arising from the same event, claim should be counted only once.

The initial step in the analysis of workers' compensation data is the development of time-specific *IR*s and *ID*s based on sound numerator and denominator data. The injury data for the numerator are discussed in section 12.2 and the exposure data for the denominator are discussed in section 12.3. In section 12.4 some techniques are presented for using these risk quantities to compare the traumatic injury risks faced by workers in various subsets of the work force.

12.2 NUMERATOR (INJURY) DATA

Reports of injuries are the usual source of information for the numerator of the risk quantities, *IR* and *ID*. The tabulation of such numerator data would be simple if a perfect injury information system existed. Unfortunately, shortcomings are found in all the injury information systems, including those based on workers' compensation claims. An appreciation of these shortcomings is essential for any analysis of injury data.

12.2.1 Occupational Injury Information System

A spatial tool for explaining injury information systems is the Venn diagram. These diagrams allow us to picture the relationship between all items in a population and those items in a subset of the total. The use of a Venn diagram for visualizing injury information is illustrated in Figure 12.1. The area within the rectangle in this illustration is defined to represent all injuries in a finite time period among all employees of a particular establishment. Included are minor injuries like paper cuts as well as injuries that occurred away from the job. The circle in Figure 12.1 represents those injuries that meet the unique criteria of a particular injury reporting system.

These criteria vary considerably, and most reporting systems utilize multiple criteria. The criteria generally concern the (1) time when the injury occurred in relation to the individual's work schedule, (2) place where the injuring event occurred in relation to where the employee's duties require him or her to be, and (3) severity of the injury as determined using a convenient yardstick like the kind of treatment required or the extent of time the employee did not work because of the injury.

One important characteristic of workers' compensation data is that the severity criteria allow for claims based on disability resulting from a work-related injury. This distinction between disability and medical criteria is illustrated by the common malady usually referred to as low back pain. Let us suppose that two individuals, a construction laborer and a librarian, each develop a painful lower back shortly after performing a work-related task. From a medical perspective, if both people experience the same level of pain, each case should be counted equally in an injury information system. But if the severity criterion of the reporting system is based on the extent of non-work days resulting from the injury, it is quite likely that the two cases will not be counted equally. The construction worker may be unable to perform his or her job with a painful back. The librarian, however, may be able to report for work and perform most tasks even with a painful back. Thus, for equally painful backs one employee may be disabled from working when the other is not.

The difference between injury information systems based on medical vis-à-vis disability criteria is depicted in the Venn diagram in Figure 12.2. The larger circle represents all occupational injuries that exceed certain medical criteria for severity—for example, all cases that require treatment by a physician.

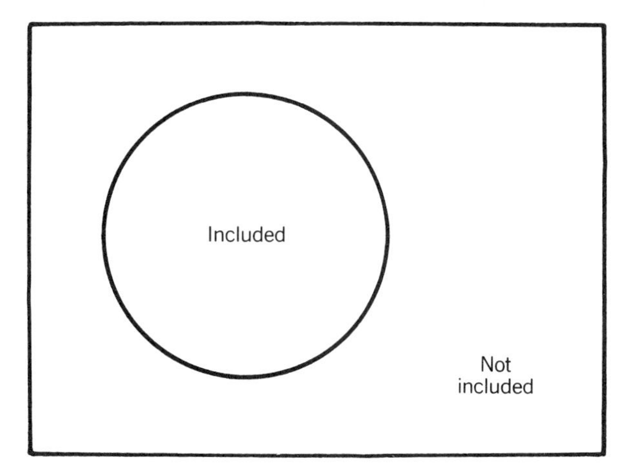

FIGURE 12.1. Depiction of all injuries (rectangle) and a subset of those included in an injury information system (circle).

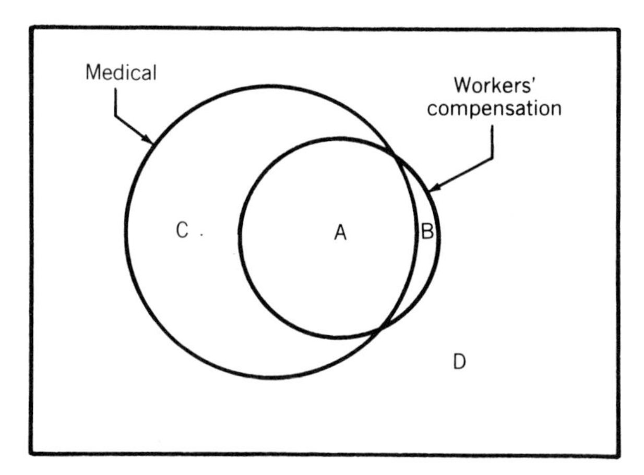

FIGURE 12.2. Injuries included in two different injury information systems, one based on medical, the other on workers' compensation (*WC*) criteria.

The smaller circle represents all occupational injuries from the same work force and time period that meet the statutory requirements for workers' compensation. Notice that not all injuries that meet the statutory compensation requirements are included among those in the medically based injury information system. An example would be a worker who calls in sick several days in a row because of low back pain, eventually reporting back to work without ever being seen by a physician. This employee could qualify for partial wage indemnification under the workers' compensation statutes of many states, although without medical testimony or reports it might be difficult for the employee to provide sufficient evidence to prove his or her eligibility for compensation.

Figure 12.2 shows four areas denoted with the letters A, B, C, and D. These areas represent the following four subsets of all injuries to a specified group of workers:

A. Injuries that meet the criteria of both injury information system (for example, the loss of a finger in a mechanical power press)
B. Injuries that meet the criteria for the workers' compensation injury information system but not the medically based system, (for example, the previously mentioned case of low back pain)
C. Injuries that meet the criteria for the medically based injury information system but not the workers' compensation system (for example, a laceration treated by the plant physician with only two hours of lost work time)
D. Injuries that do not meet the criteria for inclusion in either injury information system (for example, a paper cut or an injury sustained at home)

This same basic categorization may also be viewed using a two-by-two table (sometimes referred to as a two-by-two matrix or a fourfold table). The two rows divide the total set of injuries into two groups: those that meet and those that do not meet the criteria for inclusion in the workers' compensation information system. Similarly, the two columns divide the total set of injuries into two groups: those that meet and those that do not meet the criteria for inclusion in the medically based injury information system. Figure 12.3 shows such a table. The letters A, B, C, and D refer to the same subsets as those in the Venn diagram of Figure 12.2.

Venn diagrams and tabular matrices provide two useful tools for appreciating the relationships between all injuries sustained by a specified group of people during a specified time period (the total set), and those injuries included in one or more injury information systems. Venn diagrams are particularly useful for visualizing the meaning of the division of a complete set of injuries into smaller subsets. Tabular matrices provide an orderly mechanism for dividing a set of injuries into multiple subsets. In general, Venn diagrams are practical tools for a limited number of subsets such as those depicted in Figure 12.2. With more subsets they become more difficult to understand. Tabular matrices are advantageous when the number of subsets is larger, because of their expandability.

An important factor in any analysis of workers' compensation data is the criteria for inclusion of an injury into the recording system. The criteria used for workers' compensation systems are based on the specific and unique statute of the applicable state. Consequently, to represent the injuries included in the state compensation record systems on a Venn diagram would require 50 circles. All the circles would enclose a basic area representing those injuries that are compensable in all states. But the circles would

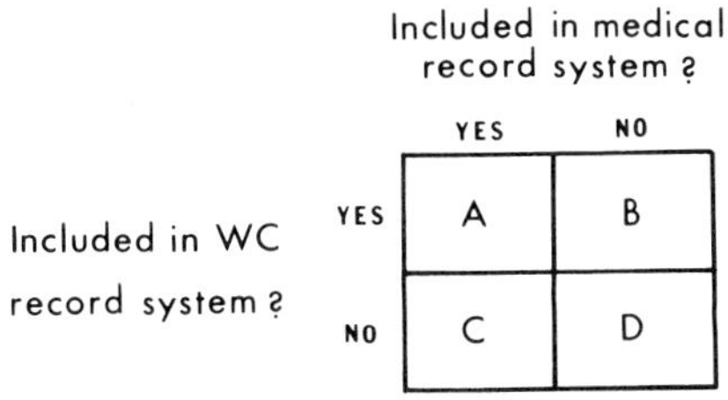

		Included in medical record system?	
		YES	NO
Included in WC record system?	YES	A	B
	NO	C	D

FIGURE 12.3. Matrix approach for subsetting the universe of injuries shown in Figure 12.2.

overlap at various points, creating small areas representing injuries that would be compensable in some states but not others. These marginal cases are the result of differences in state statutes, in the interpretation of these statutes by the state courts, and in the administrative practices of state compensation agencies. However, most injuries that are compensable in one state will be compensable in all the other states because of the fundamental similarities of workers' compensation statutes.

An understanding of the kinds of cases that qualify for compensation is important for appreciating the implications of analyses based on such data. Below is a concise outline of the fundamental requirements incorporated into state and federal workers' compensation statutes:

1. The injured individual must be an "employee" within the definition of the applicable statute. The major impact of this requirement is the exclusion of independent contractors and self-employed individuals from coverage under state compensation statutes.
2. The employer of the injured individual must have been included in the applicable compensation statute. For example, a statute may only include employers with three or more employees. State statutes also vary in their coverage of state, county, and city government employees. Furthermore, state compensation statutes do not include certain employers that have compensation systems originating from federal statutes. For example, Veterans' Administration hospitals are not included as employers subject to state workers' compensation laws because VA employees are covered by a federal compensation law. Similarly, longshore and harbor workers are covered by a special federal statute, so state compensation statutes are inapplicable.
3. The injury must arise out of and in the course of employment. Most industrial injuries clearly meet this pair of requirements if the injury occurred while the employee was doing something related to his or her job duties and during his or her work shift. A comparatively small portion of marginal cases, however, create a large volume of case law and receive a disproportionate share of attention in the popular press. When looking at all compensation cases for most categories of injury, these marginal cases comprise such a small percentage that their impact is inconsequential when comparing groups of workers in terms of their respective risk of incurring most kinds of injuries.
4. The severity of the injury must meet the applicable statutory requirements. Sufficient severity may be based on any one of three criteria: (1) if the injured employee obtains medical attention (as defined in the statute) for the injury, (2) if the injury causes the employee to be absent from work for a specified number of workdays, or (3) if the injury is specified explicitly in the statute (e.g., death or loss of specific parts of the body). The last-mentioned cases are generally referred to as "scheduled" injuries because the statute provides a formula, or schedule, for determining the minimum amount of compensation for each listed injury.

These requirements define the kind of injury cases entitled to workers' compensation. This is not quite the same as defining the kind of cases that get into the system because of various classification and coding problems.

12.2.2 Classification and Coding Problems

All injury information systems have misclassification problems. Some injuries that should be in the system fail to get included for one reason or another. And some cases that do not actually meet the inclusion criteria manage to get included. Furthermore, of those cases that are included, some will be incorrectly classified. These problems are discussed in this subsection.

The appreciation of misclassification errors involving incorrect inclusion or exclusion can be facilitated by the use of a two-by-two table. Figure 12.4 shows such a table. It depicts the four possible ways a worker's injury can be classified with respect to the workers' compensation system. The two rows divide all cases according to whether the case was or was not entered into the compensation system. This di-

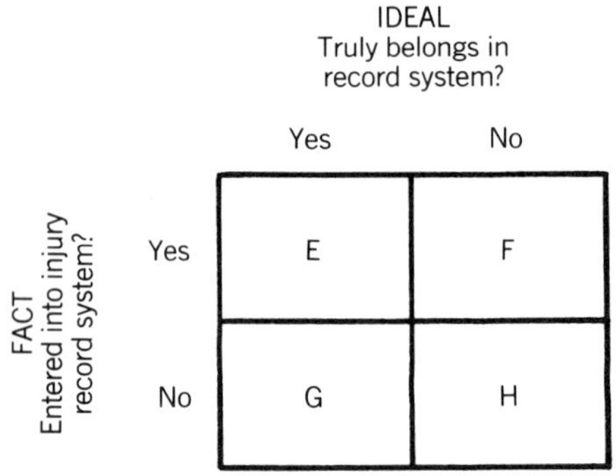

IDEAL
Truly belongs in record system?

FACT: Entered into injury record system?	Yes	No
Yes	E	F
No	G	H

FIGURE 12.4. Misclassification errors depicted in a two by two table.

mension may be regarded as the "in-fact" classification. The two columns divide each of the cases into whether or not it truly belongs in the system, assuming that everything about the case is known. This dimension may be referred to as the "ideal" classification. The cases within the four cells (denoted E, F, G, and H to avoid confusion with the letters in Figure 12.3) signify:

E—cases correctly included
F—cases incorrectly included
G—cases incorrectly excluded
H—cases correctly excluded

The erroneously classified cases are those in cells F and G. Within the workers' compensation systems, attempts to control such errors are made by those involved in the compensation system: injured workers and labor unions, employers, attorneys, and insurers. Still, some portion of qualifying injuries will not be compensated and some portion of nonqualifying cases will receive compensation.

The proportion of misclassification errors is probably dependent on the nature of the injury, although studies have not confirmed this hypothesis. The injuries most likely to be misclassified are those in which it is difficult to clearly establish a diagnosis and clearly prove or disprove that the condition arose out of and in the course of employment, (e.g., back injuries). In contrast, misclassifications involving amputations should be rare because of the clear diagnosis and the ease of establishing when and where the amputation occurred.

Another kind of error results from incorrect coding of the case descriptors. Cases are coded to facilitate computer storage and retrieval. Coding uses preestablished categories to describe the characteristics of the particular case. The coding system that is most used in the United States is derived from the American National Standards Institute "Method of Recording Basic Facts Relating to the Nature and Occurrence of Work Injuries," generally referred to as ANSI Z16.2.[8]

The Z16.2 standard is not specifically aimed at finding the causes of injuries. It is aimed at identifying facts about the circumstances that can be ascertained for a uniform recording system. This approach has the advantage of removing from the recording system some of the subjective conclusions of the individual who fills out the first report of injury. Those features of the ANSI Z16.2 standard that best exemplify the goal of reporting facts were adopted by the U.S. Bureau of Labor Statistics (BLS) when a national program was established for collecting workers' compensation case records from the states. This injury information system is called the Supplementary Data System (SDS) because its purpose is to supplement the information obtained by BLS in its annual survey of employers concerning occupational injuries and illnesses.[9]

Both the ANSI Z16.2 and the SDS system call for each injury or illness case to be described in terms of several characteristics. The most important characteristics used for the SDS system are summarized in the following outline:

1. About the injured person
   a. Age
   b. Sex
   c. Occupation

2. About the employer
   a. Principal product or service
   b. Ownership—public or private
3. About the event
   a. Source of injury or illness
   b. Type of accident or exposure
   c. Year and month of occurrence
4. About the injury or disease
   a. Nature of injury
   b. Part of body
5. About the source of the report
   a. State
   b. Reference year—the year that identifies the SDS tape where the case record is filed

Errors occur in the process of assigning case reports to coding categories due to insufficient or incorrect information on the original injury report form or due to mistakes by coding clerks. BLS has been working with those states participating in the SDS to minimize mistakes made by coding clerks. But the problems due to the basic inadequacy of reported information continue to limit the potential value of the system as an information source.

Another concern with compensation data is that there is not always a one-to-one match between claims and events. This includes instances in which a worker files two legitimate claims for injury arising from a single event. This may occur, for example, in many states when a "scheduled" injury involves an initial claim for the scheduled amount and a subsequent claim for permanent disability. There are also instances in which a single event injures more than one worker, resulting in multiple compensation claims. Thus, counting compensation claims is not precisely the same as counting injuring events.

Related to the concern for misclassification errors is a concern for loss of information due to coding. Basically, an original claim form, or employer's first report of injury, contains a certain amount of information. After being reduced to a set of coded descriptors, less information about the case is available.[10] For example, a claim form might say that a press operator was changing the die on a specific model mechanical power press when the press descended because another worker switched the power on, resulting in an amputation of the operator's index finger. After coding, the case would no longer include the discriptive information about the press, any indication that the operator was attempting to change the die, or any information about another worker overriding the safety system by turning on the power. Consequently, a subsequent search of coded injury reports to get information about injuries involving specified models of machines, the performance of setup tasks, or inadequate hazardous energy control procedures would be futile because the information was not included in the coded case data.

These various classification and coding problems mean that we should not be surprised if errors are found in the records of individual cases. The existence of errors, however, does not negate the potential of workers' compensation data for identifying common injury patterns and groups of workers who have been experiencing disproportionately high frequencies of compensable injuries. Using large numbers of cases diminishes the importance of classification and coding errors. The large number of case records contained in the SDS makes such analyses possible. For this reason, a clear understanding of the SDS coding system is essential. The following subsection lists and defines the most useful coding categories used for SDS cases.

12.2.3 SDS Coding Categories

The detailed categorization scheme used in the SDS is described in documents prepared by the Bureau of Labor Statistics and obtainable from the National Technical Information Service (NTIS), U.S. Department of Commerce, Springfield, VA 22161. A few changes are made in the coding system each year or two. Consequently, different documents are required to define the specific codes used for any particular year. Three documents are available from NTIS:

Supplementary Data System, Microdata Files User's Guide, 1976–1977. NTIS Number PB 288258.

Supplementary Data System, Microdata Files User's Guide, 1978–1979. NTIS Number 80-149982.

Supplementary Data System, Microdata Files User's Guide, 1980 Edition. NTIS Number PB 83 133 553.

Subsequent editions may be located by contacting the Office of Occupational Safety and Health, BLS, U.S. DOL, 200 Constitution Ave., N.W., Washington, DC 20006. Because of the limited availability of

TABLE 12.1. Occupational Categories Used for SDS Cases, From 1970 U.S. Bureau of Census System

Occupation Code	
	Professional, Technical, and Kindred Workers
001	Accountants
002	Architects
	Computer specialists
003	Computer programmers
004	Computer systems analysts
005	Computer specialists, NEC[a]
	Engineers
006	Aeronautical and astronautical engineers
010	Chemical engineers
011	Civil engineers
012	Electrical and electronic engineers
013	Industrial engineers
014	Mechanical engineers
015	Metallurgical and materials engineers
020	Mining engineers
021	Petroleum engineers
022	Sales engineers
023	Engineers, NEC
024	Farm management advisors
025	Foresters and conservationists
026	Home management advisors
	Lawyers and judges
030	Judges
031	Lawyers
	Librarians, archivists, and curators
032	Librarians
033	Archivists and curators
	Mathematical specialists
034	Actuaries
035	Mathematicians
036	Statisticians
	Life and physical scientists
042	Agricultural scientists
043	Atmospheric and space scientists
044	Biological scientists
045	Chemists
051	Geologists
052	Marine scientists
053	Physicists and astronomers
054	Life and physical scientists, NEC
055	Operations and systems researchers and analysts
056	Personnel and labor relations workers
	Physicians, dentists, and related practitioners
061	Chiropractors
062	Dentists
063	Optometrists
064	Pharmacists
065	Physicians, medical and osteopathic
071	Podiatrists
072	Veterinarians
073	Health practitioners, NEC
	Nurses, dietitians, and therapists
074	Dietitians
075	Registered nurses
076	Therapists

TABLE 12.1. Occupational Categories Used for SDS Cases, From 1970 U.S. Bureau of Census System *(continued)*

Occupation Code	
	Health technologists and technicians
080	Clinical laboratory technologists and technicians
081	Dental hygienists
082	Health record technologists and technicians
083	Radiologic technologists and technicians
084	Therapy assistants
085	Health technologists and technicians, NEC
	Religious workers
086	Clergymen
090	Religious workers, NEC
	Social scientists
091	Economists
092	Political scientists
093	Psychologists
094	Sociologists
095	Urban and regional planners
096	Social scientists, NEC
	Social and recreation workers
100	Social workers
101	Recreation workers
	Teachers, college and university
102	Agriculture teachers
103	Atmospheric, earth, marine, and space teachers
104	Biology teachers
105	Chemistry teachers
110	Physics teachers
111	Engineering teachers
112	Mathematics teachers
113	Health specialties teachers
114	Psychology teachers
115	Business and commerce teachers
116	Economics teachers
120	History teachers
121	Sociology teachers
122	Social science teachers, NEC
123	Art, drama, and music teachers
124	Coaches and physical education teachers
125	Education teachers
126	English teachers
130	Foreign language teachers
131	Home economics teachers
132	Law teachers
133	Theology teachers
134	Trade, industrial, and technical teachers
135	Miscellaneous teachers, college and university
140	Teachers, college and university, subject not specified
	Teachers, except college and university
141	Adult education teachers
N (142)	Elementary school teachers
143	Prekindergarten and kindergarten teachers
144	Secondary school teachers
145	Teachers except college and university, NEC
	Engineering and science technicians
150	Agriculture and biological technicians, except health
151	Chemical technicians
152	Draftsmen
153	Electrical and electronic engineering technicians
154	Industrial engineering technicians
155	Mechanical engineering technicians

TABLE 12.1. Occupational Categories Used for SDS Cases, From 1970 U.S. Bureau of Census System *(continued)*

Occupation Code	
156	Mathematical technicians
161	Surveyors
162	Engineering and science technicians, NEC
	Technicians, except health, and engineering and science
163	Airplane pilots
164	Air traffic controllers
165	Embalmers
170	Flight engineers
171	Radio operators
172	Tool programmers, numerical control
173	Technicians, NEC
174	Vocational and educational counselors
	Writers, artists, and entertainers
175	Actors
180	Athletes and kindred workers
181	Authors
182	Dancers
183	Designers
184	Editors and reporters
185	Musicians and composers
190	Painters and sculptors
191	Photographers
192	Public relations men and publicity writers
193	Radio and television announcers
194	Writers, artists, and entertainers, NEC
195	Research workers, not specified
	Managers and Administrators Except Farm
201	Assessors, controllers, and treasurers; local public administration
202	Bank officers and financial managers
203	Buyers and shippers, farm products
205	Buyers, wholesale and retail trade
210	Credit men
211	Funeral directors
212	Health administrators
213	Construction inspectors, public administration
215	Inspectors, except construction, public administration
216	Managers and superintendents, building
220	Office managers, NEC
221	Officers, pilots, and pursers; ship
222	Officials and administrators; public administration, NEC
223	Officials of lodges, societies, and unions
224	Postmasters and mail superintendents
225	Purchasing agents and buyers, NEC
226	Railroad conductors
230	Restaurant, cafeteria, and bar managers
231	Sales managers and department heads, retail trade
233	Sales managers, except retail trade
235	School administrators, college
240	School administrators, elementary and secondary
245	Managers and administrators, NEC
	Sales Workers
260	Advertising agents and salesmen
261	Auctioneers
262	Demonstrators
264	Hucksters and peddlers
265	Insurance agents, brokers, and underwriters

TABLE 12.1. Occupational Categories Used for SDS Cases, From 1970 U.S. Bureau of Census System *(continued)*

Occupation Code	
266	Newsboys
270	Real estate agents and brokers
271	Stock and bond salesmen
280	Salesmen and sales clerks, NEC
	Clerical and Kindred Workers
301	Bank tellers
303	Billing clerks
P (305)	Bookkeepers
310	Cashiers
311	Clerical assistants, social welfare
312	Clerical supervisors, NEC
313	Collectors, bill and account
314	Counter clerks, except food
315	Dispatchers and starters, vehicle
320	Enumerators and interviewers
321	Estimators and investigators, NEC
323	Expediters and production controllers
325	File clerks
326	Insurance adjusters, examiners, and investigators
330	Library attendants and assistants
331	Mail carriers, post office
332	Mail handlers, except post office
333	Messengers and office boys
334	Meter readers, utilities
	Office machine operators
341	Bookkeeping and billing machine operators
342	Calculating machine operators
343	Computer and peripheral equipment operators
344	Duplicating machine operators
345	Key punch operators
350	Tabulating machine operators
355	Office machine operators, NEC
360	Payroll and timekeeping clerks
361	Postal clerks
362	Proofreaders
363	Real estate appraisers
364	Receptionists
	Secretaries
370	Secretaries, legal
371	Secretaries, medical
Q (372)	Secretaries, NEC
374	Shipping and receiving clerks
375	Statistical clerks
376	Stenographers
381	Stock clerks and storekeepers
382	Teacher aides, exc. school monitors
383	Telegraph messengers
384	Telegraph operators
385	Telephone operators
390	Ticket, station, and express agents
391	Typists
392	Weighers
394	Miscellaneous clerical workers
395	Not specified clerical workers

TABLE 12.1. Occupational Categories Used for SDS Cases, From 1970 U.S. Bureau of Census System *(continued)*

Occupation Code	
	Craftsmen and Kindred Workers
401	Automobile accessories installers
402	Bakers
403	Blacksmiths
404	Boilermakers
405	Bookbinders
410	Brickmasons and stonemasons
411	Brickmasons and stonemasons, apprentices
412	Bulldozer operators
413	Cabinetmakers
R (415)	Carpenters
416	Carpenter apprentices
420	Carpet installers
421	Cement and concrete finishers
422	Compositors and typesetters
423	Printing trades apprentices, exc. pressmen
424	Cranemen, derrickmen, and hoistmen
425	Decorators and window dressers
426	Dental laboratory technicians
430	Electricians
431	Electrician apprentices
433	Electric power linemen and cablemen
434	Electrotypers and stereotypers
435	Engravers, exc. photoengravers
436	Excavating, grading, and road machine operators; exc. bulldozer
440	Floor layers, exc. tile setters
441	Foremen, NEC
442	Forgemen and hammermen
443	Furniture and wood finishers
444	Furriers
445	Glaziers
446	Heat treaters, annealers, and temperers
450	Inspectors, scalers, and graders; log and lumber
452	Inspectors, NEC
453	Jewelers and watchmakers
454	Job and die setters, metal
455	Locomotive engineers
456	Locomotive firemen
461	Machinists
462	Machinist apprentices
	Mechanics and repairmen
470	Air conditioning, heating, and refrigeration
471	Aircraft
472	Automobile body repairmen
S (473)	Automobile mechanics
474	Automobile mechanic apprentices
475	Data processing machine repairmen
480	Farm implement
481	Heavy equipment mechanics, incl. diesel
482	Household appliance and accessory installers and mechanics
483	Loom fixers
484	Office machine
485	Radio and television
486	Railroad and car shop
491	Mechanic, exc. auto, apprentices
492	Miscellaneous mechanics and repairmen
495	Not specified mechanics and repairmen
501	Millers; grain, flour, and feed
502	Millwrights

TABLE 12.1. Occupational Categories Used for SDS Cases, From 1970 U.S. Bureau of Census System *(continued)*

Occupation Code	
503	Molders, metal
504	Molder apprentices
505	Motion picture projectionists
506	Opticians, and lens grinders and polishers
510	Painters, construction and maintenance
511	Painter apprentices
512	Paperhangers
514	Pattern and model makers, exc. paper
515	Photoengravers and lithographers
516	Piano and organ tuners and repairmen
520	Plasterers
521	Plasterer apprentices
522	Plumbers and pipe fitters
523	Plumber and pipe fitter apprentices
525	Power station operators
530	Pressmen and plate printers, printing
531	Pressman apprentices
533	Rollers and finishers, metal
534	Roofers and slaters
535	Sheetmetal workers and tinsmiths
536	Sheetmetal apprentices
540	Shipfitters
542	Shoe repairmen
543	Sign painters and letterers
545	Stationary engineers
546	Stone cutters and stone carvers
550	Structural metal craftsmen
551	Tailors
552	Telephone installers and repairmen
554	Telephone linemen and splicers
560	Tile setters
561	Tool and die makers
562	Tool and diemaker apprentices
563	Upholsterers
571	Specified craft apprentices, NEC
572	Not specified apprentices
575	Craftsmen and kindred workers, NEC
580	Former members of the Armed Forces
	Operatives, Except Transport
601	Asbestos and insulation workers
T (602)	Assemblers
603	Blasters and powdermen
604	Bottling and canning operatives
605	Chainmen, rodmen, and axmen; surveying
610	Checkers, examiners, and inspectors; manufacturing
611	Clothing ironers and pressers
612	Cutting operatives, NEC
613	Dressmakers and seamstresses, except factory
614	Drillers, earth
615	Dry wall installers and lathers
620	Dyers
621	Filers, polishers, sanders, and buffers
622	Furnacemen, smeltermen, and pourers
423	Garage workers and gas station attendants
624	Graders and sorters, manufacturing
625	Produce graders and packers, except factory and farm
626	Heaters, metal

TABLE 12.1. Occupational Categories Used for SDS Cases, From 1970 U.S. Bureau of Census System *(continued)*

Occupation Code	
630	Laundry and dry cleaning operatives, NEC
631	Meat cutters and butchers, exc. manufacturing
633	Meat cutters and butchers, manufacturing
634	Meat wrappers, retail trade
635	Metal platers
636	Milliners
640	Mine operatives, NEC
641	Mixing operatives
642	Oilers and greasers, exc. auto
643	Packers and wrappers, except meat and produce
644	Painters, manufactured articles
645	Photographic process workers
	Precision machine operatives
650	Drill press operatives
651	Grinding machine operatives
652	Lathe and milling machine operatives
653	Precision machine operatives, NEC
656	Punch and stamping press operatives
660	Riveters and fasteners
661	Sailors and deckhands
662	Sawyers
663	Sewers and stitchers
664	Shoemaking machine operatives
665	Solderers
666	Stationary firemen
	Textile operatives
670	Carding, lapping, and combing operatives
671	Knitters, loopers, and toppers
672	Spinners, twisters, and winders
673	Weavers
674	Textile operatives, NEC
680	Welders and flame-cutters
681	Winding operatives, NEC
690	Machine operatives, miscellaneous, specified
692	Machine operatives, not specified
694	Miscellaneous operatives
695	Not specified operatives
	Transport Equipment Operatives
701	Boatmen and canalmen
703	Bus drivers
704	Conductors and motormen, urban rail transit
705	Deliverymen and routemen
706	Fork lift and tow motor operatives
710	Motormen; mine, factory, logging camp, etc.
711	Parking attendants
712	Railroad brakemen
713	Railroad switchmen
714	Taxicab drivers and chauffeurs
U (715)	Truck drivers
	Laborers, Except Farm
740	Animal caretakers, exc. farm
750	Carpenters' helpers
V (751)	Construction laborers, exc. carpenters' helpers
752	Fishermen and oystermen
753	Freight and material handlers
754	Garbage collectors

TABLE 12.1. Occupational Categories Used for SDS Cases, From 1970 U.S. Bureau of Census System *(continued)*

Occupation Code	
755	Gardeners and groundskeepers, exc. farm
760	Longshoremen and stevedores
761	Lumbermen, draftsmen, and woodchoppers
762	Stock handlers
763	Teamsters
764	Vehicle washers and equipment cleaners
770	Warehousemen, NEC
780	Miscellaneous laborers
785	Not specified laborers
	Farmers and Farm Managers
W (801)	Farmers (owners and tenants)
802	Farm managers
	Farm Laborers and Farm Foremen
821	Farm foremen
822	Farm laborers, wage workers
823	Farm laborers, unpaid family workers
824	Farm service laborers, self-employed
	Service Workers, Exc. Private Household
	Cleaning service workers
901	Chambermaids and maids, except private household
902	Cleaners and charwomen
X (903)	Janitors and sextons
	Food service workers
910	Bartenders
911	Busboys
912	Cooks, except private household
913	Dishwashers
914	Food counter and fountain workers
Y (915)	Waiters
916	Food service workers, NEC, except private household
	Health service workers
921	Dental assistants
922	Health aides, exc. nursing
923	Health trainees
924	Lay midwives
925	Nursing aides, orderlies, and attendants
926	Practical nurses
	Personal service workers
931	Airline stewardesses
932	Attendants, recreation and amusement
933	Attendants, personal service, NEC
934	Baggage porters and bellhops
935	Barbers
940	Boarding and lodging house keepers
941	Bootblacks
942	Child care workers, exc. private household
943	Elevator operators
944	Hairdressers and cosmetologists
945	Personal service apprentices
950	Housekeepers, exc. private household
952	School monitors
953	Ushers, recreation and amusement
954	Welfare service aides

TABLE 12.1. Occupational Categories Used for SDS Cases, From 1970 U.S. Bureau of Census System *(continued)*

Occupation Code	
	Protective service workers
960	Crossing guards and bridge tenders
961	Firemen, fire protection
962	Guards and watchmen
963	Marshals and constables
964	Policemen and detectives
965	Sheriffs and bailiffs
	Private Household Workers
980	Child care workers, private household
981	Cooks, private household
982	Housekeepers, private household
983	Laundresses, private household
Z (984)	Maids and servants, private household
995	*Occupation Not Reported*
	Allocation Categories
196	Professional, technical, and kindred workers—allocated
246	Managers and administrators, except farm—allocated
296	Sales workers—allocated
396	Clerical and kindred workers—allocated
586	Craftsmen and kindred workers—allocated
696	Operatives, except transport—allocated
726	Transport equipment operatives—allocated
796	Laborers, except farm—allocated
806	Farmers and farm managers—allocated
846	Farm laborers and farm foremen—allocated
976	Service workers, exc. private household—allocated
986	Private household workers—allocated

Source: U.S. Bureau of the Census, *1970 Census of Population, Alphabetical Index of Industries and Occupations,* U.S. Government Printing Office, Washington, D.C., 1971.
[a]Not Elsewhere Classified

these documents, the coding categories for the 1980 edition are presented in this section as Tables 12.1 to 12.6. Notes are included in these tables indicating changes made through 1985. Descriptions of the key coding categories follow.

Age—the chronological age of the injured at the time the injury occurred in whole year increments.

Sex—the gender of the injured.

Occupation—the occupation of the injured according to the 1970 Census Occupational Classification System[11] listed in Table 12.1. Beginning with the 1985 reporting year, BLS is using the 1980 Census Occupational Classification System (COCS).[12]

Industry—the principal business of the injured or ill worker's employer based on the 1972 SIC Manual[13] as amended by the 1977 Supplement.[14] The major (two-digit) SIC categories are listed in Table 12.2. Variations in level of coding exist for public vis-à-vis private-sector employers and between states. The User's Guide[15] specifies that (1) *private-sector* coding must be at the four-digit level for manufacturing and three-digit level, at a minimum, nonmanufacturing; and (2) *public-sector* coding at the four-digit level for manufacturing and three-digit level for nonmanufacturing are desired, but three-digit for manufacturing and two-digit for nonmanufacturing are acceptable. Public-sector industries requiring a four-digit SIC are Police Protection—SIC 9221; Correctional Institutions—SIC 9223; Fire Protection—SIC 9224; Refuse Systems—SIC 4953; and Garbage and Refuse, Collecting and Transporting Without Disposal—SIC 4212.

Ownership—the employer is classified into one of three categories: private sector, state government, or local government.

TABLE 12.2. Industrial Classification of the Employer for Two-digit Codes from the Standard Industrial Classification Manual

Division	Major Group
A. Agriculture, forestry, and fishing	01. Agricultural production—crops
	02. Agricultural production—livestock
	07. Agricultural services
	08. Forestry
	09. Fishing, hunting, and trapping
B. Mining	10. Metal mining
	11. Anthracite mining
	12. Bituminous coal and lignite mining
	13. Oil and gas extraction
	14. Mining and quarrying of nonmetallic minerals, except fuels
C. Construction	15. Building construction—general contractors and operative builders
	16. Construction other than building construction—general contractors
	17. Construction—special trade contractors
D. Manufacturing	20. Food and kindred products
	21. Tobacco manufacturers
	22. Textile mill products
	23. Apparel and other finished products made from fabrics and similar materials
	24. Lumber and wood products, except furniture
	25. Furniture and fixtures
	26. Paper and allied products
	27. Printing, publishing, and allied industries
	28. Chemicals and allied products
	29. Petroleum refining and related industries
	30. Rubber and miscellaneous plastics products
	31. Leather and leather products
	32. Stone, clay, glass, and concrete products
	33. Primary metal industries
	34. Fabricated metal products, except machinery and transportation equipment
	35. Machinery, except electrical
	36. Electrical and electronic machinery, equipment, and supplies
	37. Transportation equipment
	38. Measuring, analyzing, and controlling instruments; photographic, medical, and optical goods; watches and clocks
	39. Miscellaneous manufacturing industries
E. Transportation, communications, electric, gas, and sanitary services	40. Railroad transportation
	41. Local and suburban transit and interurban highway passenger transportation
	42. Motor freight transportation and warehousing
	43. U.S. Postal Service
	44. Water transportation
	45. Transportation by air
	46. Pipelines, except natural gas
	47. Transportation services
	48. Communication
	49. Electric, gas and sanitary services
F. Wholesale trade	50. Wholesale trade—durable goods
	51. Wholesale trade—nondurable goods
G. Retail trade	52. Building materials, hardware, garden supply, and mobile home dealers
	53. General merchandise stores
	54. Food stores
	55. Automotive dealers and gasoline service stations
	56. Apparel and accessory stores

TABLE 12.2. Industrial Classification of the Employer for Two-digit Codes from the Standard Industrial Classification Manual *(continued)*

Division	Major Group
	57. Furniture, home furnishings, and equipment stores
	58. Eating and drinking places
	59. Miscellaneous retail
H. Finance, insurance, and real estate	60. Banking
	61. Credit agencies other than banks
	62. Security and commodity brokers, dealers, exchanges, and services
	63. Insurance
	64. Insurance agents, brokers, and service
	65. Real estate
	66. Combinations of real estate, insurance, loans, law offices
	67. Holding and other investment offices
I. Services	70. Hotels, rooming houses, camps, and other lodging places
	72. Personal services
	73. Business services
	75. Automotive repair, services, and garages
	76. Miscellaneous repair services
	78. Motion pictures
	79. Amusement and recreation services, except motion pictures
	80. Health services
	81. Legal services
	82. Educational services
	83. Social services
	84. Museums, art galleries, botanical and zoological gardens
	86. Membership organizations
	88. Private households
	89. Miscellaneous services
J. Public administration	91. Executive, legislative, and general government, except finance
	92. Justice, public order, and safety
	93. Public finance, taxation, and monetary policy
	94. Administration of human resources programs
	95. Administration of environmental quality and housing programs
	96. Administration of economic programs
	97. National security and international affairs
K. Nonclassifiable establishments	99. Nonclassifiable establishments

Source: U.S. Office of Management and Budget, *Standard Industrial Classification Manual 1972,* U.S. Government Printing Office, Washington, D.C., 1980.

Nature—the nature of the injury or illness indicates the principal physical or mental characteristics of the injury or illness according to the codes listed in Table 12.3.

Part of body—the part of the injured or ill employee's body directly affected by the injury or illness according to the coding categories listed in Table 12.4.

Source—the source of the injury or illness identifies the object, substance, or bodily motion that directly produced or inflicted the injury or illness according to the categories listed in Table 12.5.

Type—the type of accident or exposure identifies the event that directly resulted in the injury or illness according to the categories listed in Table 12.6.

Year and month of occurrence—the year and month of injury or the diagnosis or onset of the illness.

The use of coded compensation records requires a strategy for selecting the cases of interest from a larger group of cases. For example, a particular analysis may call for a search of all cases in a particular state's file from reference year 1980. The first step is to find out if the state maintains such records in a coded format. Table 12.7 lists the states that participated in the SDS in 1979, 1980, 1981, or 1983. If the state is among these, it should be possible to get the state workers' compensation agency to run a search

TABLE 12.3. Nature of the Injury or Illness Used for SDS Cases, from the Bureau of Labor Statistics

Nature Code	Title and Notes
100	*Amputation or enucleation*
110	*Asphyxia, strangulation, etc.* Includes drowning and suffocation.
120	*Burn (heat)* The effect of contact with hot substances. Includes scalds and electric burns, but not electric shock. Does not include chemical burns, friction burns, effects of radiation, sunburn, or systemic disability such as heat stroke, etc.
130	*Burn (chemical)* Tissue damage resulting from the corrosive action of chemicals, chemical compounds, fumes, etc. (acids, alkalies). Includes skin, fry, anhydrous ammonia, and cement burns.
140	*Concussion* Brain, cerebral. Includes loss of consciousness as result of blow to head.
150-159	*Infective or parasitic disease*
150	Infective or parasitic disease, UNS.[a]
151	Amebiasis.
152	Anthrax.
153	Brucellosis.
154	Conjunctivitis and ophthalmia.
156	Tetanus.
157	Tuberculosis.
159	Other infective or parasitic disease (includes septicemia).
160	*Contusion, crushing, bruise* Applies to intact skin surface.
170	*Cut, laceration, puncture* An open wound. Includes injection of paint, grease, or water under pressure.
180-189	*Dermatitis:* Rash, skin or tissue inflammation, including boils, etc. Generally resulting from direct contact with irritants or sensitizing chemicals such as drugs, oils, biologic agents, plants, woods, or metals, which may be in the form of solids, pastes, liquids, or vapors and which may be contacted in the pure state or in compounds or in combination with other materials. Does not include skin or tissue damage resulting from corrosive action of chemicals, burns from contact with hot substances, effects of exposure or radiation, effects of exposure to low temperatures, or inflammation or irritation resulting from friction or impact.
180	Dermatitis, UNS.
181	Contact dermatitis (includes skin infections) (use code 185 beginning 1985).
182	Allergenic dermatitis (secondary to dermatitis) (use code 185 beginning 1985).
183	Primary infections of the skin (includes boils, carbuncles, cellulitis, abcess, lymphadenitis, impetigo, and pyoderma).
184	Other skin conditions (includes toxic erythema, lichen, and pruritus).
185	Dermatitis, allergenic or contact (category first used in 1985).
189	Skin condition, NEC.[b]
190	*Dislocation* Includes herniated or ruptured disc.
200	*Electric shock electrocution*
210	*Fracture*
220	*Effects of exposure to low temperature* Includes freezing and frostbite.
230	Hearing loss or impairment A separate injury, not the sequela of another injury.
240	*Effects of environmental heat* Includes heat stroke, sunstroke, heat cramps, and heat exhaustion. Does not include sunburn or other effects of radiation.

TABLE 12.3. **Nature of the Injury or Illness Used for SDS Cases, from the Bureau of Labor Statistics *(continued)***

Nature Code	Title and Notes
250	*Hernia rupture* Includes both inguinal and noninguinal hernias. Excludes herniated or ruptured disc (see category 190). Excludes ruptured muscles or cartilage (see category 310).
260	*Inflammation of joints, etc.* Includes bursitis, synovitis, tenosynovitis, and other conditions affecting joints, tendons, or muscles. Does not include strains, sprains, or dislocation of muscles or tendons, or their aftereffects. Select only if condition occurred over time as result of repetitive activity.
270-279	*Poisoning, systemic* A systemic morbid condition resulting from the inhalation, ingestion, or skin absorption of a toxic substance affecting the functioning of a body system. Includes chemical or drug poisoning, metal poisoning, organic diseases, and venomous reptile and insect bites. Does not include effects of radiation, pneumoconiosis, corrosive effects of chemicals, skin-surface irritations, septicemia, or infected wounds.
270	Poisoning, systemic, UNS.
271	Due to toxic materials.
272	Diseases of the blood and blood-forming organs (includes anemia and purpura).
273	Upper respiratory conditions (includes rhinitis, pharyngitis, sinusitis, and hay fever).
274	Influenza, pneumonia, etc. (includes bronchitis, asthma, pneumonitis, and emphysema).
275	Toxic hepatitis.
276	Other diseases of the gastrointestinal tract (includes gingivitis, gastroenteritis, and colitis).
279	Other toxic effects of one system only (musculoskeletal and connective tissue, e.g., osteoarthritis, rheumatism, myalgia, bursitis, tenosynovitis; circulatory system, e.g., heart, veins, arteries; nervous system, e.g., cerebral paralysis neuralgia, neuritis; the genitourinary system, e.g., nephritis, nephrosis, diseases of the urinary tract; symptoms and ill-defined conditions).
280-289	*Pneumoconiosis*
280	Pneumoconiosis, UNS.
281	Aluminosis.
282	Anthracosis.
283	Asbestosis.
284	Byssinosis.
285	Siderosis.
286	Silicosis.
287	Other pneumoconioses.
289	Pneumoconiosis with tuberculosis.
290-295	*Radiation effects* Sunburn and all forms of damage to tissue, bones, or body fluids produced by exposure to radiation.
290	Radiation effects UNS.
291	Non-ionizing radiation (includes sunburn, ultraviolet burns (not welder's flash), but not heat stroke, exhaustion, microwave radiation effects, etc.).
292	Microwave.
293	Ionizing radiation—X-ray.
294	Ionizing radiation—isotopes.
295	Welder's flash.
300	*Scratches, abrasions* Superficial wounds, including foreign body in eye even though first report indicates "no injury," and friction burns.
310	*Sprains, strains* Includes torn ligaments, ruptured muscles, and ruptured knee cartilage (in the absence of a clear statement of dislocation).
320	*Hemorrhoids.*
330	*Hepatitis (serum and infective).*
400	*Multiple injuries.*

TABLE 12.3. Nature of the Injury or Illness Used for SDS Cases, from the Bureau of Labor Statistics *(continued)*

Nature Code	Title and Notes
500	*Effects of changes in atmospheric pressure* Includes aero-otitis media and caisson disease.
510	*Cerebrovascular and other conditions of the circulatory system* Includes strokes and varicose veins (nontoxic). Excludes heart and hemorrhoids.
520	*Complications peculiar to medical care* Toxic and nontoxic.
530	*Eye, other diseases of the eye* Includes chalazion, chemical conjunctivitis, and conjunctivitis due to other toxic materials.
540	*Mental disorders* Includes acute anxiety and depression, and neurosis. Toxic and nontoxic.
550–552	*Neoplasm, tumor* Toxic and nontoxic.
550	Neoplasm, tumor UNS.
551	Malignant. Includes cancer and leukemia.
552	Benign.
560–562	*Nervous system, nontoxic conditions of*
560	Nervous system, conditions of, UNS.
561	Diseases of the central nervous system (includes cerebral paralysis and migraine).
562	Diseases of the nerves and peripheral ganglia (includes Bell's palsy and carpal tunnel syndrome).
570–572	*Respiratory system, nontoxic conditions of*
570	Respiratory system, conditions of, UNS.
571	Upper respiratory (includes rhinitis).
572	Influenza, pneumonia, etc. (includes bronchitis and asthma).
580	*Symptoms and ill-defined conditions* Nontoxic (see category 279 for toxic conditions).
900	*No injury or illness*
950	*Damage to prosthetic devices* Includes eyeglasses, false teeth, etc.
990	*Occupational disease, NEC*
991	*Heart condition (Includes heart attack)* Nontoxic.
995	*Other Injury, NEC*
999	*Nonclassifiable*

Source: U.S. Bureau of Labor Statistics, *Supplementary Data System, Microdata Files User's Guide, 1980 Edition,* Document Number PB83-133553, National Technical Information Service, Springfield, Virginia, 1982.

[a]UNS = Unspecified

[b]NEC = Not Elsewhere Classified

of their coded records according to the desired specifications. If the state does not participate in the SDS, the state workers' compensation agency will need to be contacted to find out if they have a system for storage and retrieval of case information, and if they do, how a search can be obtained. Assistance with multistate searches may be obtained by writing to the Division of Safety Research, NIOSH, 944 Chestnut Ridge Road, Morgantown, West Virginia 26505.

Whether contacting a state or NIOSH, a description of the kind of cases you want included in your search is needed. A verbal description may be used; however, experience indicates that this approach nearly always results in more computer runs and delays than necessary. The preferred approach is to specify the characteristics of the desired cases in terms of the code categories shown in Tables 12.1 to 12.6.

In addition, the request should indicate if the cases of interest must satisfy each and every characteristic provided (the "and" statement), or if satisfying one of the stated categories is sufficient (the "or"

TABLE 12.4. The Injured or Ill Part of Body Used for SDS Cases, from the Bureau of Labor Statistics

Code	Part of Body
100	Head, UNS[a]
110	Brain
120	Ear(s)
121	Ear(s), external
124	Ear(s), internal (includes hearing)
130	Eye(s) (includes optic nerve and vision)
140	Face, UNS
141	Jaw (includes chin)
144	Mouth (includes lips, teeth, tongue, throat, and taste)
146	Nose (includes nasal passages, sinus, and sense of smell)
148	Face, multiple parts (any combination of above parts)
149	Face, NEC[b]
150	Scalp
160	Skull
198	Head, multiple (any combination of above parts)
199	Head, NEC (beginning 1985 no longer used)
200	Neck, including cervical vertebrae
300	Upper extremities, UNS
310	Arm(s), UNS
311	Upper arm(s)
313	Elbow(s)
315	Forearm(s)
318	Arm(s), multiple (any combination of above parts)
319	Arm, NEC
320	Wrist
330	Hand (not wrist or fingers)
340	Finger(s)
398	Upper extremities, multiple (any combination of above parts)
399	Upper extremities, NEC (beginning 1985 no longer used)
400	Trunk, UNS
410	Abdomen (includes internal organs)
420	Back (includes back muscles, spine, and spinal cord)
430	Chest (includes ribs, breastbone, and internal organs of the chest)
440	Hips (includes pelvis, pelvic organs, and buttocks)
450	Shoulder(s)
498	Trunk, multiple (any combination of above parts)
499	Trunk, NEC (beginning 1985 no longer used)
500	Lower extremities, UNS
510	Leg(s) (above ankle)
511	Thigh(s)
513	Knee(s)
515	Lower leg(s)
518	Leg(s), multiple (any combination of above parts)
519	Leg(s), NEC
520	Ankle(s)
530	Foot/feet (not ankle or toes)
540	Toe(s)
598	Lower extremities, multiple (any combination of parts of the lower extremities)
599	Lower extremities, NEC (beginning 1985 no longer used)
700	Multiple parts (applies when more than one major body part has been affected, such as an arm and a leg)
800	Body system (applies when the functioning of an entire body system has been affected without specific injury to any other part, as in the case of poisoning, corrosive action affecting internal organs, damage to nerve centers, etc. Does not apply when the systemic damage results from an external injury affecting an external part, such as a back injury that includes damage to the nerves of the spinal cord)
801	Circulatory system (heart, blood, arteries, veins, etc.)
810	Digestive system
820	Excretory system (kidneys, bladder, intestines, etc.)

TABLE 12.4. The Injured or Ill Part of Body Used for SDS Cases, from the Bureau of Labor Statistics *(continued)*

Code	Part of Body
830	Musculoskeletal system (bones, joints, tendons, muscles, etc.)
840	Nervous system
850	Respiratory system (lungs, etc.)
880	Other body systems
900	Body parts, NEC
999	Nonclassifiable (usually due to insufficient information)

Source: U.S. Bureau of Labor Statistics, *Supplementary Data System, Microdata Files User's Guide, 1980 Edition,* Document Number PB83-133553, National Technical Information Service, Springfield, Virginia, 1982.

[a]UNS = Unspecified

[b]NEC = Not Elsewhere Classified

TABLE 12.5. Source of Injury or Illness Coding Categories Used in the SDS System, from the Bureau of Labor Statistics

Major Group	Source Code	Title
01		*Air pressure*
	0100	Air pressure, UNS[a]
	0101	High pressure
	0120	Low pressure
02		*Animals, insects, birds, reptiles*
	0200	Animals, insects, birds, reptiles, UNS
	0201	Animals
	0230	Birds
	0250	Insects
	0270	Reptiles
03		*Animal products*
	0300	Animal products, UNS
	0301	Bones
	0310	Feathers[b]
	0320	Fur, hair, wool, etc. (raw)[b]
	0330	Hides, leather
	0399	Animal products, NEC[c]
04	0400	*Bodily motion*
05		*Boilers, pressure vessels*
	0500	Boilers, pressure vessels, UNS
	0501	Boilers
	0510	Pressurized containers
	0530	Pressure lines
	0599	Pressure vessels, NEC
06		*Boxes, barrels, containers, packages*[d]
	0600	Boxes, barrels, containers, packages, UNS
	0601	Barrels, kegs, drums
	0630	Boxes, crates, cartons
	0650	Bottles, jugs, flasks, etc.
	0660	Bundles, bales
	0665	Reels, rolls
	0670	Tanks, bins, etc. (not pressurized)
	0699	Containers, NEC[d]

TABLE 12.5. Source of Injury or Illness Coding Categories Used in the SDS System, from the Bureau of Labor Statistics *(continued)*

Major Group	Source Code	Title
07		*Buildings and structures*
	0700	Buildings and structures, UNS
	0701	Buildings, office, plant, residential, etc.
	0705	Doors, gates
	0708	Windows, window frames
	0710	Bridges[e]
	0720	Dams, locks, etc.[e]
	0730	Grandstands, stadia, etc.
	0740	Scaffolds, staging, etc.
	0750	Towers, poles, etc.
	0755	Walls, fences
	0760	Wharfs, docks, etc.
	0799	Buildings and structures, NEC
08		*Ceramic items*
	0800	Ceramic items, UNS
	0801	Brick
	0810	China
	0820	Drain tile, sewer pipe, flue lining, etc.
	0830	Glazed tile, (decorative—not load bearing)
	0840	Pottery[f]
	0850	Structural tile (glazed or plain)
	0899	Ceramic items, NEC
09		*Chemicals, chemical compounds*[g]
	0900	Chemicals, chemical compounds, UNS
	0901	Acids
	0905	Alcohols
	0910	Alkalies
	0915	Aromatic compounds
	0920	Arsenic compounds
	0925	Carbon bisulphide
	0930	Carbon dioxide
	0935	Carbon monoxide
	0940	Carbon tetrachloride
	0945	Cyanides or cyanogen compounds
	0950	Halogenated compounds, NEC
	0955	Metallic compounds, NEC
	0960	Oxides of nitrogen
	0965	Paint
	0999	Chemicals and compounds, NEC
10		*Clothing, apparel, shoes*
	1000	Clothing, apparel, shoes, UNS
	1001	Boots, shoes, etc.
	1010	Gloves
	1020	Hats, head coverings
	1030	Outer coats, raincoats
	1040	Shirts, blouses, sweaters, coats
	1050	Suits, pants, coveralls, dresses
	1060	Stockings, socks
	1070	Underwear
	1099	Apparel, NEC[h]
11		*Coal and petroleum products*
	1100	Coal and petroleum products, UNS
	1101	Coal
	1110	Coke
	1120	Crude oil, fuel oil
	1130	Gasoline, liquid hydrocarbon compounds
	1140	Hydrocarbon gases
	1150	Kerosine
	1160	Lubricating and cutting oils, greases

TABLE 12.5. Source of Injury or Illness Coding Categories Used in the SDS System, from the Bureau of Labor Statistics *(continued)*

Major Group	Source Code	Title
	1170	Manufactured gases
	1180	Naptha solvents
	1190	Petroleum asphalts and road oils
	1199	Coal and petroleum products, NEC
12	1200	*Cold (atmospheric, environmental)*
13		*Conveyors*
	1300	Conveyors, UNS
	1301	Gravity conveyors
	1350	Powered conveyors
14		*Drugs and medicines*
	1400	Drugs and medicines, UNS
	1401	Biologic products
	1490	Other medicinals
15		*Electric apparatus*
	1500	Electric apparatus, UNS
	1501	Motors
	1505	Generators
	1510	Transformers, converters
	1515	Conductors
	1520	Switchboard and bus structures, switches, circuit breakers, fuses
	1530	Rheostats, starters, control apparatus, capacitators, rectifiers, storage batteries
	1540	Magnetic and electrolytic apparatus
	1599	Electric apparatus, NEC
17	1700	*Flame, fire, smoke*
18		*Food products*
	1800	Food products, UNS
	1810	Fruits and fruit products
	1820	Grains and grain products
	1840	Meat, poultry, seafood, and their products
	1850	Milk and milk products
	1870	Vegetables and vegetable products
	1890	Food products, NEC
19		*Furniture, fixtures, furnishings*
	1900	Furniture, fixtures, furnishings, UNS
	1901	Cabinets, file cases, bookcases, etc.
	1910	Chairs, benches, etc.
	1920	Counters, work benches, etc.
	1930	Desks
	1950	Floor coverings, carpets, rugs, mats, etc.
	1960	Lighting equipment, lamps, bulbs, etc.
	1970	Tables
	1999	Furniture, fixtures, furnishings, NEC
20	2000	*Glass items, NEC*
22		*Handtools, not powered*
	2200	Handtools, not powered, UNS
	2201	Axe
	2205	Blowtorch
	2210	Chisel
	2215	Crowbar, prybar
	2220	File
	2225	Fork
	2230	Hammer, sledge, mallet
	2235	Hatchet
	2240	Hoe
	2245	Knife

TABLE 12.5. Source of Injury or Illness Coding Categories Used in the SDS System, from the Bureau of Labor Statistics *(continued)*

Major Group	Source Code	Title
	2250	Pick
	2255	Plane
	2260	Pliers, tongs
	2265	Punch
	2270	Rope, chain
	2275	Saw
	2280	Scissors, shears
	2285	Screwdriver
	2290	Shovel, spade
	2295	Wrench
	2299	Handtools, not powered, NEC
23		*Handtools, powered*
	2300	Handtools, powered, UNS
	2301	Abrasive stone or wheel grinder
	2305	Buffer, polisher, waxer
	2310	Chisel
	2315	Drill
	2320	Hammer, tamper
	2325	Ironer
	2330	Knife
	2335	Powder-actuated tools
	2340	Punch
	2345	Riveter
	2350	Sandblaster
	2355	Saw
	2360	Screwdriver, bolt setter
	2370	Welding tools
	2399	Handtools, powered, NEC
24	2400	*Heat (atmospheric, environmental)*
25	2500	*Heating equipment, NEC*
26		*Hoisting apparatus*
	2600	Hoisting apparatus, UNS
	2610	Cranes, derricks
	2620	Elevators
	2630	Shovels, dredges
	2640	Other hoisting apparatus, UNS[i]
	2641	Air hoists[i]
	2642	Chain hoists, chain blocks[i]
	2643	Electric hoists[i]
	2644	Gin poles[i]
	2645	Jacks (mechanical, hydraulic, air, etc.)[i]
	2646	Jammer (logging)[i]
	2647	Mine buckets[i]
	2699	Hoisting apparatus, NEC
27	2700	*Infectious, parasitic agents, NEC*
28		*Ladders*
	2800	Ladders, UNS
	2810	Fixed
	2830	Movable, UNS
	2831	Extension ladders
	2833	Step ladders
	2835	Straight ladders
	2899	Ladders, NEC
29		*Liquids*
	2900	Liquids, UNS
	2910	Water
	2999	Other liquids, NEC

TABLE 12.5. Source of Injury or Illness Coding Categories Used in the SDS System, from the Bureau of Labor Statistics *(continued)*

Major Group	Source Code	Title
30		*Machines*
	3000	Machines, UNS
	3001	Agitators, mixers, tumblers, etc.
	3050	Agricultural machines, NEC
	3100	Buffers, polishers, sanders, grinders
	3150	Casting, forging, welding machines
	3200	Crushing, pulverizing, etc., machines
	3250	Drilling, boring, and turning machines
	3300	Earthmoving and highway construction machines, NEC
	3350	Mining and tunneling machines, NEC
	3400	Office machines
	3450	Packaging, wrapping machines
	3500	Picking, carding, and combing machines
	3550	Planers, shapers, molders
	3600	Presses (not printing)
	3650	Printing machines
	3700	Rolls
	3750	Saws
	3800	Screening, separating machines
	3850	Shears, slitters, slicers
	3900	Stitching, sewing machines
	3950	Weaving, knitting, spinning machines
	3999	Machines, NEC
40		*Mechanical power transmission apparatus*
	4000	Mechanical power transmission apparatus, UNS
	4010	Belts[j]
	4020	Chains, ropes, cables[j]
	4030	Drums, pulleys, sheaves[j]
	4040	Friction clutches[j]
	4050	Gears[j]
	4099	Mechanical power transmission apparatus, NEC[j]
		Reference Years 1976 through 1982
41		*Metal items*
	4100	Metal items, UNS
	4110	Automobile parts (metal)
	4115	Beams, bars
	4120	Bullets
	4125	Molds
	4130	Molten metal
	4135	Nails, spikes, etc.
	4140	Pipe
	4145	Screws, nuts, bolts
	4199	Metal items, NEC
		Beginning Reference Year 1983
41		*Metal items*
	4100	Metal items, UNS
	4110	Automobile parts
	4115	Structural metal
	4120	Bullets, pellets
	4125	Dies, molds, patterns
	4130	Molten metal
	4140	Pipe and fittings
	4150	Metal parts (except automobile)
	4155	Metal fasteners
	4160	Metal binders

TABLE 12.5. Source of Injury or Illness Coding Categories Used in the SDS System, from the Bureau of Labor Statistics *(continued)*

Major Group	Source Code	Title
	4165	Metal chips, splinters, particles
	4199	Metal items, NEC
42	4200	*Mineral items, metallic, NEC*
43	4300	*Mineral items, nonmetallic, NEC*
44	4400	*Noise*
45	4500	*Paper and pulp items, NEC*
46	4600	*Particles (unidentified)*
47	4700	*Plants, trees, vegetation*
48	4800	*Plastic items, NEC*
49		*Pumps and prime movers*
	4900	Pumps and prime movers, UNS
	4910	Engines
	4930	Pumps
	4950	Turbines
50		*Radiating substances and equipment*
	5000	Radiating substances and equipment, UNS
	5010	Isotopes or irradiated substances for industrial or medical use
	5020	Radium
	5030	Reactor fuel, raw or processed
	5040	Reactor wastes
	5050	Sun
	5060	Ultraviolet equipment
	5070	Welding equipment, electric arc
	5080	X-ray and fluoroscope equipment
	5099	Radiating substances and equipment, NEC
51	5100	*Soaps, detergents, cleaning compounds, NEC*[k]
52	5200	*Silica*
53	5300	*Scrap, debris, waste materials, NEC*
54	5400	*Steam*
55	5500	*Textiles items, NEC*
56		*Vehicles*
	5600	Vehicles, UNS
	5601	Animal-drawn
	5610	Aircraft
	5620	Highway vehicles, powered
		Plant or industrial vehicles
	5630	Plant or industrial vehicles, UNS
	5631	Handtrucks, dollies, and other nonpowered vehicles
	5635	Forklifts, stackers, lumber carriers, and other powered carriers
	5638	Mules, tractors, and other powered towing vehicles
	5640	Rail vehicles
	5650	Sleds, snow and ice vehicles
	5660	Water vehicles
	5699	Vehicles, NEC
57		*Wood items*
	5700	Wood items, UNS
	5710	Logs
	5720	Lumber
	5730	Skids, pallets
	5749	Wood items, NEC
58		*Working surfaces*
	5800	Working surfaces, UNS
	5801	Floor
	5810	Ground (outdoors)

TABLE 12.5. Source of Injury or Illness Coding Categories Used in the SDS System, from the Bureau of Labor Statistics *(continued)*

Major Group	Source Code	Title
	5815	Ramps
	5820	Roofs
	5825	Runways or platforms
	5830	Sidewalks, paths, walkways (outdoors)
	5840	Stairs, steps
	5845	Street, road
	5899	Working surfaces, NEC
60		*Person*
	6000	Person, UNS
	6010	Person, injured
	6020	Person, other than injured
61	6100	*Recreation and athletic equipment*
62		*Rubber products*
	6200	Rubber products, UNS
	6210	Tires
	6299	Rubber products, NEC
65	6500	*Ice, snow (not working surface)*
88	8800	*Miscellaneous, NEC*
98	9800	*Nonclassifiable*

Source: U.S. Bureau of Labor Statistics, *Supplementary Data System, Microdata Files User's Guide, 1980 Edition,* Document Number PB83-133553, National Technical Information Service, Springfield, Virginia, 1982.

[a]UNS = Unspecified.

[b]Beginning reference year 1985, codes 310 and 320 are deleted. Use code 0399.

[c]Not Elsewhere Classified.

[d]Beginning reference year 1985, new codes added: 605, bags, sacks; 610, pots, pans, dishes, trays; 615, cans; 620, pails, buckets, baskets.

[e]Beginning reference year 1985, codes 0710 and 0720 deleted. Use code 0799.

[f]Beginning reference year 1985, code 0840 deleted. Use code 0899.

[g]Beginning reference year 1985, extensive revision of codes for chemicals.

[h]Beginning reference year 1985, codes 1020, 1030, 1040, 1060, and 1070 deleted. Use code 1099.

[i]Beginning reference year 1985, codes 2640, 2641, 2642, 2643, 2644, 2645, 2646, and 2647 deleted. Replaced by 2625, hoists (air, chain, electric); 2650, jacks (mechanical, hydraulic, air, etc.).

[j]Beginning reference year 1985, codes 4010–4099 deleted. Use code 4000.

[k]Beginning reference year 1985, code 5100 deleted. Use code 0991.

statement). An easy way to organize a search strategy is to construct a statement similar to an algebraic statement. The following notation is convenient.

Symbol	Coding Dimension Represented
A	Age
G	Gender
O	Occupation
I	Industry
N	Nature
P	Part of body
S	Source
T	Type
YR	Reference year
YO	Year of occurrence
M	Month of occurrence

An example is the easiest way to explain the approach. Suppose you want to know about cases of back pain that occurred in 1980 among registered nurses (RNs). A search strategy similar to that used by

TABLE 12.6. Type of Event or Exposure Coding Categories Used for SDS Cases, from the Bureau of Labor Statistics

Major Group	Type Code	Title
01		*Struck against* Applies to cases in which the injury was produced by impact between the injured person and the source of injury, the motion producing the contact being primarily that of the person rather than of the source of injury, except when the motion of the person was generated in a fall. Includes cases of bumping into objects, stepping on objects, kicking objects, being pushed or thrown against objects, etc. Does not include cases of jumping from elevations.
	010	Struck against, UNS[a]
	011	Stationary object
	012	Moving object
02		*Struck by* Applies to cases in which the injury was produced by impact between the injured person and the source of injury, the motion producing the contact being primarily that of the source of injury rather than of the person.
	020	Struck by, UNS
	021	Falling object
	022	Flying object
	029	Struck by, NEC[b]
03		*Fall from elevation* Includes jumping from elevations. Applies to cases in which the injury was produced by impact between the injured person and the source of injury, the motion producing the contact being that of the person, under the following circumstances: (1) The motion of the person and the force of impact were generated by gravity. (2) The point of contact with the source of injury was lower than the surface supporting the person at the inception of the fall.
	030	Fall from elevation, UNS
	031	From scaffolds, walkways, platforms, etc.
	032	From ladders
	033	From piled materials
	034	From vehicles
	035	On stairs
	036	Into shafts, excavations, floor openings, etc.
	039	Fall to lower level, NEC
05		*Fall on same level* Applies to cases in which the injury was produced by impact between the injured person and an external object, the motion producing the contact being that of the person, under the following circumstances: (1) The motion of the person was generated by gravity following a loss of equilibrium and ability to maintain an upright position. (2) The point of contact with the source of injury was at the same level or above the surface supporting the person at the inception of the fall.
	050	Fall on same level, UNS
	051	Fall to the walkway, or working surface
	052	Fall onto or against objects
	059	Fall on same level, NEC
06		*Caught in, under, or between* Applies to nonimpact cases in which the injury was produced by squeezing, pinching, or crushing between a moving object and a stationary object, between two moving objects, or between parts of an object. Does not apply when the source of injury is a freely flying or falling object.
	060	Caught in, under, or between, UNS
	061	Inrunning or meshing objects
	062	A moving and a stationary object
	063	Two or more moving objects (not meshing)
	064	Collapsing materials
	069	Caught in, under, or between, NEC

TABLE 12.6. Type of Event or Exposure Coding Categories Used for SDS Cases, from the Bureau of Labor Statistics *(continued)*

Major Group	Type Code	Title
08		*Rubbed or abraded*
		Applies to nonimpact cases in which the injury was produced by pressure, vibration, or friction between the person and the source of injury.
	080	Rubbed or abraded, UNS
	081	By leaning, kneeling, or sitting on objects
	082	By objects being handled
	083	By vibrating objects
	084	By foreign matter in eyes
	085	By repetition of pressure
	089	Rubbed or abraded, NEC
10	100	*Bodily reaction*
		Applies to nonimpact cases in which the injury resulted solely from a free bodily motion which imposed stress or strain upon some part of the body. Generally applies to the occurrence of strains, sprains, ruptures, or other internal injuries resulting from the assumption of an unnatural position or involuntary motions induced by sudden noise, fright, or efforts to recover from slips or loss of balance. Includes cases involving muscular or internal injury resulting from the execution of personal movements such as walking, climbing, running, reaching, turning, bending, etc., when such movement in itself was the source of injury.
		Does not apply to overexertion cases involving the lifting, pulling, or pushing of objects, or to cases in which a bodily movement, voluntary or involuntary, results in forcible contact with an external object.
12		*Overexertion*
		Applies to nonimpact cases in which the injury resulted from excessive physical effort, as in lifting, pulling, pushing, wielding, or throwing the source of injury. Includes conditions resulting from repetitive motion in use of hand tools.
	120	Overexertion, UNS
	121	In lifting objects
	122	In pulling or pushing objects
	123	In holding, wielding, throwing, or carrying objects
	129	Overexertion, NEC
13	130	*Contact with electric current*
		Applies only to nonimpact cases in which the injury consisted of electric shock, or electrocution. Includes cases of electric burns.
15		*Contact with temperature extremes*
		Applies only to nonimpact cases in which the injury consisted of a burn, heat exhaustion, freezing, frostbite, etc., resulting from contact with, or exposure to, hot or cold objects, air, gases, vapors, or liquids.
		Does not apply in cases of injury from contact with sun's rays or other radiations, or injury by the toxic or caustic characteristics of chemicals, etc.
	150	Contact with temperature extremes, UNS
	151	General heat, atmosphere or environment
	152	General cold, atmosphere or environment
	153	Hot objects or substances
	154	Cold objects or substances
18		*Contact with radiations, caustics, toxic and noxious substances*
		Applies only to nonimpact cases in which the injury resulted from the inhalation, absorption (skin contact), or ingestion of harmful substances. Generally covers the occurrence of injuries of the following natures: drowning, poisoning, chemical burns or irritations, allergic reactions, contagious diseases, and damage to the body from exposure to the sun's rays or other radiations.
		Does not apply in cases of burns or freezing from contact with temperature extremes, or of burns from contact with electric current.
	180	Contact with radiations, caustics, toxic and noxious substances, UNS
	181	By inhalation
	182	By ingestion
	183	By absorption
	189	Contact with radiations, caustics, toxic and noxious substances, NEC

TABLE 12.6. Type of Event or Exposure Coding Categories Used for SDS Cases, from the Bureau of Labor Statistics *(continued)*

Major Group	Type Code	Title
20		*Transportation accidents, other than motor vehicle* Code for the type of vehicle in which the injured was an occupant. Applies only to cases in which the injury resulted from an accident to a transportation vehicle, other than a motor vehicle, or from the movement of such vehicle, in which the injured was an occupant, such as a train collision; an airplane crash; a ship sinking, grounding, etc. Does not apply to accidents of a personal nature which occur aboard a transportation vehicle, other than a motor vehicle, such as a fall while moving about in the vehicle or simply striking against a part of the vehicle during normal movement in the vehicle. However, if the occupant is thrown to the floor or against a part of the vehicle as a result of a sudden stopping, starting, or lurching of the vehicle, the case falls in this classification.
	200	Transportation accidents, other than motor vehicle, UNS
	201	Aircraft accident
	205	Ship or boat accident
	207	Streetcar or subway accident
	211	Train accident
	298	Transportation accidents, other than motor vehicle, NEC
30		*Motor vehicle accidents* Applies only to cases in which the injured person was an occupant in a highway-type motor vehicle at the time of injury and the injury resulted from an accident to the vehicle or from the movement of the vehicle. Code in terms of the event affecting or involving the vehicle in which the injured was an occupant. If more than one of the listed events occurred, code for the first event in the sequence. The following kinds of cases are not included in this classification: (1) Falls on or from a vehicle which do not result from an accident to the vehicle or from the motion of the vehicle. (2) Striking against or being struck by parts of a vehicle or its cargo while in or on the vehicle, unless the event resulted from an accident to the vehicle or from the movement of the vehicle. (3) Accidents occurring in the course of servicing, repairing, loading, or unloading a vehicle, unless the event resulted from an accident to the vehicle or from the movement of the vehicle. (4) Accidents in which the injured was not in or on the vehicle as operator or passenger at the time of the accident.
	300	Motor vehicle, UNS
31		Collision or sideswipe with another vehicle, both vehicles in motion
	310	Collision or sideswipe with another vehicle, both vehicles in motion, UNS
	311	With an oncoming vehicle on same road, street, or trafficway
	312	With a vehicle moving in the same direction on same road, street, or trafficway
	313	With a vehicle moving in an intersecting trafficway
32		Collision or sideswipe with a standing vehicle or stationary object
	320	Collision or sideswipe with a standing vehicle or stationary object, UNS
	321	Running into or sideswiping a standing vehicle or object in the roadway
	322	Running into or sideswiping a standing vehicle or object at side of road (not in trafficway)
	323	Struck by another vehicle while standing in roadway
	324	Struck by another vehicle while standing off roadway
33		*Noncollision accidents*
	330	Noncollision, UNS
	331	Overturned
	332	Ran off roadway (out of control)
	333	Sudden stop or start (throwing occupants out of, or against interior parts of the vehicle; or throwing contents of vehicle against occupants)
	338	Other noncollision accidents

TABLE 12.6. Type of Event or Exposure Coding Categories Used for SDS Cases, from the Bureau of Labor Statistics *(continued)*

Major Group	Type Code	Title
40		*Exposure to noise*
	400	Exposure to noise, UNS
	402	Sudden or single exposure
	405	Repeated noise
50	500	*Explosions*
		Beginning Reference Year 1985
60		*Nonhighway motor vehicle accidents* Applies only to cases in which the injured person was an occupant in a nonhighway-type motor vehicle at the time of the injury and the injury resulted from an accident to the vehicle or from the movement of the vehicle.
	600	Nonhighway motor vehicle accidents, UNS
61	610	Collision or sideswipe with another vehicle—both vehicles in motion
62	620	Collision or sideswipe with a standing vehicle or stationary object (includes injury to occupants of a standing vehicle when struck by another vehicle)
63	630	Noncollision accidents, UNS
	631	Overturned
	632	Out of control
	633	Sudden stop or start (injury resulting from bodily motion or fpom striking against interior parts of the vehicle)
	634	Struck by shifting load or falling object (occupant of vehicle struck by shifting of falling load as result of movement, including sudden stops or starts, of the vehicle)
	635	Fall from vehicle (as result of an accident to the vehicle or from the movement of the vehicle)
	636	Vehicle movement (bodily conditions incurred as a result of movement of vehicle, such as a back injury resulting from a wheel of vehicle hitting a hole)
	639	Other noncollision accidents
		Reference Years, 1977–1985
89	899	*Accident Type, NEC* Includes cases of mental disorders
99	999	*Nonclassifiable*

Source: U.S. Bureau of Labor Statistics, *Supplementary Data System, Microdata Files User's Guide, 1980 Edition,* Document Number PB83-133553, National Technical Information Service, Springfield, Virginia, 1982.
[a]UNS = Unspecified
[b]NEC = Not Elsewhere Classified

Klein, Jensen, and Sanderson[16] may be used to compare the back pain experiences of different occupational groups. An appropriate statement for such a search strategy would be:

$$\text{cases} = [YR_1 \text{ or } YR_2] \text{ and } [YO] \text{ and } [P] \text{ and } [O] \text{ and } [T_1 \text{ or } T_2 \text{ or } T_3 \text{ or } T_4 \text{ or } T_5] \text{ and } [N_1 \text{ or } N_2]$$

where:

YR_1 = Reference year: 1980
YR_2 = Reference year: 1981
YO = Year of occurrence: 1980
P = Part of body category 420: back
O = Occupation category 075: RNs
T_1 = Type of event category 120: Overexertion
T_2 = Type of event category 121: In lifting objects
T_3 = Type of event category 122: In pulling or pushing objects

TABLE 12.7. States Participating in the Supplementary Data System for Reference Years 1979, 1980, 1981, and 1983[a]

	Reference Year			
State	1979	1980	1981	1983
Alaska	Y	Y	Y	Y
Arizona	Y	Y	Y	Y
Arkansas	Y	Y	Y	Y
California	Y	Y	Y	Y
Colorado	Y	Y	Y	Y
Delaware	Y	Y	Y	Y
Hawaii	Y	Y	Y	Y
Idaho	Y	Y	Y	N
Indiana	Y	Y	Y	Y
Iowa	Y	Y	Y	Y
Kentucky	Y	Y	Y	Y
Maine	Y	Y	Y	Y
Maryland	Y	Y	Y	Y
Massachusetts	Y	Y	N	N
Michigan	Y	Y	Y	Y
Minnesota	Y	Y	Y	Y
Mississippi	N	Y	Y	Y
Missouri	Y	Y	Y	Y
Montana	Y	Y	Y	Y
Nebraska	Y	Y	Y	Y
New Jersey	Y	Y	N	N
New Mexico	Y	Y	Y	Y
New York	Y	Y	Y	Y
North Carolina	Y	Y	Y	Y
Ohio	Y	Y	Y	Y
Oregon	Y	Y	Y	Y
South Dakota	Y	N	N	N
Tennessee	Y	Y	Y	Y
Utah	Y	Y	Y	Y
Vermont	Y	Y	Y	Y
Virginia	Y	N	N	Y
Washington	Y	Y	Y	Y
Wisconsin	Y	Y	Y	Y
Wyoming	Y	Y	Y	Y

[a]Y = yes, N = no. Data for 1982 and 1984 unavailable due to restriction by the Federal Office of Management and Budget.

T_4 = Type of event category 123: In holding, wielding, throwing, or carrying objects
T_5 = Overexertion, NEC
N_1 = Nature category 310; sprains, strains
N_2 = Nature category 190: dislocation, including herniated or ruptured disc

This approach helps structure a search that will exclude cases that fail to meet a requirement in each of the six brackets. The result is a group of cases (or a computer file) that meets the particular specifications.

The contents of the file can be printed out in a variety of formats. The simplest output is a frequency table listing the categories in a specified dimension (e.g., industry or age) along with the number of cases in each respective category. Another common output is a cross-tabulation that lists the categories in one dimension as rows and the categories in a second dimension as columns. The number of cases is shown in the respective cells of the table.

These frequency tabulations provide the numbers to use in the numerator of the IRs and IDs. The next step is to obtain appropriate denominator data.

12.3 DENOMINATOR (EXPOSURE) DATA

The simple counting of cases is just the beginning step in a risk analysis. To compare the injury risk of one group of workers to another requires consideration of their respective exposures—generally quantified in units of either population or person-time (e.g., person-hours, person-shifts, or person-years)—working under conditions that could give rise to an injury that qualifies for compensation.

For example, a machine tool operation may have 10 positions for machinists. During a particular year three relevant events occurred: (1) one machinist retired and was replaced after two months, (2) another quit and was replaced after three months, and (3) two others were absent for two weeks each during the summer for military reserve duty. Thus, a total of six person-months (or one-half of a person-year) of potential exposure did not occur. As a result an estimate of the exposure time of machinists during the year would be approximately 9.5 person-years. This figure would be suitable as the denominator of an annual ID. The resulting quotient could be compared with similar quantities for other groups of workers in the establishment.

For comparisons not limited to one particular employer, an extremely useful source of employment data for the United States is the 1980 Census of the Population. It makes available impressive detail concerning people who were employed at the time the Census was taken (May, 1980). The reference rooms of most large libraries have sets of books from the U.S. Bureau of the Census describing the results.[17] It is also possible to purchase computer tapes containing even more specific information.[18] Suitable denominators can be obtained from these sources for calculating IRs for groups of workers, as well as approximate IDs for 1979 based on the number of weeks individuals said they worked during 1979.

12.4 IDENTIFYING HIGH-RISK GROUPS

Different groups of workers may be compared directly based on the magnitude of their respective IRs or IDs, provided that each group is subject to the same criteria for compensation. Otherwise, additional analyses are needed. Some potentially useful techniques based on ranks and ratios are illustrated in the next subsections.

12.4.1 Rank Methods

Techniques that utilize ranking procedures are appropriate for certain risk comparisons based on workers' compensation data. The ranking of groups of workers within a state allows for comparisons based on the same compensation statute, regulations, and administrative practices of the state's compensation agency. Ranking procedures are valuable when incidence information from multiple states is used. An appropriate procedure is illustrated in the following example.

Numerous authors have claimed that back disorders are common among nurses. Unfortunately, no existing injury or illness information system contains sufficient information to directly evaluate this claim. However, the SDS system contains pertinent information about workers' compensation cases reported as a sprain or strain involving the worker's back. An evaluation of the claim using SDS data is possible by accepting the proposition that an IR based on compensation data is positively correlated with a similar index derived from an ideal, but nonexistent, medically based injury/illness information system (per equation 12.2).

Working with this proposition, an analysis of SDS data was undertaken. The SDS occupational classification system identifies three occupations that involve the various nursing functions. Three other occupational groups were also selected for comparison. SDS records were searched to identify compensable injuries classified as back sprains or strains that occurred during 1980 in four geographically disperse states: Wisconsin (WI), Idaho (ID), New York (NY), and North Carolina (NC). Since each state has a unique compensation system, the simple addition of claims from the four states was undesirable, because states with easy compensation requirements would be overrepresented in the sum. Consequently, the IRs of each occupational group in each state were determined using SDS compensation records for the numerator and Census records for the denominator. The results of six occupational groups are shown in Table 12.8.

The occupational groups of each state were then rank ordered according to IR. To develop an overall rank order the computationally simpliest method is to take an arithmetic average of the four state-specific ranks of each occupational group. However, it is preferable to weight the ranks to account for differences in the population of workers in each group in each state.

The following formula considers the rank (R) of each occupation (i) within each state (j) to compute the weighted-average rank of each occupation:

$$\text{Weighted Average } R_i = \sum_{j=1}^{m} \frac{P_{ij}}{P_i} (R_{ij}) \qquad (12.3)$$

TABLE 12.8. Incidence Ratios for Back Strains or Sprains in 1980 in Four States

Occupational Group	State	(A) Cases	(B) Population[a]	Incidence Ratio = A/B
Registered nurses	WI	62	26,540	0.002336
Licensed practical nurses	WI	66	10,267	0.006428
Nursing aides and orderlies	WI	668	34,912	0.019134
Garbage collectors	WI	14	1,177	0.011895
Cashiers	WI	27	34,774	0.000776
Radiologic technicians	WI	3	2,455	0.001222
Registered nurses	ID	52	3,568	0.014574
Licensed practical nurses	ID	50	1,598	0.031289
Nursing aides and orderlies	ID	131	3,350	0.039104
Garbage collectors	ID	18	156	0.115385
Cashiers	ID	33	5,909	0.005585
Radiologic technicians	ID	1	387	0.002584
Registered nurses	NY	246	111,608	0.002204
Licensed practical nurses	NY	158	28,122	0.005618
Nursing aides and orderlies	NY	961	116,410	0.008255
Garbage collectors	NY	49	9,236	0.005305
Cashiers	NY	44	123,783	0.000355
Radiologic technicians	NY	12	6,490	0.001849
Registered nurses	NC	34	25,203	0.001349
Licensed practical nurses	NC	39	8,896	0.004384
Nursing aides and orderlies	NC	211	30,985	0.006810
Garbage collectors	NC	35	2,443	0.014327
Cashiers	NC	15	40,260	0.000372
Radiologic technicians	NC	7	2,112	0.003314

Source: Figures on total civilian employees from U.S. Bureau of the Census, *1980 Census of the Population, Detailed Population Characteristics,* by State, Part 51, Table 217, U.S. Department of Commerce, Washington, D.C., 1984. Figures on workers not covered by state compensation plan from U.S. Bureau of the Census, *Census of Population and Housing, 1980: Public-Use Microdata Samples Technical Documentation,* U.S. Department of Commerce, Washington, D.C., 1983.

[a]From total civilian employees reported by Bureau of Census less workers not covered by the state compensation plan (i.e., federal government civilians, self-employed individuals, and unpaid family workers) obtained from Bureau of Census computer tapes called Public-Use Microdata Samples.

where:

P_{ij} is the population of employed individuals in occupation i and state j.

P_i is the total population in the m states of employed individuals in occupation i.

R_{ij} is the rank of occupation i in state j.

The application of this formula to the data in Table 12.8 produces the weighted-average ranks shown in the right-most column of Table 12.9. These rank values indicate that two of the nursing occupational groups actually rank higher than garbage collectors, a group widely recognized for back injury problems. Note also the considerable consistency in the occupational ranks in the four states despite differences in the compensation laws and practices of the states. This consistency is not unique to these six occupational groups. An analysis of these plus 18 more occupational groups showed similar consistency.[19]

This technique for combining within-state rankings of different occupational groups may also be used for groups defined in other ways. For example, a corporation with multiple manufacturing operations could use this approach to combine the plant-specific ranks of defined groups of workers to determine a company-wide ranking. In addition, the ranking of groups facilitates the use of several statistical techniques described in nonparametric statistics books.[20,21]

12.4.2 Ratio Methods

Another way to compare groups of workers is with ratios of their respective IRs or IDs.[22] The procedure is illustrated with the data from the preceding analysis of back sprain and strain compensation cases from six occupational groups.

TABLE 12.9. Rank Order of Six Occupations in Four States According to Weighted-average Rank

Occupational Group	Rank within State				Weighted Average Rank
	WI	ID	NY	NC	
Nursing aides and orderlies	1	2	1	2	1.18
Licensed practical nurses	3	3	2	3	2.42
Garbage collectors	2	1	3	1	2.51
Registered nurses	4	4	4	5	4.15
Radiologic technicians	5	6	5	4	4.85
Cashiers	6	5	6	6	5.97

The first step is to select a reference group. It is desirable for the reference group to have a large population and either a low (but greater than zero) incidence of the injury or a minimal exposure to the hazard being investigated. In this example one of the occupational groups is used. One that has both a relatively large population in each state and a fairly low IR in each state is the cashier group. The radiologic technicians have too small a population, and the other groups have relatively high IRs. Thus, cashiers will be used for the reference group.

The remaining five occupational groups can then be compared to cashiers using a ratio of IRs:

$$\text{Ratio of incidence ratios} = \frac{IR_i}{IR_r} \tag{12.4}$$

where:

IR_i = Incidence ratio of group of interest
IR_r = Incidence ratio of reference group

For the example data this formula yields the ratio values shown in Table 12.10 for each state. The far-right column shows a weighted-average ratio computed with equation 12.3 except that state-specific ratios are used instead of state-specific ranks. These results indicate how great the IR of one group is compared to cashiers. For example, the workers' compensation records indicate that nursing aides and orderlies have an IR of back sprains and strains 22.1 times that of cashiers.

12.5 SUMMARY

Practical approaches for comparing groups of workers in terms of risk of personal injury have been presented. The procedures use workers' compensation records to obtain values for a pair of risk quantities called incidence ratio and incidence density, both of which are believed to be positively correlated with the difficult-to-measure quantity, true risk. Two procedures are described for comparing groups to each other. One procedure compares the groups in terms of their respective ranks based on incidence ratio or incidence density. The other makes comparisons based on the ratio of each group's incidence ratio or incidence density to that of a reference group.

The use of workers' compensation data for risk comparisons is in an infantile stage of development. The most common use of such data has been for assessing risk from an insurance perspective. Improvements in workers' compensation record systems during the late 1970s opened up new capabilities for risk comparisons. Many opportunities exist for improving methodology, testing assumptions, and making meaningful comparisons between groups of workers in terms of their respective risk of incurring particular types of occupational injuries.

TABLE 12.10. Ratios of the IRs of Five Occupational Groups to the IR of Cashiers

Occupational Group	Ratio within State				Weighted Average Ratio
	WI	ID	NY	NC	
Nursing aides and orderlies	24.36	7.27	22.87	18.29	22.1
Garbage collectors	14.40	13.23	14.67	38.52	19.1
Licensed practical nurses	8.32	5.54	15.40	11.69	13.0
Registered nurses	3.00	2.57	6.00	3.63	5.1
Radiologic technicians	1.58	0.49	5.17	8.82	5.0

ACKNOWLEDGMENTS

The author is indebted to Micheal Moll, Ph.D., and Patricia Cutlip for making the SDS data accessible, and the following individuals for their valuable comments on earlier drafts of this chapter: Cathe Bell; Robert Brackbill, Ph.D.; Thomas Richards, M.D.; Lyle Pleak; and Lyle Schauer.

REFERENCES

1. Employers Insurance of Wausau, *Navigating OSHA with the Vital Few,* Employers Insurance of Wausau, Wausau, Wisconsin, 1972.
2. American Society of Safety Engineers, *Dictionary of Terms used in the Safety Profession,* ASSE, Park Ridge, Illinois, 1971.
3. Kleinbaum, D. G., L. L. Kupper, and H. Morgenstern, *Epidemiologic Research: Principles and Quantitative Methods,* Lifetime Learning Publications, London, p. 99, 1982.
4. Hauer, E., "Traffic Conflicts and Exposure," *Accid. Anal. & Prev.,* Vol. 14, No. 5, 359-364, 1982.
5. Malasky, S. W., *System Safety: Technology and Application,* 2nd ed., Garland STPM Press, New York, pp. 45-46, 1982.
6. Bruhning, E., and R. Volker, "Accident Risk in Road Traffic—Characteristic Quantities and Their Statistical Treatment," *Accid. Anal. & Prev.,* Vol. 14, No. 1, 65-80, 1982.
7. Kleinbaum, D. G., L. L. Kupper, and H. Morgenstern, *Epidemiologic Research: Principles and Quantitative Methods,* Lifetime Learning Publications, London, pp. 100-103, 1982.
8. American National Standards Institute, *Method of Recording Basic Facts Relating to the Nature and Occurrence of Work Injuries, Z16.2-1962,* reaffirmed 1969, ANSI, New York, 1969.
9. Root, N., and D. Sebastian, "BLS Develops Measure of Job Risk by Occupation," *Monthly Labor Review,* Vol. 104, No. 10, 26-30, 1981.
10. Safety Sciences, *Feasibility of Securing Research-Defining Accident Statistics,* Publication Number 78-180, National Institute for Occupational Safety and Health, Cincinnati, Ohio, 1978.
11. U.S. Bureau of the Census, 1970 Census of Population, Alphabetical Index of Industries and Occupations, U.S. Government Printing Office, Washington, D.C., 1971.
12. U.S. Bureau of the Census, *1980 Census of the Population, Classified Index of Industries and Occupations,* PHC80-R4, U.S. Department of Commerce, Washington, D.C., 1980.
13. U.S. Office of Management and Budget, *Standard Industrial Classification Manual 1972,* U.S. Government Printing Office, Washington, D.C., 1980.
14. Office of Federal Statistical Policy and Standards, *Standard Industrial Classification Manual, 1977 Supplement,* U.S. Government Printing Office, Washington, D.C., 1977.
15. U.S. Bureau of Labor Statistics, *Supplementary Data System, Microdata Files User's Guide, 1980 ed.,* Document Number PB83-133553, National Technical Information Service, Springfield, Virginia, 1982.
16. Klein, B., R. Jensen, and L. Sanderson, "Assessment of Workers' Compensation Claims for Back Strains/Sprains," *J. Occup. Med.,* Vol. 26, No. 6, 443-448, 1984.
17. U.S. Bureau of the Census, *1980 Census of the Population, Detailed Population Characteristics,* by State, Part 51, Table 217, U.S. Department of Commerce, Washington, D.C., 1984.
18. U.S. Bureau of the Census, *Census of Population and Housing, 1980: Public-Use Microdata Samples Technical Documentation,* U.S. Department of Commerce, Washington, D.C., 1983.
19. Jensen, R., *"Disabling Back Injuries Among Nursing Personnel: Research Needs and Justification,"* Research in Nursing and Health, in press, 1987.
20. Hollander, M., and D. A. Wolfe, *Nonparametric Statistical Methods,* Wiley, New York, 1973.
21. Conover, W. J., *Practical Nonparametric Statistics,* 2nd ed., Wiley, New York, 1980.
22. Kleinbaum, D. G., L. L. Kupper, and H. Morgenstern, *Epidemiologic Research: Principles and Quantitative Methods,* Lifetime Learning Publications, London, p. 143, 1982.

BIBLIOGRAPHY

Chamber of Commerce of the United States, *Analysis of Workers' Compensation Laws,* published annually by Chamber of Commerce of the United States, 1615 H Street, N.W., Washington, D.C. 20062.

Harvard Law Review, Compensating victims of occupational disease, *Harvard Law Review,* Vol. 93, No. 5, 916-937, 1980.

Jensen, R., B. Klein, and L. Sanderson, Motion-related wrist disorders traced to industries, occupational groups, *Monthly Labor Review,* Vol. 106, No. 9, 13-16, 1983.

Larson, A., *Workmen's Compensation for Occupational Injuries and Death,* 2-vol. desk ed., Matthew Bender & Co., New York, 1985.

Mason, K., Accident patterns by time-of-day and day-of-week of injury occurrence, *Journal of Occupational Accidents,* Vol. 2, No. 2, 159-177, 1979.

McCaffrey, D. P., Work-related amputations by type and prevalence, *Monthly Labor Review,* Vol. 104, No. 3, 35-41, 1981.

McLennan, B. N., Product liability in the workplace: Product liability legislation and workers' compensation laws, *Labor Law Journal,* Vol. 34, No. 3, 160-171, 1983.

Root, N., and D. Sebastian, BLS develops measure of job risk by occupation, *Monthly Labor Review,* Vol. 104, No. 10, 26-30, 1981.

PART VI

OCCUPATIONAL SAFETY AND HEALTH TRAINING PROGRAMS

CHAPTER 13

How to Set Up a Training Program

A. JAMES McKNIGHT
National Public Services Research Institute
Landover, Maryland

Odell Warner was in his third day on the job at the Harrington Ice Company. One of his jobs was to empty ice storage bins. The ice was removed from the trap door on the underside of each bin. It is common, however, for ice in the bins to solidify and fail to drop through when the trap door is opened. When this "bridging" occurs, the ice has to be physically dislodged. For this purpose, Harrington constructed a wooden catwalk along the storage bins to allow employees to reach into the bins and break up the ice with a garden hoe.

Ice is carried to the storage bins by an overhead conveyer. In his first attempt at "bridging," Odell allowed the handle of the hoe to come in contact with the conveyer. Both the hoe and Odell's left arm were drawn into the conveyer with the result that Odell's arm was severed at the shoulder.

This is a true story, although the names of Odell and his employer have been changed. Odell was unaware of the danger of raising his hoe to a point where it could get caught in the overhead conveyer. No one had told him about it. The old timers knew. Eventually someone would have mentioned it. But "eventually" turned out to be too late.

13.1 WHY SAFETY TRAINING?

Cases like Odell's are common. He was one of many who discovered the hazards of their work environment the hard way. Some are lucky and get away without a scratch. Others, like Odell, aren't. While experience may be the "best teacher" for some things, job safety isn't one of them. A comprehensive and well-taught program of safety training, if not a better teacher than experience, is certainly less painful.

Odell had been trained. He had been taught how to empty ice storage bins and how to dislodge the ice when "bridging" occurred. And he had been taught to do it the right way, which prompted his supervisor to say, "If he had done it the way I had showed him, he'd never have gotten hurt."

But people make mistakes. Some mistakes are harmless. The job may not get done, but nobody is hurt or made ill. Helping workers to distinguish between harmless and harmful errors is what safety training is all about.

Giving specific training and safe job practices will not eliminate mistakes or the accidents that result from them. However, research has shown that a well-designed safety training program, aimed at avoiding specific hazards, can cut incidents of unsafe errors almost in half (Cohen and Jensen, 1984). A very small investment in training can reap great dividends in reduced injury and illness, lost work time, and workers compensation premiums.

13.2 WHAT'S IN THIS CHAPTER

This chapter will describe the activities involved in setting up a job safety program. It cannot deal with every conceivable element of the process. Rather, the chapter will focus upon those aspects of job training programs that are specific to safety. For more general information on setting up job safety programs,* a very informative publication is:

Effective Training: A Guide for the Company Instructor, International Correspondence School, Scranton, PA 18515

Developing effective job safety training programs involves answering the following questions:

Whom should you teach?
What should you teach?
When should you teach it?
How should you teach it?
What do you need to teach it?

13.3 WHOM SHOULD YOU TEACH?

Everyone! Many people, including some who are very enlightened, think that safety training is for workers alone. However, to be effective, safety training must be directed to all levels: workers, supervisors, and managers.

*To save words, the term "safety" will include prevention of both accidents and illnesses and will, therefore, serve as shorthand for both safety and health.

13.3.1 Workers

Many jobs present obvious hazards. Operating unguarded equipment, working in high places or underground, handling hazardous materials, and dealing with high voltage equipment are cases in point. The need for safety training in these jobs is fairly apparent, although the training is not always provided.

However, there are many jobs when the hazard is less obvious—jobs that require lifting, driving vehicles, or watching video terminals. Training programs dealing with these hazards are rare in industry. Yet, strained backs from incorrect lifting, eyestrain from video terminals, or injuries from automobile accidents account for a considerable amount of lost work time.

Safety training should be part of every job. Certainly, some jobs are more hazardous than others and the amount of training will vary enormously from one job to another. Yet no job should be overlooked in setting up a program of training.

13.3.2 Supervisors

Workers do not always do the things they are expected to, so a productive workforce requires supervision. The same is true of a safe and healthy workforce. It is just as much the supervisor's job to see that workers perform their jobs in the manner that is safe and healthy as it is to see that they are productive.

To fulfill their responsibilities, supervisors themselves need training. They need to be trained not only in safety but in ways of passing safety onto workers and monitoring them to make sure that they perform safely.

13.3.3 Managers

Safety training will not flourish without the active support of management. Managers all up and down the line must see that safety is important to the success of the organization. Only then will managers establish policies needed not only to maintain programs of job safety training but to make job safety a major consideration in management of the entire organization.

The enormous cost to business and industry of lost work time, workers' compensation, and damage to equipment testifies to the importance of safety to the success of any business. As has been pointed out already in this book, occupational accidents and illness cost businesses billions of dollars each year. Unfortunately, very little attention is given to the importance of safety in the schools of business that train managers. It is, therefore, up to each organization to make sure its managers recognize that safety is a critical management concern.

13.4 WHAT SHOULD YOU TEACH?

To perform safely on the job, workers must know what to do to be safe. This section will describe the general categories of performance that lead to safety, how to find out about them, and how to formulate training objectives.

13.4.1 Safe Job Performance

What do workers need to do in order to be safe? For this discussion, we'll classify job performance into the following three categories.

1. Employing safe job procedures.
2. Observing safety precautions.
3. Identifying job hazards.

Job Procedures

The most fundamental form of job safety is simply doing the work safely—lifting boxes, operating a lathe, loading a trailer, servicing a conveyor. This aspect of safety is generally included in employee training, since it is part and parcel of the job that employees are trained to do. The problem is that the safety aspects of the job are easy to neglect because they do not contribute directly to getting the work done. Indeed, workers often find that doing the job the unsafe way is faster and easier. Since most of the time nothing goes wrong, employees continue doing things the fast and easy way—until they have an accident.

Workers not only need to be taught the safe way to carry out job procedures but to be shown why the safe way is the correct way. It often helps if instructors can show, by word or example, the consequences of failure to employ safe job procedures.

Safety Precautions

Some jobs require, in addition to the normal procedures for carrying out the job safely, special "precautions" intended to ensure the safety of workers. Examples include:

- Use of safety features on operating equipment (e.g., interlocks, machine guards).
- Use of special safety equipment (e.g., fire extinguishers, resuscitators).
- Use of protective clothing or equipment (e.g., goggles, hard hats, gloves).

Safety precautions are easy to neglect since it's possible to do the job without them. Again, if workers are expected to employ safety precautions, they need to know the "why" as well as the "what." Too often job procedures and safety precautions are taught with a "do it this way because I tell you to" flavor. If that is all they are told, workers like Odell Warner may have no way of distinguishing harmless from potentially dangerous mistakes. How long would it have taken Odell's supervisor to say "If you do it the wrong way, here is what can happen."

Job Hazards

If job procedures and safety precautions were always adhered to, there would be very few accidents. However, people often make mistakes. Most job injuries and illnesses are ascribed to "human error." We are told so often that to err is human, some of us have been led to accept it as inevitable.

Human error is *not* inevitable. There are certain errors people almost never make. For example, even drivers who carelessly overrun crosswalks a lot rarely do it if there's a pedestrian there. Why? Because they see one error as essentially harmless and the other as homicidal. If they recognized the possible danger in overrunning crosswalks (forcing pedestrians into traffic), they might be more likely to avoid those errors as well. Safety training may never be successful in eliminating errors entirely. However, it can hope to succeed in preventing the most dangerous errors by acquainting employees with the hazards that are involved.

Traditionally, the route to recognizing job hazards has been experience. People get burned and learn to keep their hands away from hot surfaces. As we saw in Odell's case, experience can be a very painful teacher of job hazards. Training may help spare people the pain of learning from experience. Regrettably, very few safety training programs do as good a job of acquainting workers with the hazards of their job environment as they do in teaching job procedures and safety precautions.

13.4.2 Sources of Information

Where do we learn about safety procedures, precautions, and job hazards so we can teach others about them? Three valuable information sources are publications, job hazard analysis, and accident analysis.

Publications

A wealth of information concerning safety procedures and precautions is in technical manuals, bulletins, and safety periodicals.

Manuals. Technical manuals that accompany equipment generally provide a lot of safety information. Often, however, they are not read because workers are taught the operation or servicing of equipment and are never required to read manuals. Sometimes the manuals become separated from the equipment and, eventually, lost. A case in point is a recent accident in which the workman lost his arm in a gas dryer that failed to stop revolving when the lid was lifted. The problem was traced to a gasket that required annual servicing. However, the manual had been lost and the gasket hadn't been touched in years.

Bulletins. Manufacturers often issue bulletins describing safety precautions and noting those safety hazards that have been discovered since the equipment was manufactured. This information may be available only to purchasers who complete warranty cards and get on the mailing lists. Technical bulletins must be routinely screened when they are received so that steps can be taken to make sure that the word reaches all affected workers.

Periodicals. Periodicals that are published in various areas of business and industry often contain articles on job safety. At the same time, safety periodicals often provide articles dealing with the hazards in a particular area of business or industry. Both types of publications should be screened so that pertinent, up-to-date job safety information can be fed into training programs.

Job Hazard Analysis

Job hazard analysis is a process by which jobs are analyzed for hazards. Many textbooks in occupational safety and system safety describe methods of a job hazard analysis.

Job hazard analysis simply amounts to examining each element of job procedure to find out where hazards are most likely to arise. It is a process of going through job procedures step by step, asking the following questions at each step:

- How often is the worker called upon to perform the step?
- How likely are workers to make a mistake in performing this step?
- If workers make a mistake, how likely are they to cause damage, illness, or injury?
- How severe is any illness or injury likely to be?

The greatest potential hazards are those that rank highest on this combination of factors.

When jobs are highly repetitive, hazards are best identified by observing workers in the performance of their jobs. On the other hand, where jobs involve a great variety of tasks, it is better to interview workers to find out what they do that might be hazardous and what they view as the most hazardous aspects of their jobs.

Job hazard analysis is a time-consuming procedure. However, it can pay big dividends in alerting people to hazards before they result in accidents.

Once hazards have been identified, procedures or equipment can often be redesigned or special devices developed to remove the hazard. However, it may be impractical or impossible to remove all hazards of a particular job. If certain hazards cannot be removed, at least workers can be warned about them and be provided with procedures that allow them to minimize the risk. Job hazard analysis is therefore a valuable technique in identifying what needs to be taught in job safety training programs.

Accident Analysis

Despite efforts to prevent them, accidents will happen. If we can't completely prevent accidents, we can at least learn from those that occur and try to see that they do not happen a second time. Chapter V, "How to Conduct an Accident Investigation," described techniques for studying accidents that occur on the job.

In analyzing accidents for training purposes, the objective is to identify those human errors that led to the accidents. Knowing what errors often have led to accidents allows us to concentrate on preventing those errors. Since over 80% of job accidents are attributed to human error, preventing the recurrence of unsafe errors should have a tremendous impact on job safety.

Regrettably, accident investigation and analysis have not been as useful as they might be in identifying accident-producing worker errors. One reason is that the human errors that lead to accidents are hard to pinpoint. Often the only one who knows what happened in an accident is the victim, and even victims are typically unclear as to just what occurred ("it all happened so fast"). Accident investigators generally have to depend upon inferences drawn from the circumstances surrounding accidents. They are naturally reluctant to mix speculation about worker error with such "objective" accident data as the time of the accident, its location, equipment damage, or extent of injury, even though it may be a lot more valuable. For this reason, information on worker errors is too often missing from official accident reports.

No matter how speculative information concerning human errors may be, it is important that the information be collected so that it can be applied to safety training programs. The best way to obtain information about the errors leading to accidents is by assuring that accident reports include a narrative section in which the accident investigator describes *in detail:*

- The events surrounding the accidents, including those occurring before, during, and following the accident.
- All the errors that the investigator believes contributed to the accident (accidents rarely result from a single error).
- The steps that, had they occurred, might have prevented or lessened the severity of the accident.

A narrative description of a welding accident appears in Figure 13.1. Consider what steps might be taken in training to prevent future occurrences of this type of accident.

Many companies have job accidents reviewed by a safety panel whose duty it is to decide whether or not the accident was preventable. Disciplinary action may be taken against employees who sustain a large number of preventable accidents. These accident review panels often delve rather deeply into the workers' errors leading to accidents and therefore can be a valuable source of "lessons learned" for development of training programs. However, to encourage honesty in reporting, it might be a good idea to forgo attempts to assess blame, particularly in those cases where you are depending on the parties involved to tell you what happened.

WELDING ACCIDENT

Description of Accident

A 55-gallon drum exploded when welders Robert Guthrie and John Sterns used it as a support in welding a metal bracket. The drum had contained xylene and, having been emptied, was set aside for return to the supplier. The spigot had been removed but the bung that should have filled the opening had not been replaced. The drum was marked with a diamond shaped "Flammable Liquid" label.

The two welders had mistaken the drum for a trash can and used it to support the metal bracket they were attempting to weld. Apparently, fumes from xylene residue in the bottom of the drum leaked out through the opening where the spigot had been removed and exploded when the acetylene torch was ignited.

Errors Contributing to the Accident

The following factors contributed to the accident:

1. Removing the spigot without replacing the bung to prevent xylene fumes from escaping.
2. Leaving the drum in an area where it might be exposed to cause an open flame.
3. Operating an acetylene torch in vicinity of a container marked having flammable contents.

Recommended Preventive Action

1. Segregate drums containing any flammable residue within a fenced enclosure pending their return to suppliers.
2. Require suppliers to attach restraining wires to bungs so that they do not become separated from drums when removed.
3. Require that all containers in which flammable materials are stored be flushed out when they are emptied.
4. Instruct all employees of the hazards represented by flammable residues in containers and the importance of keeping them away from a spark or open flame.

FIGURE 13.1. Welding accident.

13.4.3 Preparing Instructional Objectives

Once you have decided what you want to teach, you need to formulate a set of instructional objectives. Instructional objectives are merely statements of what a training program is to accomplish. During development of a training program, an explicit set of objectives will help define specifically what is and isn't to be included. Objectives help to assure that the program includes everything that belongs in it and that irrelevant material doesn't stray into the program. They also help to communicate to others what the program will and will not include, by helping to guard against misconceptions.

In the rush to develop training programs against short deadlines, the temptation is strong to bypass the identification of instructional objectives and plunge directly into the development of training materials. The shortcut seldom works. As materials are prepared, differences in people's understanding as to what the program is intended to do begin to surface. Eventually, after everyone agrees on a set of objectives, materials are revised to conform with the objectives. The cost of the revision is lost time, effort, and expense.

Separate instructional objectives should be prepared for:

- Performance
- Knowledge
- Skill
- Attitudes

Performance Objectives

Performance objectives should state concisely what employees are expected to do as a result of instruction.

One category of performance objectives for job safety training involves training workers to be *able* to do something in a particular area. For example, an objective of the training program for warehousemen might be "To identify hazards involved in material handling and storage." Workers who have never handled hazardous material might not know what the dangers are. They would require instruction in *recognizing* hazards before the materials could be safely handled. These ability-oriented objectives involve instruction intended to enable employees to perform safely.

The second type of the objective deals with *making sure* that workers carry out activities that they already have the ability to perform. An example might be, "Keep storage areas free of trash." Anyone *can* pick up trash. The objective of training would be to make sure that they do. These compliance-oriented objectives are frequently overlooked in establishing objectives, since they deal with things that workers already know how to do. Yet, if they are omitted from performance objectives, the instruction needed to help assure that workers actually do what they are supposed to do may not be given.

In recent years, a lot of emphasis has been placed upon so-called measurable objectives, which refers to stating objectives in measurable terms, that is, defining precisely the necessary level of performance as well as the conditions under which the performance will be measured. This may be feasible with jobs that consist of very few tasks, performed under uniform conditions. However, it is virtually impossible where the performances occur under conditions that are as many and varied as those that make up the objectives of job safety trainng. Imagine defining the objective "keep the storage area free of trash" in measurable terms! (How much trash is there? What kind? What is there to pick it up with?)

Training objectives should be simply broad statements of what the training is intended to do. The measurement is best left to the process of testing rather than the preparation of objectives. (See Section 13.8.)

Knowledge Objectives

Once you have identified what workers must do or be able to do, the next step is to identify what they need to know in order to be able to do it. Knowledge objectives would include:

Procedures: Information about specific steps involved in carrying out a required performance, such as stacking materials or separating combustible and noncombustible materials.

Facts: Information about things and events, such as minimum clearance from sprinkler heads.

Hazards: Information about hazards such as signs of improperly stacked materials or exposed conveyer areas.

Concepts: Information about relationships such as that between distance and radiation levels.

Knowledge objectives will help to identify the subject matter that must be taught. Developing knowledge objectives systematically from performance objectives will help make certain that all important information needed to ensure safe job performance is included, and that information that is unessential—the "nice-to-know" information—is avoided.

Skill Objectives

Skills are those abilities required over and above knowledge that allow people to perform. If you tell workers how to operate a piece of equipment, and they can do it correctly the first time, the performance obviously didn't require the development of any new skill. However, if people need to practice a task in order to perform adequately, it involves some type of skill. Formulating skill objectives helps to identify those performances for which practice must be provided.

Fortunately, skill is rarely very important to the safety of most job tasks. Accidents occur more often from careless errors rather than from lack of skill. However, some safety related performances do require skill and practice, such as first aid, CPR techniques, and use of fire extinguishers. It is important that skill objectives be explicitedly formulated for these important performances to make sure the required practice is provided.

Attitude Objectives

The correlation of attitude and job safety is generally acknowledged. Many safety professionals believe that attitude is more important than knowledge or skill in achieving safe job performance.

If we wish workers to hold certain attitudes toward job safety, we must include attitudes among the objectives of a safety training program. For example, if we want workers to believe that working in a construction area without wearing a hard hat is foolhardy, formulate it as an objective. We may not succeed in reaching that objective with every trainee. However, that attitude is more likely to develop among trainees if it is made an instructional objective and the instructional program is specifically directed towards developing that attitude.

13.5 HOW SHOULD YOU TEACH?

Having established a set of training objectives, we now must decide the best way to attain these objectives. To some people, teaching is simply a matter of *telling*—getting up and describing procedures, relating facts, or explaining concepts in a lecture. The lecture is certainly an efficient way of communicating information. However, in teaching job safety, methods calling for more active *participation* by students have the following advantages:

1. Most of the students (employees) are long out of school and are not used to sitting for long periods of time without having a chance to talk.
2. The students have a wealth of experience that they can bring to bear on the problems of job safety.
3. Safety instruction is concerned as much with changing minds as with developing skills. This objective cannot be attained very well without participation of the students.

We will divide the discussion of participative training methods into the following six categories:

- Presentation
- Demonstration
- Problem solving
- Group discussion
- Practice
- Independent study

13.5.1 Presentation

Simple presentation of information will be a substantial portion of any training program. There is a lot that workers don't know about performing their job safely and they need to have the necessary information presented to them. However, this doesn't mean that they must be lectured. A skillful teacher can draw information out of students rather than feeding it to them. By a series of leading questions or prompts, instructors can get students to "discover" facts and concepts for themselves. This is particularly true of job safety instruction, much of which is common sense.

Pulling information out of students certainly takes longer than simply supplying it to them. However, involving students in the process gives the instructor feedback on how well the students are learning. Lecturers never know whether any of the information they are transmitting is really being received. Moreover, the information that students do acquire through active participation is likely to be retained longer.

13.5.2 Demonstration

Information need not be presented solely in words. Some things are better *demonstrated* than talked about. A demonstration on how to operate a piece of equipment will almost always be clearer than a description. Also, demonstrations are frequently more convincing. For example, those truck drivers who refuse to believe that a skidding wheel moves faster than a wheel that is spinning freely can be shown by taking the rear wheels of a toy vehicle and letting it run down an incline board (the rear wheels will catch up and pass the front wheels).

Demonstrations are particularly useful in showing the consequences of following or not following certain safety practices. Often used examples include:

A collection of badly dented but life-saving, hard hats.

The unsteadiness of incorrectly stacked boxes.

What can happen when machine guards are bypassed.

A factual demonstration of the consequences of hazards should not be confused with use of "scare tactics," the blood and gore movies often used in an attempt to frighten people into abiding by safety practices. Scare tactics don't provide information. Their function is to frighten rather than inform. Moreover, they don't work. Students tune out the gore and the message with it.

13.5.3 Problem Solving

Having information and being able to use it are different things. Classroom problem solving exercises give students an opportunity to actually use what they've learned. Some examples of problem solving exercises in job safety training are:

- An exercise in which students figure out how to stack supplies or load a trailer.
- An accident case study: what caused it?
- A personnel problem: how can a supervisor get a recalcitrant worker to wear a hard hat?

In addition to allowing students to apply information, problem solving exercises (1) help instructors gage how well they are getting information across, (2) identify individual students who are having difficulty learning or applying information, (3) help students to integrate the information into their own jobs, and (4) aid retention by using information.

13.5.4 Group Discussion

Where the objective of training is to change hearts and minds rather than develop abilities, fellow workers are often more influential than outside teachers. Group discussions provide a means by which influences of fellow workers can be applied. For example, some workers find hard hats or gloves uncomfortable when the temperature climbs and will take them off when nobody is watching. A discussion in which these people try to defend their position to other workers may be more successful than a teacher in convincing them that the risks of injury outweigh minor discomfort.

For a group discussion to be most effective, the following conditions need to prevail:

- The facts of the case must be laid out beforehand. Group discussion cannot flourish in an information vacuum.
- The issue must be one on which the majority of students, or at least the most respected students, support the "safe" position. Otherwise there is a risk that the discussion will only solidify opposition to the safe practice being discussed.
- The instructor needs to set up the discussion and supply information whenever requested, but otherwise be silent. While this can be difficult to do when the discussion seems to be taking a wrong turn, any attempt to impose the "company" point of view will be futile.

13.5.5 Practice

Where students are expected to develop skill in performing an activity, they must be provided an opportunity to practice. Anyone who has tried to get a fire extinguisher into action quickly while a grease fire blazes merrily on the stove can testify to the importance of practice in developing skill. The fact that many safety-related skills must be applied under emergency conditions increases the need of practice (well-practiced activities are less likely to deteriorate under emotional duress).

Simulation has often been used to develop skill in tasks that cannot be practiced in their natural environment, either because the tasks occur too infrequently or because they are dangerous. In teaching safe driving, motion picture simulators have been used to develop skills in recognizing hazards, while wetted-down pavement has been used for developing skills in handling skids.

Enterprising instructors have developed a variety of homemade simulators to allow practice in safe job performance. Often instructors will salvage wornout items of equipment and modify them so that students can operate the equipment to the extent necessary to learn safety procedures without the risk of injury. Sometimes expensive equipment has been simulated by inexpensive wooden mockups. Simulators need not look like the real thing, so long as the stimuli to which students respond and the responses they make are like the ones that are found on the job.

13.5.6 Independent Study

Workers need not convene in classrooms in order to be trained. Most literate students can obtain information as well from printed sources as they can from an instructor or training film. With the aid of printed materials, they can often practice activities by themselves and thereby develop skill as well as gain information.

Independent study can actually offer several potential advantages over classroom training:

- Students can learn at their own pace, a distinct benefit to unusually slow or unusually rapid learners.
- Instruction can be given to individual students rather than having to form classes, a particular advantage in training new employees. (This will be discussed further in Section 13.6.)
- Accompanied with periodic tests to provide a necessary learning incentive, independent study can actually lead to higher levels of achievement than classroom instruction.

Independent study requires the availability of written materials to communicate the information that would otherwise be presented by an instructor. It is therefore best employed in situations where necessary safety-related information is well documented in existing publications such as general trade publications, equipment manuals, and company policies and procedures documents.

Some company training directors are very negative about the use of independent study, claiming that "you'll never get workers to study on their own time." Experience proves otherwise, particularly when independent study is combined with more participative forms of classroom instruction. Independent study and classroom participation reinforce one another. On one hand, independent study provides the means by which students can obtain the information they need in order to participate effectively in classroom activity. On the other hand, the desire to participate effectively, and not appear as the class dummy, becomes an incentive to students to study independently.

13.6 WHEN SHOULD YOU TEACH?

The best time to teach about job hazards is before they are encountered. If we could complete all safety training before workers ever report for duty, the risk of having accidents would be reduced. However, such a heavy dose of preservice training is rarely practical. Employers naturally want workers to start working and become productive as soon as possible. Moreover, it is a rare company that is large enough to muster enough new employees at any one time to form classes for an extensive preservice training program.

Even if it were possible to provide comprehensive job safety instruction to new employees before they started working, later instruction would still be needed to accommodate changes in job safety requirements as well as to provide refresher training. For these reasons, job safety training must be given on both a *preservice* and *in-service* basis.

13.6.1 Preservice Training

The amount of instruction that can be given before workers report to the job is limited by the need for employees to be on the job and productive as soon as possible, as well as by the fact that most new workers report individually to fill individual vacancies rather than in groups large enough to form classes.

Because of the limitations imposed upon preservice training, safety subject matter must be pared down to the following categories:

General: Procedures and precautions that apply to a wide range of tasks are more suitable than those that apply only to specialized or seldom performed tasks. For example, it is more important that workers know the importance of protective clothing than the potential hazard of some specialized tool.

Critical: Things that are likely to cause severe accident or illness are more important to know about than those that involve low risk. Priority should be given to preventing those errors that are most common, most likely to result in an accident or illness, and most likely to produce the most severe consequences. For example, in view of the high incidence of back sprain among warehousemen, it's probably more important that they get early instruction in lifting techniques than in the hazards of forklift operations.

Simple: Preservice instruction is given at a time when employees are likely to be unfamiliar with job procedures. It therefore makes sense to concentrate on those things that are simple to teach and to leave more complicated instruction to in-service training.

If at all possible, preservice safety instruction should be integrated with the normal job training given to new employees. Safety procedures and precautions should be viewed as being part of the right way to do the job, not as something done solely out of concern for safety.

Integrating safety with regular job instruction has an added benefit. If the supervisors and coworkers who conduct regular job training teach their employees safe performance long enough, they may start doing it themselves. While this may seem like a "backdoor" method for improving the safety of experienced workers, it is a fact that many people learn by teaching and there is no reason why safety shouldn't benefit from this phenomenon.

To assure that safety is made a part of regular job instruction, the procedures that new employees are taught to perform and any materials that go with them should be examined closely to make sure that the job procedures themselves are safe, that associated safety precautions are taught, and that critical hazards are identified. Some elements of safety that should be part of preservice instruction cannot be readily integrated into regular job instruction. For example, special emergency procedures (fire, chemical leaks, evacuation, etc.) may be highly important, but are not easy to integrate into ordinary job instruction. It is still possible to require new employees to read a pamphlet on the subject or to watch a special video dealing with it.

13.6.2 In-Service Training

Where the number of new employees starting at any one time is rather small, classes of sufficient size to make for efficient job safety training can only be mustered from across the entire workforce on an in-service basis. For purposes of discussion, we'll divide employee in-service programs into three categories:

1. Total job training.
2. Specialized training.
3. Refresher training.

Job Training

At some time, all employees should be given a total job safety course, a comprehensive and intensive program intended to instruct them in all of the safety procedures, precautions, and hazards of their jobs. It can be taught on an annual, semiannual, or monthly basis, depending upon the size of the organization, to all workers employed in the time since it was last given.

Job safety training programs should be tailored as much as possible to individual jobs. Some companies try to get by with a single "one size fits all" program. Such programs typically treat safety in only the most general terms and do not provide the instruction in specific safety precautions and job hazards that is needed to have a truly significant impact upon occasional accidents and illnesses.

Where there are many individual jobs, instruction should be tailored to groups of jobs having similar instructional objectives.

Specialized Training

Even the most comprehensive general safety program cannot anticipate all the problems that employees may encounter. As conditions change or we learn new things, we need to inform all employees. This need can be met through specialized training program focusing on specific procedures, precautions, and hazards arising from such occurrences as:

- Purchase of new equipment.
- Changes in job procedures.
- A sudden rash of accidents or illnesses stemming from a particular situation.
- Manufacturer's bulletin advising of a particular hazard.

Specialized courses can range in duration from just a few minutes to several hours depending upon the complexity of the problem they are designed to deal with. The employees involved can range from just a few people working in one section to an entire work force. What is important is that when a condition arises, those affected by it should be quickly informed through an instructional program that not only gets the message across effectively but *makes sure that it is understood.* One reads too frequently of tragic accidents that stem from conditions identified by management but were handled through memos or signs rather than through a formal training program.

Refresher Training

No matter how well informed or well intentioned workers may be, their adherence to safety procedures will occasionally flag. Periodic refresher programs can serve as reminders and thereby help to maintain a high level of safety consciousness.

To be effective, refresher training must focus upon specific aspects of the job. Sending employees to the same general safety course year after year is not refresher training. People don't forget everything, and forcing them to sit through instruction about items they already know is an enormous waste in time. Worse than that, it is seen as a joke, with the result that even those portions of the course that could provide useful refresher instruction are tuned out.

13.7 WHAT RESOURCES DO YOU NEED?

You have decided who and what you need to teach, along with how and when you need to do it. The next question is, what will it take to bring it all off?

To run an effective job training program, you need:

- Instructors
- Materials
- Equipment
- Facilities
- Students

13.7.1 Instructors

Of the various elements of a job safety training program, the instructor is by far the most important. A good instructor can teach without elaborate instructional aids, training equipment, or facilities. On the other hand, the best aids, equipment, and facilities in the world will produce nothing without a competent instructor.

What does it take to make a competent instructor?

- Technical know-how
- Teaching ability
- Credibility

Technical Know-How

We have all known teachers whose knowledge of the subject they are teaching barely allows them to stay one page in front of their students. They may be all right as long as they stick to what is written in the book, but let one student ask a question and they are sunk.

To teach effectively, instructors must be thoroughly grounded in those aspects of occupational safety and health that apply to the course they are teaching. Organizations that are fortunate enough to have one or more qualified safety professionals on the staff have technical competence at hand. Those organizations that lack safety professionals will have to find people who can teach and provide them with the subject matter expertise. Instructors can be obtained from the ranks of supervisors, teachers, or experienced employees who have what it takes to teach other employees, provided they can learn the subject matter beforehand.

A variety of means exists by which people can acquire sufficient knowledge of job safety and health to be able to teach it.

- Many community and junior colleges offer courses in occupational safety and health. Check local college catalogues.
- Many universities offer special evening or weekend courses for adults. Call the university department that handles occupational safety instruction.
- A number of nonprofit and profit-making organizations put on packaged seminars or short courses in aspects of job safety. Safety and health periodicals often advertise such courses.

Local safety councils should also be able to provide information on where occupational safety and health courses are being taught.

Teaching Ability

The ability to teach may be a little harder to come by than technical know-how. While almost anyone can learn occupational safety and health, not everyone can learn to teach well.

If a training department exists, it may be able to provide qualified instructors. Be careful, though; people who set up and administer training programs can't always teach. Also beware of lecturers. In the highly interactive form of instruction needed to teach job safety to employees, skill in public speaking and rhetoric is probably less important than:

- Ability to relate to people as individuals.
- Sincere interest in, and satisfaction from, helping people to learn.
- Patience and understanding with people who are slow to grasp what is being taught at the moment.
- Dedication to the subject matter and its ultimate goal, the protection of employee's protection and health.

There may be any number of "natural" instructors among the supervisors and the work force. Employees who are good at their jobs and have shown a flair for teaching job skills should be considered candidates for teaching job safety.

Credibility

Students must believe that the teacher knows the subject, so a thorough knowledge of both the job and job safety is essential to credibility. However, credibility is also bolstered if teachers are seen as people who *practice* what they teach. An experienced, skilled, and respected supervisor or a worker who is known to follow safety practices religiously can serve as a role model as well as an instructor. What must not be allowed to happen is for someone who is obviously not safety conscious—someone who has a poor safety record—to attempt to teach job safety. In a subject so dependent upon attitude and motivation as job safety and health, having a poor role model as a teacher dooms the course from the start.

13.7.2 Instructional Materials

Well-designed instructional materials can greatly improve the effectiveness of a training program. Generally speaking, the contribution of instructional materials is due to their content and the way they are organized rather than their appearance. Business and industry are being constantly bombarded by purveyors of "slick" materials—books, films, and so on—that are enormously pleasing to the eye but offer very little substance.

Anyone considering materials for use in a job safety and health program must go beyond the physical appearance of the materials to the content and organization. The two most important questions to ask about any book, film, or other item of material are:

1. Does it provide the information that users need to have?
2. Is it organized in such a way as to provide the information when it is needed?

Some of the most useful training materials are the home-grown type, developed within a company itself. They may be simply typewritten and photocopied, but they contain information about procedures, precautions, and hazards specific to the company that would never be found in materials prepared by commercial publishers.

Materials that can help to contribute to the effectiveness of job safety programs include:

- Instructor materials.
- Student materials.
- Instructional aids.

Instructor Materials

In job safety and health programs, most of what employees learn will come from the instructor. Materials that can help the instructor to teach effectively include:

- Objectives
- Lesson plans
- Support materials

Objectives. No two people will share exactly the same view as to what should be taught in a job safety training program. Once instructional objectives have been developed, they should be made available to all instructors so there is a clear understanding of what students are expected to do and know as well as what skills and attitudes they are expected to possess.

Lesson Plans. Lesson plans are simply sets of notes for teachers. A comprehensive and detailed lesson plan can serve several purposes.

- It helps teachers through a lesson, prompting them as to what happens next and making sure that everything that should be taught is included.
- It helps keep instructors on track, restraining them from "war stories" or other digressions.
- It provides a description of the course that can be made available to management, governmental agencies, or others who have the desire or need to know.

The primary function of a lesson plan is to *guide* instructors; it isn't intended to teach the subject matter. The level of guidance in lesson plans varies greatly. Some lesson plans take the form of verbatim scripts that instructors are expected to read. A script may be a necessity for an instructor who knows absolutely nothing about the subject or teachers who teach so seldom that they hardly remember anything. Most people in training agree, however, that a script leads to poor teaching, since it keeps the instructor's nose buried in a lesson plan, making it impossible to see and react to students. In addition, it generally leads to dull, monotonous presentation and prevents the interaction needed for effective instruction.

A lesson plan should be viewed as a guide that simply helps to prompt truly knowledgeable instructors. If the instructors do not know the subject matter, they need a textbook or a teacher preparation program, not a lesson plan.

The level of detail presented by a lesson plan will vary upon how frequently instructors are expected to teach. If they teach very frequently, say every week, they may need only a list of topics. On the other hand, if they teach only once every few months, they will probably need a more detailed list of points to be covered under each topic.

Some lesson plans provide two levels of detail. The body of the lesson plan contains detailed guidance, listing just every point to be covered. A more general set of guides is also provided, either in the form of highlighted headings or in a separate column. The idea is that instructors who are not very familiar with the course will use the detailed notes. As they become more familiar with the lesson plan, they become less dependent upon the details and use the more general notes. If they forget something, it is an easy matter to refer to the more detailed guidance.

An illustrative example of a lesson plan employing two levels of detail is shown in Figure 13.2.

Support Materials. In addition to objectives and lesson plans, a variety of materials can be made available for use of instructors outside of the lesson itself. Such support might include:

Text. While a text or other detailed discussion of subject matter is unwelcome within the lesson plan, it may be provided among support materials in order to:

- Allow instructors who are somewhat weak in the subject matter to become more knowledgeable.
- To provide knowledgeable instructors with a last-minute "brush up" before a lesson.
- As a resource to aid in answering questions raised during a lesson.

Possible sources of technical information will be mentioned shortly in discussion of independent study materials.

Tests. When tests are to be given to students, copies of the test should be provided to the instructor, along with an answer key. It is important that instructors review tests in detail before teaching a course in order to make sure that (1) specific items in the test are covered during the course, (2) the information given by the instructor agrees with the answers to the questions, and (3) the information in tests is given in a form that enables students to recognize the correct answer. Instructors should not highlight any of the specific information taught in class as "being on the test." The information called for on any test is merely a sample of all that students are expected to know. The test results cannot be expected to give a

LESSON PLAN

Notes	Detail
Visual No. 4	Show Visual No. 4 "Examples of Carbon Dioxide Fire Extinguishers" Review the key points that workers must know about fire extinguishers.
Parts of the fire extinguisher	Review the parts of the fire extinguisher.
Locking pin	Explain the locking pin. The locking pin presents accidental triggering of the fire extinguisher and discharge of the liquid carbon dioxide. The locking pin must be pulled out before the fire extinguisher can be used.
Squeeze grip	Squeeze grip Squeezing the grip causes the carbon dioxide to be discharged from the fire extinguisher. Be careful not to squeeze the grip until the fire extinguisher is aimed at the fire.
Carrying handle	Carrying handle The carrying handle is to be used when transporting the fire extinguisher from one place to another. Always use carrying handle when transporting the fire extinguisher. Holding fire extinguisher by the carrying handle minimizes the chances of accidently dropping it and damaging it.
Cylinder	Cylinder The cylinder holds the liquid carbon dioxide.
Siphon tube	Siphon tube Within the cylinder, a siphon tube carried the liquid carbon dioxide to the hose.
Hose	Hose The hose is fastened to the top of the fire extinguisher and allows the liquid carbon dioxide to be pointed at the fire without having to turn the entire fire extinguisher.

FIGURE 13.2. Illustrative example of a lesson plan.

valid appraisal of the students' total knowledge of the subject if they have an opportunity to bone up on specific questions.

Visuals. It is desirable to provide instructors with hard copy of any visuals used in instruction so they may be previewed along with the lesson plan. Some instructors like to insert hard copy of any visuals in the body of the lesson plans in order to see the visual during class without having to turn away from the students to look at a screen.

Narrations. Use of slide-cassette presentations as an instructional aid will be discussed in a moment. Where they are used, a copy of the narrations will enable the instructor to:

- Manually advance the slides if no synchronizer is available, or if the slides get out of sync.
- Deliver the narration orally if cassette player or cassette are unavailable.
- Vary the narration to adapt the presentation to local needs.

Packaging. It is beneficial to the instructor to put instructor materials—objectives, lesson plans and support materials—under one cover. Thus they will all be in one place and the instructor can easily refer back and forth between them. A three-ring binder is ideal in that it allows instructors to reorganize the material to suit their own preferences and insert material as needed to keep the course updated.

STUDENT MATERIALS

A variety of materials can be provided to students to help them obtain course objectives. We can classify them according to their purpose:

- Independent study materials.
- Classroom materials.
- Job materials.

Independent study materials. Students who are expected to study by themselves must be provided the materials with which to do it. A compilation of safety procedures, precautions, and hazards in the form of a textbook would certainly meet the needs. However, few employers have the resources to prepare a job safety text for their employees. Fortunately, the need for study material can be handled simply by having available:

- Sections of safety and health publications dealing with materials relevant to the particular job.
- Brochures and pamphlets dealing with related safety concerns.
- Sections of equipment manuals dealing with safety.

If materials are reproduced for this purpose, they must either be free of copyright or releases must be obtained from the publishers. A great volume of safety and health material in the public domain can be obtained from the National Institute of Occupational Safety and Health. Write to:

National Institute of Occupational Safety and Health
4676 Columbia Parkway
Cincinnati, OH 45226

Classroom Materials. Certain classroom activities call for handouts to be available to students. These include:

- Case studies to serve as a basis for problem solving exercises and group discussions.
- Materials to be filled out during class, such as work sheets for calculations, administrative forms, and equipment checklists.
- Technical reference materials supplying data to be used in classroom exercises.

Job Materials. Materials provided to students in training can go beyond supporting instruction to actually helping them perform their jobs more safely. Examples are:

- Equipment operating manual and lists.
- Warnings that can be posted on hazardous equipment or in high hazard areas.
- Data sheets listing maximum safe levels of various substances.
- Company safety policies and procedures.

Instructional Aids. Several different types of training materials can be provided to help instructors teach more effectively. The most common types are:

- Visuals
- Audiovisual presentations
- Videocassette recording

Visuals. The use of visuals allows student to use their eyes as well as their ears to learn. The two most common types of visuals are:

- 2 × 2 slides
- 8½ × 11 transparency

Slides are easier to handle than transparencies and generally less expensive to produce and reproduce. Photographic slides are particularly useful in helping workers to identify various job safety hazards. Keep a 35 mm camera at hand to shoot pictures of hazards as they are encountered in the work situation. Text slides are also relatively easy to prepare by photographing printed materials. Pamphlets describing techniques for preparing very attractive slides can be found in most photography stores.

Overhead transparencies are most effectively used where a lot of detail must be presented (equipment diagram, detailed checklist) or where the visual must be manipulated during instruction (showing how to fill in a form). Transparencies also have the virtue of being quickly prepared from hard copy with present-day reproducing machines.

A-V Presentations. Films and slide-cassette presentations have been widely used in job safety training. Unfortunately, they have also been widely *misused.* Films and slide-cassette presentations have the ability to present a variety of types of information in a particularly vivid form. Regrettably, the training films sold do not always make use of these potential benefits. Too often they are intended to entertain more than enlighten. And too often, those who buy them on behalf of job safety training are not competent to judge whether the presentations have any training value or not.

Before purchasing any audio visual presentation, examine it closely and ask the following questions:

What does it say? What information does it present? How much of the information is job relevant? How much of the content is devoted to irrelevant information or some "story line"?

How well does it communicate? Is the relevant information communicated in such a way that it can be understood?

Is it well organized? Many A-V producers have an aversion to structure, and deliberately present information in a free-form manner that can be difficult for students to follow and organize in their own minds.

Is it up to date? A-V presentations may be marketed for many years in order to recoup production costs. There is nothing wrong with old materials as long as they don't supply out-of-date information or have an obviously dated appearance.

Is the medium fully exploited? To be worth its cost and the time it takes to show, an audio-visual presentation should provide something that the instructor cannot. Be wary of films that show little more than "talking heads."

Videocassette Recordings. Videocassette recorders (VCRs) have an enormous potential for employee training. Videos depicting specific safety procedures and job hazards can be prepared quickly and inexpensively. The person using a video can see immediately what has been recorded and reshoot it if necessary without additional material costs. Videocassette recording allows very effective presentations to be prepared with little training or experience. In addition to providing video presentations, VCRs can be used by students to watch their own performance during training and spot things they are doing wrong.

One minor limitation of video is that it cannot be shown to large groups without using very expensive monitors or projection equipment. This limitation is no handicap in most job safety training programs, where classes are quite small. On the plus side is the fact that video monitors can be brought to almost any place within a building very easily and can be shown without dimming the lights.

13.7.3 Equipment

The equipment used in training can be divided into three categories:

- Training equipment
- Training devices
- Operational equipment

TRAINING EQUIPMENT

Training *equipment* is that equipment designed for general training purposes, such as projectors, videorecorders, chalkboards, easels, and so on. The types of training equipment required for any course will be dictated by the instructional methods and materials that have been selected for use in instruction. Local audiovisual dealers can provide information on the advantages and disadvantages of specific models and grants. A useful reference providing information on the effective use of audiovisual equipment is: *Media Equipment: A Guide and Dictionary,* by K. C. Rosenberg and J. S. Doskey, Littleton Co. Libraries Unlimited, 1976.

TRAINING DEVICESS

Training *devices* are items of equipment that have been designed for specific training purposes, including simulators, working models, and mockups.

Training devices generally can be used to provide demonstrations and allow practice of job tasks where use of operational equipment is (1) too expensive, (2) too dangerous, or (3) unable to present job tasks on cue. Because of their limited scope of use, training devices are not mass produced. However, enterprising instructors often build training devices by salvaging wornout pieces of operational equipment and modifying them for training purposes. Examples include:

- "Bugging" equipment to demonstrate hazards and how to prevent them.
- Cutting equipment open to reveal internal hazards.
- Modifying equipment electrically or mechanically to allow demonstration of operating hazards without students being shocked, burned, cut, pinched, and so on.

Where old equipment is not available or serviceable, many instructors have actually constructed models, in full or reduced scale, to permit demonstration or practice of safety procedures.

OPERATIONAL EQUIPMENT

In demonstrating many aspects of job safety, there is nothing like the actual equipment for effective instruction. One advantage that industrial job safety programs enjoy over many vocational school programs is that the equipment is readily at hand. This includes both production equipment (lathe, forklift, hammers, etc.) and safety equipment (respirators, fire extinguishers, emergency lights, etc.). In large companies with extensive training programs, it may be feasible to allocate units of operational equipment specifically to safety training purposes and put them in the classroom where they can be used easily by students without interfering with normal job operations. Where this isn't feasible, students can be taken to the equipment in order to provide demonstrations and practice of critical tasks.

13.7.4 Facilities

While many books on training discuss them at great length, facilities are probably the least important aspect of job safety training. If training has the right objectives, employs the right methods to achieve them, and uses qualified instructors, it can be given in a barn. This is fortunate, since the facilities for employee training are rather limited in most industrial and business organizations. Where elaborate facilities are available, they are more likely to be at corporate headquarters than in the plants and offices where most employees work.

All that is needed for job safety instruction is a room that will accommodate about 12–18 people. It is rarely efficient to muster more than this number at any one time for a true safe job safety training program, particularly if instruction is to involve active participation by students. Where instruction involves working with heavy equipment, it is usually better to bring the students to the equipment than to bring the equipment to the students.

Arrangements of classroom equipment and facilities are described in training textbooks. The arrangement most frequently used is one in which students sit behind tables on which they can keep books and classroom materials and take notes. Two pairs of 3-person tables, 1 on each side of the room, would accommodate 12 students. For a class of 25, 3 sets of 4-person tables can be placed on each side of the room. The tables and chairs should be turned slightly inward so that all face front-center, where the instructor and video monitor or projection screen will be located.

13.7.5 Students

Training programs can be successful without elaborate training aids and devices, without special facilities and even without instructors. But there is one thing that a training program cannot do without and that is *students.* Many employees have set up job safety programs to protect their employees from injury or illness and then have been unwilling to provide time for employees to participate. The principle, you don't get something for nothing, applies as much to job safety as to anything else. Management must be willing to provide ample time for workers to attend job safety training on a continuing basis if they expect to reap the benefits.

13.8 HOW CAN YOU EVALUATE TRAINING?

Two things must be evaluated in training: the students and the program. For the most part, these aspects of evaluation can be handled by the same methods and measures. To evaluate students, you look at individual test scores; to evaluate the program, you look at overall results. For example, if a few students do poorly on a knowledge test, you interpret the results as meaning that those students either were not paying attention or were slow to grasp the information. If overall results are poor, it means that the methods and materials used in training are not effective.

13.8.1 The Scope of Evaluation

The scope of any training evaluation is defined by the objectives of the training program. Because of the great number of specific performances that make up the objectives of job safety training, it is not possible to include all of them in the evaluation. Tests used in evaluation can only *sample* from the specific performances, knowledge, skills, and attitudes that make up objectives. The items on a test are a sample of what workers are responsible for. If the items are correctly chosen, the sample will provide an accurate estimate of what people know, can do, and believe. To provide an accurate sample, the items making up a test must meet two conditions:

1. They must be drawn from all of the performances, knowledge, skills, and attitudes making up safety on that particular job. They cannot concentrate on only one aspect of the job.
2. Students can't know what specific items are going to be on the test. Telling them lets them prepare for those specific items and tests results no longer reflect attainment of the overall objectives.

The four types of measures that have been used to evaluate students and programs are:

- Performance measures
- Knowledge tests
- Skill tests
- Attitude surveys

13.8.2 Performance Measures

A performance measure is simply a test in which the workers are required to demonstrate performance of some aspect of their jobs. Performance measures can be divided into three categories:

- Tests
- Observations
- Records

Performance Tests

A performance test is a standardized system for measuring performance of some job tasks, for example:

- Stacking boxes with a forklift.
- Backing a truck into an alley.
- Feeding material into a drill press.

In a performance test, every aspect of the task to be performed, the conditions under which the performance is to occur, and the criteria for measuring the level of performance must be specified in detail. Such detailed specifications are the only way to make sure that the test is given the same way every time and therefore can be used to evaluate courses.

One limitation of performance tests in assessing safety of performance is that they assess only what people can do not what they normally do. Much of job safety training involves getting workers to comply with the safety procedures and precautions that they already know how to perform, such as wearing hard hats, deactivating equipment before working on it, and so on. Workers will, of course, employ safety practices when they know they are being tested, whether they do so on the job or not. Therefore, performance tests are best used in testing the worker's ability to meet objectives that involve special knowledge or skill, such as the use of respirators or fire extinguishers.

Performance Observation

Those performances that cannot be validly measured through performance tests can often be assessed through observations of workers in their everyday job performance. The workers should not know that they are being observed, or at least not for the specific purposes of evaluating how safely they are performing. The observations can be made as a part of a supervisor's routine daily monitoring of performance.

To use performance observations for evaluation, you need to:

- Prepare objectives identifying the specific behavior to be observed.
- Tally each instance of an correct behavior and each error. (Don't just count errors, since that doesn't account for differences in the amount of work.)
- Make all observations of a particular worker at the same time of day and day of week so that observations aren't affected by differences in conditions.
- Compare the rate of error before and after training to see if there is any improvement.

PERFORMANCE RECORDS

Many of the records kept by companies can be useful in evaluating, training and in identifying areas where improvement is needed.

Accident records: In addition to helping to determine what should go into training, accident records can help show how effective training has been in improving job safety.

Equipment records: Accidents that result only in property damage may not show up in accident records. Equipment repair records may serve as an indication of how training has affected minor accidents.

No attempt should be made to use accident or repair records unless there are enough of them to produce a reliable sample of incidents. As a rough rule of thumb, there should be at least 1000 accidents or equipment breakdowns in periods being compared (before and after training) to give confidence in the results.

13.8.3 Knowledge Tests

When it comes to assessing what people know about safety of job performance, written knowledge tests are much more efficient than performance tests and better at diagnosing specific knowledge deficiencies. Where only a few people are being treated, subjective tests can be used, that is, students can simply be given questions and asked to write out their answers, for instance What are the biggest hazards in working with the XYZ equipment?

If a training program is to be given many times and large number of students are involved, objective, multiple-choice tests take less time to administer and score. Here are some guidelines to observe in writing multiple choice items for safety training programs:

1. Use simple language; the test is to measure safety, not vocabulary or reading ability.
2. Try to focus upon information people *need* in order to behave safely; ignore "nice-to-know" information such as history.
3. Avoid asking for definitions unless they are essential to safety.
4. Avoid asking for information that workers would normally look up when it's needed.
5. Always ask for the *correct* answer, not the "best" answer.
6. Make all alternative responses deal with the same piece of information.
7. Avoid *true-false* questions; few things are completely true or false.
8. Avoid "good" words like *safe* or *cautious.*
9. Keep all alternatives about the same length.
10. Avoid negative questions, such as "Which of the following is not. . . ." People tend to forget that are looking for the wrong answer.
11. Avoid *all of the above* items; people may pick the first correct answer recognizing that they are all correct.

13.8.4 Skill Tests

While the most valid way to measure job safety skills is through performance tests, it is often quicker, cheaper, or safer to develop special tests of particular job skills. Two examples of such tests are:

- "Road-eo" exercises to test skill in handling vehicles.
- A set of photographic slides to test the ability to recognize hazardous situations.

It is important to make sure that the skills being tested are the same ones required on the job. The people who score highest on the test should also be rated as most skillful on the job.

13.8.5 Attitude Surveys

While attitudes are among the most important determiners of safe job performance, they are also among the most difficult to measure. People tend to be rather guarded in revealing their true feelings about things having to do with safety. However, if attitudes are surveyed under conditions in which anonymity is guaranteed, the results can be illuminating. Attitude measures can help in identifying the specific attitudes most in need of improvement and in evaluating the effectiveness of instruction designed to bring about the improvement.

The most common way to measure attitudes toward job safety is through surveys in which workers are asked to state their opinion relative to various safety issues. An item from such a survey might be: Workers in construction zones should wear hard hats:

a. All the time.
b. When there is danger of falling objects.
c. When they want to.

One would hope that workers who are trained in job safety would believe that situations creating hazards to the head cannot always be anticipated would be more inclined to choose **a.** If they didn't, it may be that the training in the importance of hard hats needs improvement. Of course, if everyone accepted **a** at the outset, there might not be a need for training on this issue at all.

13.9 SUMMARY

An effective job safety program can be established with surprisingly little cost or technical expertise if you go about it in a thoughtfully and systematically way. Anyone who is knowledgeable and dedicated to job safety can set up an effective program using the guidelines presented in this chapter. Here are some questions you can ask yourself as a final check:

WHOM TO TEACH

Workers: Do I have a program for all the workers who need it, not just those who operate obviously dangerous equipment?

Supervisors: Do I have a special program for supervisors, to help them teach and monitor safety as well as practice it?

Management: Do I have a means of convincing management that an effective job safety program can pay off for the company?

WHAT TO TEACH

Have I made the best possible use of job hazard analysis, available publications, and accident information to identify the most important job safety procedures, special precautions, and hazardous conditions?

Objectives. Have I written down what I want to accomplish, what I want people to do, know, believe, and be skillful at?

HOW TO TEACH

Am I using the right method to teach each subject?

Am I allowing the workers enough opportunity to interact with the teachers and with one another?

Have I given students a chance to study on their own—and a good reason for doing it?

WHEN TO TEACH

Am I using the right mixture of:

Preservice training: To teach workers the most critical aspects of job safety before they are exposed to job hazards?

In-Service training: To ultimately teach all aspects of job safety; to cover special hazards and to provide occasional reminders.

WHAT YOU NEED TO TEACH

Instructors: Do the instructors have the technical know-how, teaching ability, and credibility to be good teachers?

Instructor materials: Do I have a detailed set of lesson plans to make sure instructors are teaching the right things?

Student materials: Have I provided those materials (and only those materials) that students need in order to learn and perform on the job?

Instructional aids: Are the films and other instructional aids I am using appropriate to what I am trying to teach?

Equipment: Do I have access to the equipment needed for training?

Students: Can I be sure that students will get enough time to complete the program?

HOW TO EVALUATE

Evaluating needs: Have I evaluated the current state of affairs to know where training is needed?

Evaluating students: Do I have a system for evaluating student progress and final achievement?

Program evaluation: Do I have a system for evaluating the effectiveness of the program in proving job safety company-wide?

REFERENCE

Cohen, H. H. and Jensen, R. C., Measuring the Effectiveness of an Industrial Lift Truck Safety Training Program, *Journal of Safety Research,* Vol. 15, pp. 125–135, 1984.

CHAPTER 14

How to Select Training Methods and Media

MARGARET C. SAMWAYS
NUS Corporation
Pittsburgh, Pennsylvania

14.1 OVERVIEW OF APPROACH

Historically, the selection of methods and media for occupational safety and health (OSH) training has put the cart before the horse. Too often, the decision-maker is guided more by the availability of certain media techniques than by the instructional job that needs to be done. Paradoxically, the packaging and marketing of OSH training programs simplifies decision making, yet at the same time makes it more difficult. Attractively packaged and cleverly produced audiovisual programs can answer real instructional needs; on the other hand, the instructional intent of shaping or changing work behaviors can be lost.

When one considers that the human factor accounts for an estimated 90% of the causation of occupational illnesses and injuries, and that a healthy workforce is a major factor in productivity and profitability, the importance of effective training cannot be overestimated.

Reaching the goal of an informed workforce presents a formidable training problem. Workers' appraisal of workplace risk is not always rational and is influenced by societal factors outside the control of the trainer. For example, a worker who is diligent in following safe work procedures with certain chemicals may be lax when working with the same chemicals in his home hobby shop. Conversely, workers may reject health measures in the workplace because of peer pressure to appear "macho." Serious health risks, such as cigarette smoking, are voluntarily continued off the job, while at the same time a great concern may be felt about workplace exposures to low levels of chemical agents that have no known ill effects. More critical still, it is known that people tend to reject or not to recall "bad news" messages, such as an unfavorable prognosis given by a physician. Philip Ley (1966) began work in the area of physician/patient communication due to his early finding that "patients tend to have particularly poor memories for medical information." Ley found, in fact, that persons receiving information about their health forget over one third of what they are told. Gillum and Barsky (1974) found that both high and low levels of anxiety appear to have an adverse effect on the retention of information. In the workplace, Leventhal found that fear arousal in messages caused by vivid descriptions of hazards can evoke desired concern, but that the same message must also contain instructions for controlling the danger so as to reduce the feeling of helplessness (Leventhal, 1970, 1967). The messages are clearer when they refer to obvious hazards with immediately harmful effects, such as the flammability of gasoline vapors. Less understandable are messages about the statistical probability of long-term harmful effects resulting from exposure to chemicals to which workers are exposed day after day. These issues, which lie at the heart of communicating hazards effectively to workers, are addressed in the *Guidelines to Effective Worker Communications,* issued by the National Institute of Occupational Safety and Health (NIOSH), in 1985. All hazard communication training programs, and indeed any other occupational safety and health training programs, should recognize these factors, which run counter to the training effort. The "performance" nature of the OSHA Hazard Communication Standard has made it necessary to be able to demonstrate that acceptable levels of performance have been reached. Throughout this chapter, the importance of evaluating the effectiveness of training is stressed, and Section 14.7 is devoted to this issue. As pertains to the Hazard Communication Standard, the use of tests to measure and document worker understanding is essential. Without such documentation, it will be impossible to show the visiting compliance officer or company auditor that the transfer of knowledge has occurred. Potential losses resulting from a deficit of testing documentation are even greater in the area of "failure to inform" discussed in the section called *Legal Liability Issues.*

To achieve effective training at the lowest possible cost, a decision process should be followed, involving many diverse factors. The first step, Needs Assessment, has to do with assessing the needs and the forces, external and internal, that trigger the needs. The second, Task Characteristics, concerns the characteristics of the desired training outcomes, such as skills, knowledge, or attitudes. The third, Population Characteristics, involves the analysis of the target workforce, including such factors as age and educational level. The fourth, Resource Characteristics, covers such items as budgets, the availability of training facilities, and the options to purchase programs and services. Only after all these items are considered is it possible to make an informed selection among the many method and media choices.

14.2 NEEDS ASSESSMENT FACTORS

14.2.1 External Factors

Regulatory

Of all the external forces that bring about a need for training, the easiest to identify are regulatory requirements. The federal government has generally recognized the need for worker education and knowledge and has established requirements concerning various industrial hazards.

The Occupational Safety and Health Administration (OSHA), following the passage of the Occupational Safety and Health Act in 1970, has taken the lead with respect to the design of effective worker training. OSHA has also established rules concerning specific chemicals, personal protection devices and procedures, and hazard communications. Many OSHA regulations are derived from national consensus standards. Recently, OSHA has issued Voluntary Training Guidelines (Federal Register, Vol. 49, No. 146) that describe the steps that should be followed to develop an effective program. The Mine Safety and Health Administration (MSHA), now administered with, OSHA by the Department of Labor, has also developed prescribed training for miners, and is in the process of refining its training requirements. Training is also required under the Resource Conservation and Recovery Act (RCRA) and under Title 10, 19.12, by the Nuclear Regulatory Commission.

State and local governments have, in the last few years, issued a number of laws and ordinances that involve training. A brief summary of some major regulatory requirements follows.

OSHA General Duty Clause

The General Duty Clause of the OSH Act (Public Law 91-596) requires that each employer:

1. Shall furnish to each of his employees employment and a place of employment free from *recognized* hazards that are causing or are likely to cause death or serious physical harm to his employees.
2. Shall comply with occupational safety and health standards promulgated under this Act.

It is generally understood that the requirement to provide training and information is implicit in these statements. More specific implications are in the description of the responsibilities of the employee who

> shall comply with occupational safety and health standards and all rules, regulations, and orders pursuant to this Act, which are applicable to his own actions and conduct.

Obviously, compliance with rules and safe work actions can only be achieved through effective training. The employee also has numerous rights under each standard promulgated following the OSH Act, one of which is the "right to be made aware of the hazards found in the workplace." For example, what hazardous physical or chemical agents are used or stored in the workplace? What health problems are associated with exposures to such agents? What methods can be employed to minimize exposure? What type of protective devices should be used, and how should they be used? Employees should also know the appropriate response to all warning signs and signals and should be told to report to their supervisor any condition that could promote injury or illness. They also are entitled to know the magnitude and severity of any or all of their exposures to physical and chemical agents and may, if they so desire, obtain copies of medical evaluations made for the purpose of managing such exposures. This knowledge can only come about by means of systematic education and training.

The implication for training of the General Duty Clause and the standards that have been promulgated are immense. The Hazard Communication Standard and state/local Right-to-Know Laws, discussed in detail in Section 14.2.1, have only served to focus even more specifically on the essential role of worker communication and training. It is apparent that training is the point of delivery for all industry's efforts to prevent occupational injuries and illness, and is therefore the key to an effective program.

Specific Vertical Standards

In contrast to the general injunction to train that is implicit in the General Duty Clause, there have been a number of "vertical" standards issued by OSHA that apply to specific chemicals or operations. All such standards issued in recent years contain specific training requirements. These requirements can be found at the end of each standard, since it is felt that training is the last function to be performed once all other controls are in place. The requirements of the Lead Standard (29 CFR Part 1910.1025) are typical.

The requirements are:

LEAD

1910.1025 (1) (1) and (2)

(1) Employee information and training.

(1) Training program.

(i) Each employer who has a workplace in which there is a potential exposure to airborn lead at any level shall inform employees of the content of Appendixes A and B of this regulation.

(ii) The employer shall institute a training program for and assure the participation of all employees who are subject to exposure to lead at or above the action level or for whom the possibility of skin or eye irritation exists.

(iii) The employer shall provide initial training by 180 days from the effective date [Editor's note: OSHA's lead standard became effective February 1, 1979.] for those employees covered by paragraph (1) (1) (ii) on the standard's effective date and prior to the time of initial job assignment for those employees subsequently covered by this paragraph.

(iv) The training program shall be repeated at least annually for each employee.

(v) The employer shall assure that each employee is informed of the following:

(A) The content of this standard and its appendixes;

(B) The specific nature of the operations which could result in exposure to lead above the action level;

(C) The purpose, proper selection, fitting, use, and limitation of respirators;
(D) The purpose and a description of the medical surveillance program, and the medical removal protection program including information concerning the adverse health effects associated with excessive exposure to lead (with particular attention to the adverse reproductive effects on both males and females);
(E) The engineering controls and work practices associated with the employee's job assignment;
(F) The contents of any compliance plan in effect; and
(G) Instructions to employees that chelating agents should not routinely be used to remove lead from their bodies and should not be used at all except under the direction of a licensed physician;
(2) Access to information and training materials
(i) The employer shall make readily available to all affected employees a copy of this standard and its appendixes.
(ii) The employer shall provide, upon request, all materials relating to the employee information and training program to the Assistant Secretary and the Director.
(iii) In addition to the information required by paragraph (1) (1) (v), the employer shall include as part of the training program, and shall distribute to employees, any materials pertaining to the Occupational Safety and Health Act, the Regulations issued pursuant to the Act, and this lead standard which are made available to the employer by the Assistant Secretary.

These and other training requirements of OSHA Standards can be found in the OSHA publication *Training Requirements in OSHA Standards and Training Guideline* (1985).

MSHA

The Bureau of Mines was established in the Department of the Interior in 1910, as the result of a series of coal mine disasters. In 1973, the Mine Enforcement Safety Administration was created from the health and safety division of the Bureau, and charged with administering the Metal and Non-Metal Mining Act of 1966 and the Coal Mine Safety and Health Act of 1969. In 1977, the Federal Mine Safety and Health Act (PL 91-173, as amended by PL 95-164) moved enforcement from the Department of the Interior to the Department of Labor and created the Mine Safety and Health Administration (MSHA). MSHA now shares an equal position with OSHA under the Secretary of Labor.

Throughout its history, the Bureau emphasized the importance of training for miners; this emphasis has been continued in MESA and MSHA. The requirements are set out in Title 30 of the Code of Federal Regulations for both underground and surface mines. The training is expressly prescribed; for example, each new underground miner "shall receive no less than 40 hours of training as prescribed in this section before such miner is assigned to work duties." The prescription for each section of the course then follows. MSHA training is thus time based rather than performance based, and the tacit assumption is made that all workers learn the same amount at an equal rate. Since there is no requirement to find out if the training is understood, the training certificate that each miner receives certifies only that he or she has attended the prescribed courses for the stated number of hours. This approach is in contrast to the performance nature of the OSHA Hazard Communication Standard.

EPA

The Clean Air Act was passed and the Environmental Protection Agency established in 1970; this had the effect of consolidating all federal pollution control agencies into one organization. In 1976, the Toxic Substances Control Act (TSCA) gave the EPA administrative and enforcement powers over new and existing chemicals entering the environment by means of premanufacturing notification, screening, testing, manufacturing bans, and penalties.

Section 8(e) of TSCA requires that workers be informed annually of their right to report to the EPA any information that reasonably supports a conclusion that a chemical substance or mixture presents a substantial risk of injury to health or the environment. Similarly, under the 1976 Resources Conservation and Recovery Act (RCRA), workers should be informed of their right to report to the EPA concerns regarding the handling and disposal of hazardous and other wastes.

These worker "informing" requirements of TSCA and RCRA, which can also be found in OSHA standards, are less onerous to the trainer than the actual training requirements mandated in OSHA and MSHA standards. They can be met through simple informational means, such as posters, explanations in the company newsletter, or safety meeting talks with handouts. Some inexpensive slide/tape programs covering the TSCA 8(e) and RCRA requirements are available from vendors.

DOT

Several sections of CFR Title 49 relate to the shipment of hazardous materials, which requires training. Section 173.1(6) states that: "It is the duty of each person who offers hazardous materials for transportation to instruct each of his officers, agents and employees having any responsibility for preparing hazardous materials for shipment as to the applicable regulations." Training is also required of Carriers. Sections 174.7 (rail); 175.20 (air); 176.13 (water); 177.800 and Parts 390-397, excluding Sections 397.3 and 397.9 (highway), all state that "It is the duty of each carrier to make the prescribed regulations effective and to thoroughly instruct each of his officers, agents and employees in relation thereto."

These training requirements have prompted organizations such as the American Trucking Association to offer seminars at regular intervals across the country. Some instructional packages are also available.

Hazard Communication Standard and State and Local Right-To-Know Laws

The OSHA Hazard Communication Standard (29 CFR 1910.1200) has been termed the most significant landmark in the history of occupational safety and health regulatory activity. The standard is unique because it is a "performance" rule; that is, industry has great latitude in the manner of compliance, as long as the intent of the standard is met. Since the intent is to communicate hazards to workers in an effective manner, training is the key requirement that determines the success of all the other requirements pertaining to hazard assessment, labeling, material safety data sheet distribution, and so on. This standard alone has caused a quantum leap in the visibility of training activity, and will have a considerable impact on the availability of budgetary and other resources to the trainer.

Because of its importance, the training and information requirements are summarized below: The employer must:

- Establish a training and information program for employees exposed to hazardous chemicals.
- Provide such training at the time of initial assignment.
- Provide training whenever a new hazard is introduced into the work area.
- Train employees who have not received initial training.

1. The employer must inform employees of:
   - The requirements of this standard.
   - Operations in their work area where chemical hazards are present.
   - Where the facility is keeping the written materials required by the standard.
     a. Written hazard evaluation procedures.
     b. Written program.
     c. Lists of hazardous chemicals.
     d. MSDS collection.
2. The employer must train employees in:
   - Methods and observations they may use to detect the presence of a hazardous chemical in their work area, e.g., visual presence, smell.
   - The physical and health hazards of chemicals in their work area (may be general to classes of chemicals).
   - Specific hazardous chemical handling information available through the MSDS.
   - The protective measures employees can take to protect themselves from hazardous chemicals.
   - The specific procedures initiated to provide protection, including:
     a. Work practices.
     b. Personal protective equipment.
     c. Emergency procedures.
     d. How the hazard communication program is implemented in the workplace.
     e. How to read and interpret information on labels.
     f. How to read and interpret material safety data sheets.
     g. How to obtain available hazard information.
     h. How to use available hazard information.

The compliance date for training and information was May 25, 1986.

The standard makes it clear that much of the information and training must be workplace specific. This means that purchased "generic" training programs will not meet all the requirements, although they may be useful for orientation to such topics as general categories of physical and chemical hazards, understanding the material safety data sheet, and understanding the purpose and method of operation of personal protection equipment. The onus of developing, implementing, and documenting training will fall heavily on the field or location coordinator, who is the only person fully acquainted with the specific characteristics of each workplace. The Compliance Directives, issued in August 1985, have provided further indications of what constitutes "compliance" in the area of training. Some guidance has been given by OSHA in the *Training Guidelines,* mentioned previously. Although these guidelines are intended for all occupational safety and health training, they were issued at a time that made them particularly relevant to the hazard communications training problem. There are obviously many ways in which the training could be approached, varying for all the reasons that are described later in this chapter. A general approach is described below.

Steps to take to meet the training/information requirements of the Federal Hazard Communication Standard

I. Evaluate training needs
   A. Consider previous training history: What training has been conducted; how often? What was the content? How effective was it?
   B. Use existing company records to define exposure groups
      - Inventory of chemicals used, stored or otherwise present in the workplace; serves as an index of MSDS sheets for hazardous materials present in each workplace.
      - Description of job tasks performed for each job classification; gives information on potential exposures to hazardous chemicals.
      - Further define exposure groups by reviewing available industrial hygiene monitoring data.

II. Formalize/standardize existing training programs to fulfill requirements
   A. Program criteria
      1. Information about the Hazard Communication Standard
         a. Information on the requirements
            - Provide written information or verbal instruction
      2. Program training
         a. Hazardous substance list
            - Provide information on the location and availability of the list
            - Provide access
         b. MSDS
            - Explain purpose and content of MSDS
            - Specify location and availability of MSDSs
            - Provide MSDS Users Guide for reference and periodic review
            - Store MSDSs for each workplace in a binder that is retained in the workplace
            - For employees working in remote locations, compile binder of MSDSs to be carried in their vehicle
         c. Labeling
            - Describe workplace labeling program
            - Explain hazard warning terms
            - Where chemicals in unlabeled pipes or vessels are identified using process sheets, batch tickets, etc., train employees to link this information to specific chemical identity and hazard information (MSDS)
      3. Chemical training
         a. Tailor chemical hazard training to the exposure groups previously identified
         b. Where appropriate, group chemicals into classes with similar hazards and precautionary handling procedures
         c. Provide specific training for high hazard chemicals
         d. Provide additional training when new chemicals are introduced into the workplace, or when employees are reassigned to another work area involving different chemical exposures
   B. Performance evaluation
      - Provide means of assessing the effectiveness of existing training programs
   C. Documentation
      - Employee
      - Training received
      - Frequency
      - Follow-up

III. Develop new training programs where needed
   A. Track the use of new elements in the workplace that might necessitate the development or purchase of new programs
      1. New chemicals
      2. New work procedures
      3. New policies and administrative procedures
      4. New information sources, such as changes in the MSDS
   B. Review and evaluate ongoing programs
      1. Replace ineffective programs
      2. Upgrade programs by the addition of new supplementary materials
   C. Search for alternative programs for retraining purposes
      1. Review new vendor programs as possible candidates
      2. Interact with peers through professional and trade associations to find out new and improved training methods and media

The state and local right-to-know laws and ordinances are different in intent from the Federal standard. First, they are not "performance" oriented, but are concerned more with the disclosure of chemical information by industry. Secondly, they reach beyond the workplace to the community, an area not touched by OSHA. At the time of writing, all except three of the state laws contain training requirements; two of these three plan to follow the federal requirement. One difference to be found in some states laws is an annual retraining requirement. It is possible that OSHA may eventually "retrofit" its requirements in this regard to be consistent with these states. The lack of a "performance" orientation in the state laws is a significant factor, and there is no requirement, either explicit or implicit, that training has to be effective. Since the state and local laws may eventually be preempted by the federal requirement, it is advisable to work towards the same goal as the OSHA requirement, which is an informed workforce knowledgeable of the hazards and skilled in self-protection measures.

Consensus Standards

After the Occupational Safety and Health Act was passed in 1970, the initial standards to appear consisted of existing mandatory national consensus standards and of existing federal standards promulgated under five acts concerning construction work, ship repairing, shipbuilding, shipbreaking, and longshoring. The standards became effective August 27, 1971. The major sources of the national consensus standards adopted in 1971 were the American National Standards Institute and the National Fire Protection Association (NFPA). Other bodies that had material approved for incorporation by reference included the American Conference of Governmental Industrial Hygienists (ACGIH), the American Society for Testing Materials (ASTM) and the Underwriters Laboratories (UL). However, any organization that deemed itself a producer of national consensus standards was invited to write to the assistant secretary of labor before February 1983 to determine whether any of its standards met the criteria defined in the Act. This idea of using standards derived from private organizations with industry representation has enabled OSHA and industry to keep in step better than if OSHA had imposed totally new requirements. It also means that new consensus standards must be carefully watched for their regulatory impact. For example, an upcoming ANSI Standard (Z-135) that may have some impact on hazard communication training deals with hazard warning signs. Adoption of these signs might necessitate changes in the training program.

Legal Liability Issues

Before the mid-1970s, responsibility for the occupational health and safety area resided principally with the occupational physician, the industrial hygienist, the safety engineer, and other medical or scientific personnel. The lawyer was consulted on occasion to interpret a new regulation or to review policies and procedures prior to publication. Since 1973, however, a revolution in the legal arena has forced industry to focus its attention on the legal issues associated with toxic substances. The case viewed as the landmark was *Borel v. Fibreboard Paper Products Corporation et al.*, 93 F.2d 1076 (1973), in which Clarence Borel, an industrial insulation worker, sued the manufacturers of insulation materials for failure to warn of the dangers involved in handling asbestos. As a result of this and subsequent cases, personal injury suits associated with alleged exposure to toxic substances are having, and will continue to have, a dramatic economic impact upon the industrial community. Known as "toxic torts," such suits have been initiated by the manufacturer's own employees, the employees of other companies to whom a company may supply raw materials or other products, consumers, contractors, and other persons entering a manufacturing facility, and persons who may be exposed to emissions or wastes generated during manufacturing operations. Although there have been very few verdicts, many settlements of millions of dollars have been made in toxic tort cases, and there are billions of dollars in outstanding claims. The first class of potential plaintiffs, the manufacturer's own employees, is the group of most interest to this discussion. From 1908, when the first workers' compensation law that successfully passed the test of constitutionality was adopted by the federal government, the concept of compensation for any disease or injury occurring at work or caused by the work environment became universal. Although the insurance coverage varies from state to state, acceptance by the worker of workers' compensation payments precludes recovery of additional compensation by a legal suit. However, most state workers' compensation laws contain exceptions that will allow employees, under certain circumstances, to sue their employers directly. The most common exception involves damages incurred as a result of intentional actions that may make the employer subject to punitive or exemplary damages beyond the compensatory damages suffered by the plaintiff. The most common basis for such suits is the employer's duty to warn workers if he knows or has reason to believe that products or materials may be toxic at concentrations at which exposures may occur.

The scientific data regarding the toxicity of a substance need not be conclusive for a duty to warn to exist; a legal obligation to warn arises if enough evidence exists that a "reasonable man" would want to be warned so that exposure could be avoided. As was stated in the Borel case, "an insulation worker, no less than any other product user, has a right to decide whether to expose himself to the risk." The duty to warn not only creates a number of difficult legal problems for the lawyers, but some practical questions for the health and safety staff, such as: What information must be transmitted? How must it be transmitted? How must it be transmitted in order to warn adequately? How can one tell that the information transmission was performed adequately?

These questions fall into the lap of the safety officer or supervisor with training responsibilities, who must plan and budget for a high quality program using the most effective methods and media. Documentation of the program becomes of paramount importance in view of the delayed nature of some occupationally induced diseases, particularly cancer. Although the law in the area of toxic tort is far from settled, it is clear that effective and documented training and education, particularly in the area of chemical hazards, can play a leading role in eliminating a substantial legal liability, which may have a significant impact upon the company's balance sheet.

Labor

The role of labor in occupational safety and health has increased dramatically since the inception of the OSH Act. Participation in occupational health decisions has been included in recent large contracts, including that signed in 1984 by the United Auto Workers. Field studies are conducted, seminars are offered, and newsletters are distributed by national and local labor groups. Where worker training is concerned, the "New Directions" grants issued by OSHA for the development of training expertise in nonprofit organizations have significantly affected the training focus of unions; examination of the lists of grantees reveal that a great deal of this money has been granted to labor. Products of the grants range from slide/tape packages dealing with specific hazards or conditions to surveys, personnel development, and workshops. Very recently a new trend has emerged, that of the joint development of training programs by unions and industry trade associations. There is no doubt that labor's interest in the critical area of worker hazard communication and training, always strong, it continuing to grow.

OSH Committees

At its worst, the Labor–Management Safety and Health Committee can be an adverserial and unproductive battleground. At its best, the committee can anticipate health and safety problems and respond quickly and effectively when problems occur. Thousands of labor contracts throughout the country call for an OSH committee; unfortunately, too many of these committees lack authority, resources, and training, and have no formal or meaningful relationship with line management. The successful committees are remarkably cost-effective because they focus their efforts on controlling factors that often elude managements' control. As Kevin M. Sweeney says in his book, *Building an Effective Labor-Management Safety and Health Committee, A Practitioner Manual,*

> not only will an effective committee cut costs (e.g. OSHA fines, workmen's compensation, legal expenses, litigation, liability insurance premiums), it will develop and disseminate information, actively address safety and health situations before they become problems, participate in the design of policies and procedures and resolve problems.

The OSH Committee interest in training ideally should extend to its own training. The training for a committee should include not only instruction in occupational safety and health matters, but training in what are called "group process skills." Without such skills the group will have trouble functioning productively. Once a climate of cooperative interaction is established, new members rotating onto the committee will learn to model their behavior to be consistent with that of the group. Periodic retraining is recommended.

Union OSH Groups

As a result of the emphasis on worker health and safety, the larger unions have systematically established centralized health and safety staffs, including such disciplines as occupational medicine, industrial hygiene, and toxicology. In addition to providing technical guidance to the union, these groups review regulatory activity and call attention to their position on major issues. Strong challenges have been made to OSHA standards, such as the benzene, noise, and hazard communication standards. When they feel it is in the best interests of their membership, the OSH group also participates in litigation. As noted earlier, much of the litigation involves alleged failure to inform; this activity is thus of great relevance to the training effort.

14.2.2 Internal Factors

In contrast to the external factors, the internal factors that drive the occupational health and safety effort are specific to each company's products and processes, existing data collection and analysis systems, and management style. Although the most obvious internal factor is company policy, policy does not occur in a vacuum, but is triggered by other factors such as an increase in incidence or severity rates. As health costs mount, the tendency to use training as a solution to many different kinds of problems will increase the pressure on the field health or safety practitioner to supply a quick training "fix." In this section, we will look briefly at some of the different kinds of data that might indicate the existence of a problem; whether training is an appropriate solution will be discussed in Section 14.3.1.

Morbidity and Mortality Data

Not many companies are engaged in formal retrospective and prospective epidemiology studies, although some may be participating in such studies under the umbrella of their trade association. Industry studies will tend to identify groups of workers who have been or may be at a greater risk of disease than the general population because of their exposure to specific hazardous materials. Results are reported in terms of statistical probability. From the trainer's viewpoint, this is probably the least useful of all the internal triggers because of its lack of immediacy and specificity.

Incidence and Severity Rates

The investigation that follows an incident resulting in loss of life, severe injury, or property damage may reveal an urgent training need. Data kept over a period of time may also reveal such a need, particularly when an undesirable trend is seen to be developing. For example, an increase in the number of forklift truck incidents may show upon examination that training is hard to understand, that there is a language problem, or that not enough guided practice has been given.

Medical Surveillance Data

Several OSHA standards require regular medical surveillance of workers exposed to identified physical, chemical, or biological hazards. The occupational physician may also note an unexplained increase in visits to the medical department for skin or eye irritation. Abnormal findings or increased visits may indicate a need for changes in engineering controls, administrative controls, or personal protection controls. Training, generally considered an administrative control, may be involved in any or all of these.

Industrial Hygiene Monitoring Data

The data generated by the industrial hygienist can trigger immediate actions. For example, noise measurements may show that a hearing conservation program, including training, is necessary. Changed levels of contaminants may indicate the need for a more comprehensive respiratory protection program, which will require training. Since it is now required to make industrial hygiene monitoring data available to the affected employees (29 CFR 1910.20), a training program on the role of the industrial hygienist and the interpretation of findings would be timely.

Sickness and Absenteeism Costs

Sickness and absenteeism costs, like the epidemiologic data, tend to be tracked over a long period of time and are of little immediate use to the trainer. One thing they may reveal, however is the need for training and information on off-the-job hazards. Off-the-job accidents are three to four times more frequent than on-the-job accidents, and can result in great financial losses to the company. Any reduction in such losses by helping workers to transfer concepts of safety from the workplace to the home will have a positive effect on the company's bottom line. Management may also ask the person responsible for training to organize and arrange a workplace health promotion program covering such topics as stress management. Resources are available in most communities to advise on or present programs of this sort.

Workers' Compensation Costs

Like the other indices that are tracked over a period of time, Workers' Compensation costs will not provide an immediate training trigger. Trends may emerge, however. For example, a number of claims for low back injury may indicate the need for a review of lifting tasks and practices from an ergonomics point of view, followed by training in safe lifting practices.

Company Policies

The riskier the industry, the more likely it is that a company has well-established health and safety policies backed up by compliance audits. Few companies have a policy that focuses specifically on training and education, although it is possible that such policies will be generated as a result of some of the regulatory and legal liability issues discussed earlier.

14.3 TASK CHARACTERISTICS

The determination of an instructional need, triggered either by external or internal factors, leads to an analysis of the training task. The assumption that an instructional need does exist should never be made lightly; training programs are doomed to failure if they are offered as solutions to what are actually management problems. The following sections discuss the diagnosis of a training problem and the characterization of the types of learning that are desired. Selection of methods and media should be influenced more by these task characteristics than by any other factors.

14.3.1 Diagnosis of a Training Problem

In their book *Analyzing Performance Problems,* Robert Mager and Peter Pipe (1970) point out that the performance discrepancy must be analyzed before solutions such as training are prescribed. Because of management's reluctance to acknowledge that poor workplace health and safety performance rests anywhere but with the worker, there is usually a strong desire to solve such problems by means of training. Training is also regarded as a comparatively low-cost solution, to be offered in the form of a one-time presentation of a purchased movie or slide show. If the accidents or hazardous exposures continue, the worker is blamed for not paying attention to the presentation or the trainer is blamed for failing to select a sufficiently motivational film. Sometimes, management feels that the workers' poor attitude is at fault. Although it is true, as described in the introduction to this chapter, that roadblocks do exist in the effective communication of risks to workers, pointing the finger of blame serves only to obscure the fundamental issue of the nature of the problem.

What, then, can truly be called a training problem? A training problem exists when an undesirable workplace health or safety condition is due to a lack of knowledge or skill on the part of the worker. An upgrading of knowledge or skill should thus result in a measurable improvement. Mager and Pipe's flow diagram, which tracks the universe of possible reasons for performance discrepancies, is shown in Figure 14.1. It can be seen that formal training is only one of several alternatives to solving the perceived problem, which is often seen as one of "poor attitudes." Other alternatives, such as removing obstacles to satisfactory performance, should be considered.

As an example in the health and safety area, the use of some personal protective equipment can be considered "punishing" when it is highly uncomfortable; workers may wear it incorrectly or not wear it at all. Modification or change of the equipment may "remove the punishment" and obviate the need for training, which will not be effective as long as the discomfort factor remains. Training is definitely required, however, when the frontiers of workplace hazards are extended; for example, when a potentially hazardous new chemical process is introduced or when a worker is assigned to a new potentially hazardous task. Although the question of "attitudes" will be discussed briefly, attitudes should be regarded as an outcome of the ways in which management makes known its commitment to worker safety and health.

One of these ways, of course, is the conduct of effective knowledge and skills training in situations where training is appropriate.

A host of problems exist that do not call for training solutions. Among these are lack of clear policies, ambiguity regarding safe work procedures, engineering or ergonomic deficits, the imposition of unrealistic and stressful production goals, confusing or obscure warning signals or signage, and an absence of feedback to workers on the setting and achievement of safety goals. The application of training as a solution to any of these or other issues unrelated to worker knowledge and skills will result in a spiral of failure and continuing blame.

Let us look now at the different types of learning output and the effect these will have on the selection of methods and media.

14.3.2 Type of Learning Output

KNOWLEDGE

A core of health and safety knowledge essential to the safe performance of any job, as well as pieces of knowledge that are specific to unique situations, processes, or hazards is in existence. It is important, for example, to know what potential hazards are associated with tasks such as confined space entry, to know the safe response to emergency situations, to be familiar with correct procedures, and so on. Knowledge is best gained when:

- Learning objectives are made clear to the worker.
- The facts to be learned are broken down into a logical sequence.
- Examples are drawn from the worker's own experience.
- The worker has the opportunity to participate or respond.
- The worker receives feedback on his or her progress by means of quizzes or tests, and enough repetition is built in to provide "overlearning" and thus to delay the process of forgetting.

In most workplace situations, immediate practice is provided by the daily performance of the job. However, there are some situations, such as emergency response, that are, by their very nature, infrequent. For these, it is important to retrain periodically. In addition, knowledge is acquired more easily if several channels of communication are used. For example, it is more effective to present a lecture with audiovisuals than a lecture alone. The addition of other learning aids, such as small group exercises using flipcharts or handouts containing a summary of the material, will add even more to the effectiveness of the learning experience.

The term "knowledge" embraces several different levels of complexity of learning. There is obviously a difference, for example, between having to memorize a list of facts and having a general understanding of what those facts mean. At a more superficial level, it may be necessary only to know where the facts are kept and how to gain access to them. The decision as to what kind or level of knowledge is needed will, of course, have an impact on the cost of training and also on the selection of methods and media. Some comments on different levels or types of knowledge follow.

Recognition. A good example of a situation in which recognition is an appropriate level of knowledge is the interpretation of posted warning signs. It would be absurd and counterproductive to require a worker to memorize the words on all warning signs and to be able to reproduce them, correctly spelled, in a quiz. Nor would we expect a worker to be able to tell us what the dimensions of the signs are, or of

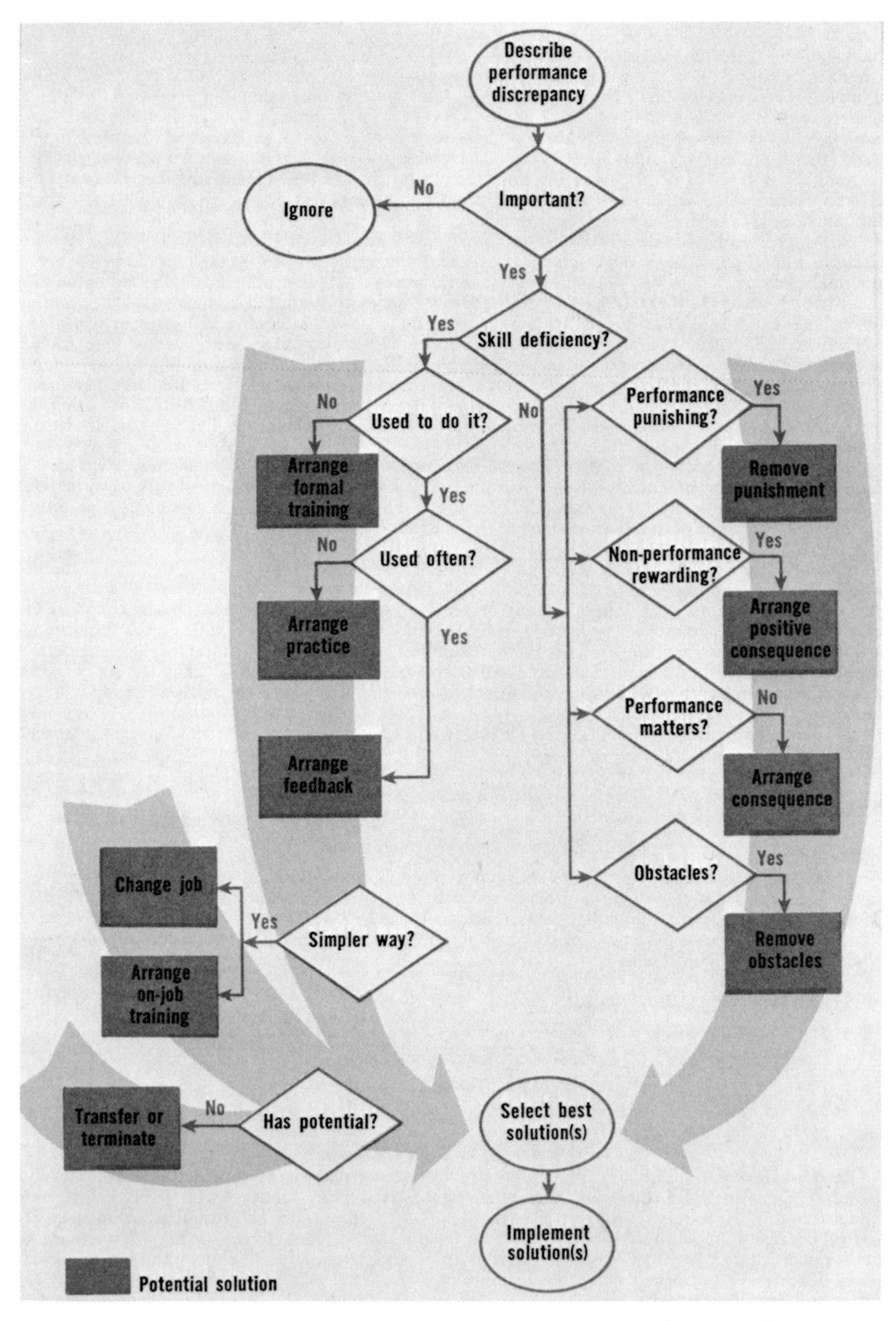

FIGURE 14.1. Analyzing performance problems. (From Mager, R. F., and Pipe, P. *Analyzing Performance Problems.* Reproduced courtesy of David S. Lake Publishers, 1984.)

what material they are made. It is sufficient if the worker recognizes the signs saying Danger or Noisy Area: Hearing Protection Must Be Worn and takes the appropriate action. The signs present images through the use of shapes, colors, and signal words that are meant to trigger certain actions, such as donning safety glasses or hearing protection. Training for recognition is simple to conduct once the purpose of the training is clearly identified. An explanation of signs using color slides or a pamphlet containing pictures and explanations will probably be sufficient.

Discrimination. At a different level from recognition, discrimination requires a worker to know what a thing is or is not in comparison to other objects or situations. For example, most facilities have a system of emergency whistles or hooters that indicate various situations ranging in meaning from *gather at a designated place* to *emergency evacuation.* Each signal is meaningful only in the context of the others. Similarly, yet in response to a quite different need, discrimination is required in selecting the correct respirator for different kinds of hazards. The worker needs to know all possible choices before deciding that in a suspected H_2S environment self-contained breathing apparatus (SCBA) will be necessary. The training in these instances should present the range of choices and give practice at contrasting and comparing them. Job aids such as charts or tables in which the characteristics or pros and cons are listed are useful tools for teaching discrimination. True/false or multiple-choice quizzes will test the worker's ability to select the correct option or action.

Understanding. The term "understanding" has been so misused in the training field that it has to be approached warily. A training objective stating that at the end of training the "worker must understand the Material Safety Data Sheet" could mean any one of a number of different things. For example, it could mean that the worker should have an understanding of terminology equal to that of a professional toxicologist. It could also mean that the worker should understand the origin and history of the MSDS, or should be able to reproduce one by rote memory. Such levels of understanding would require a mas-

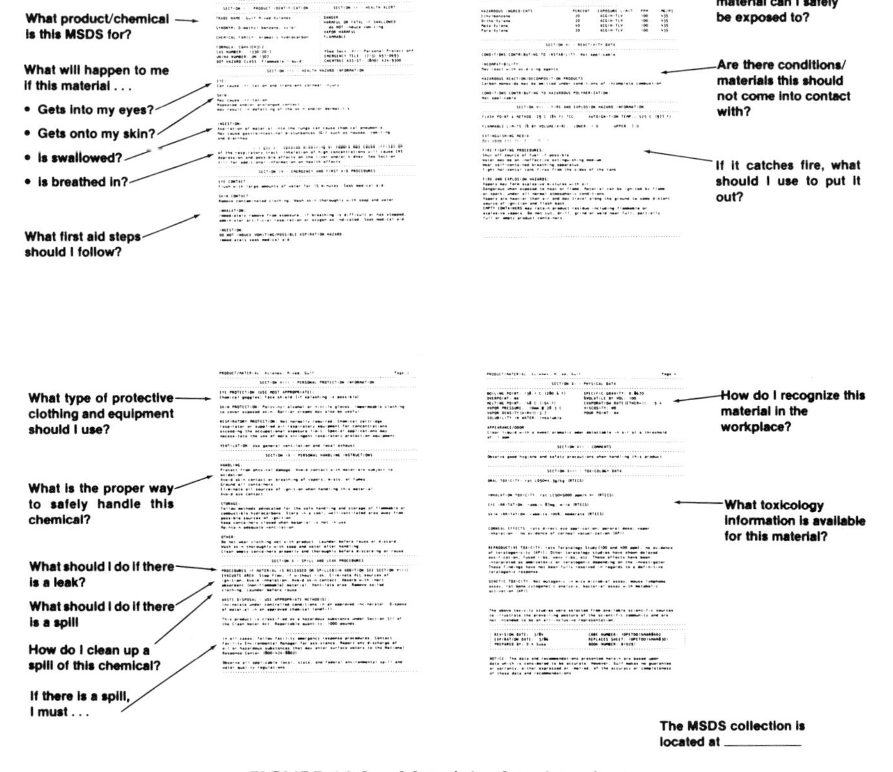

FIGURE 14.2. Material safety data sheets.

sive training effort, far beyond the scope and budget of the trainer. They also would be counterproductive to the usefulness of the MSDS as an informational tool. What, then, is meant by "understanding" in this context? Figure 14.2 is a workplace poster designed by Gulf Oil Corporation that provides the first step in understanding the MSDS. The different sections of the MSDS are identified by means of such simple questions as "What is the proper way to safely handle this chemical?" A series of slide/tape programs explaining the purpose and terminology of the MSDS goes into greater detail. In addition, a *User's Guide and Glossary of Terms* is housed with the MSDS collection. This combination of poster, training sessions, and reference guide provides a level of "understanding" that enables the worker to interpret important health and safety protection information. In other contexts, and for other tasks, the need to "understand" might require a totally different kind of training activity. For example, if we wanted a worker to understand the correct extinguishing agents to use on different types of fire, we might decide on closer examination that this is a discrimination task such as those described in the previous section. When the desired outcomes of training are clearly defined, the appropriate training methods will also become clear.

Skills. Skills training involves the worker "doing" something rather than "knowing" something, although nearly all skills training involves some element of knowledge. Since a general rule of thumb in all training is to simulate the real-world task as closely as possible, it follows that skills training involves the actual performance, with guidance, of the target skill. What is not always recognized is that skills training is facilitated if the task is broken down into a series of steps, so that each step is mastered before the entire task is attempted. Safe procedures should be built into the training; the muscles are not as forgetful as the mind, and the safe way of performing a skill should become habitual. Skills are acquired most easily if:

- The trainer gives an overview of the entire task, if necessary, teaching essential nomenclature.
- The trainer performs the task, describing the sequence as he or she progresses.
- The worker performs, with guidance and reminders.
- The worker practices alone.
- Recognition is given for skill mastery.

The last item, which should include praise for the mastery of each step as well as such rewards as badges, hard-hat decals, or certificates, leads to the next area, that of attitudes.

Attitudes. If "understanding" is a fuzzy training concept, "attitudes" are generally fuzzier. As mentioned earlier, poor safety performance is often attributed to poor worker attitudes, and training problems are sometimes prescribed as a solution. It is important to remember that training is *not* a solution for problems in which no training deficit exists. Other factors, often not accessible to the trainer, may lie at the root of the workers' undesirable behaviors, which seem to reflect a "poor attitude." How, then, can training have any impact on attitudes?

Workers generally tend to react favorably when they are given some "ownership" of the health and safety effort. (This can be achieved in a number of different ways; for example, by soliciting their ideas through such mechanisms as Quality Circles or suggestion box systems).

Some ways to produce this effect are:

- Their training includes a clear delineation of their own, as well as management's responsibilities.
- They are rewarded for good safety performance. An immediate pat on the back to a new worker for remembering a safe work procedure is more effective than a certificate issued at the end of the year.
- They are not punished or humiliated for honest mistakes that could result in an unsafe incident. Instead, the alert supervisor should correct the mistake and show how the task should be performed.
- Most of all, the commitment of management to safety is observable and real.

14.4 POPULATION CHARACTERISTICS

14.4.1 Demographics

If the determination of the appropriate training methods and media is made only after consideration of task characteristics, population characteristics, and resource characteristics, assessment of the worker target group is the second leg of this three-legged stool. Training that is not appropriate to the worker population can be wildly off target. Factors such as average age, educational level, and length of service will influence the design of the training and the kinds of examples that are used.

AGE

An aging workforce has characteristics that can present special training problems. The original training may have been conducted many years ago, and may not even be relevant today. Often, on-the-job training has consisted of watching an experienced worker perform the job, and as a consequence picking up his or her poor safety habits. In fact, an environment of disregard for health and safety can develop from a nucleus of workers with such habits. On the other hand, experienced older workers are less likely to make the mistakes made by new or newly assigned workers. Experienced older workers generally have a lower absenteeism rate, but when they do get sick, they have illnesses of greater duration and severity. They may be more attuned to written manuals and instructions than younger workers who have grown up in the electronic age. Special concerns related to hearing loss, low back problems, or other disorders may dictate a slightly different training approach.

Length of Service

The previous remarks apply also to length of service. Workers who have worked safely in jobs for a long time should be considered as resources to the trainer. They can contribute examples of good and poor safety practices, and can be helpful in orienting new workers. It has become the practice in many high-risk working environments to identify the new worker by a different colored hard hat, so that the more experienced workers can watch out for them and guide them.

The "New Workforce"

Mention should be made of the "new workforce," consisting of younger people who appear to have a different and more self-centered attitude to work.

A report by Yankelovich and Immerwahr (1983), *Putting the Work Ethic to Work,* noted that changes have occurred in the composition of the workforce, and in the social values and realities that shape people's working lives. Increase in education levels has resulted in a desire for more job discretion, autonomy, and freedom. The old "carrot and stick" system of the prospect of high pay vs. fear of unemployment has also lost much of its impact, according to the authors, because social changes have provided cushioning effects. Finally, attitudes toward authority have changed, and the younger and better educated workers have learned to question management decisions. All these factors have an impact on the design and delivery of training.

Reportedly, members of the "new workforce" tend to move from job to job seeking immediate rewards rather than waiting, as older workers have traditionally done, for the delayed rewards of long-time service. This characteristic can result in high turnover and, as a consequence, causes an ongoing training problem.

The prevalence of drug usage, particularly marihuana abuse, has been suspected of having an impact on safety performance and on job performance generally, and is a factor to be considered. In his publication *Marihuana—The Other Side,* Ronald Kuest points out that because the active ingredient THC is taken up and stored in body fat, the chronic user never escapes from the chemical effects of THC acting on the brain. He or she is always affected to some degree in areas very important to safe work performance: concentration, judgment, short-term memory, decision-making, and motivation. Since it is estimated that 62% of all adults 22–25 years old or 38% percent of all males 20–30 years old have used marihuana, the trainer should be alert to the problem.

Rosabeth Moss Kanter, in her book *The Change Masters,* speaks of another more positive aspect of the new workforce. While forty years ago only a small percentage of the work force were college graduates, today 25% of all people who work and 20% of all crafts workers are graduates. For this reason, it may be easier for the trainer to communicate health and safety messages.

Where preferences for training methods or media are concerned, younger workers have had years of exposure to TV, and tend to respond well to video, interactive video or computer assisted instruction.

Gender

So far, the discussion has not focused on either sex as having particular training needs. However, in the field of occupational health and safety there are some very definite gender-related problems. For example, in a workplace where there is a potential for exposure to teratogens or fetotoxins, all fertile females should be informed at the time they are hired of the potential for damage to the fetus. Radiation workers are another group who should be informed about the potentially mutagenic effects for both males and females.

Realistically, some jobs tend to be sex-stereotyped. Most Visual Display Unit (VDU) operators are female, for example, and the training classes concerned with ways to combat vision and other VDU discomforts will be directed principally to females. Most workers engaged in heavy materials handling, on the other hand, are male. The macho attitude that causes back and other injuries will be found in these male groups, and such attitudes should be addressed by the trainer.

Educational Level

Of all the population characteristics, educational level will have the most impact on the methods and media selected for training. In this category we can include language differences and present level of job-specific knowledge.

Instructional designers have encountered the two extremes of educational level. On the one hand, there exist sophisticated groups of workers, such as researchers in a laboratory, who are bored and turned off by simple explanations of basic material. On the other hand, many groups have low-level reading and writing skills and limited vocabularies. Another common experience is to design a training program, only to find on pretesting the materials in a field tryout, that most of the workers know most of the material before the training begins.

The lesson learned from such experiences is to gather as much information as possible about the target population before making any assumptions about the level of complexity, the method of presentation, and so on. In 1980, the American Petroleum Institute conducted a study of the readability of refinery operator training materials prepared by a contractor, Howell Training Company (Cornelius and Krager, 1980). It was found that the average reading level of the materials was between the 9th and 10th grade level, although some of the materials were written at a higher level of difficulty. This followed an effort by the Subcommittee on Personnel Selection to ascertain the educational level of petroleum and petrochemical workers that revealed that, although the educational level of workers in the sample averaged grade 12.7, the distribution of reading level test scores approximated the grade 9 on national public school norms. On the average, the reading skills appeared to match the reading difficulty of the API-sponsored training materials. The information gave valuable feedback to API and its training materials contractor, and paved the way for further development of materials appropriate to the target workforce.

In groups of workers where there is a great range of educational levels, self-paced methods such as programmed instruction or "branched" methods such as interactive video or computer assisted instruction should be considered. Pretesting will help to determine what does or does not need to be taught, and preparatory modules of basic instruction will sometimes help to bring the less knowledgeable learner up to speed. Many packaged training programs are now sold in more than one language, and it is not difficult to find translators to prepare foreign-language versions of slide/tape or video scripts.

14.4.2 Group Size

The format of training will also be dictated by the size of the groups of workers who need to be trained. In large operations, where there are across-the-board health and safety problems such as housekeeping, large groups are often convened. In small operations, or where only a few workers are affected, training is conducted in small groups.

Large Groups

A large group presents a challenge to the trainer; more than 20 or 30 people tend to coagulate into an unresponsive herd. It is difficult to tell how many of the participants understand the topic, and the burden of responding to the training usually falls on a few interested people. Lectures, movies, and slide shows, or multiple video monitors are the only methods big enough to embrace the whole group. Several techniques are available, however, to break up the large group into small groups or individuals. Small subgroups can be given problems to work on, individuals can be selected to participate in role play, or individual self-paced training programs can be handed out for participants to complete at their own speed. Paper and pencil exercises can be handed out and discussions held on the results. The basic formula for effective instruction, which is telling, showing, and doing, can be applied in large groups with sufficient planning.

Small Groups

Small groups are so flexible that training can be conducted in the training room, at the worksite, or on an ad hoc basis when there is a point to be made. Any group of less than 20 individuals tends to be responsive and participatory. It is possible for workers to gather round to observe demonstrations of skills, such as donning a respirator, or to participate in guided discussion. Chalkboard or flipchart illustrations can be seen easily. Interruptions and questions can be encouraged without loss of control. A feeling of partnership can be developed in workers who undergo training together.

Individual Workers

In many instances, it is necessary to train individual workers. For example, in an operation with a low turnover rate, it is not possible to indoctrinate new hires in groups. If work with hazardous physical or chemical agents is involved, the workers have to be trained "at the time of initial assignment" (OSHA 1910.1200). Assignment changes or process changes will also necessitate immediate training for experienced workers. This situation could become so labor-intensive for the trainer that little time would remain for other activities. For individuals, consideration should be given to the use of methods that release the trainer, such as self-paced programmed instruction texts, interactive video, slide/sound projectors, or computer assisted instruction. In a training room containing simulated machinery or systems, several individuals can pursue different courses under the eye of one trainer. It is important, when using individualized methods or media, to allocate some trainer time and also some group interaction for the purposes of encouragement, discussion, and questions.

14.4.3 Geographic Factors

The training solutions that can be applied in large centralized operations are not available to companies that have many small decentralized locations.

Centralized Operations

Centralized operations generally have the luxury of a training staff, a training room, sophisticated equipment, and a controlled training schedule. Records are kept in a central location and the entire program is managed in a consistent manner. Rewards such as training program completion certificates, badges, and hard hat decals are recognized by everybody as representing the attainment of a certain level of achievement. The volume of training permits the purchase of the latest audiovisual aids; no one has to sit through the same film over and over again.

Decentralized Operations

The picture in small decentralized operations is entirely different. The trainer is often a person who already wears several "hats" and has many other responsibilities. When there is no training room, sessions may have to be conducted in the cafeteria or in rented rooms at the local hotel. Budgets may not permit the purchase of up-to-date equipment or audiovisual aids. Often, the training of new hires is assigned to an experienced worker. There is a lack of consistency among decentralized operations; each facility tends to follow its own path.

Solutions to these problems can be found. For example, it is possible to convene the trainers from scattered locations for a "train-the-trainer" course, in which they learn how to plan, conduct, and evaluate training. Guidelines and manuals can be provided to these individuals to insure consistent methods of implementing and documenting training programs. A centralized lending library of slide/tape programs, films, and other aids can be set up and the availability of new programs made known to location trainers. In instances where it is desirable to have training of the highest quality and consistency, such as fire-brigade training, it is possible to send selected workers to such professionally operated centers as fire schools. In some instances, workshops are offered in cities across the country by such groups as the American Trucking Association, which conducts workshops on the DOT Transportation of Hazardous Materials training requirements. All these solutions involve some centralized leadership, either by an individual, a department, or a planning committee.

14.5 RESOURCE CHARACTERISTICS

Occupational health and safety training programs generally operate under severe constraints. The long-range preventive nature of the activity makes it difficult for management to perceive a quick payoff. Until the advent of the Hazard Communications Standard and right-to-know laws, training has been the Cinderella in comparison with the more costly preventive programs such as fire protection, toxicology, and medical surveillance. The trainer's slogan, "You pay for training whether you do any or not," has tended to fall on deaf ears. It has not always been appreciated that poorly planned and ineffective training is the most expensive of all, particularly when such training fails to halt incidents that result in loss of life, injuries, or property damage.

14.5.1 Management Support

The most important and significant resource is, of course, management support. Grudging support that allows only "letter-of-the-law" training under adverse conditions undermines the entire training effort. Enthusiastic lip service without the provision of material resources also has a negative effect. Management can best show support by supplying necessary resources of manpower, budget, and training time.

Budgets

It is unusual for a separate budget to be established for occupational health and safety training. Traditionally, funds are made available from the Safety, Industrial Hygiene or Human Resources department budgets. This is partly due to the fact that there has rarely been an entity called "the health and safety trainer"; instead, someone from one of those groups is assigned an extra area of responsibility.

In small operations the training "hat" is worn, along with several other hats, by the plant manager or operations manager. Training is, in fact, often delegated to the first-line supervisor. With the advent of the OSHA Hazard Communications Standard and state right-to-know laws, this picture is beginning to change. Estimates of how much it will cost to train at-risk workers vary widely; however, it is clear that the costs will be highly visible. OSHA has calculated that the training portion of the standard will cost only a few dollars per worker. Industry estimates set the figure much higher. At a Cincinnati Chamber of Commerce seminar, as reported in the February 7, 1985 issue of the *Occupational Safety and Health Reporter* (BNA, Volume 14, Number 35), a chemical industry representative stated that "training is really a big ticket item," which could cost $22.50 for each of the 14 million workers affected nationwide, or a total of $316 million. It is not known what extra costs will be incurred in starting up this massive training effort, including the development of documentation systems that will allow companies to track workers as they start new assignments or are exposed to new processes or chemicals.

Since these costs are additional to the costs of existing safety training, such as forklift truck operator training or hearing conservation education, it can be seen that a fairly substantial operating budget is needed. Some of the budget should cover the education and the time of the person(s) who develop, present and document the training, or outside consultants to assist with these functions. Some should be assigned for manuals, handouts, purchased audiovisual packages, and other materials. Capital cost items, such as necessary space, equipment, training aids, simulators and record-keeping systems should be budgeted. In summary, training will make a leap from a relatively low-cost item to a position of budgetary visibility.

On-the-Job Training Time

Even when management is fully supportive, time off the job for training can become a problem, and companies have come up with some innovative ideas to reduce it. For example, in a plant covering a wide area, one company uses a trailer as a mobile training room. The trailer is parked in each major work area in turn, so that workers do not have to walk far from their job for training sessions.

Another common practice is to "piggyback" the training onto scheduled safety meetings; that is, to design the training in modules, so that each module can be presented within the time-frame of a safety meeting. Modules of 20 minutes long will allow time for questions and discussion in a 30–40 minute meeting.

A third option is to ask workers to do a little homework before a scheduled group meeting, so that they all enter the meeting with some common knowledge of basic terms or concepts. If the "homework" takes the form of an attractively illustrated booklet, and it is known that a quiz will be given at the beginning of the meeting, this idea may work well. For some of the programs at Gulf Oil, take-home booklets were designed in a comic book format. These serve two purposes: they outline major points to be covered in the training, and, because of the attractive format, are likely to be shared with the family. The inclusion of the family has the positive effect of bringing their interest and influence to bear upon workplace health and safety and the importance of compliance with safe work practices.

Availability of Trained Trainer

It is apparent that decisions regarding training are complex, and that an untrained person would have a difficult time setting up and delivering effective programs. The third manifestation of management support, then, is to make available to the field operations a trained "trainer." This does not mean that a fully trained professional trainer should be hired for every plant location; in many cases, less expensive options exist. For example, a centralized core of training expertise can be established, to be made available to field locations as needed. Another attractive alternative is to set up "train-the-trainer" workshops, to be administered either by a centralized training function or by a contractor. It is possible also to send trainers to generic workshops run by vendors and by such bodies as the American Society for Training and Development. The advantages of sponsoring this kind of training are as follows:

- A cadre of trained trainers can be developed who are able to implement programs on an on-going basis.
- The trainer, usually the safety officer or some other person wearing another "hat," can act as the focal point for the transmission of new programs and materials to the plant.
- Training records and documentation of program effectiveness can reside with this person.
- A beneficial relationship is established between the trainer and the centralized training professional or consultant.

Most workshops are approximately a week in length and contain the same basic elements. These are:

- Needs assessment, theory, and practice.
- Practice at defining performance objectives.
- Selecting appropriate methods and media.
- Developing the content of a program.
- Practicing presentation skills.
- Documentating and evaluating.
- General program management.

Personnel to be trained as trainers must be carefully selected; not everyone has either the inclination or the people-oriented attributes that make a good trainer. Management support is necessary in the selection process so that people are not identified simply on the basis of having little else to do.

14.5.2 Location Considerations and Constraints

Most industrial training in the real world has to be planned and conducted in the face of many physical constraints. Some of these have been discussed previously. The type of location, the presence or absence of training facilities, and the availability of equipment are all important factors.

Type of Location

As an example of a "worst case" situation, consider the quandary of a trainer whose trainees are strung out in very small groups along a pipeline or in small marketing operations. Or think of the logistical problems presented when workers are captive on offshore oil drilling platforms for one or two weeks at a time, and are allowed to go home to towns all over the country on their off-duty weeks. Such problems are faced by the oil industry, where there is a great deal of mandatory training to be presented. An entirely different scenario is seen in large manufacturing operations, where many workers are doing similar jobs and can be trained in groups.

To return to the drilling platform example, it is interesting to see what the industry has done to overcome the problems. One alternative is to set up a regional or local training center, and to bring workers in from offshore for scheduled training. Another is to turn to techniques of individualized instruction, such as interactive video, computer-based training, or self-paced paper-driven texts. Workers can train at mini- or microcomputer terminals located on the platforms in their off-duty time. Most remote locations have video playback equipment on which to show movies, and it is possible to "piggyback" safety training videocassettes onto this system. A third option is to appoint a traveling training coordinator who, like a traveling preacher in the Old West, appears periodically to conduct all necessary training. There are, doubtless, many other methods used in the industry in combination with one or more of the three above.

Training Facilities

Training is best conducted in a dedicated training facility away from plant noise and distractions. Good lighting, comfortable chairs with a writing arm (or simply good tables and chairs), and blinds to screen out light during audiovisual presentations are assets.

Audiovisual Equipment Availability. In a study conducted by the Worker Training Task Group of the Chemical Manufacturers Association (1984), it was found that the most common pieces of equipment are the slide projector and tape recorder. Running second to that are video playback equipment and movie projectors. The humble overhead projector is also commonly found. Often the slide projector is a "slide/sound" model, on which the slide changes are triggered by the audiotape. These pieces of equipment are very basic and can easily be rented. More sophisticated machines, such as equipment for interactive video, are rarely found. Some caution is needed in purchasing vendor training packages that can only be displayed on one type of equipment, since not all the locations in the company will necessarily have compatible machines.

14.5.3 Purchased Programs and Services

Purchased programs and services have become a great deal easier to find over the last few years. From only a handful of vendors and consulting companies, the field has grown and expanded. Because of the passage of various right-to-know laws and the Federal Hazard Communication Standard, this particular area has recently grown faster than that of traditional safety training program assistance. As alternatives to the hiring of in-house personnel, purchased programs and services offer some attractive options to management. Money will be squandered, however, if selections are made more on the basis of the attractiveness of the materials than on their suitability in meeting your specific needs.

SELECTION

The same selection factors apply to purchases as to in-house program development. The first question to ask is whether the perceived need does indeed reflect a training problem. Even the highest quality purchased program will do nothing to solve engineering, administrative, or other problems associated with the commitment and decisions of management. If the problem is truly a deficit in knowledge or skills or a need to acquire new knowledge or skills, a purchased program may provide all or part of the solution. The next step is to define the desired performance outcomes in such a way that the most appropriate program or service is selected. Upon completion of this step, it is a good idea to review the existing programs at your location to see if you can save reinventing the wheel by including them in your overall scheme. An easy way to do this is to construct a matrix. Along the top you can list all the desired performance outcomes, such as "worker can list the correct actions to take in the event of a spill" (knowledge), or "worker will perform the designated actions correctly in the event of a spill" (skills). Down the left-hand side of the matrix you can list all the manuals, guidelines, and training programs that you now possess. If any of them satisfy one or more of the outcomes listed along the top, put a checkmark in the square of the matrix where demand and supply intersect. In this manner, the remaining empty squares can be identified as presenting a need for purchased programs or services.

The next step is to eliminate from all available programs those that will be useless to you because of their hardware demands or any other factors. For example, there is no point in purchasing videocassettes that are not compatible with your equipment. A less obvious example is the "hit and run" type of service; this is typified by the groups who charge a high fee for putting on a one-time-only training program or presentation. When new hires have to be trained or annual retraining is needed, there is no follow-up mechanism or program in place to train them, and the consultant group must be brought in again. Selection among the remaining programs should be made on the basis of answers to the following questions:

- Does this program/service meet precisely the needs I have identified?
- If not, is there some built-in flexibility that will allow me (or the contractor) to make appropriate modifications?
- Are there some data I can see that indicate that this program has been effective with groups of workers similar to mine?
- Is the vendor or consultant willing to give me the names of previous clients, so that I can find out how well the program worked in a similar operating environment to mine?
- Can I preview all or part of the program?
- Is there a leader's guide that will help me to administer the program?
- Are there tests that will allow me to document my workers' understanding of the content of the program?
- Are there any student handouts or job aids that will be useful to workers as continuing reminders when they get back on the job?
- Are the examples used appropriate to my kinds of workers, our part of the country, and so on?

You may wish to add more questions to this list, depending upon the special characteristics of your workplace or workers. For example, it may be important to find out whether a Spanish version of the program is available. If you are purchasing "train-the-trainer" services, you might wish to take the whole workshop yourself before making up your mind. If a contractor is going to design a custom-made program for you, you will have questions about the stability of his or her business, since it would be disastrous to end up with a half developed program.

EVALUATION

If the purchased program has supporting data that show its effectiveness with groups of similar workers, you can predict that it will probably work well with your workers. Field-tested programs are still rare, however, and you may wish to evaluate the program you have purchased before making a decision to continue or to look for some other alternative. The techniques described in a later section on evaluation and documentation will be useful for this purpose.

14.6 SELECTION OPTIONS

Up to this point, the varying task, population and resource characteristics that help to shape the method/media decision have been reviewed. There are also intrinsic factors that characterize each teaching method and make it more suitable in some contexts than in others. In this section a variety of options will be described briefly, with some of the pro's and con's discussed. A summary of steps to follow in media selection is presented. Finally, a matrix that shows the applicability of several common media to a variety of learner and task requirements is shown. More detailed information can be found in the Training and Development Handbook (ASTD, 1976).

14.6.1 Lecture/Discussion

Despite the availability of many instructional methods, the lecture remains the most common method of communicating information. The CMA survey of member companies found that audiovisual aids and on-the-job training were most commonly used, but the audiovisual aids were most probably accompanied by some type of lecture. The following analysis of pros and cons is in the form of charts derived from an overview of instructional methods prepared by Marilyn C. Tirrell.*

Method	Pros	Cons
Lecture	Can cover a large amount of information. Can be used with any group, large or small. Allows total control of information by lecturer. Can be effective for teaching knowledge(s).	Does not encourage active participation/ interaction. One-way communication limits understanding of learner needs. Effective, attention-holding lecturing requires sophisticated verbal skills. Not appropriate for teaching skills.

The lack of student participation in the lecture environment is probably its least desirable feature. In fact, a lecture, at its worst, is a situation in which the knowledgeable person (the lecturer) practices talking to himself about what he already knows. Students have been known to sleep through lectures, or spend the time in thinking about other things. At its best, especially when combined with other methods, the lecture can be stimulating and interesting.

A discussion involves greater risk than a lecture. Control of the class can be lost unless the discussion is guided very carefully. The pro's and con's are analyzed below.

Method	Pros	Cons
Discussion (Uses open-ended questioning technique)	Involves learners *actively* in the learning. Instructor gets valuable feedback on learner needs/ attitudes/knowledge(s). Issues and assumptions can be explored and clarified. Can be leader-guided for large groups or can be small-group oriented. Learners can *discover* new concepts. Learning climate is more relaxed, i.e., less formal than lecture. Can be effective for teaching knowledge(s).	Discussions can go off on irrelevant tangents. Requires skill in maintaining class control. Open-ended questions must be structured in logical progression for "discovery" to occur. When used during large group breakdown into small groups, facilitators are needed to assist instructor.

*Some Pro and Con summary portions of this chart reproduced by courtesy of Tirrell, M. C., *The Learning Curve: Train the Trainer,* Visucom Productions, Inc., Redwood City, CA, 1985.

14.6.2 Demonstration

In skills learning, such as learning to operate self-contained breathing apparatus, a demonstration is an essential part of the learning sequence. For critically important tasks, the demonstration should be live rather than filmed. It should not be attempted for large groups, unless the objective is simply to familiarize students with the general sequence of a task. Pros and Cons are summarized below.

Method	Pros	Cons
Demonstration	Relates information to the real world. Attention-getting. Flexible pace that can be geared to learners' needs/ capabilities. Can be repeated over and over, if necessary to accomplish learning. Excellent for teaching skills when followed by learner practice.	Requires *thorough* preparation. Should be limited to small groups or one-on-one. (Closed circuit TV can be used for some large group demos depending on the level of skill(s) complexity.)

14.6.3 Small Group Practice (Case Study, Role Play, Games, Debate)

An effective way of dealing with a large, unwieldy group is to break it down into small groups. Each subgroup is assigned a task or project and usually has a recorder, to record decisions on an easel chart, and a reporter, to report back to the large group at the end of the exercise. With careful preplanning and organization of materials, it is possible for one instructor to have as many as four small groups working in different corners of the room. The instructor usually circulates to provide guidance and answer questions. Alternatively, a few participants at a time can be drawn from a large group to perform in role play situations. A summary appears below.

Method	Pros	Cons
Small group practice Case study Role play Games Debate	Can be used in large groups as a less threatening way for all learners to participate. Can build group rapport by pooling individual talents to achieve a common goal. Provides an excellent opportunity to apply new knowledge. Can be used to simulate a variety of real-world problems/situations. While not optimum for applying all skills, it can be valuable for practicing interpersonal/ decision-making skills. Can contribute to development of new/changed attitudes. Learners have the opportunity to learn from each other.	Practice activities must be structured *in detail* to assure that learning will occur. Dominant personalities in the groups may overwhelm less aggressive members. Can result in unpleasant arguments within groups made up of diverse personalities/background. Facilitators to monitor the progress of each group are needed when more than 4 groups are involved concurrently. Group size may be restrictive; should be limited to from 3–7 learners per small group. Does not allow for specific feedback on performance of/to the individual.

14.6.4 On-the-Job Training

In many operations, large and small, on-the-job training is still the common way to introduce new or newly assigned workers to the job. At its best, it allows the neophyte to observe a skilled worker and to benefit from his or her years of experience. At its worst, it exposes the new worker to risky shortcuts, poor safety practices, and a confusing exposition of facts and work practices. To the experienced worker, the task has become so habitual that many steps are not described in a logical sequence, or not described at all. If on-the-job training is to be given, some support must be offered by a trainer who can monitor and advise the experienced worker.

14.6.5 Simulation

In many occupational health and safety learning situations, practice in the "real world" could be dangerous and/or expensive. Airline pilots, nuclear plant operators, and deep-sea divers are extreme and dramatic examples of students who cannot receive their initial training on the job. Simulators for some of these jobs are in themselves very expensive, often costing thousands of dollars. Less dramatic situations exist, however, where simulators are useful. Training for jobs that tend to be widely scattered, such as maintenance mechanics or off-shore rig roustabouts, can often be conducted economically by keeping the students under the eye of the instructor in a classroom containing simulated tasks and machinery. This method has the virtue of allowing the instructor to set up certain conditions to teach specific points. It generally is combined with lectures, demonstration, and other methods. Simulation is not economical if there are few students to be trained or if the task is transitory in nature.

14.6.6 Individualized Instruction

The educational levels, experience, and abilities of the worker group may be so diverse that individualized instruction is the only answer. Even when workers are fairly homogeneous, individualization is useful as a tool for training new employees. Rather than allow these people to wait until the new employee group is sufficiently large, immediate training can be made available. This consideration has become important for two reasons: the Hazard Communication Standard requires training "at the time of initial assignment"; and, from a worker protection viewpoint, it is not a good idea to allow untrained workers to start new jobs, particularly since they are vulnerable to picking up poor safety habits. Individualized instruction ranges in complexity from paper and pencil independent study materials to slide/sound projectors, computer assisted instruction, and interactive video systems. Most systems have some mechanism for student response; in computerized systems, the responses are recorded automatically. The documentation of responses has obvious advantages for legal liability protection purposes, because it is possible to see exactly what each student has learned.

A second advantage is that it is possible to evaluate student responses for the purpose of assessing retraining needs. Some pros and cons follow.

Method	Pros	Cons
Independent study/ individualized instruction	Learners can proceed at their own speed. Learner receives feedback on level of mastery/performance at periodic intervals. Opportunity to go back over material until it is learned. Eliminates the sometimes negative effect of peer pressure. Forces learner to be responsible for the learning. Is an effective supplemental method when combined with other instructional methods. Allows for identification of remedial training needs.	Highly dependent on quality of written/printed materials or other media. Requires highly structured planning. Absence of human interaction, i.e., sharing of ideas, debating concepts, can limit the scope of learning and can take the fun out of learning. More effective for teaching knowledge(s) than for teaching skills.

Programmed Instruction

The genesis of programmed instruction, in the early 1960s, occurred at a time when sophisticated electronic delivery systems were not available. As a result, programmed instruction packages, both linear and branched, are inevitably associated with paper and pencil. This is unfortunate; the concepts of active self-paced learning, with immediate feedback to the student as to the correctness of the response, are equally applicable to electronic linear and branched systems. The idea that the subject should be broken down in a logical manner into small steps and that each sequence of steps be reviewed and practiced is perfectly valid in any medium. The paper-and-pencil programs are still used because of their economy and flexibility. For example, it is possible to give students a programmed text to study on their own time, and then to convene a group meeting to test their knowledge and conduct a discussion. The individual study prepares students for the discussion and brings everyone to the meeting with the same knowledge base.

Computer Assisted Instruction

The terms "computer managed" and "computer based" instruction fall under the general umbrella of computer assisted instruction. At a minimum, the computer can be used to record the fact that training occurred and to maintain a database of each workers' training. At the maximum, the computer acts as the instructor, and the program that each student sees is modified on the basis of his or her unique pattern of correct and incorrect responses. The rate of expansion and change is so rapid that any description of hardware and software options and costs would quickly become obsolete. However, it is recommended that a thorough study be conducted before expensive equipment and programs are acquired. A particular word of caution relates to the anticipated "life" of the technical content. For example, if you are aware that the hazard status of a chemical is subject to change due to ongoing toxicological testing, it would not make sense to purchase an expensive program that will be useless within three months. Some advantages and disadvantages are:

Method	Pros	Cons
CAI	Adapts to various levels and speed of learners. Can provide simulation and demonstrate complex concepts. Can be remote, yet connected to a central source. Can interact without a keyboard (touch sensitive). Has great capacity and capabilities.	Hardware expensive, though prices falling. Hardware and software can become outdated quickly. Requires extensive development time. Little student-to-student interaction. Scheduling students a problem if few machines available. Security problems for equipment.

INTERACTIVE VIDEO

Many of the advantages of computer assisted instruction are true for interactive video. Users create unique lesson sequences based on their own information requirements and learning styles, by selecting from many preplanned options. Each response triggers appropriate system feedback, so that learners are provided with a unique path through the program. A disadvantage of videotapes is the time, following a learner response, that is taken to search the tape for the next appropriate statement or question. The most advanced technology is seen in laser discs, which are encoded in modular rather than linear form; the search time is reduced to almost nothing. As with any individualized instruction, the cost of this technology may be justified in a situation where there is high turnover in the workforce, so that newly hired workers can immediately receive training. Some advantages and disadvantages of interactive video are:

Method	Pros	Cons
Interactive video	Adds realistic pictures and movement to all the advantages shown for CAI. Many options and learning tracks possible. Videodiscs search rapidly, are easily stored. Fast developing technology—many exciting developments taking place.	Same as for CAI. If videotape, search time is slow after student responds. As with CAI, fast moving technology can leave you behind.

14.6.7 Audiovisual Techniques*

Many audiovisual tools are available to the trainer, and some of them are extremely low in cost. Sometimes the simple and economical tools do a more effective job than the complex presentation systems.

BLACKBOARD AND CHART PAD

These media can highlight important points and give emphasis. Conference pads can be prepared before the training session so that the material is ready when needed for teaching or review. They can also provide a record of the session by being taped to the walls or placed on file.

Method	Pros	Cons
Blackboard/ chart pad	Very flexible for controlling discussions. Excellent for emphasis. Can be used in almost any situation. Requires no special skill. Inexpensive.	Not good for complicated topics. Blackboard not good when a permanent record is needed. Possible loss of consistency from one group to the next.

*Some of the following information was derived from a compilation prepared by Tom Jamrose.

Charts and Flannel Boards

When a simple method of depicting a sequence, growth, contrast, or facts in chronological order is desired, these media are useful. They focus attention on key points and, once prepared, can be used again. The pros and cons are similar to those of the blackboard and chart pad.

Pictures and Posters

Posters are useful in defining very important topics or messages. Since they are quite expensive, posters should not be used for topics that affect very few workers and/or are of transient interest. Discretion should be used in the use of cartoons; it is easy to overstep the bounds of good taste in depicting matters having to do with health and safety.

Overhead Transparencies

There are good reasons why the overhead projector is used so frequently for management presentations in executive boardrooms. It is the most reliable and also the most versatile of all the techniques for projecting an image. A spare bulb and an extension cord are all the maintenance equipment needed.

Method	Pros	Cons
Overhead transparencies	Very versatile. Projected image can be manipulated. Easy to produce on copy machine. Useful for systematic, sequence presentation. Projector is simple to operate. Presentation easily controlled by instructor. Useful with large or small groups. Inexpensive.	Transparencies are large, require special storage arrangements.

35mm Slides and Slide/Tapes

Slide presentations and synchronized slide/tape programs have been the workhorse of projected media for years. Even the smallest plant either owns or can rent a slide projector, and everyone has a tape recorder. The slide/sound projectors that incorporate the slide carousel and tape cassette in one neat package are particularly good for individual and small group presentations. Some are also designed to project at a distance for large groups. Programs can be written and assembled in-house, using local management and workers as photographic models. When new processes and problems are encountered, changes are easy to make.

Method	Pros	Cons
35mm slides and slide/tape	Easily handled, stored and arranged in sequence. Good for individuals and groups of any size. Flexible/adaptable. Large number of commercially available programs. Can be combined with taped narration for repeatability. Fairly economical.	Loose slides easily disorganized. Requires some photographic skills and lab assistance. Need special equipment for close-up copying.

Motion Picture Film

Movies, usually 16 mm, are widely used for safety training. There are literally hundreds of commercially available films dealing with almost every conceivable safety topic. Most films are of high quality, and are excellent for dramatizing events and commanding attention. Films are expensive, though, and should not be developed for topics that are of a transient nature. Outmoded cars and clothes also look very conspicuous in movies and can detract from the message. Since workers do not like to see the same film twice, it is wise to consider a shared arrangement with other company locations or through trade associations.

Method	Pros	Cons
Motion picture film	Film size is standardized. Many commercially available materials. Show motion and relationships. Good for groups or individuals. Can easily be transferred to videotape.	Expensive to prepare, repair, and rent. Because of expense, needs careful professional planning. Useless when material is out-of-date.

VIDEO (INCLUDING VIDEOTAPE, TELEVISED INSTRUCTION/CONFERENCING)
Our society has become so attuned to television that workers feel very comfortable in front of a TV monitor. The use of videocassettes has become so widespread that many 16 mm movies are now available in video formats. Although large-screen video projection systems can be found, most programs are shown on small monitors. This means that groups must be limited to 15 people, or multiple monitors must be purchased or rented. Companies do not always have consistent playback equipment, so that videocassettes bought in one of the three formats (three-quarter inch, betamax, VHS) will not be playable on all machines.

Method	Pros	Cons
Video	Permits same image to be transmitted to large numbers of people in many locations. Easy to play back. Allows for use of slides, overhead, or other media within the video program. Videotape is reusable.	Image display limited to size of monitor. High cost of production. Equipment standards not uniform worldwide. Tape/cassette standards not uniform—can be problematic for distribution of completed program.

14.6.8 Summary of Steps in Media Selection

Each learning topic or situation is unique, and for each situation selection of an appropriate medium can be made. No cookbook prescription exists; there are too many task and population variables. It is not possible to make mechanical rules based on the subject matter involved or the instructional methodology that happens to be available. The following steps are derived from Briggs *Handbook for Procedures for the Design of Instruction* (1970).

1. *Define the resources:* Note any limiting conditions for both development and implementation of the training program in terms of time, costs and resources available.
2. *Define the size of the group:* Decide whether you intend, or are able, to conduct individual, small group, or large group instruction.
3. *Define population characteristics:* Note educational levels, language preferences, or any other unique characteristics of learners.
4. *Define task characteristics:* Note the types of desired learning outcomes (knowledge or skills).
5. *List the general and specific instructional events* for the type of learning the performance outcomes represent.
6. *Arrange the instructional events in the desired order* and consider whether more than one application of each event is needed.
7. *List the media options* from which a choice is to be made for each event.
8. *Make a tentative media selection* for each event from among the options recorded. Check the advantages or disadvantages, of each type of media for group or individual use.
9. *Review the entire series of tentative media choices, seeking optimum "packaging."*
10. *Make final media choices* for "package" units.
11. *Write guidelines to the instructor* for presenting the units, including a prescription for instructional events, such as discussion sessions, not provided by the media.
12. *Prepare guidelines for the students,* if necessary, to assist them in using the media correctly. This is particularly applicable for individual instruction.

Although the unique nature of each learning task has been emphasized, it is apparent from the previous analysis of different methods and media that some applications are clearly suitable or unsuitable in a given set of conditions. The following matrix (Figure 14.3), designed by John G. Wilshwen, Jr. with the assistance of Dr. Richard Stowe, and reproduced in *Briggs Handbook* (1970), brings together an enor-

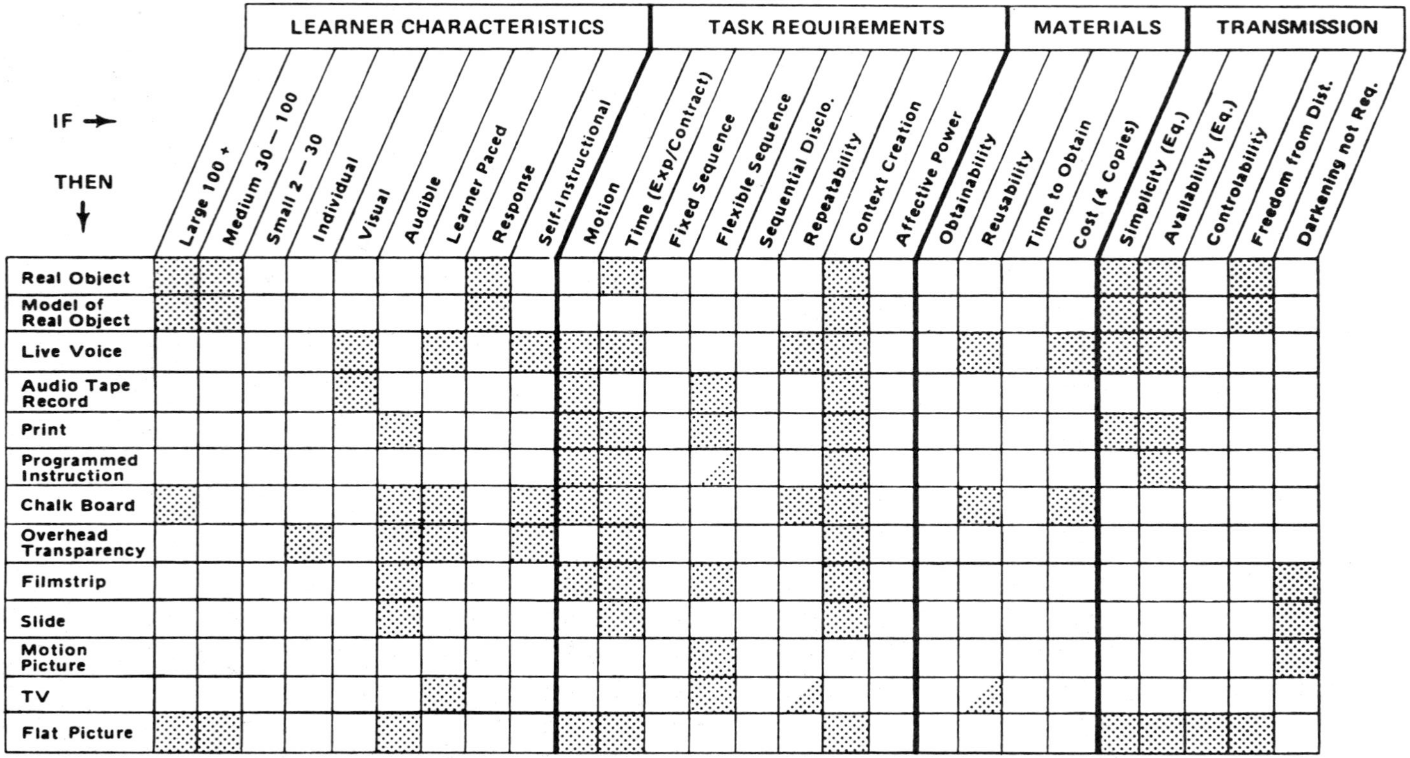

IF →	LEARNER CHARACTERISTICS									TASK REQUIREMENTS								MATERIALS				TRANSMISSION				
THEN ↓	Large 100 +	Medium 30 — 100	Small 2 — 30	Individual	Visual	Audible	Learner Paced	Response	Self-Instructional	Motion	Time (Exp/Contract)	Fixed Sequence	Flexible Sequence	Sequential Disclo.	Repeatability	Context Creation	Affective Power	Obtainability	Reusability	Time to Obtain	Cost (4 Copies)	Simplicity (Eq.)	Availability (Eq.)	Controlability	Freedom from Dist.	Darkening not Req.
Real Object																										
Model of Real Object																										
Live Voice																										
Audio Tape Record																										
Print																										
Programmed Instruction																										
Chalk Board																										
Overhead Transparency																										
Filmstrip																										
Slide																										
Motion Picture																										
TV																										
Flat Picture																										

FIGURE 14.3. Indications/contraindications for various methods and media. (From Briggs, L., *Handbook for Procedures for the Design of Instruction,* 1970. Reproduced courtesy of the American Institutes for Research, Wash., D.C.)

mous amount of information in a limited space. In the matrix, solid shading means "*not* applicable"; partial shading means "partially applicable," and empty cells means "applicable." An explanation of terms is given following the matrix.

Explanation of Terms

Learner Characteristics

Large, medium, small, individual: Refer to sizes of groups of learners.

Visual: Learner characteristics dictate that the stimulus material be visual.

Audible: Learner characteristics dictate that the stimulus material be audible.

Learner paced: Learner characteristics dictate that the rate of presentation be controlled by the learner.

Response: The medium contains provision for incorporating demand for learner response.

Self-instruction: Learner characteristics dictate that stimulus materials be designed so that learner is able to use them with little or no supervision.

Task Requirements

Motion: Task requirements indicate that motion must be depicted.

Time (exp/contract): Refers to the possibility of expanding or contracting length of presentation as compared with real-time experience of same phenomena: e.g., slow motion or speeded motion pictures, compressed or expanded speech devices.

Fixed sequence: Refers to characteristic of medium that does not permit change in sequence of presentation beyond forward or reverse.

Flexible sequence: Medium permits change in order of presentation of stimuli.

Sequential disclosure: Medium permits revelation of material bit by bit and allows retention of prior bits as further bits are revealed.

Repeatability: Medium allows complete or partial redisplay.

Context creation: Refers to capability of media to transport learner from awareness of real world to context artifically contrived. Motion pictures are an obvious example, but it is our contention that all media have this capability to some degree. A book has it, for example.

Affective power: All media have the power to move people emotionally to some degree.

Materials. We feel that items in this group are reasonably clear.

Transmission

Simplicity: How simple is the equipment to operate?

Availability: How readily available is the equipment required to display the stimulus materials?

Controllability: How much control over the transmission can be exercised by the instructor? (Start/stop, slower/faster, freeze frame, volume change, forward/reverse, repeat, switch to different medium.)

Freedom from distraction: To what extent does the equipment distract the learners from the intended stimuli?

Darkening not required: Medium can be presented without necessity of darkening learner environment.

14.7 EVALUATION AND DOCUMENTATION

The last step in the instructional design process is probably the most important and the least often performed. Logically, when valuable manpower, time, and resources have been devoted to the development of a training program, it makes sense to find out whether the program is effective in meeting its knowledge and/or skills objectives. No one can predict with certainty how effective a program will be, and even the most carefully developed training benefits from "fine-tuning" based on field test results. When the health and safety of workers is at stake, evaluation becomes even more essential.

Documentation of training has been developed from a simple roster of attendees to complex systems that track the unique path of each worker as he or she is assigned to new jobs that require additional training. Throughout this chapter the importance of legal liability protection has been stressed; documentation is the most important protection of all. When evaluative procedures show the training to be effective, and these data are properly documented, the training process has come full cycle and the training need is demonstrably satisfied. However, the system is not static; needs must be continually assessed on the basis of continuing evaluation.

14.7.1 Evaluation Methods

A distinction should be made between *formative* evaluation, or the testing of the effectiveness of the training program as it evolves, and *summative* evaluation, which gives information on the long-term impact of training on worker knowledge, skills, and behaviors. Most occupational safety and health training programs are not evaluated at all. Very few formative evaluations, in which programs are tried out on samples of the target population, are performed. Those formative and summative evaluations that are performed generally consist of paper-and-pencil tests of comprehension; although these are valuable for documenting comprehension, they do not test transfer of the knowledge or skills to the job. An element working in favor of the trainer is the daily practice of job knowledge and skills. Safe work practices, once learned, should become habitual. Knowledge and skills that are used infrequently, however, will not be retained. Emergency procedure skills, for example, should periodically be reevaluated, and retraining conducted as needed. Another element to be considered is the level of mastery to be set as a criterion. A commonly accepted rule of thumb is that training is effective if 90% or more of the trainees achieve 90% or better on a posttest that reflects the original instructional objectives. Compared with the normal distribution of achievement reflected in most educational grading systems, "90 for 90" is a phenomenal figure. In the world of workers' safety and health, however, it could be less than satisfactory. An extreme example is that of the 10 diving trainees—is it satisfactory if only 9 rise to the surface after the posttest? Similar questions could arise in failure-to-inform suits if the individual claims to be one of the small minority who did not understand the hazard communication training. The answer to these considerations is not to discontinue and bury the evaluation process, but to revise, customize, or repeat the training as necessary. The ultimate goal of having all workers aware of and responsive to workplace hazards will allow no other solution.

The following guide to testing and evaluation prepared by the Health and Safety Education Department of Gulf Oil Corporation should be helpful in the selection and use of evaluation methods for tests of immediate comprehension following training.

EFFECTIVE EXAMINATIONS AND EVALUATIONS

I. Why test and evaluate? And what's the difference anyway!
   A. The difference between evaluations and examinations.
      1. Examinations or tests assess the change in behavior/attitude or the acquisition of knowledge/skill by the students.
      2. Evaluations are analyses of the training personnel, methods, and materials.
      3. Follow-up evaluations determine whether or not the training program(s) has contributed to the overall goals and objectives of the business/corporation.
      4. Examination (testing) is a component of evaluation.
   B. The value of tests to trainees.
      1. Provide feedback on progress.
      2. Provide information on the most important points of the training course.
      3. Provide an opportunity to apply learning.
      4. Influence trainees to review the work covered/pay attention to the material being presented.
   C. The value of tests to the instructor.
      1. Indicate whether or not objectives have been accomplished.
      2. Highlight those points the instructor did not clarify.
      3. Identify those students who require individual/extra instruction.
      4. Provide proof of training effectiveness.
   D. The value of tests to management.
      1. Assist in reporting on the effectiveness of training programs.
      2. Provide documentation of training and information on employee competency levels.
      3. Can be directly related to improved performance in follow-up evaluations.
   E. Characteristics of an adequate training evaluation plan.
      1. *Must* measure a broad range of objectives by a variety of techniques which are based on the kinds of learning to be accomplished.
      2. *Must* be systematic and an integral part of the training program development.
         a. Learning objectives must be determined in measurable terms.
         b. Testing should be planned during as well as at the end of training.
         c. Testing can be part of the teaching methodology.
      3. *Must* focus on changes in the student's behavior.
      4. *Must* provide for measure of all areas encompassed by the training program.

5. *Must* involve the widest possible participation of company personnel.
   a. Management support.
   b. Supervisory endorsement and follow-up.
   c. Public affairs: internal and external visibility/credibility.

II. Basic principles of test construction.
  A. It is important to follow the major steps in test construction.
    1. Determine the scope of the test.
    2. Determine what is to be measured.
    3. Select the test items.
    4. Select the technique.
    5. Fix the length of the text.
    6. Select the final items.
    7. Arrange the items in final form.
    8. Prepare directions for the test.
    9. Prepare a scoring device.
    10. Question the question.
  B. Do's and don'ts of test item development.
    1. Keep the language and punctuation clear.
    2. State each item in the working language of the employee.
    3. Avoid asking questions on trivial details.
    4. Use plausible choices.
    5. Keep weighting of items to a minimum.
    6. Increase the number of test items to increase reliability.
    7. Include clear directions for completing each type of test item.
    8. Avoid trick questions.
  C. Pretesting and posttesting as an evaluation tool. Pretesting allows prior knowledge to be measured, and provides a baseline for the measurement of gain.
  D. Subjective tests vs. objective tests.
    1. Advantages and disadvantages of the essay test and oral test.
    2. Objective tests include multiple choice, true-false, completion, matching test items, and performance testing.

III. Guidelines for development of seven types of test items.
  A. Essay test items are effective when the training stresses a "whole" or *gestalt* concept and measure the student's ability to:
    1. Organize facts/ideas.
    2. Reason with or from the knowledge gained.
    3. Express thoughts clearly.
  B. Oral questioning is both a testing and a teaching technique.
    1. Direct questioning.
    2. The overhead question.
    3. The reworded question.
    4. The reverse question.
    5. The rhetorical question.
  C. True-false items are easy to score and can be useful when worded with care.
    1. Limit each statement to a single idea.
    2. Avoid the use of specific determiners, e.g., all, never, frequently, sometimes, etc.
    3. Make sure the statements are unequivocally true or false.
    4. Quote the authority/source when a controversial statement is used.
    5. Avoid the use of negative statements.
    6. Try to use approximately equal number of true and false statements.
  D. Multiple-choice items either begin with an incomplete sentence or are a question.
    1. The problem should clearly point to the theme of the correct alternative answer.
    2. Incorrect alternatives should be plausibly related to the problem.
    3. Correct alternatives should not be consistently different in appearance from incorrect alternatives.
    4. All alternatives in the item must be grammatically consistent.
    5. Avoid grammatical cues and sentence structures that give away the correct alternative.
    6. Avoid employing such alternatives as "none of the above," "both a and b above," "all of the above."
    7. All choices should be plausible.
    8. Avoid including material unrelated to the theme of the intended response in the problem.
    9. Use negatives sparingly in the premise statements.
    10. Avoid irrelevant sources of difficulty in the statement of the problem or in the alternatives.

11. Make each item independent of every other item.
12. Alternatives within an item should not overlap or be synonymous with one another; they should be different but related to the premise.
13. Alternatives should be randomly ordered for each item.
14. Avoid confusing the multiple-choice testing style with other testing techniques.
15. Ensure that item content relates to important aspects of the subject matter.

E. Matching items are most useful for measuring selection or recognition.
1. The list of options should be at least 50% longer than the list of items.
2. The list of premise items should be relatively short.
3. When a matching item consists of a word or phrase having many associations to be paired with a more specific word or phrase, the more specific term should be placed in the stem column.
4. All the entries in each list should relate to the same central theme.
5. Premises and response sets should be arranged alphabetically, chronologically, or numerically.
6. All items of the matching set must be included on the same page.
7. Test instructions should clearly state how the matching is to be performed.

F. In responding to completion or fill-in questions, learners must recall, compute, or otherwise create the answers.
1. Only significant words should be omitted.
2. As a rule, use only one or two blank spaces.
3. Make sure that only one term will sensibly complete the statement.
4. When omitting words to make incomplete statements, leave enough clues.
5. Place the blank space near the end of the sentence to make the task easier and to avoid the trainees having to work backward for the answer.
6. Avoid repeating textbook phrasing word for word.
7. Avoid grammatical clues to the correct answer.
8. Use completion-type item only when appropriate.

G. The primary advantage of performance tests is that they are one of the best means of improving skills, because failures/weaknesses can be observed and used as the basis for additional practice.
1. Motivate trainees to do.
2. Insure that practice makes perfect.
3. Keep achievement standards progressive.
4. Make the test conditions realistic.
5. Apply and test the material as taught.
6. Make each step move to the next.

IV. Administration and documentation of tests and evaluations.

A. One of the characteristics of a good test is the formulation of a good set of test directions.

B. In determining the effectiveness of tests, there are many aspects to consider.
1. *Validity:* Did the test measure what it was designed to measure? Check the specific objectives.
2. *Reliability:* Did each test item measure with accuracy and consistency? Follow up on-the-job assists to determine this.
3. *Mechanical simplicity:* Was the test easy to give, easy to take, and to score? Instructors and trainees are working too hard if the answers are no.
4. *Discrimination:* Was there a reasonable distribution of scores (or items missed)?
5. *Comprehensiveness:* Was there at least one test item on each key point covered in the instruction?
6. *Range of difficulty:* Did the test include some easy items, some items of medium difficulty, and some that were fairly difficult?
7. *Objectivity:* Was scoring affected by the opinions of the scorer? This is especially true for essay test items.

C. Maintaining adequate training records is essential to provide proof that effective training and educational programs have been presented.
1. Keep one complete set of materials for each unit of instruction.
2. Maintain information on courses taken in employee file.
3. Maintain testing/performance data, when applicable.

14.7.2 Documentation Systems

Documentation is the heart of a training system; training cannot be proved or disproved without documentation. The driving force behind documentation, particularly in the area of hazard communication, is the chemical inventory. Once the chemicals in the workplace are inventoried, it is possible to identify the workers exposed to those chemicals and plan and record the training accordingly. It is possible to purchase software compatible with commonly used personal computers or with mainframes that will assist in organizing this information.

HARD COPY SYSTEMS

The traditional hard copy record of training is the sign-off sheet passed around at the training meeting. This sheet should be interpreted as meaning only that people were or were not present at the meeting; the fact that they were asleep, were daydreaming, or did not understand is not recorded. A step up from this system is the individual record; for example, a record in each worker's file will indicate which training courses he or she has attended. A quantum leap is taken if a record of each individual's understanding is retained. This is generally achieved by keeping test scores on tests that were designed to measure comprehension. Lastly, a record of safe performance on the job, or what we have called summative evaluation, can be kept. Planned observations of workers on the job can be made. If the worker is new at the job, a formal review can be made with the worker's knowledge to see if he or she knows the correct work procedures. For experienced workers, it is possible to document the fact that the job is or is not being performed in a safe manner without so informing the worker. Such observations, made solely for the purpose of correcting and guiding the worker, can be documented.

COMPUTER MANAGED SYSTEMS

A record of the documentation steps outlined above can of course be managed by a computer. In a system that is managed totally by the computer, the training itself can be computer based. The responses of workers to each step of the training program will trigger the presentation of supplementary or remedial training steps or an advance to more difficult steps. When used simply as a device for recording training, the computer is still a useful tool. For example, educational institutions commonly use software packages to record grades, class attendance, missed tests, and so on. In the workplace, it is possible for the computer to record industrial hygiene monitoring data for all the chemicals to which each worker is potentially exposed, with the inclusion of data on training and scores on tests. It is suggested that computer managed options should be explored before a decision is made on how best to document training.

14.7.3 Cost-Effectiveness

Eventually, the questions relating to the evaluation of training include the cost factor. Typical questions are: "Could we have done an equally good job for less money?" or "Could we have done better if we had spent more?" Most typically, the question arises as to whether anything of any value can be done with severely restricted resources. As noted earlier, really effective training never comes cheaply; it requires expenditures of time and effort that can be appreciated only by those who have gone through the entire experience. However, some methods and media are inherently "Cadillacs," and should not be undertaken on a compact-car budget. Some of these were identified in the earlier discussion of media options.

Two rules of thumb should be followed in attempting to develop an effective program at the lowest possible cost. One is to avoid costly revisions by following a carefully prescribed sequence of actions. The other is to design a "lean" program that can be expanded on the basis of field test results. The format shown in Figure 14.4 is suggested as a useful sequence to follow.

The process does not include the evaluation steps, and is essentially the same whether or not an outside contractor is involved. The critical points in the process are outline approval, script approval, story board approval, final review conference, and distribution. Each of these points represent a "go-no go" decision that allows revisions to be made at least possible expense. The sequence keeps second-guessing to a minimum, although it cannot prevent last minute sabotage of an expensive program by senior management.

The concept of a "lean" program is that it is difficult to prune a program that is initially fat. The flow of the training sequence is such that removal of material may cause awkward gaps. Field testing of a lean program, on the other hand, will show which portions can effectively remain lean and which need additional material to assist in learner acquisition of skills or knowledge.

14.8 RESOURCE MATERIALS

Developments in the field of occupational health and safety training have caused a surge in the availability of resources. There is no way of keeping such a list current, but access to one resource often yields information about others (see Chapters 7 and 11).

14.8.1 Resources for Assistance

Since health and safety training programs are difficult and expensive to develop, it is helpful to take advantage of assistance available from a variety of professional, governmental, and trade organizations

GULF HEALTH AND SAFETY EDUCATION DEPARTMENT

TRAINING PROGRAM DEVELOPMENT
AND REVIEW PROCEDURES

* If contractor is used.

DATE(S)	ACTIVITY/TASK	H&SE STAFF	TECHNICAL AND LEGAL SUPPORT	CLIENT REP.	CONTRACTOR STAFF
	Pre-Production Conference(s)				
	Statement of Objectives				
	*Proposal(s) Review				
	*Contract Award				
	Strategy and Design Conference				
	Subject Matter Input				
	Outline Write				
	Outline Approval				
	Script Write, including tests				
	1st Script Conference				
	Script Re-write				
	Script Approval				
	Preparation of Story Board				
	1st Story Board Conference				
	Story Board Approval				
	Develop Art/Manuals/Refine Tests				
	Arrange Site Visits				
	Photography				
	Visual Review				
	Revise Visuals				
	Final Review Conference				
	Production of Slides/Video				
	Sound Studio/Sound Approval				
	Assemble and Deliver				
	Distribute to Client(s)				

Program Topic: ______________________ Form No. __________

Outside Contractor: ______________________

FIGURE 14.4. Sample development form.

that serve as forums for the exchange of information. Some of these groups have also developed training programs that are useful to other organizations, and generally make these available. The following list is not exhaustive, but includes many of the most useful sources of assistance. It was derived as part of a project conducted by the Task Group on Communication and Education of the Occupational Safety and Health Committee of the Chemical Manufacturers Association.

I. Trade Associations

1. *American Petroleum Institute (API)*
   API has produced several training programs. The *API Publications and Materials Catalog for 1983* contains training programs on fire fighting, accident control techniques, and safe handling practices for petroleum industry workers. API addresses specific topics such as confined space entry, equipment lockout, industrial hygiene orientation and fire protection.
   Some audiovisual materials are included in the Catalog.
   American Petroleum Institute
   1220 L Street N.W.
   Washington, DC 20005
   (202) 682-8000
2. *National Agricultural Chemicals Association (NACA)*
   NACA, through its OSHA Committee, has developed and adopted a set of *Good Workplace Practices (GWPs) for the Manufacture and Formulation of Pesticides.* These GWPs are available in booklet form and describe recommended industrial hygiene program components based on the best practical combination of administrative and engineering controls, work practices, personal protective equipment, and control monitoring.
   NACA and OSHA recently developed a slide/tape program based on the GWPs. The slide/tape program as well as the GWP booklet cover recommendations on hazard awareness, employee indoctrination and training, facility and equipment design, operating procedures, housekeeping, hygiene facilities, personal protective equipment, and medical surveillance. The slide/tape program is intended to provide information to employers in the industry and to be of use in worker education and training programs. The GWP booklet and the slide/tape program are available from NACA.
   The National Agricultural Chemicals Association
   1155 15th Street, N.W.
   Washington, DC 20005
   (202) 296-1585
3. *The Society of the Plastics Industry, Inc. (SPI)*
   SPI is presently involved in the second year of an OSHA New Directions Program Grant project aimed at developing worker safety and health training programs for the plastics processing industry and is working with a consultant on this project to produce a manual for supervisors and a series of eight 15-minute slide/tape training modules.
   The manual, designed for supervisors, has been completed and is entitled *Safety & Health Reference Guide for the Plastics Processor.* It is issued as an $8\frac{1}{2} \times 11$ looseleaf binder so that changes and updates can be made easily.
   One slide/tape module entitled: "Accident Prevention, Investigation and Response," is aimed at the supervisor, and the remaining seven are designed for workers in the plastics processing industry. Titles include Mechanical Materials Handling, Plastics Machinery Safety, Noise, and Electrical Safety. Four of the modules were completed in 1982, and the remaining four were completed in 1983. SPI is planning "train the trainer" seminars for member companies on the operation of good plant safety programs through the use of available SPI training materials. The manual and slide/tape modules will be available for purchase from SPI. Further information can be obtained by calling SPI's Committee on Occupational Safety and Health, in New York.
   Society of the Plastics Industries, Inc.
   355 Lexington Avenue
   New York, NY 10017
   (212) 503-0600
4. *National Paint and Coatings Association (NPCA)*
   NPCA has developed a hazardous materials identification system (HMIS) that contains training ideas.
   National Paint and Coatings Association
   1500 Rhode Island Ave., N.W.
   Washington, DC 20005
   (202) 462-6272

5. *Synthetic Organic Chemical Manufacturers Association (SOCMA)*
   Synthetic Organ Chemical Manufacturers Ass'n.
   1330 Connecticut Ave. Suite 300
   Washington, DC 20036
   (202) 659-0060
6. *Chemical Specialties Manufacturers Assn. (CSMA)*
   1001 Connecticut Avenue, N.W.
   Washington, DC 20036
   (202) 872-8110

II. Federal Agencies

1. *Occupational Safety and Health Administration (OSHA)*
   OSHA has produced several worker training programs that are in the public domain and can be modified or copied to suit the needs of the individual company. Lists of currently available programs can be obtained from the nearest OSHA Regional Office in your region.

   OSHA has also developed extensive programs for compliance officer training. These provide inexpensive sources of slides and scripts, again in the public domain, and useful in building individual company programs. Information on these programs can also be obtained from your regional OSHA office.

   The OSHA Training Institute in Des Plaines, Illinois, offers many courses for private sector employees as well as federal and state personnel. The courses are excellent and inexpensive.

   A useful reference for the development of worker health and safety training is OSHA's booklet *Teaching Safety and Health in the Workplace: An Instructor's Guide.*

   Periodically, an *OSHA Publications and Training Materials* listing is issued (OSHA 2019). For information, contact the regional OSHA office listed in your local telephone directory.
2. *National Institute for Occupational Safety and Health (NIOSH)*
   In October, 1981, NIOSH issued a comprehensive guide, entitled *Audiovisual Resources in Occupational Safety and Health: An Evaluation Guide* (DHHS-NIOSH Publication No. 82-102), which is a compilation of critical reviews of training materials.
   Division of Training and Manpower Development
   U.S. Department of Health and Human Services
   CDC–NIOSH
   4676 Columbia Parkway
   Cincinnati, OH 45226
   (513) 533-8236
3. *National AudioVisual Center (NAVC)*
   NAVC is the central source for federal audiovisual materials, and periodically issues catalogs of worker health and safety training programs. Catalog orders and inquiries should be addressed to:
   National AudioVisual Center
   Order Section MS
   8700 Edgeworth Drive
   Capital Heights, Maryland 20743-3701
   (301) 763-8236
4. *U.S. Department of Commerce, National Bureau of Standards (NBS)*
   NBS has published documents on effective hazard warning symbols and signs through their National Engineering Laboratory.
   U.S. Department of Commerce
   National Bureau of Standards
   Washington, DC 20234
   (301) 921-2000

III. Service Organizations

1. *The National Safety Council (NSC)*
   The National Safety Council has several safety and health education and training programs available. Catalogs of programs can be obtained by contacting your local safety council or the National Council in Chicago. New programs are being issued continually.
   National Safety Council
   444 North Michigan Avenue
   Chicago, IL 60611
   (312) 527-4800

2. *National Association of Manufacturers (NAM)*
   NAM has a task group which is presently working on the production of two manuals dealing with:
   a. Health and safety training films made by NAM member companies.
   b. Health and Safety training programs made by NAM.
      National Association of Manufacturers
      1331 Pennsylvania Ave., N.W.
      Suite 1500 North
      Washington, DC 20004-1703
      (202) 637-3000
3. *Organization Resources Counselors, Inc. (ORC)*
   ORC is looking into the development of audiovisual training programs of common interest to a broad spectrum of companies.
   Organization Resources Counselors, Inc. (ORC)
   1331 Pennsylvania Avenue N.W.
   Washington, DC 20004
   (202) 737-6330

IV. Professional Societies

1. *American Society for Training and Development–Occupational Safety and Health Special Interest Group (ASTD-OSHSIG)*
   The ASTD–OSHSIG is composed of widely diverse industrial interests, such as chemical process safety, mining safety, and transportation safety. It meets annually for a half-day session and addresses broad topics in health and safety. Information can be obtained by contacting the Special Interest Group Coordinator at ASTD Headquarters:
   American Society for Training and Development
   1630 Duke St.
   P.O. Box 1443
   Alexandria, VA 22313
   (703) 683-8100
2. *American Society of Safety Engineers (ASSE)*
   ASSE sponsors professional development courses and seminars for the benefit of their members.
   American Society of Safety Engineers
   1800 East Oakton St.
   Des Plaines, Illinois 60018
   (312) 692-4121
3. *American Association of Occupational Health Nurses*
   3500 Piedmont Road
   NE Suite 400
   Atlanta, GA 30305
   (404) 262-1162
4. *American Conference of Governmental Industrial Hygienists (ACGIH)*
   American Conference of Governmental Industrial Hygienists
   6500 Glenway Ave.
   Building D-7
   Cincinnati, OH 45211
   (513) 661-7881
5. *American Industrial Hygiene Association (AIHA)*
   AIHA is actively involved in health and safety education. The Training, Education and Communications Committee sponsors a half-day session on training and hazard communication as well as professional development courses at the annual conference.
   American Industrial Hygiene Association
   475 Wolf Ledger Parkway
   Akron, OH 44311-1087
   (216) 762-7294
6. *American Occupational Medical Association*
   2340 South Arlington Heights Rd.
   Suite 400
   Arlington Heights, IL 60005
   (312) 228-6850
7. *American Medical Association (AMA)*
   American Medical Association
   535 N. Dearborn Street
   Chicago, IL 60610
   (312) 645-5000

8. *American Chemical Society (ACS)*
ACS sponsors professional development courses, seminars and symposia for the benefit of their members. Their available safety-related training programs are directed toward educational institutions such as colleges and universities (laboratory safety), and not toward the chemical worker.
American Chemical Society
1155 16th Street, N.W.
Washington, DC 20036
(202) 872-4600

V. Standards and Specification Groups
1. *American National Standards Institute (ANSI)*
ANSI has issued a Standard, ANSI Z 129.1 entitled: *Precautionary Labeling of Hazardous Industrial Chemicals,* which is available for purchase and can be used as a guide in developing hazard communications for workers.
American National Standards Institute
1430 Broadway
New York, NY 10018
(212) 354-3300
2. *American Society for Testing and Materials (ASTM)*
ASTM Committee E-34 on Occupational Health and Safety
1916 Race Street
Philadelphia, PA 19103
(215) 299-5483
3. *Underwriters Laboratories, Inc. (UL)*
Underwriters Laboratories, Inc.
333-T Pfingsten Road
Northbrook, IL 60062
(312) 272-8129

VI. Insurance Companies
Insurance companies that deal with occupational health and safety in industry have developed training programs for their clients. It is recommended that you check with your company's insurance carrier for information or contact one or both of the following organizations:
Alliance of American Insurers
1501 Woodfield Rd.
Suite 400 W
Shomberg, IL 60173
(312) 490-8500
American Insurance Association
85 John Street
New York, NY 10038
(Washington, DC 202-293-3010)

VII. Fire Protection Associations
1. *National Fire Protection Association (NFPA)*
NFPA develops specific resource materials on fire safety and publishes a catalog.
National Fire Protection Association
Batterymarch Park
Quincy, MA 02269
(617) 770-3000

REFERENCES

Cornelius, E. T. III and Kraiger, K., *Analysis of the Reading Level of American Petroleum Institute Training Materials,* Report submitted by the Ohio State University Research Foundation, March 10, 1980.

Kanter, R. M., *The Change Masters,* Simon & Schuster, New York, 1983.

Kuest, R. D., *Marihuana—The Other Side,* Management by God, Olympia, WA, 1980.

Mager, R. F., and Pipe, P., *Analyzing Performance Problems,* Fearon-Pittman, 1970.

Sweeney, K. M., *Building an Effective Labor-Management Safety and Health Committee, A Practitioner Manual.* Produced under a grant from OSHA to the American Center for the Quality of Work Life, 1411 K Street, N.W., Washington, D.C. 20005, 1984.

Worker Training Task Group of the Chemical Manufacturers Association, *Worker Health and Safety Education in the Chemical Industry,* CMA, Washington, D.C. 1984.

BIBLIOGRAPHY

American Society for Training and Development, *Be a Better Task Analyst*, ASTD, 1630 Duke Street, P.O. Box 1443, Alexandria, VA 22313.

American Society for Training and Development, *Training and Development Handbook* (R. L. Craig, ed.) 2nd Ed., New York: McGraw-Hill, 1976.

Briggs, Leslie J., *Handbook of Procedures for the Design of Instruction*, American Institutes for Research, Monograph No. 4, 1970.

Chemical Manufacturers Association, *Worker Health and Safety Education in the Chemical Industry*, April, 1984.

Cohen, A., et al., Self-Protective Measures Against Workplace Hazards, *Journal of Safety Research*, 1979, **11**(13):121-131.

Davies, I. K., *Instructional Technique*, New York: McGraw-Hill, 1980.

Denova, C. C., *Test Construction for Training Evaluation*, New York: Van Nostrand Reinhold Company, 1979.

Hamrick, M. H., Anspaugh, D. J., and Smith, D. L., Decision-Making and the Behavior Gap, *The Journal of School Health*, October, 1980, 455-458.

Heath, E. D., Worker Training and Education in Occupational Safety and Health: A Report on Practice in Six Industrialized Western Nations, *Journal of Safety Research*, **13**(2):1982.

Kemp, J. E., *Planning and Producing Audio-Visual Materials*, New York: Thomas Y. Crowell, 1975.

Knowles, M. S., *The Modern Practice of Adult Education*, Chicago: Association Press, 1980.

Mager, R. F., *Preparing Instructional Objectives*, 2nd Ed., Fearon-Pitman, 1975.

Mager, R. F., *Measuring Instructional Intent*, Fearon-Pitman, 1973.

Manning, D. T., Writing Readable Health Messages, *Public Health Reports*, 1981, **96**(5):31-34.

Manning, D. T., Writing to Promote Health and Safety Behavior in Occupational Settings, *Journal of Technical Writing and Communication*, 1982, **12**(4):301-306.

Margolis, F. H., *Training By Objectives: A Participant Oriented Approach*, Cambridge, MA.: McBer, 1970.

McCormick, E. J., *Job Analysis: Methods and Applications*, New York: American Management Association, 1979.

National Institutes for Occupational Safety and Health *Audiovisual Resources in Occupational Safety and Health*, NIOSH, Cincinnati, 1981.

Pipe, P., *Objectives-Tool For Change*, Belmont, CA: Fearon Publishers, 1975.

Samways, M. C., Cost-Effective Occupational Health and Safety Training, *American Industrial Hygiene Association Journal*, (44)1983.

Yankelovich, D. and Immerwahr, J., *Putting the Work Ethic to Work: A Public Agenda Report on Restoring America's Competitive Vitality*, New York: The Public Agenda Foundation, 1983.

CHAPTER 15

How to Measure Training Effectiveness

Jack B. Re Velle
Hughes Aircraft Company
Los Angeles, California

15.1 INTRODUCTION

The cost of providing OSH training in an organization is increasing, and as organizations scramble to fund their budgets, it is imperative that they know where the money is spent and for what purposes. This chapter explores in detail 10 statistically oriented methodologies used to ascertain safety training effectiveness. Although the methods must be tailored to the specific situation, a number of fundamental principles and guidelines can be useful for any situation.

15.2 REASONS FOR MEASURING TRAINING EFFECTIVENESS

This discussion will not include the use of such conventional measures as instructor evaluations. There are two reasons for this:

1. Literature on the subject indicates that students are good observers of instructor performance, but are poor evaluators of class content.
2. Job performance is the final measure of any training program. No matter how well or how poorly students rate an instructor or a course, what really matters is the extent of desired change in the students' behaviors.

15.3 CHAPTER ORIENTATION

We will use five measures of training effectiveness as proxies for student behavior:

- Accident rates.
- Hazard reporting rates.
- Injury costs.
- Personnel evaluations.
- Workers' compensation premiums.

15.3.1 Accident Rates

Accident rates are a commonly used measure of work performance. While exact values are relative to a variety of environmental elements, differences between time periods, between shifts for the same time period, between industries, etc., are excellent indicators of change. One such change could be the difference between two departments, one that has received safety training and one that has not. Another change could be the difference between the accident rates of a department that was exposed to one type of safety training program and one that received a different type of training. Thus, accident rates become measurable indicators of the effectiveness of safety training.

15.3.2 Hazard-Reporting Rates

Hazard-reporting rates are not used as commonly as accident rates to help measure the quality of safety training, but they are excellent internal indicators nonetheless. The hazard reporting rate is a function of at least five contributing factors: (1) number of physical hazards that require reporting, (2) number of personnel hazards that require reporting, (3) motivation of personnel to report hazards that have been detected, (4) capacity of personnel to discern the presence of hazards, and (5) the organizational climate, which may or may not be supportive of a high hazard-reporting rate. Comprehensive safety training substantially influences the last three factors, and so hazard-reporting rates are used to measure safety training effectiveness.

15.3.3 Injury Costs

Injury costs are direct measures of how well or how poorly a safety training program has impacted costs of operation. Depending upon a variety of interrelated factors, injury costs can be used to demonstrate the cost benefits associated with safety training types, delivery modes, durations, and so on.

15.3.4 Personnel Evaluations

Personnel evaluations are traditionally rendered on each employee at least annually. No matter what scoring system is used, they are easily adapted to measuring safety training effectiveness. Employees who have internalized their safety training are expected to perform better, and thus, receive more points on their performance evaluations. The better the training, the greater the internalization, the better the performance, and therefore, the higher the evaluation.

15.3.5 Workers' Compensation Premiums

Workers' compensation premiums are paid by employers to their insurance carriers. The better a company's safety performance, that is, the lower the company's accident frequency and severity rates, the lower the premiums. And as noted earlier, accident rates are commonly used proxies for safe work performance. Safety training determined to be effective should be expected to contribute to reduction of workers' compensation premiums.

15.3.6 Methods of Inference

We will also demonstrate the applicability of ten methods of statistical inference to measuring safety training effectiveness. These methods are:

- Probabilistic analysis (using Bayes theorem).
- Chi-square analysis.
- Statistical control charts.
- Analysis of variance.
- Regression analysis.
- Student's *t*.
- Sampling/surveying.
- Hypothesis testing.
- Experimental design.
- Correlation analysis.

Table 15.1 presents a measures–methods matrix that outlines the ten examples that make up the remainder of this chapter.

15.4 OUTSIDE REFERENCES

If necessary, the reader is urged to make extensive use of several of the many available texts on statistics, statistical quality control, and experimental design. In the event that the explanations in the examples that follow are insufficient, the reader with limited previous exposure to statistical analysis should expect to have to devote some time and effort in this area.

15.5 EXAMPLES

The remainder of this chapter is devoted to examples of the five measures of training effectiveness using ten measures of statistical inference as identified in Table 15.1.

15.5.1 Example: Accident Rates and Probabilistic Analysis

PROBLEM

The probability that an employee will have an acceptable accident history (however a company may choose to define it) is 60%. The mathematical statement is written as:

$$P(A) = 0.60$$

Each employee completes a safety orientation course and is tested at the conclusion of the course. Among current employees, company records indicate that 80% of the employees who ultimately have acceptable accident histories pass the test, but only 40% of those who have unacceptable accident histories pass. Let P mean *pass* and A' mean *not acceptable*, then:

$$P(P/A) = 0.80$$

and

$$P(P/A') = 0.40$$

Suppose a new employee takes the safety course and passes the test. Clearly, the probability, $P(A/P)$, that the employee will have an acceptable accident history given that he or she passes the test, is greater than 0.60, because this employee is no longer one of the general group whose probability is 0.60,

TABLE 15.1. Measures–Methods Matrix

Pre/Post Rates and Ratings: Measures of Training Effectiveness	Test Statistics: Methods of Inference[a]									
	Probabilistic Analysis	Chi-Square Analysis	Statistical Control Charts	Analysis of Variance	Regression Analysis	Student's t	Sampling/ Surveying	Hypothesis Testing	Experimental Design	Correlation Analysis
Accident rates	1	o	o	o	o	6	o	o	o	o
Hazard reporting rates	o	2	o	o	o	o	7	o	o	o
Injury costs	o	o	3	o	o	o	o	8	o	o
Personnel evaluations	o	o	o	4	o	o	o	o	9	o
Workers' compensation premiums	o	o	o	o	5	o	o	o	o	10

[a] o indicates compatibility of measures–methods interface; x indicates example problem number.

but is, instead, one who has passed the test. *Before* he or she took the test, the probability was 0.60; this is called the *prior* probability.

Our interest is in computing the *posterior* probability, the probability that the employee will have an acceptable accident history after having taken the safety course and passed the test.

Solution

The calculations rest upon the definition of *conditional probability,* in particular upon the statement that the joint probability of two conditions, X and Y, is equal to the product of the conditional probability of one, given the other, times the marginal probability of the other. Mathematically, this is written as:

$$P(X, Y) = P(X/Y) \cdot P(Y) = P(Y/X) \cdot P(X) \tag{15.1}$$

In this problem, we know that:

$$P(A) = 0.60$$
$$P(P/A) = 0.80$$
$$P(P/A') = 0.40$$

From the definition of conditional probability, we compute:

$$P(P, A) = P(P/A) \cdot P(A) = (0.80)(0.60) = 0.48$$
$$P(P, A') = P(P/A') \cdot P(A') = (0.40)(0.40) = 0.16$$

Note: $P(A) + P(A') = 1.00$, since the two events, acceptable accident history and unacceptable accident history, are mutualy exclusive and collectively exhaustive. Knowing that $P(A) = 0.60$, a simple subtraction exercise produces $P(A') = 0.40$.

Now we have:

	P	P'	Total
A	0.48	v	0.60
A'	0.16	w	y
Total	0.64	x	z

We can now begin to complete this probability table:

$z = 1.00$, since it is the sum of both the marginal probabilities:

$$P(A) + P(A') = 1.00$$

and

$$P(P) + P(P') = 1.00$$

Since we know $P(A) = 0.60$:

$$P(A) + y = 1.00$$
$$0.60 = y = 1.00$$
$$y = 0.40$$

Since we know $P(P) = 0.64$:

$$P(P) + P(P') = 1.00$$
$$0.64 + x = 1.00$$
$$x = 0.36$$

The probability table now looks like:

	P	P'	Total
A	0.48	v	0.60
A'	0.16	w	0.40
Total	0.64	0.36	1.00

Now we can complete the probability table:

$$0.48 + v = 0.60$$
$$v = 0.12$$
$$0.16 + w = 0.40$$
$$w = 0.24$$

Note: A quick check of the calculations to assure accuracy is in order.

Since we know that

$$v + w = 0.36$$

we make the necessary substitutions to verify the values of v and w:

$$0.12 + 0.24 = 0.36$$
$$0.36 = 0.36$$

We have a check.

The probability table is now complete.

	P	P'	Total
A	0.48	0.12	0.60
A'	0.16	0.24	0.40
	0.64	0.36	1.00

At this point we can answer any number of questions from the completed probability table. For example, what is the probability of an employee having an acceptable accident history given that he or she passed the test at the conclusion of the safety course, that is, what is the value of $P(A/P)$?

From the definition of conditional probability, we know that our mathematical statement is

$$P(A, P) = P(A/P) \cdot P(P) \tag{15.2}$$

We want the value of $P(A/P)$. The values of $P(A, P)$ and $P(A/P)$ and $P(P)$ are extracted from the probability table. $P(A,P)$, the joint probability of an acceptable accident history *and* passing the test at the conclusion of the safety course, is 0.48. $P(P)$, the marginal probability of passing the test, is 0.64. Making the necessary substitutions we have

$$0.48 = P(A/P) \cdot (0.64)$$
$$P(A/P) = 0.48/0.64$$
$$= 0.75$$

Without mathematical symbols, this means that the probability is 0.75 that an employee, selected at random, who has passed the test will have an acceptable accident history.

Let's try a second example. What is the value of $P(A'/P')$? We proceed as before.

$$P(A', P') = P(A'/P') \cdot P(P')$$
$$0.24 = P(A'/P') \cdot (0.36)$$
$$P(A'/P') = 0.24/0.36$$
$$= 0.67$$

Without symbols, this means that the probability is 0.67 that an employee, selected at random, who has *not* passed the test will have an unacceptable accident history.

Another approach to the solution to problems of this type is to use Bayes theorem, which states, in general, that

$$P(Y/X) = \frac{P(X/Y) \cdot P(Y)}{P(X/Y) \cdot P(Y) + P(X/Y') \cdot P(Y')} \tag{15.3}$$

Restating this in terms of the first example, we have

$$P(A/P) = \frac{P(P/A) \cdot P(A)}{P(P/A) \cdot P(A) + P(P/A') \cdot P(A')}$$

Making the necessary substitutions in Bayes theorem from the original problem and the completed probability table, we have

$$P(A/P) = \frac{(0.80)\,(0.60)}{(0.80)\,(0.60) + (0.40)\,(0.40)}$$

$$= \frac{0.48}{0.48 + 0.16} = \frac{0.48}{0.64}$$

$$= 0.75$$

which of course is the conclusion we reached earlier.

15.5.2 Example: Hazard-Reporting Rates and Chi-Square Analysis

PROBLEM

It is claimed that an equal number of employees who have and have not completed a specific safety training course report hazards that they spot while on the job. A random sample of 40 employees are observed, and of these, 25 have completed the course and 15 have not. Test the null hypothesis that the overall number of employees reporting hazards is equal by applying the chi-square test and using the 5% level of significance.

Solution

We begin by developing a table of information already available.

	Employees		
	Completed Training	Did not Complete Training	Total
Number in Sample, f_o	25	15	40
Number Expected, f_e	20	20	40

Next, we state our null and alternate hypotheses, which are, respectively:

H_0: The number of employees who completed the safety training course is equal to the number who did not.

H_1: The number of employees who completed the safety training course is not equal to the number who did not.

If we let

x = the number of employees who completed the safety training course

and y = the number of employees who did not complete the safety training course then we can restate the null and alternate hypotheses in quantitative terms:

$$H_0: x = y$$

$$H_1: x \neq y$$

or

$$H_0: x - y = 0 \text{ (thus, the null hypothesis)}$$

$$H_1: x - y \neq 0$$

Next, we search out the critical chi-square value from a chi-square table. To do this we first need two bits of information: the number of degrees of freedom (df) and the level of significance desired for protection.

The df value is determined by

$$df = k - m - l \tag{15.4}$$

where k is the number of observed categories, in this case $k = 2$,
and m is the number of population parameters estimated, in this case $m = 0$.

Therefore,

$$df = 2 - 0 - 1 = 1$$

The level of significance, which is the complement of the confidence level, was given in the problem statement as 5%, therefore alpha (α, the Greek letter used to symbolize the level of significance) equals 0.05.

With $df = 1$ and $\alpha = 0.05$, we can find the critical chi-square value for this problem:

$$\text{critical } \chi^2 \, (df = 1, \alpha = 0.05) = 3.84$$

The next step is to calculate the actual chi-share value for this problem. The χ^2 value is determined by

$$\chi^2 = \sum \frac{(|f_o - f_e| - 0.5)^2}{f_e} \tag{15.5}$$

where f_o is the frequency of occurrence of each observed category,
and f_e is the frequency of occurrence of each expected category.
$|f_o - f_e|$ is the absolute value (without regard to sign) of the difference between the two frequencies.
0.5 is the Yates factor to correct for continuity (this is discussed in most basic statistics texts).
Therefore,

$$\chi^2 = \frac{(|25 - 20| - 0.5)^2}{20} + \frac{(|15 - 20| - 0.5)^2}{20}$$

$$\chi^2 = \frac{(5 - 0.5)^2}{20} + \frac{(5 - 0.5)^2}{20}$$

$$\chi^2 = \frac{(4.5)^2}{20} + \frac{(4.5)^2}{20}$$

$$\chi^2 = 2.02$$

The solution to the problem is concluded by comparing the calculated value of χ^2 to the critical value of χ^2. If the calculated value is equal to or greater than the critical value we cannot accept the null hypothesis, i.e., we must accept the alternate hypothesis.

In this case we have

$$\text{calculated} \quad \chi^2 = 2.02$$

$$\text{while critical} \quad \chi^2 = 3.84$$

Since the calculated χ^2 is less than the critical χ^2, we cannot reject the null hypothesis at the 5% level of significance. Said another way, when $\alpha = .05$ we have no reason not to believe the null hypothesis that safety training has had no impact on the hazard reporting rate.

15.5.3 Example: Injury Costs and Statistical Control Charts

PROBLEM

At this point we turn our attention to injury costs. Our concern is whether they are increasing, decreasing, or remaining unchanged. If the cost comparison is to include time frames greater than a year, then it is appropriate to use an inflation or deflation factor to keep the time value of money constant. This problem treats the various costs as if one or the other of these two factors has already been appropriately applied.

To determine if injury costs are "in control," we will apply the concept of statistical control charts, which are most commonly used in quality control/defect analysis.

Two types of statistical control charts exist: attributes and variables. *Attributes* charts are used when the data being collected can only be counted, that is, measurement is not possible. An example of this is the number (or frequency) of accidents each month in a given department. Variables charts, on the other hand, are used when the data being collected can be measured. Examples of this type of charting include accident rates and injury costs. The problem that follows is an application of variables control charting to the control of injury costs. Specifically, we will apply the most commonly used, most versatile of all control charts, the x-bar and R Chart.

An x-bar (usually written as $\bar{x}$ and read as "x-bar") and R control chart is one that shows the mean value, $\bar{x}$, and the range, R. The $\bar{x}$ portion of the chart is used primarily to show any changes in the mean (arithmetic average) of a process, while the R portion shows any changes in process dispersion or variation around the mean. This chart is particularly useful because it shows changes in both mean and dispersion values simultaneously. Thus, the $\bar{x}$–R control chart is quite effective for detecting process abnormalities. Table 15.2 includes sample injury cost data for a manufacturing firm for a 25-month period. The data has been coded from the original injury cost figures, for example, the first sample cost in the first month is shown in Table 15.2 as 14.0. The actual injury cost was $1400, but to ease the necessary subsequent computations, this was coded to 14.0. The second sample in the first month is presented as 12.6; this is the coded value from a injury cost of $1260. Each of the five sample values for all 25 months have been generated the same way: coded by dividing the original value by 100. Other problems may also lend themselves to coding of the actual data to produce data which is more easily computed.

Table 15.2 also includes an $\bar{x}$ and R value for each of the 25 months observed.

Solution

The 11 steps necessary to produce an $\bar{x}$–R chart are:

1. Collect the data. It is best to have at least 100 samples taken from recent data from a process similar to one which will be used in the future. (In this example, we have 25 months of injury cost experience for a manufacturing firm.)
2. Place the data into logical subgroups. These subgroups can be according to measurement or in order of collection, and should include from two to five samples each. The data should be divided into subgroups in keeping with the following conditions:

   a. Data obtained under the same technical conditions should be used to form a subgroup.

   b. A subgroup should not include data from a different time period or of a different nature.

   Note: For the reasons above, data are usually divided into subgroups according to date, time, lot, and so on. The number of examples in a subgroup determines the size of the subgroup and is represented by n; the number of subgroups is represented by k. (In this example, we have selected 5 samples from each of the 25 months. Thus, $n = 5$ and $k = 25$.)
3. Record the data on a data sheet. It should be designed to facilitate computation of $\bar{x}$ and R values for each sub-group. Figure 15.1 is a frequently used format to collect and plot $\bar{x}$ and R chart values.
4. Find the mean value, $\bar{x}$. Use equation 15.6 to calculate $\bar{x}$ for each subgroup. Compute the mean value $\bar{x}$ to one decimal place beyond that of the original measurement value. (In this example, a coded sample value of 14.0 was drawn from an uncoded sample value of 1400.)

$$\bar{x} = \frac{x_1 + x_2 + x_3 + \ldots + x_n}{n} \tag{15.6}$$

TABLE 15.2. $\bar{x}$–R Control Chart

Month	Sample No.						
No.	1	2	3	4	5	$\bar{x}$	R
1	14.0	12.6	13.2	13.1	12.1	13.00	1.9
2	13.2	13.3	12.7	13.4	12.1	12.94	1.3
3	13.5	12.8	13.0	12.8	12.4	12.90	1.1
4	13.9	12.4	13.3	13.1	13.2	13.18	1.5
5	13.0	13.0	12.1	12.2	13.3	12.72	1.2
6	13.7	12.0	12.5	12.4	12.4	12.60	1.7
7	13.9	12.1	12.7	13.4	13.0	13.02	1.8
8	13.4	13.6	13.0	12.4	13.5	13.18	1.2
9	14.4	12.4	12.2	12.4	12.5	12.78	2.2
10	13.3	12.4	12.6	12.9	12.8	12.80	0.9
11	13.3	12.8	13.0	13.0	13.1	13.04	0.5
12	13.6	12.5	13.3	13.5	12.8	13.14	1.1
13	13.4	13.3	12.0	13.0	13.1	12.96	1.4
14	13.9	13.1	13.5	12.6	12.8	13.18	1.3
15	14.2	12.7	12.9	12.9	12.5	13.04	1.7
16	13.6	12.6	12.4	12.5	12.2	12.66	1.4
17	14.0	13.2	12.4	13.0	13.0	13.12	1.6
18	13.1	12.9	13.5	12.3	12.8	12.92	1.2
19	14.6	13.7	13.4	12.2	12.5	13.28	2.4
20	13.9	13.0	13.0	13.2	12.6	13.14	1.3
21	13.3	12.7	12.6	12.8	12.7	12.82	0.7
22	13.9	12.4	12.7	12.4	12.8	12.84	1.5
23	13.2	12.3	12.6	13.1	12.7	12.78	0.9
24	13.2	12.8	12.8	12.3	12.6	12.74	0.9
25	13.3	12.8	12.0	12.3	12.2	12.72	1.1
						$\Sigma\bar{x} = 323.50$	$\Sigma R = 33.8$
						$\bar{\bar{x}} = 12.940$	$\bar{R} = 1.35$

For data from the first month, the calculation is:

$$\bar{x} = \frac{14.0 + 12.6 + 13.2 + 13.1 + 12.1}{5}$$

$$\bar{x} = \frac{65.0}{5} = 13.00$$

And for the second month,

$$\bar{x} = \frac{13.2 + 13.3 + 12.7 + 13.4 + 12.1}{5}$$

$$\bar{x} = \frac{64.7}{5} = 12.94$$

5. Find the range, R. Use equation 15.7 to compute R for each subgroup.

$$R = x_{\text{(largest value)}} - x_{\text{(smallest value)}} \qquad (15.7)$$

For the first two months in Table 15.2, R is:

$$R = 14.0 - 12.1 = 1.9$$

$$R = 13.4 - 12.1 = 1.3$$

430813-7

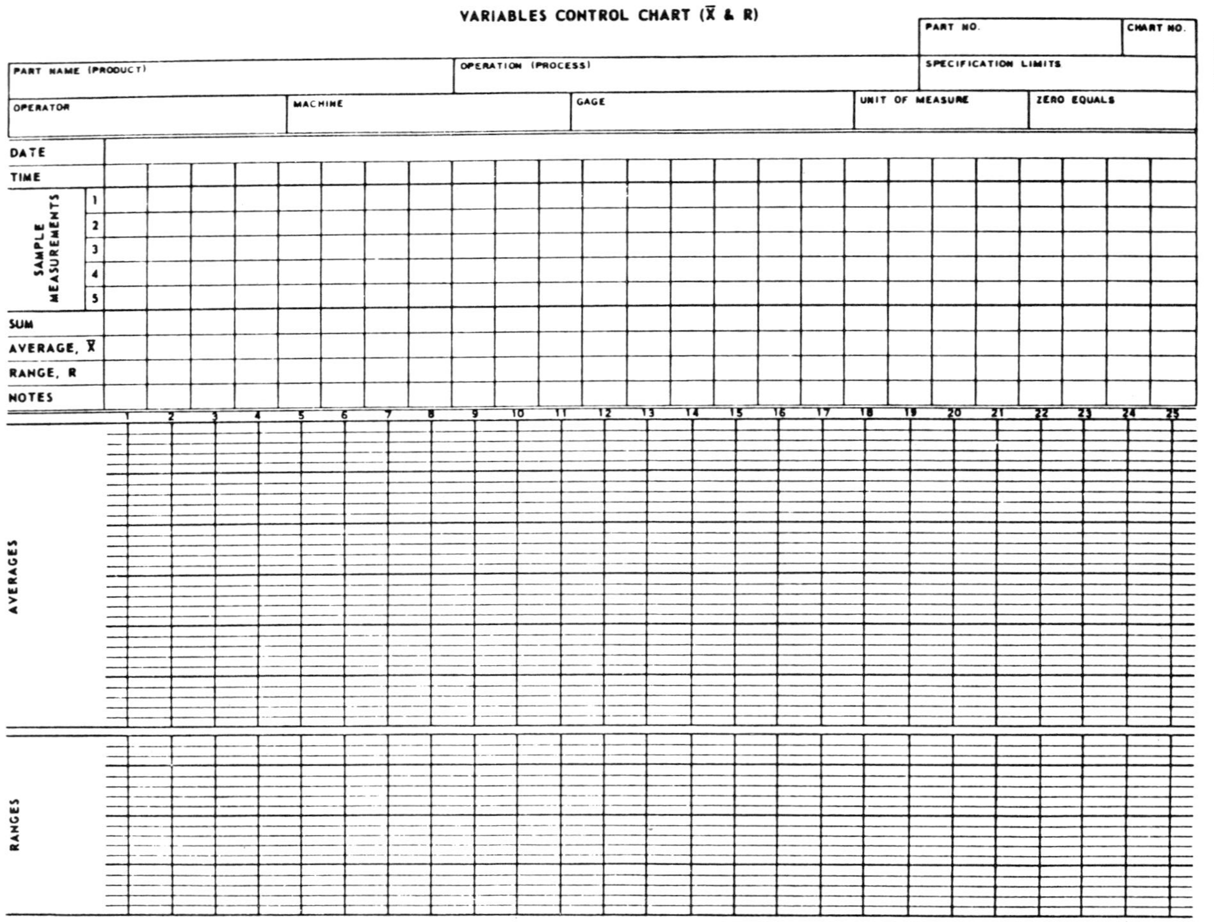

VARIABLES CONTROL CHART ($\bar{X}$ & R)

PART NO.
CHART NO.
PART NAME (PRODUCT)
OPERATION (PROCESS)
SPECIFICATION LIMITS
OPERATOR
MACHINE
GAGE
UNIT OF MEASURE
ZERO EQUALS

DATE
TIME
SAMPLE MEASUREMENTS 1 2 3 4 5
SUM
AVERAGE, $\bar{X}$
RANGE, R
NOTES

1 2 3 4 5 6 7 8 9 10 11 12 13 14 15 16 17 18 19 20 21 22 23 24 25

AVERAGES

RANGES

FIGURE 15.1. $\bar{x}$ and R chart, blank.

6. Find the overall mean, $\bar{\bar{x}}$. (This is usually read as "x-double bar.") Use equation 15.8 to total all the mean values, $\bar{x}$, for each subgroup and divide by the number of subgroups, k.

$$\bar{\bar{x}} = \frac{\bar{x}_1 + \bar{x}_2 + \bar{x}_3 + \ldots + \bar{x}_n}{k} \tag{15.8}$$

Compute the overall mean value, $\bar{\bar{x}}$, to two decimal places beyond that of the original measurement value. For the data in Table 15.2, we have:

$$\bar{\bar{x}} = \frac{13.00 + 12.94 + 12.90 + \ldots + 12.72}{25}$$

$$\bar{\bar{x}} = \frac{323.50}{25} = 12.940$$

7. Determine the average value of the range, $\bar{R}$. Using equation 15.9, total R for all the subgroups and divide by the number of subgroups, k.

$$\bar{R} = \frac{R_1 + R_2 + R_3 + \ldots + R_k}{k} \tag{15.9}$$

Compute the average range value, $\bar{R}$, to one decimal place beyond that of R. For the data in Table 15.2, we have:

$$\bar{R} = \frac{1.9 + 1.3 + 1.1 + \ldots + 1.1}{25}$$

$$\bar{R} = \frac{33.8}{25} = 1.35$$

8. Compute the control limit lines using the coefficient values of Table 15.3 and equations 15.10a, b, and c for $\bar{x}$ control charts, and 15.11a, b, and c for R control charts.

$\bar{x}$ Control Charts

$$\text{Center Line} = CL = \bar{x} \tag{15.10a}$$

$$\text{Upper Control Limit} + UCL$$
$$UCL = \bar{\bar{x}} + A_2 \bar{R} \tag{15.10b}$$

$$\text{Lower Control Limit} = LCL$$
$$LCL = \bar{\bar{x}} - A_2 \bar{R} \tag{15.10c}$$

R Control Charts

$$\text{Center Line} = CL = \bar{R} \tag{15.11a}$$

$$\text{Upper Control Limit} = UCL$$
$$UCL = D_4 \bar{R} \tag{15.11b}$$

$$\text{Lower Control Limit} = LCL$$
$$LCL = D_3 \bar{R} \tag{15.11c}$$

TABLE 15.3. Table of Factors for $\bar{x}$-R Charts

n	A_2	D_4	D_3	
2	1.880	3.267		
3	1.023	2.575		
4	0.729	2.282	}	Do not apply
5	0.577	2.115		
6	0.483	2.004		
7	0.419	1.924	0.076	

For the data in Table 15.2, we have:

***x̄* Control Chart**

$$
\begin{aligned}
CL &= \bar{\bar{x}} = 12.940 \\
UCL &= \bar{\bar{x}} + A_2 \bar{R} \\
&= 12.940 + (0.577)(1.35) \\
&= 12.940 + 0.779 \\
&= 13.719 \\
LCL &= \bar{\bar{x}} - A_2 \bar{R} \\
&= 12.940 - (0.577)(1.35) \\
&= 12.940 - 0.779 \\
&= 12.161
\end{aligned}
$$

***R* Control Chart**

$$
\begin{aligned}
CL &= \bar{R} = 1.35 \\
UCL &= D_4 \bar{R} \\
&= (2.115)(1.35) \\
&= 2.86 \\
LCL &= D_3 \bar{R} \\
&= (0)(1.35) \\
&= 0
\end{aligned}
$$

9. Construct the control chart. Using either control chart paper such as Figure 15.1 or graph paper, set the vertical index so that the *UCL* and *LCL* are separated by 1–1½ in. (20 to 30 mm). Draw in the three control lines and the necessary numerical values. Traditionally, the center line is a solid line. For process *analysis*, the *UCL* and *LCL* are broken (dashed) lines, while for process *control*, they are dotted lines.
10. Plot the $\bar{x}$ and R points computed for each subgroup (25 months in this problem) on the same vertical line (each line is one month's sample). For the $\bar{x}$ values, use a dot or period, and for the R values use a small x. Circle all points that exceed the control limit lines to distinguish them from the others. The dots and x's should be ⅛–¼ in. (2 to 5 mm) apart. Figure 15.2 shows a completed control chart based on the data in Table 15.2.
11. Write in the necessary information. On the left edge of the control chart, write $\bar{x}$ and R, and on the upper left of an $\bar{x}$ control chart, write the n value. Also, indicate the nature of the date, the period when it was collected, and instruments used, the person(s) responsible, etc.

15.5.4 Example: Personnel Evaluations and Analysis of Variance

PROBLEM

Fifteen trainees are directed to complete a safety training program and are randomly assigned to three different types of instructional approaches. Each of these three methodologies is concerned with developing a specific level of knowledge in the safe performance of the trainees' jobs. The achievement test scores at the conclusion of the instructional unit are reported in Table 15.4, along with the mean performance score associated with each instructional approach.

Use the analysis of variance procedure (also known as ANOVA or AOV) to test the null hypothesis that the three sample means were obtained from the same population at the 5% level of significance.

Solution

As in most applications of statistical inference, there is a step-by-step procedure that, if closely adhered to, will produce easily understood results. In this solution ANOVA, which was first developed by R. A. Fisher and in whose honor the F distribution was named, will be applied to ascertain the preferred instructional approach.

1. Compute the mean for each sample group and then determine the standard error of the mean, $s_{\bar{x}}$, *based only on the several sample means.* Computationally, this is the standard deviation of these several mean values.
2. Now, given the formula $s_{\bar{x}} = s/\sqrt{n}$, it follows that $s = \sqrt{n}\, s_{\bar{x}}$ and that $s^2 = ns_{\bar{x}}^2$. Therefore, the standard error of the mean computed in (1) can be used to estimate the variance of the (common)

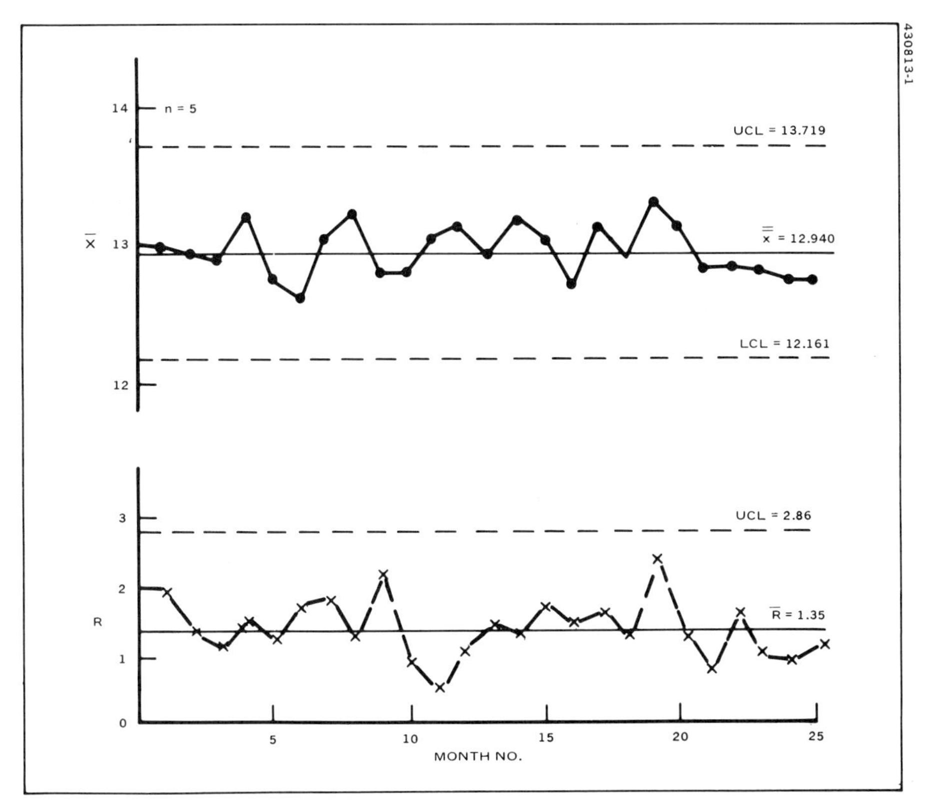

FIGURE 15.2. $\bar{x}$ and R chart for Table 15.2.

TABLE 15.4. Achievement Test Scores of Trainees Under Three Methods of Instruction

Instructional Method			Test Scores			Total Test Scores	Mean Test Scores
A_1	86	79	81	70	84	400	80
A_2	90	76	88	82	89	425	85
A_3	82	68	73	71	81	375	75

population from which the several samples were obtained. This estimate of the population variance is called the *mean square between groups (MSB).*

3. Compute the variance separately within each sample group and in respect to each group mean. Then pool these variance values by weighting them according to $n - 1$ for each sample. This weighting procedure for the variance is an extension of the procedure for combining and weighting two sample variances. The resulting estimate of the population variance is called the *mean square within groups (MSW).*
4. If the null hypothesis that $\mu_1 = \mu_2 = \mu_3 = \ldots = \mu_k$ is true, then it follows that the two mean squares obtained in (2) and (3) are unbiased and independent estimators of the same population variance, σ^2. However, if the null hypothesis is false, then the expected value of the *MSB* is larger than the *MSW*. Essentially, any differences among the population means will inflate *MSB* while having no effect on *MSW*.
5. Based on the observation in (4), the F distribution can be used to test the difference between the

two variances. A one-tail test is involved, and the general form of the F test in analysis of variance is

$$F_{df_1,\, df_2} = \frac{MSB}{MSW}$$

If the F ratio is in the region of rejection for the specified level of significance, then the hypothesis that the several sample means came from the same population is rejected.

This problem illustrates the application of these five steps to a hypothesis-testing problem involving the difference among three means.

From the hypothesis H_0: $\mu_1 = \mu_2 = \mu_3$ and H_1: The means are not mutually equal.

1. The overall mean of all 15 test scores is

$$\bar{x}_T = \frac{\Sigma x}{n} = \frac{1200}{15} = 80$$

The standard error of the mean, based on the three sample means reported, is

$$s_{\bar{x}} = \sqrt{\frac{(\bar{x} - \bar{x}_T)^2}{\text{No. means} - 1}} = \sqrt{\frac{(80 - 80)^2 + (85 - 80)^2 + (75 - 80)^2}{3 - 1}}$$

$$= \sqrt{\frac{50}{2}} = 5.0$$

2. $MSB = ns_{\bar{x}}^2 = 5(5.0)^2 = 5(25) = 125$
3. From the general formula:

$$s^2 = \frac{\Sigma\,(x - \bar{x}\,)^2}{n - 1}$$

the variance for each of the three samples is

$$s_1^2 = \frac{(86 - 80)^2 + (79 - 80)^2 + (81 - 80)^2 + (70 - 80)^2 + (84 - 80)^2}{5 - 1}$$

$$= \frac{154}{4} = 38.5$$

$$s_2^2 = \frac{(90 - 85)^2 + (76 - 85)^2 + (88 - 85)^2 + (82 - 85)^2 + (89 - 85)^2}{5 - 1}$$

$$= \frac{140}{4} = 35.0$$

$$s_3^2 = \frac{(82 - 75)^2 + (68 - 75)^2 + (73 - 75)^2 + (71 - 75)^2 + (81 - 75)^2}{5 - 1}$$

$$= \frac{154}{4} = 38.5$$

Then,

$$\hat{\sigma}^2\,(\text{pooled}) = \frac{(n_1 - 1)\,s_1^2 + (n_2 - 1)\,s_2^2 + (n_3 - 1)\,s_3^2}{n_1 + n_2 + n_3 - 3}$$

$$= \frac{(4)\,(38.5) + 4(35.0) + 4(38.5)}{5 + 5 + 5 - 3}$$

$$= \frac{448}{12} = 37.3$$

Therefore, $MSW = 37.3$

4. Since MSB is larger than MSW, a test of the null hypothesis is appropriate.

Critical $F(df = k - 1,\ kn - k;\ \alpha = 0.05) = F(2,\ 12;\ \alpha = 0.05) = 3.88$

5. $F = MSB/MSW = 125/37.3 = 3.35$

Since this F ratio is not in the region of rejection of the null hypothesis at the designated 5% level of significance, the hypothesis of no difference among the sample means cannot be rejected.

In less mathematical jargon, this means that since the F ratio just calculated is less than the tabulated value of F for the appropriate degrees of freedom, we don't have sufficient statistical evidence to reject the null hypothesis. In the absence of any other information to the contrary, we must conclude that no one of the instructional approaches is any better than the other two.

15.5.5 Example: Workers' Compensation Premiums and Regression Analysis

PROBLEM

In this example we will apply linear regression (least squares method) to establish a predictive relationship between safety training and the cost of workers' compensation premiums (WCP) to employers.

Let's review the training and financial records for a given company for a 16-year period. We need to compare the results of whatever safety training was provided with the company's cost of workers' compensation premiums. Table 15.5 provides a summary of the training and financial records from 1970 (the same year that OSHA was created) to 1985. Because workers' compensation premiums are predicated on a company's most recent accident history, a lead-lag relationship exists between them. Therefore, it will be necessary to ascertain the extent and timing of the relationship as a part of the solution to this problem. Table 15.5 presents three relationships:

Relationship Number	**Independent Variable**		**Dependent Variable**
	Average training hours per employee		WCP: \$ per \$1000
1	x	and	y_1 = same year
2	x	and	y_2 = next year
3	x	and	y_3 = two years later

As in any bivariate analysis, a search for a solution methodology is expedited by starting with a picture of the process. In this case we begin with a scattergram (an x-y graphical plot of both variables) to determine if any visual evidence of a relationship exists. Figure 15.3 is a scattergram of the data in Table 15.5. However, the mass of data presented in Figure 15.3 is not easily deciphered. Therefore, the three relationships are plotted individually in Figure 15.4

TABLE 15.5. Workers' Compensation Premium (WCP) and Safety Training Records

Year	Average Training Hours per Employee (x)	WCP: Dollars per $10000		
		Same Year (y_1)	Next Year (y_2)	Two Years (y_3)
1970	5	70	75	80
1971	10	75	80	70
1972	15	80	70	70
1973	15	70	70	65
1974	20	70	65	55
1975	30	65	55	50
1976	20	55	50	60
1977	20	50	60	65
1978	30	60	65	55
1979	35	65	55	50
1980	40	55	50	40
1981	30	50	40	45
1982	40	40	45	45
1983	40	45	45	40
1984	35	45	40	35
1985	40	40	35	30

All dollar figures are stated in 1985 dollars.

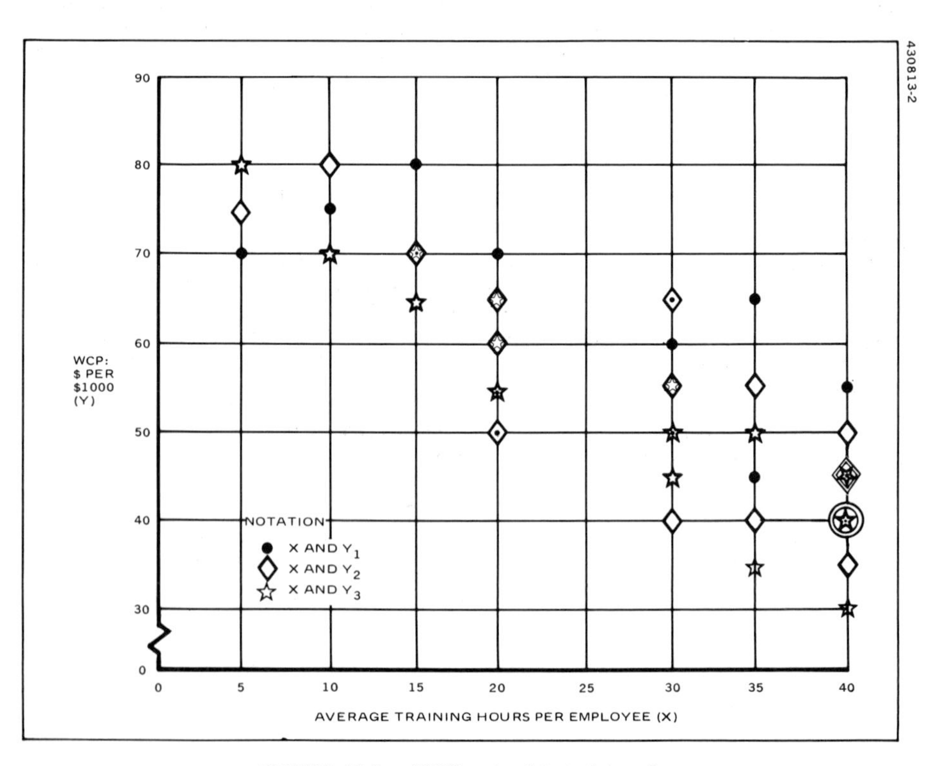

FIGURE 15.3. WCP and safety training: I.

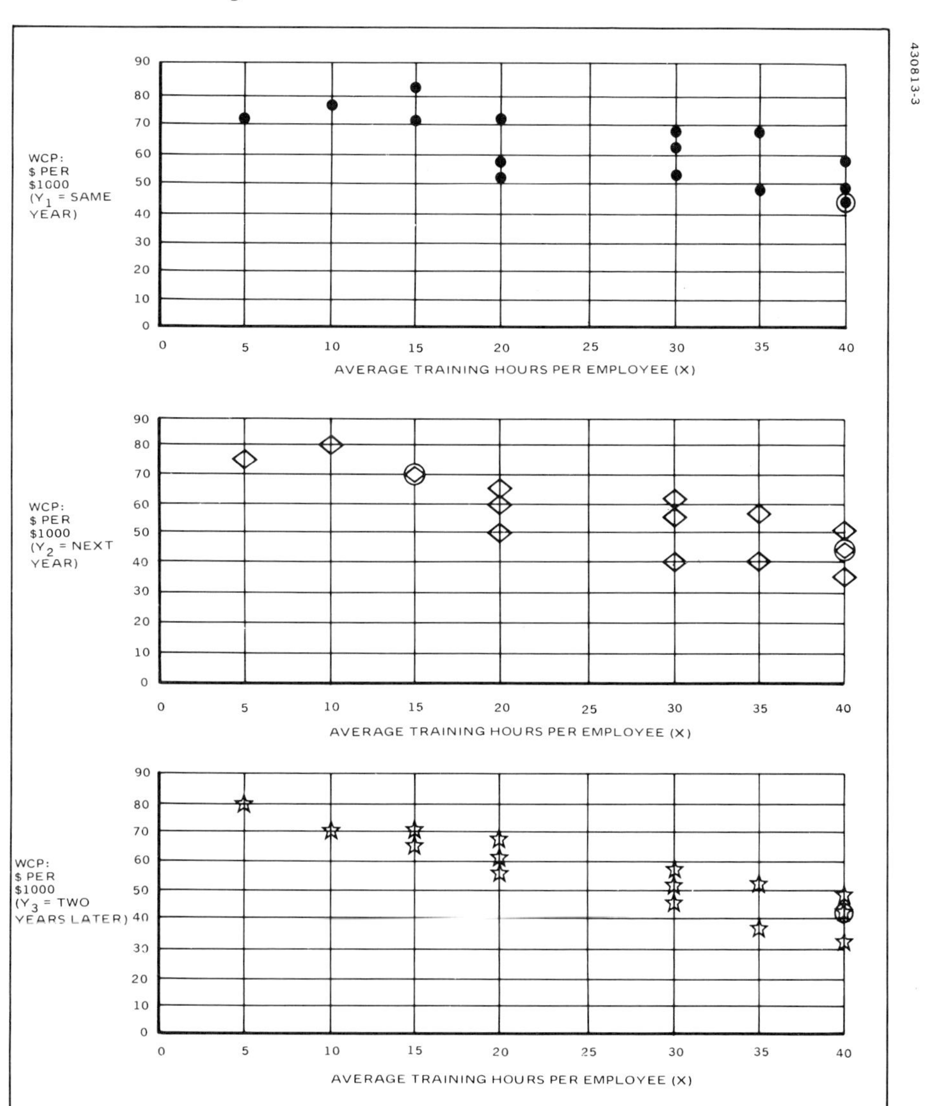

FIGURE 15.4. WCP and safety training: II.

Solution

We have three mathematical equations to use in regression analysis. They are:

$$\bar{y} = a + bx \tag{15.13}$$

where

$\bar{y}$ = the expected WCP cost per average hour of training
a = the value of the y-intercept (when x is zero)
b = the slope of the regression line
x = the average training hours per employee

$$b = \frac{\Sigma xy - n\overline{xy}}{\Sigma x^2 - n\overline{x}^2} \tag{15.14}$$

where

Σxy = the sum of n cross-products of xy
n = the number of bivariate data points
$\bar{x}$ = the expected training hours per employee
Σx^2 = the sum of n values of x squared
$\bar{x}^2$ = the square of the expected training hours per employee

$$a = \bar{y} - b\bar{x} \qquad (15.15)$$

where all symbols have been previously defined.

Using equation 15.14 first, we calculate the slopes of the regression lines for all three relationships. Relationship 1 data is drawn from Table 15.6, relationship 2 data from Table 15.7 and 3 from Table 15.8.

Relationship 1

$$b_1 = \frac{\Sigma xy_1 - n\overline{xy}_1}{\Sigma x^2 - n\bar{x}^2}$$

$$= \frac{23{,}150 - (16)(26.56)(58.44)}{13{,}325 - (16)(26.56)^2}$$

$$= \frac{-1684.66}{2038.06}$$

$$= -0.827$$

TABLE 15.6. Relationship 1 Regression Calculations

Average Training Hours per Employee (x)	WCP: Dollars per \$1000 Same Year (y_1)	xy_1	x^2	y_1^2
5	70	350	25	4900
10	75	750	100	5625
15	80	1200	225	6400
15	70	1050	225	4900
20	70	1400	400	4900
30	65	1950	900	4225
20	55	1100	400	3025
20	50	1000	400	2500
30	60	1800	900	3600
35	65	2275	1225	4225
40	55	2200	1600	3025
30	50	1500	900	2500
40	40	1600	1600	1600
40	45	1800	1600	2025
35	45	1575	1225	2025
40	40	1600	1600	1600
Total 425	935	23,150	13,325	57,075
Mean 26.56	58.44			
Symbol $\bar{x}$	$\bar{y}_1$	Σxy_1	Σx^2	Σy_1^2

TABLE 15.7. Relationship 2 Regression Calculations

Average Training Hours per Employee (x)	WCP: Dollars per $1000 Next Year ($y_2$)	xy_2	x^2	y_2^2
5	75	375	25	5625
10	80	800	100	6400
15	70	1050	225	4900
15	70	1050	225	4900
20	65	1300	400	4225
30	55	1650	900	3025
20	50	1000	400	2500
20	60	1200	400	3600
30	65	1950	900	4225
35	55	1925	1225	3025
40	50	2000	1600	2500
30	40	1200	900	1600
40	45	1800	1600	2025
40	45	1800	1600	2025
35	40	1400	1225	1600
40	30	1200	1600	900
Total 425	895	21,700	13,325	53,075
Mean 26.56	55.94			
Symbol $\bar{x}$	$\bar{y}_2$	Σxy_2	Σx^2	Σy_2^2

TABLE 15.8. Relationship 3 Regression Calculations

Average Training Hours per Employee (x)	WCP: Dollars per $1000 Two Years ($y_3$)	xy_3	x^2	y_3^2
5	80	400	25	6400
10	70	700	100	4900
15	70	1050	225	4900
15	65	975	225	4225
20	55	1100	400	3025
30	50	1500	900	2500
20	60	1200	400	3600
20	65	1300	400	4225
30	55	1650	900	3025
35	50	1750	1225	2500
40	40	1600	1600	1600
30	45	1350	900	2025
40	45	1800	1600	2025
40	40	1600	1600	1600
35	35	1225	1225	1225
40	30	1200	1600	900
Total 425	855	20,400	13,325	48,675
Mean 26.56	53.44			
Symbol $\bar{x}$	$\bar{y}_3$	Σxy_3	Σx^2	Σy_3^2

Relationship 2

$$b_2 = \frac{21{,}700 - (16)(26.56)(55.94)}{13{,}325 - (16)(26.56)^2}$$

$$= \frac{-2072.26}{2038.06}$$

$$= -1.017$$

Relationship 3

$$b_3 = \frac{20{,}400 - (16)(26.56)(53.44)}{13{,}325 - (16)(26.56)^2}$$

$$= \frac{-2309.86}{2038.06}$$

$$= -1.133$$

Using equation 15.15 and the results derived from use of equation 15.14, the next step is to calculate the y-intercepts for all three relationships.

Relationship 1

$$\begin{aligned} a_1 &= \bar{y}_1 - b_1\bar{x}_1 \\ &= 58.44 - (-.827)(26.56) \\ &= 58.44 + 21.97 \\ &= 80.41 \end{aligned}$$

Relationship 2

$$\begin{aligned} a_2 &= 55.94 - (-1.017)(26.56) \\ &= 55.94 + 27.01 \\ &= 82.95 \end{aligned}$$

Relationship 3

$$\begin{aligned} a_3 &= 53.44 - (-1.133)(26.56) \\ &= 53.44 + 30.09 \\ &= 83.53 \end{aligned}$$

Now we can summarize the results of the foregoing calculations, as follows:

Relationship Number	b	a	Equation 15.13: $\bar{y} = a + bx$
1	$-.827$	80.41	$\bar{y} = 80.41 - .827x$
2	-1.017	82.95	$\bar{y} = 82.95 - 1.017x$
3	-1.133	83.53	$\bar{y} = 83.53 - 1.133x$

These regression lines for Relationships 1, 2, and 3 are plotted in Figure 15.5. Examination of these lines indicates that when the average safety training hours per employee is less than 10, estimates of the workers' compensation premium costs will be lowest using Relationship 1. However, when training hours are over 10, workers' compensation premium estimates will be lowest using Relationship 3.

15.5.6 Example: Accident Rates and Student's *t*

PROBLEM

Accident rates for two different operating locations (Plant A and Plant B), each with nine departments, have been compiled for one year following completion of safety training at both plants. Plant A personnel were trained using a series of conventional classes, while their counterparts at Plant B were trained

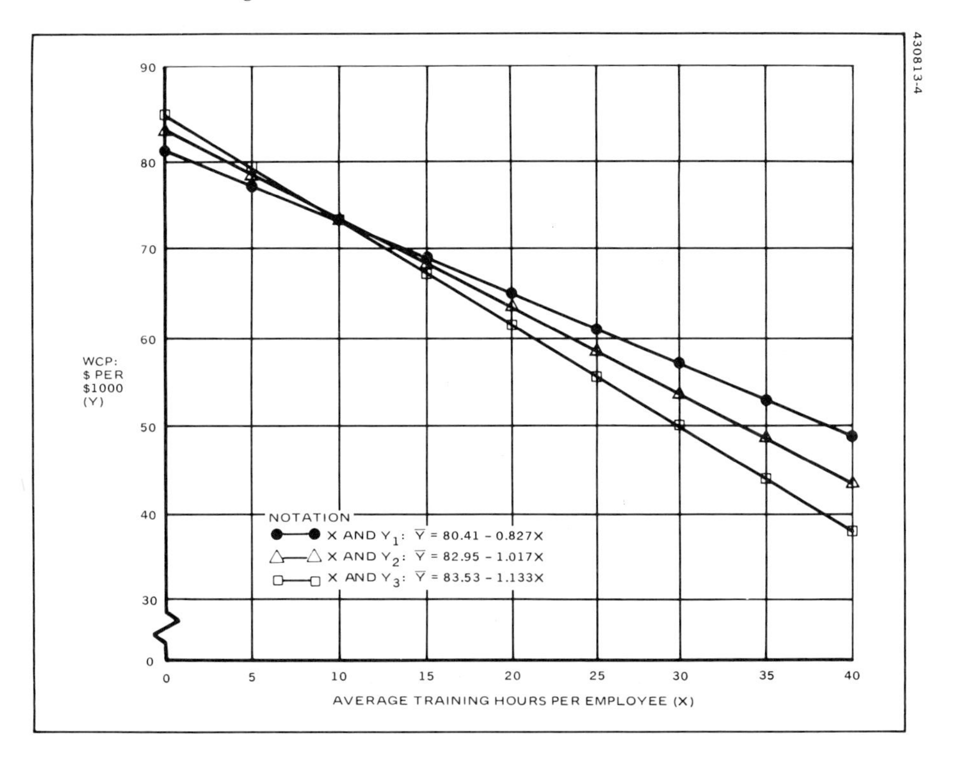

FIGURE 15.5. WCP and safety training: Regression relationships.

using a newly purchased set of interactive video disks designed for use on an individual basis. The resulting accident rates are as follows:

Plant A	35	31	29	25	34	40	27	32	31
Plant B	41	35	31	28	35	44	32	37	34

Do the data provide sufficient evidence to indicate that the mean accident rate is different for the two training methods? In this case, we will use the level of significance as 5%, that is, the probability of a Type I error is 5% ($\alpha = 5\%$).

Solution

This experiment employed an independent sampling design to compare the two safety training methods, since different experimental units (departments) were used for the two methods and no effort was made to pair-match the departments. The basic elements of the experiment are:

- Experimental units: nine departments in each of two plants.
- Response: accident rates.
- Treatments: the two types of training methods, conventional classes and interactive video disks.

Since we wish to make a parametric comparison of the two treatment (population) means, let μ_1 and μ_2 represent the mean accident rates for the conventional class and interactive video disk populations, respectively. The small-sample parametric test setup for testing the research hypothesis that these mean scores differ is as follows:

- *Null hypothesis*, H_0: $(\mu_1 - \mu_2) = 0$
- *Alternate hypothesis*, H_A: $(\mu_1 - \mu_2) \neq 0$
- *Test statistic*, t:

$$t = \frac{(\bar{y}_1 - \bar{y}_2) - 0}{s_p \sqrt{\frac{1}{n_1} + \frac{1}{n_2}}} \tag{15.16}$$

where

$\bar{y}_1$ and $\bar{y}_2$ are sample means
s_p is the pooled estimate of the common standard deviation
n_1 and n_2 are the sample sizes

- *Assumptions:*

1. The two populations of accident rates are approximately normal.
2. The variances of the populations of accident rates are equal.

- *Rejection region:*

The alternate hypothesis is two-tailed. For $\alpha = 0.05$ and degrees of freedom (df) equal to $(n_1 + n_2 - 2) = (9 + 9 - 2) = 16$, the critical value for $t_{\alpha/2}$ is 2.120. This value is located in the Student's t table, which can be found in any basic text on statistics. Since $t_{.05/2} = t_{.025} = 2.120$, the rejection region for H_0 is:

$t < -2.120$ or $t > 2.120$, which is also written as

$-2.120 > t > 2.120$. This is presented in Figure 15.6.

Before we can calculate the test statistic, t, we need the sample means and the sum of the squares of the deviations about the means. Thus,

$$\bar{y}_1 = 31.56$$

$$\sum_{i=1}^{9} (y_i - \bar{y}_1)^2 = 160.22$$

$$\bar{y}_2 = 35.22$$

$$\sum_{i=1}^{9} (y_i - \bar{y}_2)^2 = 195.56$$

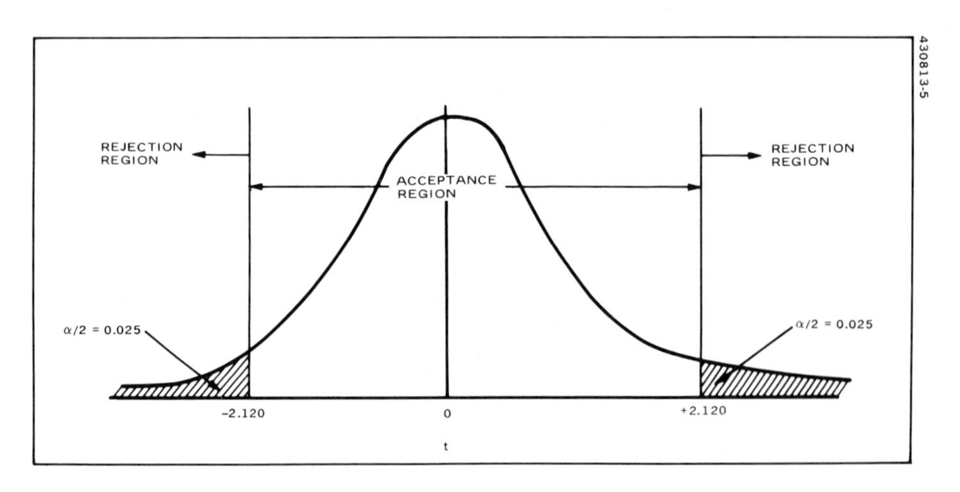

FIGURE 15.6. Two-tailed rejection region.

Then, the pooled estimate of the common variance is:

$$s_p^2 = \frac{\sum_{i=1}^{9} (y_i - \bar{y}_1)^2 + \sum_{i=1}^{9} (y_i - \bar{y}_2)^2}{n_1 + n_2 - 2}$$

$$= \frac{160.22 + 195.56}{16}$$

$$= 22.24$$

Thus, the standard deviation is:

$$s_p = 4.72$$

Now we calculate the test statistic, t:

$$t = \frac{(\bar{y}_1 - \bar{y}_2)}{s_p\sqrt{\frac{1}{n_1} + \frac{1}{n_2}}} = \frac{31.56 - 35.22}{4.72\sqrt{\frac{1}{9} + \frac{1}{9}}}$$

$$= \frac{-3.66}{(4.72)(.471)} = \frac{-3.66}{2.225}$$

$$= -1.65$$

Since this value is not in the rejection region displayed in Figure 15.6, we must conclude that these data do not support the alternate hypothesis at the 0.05 level of significance. In other words, there is insufficient evidence to conclude that the mean accident rates differ for the two training delivery populations.

15.5.7 Example: Hazard-Reporting Rates and Sampling/Surveying

PROBLEM

Suppose a need exists to determine average hazard-reporting rates within an organization. To generate a valid statistical estimate of the average rate, we must first determine the required sample size.

In this example we will assume that the desired size of the confidence interval and the associated degree of confidence are known. If the population standard deviation can be estimated, such as from the results of earlier studies, the required sample size for survey based on the use of the normal distribution is:

$$n = \left(\frac{z\sigma}{E}\right)^2 \tag{15.17}$$

where

n = the desired sample size
z = the number of standard deviations associated with the specified degree of confidence
σ = the standard deviation of the population (or an estimate thereof)
E = the "plus and minus" error factor allowed in the confidence interval (E is always one-half the total confidence interval)

In this example we want to estimate the average hazard reporting rate within an organization so that it can be compared either to previous rates in the same organization or those of other organizations. The mean rate needs to be determined within 3 reports per month (plus or minus) and with a 90% confidence. Based on data collected a year ago, we estimate the hazard reporting rate standard deviation to be σ = 20 reports per month.

Solution
Thus, the minimum required sample size is:

$$n = \left(\frac{z\sigma}{E}\right)^2$$
$$= \left(\frac{1.65 \times 20}{3}\right)^2$$
$$= (33/3)^2 = 11^2$$
$$= 121$$

In this case a z value of 1.65 was used to correspond to a desired confidence level of 90%. The following list of two-tail equivalencies can be used in this way.*

Confidence Level	z
90%	1.65
95%	1.96
98%	2.33
99%	2.58

15.5.8 Example: Injury Costs and Hypothesis Testing

PROBLEM
A distributor of a commercial safety training program tells a prospective purchaser of the program that other companies in their industry that use the program have reduced their average injury costs to $15,000 each. For the purposes of this example, let's assume that injury costs in this industry are approximately normally distributed, and that based on a recent insurance company survey the standard deviation can be accepted as being about $2,000.

To determine the validity of the distributor's claim, a survey is conducted of a random sample of n = 15 program users. The mean injury cost reported by the program users is $\bar{x}$ = $14,000. Test the null hypothesis, H_0, that μ = $15,000 by establishing critical limits of the sample mean in terms of dollars, using the 5% level of significance. To verify the results, test H_0 by using the standard normal variable, z, as the test statistic.

Solution
It should be noted that the normal probability distribution can be used even though the sample is small enough to use Student's t distribution, because the population is assumed to be normally distributed and the standard deviation is known.

Since $H_0 : \mu = \$15{,}000$
and $H_1 : \mu \neq \$15{,}000$

the critical limits of $\bar{x}$ when $\alpha = 0.05$ are:

$$\bar{x}_{CR} = \mu_0 \pm z\,\sigma_{\bar{x}} = \mu_0 \pm z\left(\frac{\sigma}{\sqrt{n}}\right) \quad (15.18)$$
$$= 15{,}000 \pm 1.96\left(\frac{2000}{\sqrt{15}}\right)$$
$$= 15{,}000 \pm 1.96\left(\frac{2000}{3.87}\right)$$
$$= 15{,}000 \pm 1.96\,(516.80)$$
$$= 15{,}000 \pm 1013$$
$$= \$13{,}987 \text{ and } \$16{,}013$$

*This is drawn from a standard normal distribution table, which can be found in any basic statistics text.

Since the sample mean, $\bar{x} = \$14{,}000$, is in the region of acceptance (see Figure 15.7) of the null hypothesis, the claim regarding the results of using the safety training program made by its distributors cannot be rejected at the 5% level of significance. See Figure 15.7.

To verify these results, we must first determine the critical z value.

Critical z (with $\alpha = 0.05 = \pm\ 1.96$)*

Next, the standard error of the mean is calculated:

$$\sigma_{\bar{x}} = \frac{\sigma}{\sqrt{n}} = \frac{2000}{\sqrt{15}} = \frac{2000}{3.87} = \$516.80$$

Then we calculate z for this situation:

$$z = \frac{\bar{x} - \mu}{\sigma_{\bar{x}}} = \frac{14{,}000 - 15{,}000}{516.80} = \frac{-1000}{516.80}$$
$$= -1.93$$

Since the computed z value of -1.93 is in the region of acceptance of the null hypothesis, the distributor's claim cannot be rejected when the level of significance is five percent. Figure 15.7 presents this situation as well.

15.5.9 Example: Personnel Evaluations and Experimental Design

PROBLEM

In this example we have a collection of personnel evaluations rendered approximately one month following completion of a safety training program. At the time of the program, 15 trainees were exposed to three different training delivery methods: A_1, A_2, and A_3. For purposes of this example, let's say that method A_1 was attendance at a conventional classroom presentation, A_2 was use of a programmed learning text, and A_3 was a combination of the first two. Before beginning the training program, the trainees were categorized by their supervisors according to their perceived levels of ability. Table 15.9 presents the results of the evaluations within the context of the training delivery methods and trainee abilities.

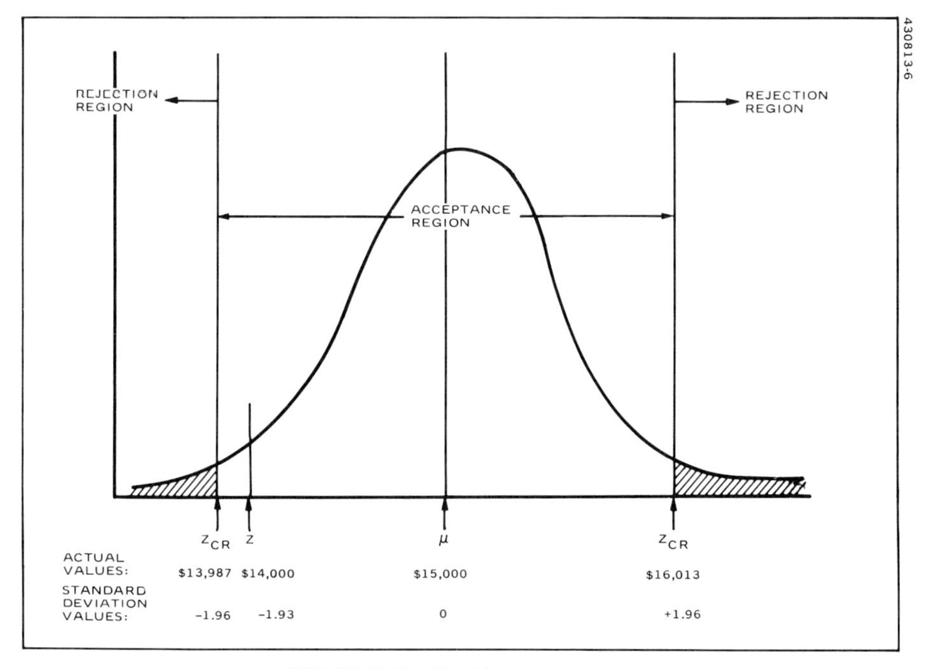

FIGURE 15.7. Decision parameters.

*This is drawn from a standard normal distribution table, which can be found in any basic statistics text.

TABLE 15.9. Trainee Evaluations

Level of Ability	Method of Instruction[a]		
	A_1	A_2	A_3
B_1	86	90	82
B_2	84	89	81
B_3	81	88	73
B_4	79	76	68
B_5	70	82	71

[a]Minimum = 10, Maximum = 100.

The personnel evaluation data in Table 15.9 is organized in an experimental design described by statisticians as a *randomized block design.* This and other experimental designs are fully described in a number of texts on statistics and experimental design.

In this example we will test the null hypothesis that there is no difference in the mean performance evaluation of the trainees among the three methods of instruction, using the five percent level of significance.

Solution

We begin as is common in statistical analysis by stating the null and alternate hypotheses. In this example there are two of each:

$A \quad H_0: \alpha_k = 0$ for all columns
$\quad\; H_A: \alpha_k \neq 0$ for all columns

$B \quad H_0: \alpha_j = 0$ for all rows
$\quad\; H_A: \alpha_j \neq 0$ for all rows

Next, we need to expand the randomized block design of Table 15.9 to provide data necessary to perform an analysis of variance. This expansion is presented in Table 15.10.

The various quantities required for Table 15.11, the analysis of variance (ANOVA or AOV) table, are:

For j (The Subscript Row Identifiers)

T_j: $T_1 = 258$ $\quad T_2 = 254$ $\quad T_3 = 242$ $\quad T_4 = 223$ $\quad T_5 = 223$

T_j^2: $T_1^2 = 66{,}564$ $\quad T_2^2 = 64{,}516$ $\quad T_3^2 = 58{,}564$ $\quad T_4^2 = 49{,}729$ $\quad T_5^2 = 49{,}729$

For k (The Subscript Column Identifiers)

T_k: $T_1 = 400$ $\quad T_2 = 425$ $\quad T_3 = 375$

T_k^2: $T_1^2 = 160{,}000$ $\quad T_2^2 = 180{,}625$ $\quad T_3^2 = 140{,}625$

where: $n_1 = 5$ $\quad n_2 = 5$ $\quad n_3 = 5$

Overall: $T = 1200$ $\quad T^2 = 1{,}440{,}000$ $\quad N = 15$

TABLE 15.10. Analysis of Variance Factors

Level of Ability	Method of Instruction			Total	Mean
	A_1	A_2	A_3	T_j	$\bar{x}_j$
B_1	86	90	82	258	86.0
B_2	84	89	81	254	84.7
B_3	81	88	73	242	80.7
B_4	79	76	68	223	74.3
B_5	70	82	71	223	74.3
				Grand Total	
Total, T_k	400	425	375	$T = 1200$	Grand Mean
Mean, $\bar{x}_k$	80	85	75		$\bar{x} = 80.0$

TABLE 15.11. ANOVA Table for Analysis of Three Methods of Instruction According to Ability Level

Source of Variation	Sum of Squares SS	Degrees of Freedom df	Mean Square MS	F Ratio
Between treatment groups, A (method)	250.0	$3 - 1 = 2$	$\frac{250.0}{2} = 125.0$	$\frac{125.0}{10.1} = 12.4$
Between blocks, B (ability level)	367.3	$5 - 1 = 4$	$\frac{367.3}{4} = 91.8$	$\frac{91.8}{10.1} = 9.1$
Sampling error, E	80.7	$(5-1)(3-1) = 8$	$\frac{80.7}{8} = 10.1$	
Total, T	698.0	$15 - 1 = 14$		

Therefore,

$$\frac{T^2}{N} = \frac{1,440,000}{15} = 96,000$$

and

$$\sum_{j=1}^{J} \sum_{k=1}^{K} x^2 = 86^2 + 84^2 + 81^2 + \cdots + 71^2 = 96,698$$

The sum of squares between *methods of instruction*, that is, between treatment groups, is designated as SSA:

$$SSA = \sum_{k=1}^{K} \frac{T_k^2}{n_k} - \frac{T^2}{N}$$

$$= \frac{160,000}{5} + \frac{180,625}{5} + \frac{140,625}{5} - 96,000$$

$$= 250$$

The sum of squares between *ability levels*, that is, between blocks, is designated as SSB:

$$SSB = \frac{1}{K} \sum_{j=1}^{J} T_j^2 - \frac{T^2}{N}$$

$$= \frac{1}{3}(66,564 + 64,516 + 58,564 + 49,729 + 49,729) - 96,000$$

$$= 367.3$$

The sum of squares *total* is designated as SST:

$$SST = \sum_{j=1}^{J} \sum_{k=1}^{K} x^2 - \frac{T^2}{N}$$

$$= 96,698 - 96,000$$

$$= 698$$

The sum of squares resulting from *experimental error* is designated as SSE:

$$\begin{aligned} SSE &= SST - SSA - SSB \\ &= 698 - 250 - 367.3 \\ &= 80.7 \end{aligned}$$

Table 15.11 presents the ANOVA for this data. With respect to the linear equation representing this model, the two F ratios in Table 15.11 are concerned with the tests of the null and alternate hypotheses stated earlier.

In terms of practical implications, the first null hypothesis is concerned with testing the difference among the column means, *which is the basic purpose of the analysis.* The second null hypothesis is concerned with testing the difference among the row means. This is done to identify and control this source of variability due to individual differences and thereby reduce the variability attributed to sampling error.

Using the 5% level of significance, the required F ratio for the rejection of the first null hypothesis ($df = 2,8$) = 4.46, while the required F for the second null hypothesis ($df = 4,8$) = 3.84. Thus, both of the calculated F ratios in Table 15.11 are in the region of rejection of the null hypothesis.

We conclude that there is a significant difference in personnel evaluations for the different methods of instruction, and also that there is a significant difference in evaluations for the different levels of ability.

15.5.10 Example: Workers' Compensation Premiums and Correlation Analysis

PROBLEM

This example is an extension of the analysis of the workers' compensation premium problem, Problem 5. In an effort to gain further understanding of the three relationships described and analyzed in that problem, we will first determine the coefficient of determination and then the coefficient of correlation.

The coefficient of determination indicates the proportion of the variance in the dependent variable (workers' compensation premium), which is statistically explained by the regression equation, that is, by knowledge of the associated independent variable (average safety training hours). Equation 15.19 symbolically defines the coefficient of determination, r^2:

$$r^2 = \frac{a\Sigma y + b\Sigma xy - n\bar{y}^2}{\Sigma y^2 - n\bar{y}^2} \tag{15.19}$$

where

a = the value of the y-intercept (when x is zero)
Σy = the sum of all n values of y
b = the slope of the regression line
Σxy = the sum of n cross-products of xy
n = the number of bivariate data points
$\bar{y}^2$ = the square of the expected WCP cost per average hour of training
Σy^2 = the sum of n values of y squared

The coefficient of correlation is the square root of the coefficient of determination and is a measure of the linear relationship between the two variables, x and y. Equation 15.20 symbolically defines the coefficient of correlation, r:

$$r = \sqrt{r^2} \tag{15.20}$$

where r^2 = coefficient of determination.

Using equations 15.19 and 15.20 in conjunction with data from Section 15.5.5 (Problem 5), determine r^2 and r for all three x-y relationships and interpret the results.

Solution

Relationship 1

$$r^2 = \frac{(80.41)(935) + (-.827)(23{,}150) - (16)(58.44)^2}{57{,}075 - (16)(58.44)^2}$$

$$= \frac{75{,}183.35 - 19{,}145.05 - (16)(3415.23)}{57{,}075 - (16)(3415.23)}$$

$$= \frac{75{,}183.35 - (19{,}145.05 + 54{,}643.74)}{57{,}075 - 54{,}643.74}$$

$$= \frac{1394.56}{2431.26}$$

$$= 0.574$$

Therefore,

$$r = \sqrt{0.574}$$
$$= -0.757$$

Note: The sign, plus or minus, given to r is drawn from the sign associated with the slope of the regression line. In this case the sign is minus.

Relationship 2

$$r^2 = \frac{(82.95)(895) + (-1.017)(21{,}700) - (16)(55.94)^2}{53{,}075 - (16)(55.94)^2}$$

$$= \frac{74{,}240.25 - 22{,}068.90 - (16)(3129.28)}{53{,}075 - (16)(3129.28)}$$

$$= \frac{74{,}240.25 - (22{,}068.90 + 50{,}068.54)}{53{,}075 - 50.068.54}$$

$$= \frac{2102.81}{3006.46}$$

$$= 0.699$$

Therefore,

$$r = \sqrt{0.699}$$
$$= -0.836$$

As in Relationship 1, the sign is minus since the slope of the associated regression line is negative.

Relationship 3

$$r^2 = \frac{(83.53)(855) + (-1.133)(20{,}400) - (16)(53.44)^2}{48{,}675 - (16)(53.44)^2}$$

$$= \frac{71{,}418.15 - 23{,}113.2 - (16)(2855.83)}{48{,}675 - (16)(2855.83)}$$

$$= \frac{71{,}418.15 - (23{,}113.2 + 45{,}693.34)}{48{,}675 - 45{,}693.34}$$

$$= \frac{2611.61}{2981.66}$$

$$= 0.876$$

Therefore,

$$r = \sqrt{0.876}$$
$$= -0.936$$

Note: the sign continues to be minus.

In summary, we have:

Relationship Number	Coefficient of Determination	Coefficient of Correlation
1	0.574	−0.757
2	0.699	−0.836
3	0.876	−0.936

Thus, it can be easily seen that recognition and application of the lead-lag relationship between safety training hours and workers' compensation premiums provides greater insight into the effectiveness of safety training. Relationship 3, where workers' compensation premiums are offset by two years from the average safety training hours, offers the greatest explanation of the WCP variance as well as the greatest measure of the linear relationship between the independent and dependent variables.

BIBLIOGRAPHY

Johnston, W. L. and T. R. Rogers, Measuring Safety Performance, *Industrial Engineering,* December 1975.

Peterson, D., *Analyzing Safety Performance,* Garland STPM Press, New York, 1980.

Peterson, D., *Techniques of Safety Management,* 2nd Ed., McGraw-Hill, New York, 1978.

Re Velle, J. B., *Safety Training Methods,* Wiley, New York, 1980.

Shafai-Sahrai, Y., *Determinants of Occupational Injury Experience,* Michigan State University Division of Research, East Lansing, 1973.

Tarrants, W. E., *The Measurement of Safety Performance,* Garland STPM Press, New York, 1980.

CHAPTER 16

How to Achieve Cooperation and Compliance from an Employee Occupational Safety and Health Training Program

VICTORIA COOPER MUSSELMAN
Carnow, Conibear & Associates, Ltd.
Chicago, Illinois

16.1 INTRODUCTION

Training as a component of the solution to workplace environmental and occupational health problems continues to grow in importance. Over the last few years, legislative and regulatory activity has continued to demonstrate this trend. More than 20 states have passed employee right-to-know laws requiring training of employees (Musselman, 1984). Many other state legislatures are currently considering similar action. The Federal Occupational Safety and Health Administration has also issued its *Hazard Communication Standard* (1984). The Environmental Protection Agency has issued training requirements under Toxic Substance Control Act regulation (EPA, 1983, 1984). Increasingly, occupational health and safety professionals are being called upon to provide employee training in proper work practices, including health hazards of industrial chemicals, safety procedures, and the fit, use, and maintenance of protective equipment.

These increasing regulatory and legislative requirements for training pose a difficult problem for the business community. Both the direct and indirect costs of training can be high, yet as a result of these regulatory and legislative initiatives, training has become mandatory for more and more employers (OSHA, 1977).

Awareness is growing, too, among the general population about the potential hazards of chemicals in the environment. Companies are also becoming more sensitive about potential health hazards related to toxic exposures. This is due in part to the legal ramifications of employee exposure to hazardous chemicals and cancer-causing agents in the workplace environment. Employer trade groups and employee unions are also putting pressure on industry to increase the amount of health and safety training provided to employees (*Chemical Week*, January 23, 1985).

16.2 ELEMENTS OF A SUCCESSFUL PROGRAM

The purpose of this chapter is to assist those involved in or responsible for health and safety training to develop effective programs. The text includes description of the basic requirements of a successful program and a discussion of the step-by-step procedures required to set up such a program. Included in this chapter is a description of the methods used to evaluate your training efforts.

It is important to realize however, that there is no cookbook method for developing training that allows you to serve up a perfect program every time. You can realistically expect that your program should communicate information, teach skills, and influence attitudes. It can also help foster cooperation and establish corporate credibility in the eyes of the staff, employees, and the public. A training program, however, *cannot* substitute for a coherent and comprehensive corporate health and safety policy. Training communicates information and ideas that are developed and formulated prior to training. The strength of the overall corporate commitment to a safe and healthy work environment will be reflected in the training program developed to implement the policy and procedures. Though the training can be well organized and well presented, it can never be better than the policy it represents.

Many companies find it difficult to implement effective training programs. But other firms have instituted successful programs. What do successful programs look like? How can you tell if a program is successful? A program must be considered successful if it meets its specific program objectives and helps to reduce accidents, injuries, and illnesses due to exposure to workplace safety and health hazards. In order for such a program to meet the success goal, it must meet the criteria that you have established previously. First, it must be a cooperative effort among all staff, including management and hourly personnel. In addition, there must be sufficient information and resources available to both management and hourly employees to implement the program. Next, the program needs clear-cut and achievable goals and objectives. Last, there must be an organized and consistent feedback system for safe behavior that is administered in an evenhanded way.

16.2.1 Cooperation

Cooperation may seem like an obvious requirement of a teaching–learning situation. However, in the workplace setting, a true cooperative effort is often difficult to achieve. Such an effort requires employees at all levels of the organization and with varying goals and responsibilities to work together.

On the management side, executive level personnel must support cooperative health and safety efforts with resources and a clear mandate. Middle-level management must distribute the mandate and resources in a way that will provide the most effective cooperative effort. Line supervisors must implement the program on a day-to-day basis.

Hourly employees and their representatives must also feel a part of this cooperative effort if they are to comply with health and safety regulations in the workplace. They need to believe that they can have meaningful input when implementing the program. It is also necessary for employees to be educated about the advantages to themselves when they follow health and safety guidelines in the workplace. These advantages can include such things as personal health, lowered risk of accidents and diseases, as well as rewards and incentives provided by the company (Becher, 1974).

The corporate commitment of management to an occupational safety and health program must be obvious and consistent for a successful program. Implementing training programs is a successful method for communicating overall commitment to company employees, both management and hourly.

16.2.2 Information On Resources

Another element of an effective program requires the corporation to provide adequate information for program implementation to both management and employees. An essential component is information about how the company plans to solve health and safety problems on the job. Though training alone cannot solve health and safety problems, it can demonstrate that appropriate resources are available for problem solving and can explain to employees how to use the resources. Training programs provide information, teach skills, and encourage awareness. But, as stated before, they can never be a substitute for an overall health and safety program and must be seen as just one part of a comprehensive program.

The role of training is to communicate information. Providing information does not guarantee that it will be learned or acted upon. The link between knowing what to do and doing it, however, is not always clear. An obvious example of this are those people who understand the health hazards associated with smoking cigarettes yet continue to smoke.

An individual may or may not follow known safety and health guidelines, but can't possibly follow appropriate procedures when he or she doesn't know them. Therefore, a thorough and comprehensive training program must provide the information baseline necessary for implementing a health and safety program.

16.2.3 Goals And Objectives

Another essential element of an effective health and safety program and its training component is a written program and objectives. Many regulations, such as the OSHA respirator standard and the OSHA hazard communication standard require written training programs. The objectives developed for these programs provide an essential benchmark for judging the success or failure of the training program itself evaluating the effectiveness of your efforts.

Formulating goals and objectives that are realistic and attainable as well as measurable will be discussed in much greater detail later on. However, it is important to consider these goals as part of the information that needs to be communicated to employees. When all participants are knowledgeable about the final destination, the more likely it is that everyone will get there.

16.2.4 Feedback

The final essential element of an effective health and safety program must be feedback: continuous reinforcement that motivates all, employers and employees alike, to behave in a safe and healthy manner. Training can be more or less formal depending upon the circumstances. The classic image of training and education is of a classroom with a teacher in front of a quiet audience. However, feedback on a day-to-day or job-by-job basis, as well as rewards for successful completion of projects or meeting objectives as well as incentives that motivate, must all be seen as part of a successful training program.

16.3 HISTORY OF HEALTH AND SAFETY TRAINING

Safety training designed to prevent injury and accidents on the job are the historic roots for the health and safety training we know today. Such training was and still is particularly important in industries such as mining, construction, and manufacturing. Organizations such as the National Safety Council and its state affiliates have developed employee educational programs that specifically address accident prevention techniques. Among the topics covered by such programs are slips and falls, ladder safety, mine explosion prevention, and machine guarding. The motivation for industry to implement these programs comes mainly from escalating workers' compensation costs over the last decades. Since rates for this type of insurance are often set based on the number of lost time accident days experienced by a company, prevention programs in the safety area can have a direct impact on total insurance costs.

Job safety analysis techniques are used in developing these programs. The purpose here is to review every aspect of a particular job, and based on that review, identify those elements of the job that are particularly dangerous. Based on this analysis, appropriate engineering controls are implemented where possible and where not possible, employees are trained about the potential dangers and how to prevent them. A properly implemented job safety analysis program involves the person performing the operation as well as management, supervisory, and safety staff. Members of this team participate in the process of making a job safety analysis of the particular situation. If implemented correctly, it is a "bottom-up" approach to safety programs (Re Velle, 1980).

An alternative to this type of program is the company-wide motivational program, such as Dupont's Stop Program. These programs feature company-wide participation from top management to hourly employees and can be characterized by their teamwork approach to safety. Some of these programs encourage competition between different departments within companies to reduce the number of lost work days. Large signs often seen at the entrance gates of factories herald the number of lost time accident free days a company has had. Such programs often use rewards and incentives to encourage safety performance.

Job safety analysis and company-wide motivational programs for safety have been around for many years. For the last ten years, however, there has been a gradual but ever-increasing emphasis on a different type of safety training. While the kinds of programs described above emphasize accident and injury prevention, particularly such acute, unsafe events as slips, trips, and falls, more recently emphasis has been placed on health hazard training. Of particular concern here is minimizing the long-term exposure to chemical hazards. Such training requires a much more sophisticated approach. Complex concepts such as relative risk; routes of entry such as inhalation, ingestion, and skin contact; environmental control technology such as ventilation systems; and the use of protective equipment required must be communicated as part of training programs regarding health hazards (Cohen, et al., 1985).

The initial OSHA regulations did not require any health hazard training. As more and more information about toxicity and hazards associated with substances in the workplace became known, there has been a shift in policy leading to an increasing emphasis on health hazard training.

The competing forces on this issue have been, on the one side, employee groups and their representatives pressuring for access to information about hazardous chemicals in the workplace. At the other end of the spectrum, industry has emphasized that proprietary and trade-secret requirements prevented full disclosure regarding hazardous substances in the workplace.

These extreme positions have moved closer together over the years. The incident in Bohpal, India has provided strong motivation for the chemical industry to see that increasing access to information by employees and their representatives is an important way of preventing accidents and minimizing future liability claims. Employee groups are also coming to grips with the importance of individual behavior in preventing long-term exposure to hazardous chemicals in the workplace.

There has also been an increased emphasis in America on the health status of the general population. Concerns about the relationship between smoking and disease as well as diet and exercise have grown parallel to the concern about health hazards in the workplace.

These dual concerns have motivated the passage of new laws and the promulgation of regulations that emphasize the transfer of information from employer to employee about the hazards of workplace substances. In general, these rules fix responsibility with the employer for assembling and disseminating this information.

Since 1978, when OSHA issued its first training requirement for working with a regulated substance, there have been a myriad of new laws and regulations that cover this issue. The most prominent among them is the OSHA *Hazard Communication Standard,* first issued in 1981, then withdrawn and reissued in 1984. In addition, there have been many state right-to-know laws passed as well as other laws with training requirements including the Toxic Substance Control Act (TSCA), the Resource Conservation and Recovery Act (RCRA) and the Federal Insecticide, Fungicide, and Rodenticide Act (FIFRA).

The underlying belief that motivates those who support these laws, rules, and regulations is that there is a link, both psychological and behavioral, between knowing about the hazards of a chemical and doing something to protect oneself. That link has not been clearly established scientifically. Though there is conflicting evidence about this assumption, it is logical to assume that if you don't know about the hazards related to a chemical you work with, then you are unlikely to be able to protect yourself from the hazards (Cohen, 1982).

16.4 BENEFITS FROM HEALTH AND SAFETY TRAINING PROGRAMS

Though the regulatory requirements for training in occupational health and safety are increasing, successful programs are those that go beyond the letter of the law. The company that puts emphasis on training regarding safety and health practices for employees and demonstrates concern through these programs is able to achieve cooperation and compliance for healthy and safe behavior.

Mangers often find it difficult to justify the cost of serious and consistent training efforts. Thorough and comprehensive programs add quantifiable cost to the bottom line in the short term. So when budget problems occur, it is often training that suffers.

A more long-term approach to understanding the importance of health and safety training is needed to build a truly successful program. The long-term benefits of such a program, however, are not as easily quantifiable as the short-term costs.

Part of the difficulty associated with quantifying these benefits has to do with the difficulty in measuring the outcome. The goal of most health and safety training programs is to *prevent* accidents, injuries, and diseases. Performing cost-benefit analysis to indicate effect over the long term is expensive as well as methodologically difficult.

Even though many of the long-term advantages of preventive training programs are hard to quantify, it is important to understand their value. Even if they are not easy to quantify, they will have an impact on the long term financial health of a company and the personal health of its employees (Zofar, 1980).

16.4.1 Regulatory Compliance

The most obvious advantage of instituting health and safety training programs is that the company will fulfill its regulatory requirements. This is clearly a short-term goal, but important nonetheless. It is not difficult to see that if you must put resources into providing some type of training program it might as well be one that meets the basic requirements. Complying with regulatory requirements can sometimes be quite complex. For example, CFR 1910.134, the OSHA *Respirator Standard,* applies to facilities where engineering controls are not feasible for protecting employees from inhalation hazards. To meet the requirements of this standard, an employee must complete eleven steps specified by the standard. Included among them are: medical screening of employees, respirator-fit testing, training programs for respirator users, and a formal written respirator plan.

Employee training implemented as part of a respiratory protection program provides the forum where employees learn about the company's attitude and willingness to protect employees from occupational hazards. How vigorously the company implements regulatory compliance programs says a lot about the company's level of concern. The more obviously concerned about the matter a company is, the more likely employees will be to implement the program.

16.4.2 Reduced Health Care Cost

The results of a successful health and safety training program can include reduced accident rates, injuries, and diseases associated with exposure to unsafe situations or hazardous chemicals. The health care and insurance costs associated with work related accidents, injuries, and diseases are considered in most situations as part of the total costs of health and safety. The higher these costs, the more of a negative impact they have on the corporate bottom line.

Included among these costs are insurance, workers' compensation, and the cost of fringe benefits associated with employment such as sick time and disability pay. The cost of these items is directly related to accident and injury rates at any facility. If meaningful training program that educates employees on how to reduce the risk of accidents and injuries does reduce these occurrences, the total cost of the program can be lowered. Of course, the training program must be based on a strong corporate safety and health program that is designed to reduce accidents and illnesses.

16.4.3 Employee Relations

Another advantage of a serious health and safety education program in a facility concerns employee relations. Much has been written and said regarding Japanese management techniques that emphasize the overall advantages to a firm that implements policies based on employee/employer cooperative relationships. A similar impact can be realized through a thorough and consistent health and safety program that includes a strong training component. When such a program is instituted, it can have a positive impact on employee relations and motivation. Employees feel more positive about an employer who spends time and resources teaching them how to protect themselves from physical and health hazards associated with their job. This demonstrated concern for employee well-being can have a positive impact on the productivity of the operation.

16.4.4 Future Liability

One of the most important long-term effects of a comprehensive employee health and safety training program is a defensive one. Such programs can play a very large role in preventing possible future legal liability. Informing employees about the safety and health hazards associated with their jobs and documenting that this training in fact took place, as well as informing employees about the actions being taken by the company to prevent future health and safety problems, can be used to defend against future liability claims. You can never prevent an employee from claiming in the future that you have failed to warn him or her about the hazards associated with the job, but by providing information you can limit the employer's special liability that is associated with the failure to warn.

The general concept of failure to warn is not a new one. It is being used more and more in toxic tort litigation, whether it be consumer oriented or job related. The legal principle that failure to warn leads to special liability has become quite well established. In some cases, claims of failure to warn have lead to plaintiffs circumventing workers' compensation law. In these circumstances, employees sue employers directly, which is prevented under the regular workers' comp system.

By providing information to employees regarding the hazards in the workplace, a company can make their employees partners in prevention of accidents and illnesses. By involving them now in such a program, future liability will be limited. Employee right-to-know laws formalize this requirement. However, just by following the letter of the law and not the spirit of full disclosure, you may risk future liability. This may seem a "far out" and unlikely approach by industry, but such groups as the Chemical Manufacturers Association have gone on public record supporting this approach.

16.5 KNOWLEDGE AND MOTIVATION THE KEY TO COOPERATION AND COMPLIANCE

The overall objective of a health and safety education program is to make it possible for employees to work more safely on the job. In order to achieve the desired level of cooperation and compliance, an employee must have the knowledge as well as the motivation to work safely in the workplace environment.

16.5.1 Knowledge

To achieve this goal, three types of learning must take place. First, an individual must have the information (knowledge) so he or she is able to work safely. Educators call this the *cognitive domain.* Typically, the knowledge component of a chemical hazards education program teaches an employee the difference between the three possible routes of entry when working with hazardous chemicals: inhalation, ingestion, and skin penetration. In the safety area, a typical knowledge requirement would include information about proper machine guarding, confined space entry hazards, and the types of grounding to be used when working with electrical equipment.

Depending on the nature of the operation, the knowledge, or information, can be more or less complex. For example, plating operations usually are well understood and the hazards associated with plating can be controlled by ventilation. The information required to communicate this knowledge is limited and understandable by the average person.

Chemical plant operations are not always as easy to understand. Many chemical operations require multiple raw materials, sophisticated processing equipment, and complex piping, pumping, and control facilities. Learning about the hazards associated with a facility of this type can require lengthy training for those employed at these facilities. Included in such programs is information regarding safety procedures, preventive maintenance, and the potential health effects of multiple chemicals and mixtures of chemicals.

Even more complex than chemical operations are such facilities as nuclear power plants. The operators must know and understand very complex equipment. As the complexity of a facility increases, the amount of knowledge required obviously increases, as does the amount of time required to communicate this knowledge, and the amount of time it will take for the operator to learn it (Feuer, 1985).

16.5.2 Behavioral Skills

Knowing about potential health and safety hazards and their control is just the first step. An individual must also have the skills necessary to implement what they have learned in the cognitive realm. These skills are often considered by educators to be in the *psychomotor domain.* This includes the actual physical behaviors an employee is expected to carry out in order to work safely in a potentially hazardous situation. Such learning is clearly performance oriented and covers the actual procedures or behaviors expected for safety to be maintained.

For example, there are prescribed steps to be followed when confined space entry is required. Such a training program would cover the actual skills of donning protective equipment, such as shoring appropriate air lines into the environment, and buddy systems. In the area of health hazards, the psychomotor skills required include proper use of protective equipment when cleaning up spills or leaks, and the demonstrated mastery for carrying out those procedures.

As the complexity of the operation increases, required safety behaviors go beyond skills for personal protection. The skills required include the total facility operation. Sophisticated skills learning can include such things as computer accident simulation, disaster drills, and procedure demonstrations on a repeated basis (Fitch, et al., 1976).

16.5.3 Attitudes

Knowledge can be defined as knowing *what* to do; skills can be defined as knowing *how* to do it. The last area of learning that must take place for an educational program to be effective is the encouragement of appropriate attitudes about a safe work environment. Educators call this the *affective domain*. Learning is much more difficult to achieve in this area and the inappropriateness of employee attitudes is a common complaint by many who are responsible for implementing health and safety training programs. Though employees know they should wear earplugs when exposed to excessive noise, that does not mean that they always will. Cavalier attitudes about working with caustics and acids have been a problem with older workers for years.

The importance that attitudes play in health and safety behavior has only recently been recognized as it applies to occupational health and safety concerns. The complexity of the process of formulating attitudes about health risks has been the focus of study by cognitive, organizational, and social psychologists. A model developed in the early 1970s, called the Health Belief Model, has been the starting point for those involved in developing methods of effectively communicating health hazard information (Janz and Becher, 1984).

As the concept of employee right to know has developed, health risk information and attitudes about health risk have become an important component of any safety training program. How this information is communicated is influenced by the attitudes of the communicator and the receptiveness of the learner. This process is affected by such concerns as legal liability and the level of trust between the giver and recipient of the information. This relationship and the methods for influencing outcomes has only recently begun to be presented in the professional literature (Hopkins, 1982).

Carrying out a successful program in industry is a difficult task. In a company where health and safety matters are taken seriously, management may provide the time necessary to learn about hazards and to specify the amount of information and knowledge transferred. Such a company can and also will provide adequate training to ensure that an employee has the skills necessary to carry out the procedures.

However, influencing attitudes is not as easy. We learn our attitudes and habits from the world outside our work as well as at our workplace. Our attitudes are affected by those around us. If an hourly employee is provided the knowledge and skills necessary to carry out a task, yet the foreman or supervisor minimizes the importance of performing the task properly, the employee's attitude about carrying through on safe behaviors will be influenced by the attitude of the supervisor.

It is sometimes the case that the attitude of corporate management is different and more positive towards health and safety problems than the direct supervisor, whose main objective is production. Mixed messages about implementing health and safety procedures can have a very negative impact on an employee's attitude.

16.5.4 How Are People Motivated?

It is widely accepted that such a thing as corporate culture exists. The health and safety climate is part of that culture. In companies where there is a positive climate for safe and healthy behavior, employees can be motivated to perform operations safely. But it is important to recognize that these attitudes must be present at *all* levels of an organization for effective results that motivate employees.

A direct link exists between an individual's attitude and his or her motivation. More positive attitudes about possibilities and potential outcomes lead to stronger motivation. Employees who think or believe that they can protect themselves and their fellow employees by performing their jobs in a safe and health promoting way will be more motivated to do so. Attitudes are more easily understood if they are seen as the level of confidence a person has about his knowledge and skills.

Employees who believe that the health and safety program information describing the potential hazards of working with a hazardous substance are correct will be motivated, by their own self-interest in a long and healthy life, to carry out the activities or perform the skills that are required to make the job safe.

Besides self-interest and self-motivation, the examples of others provide strong motivation as well. The supervisor who requires strict adherence to health and safety procedures is a strong motivator for the employees to do so themselves. The motivation of the supervisor is also important to consider. If the safety and health record of a supervisor is reviewed as part of his performance appraisal, and therefore has an impact on his salary and promotion, strong self-interest-based motivation exists.

Reinforcement for positive actions is one of the strongest motivational tools. There is no better way to get someone to repeat a particular procedure than by positively praising them or rewarding them for carrying out the procedure. This is a basic tenet of human psychology. We repeat those actions that elicit positive "strokes" and eliminate those that don't. Elaborate reward and incentive programs, such as the games and contests maintained earlier, are not necessarily the best way to reinforce for positive actions. Sometimes these programs can seem to play favorites. Often a word of encouragement by a supervisor or coworker and compensation review based on evaluation of safety performance in the department provides very strong reinforcement for positive actions (Smith et al., 1978).

16.6 TENS STEPS TO EFFECTIVE HEALTH AND SAFETY TRAINING PROGRAMS

The first section of this chapter has described the appropriate context for health and safety training. It was meant to explore the complex set of variables that influence health and safety training programs. In spite of the possible problems and potential pitfalls that anyone charged with the training task may encounter, clear-cut procedures exist that can be utilized to accomplish one's goals.

In order to implement an effective program that will lead to cooperation and compliance, ten basic steps can be identified. These ten steps are as follows:

1. Perform needs assessment.
2. Set objectives.
3. Evaluate and review previous training efforts.
4. Determine content of training.
5. Assemble available resources for implementing program.
6. Implement the program.
7. Determine training environment.
8. Plan methods for documenting and evaluating the program.
9. Specify feedback measures to be used.
10. Write a formal training plan.

These ten steps are based on standard techniques developed by educators implementing any type of educational program. The details associated with each of these steps will be discussed one at a time. It is important to remember that implementing this program will differ from location to location. This in not a cookbook formula. Implementation will depend upon how complex the situation is upon the total number of potential hazards identified (Smith et al., 1978).

16.6.1 Education Needs Assessment

The purpose of a needs assessment is to establish what your training requirements are. How you conduct and who participates in the needs assessment process can have a large influence on the success or failure of your training program. The intent of the needs assessment is to survey all areas of health and safety compliance activity to determine where employee and employer participation in a training program would enhance the program (Smith et al., 1978).

The first and most obvious area to look at is your regulatory requirements. If your firm works with OSHA-regulated substances that require special training, this is an obvious area of training need. Such substances as benzidine, inorganic arsenic, vinyl and polyvinyl chloride, as well as lead and benzene are among substances requiring special training under OSHA requirements (OSHA, 1985).

In addition, many states have specific laws regarding training people who work with asbestos. The OSHA *Hazard Communication Standard* as well as the state right-to-know laws also require employee training about potential health effects of various hazardous chemicals.

Regulatory requirements for training are also included in laws associated with the control of hazardous waste. Generator, haulers, and storers of hazardous waste must provide information and teach skills that ensure that employees handle these dangerous substances properly.

In the safety area, though there are few regulatory requirements, a review of insurance claims as well as lost-time and non-lost-time workplace accidents will help you pinpoint jobs with high accident rates and/or large numbers of insurance claims. Review of insurance and health records should certainly be a part of your needs assessment.

Any corporate health and safety policies that exist in your firm should also be carefully reviewed to determine if there are any specific training needs required by these policies. Since it is the procedures and practices outlined in your organization's health and safety policy that will make up a significant portion of the information to be communicated in your training program, a careful review of these policies and procedures is essential.

If no policies are in existence or your policies are weak in areas where your accident and illness incidence rates are high, you should consider some additional steps that you might want to add to the process. Since the purpose of training is to communicate information and motivate individuals, policy deficiencies must be corrected prior to developing training programs. This task has been described in detail in other sections of this book. If policies are weak or nonexistent, it will make your task of developing training content impossible. Therefore, a thorough review of policies and procedures will show you where you may encounter problems.

Another source of information for your needs assessment is direct communication with employees at your company. This can play an important role in the motivational process, because, by requesting information and feedback from managers, supervisors, employees, and their representatives, you indicate to them your seriousness about establishing a safe and healthy workplace. This process can be carried out through personal interviews, surveys, or mail drops for anonymous comments.

On the one hand, you want to make sure that all the people involved feel that they are participating in the process. A reverse problem you may face is a rising level of expectation among your target audience regarding what they might expect from the training program you are developing. Matching perceived needs of the chief operating officer, plant manager, or union president may be difficult. It is the responsibility of the person implementing the training program to communicate to all staff the importance being placed on this process, yet, at the same time, not raise expectations to an unattainable level.

As a result of your needs assessment, you will have created a long "laundry list" of possible areas for training. Using this list as the basis of all possible training, you must begin setting priorities. This is the task of step 2, setting objectives (Votjecky, 1985).

16.6.2 Setting Objectives

Now that you have a complete list of all possible areas where training might be used to help prevent health and safety problems from occurring, you are ready to make some very hard choices about what you can accomplish given your resources and authority.

Since objectives provide the framework for implementing your health and safety training program, it is important at this step to get top management approval and active support. Only with their sanction will you have the authority needed to implement these programs. If you work in a setting where the employees are represented by a union or an employe organization, their cooperation and agreement with training objectives may also be important.

Recently negotiated union agreements in the auto industry have established procedures for union representatives to become part of the planning process for occupational health and safety training. The logic behind this development is that health and safety program implementation requires a cooperative effort. The earlier the cooperation begins the more effective it will be.

Defining your objectives often turns out to be a negotiations process among all concerned. Whether or not you can pinpoint achievable objectives and goals will have a major impact on gaining cooperation and compliance in your safety and health program.

Management Objectives

Four types of objectives must be considered when formulating your training program. They are: (1) management objectives, (2) knowledge objectives, (3) behavioral objectives, and (4) attitude objectives. Management objectives reflect corporate policy regarding health and safety training. Such objectives might include only meeting regulatory requirements. Making compliance with legal requirements your only management objective is shortsighted, however. Management objectives provide the organizational context for your training program. Consider expanding them to include such items as: how frequently training should occur, say annually or monthly; specifying resource allocation for training; and delineating responsibility for implementation. Some possible management objectives to consider are:

- All employees will spend a minimum of ten hours a year in health and safety training.
- At least 5% of all department operating budgets will be dedicated to health and safety training.
- Corporate executives, plant managers, and supervisory staff will participate directly in the training process.

These management objectives really provide goals and a framework for your training program. The other types of objectives provide more specific guidelines for your training program.

Behavioral Objectives

Behavioral objectives, expressed in measurable terms, are the core of your training goals. These behavioral objectives focus on employee behavior (Chhokar and Wallin, 1984), and their success can be measured. In this case you are trying to teach an employee how to perform a task safely. You must state your objectives in terms that every participant can understand. An example of a behavioral objective is one that sets out specific performance expectations, for example: at the end of this training program employees will be able to understand how a respirator works, clean their respirators, and check the fit of their respirators.

Knowledge Objectives

Knowledge objectives are another key element in the objective-setting process. These objectives define the information you plan to impart to your employees. State right-to-know laws, the federal OSHA *Hazard Communication Standard,* and the specific training requirements for designated substances require that you provide certain types of information about chemicals. A typical knowledge objective is: the employees will understand their legal right to specific information, will be able to interpret material safety data sheets, and will know how chemicals are used in our facility.

Another type of knowledge is the one we commonly see in the health and safety standards, which states that an employee will be informed about the potential health problems related to working with a specific chemical. This objective does not require the employee to do anything, as is the case with a behavioral objective; however, the knowledge of specific information about a product, process, or procedure is a requirement that must be met in order to be able to fulfill the behavioral objectives.

ATTITUDE OBJECTIVES
You must also consider attitude objectives. As described earlier, employee attitudes are influenced by the safety and health environment established by corporate management. The stronger corporate commitment toward improving safety and health is, the more likely employees will be to change behavior when working with hazardous chemicals. An example of an attitude objective is: employees will have a concept of relative risk, they will understand why working with one chemical involves more risk than working with another; they will understand why it is important to work safely with chemicals. The specific attitude about a health and safety program in general is communicated at all levels of management.

All types of objectives must be realistic if they are to be met. It is often the case that though an employer's objectives appear laudable and appropriate on paper, they are unachievable in reality. It is the actual day-to-day operation that determine how effective your health and safety program is. Contradictions between how your program is implemented and its overall goals will have a direct impact on the trust between you and the employees. If no clear-cut correlation between objectives and management attitudes about implementation exists, your entire program may be sabotaged. The attitudes of management are clearly communicated through management behavior. When day-to-day attitudes are contradictory to "official" policy, a hidden message is being communicated loud and clear: don't do as I say, do as I do!

16.6.3 Evaluating And Reviewing Previous Training Efforts

Now that you have determined training needs and developed a series of goals and objectives that will satisfy these needs, you are, at this point, ready to begin to develop a training program. It will be worth the time it takes to investigate previous efforts in health and safety training. The purpose here is to see if you can borrow resources and ideas from previous efforts so that you can limit the amount of new program development you must initiate.

The person with overall responsibility for the training program should not consider himself or herself a "one-man show." There are several places in which to look for help. You need to consider who else on the staff may have experience with health and safety training. If you are with the training department in a larger facility but without a health and safety background, you will have to get help from people in the plant who are knowledgeable about health hazards, plant operations, or safety concerns. If you are a health and safety expert, you may want to get help from the training department or the human resource development group in your company. They are trained to develop motivational programs, and may strengthen your efforts considerably.

You might also find resources in other locations where your company offers training programs on the same hazards. These locations may have written programs and training materials available for review and possible use. You should also consider the use of consultants knowledgeable in the specialized field of occupational health and safety training who could assist you in the program. The actual expense may be less if you use a consultant to assist you than the cost in time and materials for you to develop the program alone.

By looking at and evaluating previous training efforts and talking to people who have participated in those efforts, you can probably avoid mistakes that were made in the past and can avoid reinventing the wheel when it may not be necessary.

16.6.4 Determining Content Of Training

Now that you have established your objectives, or, in other words, established the goals that can help you evaluate the outcome of the training process, it is time to decide what information and skills you need to teach in order to meet these objectives.

The training program content, or what is taught, is often called the program curriculum. As you develop the content outline, it is very important to make sure that your curriculum provides enough information. It is also important to ensure that adequate time is spent on the tasks designed for skills development. Employees need adequate time to learn what is necessary to fulfill the objectives you have set. For example, if one of your objectives is the proper use of respirators, you will have to include some hands-on experience with respirators as part of your curriculum.

In order to ensure that you do provide information and skills development that meet your objectives, it is often useful to use these objectives as a guidepost for outlining your program content. Since one of the key elements of a program that fosters cooperation and compliance is group participation, including the actual objectives as part of the curriculum for the training program can be useful. When explaining your objectives becomes a curriculum element in the program, the employee will be aware of the full scope of the company's health and safety program as well as understand what is expected of him or her in relation to that program.

Many of the federal and state regulatory requirements related to training clearly specify specific curriculum items that must be included in the program. You should carefully review these regulations and laws to ensure that your curriculum covers the points that are required under these rules. Often your objectives provide overall guidelines for your program. Repeated review of regulatory requirements can help you stay on track throughout the program development process.

Since the overall purpose of health and safety programs is to prevent accidents, illnesses, and injuries, it is important to consider the appropriate amount of general information that should be communicated in your health and safety program. Two things should be considered when making these decisions.

First of all, there is a clear need-to-know criterion. This would include the information an employee needs to know to be able to carry out a procedure safely. The second criterion has to do with employee motivation. It is important to consider how much information a person needs to understand the potential hazard of unsafe practices.

Obviously, when training for work with asbestos, just saying that asbestos can harm your lungs is inadequate (and in many states illegal). To motivate someone to work safely with asbestos it is necessary to explain that asbestos is a cancer-causing agent, that exposure to asbestos can lead to fatal lung diseases, and that every precaution must be taken to prevent exposure. In addition, the link between the hazards of asbestos and smoking must be communicated.

Asbestos provides a clear-cut case of how curriculum information about a substance can be a motivating factor in encouraging safe behavior. With other hazardous substances and unsafe situations this link is not always clear. But it is important not to overly alarm employees about hazards because it can lead to such erroneous beliefs as "everything causes cancer." This can lead to an employee not taking the warning seriously. The other side of this trap is that *minimizing* potential hazards may create a future liability for you. Therefore, it is important to assess a potential hazard accurately and communicate information in your training program that reflects that assessment.

The OSHA Standards for regulated substances contain common curriculum elements. They are listed below in order to provide a useful framework in which to think about specific training program content. They can be adapted for use with physical hazards or safety problems, as well. The program elements include:

- Nature of the hazard, toxicity.
- Review of the relevant government standards.
- Specific nature of the operations likely to release the regulated substance.
- Purpose of medical surveillance when required.
- Emergency procedures.
- Engineering controls.
- Respirator and other personal protective equipment usage.
- First aid.

An additional program element to consider is an overview of company health and safety policy. Using a motivational message from corporate officers, either written or verbal, can provide strong reinforcement for your program.

16.6.5 Assembling Resources

Now that you have decided what you will teach, identifying the resources that can help you actually implement the direct training is the next step. You may have already identified other staff at your facility or at the corporate level who are knowledgeable in health hazards, plant operations, or employee training. Now is the time to figure out ways to involve these individuals in the actual teaching process. They may be available to write sections of a program to be delivered by someone else or they actually may be able to teach in your program (Vojtecky, 1985).

Who should teach in a training program can often be a difficult problem. The teacher is often determined by policies over which you have no influence. For example, your firm may have established a policy that first-line supervisors provide employee training. Alternatively, the health and safety officer may be designated as the trainer. These individuals may be charged with this teaching responsibility but not be adequately trained themselves to carry it out. Teaching skills can be learned, however. Many "train the trainer" programs are available. If you encounter the situation described above, try to find the resources necessary to improve the trainers' teaching skills.

If you have more flexibility in assigning instructors, you should choose teachers carefully. Some people are natural teachers. These people are easily identifiable because they communicate information with ease, are patient with learners, and like to teach. When possible, such individuals should be used as instructors in your program.

You also should already have determined whether there are any existing programs available from other locations or past training efforts. This is the time to review them and decide what material to use and how to use it.

You should also review the off-the-shelf training products available to see if they can supplement your specific material. Consulting firms and services are another resource that can help you develop your program.

In addition to human resources, you must also consider supplies and equipment resources. Make sure to determine what audiovisual equipment is available and what you need to use the existing programs. You should also consider what handouts you would like to give participants, what slides you need to make, and how much all of this will cost.

Audiovisual aids and handouts provide a useful way of communicating information. Some people learn best by hearing information. Others learn best when they see or read information. You should, therefore, include all three forms of communication—verbal, written, and visual—for an effective program.

Another resource you will need is equipment for hands-on training. If you are training people to use respirators, for example, a full set of respirators and respirator packs will be necessary for each student.

After you have decided what equipment and supplies you will need, it is advisable to develop an equipment checklist, such as the one in Table 16.1. This provides a reminder for you each time you present your program. This checklist should be very detailed to ensure that you include all the equipment that you need. A sample training program checklist used for an asbestos hazard training program follows. It will give you an idea of how you can structure a checklist for your own program.

16.6.6 Implementing The Program

Involve as many people as practical in implementing your program. It is important for compliance and cooperation that you create a sense of team spirit.

To do so, there are several things that you can do. First of all, you should have your peers and experts review the program as you have designed it so far. It can be very embarrassing if you develop and implement a training program in a particular area of the facility operation but have not involved the supervisor, and when you present the program, you find that there are some gross inconsistencies between what you are presenting and what is the reality of the situation. This kind of embarrassment can have a serious effect on the credibility of your program. Therefore, it is very important that your program be reviewed by the people who are responsible for implementing the safety procedures your program covers.

Another way to create team spirit and to make this a cooperative effort is to involve key people in your firm in support of the program. Written support from top executives is at least a strong indication of the company's intent. Visible support and compliance by first-line supervisory personnel is also essential. If employees are expected to wear respirators or earplugs in a noisy environment, it is essential that management and supervisory personnel do the same. In a unionized setting, participation and motivation by union officials as an encouragement to your employees can also be very effective.

16.6.7 Determining Training Environment

Now you're ready to implement your training program. The actual environment in which training takes place includes such considerations as class size, location, length of instruction, and number of sessions.

Your class size should be small enough so that everyone can have an opportunity to ask questions and participate. A maximum for specific training programs is between 25 and 35 people. General information can be given to a much larger group as long as there are also specific small-group meetings where people can speak up and ask questions. In general, people working together should learn together.

The amount of time spent on training will depend on the amount of information you need to communicate. Training time away from production can be expensive. Keep this in mind as you develop your program.

You should also make sure that the length of time and number of sessions required allows each presented unit to contain a learnable and understandable amount of information. Repeat training about the same information is often more effective than detailed information about a topic only once. In other words, it makes sense to schedule two sessions of one hour or one half-hour session every month rather than a full day seminar once a year. In this way people can absorb smaller bits of information that they can process, implement, and review later.

Location and classroom size also must be considered. The more conducive to learning your classroom is, the more likely it is you will get your information across. The optimal space is a well-lit, quiet classroom away from any interruption and disturbances. However, this ideal is not always easily achieved. A classroom that is temporarily set up in the middle of a production operation clearly does not indicate a strong commitment on the part of the company to implement these programs. Therefore, it is important and advisable to consider the best possible location for your classroom presentations.

TABLE 16.1. Equipment Checklist for Asbestos Training Sessions

The following checklist will provide a guide for those materials necessary to put on an asbestos training session with respirator fit testing. The quantity of some of the items may change according to the number of students or the target audience.

Type	Quantity
Respirators	
MSA, half facepiece	6 large 4 medium 1 small
Scott half facepiece respirator	1
Scott full facepiece twin cartridge respirator	1
MSA full facepiece twin cartridge respirator	1
#3M8710 respirator	1
#3M9920 respirator	1
Sesco half facepiece respirator	2
Organic vapor cartridges	An appropriate number to be used during fit testing

Type	Quantity
Personal Protective Equipment	
Asbestos cartridges or filter pads	An appropriate number to be used during PPE familiarization
Disposable boots	2
Disposable suits	4
Disposable hairnet	1
Disposable hoods	2
Ear muffs	1
Chemical splash goggles	1
Ear plugs	5

Type	Quantity
Respirator Fit Testing Equipment	
Fit testing support mechanism (tripod)	1
Plastic garment bag	1
Plastic hanger	1
Banana oil in a bottle	1
Cotton balls	2
Pipe cleaners	2

Type	Quantity
Sampling Equipment	
Sampling pumps	2
Cassette filters	4 (2 each of 37 mm and 2 each of 25 mm)
Tygon tubing	2

TABLE 16.1. Equipment Checklist for Asbestos Training Sessions *(continued)*

Type	Quantity
Removal Equipment	
Glove bag	1
Handout Material	
Number of handouts will depend on the number of students anticipated. These handout materials include:	
Asbestos training tests	
Material safety data sheets (MSDS)	
Containment area handout	
Quick reference handout	
Audiovisual Equipment	
35 mm projector with remote control	
Projector screen	
16 mm movie projector	
Overhead transparency projector	
16 mm film "Doin' It Right"	
3 slide trays with selected slides	
Electric pointer	
Mechanical pointer	
Miscellaneous Equipment	
Two sets of 5 asbestos bulk samples	
______________________________	
______________________________	
______________________________	
______________________________	

16.6.8 Planning Methods For Documenting And Evaluating Your Program

There are several reasons why your training program should be well documented. It will be important in the future to review what you have done previously so that you may improve and alter your program as needed. To do this you should keep your content outlines, training supplies, equipment checklist, and any other information about the mechanics of setting up your program.

It is also important to keep a record that training has taken place. Sign-in sheets are an appropriate method of achieving this goal. In larger organizations it may be possible to computerize the records of who attended what sessions and document the training objectives and outlines for each department. By using this kind of documentation, you can make sure that all people have received the appropriate training and can schedule repeat sessions for those who missed. Keeping accurate and complete records of your training program will make it much easier to implement it again or to change it as the need arises.

Another reason for documenting your program is to determine if your training has been effective. This requires training evaluation. Simply stated, a training evaluation compares your stated objectives with your results. Based on this comparison you can make judgments about the effectiveness of your program (Cohen and Jensen).

Different evaluation techniques can measure both the success of your presentation and changes in employee behavior as well as help you to review procedures. Such procedures include skills testing, observation, and proxy measures for behavior evaluation. Evaluation serves as a guide to improving your program. Since cooperation and compliance are the primary goal and purpose of designing a training program, it is also useful to use participant surveys or questionnaires to get their opinion of the program. Such evaluation procedures can help you improve your program as well as help the participants understand that the intent of the program is to develop an effective and cooperative health and safety climate. In order to make the evaluation process clearer, two tables have been included, each of which will be discussed in detail.

EVALUATION METHODS AND DOCUMENTATION

The first type of evaluation, shown in Table 16.2 is "content review." The objective here is to make judgments regarding the level of understanding, completeness, and accuracy of the information presented. Prior to training, others in the organization who are familiar with the objectives should review the plans. Based on their experience and knowledge, they can identify problems with the program. This review can be as formal or informal as required. As stated before, this audit of your program should include a review of previous training programs. This will help identify problems or mistakes encountered in the past, which can be corrected when planning for the future.

The second step of content review is evaluation by the employees being trained. The most common method is to provide each participant with an evaluation form that is appropriate for his or her level of education, such as the one shown in Table 16.3. The form should include questions regarding relevance, understanding, completeness, and accuracy of the material presented. The participants should be allowed to hand these in anonymously so they do not feel obligated to respond positively. Though external influences such as lack of enthusiasm for the program and conflict with production incentives may influence responses on such a questionnaire, it is important to give employees a chance to comment on the value of the program.

The second type of evaluation to be included in this category is audit data, or documentation. How you plan to use this data will determine how much information you need to collect. For example, who attended the programs can be described according to the department the individual works in, his or her

TABLE 16.2. Possible Evaluation Techniques

Type	Database	Results
	PROGRAM EVALUATION	
Documentation	Who, what, where, etc.	Audit of program Lists of no-shows % of participation
Content review by peers and experts	Review in planning phase by peers and experts	Judgments about completeness and accuracy
Content review by participants	Review after program by participants	Judgments about understanding and relevance
Qualitative evaluation	Documented impressions, feelings, observations by peers, participants, and presentors	With above info, a complete review of program

TABLE 16.3. Hazardous Chemical Training Evaluation

Date: ____________________

1. Which parts of this program were most useful to you?
   Explain: ____________________

2. Which parts were least useful?
   Explain: ____________________

3. What did you learn about working with hazardous chemicals that you did not know before? ____________________

4. Do you have any recommendations for topics that should be added? ____________________

5. How could the program be improved?
   ____

position in the organization and/or job responsibilities. Compiling lists of absentees and new hires will make it easier to follow up and encourage participation in later sessions. What sessions or types of training were offered is another important piece of audit data. When, where, who taught, and how long the programs were is also important. This information can be gathered at the time of the events and should be compared to a preplanned training schedule and preestablished criteria and expectations of participation. These data can be prepared immediately after the presentation of a program. The final task of audit and documentation is the preparation of a written health training program and evaluation plan. The information collected as described above will be used to prepare and update this plan.

A third important area is qualitative evaluation. This evaluation does not rely on facts and figures, but is based on the evaluator's impressions, feelings, and observations of the group process. Peers, employees, and instructors should be given an opportunity to provide a qualitative impression of the educational program. When this information is compared to the hard data of questionnaires, peer review, and audit data, one can get a full picture of the impact of the program.

Evaluating Changes In Employee Behavior

The described types of evaluation are short-term ways to review the success or failure of a program. However, it is the goal of a health education program to influence behavior of employees in the long term as well. You want to determine, for example, whether changes in work practices or use of such protective equipment as respirators reduces health risks in the future. It is important to plan ahead how you will implement these evaluation techniques so as to achieve these goals. Refer to Table 16.4 for a complete listing of these techniques.

The first and most obvious method for making such an evaluation is to monitor and observe the level of compliance on the part of employees. This can be done by using recorded spot checks made by safety and health personnel or first-line supervisors. This methodology, however, has many problems. Such activity is often perceived by the employee as punitive. Another problem to consider is that employees' awareness that this type of evaluation is taking place may itself affect behavior and, therefore, not be representative of the overall practices of the employee or the department.

A second type of evaluation to estimate changes in employee behavior is skills testing, which is most commonly done in a pretest, posttest format. Employees must be pretested to determine a base line measure of understanding of required behavior. Once the training is completed, a posttest is then administered to determine if understanding of requirements for proper compliance has increased. Once again, this method has problems. The most important one is that the pretest posttest format measures the level of understanding of a problem but *not* the performance. Employees might very well know that they are to use a respirator in a given circumstance; however, when the time comes to use it, they may be unable to, for a variety of reasons. As mentioned before, availability of equipment and attitudes of foremen and others in the facility are just two of the impediments to proper behavior.

Another evaluation technique uses proxy or unobtrusive measures to determine changes in employee behavior and compliance with proper work practices. These measures are intended to identify activities that represent changes in behavior but do not require direct observation or testing of the behaviors themselves. An example of a proxy measure, as shown in Table 16.4, would be the use of personal protective equipment. This type of measure works in the following way. In a department where respirator usage was required, the use of disposable pieces of equipment, such as filters, would be counted before and after training. If there was an increased use of filters, someone would be able to document, by proxy, that change in behavior had occurred. A similar example is the use of disposable earplugs. If the use of hearing protection was the behavior that the program was trying to influence, then increased use of disposable plugs would indicate increased compliance.

Another type of proxy measure often used in industry is physical measures such as blood lead levels, pulmonary function testing, and allergy testing. When coordinated with a medical surveillance pro-

TABLE 16.4. Possible Evaluation Techniques

Type	Database	Results
	BEHAVIORAL CHANGES EVALUATION	
Observation	Recorded spot checks of employee behavior	Level of compliance
Skills testing	Pretest (baseline) Posttest (changes)	Level of knowledge of proper procedures
Proxy measures	Pre- and post-training measures of: Use of protective equipment Physical measures such as blood lead levels	Indirect measure of behavior

gram, such measures can be of use in evaluation. Because this type of testing is an indirect measure of behavior, the tests must be done periodically. When done in this manner, test results can monitor decreases in abnormal physical changes after education takes place. This, of course, assumes that baseline measurements have been established prior to training (Musselman, January, 1984).

16.6.9 Specifying Feedback Measures To Be Used

In addition to evaluating the success of your program and eliciting opinions about it from the participants, it is also important to understand that training alone is not adequate to establish an effective health and safety program. Feedback regarding proper practices and/or rewards and incentives for preventive behavior must be an integral part of any kind of health and safety training program.

At a management level, the most effective programs are ones where the judgments about the success and failure of health and safety programs become part of the performance appraisal and salary review process. For a manager or supervisor, health and safety programs will be taken most seriously when they become part of the bottom line. In other words, persons who know that future raises and promotions will be based in part on how they implement safety and health programs in their department are more likely to implement them in a serious and consistent fashion.

Employee cooperation and compliance can also be influenced by feedback measures. Repeat training is a strong feedback measure, because it allows time for employees to discuss the problems that they have had implementing the information and skills they have learned. Verbal and written recognition for positive departmental performance is a traditional method of feedback used in industry. Prizes and rewards for individuals or departments has also been used as a feedback measure. These programs can be successful if they are administered fairly and consistently. However, the best feedback measures are ones that take a problem solving approach. When an employee identifies a problem, quick response by management will serve as very strong reinforcement and feedback that the company takes health and safety seriously. When an employee feels the company is ignoring a problem rather than solving it, the clear message is that the employer is not interested in health and safety.

16.6.10 Writing A Formal Training Plan

Under the new OSHA *Hazard Communication Standard,* a formal document called the Hazard Communication Plan is required for regulatory compliance. It is wise to create such a document for your records that covers *all* of your health and safety training. This document must be available for employees and managers, as required by the *Hazard Communication Standard,* and also can be used when the OSHA inspector or the State Department of Labor inspector or your insurance carrier wants to know how you are training your employees. Since training is becoming more and more of a requirement, such a document will serve as a handy reference.

16.7 SUMMARY AND CONCLUSION

In addition to health and safety programs, you might want to consider some additional programs. Wellness and employee assistance programs also provide employees with an indication of the seriousness of your company about protecting their health and safety. Some smaller companies may not have such

TABLE 16.5. Requirements of an Effective Training Program

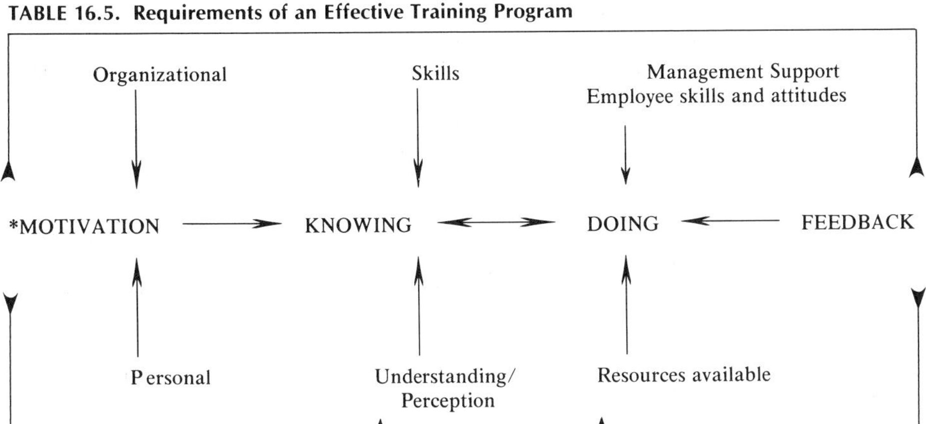

program or there may be no resources to implement them. However, local companies can team together to provide wellness and employee assistance programs for their employees through local agencies such as the YMCA.

In closing, the most important thing to understand is that motivation, cooperation, and compliance are directly related to the attitudes of all company personnel from the president on down. Table 16.5 describes this relationship. If there is no commitment to a strong program, no amount of training will generate cooperation and compliance. If that commitment exists, and the resources to implement it are made available, a strong health and safety program will foster cooperation and compliance at your location. If the employees, either salaried or hourly, understand that the company cares about the health of its employees and has an interest in protecting their health, you will have gone a long way to implementing an effective program.

REFERENCES

Becher, M. H., Ed., The Health Belief Model and Personal Health Behavior, *Health Education Monographs,* Vol. 2, No. 4, Winter, 1974.

Chhokar, J. S., and Wallin, J. A., Improving Safety Through Applied Behavior Analysis, *Journal of Safety Research,* Vol. 15, pp. 141–151, 1984.

Cohen, A., Behavioral Approaches to Personal Protective Equipment Usage, *Symposium on Occupational Safety Research and Education,* U.S. Dept. of Health and Human Services, Public Health Service, Centers for Disease Control, NIOSH, February, 1982.

Cohen, A., Colligan, M. J., and Berger, P., Psychology in Health Risk Messages for Workers, *Journal of Occupational Medicine,* Vol. 27, No. 8, 1985.

Cohen, H. H. and Jensen, R. C., Measuring the Effectiveness of an Industrial Lift Truck Safety Training Program, *Journal of Safety Research,* Vol. 15, No. 3.

EPA, *Significant New Use Rule on Premanufacturing,* Notice 83-394, December 19, 1983, 48IR57441.

EPA, *Significant New Use Rule on Premanufacturing,* Notice 83-23 and 83-24, 83-75, January 1984, 49FR99.

Feuer, D., Training at Three Mile Island: Six Years Later, *Training,* March, 1985.

Fitch, H. G., Hermann, J., and Hopkins, B. L., Safe and Unsafe Behavior and Its Modification, *Journal of Occupational Medicine,* Vol. 18, No. 9, September, 1976.

Hopkins, B. L., *Behavioral Procedures for Reducing Worker Exposure to Carcinogens,* National Institute for Occupational Safety and Health, Cincinnati, OH, 1982.

Janz, N. K. and Becher, M., The Health Belief Model: A Decade Later, *Health Education Quarterly,* Vol. 11, No. 1, 1984.

Komaki, J., Barwick, K. D., and Scott, L. R., A Behavioral Approach to Occupational Safety: Pinpointing and Reinforcing Safe Performance in a Food Manufacturing Plant, *Journal of Applied Psychology,* Vol. 63, No. 4, pp. 434–445.

Michalak, D. F. and Yager, E. G., *Making the Training Process Work,* Harper and Row, New York, 1979.

Musselman, V. C., Health and Safety Training: Impact on the Behavioral Bottom Line, *National Safety News,* January, 1984.

Musselman, V. C. Do Your Employees Have the Right-to-Know? *National Safety News,* June, 1984.

OSHA, *Hazard Communication Standard,* Federal Register, November 27, 1984.

OSHA, *Training Requirements in OSHA Standards and Training Guidelines,* Washington, D.C. OSHA, 1985.

Re Velle, J. B., Safety Training Methods, Wiley, New York, 1980.

Smith, M., Cohen, H., Cohen, A., and Cleveland, R., Characteristics of Successful Safety Programs, *Journal of Safety Research,* Vol. 10, No. 1, Spring, 1978.

UAW's Right-to-Know Breakthrough, *Chemical Week,* January 23, 1985.

Vojtecky, M. A., Workplace Health Education, Principles in Practice, *Journal of Occupational Medicine,* Vol. 27, No. 1, January, 1985.

Vojtecky, M. A., Workplace Health Education: Who Should Educate?, *American Journal of Industrial Medicine,* 1985, Vol. 7, pp. 81–85.

Voluntary Training Guidelines, Notice, Federal Register, Friday, July 27, 1984.

Zofar, D. Safety Climate in Industrial Organizations: Theoretical and Applied Implications, *Journal of Applied Psychology,* 1980, Vol. 65, No. 1, 96-102.

PART VII

ANALYTICAL TOOLS

CHAPTER 17

How to Apply Simulation Modeling and Analysis to Occupational Safety and Health

MAHMOUD A. AYOUB

North Carolina State University
Raleigh, North Carolina

17.1 INTRODUCTION

Safety management can be described as an approach to hazard control that emphasizes behavioral change. This traditional definition is based on the questionable assumption that hazard control should be achieved at any cost and by any means, and so is in clear defiance of the compelling limitations of resources and technology. In an era marked by increasing concern for efficient utilization of resources, coupled with a cost-conscious industry and public, hazard control has to be predicted on optimum allocation of resources. Methods and techniques of operations research and management science can and should be used to deal with the multifaceted problems of allocation and scheduling of resources and evaluation of competitive alternatives.

Most problems of allocating safety resources can be formulated as optimization models of one type or another—linear and nonlinear models, inventory models, queueing models, economic decision models, and so on. All optimization models share common characteristics in that the form of each is well known and basically independent of the type of problem or situation being studied. In contrast, the content (decision variables, parameters, constants, etc.) of each model varies from application to application, from modeler to modeler. Optimization models have been in use in a variety of situations for years, and there is now available a collection of efficient, as well as sufficient, computerized algorithms that can be used by those whose interests lie in the utilization of the approach rather than in the development and enhancement of algorithms. For illustrative applications of some of these models, see Ayoub (1979).

The counterpart of optimization models, simulation modeling, is an abstraction of reality, either in every aspect or in a finite number of selected dimensions. Each model is highly dependent on the purpose of the simulation study and the availability and quality of the data used in building the model. Simulation studies use models for analyses and for making predictions. There are two types of simulations: physical simulation and computer simulation. The former has been around for years and has its roots in many engineering design and research projects, for example, wind tunnels, detailed models of dams, and irrigation systems. Computer simulation is a relatively new development in the state of the art, and its history parallels that of computer technology. The concept of simulation (namely, Monte Carlo) has been around for some time, but has become a viable tool for research only through the availability of computers and simulation languages.

Simulation models are usually hybrids of mathematical expressions and decision rules that prescribe the course of action to be taken upon the occurrence of certain events. In some cases, simulation models can be built in such a way as to utilize an optimization algorithm as an integral part of the model. The presence of such an algorithm can assure that every time a decision has to be made within the simulation environment, an optimum decision will be reached. Simulation, by its nature, is more amenable to online interactive modeling approaches. A limiting factor in using modeling for studying systems heavily dominated by behavioral components rests with the difficulty of quantifying the decision-making process. Many worthwhile modeling studies were aborted because of this. Current simulation technology makes it possible to overcome this drawback by incorporating the user as an "active component of the model," with no attempt to describe what a person (user) will do under this or that condition. Instead, the person is presented with a summary or global view of the prevailing conditions in the system, and then is asked to suggest a course of action or actions to be followed during the next time period. In this fashion, several decision makers can be incorporated in the simulation model.

17.2 SIMULATION MODELING AND ANALYSIS

The development and implementation of a large-scale simulation model is a process that encompasses a set of well-defined and interrelated activities: (1) model specification, (2) model design and architecture, (3) data gatherizing and categorization, (4) model algorithms and programming, (5) model testing, (6) model documentation and (7) model implementation. Each of these is briefly described below. Various aspects of coordinating and managing the simulation project are also given. Finally, the design of simulation experiments is introduced and discussed. For detailed coverage of these as well as other related topics, Emshoft and Sisson (1970), Shannon (1975), Fishman (1978), and Pritsker and Pegden (1979) are quite useful.

17.2.1 Model Specification

A simulation model, regardless of its level of sophistication, will ultimately be used to serve management; thus, management's support and needs have to be established at the outset and nursed throughout the development process. Many excellent and worthwhile models have been shelved because management failed to identify with them or to perceive their true value and potential. Management can be offered an active role in the process of model development through a number of mechanisms: personal interviews, reviews and surveys of typical management decision problems, and workshops where illustrative examples and successful simulation applications are presented and managers' reactions and comments are solicited. Several iterations may be required before management can reach a consensus on what is expected from the model and on the operational requirements for future maintenance and expansion of the model. Management involvement should not end with the definition of model objectives and limitations but rather should be extended and applied at various levels to other development activities.

An organization may elect to employ a consultant or group of consultants to carry out the development process. In this context, the organization assumes the role of an observer. Accordingly, when the model is developed and implemented, the organization may not have the in-house capability to work with model expansion and modification—events that are certain to occur. If staff expertise to respond quickly to management requests is lacking, the credibility and usefulness of the simulation model will be affected, and it may be abandoned. To overcome some of these difficulties, the organization may elect to build in-house expertise in simulation modeling. In addition, consultants may be utilized to assure that the model will be developed on a sound basis and in accordance with the best technical standards currently available. The point to remember is that a simulation model is not developed once and for all, but rather is continually modified and changed to meet demands of the changing environment and the characteristics of the system it portrays. For this, an in-house capability to maintain and modify the system is a must.

In many applications, a single simulation model may be too restrictive to meet the demands of varying degrees of details and exactness. Because of this, it is not uncommon to develop a family of models to deal with various scenarios, or "what if" situations. For example, top management interested in defining general policies and procedures would need the model to represent general characteristics of the organization. On the other hand, for an operational manager concerned about meeting schedules and targeted delivery dates, a model reflecting the detailed movement of products, status of machines and work schedules would be required. These two models would consider the organization from two different angles; however, they have one thing in common: both use the same database and reflect the same organizational structure and characteristics.

How many models should an organization have? The answer can be obvious and straightforward, for instance, one or two. This is especially true in cases where the decision needs of the organization are well defined and clearly stated. In cases where it is difficult to delineate the role to be assumed for the potential model, the developer may be forced to perform a feasibility study and develop some prototype models. These models would then be used to arrive at the specific family of models suitable for the overall needs of the organization. In any event, in the course of defining the architecture of the simulation model or models, the developer should design models that can meet existing demands and yet be flexible enough to meet future demands as well.

17.2.2 Model Design and Architecture

A simulation model is a collection of components woven together through a host of relationships and decision rules. Each component is assumed to represent a major function or activity of the system being studied. In the model, a component is defined in terms of input, output, environment, and decision rules. Input is simply the flow the component receives from other components within the model. The flow might be information, data, or resources. Output is the response of the component to the input given. This might be initiation of an activity (task), termination or cancellation of activities in progress, or change of certain operating rules of some other components of the model. Environment is taken to mean all the internal and external conditions that might influence the functioning of the component, that is, responding to a given level and type of input. The environment for a given component is generally defined as a combination of two things: (1) rules and policies that can be specified a priori and, as such, assumed to be independent of the model performance, and (2) events that can occur due to interaction among model components. The first remains fixed throughout a simulation run, while the second requires a choice of action(s) to be taken upon the occurrence of certain events.

Relating input, output, and environment can give us the decision rule for the component. The environment of a given component can be described by decision trees when the number of conditions to be considered is relatively small, or by decision tables when there are several conditions to be dealt with. To define model components, three primary areas of the system should be investigated in detail: organizational structure; characteristics of each function (input, output, resources, information flow, interactions with other functions); and decision points that control the flow (information, resources, materials, etc.) within the organization. Traditional techniques such as task analysis and studies of process flow (flow charts) can be used effectively to develop the model architecture (Gane and Sarson, 1979).

Architecture can be defined in terms of either assessment models or decision models. Assessment models require no decisions to be made within the simulation environment. The simulation is used simply to create random events, and then to manipulate the resulting individual values to compute an overall response for each time period. Decision models can be viewed as a set of relationships and decision rules that direct flow through model components. Decision models vary widely in the methods by which decisions concerning the flow are made. For example, the flow may be controlled by a schedule (given a priori) that details what actions should be taken upon the occurrence of certain events. In contrast, for some models decisions may be reached based on the global status of the systems they portray. Thus the flow is no longer controlled by just the type and nature of events occurring at a given point, but by events within and among all model components. In this context, success or failure of past decisions will influence future action and choices; this is an adaptive model. Furthermore, decisions at any point can be influenced to a large extent by events and happenings outside the defined system boundaries and domain—government regulations, pressure of labor unions, special interest groups, state of the economy, and so on.

For some applications, the question of resource availability will dictate the actions to be taken. For example, in a model reported for fire-fighting studies (Rider et al., 1975), responding to fire calls is a decision based on the status of the system (number of units unavailable—out on call or out of service) and the type and size of reported fires. If no units are available, no action will be implemented and no decision can be reached. Delays in reaching decisions can have a severe impact on system performance and its overall effectiveness. Such a decision model, which incorporates resource limitations, is by far the most complex and, perhaps, the one that comes closest to depicting real systems.

In some cases, simulation models are used to test system designs under dynamic conditions. Consider the case of designing large control panels typical of chemical plants, oil refineries, and nuclear reactors. Any of these panels carries hundreds or perhaps thousands of instruments (e.g., displays) used to monitor and control the performance of various processes in a given plant. The safe performance of a plant depends on the ability of the operator (or operators) to detect changes in system status and to initiate an appropriate and timely response to the perceived signal. To assure safe performance, the operator's response is controlled through design considerations and work assignments. However, basically panels are designed on the assumption that each operator will be working under normal operating conditions, and accordingly, no consideration (at least in an explicit sense) is given as to how the operator's performance will be affected under stress (e.g., the fear of a melt-down in a nuclear reactor). Consequently, costly mistakes and omissions in panel designs may be avoided through simulation modeling and analysis. Through simulation, a design configuration can be tested and evaluated under various levels of stressful situations and emergency conditions (for an example see Pulat, 1980). The results of such tests would define design changes to be implemented to assure that operators' performance will be within acceptable (safe) limits.

17.2.3 User Interface with the Model

In cases where more than one model is required, attempts should be made to have the users interface with the models through a single input routine. This routine would aid the user in selecting the appropriate model and prepare the user's data (given unstructured) to match the input specifications of the selected model. To facilitate the production of custom-tailored reports and one-of-a-kind analyses, the user should have a special output routine. This can simply be a report writer that can extract from standard simulation output data the special reports and analyses required by the user. Some special applications may call for the use of an optimization model in concert with, as well as in support of, the simulation model. In such cases, plans should be drawn up specifying how the models will be linked together, how the information will flow between the models, and what control the user would exercise over the two models.

17.2.4 Data Gathering and Classification

Data required to define model structure and interrelationships among model components can be collected in a number of ways. It can be extracted from the existing organization database, obtained by monitoring the performance of some key operations and functions over a reasonable period of time, or by interviewing personnel of various functions to obtain their best estimates of various types of performance data and hierachy of operations. When data are obtained through interviews and personal estimates, a technique like the Delphi method (Linstone and Turoff, 1975) can be used to add some uniformity and perhaps credibility to the outcome.

Following collection, data have to be categorized and described in terms of some appropriate probability distributions. The distribution that best fits the data should be selected. Standard goodness of fit tests can be used for such purposes (Phillips, 1972). If the standard distributions (e.g., normal, log normal, and Poisson) are not suitable, then empirical table-lookups and the like can be used to define model data.

In addition to performance data, several other pieces of data will be required to define the model completely. These might include work and information flow procedures, physical and technological requirements for various production functions, and overall regulatory standards that must be met by the organization.

The data types discussed above can be considered as generic to the simulation model, and accordingly, have to be provided independent of model users. Some of the data may be considered as fixed, while others may be subject to change due such factors as modification of the operational environment, market expansion, introduction of new regulatory requirements, and so on. Plans and procedures should be developed to assure that model generic data will be up to date as well as reflective of the organization environment.

17.2.5 Model Algorithm

Once the architecture of the simulation model has been defined and detailed, attention is next focused on the task of selecting a simulation algorithm for model implementation. A simulation algorithm should entail the following: (1) generation of random numbers, (2) sampling from probability as well as empirical distributions, (3) keeping a simulation clock, (4) scheduling of model tasks and activities, and (5) collecting statistical data on key model events, resources, and activities.

In selecting an algorithm, one of two alternatives can be pursued: custom tailoring an algorithm using one of the available high level computer languages such as FORTRAN, PL/I, or APL; or using one of the existing simulation packages such as GPSS, SIMSCRIPT, GASP, GERT, or DYNAMO. Each of these alternatives has its pluses and minuses with regard to development time, skills required, and reliability of the algorithm developed. However, using a simulation package may prove to be the best course of action on the basis of cost and development time. In deciding among competing packages, the following should be taken into consideration: (1) error checking capability, (2) level of skill required before it can be effectively utilized, (3) ease of adaptation to meet the specifications defined for the model, (4) hardware requirements for package implementation, (5) degree of maintenance and support provided by the package developers, and (6) type and level of documentation given for the package.

In addition to the technical considerations above, the cost and the performance history of the package should bear on the final decision. Shannon (1975) gives a rather detailed and excellent review of some of the primary packages and languages. He also details an approach that can be used to select the package (language) that best suits a given application.

17.2.6 Testing

A large scale simulation model encompasses many components, data files, and decision rules. It represents the designer's abstraction of the important and key aspects of the system under consideration. The value of such a model lies in its ability to predict accurately the behavior of the system it portrays. Testing a simulation model may be taken as determining how close the model comes to predicting known performance. This can be accomplished by examining model performance in dealing with diverse decision problems encountered in the past. Testing the model in its entirety—components, data elements, structure, and so on—is usually presented in the literature as a criterion-related validity test. It measures how accurately the model predicts the performance, as well as the response of the system for a given set of input data and policy conditions. In using criterion related validity, no attempt is made to establish or define the accuracy of any specific component of the model, but rather to define the accuracy of total performance and output.

Other types of tests include content validity tests and face validity tests. A content test seeks to establish that the model components, structure, relationships, and so on are valid and accurate presentations of the system modeled. On the other hand, face validity tests deal with the question of relevancy: Do the model and its output show a high degree of association with the system and its operation? For the model to have validity, the potential user must be able to associate model performance with expected system performance. To define content and criterion related validity tests, a host of statistical tests can be used, test of means, test of variances, chi-square tests, and so forth.

17.2.7 Documentation

One of the important tasks of a simulation project is that of model documentation and development of user manuals. Unfortunately, in many studies, it is perhaps the one task that receives little attention and, in many cases, is done in haste and as an afterthought. If a simulation model is to be successful, it has to be understood and has to be reached by the potential users. For such reasons, proper and sufficient documentation has to be provided. Two types of documentation accompany a simulation model: (1) design manuals to cover model architecture, algorithms, general data, and so soon; (2) manuals to allow potential users to utilize the model effectively, to detail and interpret the output reports, and to provide the user with a range of illustrative examples of possible model applications.

These manuals should be designed and developed as the simulation project unfolds, and certainly should not be worked on until the model is developed and tested. For user manuals, the background and skill of the potential users will definitely dictate the level and type of coverage, the languages used, and so on. By planning for the manuals early in the development process, user needs will be properly accounted for and reflected in model specification and design, thus eliminating last minute surprises and perhaps confusion and disappointment.

17.2.8 Implementation

With the model fully developed, documented, and proved accurate, implementation is the next and final task of the simulation project. Implementation encompasses two major activities: (1) introducing the model and its supporting manuals to the potential users, and (2) monitoring model performance and user reaction. The model can best be introduced through workshops and demonstration projects geared to the needs of various departments or management groups within the organization. A model can be excellent in its design and documentation, but it may fail to gain user acceptance and support. Records should be kept on user interaction with the model, problems encountered either in preparation of input data or interpretation of output reports, and circumstances where the model failed to respond to specific needs. Examination of these records will yield some insight as to what changes and modifications, if any, might be necessary to enhance the user's interface with the model.

17.2.9 Project Management

The various tasks associated with a simuation project have to be coordinated and scheduled in a manner that will best utilize the organization resources and assure completion within a given time frame. The failure of many simulation projects is due to the fact that they tend to be treated as academic exercises where most of the research work is carried out by precious few. Thus, personal biases tend to color the direction of the research (the tasks to be started, the time allocated to any given task, sequencing of different tasks, etc.).

For industry or large-scale studies, a structured approach is a must. A project manager may be assigned the responsibility of overseeing and coordinating the work of a number of researchers (system designers, programmers, manual writers, and workshop instructors, outside consultants, etc.) assigned to the different tasks. The project manager, in turn, reports to a steering committee assembled from the various potential user groups within the organization. Having such a committee will assure that user interests are protected throughout the various phases of model development and implementation. Furthermore, it will eliminate the climate of apprehension and suspicion usually accorded new projects or undertakings by making the users part of the process itself. Upon implementation, model maintenance and upkeep should be assigned to an individual who is thoroughly familiar with the intricacies and details of the project. The selected individual will, in effect, act as the model manager and will be in a position to respond to user demand and needs.

For small applications and feasibility studies, other project management approaches can be used. However, regardless of the approach used, some degree of control and accountability over the project should be exercised; otherwise, the outcome is likely to be less than acceptable, if not disastrous.

17.2.10 Simulation Experiments

A fully implemented simulation model offers the decision maker the laboratory by which a host of studies can be performed. Any decision contemplated by the organization might be carried out by changes in utilization of resources, adoption of new policies and procedures, or restructuring some functions within the organization. All of these can be manipulated and defined by the decision maker at will; the question is which one (or ones) would have a significant impact on the overall performance.

Suppose that the performance of the system (organization) is lagging behind expectation in one or more areas. Relying on knowledge of and past experience with the system, a set of corrective actions can be defined separately or collectively to bring the system back to an acceptable performance level. Now the questions are: What corrective actions should be implemented? At what level should those selected be implemented? To answer these effectively, an experiment is needed to identify the action(s) that will significantly improve the system performance. In carrying out an experiment in accordance with a given design and structure, the steps involved are:

1. Select the independent variables to be considered, that is, the corrective actions.
2. For each variable, determine the levels to be assigned during the experiment. A variable such as age may assume an infinite number of levels. However, in the interest of economy, only those levels that would be practical as well as meaningful are considered (e.g., three age groups may be sufficient). In general, two to three levels for each variable are adequate.
3. Design the experiment. There are two basic design alternatives. *Complete factorial design* means that the performance of the system (response) is measured for all possible combinations of variable levels. In contrast, *response surface design* is basically a search technique that attempts to locate the levels that produce maximum system response using only a few combinations of variable levels. Within each design type, there exist many variations and alternatives for conducting the experiment and the corresponding data analysis. Choosing one design over another is basically a function of the number of possible combinations that will be tested and the statistical method that will be used to categorize and analyze the experimental data. Both of these will influence greatly the computer time required for running the experiment.

4. Use the simulation model to predict the system response for each experimental condition (specific combinations of variable levels). Responses corresponding to all experimental conditions constitute the data to be analyzed. In collecting the data, issues of sample size and number of replications have to be considered.
5. Finally, analyze the experimental data to arrive at the proper conclusions concerning the variables and their contributions toward the overall performance of the system. To this end, many computerized statistical systems do exist and, indeed, offer a host of analysis techniques to meet a multitude of experimental designs, SAS (1981) and SPSS/PC (1984) are just two examples of widely used systems. In addition, many statistical textbooks offering full treatment of the subject can be consulted for further details (e.g., Anderson and McLean, 1974).

17.2.11 Cost of Simulation

Large-scale simulation models can be quite expensive, since they require a large financial commitment, first for their development and then for their maintenance. Before commencing a simulation study, its cost should be clearly defined. The estimated cost of the study should include development of software (model specification, model design, programming, and testing); hardware requirements (acquisition of special equipment, modification of existing systems, etc.) and staffing for model maintenance and user services.

The total cost of the simulation model should be contrasted with the cost the organization is willing to incur in finding out the implications and consequences of various strategic and planning decisions, that is, the value the organization places on the information to be generated by the model for a given decision scenario. In almost all cases, the benefits of having the simulation model offset its cost many times over.

Starting a simulation study does not necessarily mean it will be concluded successfully—the model fully developed and implemented. The risk associated with the development process must be appreciated and accepted. It is possible that after the specifications are defined, it may be found infeasible (for technical, political, or other reasons) to pursue the simulation further. The chance of this happening is quite good, especially in the early stages of development. Fortunately, at that point the cost is typically not very great and can be sustained without sending shock waves through an organization's accounting system. However, it takes a disciplined and experienced modeler to terminate the simulation at the right time; waiting too long may make it difficult to scrap the simulation project and to write it off as a small loss to the organization. Therefore, as pointed out previously, the need for detailed and periodic reviews of all aspects of the simulation study cannot be overemphasized.

17.3 SIMULATION BASICS

Simulation has been defined as the process of recreating data. It has also been said that a simulation model is a collection of components possessing various relationships. Therefore, as the simulation proceeds, many data points will be generated. Individually, each point can be attributed to a specific source with known characteristics (e.g., a well-defined probability distribution). However, at any time and for a given distribution, the specific data point to be recreated cannot be predicted (i.e., its occurrence is totally random). It follows that when several data points are recreated, each from its own source, a host of data profiles may be realized. Each of these profiles, in turn, would have its own impact (positive or negative) on the system being modeled. In sum, although the sources and the basic simulation data points are known fully, the values obtained from them (individually or collectively) are random; accordingly, elicited responses from the system cannot be given a priori. In this last statement lies the power of simulation modeling—its basic premise of starting with the known and then being able to project what will be!

17.3.1 Monte Carlo Sampling

Recreation of data is fundamental to all simulation algorithms, as well as to fully developed simulation languages. Traditionally, the process is given the catchall name of "Monte Carlo." In less exotic words, it can be simply called "random sampling" (simulated from theoretical and empirical distributions). The process is rather simple and straightforward, and involves the components in the next sections.

17.3.2 Generation of Random Numbers

A sequence of numbers, each of which has an equal chance of occuring at any place in the sequence, is called random. In addition, successive numbers in the sequence are independent. The roulette wheel in a gambling casino produces such numbers, assuming of course, that the wheel in honest. Each number on the wheel has the same chance of being selected as the winning number; the occurrence of any number does not change or influence the outcome of successive spins. Random numbers have many uses and can be found in practically every textbook on statistics, management science, and operations research. Table 17.1 is an example of a collection of four-digit random numbers.

For simulation modeling, random numbers are generated by employing mathematical expressions known as pseudonumbers (Graybeal and Pooch, 1980). These are generated by starting with a first number (usually referred to as a random number seed) and then through mathematical manipulation (squaring, a combination of multiplication and addition, etc.) a next random number is generated. The new number is then used as its predecessor to generate the next random number. The process continues in this fashion; each repetition yields a new number.

17.3.3 Data Categorization: Cumulative Probability Distribution Function

Data describing various aspects of a system and its components have to be categorized somewhat descriptively. Histograms and frequency tables are two examples of data categorization and summarization. If feasible, another type of data categorization is achieved through the use of such standard probability distributions as normal, Poisson, Erland, triangular. Choosing from among these distributions depends on the data and its characteristics.

Goodness of fit tests can be used to aid in selecting the proper distribution.

Following data categorization, cumulative distributions are then developed. Each cumulative distribution is basically a plot of the random variable X vs. the probability of the variable being X or less. The cumulative distribution function is usually denoted $F(X)$.

EXAMPLE 17.1

Assume that for a random variable, X, we have

X	$P(X)$
1	.1
2	.2
3	.3
4	.15
5	.05

Cumulative probability $F(X)$ is obtained by successively adding the probability figures $P(X)$ preceding each of the given values. The result would be as follows:

X	$P(X = a)$	$F(X) \leq P(X)$
1	.10	.10
2	.20	.30
3	.30	.60
4	.15	.75
5	.25	1.0

A plot of the cumulative distribution above is shown in Figure 17.1

The cumulative function given represents a discrete distribution, for the values of the random variable X change at discrete points. The counterparts of discrete functions are then continuous distributions similar to that of the normal probability function (see Figure 17.2).

TABLE 17.1. Four-Digit Random Numbers.[a]

5269	6867	3781	0884
5173	0108	4089	3558
1169	3049	7772	9190
1644	0384	7509	1596
9747	2344	6337	6919
6403	3730	4748	4918
7840	6510	8497	2388
4341	6063	7311	9291
5716	7602	4171	6603
0979	5202	2397	7564

[a]Table can be entered at any point. However, once an entry point is selected, it is recommended that successive numbers (especially for the same distribution) be selected in the order given.

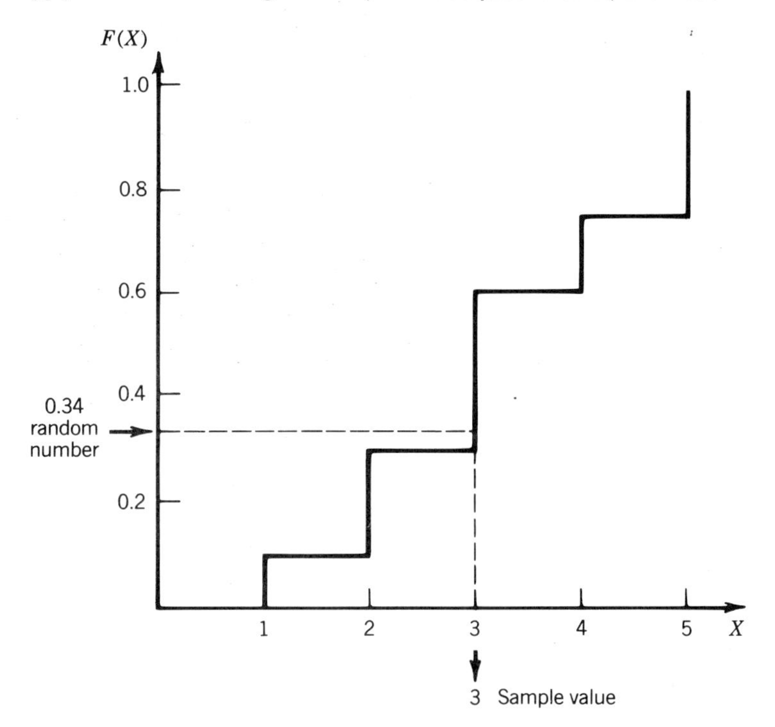

FIGURE 17.1. Cumulative distribution of a discrete function.

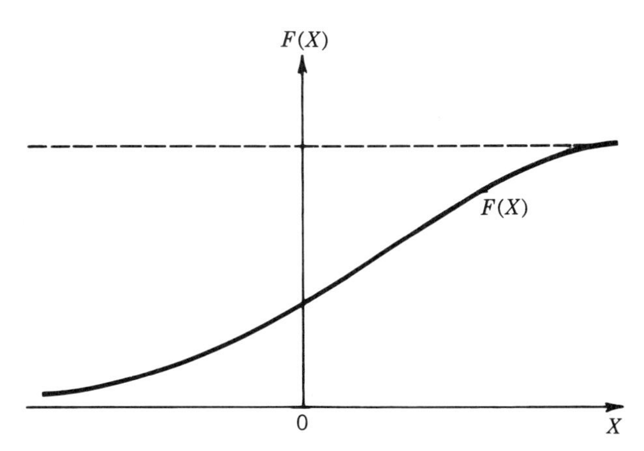

FIGURE 17.2. Cumulative distribution of a continuous function.

17.3.4 Drawing Sample Values

Having obtained some random numbers and developed the cumulation distribution function, we next proceed to draw samples of the random variable. This is done by finding the values that correspond to the given random numbers. Each value can be obtained graphically by projecting a line horizontally at the point on the ordinate $F(X)$ corresponding to the value of the random number until the line intersects the curve (for continuous distributions) or one of the staircase segments (for discrete distributions). The value of the abscissa corresponds to this intersection.

EXAMPLE 17.2

Suppose we have generated the sequence of random numbers

0.34 0.24 0.23 0.38 0.64 0.36 0.35

Entering each of these numbers on the ordinate of Figure 17.1, we generate the following set of values for the random variable X:

Random Number	Sample Values
0.34	3
0.24	2
0.23	2
0.38	3
0.64	4
0.36	3
0.35	3

The same set of sample values can also be obtained by simply writing some decision rules specifying the value (V) that corresponds to each range of random numbers (RN). Specially, for the example at hand, we can write

$$\begin{aligned}
&\text{If } && RN \leq 0.10,\ V = 1\\
&\text{If } .10 < && RN \leq 0.30,\ V = 2\\
&\text{If } .30 < && RN \leq 0.60,\ V = 3\\
&\text{If } .60 < && RN \leq 0.75,\ V = 4\\
&\text{If } && RN \leq 0.75,\ V = 5
\end{aligned}$$

Again, the two methods (graphical and decision rules) are the same; however, the latter is rather easy to implement and quite suitable for computer algorithms.

Simulated sampling from probability distributions can be extended to encompass more than one distribution. Here, we consider situations where there is a need to sample from several distributions; however, the sample obtained from one will determine how many samples to be drawn from the others.

EXAMPLE 17.3

Consider the case of simulating accidents and determinating the corresponding costs. Accidents occur with the following probability:

Accident Type	Probability
First aid	0.6
Lost time	0.1
Equipment damage	0.3

The costs associated with the various accident types are given by

1. First Aid		2. Lost Time		3. Equipment Damage	
Cost $	Probability	Cost $	Probability	Cost $	Probability
25	0.3	500	0.2	100	0.5
50	0.5	1000	0.5	200	0.3
100	0.2	2500	0.3	500	0.2

Based on past history, we anticipate that a total of 20 accidents will occur during the next period. Now, what would be the total cost for all accidents? To answer this question we proceed as before and use Monte Carlo sampling as follows:

1. Obtain sample from the accident-type distribution. This gives a particular accident type (i.e., first aid, lost time, or equipment damage).
2. For the type of accident determined in (1), compute an associated cost by sampling from the appropriate distribution.
3. Repeat (1) and (2) until a total of 20 accidents has been generated. Adding the individual costs of all accidents, obtain the total accident cost for the next period.
4. Obtaining the total cost of 20 accidents constitutes but one sample point. To have a somewhat realistic estimate, repeat the process (Steps 1–3) several times. From the estimate obtained, mean and standard deviation of the total accident cost are then computed.

Figures 17.3, 17.4, 17.5 and 17.6 give the cumulative distributions for the accident types and the three cost categories. Results of sampling from the distributions are summarized in Table 17.2. Table 17.3

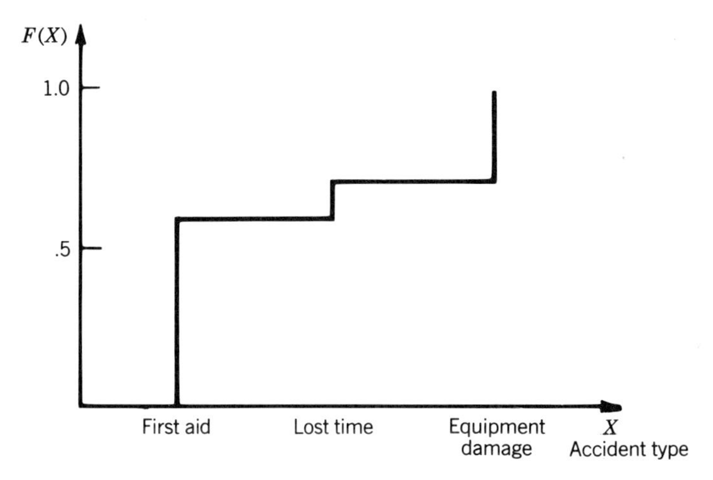

FIGURE 17.3. Distribution of accidents.

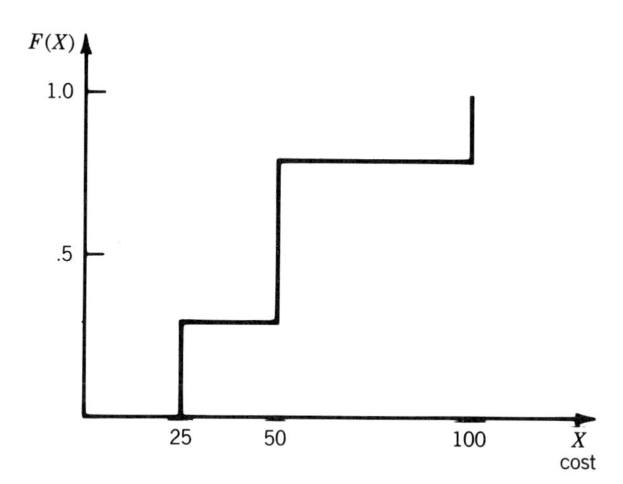

FIGURE 17.4. Distribution of first-aid costs.

shows the estimated cost for ten trials (repetitions). From the data presented, we determine that mean cost = \$7072.50 and standard deviation = \$175.74.

In a similar manner, Monte Carlo sampling can be extended to span many interrelated distributions.

By virtue of their well-defined functions, standard distributions are well suited for producing sample values through direct computation instead of graphically. For example, to obtain a sample X from a normal distribution with mean μ and standard deviation σ, we use

$$X = \mu + \sigma\,(RNN)$$

The random number (RNN) should be obtained from a table of normally distributed numbers (see Table 17.4). In the case of exponential distribution, the random sample X is given by

$$X = -\mu \ln\,(RNN)$$

where μ = mean.

RN = random number (uniformly distributed, see Table 17.1). Similar expressions can be developed for other distributions.

The procedure just described is the essence of Monte Carlo sampling. Reflection on the method shows that we have done nothing more than use data categorization in reverse. In other words, when studying a phenomenon such as the number of accidents vs. day of week, we usually start by determining

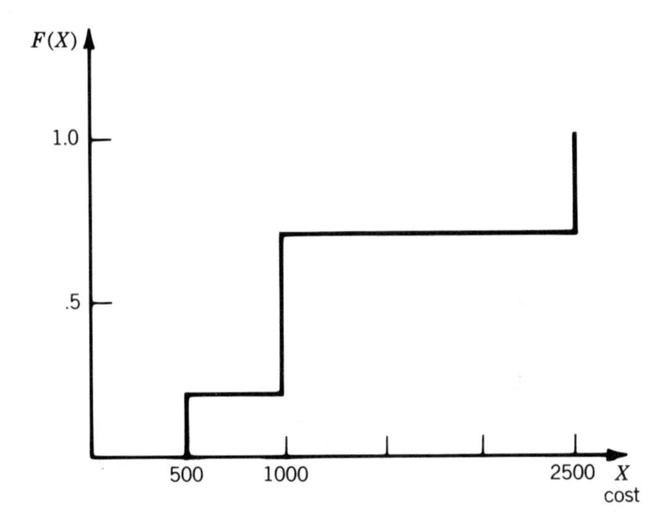

FIGURE 17.5. Distribution of lost time costs.

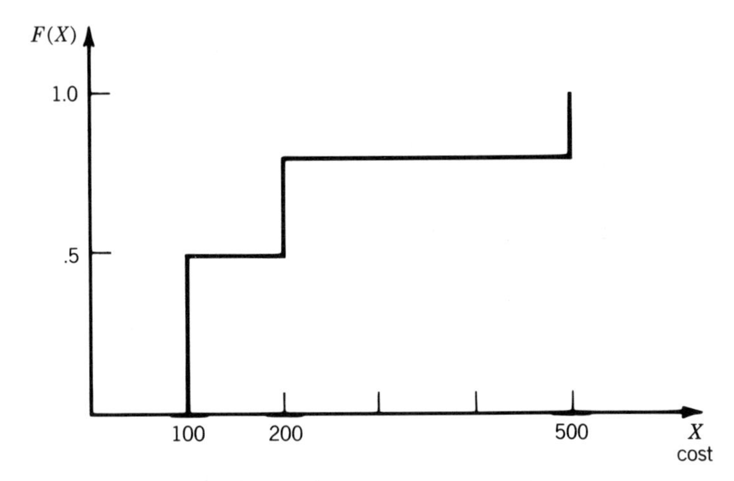

FIGURE 17.6. Distribution of equipment damage costs.

(through observations, surveys, etc.) the frequency with which a certain number of accidents occurs. From this, an appropriate probability distribution is developed.

In a simulated sampling, we reverse the roles and assume that we have a population of random numbers from which we draw a sample. To every number in the sample we assign a value. The nice thing about all of this is that to draw a sample of random numbers, we need only a table or a mathematical expression! Below are four additional examples that demonstrate how useful Monte Carlo simulation can be in dealing with some typical safety management problems.

EXAMPLE 17.4

An OSHA compliance officer is scheduled to inspect the Buoya Manufacturing Company for possible violation of the noise standard. The overall noise level in the Buoya Plant varies from hour to hour and from day to day. In general, the noise level varies with the number of machines operating simultane-

TABLE 17.2 Accident Costs for the First Simulation

	1	2	3	4	5	6	7	8	9	10	11	12	13	14	15	16	17	18	19	20
First random number	.29	.25	.26	.93	.88	.08	35	.91	.15	.69	.49	.23	.92	.66	.75	.61	.31	.37	.49	.59
Accident type[a]	1	1	1	3	3	1	1	3	1	2	1	1	3	2	3	2	1	1	1	1
Second Random number	.62	.72	.96	.71	.05	.22	.50	.68	.72	.26	.68	.81	.68	.99	.30	.75	.75	.72	.43	.84
Cost ($)	50	50	100	200	100	25	50	200	50	1000	50	100	200	2500	100	2500	50	50	50	100

[a]1 = First aid.
2 = Lost time.
3 = Equipment only.

Total cost = $7,525
Average cost = $7,025/20 = $376.25
Standard deviation = $802.07

TABLE 17.3. Total Accident Costs for Ten Simulations

1	2	3	4	5	6	7	8	9	10
7025	5000	8100	7500	6100	9000	7700	6500	7000	6800

Average total cost = \$7072.50.
Standard deviation = \$175.74.

TABLE 17.4 Random Normal Numbers $\mu = 0$, $\sigma = 1$.

−0.98	0.76	−1.22	0.99	−1.29
−0.91	0.43	−1.11	0.04	−1.96
−0.67	0.92	−0.72	−0.47	0.15
−0.79	0.66	1.70	−1.02	1.82
−0.55	−0.07	−2.08	−0.35	−0.17
−0.42	0.46	−0.61	−0.78	0.77
0.10	−0.33	−0.52	0.18	0.38
0.11	−0.65	0.30	−0.36	0.54
−1.42	1.01	0.10	−0.20	0.71
0.17	1.75	−0.86	0.93	−0.98

ously and with the type of materials being processed. From previous sound surveys, the following data are given:

Overall Noise Level dBA	Probability
88	0.4
90	0.3
92	0.2
95	0.1

What is the probability that the inspector finds the company in compliance? The inspector will base his (her) decision on the results of ten sound level readings taken throughout the plant.

Table 17.5 is used to simulate the OSHA inspector's ten noise-level readings. The assumption here is that the sound readings will be taken randomly with respect to both time and space. To be in compliance, the noise level must average less than or equal to 90 dBA.

Results of the first simulation run are shown in Table 17.6. The average values for the first five runs are 89.3, 90.1, 90.5, 90.1, and 89.5. Since two of the values are 90.0 or less, the company stands a 40% chance of being found in compliance. The standard deviation after five runs is 0.49.

After five more runs are completed, values of 91.2, 90.4, 90.5, 89.8, and 91.0 are obtained. Now there are three figures corresponding to compliance; the company's chances are 30%. The standard deviation for the ten runs is 0.61. The next five runs produce values of 89.7, 90.5, 89.8, 91.0, and 9.1.5. With a total of five values out of fifteen runs indicating compliance, our probability is now 33%. The standard deviation increased to only 0.65. Therefore, we can be fairly confident that will the company be found in compliance only about one-third of time.

EXAMPLE 17.5

A state OSHA director would like to determine an appropriate level of staffing for his inspection and compliance force. He gives the following estimates for inspection time per plant:

Time (h)	Probability
6	0.4
8	0.3
16	0.2
24	0.1

In some cases, the safety inspection is followed by an industrial hygiene survey. The probability of a plant getting both safety and hygiene inspections is 0.25. The time required to perform an industrial hygiene inspection is assumed to be normally distributed with mean = 5 hours and standard deviation = 1.5 hours.

TABLE 17.5. Basis for Noise-Level Determination

Noise Level dBA	Probability	Random Numbers
88	0.4	0–39
90	0.3	40–69
92	0.2	70–89
95	0.1	90–99

TABLE 17.6. Results of the First Simulation

Sample	Random Number	Noise Level dBA
1	50	90
2	48	90
3	28	88
4	27	88
5	34	88
6	62	90
7	27	88
8	17	88
9	94	95
10	33	88
		Average = 89.3

The OSHA director is planning to inspect 200 plants in the next six months. Existing policies limit each inspector to a maximum of 20 hours of field work per week. How many safety inspectors should be hired? How many industrial hygienists would be needed? What is the average work load per inspector?

Since the problem calls for answers based on a weekly work load, let us assume that of the 200 plants to be inspected, no more than 8 will be attempted in the 26 weeks. We can simulate this 8-plant work week until we feel we have a range of values for the required manpower.

Eight simulations were made in accordance with the procedure shown in Figure 17.7. Details of the first two runs are given in Table 17.7. The overall results of eight simulations are:

Hours/Week for Safety Inspections	Hours/Week for Industrial Hygiene Inspections
70	0
82	0
96	14
94	14
52	13
82	17
60	21
106	5

Even without further simulation runs, it seems that five safety inspections and one industrial hygiene inspector will be sufficient at 20 hours/week each. In only one of eight cases does the work load exceed this manpower level. This deficit can be countered by spare time remaining in a slack week. With this manpower allocation, the average work load can be found by averaging the simulated weekly requirements and dividing by the number of inspectors. For safety inspectors, the load is $80.25/5 = 16.05$ hours/week; the industrial hygienist will put in 11.62 inspection hours/week. The slack time could be lessened if one inspector certified for both safety and industrial hygiene was employed, leaving only five total inspectors.

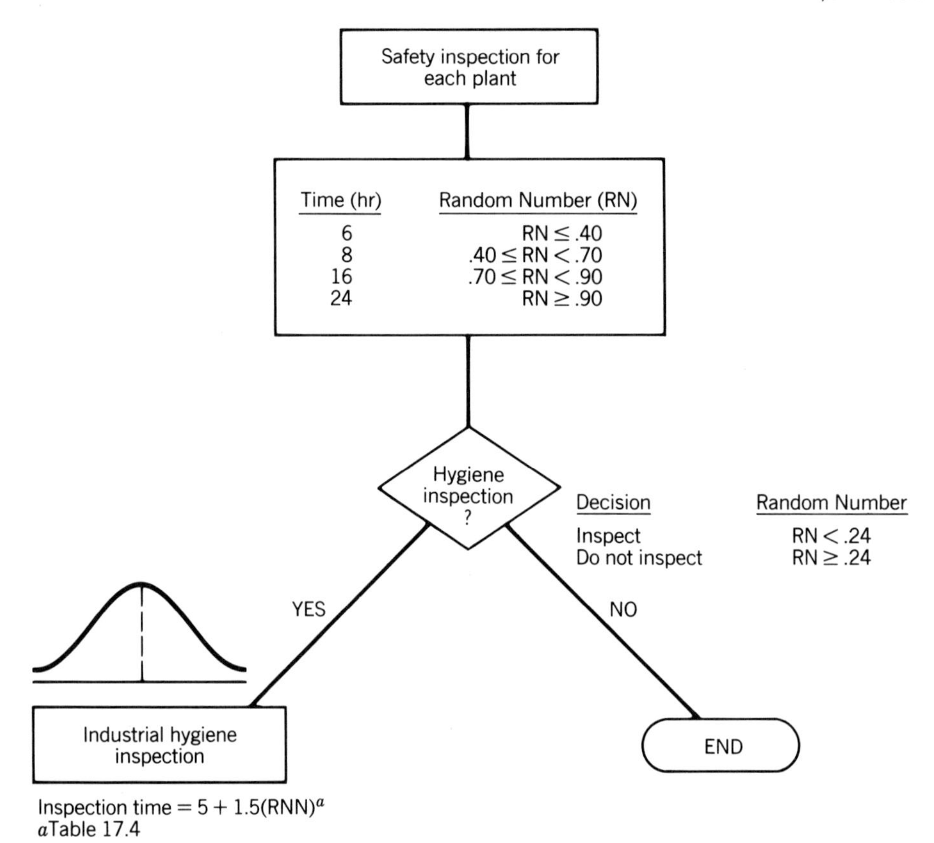

FIGURE 17.7. OSHA manpower determination.

EXAMPLE 17.6

Figure 17.8 gives distribution of noise levels throughout a manufacturing plant. Because of various work assignments, some employees are exposed to a wide variety of noise levels during any given work shift. A time study of those employees over several days provides the following:

Area	Time (h)
I	2
II	1
III	3
IV	1
Total	9

Based on the given data, what is the probability that the company can be found in compliance with the OSHA noise standard?

From the data provided, we can see how long, on the average, a worker spends in each of the areas each day. Within each area the noise levels are different. Probabilities can be assigned to each noise level in proportion to its occurrence in each of the four areas. If we assume that over a period of time a worker will stay in each noise-level area for a length of time proportional to its fractional floor space, we can perform a simulation of expected noise exposure. Estimated percentages of each noise level and their associated random numbers are shown in Table 17.8.

A worker's daily exposure to noise may be simulated by generating two random numbers each for Areas I and II, three numbers for Area III and one for Area IV.

In this manner, the noise levels are determined at intervals of one hour each. Results of the first

TABLE 17.7. Results of the First Two Simulations

Safety Inspection Plants	Random Number	Time (h)	Random Number	Decision	Random Number	Industrial Hygiene Inspection Time (h)
First Simulation Run						
1	.63	8	.44	No		
2	.58	8	.63	No		
3	.65	8	.76	No		
4	.12	6	.50	No		
5	.47	8	.49	No		
6	.40	8	.63	No		
7	.69	8	.13	Yes	0	5
8	.82	16	.14	Yes	0.66	4
Totals		70				9
Second Simulation Run						
1	.72	16	.33	No		
2	.39	6	.86	No		
3	.85	16	.37	No		
4	.30	6	.48	No		
5	.31	6	.97	No		
6	.40	8	.91	No		
7	.73	16	.33	No		
8	.53	8	.55	No		
		82				0

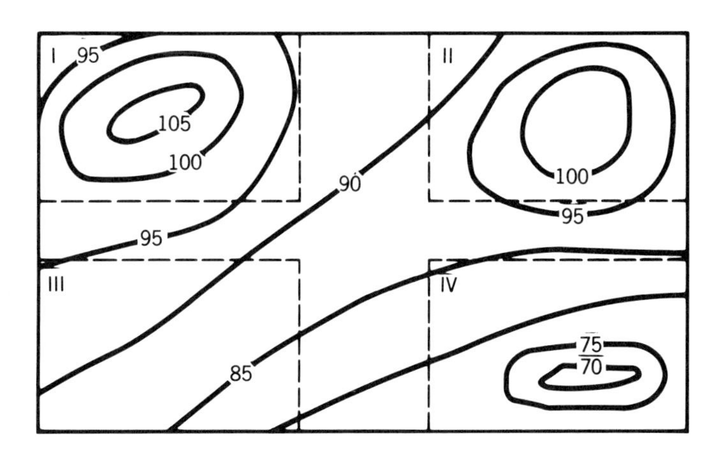

FIGURE 17.8. Noise levels in the plant, dB A.

TABLE 17.8. Estimated Occurrence of Noise Levels Within Each Area

Noise Level dBA	%	Random Number	Noise Level dBA	%	Random Number
	Area I			Area II	
90–95	20	0–19	90–95	25	0–24
95–100	30	20–49	95–100	40	25–64
100–105	35	50–84	100–105	35	65–99
>105	15	85–99			
	Area III			Area IV	
80–85	25	0–24	<70	10	0–9
85–90	40	25–64	70–75	15	10–24
90–95	35	65–99	75–80	50	25–74
			80–85	25	75–99

TABLE 17.9. Results of the First Simulation

Random Number	Noise Level Range dBA	Assumed Value dBA
	Area I	
99	>105	105
07	90–95	92
	Area II	
26	95–100	97
82	100–105	102
	Area III	
84	90–95	92
41	85–90	87
48	85–90	87
	Area IV	
46	75–80	77

simulation are given in Table 17.9. For a state of compliance to exist, the following condition should be met:

$$\Sigma \frac{C_i}{T_i} \leq 1$$

where

C_i = actual exposure time at the ith noise level.
T_i = the maximum permissible exposure time for the ith noise level. Table 17.10 defines the limiting times for various noise levels.

From Table 17.9, it is clear that the time-weighted exposure is indeed greater than 1; thus compliance is lacking. Of a total of 20 simulations, only 2 runs were found to yield a weighted exposure of less than 1. Therefore, it is safe to conclude that the probability that the company will be found in a state of compliance is 0.10—a very low figure, which should entice management into taking some drastic steps for instituting corrective measures.

EXAMPLE 17.7

An insurance company is interested in defining the premium it should charge its industrial customers for workers' compensation coverage. The company estimates that the value of all the claims to be processed in a year in normally distributed with a mean of $750,000 and standard deviation of $50,000. The company will insure 1000 clients. The overhead cost (service charge) per claim is as follows:

Cost per Claim	Probability
50% of the claim	0.60
25% of the claim	0.40

What insurance premium should the company charge each of its clients?

The basic solution approach for the problem at hand is shown in Figure 17.9. We will demonstrate the procedure by carrying out several simulation runs of ten trials each. Results of the first run are given in Figure 17.10. The total claims value would be approximately $1,053,000 annually. Each of the 1000 clients should be charged $1053.

EXAMPLE 17.8

OSHA estimates that the canning industry will incur $150 per plant in order to comply with a new standard regulating conveyors. However, because some degree of compliance with the standard can be attributed to standard industry practices, the estimated cost has to be somewhat modified. From industry data and known practices, the following can be assumed:

Existing Level of Compliance	Probability
10%	0.35
25%	0.25
75%	0.25
80%	0.25
90%	0.10

In computing the annualized compliance cost per plant, OSHA assumes an interest charge (i) of 10% with an economic life (n) of 15 years. There are many reasons for i and n to vary, for instance, the state of the economy (growth, decline, etc.). Economic forecasts provide the following:

n (years)	Probability	i(%)	Probability
10	0.5	10	0.2
15	0.3	15	0.5
20	0.2	20	0.3

What is the most likely annualized cost for the industry?

The cost of compliance depends on the existing level of compliance and its corresponding percentage of occurrence. Hence, the cost will be .90 ($150) = $135 in 35% of the plants; .75 ($150) = $112.50 in 25% of the plants, and so on.

The annualized cost is determined from the nine different combinations of i and n, which yield the capital recovery factor (A/P, $i\%$, n) and the cost of compliance (present cost, P):

$$A = P(A/P, i\%, n)$$

The three stochastic functions (cost of compliance, i, and n) can be represented as random number strings, as in the preceding examples. Three random numbers will be required to generate one value for

TABLE 17.10. Permissible Noise Exposure[a]

Duration per day (h)	Noise Level dBA
8	90
6	92
4	95
3	97
2	100
1 1/2	102
1	105
1/2	110
1/4 or less	115

[a]OSHA general industry standards (1910.95).

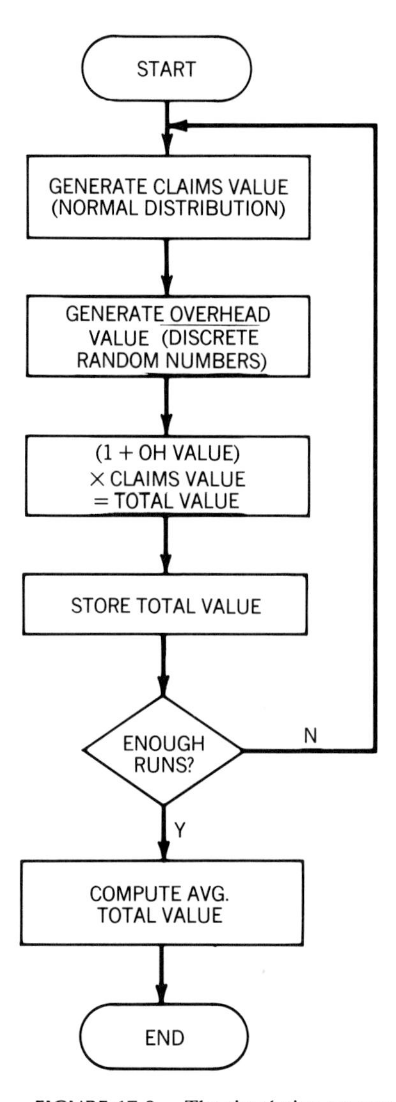

FIGURE 17.9. The simulation process.

the annualized cost of compliance. The problem model is presented in Figure 17.11. A few manual runs were performed to illustrate typical simulation results. The average annualized cost after 15 runs of $19.62 may be used as a rough estimate of the probable yearly cost to the industry.

EXAMPLE 17.9

In the maintenance department, a utility worker is assigned an area that is about 20% of the total shop area. The worker performs a variety of jobs in the assigned area: loading and unloading, cleaning, helping other workers, and so on. Throughout the shop, an overhead (bridge) crane travels from end to end for handling heavy pieces of equipment. Movement of the crane is random and in a sense cannot be predicted, for its position is controlled by the needs of the different work stations in the maintenance area. The crane is rather old and has not been kept in good working order due to poor preventive maintenance. The probability of the crane dropping its load due to chain breakage or improper loading is 0.10. What is the probability that the utility worker will be involved in a crane accident in a period of eight hours?

Normal Random Number (RNN)	Claims Value[a] (dollars)	Random Number (RN)	OH Value	Total Value
.5	775,000	.96	.25	968,750
0	750,000	.72	.25	937,500
−1.0	700,000	.38	.50	1,050,000
2.5	825,000	.01	.50	1,237,500
− .5	725,000	.36	.50	1,087,500
.5	775,000	.97	.25	968,750
−2.0	650,000	.94	.25	812,500
1.0	800,000	.28	.50	1,200,000
.5	775,000	.67	.25	968,750
− .5	725,000	.38	.50	1,087,500

[a] = 750,000 + 50,000 (RNN) Average = 1,031,875

The simulation was repeated 14 more times; the corresponding average values of the claims are:

\$1,031,875	\$1,137,500	\$1,049,375
1,100,750	1,007,500	1,086,250
1,072,000	1,002,962	1,051,250
1,050,875	1,056,250	1,023,750
1,063,750	1,070,625	1,000,000

FIGURE 17.10. Average cost of claims.

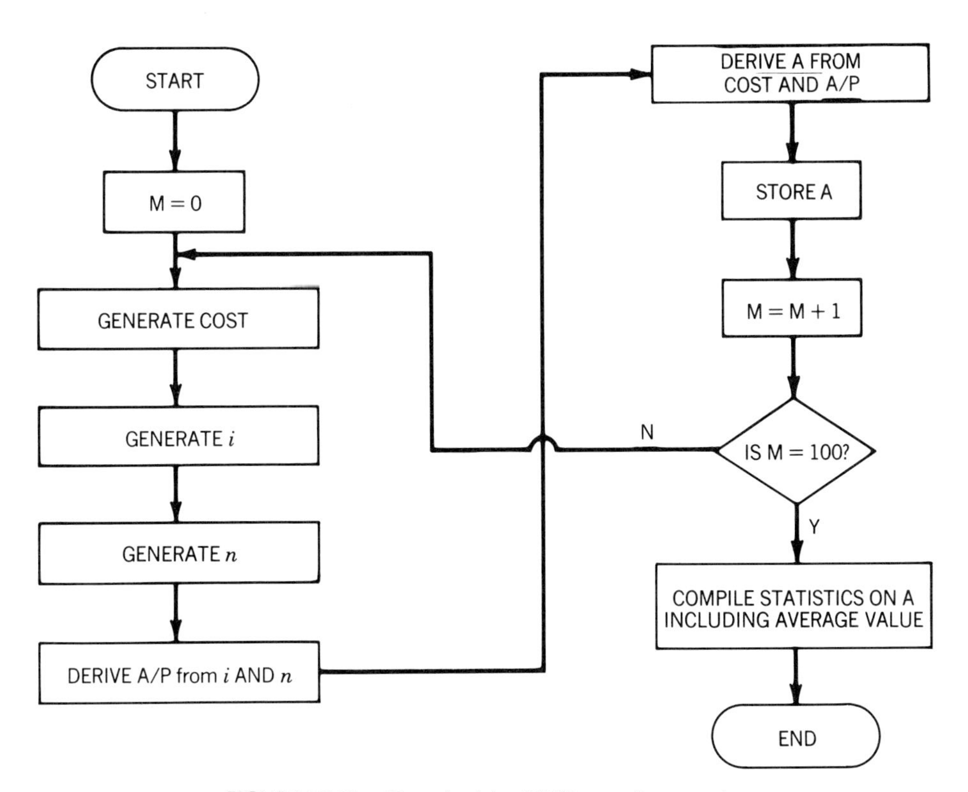

FIGURE 17.11. Flow chart for OSHA compliance cost.

In order to work this example, three assumptions must be made:

1. If the crane drops its load over the worker's assigned area (20% of the total shop area), he will be involved in an accident.
2. The worker's area will be arbitrarily assigned random numbers 30–49 (20% of the random numbers).
3. The action of the crane dropping its load will be assigned to random numbers 0–9.

Since the crane is above the worker's area 20% of the day, the appearance of any of the random numbers 30–49 will represent the hazardous condition. Then, if any of the random numbers 0–9 appears next, a simulated accident occurs.

Three hundred trials were performed, and five accidents occurred. Therefore, the probability that the worker will be involved in an accident is $5/300 = 1.7\%$.

EXAMPLE 17.10

A large company is evaluating a preemployment screening program, which is claimed to have been developed in accordance with the Equal Employment and Opportunity Commission (EEOC) guidelines and recommendations. The company implemented the program for a pilot test at one of its plants. The screening scores for those persons examined are given below:

Men: Scores are normally distributed with mean = 80 and standard deviation = 15.
Women: Scores are similarly distributed with mean = 60 and standard deviation = 20.

The company will hire a labor market equally divided between men and women. The company plans to hire 30 workers per year for the next five years. Minimum passing test score for both men and women is 70.

What will be the composition of the company labor force by the end of the fifth year? Will the company be in compliance with the EEOC rule that demands that the workforce should be representative of the labor market accessible to the company?

A simulation model, shown in Figure 17.12, is designed to count the number of males (M) and females (F) that would be employed out of the next 150 applicants (N). The history of test scores indicates that there will be some favor given to males under the screening program. The simulation will show how much difference this bias will make as far as the total workforce is concerned. If the company cannot comply with the EEOC rule, experiments may be conducted with the model to find the minimum reduction in job requirements to make the job more accessible to females.

EXAMPLE 17.11

An operator works on a small punch press, feeding the press with small parts to be stamped. Operating the press involves two distinct steps. First, the operator loads the metal piece into the press. Second, the press moves the die downward (down cycle) to complete stamping. Therefore, in this situation, we have two patterns synchronized with each other in such a way that the operator manages to pull his hands out of the die area before the press die reaches the metal piece. The two patterns can be depicted as shown in Figure 17.13.

The first observation about this problem might be that working on the press in question should certainly be avoided. However, the high accident rate is caused by a couple of biases in the solution. For simplicity, an accident was assumed to occur if the cumulative loading time (CUMTO) equalled or exceeded the press cycle time (TM) when, in fact, an accident could not occur unless the two times coincided. Also, the fact that negative (and some low) loading times cannot be used suggests that the normal distribution should be altered slightly in order to reflect the loading time function accurately. Correction of these two factors will result in a more accurate simulation.

Timing of press action (cycle) is uniform and can be determined in advance. In contrast, for many reasons (fatigue, distraction, poor eye–hand coordination, etc.) the operator action can occur at somewhat irregular intervals. An accident results when the sequence of loading actions is disrupted to the extent that it coincides with the press cycle: the operator's hand will be caught in the press die.

Suppose that the press cycles once every minute, $t_m = 1$. The time interval between successive loadings (t_o) is normally distributed with mean (μ) = 1 and standard deviation (σ) = 0.5 minutes. What is the probability that a press-related injury will occur in an eight-hour period?

We may safely assume that the operator will not try to load a part before the preceding part has been stamped. Let us also assume that the operator begins the sequence at $t = 0.5$ minutes. Therefore, the first possible accident would occur at $t = 2$ if $t_o \geq 1.5$ minutes.

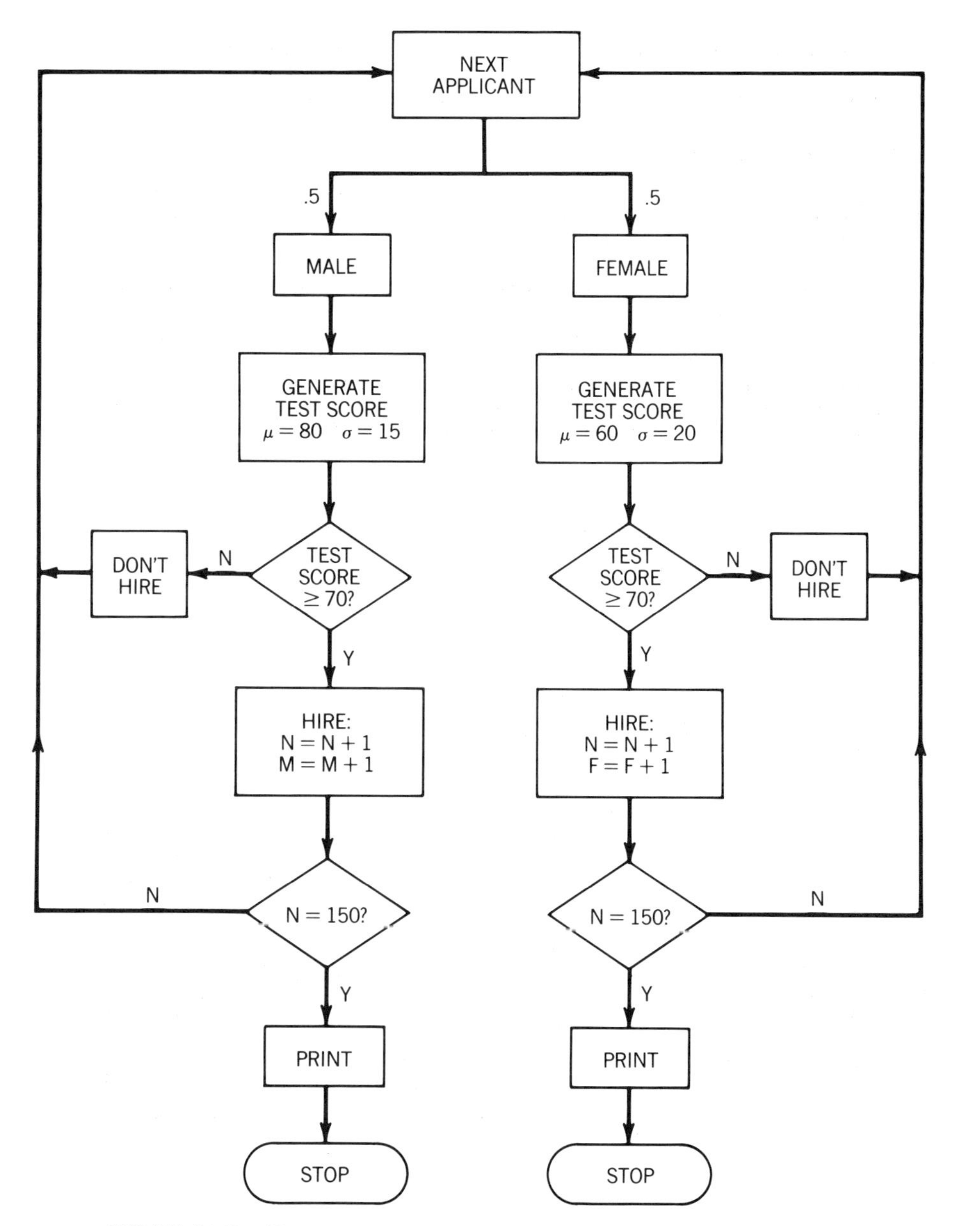

FIGURE 17.12. Flow chart of EEOC discrimination rule compliance simulation.

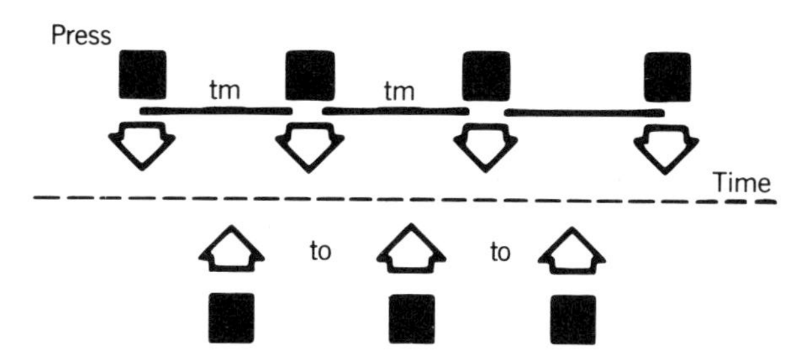

FIGURE 17.13. Press operation sequence.

In the computer simulation model (Figure 17.14), *TO* will be generated from a normal distribution random number routine.

Initial simulation runs produced an accident in four of thirteen cycles.

17.3.5 Simulation Algorithms

Monte Carlo sampling makes no provisions for expenditure of time or scheduling of activities in cases where limitations of resources may introduce delays and the formation of queues. In this context, Monte Carlo implies that events proceed (occur) with no time lost between successive events. This is not always true, especially when one is modeling complex systems and their related activities. With queues, issues concerning the selection of which activities (tasks) to be started next, how resources will be assigned to competing activities, and a host of other issues, have to be dealt with. Dealing with all these issues requires an algorithm that offers more than just sampling. The algorithm, among other things, should be able to sample from theoretical as well as empirical distributions and should maintain files of model activities and allocation of resources in accordance with some prescribed rules. In summary, these requirements describe the characteristics of any simulation algorithm. It follows that we view simulation as expanded Monte Carlo sampling, allowing for the expenditure of time and the occurrence of delays.

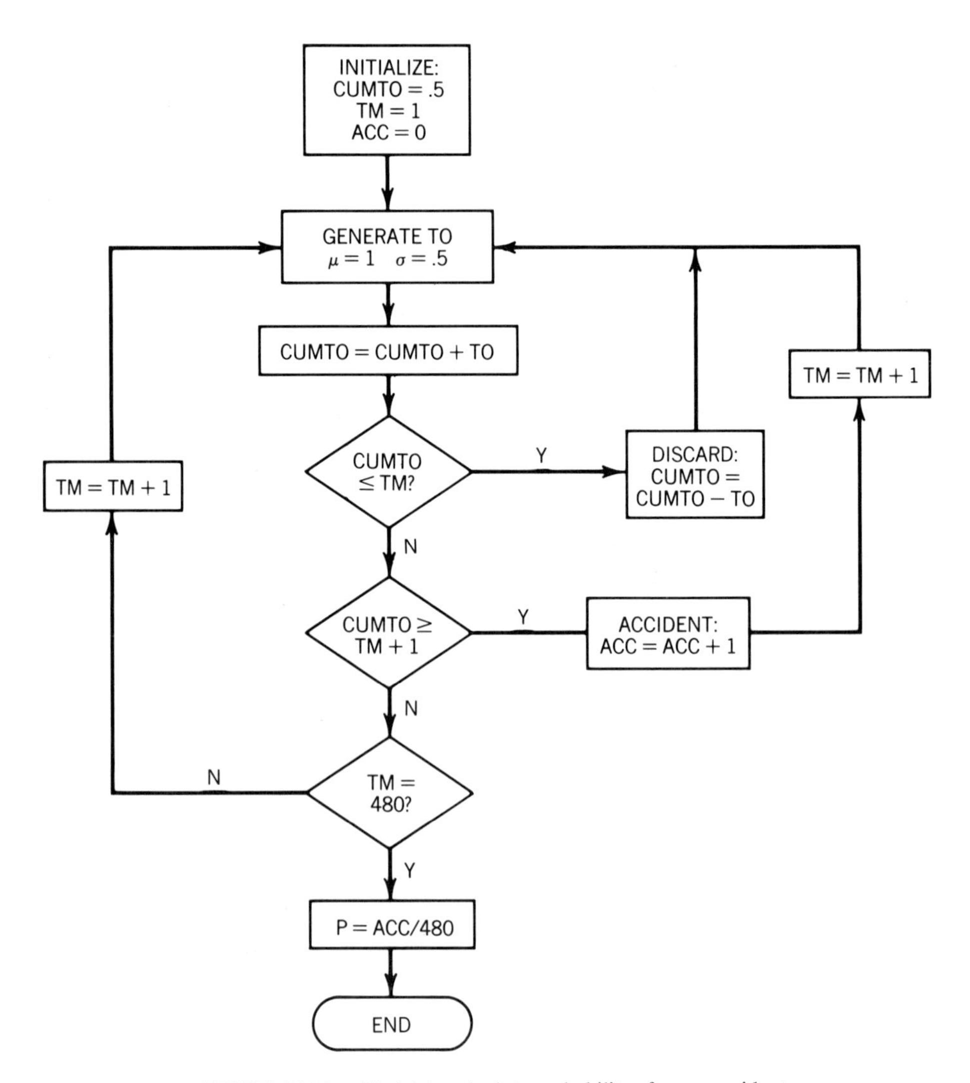

FIGURE 17.14. Model to calculate probability of press accident.

The following is an example of how this is achieved.

EXAMPLE 17.12

A state OSHA program has a compliance officer assigned on a full-time basis to conduct industrial hygiene investigations. Requests for such investigations are made on the basis of the results of the general OSHA inspections and compliance surveys. In any given week, the demand for industrial hygiene investigations can be given by

Number of Requested Investigations	Probability
0	0.5
1	0.25
2	0.25

Time per investigation is assumed to be normally distributed with mean = 2.5 days and standard deviation = 0.5 days. Investigations are carried out on a first-come first-served basis. Accordingly, requests received while the officer is busy are placed on the list of future investigations. From this list, requests are handled in the order in which they were filed. All requests are assumed to be made at the beginning of the week. The officer works only five days a week, with no provision for overtime. Therefore, investigations that cannot be finished in the week in which they started are carried forward to the following week. Determine the number of investigations the officer would be able to handle in a ten-week period.

To deal with the problem at hand, we would proceed as follows:

1. Generate the requests for a week. The number of industrial hygiene investigations to be conducted is obtained by sampling, using the demand data given.
2. If the compliance officer is busy, the requests have to be added to the others that have not yet been processed. Otherwise, the officer is immediately assigned to start a requested investigation. Time (days) of the investigation is assigned using the formula for sampling from normal distribution: $2.5 = 0.5\,(RNN)$; RNN = random number obtained from Table 17.4; 2.5 and 0.5 are the mean and standard deviation of the distribution respectively. The officer will remain busy for this amount of time.
3. When the officer becomes free, the requests, if any, at the top of the waiting list will be considered next.
4. The process of creating requests and carrying out the corresponding investigations continue until the total time assigned to the simulation expires—ten weeks, in our example.

The steps above are carried out in a format similar to that of Table 17.11. Solid lines in Table 17.11 represent investigation times, while dashed lines portray time before the investigation can be commenced for a given request. Circled numbers are successive requests generated during the simulation period. Dashed lines in Table 17.11 give the total waiting times before investigations can be started. This gives an average of two days of waiting per request for investigation. From the table we also can determine the utilization level of the industrial hygienist, the longest number of periods he remained busy, remained idle, and so forth.

The preceding adaptation of Monte Carlo sampling to handle simulations of models with resources and time constraints is a simple exercise that is characterized by very few events and activities. It should not be difficult however, to appreciate how cumbersome and time consuming the procedure gets when we start dealing with several events coupled with a hierarchy of activities. In other words, it would no longer be the case of scheduling one activity, such as hygiene investigation, but rather the case that the start or completion of one or more activities will dictate the events to be scheduled. This brings us a bit closer to requiring a computerized algorithm for handling our simulation.

17.3.6 Computer Algorithms

The two general classes of simulation algorithms are discrete and continuous. *Discrete* means that we are interested in monitoring the system of interest only at some finite points in time. Accordingly, the simulation is a sequence of "snapshots," each of which is taken at some given time. Each of these shots gives information on the status of the system and its activities. Because of this, simulation time is incremented in steps, (see Figure 17.15) each of which corresponds to duration of an activity, delay time spent in a queue, and so forth.

Discrete algorithms are applicable to the class of simulation models where decisions concerning starting of activities or assignment of resources are made following the occurrence of some events. During the time between events we assume that the status of the system being modeled is irrelevant to our analysis.

TABLE 17.11. Simulation Data Sheet

	Week									
	1	2	3	4	5	6	7	8	9	10
Random number for generating requests	.82	.12	.84	.98	.46	.70	.08	.76	.77	.82
Number of requests	2	0	2	2	0	1	0	2	2	2
Random numbers for investigation times	1.57 −0.23	0	1.44 0.05	−2.56 2.23	—	0.81	—	0.17 −1.06	2.39 0.25	−1.17 0.08
Investigation time 2.5 ± 0.5 (*RNN*)	3.28 2.38	0	3.72 2.75	1.22 3.66	0	2.9	0	2.56 1.97	3.6 2.62	1.92 2.54

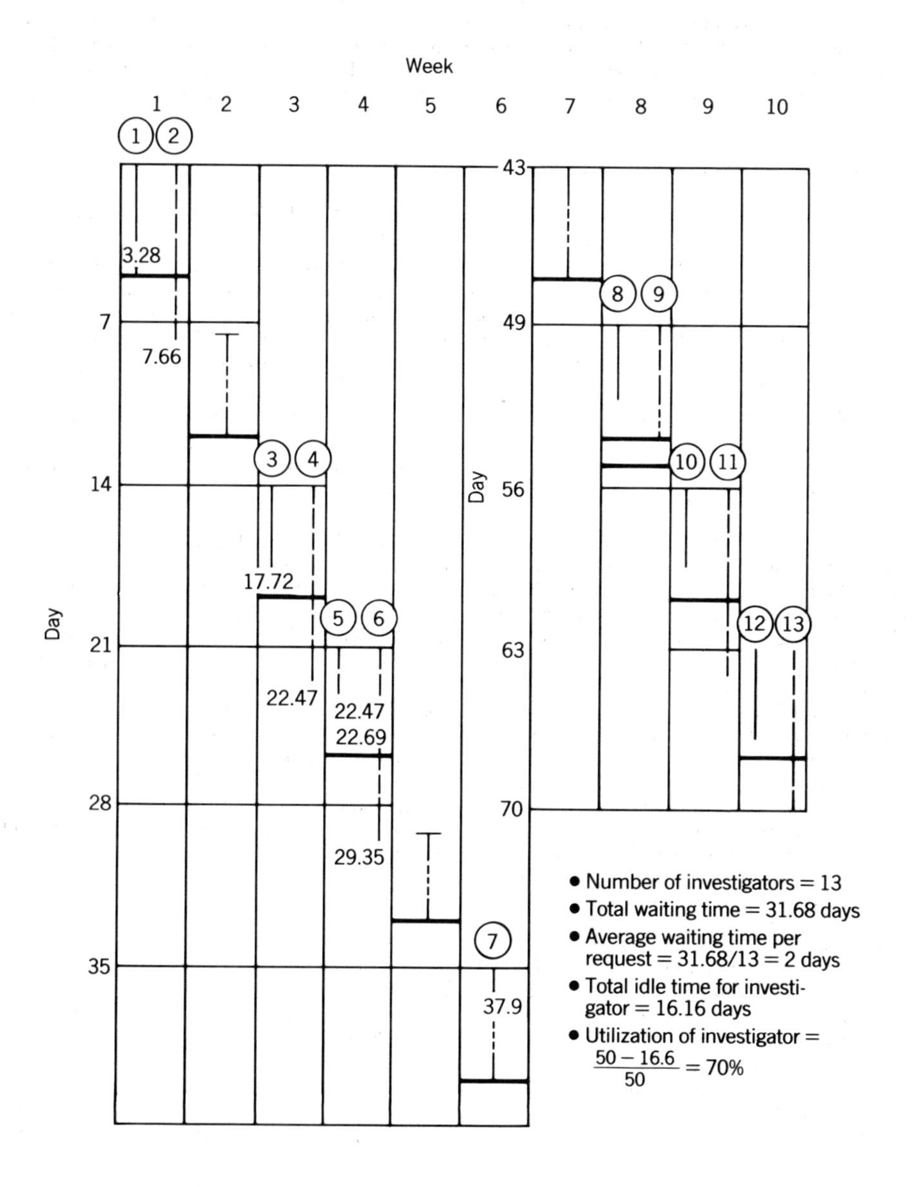

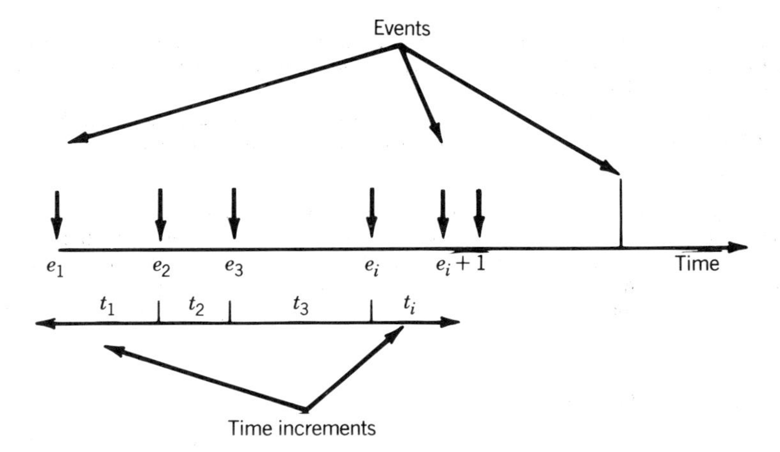

FIGURE 17.15. Increments of simulation time.

As an example, consider the OSHA inspection situation of the preceding example. To have an inspection, we need (1) an inspector who is free and (2) a site where the inspection will be carried out. After the inspection is commenced, we do not have much to contribute until the inspector becomes free or a new request for inspection is received. Either of these (free inspector or request for inspection) constitutes an event that will trigger some response from us. That is precisely the way we carried out the simulation for the OSHA example.

Continuous algorithms are utilized for models whose variables vary in a continuous fashion. At any point, the value of a variable depends on, among other things, its previous value. It is customary for continuous algorithms to represent the relationships among successive values by using different equations (approximation of differential equations). Coupled with this, the simulation time is represented as a sequence of equal intervals, each of a relatively short duration. Economic, biological, and political simulation models are highly suited for continuous algorithms. For example, the Continuous Systems Modeling Program (CSMP) has been used to simulate the cardiovascular system. As another example, consider the safety management model developed by Diehl, 1974. Utilizing the approach of systems dynamics, Diehl's model is an attempt at defining the impact of (1) organizational financial posture, (2) management attitude, (3) type of production technology utilized, and (4) investment in accident (disease) prevention measures on the total safety and health performance of the firm. The model reportedly utilized complex feedback loops and relationships that involve people, materials, equipment, money, and production orders.

Short of developing a full-fledged simulation algorithm, we may consider the use of one of the many existing languages for discrete and continuous simulations.

Available Languages

Discrete simulation languages may be classified into two distinct groups: structured and semistructured languages. *Structured languages* present the modeler with a menu of building blocks and functional components with which a simulation model can be structured and coded for computer processing. General Purpose Simulation System (GPSS) and Graphical Evaluation and Review Techniques (GERT) are two structured languages that are primarily used for modeling systems characterized by queues and limited resources. In contrast, *semistructured languages* present the modeler with a collection of routines that can be used for carrying out many of the standard tasks typical of simulation languages (scheduling of events and activities, collection of statistics, maintaining the simulation clock, etc.). These languages give the user the sole responsibility for translating the simulation model into logically and functionally correct computer codes. General Activity Simulation Program (GASP) and SIMSCRIPT fall into the the category of semistructured languages.

The fundamental difference between structured and semistructured languages lies in the level of programming involvement expected from the modeler. Structured languages require little or no experience with computer programming or with the mechanics of the simulation process itself. On the other hand, the use of semistructured languages mandates proficiency on the part of the modeler in both programming and simulation algorithms.

In general, structured languages offer the following advantages:

1. They are self-documenting. Through the use of standard symbols and graphics, the structure and basis of the simulation model (components, data, relationships, etc.) are easily communicated to others who may not have been involved with the model development process.

2. They free the modeler from many of the time-consuming and error-prone tasks needed to structure and translate simulation models into appropriate computer codes. In this respect, the modeler is involved only with the modeling process.
3. The modeler does not need to be proficient in any programming language, (e.g., FORTRAN or SIMSCRIPT). He or she only needs to understand how to translate the model into the input data (building blocks) defined by the language.

Structured languages, however, have the drawback of requiring a somewhat rigid structure, which makes modeling certain phenomena quite a chore and, at times, simply impossible. Indeed, for some large-scale applications, the use of semistructured languages is a must. To the disappointment of many transient simulation modelers, mastering one of the semistructured languages such as SIMSCRIPT requires time and hands-on experience—two requisites that cannot be justified for minor one-time applications.

17.3.7 Games and Gaming

Simulation is traditionally used to evaluate a given system design or action plan. As discussed previously, simulation gives the user a paper-and-pencil laboratory to study various what-if questions concerning a system and its performance. Another use of simulation is in the area of management gaming. A gaming situation results when (1) a decision has to be made by a person and (2) a consequence of this decision results. The decision outcome may be positive in that the person has perceived the correct situation and, accordingly, the system has performed as expected. Conversely, failure of the decision-maker to comprehend the prevailing conditions of the system can lead to erroneous decisions that may seriously affect system performance. By way of feedback, a person can be trained to make decisions, to integrate several pieces of information in order to define a prevailing condition, and to make use of some "rules of thumb" gained from continuous experimentation. Such training is usually given by on-the-job experience, a process that may consume years before an acceptable level of training is reached. To reduce somewhat the normal time span required for management decision-making, carefully structured computer games are widely utilized. The essence of any of the games is simple: A person is presented with a scenario and then asked to make a decision based on the information provided and his or her understanding of the situation. Next, the person is given some kind of feedback on the behavior of the system following implementation of the decision. At this point, the person may introduce some changes to the system or may elect to do nothing. The specific action will depend on the feedback and the judgment of the person involved. The process repeats itself for many cycles. As the person goes through the game cycles, his or her ability to make and reach the right decisions should be improved. In some situations, the person may capture the characteristics of the situation under consideration to the extent of developing some rules for making decisions. These rules can vary from defining some simple conditions for each decision to rules that would involve the use of highly developed decision models such as linear programming. In any event, the outcome of using a game, computerized or otherwise, will have its profound impact (in a positive sense) on the person's ability to make decisions and to "size up" given situations.

17.4 APPLICATIONS

The potential of simulation as a tool for safety and health research and applications is basically unlimited. It is readily applicable to a host of safety and health problems; the following is a partial list:

1. Study of the accident phenomenon, its underlying mechanism, its contributing variables (human, machine and environment).
2. Evaluation of the effectiveness of various hazard control measures: control of physical hazards (alternatives for safeguarding), control of environmental hazards (selection of noise control approach), control of human behavior (training selection, enforcement).
3. Manpower planning for the safety and health organization based on organizational structure and available resources.
4. Inflationary impact of safety and health regulations given current state of compliance in industry, compliance alternatives, and potential compliance costs.
5. Evaluation of various modes for safety and health training—on-the-job retraining and retrofitting vs. formal (college) training.
6. Analysis of the standard-setting process and its driving and restraining forces—politics, media campaigns, pressure groups, degree of public awareness, and comprehension of industrial hazards and economy.

7. Evaluation of alternative sites for storage of nuclear waste based on site topography, population at risk, and quality of transportation links between sites and waste producers.
8. Evaluation of land, sea, and air transportation systems for handling hazardous materials. Considerations should be given to characteristics (physical, chemical, etc.) of materials, reliability of transportation system employed, population exposed, feasibility and effectiveness of hazard containments following spills and regulatory constraints.
9. Effectiveness of safety communication within multiplant corporations. The simulation should consider: degree and extent of safety inspection and accident investigation practiced, the place of the safety and health departments within the organization, characteristics of data handling system, training and expertise of the safety and health staff.
10. Design of safety performance aids (safety signs, safe operating procedures, accident investigation manuals, etc.).
11. Evaluations of different patterns of materials flow within a given plant, taking into account characteristics of materials handled, materials handling equipment used, plant layout, regulatory requirements, and variations in product demands.
12. Impact of various work schedules (shift work, four-day work week) on safety performance.
13. Development of management games for training safety and health professionals to handle effectively issues concerning the allocation and scheduling of resources; formulation of policies and procedures.
14. Assessment of the impact of various preemployment screening program on safety performance and the regulatory requirements of Equal Employment Opportunity Commission (EEOC).
15. Prediction of health profiles of employees exposed to a mix of industrial hazards (with varying intensities and durations) throughout ten or more years of employment.

Two examples given below will illustrate some of the problem areas outlined above.

17.4.1 Effectiveness of the Safety Organization

To develop a computerized approach for studying the effects of different management systems on the overall safety performance of the industrial firm, management gaming and simulation are used in the following model. (For a good review of gaming literature, see Belch, 1974; Gibbs, 1974). The proposed model simulates the interaction between the accident process and the management process. The *accident* process describes the steps involved in producing accidents: hazard generation and transformation of the generated hazards into accident-producing situations. The *management* process deals with managerial decisions and acts that span the three phases of the accident control process; hazard recognition, hazard analysis, and hazard control.

Management and accident processes are time dependent. For instance, in the case of the accident process, time will elapse before a hazard is generated; following this, the hazard may be queued for some time before it produces an accident. On the other hand, the time behavior of the management process is dependent upon such items as the structure of the organization, existing safety policies, and the availability of the safety program.

Using time as the control variable, the model simulates a race against time by management to eliminate hazards capable of causing either an accident or near accident. Figure 17.16 is intended to portray the basic philosophy behind the model: an accident or an incident occurs when protection normally provided by a safety program is absent. This protection could be in the form of personal protective equipment, machine guarding, isolation devices, substitution devices, or other equipment normally used to protect the worker from the hazards of the industrial environment. The concept, exemplified in Figure 17.17, follows from what Kibbee (1972) calls the "teeter-totter" principle. Kibbee relates his explanation for this principle to the field of economics, postulating the counterbalancing effect of a firm that distributes stock profits vs. its working capital. Worker protection costs management an initial capital outlay that either may outweigh its usefulness, or prove a wise investment; however, the unprotected worker (or system) will be involved in an accident or an incident only while unprotected (the bottom section of the figure).

Hazards are assumed to be generated on a time basis; their occurrence is a discrete event, but their lingering presence is a continuous phenomenon. Each firm, institution, or others using the model must express the hazard generation rate at a daily function of its (OSHA) incidence rate. One way to be obtain this note is by using Heinrich's 1-29-300 classic pyramidal formula and extending the base by approximately the same proportion for one level. The ratio is then 1:29-300 = 3000, with the latter (Figure 17.18) representing the number of hazards contributing to the top parts of the pyramid. By the use of proportion, the mean time between hazardous occurrences can be estimated for a given incidence rate—a rate based on the data included in the first two levels of the pyramid. As shown in Figure 17.18, other formulae may also be used (Kann, 1974).

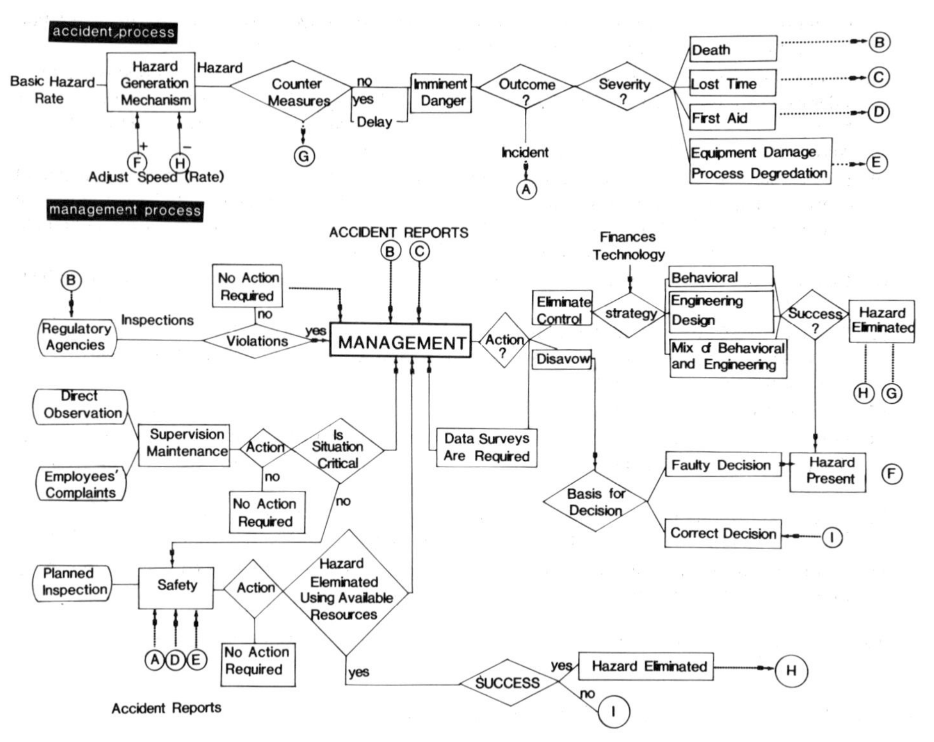

FIGURE 17.16. Flow chart of accident management model.

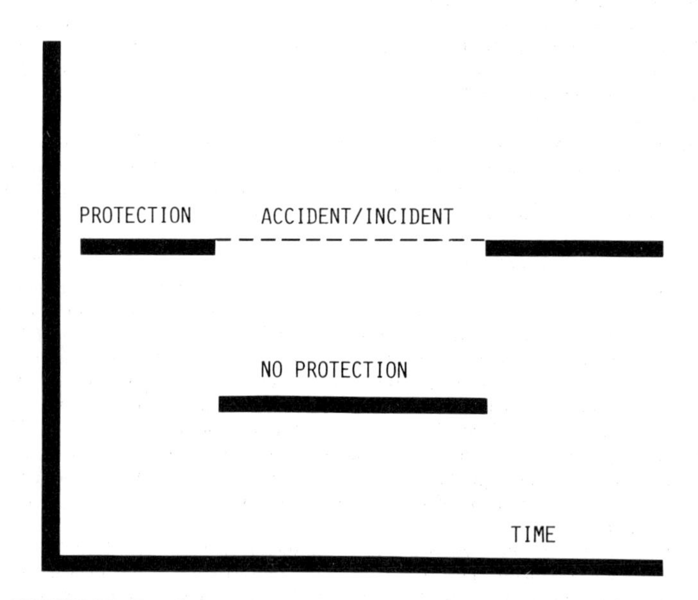

FIGURE 17.17. Control sequence for avoidance or generation of an accident.

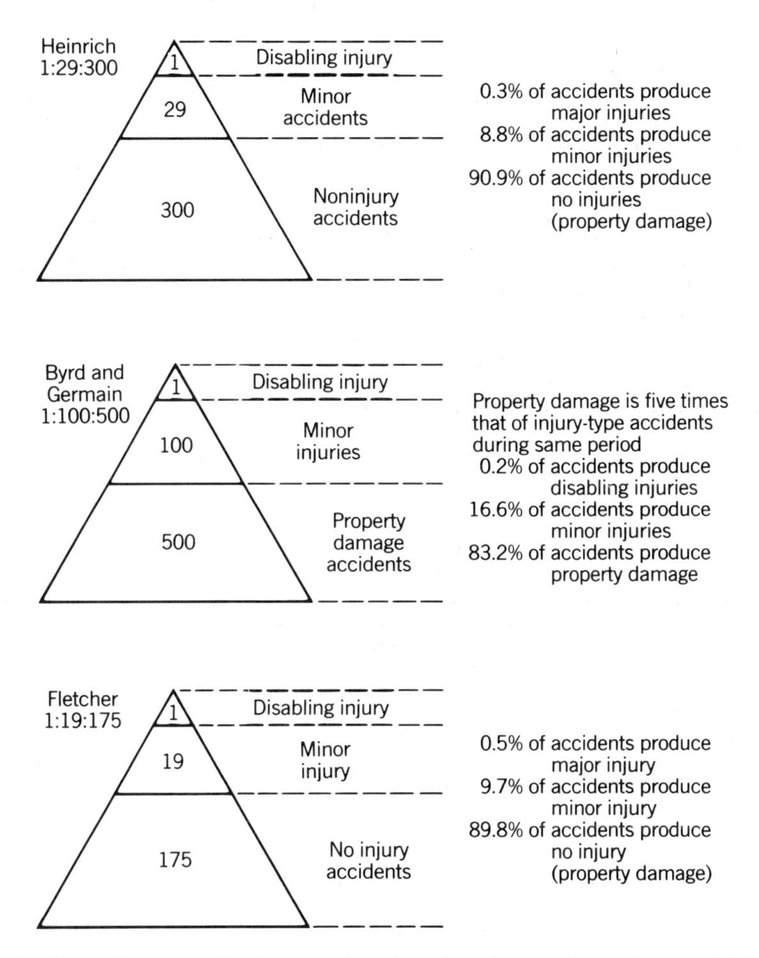

FIGURE 17.18. Pyramidal formulas for calculation of accident rates (Kanon, 1974).

As the hazards are generated and subsequently queued for potential accident/incident cases, some fade and do not actually cause accidents or incidents (personnel obviously being keyed to avoid the hazard). Hazards culminating in either an accident or an incident can lead to one of four events: first aid, lost time, death, or process or equipment degradation, the latter being a result in which personnel are not injured. Based on Heinrich's formula, the probability of each of the four possible events can be derived.

While a host of information inputs and events trigger the communication processes that ultimately lead to hazard elimination, the enumeration of hazards in terms of specific type, nature, and level is considered to be superfluous to the intent of the model. This philosphy is similar to that of Howard (1968); he indicates a growing consensus of safety specialists is that "it is infeasible to initiate system development by modeling a total organization's information system." Howard reports that such models have invariably failed, whereas others attempting to model smaller segments of the organization have been successful.

Communication is shown to begin either with the various personnel within the organization or from those outside the organization. It follows seven possible channels: inspections by regulatory agencies; direct observation by management or supervision; employee complaints; a rash of first aid cases; substantial losses due to equipment or process degradation; routine inspections from within the organization; and an accumulation of near accidents. Regulatory agency inspections are communicated directly to management, direct observations and employee complaints are passed on to supervision or maintenance, and all other information inputs (primarily generated from accident reports) are communicated to the safety function. Each communication channel is given a time value to indicate how often the flow of information can be expected to occur and also is given a regenerating loop to indicate the time interval between successive inputs.

Management, made aware of the hazardous condition, can be expected to do one of three things: (1) recognize the condition as hazardous and initiate hazard elimination procedures; (2) ask for further surveys and analyses before making a decision; or (3) conclude that the condition or practice reported to be hazardous can be accepted as a fair risk. Further surveys could result in no action or implementation of hazard control measures. No action, after a selected time interval, could lead to reevaluation by management, maintenance, or supervision. Probabilities are assigned for the three possible outcomes of the management decisions; time values are also assigned to indicate the amount of time required to perform the indicated activities and are expressed as slow, usual, and fast responses. If management decides to provide interim protection (such as personal protective equipment), then the accident/incident process will be slowed down. If none is provided, the process is speeded up or continues to operate at the same speed. The increase or decrease in speed of the model is represented by the section of the model under the accident process, labeled "countermeasures."

To summarize, the intensity of management response will depend upon the value of the information received. For instance, management receiving an OSHA citation will undoubtedly move quickly and, perhaps, at considerable expense, to correct the hazards cited. In contrast, information contained in some inspection reports may not produce any immediate hazard-control actions.

Hazard elimination requires the availability of technology, finances, and time. First, we should be able to define an approach that is technically and otherwise feasible for controlling the recognized hazards. Second, cost of the selected control approach must be within reach, considering the organization resources and financial posture. Finally, it takes time to implement the selected control approach and to bring the recognized hazards effectively under control.

In general, regardless of hazard type and characteristics, control is achieved by using either behavioral or engineering approaches; in some situations, using a mix of the two can be justified. Behavioral approaches encompass such things as training, selection, and enforcement. On the other hand, engineering emphasizes the technical changes to be introduced in the workplace and work practices, for example, safeguarding of machines. Behavioral approaches can be implemented in a relatively short time. In contrast, their reliability (i.e., remaining effective in the long run) is rather low. Behavioral approaches usually cause a surge in a safety and health activities, the result being the elimination or control of most workplace hazards. However, this does not last long, and with time, the sharpened interest in safety activities will dissipate and eventually control of the hazards will be all but lost. Nor are engineering approaches without shortcomings. Due to a host of factors (poor maintenance, wear and tear, etc.), hazard control through engineering approaches my not be successful under all circumstances and for all situations.

Therefore, any type of hazard elimination (control) is subject to either success or failure. If success (abatement) is obtained, then the accident/incident process is slowed down proportionately; failure increases the speed of the process. Information concerning failure to control the hazards—after a time interval necessary for detection—can trigger either a reevaluation through maintenance and supervision, or an initiation of a new hazard-control attempt. As pointed out by Firenze (1973), hazard control is neither total failure nor total success. Stated differently and in light of the discussion above, hazard-control approaches (behavioral or engineering) do have a limited service life, after which their effects will be lost and hazards will start to accumulate. When this point is reached, the system is changed once more to reflect the new speed of the process.

The steps involved in the simulation (running the model) are as follows:

Step 1. Define performance and decision parameters for various model components. In each case, three pieces of data are sought: the time it takes before a decision is reached, the type and number of inputs required to trigger a response, and the frequency with which each input is received. Whenever applicable, time parameters should be described using appropriate probability distributions.

Step 2. According to the organizational structure and existing policies and regulations, determine the different activities of the management process that will be active during the model simulation. For example, if the particular company under consideration does not have an active safety program, all information inputs to be generated by the safety department will not be considered during simulation.

Step 3. Simulate the model using input data generated from Steps 1 and 2.

Step 4. After examining the simulation results of Step 3, some policy and/or procedure changes may be introduced. The effects of these changes upon the overall safety performance of the organization can be ascertained by simulating the model several more times. This process of introducing or making changes, and then simulating, can be repeated until all possible policies have been assessed.

The model described above was coded using QGERT and then tested by studying the effects of three management policies on the overall safety performance of a hypothetical firm. The three policies used were:

1. An ongoing safety program does exist.
2. No safety program is available.
3. No safety program is available; management is not always supportive of safety activities and expenditures.

Data for the various components of the model were estimated by conducting a survey of ten furniture plants in North Carolina. Summary of ten simulations (each representing a full year of performance) is shown in Table 17.12. The results were in the expected vein; that is, the firm with the safety program and the firm with no safety program fared better than the firm with no safety program and in which management always asked for further surveys (considering the lowest number of observances as the better performance).

To test accurately the impact of various organizational structures and policies, one needs to go a little bit further than what has been presented thus far. That is, a full-fledged experiment should be designed and carried out using the proposed model; such an experiment in itself would constitute a worthwhile study.

17.4.2 Evaluation of Maintenance Manuals

Preventive and corrective maintenance activities do constitute an important and integral part of any trustworthy hazard-control as well as accident-prevention program. Any practical system can fail from time to time, no matter how remote the chance is kept through system design (reliability) and preventive maintenance. To fix a system, the failing components have to be identified, then repaired or replaced with equivalent components (spares). The process of identifying the failing components is called troubleshooting—an important concept in corrective maintenance.

Successful troubleshooting requires knowing the system and its operation; this was relatively easy when systems were simple and small. However, as systems grew larger and more complex, it became difficult, time-consuming, and sometimes even impossible for a single person to acquire all the knowledge needed to maintain such systems (Elliott, 1967). To overcome this difficulty, fully proceduralized troubleshooting guides are used. These are step-by-step procedures that guide maintenance personnel through the troubleshooting process. A guide tells the maintenance person where to check first, how to interpret the check result, and then what to do next. The purpose of using fully proceduralized guides is to minimize the number of decisions to be made by the maintenance person.

Several studies have demonstrated the effectiveness of fully proceduralized guides in performing maintenance activities (Elliott, 1967; Post and Price, 1973; Foley and Camm, 1972; Joyce and Chenzoff, 1973; Smillie, 1977). Fewer errors and shorter maintenance time can be expected when troubleshooting guides are used. Furthermore, the use of guides can practically eliminate the need for training or reduce it to as little as a few dozen hours. Troubleshooting guides, however, have some limitations. Some of these are:

1. In designing a troubleshooting guide, it is practically impossible to deal with every potential system problem; e.g., a conductive metal can drop into electronic circuits and short adjacent circuits. This could cause a malfunction of the system. Since the short can occur in any place, it would be impossible to design a guide for isolating and repairing this type of fault.
2. Fully proceduralized guides do not promote opportunities for logical reasoning and learning. This can have a serious impact on the attitude and motivation of maintenance personnel.
3. The guides are designed on the implicit assumption that when a system fails, the initiating causes (troubles) will remain in a permanent state—the trouble will appear every time a check is made.

TABLE 17.12. Result of executed simulations.

	Accident Type			
Condition	First Aid	Process or Equipment Degradation	Lost Time	Death
Safety program	36	405	0	0
No safety program	41	414	0	1
No safety program; further surveys required	54	500	2	1

This is not always the case for many systems. Indeed, troubles are known to appear intermittently. Therefore, using fully proceduralized guides to deal with intermittent troubles may not be very effective, and in some instances, may lead to erroneous conclusions.

The degree to which trouble intermittency appears varies a great deal within and among system components. Some troubles may appear once in three or four tests; some may appear once in several hours of operation or perhaps eery month or two. As long as the trouble appears relatively frequently, it will not too severely impact effectiveness of the guide. On the other hand, if the trouble occurs rarely, say once a month, it would be almost impossible to pin down the failing component on the first attempt. The issue then becomes one of identifying under what conditions of trouble intermittency a fully proceduralized guide would lose its effectiveness. For this and similar issues, the following model can provide a starting point.

17.4.3 A Troubleshooting Model

We consider any troubleshooting process as a sequence of the following steps (see Figure 17.19):

1. Intermittent trouble occurs and a maintenance person is called in.
2. He makes preparation for troubleshooting (gets symptom information from the operator, removes cover, etc.).
3. He runs a test.
4. If the trouble appears, he proceeds to the next step. If not, he skips to Step 6.
5. He follows the subsequent instruction steps until he is directed to the failing component or comes to the next test run. When the failing component is found, the troubleshooting process is completed. Otherwise, he goes back to Step 3.
6. He may or may not notice that the trouble did not show up during the test. If he notices, he would repeat the test within some time limits or for a certain number of trials. In this case, he goes back to Step 3. If repeating the test does not identify the failing component, he would quit further troubleshooting, assuming that the trouble is too intermittent. In this case, the system is not fixed and most likely will fail later.
7. If he fails to recognize that the trouble did not occur, then he will follow a wrong path in the guide. And after following some instruction steps, he will conclude the test without locating the malfunctioning component.
8. He may or may not notice that he was on a wrong track. If he does not, the trouble is not really fixed. If he does, there are two possible consequences. One is to quit right there, while the other is to repeat the whole process from the beginning, that is, go back to Step 3.

The troubleshooting process described above was simulated using QGERT with the following specifications and parameters.

1. Availability of a minicomputer with a built-in exerciser for logic testing is assumed. This test takes three minutes to run.
2. Time to perform instruction steps has a beta distribution.
3. Troubles appear intermittently. If the trouble is not encountered during a test run, the test will be repeated up to 5 times.
4. On the average, 2.5 decision steps will be used in any given troubleshooting session. The distribution of the number of decision steps is assumed to be normal, with standard deviation equal to 0.8.
5. If the maintenance person notices that he was not following the right sequence of instructions, he may repeat the process only once.

The results of 500 simulations are given in Table 17.13 and Figure 17.20. As shown in Figure 17.20, the effectiveness of the troubleshooting guide drops very rapidly as the trouble intermittency decreases (trouble appears less and less). When trouble appears very intermittently (probability less than 0.1), the troubleshooting guide is almost useless (Table 17.13). This becomes significant if we consider an extreme case:
A problem appears, say once a month. Suppose we could tell whether the trouble appears or not, and we could see the trouble at the first decision point in the guide. To proceed further, we have to wait at the next decision point until the trouble appears again, which may be a month ahead.

Generally speaking, by increasing the number of test repetitions, the effectiveness (the probability of locating the troubles) will be improved. However, it will be at the expense of more troubleshooting

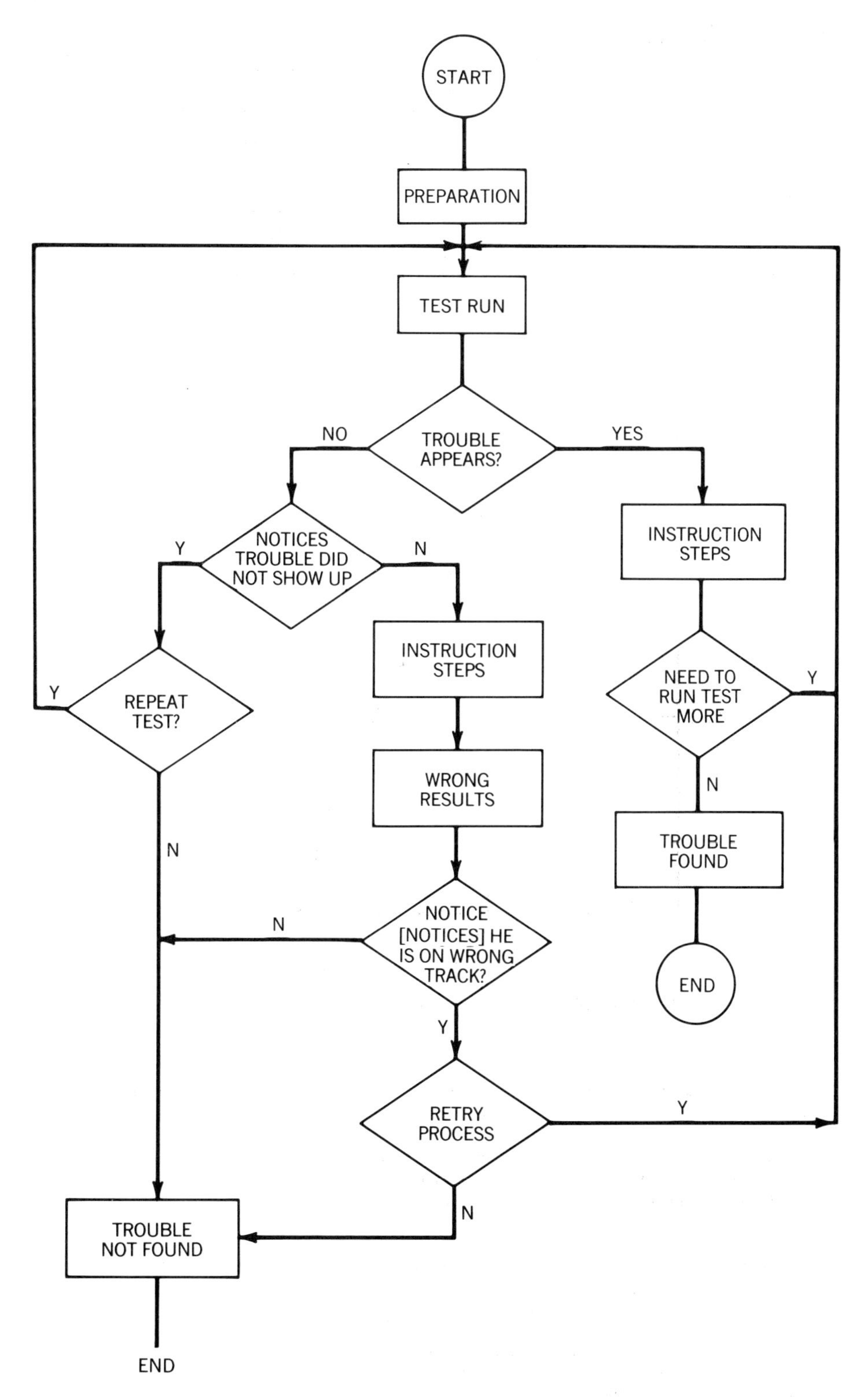

FIGURE 17.19. Troubleshooting flowchart.

TABLE 17.13 Guide Effectiveness

Trouble Intermittency	Guide Effectiveness	Average Diagnostic Hour
.9	.98	81.55
.7	.93	101.34
.6	.83	113.16
.5	.65	122.53
.4	.47	122.05
.3	.25	122.20
.1	.01	83.60

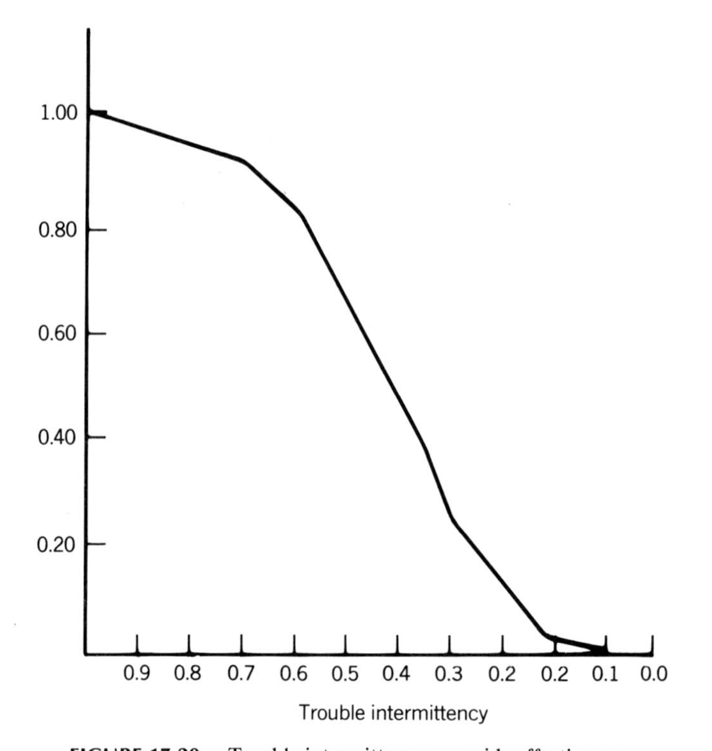

FIGURE 17.20. Trouble intermittency vs. guide effectiveness.

hours. The fully proceduralized troubleshooting guides can be very ineffective on intermittent failures. It will be necessary to improve the guides in this respect or to develop a new maintenance concept to deal with intermittent failures. Post and Price (1973) conducted a study on maintenance documentation comparing the traditional maintenance documents (deductive maintenance approach) and the proceduralized troubleshooting guide (directive maintenance approach). They suggest an integrated approach which takes advantage of both approaches.

Extension

The model just described may be used to explain how, under certain conditions, accident investigation can be misleading, even with the best intentions and availability of seasoned investigators. For example, consider the following hypothetical situation.

1. A traffic intersection was controlled with 4-way stop signs. Following a rash of serious accidents, it was decided to replace the stop signs by a traffic light. Here most of the accidents at the intersection were attributable to poor traffic control. Indeed, after the traffic light was installed, serious accidents at the same intersection were eliminated. In other words, reviewing some of the

accident cases led the investigator to define the absence of a traffic light as the primary cause of the problem. Since the intersection had no traffic light, that is, the trouble appeared permanently; accordingly, it was not difficult for the investigator to reach the proper conclusion.

2. Consider the same traffic intersection, now equipped with a traffic light. A few months after the installation of the traffic light, a sequence of serious accidents occurred at the intersection. Careful analysis of these accidents led to the inevitable conclusion that identified human error was the primary cause, for every time an investigation was conducted, each aspect of the intersection, including the traffic light, was according to standards and specifications. However, the number of accidents did not decrease and persisted even after additional warning signs were posted at the intersection. With this as background, a special study of the intersection and its traffic was undertaken for a period of several months. This lengthy investigation revealed that the traffic light did not function properly all the time, and in many instances it gave the right of way to both traffic directions simultaneously—an open invitation to accidents.

In this case, the trouble (malfunctioning traffic light) was intermittent to the extent that every time an investigation was undertaken (following an accident) it did not appear; every investigation concluded that the traffic light was functioning properly. When it was discovered to be otherwise, it was too late as well as being costly.

The essence of the traffic case just described can be found (at least conceptually) in many accident-producing situations, for the simple reason that we deal with people who can change their predictable responses as the characteristics and nature of their working environments change. These changes can be periodic and might be difficult to trace following an accident—an important prerequisite for accident reconstruction.

REFERENCES

Anderson, V. L., and McLean, R. A., *Design of Experiments: A Realistic Approach,* Marcel Dekker, New York, 1974.

Ayoub, M. A., Integrated Safety Management Information System; Part II: Allocation of Resources, *Journal of Occupational Accidents,* 2: 135-157, 1979.

Ayoub, M. A., Integrated Safety Management Information System; Part II: Allocation of Resources, *J. Occupational Accidents,* 2: 135-157, 1979.

Belch, J., *Contemporary Games: A Directory and Bibliography Covering Games and Play Situations Used for Instruction and Training by Schools, Colleges and Universities, Government, Business and Management,* Gale Research Company, Detroit, 1973.

Diehl, A. E., *A System Dynamics Simulation Model of Occupational Safety and Health Phenomena,* Unpublished Ph.D. Dissertation, North Carolina State University, Raleigh, NC, 1974.

Elliott, T. K., *Development of Fully Proceduralized Troubleshooting Routines,* Aerospace Medical Research Laboratories Technical Report 152, November, 1967.

Emshoft, J. R., and Sison, R. L.; *Design and Use of Computer Simulation Models,* Macmillan, New Yor, 1970.

Firenze, R. J., The Logic of Hazard Control Management, in: Widner, J. T., *Selected Readings in Safety,* Academy Press, Macon, GA, 1973.

Fishman, G. S., *Principles of Discrete Event Simulation,* Wiley, New York, 1978.

Foley, J. P., J., and Camm, W. B., *Job Performance Aids Research Summary,* Air Force Systems Command, August, 1972.

Gane, C., and Sarson, T., *Structured Systems Analysis: Tools and Techniques,* Prentice-Hall, Englewood Cliffs, NJ, 1979.

Gibbs, G. I., ed., *Handbook of Games and Simulation Exercises,* Sage Publications, Beverly Hills, California, 1974.

Graybeal, W. and Pooch, V. W., *Simulation: Principles and Methods,* Winthrop Publishers, Cambridge, MA, 1980.

Howard, J. A., and Morgenroth, W. M., Information Processing Model of Executive Decision, *Management Science,* 14, (March 1968) 416-428.

Joyce, R. P., Chenzoff, A. P., Mulligan, J. F., and Mallory, W. J., *Fully Proceduralized Job Performance Aids* (Three Books), Air Force Systems Command, December, 1973.

Kanon, J. C., Safety Standards—The Systematical Approach and Methodical Basis for Personal Safety and Damage Control, in: *Symposium on Working Place Safety,* 22-26 July, 1974, Bad Grund, FRG, 1974.

Kibbee, J., Kraft, C., and Nanua, B., Management Games, in: Carlson, J.G.H., and Misshauk, M.J., *Introduction to Gaming: Management Decision Simulations,* John Wiley, New York, 1972.

Linstone, H. A., and Turnoff, M., *The Delphi Method: Tehcniques and Application,* Addison-Wesley, Reading, MA, 1975.

Post, J. T., and Price, H. E., *Development of Optimum Performance Aids for Troubleshooting,* Bio Technology, Virginia, April 1973.

Phillips, D. T., *Applied Goodness of Fit Testing,* AIIE Monograph Series, AIIE-OR-72-1, Atlanta, GA, 1972.

Pritsker, A.A.B., and Pegden, C. B., *Introduction to Simulation and SLAM,* Halstead Press and Systems, New York, 1979.

Pulat, B. M., *Computer Aided Panel Design and Evaluation System—CAPADES,* Ph.D. Dissertation, Dept. of Industrial Engineering, North Carolina State University, Raleigh, NC, 1980.

Rider, K., Hausner, J., Shortell, R., Bligh, J., and Candeloro, T., *An Analysis of the Deployment of Fire-fighting Resources in Jersey City, New Jersey,* The Rand Corporation, R-1566/4-HUD, August 1975.

Shannon, R. E., *Systems Simulation: The Art and Science,* Prentice-Hall, Englewood Cliffs, NJ, 1975.

SAS User's Guide, SAS Institute, Inc., Raleigh, NC, 1979.

Smillie, R. J., and Ayoub, M. A., Accident Causation Theories: A Simulation Approach, *Journal of Occupational Accidents,* 1 (1976), 47–68.

SPSS for the IBM PC, Chicago, 1984.

CHAPTER 18

How to Apply System Safety to Occupational Safety and Health

LEWIS BASS
The KAIROS Company
Mountain View, California

18.1 INTRODUCTION

This chapter addresses the application of the system safety approach to occupational health problems. It contrasts and compares traditional industrial safety methods with those of system safety. Actual system safety analyses of occupational hazards are used as examples.

Traditionally, system safety professionals have participated in planning and evaluating product safety, whereas industrial health and safety professionals were concerned with safety during production. Now, a growing number of manufacturing companies are successfully integrating product and industrial safety in their organizations to reduce accidents and improve profitability. System safety techniques, the tools of the safety professional, are brought to bear on the problem of production safety from the earliest stages of design through the shipment of products and disposal of wastes. Safety concerns are documented and tracked throughout this process.

System safety techniques allow assessment of the interaction of processes, equipment, and workers in an industrial facility. Applied at the concept, design, and pilot-plant stage of an industrial process, these techniques provide a method for early introduction of occupational health concerns. They generate requirements for designing safe procedures and environmental control and monitoring programs. Furthermore, they set the stage for management support of the safety effort.

18.1.1 The System Safety Approach

The system safety approach begins with defining the system under study. For industrial products, the system consists of the operator, the equipment and tools being used, the facility, environmental conditions, and other equipment at the site. The system safety approach evaluates the safety of the system throughout its life cycle of modes of operation, considering for each life cycle mode foreseeable situations that may arise due to equipment failure, human error, and environmental conditions. In the case of a degreasing machine, for example, the life cycle modes at the workplace are selection by the purchasing department, shipment by the equipment manufacturer, installation, operation, repair, maintenance, and disposal.

System safety techniques provide a thorough, systematic approach with which to address workplace hazards. These techniques are as effective in analyzing the workplace as they are for assessing products and systems. The use of industrial safety techniques in conjunction with system safety techniques can isolate hazards in the workplace more quickly and effectively than the use of either type of technique alone. Hazard analysis, failure modes and effects analysis, operational hazard analysis, fault tree analysis, software safety analysis, and robotic safety analysis are techniques used to identify the hazards of the worksite, to determine how accidents could occur, and to design equipment and develop facility and procedural safeguards to reduce risk.

An understanding of how things fail is basic to the application of the system safety approach. Accidents happen when physical systems fail, when unforeseen events occur, or when the humans coming into contact with systems make errors. Possible equipment failures and human errors must be anticipated. Safety professionals weigh the consequences of failures and establish priorities among the hazards present. They also recommend methods to reduce the likelihood and/or consequences of failure.

Many safety problems are first encountered in response to an emergency of some type. Once the emergency is over, the problem may be considered solved—until another emergency occurs. Solving problems once they have occurred is an important part of the responsibilities of the safety professional, but the true role of the safety professional is to prevent problems from arising in the first place.

18.2 EVALUATING WORKPLACE HAZARDS

Hazards must be identified before they can be prevented and controlled. The earlier in the design process hazards are recognized, the easier it is to make changes. If the hazard analysis process is performed prior to design, designers can be alerted to the problems and can concentrate on eliminating or controlling them. Hazards may occur in an occupational setting because management and workers are not aware of their potential.

Most hazards fall into one of four categories: physical, chemical, biological, and ergonomic.

A physical hazard is an energy hazard that directly affects the worker. Some examples are radiation, being struck by an object, falling, or being subjected to heat, stress, or noise. The function of industrial safety equipment, for instance, is to prevent injuries to workers. If this function is not performed, the user may be injured. Less obvious is the possibility that the safety equipment will perform "correctly" (i.e., perform its intended function) but still fail to protect the worker from injury. An example of this might be a fall arrest system (life line, rope grab, lanyard, and safety belt) that successfully arrests the fall of the user but does so in such a manner that the deceleration forces cause internal injuries.

A chemical hazard is a hazard caused directly or indirectly by a specific nonliving material substance. Of the traditional categories of hazards, the chemical hazard is one of the most difficult to deal with. A knowledge of chemistry, biology, and toxicology is required to control such problems. Even when the safety professional has this background, the specific hazards involved in a production process may be difficult to determine. Polychlorinated biphenyls (PCBs) are an excellent example of a recently discovered chemical hazard that has been implicated in liver disease and cancer. Thousands of pounds of PCBs have been released into the environment and pose a significant environmental threat if they are not eliminated. Many other chemicals have also been used for years before they were suspected of being hazardous. Carbon tetrachloride, asbestos, and formaldehyde, for example, were once considered harmless but were later found to cause serious illness.

A biological hazard is a living organism that causes a health problem for the worker. Examples are yeasts, fungi, and infectious bacteria. Grain handlers who inhale grain dust can contract a condition called "farmer's lung" caused by a fungus sometimes contained in grain. Exposure to dust from vegetable fibers such as cotton or flax can cause upper respiratory tract infections and allergic reactions in some people. Slaughterhouse workers and fish handlers are at special risk from bacterial infections.

An ergonomic hazard arises from the position or stress the worker is exposed to in the course of the job. Examples include the position the person must assume, the weight that must be lifted, the angle at which the worker must hold his or her head, and the illumination level of his or her work station. An important safety task relating to this type of hazard is called "defining the operator," which involves defining the operator's capabilities. Human factors engineering principles must be applied to reduce the probability that the operation will be unsafe. The operator's expectations of how equipment operates and his or her skills and capability are critical safety factors. The operator must have the skill to understand the instructional material that comes with new equipment, or the company must create appropriate training courses and materials. Special training may be required to assure safe and effective operation. Human factors such as the reluctance of operators to follow safety precautions if the precautions are perceived to interfere with their productivity must also be considered.

18.2.1 Evaluating Risk

When considering the "safety" of a system or process, the level of risk associated with the use must be evaluated. Is the risk acceptable? Can the risk be reduced? At what cost? Based on the risk presented, what are the economic and utility trade-offs when considering alternatives to reduce the risk?

For each of the major hazards, risks can be estimated from tests, experience, and medical and engineering judgment. This allows the analyst to evaluate the acceptability of the risks associated with the system. Hazard severity and probability levels, defined in Table 18.1, are valuable tools for possible risk reduction. The probability and severity of some of the events can be reduced through design—adding guards, interlocks, and warnings, for example. Other hazards result from operator behavior. The capacity of the employer to control user behavior and to block accident scenarios varies. User populations can be restricted to trained professionals. Conditions of acceptable use can be defined in instruction manuals and operator training programs.

18.2.2 The System Safety Analysis Process

In actual practice, the steps involved in the system safety analysis process are not necessarily performed in the order we indicate. Potential hazards, for example, are documented whenever they are recognized. This may be before the formal analysis has begun, or after what seems to be the completion of the analysis. Because the process is iterative, the analyst may find that he or she performs a number of steps, evaluates the findings, and then repeats portions of the analysis from a different point of view. There are no set rules to be followed when performing an analysis. The following are guidelines to assist the analyst in performing a systematic and adequate review.

18.2.3 Safety Task Description

Prior to beginning the analysis, the scope and depth of the analysis are determined and agreed to. The scope is defined by a number of interrelated factors and requirements, including the projected cost of the analysis, schedule deadlines, available manpower, and corporate risk if an accident occurs. The safety professional must also define who does what, how much control over the design he or she has, and of what an adequate, comprehensive analysis would consist. The definition of an adequate, comprehensive analysis is influenced by the magnitude of the potential risks and by the resources available.

The potential risk that an accident or mishap will occur can be weighed against the cost of the analysis and the corrective action required. The analyst may accept the certification of the project engi-

TABLE 18.1. Hazard Assessment

Hazard Severity

Category I, Catastrophic—may cause death or system loss.

Category II, Critical—may cause severe injury, severe occupational illness, or major system damage.

Category III, Marginal—may cause minor injury, minor occupational illness, or minor system damage.

Category IV, Negligible—will not result in injury, occupational illness, or system damage.

Hazard Probability

Descriptive Word		Specific Individual Installation	All Installations
Frequent	(A)	Likely to occur frequently	Continuously experienced
Probable	(B)	Will occur several times in life of installation	Will occur frequently
Occasional	(C)	Likely to occur sometime in life of installation	Will occur several times
Remote	(D)	Unlikely but possible to occur in life of installation	Unlikely but can reasonably be expected to occur
Improbable	(E)	So unlikely, it can be assumed occurence may not be experienced	Unlikely to occur, but ossible

Adapted from Department of Defense, Military Standard 882B, System Safety Program Requirements (March 30, 1984).

neer that there are no excessive risks associated with the process, or the analyst may verify the safety of the system by independently reviewing facility drawings and engineering analyses and calculations.

After the scope of the analysis has been determined and a definition of an adequate and comprehensive review has been agreed upon, the analysis is begun. The analysis continues until the safety-related problems that are exposed are satisfactorily resolved. The system safety analysis process is iterative and considers the entire system throughout its life cycle. It begins as an overview and becomes more detailed as specific areas of concern are recognized. The logic of each safety analysis process is the same, even though the emphasis or method of presenting the results may create apparent differences.

18.2.4 Understanding the System

First, an understanding of the System must be developed. This understanding must be complete enough to allow the analyst to prepare a written response to the following questions:

1. What does the system do? What is its intended function?
2. How does the system perform its intended function?
3. What raw materials are used?
4. What is the process, or what are the manufacturing steps?
5. What energy sources are involved and how?
6. What by-products are produced by the system?
7. Who is affected by the system, and what are the capabilities and limitations of these individuals?
8. How is the system built, used, maintained, repaired, and disposed of at the end of its useful life?
9. What are the interfaces with other systems or objects?

This list is illustrative only. A given system may not include all of the items, or it may include other items. The analyst must have a clear understanding of the entire system before the analysis can be completed. During this process the analyst documents hazards that are to be investigated later in greater detail.

Hazard Identification

After the system is well understood, general areas of possible hazard can be determined. At the beginning of the analysis, a hazard checklist can be helpful to ensure that all areas of concern have been evaluated. Hazards found on a typical checklist include:

1. Electrical—does the system use potentially hazardous voltages and currents?
2. Toxic materials—does the system use materials that are toxic or may become toxic (as in a fire)?
3. Fire—could the system present a fire hazard?
4. Acceleration—does the system have moving parts, or parts that might impact a person (springs, fluids, etc.)?
5. Explosion—are explosive materials or pressurized systems involved?
6. Chemical reactions—are potentially reactive materials used?
7. Heat and temperature—are hot or cold temperatures involved?
8. Mechanical—are rollers, gears, cutters, or similar items involved?
9. Pressure—are potentially hazardous pressures involved?
10. Radiation—are hazardous levels of radiation (ionizing, non-ionizing, infrared, visible, ultraviolet, microwave) used?
11. Vibration and noise—can the system cause injury or be damaged due to vibration and noise?
12. Miscellaneous—are other items such as biological hazards and environmental pollution involved?

After the categories of potential hazards have been determined, the system is analyzed to determine if the hazard is "credible." A credible hazard is one that can reasonably be anticipated to occur. Next, the level of risk is assessed to see if it is great enough to warrant corrective measures.

18.2.5 Life Cycle/Operational Modes

After it is determined which hazards are to be evaluated in further detail, the analyst must review the design of the system to determine how each of the specific hazards is involved during the system's life cycle. The analytical tools of system safety are used during this phase of the analysis. Depending upon the particular system and the scope of the project, it may be appropriate to use hazard analysis (HA), fault tree analysis (FTA), failure modes and effects analysis (FMEA), operating hazard analysis (OHA), or any other techniques that determines how specific hazards could result in an injury. These tools are also useful for documenting the analyst's activities and communicating this information to others.

As each specific hazard is analyzed, it is evaluated from the point of view of the entire system during each phase of the life cycle of the system.
Typical life cycle phases include:

- the *pre-use phase*, during which the concerns surround procurement, assembly, and installation of the system;
- the *use phase*, during which the system is operated;
- the *repair and maintenance phase*, during which operational safety systems such as interlocks and covers are defeated or removed during repair or maintenance activities;
- the *disposal phase*, during which long-term environmental safety can be a primary concern.

Possible exposure to a particular hazard may be minimal during normal operation but severe during maintenance operations. Sometimes it is useful to look at each phase of the life cycle separately. At other times it may be easier to follow a particular hazard through the series of phases. There is no hard-and-fast rule for determining which approach is preferable. It depends upon the system, the analyst, and the objectives of management. A number of approaches may be used, as necessary, at different stages of the analysis.

18.2.6 Corrective Action

As the analysis proceeds, the safety analyst looks for safety problems in the hardware, in human factors, and in the interaction of the machine with the rest of the building or plant. Historical data can be particularly useful in identifying potential problems.

It is necessary at some early point to determine which hazards to analyze for risk reductions. The original definition of the scope of the study will help to determine the relative priority of the various hazards.

Some types of hazards are more likely to be resolved than others. Minor hazards that involve a low level of risk and are fairly obvious can easily be resolved. Such hazards are normally corrected as a matter of course after they are identified. Another important category involves hazards that are specified in the regulations of administrative agencies such as the Occupational Safety and Health Administration and the Department of Transportation. These must almost always be resolved, as the alternative can be a morass of legal difficulties. Hazards specified in company policies or documents such as customer purchase specifications must also be given priority. The difficult part of the hazard analysis process concerns the "derived" requirements, or hazards that are not included in laws or specifications. These hazards exist because of the specific system's design and use, and must be considered.

After the system has been analyzed and a list of hazards generated, the analyst must assure that satisfactory corrective action is accomplished. The simple solutions and those dictated by law or specification are usually instituted after management is made aware of their existence. The safety analyst may be asked to suggest ways to comply with the hazard control requirements. The analyst must be assured that proposed corrective actions will solve the problem without introducing additional unacceptable hazards into the system.

The challenge for the safety analyst is in resolving the derived hazards. He or she must be able to demonstrate to management that the hazard actually exists and that the risks involved are serious enough to warrant corrective action. He or she will probably be asked what corrective action is required. Sometimes the problem is easily understood and specific recommendations can be presented. Usually the safety analyst is limited to making suggestions. It is up to management to determine which corrective actions to accept and how to implement them. The safety analyst must determine possible solutions, suggest approaches, and be flexible enough to accept alternative solutions that effectively meet the hazard control requirements.

Hierarchy of Corrective Action

There is a group of activities for reducing risk known as the hierarchy of corrective action. The use of all of these actions together to reduce a hazard provides the most protection. Often, only one or two of the precautions are necessary or practical.

The first effort to reduce risk is to evaluate the design. When possible, it is always preferable to reduce risks through changing the design of the workplace or selecting safer equipment and materials. The early involvement of people familiar with workplace abatement of hazards can alert process designers to the occupational safety implications of alternatives. The process can then be modified to reduce or eliminate the hazard. Design solutions include the use of the safest process chemicals and of fail-safe or redundant controls.

If the process or equipment is already installed and in operation, the problem of hazard abatement becomes more difficult. There is the additional cost of retrofitting hardware and changing the design and design drawings.

After the equipment and materials are chosen, hazardous parts, energy sources, or materials can be isolated from human contact by adding interlocks, guards, and other protective equipment.

When it is impractical to completely isolate a hazard, it is usually possible to have people subject to the danger wear protective equipment. Protective equipment for personnel can include eye protection, hearing protection, respirators, and special clothing.

When operator intervention is required, indicator lights or signals may be used to avoid more serious situations. Warning lights and alarms can be installed to signal imminent danger of accident. Personnel can then take immediate countermeasures or evacuate.

Labels and manuals providing special instructions or warnings are an important element in the effort to reduce risk. Warnings alone, however, are usually insufficient to reduce the likelihood or severity of accidents. When used, warnings must be easily understood by the persons they are intended to alert.

An additional measure of protection can be gained by requiring employees to follow special procedures when they are subject to hazards. This may require special training. Refresher courses may be required to reduce the possibility that they will forget instructions or become complacent.

Finally, the severity of damage to people and property in the event of accident can be reduced by installing sprinkler systems and other control equipment. Rescue or survival devices such as breathing apparatus and fire-resistant clothing may also be required.

Corrective Action Implementation

The corrective action chosen will be the most effective and realistic solution available, given the variety of requirements of the corporation.

Management will certainly be concerned with expense. The initial direct costs of proposed changes are usually highest for the design option and lowest for warnings. The level of effectiveness is directly related to costs, with redesign being the most effective and warnings the least effective.

The safety analyst typically interacts with the level of management directly involved in operating the system. Since costs the company may have to pay because of injuries are generally not directly chargeable to the manufacturing department, the safety analyst must present recommendations in a manner that makes line and top management understand their responsibility for future accidents if the recommendations are not followed. Any leverage that the safety analyst has is based on the fact that by not accepting a recommendation, management is accepting responsibility for the consequences.

VERIFICATION AND DOCUMENTATION
Once a method of resolution has been agreed upon, it is the safety analyst's responsibility to verify that modifications are performed in a manner that will decrease the risk to an acceptable level without introducing other unacceptable risks.

Finally, the accomplishments of the system safety analysis process must be documented. The documentation includes the scope of the analysis, a description of the analyses performed and the hazards that were found, the rationale for determining that the hazards presented risks that required action, a description of what was done to reduce the risks, and an explanation of the method and rationale used to verify that the risks were actually reduced.

18.3 INTERACTION OF SAFETY AND THE CORPORATION

A system safety perspective allows the creation of a comprehensive safety program encompassing the entire industrial enterprise. Most corporate departments affect the safe operation of the plant and must participate in the safety effort.

The safety professional rarely personally reduces accidents. Every aspect of the safety program requires the commitment of corporate management, equipment designers, process engineers, facility planners, manufacturing personnel, and the personnel, medical, and security departments. Safety professionals motivate and coordinate the commitment to safety in cooperation with a diverse group of people.

18.3.1 Interaction With Top Management

Top management is ultimately responsible for judging acceptable levels of occupational and product risk, since they are responsible for the firm's reputation and liability exposure. Failure to achieve the required level of safety may result in economic damages, regulatory sanctions, or criminal prosecution and jail terms in some cases. Also, top management has increasingly realized that corporate profits are related to the success of the total safety program. Studies have shown that an unsafe plant produces greater quantities of defective end products. Productivity increases as the safety of the workplace increases, and workers produce more and higher quality products. Corporate excellence in manufacturing increases customer acceptance, market share, and profits. The safety professional is the person delegated by top management to assure that safety *is* a part of corporate excellence.

18.3.2 Interaction With Product Designers

Product and process engineering have traditionally been concerned with end product quality, safety, and performance. Little thought is generally given during design to the hazards involved in manufacturing the product. Safety people must work with product design personnel to assure that new products incorporate safety features that protect not only the end user but also the workers who produce the product. Many workplace injuries can be avoided if safety professionals inspire end product designers at this crucial stage of the design process. When plant safety is taken into account during product design, a different set of design alternatives is often the result. These changes may have a big impact on the safety and ergonomics of production, with little or no change in product performance or cost.

18.3.3 Interaction With Manufacturing and Process Engineers

The safety professional works with manufacturing and process engineers to subject all proposed equipment and processes to system safety analyses. System safety techniques are used to create manufacturing and process specifications. The results of these exercises guide process design decisions and highlight areas for special training and procedural cautions. When creating a manufacturing or process specification, the entire sequence of operations during the process must be considered for safety. Material handling, production, maintenance and repair of production equipment, and emergencies and disasters must all be considered.

18.3.4 Interaction with Plant and Facility Engineers

The interaction of activities in the plant may pose hazards that an analysis of an individual process or work station would not disclose. The safety professional must work with the plant engineers who design new or modify existing facilities. The working environment must not pose unusual hazards to company employees, visitors, or the community. By-products or wastes from the operation must be handled so that the community and workers are not endangered. Changes in facility design at inception may avoid the need for costly retrofitting or, worse still, loss of morale, production, and workers' compensation costs resulting from employee injury. For the facility to be safe, a system approach to detecting hazards reducing risks is necessary. Environmental controls such as hazardous material management, air monitoring, and ventilation requirements are best considered at this stage.

18.3.5 Interaction with Plant Maintenance

Proper maintenance of equipment and facilities is critical to continued safe operations. The safety specialist must ensure that maintenance personnel are committed to the goals of the safety program. They should be alerted to hardware components that are crucial to safety, and they must observe safety precautions when working on equipment. Lockout procedures and other protective measures must be enforced.

18.3.6 Interaction with Purchasing

Procurement personnel should be advised by safety people about the safety requirements of equipment and materials purchased by the firm. A system safety approach to the firm's activities will reveal which features of equipment purchased for company use are crucial to safety. The interaction of different pieces of equipment and materials may pose hazards that are not obvious when each piece of equipment is considered separately. The acquisition of equipment with documented safety margins and effective warnings, instructions, and safeguards increases the likelihood of long productive use and reduces risks to company personnel.

18.3.7 Interaction with the Personnel Department

Safety professionals can help the personnel department select workers who will not endanger themselves or others on the job. Minimum physical requirements for jobs crucial to safety must be established, and a safety training program must be developed and maintained. Safety training should address normal, safe practices as well as procedures during emergencies such as fires, injuries, or natural disasters. Employee training in standard operations is also the concern of the safety professional. Proper training in lifting to avoid back injuries can have as much impact on the risk of accident as the safe design of production equipment. The adequacy of training must be carefully monitored by the safety specialist. Increased training may be recommended where problems are identified.

18.3.8 Interaction with the Medical Department

Safety specialists must work with medical personnel to ensure that workers are screened for work-induced health problems. Close monitoring of workers' health can provide early warning of undetected hazards in the workplace. Workers' health must be monitored to prevent ill or injured workers from compromising the safety of themselves or others.

18.3.9 Interaction with the Legal Department

The legal department may draw upon the skills of the safety professional to assist in regulatory compliance and reporting. This will involve OSHA, EPA, and other federal, state, and local agencies. The safety professional also provides product liability and workers' compensation litigation support and can assist counsel in accident investigation and discovery.

18.4 SYSTEM SAFETY TECHNIQUES

18.4.1 Hazard Analysis

The hazard analysis technique is used to identify hazards present in equipment and systems during installation, production, and maintenance operations and to describe how they occur in an intended and foreseeable use environment. This technique identifies how a hazard can result in personal injury or property damage. The technique organizes safety-related concerns into a common framework, listing each major type of hazard, the potential causes, and preventive measures. The hazard analysis is useful for highlighting hazards that could become the subject of more detailed study, testing, or sampling.

Certain aspects of safety such as guards, interlocks, safety margins, and warnings affect the *likelihood* of an accident. Other aspects, including personal protective equipment such as roll bars and seat belts for tractors and front-end loaders, reduce the *consequences* of an accident. Reasons for component and procedure failures of the system involved are characterized.

Hazard analysis considers two basic items. First, factors that can prevent the "correct" functioning of the subject equipment are identified. Second, other safety-related problems that may be created by the equipment are identified. In both cases, the analysis considers the relative likelihood and severity of the various problems.

The hazard analysis technique considers, among others, the following aspects of workplace safety:

1. Potentially hazardous components
2. The interface between critical components and between several pieces of equipment
3. Possibly hazardous operations and emergency conditions
4. Alternative design approaches to potentially hazardous conditions
5. Operating and environmental factors that may contribute to an accident
6. The impact of government and industry standards

Hazard analysis is based on a thorough understanding of the function, design, and application of the system. Information sources include design drawings and engineering analyses; functional flow diagrams and product literature; government or industry studies and specifications; and the personnel who design, manufacture, test, sell, use, and regulate the product.

All potential safety problems must be considered. For problems judged likely to occur or that have severe consequences, "preventive measures" must be addressed in the analysis. Preventive measures are actions taken to prevent the occurrence of safety problems. These measures include, but are not limited to, equipment design modification, incoming inspection to ensure receipt of defect-free equipment, and adequate training of users.

Hazard analysis considers (1) hazard description; (2) potential problem causes; (3) failure mecha-

TABLE 18.2. Potential Hazards

Kinetics	*Pressure*
Inadvertent motion	Compressed gas
Movement of loose objects	Accidental release
Impacts (sudden stops)	Container rupture
Falls	Overpressurization
Falling objects	
	Leak of Material that Is:
Chemical (Nonfire)	Flammable
Corrosion	Toxic
	Reactive
Electrical	Corrosive
Short circuit	Slippery
Overheating	Asphyxiating
Electrocution	
	Radiation
Contamination	Non-ionizing
Harmful matter	Ionizing
Incompatible matter	
	Toxicity
Explosives	Asphyxiating
	Irritant
Flammability	Poison
	Carcinogen
Temperature	
Hot surface	*Vibration and Noise*
Cold surface	Noise source
Increased/decreased gas pressure	Metal fatigue
Flammability	Loosening of connectors
Volatility	"White fingers"
Reactivity	
Size changes causing damage to equipment	
Mechanial	
Shape (edges or points)	
Moving equipment	
Pinch points	
Stability/toppling tendency	
Ejected parts or fragments	

Source: Adapted by P. Schroeder, Zurich Insurance Company, from W. Hammer, *Product Safety Management and Engineering*, Prentice-Hall (1980).

nisms or scenarios; (4) the probability of their occurrence; (5) the effects of their occurrence; and (6) measures to be taken to prevent their occurrence. The analysis can be guided by a checklist and engineering model. Details of the checklist and model will vary from system to system, as will the depth and scope of the analysis being performed.

Potential hazards are covered in Table 18.2. Potential consequences are listed in Table 18.3.

Example Analysis—Hazard Analysis Technique

Hazard analyses of semiconductor processing equipment have been performed to assess the interaction of personnel and toxic gases, high voltages, and microwave and radio frequency radiation. The hazard analysis techniques used in this and later examples can be applied in many different kinds of situations and are not limited to the specific types of circumstances and equipment described.

A high-pressure oxidation system was assessed using the hazard analysis technique (see Table 18.4). This type of system is used to perform high-volume wafer oxidation, which is one of the steps in the production of semiconductor wafers. The objective of the device is to produce the maximum quantity of high quality defect-free wafers.

Because hydrogen and oxygen are used in this system, the potential for an explosion was evaluated. Gaseous hydrogen and oxygen are present in a pressurized vessel at temperatures approaching 1000°C. To guard against a detonation capable of causing the pressure vessel to rupture, several safety devices are used (such as pressure relief, gas sensing, and pressure sensing). To produce a catastrophic explosion, a combination of several failures, both human and mechanical, is necessary. A "worst case" situation could possibly cause the vessel to rupture, but this is considered a highly remote possibility for normal use with trained operators. However, the possibility of an explosion is increased by nonstandard use (such as checkout, nonstandard procedures, maintenance, and training) and by the use of untrained or unqualified operators.

Other hazards, such as exposure to high voltage and thermal burns, were also considered as possible causes of accidents involving the system. One major weakness uncovered during the hazard analysis process was that documentation involving the device was incomplete and inconsistent. Because these units are sometimes custom designed, one device may differ from another, and a given device may not be identical to the one described in the instruction manual. Such discrepancies, and lack of adequate information, can potentially result in human errors during installation, maintenance, and operation.

Based on the analysis, improvements were suggested to increase the safety of the system. Most important was a lockout device to prevent access to hazardous parts of the equipment during operation. On the original system an operator could remove the back panels with a screwdriver while the machine was in use. Removal of the back panels provides access to high-voltage equipment, gas pressure settings and regulators, and other sensing and logic devices. It was suggested that these items be safeguarded to prevent access and "adjustments" while the unit is in operation. The need for better "human engineer-

TABLE 18.3. Potential Consequences

Mechanical Impact	*Chemical Impact*
Cuts	Burn
Punctures	Asphyxiation
Bruises	Organ damage
Broken Bones	Respiratory system damage
Particles in eyes	Circulatory system damage
Crushing	Dermatosis
Strains	Cancer
Noise and vibration	Nervous system damage
Shock (pressure)	Teratogenic effect
Drowning	Mutagenic effect
Thermal Impact	*Biological Impact*
Burns (heat)	Infection
Burns (cold)	Intoxication
Cold immersion	Poisoning
Heat exhaustion	
Wind chill	
Electrical Impact	*Radiation Impact*
Shock	Burns
Burns	Genetic

Source: Adapted by P. Schroeder, Zurich Insurance Company, from W. Hammer, *Product Safety Management and Engineering*, Prentice-Hall (1980).

TABLE 18.4. Hazard Analysis—High-Pressure Oxidation System

Category	Consequence	Cause	Preventive Measures
1. Rupture of pressure vessel	Serious injury or death	A. Mechanical failure of pressure vessel operating within normal pressure range (less than 400 psi or 25 atmospheres).	• Each vessel is hydrostatically tested to 680 psi or 45 atmospheres per applicable ASME code provisions. • Vessel has burst pressure of 10 times maximum rated operating pressure. • Periodic inspection for corrosion of components (such as resulting from HCl and water purging).
		B. Mechanical failure of pressure vessel operating above maximum certified pressure level.	• Pressure regulator. • Pressure relief/check valve in gas tubing set below maximum operating pressure. • Vessel and tube overpressure switch to vent at maximum operating pressure. • Relief-valve set at, or just above, maximum operating pressure. • Gas tubing is used with burst pressure below burst pressure of vessel.
		C. Mechanical failure of pressure vessel due to detonation of H_2O_2 gases	• Hydrogen sensors to detect leaks or to detect hydrogen-rich condition. • Fail-safe valving to shutdown process gas flow in the event of power loss or abort. • Interlocks to prevent H_2 gas flow before there is O_2 gas flow (thus preventing hydrogen-rich condition). • High-pressure H_2 flow and low-pressure flow detectors. • Temperature sensor indicating operation below 600°C (hydrogen ingnition temperature).
2. Rapid decompression of pressure vessel	Serious injury or death	A. Opening of loading door when unit is pressurized.	• The load door is equipped with a microswitch that only permits the door to be actuated if the pressure drops below 1 psi.
		B. Undoing bolts(s) of pressure vessel when unit is pressurized.	• Provide some indication (oral or visual or both) that unit is under pressure. • Label/warning affixed to potential danger points. • Training of operating/maintenance personnel.

TABLE 18.4. Hazard Analysis—High-Pressure Oxidation System *(continued)*

Category	Consequence	Cause	Preventive Measures
3. Toxicity	Minor to serious injury	Release of HCl gas (normally used to flush the system) due to:	
		A. Ruptured plumbing or leak at fittings.	• HCl pressure regulator and inspections.
		B. Inadvertent release by operator.	• Proper training of personnel.
		C. Improper maintenance procedures.	• Adequate documentation and training of personnel.
4. Electrical shock	Minor to serious injury	A. Contact with high voltage on inside of unit.	• Lockout mechanism that prevents access when power is turned on.
		B. Shorting of high voltage line to chasis of unit.	• Proper grounding of unit, inspection checks, and proper maintenance training.
5. Excessive heat/ burns	Minor to serious injury.	A. Loss of water cooling.	• Water flow meter with indicator. • Use only D.I. water to prevent plugging by scale. • Temperature sensor to indicate overheating and to abort operating.
		B. Contact with hot vessel.	• Operator will wear thermal gloves.
		C. Contact with hot wafers or wafer boat.	• Allow sufficient time for cooling to safe temperature after operating (proper training).
6. Flammability	Serious injury to death	Presence of hydrogen as a process gas.	• Proper containment and handling during installation, maintenance, and operation. • Ground all lines and equipment.

ing" of the operator control panel was also crucial. The combination of lights, push lights, toggle switches, dials, and meters was such that an operator could not quickly comprehend which indicators and controls were related, and therefore could not readily prioritize and utilize the information presented. An emphasis on human engineering is especially important in providing sufficient information for effective responses to emergency situations.

18.4.2 Failure Modes and Effects Analysis

Failure modes and effects analysis (FMEA) is a system safety engineering technique used to assess each type of safety-related failure associated with equipment. The technique requires a detailed evaluation of the equipment being assessed. It lists each "component," the potential failure modes, and their possible effects.

FMEAs are performed at the component level to determine possible ways that equipment can fail and to determine the effect of such failures on the system. The FMEA is used to assure that component failure modes and their effects have been considered and either eliminated or controlled; that information for maintenance and operational manuals has been provided; and that input to other safety analyses has been generated.

The FMEA provides early evaluation of an installation, with recommendations for improvements. It documents possible failure modes in the system's design; determines the effect of each identified failure mode on system operation; identifies items critical to workplace safety; assists in the evaluation of equipment design features such as redundancy; generates recommendations for equipment design improvements or contingency options; allows priorities to be set for receiving, testing, inspection, and maintenance; provides information that can be used to determine the causes of accidents; classifies each potential failure to a criticality category and assigns a probability of occurrence of potential failure; provides a basis for comparison in the event of subsequent failures and corrective action; provides information for more sophisticated system safety analyses; and provides information for operation and maintenance manuals, handbooks, and training programs.

FMEAs allow evaluation of failure sources by predicting the effect of the failure on safety. A failure is defined as the inability of an item to perform its required functions safely. (Defects and discrepancies that do not adversely affect safety are *not* considered failures). A failure mode is the physical description of the manner in which a failure occurs (e.g., open circuit, leak structural failure).

A functional statement is prepared for each system and subsystem being analyzed. These statements give a narrative description of the operation of each item for each life cycle mode. Potential failure modes for each item are analyzed and their effects on the system determined. The analysis determines the effect of the failure on the operation of the subsystem and complete system, the criticality of each failure mode, the frequency of each failure mode, and the preventive measures used to eliminate or reduce failure mode frequency and/or criticality.

The criticality of the effect of each failure is categorized. A frequency (or probability) of loss is assigned for each failure mode.

Once the FMEA is complete, a critical item list (CIL) can be prepared. The CIL identifies failures whose effects exceed a predetermined threshold of risk. The CIL serves as input to further safety analysis efforts. Each item on the CIL is examined in detail to determine the nature and extent of the safety hazards associated with the item. This analysis aims at the elimination or reduction and control of the hazards.

Many FMEAs are prepared as a cooperative effort between safety, manufacturing, and equipment-design engineers. Manufacturing and equipment-design engineers are generally best qualified to identify failure modes and establish the effects. Safety engineers' expertise is most useful in determining the expected frequency and criticality of occurrence and in establishing the seriousness of the failure.

The FMEA can be used to expand on particular safety concerns that surface during the hazard analysis. Highly probable events with severe consequences can be identified for corrective action. Different equipment can be purchased, or components with lower failure rates can be selected.

Example Analysis—Failure Modes and Effects Analysis Technique

A failure modes and effects analysis of an ion implanter is shown in Table 18.5 and Table 18.6 for the implanting and bottle-changing operations. This analysis showed that alterations in existing ventilation systems and revision of emergency procedures were required.

An ion implanter is a device designed to implant source material during the semiconductor production process to produce particular electrical characteristics in the completed wafer. The first requirement of an ion implantation system is the ability to generate ions of the desired atomic species. A gaseous or solid source is used, generating ions by boiling them off in the ion source chamber. The correct ions are separated from any others that may be present by bending them through a preset angle using an electromagnetic field. X-rays are generated as a by-product when the ion beam strikes any dense material. The selected ions are then accelerated using radio frequency radiation; a wave guide directs the ions to the target wafer. These systems use high voltages and extremely toxic source materials such as arsenic. They produce toxic by-products that must be prevented from contacting personnel or polluting the environment.

Exposure guidelines have been developed by the American Conference of Governmental Industrial Hygienists for the various source gases. Threshold Limit Values (TLVs) for the source gases normally used in the ion implanter have been developed. Periodic sampling of the workplace was recommended to assure that exposures were within acceptable levels.

18.4.3 Job Safety Analysis/Operating Hazard Analysis

Job safety analysis and the operating hazard analysis are techniques for reviewing a work activity for inherent or potential hazards. The job is broken down into its individual actions and each activity is described in detail. Information about the job is collected either through direct observation of the employee or discussion of the job with one or more of the people who perform it. The potential hazards of each element of the task are listed, along with ways to eliminate or reduce the effects of the hazard.

Example Analysis—Job Safety Analysis/Operating Hazard Analysis Technique

In the case of an industrial ultrasonic cleaner used to clean parts in preparation for later processing, one hazard is the danger of using a toxic or flammable solvent. The process should be designed such that a nontoxic and nonflammable solvent will do the job, though this solution may involve altering the manufacturing process so that the need for a strong cleaner is eliminated. The operational steps for an ultrasonic cleaner are installation; operator training; filling the system with chemicals; operation (open top,

TABLE 18.5. Failure Modes and Effects Analysis, Ion Implanter, Toxicity Hazard (Implanting Operation)

System	Component	Failure Mode	Failure Effect
Exhaust	Pumps	Exhaust air flow less than 1200 cfm. Ambient air inlet less 200 cfm. System shut down due to non-ion implanter emergency.	Inadequate dilution of toxic gases.
Scrubber	Water pump	Inadequate water flow.	Increased environmental pollution.
		Inadequate water spray.	Excessive gaseous concentrations.
Alarm/leak sensor	Pressure differential sensor Leak detector		Lack of warning of presence of toxic gases.
Ion implanter gas delivery system	Swage fittings	Inadequate seal	Emission of toxic gases.
	Gas bottles	A. Leak at fitting. B. Improper operation by personnel.	Emission of toxic gases.
	Plumbing	Hole in plumbing.	Emission of toxic gases.
	Breathing apparatus	A. Unit malfunctions or is unavailable in event of gas leak. B. Improper fit. C. Improper operation by personnel.	Inability to evacuate/rescue safely.
Electrical system		Loss of all electrical power to implanter, rendering internal venting and pumping system inoperative. Gas feed manifold interlock fails to shut down gas feed in event of external system malfunction.	Accumulation of toxic gases in implanter.

put part in basket, close top, turn on, remove part); maintenance; and disposal of by-products. Hazards or failure modes and their causes and effects are assessed for each step. Safeguards that were recommended for the most serious risks included interlocks, protective equipment, facility requirements, instructions, and warnings.

For the ion implanter discussed earlier an evaluation determined the operational modes to be as follows:

Operational Modes	Hazard Control Requirements
I. Initial installation of equipment	Effective training for operator of scrubber and exhaust system.
II. Implanting	Operator training.
III. Changing gas bottles	Proper procedure and breathing apparatus.
IV. Emergency/rescue situations	Breathing apparatus and proper training.
V. Disposal	Disposal of gas bottles (usually returned to gas supplier) and process by-products.

One of the higher risk operations is changing source gas bottles. The operation must be performed by a knowledgable operator wearing (or having available) a self-contained breathing unit. The exhaust system must be operational. The proper selection, operation, and maintenance of the gas supply, exhaust system, and scrubber are the responsibility of the user. Guidelines for appropriate ventilation are supplied by the equipment manufacturer.

TABLE 18.6. Failure Modes and Effects Analysis, Ion Implanter, Toxicity Hazard (Bottle-Changing Operation)

System	Component	Failure Mode	Failure Effect
Exhaust	Pumps	Air flow less than 1200 cfm. Ambient air inlet less than 200 cfm. System shut down due to non-ion implanter emergency.	Inadequate dilution of toxic gases.
Scrubber	Water pump	Inadequate water flow.	Excessive concentration of dissolved gases.
		Inadequate water spray.	Excessive gaseous concentrations.
Alarm/leak sensor	Pressure differential sensor Leak detector		Lack of warning of presence of toxic gases.
Ion implanter gas delivery system	Swage fittings	Adequate seal under vacuum but inadequate seal under pressurized conditions.	Emission of toxic gases.
	Gas bottles	A. Fracture of bottle or cap.	Excessive gas concentrations.
		B. Failure to keep gas supply bottle under exhaust hood when changing.	Excessive gas concentrations.
		C. Regulator malfunction.	Improper gas concentrations.
	Breathing apparatus	A. Failure to use during bottle-changing operations.	Possible exposure to excessive gas concentrations.

Exposure to electrical shock can be minimized by the use of removeable key-operated safety interlocks and circuit breakers.

The interface between the ion implanter, facility electrical system, and exhaust system can be a source of safety problems. Emergency evacuation procedures were developed for situations when the exhaust system was not functioning.

If the exhaust system is not functioning there will be inadequate ventilation at the implanter in the event of a toxic leak. If there is no toxic gas detection system, the operator should use breathing equipment and evacuate the area immediately. The surrounding area in the building should also be evacuated until the exhaust system is again operative.

If there is a gas detection system, the operator may complete the current run. He or she should not begin a new run until the exhaust system is functioning.

18.4.4 Fault Tree Analysis

Fault tree analysis (FTA) is the technique used to assess the possibility that a specific hazardous event will occur. It visually depicts the interrelationship of combinations of hardware failure, environmental conditions, and human error that can result in the occurrence of a specified accident. The accident under investigation is defined as the top event of the fault tree.

The fault tree is used to provide documented evidence of compliance with safety requirements, to identify the ways the system can fail, to isolate specific failures, and to assess the impact of design changes and alternatives. It is used to improve installation design. One advantage of this logic method is its ability to enhance communication between safety, engineering, management, and other participating departments.

There are two phases of the fault tree method. First, the system is modeled by developing a logical representation of the system's hazard and failure characteristics. Second, the model is analyzed to determine the qualitative and quantitative implications of the logic diagram.

The fault tree is so structured that the undesired event appears as the top event of the fault tree. This top event defines the scope of the logic model, the bounds of the analysis, and the resources, time, and money that will be required.

Below the top undesired event are the sub-undesired events. These include the potential hazards and failures that are the immediate causes of the top event. These events may be inadvertent or unauthorized functions or the failure of the system to perform a required operation. They are determined from preliminary analysis of specific safety considerations. As the analysis proceeds to completion, engineering changes and increased knowledge about the system will require that modifications be made to the tree structure.

At the second level in the logic tree, the major systems that can cause the sub-undesired events are depicted. Analysis of flow diagrams, logic diagrams, and schematics of the system describe the interrelationship between the major system level and the subsystem and detailed hardware level. At this third level, contributing events and failures are diagrammed systematically to show their relation to each other and to the undesired event being investigated.

The analysis process begins with the event that could directly cause the undesired event and works down through the system to determine combinations of events and component failures that could bring about the accident. The resulting diagrams are called "fault trees." When more than one event is involved, the tree indicates whether they must all occur in combination or whether they may occur alone. The circumstances under which each of the events in the logic tree could occur are determined. This involves examining each component or subsystem capable of producing an event in the fault tree and determining how its failure can contribute to the undesired event.

Analyzing the Logic Tree

After the logic tree has been constructed, suitable mathematical expressions are developed using Boolean algebra. These expressions are the mathematical statement of the "and," "or," and inhibit relationships, and are evaluated by hand or with suitable digital computer algorithms.

A qualitative analysis is performed to identify the minimal combinations of conditions that will cause the accident to occur. These combinations are known as the minimal cut set. They are important because they give the unique modes by which the top event can occur and represent the failure scenarios of the system. If a quantitative analysis is to be performed, the probability of occurrence, or failure rate, of each component must be provided. These rates may be known or may be obtained from supplier's test data, comparison with similar equipment, experimental data, or engineering analysis. Quantitative analysis can then be performed to determine the likelihood of the intermediate top events, the probabilities of the critical paths, and the importance ranking of the component failure modes and minimal cut sets.

At this point, changes in the configuration of the logic tree may be made and the analysis rerun for the purpose of trade-offs. Also, optimization studies may be performed or failure rate data altered for sensitivity evaluations. The results of the analysis are next evaluated and the model reanalyzed until final results and recommendations are produced.

Fault Tree Construction

The first step in the construction of a fault tree is to define the top undesired fault event to be considered in the analysis. The top event can be expressed in terms such as "explosion" or "personnel exposure to toxic substances." An effective approach is to consider undesired events that are of primary concern. Then the top structure of the fault tree can be depicted as a combination of the top undesired event and sub-undesired events, including potential accidents and hazardous conditions that are the immediate causes of the top event.

Next, an understanding of the system and its intended use must be acquired. Any entity can be labeled a "system," but it should be accurately defined. The boundaries of the system and its elements must be defined as early as possible and revised as required during the course of the analysis. The definition of the system should include operating conditions as well as environmental situations and the human role in system operation. Such delineation establishes the limits for succeeding steps in the analysis process and often reduces complex systems to manageable parts.

Once the top event has been defined and the system is clearly identified, the system is analyzed to determine the higher order events that can cause the top event to occur.

The next level in the fault tree is the major system level. The analyst divides the operation of the system into phases and subphases. During this step, the analyst determines the logical interrelationship of lower events that can cause the higher order events.

Finally, the logical relationships among fault events are developed. A fault event description must include *what* the system or component fault state is and *when* that system is in the fault state. A fault is a condition (not necessarily a failure) of a system, subsystem, or component that contributes to the possible occurrence of the undesired event. It is important to be explicit in defining each fault event in order to eliminate ambiguities that could arise in subsequent development of the tree.

Fault events can be divided into two categories: basic events and gate events. Basic events are events that are related to a specific failure rate and fault duration time. They represent the logical transition of system elements (usually at the component level) from an unfailed state to a failed state. These events are used only as inputs to a logic gate (never as outputs). In a fault tree diagram, basic events are usually depicted by circles and diamonds, depending upon the information and status of each of these events. On the other hand, a gate event is an event (such as system failure) that results from output of a logic gate. Since the gate event is dependent upon the input events and the type of logic gate function, it is a dependent event. The gate event is not the logic gate itself, but the *result* of the logic gate function and the input events. The gate event is depicted by a rectangle above the logic gate. As development of the fault tree progresses, gate events on one level naturally become outputs of gate events on the next lower level.

Fault events can be characterized as being "state-of-component" or "state-of-system" faults. If it is possible for a failure of the subject of the fault event to be the complete cause of the event, then the fault event is a "state-of-component" fault. If the fault event cannot be triggered by a simple component failure, it is a "state-of-system" fault.

A primary failure is the failure of a component within the design envelope (i.e., failure due to inherent characteristics of the system element under consideration). It is assumed that a primary failure of one component does not contribute to the primary failure of another component.

A secondary failure is the failure of a component outside the design envelope (i.e., failure due to excessive environmental or operational stresses being placed on the component). The development of a secondary failure usually requires extensive knowledge of the system. The analysis can become quite involved since significant fault interrelationships must be developed. A secondary failure, unlike a primary failure, can be caused by a primary or secondary failure of another component or set of components in the system. A secondary failure is due to either an operational stress in the component or an environmental stress placed on the component. Operational stress of an individual component may result from out-of-tolerance conditions of amplitude, frequency, duration, or polarity. These conditions may be considered singly or in combination. Environmental stress is an energy input that the component may experience from system energy sources.

Command Fault Event

A command fault is the inadvertent operation of the component due to failure of a control element. It represents signal flow through the system along a single branch of the fault tree. It terminates in primary and/or secondary events. An example of a command fault is the inadvertent application of power to a relay coil. The closing of the relay contacts is then considered to be a command fault. A command fault is always a "state-of-system" fault. A command fault event creates an orderly and logical extension of the analysis at each level of the fault tree.

Example Analysis—Fault Tree Analysis Technique

A fault tree analysis was performed on the high pressure oxidation system described earlier to evaluate how a rupture of the pressure vessel could result. An excerpt from the tree showing the upper logic levels is contained in Figure 18.1. The definitions for the symbols used in the tree are shown in Figure 18.2.

18.4.5 Software Safety Analysis

Many industrial processes are now developed to be controlled by software. The introduction of this new technology presents unique safety problems. Some systems rely solely on software safety interlocks, so it has been necessary to develop hazard controls unique to software-controlled systems that will prevent unauthorized personnel from defeating these interlocks.

Customers usually specify a particular process when they purchase a machine and use the machine according to parameters specified by its manufacturer. Misuse of the equipment falls within one or more of the following three categories:

1. The use of processes other than those specified or approved by the manufacturer
2. The use of incorrect interfaces between the device and the environment
3. Modification of the hardware or software

Software Hazards and Hazard Control Requirements

Possible hazardous situations and their prevention and/or resolution must be documented. Processes and the local environment must be monitored by the software to determine when unsafe conditions exist or when processes are out of tolerance and could produce an unsafe condition. If an unsafe condition is found to exist, the software will activate an alarm, give the operator information concerning the problem, advise the operator about what to do next, and if appropriate, take steps to make the system safe in the best manner possible given the information that it has at that time. The software will continue to monitor the system to make corrections to the procedure for making the system safe if information becomes available that might change that procedure.

It is crucial that the operator not be able to input out of sequence or to input incorrect commands. Unnecessary controls should not be capable of being energized, and the control panel must be designed so that only the active controls are labeled with their current function.

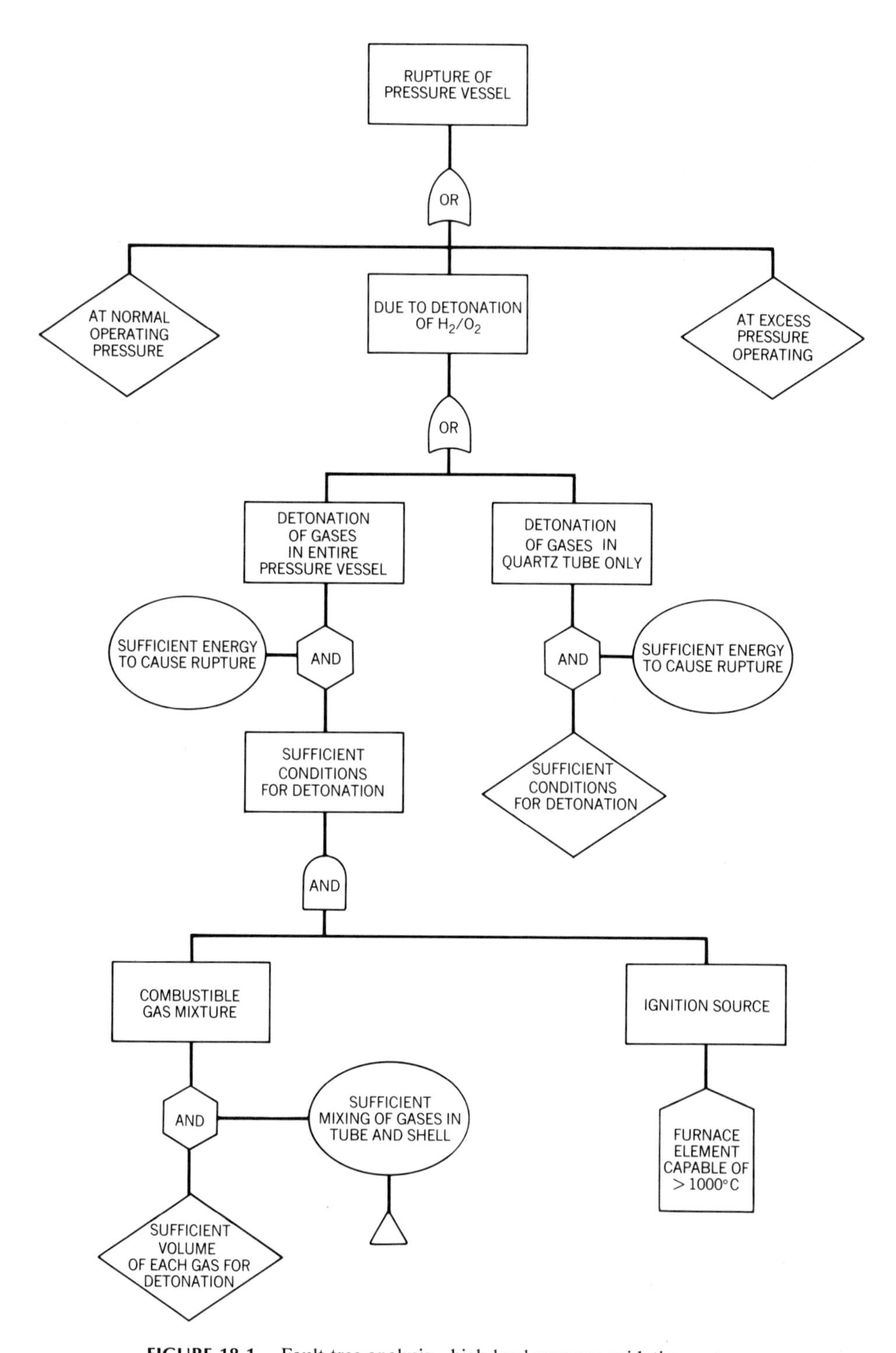

FIGURE 18.1. Fault tree analysis—high-level pressure oxidation system.

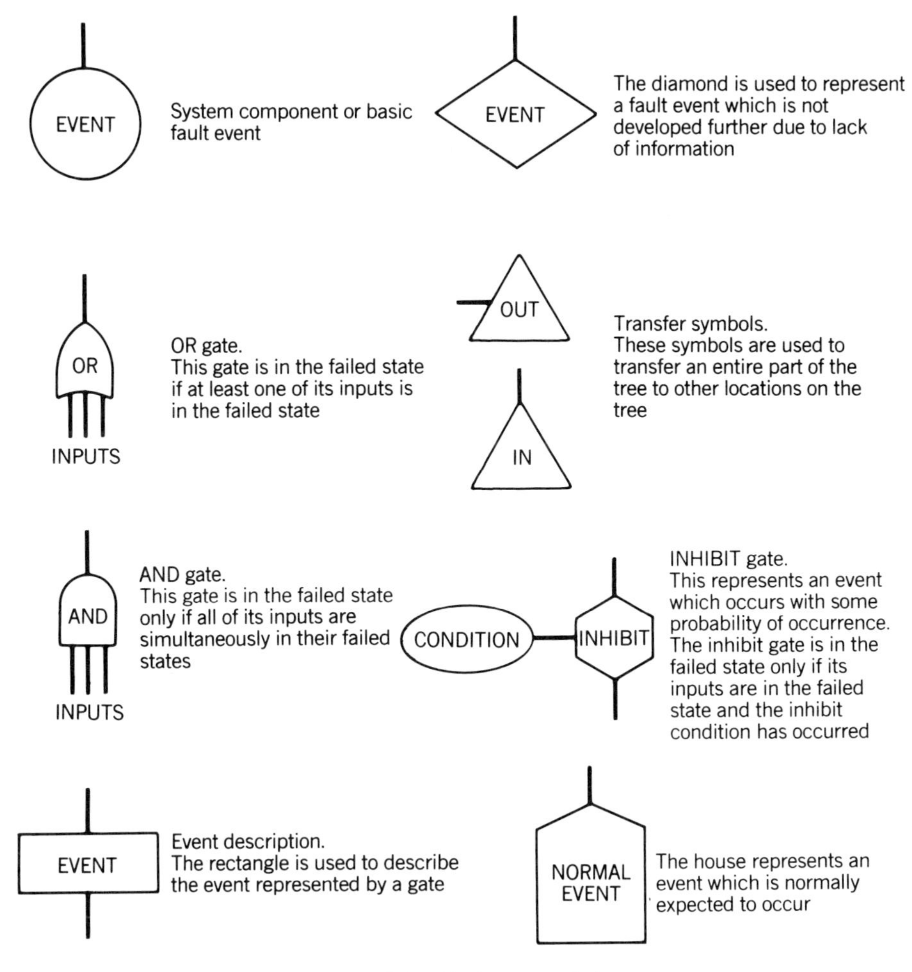

FIGURE 18.2. Fault tree symbology.

The software must check for parity before using any stored values. Parity is initially set by causing every data word to be odd or even. To make this possible, a separate bit is allowed for each word. If a bit in the word changes for some reason, then the word will change from odd to even or vice versa, and the parity check will fail. Loss of parity indicates that some of the information has been lost or changed and thus that the safety of the system can no longer be assured. If parity is lost an alarm must be given and the system made safe until the problem can be corrected.

It is important that the software be able to determine its position in the programming sequence. The proper steps must be verifiable at any given point within the program. This requirement is included to assure that a software jump, which could put the program into any location at any time, has not occurred. If a software jump does occur and is not detected, almost anything under software control could happen.

A list of known hazardous situations must be maintained in the software. The software must be able to pause the operation, activate an alarm, notify the operator of the problem, advise the operator about appropriate responses, and allow the required procedures to be carried out when needed.

If an error or an out-of-tolerance parameter is measured, an alarm must be activated. The system must be placed into a safety mode, the operator must be advised about the problem, and appropriate responses must be indicated. The software must allow the operator to carry out the required procedures.

There must be an automatic safe shutdown procedure from any point within the procedure. The software should notify the operator of the condition and then determine the best way to make the system safe with the available information. The operator should be notified prior to each shutdown step to allow his or her intervention. If the operator's response does not agree with the software's evaluation, the operator should be notified of the difference, told about the possible outcome of the intervention, and asked to verify the command. It is possible that in the event of a failure or error the software will not have access to the information necessary to safely shut down the system. It is also possible that it will indicate the correct procedure, but that the operator will try to do something hazardous. Human operators must have ultimate control, but they must be given as much information as possible to help them make the correct decision, and they must be checked so that an inadvertent response does not create a hazardous situation.

The system must permit manual override of the controls. This is required for maintenance and for emergency situations. There must be ways to physically override the system, such as the use of hand-controlled valves, and also a way to override the software control. This override capability must have safeguards so that the operator doesn't accidentally activate the manual override.

The software must be able to tell if the sensors are giving valid information. If a sensor fails for some reason, the system must be able to detect that failure so that it doesn't proceed using incorrect information.

The system must be able to tell if it is sending the intended information to the operator. There must be a feedback between the program and the display devices to make sure that the displayed information is the same as the intended information.

Output messages to the operator should not be in an abbreviated form, unless that form is more easily recognized than the full text. It is possible for the operator to misinterpret the information if abbreviations are used, especially when error messages infrequently seen by the operator are sent. An example of an abbreviation that probably *would* be allowed is the term "laser," which would be much more likely to be understood than "light amplification by stimulated emission of radiation."

The system should take into account specific situations to determine if an emergency shutdown is appropriate. There are situations where the safest shutdown procedure is not to shut everything off. If situations might exist that would require some other procedure, they must be taken into account by the software.

Especially hazardous operations require a voting system to determine what to do next. If the situation warrants it, several sensors may be interrogated and the one that is different can be assumed to be incorrect. If this happens, the operator must be informed so that the failed sensor can be replaced.

The software must be able to keep track of hazardous locations and operate safety interlocks. Not only must the interlocks be activated, but the fact that an interlock is being used or overridden must be indicated. Two sensors may be required at the interlocked locations, one to control the interlock and the other to indicate when the interlock has been overridden or is not operating. A combination of physical and software interlocks may be required for the system, with the more hazardous situations requiring physical interlocks and/or lockouts.

The main safety requirement is to prevent the operator and the software from causing a hazardous situation by performing an incorrect step or series of steps. The best way to do this is to make sure that the information used is correct, that assumed conditions exist, and that the correct procedures to use with these conditions are understood and performed. Compliance with the safety requirements assures that these procedures will be performed correctly during normal operations and when a failure or unexpected condition occurs.

Example Analysis—Software Safety Analysis Technique

This analysis concerns the safety requirements of the software for the microprocessor that controls a semiconductor wafer etcher. These systems are designed for fully automatic operation, with an operator in the area to load and unload wafers while monitoring the process. A system of load locks and robots may be incorporated into the design to move the wafers from the clean room into and out of the process chamber.

Because the system is automatic, a microprocessor controls all of the parameters necessary to process the wafers. This includes control of the gas supply valves, pumping systems, RF power supply, end-point detection devices, and the load locks and robots.

The microprocessor must have access to a large number of measurements to properly control the system. These include:

1. Gas flow rate into the process chamber
2. Pressure at the following locations
    a. the gas supplies
    b. the process chamber
    c. the load-lock chambers
    d. the pumps
    e. the effluent lines
    f. the interior of the cabinets and ambient (the pressure difference is used to check that a positive air flow is occurring within the cabinets)

3. Temperature at the following locations:
   a. within the process chamber
   b. ambient
   c. within motor windings
   d. within the electronic systems, including the microprocessor
   e. in the cryogenic refrigeration system
4. RF power (frequency, power level, and reflected power)
5. End-point detection systems, which may be one or more of the following types:
   a. laser
   b. spectrographic
   c. voltage drop between the cathode and the anode
6. Position and state of gas valves
7. Gas detectors (leak detection)
8. Position of door (open or closed)
9. Position and action of robots and wafers
10. Gas flow rates out of the system
11. When protective covers have been removed
12. Time (to determine when processes have taken too long)

Operators of these systems usually have some formal training in the use of the system. The operation of the equipment is quite repetitious and not very interesting. Therefore, personnel are less likely to be able to respond to unusual situations or emergencies without specific detailed instructions and training.

The users of this system are manufacturers of integrated circuits for use in electronic devices. The manufacturers are located worldwide, so the system must meet a wide variety of codes and requirements. The users of this equipment can be expected to be quite technologically sophisticated. By necessity they hire process engineers who are familiar with this type of equipment.

18.4.6 Robotic System Safety Analysis

Robots can perform in environments that pose hazards for humans. They are designed to work in a defined area, called an envelope, doing predictable, routine jobs. Their use in factories is accelerating rapidly. Because their widespread use is a recent development, robot users and manufacturers need to be alerted to their particular hazards, including unpredictable action patterns, the ability to move in free space, and the possibility of unexpected reconfiguration. Specific potential dangers are the possibilities that a worker will be pinned, crushed, or burned.

General Robotic Safety Requirements

Guarding requirements for the working envelope of the robot are necessarily complex. The robot cannot see or reason. Any obstacle placed in its programmed path will be struck. During a failure mode, the robot can move to locations outside its programmed area. Programmed nonmovement periods make it unsafe to assume that the robot is not about to move because it is not in motion.

Safety interlocks can provide reasonable protection for parts and equipment, but for the safety of personnel the complete working area must be enclosed by a fence. The fence should be high enough to physically prevent personnel from inadvertently walking into the reach of the robot. Clearance at the bottom should be sufficient to allow for maintenance and cleaning of the ground, but small enough to prevent entry. Warning signs must be placed on the fencing and indicate that access is for authorized personnel only. The floor surrounding the robot must be marked to indicate the complete possible envelope of movement of the robot.

An overlap of work location arises when the robot is welding at one station while the manned setup is performed at an adjacent station. A physical barricade must be erected that can stop the robot's motion and ensure segregation of persons and the robot. Since access is required for teaching, fencing and barricading must not present a trapping hazard. A minimum one-foot clear space must exist between the extreme movements of the robot and the fence. Clear space should not be occupied by toolboxes, lockers, racks, parts, or portable equipment.

To ensure that persons and robot remain segregated, the work envelope of the robot should contain trip devices, such as pressure-sensitive mats connected to the robot control unit and photoelectric systems. Entry into the working envelope of an operating work station must inactivate the robot. It must not be possible to circumvent or step over a pressure mat into a position of danger. A photoelectric system is needed to control a light curtain that prevents the robot from entering a work station occupied by a person. Activating the pressure mat activates the light curtain surrounding the robot arm entry route, which in turn inactivates the robot.

Additional precautions include interlocks on the entry gates to prevent entry into an operating work station: reset buttons located outside the working envelope of the robot for programming the robot to stop when finished and to proceed into the next work station only upon command from a start button located outside the work envelope: a red "mushroom-head" emergency stop button at each station with restart possible only at the main control; and an emergency pressure-sensitive stop device on the welding head that activates to prevent damage to the robot and/or persons when the robot arm welding head strikes an object.

Example Analysis—Specific Robotic Safety Requirements

A system safety analysis was performed on a robot used for welding. Welding liberates light in the ultraviolet and infrared (UV/IR) regions of the spectrum. Welding robots can operate in such an environment without visual hazard, but UV/IR light can produce acute and/or chronic adverse health effects in a human operator. Therefore, all welding stations must use spectra (transparent) welding curtains on all four sides. These curtains may be attached to the fences and barricades and provide a UV/IR light-tight barrier to all other areas.

Irregularities in the power supply should not cause unwanted movements of the robot. All electrical equipment should comply with the safety requirements of ANSI standards for electrical equipment of machine tools. Power cables must be protected, firmly secured, and located to avoid tripping and snagging. The electrical power supply must be adequate quality to prevent hazards, and filters or separate cables from the transformer must be installed or used to give protection from main line interferences. Magnetic and other interferences must be eliminated from the plant area.

Hazards can arise from stored energy systems, including hydraulic pneumatic accumulators, springs, spring guards, weights, counterweights, electric motors, and flywheels. To protect against these potential hazards, a facility is needed to dump pressure from accumulators; protection must be maintained for all parts against overpressure; flexible hoses must be restrained or shielded in case of failure; check and relief valves must be fitted to prevent the robot arm from dropping suddenly due to power loss; a shot pin cylinder mechanism should support the shoulder when the hydraulics are turned off; and a captive key system designed to function only when power is off is needed to lock the doors of the robot's cabinet and power systems.

Routine Precautions in Robotic Safety

Human exposure to a robot within the robot enclosure is necessary during programming and routine production and maintenance activities. The risk that a human operator will mishear, misinterpret, or forget instructions, requests, or promises is great. Documentation and access procedures can minimize this hazard.

A "permit to work" system should be required for maintenance, diagnostics, or programming work within the enclosure. This permit must clearly identify the work to be done, who will supervise, who carries out each section of work, the safety precautions taken, and a time limit. Access to the enclosure is denied at all times when the robot is in automatic cycle, and reduced speed and torque, used during the "teaching" or programming process, must be selected if the robot is to be operated manually within the fence. Fault diagnosis, maintenance, and programming must always be performed with the hand-held pendant, and the pendant must be controlled exclusively by "dead man" switches that deactivate when pressure is released.

Robotic safety is in its infancy. As robots become more "intelligent," the associated workplace hazards will increase since greater human involvement will be necessary. System safety techniques are tools that can be utilized to predict the likelihood of these new hazards, pinpoint their location, and provide methods of preventing them.

18.5 SAFETY CONTRACT

A useful way to involve personnel in the process of achieving safety goals is to develop a contract with each department head covering safety activities and goals for the next period of time. Since safety specialists cannot directly act in ways that reduce accidents, they must assure that others do. Formal agreements or "contracts" should precisely and clearly define the required activities and results. They must be based on realistic objectives that can be achieved with the available resources of the company.

The contract may also require actions of the safety specialist. It may be agreed that additional training, equipment, or facilities are required and that it is the safety person's responsibility to arrange for such help. In some cases, supporting evidence from sampling and analysis is needed to reach agreement with department heads. Safety people are responsible for providing the information required to persuade managers of the need to act. Conformance to the contract by all parties based on understanding and active commitment to its objectives is the safety professional's goal.

The contract should contain unambiguous criteria for measuring success and specify a time for completion and evaluation. The contract is negotiated between the safety professional and the department head, but it is backed up by authority delegated from top management. In some firms, contributions to the safety of a department are considered when promotions and raises are determined. The greater the commitment of top management, the more binding the safety contract will be.

18.5.1 Safety Audits

An audit procedure is important to assure that the plan is having the desired effect. If no effect is measured, the implementation of the plan should be reevaluated. The action "contracts" of the safety program should be audited at regular intervals. Procedures, equipment, and accidents records must be reviewed to determine if failure to achieve a goal requires revision of the contract or stricter enforcement of existing procedures. Problems and deficiencies in the existing safety program can be met by existing plans of action with department heads. A good safety program is responsive to the changes in circumstances and needs of the company. Safety audits performed at regular intervals are an important part of a functioning safety program.

18.6 CONCLUSION

The system safety engineer's ultimate objective is to reduce the number of preventable accidents. A safety professional is paid to get results. To reduce accidents, people must be motivated to behave in ways that reduce the risk of injury or damage. Audits of safety performance are basic to the success of a safety program. The safety professional must follow through to see that the plan of action is implemented.

Early application of the techniques of hazard analysis, failure modes and effects analysis, job safety/operating hazard analysis, fault tree analysis, software safety analysis, and robotic safety analysis is the most effective combination of actions to reduce occupational hazards. Safety professionals work most effectively when they serve as coordinators between management and the process or product. System safety tools provide a common "language," ensuring clear communication and understanding between managment, production/process team members, and the safety professional in achieving safety goals.

ACKNOWLEDGMENTS

The author acknowledges the contributions of Charles Hoes, P.E., system safety engineer, Frank Herzog, P.E., and Jan Cummins, who assisted in the preparation of this chapter.

BIBLIOGRAPHY

Bass, L., *Products Liability: Design and Manufacturing Defects*, Shephard's McGraw Hill Book Co., 1986.

Hammer, W., *Product Safety Management and Engineering*, Prentice-Hall, 1980.

Koren, H., *Handbook of Environmental Health and Safety*, vol. I, Pergamon Press, 1980.

Practicing Law Institute, *Product Liability of Manufacturers: Prevention and Defense 1984*, 1984.

U.S. Department of Health, Education, and Welfare, P.H.S., CDC, NIOSH, *Occupational Diseases: a Guide to Their Recognition*. Rev. June 1977. DHEW (NIOSH) pub no. 77-181.

CHAPTER 19

How to Apply the Microcomputer to Occupational Safety

BRAD T. GARBER
University of New Haven
New Haven, Connecticut

19.1 INTRODUCTION

With each passing year societal pressures for improved safety performance increase. In order to respond to this challenge the safety professional must continually develop new and innovative means of improving safety programs. Many types of technological advances have eased this task. Of particular importance are those in the area of computer science. One such advance, the development of the microcomputer, has great significance to the safety professional.

The practice of safety involves expending a considerable amount of effort on information handling and retrieval. It is in these areas that the safety professional can derive the greatest benefit from microcomputers. They permit tasks, which when done by hand are time-consuming and tedious, to be performed quickly, efficiently, and inexpensively. The resulting increases in productivity are directly translatable into better safety programs.

19.2 SAFETY STATISTICS

Measuring performance is of inestimable importance in the field of safety. It is only by keeping accurate and meaningful accident statistics that one can determine the effectiveness of safety programs. Since doing the calculations involved in statistical analyses is often time-consuming, difficult, and uninteresting, a tool such as the microcomputer that can automate the process is of great value. Not only can the microcomputer perform mathematical computations at high speed, but it can record results, produce reports, and prepare charts and graphs. As an added benefit, the likelihood of error is dramatically reduced.

An example of how a microcomputer can be used for handling safety statistics follows:

A hypothetical company has six plants and a corporate headquarters. At the end of every month each location sends data on their employee hours and OSHA-recordable injuries and illnesses to the corporate safety department. The injuries and illnesses are broken down into four categories: fatalities, lost workday cases involving days away from work, lost workday cases involving restricted activity, and non-lost workday cases involving medical treatment. Total recordable injuries and illnesses are computed by summing the categories. Incidence rates are calculated as follows:

$$\text{Incidence rate} = \frac{\text{number of injuries or illnesses} \times 200{,}000 \text{ hours}}{\text{employee hours of exposure}}$$

A monthly report on safety performance is prepared for management.

A microcomputer is used to do the following: (1) record the incoming data, (2) compile statistics and

```
100 DIM D(7,12,5),M$(12)
110 FOR N = 1 TO 7
120 READ L$(N)
130 NEXT N
140 DATA "HEADQUARTERS","ATLANTA","DETROIT","CLEVELAND","NEWARK","PHILADELPHIA",
"WILMINGTON"
190 ON ERROR GOTO 7000
200  OPEN "SDATA.TXT" FOR INPUT AS # 1
205 ON ERROR GOTO 0
210  FOR N1 = 1 TO 7
220  FOR N2 = 1 TO 12
230  FOR N3 = 1 TO 5
250  INPUT # 1,D(N1,N2,N3)
260  NEXT N3
270  NEXT N2
280  NEXT N1
290 CLOSE
402  FOR N = 1 TO 12 : READ M$(N) : NEXT N
403  DATA JANUARY,FEBRUARY,MARCH,APRIL,MAY,JUNE,JULY,AUGUST,SEPTEMBER,OCTOBER,NO
VEMBER,DECEMBER
500 CLS
510 PRINT "                                ABC CORPORATION"
520 PRINT "                           SAFETY STATISTICS PROGRAM"
530 PRINT:PRINT:PRINT:PRINT:PRINT:PRINT
540 PRINT "               ENTER <1> TO ENTER DATA FOR MONTH"
550 PRINT "               ENTER <2> TO VIEW DATA"
560 PRINT "               ENTER <3> TO CORRECT DATA"
570 PRINT "               ENTER <4> TO GENERATE TABLES"
580 PRINT "               ENTER <5> TO SET UP DATABASE FOR NEW YEAR"
590 PRINT "               ENTER <6> TO END PROGRAM"
595 PRINT
600 INPUT CHOICE
605 IF CHOICE > 6 THEN GOTO 500
610 CLS
620 ON CHOICE GOTO 1000,2000,3000,4000,5000,6000
1000 INPUT "ENTER NUMBER OF MONTH (EX. 7 FOR JULY)  ";MONTH
1010 CLS
1020 FOR N = 1 TO 7
1030 PRINT "                             ENTER DATA FOR ";L$(N)
1035 PRINT :PRINT
1040 INPUT "ENTER NUMBER OF FATALITIES ";D(N,MONTH,1)
1050 INPUT "ENTER NUMBER OF LOST WORKDAY AWAY FROM WORK CASES ";D(N,MONTH,2)
1060 INPUT "ENTER NUMBER OF LOST WORKDAY RESTRICTED CASES ";D(N,MONTH,3)
1070 INPUT "ENTER NUMBER NON LOST WORKDAY MEDICAL CASES ";D(N,MONTH,4)
1075 INPUT "ENTER NUMBER OF HOURS FOR THE MONTH ";D(N,MONTH,5)
1080 CLS
1100 NEXT N
1110 OPEN "SDATA.TXT" FOR OUTPUT AS # 1
1120 FOR N1 = 1 TO 7
1130 FOR N2 = 1 TO 12
1140 FOR N3 = 1 TO 5
1165 PRINT # 1,D(N1,N2,N3)
1170 NEXT N3
1180 NEXT N2
1190 NEXT N1
1200 CLOSE
1210 GOTO 500
```

FIGURE 19.1. A safety statistics program written in the IBM PC version of BASIC.

```
2000 CLS
2010 INPUT "ENTER MONTH NUMBER ";M1
2015 PRINT
2020 PRINT
2030 PRINT"LOCATION             FATAL      LWCAFW     LWCR       NLWC        HOURS"
2035 PRINT
2040 FOR N = 1 TO 7
2050 PRINT L$(N) TAB(20) D(N,M1,1) TAB(30) D(N,M1,2) TAB(40) D(N,M1,3) TAB(50)
 D(N,M1,4) TAB(60) D(N,M1,5)
2065 PRINT
2070 NEXT N
2075 PRINT
2080 INPUT "PRESS RETURN TO CONTINUE ";A$
2090 GOTO 500
3000 CLS
3010 INPUT "ENTER MONTH OF INCORRECT DATA ";M1
3015 PRINT "ENTER NUMBER OF PLANT "
3020 PRINT : FOR N = 1 TO 7 : PRINT L$(N) ; " = " ; N : NEXT N
3025 PRINT
3030 INPUT NUM
3035 PRINT
3040 INPUT "ENTER NUMBER OF FATALITIES ";D(NUM,M1,1)
3050 INPUT "ENTER NUMBER OF LOST WORKDAY AWAY FROM WORK CASES ";D(NUM,M1,2)
3060 INPUT "ENTER NUMBER OF LOST WORKDAY RESTRICTED CASES ";D(NUM,M1,3)
3070 INPUT "ENTER NUMBER NON LOST WORKDAY MEDICAL CASES ";D(NUM,M1,4)
3075 INPUT "ENTER NUMBER OF HOURS FOR MONTH ";D(NUM,M1,5)
3080 OPEN "SDATA.TXT" FOR OUTPUT AS # 1
3090 FOR N1 = 1 TO 7
3100 FOR N2 = 1 TO 12
3110 FOR N3 = 1 TO 5
3120 PRINT # 1,D(N1,N2,N3)
3130 NEXT N3
3140 NEXT N2
3150 NEXT N1
3160 CLOSE
3170 GOTO 500
4000 CLS
4010 INPUT "GENERATE TABLES FOR WHICH MONTH NUMBER ";M2
4025 IF D(1,M2,5) <> 99 THEN GOTO 4035
4026 PRINT
4027 PRINT "NO DATA EXISTS FOR THE MONTH OF ";M$(M2)
4028 PRINT
4029 INPUT "PRESS RETURN TO CONTINUE"; A$
4031 GOTO 500
4035 CLS
4036 PRINT "                                   TABLE I"
4040 PRINT "                               ABC CORPORATION"
4050 PRINT "                          SAFETY REPORT FOR ";M$(M2)
4060 PRINT
4070 PRINT "                        YEAR TO DATE INCIDENCE RATES"
4073 TRC = 0 : HRC = 0 : FC = 0 : LAC = 0 : LRC = 0 : MC = 0
4074 FOR N = 1 TO 8
4075 F(N) = 0 : LA(N) = 0 : LR(N) = 0 : MC(N) = 0 : HR(N) = 0
4076 NEXT N
4080 FOR P = 1 TO 7
4090 FOR M4 = 1 TO M2
4100 F(P)=F(P) + D(P,M4,1)
4101 LA(P)=LA(P) +D(P,M4,2)
4102 LR(P)=LR(P) +D(P,M4,3)
4103 MC(P)=MC(P) +D(P,M4,4)
4105 HR(P)=HR(P) +D(P,M4,5)
4110 NEXT M4
4115 TR(P) = F(P) + LA(P) + LR(P) + MC(P)
4120 NEXT P
4130 PRINT:PRINT:PRINT
4140 FOR N = 1 TO 7
4150 FC = FC + F(N)
4160 LAC = LAC + LA(N)
4170 LRC = LRC + LR(N)
4180 MC = MC + MC(N)
4190 HRC = HRC + HR(N)
4195 TRC = TRC + TR(N)
4200 NEXT N
4300 PRINT "LOCATION              TOTAL RECORDABLES INCIDENCE     AWAY FROM WORK I
NCIDENCE"
4305 PRINT "                                 RATE                          RATE"
4306 PRINT
4310 FOR N = 1 TO 7
4320 PRINT L$(N) TAB(30);
```

FIGURE 19.1. *(Continued).*

```
4321 PRINT USING "####.##";TR(N)*200000!/HR(N);
4322 PRINT TAB(60);
4323 PRINT USING "####.##";LA(N)*200000!/HR(N)
4325 NEXT N
4330 PRINT : PRINT
4340 PRINT "CORPORATION" TAB(30)
4341 PRINT USING "####.##" ;TRC*200000!/HRC;
4342 PRINT TAB(60);
4343 PRINT USING "####.##" ; LAC*200000!/HRC
4345 PRINT
4350 INPUT "PRESS RETURN TO CONTINUE"; A$
4360 CLS
4436 PRINT "                                        TABLE II"
4440 PRINT "                                     ABC CORPORATION"
4450 PRINT "                                 SAFETY REPORT FOR ";M$(M2)
4460 PRINT
4470 PRINT "                                NUMBERS OF INJURIES BY PLANT"
4471 PRINT "                                       YEAR TO DATE"
4472 PRINT : PRINT
4475 PRINT "LOCATION           FATAL       LOST-AWAY       LOST REST    MEDICAL
TOTAL"
4476 PRINT : PRINT
4490 FOR N = 1 TO 7
4500 PRINT L$(N) TAB(20) F(N) TAB(33) LA(N) TAB(48)  LR(N) TAB(60) MC(N) TAB(70)
TR(N)
4510 NEXT N
4520 PRINT : PRINT
4530 PRINT  "CORPORATION" TAB(20) FC TAB(33) LAC TAB(48) LRC TAB(60) MC TAB(70)
TRC
4531 PRINT
4550 INPUT "PRESS RETURN TO CONTINUE"; A$
4551 CLS
4636 PRINT "                                        TABLE III"
4640 PRINT "                                     ABC CORPORATION"
4650 PRINT "                                 SAFETY REPORT FOR ";M$(M2)
4660 PRINT
4670 PRINT "                            YEAR TO DATE INCIDENCE RATES BY MONTH"
4671 PRINT "                                   TOTAL CORPORATION"
4672 PRINT : PRINT
4680 PRINT "MONTH              TOTAL RECORDABLES INCIDENCE     AWAY FROM WORK IN
CIDENCE"
4681 PRINT "                                 RATE                          RATE"
4682 PRINT
4872 FOR Q = 1 TO M2
4873 TRC = 0 : HRC = 0 : FC = 0 : LAC = 0 : LAR = 0 : MC = 0
4874 FOR N = 1 TO 8
4875 F(N) = 0 : LA(N) = 0 : LR(N) = 0 : MC(N) = 0 : HR(N) = 0
4876 NEXT N
4880 FOR P = 1 TO 7
4890 FOR M4 = 1 TO Q
4900 F(P)=F(P) + D(P,M4,1)
4901 LA(P)=LA(P) +D(P,M4,2)
4902 LR(P)=LR(P) +D(P,M4,3)
4903 MC(P)=MC(P) +D(P,M4,4)
4905 HR(P)=HR(P) +D(P,M4,5)
4910 NEXT M4
4915 TR(P) = F(P) + LA(P) + LR(P) + MC(P)
4920 NEXT P
4940 FOR N = 1 TO 7
4950 FC = FC + F(N)
4960 LAC = LAC + LA(N)
4970 LRC = LRC + LR(N)
4980 MC = MC + MC(N)
4990 HRC = HRC + HR(N)
4991 TRC = TRC + TR(N)
4992 NEXT N
4993 PRINT M$(Q) TAB(30);: PRINT USING "####.##"; TRC*200000!/HRC;
4994 PRINT TAB(60);:PRINT USING "####.##"; LAC*200000!/HRC
4996 NEXT Q
4997 PRINT
4998 INPUT "PRESS RETURN TO CONTINUE"; A$
4999 GOTO 500
5000 OPEN "SDATA.TXT" FOR OUTPUT AS # 1
5010 FOR N1 = 1 TO 7
5020 FOR N2 = 1 TO 12
5030 FOR N3 = 1 TO 5
5040 D(N1,N2,N3) = 99
5045 PRINT # 1,D(N1,N2,N3)
5050 NEXT N3
```

FIGURE 19.1. *(Continued).*

```
5Ø6Ø NEXT N2
5Ø7Ø NEXT N1
5Ø8Ø GOTO 5ØØ
6ØØØ CLOSE
6Ø1Ø END
7ØØØ CLS
7Ø1Ø PRINT "NO DATA FILE EXISTS, RUN FUNCTION 5 "
7Ø15 INPUT "PRESS RETURN TO CONTINUE ";A$
7Ø2Ø GOTO 5ØØ
```

FIGURE 19.1. *(Continued).*

compute incidence rates, (3) generate tables, and (4) prepare graphics. A program written in BASIC is used for the first three tasks and a commercial program for the fourth.

The BASIC program (Fig. 19.1). was designed to be easy to use. It is menu-driven, which means that when running the computer the operator can tell it what to do by selecting from a group of choices that are displayed in plain English on the screen. The starting menu is shown in Figure 19.2.

To enter data, function 1 is chosen from the starting menu (Fig. 19.3). Function 2 permits the viewing of data (Fig. 19.4), and function 3 allows corrections to be made. Tables are generated by function 4 (Fig. 19.5). At the beginning of the year it is necessary to run function 5 to initialize the data base.

Graphs and charts used in the monthly report are produced using a commercially available program called Chartmaster (Figs. 19.6 and 19.7). The program has facilities for producing bar, pie, and line charts as well as scatter diagrams. It can output either to a graphics printer for routine applications or to a plotter if high-quality graphics are desired (Figs. 19.8 and 19.9).

```
                    ABC CORPORATION
               SAFETY STATISTICS PROGRAM


ENTER <1> TO ENTER DATA FOR MONTH
ENTER <2> TO VIEW DATA
ENTER <3> TO CORRECT DATA
ENTER <4> TO GENERATE TABLES
ENTER <5> TO SET UP DATABASE FOR NEW YEAR
ENTER <6> TO END PROGRAM
```

FIGURE 19.2. Starting menu generated by safety statistics program.

```
                              ABC CORPORATION
                         SAFETY STATISTICS PROGRAM


           ENTER <1> TO ENTER DATA FOR MONTH
           ENTER <2> TO VIEW DATA
           ENTER <3> TO CORRECT DATA
           ENTER <4> TO GENERATE TABLES
           ENTER <5> TO SET UP DATABASE FOR NEW YEAR
           ENTER <6> TO END PROGRAM
? 1
ENTER NUMBER OF MONTH (EX. 7 FOR JULY)  ? 3
                              ENTER DATA FOR HEADQUARTERS

ENTER NUMBER OF FATALITIES ? Ø
ENTER NUMBER OF LOST WORKDAY AWAY FROM WORK CASES ? Ø
ENTER NUMBER OF LOST WORKDAY RESTRICTED CASES ? Ø
ENTER NUMBER NON LOST WORKDAY MEDICAL CASES ? 1
ENTER NUMBER OF HOURS FOR THE MONTH ? 293845
```

FIGURE 19.3. Data entry format for safety statistics program.

```
ENTER MONTH NUMBER ? 1
```

LOCATION	FATAL	LWCAFW	LWCR	NLWC	HOURS
HEADQUARTERS	Ø	1	Ø	2	123543
ATLANTA	Ø	Ø	Ø	1	434312
DETROIT	Ø	Ø	Ø	Ø	333214
CLEVELAND	Ø	Ø	1	Ø	665432
NEWARK	Ø	Ø	Ø	Ø	123298
PHILADELPHIA	Ø	Ø	Ø	1	126543
WILMINGTON	Ø	1	Ø	Ø	654879

FIGURE 19.4. Table generated by view function.

19.3 SPREADSHEETS

As an alternative to employing BASIC for handling safety statistics, a type of software package known as a "spreadsheet" can be used. These programs generate tables that consist of a grid having a user-defined number of rows and columns (Fig. 19.10). After raw data is entered into cells on the grid, the user enters a series of simple instructions that causes calculations to be made and the results to be recorded in unoccupied cells. Once the instructions have been entered they are permanently stored. If the raw data is changed, the spreadsheet is automatically retabulated. An example of a spreadsheet used for compiling safety data is shown in Figure 19.11.

The major advantage of using a spreadsheet as opposed to writing a standard computer program for safety statistics is that a record-keeping system can be set up much more rapidly and easily. The major disadvantage is that spreadsheets lack the flexibility provided by standard programming languages. As a result, the types of calculations that can be performed and the ways in which output can be formatted are severely limited.

19.4 OTHER SAFETY APPLICATIONS OF MICROCOMPUTERS

In addition to providing an excellent tool for generating safety statistics reports, microcomputers can be used by the safety professional for a variety of other purposes. These include word processing, database management, communicating with on-line literature searching services, and doing repetitive or complicated calculations. Each of these applications will be discussed separately.

19.5 WORD PROCESSING

Perhaps one of the most useful applications of microcomputers is word processing. Since the safety professional often communicates findings in the form of written reports, word processors can be of considerable value. A number of full-featured word processing programs are available for microcomputers. They have essentially all of the features found in dedicated word processing units.

Word-processors are electronic document-producing devices. They differ from normal typewriters in that letters typed on the keyboard are displayed on a video screen and stored in computer memory before they are committed to paper. As a result, changes can be made, errors corrected, and sentences or paragraphs moved at electronic speed with little inconvenience to the typist. It has been estimated that word processors can increase the productivity of typists by a factor of two or more.

In addition to these capabilities, many word processors can perform advanced functions such as checking spelling, merging form letters with mailing lists, performing calculations on data within documents, and merging text from several documents.

The effectiveness of safety programs is due in large part to the quality of the reports produced by safety professionals. The ability of the word processor to facilitate production of high-quality, attractive reports therefore makes it a valuable resource in the safety office.

TABLE I
ABC CORPORATION
SAFETY REPORT FOR FEBRUARY

YEAR TO DATE INCIDENCE RATES

LOCATION	TOTAL RECORDABLES INCIDENCE RATE	AWAY FROM WORK INCIDENCE RATE
HEADQUARTERS	3.24	Ø.81
ATLANTA	Ø.46	Ø.ØØ
DETROIT	Ø.3Ø	Ø.ØØ
CLEVELAND	Ø.15	Ø.ØØ
NEWARK	Ø.8Ø	Ø.8Ø
PHILADELPHIA	1.58	Ø.ØØ
WILMINGTON	Ø.15	Ø.15
CORPORATION	Ø.49	Ø.12

TABLE II
ABC CORPORATION
SAFETY REPORT FOR FEBRUARY

NUMBERS OF INJURIES BY PLANT
YEAR TO DATE

LOCATION	FATAL	LOST-AWAY	LOST REST	MEDICAL	TOTAL
HEADQUARTERS	Ø	1	Ø	3	4
ATLANTA	Ø	Ø	Ø	2	2
DETROIT	Ø	Ø	Ø	1	1
CLEVELAND	Ø	Ø	1	Ø	1
NEWARK	Ø	1	Ø	Ø	1
PHILADELPHIA	Ø	Ø	Ø	2	2
WILMINGTON	Ø	1	Ø	Ø	1
CORPORATION	Ø	3	1	8	12

TABLE III
ABC CORPORATION
SAFETY REPORT FOR FEBRUARY

YEAR TO DATE INCIDENCE RATES BY MONTH
TOTAL CORPORATION

MONTH	TOTAL RECORDABLES INCIDENCE RATE	AWAY FROM WORK INCIDENCE RATE
JANUARY	Ø.57	Ø.16
FEBRUARY	Ø.49	Ø.12

FIGURE 19.5. Tables generated by safety statistics program.

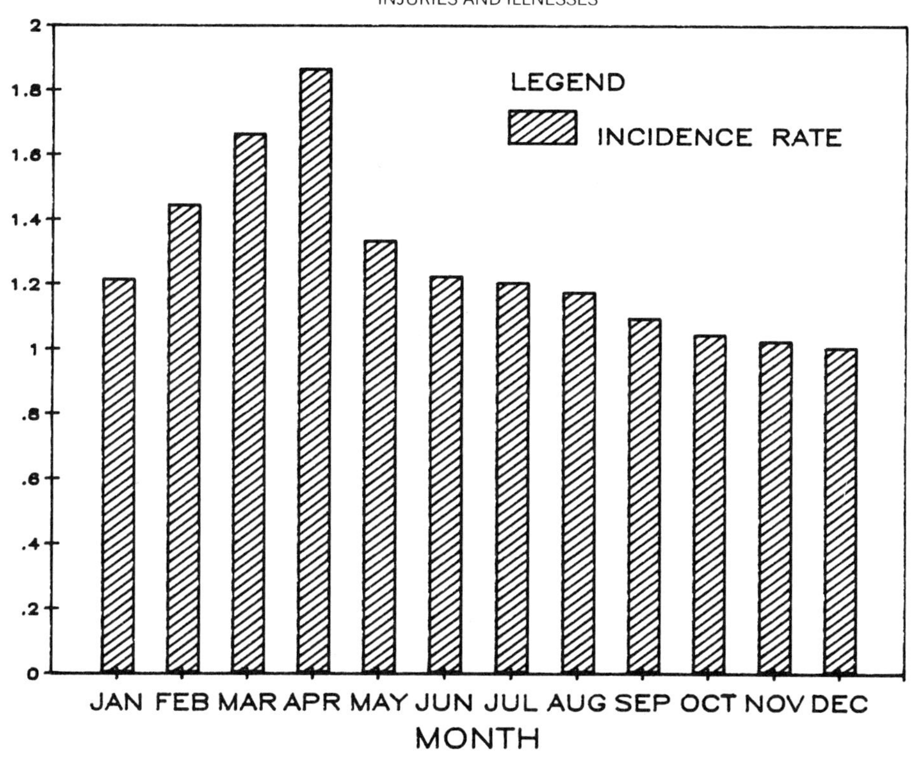

FIGURE 19.6. Bar chart produced by commercially available graphics software.

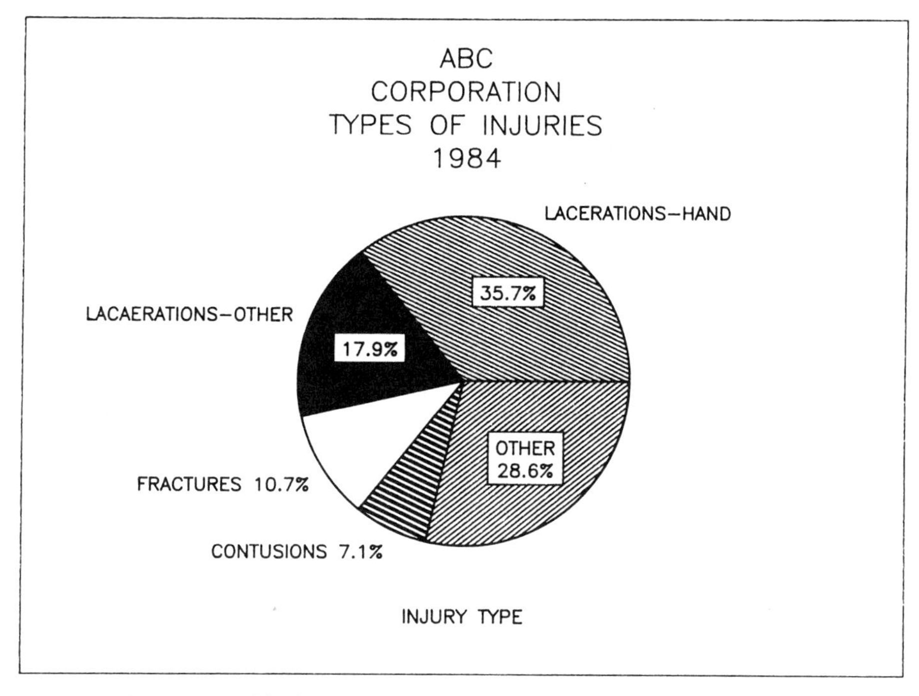

FIGURE 19.7. Pie chart produced by commercially available graphics software.

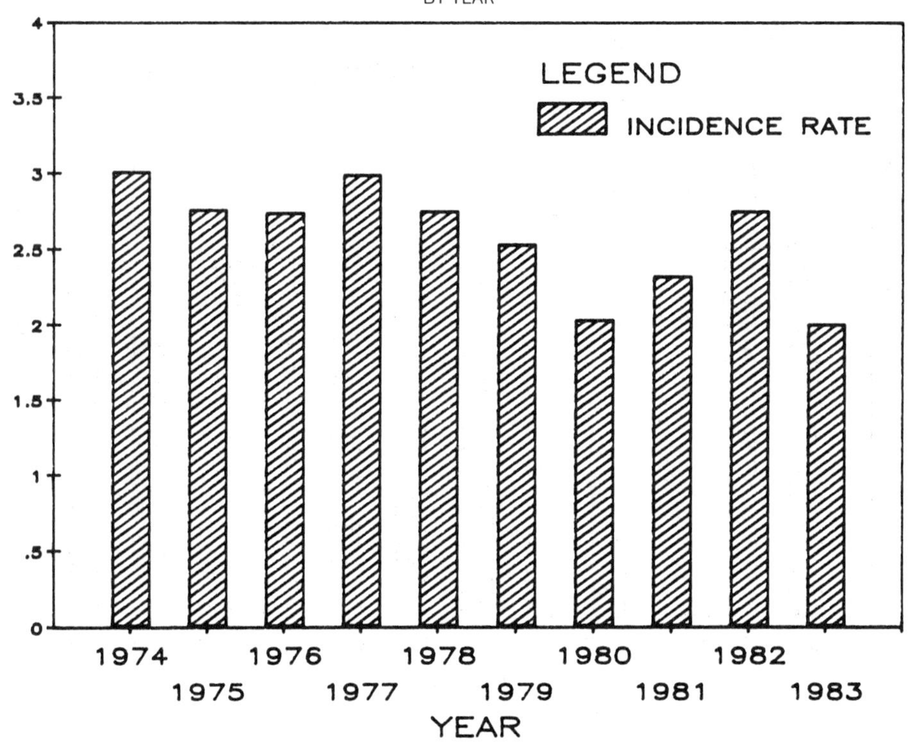

FIGURE 19.8. A bar chart produced on a graphics plotter.

19.6 DATABASE MANAGEMENT

Database management software packages are programs that store and retrieve records. There are many types of programs available, but they perform basically the same functions. They can create a record structure, add records, delete records, modify records, and generate reports.

An example of using a database management program for a safety application follows:

A company that manufactures explosive materials requires that all employees receive a refresher course on explosives safety every three months. The course is given on a monthly basis and the safety department must determine who is required to attend.

A microcomputer data base is set up using DBASEII, a popular database management program. Each record contains the employee's name, social security number, and the date that he or she last attended the course. A routine written in DBASEII language produces a starting menu (Fig. 19.12). This allows for easy use of the system. Each month after the course is completed, function 2 is employed to record attendance data. Function 1 is run before the course and generates a list of employees whose attendance is required. Examples of program input and output are shown in Figures 19.13 and 19.14.

19.7 LITERATURE SEARCHING SERVICES

A number of dial-up on-line literature searching services of interest to safety professionals are available, and microcomputers can be used as terminals to access them. These services allow the user to determine if any journal articles exist on a specific topic of interest. For instance, if one wanted to find out what is known about factors that influence gastrointestinal absorption of lead, one could dial up the National Library of Medicine's on-line search computer and direct it to provide data on relevant journal articles. The information available includes titles, authors, sources, and abstracts (Fig. 19.15). A list of some of

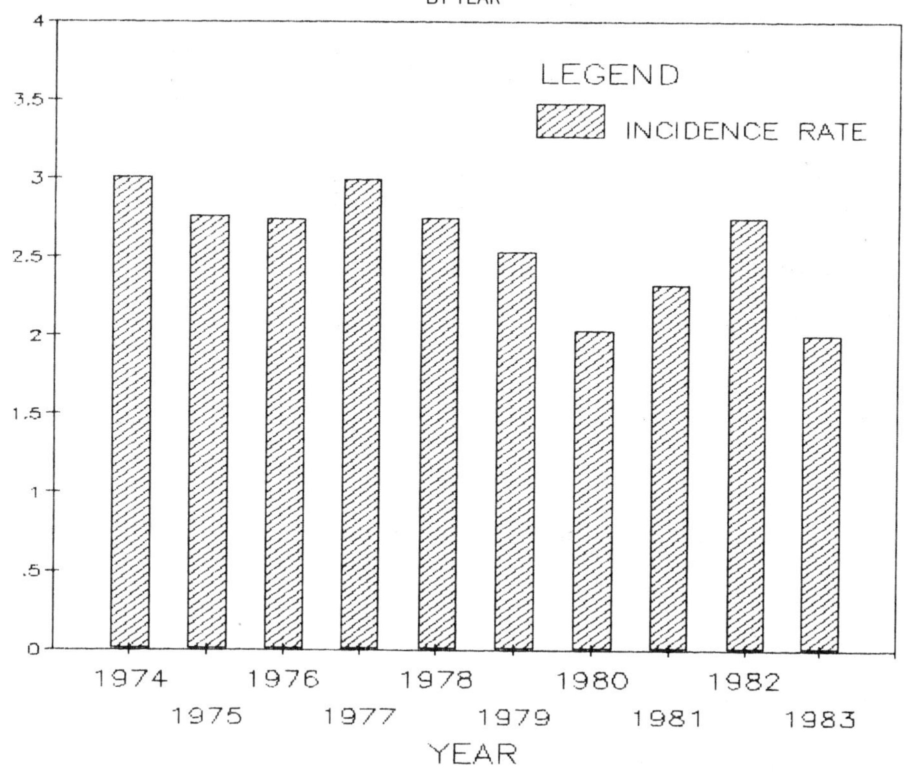

FIGURE 19.9. The same bar chart as shown in Figure 19.8 produced on a graphics printer.

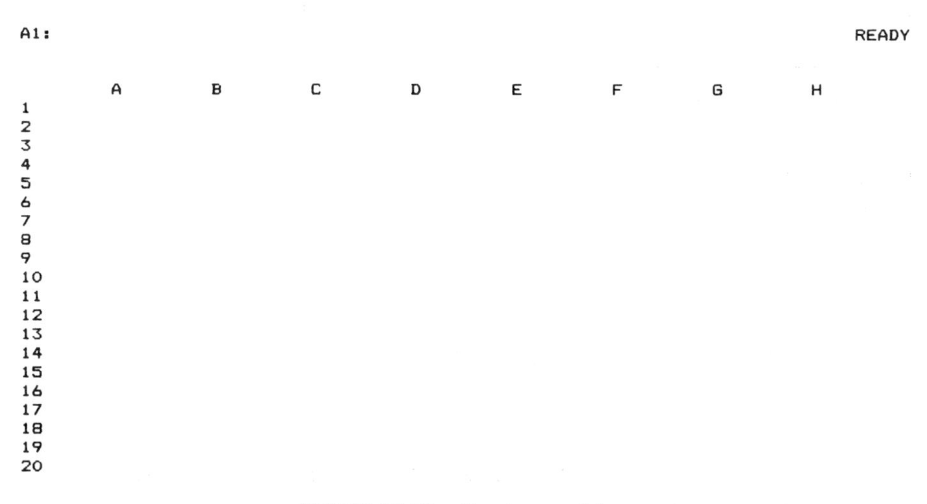

FIGURE 19.10. Empty spreadsheet.

A1: (FO) READY

	A	B	C	D	E	F	G	H
1			ABC CORPORATION					
2			INCIDENCE RATES					
3								
4								
5	LOCATION	FATAL	LOST	LOST	NON LOST	HOURS	TOTAL	TOTAL
6			WORKDAY	WORKDAY	WORKDAY		LOST	RECORDABLES
7			AWAY	RESTRICTED			WORKDAY	I.R.
8							I.R.	
9								
10								
11	HEADQUARTERS	0	1	0	2	123543	1.62	4.86
12	ATLANTA	0	0	0	1	434312	0.00	0.46
13	DETROIT	0	0	0	0	333214	0.00	0.00
14	CLEVELAND	0	0	1	0	665432	0.30	0.30
15	NEWARK	0	0	0	0	123298	0.00	0.00
16	PHILADELPHIA	0	0	0	1	126543	0.00	1.58
17	WILMINGTON	0	1	0	0	654879	0.31	0.31
18								
19	CORPORATION	0	2	1	4	2461221	0.24	0.57
20								

FIGURE 19.11. Spreadsheet set up for safety statistics. Columns G and H are calculated automatically from raw data in columns A through F.

```
*********************COURSÈ ATTENDANCE SYSTEM********************

CHOICES:

<1>  LIST EMPLOYEES WHO NEED TO ATTEND COURSE
<2>  ENTER COURSE ATTENDEES
<3>  ENTER NEW EMPLOYEE
<4>  DELETE EMPLOYEE
<5>  EXIT SYSTEM


SELECT CHOICE:
WAITING
```

FIGURE 19.12. Starting menu for DBASEII safety course attendance system.

```
*********************COURSE ATTENDANCE SYSTEM********************

CHOICES:

<1>  LIST EMPLOYEES WHO NEED TO ATTEND COURSE
<2>  ENTER COURSE ATTENDEES
<3>  ENTER NEW EMPLOYEE
<4>  DELETE EMPLOYEE
<5>  EXIT SYSTEM


SELECT CHOICE:
WAITING 2



          ENTER DATE OF COURSE:Ø22785:



          ENTER EMPLOYEES SOCIAL SECURITY NUMBER:123456789:
```

FIGURE 19.13. Data entry format for safety course attendance system.

```
**********************COURSE ATTENDANCE SYSTEM********************

 CHOICES:

 <1>  LIST EMPLOYEES WHO NEED TO ATTEND COURSE
 <2>  ENTER COURSE ATTENDEES
 <3>  ENTER NEW EMPLOYEE
 <4>  DELETE EMPLOYEE
 <5>  EXIT SYSTEM


 SELECT CHOICE:
WAITING 1



           ENTER DATE OF COURSE:Ø22785:


              EMPLOYEES REQUIRED TO ATTEND COURSE

Ø22785

CARL CANNON            1Ø2884  345678912
DONALD DICKSON         Ø92884  456789123
EARL ERIKSON           Ø92784  567891234
```

FIGURE 19.14. Report listing employees who are required to attend the course given on inputted date.

```
SS 5 /C?
USER:
PRINT FULL 1 INDENTED SKIP 17
PROG:

18
AUTHOR                    GARBER BT
AUTHOR                    WEI E
TITLE                     Influence of dietary factors of the
                          gastrontestinal absorption of lead.
SECONDARY SOURCE ID       HEEP/74/09469
SOURCE                    TOXICOL APPL PHARMACOL; 27 (3). 1974 685-691
ABSTRACT                  HEEP  COPYRIGHT: BIOL ABS.  Gastrointestinal
                          absorption of Pb was investigated in mice
                          after oral administration of lead acetate
                          labeled with 210-Pb.  When doses of 0.2, 2
                          and 20 mg of Pb/kg were given, the magnitude
                          of the dose did not appear to affect
                          significantly the percent absorbed.  The
                          presence of food in the gastrointestinal
                          tract reduced Pb absorption when a tracer
                          dose was administered but did not affect
                          absorption after 2 mg of Pb/kg orally.  The
                          chelators nitrilotriacetic acid and sodium
                          citrate increased absorption of Pb, as did
                          orange juice, a source of citric acid.  Milk
                          and the chelating agents, EDTA and
                          diethylenetriaminepentaacetic acid, did not
                          affect significantly Pb absorption.
```

FIGURE 19.15. Part of the output from a National Library of Medicine Literature Search on factors influencing the gastrointestinal absorption of lead.

the organizations that maintain literature-searching databases of interest to safety professionals is given in Fig. 19.16.

19.8 REPETITIVE OR COMPLICATED CALCULATIONS

The microcomputer is an excellent tool for doing repetitive or complicated calculations. It provides a number of advantages over using hand calculators or other manual methods. In addition to the obvious speed advantages, the microcomputer allows the user to format output into an easily understandable form, record the raw data entered as well as the results, avoid the use of paper, and easily determine the source of errors.

An example of a computer program to calculate incidence rates from numbers of injuries and employee hours is shown in Figure 19.17. The output (Fig. 19.18) is formatted so that both the raw data and the results can be viewed. Since the only significant source of error in this example is entering incorrect data, this output format facilitates ensuring accurate results.

DATA BASE VENDOR	ADDRESS	PHONE NUMBER
System Development Corporation (SDC)	2500 Colorado Avenue Santa Monica, CA 90406	(800) 421-7229 (CA: 800-352-6689)
Dialog	3460 Hillview Avenue Palo Alto, CA 94304	(800) 227-1960 (CA: 800-982-5844)
Bibliographic Retrieval Services, Inc.	1200 Route 7 Latham, NY 12110	(800) 833-4707 (NY: 800-553-5566)
National Library of Medicine Medlars Management Section	8600 Rockville Pike Bethesda, MD 20209	(301) 496-6193

FIGURE 19.16. Vendors who provide on-line databases of interest to safety professionals.

```
10 DIM P$(500),H(500),L(500),R(500),LIR(500),RIR(500),H$(500)
20 C = C + 1
30 INPUT "ENTER PLANT NAME ";P$(C)
40 INPUT "ENTER EMPLOYEE HOURS ";H$(C)
45 H(C) = VAL(H$(C))
50 INPUT "ENTER NUMBER OF LOST WORKDAY CASES ";L(C)
60 INPUT "ENTER TOTAL NUMBER OF RECORDABLES ";R(C)
70 INPUT "IS THIS THE LAST PLANT ";B$
90 LIR(C) = L(C) * 200000! / H(C)
100 RIR(C) = R(C) * 200000! / H(C)
110 IF LEFT$(B$,1) <> "Y" THEN GOTO 20
130  PRINT "                    INCIDENCE RATE CALCULATIONS"
140 PRINT : PRINT
150 PRINT"                RAW DATA                                   CALCULATED DAT
A"
160 PRINT"-------------------------------------------------   ------------------
------"
170 PRINT"PLANT                   EMPLOYEE   LOST      TOTAL         LOST          TOTAL
      "
180 PRINT"NAME                    HOURS      WORKDAY   RECORDABLE    WORKDAY       RECOR
DABLE "
190 PRINT"                                   CASES     CASES         I.R.          I.R.
      "
200 PRINT"_____________   __________  _____     _____        _________     _____
      "
205 PRINT
210 FOR N = 1 TO C
220 PRINT LEFT$(P$(N),14) TAB(20) H$(N) TAB(31) L(N) TAB(40) R(N) TAB(53);
230 PRINT USING "####.##"; LIR(N);
235 PRINT TAB(66) USING "####.##"; RIR(N)
250 NEXT N
260 END
```

FIGURE 19.17. BASIC (IBM PC version) program that computes incidence rates.

INCIDENCE RATE CALCULATIONS

RAW DATA				CALCULATED DATA	
PLANT NAME	EMPLOYEE HOURS	LOST WORKDAY CASES	TOTAL RECORDABLE CASES	LOST WORKDAY I.R.	TOTAL RECORDABLE I.R.
HEADQUARTERS	123543	1	3	1.62	4.86
ATLANTA	434312	Ø	1	Ø.ØØ	Ø.46
DETROIT	333214	Ø	Ø	Ø.ØØ	Ø.ØØ
CLEVELAND	665432	1	1	Ø.3Ø	Ø.3Ø
NEWARK	123298	Ø	Ø	Ø.ØØ	Ø.ØØ
PHILADELPHIA	126543	Ø	1	Ø.ØØ	1.58
WILMINGTON	654879	1	1	Ø.31	Ø.31

FIGURE 19.18. Output from incidence rate calculating program.

19.9 SUMMARY

The microcomputer is a tool that can be used to great advantage by the safety professional. Technological advances have lowered the cost of hardware and have improved the quality of software. A number of commercial programs specifically designed for safety record-keeping are now available. The state of the art has reached the point where neither cost nor complexity are barriers for most potential users of microcomputers.

BIBLIOGRAPHY

Garber, B. T., A microcomputer approach to safety recordkeeping, *Occ. Hlth and Safety,* January 1983.

Greenburg, L., The practice of occupational safety and health—facing the twenty-first century, *Am. Ind. Hyg. Assoc. J.,* Vol. 43, No. 9, 1982.

Greenburg, L., Enhancing the usability of safety software for microcomputers, *Am. Ind. Hyg. Assoc. J.,* Vol. 45, No. 7, 1984.

Whyte, A. A., A guide to occupational health information systems, *Occ. Hlth and Safety,* August 1982.

CHAPTER 20

How to Apply Economic/Decision Analysis Techniques to Occupational Safety and Health

MAHMOUD A. AYOUB

JOHN R. CANADA

North Carolina State University
Raleigh, North Carolina

20.1 COST CONCEPTS

Many, if not most, safety and health decisions involve trade-offs with cost considerations. This chapter will summarize and illustrate the use of some common tools and techniques for considering economics to facilitate rational decisions. In general, higher levels of safety and health protection involve higher costs of some types—such as for equipment, supplies, and personnel procedures to make that safety and health protection level possible.

Fundamental to the making of sound economic analyses is the proper consideration of relevant costs and savings. The following are some major concepts and definitions:

1. Only those costs/savings that differ between the alternatives under consideration need to be considered. These can be called costs/savings that are *incremental* between at least two alternatives. All other costs/savings are said to be *fixed*—that is, because they are constant across all alternatives they do not affect the choice among alternatives and hence can be ignored in making that choice.

EXAMPLE
Two types of radiation protection may involve differing supplies costs and personnel time savings, but require the same equipment costs for the same time period. In this case, an economic comparison should consider the *incremental* supplies and personnel time savings for the two alternatives, but can ignore the *fixed* equipment cost.

2. One may well need to consider *opportunity* costs/savings in addition to the obvious (usually cash) costs/savings. An *opportunity* cost occurs when the use of a limited resource (such as time, material, or money) results in *foregoing* (or giving up) the chance to use that resource to advantage elsewhere.

EXAMPLE
Suppose it is being decided whether the only available safety inspector (costing $100/day) should be assigned to Building A, where it is presumed he can cause savings of $125/day. That same inspector, if assigned to Building B, can cause savings of $175 per day. That $175 is the *opportunity* cost of assigning the inspector to Building A rather than B. Comparison of the $175 with the $125 savings says that he should not be assigned to A, even though the $125 savings is greater than the $100 costs of the inspector.*

3. Just as *opportunity* costs/savings should be identified and considered, the opposite counterpart, *sunk* costs/savings, should be ignored. *Sunk* costs/savings are any that are the result of past decisions or commitments. They are *fixed* with respect to future alternatives and thus have no effect on a choice between those alternatives.

EXAMPLE
$1 million was recently spent on Nuclear Safety System A, and it cost $100,000/year to operate. An alternative System B would scrap System A at a cost of $20,000 and then cost $70,000/year to operate. Despite the psychological difficulties of spending $20,000 to scrap the $1 million System A, one should be willing to do so to save $30,000/year.

20.2 IMPORTANCE OF CONSIDERING ALL ALTERNATIVES

It is easy to state, but often overlooked, that all feasible alternatives should be considered to ensure that the best decision is reached. There is a tendency to identify two or more alternatives and to study them until one is very confident which is the best decision. However, if there exists some other alternative that is in fact better than the one chosen, then the choice is suboptimal—the opportunity to choose the best has been foregone.

*If budget permits and the numbers are valid, one would want inspectors for both A and B.

20.2.1 Cost/Savings Table

One fundamental tool for displaying alternatives and outcomes to aid in decision making is a cost/savings table (sometimes called a "Payoff Tableau") as shown in Table 20.1. The best choice might become obvious from inspecting this table. If not, other decision aids such as those described below may be useful. The first one, aspiration level, makes no assumptions regarding probabilities of the various outcomes.

20.2.2 Aspiration Level

The "aspiration" or "maximum acceptable cost" method involves possible elimination of one or more alternatives, perhaps to the point that the best choice becomes clear.

EXAMPLE
In Table 20.1, if the firm were unwilling to accept any alternative that involved an outcome costing more than, say, $160,000, then alternatives A and B are eliminated and perhaps one can make a choice between alternatives C and D. If, on the other hand, the $160,000 were reduced to $100,000, then the only choice is alternative D.

20.2.3 Mimimax Rule

The minimax rule is very conservative because it suggests that the decision maker find the maximum cost associated with each alternative and select the alternative that has the minimum of those maximum costs. (Its counterpart in case of savings is called the "maximin rule.")

EXAMPLE
The maximum costs for each alternative happens to be in the "Severe Accident" column of Table 20.1. The minimum of these is 95, which would make alternative D the choice using this rule.

20.2.4 Minimin Rule

The minimin rule is the most optimistic or nonconservative because it suggests that the decision maker find the minimum cost associated with each alternative and select the alternative that has the minimum of these minimum costs. (Its counterpart in case of savings is called the "maximax rule.")

EXAMPLE
The minimum costs for each of the alternatives are as follows: A, 0; B, 3; C, 10; and D, 13. Of these the minimum is 0, which makes alternative A the choice using this rule.

20.2.5 Hurwicz Rule

The Hurwicz rule merely provides for a weighting of the optimistic vs. conservative outcomes for each of the alternatives and then selecting the best alternative as follows: Choose an index of optimism, a, such that $0 \le a \le 1$. For each alternative, compute the weighted outcome using the formula $a \times$ (optimistic outcome) $+ (1 - a) \times$ (conservative outcome), and then select the alternative with the best weighted outcome.

TABLE 20.1. Example Table of Alternatives and Outcomes (Costs in $000).

Alternative		State of Nature		
Insurance	Protective Device	No Accident	Moderate Accident	Severe Accident
(A) No	No	0	50	500
(B) No	Yes	3	40	200
(C) Yes	No	10	15	110
(D) Yes	Yes	13	13	95

EXAMPLE
For the problem in Table 20.1, and assuming a is 0.9, the calculation of weighted outcomes is as follows:

Alt	
A	0.9(0) + 0.1 (500) = 50
B	0.9(3) + 0.1 (200) = 22.7
C	0.9(10) + 0.1 (110) = 20
D	0.9(13) + 0.1 (95) = 21.2

The lowest of the above weighted costs is the 20 for alternative C, which would make it the choice for this particular weighting of a.

20.2.6 LaPlace Rule

The LaPlace rule merely assumes that all outcomes are equally likely and chooses the best alternative based on calculation of the expected outcomes of each. The expected average outcome of an alternative with equal probability is merely the sum of the different outcome values divided by the number of outcomes.

EXAMPLE
Applied to the alternatives in Table 20.1, the calculations for this rule are as follows:

Alt	Expected Outcome
A	(0 + 50 + 500)/3 = 183.3
B	(3 + 40 + 200)/3 = 81
C	(10 + 15 + 110)/3 = 45
D	(13 + 13 + 95)/3 = 40.3

The lowest of the above expected costs is 40.3, and thus alternative D would be the best choice using this rule.

20.2.7 Probabilities—Expected Values and Standard Deviations

Whenever one can assign probabilities to all possible outcomes, the most commonly used measures that can be calculated for each alternative are:

$$\text{Expected Value (Mean): } \bar{x} = \sum_{i=1}^{n} x_i P(x_i)$$

$$\text{Variance: } \sigma^2 = \sum_{i=1}^{n} x_i^2 P(x_i) - (\bar{x})^2$$

$$\text{Standard Deviation: } \sigma = \sqrt{\text{variance}}$$

Where

x_i = the ith outcome
$P(x_i)$ = probability of the ith outcome (which must sum to 1.00 for all outcomes)
n = number of outcomes for an alternative

EXAMPLE
Using the example in Table 20.1, let the probabilities of each of the accident outcomes (rather than being equal as for the LaPlace Rule) be as follows:

No	Moderate	Severe
0.5	0.4	0.1

The calculation of expected outcomes is:

Alternative	
A	$0(0.5) + 50(0.4) + 500(0.1) = 70$
B	$3(0.5) + 40(0.4) + 200(0.1) = 37.5$
C	$10(0.5) + 15(0.4) + 110(0.1) = 22$
D	$13(0.5) + 13(0.4) + 95(0.1) = 21.2$

If the criterion is to minimize the expected outcome (cost), then alternative D is best. The calculation of variance of outcomes is:

Alternative	
A:	$(0)^2(0.5) + (50)^2(0.4) + (500)^2(0.1) - (70)^2 = 21{,}100$
B:	$(3)^2(0.5) + (40)^2(0.4) + (200)^2(0.1) - (37.5)^2 = 3{,}238$
C:	$(10)^2(0.5) + (15)^2(0.4) + (110)^2(0.1) - (22)^2 = 866$
D:	$(13)^2(0.5) + (13)^2(0.4) + (95)^2(0.1) - (21.2)^2 = 605$

Thus the calculation of standard deviations of outcomes is:

Alternatives	
A	$\sqrt{21{,}100} = 145.3$
B	$\sqrt{3{,}238} = 56.9$
C	$\sqrt{866} = 29.4$
D	$\sqrt{605} = 24.6$

If the criterion is to minimize the standard deviation then alternative D is best.*

20.2.8 Expectation-Variability

Sometimes it is very useful to decide between alternatives based on expected values adjusted by a function of some measure of variability such as variance, standard deviation, or even range for each alternative.

EXAMPLE
Suppose it is judged that for the example in Table 20.1, all expected values (costs) should be increased by 0.2 of the respective standard deviations. The resulting values for decision-making purposes are:

Alternative	
A	$70 + 0.2\,(145.3) = 99.1$
B	$37.5 + 0.2\,(56.9) = 48.9$
C	$22 + 0.2\,(29.4) = 27.9$
D	$21.2 + 0.2\,(24.6) = 26.1$

Thus, alternative D has the lowest "adjusted" cost value and is the best by a slight margin.

20.2.9 Certainty Equivalence

A technique highly related to the expectation-variability approach is to ask the decision maker questions such as "What amount for certain would you be willing to pay (or receive) in order to give up the various possible costs (or savings)?" These various possible costs or savings are normally those associated with a given alternative, and one may or may not have probabilities associated with each possible estimate.

EXAMPLE
For the decision problem in Table 20.1, the question for alternative A might be stated as follows: "What certain cost would be just as good (or bad) in your mind as taking a chance of losing $0 if there is no accident, or $50,000 if there is a moderate accident, or $500,000 if there is a severe accident?" The answer to this is subjective, depending on the decision maker's attitude toward risk—that is, toward the consequences of the various possible outcomes. If the decision maker is "risk-neutral," he would base his certainty equivalence on the expected (long-run average) cost. Depending on his perception of the

*It is not normally experienced that the alternative that has the best expected outcome also has the lowest variability of outcomes.

probabilities, that could be anything between \$0 and \$500,000. Let's assume the probabilities given earlier; thus, the certainty equivalent would be \$70,000. On the other hand, the more normal "risk-averse" decision maker would probably have a certainty equivalent of much more than the \$70,000, particularly if he viewed the possible \$500,000 loss as catastrophic.

For the sake of illustration, let's assume that a moderately risk-averse decision maker would answer as follows:

Alternative	One Decision Maker's Certainty Equivalent
A	140
B	50
C	38
D	40

The lowest of these certainty equivalent costs is 38, so alternative C would be the choice by a slight margin.

20.3 TIME VALUE OF MONEY FUNDAMENTALS

Analyses involving significantly differing amounts of monies over time should include provision for weighting those monies according to timing. Figure 20.1 shows a standard diagram for the relative placement of the various amounts that can be converted from one to the other at some rate of interest or earning power of the money. Four formulae—for finding the present worth of a future amount and uniform series equivalents given other known amounts—for discrete compounding (together with factor symbols and examples) are summarized in Table 20.2.

To illustrate, consider the first example on the right side of Table 20.2. The formula that applies is

$$P = F[1/(1 + i)^{10}] = \$4045\ [1/(1 + .15)^{10}] = \$1000$$

Using the functional symbols system and tabled factors, the same problem can be set up and solved as

$$P = F(P/F,\ i\%,\ N) = \$4045\ (P/F,\ 15\%,\ 10)$$
$$= \$4045(0.2472) = \$1000$$

where the (P/F) factor value of 0.2472 can be obtained from the very abbreviated Table 20.3. (Note: much more extensive tables are available in engineering economics texts.)

It should be noted that one can compute F given P, or P, F, or G, given A, by using the reciprocal of

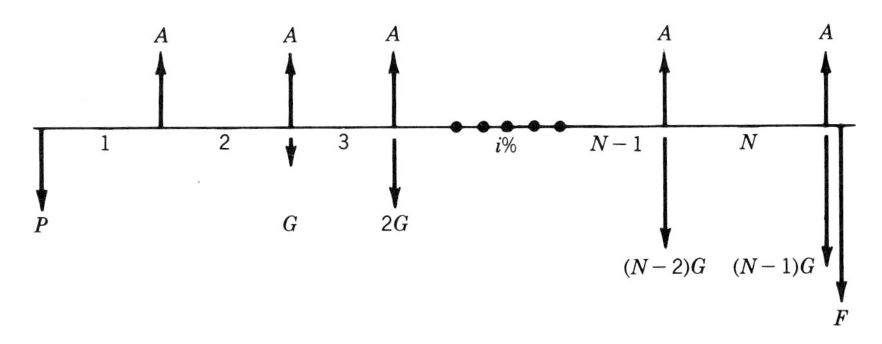

FIGURE 20.1. Standard time diagram for uniform series related to present worth, future worth, and uniform gradient.

TABLE 20.2. Summarization of Discrete Compound Interest Factors and Symbols for Finding Present and Uniform Series Equivalents.[a]

To find[b]	Given	Multiply "Given" by Factor Below	Factor's Name	Factor Functional Symbol	Example (answer for $i = 15\%$) (Note: All uniform series problems assume end-of-period payments.)
P	F	$\frac{1}{(1+i)^N}$	Single sum present worth	$(P/F, i\%, N)$	A company desires to have \$4045 10 years from now. What amount is needed now to provide for it? Ans.: \$1000
A	P	$\frac{i(1+i)^N}{(1+i)^N - 1}$	Capital recovery	$(A/P, i\%, N)$	What is the size of 10 equal end-of-year payments equivalent to an initial cost of \$1000? Ans.: \$199.30
A	F	$\frac{i}{(1+i)^N - 1}$	Sinking fund	$(A/F, i\%, N)$	What equal end-of-year amounts for 10 years are equivalent to \$4045 at the end of the 10th year? Ans.: \$199.30
A	G	$\frac{1}{i} - \left(\frac{N}{(1+i)^N - 1}\right)$	Gradient to Uniform Series	$(A/G, i\%, N)$	What equal annual saving is equivalent to \$58.90 at the end of year 2, and then increasing by \$58.90 at the end of each year thereafter until the end of the year 10? Ans.: \$199.30

[a]Key: i = Interest rate per interest period; N = Number of interest periods; P = Present worth; A = Uniform series amount; F = Future Worth; G = Uniform gradient

[b]NOTE: P, given F; or P, F, or G, given A, can be computed as the reciprocal of the respective formulas given above.

TABLE 20.3. Abbreviated Table of Discrete Compound Interest Factors.

	Single Sum Present Worth Factor (P/F, i%, N)				Uniform Series Capital Recovery Factor (A/P, i%, N)			
$N\backslash i$% →	5%	15%	25%	50%	5%	15%	25%	50%
1	0.9524	0.8696	0.8000	0.6667	1.0500	1.1500	1.2500	1.5000
5	0.7835	0.4972	0.3277	0.1317	0.2310	0.2983	0.3718	0.5758
10	0.6139	0.2472	0.1074	0.0173	0.1295	0.1993	0.2801	0.5088
15	0.4810	0.1229	0.0352	0.0023	0.0963	0.1710	0.2591	0.5011
20	0.3769	0.0611	0.0115	0.0003	0.0802	0.1598	0.2529	0.5002
40	0.1420	0.0037	0.0001	0.0000	0.0583	0.1506	0.2500	0.5000
	Uniform Series Sinking Fund Factor (A/F, i%, N)				**Gradient to Uniform Series Factor (A/G, i%, N)**			
$N\backslash i$% →	5%	15%	25%	50%	5%	15%	25%	50%
1	1.0000	1.0000	1.0000	1.0000	0.000	0.000	0.000	0.000
5	0.1810	0.1483	0.1218	0.0758	1.903	1.723	1.563	1.242
10	0.0795	0.0493	0.0301	0.0088	4.099	3.383	2.797	1.824
15	0.0463	0.0210	0.0091	0.0011	6.097	4.565	3.453	1.966
20	0.0302	0.0098	0.0029	0.0002	7.903	5.365	3.767	1.994
40	0.0083	0.0006	0.0000	0.0000	13.377	6.517	3.995	2.000

the respective formulas given above. For example, the factor for finding a future equivalent, given the present equivalent, $(F/P, i\%, N) = 1/(P/F, i\%, N)$.

One can use combinations of the above to find the present, annual or future equivalent of any pattern of cash flows.

EXAMPLE

Suppose one wants to find (1) the equivalent annual worth and (2) the equivalent worth at the end of 10 years of a present amount of \$1000 plus a gradient of \$100/year beginning at the end of year 2 with interest at 15%/yr. This can be calculated as

$$\begin{aligned}(1)\ A &= \$1000\,(A/P,\,15\%,\,10) + \$100\,(A/G,\,15\%,\,10)\\ &= 1000\,(0.1993) + \$100\,(3.3830) = \$537.6\end{aligned}$$

$$\begin{aligned}(2)\ F &= A(F/A,\,15\%,\,10) = A\,(1/(A/F,\,15\%,\,10))\\ &= \$537.6\,(1/.0493) = \$10{,}905\end{aligned}$$

20.4 ECONOMIC ANALYSIS METHODS

The most commonly used economic analysis methods to aid in deciding whether a given safety/health project is justified are (1) annual worth, (2) present worth, (3) benefit-cost ratio, and (4) payback period. Each will be illustrated in before-income-tax analyses for a sample project involving uniform annual cash flows.

EXAMPLE

A proposed safety project would involve an investment of \$20,000 and is expected to have annual cash benefits (savings) of \$10,000 and annual cash disbursements of \$4,000 over a life of 5 years, at which time it would have a salvage value of \$6,000. It is desired to determine if the project should be recommended if the firm has a minimum before-tax attractive rate of return (hurdle rate) of 15%.

(1) *Annual Worth*

Cash Savings		\$10,000
— Cash Disb:	−	4000
— Investment: \$20,000 (A/P, 15%, 5)		
= 20,000 (0.2983) =	−	5966
+ Salv.: \$6000 (A/F, 15%, 5)		
= 6000 (0.1483)	+	889
Net Annual Worth =	+\$	923

(+ \$923 > 0; ∴ Project is good)

(2) *Present Worth*

Each of the cash flows can be converted to present equivalents and the result summed to find the net present worth. A shorthand way is to use the above net annual worth as follows:

$$\begin{aligned}\text{Net Present Worth} &= \text{Net Annual Worth} \times (1/[A/P,\,15\%,\,5])\\ &= +\ \$923\,(1/0.2983)\\ &= +\ \$3094\end{aligned}$$

(+ \$3094 > \$0; ∴ Project is again shown to be good)

(3) *Benefit-Cost Ratio*

This ratio can be computed by many different formulas according to what is called a benefit and what is called a cost. In any case, all cash flows have to be converted into the same equivalents (such as annual). Using the plus and minus signs in the above calculation of net annual worth to define what are "benefits" and "costs," respectively, and using the annual equivalents:

$$\frac{\text{Benefits}}{\text{Costs}} = \frac{\text{Savings} + \text{Salvage}}{\text{Disbursements} + \text{Investment}}$$

$$= \frac{\$10{,}000 + \$889}{4000 + 5966} = 1.09$$

(1.09 > 1.00; ∴ Project is again shown to be good)

(4) *Payback (Payout) Period*
This method merely calculates the number of years for the net cash flows (savings-disbursements) to just equal the investment. If, as is often done, the salvage value is ignored, for the above example involving uniform cash flows:

$$\text{Payback Period} = \frac{\text{Investment}}{\text{Annual Cash Savings} - \text{Disbursements}}$$

$$= \frac{\$20{,}000}{\$10{,}000 - \$4000} = 3.33 \text{ years}$$

If salvage value is recognized,

$$\text{Payback Period} = \frac{\text{Investment} - \text{Salvage}}{\text{Annual Cash Savings} - \text{Disbursements}}$$

$$= \frac{\$20{,}000 - \$6000}{\$10{,}000 - \$4000} = 2.33 \text{ years}$$

Either of the above can be compared against an arbitrary standard depending on the firm's availability of capital and perceived risk for the particular class of projects. A typical firm generally has payout period cut-offs from 1 years to 5 years. The above project, even though it is shown to be good by the other methods (which give consistent "good" or "not good" answers), may not be "good" by the payback period method.

20.4.1 Break-even Method

This method is easy and useful when one is particularly uncertain about one of the key parameters (variables) needed for the economy study. All one needs to do is to solve for the value of that parameter at which the project is barely justified. Then, one hopes to be able to judge if the parameter is likely to turn out higher or lower than that break-even value and thus whether the project is "good" or "not good."

EXAMPLE
Let us use the same data for the above \$20,000 safety project except assume that annual benefits (savings) are particularly hard to estimate (i.e., ignore the prior \$10,000 estimate). Using annual equivalents, the project will be just on the borderline if the net annual worth = 0. Thus, again using annual equivalents:

$$\text{Cash Savings} - \text{Cash Disbursements} - \text{Investment} + \text{Salvage} = 0.$$

$$\text{Cash Savings} - \$4000 - \$20{,}000\,(A/P,\ 15\%,\ 5) + \$6000\,(A/F,\ 15\%,\ 5) = 0$$

$$\text{Cash Savings} - \$4000 - \$5966 + \$899 = 0$$

$$\text{Cash Savings} = \$9077$$

If one thought cash savings were going to turn out to be greater than \$9077, the project is "good." Otherwise, the project is "not good."

20.5 DECISION TREES FOR CONSIDERING SEQUENTIAL ALTERNATIVES AND/OR MULTIPLE OUTCOMES

The decision tree technique involves graphic portrayal of the alternatives and outcomes for a decision problem. Figure 20.2 shows such a portrayal for the problem of Table 20.1, including (in parentheses) the probabilities of the various accident outcomes as assumed in the section on expected values.

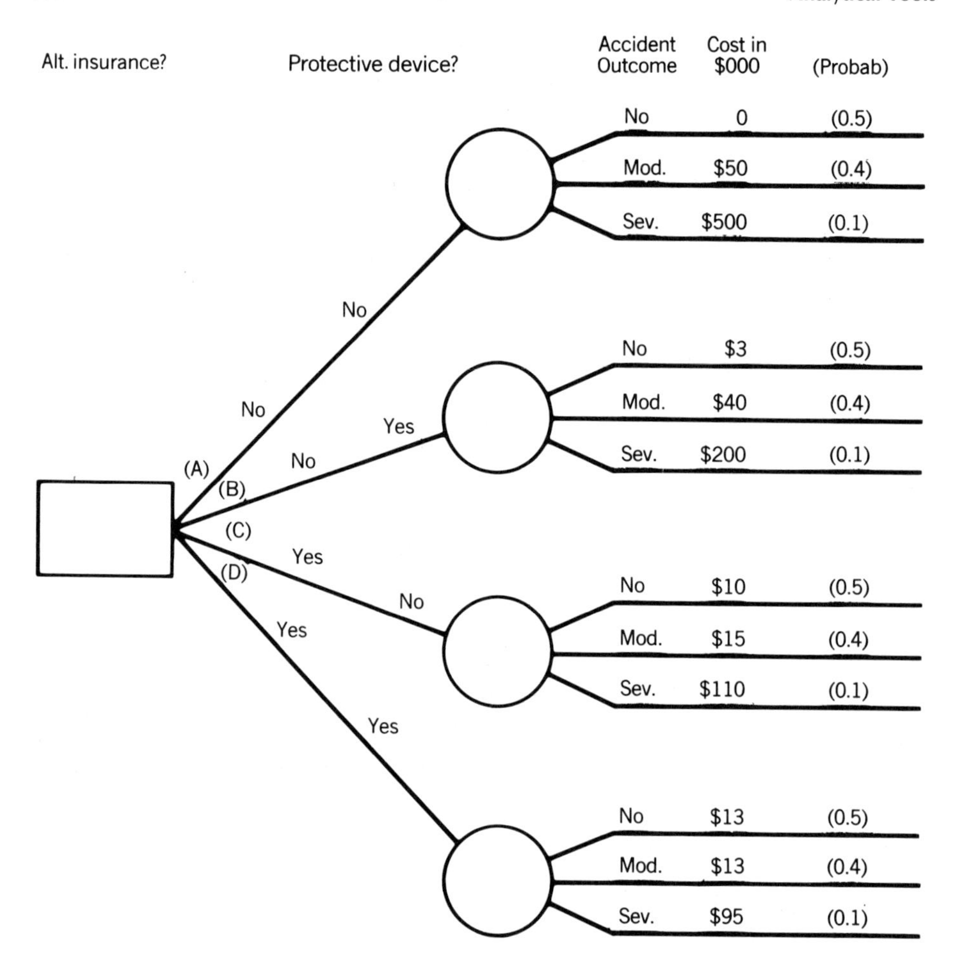

FIGURE 20.2. Example—decision tree problem.

The decision tree technique is most powerful for describing and analyzing problems involving a sequence of alternatives and/or outcomes over time. The following example involves only multiple alternatives.

EXAMPLE—Evaluating Noise Control Methods
The machine shop of a small manufacturing company was found to be in violation of the OSHA noise standard; the overall noise level was determined (by OSHA compliance officer) to be 115 db(A) for an eight-hour exposure. The company has been given 30 calendar days to bring the machine shop into compliance with the noise standard, 90 db(A) for eight hours of exposure. Failure to comply with the noise standard would make the company liable to citation for a willful violation that normally carries a $10,000 fine. Management, concerned about the problem and grave consequences that the company could face, has directed the safety engineer to take the necessary steps to meet OSHA requirements. In addition, management has made it clear that the noise reduction task should be accomplished at the minimum cost possible: Be imaginative, consider all possible alternatives which can be used to effectively combat our noise problem, then decide on the technique or techniques that could be used to put the company in compliance at minimum cost."

After careful study of the problem and a review of most of the available noise control techniques, the safety engineer chose to closely examine the following: (1) complete enclosure of all noise sources in the shop, (2) increasing the distance between noise source and those employees who do not have to work closely with the machines, (3) providing employees (machine operators as well as others) with personal protective equipment (e.g., ear muffs), and (4) reducing exposure time for all employees to the limits specified by the noise standard.

Further study led to the identification of several promising alternative plans and techniques for the enclosure approach, that is, control of noise at source. Tables 20.4 through 20.6 show the feasible controls, costs, and estimated reductions in noise levels for the machine shop. The engineer's problem is to choose the alternative or approach that will most economically reduce the overall noise level to no more than 90 db(A) for an exposure of eight hours.

Because of the nature of this problem (24 possible plans or combinations of alternatives), a decision diagram, as shown in Figure 20.3, is useful for visually displaying all alternatives and studying solution sensitivity.

TABLE 20.4. Reduction in Noise Level as a Function of Enclosure Type, and Related Costs

	Enclosure Type		
Factor	Type 1 Partial	Type 2 Total	Type 3 Barrier
Reduction in db(A)	−10	−15	−8
$ Cost	2000	4500	1500

TABLE 20.5 Additional (and Cumulative) Reduction in Noise Level when Vibration Isolation is Combined with Enclosure, and Related Costs

	Vibration Isolation Type			
	Type 1 (Vibration Mount)		Type 2 (Vibration Mount, Surface Damping)	
Enclosure Type	Reduction in db(A)	$ Cost	Reduction in db(A)	$ Cost
Type 1	−7	1000	−11	1500
	(−17)[a]	(3000)	(−21)	(3500)
Type 2	−10	1000	−13	1500
	(−25)	(5500)	(−28)	(6000)
Type 3	−8	1000	−12	1500
	(−16)	(2500)	(−20)	(3000)

[a]The figures in parentheses represent the cumulative noise reduction effects and costs of the combined vibration isolation and enclosure types.

TABLE 20.6. Additional (and Cumulative) Reduction in Noise Level when Absorption Methods are Combined with Vibration Isolation and Enclosure, and Related Costs

	Absorption Type					
	Type 1 (Walls)		Type 2 (Walls and Ceiling)		Type 3 (Walls, Ceiling, Floor)	
Vibration Isolation Type (VI) + Enclosure (E) Type	Reduction in db(A)	$ Cost	Reduction in db(A)	$ Cost	Reduction in db(A)	$ Cost
VI type 1	−4	1400	−7	2800	−3	3200
+E type 1	(−21)[a]	(4400)	(−24)	(5800)	(−20)	(6200)
VI type 1	−4	1400	−7	2800	−3	3200
+E type 2	(−29)	(6900)	(−32)	(8300)	(−28)	(8700)
VI type 1	−4	1400	−7	2800	−3	3200
+E type 3	(−20)	(3900)	(−23)	(5300)	(−19)	(5700)
VI type 2	−3	1400	−9	2800	−6	3200
+E type 1	(−24)	(4900)	(−30)	(6300)	(−27)	(6700)
VI type 2	−3	1400	−9	2800	−6	3200
+E type 2	(−31)	(7400)	(−37)	(8800)	(−34)	(9200)
VI type 2	−3	1400	−9	2800	−6	3200
+E type 3	(−23)	(4400)	(−29)	(5800)	(−26)	(6200)

[a]The figures in parentheses represent the cumulative noise reduction effects and costs of the combined absorption, vibration isolation, and enclosure types.

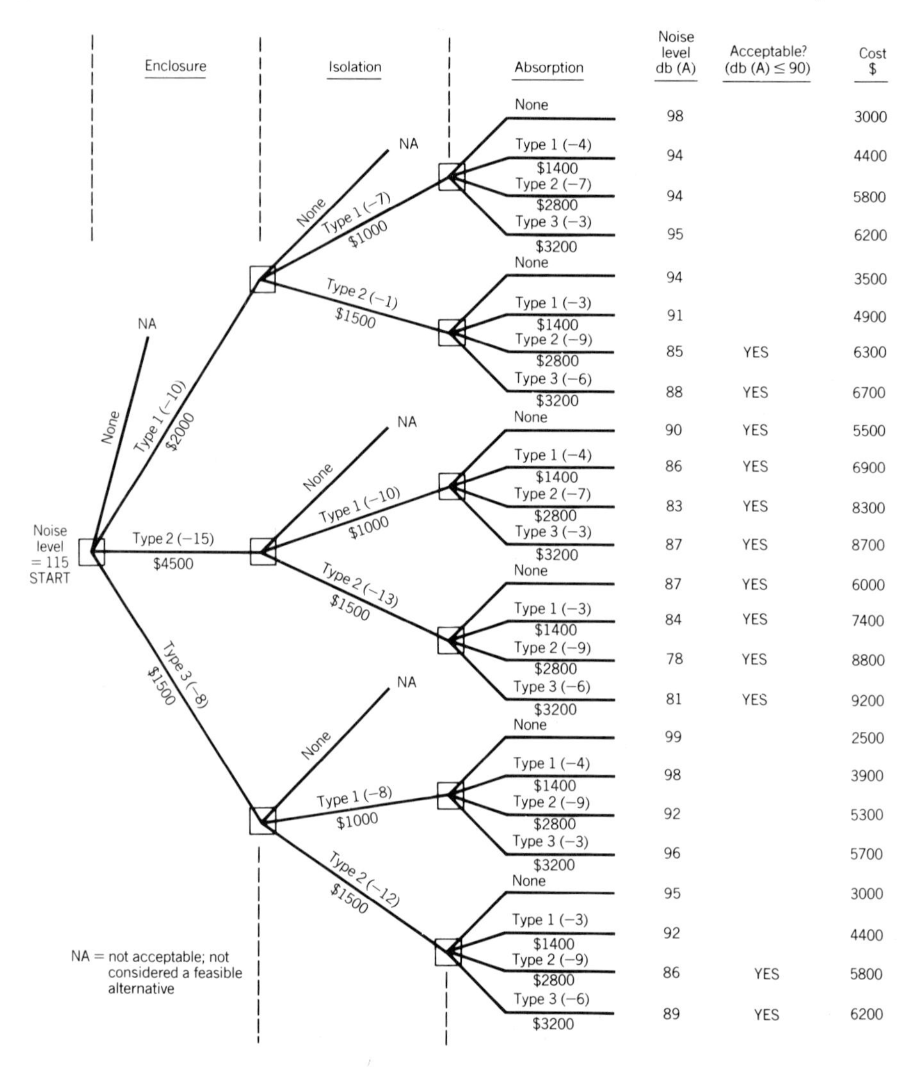

FIGURE 20.3. Decision diagram for the noise control methods problem; some type of enclosure and some type of isolation are predetermined to be necessary to meet the noise specification of 90 db (A).

From Figure 20.3 the best plan for noise compliance is seen to be one that utilizes enclosure type 2, isolation type 1, and no absorption type. This plan would result in reducing existing noise level to 90 db(A) at a total cost of $5500. Closer examination of Figure 20.3 shows, for example, that while 90 db(A) can be achieved for $5000, a further reduction of 4 db(A) can be achieved for only $300 added cost.

20.6 MATHEMATICAL MODELING APPROACHES

The above economic analysis methods are primarily basic quantitative techniques, often combined with qualitative considerations, which are thought to be most readily useable for analyzing the economics of safety problems. There are numerous mathematical modeling approaches, often identified as part of the field of operations research, that have potential for providing powerful analysis insights into safety problems. The following are two examples of such models.

Example—Determining Optimal Inspection Intervals
A manufacturing company accumulates two million man-hours of labor exposure every year. During the last three years, the company maintained an average frequency rate of 15 lost-time injuries per one million man-hours of exposure. The company safety records result in projections of the following average cost data for various accident categories:

1. Lost-time cases	\$1200/case
2. Minor-injury cases	\$ 100/case
3. No-injury cases (material losses)	\$ 50/case

The safety manager can inspect each department and area of the company periodically and issue inspection reports to all concerned. Hazards cited in these reports are assumed to be corrected (eliminated or brought under control) before the next scheduled inspection, but new hazards are assumed to be initiating as a linear function of the time since the last inspection. If each inspection costs the company \$2,000 what is the optimal time between inspections, and what is the associated total cost/year for accidents and inspections?

Safety inspections can accomplish two things. They can (1) identify safety and health hazards as well as deficiencies in existing safety programs; and (2) reinforce or enhance interest in safety and health practices. Following each inspection, an impact will be felt for a period of time, during which a special effort will be made to combat safety and health hazards. The immediate result of this effort should be a substantial reduction in the number of hazards at the workplace inspected.

In most instances, however, this would not last for long, and with time the sharpened interest in safety activities will dissipate and eventually fall to the level existing prior to the inspection. In other words, performing a safety inspection could cause a surge in safety and health activities that would usually result in the elimination or control of most workplace hazards. The problem of inspection, therefore, may be presented as a situation in which hazards are allowed to accumulate between inspections, and following each inspection the majority of these hazards would be eliminated in a short span of time relative to the time interval between successive inspections.

From the preceding discussion, we can make the following assumptions:

1. Hazards are accumulated between inspections at a linear rate with the maximum level of accumulation reached immediately prior to inspection and correction.
2. The more frequent the safety inspection, the fewer hazards will be accumulated between successive inspections.
3. The fewer the number of hazards accumulated, the less the chance for causing accidents.
4. Elimination of hazards following an inspection does not preclude or lessen their occurrence in subsequent periods.

With these assumptions in mind and from a safety viewpoint, it would be desirable to maintain low levels of hazard accumulation, that is, to perform as many inspections as practically feasible. However, from an economic and resource viewpoint, a large number of inspections may not be an ideal solution, since as frequency of inspection increases, inspection cost increases. On the other hand, when the frequency of inspection is very low (or no inspections at all are made), the number of hazards accumulated will be large, and consequently the potential for accidents and associated economic losses will be substantial.

The problem, then, is to find a frequency of inspection that will minimize the total cost (cost of inspection plus cost of accumulating hazards). To accomplish this, concepts of inventory theory for the determination of economic order quantities could be utilized. In using this approach, one can assume that hazards that accumulate between inspections are equivalent to units in stock, and that each inspector is analogous to placing an order.

The cost associated with hazard accumulation can be estimated by applying the loss potential triangle shown in Figure 20.4. In this case, one can assume that there is a direct relationship between the number of accidents and the level of hazard accumulation. Therefore, for a frequency rate of 15 and exposure time of two million hours, we compute the following:

1. Lost-time cases = (15/1 million) × 2 million = 30/yr.
2. Minor-injury cases = (30 minor/1 lost-time) × 30 lost-time = 900/yr.
3. No-injury cases = (300 non-injury/1 lost-time) × 30 lost-time = 9000/yr.
4. Total number of hazards = (3000 hazards/1 lost-time) × 30 lost-time = 90,000/yr.

The total cost of all accidents can be computed (based on data given): 30 (\$1200/lost-time) + 900 (\$100/minor) + 9000 (\$50/no-injury) = \$576,000/yr.

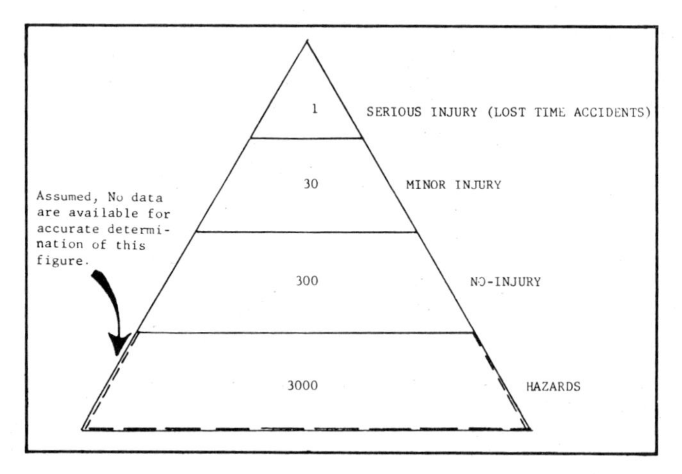

FIGURE 20.4. The loss potential triangle, showing relative frequencies of various types of injuries, as suggested in safety literature.

For this total cost and number of hazards estimated, the average cost per hazard is \$576,000/90,000 = \$6.40.

Next, adapting a classic economic order quantity model such as in Starr and Miller[1] we can write (see Table 20.7 for definition of symbols):

$$H = \sqrt{\frac{2H_T C_I}{C_H}}$$

$$t_I = \sqrt{\frac{288\ C_I}{C_H H_T}}$$

$$TC_t = \sqrt{\frac{12\ C_I}{t_I} + \frac{t_I H_T C_H}{24}}$$

where:

Substituting $H_T = 90{,}000$, $C_H = \$6.40$, and $C_I = \$2000$ yields the following optimum solution:

$$H_I = 7500$$

$$t_I = \text{one month}$$

$$TC_t = \$26{,}000$$

The optimum solution implies that safety inspection should be performed once every month, and this will be accomplished at a total cost of \$26,000.

In the solution presented above, the assumption was made that all hazards are equal in their potential contribution to accident costs and that all hazards can be eliminated when an inspection is made. It was also assumed that costs of safety inspections were not hazard-dependent; that is, the cost was merely a function of making the inspection or visit and was not necessarily related to the nature of type of hazards at the workplace.

These two assumptions could be relaxed, and a more realistic model of the problem could be developed. For instance, one may want to treat all hazards on an individual basis and compute a corresponding inspection cost for each hazard. Here it is assumed that inspection cost will be a function of the type and nature of the hazard at hand. For example, recognizing and assessing a noise hazard could involve more work (time, special skills, etc.) than recognizing housekeeping hazards (e.g., an extension cord on the floor).

To deal with the inspection problem under these conditions, the use of inventory models for "many items from one supplier" would be appropriate. These models could be studied under various policy restrictions such as maximum number of inspections and tolerable (acceptable) levels of accident costs.

TABLE 20.7 Definition of Parameters for Inspection Problem

For Inspection Interval Problem	For Analogous Order Quantity Problem
H = Optimum level of hazard accumulation between successive inspections	Optimum order quantity
H = Total number of hazards that is T expected to be accumulated during the time period study, a year in this example	Total demand during period
C = Cost of performing a safety and health I inspection	Cost of placing an order
C = Cost of exposure to a hazard per year, H	Cost of carrying inventory
t = Optimum time interval (months) between I inspections	Optimum time between orders
TC = Total optimum cost of inspection t hazards/month	Total optimum cost ordering and carrying

Example—The Inspection Games

The enforcement of and compliance with OSHA standards have all the makings of a competitive game. There are two or more competitors (industry and OSHA) and a conflict situation (e.g., compliance or no compliance). Decisions are made simultaneously by the competitors (competitors do not know a priori how their opponents will make their decisions or moves), each decision yields on outcome to each of the participants (gain or loss), and each competitor is motivated to either maximize gain or minimize losses.

From the OSHA standpoint, all occupational workplaces should be checked for compliance with the standards. Due to resource and technical limitations, however, OSHA has to adopt an inspection strategy or policy that calls for inspecting a selected number of workplaces. Such strategy might be to inspect all workplaces employing 50 or more persons for safety violations, to inspect 10 percent of all workplaces for health violations, or to temporarily postpone inspections for a selected number of businesses. The last strategy might be justified on the ground that some businesses have a safety performance that is above average or excellent and their exclusion from inspection would constitute an acceptable risk.

For industry, there are three basic options available when it comes to putting a workplace into compliance. First, it can ignore the standards and risk strict financial penalties and legal problems at a later date. Second, it can institute a cosmetic compliance program limited only to obvious safety hazards (e.g., housekeeping hazards). Third, it can accept that full compliance with the applicable standards (safety and health) may be the only feasible and economical option in the long run. Associated with each of the three options is a certain payoff, which is contingent upon the enforcement strategy adopted by OSHA. For example, if industry adopts the option of no compliance while OSHA opts for full inspection, a considerable financial loss could be incurred by the industry. Other losses or gains can be determined for all possible combinations of enforcement-compliance options or strategies. Such data is commonly presented in the form of payoff matrices—one for each competitor (i.e., OSHA or industry).

For industry, entries of the payoff matrix reflect the sum of the two cost components: (1) the cost of implementing the chosen option or strategy, and (2) the fine that might be assessed by OSHA if the industry was inspected when OSHA operated under enforcement policy strategy i. In determining P_{ij}, we accept the pessimistic view that every OSHA citation will yield a financial penalty to industry, the magnitude of which will depend on the level of compliance maintained by the industry. For OSHA, the payoff matrix is the negative of the industry matrix. In computing P_{ij}, actual or exact figures need not be precise, for the relative values are sufficient to obtain a solution.

The OSHA enforcement-compliance games (as defined in the previous discussion) can be solved by using one of the game theory methods, or perhaps linear programming. For our purposes here, we will formulate the game as an LP model (for other methods of game theory, see Ackoff and Sasieni, 1968).

Let X_i, $i = 1, 2, 3$, be the ith compliance strategy for the industry group under consideration; Y_j, $j = 1, 2, 3$, is the jth enforcement strategy of OSHA for this industry group. The industry follows an optimal compliance policy (X_i, $i = 1, 2, 3$), if

$$\sum_{j=1}^{3} \sum_{i=1}^{3} P_{ij} X_i Y_i \geq V$$

for all strategies of OSHA, Y_j, j = 1, 2, 3. Similarly, OSHA enforcement policy is optimal if

$$\sum_{i=1}^{3} \sum_{j=1}^{3} P_{ij} Y_j X_i \leq V$$

for all strategies (X_1, X_2, X_3) of the industry group.

$$\sum_{i=1}^{3} X_i = 1$$

proportions of time spent on using (playing) all strategies should sum to unity

$$\sum_{j=1}^{3} Y_j = 1$$

where

V = value of the game, which remains unknown to both competitors until the completion of the game.

In terms of a linear programming model, the above equations can be written as

Maximize V

subject to

$$\sum_{i=1}^{3} P_{ij} X_i \geq V \qquad J = 1, 2, 3$$

$$\sum_{i=1}^{3} X_i = 1$$

$$X_i \geq 0$$

For OSHA, the formulation can be written as

Minimize V

subject to

$$\sum_{j=1}^{3} P_{ij} Y_{ij} \leq V \qquad i = 1, 2, 3$$

$$\sum_{j=1}^{3} Y_j = 1$$

$$Y_j \geq 0$$

In formulating the linear programming models, we are accepting the premise that industry will always attempt to maximize its gain and OSHA will seek to minimize the losses that may result from adopting the wrong enforcement policy. Having formulated the appropriate models, the next step is to define the data (payoff matrices) and then proceed to solve the models by using a computerized LP algorithm.

For the purpose of illustration, consider the hypothetical payoff matrix for an industry given in Table 20.8. For a given compliance strategy, each entry in the payoff matrix is the utility (positive or negative) that would be expected by the industry group if OSHA opted for the various inspection policies. Of course, the OSHA payoff matrix is the opposite of the industry matrix. Applying the LP approach (formulated above) for solving the game with the given payoff matrices, the optimum strategy for both OSHA and industry during a finite period of time (e.g., a year) would be as follows:

1. Optimum compliance strategy for the industry group
   a. Opt for no compliance 14 percent of the time.
   b. Comply with safety standards 20 percent of the time.
   c. Choose to comply with both safety and health standards 66 percent of the time.
2. Optimum enforcement strategy for OSHA
   a. Not inspect this industry group 65 percent of the time.
   b. Perform safety inspection 20 percent of the time.
   c. Perform full safety and health inspection 75 percent of the time.

TABLE 20.8 Payoff Matrix for the Industry Group

Industry Compliance Strategy	OSHA Inspection Strategy for This Industry Group		
	No Inspection	Safety Inspection Only	Full Safety and Health Inspection
No compliance	100	−50	−100
Compliance with safety standards, i.e., elimination of obvious hazards, including hazards of housekeeping, machine guarding, means of egress, etc.	50	75	−50
Full safety and health compliance	25	50	100

The implication of the optimum solution is simply that in the long run it would be beneficial for the industry to comply with OSHA standards. On the other hand, insofar as OSHA is concerned, a full-fledged inspection program is not warranted for this industry group. Instead, a few inspection visits should be made to some members of the industry group in order to establish an awareness of OSHA and its activities. In other words, this industry group should not be considered a candidate for the OSHA target industry program.

The compliance and inspection strategies presented above are valid only for the payoff matrix given. Any variations in the entries of the matrix would be expected to yield other strategies or solutions.

20.7 ECONOMIC IMPACT OF STANDARDS

OSHA bears direct responsibility for developing and promulgating occupational safety and health standards and for issuing regulations. These standards and regulations are at the heart of the Occupational Safety and Health Act and are the primary tools used in the effort to reduce accidents.

Before a standard can be implemented by an industry, many checks are performed on the standard to ascertain the extent to which it will be effective and efficient. These checks are concerned not only with a standard's ability to reduce the number of accidents in an establishment but also with the ability of an employer to come into compliance with a standard's requirements. In some situations, the costs of compliance may be high and may result in adverse effects on the establishment. These adverse effects may be so severe that they effectively remove the business from the economy at the cost of an untold number of jobs. In this case it may be argued that the standard has accomplished its purpose of reducing accidents. If the job is nonexistent, so too will be any accidents associated with it. Obviously this is not the goal of OSHA. Therefore, some attention must be directed toward determining what the economic impact of a standard will be before it is fully implemented. Further, a standard should be scrutinized not solely in isolation but also in conjunction with all of the other standards it will join after adoption. This is necessary to discover any appreciable costs that will only become apparent if the standard is implemented in association with other standards.

20.7.1 Determining Economic Impact

The impact a standard has upon a particular industry is determined by how much that industry must change to meet the requirements of the standard. Two primary types of change affect an industrial environment: (1) physical change, and (2) behavior change.

PHYSICAL CHANGE

Physical change is concerned with the addition or modification of equipment used by the industry.

Machinery and equipment design is an ongoing, dynamic, evolutionary process. Any changes made to this aspect of an industry's operation must, therefore, be constantly evaluated and assessed. Design and purchase policies should ensure that equipment will meet the requirements of a standard without necessitating the later addition of safety devices. Retrofits are usually expensive and not as effective.

Behavioral Change

Behavioral changes are directed toward the modification of the actions of employees performing a particular task or working in a specific environment. Heinrich, in an often cited document (see Bibliography), purports that people—not things—cause accidents. His studies indicate that 88 percent of all accidents are caused by people and only 10 percent are caused by things. This means that if people observe safety practices then accidents may be reduced.

A problem traditionally noted in association with attempts to alter human behavior concerns the longevity of the behavioral change. It has been argued that humans, by their very nature, have a tendency to return to old habits. Therefore, when adherence to new habits is taught, the consequent behavioral change, if any, is short-lived. If this argument is at all sound, it suggests that if behavioral changes are to be relied upon as a primary method of accident reduction, then reliable training procedures have to be developed and utilized.

Monetary Cost of Physical and Behavioral Change

In considering the implementation of a type of change in an industrial environment (whether that change be physical of behavioral), one factor that must be addressed is the monetary cost of the proposed change.

The cost of physical change is often thought of as being determined merely by the purchase price of some equipment. This is not the case in a vast majority of situations, however, because installation costs and associated training adaptation costs must also be considered. In some instances the cost of installation and training may be greater than the cost of the equipment itself. Maintenance costs may also need to be considered.

The costs associated with behavioral changes include the cash costs of providing an instructor, books, and hourly wages to trainees, plus the opportunity cost of lost production time resulting from the removal of the employee from the job so that training sessions may be attended.

Deciding on Physical or Behavioral Change

There is usually not one clear-cut type of change available to industry to come into compliance with a particular standard's provisions. Several alternative types of physical and/or behavioral changes must usually be considered. The type or combination of alternatives selected ultimately determines the monetary impact of the standard on an industry. In some situations, the provisions of a standard dictate whether the change associated with compliance will be physical or behavioral. In others, it is left to industry to determine its own method of compliance. In either case, however, the objective of industry is to satisfy the requirements of a standard while keeping total compliance costs to a minimum. Because there is usually more than one alternative available, each must be individually examined and assessed in terms of its potential benefits and costs.

Physical changes typically have a higher initial cost, but in most situations they also have a proportionally higher life expectancy. The long-term effectiveness of behavioral changes, on the other hand, is often suspect. Also of concern with behavioral change is the fact that many small establishments may be forced to enlist the help of outside consultants. This method of implementing change may, therefore, become inordinately expensive. It should be pointed out that the alternative selection procedure must be carried out for each provision and standard applicable to the firm.

20.7.2 Compliance Cost

To estimate compliance cost for an industry, a specific evaluation procedure may be utilized. The procedure consists of two parts. The first part seeks to develop a valid baseline from which, in part two, compliance costs can be estimated.

To establish the baseline, the industry is categorized into representative groups of establishments. The equipment used by each of the categories is listed, and the safety practices and features currently in use are determined. Then, a number of representative model plants and process flows deemed to be representative of the industry are constructed. Compliance cost is then estimated directly from the baseline data. The compliance cost estimates can only be as accurate as the baseline data are representative of the industry at large. This fact points to the desirability of obtaining valid and reliable baseline data.

20.7.3 Methods Available for Determining a Baseline

The primary methods of obtaining baseline data include: (1) plant visits, (2) literature searches, (3) surveys and questionnaires, and (4) some combination of these methods.

Plant Visits

Establishments are first selected and then visited. During the visit, process flows, types of equipment in use, material handling systems, and layouts utilized are noted and recorded. By visiting a large number of facilities and determining their commonalities, it would be possible to construct a representative model plant for the industry at hand. The obvious hazard of this approach is the plants that have been visited may not be representative or may be in some manner unique. The representativeness of a facility may be difficult to judge, especially to an inexperienced eye. In some industries (printing, for example) equipment obsolescence takes place so rapidly that no two plants could easily be compared.

A further limitation of the plant visit method is the time required to conduct plant tours. This factor may become especially prohibitive if there is not an abundance of target facilities clustered in close proximity and long travel distances are therefore involved.

Literature Search

This method involves examining textbooks, trade journals, production process manuals, and the like in an effort to gain an insight into the workings of a particular industry. A convenient source of such information is provided by the U.S. government in the *Census of Manufacturers.* These reports put forth data in terms of major groups engaged in a particular facet of the industry, and then give detailed information about various subgroups (so-called SIC codes). The primary problem associated with using literature as a source of information concerns the necessity of having current accounts of an industrial situation. Such accounts may be difficult to obtain because the rapid changes that occur in some industries may effectively outmode a textbook soon after its publication. This is also true of the *Census of Manufacturers* report, which is completed once per year and requires four to five months to reach the shelves after completion. Also, the latest procedures and equipment may not be available to the general public due to company policy and restrictions concerning trade secrets.

Surveys

This procedure involves making inquiries using either questionnaires or telephone interviews. The shortcomings of this method are primarily due to the inordinate amount of time necessary to prepare, send, and receive back questionnaire forms from a representative sample. Difficulties may also be encountered when explaining and interpreting difficult processes and complicated equipment over the phone or by mail. Finally, some companies are reluctant to divulge information concerning their operations when they are requested to respond through such an informal medium.

Combination Method

Because of the limitations associated with each of the individual methods of determining a baseline and the importance of obtaining a valid baseline, it is usually necessary to employ a combination of the procedures discussed above. By using a combination of methods, checks can be performed on each procedure to determine that procedure's validity and, it warranted, to adjust the results obtained by the procedure. For example, a literature search could be conducted first to determine as nearly as possible the type of processes, equipment, layouts, etc., found in a specific industry. Following this, a limited number of plant tours of representative facilities could be taken to determine if the data obtained in the literature search are accurate. Finally, a limited number of surveys could be completed to verify the information obtained by the first two methods. In this final check, specific (rather than open-ended) questions based on the information obtained by the first two methods would be asked to simplify the survey procedure as much as possible.

Baseline Data—General or Specific?

A question that must be addressed at this time concerns how detailed and specific (as opposed to general and encompassing) the baseline data generated should be. The answer to this inquiry depends largely upon the existence (or nonexistence) of a generic family of establishments in the industry being scrutinized. If such generic families exist, they could be delineated to form the baseline categories. A model plant from each of the generic families could then be constructed to represent that entire generic family; the compliance costs associated with it would be considered also representative.

Generic Industrial Families

As used in the preceding paragraph, a generic family is defined as those establishments in industry that have commonalities in terms of processes, safety practices, equipment, and the like, sufficient to render them very similar to one another.

It has been held that an industry is defined by its input, the raw materials that enter it, and by its output, the product and/or service that leaves it. By this notion, if input and output are similar, then the processes associated with the throughput are also similar. Following this logic, if some number of industries have similar inputs and outputs, then they also have similar throughput processes and can therefore be called a generic family. Obviously, however, this need not always be the case, for factors such as degree of automation and recency of equipment may contribute to throughput differences.

If generic industrial families do, in fact, exist, then the task of arriving at valid compliance costs may be simplified. Such families would make the industrial categories highly similar with respect to their constituents and thereby allow for accurate industrial baseline data. If no such clear-cut generic families are found to exist, then the compliance cost estimates will be proportionally affected. It follows that the baseline data should be as exhaustive and exclusive as possible.

The goal in determining a baseline is, then, to divide establishments into groups that are representative of portions of an industry that will bear similar compliance costs. This should be accomplished using generic families wherever the industrial situation permits. A model plant from each group is then constructed and compared to a standard to ascertain the compliance costs it will bear. This compliance cost is then held as being representative of the cost to be absorbed by every other establishment in that group. Once the compliance costs of the groups have been established, it is possible to aggregate them so as to reflect the cost to the entire industry.

20.7.4 Unit Costs

Once an industry baseline has been determined, it is necessary to evaluate the costs associated with each provision of a standard. These costs are termed "unit costs."

Unit costs are the costs attached to the various methods and options available to an establishment for it to come into compliance with a particular standard. In many situations, besides having a choice between physical and/or behavioral change, a company also has to select between different *types* of physical and/or behavioral changes. Each of these types would have its own peculiar benefits and costs. Therefore, for the compliance cost estimate to be accurate, each method should be considered separately in making the estimates. If the procedure of taking every potential combination of compliance methods into consideration were to be performed by hand for each provision of a standard, it would, at best, be extremely time-consuming. For this reason, the aid of a computer and of simulation techniques can be a great asset.

Simulation procedures could be used to construct model plants, using the baseline data accumulated for all the categories. Different compliance methods could be incorporated into the model plants. In this manner an array of costs could be generated along with their respective probabilities of occurrence. From these data the probability of an establishment having to absorb a particular cost of compliance with a standard could be ascertained. Also, "most likely" costs associated with a specific generic family or industrial category could be computed.

Finally, after the individual costs for meeting each provision have been determined, it would be possible to determine where the largest costs to industry exist (e.g., costs of machine guarding, training procedures, etc.). These most significant costs could then be compared to the most significant accident-reducing provisions, and it is thereby possible to determine the "most important" provisions within those standards. (For a detailed treatment of simulation, see Chapter 17.)

Incremental Cost

Before moving to a demonstration of some of the procedures and techniques discussed, it would be beneficial to mention the advantage of using incremental costs for compliance cost estimation. The basis for using incremental costs is that some of the provisions of a proposed standard are similar (if not identical) to those detailed by other standards already in effect. Therefore, it would be redundant and inaccurate to calculate an additional cost to the industry since the industry, by law, would already be in compliance with the provision in question. For this reason, incremental costs should be calculated and utilized whenever possible.

20.7.5 Compliance Cost Determination—An Example

This example illustrates how compliance costs for a standard regulating materials-handling equipment can be determined. Two sample industries that are completely distinguishable in terms of technology, product, and structure were considered.

To compute compliance costs, the two major industries were categorized into small, medium, and large establishments on the basis of number of employees. Next, typical process flows associated with the respective products of each industry were described. The proposed standard was then applied to each category, with the result being the compliance cost of that category. Through various computations, the results obtained for each category were then obtained. The annualized costs of capital expenditures are given by:

$$IC\,(A/P,\ i\%,\ N)$$

where

IC = installed capital cost (the basic price of the capital and all costs and fees incident to its installation),
N = lifetime of the assets in years, and
i = the interest rate.

The training cost (recurring) is given by

$$(A)\,(P)\,(H)\,(W)$$

where

A = the new hire accession rate
P = the number of production workers exposed to equipment hazards
H = the incremental time in hours needed for safety training
W = the entry level hourly wage + direct benefits.

Prior to categorization of each industry, an in-depth field survey of six plants (primarily small and medium) was undertaken. In addition, a phone survey of several companies and of equipment manufacturers was conducted to refine and somewhat validate the results and data obained from the field study. In the interest of defining and categorizing various types of production processes in each industry, textbooks, technical journals, research reports, and industry data were used extensively.

In developing model plants for each industry, two approaches similar in concept but different in implementation were used. For the first industry, the industry was abstracted via one basic model representing a small plant (a borderline case between small and medium plants). The data for this small model plant were then adjusted to reflect the processing and handling characteristics of medium and large plants. This is possible because of the simple fact that the processing technology and the equipment used in the industry were considered fixed among all plants, small and large. In the case of the second industry, a set of plant models was used to describe the three primary products of the industry.

Based on the data gathered through field and phone surveys, literature searches, and the personal experience of the project staff, compliance methods with the proposed standard and the coresponding practice in each industry were obtained, complete with unit cost for each compliance activity.

Using the models developed and the unit cost for each compliance activity, the total compliance cost for each industry was then computed and summarized. Compliance cost summaries by compliance activity for each of the two industries are given in Tables 20.9 and 20.10. Some general descriptive statistics concerning the distribution of compliance costs for each industry are given in Tables 20.11 and 20.12. In all tables, two compliances cost figures are given: (1) conservative, assuming that the industry is currently not in compliance with the proposed standard, and (b) most likely, assuming that some degree of compliance with the standard is now the practice of the industry.

All compliance costs given represent *incremental costs* in that they reflect only the additional cost to be incurred by the industry in the course of compliance with the standard. The implication here should be clear and straightforward—some of the provisions of the standard required compliance practices and features similar, if not identical, to those detailed by other OSHA standards. For this reason, compliance with some of the provisions should not constitute an additional cost to the employer.

For the first industry, the compliance cost per plant varied from \$565 to \$1115; for the second, it varied from \$221 to \$382. These cost figures reflect the differences in technology and in plant size distribution between the two industries. The second industry shows a high concentration of small plants, so the average cost per plant is biased (skewed) toward the cost for small plants. In contrast, the first industry is dominated by medium and large plants, which tends to push the industry average somewhat closer to the cost for medium plants.

Provisions of the standard calling for the use of warning devices account for the largest percentage of the total compliance cost (30 percent or more). Next are work practices described in terms of training and maintenance; these account for 20 to 30 percent of the total cost. The remaining 40 percent of the total cost is distributed among three or four provisions of the standard.

20.8 SUMMARY AND PERSPECTIVE

This has been a brief review of a wide range of approaches and techniques for analyzing the economics of safety problems. Several illustrative examples are included. The potential for applying these methods to safety problems is growing rapidly.

Economic analyses based on statistical data have their limitations, the most significant of which is the accuracy of the input estimates for accident probability and accident cost. Statistical accuracy requires a large sample size. On a local scale, similar accidents occurring on similar equipment are far too infrequent to yield accurate estimates of probability. Accident costs may likewise vary.

Such limitations would not prohibit the use of quantitative economic analyses techniques, however. For those situations that cannot be handled with sufficient statistical accuracy, subjective value judgements based on professional experiences can be utilized. When dealing with subjective input estimates, it is essential that the rationale for obtaining such estimates be acceptable to and understood by management.

Accident costs may also be estimated in a number of different ways. Good sources that deal with the subject of accident cost determination are Simonds and Grimaldi,[2] Calabresi,[3] and Peterson.[4]

Because most safety decisions must ultimately relate to the economics of limited resources, the safety professional should strive to objectively analyze the economics of alternatives and then to present recommendations in a manner acceptable to the management community. To the extent that the analysis can

TABLE 20.9 Total Compliance Cost for Industry No. 1

Compliance Activity	Installed Capital Cost ($ Millions)	Annual Capital Charge ($ Millions)	Annual Operating Cost ($ Millions)	Total Conservative Annualized Cost ($ Millions)	Total Likely Annualized Cost ($ Millions)
Pretesting procedures	—	—	—	—	—
Restrictions on guard removal	—	—	—	—	—
Restrictions on equipment use	—	—	—	—	—
Housekeeping procedures	—	—	—	—	—
Repair and maintenance procedures	.939	.186	0	.186	.105
Training	—	—	.313	.313	.282
Crossovers	—	—	—	—	—
Drip pans	—	—	—	—	—
Power loss safeguards	—	—	—	—	—
Guarding	2.156	.337	—	.337	.094
Emergency stops	2.256	.375	—	.375	.282
Warning devices	2.410	.401	.490	.891	.417
Guiding and centering devices	—	—	—	—	—
Lock-out devices	—	—	—	—	—
Brake systems	4.106	.556	—	.556	.167
TOTAL	11.867	1.855	.803	2.658	1.347

TABLE 20.10. Compliance Cost for Industry No. 2

Compliance Activity	Installed Capital Cost ($ Millions)	Annual Capital Charge ($ Millions)	Annual Operating Cost ($ Millions)	Total Conservative Annualized Cost ($ Millions)	Total Likely Annualized Cost ($ Millions)
Pretesting procedures	—	—	—	—	—
Restrictions on guard removal	—	—	—	—	—
Restrictions	—	—	—	—	—
Housekeeping procedures	—	—	—	—	—
Repair and maintenance procedures	1.376	.229	0	.229	.114
Training	—	—	.417	.417	.376
Crossovers	.409	.063	0	.063	.006
Drip pans	—	—	—	—	—
Power loss safeguards	—	—	—	—	—
Guarding	1.640	.273	0	.273	.074
Emergency stops	1.724	.287	0	.287	.215
Warning devices	—	—	1.049	1.049	.577
Guiding and centering devices	.174	.029	—	.029	.001
Lock-out devices	—	—	—	—	—
Brake systems	—	—	—	—	—
TOTAL	5.323	.881	1.466	2.347	1.363

TABLE 20.11. Contribution of Each Compliance Activity Toward Total Compliance Cost: Industry No. 1

Activity	Total Conservative Annualized Cost (million dollars)	Percent of Total Cost
Warning devices	.891	33.5
Brake systems	.556	20.9
Emergency stops	.375	14.1
Guarding	.337	12.6
Training	.313	11.7
Repair and maintenance	.186	6.99
TOTAL	2.658	100.
Activity	**Most Likely Cost (million dollars)**	**Percent of Total**
Warning devices	.417	30.95
Training	.282	20.94
Emergency stops	.282	20.94
Brake systems	.167	12.39
Repair and maintenance	.105	7.79
Guarding	.094	6.97
TOTAL	1.347	100.

TABLE 20.12. Contribution of Each Compliance Activity Toward Total Cost: Industry No. 2

Activity	Total Conservative Annualized Cost (million dollars)	Percent of Total Cost
Warning devices	1.049	50.92
Training	.417	20.24
Guarding	.273	13.25
Repair and maintenance	.229	11.12
Crossovers	.063	3.06
Guiding and centering devices	.029	1.41
TOTAL	20.6	100.
Activity	**Total Likely Annualized Cost (million dollars)**	**Percent of Total**
Warning devices	.577	50.26
Training	.376	32.75
Repair and maintenance	.114	9.93
Guarding	.074	6.45
Crossovers	.006	.52
Guiding and centering devices	.001	.09
TOTAL	1.148	100.

be expressed quantitatively, one can be relieved of the subjectivity caused by lengthy descriptions, and thereby increase the chances that the best safety alternatives will be considered, recommended, and chosen.

ACKNOWLEDGMENT

Some of the examples presented in this chapter are taken from other papers by the authors published in the *Journal of Occupational Accidents* and *Professional Safety*. The section on the economic impact of standards relies heavily on research performed by Mr. Andrew Matuszak in the course of his graduate studies at North Carolina State University. We are thankful for all the help we received from our students.

Parallel thoughts were adapted from John R. Canada and John A. White, Jr., *Capital Investment Decision Analysis and Engineering*, (c) 1980, by permission of Prentice-Hall, Inc., Englewood Cliffs, N.J.

REFERENCES

1. Starr, M. K., and Miller, *Inventory Control: Theory and Practice*, Prentice-Hall, 1962.
2. Simonds, R. H., and J. V. Grimaldi, *Safety Management*, rev. ed., Irwin, 1963.
3. Calabresi, G., *The Costs of Accidents: A Legal and Economic Analysis*, Yale University Press, 1970.
4. Peterson, D., *Techniques of Safety Management*, McGraw-Hill, 1971.

BIBLIOGRAPHY

Ayoub, M. A., Integrated safety management information system part II: Allocation of resources, *Journal of Occupational Accidents*, 2(1979), 191–208.

Canada, J. R., and M. A. Ayoub, The economics of safety: A review of the literature and perspective, *Professional Safety*, Vol. 22, No. 12, December 1977.

Canada, J. R., and J. A. White, *Capital Investment Decision Analysis for Management and Engineering*, Prentice-Hall 1984.

Heinrich, H. W., *Industrial Accident Prevention*, New York: McGraw-Hill, 1st ed., 1931.

PART VIII

OCCUPATIONAL SAFETY AND HEALTH ECONOMICS

CHAPTER 21

How to Consider Cost-Benefit Analysis in Occupational Safety Practice

MICHAEL F. BIANCARDI
Wausau, Wisconsin

21.1 SCOPE

Considerable controversy has accompanied the application of cost-benefit analysis to situations impacting directly on the safety and well-being of people. Cost-benefit analysis is a money-oriented process. Activities that attempt to evaluate the health and safety of people in dollars and cents, which may be interpreted by some to pit dollars against people, are suspect in the minds and consciences of many. Notwithstanding, cost-benefit analysis has been in considerable vogue in the evaluation of health and safety circumstances affecting large population segments or requiring extensive investments of time and money—particularly in the evaluation of governmental programs or in the development of national health and safety policy.

Here our concern is the practice of occupational health and safety in an industrial or commercial operation as seen from the viewpoint of the responsible manager. While there must be significant concern in these activities for the protection of property and the productivity and continuity of operations, the core of such pursuits is the well-being of people. Thus, the existing differences of opinion concerning the propriety of cost-benefit analysis for national safety policy and programs may well follow into the business scene.

Our focus, then, on this cost-oriented procedure must be, not with a wide angle lens that brings matters of governmental policy and programs into view, but with a lens that concentrates on an organization's employee safety, fire safety, product safety, and the environmental, highway, and public incidents that may have liability exposures for the organization. Such exposures may be included in the responsibilities of the staff person directing the health and safety activities of an enterprise. Further, our discussion of this technique must reflect on its possible use in businesses and other endeavors of all sizes and types, and to all kinds of health and safety practitioners who impact on business: those responsible for industrial and commercial programs, private consultants, insurance consultants, government specialists, and academic instructors and researchers.

21.2 History and Examples

In his book *Essentials of Engineering Economics,* James L. Riggs traces the origin of benefit analysis back to 1844. In this country, the Rivers and Harbors Act of 1902 required the Army Corps of Engineers to evaluate the merits of the river and harbor projects, including the amount that commerce would benefit as compared to the estimated cost. Riggs points out that government participation in public projects was extended by the Flood Control Act of 1936, which called for benefit evaluation of flood control projects.

Two engineering uses of cost-benefit evaluation are illustrated by examples used by Professor Riggs. The first deals with alternatives for a single project. The second, provided here, shows factors that make up the costs and benefits of the structures under consideration, none of which prove to be economically feasible.

21.2.1 Engineering Examples

*Example 1**

J. L. Riggs, *Essentials of Engineering Economics,* McGraw-Hill, New York, 1982.

Benefit-to-Cost Criteria

In comparing benefit (B) to cost (C), several different perspectives are reasonable. Consider the simplified data in Table 21.1 that describe the alternatives for a small flood-control project. The current average annual damage from floodings is $200,000. Three feasible options are available to reduce the damages: Each larger investment of public funds provides greater protection.

The following criteria indicate different plausible preferences among the alternatives:

1. *Minimum investment:* Choose alternative A. If funds are severely limited, this may be the only possible choice.
2. *Maximum benefit:* Choose alternative D. Flooding would occur only during extremely wet seasons.
3. *Aspiration level:* Depends on the threshold set for cost or benefit. If, for instance, the aspiration level is to reduce flood damage by 75 percent, alternative C should be chosen because it meets the aspiration with a lower cost than alternative D. Similarly, an annual cost threshold of $100,000 indicates a preference for alternative B.

**Source:* Riggs, J. L., *Essentials of Engineering Economics,* McGraw-Hill, New York, 1982. Reprinted with permission.

TABLE 21.1. Annual Costs and Benefits From Different Levels of Investment in a Small Flood-Control Project

Alternative	Equivalent Annual Cost of Project	Average Annual Flood Damage	Annual Benefit
A: No flood control	$ 0	$200,000	$ 0
B: Construct levees	40,000	130,000	70,000
C: Small reservoir	120,000	40,000	160,000
D: Large reservoir	160,000	10,000	190,000

The three projects and the do-nothing alternative are mutually exclusive. Data for selecting the most attractive are shown below, where the figures are in thousands of dollars.

			Total		Incremental			
Alternative	Annual Benefit, B	Annual Cost, C	B/C	B-C	ΔB	ΔC	ΔB/ΔC	ΔB-ΔC
A	$ 0	$ 0	0	$ 0				
B	70	40	1.75	30	$70	$40	1.75	$30
C	160	120	1.33	40	90	80	1.125	10
D	190	160	1.19	30	30	40	0.75	−10

Source: Riggs, J. L., *Essentials of Engineering Economics,* McGraw-Hill, New York, 1982.

4. *Maximum advantage of benefits over cost (B−C):* Choose alternative C.
5. *Highest benefit-to-cost ratio (B/C):* Choose alternative B.
6. *Largest investment that has a benefit-to-cost ratio greater than 1.0:* Choose alternative D.
7. *Maximum incremental advantage of benefit over cost* (ΔB−ΔC): Choose alternative B.
8. *Maximum incremental benefit-to-cost ratio (ΔB/ΔC):* Choose alternative B.
9. *Largest investment that has an incremental B/C ratio greater than 1.0:* Choose alternative C.

*Example 2**

FEASIBILITY INVESTIGATION OF A WATER RESOURCE PROJECT

The Corps of Engineers was directed by the Congress of the United States to make a study ot Marys River Basin. The purpose of the study was to determine what could be done to reduce flood damages and to conserve, use, or develop the basin's water resource.

Six alternatives were investigated: (1) floodproofing of individual structures, (2) building levees, (3) improving river channels, (4) instituting a system of landowner-constructed dams, (5) implementing a system of small tributary reservoirs, and (6) erecting multipurpose reservoirs. The costs of floodproofing, levees, and channel improvements were found to be much greater than the benefits, mainly because of the lack of secondary benefits. Landowner-constructed dams built with government financial support would have some localized benefits, but would require an unreasonable amount of land for the storage obtained, and problems of balancing outflows made the alternative impractical.

**Source:* J. L. Riggs, *Essentials of Engineering Economics,* McGraw-Hill, New York, 1982, and Summary Report on Marys River Basin Water Resource Study, U.S. Corp of Engineers, 1975.

The last two alternatives also proved to be economically infeasible, as shown in Table 21.2.

21.2.2 Health and Safety Examples

Cost-benefit analysis (CBA) is a relatively new technique for health and safety decision-making. The planning, programming, and budgeting system initiated by the Johnson Administration in 1965 gave impetus to such use. In what was then the Department of Health, Education and Welfare, the task of conforming with a new governmental procedure fell on transferees from the Department of Defense who, although long experienced in systems analysis, were now obliged to make monetary estimates of the effects of competing programs.

Research in the literature shows that during the intervening years cost-benefit analysis has been applied by many government departments and other analysts and researchers to help make health and safety decisions. Since actual use of cost-benefit analysis in industrial safety appears to be nonexistent,

TABLE 21.2. Benefit-Cost Analyses for Alternative Reservoir Projects

	Noon Dam	Wren Dam	Tumtum Dam	Tributary Dam System
Costs, total				
Construction	$51,352,000	$52,944,000	$37,024,000	$19,090,000
Investment[a]	57,386,000	59,165,000	41,374,000	21,333,000
Costs, annual				
Interest and amortization[b]	3,380,000	3,485,000	2,437,000	1,257,000
Operation, maintenance and replacement	210,000	180,000	170,000	135,000
Total average annual cost	$ 3,590,000	$ 3,665,000	$ 2,607,000	$ 1,392,000
Benefits, average annual				
Flood control	$ 945,400	$483,100	$258,500	$126,300
Recreation	573,000	345,000	401,000	163,000
Irrigation water supply	85,000	85,000	85,000	85,000
Municipal and industrial water supply	90,600	63,400	8,800	0
Fish and wildlife enhancement	0	0	0	0
Total average annual benefits	$1,694,000	$976,500	$753,300	$374,300
Benefit-cost ratio	0.47:1	0.27:1	0.29:1	0.27:1

Source: Riggs, J. L., *Essentials of Engineering Economics*, McGraw-Hill, New York, 1982. Summary Report on Marys River Basin Water Resource Study, U.S. Corps of Engineers, 1975
[a]Construction cost plus interest during construction.
[b]The annual cost equivalent to the total cost spread over a 100-year period computed at 5 7/8% interest rate.

or at best, highly limited, two examples outside of business safety practice are cited here to help develop a uniform point of departure. Also, they should be useful in relating this evaluation technique to occupational safety practice.

*Example 3**

Abstract

This study develops a decision method for evaluating the social acceptability of industrial controls on hazardous materials. Decisions are based on a "multiple criteria approach" that jointly considers measures such as risk benefit trade-off, minimum reducible health risk, maximum acceptable cost, and implicit value of human life. Health risks are calculated by combining separate estimates of production and usage patterns, emissions to air and water, effectiveness of controls, pollutant dispersion and human susceptibility. Economic benefits consider employment, trade and consumer impacts, as well as direct cost of controls. The analysis focuses on asbestos as an example hazard. Relative values of hazard reduction alternatives are examined for asbestos manufacturing exhaust filters and for asbestos substitutions in brake linings. Preliminary calculations indicate risk reductions of these alternatives cannot justify their social costs.

**Source:* Moll, K. D., and D. P. Tihansky, Risk Benefit Analysis for Industrial and Social Needs, *American Industrial Hygiene Association Journal* (38) 4/77. Reprinted with permission of American Industrial Hygiene Association.

Moll and Tihansky conclude, in part

> If one wishes to assign a value to saving lives, the chart shows that the factory filter alternative costs less than $10 million per life saved whereas brake substitution costs about $100 million per life. Neither alternative comes close to the $300,000 valuation that workers in hazardous occupations implicitly give to their own lives.

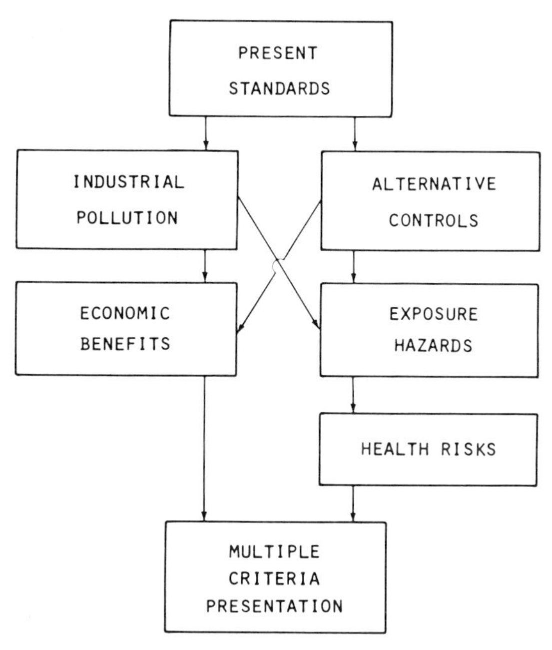

FIGURE 21.1. Risk/benefit methodology: analyticaly steps. *Source:* Risk Benefit Analysis for Industrial and Social Needs, Moll, K. D., and Tihansky, D. P., *American Industrial Hygiene Association Journal,* 4/77. Reprinted with permission of the American Industrial Hygiene Association.

EMISSION SOURCE:	MANUFACTURING FACILITIES	AUTOMOBILE BRAKES
CONTROL METHOD:	FABRIC FILTERS	SUBSTITUTE MATERIAL
ORIGINAL EMISSIONS:	547 METRIC TONS	129 METRIC TONS
CONTROL EFFECTIVENESS:	96%	100%
NATIONAL COST:		
LOW	$2.5 MILLION	$52 MILLION
MEAN	$3	$65
HIGH	$3.6	$81

FIGURE 21.2. Asbestos control costs and effectiveness. *Source:* Risk Benefit Analysis for Industrial and Social Needs, Moll, K. D., and Tihansky, D. P., *American Industrial Hygiene Association Journal,* 4/77. Reprinted with permission of the American Industrial Hygiene Association.

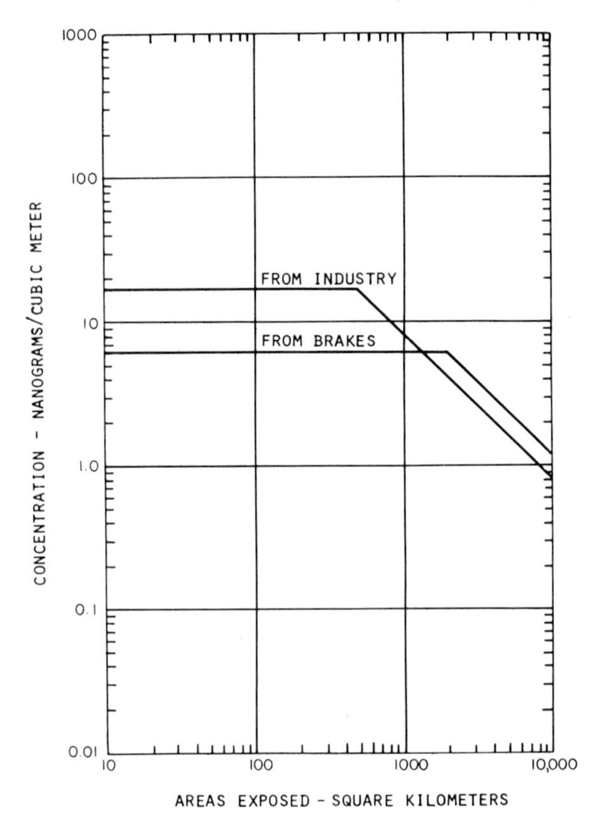

FIGURE 21.3. Asbestos air concentrations. *Source:* Risk Benefit Analysis for Industrial and Social Needs, Moll, K. D., and Tihansky, D. P., *American Industrial Hygiene Association Journal,* 4/77. Reprinted with permission of the American Industrial Hygiene Association.

*Example 4**

ABSTRACT

Cost-benefit analysis is applied to an evaluation of two strategies for reducing risk from pre-1977 model year Pintos. The first strategy is a design modification which might have been undertaken by Ford in 1970 when the Pinto was first produced. The second strategy is the recall which was undertaken by Ford in 1978. The analyses of the two strategies were based on the costs of strategy implementation and the benefits from the reduction in accidents involving pre-1977 model year Pintos. The results indicated that the recall strategy was more cost-ineffective than the design modification strategy. Consideration was also given to a third risk response strategy which included no corrective action.

**Source:* Dardis, R. and Zent, C., The Economics of Pinto Recall, *Journal of Consumer Affairs,* Winter 1982. Copyright permission given by the American Council on Consumer Interests.

Dardis and Zent conclude

> Though the recall is shown to be inefficient, there are several reasons it was implemented. The NHTSA (National Highway Traffic Safety Administration) investigation revealed that the problem did exist. Evidence indicated that the Ford new-product design process had put the Pinto into production before sufficient testing of the design. The NHTSA investigation, subsequent to Dowie's* claims, revealed that in fact the Pinto, as it was finally produced, was not pretested; a

*Dowie, M., Pinto Madness, *Mother Jones Magazine,* September/October 1977.

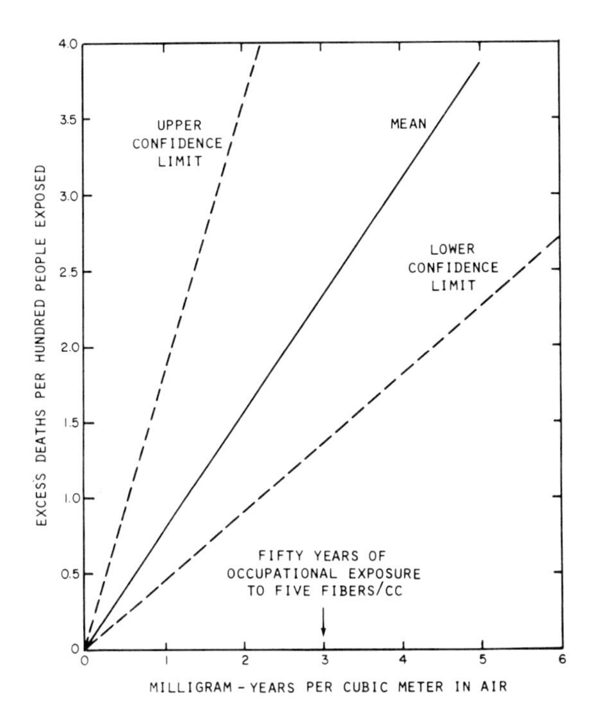

FIGURE 21.4. Asbestos dose-response: total. *Source:* Risk Benefit Analysis for Industrial and Social Needs, Moll, K. D., and Tihansky, D. P., *American Industrial Hygiene Association Journal,* 4/77. Reprinted with permission of the American Industrial Hygiene Association.

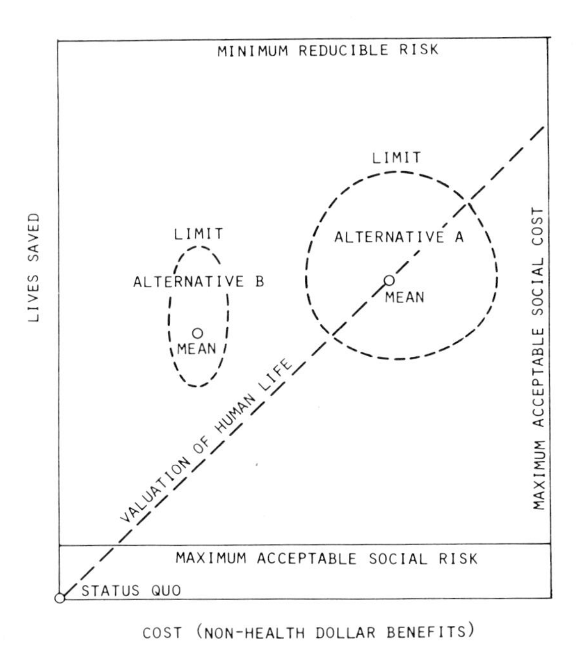

FIGURE 21.5. Multiple criteria comparison method. *Source:* Risk Benefit Analysis for Industrial and Social Needs, Moll, K. D., and Tihansky, D. P., *American Industrial Hygiene Association Journal,* 4/77. Reprinted with permission of the American Industrial Hygiene Association.

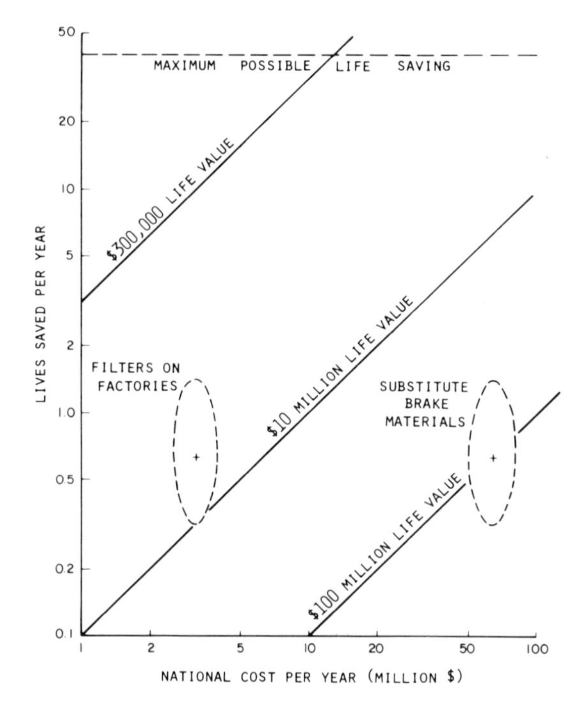

FIGURE 21.6. Asbestos pollution control alternatives. *Source:* Risk Benefit Analysis for Industrial and Social Needs, Moll, K. D., and Tihansky, D. P., *American Industrial Hygiene Association Journal,* 4/77. Reprinted with permission of the American Industrial Hygiene Association.

TABLE 21.3. Total Estimated Accident Costs From Pre-1977 Model Year Pintos ($ million)

Period	Costs of Deaths	Costs of Injuries	Accident Costs
1970-78	6.013	2.345	8.359
1979-88	1.597	0.623	2.219
1970-88	7.610	2.968	10.578

Source: Dardis, R. and Zent, C., "The Economics of the Pinto Recall", *Journal of Consumer Affairs,* Winter 1982. Copyright permission given by the American Council on Consumer Interests.

special Maverick was used for the Pinto pretests. Public pressure demanded that something be done. The lives which might be lost by inaction became important since they were identified lives and could be saved by policy decision. Though lost lives could not be resurrected, existing lives might be protected by the recall strategy. Thus, political pressure resulted in a recall of the pre-1977 model year stock. The recall strategy may be viewed as a government effort to (1) avert future losses due to the defective Pintos and (2) compensate society for past losses. It was not, however, an efficient use of society's limited resources since the study findings suggest that the cost of the recall greatly exceeded the benefits. Thus, it can be argued from an economic efficiency perspective that once it was too late for the design modification strategy, it would have been better to have taken no corrective action.

These two examples give some insight into current usage of cost-benefit analysis in the national health and safety arena. They illustrate the kinds of problems evaluated and the people and organizations using this technique. They show only the tip of the iceberg on the complexity of the method, the volume of factors that need to be considered, and the uncertainties that may exist. The examples are to help give a prefatory view of the process.

TABLE 21.4. Estimated Unit Costs of the 1978 Recall

Recall Component	Year	Unit Cost Estimate (Current Dollars)	Unit Cost Estimate (1975 Dollars)
Mailing notices	1978	0.10	0.082
	1979	0.13	0.096
Parts[a]	1978	15.00-20.00	13.17-17.56
Labor	1977	15.50[b]	13.24-15.47
	1979	21.25[c]	
Opportunity cost of time[d] (2 hours)	1975	9.08	9.08

Source: Dardis and Kent, op. cit. Copyright Permission given by the American Council on Consumer Interests.
[a]Jake Kalderman.
[b]National Automobile Dealers Association.
[c]Weighted average derived for this study is from information provided by the Ford Motor Company and the National Automobile Dealers Association, 1977 Advertising Questionnaire.
[d]U.S. Department of Commerce.

TABLE 21.5 Total Estimated Costs of the 1978 Recall

Component	Low Cost ($ million)	High Cost ($ million)
Mailing notices	0.087	0.087
Parts	4.435	5.913
Labor	4.458	5.209
Opportunity cost of time	—	3.058
TOTAL	8.980	14.267

Source: Dardis and Kent, op. cit. Copyright permission given by the American Council on Consumer Interests.

TABLE 21.6. Cost-Benefit Analysis of the 1978 Recall

Estimated Costs ($ million)	Estimated Benefits ($ million)	$\frac{C}{B}$	Implicit Value of a Life ($ million)	Implicit Value of a Discounted Life Year ($1,000)
8.980	1.465	6.13	1.272	129
14.267	1.465	9.74	2.055	209

Source: Dardis and Kent, op. cit. Copyright permission given by the American Council on Consumer Interests.

TABLE 21.7. Costs of the 1970 Potential Design Modification Strategy: Low Estimate

Calendar Year	Sales[a] (1,000)	Estimated Unit Cost ($)	Estimated Total Cost ($ million)	Present Value of Estimated Costs in 1970 ($ million)
1970	76	8.00	.608	.608
1971	328	8.00	2.624	2.385
1972	287	8.00	2.296	1.897
1973	268	8.00	2.144	1.611
1974	192	8.00	1.536	1.049
1975	170	8.00	1.360	.844
1976	106	8.00	.848	.479
TOTAL				8.873

Source: Dardis and Kent, op. cit. Copyright permission given by the American Council on Consumer Interests.
[a]Automotive News.

TABLE 21.8. Costs of the 1970 Potential Design Modification Strategy: High Estimate

Calendar Year	Sales[a] (1,000)	Estimated Unit Cost ($)	Estimated Total Cost ($ million)	Present Value of Estimated Costs in 1970 ($ million)
1970	76	18.66	1.418	1.418
1971	328	18.66	6.120	5.564
1972	287	18.66	5.355	4.425
1973	268	18.66	5.001	3.757
1974	192	18.66	3.583	2.447
1975	170	18.66	3.172	1.969
1976	106	18.66	1.978	1.116
TOTAL				20.696

Source: Dardis and Kent, op. cit. Copyright permission given by the American Council on Consumer Interests.
[a]Automotive News.

TABLE 21.9. Cost-Benefit Analysis of the 1970 Potential Design Modification

Costs ($ million)	Benefits ($ million)	$\frac{C}{B}$	Implicit Value of a Life ($ million)	Implicit Value of a Discounted Life Year ($1,000)
8.873	10.578	0.84	0.121	12
20.696	10.578	1.96	0.364	37

Source: Dardis and Kent, op. cit. Copyright permission given by the American Council on Consumer Interests.

TABLE 21.10. Comparison of Three Risk Response Strategies ($ million)

Strategy	Accident Prevention Costs	Accident Costs	Total Costs
Design modification (1970)			
Low cost	8.873	0	8.873
High cost	20.696	0	20.696
Recall (1978)			
Low cost	8.980	9.113	18.093
High cost	14.267	9.113	23.380
No corrective action	0	10.578	10.578

Source: Dardis and Kent, op. cit. Copyright permission given by the American Council on Consumer Interests.

21.3 DESCRIPTIONS AND DEFINITIONS

Much has been written describing cost-benefit analysis and characterizing its essentials. Many definitions have evolved. Phrases characterizing cost-benefit analysis include: "compare solutions", "general interest of the community", "consistency of objectives", and "excess social benefit over cost." Users of the technique have offered a number of descriptions and definitions. Some of the descriptions have confined themselves to cost-benefit analysis, while others have included or referred to cost-effectiveness analysis as well.

21.3.1 Cost-Benefit Analysis vs. Cost-Effectiveness Analysis

Steven Kelman (1981) suggests that "At the broadest and vaguest level, cost-benefit analysis may be regarded simply as systematic thinking about decision-making."

The late Jack Recht, when he was head of statistics and research at the National Safety Council, defined cost-benefit analysis as "a method of evaluating alternative proposals for accomplishing a specific objective. The name itself describes a method. First, the costs and the expected benefit (or effectiveness) of each proposal are estimated; then, the proposals are compared on the basis of their cost/benefit ratios" (Recht, 1966). Note that this definition includes the word *effectiveness;* it also implies that alternatives are compared and confines use to a single objective.

A different definition has been provided by Brian O'Neill and A. D. Kelley of the Insurance Institute for Highway Safety. They report that "cost/benefit analysis measures a planned program's cost against its expected benefits, using identical monetary units of measurement—most often dollars—on both sides of the ledger."* Here the framework for comparison appears to be the costs and benefits of a single specific program, apparently to determine if the benefits of that program exceed its costs. An inference here might be that the relative monetary values would not only indicate the economics involved in applying a control but, also, whether the risk should be controlled.

Table 21.11 shows such a comparison in simple form. It is reported to be from an internal memorandum of an automobile company as part of a gas tank safety evaluation. Here the costs outweigh the benefit dollars. One might infer that this analysis had strong bearing on the decision not to change the gas tank design.

Although the terms "cost-benefit analysis" (CBA) and "cost-effectiveness analysis" (CEA) have been used interchangably by some, O'Neill and Kelley differentiate between the two, the differentiation being the comparison of alternatives for a uniform goal for cost-effectiveness analysis. Other differences have been expressed. Some of these are captured in the following descriptions and lead us to what appears, currently, to be an accepted differentiation between the two processes.

*O'Neill and Kelley, 1975. Reprinted with permission. Society of Automotive Engineers, Inc.

C. C. Ossler defines cost-benefit analysis as

> an evaluative technique for identifying, assessing and weighing the costs and benefits of a program or service. (It is) used to evaluate the financial and economic merits of conducting a program. All costs and benefits are quantified in common monetary terms. (It) loses power when important costs and benefits cannot be expressed in monetary terms. She defines cost-effectiveness analysis as "an evaluative technique for identifying, assessing and weighing the costs of monetary effects of a program. (It is) used to compare alternative methods for achieving specific program objectives. (The) purpose of cost-effectiveness analysis is to determine which alternative will attain an objective for the least cost.
>
> *Occupational Health Nursing,* January 1984

Herbert E. Klarman, (1982) showing concern about valuations of human life states, "These doubts have led me, over the years, to turn to cost-effectiveness analysis, which is a truncated form of cost-benefit analysis that stops short of putting an economic value on the health status outcomes of programs."

TABLE 21.11. Gas Tank Evaluation

Benefits

Savings: 180 burn deaths, 180 serious burn injuries, 2,100 burned vehicles.
Unit Cost: $200,000 per death, $67,000 per injury, $700 per vehicle.
Total Benefit: 180 × ($200,000) + 180 × ($67,000) + 2,100 × ($700) = $49.5 million.

Costs

Sales: 11 million cars, 1.5 million light trucks.
Unit Cost: $11 per car, $11 per truck.
Total Cost: 11,000,000 × ($11) + 1,500,000 × ($11) = $137 million.

Source: Dowie, M., *Business and Society Review,* Fall 1977, and *Mother Jones* Magazine.

Thus CBA and CEA are identical up to placing monetary values on outcomes. CEA is simpler and less time-consuming, but more limited in use. Cost-effectiveness analysis must be confined to programs aimed at identical objectives—a meaningful limitation necessitated by the fact that no common evaluator, such as money, is used to enable comparisons among dissimilar objectives or outcomes, e.g., different kinds and severity of injuries, injuries vs. disease, health or safety vs. other kinds of public benefits such as housing, education, etc.

21.3.2 Summary of Uses

Thus appropriate uses of cost-benefit analysis and cost-effectiveness analysis are:

Cost-Effectiveness Analysis: Use where identical outcomes exist to help select among alternatives for achieving that outcome.

Cost-Benefit Analysis: Use where there are projects with differing objectives or outcomes to help choose among the alternatives, or use for an actual monetary cost vs. monetary benefit evaluation of a single program/project.

When used properly, both processes can be of assistance to:

1. Make choices among alternative courses of action.
2. Establish a framework for measuring program performance after implementation.

21.4 MAKING COST-BENEFIT ANALYSES

21.4.1 Broad Requisites

Since the steps in making CBA and CEA are identical except for benefit evaluation, this section will concentrate primarily on making cost-benefit analyses. Making cost-benefit analyses includes requisites captured by R. A. Donvito (1978) in providing guidance for the use of the government's planning, programming, budgeting system. These are:

- Objectives of the program.
- Financial requirements.
- Benefits expected to result.
- Uncertainties involving both cost and benefit estimates.
- The time when the costs are expected to be incurred and the benefits expected to result.

The scope or range of analysis and the basic question to be resolved may vary. Donvito outlines two specific approaches. One is a *fixed budget approach,* which assumes the availability of a given level of funds. The question involved here is: With a fixed level of resources available, what course of action should be taken to achieve maximum benefits? The objective is to weigh alternatives in terms of costs, benefits, risks, and timing to select the measures that would do most toward achieving goals, while still remaining within the constraints of available funds. The other method, the *fixed effectiveness approach,* assumes that some specific objective is to be accomplished. The question here is: Considering that some specific objectives must be accomplished, what courses of action should be taken to minimize the cost of achieving these objectives?

TABLE 21.12. Steps Performed in Cost-Benefit and Cost-Effectiveness Analysis

Step I	Identify the stimulus or problem for conducting economic analysis.
Step II	Specify the objectives or goals to be achieved by the program under study.
Step III	Determine alternative means of attaining those objectives or goals.
Step IV	Enumerate the costs and benefits or effects for each alternative.
Step V	Assign monetary values to the costs and benefits or assign units of effectiveness.
Step VI	Perform discounting.
Step VII	Calculate the cost-benefit or cost-effectiveness ratio.
Step VIII	Compare these ratios by use of a decision matrix and feed this information into the decision-making process.

Source: Ossler, C. C., Cost-Benefit and Cost-Effectiveness Analysis in Occupational Health, *Occupational Health Nursing,* January 1984.

21.4.2 Factors/Elements in Making Evaluations

C. C. Ossler in her article, Cost-Benefit and Cost-Effectiveness Analysis in Occupational Health, details the cost-benefit analysis process identifying the steps and giving examples of factors. These are shown in Tables 21.12 and 21.13. Additionally, she provides examples of effectiveness measures (Table 21.14), a decision matrix (Table 21.15), examples of costs and benefits related to a specific program (Tables 21.16 and 21.17) and some definitions.

21.4.3 Definitions*

DISCOUNTING
Reduces future monetary costs and benefits to their equivalent present worth. Adjusts for the fluctuation in interest and inflation rates.

DIRECT COSTS
Identifiable monetary costs (or costs that have monetary equivalence) involved in providing a specific service. Usually include capital and operating expenditures that can be attributed to one program or service.

TABLE 21.13. Cost Centers Used in Economic Analysis of Occupational Health Programs

Personnel	Professional, clerical, administrative salaries, and fringe benefits.
Supplies	Records, printed materials, desk supplies, clinical supplies.
Capital expenditures	Sound booth, spirometer.
Facility	Space rental, utilities.
Lost production time	

Source: Ossler, op. cit.

TABLE 21.14. Units of Effectiveness Used in Economic Analysis of Occupational Health Programs

Absenteeism rate
Health care services utilization
Lifestyle habits: smoking, obesity, seatbelt use, alcohol ingestion
Incidence, prevalence of work- and non-work-related injuries and illnesses
Worker compensation claims: number, severity
Number of accidents
Productivity

Source: Ossler, op. cit.

TABLE 21.15. Cost-Benefit/Effectiveness Decision Matrix for Two Alternatives: A and B

Benefit/Effectiveness of Alternative A Compared to Alternative B	Cost of Alternative A Compared to Alternative B		
	Less	Equal	More
Less	?	Select B	Select B
Equal	Select A	Either A or B	Select B
More	Select A	Select A	?

Source: Ossler, op. cit.

*Ossler, C. C., 1984.

TABLE 21.16. Cost Centers and Benefits Used in the 1980 A. D. Little Study

Costs	
Physical facility, utility services, capital equipment and fixtures, depreciation	
Supplies, journals, printed materials	
Personnel salary and fringe benefits	
Training costs	
Employee time lost by utilizing services	
Benefits	
Direct:	Decreased cost of physical examinations
	Salary savings from decreased lost worktime for onsite (versus offsite) health care
	Savings on workers' compensation claims
	Savings based on OSHA recordable events
	Savings based on first aid log data
	Reduced nonoccupational health care costs
Indirect:	Reduced absenteeism
	Reduced labor turnover
	Increased worker productivity

Source: Ossler, op. cit.

TABLE 21.17. Cost Centers and Benefits Used in the 1980 Worksite Hypertension Treatment Study (NIH)

Costs
Laboratory tests
Drugs supplied by the clinics
Clinical supplies
General supplies and operating expenses, records and computer processing
Equipment use and depreciation
Facility rent
Personnel (clinical and administrative)
Benefits
Reduced revisits
Reduced drug utilization and costs
Reduced cost of treatment
Blood pressure control

Source: Ossler, op. cit.

INDIRECT COSTS

Opportunity costs, or the costs of not being able to use resources for a different purpose. Any costs that cannot be directly linked with the program under study.

DIRECT BENEFITS

The tangible positive effects produced by a program. Avoided costs such as prevention of disease or premature death, prevention of absenteeism, or a work-related injury.

INDIRECT BENEFITS

Estimated values of secondary, often less tangible, effects such as increased productivity.

21.4.4 Life Valuation

Computations on value of life are inherent (and controversial) in making cost-benefit analyses. Rachel Dardis provides insight into the several approaches to making valuations in her paper "A Critical Evaluation of Current Approaches to Life Valuation in Cost-Benefit Analysis," including "societal valuation," "human capital" and "willingness to pay."

Tables 21.18 and 21.19 provide average costs for fatalities and injuries as prepared by the National Highway Traffic Safety Administration (NHTSA); no values are included for pain and suffering.

TABLE 21.18. Average Cost of Traffic Fatalitites

Factor	Average Cost
Medical	$ 1,370
Productivity	236,865
Property damage	3,406
Legal and court	13,394
Coroner, medical examiner	168
Emergency	290
Insurance	12,523
Public Assistance Administration	576
Government programs	135
Average cost per traffic fatality	$268,727

Source: The Economic Cost to Society of Motor Vehicle Accidents, Reference Number DOT HS806346, January 1983.

TABLE 21.19. Average Economic Costs For Motor Vehicle Injuries

Abbreviated Injury Scale	Average Cost
5	$190,010: Most serious injuries, e.g., spinal cord damage.
4	43,729
3	10,260
2	4,592
1	2,276: Least serious injuries, e.g., bumps, bruises, cuts.

Source: The Economic Cost to Society of Motor Vehicle Accidents, Reference Number DOT HS806346, January 1983.

Tables 21.1 through 21.10 illustrate the kind of cost and benefit factors and numbers developed in making cost-benefit analyses.

21.5 PROBLEMS/CONTROVERSY

Although the factors to be included in a cost-benefit analysis can be stated specifically, many impediments to the process of developing realistic, meaningful, numbers exist. Controversy over the values developed and imprecisions in the numbers used may arise from:

1. The lack of reliable data needed for many of the cost calculations.
2. Controversy as to whether or not monetary values should be placed on human life, pain, suffering, and well-being; and if so, what values might be used.

21.5.1 Data Problems

To begin with, little knowledge exists about the typical functional relationship between safety programs and reductions in death, injury, and property damage attributable to specific safety controls. Unfortunately there is no widespread data network for gathering information relating causation and control. Additionally, pertinent research has been scant.

The variability of data or estimates is clearly illustrated in the following (Harvey, 1983):

> It soon became clear that a single source of error vastly overshadowed all others: the degree of seriousness of the nonfatal injuries to be prevented by a D.R.L. (daylight running lights) regulation. It was not known how many of these injuries lead to such tragedy as blindness or paraplegia and how many are curable. D.O.T. (Department of Transportation) officials assessed a societal trade-off value between fatalities prevented and serious injuries prevented that range from 1:20 to 1:100, a relative difference of 500 percent.
>
> *Journal of the Operational Research Society,* Vol. 43 No. 1, 1983. Reprinted with permission.

21.5.2 Intangibles

If intangible benefits are to be considered, what monetary values should be assigned? The general well-being of a population, better interpersonal and working relationships, improved personal capabilities (physical, mental, emotional) are much-sought-after prizes of safety and health projects. Can these be assumed to have zero economic value and left out of valuations? After all, in evaluating such items as computer installations, attempts are made to include economic credit for such items as better productivity and better decision-making. Where is one to account for the economic benefits of safety programs which arise from better productivity, the discovery of production snags, lower absenteeism, fewer production interruptions, improved relationships, etc., if not in a comprehensive economic analysis? But some of these are the kind of intangibles that Grimaldi and Simonds, *Safety Management,* suggest are not well enough documented to be included in believable uninsured costs of occupational accidents and injuries. Yet, to omit such values would produce incomplete benefit numbers.

A consensus process for valuation of intangibles is offered by R. Litecky in his article, "Intangibles in Cost-Benefit Analysis." He describes a method for developing, in concert with operating management, consensus values for intangible benefits attributable to the installation of computers. There is a question, however, as to whether people realize and appreciate the intangible benefits of safety and health activities enough to allow adaptation of a similar consensus process.

21.5.3 Safety Perceptions/Risk Acceptance

Another problem: how much of a factor is economics to safety decision-making, in general, and to the particular safety decision at hand? There is an ambivalence about safety that makes the importance of economics a variable factor. Chauncey Starr, who has written much on the acceptability of risk, reveals that the public does not place the same value on all kinds of risks. People generally accept greater risk for voluntary than for involuntary activities. For example, a worker may have no compunction about chain-smoking cigarettes, but objects violently to cutting oil smoke that may be generated by his machine.

The duality of people in their valuation of safety and the difference in attitudes toward voluntary and involuntary situations is reinforced by the situation described in a *Wall Street Journal* article, Why the Pinto Jury Felt Ford Deserved $125 Million Penalty. When justifying the verdict, one jury member is reported to have said, "Ford knew people would be killed." The same person is the only member of the jury who drives a Pinto. Yet, he reported, he will keep on driving it.

In his book, *Of Acceptable Risk,* William Lowrance explores the uncertainties of risk acceptance and safety, and suggests that judging safety is a matter of personal and social value judgment. He suggests, also, that perceptions of safety are not constant from person to person, group to group, time to time, and situation to situation.

Thus the uncertainties about safety that may counterbalance cost-benefit findings are:

1. Risk acceptance is variable among individuals and groups of people.
2. Safety is highly subject to social values.

21.5.4 Moral Considerations

Steven Kelman (1981) in "Cost-Benefit Analysis, An Ethical Critique," adds these thoughts about the cost-benefit process:

1. In areas of environmental, safety, and health regulation, there may be many instances where a certain decision might be right even though its benefits don't outweigh its costs.
2. There are good reasons to oppose efforts to put dollar values on nonmarketed benefits and costs.
3. Given the relative frequency of occasions in the areas of environmental, safety, and health regulation where one would not wish to use a benfits-outweigh-costs test as a decision rule, and given the reasons to oppose the monetizing of nonmarketed benefits or costs that is a prerequisite for cost-benefit analysis, it is not justifiable to devote major resources to the generation of data for cost-benefit calculations or to undertake effort to "spread the gospel" of cost-benefit analysis further.

> We believe that some acts whose costs are greater than their benefits may be morally right, and contrary-wise, some acts whose benefits are greater than their costs may be morally wrong.
>
> We do not do cost-benefit analysis of freedom of speech or trial by jury—or, indeed, any element of the Bill of Rights.
>
> *Regulation,* Jan./Feb. 1981. American Enterprise Institute.

Certainly, there are risk/hazard situations that should not be put to economic scrutiny, and doing so may introduce unnecessary problems and confrontations.

21.5.5 Misuse

Besides the potential imprecisions of cost-benefit analysis as well as the controversy that accompanies human valuation and the general worth of health and safety activities, there are misuse problems:

1. Using the process as the only or major determinant in making safety decisions.
2. Using the process to determine the need for exposure control.

Cost-benefit analysis provides the framework for comparison of the costs and benefits of a single specific program to determine if the benefits of that program exceed its costs. Where the costs outweigh the benefit values, an inference could be made that the monetary values evolved evaluate not only the economics of applying a control but also whether the hazard deserves to be controlled. Performing economic valuations that seem to produce hard numbers may cause neglect of other significant decision factors. Factors other than economics (social, political, legal) may be the most significant decision determinants. See the gas tank evaluation (Table 21.11) and Example 4.

The economics of a choice must never be overlooked, or should it be the sole or major determinant, either in business safety practice or in national policy and program issues. True, the economic bottom line must be the immediate and continuing objective in the survival of a business. But, generally, it is not looked upon as the only purpose of the business. Purposes are many and are directed toward owners, shareholders, customers, suppliers, employers, the community, and government. Thus, management's motivations may be multiple, and for many safety projects, employee relations, union relations, or legal compliance may be the governing factor. To assume that sheer economics is the only driving force for safety decisions promotes unnecessary fiscal gymnastics and constructs a motivational straitjacket for the support of worthwhile projects.

Social, political, and legal reasons may prove, at times, more eye-catching than a few dollars gained, particularly when such dollars may well be overshadowed by sizable fiscal factors arising from other aspects of a business or from the national economy. The attitudes and perceptions of those to be motivated may well preclude extensive monetary emphasis. Consider these two business examples:

> The president of a company who said, when the economics of safety were being discussed, "We're in business to make a profit, but we make it from our products, not our people."
>
> The plant manager, who while being given cost alternatives to control a crane runway repair hazard, interrupted the presentation and said, "What's with all these options? Let's assume our sons are to be working on the crane runways, which alternative would we want?"

21.6 OCCUPATIONAL SAFETY PRACTICE

21.6.1 Lack of Use

The application of cost-benefit analysis to occupational safety appears to be nonexistent, perhaps because the discipline is still in its formative stages. Its application and importance to safety and health at the plant or company level has not been documented through trial, use, and analysis. Replying to a request for specific examples illustrating the use of cost-benefit evaluation in industry, Allen F. Hoskin, manager, Statistics Department, National Safety Council, provided this information:

> A few years ago, the Council asked a few of its largest members to provide information about the use of cost-benefit analysis in their operations. We received a few replies that described programs that turned out to be cost-beneficial but a cost-benefit analysis was not performed before implementation. Their replies were not published.
>
> Since we talked, I remembered a study the Council performed in 1972 for OSHA on the feasibility of performing cost-benefit analysis at the establishment level. The conclusion was that it was not feasible because very few establishments could identify the indirect underlying costs of accidents, damage to equipment, total costs, and cost of time lost by injured workers.

21.6.2 Data Problems

It is very easy to visualize that problems do not diminish when considering cost-benefit analysis in occupational safety and health practice. If anything, they become greater. First, garnering cost figures is not an easier task in the occupational situation. Generally the costs of injury and damage resulting from occupational exposures are termed "insured" and "uninsured" (Grimaldi and Simonds, *Safety Man-*

agement). But ordinary usage seems to associate these terms with on-the-job incidents. Similar losses may accrue from off-the-job injuries, product liability, and other third-party liability situations. And none of these cost groupings normally consider the broad societal costs that, by definition, cost-benefit analyses should include. Yet, even if such societal costs (and benefits) were available, their importance, to and impact on some business managements might be conjectural.

So potential cost-benefit analysts are faced with an extreme paucity of data, even when analyses for single establishments are considered—particularly in the area of uninsured (indirect) costs. Unfortunately, cost data available through an organization's normal accounting system may not be directly applicable and may be difficult to adapt, since such information is gathered for financial control purposes (rather than health and safety uses).

21.6.3 Economics as a Decision Factor

The engineering profession applies scientific principles for the practical benefit of people—society. In such practice, engineers are constantly confronted with economic choices. Professor James L. Riggs, in *Essentials of Engineering Economics* (1982), discusses "engineering decision-makers":

> Which one of several competing designs should be selected?
>
> Should the machine in use be replaced with a new one?
>
> With limited capital available, which investment alternative should be funded?
>
> Would it be preferable to pursue a safe, conservative course of action or follow a riskier one that offers higher potential returns?
>
> Among several proposals for funding that yield substantially equivalent worthwhile results but have different cash flow patterns, which is preferable?
>
> Are the benefits expected from a public-service project large enough to make its implementation costs acceptable?
>
> Two characteristics of the above questions should be apparent. The first is that each deals with a choice among alternatives, and the second is that all of them involve economic considerations. Less obvious are the requirements of adequate data and an awareness of technological constraints, to define the problem and to identify legitimate solutions. These considerations are embodied in the decision-making role of engineering economists to
>
> 1. Identify alternative uses for limited resources and obtain appropriate data.
> 2. Analyze the data to determine the preferred alternative.

Occupational safety specialists face similar decision situations, which will require that future occupational safety efforts focus on the formation of a better database for costs, causes, and effectiveness of controls. In the meantime, using the little general information available and estimating information needed for the project at hand, economic consideration should become a more intrinsic element in occupational safety decision-making and program evaluation. Heads of health and safety functions face many planning and programming decisions involving such specific questions as:

1. Will the program pay for itself? In what length of time?
2. Is control of the hazard affordable, that is, how much will correction cost vs. available funds?
3. Which hazard produces the most serious exposure?
4. Which corrective measure will be most efficient in reducing the exposure?
5. Which corrective measure will produce the most exposure reduction?
6. Which corrective measure under consideration costs the least? the most?
7. Which exposure problem is most urgent?
8. Which corrective measure will be most acceptable to all of the audiences concerned?

Questions such as these are not always subject to resolution via cost analyses. But, where uncertainties exist with respect to *priorities, costs,* and *efficiencies,* there is obvious need to consider cost evaluations and thus the possible use of cost-benefit analysis. Whether or not cost-benefit analysis is applicable will depend on the degree of uncertainty, the level of expenditures involved, and the amount of time needed to make the evaluation.

Chris Whipple (1977), discussing risk analysis, proposes that in the majority of cases in which safety engineering is applied, and particularly where consumer product risks and occupational safety issues are involved, normal everyday practice will suffice, and using formal risk analysis will be burdensome in time and cost. Risk analysis, he concludes, *is not intended to contribute to the management of the most common risks.* It would follow that this logic applies equally to the extensively comprehensive, complex process of cost-benefit analysis. For the general run of occupational health and safety situations, where a single objective is to be achieved, many of the factors (e.g., societal benefits, cost reduction or increase, fewer production interruptions) may be common to all alternatives and factor out. Thus the economic evaluation may boil down, simply, to a comparison of implementation costs.

In spite of its humanitarian objective, safety must be subjected more and more to cost evaluations—*before* and *after* implementation. The efficient use of money for the support of safety programming (as contrasted to its use for competing non-health-and-safety activities), the efficient selection among competing safety ventures, and the selection of the most efficient way to achieve a given safety result demand such evaluation. It is unfortunate that reliable data/information has not been captured, particularly for uninsured (tangible and intangible) costs and benefits. Aggressive use of cost evaluation and cost justification techniques, even with the current lack of data, should provide support for proposed safety programs and should promote the generation of data for an improved safety data bank. A more comprehensive set of standards for the collection of such data would indeed be helpful.

21.6.4 Uses and Adaptations

Proponents of cost-benefit analysis claim that this technique can be *adapted* for many different levels of programming, extending from a control directed at a specific hazard to the spending of a total operating budget of a safety organization—or, perhaps, even helping to determine how large that safety budget should be, as compared to budgets for other company functions.

William Fine (1973) adapts cost-benefit techniques and develops a formula for "risk score" and one for "justification" that he applies to fairly straightforward industrial situations. One of his examples involves controls for unsupported compressed oxygen cylinders; another, the safety of a 12,000-gallon propane storage tank.

The higher the number of dollars, the greater the requisite to consider cost-benefit analysis in occupational safety practice. Certainly, the establishment of a budget aimed at achieving a continuing improvement in safety performance (and its attendant benefits) might be the ultimate potential use of cost-benefit analysis, particularly when costs and benefits can include those related to on-the-job incidents; off-the-job safety; workers' compensation insurance and health and accident insurance; product safety and public liability matters; fire prevention and protection; disaster control; the safety aspects of employee relations, union relations, public relations, community relations and government relations (including OSHA compliance) matters; production interruptions; and discovery of production inefficiencies. Although hard data does not exist for most of these factors, estimates might be attempted using a consensus process.

One large chemical company has made such an evaluation. Figures were developed, using a consensus process that involved operating management, to evaluate the impact of its safety program on the company's profitability and to substantiate the maxim, Safety Pays. Totals for expenses and savings were developed. Savings included those effected in the area of workers' compensation insurance, replacement personnel, benefits resulting from fewer off-the-job injuries, property insurance, boiler and machinery insurance, and leased property insurance. Costs included staff safety personnel, and expenses required for additional personnel for training, plant project reviews, various inspections and audits, and additionally required capital facilities. The results confirmed the efficacy of safety expenditures, and could form the basis for future planning and budgeting for safety.

Another example—not cost-benefit analysis—but, rather, an example of the use of factors that might be part of a cost-benefit analysis. A particular company had some years of economic hardship. As expense controls were put into effect, moneys allocated to staff safety activities suffered. At the same time, injuries increased. Simple cost comparisons were made for identifiable safety program expenses and direct dollar costs for injuries. These showed clearly that the injury costs rose more than the safety dollars were being reduced, producing a total increase in operating cost. This direct comparison of the two related costs was enough to reopen the purse strings for safety. Figures 21.7, 21.8, and 21.9 show the cost relationships graphically.

Now, let us look at an example of a postprogram justification. At a large company, a lighting and painting program of the large-machinery erection shop was undertaken—ostensibly for safety and other benefits. The shop was 1250 ft long, about 90 ft high with two tiers of overhead traveling cranes, and about 75 ft wide. Hundreds of people worked in this shop. Extensive before and after figures were kept on illumination levels and injuries. The figures showed that there was a significant improvement in injury rates after the lighting and painting. Management was satisfied that the program was worthwhile and that the program "paid." (See Figure 21.10.) But, to what extent did the reduction in injuries offset, monetarily, the cost of the painting and lighting improvements? Did the reduction in injuries justify the hundreds of thousands of dollars of cost? A detailed cost evaluation before and after the project would have been even more meaningful.

Another approach to an evaluation of costs and benefits is provided by Frank Gagne, Jr. (October 1979). Two nomographs such as these shown in Figures 21.11 and 21.12 are used. An illustrative example is provided: a mandatory eye protection program for a 3000-person shop where drilling, shaving, and cutting operations account for 60% of total operating time. The steps in using the nomographs are:

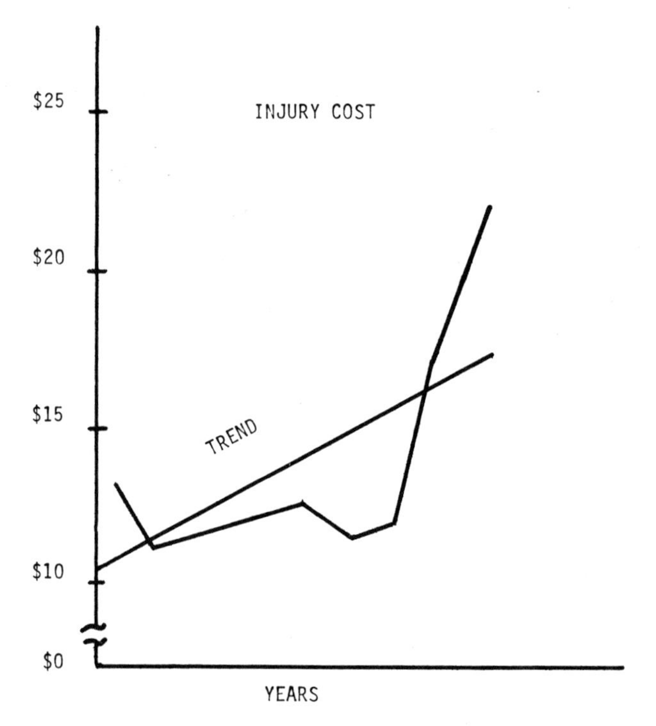

FIGURE 21.7. Injury expense per employee.

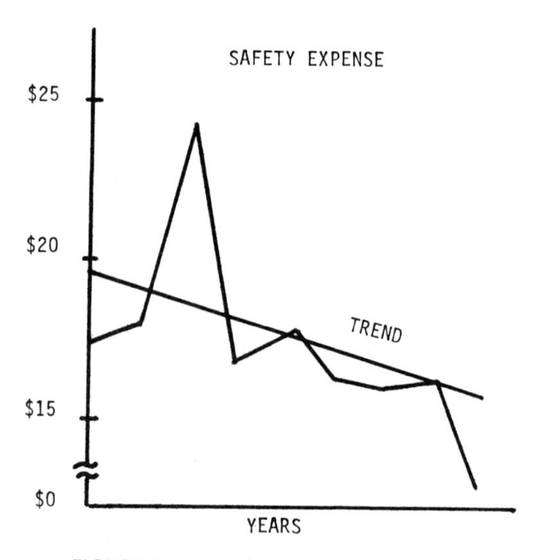

FIGURE 21.8. Safety expense per employee.

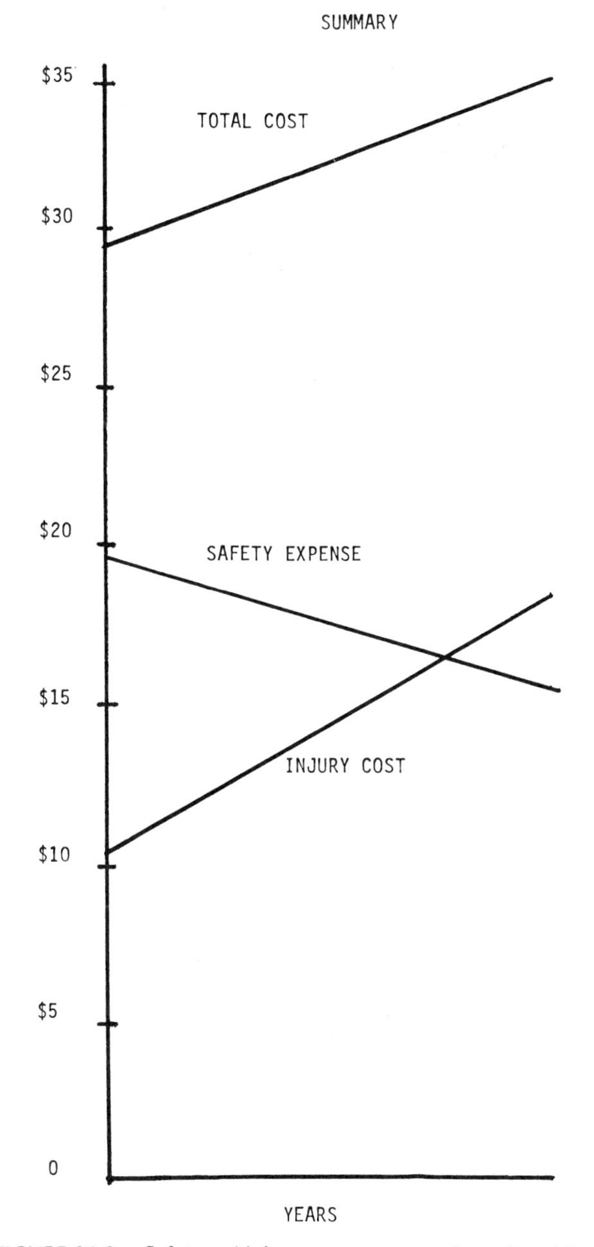

FIGURE 21.9. Safety and injury expense per employee (trends).

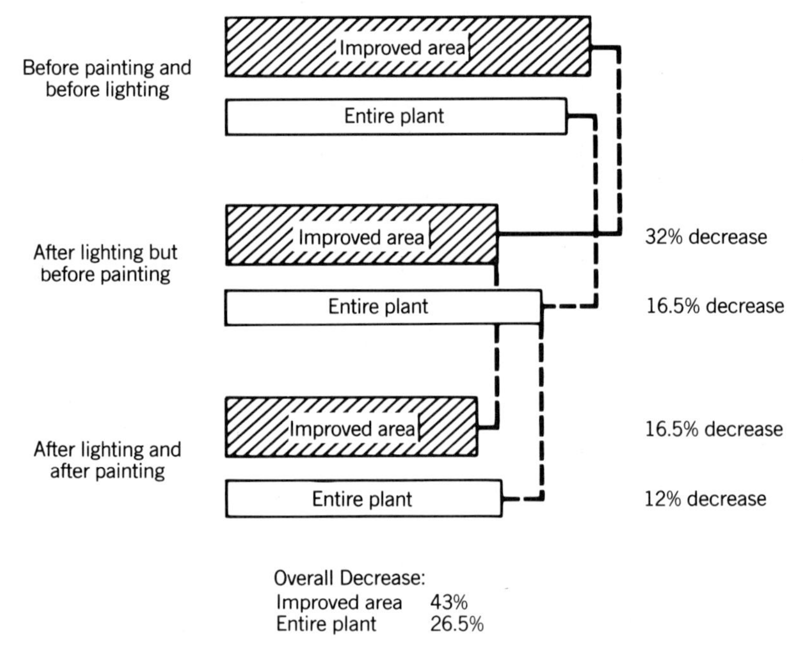

FIGURE 21.10. Injury frequency comparisons.

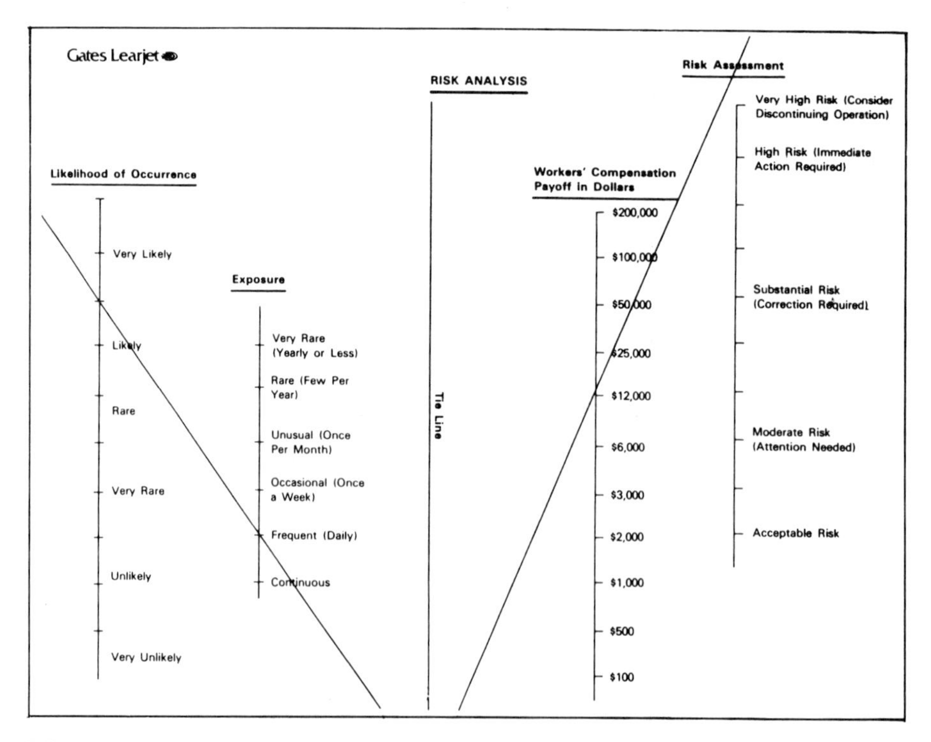

FIGURE 21.11. Risk analysis nomograph. *Source: National Safety Council Aerospace Newsletter,* 10/79.

RISK ASSESSMENT (SEE FIGURE 21.11)

Step	*Estimate*
1. Determine likelihood of injury.	Very likely.
2. Determine exposure rating.	Frequent (daily).
3. Locate intersection with tie line (extend beyond chart, if necessary).	
4. Assign monetary value of injury.	Possible loss of an eye ($15,492 Workers' Compensation award).
5. From nomograph determine order of risk.	Very high (consider discontinuing operation).

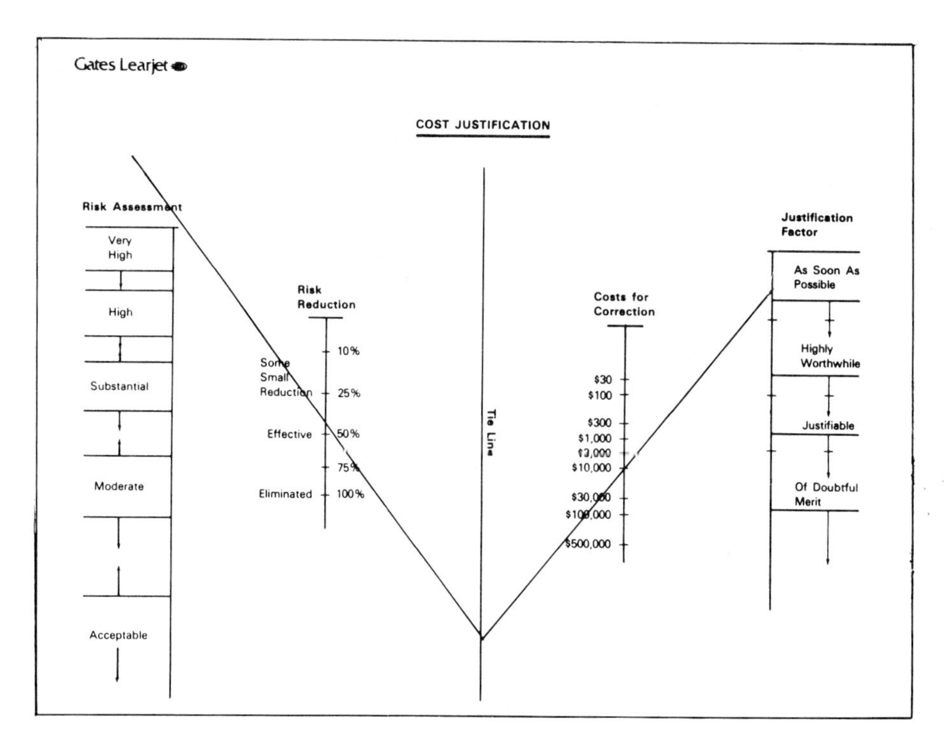

FIGURE 21.12. Cost justification nomograph. *Source: National Safety Council Aerospace Newsletter,* 10/79.

JUSTIFICATION FACTOR (SEE FIGURE 21.12)

Step	*Estimate*
6. Mark "order of risk" on nomograph.	
7. Estimate effectiveness of control.	40% reduction.
8. Intersect tie line.	
9. Estimate implementation cost.	$10,000.
10. From nomograph determine justification factor.	As soon as possible.

21.7 SUMMARY AND CONCLUSION

Cost-benefit analyses can be of assistance to the decision maker when major uncertainties as to a course of action exist—uncertainties related to priorities, costs, or efficiencies of alternatives.

The uncertainties within the process itself must be recognized, too. It is not precise in spite of the fact that it is quantitative. Often, values must be constructed or estimated—value judgments are ever present. Relationships between factors being evaluated and the objectives to be achieved may not have been determined reliably. Disagreement may exist with regard to the values assigned to tangible and intangible factors among the various audiences concerned with the objective—management, labor, government.

Before and after the fact, cost justifications are needed to help occupational safety and health decision-making and to systematically document and quantify the various economic and control aspects of safety and health. Cost-benefit analysis can be used:

When expenditures are large enough and time needed can be justified.

When proper choice of alternatives is not obvious or cannot be selected using simpler, more abbreviated analyses. For these simpler analyses, cost-benefit analysis provides a menu of factors from which to choose in making such truncated cost evaluations.

Finally, identification of exposure and risk evaluations concerning potential human and property losses must be the vehicles for determining needed controls. Other considerations such as economics, political (relationships in the broadest sense), and ancillary benefits should assist only in determining priorities and helping make selections among control methods.

REFERENCES

Biancardi, M. F., The Cost/Benefit Factor in Safety Decisions, *Professional Safety,* November 1978.

Dardis, R., A Critical Evaluation of Current Approaches to Life Valuation in Cost-Benefit Analysis, *The Journal of Consumer Affairs,* Summer 1981, Vol. 15, No. 1.

Dardis, R. and Zent, C., The Economics of the Pinto Recall, *The Journal of Consumer Affairs,* Winter, 1982, Vol. 16, No. 2.

Donvito, P. A., The Essentials of a Planning-Programming-Budgeting System, U.S. Department of Commerce/National Bureau of Standards, July 1978.

Dowie, M., How Ford Put 2,000,000 Fire Traps on Wheels, *Business and Society Review,* Fall 1977, No. 23.

Fine, W. T., Mathematical Evaluations for Controlling Hazards, National Safety Congress, 1973.

Gagne, F. Jr., Risk Analysis Valuable Approach to Cost Assessment, *National Safety Council Aerospace Newsletter,* October 1979.

Grimaldi, J. V. and Simonds, R. H., *Safety Management,* Richard D. Irwin, Homewood, IL, 1975.

Harvey, C. M., Cost-Benefit Study of a Proposed Daylight Running Lights Safety Programme, *Journal of Operational Research Society,* Vol. 34, No. 1, 1983.

Kelman, S., Cost-Benefit Analysis, An Ethical Critique, *Across the Board,* July/August 1981.

Klarman, H. E., The Road to Cost-Effectiveness Analysis, *Milbank Memorial Fund Quarterly,* Fall 1982.

Litecky, C. R., Intangibles in Cost-Benefit Analysis, *Journal of Systems Management,* February 1981.

Lowrance, W. H., *Of Acceptable Risk,* Wm. Kaufmann, 1976.

Moll, K. D. and Tihansky, D. P., Risk-Benefit Analysis for Industrial and Social Needs, *American Industrial Hygiene Association Journal,* April 1977.

O'Neill, B. and Kelley, A. B., Costs, Benefits, Effectiveness, and Safety: Setting the Record Straight, *Professional Safety,* August 1975.

Ossler, C. C., Cost-Benefit and Cost-Effectiveness Analysis in Occupational Health, *Occupational Health Nursing,* January 1984.

Recht, J. L., How To Do A Cost/Benefit Analysis of Motor Vehicle Accident Countermeasures, National Safety Council, September 1966.

Riggs, J. L., *Essentials of Engineering Economics,* McGraw-Hill, New York, 1982.

Wall Street Journal, Why the Pinto Jury Felt Ford Deserved $125,000,000 Penalty, February 14, 1978.

Whipple, C., Risk Analysis, National Safety Congress, 1977.

CHAPTER 22

How to Reduce Work Injuries in a Cost-Effective Way

FOSTER C. RINEFORT
Eastern Illinois University
Charleston, Illinois

22.1 INTRODUCTION

Peter F. Drucker, in his 1980 book, *Managing in Turbulent Times,* states that "rarely has a new social institution, a new social function emerged as fast as management is this century. . . . But rarely also has a new institution, a new leadership group, faced as demanding, as challenging, as exciting a test as the one that managing in turbulent times now poses to the managements of businesses and non-business pubic service institutions alike." This challenge, making our various institutions more productive, is a continuing theme in these increasingly demanding times. One area in which this increased effectiveness is often called for is occupational safety and health.

As some of our resources diminish and we face an increasing need to use available resources more effectively, the waste resulting from work injuries and occupational diseases continues to be an important concern in our society. Such incidents represent a source of pain, concern, and real financial cost to those who are injured or who sustain an occupational disease. Such events significantly affect employers and can result in sizable monetary expense, decreased employee morale, and poorer industrial relations and public relations. Society also loses when the productivity and quality of life of its most important asset, its people, is reduced.

Employers have reduced work injuries through the years by means of safety programs consisting both of activities to control unsafe physical conditions and of activities designed to orient, train, and motivate employees to work safely. Occupational diseases have most effectively been reduced by the recognition, evaluation, and proper control of those environmental factors and stresses that occur at work.

The purpose of this chapter is to provide some information about cost-benefit and cost-effectiveness analysis, about the state of the art, and about the results of an extensive cost-effectiveness study of safety and health covering Texas manufacturing firms performed by the author.

In simplest terms, cost-benefit analysis is a measurement of the cost of performing a certain action or actions, a measurement of the benefits that occur as a result of these actions, and, normally, some calculation of the relationships between these two factors. The approach is based on the concept of the potential Pareto improvement, which is defined as a costless economic rearrangement in which the gains can be distributed so as to make everyone in the community better off. The purpose is to maximize benefits within the constraints of a given set of costs.

Relationships between values obtained can be estimated in two ways. In one approach, a ratio of one value to the other can be computed, providing a benefit-cost ratio. In the second approach, costs can be subtracted from benefits, providing a net benefit value.

An important variation on cost-benefit analysis is labeled cost-effectiveness analysis. In this approach, the costs of alternative means of achieving a goal are compared to each other. The alternative approaches can be different programs, devices, technologies, or a combination of approaches. The problem then is to minimize costs to achieve a desired level of benefits.

Cost-benefit analysis and cost-effectiveness analysis have been used by economists, engineers, scientists, and businessmen for a number of years to evaluate a wide variety of problems with varying degrees of precision. The approach was refined, reported upon, and utilized in a number of instances in France and England in the 19th century. It has been extensively developed and used in the United States, other Western countries, and in the more rapidly developing countries in this century, particularly since 1945.

During the 1940s, 1950s, and 1960s, the technique was used to analyze problems as diverse as irrigation projects, hydroelectric power schemes, road construction, railway construction, inland waterways, urban renewal, recreation facilities, land reclamation, health problems, public works projects, educational activities, governmental programs, research and development activities, and national defense.

During the 1970s, some of the earlier problems with the technique were mitigated and cost-benefit analysis and cost-effectiveness analysis were used more extensively than ever before to deal with an even wider variety of problems. Questions about the precision of some earlier studies have been at least partially resolved by the use of computers to perform calculations and by increasing rigor in the determination and calculation of values. Some more recent applications of these techniques include evaluations of environmental preservation programs, the effects of government training programs, governmental income maintenance programs, manpower planning, the effects of activities which together deal with organization social responsibilities, multinational corporation activities, air and water pollution, corporate fund raising, employee physical examinations, job enrichment programs, and occupational safety and health. Summarizing, the technique has been utilized with increasing degrees of precision and completeness to deal with many of the complex issues and interdependencies existing in our society.

The technique of cost-benefit analysis or cost-effectiveness analysis can help governmental entities and agencies, trade associations, and individual organizations deal with occupational safety and health by its systematic approach to problem solving. The technique, like a number of other quantitative tools, first requires that objectives be formulated carefully, that aspects of the problem be analyzed, and that cause-effect relationships be reviewed, thus helping clarify formally the possible courses of action that can be pursued. Next, estimates of externalities or side effects or spillover effects can be made to help guide organizations by allowing a systematic investigation of the effects of work injuries and occupational diseases upon immediate family members, other institutions, and society. This is the least understood aspect of cost-benefit analysis and is often difficult to verbalize and is quite difficult to measure quantitatively. Then the effects of time can be investigated, in order to quantify results through a time horizon. However, the "snapshot approach," or the estimated costs and benefits at one point in time, will normally be a practical way to deal with this issue of time and can provide reasonably complete information about relationships. Last, estimates of the likelihood, or probability, of various values and relationships can be made. The provision of this data can add important credence to the results of such a study.

Cost-benefit or cost-effectiveness analysis can thus be a useful tool to better control work injuries and occupational diseases. Cost-benefit analysis is probably most useful in estimating economic relationships between various proposed actions and the expected benefits of such actions. Cost-effectiveness analysis is a more complex approach, which can be used to estimate the combination of activities that together achieve a goal, as reduced work injuries or fewer occupational diseases, at lowest cost.

In spite of limitations of both data and methodologies, cost-benefit and cost-effectiveness analysis can be useful to many of the institutions in our society. Studies can provide affected parties with a great deal of information at fairly reasonable costs. When done by individual organizations to assist in solving a specific problem, the analysis can be performed using a common-sense approach and obtaining information up to the point of diminishing returns. Finally, of greatest importance, such studies can help make the best use of scarce resources and can increase the quality of decision-making in an increasingly uncertain world.

The article entitled "Ten Practical Principles for Policy and Program Analysis" by A. C. Enthoven, which appeared in *Benefit-Cost and Policy Analysis,* 1974, provides the following practical principles for performing an effective cost-benefit or cost-effectiveness analysis:

1. Good analysis must be the servant of judgment and not a substitute for it.
2. Analysis should be open and explicit and should provide a framework for constructive debate and discussion.

3. There usually is no single best answer, so problems should not be forced into an optimizing model. Instead, look for various satisfactory solutions and for bad solutions that should be avoided.
4. Keep the conclusions simple.
5. It is better to be roughly right than exactly wrong. Therefore, getting the facts approximately right and framing properly the correct questions to be asked is an important, difficult, and underrated problem.
6. Always start by getting an overview of the problem and then proceed to the detail.
7. Consider the basis upon which current decisions are actually being made and then make realistic estimates about whether you can improve on that situation.
8. Determine whether questions basic to the problem under study have been sufficiently dealt with so you can build on this information to analyze the problem being studied in a satisfactory manner.
9. Do not overemphasize the quantitative aspects and do not ignore nonquantifiable factors that may be of decisive importance.
10. Think critically and realistically about the prospects for implementation, remembering what Robert Burns said about the best-laid plans of mice and men.

Cost-benefit analysis and cost-effectiveness analysis can be useful to help our society deal with difficult questions. These techniques are and should be tools or aids to provide additional information to decision makers and not ends in themselves. Like other tools, this approach has both strengths and weaknesses that should be understood. Some strengths are its disciplined approach to problem solving and its ability to provide quantitative estimates of a wide variety of relationships. Limitations include the difficulty of quantifying some variables, the temptation to use the technique to reach a predetermined desired course of action, and the problem of describing and measuring externalities, spillover, or third-party effects. As with any other tool, its true value can probably best be determined by evaluating the results of decisions made utilizing the technique.

22.2 OVERVIEW OF ECONOMIC APPROACHES

A review of general studies on the economics of safety properly should begin with the 1970 book by Guido Calabresi entitled, *The Cost of Accidents: A Legal and Economic Analysis.* This book discusses various means of achieving accident cost reduction through economic modeling and concludes that the most effective approach is one that seeks to minimize the sum of accident costs and accident prevention costs. Walter Oi's 1974 article, "On the Economics of Industrial Safety," provides a number of economic models of occupational safety and summarizes some of the results of his studies relating variations in work injuries to industry, labor turnover, the age and sex of employees, and the varying sizes of firms as measured by numbers of employees. In his book, *Of Acceptable Risk: Science and the Determinants of Safety,* published in 1976, William W. Lowrance discusses some of the ways that science can define and control social risk. Major topics covered are criteria for deciding how much risk is acceptable, risk assessment and reduction, the problem of establishing safety policies, and the role of regulation.

Another book published that same year, *Crisis in the Workplace: Occupational Disease and Injury*, by Nicholas Ashford, is an extensive discussion and analysis of the technical, legal, political, and economic problems regarding occupational health and safety. It also provides a review of practices in western European countries, extensive references to much of the literature, and suggests a coherent approach to the subject of occupational disease and work injury reduction.

The economic valuation of human lives is a difficult subject to deal with because to an individual, the value of his own life is infinite, even though the value to society or to associates may be a sum that can be estimated. Dorothy Rice and Barbara Cooper, in an article in the *American Journal of Public Health* entitled "The Economic Value of Human Life," provide estimates of the present value of lifetime earnings of individuals classified by sex, age, race, and years of schooling completed.

More current estimates are made by Richard Thaler and Sherwin Rosen (1976), who utilize the risks of death in various occupations to estimate the value of saving a life. In their article, "The Value of Saving a Life: Evidence from the Labor Market," these two authors provide values of between $176,000 and $260,000 per life. Other information about the valuation of life is provided in articles by Henderson (1975), Zeckhauser (1975), and Conley (1976), and by Jones-Lee in his book entitled *The Value of Life: An Economic Analysis* (1976).

The most extensive recent review of the effects of workers' compensation insurance costs upon work injury experience is contained in the *Report of the National Commission on State Workmen's Compensation Laws.* One of the major conclusions of this study was that there is no clear evidence to show that accident rates decline as experience rating becomes more powerful, suggesting that other forces influencing safety records are stronger than potential savings in workers' compensation insurance premiums resulting from a superior safety record. Important articles on this subject are found in Volumes I and III of *Supplemental Studies for the National Commission on State Workmen's Compensation Laws.* These articles were written by Walter Oi, Robert S. Smith, Louise Russell, and James Chelius. Additional information on this subject is provided by Bruce Boals (1969), Melvin Berkowitz (1971), and Daniel Kasper (1977).

Another way of evaluating the economics of occupational safety is to estimate wage differentials paid to workers in more hazardous occupations. Richard Thaler and Sherwin Rosen (1973) have utilized insurance company data on the risk of death in various occupations and have made estimates of the additional wages paid in more dangerous occupations. Robert W. Smith has also estimated wage differentials for more dangerous job classifications, estimating that they can be calculated for risks of accidental death at acceptable levels of confidence. In addition, Kip Viscusi (1976) determined that workers in occupations perceived to entail higher work injury risks are paid a wage differential of approximately $375–$420 per year for exposure to that risk.

While a great deal has been written regarding governmental intervention in the field of occupational safety, there have been relatively few research studies regarding the results of such government activities. In 1975, Aldona DiPietro estimated the effects of federal government inspections upon 18 different manufacturing industries and determined that such inspections correlated with higher work injury rates. In that same year Cooper and Company evaluated the effects of OSHA upon the longshoring industry and found that best results can occur by stimulating changes in technology, stressing cooperation between all affected parties, and demonstrating the benefits of safer operations to management. A report prepared by John Mendeloff, completed in 1976, estimated that a fully effective OSHA inspection program could eliminate 18% of a series of work injuries analyzed by a panel of California safety engineers.

In his book, *The Occupational Safety and Health Act,* Robert W. Smith reviews the results of a number of studies conducted by the author and suggests that an injury tax on employers in place of governmental regulation would be the most efficient way to encourage employers to further emphasize injury prevention. An article, "Safety at Any Price?," by Walter Oi evaluates the regulatory approaches of information dissemination, imposed penalties, and enforced standards. He suggests that society should be willing to pay for safety when the benefits received are less than the price paid for them.

The Impact of OHSA: A Study of the Effects of the Occupational Safety and Health Act on Three Key Industries—Aerospace, Chemicals and Textiles by Herbert Northrup, Charles Perry, and Richard Rowan reports an extensive study of these industries and concludes that current federal efforts do not promote optimal resource allocation, that there is a substantial overestimate of the need for the regulation of occupational safety and a similar underestimate of the effectiveness of market forces regarding occupational risks. Richard Zeckhauser and Albert Nichols have provided information to the Senate Committee on Governmental Affairs, which broadly reviews the subject of the economics of safety, including appropriate levels of safety, market imperfections, regulatory intervention, and government involvement in safety and health. An unpublished paper by Nicholas Ashford, entitled "Economic Issues in Occupational Health and Safety," discusses the appropriateness of economic analysis in this area and the use of economic incentives as an alternative to regulation.

To summarize, there have been a few studies of the economics of government intervention regarding occupational safety, which generally agree that current approachs are not as effective as they might be. They propose various solutions ranging from doing away with current regulatory activities and substituting an injury tax on employers to essentially continuing current activities without major change.

A number of nonmonetary studies of occupational safety or studies provide qualitative information and insights about the relative effectiveness of occupational safety activities and approaches; they are a part of this literature review. Some of these studies are summarized and discussed in an article entitled "Factors in Successful Occupational Safety Programs," written in 1977 by Alexander Cohen. One of these studies reviews the programs and practices of twelve safety-award-winning coal mining companies as analyzed by R. T. Davis and R. W. Stahl. In 1971 Yaghoub Shafai-Sharai surveyed eleven matched pairs of firms in eleven different Michigan industries and concluded that top management interest, more supervision, better workplace conditions, more safety devices on machinery, better accident recordkeeping, and an older, long-service labor force of married employees correlated with lower accident rates.

In 1974 William H. Mobley surveyed safety specialists and line managers in 43 glass and ceramics industry plants. He concluded that the most important behavioral and motivational factors were safety program support by top management, supervisors who set good safety examples, accidents seen as the result of unsafe acts, not unsafe conditions, supervisors with high safety expectations, and an organization seen by employees as safety conscious. The last of these selected studies is the 1975 research by Alexander Cohen, Michael Smith, and Harvey Cohen, *Safety Program Practices in High versus Low Accident Rate Companies.* Based upon mailed survey data supplied by 42 matched pairs of Wisconsin companies, it was found that the following tend to lead to improved safety performance: greater staff stature and commitment to safety, greater utilization of outside resources, greater variety of safety incentives, more job safety training, a more stable work force, and a safety program that emphasizes a better balance between the engineering and nonengineering approaches toward accident prevention.

22.3 THE TEXAS COST-EFFECTIVENESS STUDY

The study consisted of a detailed review of the occupational safety activities and the safety performance of 140 Texas chemical, paper, and wood product manufacturing firms and a statistical analysis of the information obtained. The manufacturing sector was selected because of the relatively large number of occupational injuries that occur here, because practices in this sector are often more settled, and because information about this industry could be secured reasonably easily.

Three specific manufacturing industries were selected for the study. One was the chemical and allied products industry, which was selected to be representative of manufacturers that have better than average safety performance as measured by the lost-time injury frequency rate. The second industry selected was the paper and allied products industry, because its experience, according to the same measure of safety performance, was close to the average for the manufacturing sector. The third industry was lumber and wood products, which over the years, has sustained poorer than average work-injury experience as measured by lost-time rates.

It was estimated that many of the differences between work-injury rates of firms in these three industries occurred because of basic differences between the nature of the three industries and because of differences caused by the different sizes of firms. Therefore, information was obtained from firms classified as small, medium-sized, and large in each of the three industries studied. Small firms had fewer than 50 full-time employees, medium-sized firms employed between 50 and 200 full-time people, and large firms employed more than 200 people.

A questionnaire was prepared so that specific information could be gathered. It asked detailed questions about safety activities during 1974, the safety record or performance during 1972, 1973, and 1974, and about other related activities. Specifically, questions were asked about top management safety activities, safety and health staff, new employee orientation and training, safety rules, activities to maintain interest in safety, safety meetings, safety inspections, the provision of personal protective equipment, the correction of unsafe physical conditions, physical examinations, injury treatment facilities and staff, off-the-job safety activities, safety training for experienced employees, and safety record-keeping activities.

The questionnaire also asked respondents to rank the relative effectiveness of these various activities. Another portion of the questionnaire asked about the interest of top management, the design and layout of the facility, and housekeeping at the location.

Measures of the safety record or performance were obtained by questions about the number of fatalities, lost-time injuries, days lost, medical cases, first-aid cases, man-hours worked, injury frequency rates, injury severity rates, and dollar costs of work injuries.

To obtain further information about the business firm surveyed, the last section of the questionnaire asked about the products produced, the number and average annual salary of employees by major groupings, the age of the facility, the name of the workers' compensation insurance carrier, and the names of the labor union, if any.

The questionnaire was used to obtain information from the Texas manufacturing firms in two ways. First, the survey forms were mailed to representative firms and 86 usable responses were secured. Second, 54 interviews were conducted throughout the state of Texas during the spring and early summer of 1975. A total of 140 completed questionnaires were obtained. This represented 8% of all such firms in Texas and 29% of the employees in the industries selected for study.

The monetary costs of the safety activities of each firm in the survey were calculated first. These costs were the total cost of each of the various safety activities performed during 1974, stated in terms of dollars per employee so that they could be compared with each other and with the estimated monetary cost of work injuries. These figures can be converted to the total dollar costs to an organization by being multiplied by the number of employees in the firm.

Safety staff costs were calculated by adding the reported annual salaries of full-time safety personnel and the correct proportion of the salary of part-time safety personnel. The cost of top management safety activities was computed by estimating the monetary value of the time spent on safety by the top manager at the location. New employee orientation costs consisted of the value of the estimated time spent on such orientation by both trainers and new employees. Safety rule costs were the costs of preparing and distributing these rules during 1974. The cost of activities designed to maintain employee interest in safety included the cost of safety contests, publications, posters, bulletin boards, incentives, and special programs. Safety meeting costs were the cost of department or employee, supervisory, management, labor-management, and other safety meetings. Safety inspection costs consisted of the cost of inspections by safety staff, supervisors, other management employees, headquarters personnel, insurance carriers, government employees, and consultants.

The total costs per employee of safety activities for Texas firms grouped by size and industry are provided in Table 22.1 Review of this table reveals an interesting finding of this study. In seven of the nine groupings of firms by size and industry, those firms with lower work-injury frequency rates spent less, on the average, than those similarly classified firms that had poorer safety records. One possible explanation is that good safety performance is more dependent upon the correct mix of safety activities than upon the total amount of money spent.

The estimated monetary cost of work injuries to the Texas firms was also calculated. This consisted of the direct or insured cost of work injuries, an estimate of the monies paid beyond these direct costs to Texas workers' compensation insurance carriers (in Texas employers are required to purchase workers' compensation insurance), and an estimate of the indirect or noninsured costs of work injuries.

The direct cost of work injuries was calculated by adding together the estimated cost of fatal, lost-time, and medical or doctor cases. Based upon information provided by the Texas State Board of Insurance, the average direct cost of fatal work injuries, primarily consisting of death benefits paid to survivors, was estimated to be $22,700. The monetary cost of disabling or lost-time work injuries, as measured by the American National Standards Institute Standard Z16.1, was estimated from information supplied by those surveyed.

As detailed by the Texas Workers' Compensation Act, these costs consist of the payment of medical and hospital expenses, weekly indemnity paid to injured employees during a portion of the time they are unable to work, and lump sum payments to compensate for the loss or loss of use of a portion of the body.

The average cost of such work injury, estimated to be $1910, represented the greatest proportion of direct injury costs. Medical or doctor case costs were estimated to be $30, again based upon information supplied by those surveyed. The total direct cost of work injuries for each respondent was calculated to be the average number of fatalities, if any, occurring in 1972, 1973, and 1974, times $22,700 per fatality, plus the average number of disabling or lost-time work injuries reported in 1972, 1973, and 1974, times $1,910 per such injury, plus the average number of medical cases reported during 1972, 1973, and 1974, times $30 per medical case.

Estimates of the monies paid beyond direct losses to Texas workers' compensation insurance carriers were made next. The Texas State Board of Insurance indicated that the cost of losses represented 56% of the cost of Texas workers' compensation insurance. Therefore, total insurance costs were estimated for each firm surveyed by multiplying the direct cost of work injuries by 1.79.

The noninsured or indirect cost of work injuries was also estimated. Estimates of these indirect, noninsured costs were requested from those firms surveyed, but none of the firms were able to provide this information. Therefore, following the procedures recommended by Rollin H. Simonds based upon his research regarding these indirect costs (Simonds, 1949; Simonds and Grimaldi, 1975), these costs were estimated to be equal to the direct or insured cost of work injuries.

The total cost of work injuries was then calculated to be the sum of the direct costs of fatalities, lost-time injuries, and medical cases, plus an estimate of additional monies paid to workers' compensation insurance companies, plus the estimated indirect, noninsured cost of such injuries.

This information, stated in terms of these costs per employee, is provided in Table 22.2 which shows that there were large differences between the average costs of work injuries for firms with low work-injury frequency and high work-injury frequency rates. It also shows that there were sizable differences between the cost of work injuries occurring in the three industries studied.

The data obtained about the cost of occupational safety activities and about the monetary cost of

TABLE 22.1. Cost of Safety Activities Per Employee in 1981 Dollars

	Chemicals		Paper Products		Wood Products	
	Low Injury Freq. Rate	High Injury Freq. Rate	Low Injury Freq. Rate	High Injury Freq. Rate	Low Injury Freq. Rate	High Injury Freq. Rate
Small firms	$ 662	$1032	$448	$554	$242	$456
Medium-sized firms	789	998	299	250	178	354
Large firms	1211	1152	325	840	234	355

Notes: 1. Low injury frequency rate refers to firms that have work-injury frequency rates lower than the average rate for the industry.
2. High injury frequency rate refers to firms that have work-injury rates higher than the average rate for industry.
3. Small firms employed fewer than 50 full-time employees. Medium-sized firms employed between 50 and 199 full-time equivalent employees. Large firms employed 200 or more full-time equivalent employees.

TABLE 22.2. Cost of Work Injuries Per Employee in 1981 Dollars

	Chemicals		Paper Products		Wood Products	
	Low Injury Freq. Rate	High Injury Freq. Rate	Low Injury Freq. Rate	High Injury Freq. Rate	Low Injury Freq. Rate	High Injury Freq. Rate
Small firms	$26	$760	$200	$ 696	$ 58	$1907
Medium-sized firms	82	536	230	1010	621	1870
Large firms	53	509	160	533	283	1350

Notes: 1. Low injury frequency rate refers to firms that experienced work-injury frequency rates lower than the average for the industry.
2. High injury frequency rate refers to firms that experienced work-injury frequency rates higher than the average rate for the industry.
3. Small firms employed fewer than 50 full-time employees. Medium-sized firms employed between 50 and 199 full-time equivalent employees. Large firms employed 200 or more full-time equivalent employees.

work injuries was analyzed in three ways. First, graphic comparisons were made. Second, multiple-linear regressions were computed. Third, costs were calculated as a percent of the average annual hourly payroll.

Figure 22.1 is a bar graph providing a comparison of the average cost of safety activities per employee for the chemical, paper and wood product industries. This figure shows that the Texas chemical industry spent proportionately more for safety activities and experienced relatively low losses. Therefore, there generally appears to be a cause-and-effect relationship between expenditures to prevent work injuries and work injuries as measured by their monetary costs. However, as previously indicated, the data provided in Tables 22.1 and 22.2 show that these relationships are more complex than expected.

To study these relationships further, the data were analyzed by the multiple linear regression statistical technique. This analysis represents the heart of the study. The technique provides a way of estimating the effect of various safety activities and other variables upon the work injuries that occurred, both measured in terms of dollar costs per employee. The specific technique used was the James H. Goodnight Maximum R^2 improvement multiple linear regression program of the North Carolina State University Statistical Analysis System.

This procedure calculates which single safety activity or variable explains the greatest amount of the difference between the cost of work injuries for firms with low-injury costs and those firms with high injury costs. Then it calculates which two safety activities or variables working in combination best explain differences between low and high injury costs.

In similar fashion, the program calculates which three, four, five, and so on variables, again working in combination, best explain differences between low and high injury costs. Because both the safety activities, or variables and the cost of work injuries were stated in terms of monetary costs per employee, this procedure provided an estimate of the effect of spending a given amount, as one dollar per employee, for a certain safety activity such as safety meetings, upon the cost of work injuries per employee.

One possible outcome of such an expenditure was a large decrease in work injury costs, indicating a cost-effective activity. A second possible outcome of the expenditure was a decrease in work injury costs to the employer, but less than the cost of the activity or between $.01 ad $1.00 per dollar per employee expended upon the safety activity. This outcome was labeled partially cost-effective. A third possible outcome was an increase in work injury costs, indicating that firms that spent more for such activities also experienced higher work injury costs. This outcome was labeled cost-ineffective.

The combination of variables utilized to estimate differences between firms with low work-injury costs and those with high work-injury costs were:

Management activities regarding safety.
Safety and health staff.
Safety orientation of new employees.
Safety rules.
Activities designed to maintain employee interest in safety.
Safety meetings.
Safety inspections.
Personal protective equipment.

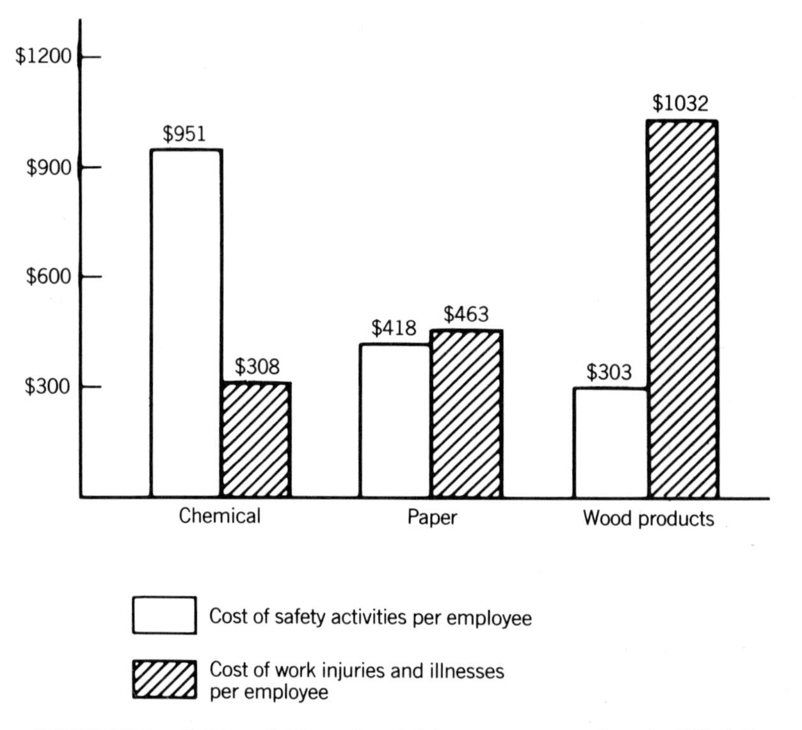

FIGURE 22.1. Safety activities and work injury costs per employee in 1981 dollars.

Guarding or the correction of unsafe physical conditions.
Physical examinations.
Medical or injury treatment facilities and staff.
Off-the-job safety activities.
Safety training for experienced employees.
Safety record keeping activities.
The number of full-time equivalent employees.
Span of control defined in this study as the cost of supervision per employee.

All of the these variables except number of employees provided monetary estimates of the effect of a given activity upon work-injury costs. In this case, the results of an addition or reduction in the number of employees was related to injury costs.

Calculations were performed for all firms in the same industry, all firms of the same size grouping, and for each of the nine groupings of firms of the same size in the same industry. The calculations indicated that different combinations or mixes of safety activities best explained differences in work-injury costs in organizations of different size groupings or in small, medium-sized, and large firms.

The analysis also indicated that those activities, which together make up a safety program, explained approximately 60% of the difference between the cost of work injuries in large or medium-sized firms or in firms with more than 50 employees. Approximately 27% more of this difference was explained by the size of the firm, as measured by the number of employees and by span of control or employees per supervisor, measured here in terms of the cost of such supervision per employee. Therefore, for firms of these sizes, appropriate safety programs can reduce work injury costs, but only up to a point. In small firms, or those with fewer than 50 employees, safety program activities explained 75–95% of the difference between the cost of work injuries for firms with low injury costs and those with high injury costs. A summary of this information expressed in terms of R^2, or the proportion of the variation between low and high work-injury costs per employee explained by various activities or variables, is provided in Table 22.3.

Next, a review of those specific activities or variables which, interacting with each other, best explained differences between the cost of work injuries, is provided in Table 22.4. This table shows the R^2,

TABLE 22.3. Analysis of Variables Explaining Differences Between Low and High Work Injury Costs

Size of Firm	Chemical		Paper		Wood Products	
	Variable	R^2	Variable	R^2	Variable	R^2
Small	Safety program	78%	Safety program	95%	Safety program	75%
	Not explained	22	Employee number	01	Span control	17
			Not explained	04	Employee number	02
					Not explained	06
	Total	100%	Total	100%	Total	100%
Medium	Safety program	55%	Safety program	58%	Safety program	66%
	Employee number	18	Employee number	25	Employee number	32
	Span control	03	Not explained	17	Not explained	01
	Not explained	24				
	Total	100%	Total	100%	Total	100%
Large	Safety program	51%	Safety program	76%	Safety program	65%
	Span control	19	Employee number	15	Span control	23
	Employee number	13	Span control	05	Employee number	12
	Not explained	17	Not explained	04	Not explained	01
	Total	100%	Total	100%	Total	100%

Note: R^2 is the proportion of the variations between low and high work-injury costs per employee explained by variables.

or the proportion of the variation between low and high work-injury costs explained by each of the variables, as well as the sign of each of the coefficients (multipliers) of the variables. A positive sign indicates that an increase in expenditures for the activity helped explain increased work-injury costs. A negative sign indicates that larger monetary expenditures resulted in decreased costs of work injuries. Therefore, for the large chemical firms, greater expenditures for off-the-job safety, medical facilities and staff, safety training, and for additional supervision, or span of control, decreased work-injury costs.

Similarly, additional expenditures for guarding or the correction of unsafe physical conditions, safety staff, orientation, and for safety records, correlated with higher work-injury costs. In similar fashion, somewhat different combinations of the various safety activities and other variables best explained differences between firms with low and high work-injury costs in other industries and for medium-sized and small chemical firms.

To provide still more information about the estimated effect of the various safety activities upon the cost of work injuries, some further results of this study are presented in Table 22.5. This table provides a listing in decreasing rank order of the effect of the various activities upon the cost of work injuries per employee for the firms studied. Those activities located above the horizontal line in each grouping of data by industry and size were cost-effective, or caused a reduction in work injuries greater than the expenditure for the activity. Those activities listed below the horizontal line were not cost-effective, or were related to high work-injury costs. All of the activities or variables that entered the multiple linear regression analysis at satisfactory probabilities are listed, in order to provide a more complete estimate of the relative effect of these activities.

A review of the data presented indicates that the 14 safety activities, the span of control or available supervision, and the firm size as measured by the number of employees interact with each other in complex, subtle ways to explain differences between those firms with low work-injury costs and those with high injury costs. The study further shows that there are no easy, simple answers to the question of how best to reduce work-injury costs, but that relationships between the cost of safety activities and work-injury costs do exist and that they can be estimated.

One last evaluation of the relative effectiveness of the variables was made in order to estimate relationships, realizing both the complex nature of these relationships and the fact that different combinations of these variables explain differences in firms in different industries and in firms of different sizes. These estimates are based upon an averaging for the nine classifications of firms by industry and by size of both the coefficients or values that maximize the effect of the variables and the R^2 value or proportion of the cost of work injuries explained by the various activities such as safety meetings and safety record

TABLE 22.4. Comparison of Effect of Selected Variables on Cost of Work Injuries per Employee Measured by Multiple Linear Regression Technique

Size of Firm	Chemical			Paper			Wood Products		
	Variable	Sign	R^2	Variable	Sign	R^2	Variable	Sign	R^2
Small	Staff	+	30%	Guarding	+	32%	Physicals	+	25%
	Orientation	+	21	Records	+	27	Span control	−	17
	Records	−	10	Off-the-job	+	16	Meetings	−	16
	Inspections	−	09	Physicals	−	11	Interest	+	11
	Rules	−	08	Inspections	+	05	Training	−	11
				Staff	−	04	Orientation	−	12
				Employee no.	+	01	Employee no.	−	02
	Total		78%	Total		96%	Total		94%
Medium	Employee no.	−	18%	Employee no.	+	25%	Orientation	−	50%
	Meetings	−	15	Rules	+	18	Employee no.	−	32
	Interest	+	13	Guarding	+	09	Physicals	+	09
	Management	−	11	Orientation	−	09	Rules	−	03
	Training	−	06	Training	−	07	Records	+	03
	Equipment	−	03	Management	+	07	Training	+	03
	Orientation	+	03	Interest	−	06	Management	−	01
	Span control	−	03						
	Off-the-job	−	04						
	Total		76%	Total		83%	Total		99%
Large	Span control	−	19%	Off-the-job	−	31%	Physicals	+	25%
	Medical	−	16	Management	+	21	Span control	−	23
	Employee no.	−	13	Employee no.	−	15	Medical	−	23
	Staff	+	10	Span control	−	15	Staff	+	16
	Training	−	10	Interest	−	14	Employee no.	−	12
	Records	+	04						
	Guarding	+	04						
	Orientation	+	04						
	Off-the-job	−	03						
	Total		83%	Total		96%	Total		99%

Note: R^2 is the proportion of the variations between low and high work-injury costs per employee explained by a variable.

keeping expenditures. Based upon these criterion, those variables that were cost-effective, listed in decreasing order of cost-effectiveness, were as follows:

Safety rules.
Off-the-job safety activities.
Safety training.
Safety orientation.
Safety meetings.
Medical facilities, staff, and supplies.

Based upon these same criteria, the following were found to be partially cost-effective:

Span of control or cost of supervision.
Personal protective equipment.
Safety inspections.

Estimated in the same way and ranked in increasing order of cost-effectiveness, the following were found in this study not to be cost-effective:

Guarding or the correction of unsafe physical conditions.

Safety staff.
Activities to maintain interest.
Physical examinations.
Safety records.

This means that expenditures for guarding or the correction of unsafe conditions were somewhat cost-ineffective in firms studied and that record keeping activities were highly cost-ineffective.

This study does not conclude that cost-ineffective activities are undesirable. It does suggest that *current levels* of expenditures for those activities are sometimes not as cost-effective as expenditures for the other activities previously described. It further indicates that at current levels of spending, further reductions of work-injury costs in the industries studied may be achieved best by further expenditures for the most cost-effective variables.

Let us try to assess why some of the activities were cost-effective and some were not, for the firms studied. Safety rules were the most cost-effective activity, possibly because firms that spent larger amounts of money to develop and distribute safety rules evaluated the hazards of many jobs and communicated much of this information in a useful form to employees.

TABLE 22.5. Decreasing Rank Order of the Cost-Effectiveness of Selected Variables

Size of Firm	Chemical	Paper	Wood Products
Small	Off-the-job	Off-the-job	Orientation
	Records	Physicals	Training
	Training	Orientation	Meetings
	Rules	Staff	Management
	Inspections	Span control	Span control
	Guarding	Guarding	Rules
	Management	Inspections	Equipment
	Staff	Meetings	Physicals
	Orientation	Management	Interest
	Equipment	Medical	
	Physicals	Records	
Med.	Management	Medical	Rules
	Off-the-job	Training	Training
	Rules	Equipment	Management
	Medical	Interest	Medical
	Equipment	Orientation	Staff
	Training	Span Control	Interest
	Meetings	Guarding	Guarding
	Span control	Inspections	Meetings
	Orientation	Physicals	Orientation
	Interest	Management	Physicals
	Physicals	Rules	Records
	Records		
Large	Off-the-job	Off-the-job	Medical
	Physicals	Records	Records
	Inspections	Interest	Interest
	Medical	Inspections	Span control
	Training	Meetings	Meetings
	Span control	Training	Guarding
	Guarding	Span control	Physicals
	Staff	Staff	Orientation
	Orientation	Orientation	Staff
	Records	Management	Training
	Interest		Equipment
	Management		Rules
			Management

Note: The activities listed above the horizontal line in each grouping were cost effective. The activities below the horizontal line were cost-ineffective.

Possibly more extensive off-the-job safety activities partially explain the lower injury costs of these firms, because such activities help convince employees that the companies are truly interested in the safety of their employees.

Larger expenditures for safety orientation, safety training, and safety meetings also effectively reduced work-injury costs, possibly because these activities help teach employees the safe way to work. More extensive medical facilities, staff, and supplies may also both provide a valued service and communicate the interest of the firm in the safety of its employees. In addition, a good full- or part-time industrial nurse often provides useful support for a safety program.

Expenditures for span of control, defined in this study as the cost of supervision per employee, were partially cost effective. This means that the cost of additional supervision correlated with lower work-injury costs, but not in an amount equal to the full cost of such additional supervision. However, if the cost of additional supervision could be partially justified for other reasons, such as improvements in productivity or the improved quality of products produced, then these factors, coupled with a partial justification in terms of some reduction in the cost of work injuries, might make the provision of additional supervision desirable.

Safety inspections, which can be both a training device and a means of finding unsafe physical conditions, were also partially cost-effective. Similarly, personal protective equipment was partially cost-effective, possibly reflecting differing degrees of control over the issuance of such devices at surveyed locations.

Another series of activities was found to be cost-ineffective. Guarding, or the correction of unsafe physical conditions, was frequently not cost-effective in this study, possibly because greater emphasis on this important part of occupational safety work diminished emphasis upon the more cost-effective training activities, and possibly because the large amounts of money being spent on such activities are currently achieving diminishing returns. Similarly, expenditures for larger safety staffs were frequently not as cost-effective as the smaller, leaner safety staffs found in some firms.

Activities to maintain employee interest in safety may not be cost-effective, because such techniques as safety contests and safety posters may be relatively poorer ways to encourage employee safety. Physical examinations often related to higher work-injury costs, possibly because such examinations correlate with higher employee turnover and the necessity of hiring new, inexperienced employees. As expected, increased costs of maintaining safety records related to higher work-injury costs, as those firms with greater numbers of injuries spent more money maintaining records of these occurrences.

Management, or the estimated value of the time the top manager at the location spent on safety, was frequently not cost-effective in this study. No special significance is attached to this, because three other measures of this variable in this study indicated that the support and interest of top management was an important part of an effective safety program.

One possible explanation of the results of this study is that the firms surveyed approach occupational safety in two ways. One group of firms, those with lower work-injury costs and with slightly lower safety program costs, may emphasize injury prevention and those activities that they have found over time best reduce their work injuries. Another group of firms, those with higher than average work-injury costs, which spend slightly more on their safety programs, may emphasize compliance with governmental standards and regulations rather than injury prevention.

The third and last way these data have been reviewed is by comparing the costs of both safety activities per employee and of work injuries per employee to hourly-employee average annual wages, as shown in Table 22.6. The table shows that these costs represented 8.4% of average hourly wages for chemical firm respondents, 7.1% for paper manufacturers, and 13.7% for wood product firms. This evaluation indicates that the proportion of employee safety costs varies significantly in the manufacturing firms surveyed. Therefore, *this reason,* coupled with other factors such as humanitarian considerations and the central investment per employee, may explain why safety practices vary so widely from industry to industry.

Essentially, this study found that subtle combinations of highly interrelated safety activities best

TABLE 22.6. Cost of Safety Activities and Work Injuries by Percent of Hourly Employee Average Annual Wages

	Chemical	Paper	Wood Products
Hourly employee annual wages	$16,150	$13,010	$10,040
Safety program costs as percent of wages	6.1%	3.3%	3.1%
Work injury costs as percent of wages	2.3%	3.8%	10.6%
Both safety program and injury costs as percent of wages	8.4%	7.1%	13.7%

Note: Annual wages are in 1981 dollars.

explained the difference between Texas firms with different work-injury costs. This combination was quite different for firms of different sizes, and was somewhat different for firms in different industries. It was found that a better combination, or mix, of these various safety activities, rather than greater monetary expenditures for some of them, was frequently the best way to reduce work-injury costs. This better combination seemed to consist of sufficient expenditure for safety rules, off-the-job safety, safety training, safety orientation, safety meetings, medical facilities and staff, and of good top management interest and participation in safety. The most effective mix, therefore, would seem to be a balanced approach, which combines both engineering and nonengineering and, probably, places more emphasis upon the nonengineering aspects. Individual locations or firms may find that they can best reduce work injuries by reevaluating their current activities, taking this information into consideration but always tempering it with the results of their own past experiences.

Last, this study indicates that the relatively straightforward question on how best to reduce work injuries is not answered easily. Some tentative answers, which can be of practical value to manufacturing firms and which, because of the lack of other studies, may also be useful to organizations in other sectors of the economy, have been provided by this study.

22.4 IMPLICATIONS FOR GOVERNMENT POLICY

The Occupational Safety and Health Act of 1970 has affected U.S. employers as much as any piece of legislation. It has required them to spend large amounts of money in order to meet extensive requirements regarding the physical condition of their structures, equipment, and materials.

The Occupational Safety and Health Act represents only the most recent major event in the long course of worker safety in this country. The industrial revolution and grouping of employees in factories first created the conditions in which occupational injuries and diseases could occur and be recognized. Legal decisions holding employers liable for worker injuries, state workers' compensation statutes, and concern for worker safety caused employers to begin to engage in activities designed to reduce work injuries. The first activities, primarily the correction of unsafe physical conditions, resulted in significant reductions in both the numbers of work injuries and the seriousness of many of those injuries that occurred. This approach represents both the starting point for accident prevention activities and an important aspect of such programs.

Facing increased demands for labor, both during and after World War II, and influenced by the work of such pioneers in the field as Heinrich and the National Safety Council, employers began to focus their attention on the unsafe acts of workers by orienting, training, and motivating their employees to work safely. This approach, coupled with existing activities, resulted in still further reductions in the number and seriousness of work injuries.

In 1970, in response to increased public attention to the quality of life, the interests of organized labor, and an increase in the number of reported work injuries, the Occupational Safety and Health Act was passed. Following the approach of existing state and federal regulations in this area and lacking evidence that an alternative approach would be more effective, the primary approach taken, as indicated, was the establishment of standards regulating the physical condition of workplaces, inspection

TABLE 22.7. Comparison of the Relative Effectiveness of Variables in the Cost-Benefit Analysis of Selected Texas Manufacturers

Cost-Effective		Partially Cost-Effective		Cost-Ineffective	
Variable	Effect of $1 Expenditure Upon Injury Costs	Variable	Effect of $1 Expenditure Upon Injury Costs	Variable	Effect of $1 Expenditure Upon Injury Costs
Rules	−$58	Inspections	−$.81	Guarding	+$ .64
Off-the-job	−$53	Meetings	−$.38	Staff	+$ .80
Training	−$40	Span Control	−$.32	Equipment	+$ 6.00
Medical	−$14			Interest	+$10.00
Orientation	−$ 4			Physicals	+$23.00
				Records	+$32.00

Note: These values are an average of the initial B values (coefficients or multipliers of safety activity variables) in multiple linear regression equations for each group of survey respondents for which the value appeared. They are somewhat different than other reported replationships, which are based on both B Values and R^2 values. (R^2 is the proportion of the variations between low and high work-injury costs per employee explained by a variable.)

TABLE 22.8. Proportion of Monies Spent by Texas Manufacturers for Employee Training, for Correction of Physical Conditions and for Both Activities

	Percent of Safety Program Costs		
Classification by Size and Injury Experience	Unsafe Acts	Physical Conditions	Both
Large—low injuries	36%	53%	11%
Large—high injuries	30%	58%	12%
Medium—low injuries	31%	49%	20%
Medium—high injuries	21%	65%	14%
Small—low injuries	24%	62%	14%
Small—high injuries	17%	71%	12%

Notes: 1. Large employers have more than 200 employees.
2. Medium-sized employers have between 50–199 employees.
3. Small employers have less than 50 employees.
4. Safety program activities oriented toward unsafe acts includes meetings, employee orientation, training, rules, interest, off-the-job safety, medical and physicals.
5. Safety program activities oriented toward physical conditions include guarding, inspections and personal protective equipment.
6. Safety program activities oriented toward both covers staff, management and records keeping costs.

for violations, and the assessment of fines. The results have been major increases in expenditures by U.S. firms to comply with OSHA requirements, dissatisfaction on the part of many businessmen because of this further regulation of their activities, and some reduction in the number of deaths, primarily in more hazardous occupations, as some of the more serious physical hazards were eliminated. No similar significant decrease in either the number of reported work injuries or their seriousness has occurred. Finally, to date, the results of the Act have not been perceived by many to be proportionate to the efforts and monies expended.

There seem to be several implications of the Texas study for public policy. Tables 22.7 and 22.8 highlight important conclusions of the study, which lead to the following recommendations for public policy. First, this research study provides increased understanding of the problems of the Occupational Safety and Health Administration because it concludes that employers have often been required to perform activities that, at required levels of expenditures, were not cost-effective or, at these levels of expenditures, did not clearly contribute to the reduction of work injuries. Second, the study quantifies and measures those activities that are estimated to be cost-effective. Thus the Occupational Safety and Health Administration and other appropriate governmental agencies can have a basis for encouraging those activities by employers and employees that are normally considered a complete safety program. Third, a number of recent legal cases have seemed to say that employers may comply with governmental requirements in the most cost-effective way. This study would then indicate a fairly extensive redirection of such governmental programs.

REFERENCES

Amchan, A. J., "The Future of OSHA," *Labor Law Journal,* September 1984, p. 547.

American National Standards Institute, *Method of Recording and Measuring Work Injury Experience,* New York: American National Standards Institute, 1969.

Ashford, N., *Crisis in the Work Place: Occupational Disease and Injury,* Cambridge, MA, M.I.T. Press, 1976.

Ashford, Nicholas, "Economic Issues on Occupational Health and Safety," unpublished paper, Massachusetts Institute of Technology, 1979.

Ashford, N., The Role of Risk Assessment and Cost-Benefit Analysis in Decisions Concerning Safety and the Environment, FDA Symposium on Risk Benefit Decisions and the Public Health, Colorado Springs, CO, 1978.

Barr, A. J. and Goodnight, J. H., *A User's Guide to the Statistical Analysis System,* Raleigh, NC: North Carolina State University, 1974.

Bequele, A., The Costs and Benefits of Protecting and Saving Lives at Work: Some Issues, *International Labour Review,* Jan.-Feb. 1984.

Berkowitz, M., Allocation Effects of Workmen's Compensation, *Proceedings of the 24th Annual Meeting of the Industrial Relations Research Association,* Dec. 1971.

Biancardi, M., The Cost-Benefit Factor in Safety Decisions, *Professional Safety,* November 1978.

Boals, B. R., *Some Economic Aspects of Workmen's Compensation in Florida,* Ph.D. dissertation, University of Florida, 1969.

Calabresi, G., *The Cost of Accidents: A Legal and Economic Analysis,* New Haven, CO, Yale University Press, 1970.

Cannon, J. A., Economic Analysis of Hazards, *Journal of Safety Research,* Winter 1974.

Carlin, D. and Planek, T. M., Risk Evaluation in Industry: Methods and Practice, Part I and Part II, *Professional Safety,* March and April 1980.

Chamber of Commerce of the United States, *Analysis of Workmens Compensation Laws,* Washington, D.C.; Chamber of Commerce of the United States, 1974.

Cohen, A., Cohen, H., and Smith, M., Characteristics of Successful Safety Programs, *Journal of Safety Research,* Spring 1978.

Cohen, A., Factors in Successful Occupational Safety Programs, *Journal of Safety Research,* December 1977.

Cohen, A., Smith, M., and Cohen, H., *Safety Program Practices in High versus Low Accident Rate Companies,* Cincinnati, OH: U.S. Department of Health, Education and Welfare, 1975.

Conley, B. C., The Value of Human Life in the Demand for Safety, *The American Economic Review,* March 1976.

Cooke, W. and Gantschi F., Plant Safety Programs and Injury Reduction, *Industrial Relations,* Fall 1981, p. 245.

Cooper and Company, *A Casual Study of Accidents in the Longshoring Industry and OSHA's Effectiveness: Final Repor,* Stamford, CT, Cooper and Company, 1975.

Cordtz, D., Safety on the Job Becomes Major Job for Management, *Fortune,* November 1972.

Dasgupta, A. K. and Pearce, D., *Cost-Benefit Analysis,* London: Macmillan, 1972.

Davis, R. T. and Stahl, R. W., Safety Organizations and Activities of Award Winning Companies in Coal Mining, Information Circular 8224, Pittsburgh, PA: U.S. Department of Interior, Bureau of Mines, 1964.

Dewhurst, R. F., *Business Cost-Benefit Analysis,* New York: McGraw-Hill, 1972.

Diehl, A. E., *A Systems Dynamics Simulation Model of the Occupational Safety and Health Phenomenon,* Ph.D. dissertation, North Carolina State University, 1974.

DiPietro, A., *Data Needs for the Evaluation of OSHA's Net Impact,* Office of Evaluation, Office of the Assistant Secretary for Policy, Evaluation and Research, U.S. Department of Labor, August 1975.

Drucker, P. F., *Managing in Turbulent Times,* New York; Harper & Row, 1980.

Drucker, P. F., *Management: Tasks, Responsibilities, and Practices,* New York: Harper & Row, 1974.

Eads, G., The Benefits of Better Benefits Estimation, A. Ferguson, ed., *Benefits of Safety and Health Regulations,* Cambridge, MA: Ballinger, 1981.

Eckstein, O., A Survey of the Theory of Public Expenditure Criteria, *Public Finances: Needs, Sources and Utilization,* Princeton, NJ: Princeton University Press, 1961.

Editors, A Bicentennial Look at Safety, *Professional Safety,* **21,** April 1976.

Editors, *Occupational Safety & Health Cases, Vol. 4,* Washington, D.C., Bureau of National Affairs, Inc., 1977.

Enthoven, A. C., Ten Practical Principles for Policy and Programs Analysis, *Benefit-Cost and Policy Analysis,* 1974, Zeckhauser, R. et al., ed., Chicago: Aldine Publishing, 1975.

Ferguson, A. and LeVeen, P., *The Benefits of Health and Safety Regulations,* Cambridge, MA, Ballinger, 1981.

Ferry, T. S., The Price of Safety Demands More Than Safety Expertise, *Professional Engineer,* March 1981, pp. 15–17.

Goldman, T. A., ed., *Cost-Effectiveness Analysis,* New York: Praeger Press, 1967.

Green, M. and Waitzman, N., *Business War on the Law,* Learning Research Project, Washington, D.C., 1980.

Guzzardi, W., Reagan's Reluctant Deregulators, *Fortune,* March 8, 1982.

Guzzardi, W., The Mindless Pursuit of Safety, *Fortune,* April 9, 1979.

Harberger, A. C., Haveman, R., Margoulis, J., Niskanen, W., Turvey, R., and Zeckhauser, R., *Benefit-Cost and Policy Analysis,* Chicago: Aldine Press, 1972.

Haveman, R. M., Harberger, A. C., Lynn, L. E. Jr., Niskanen, W., Turvey, R., Zeckhauser, R., *Benefit-Cost and Policy Analysis,* 1973, Chicago: Aldine, 1974.

Henderson, M., The Value of Human Life, *Search,* Jan.-Feb. 1975.

Hinrichs, H. M., and Taylor, G. M., *Program Budgeting and Benefit-Cost Analysis,* Pacific Palisades, CA: Goodyear, 1969.

Irvin, G., *Modern Cost-Benefit Methods,* New York: Harper & Row, 1978.

Jones-Lee, M. W., *The Value of Life: An Economic Analysis,* Chicago: University of Chicago Press, 1976.

Kasper, D. M., For a Better Workers Compensation System, *Harvard Business Review,* March-April 1977.

Klein, T. A., *Social Costs and Benefits of Business,* Englewood Cliffs, NJ: Prentice Hall, 1977.

Krikorian, M., An Analysis of the Occupational Safety and Health Act of 1970, ASSE *Journal,* 18, April 1973.

Levin, M., Politics and Polarity: The Limits of OSHA Reform, *Regulation,* Nov.-Dec. 1979.

Lowrance, W. W., *Of Acceptable Risk: Science and the Determinants of Safety,* Los Altos, CA: William Kaufman, 1976.

MacLaury, J., The Job Safety Law of 1970: Its Passage was Perilous, *Monthly Labor Review,* March 1981, pp. 18-24.

Marlow, M. L., The Economics of Enforcement: The Case of OSHA, *Journal of Economics & Business 1982,* **34,** p. 165.

Marx, T. G., The Cost of Living: Life, Liberty and Cost-Benefit Analysis, *Policy Review,* Su 1983, pp. 53-58.

McKean, R., Cost-Benefit Analysis, in *Managerial Economics and Operations Research,* E. Mansfield, ed., New York: W. Norton, 1980.

Mendeloff, J., *An Evaluation of the OSHA Program's Effect on Workplace Injuries: Evidence from California through 1974,* Washington, D.C., U.S. Department of Labor, 1976.

Mendeloff, J., *Regulating Safety: An Economic and Political Analysis of Occupational Safety and Health Policy,* Cambridge, MA, MIT Press, 1979.

Mishan, E. J., *Cost-Benefit Analysis: An Informal Introduction,* New York: Praeger Press, 1975.

Mishan, E. J., *Economic Efficiency and Social Welfare,* London: Allen & Unwin, 1981.

Mobley, W. H., Managerial Evaluations of Safety Motivation and Behavioral Hypotheses, unpublished paper, University of South Carolina, 1974.

Niskanen, W. A., Harberger, A. C., Haveman, R. H., Turvey, R., and Zeckhauser, R., *Benefit-Cost and Policy Analysis,* 1972, Chicago: Aldine, 1973.

Northrup, H., Charles, P. E., Rowan, R., *The Impact of OSHA: A Study of the Effects of the Occupational Safety and Health Act on Three Key Industries—Aerospace, Chemicals,* and *Textiles,* Philadelphia, PA: The Wharton School, University of Pennsylvania, 1978.

Oi, W. Y., On the Economics of Industrial Safety, *Law and Contemporary Problems,* Summer-Autumn, 1974.

Oi, W. Y., "Safety at Any Price?," *Regulation,* November-December 1977.

O'Neill, B. and Kelley, A. B., Costs, Benefits, Effectiveness and Safety: Setting the Record Straight, *Professional Safety,* August 1975.

Prest, A. and Turvey, R. P., "Applications of Cost-Benefit Analysis," *Managerial Economics & Operations Research,* Edwin Mansfield, ed., New York: W. Norton & Co., 1980.

Ray, A., *Cost-Benefit Analysis: Issues and Methodologies,* Baltimore, MD: Johns Hopkins Press, 1984.

Recht, J. C., *How To Do a Cost-Benefit Analysis of Motor Vehicle Accident Counter Measures,* Chicago: National Safety Council, 1966.

Rice, Dorothy and Barbara Cooper, "The Economic Value of Human Life," *American Journal of Public Health,* Nov. 1967.

Rinefort, F. C., *The Need for Occupational Safety and Health Technicians in California,* MBA thesis, California State University, San Francisco, 1972.

Rinefort, F. C., Some Aspects of the Occupational Safety and Health Act of 1970, unpublished paper, Texas A & M University, 1974.

Rinefort, F. C., *A Study of Some of the Costs and Benefits of Occupational Safety and Health in Selected Texas Industries,* Doctoral Dissertation, Texas A & M University, 1976.

Rinefort, F. C., "A New Look at Occupational Safety," *Personnel Administrator,* November 1977.

Rinefort, F. C., "The Economics of Safety," *Professional Safety,* July 1979.

Rinefort, F. C., "A New Direction for OSHA?", *Professional Safety*, November 1978.

Shafai-Sharai, Y., *An Inquiry into Factors that Might Explain Differences in Occupational Accident Experience of Similar Size Firms in the Same Industry*, Ph.D. dissertation, Michigan State University, 1971.

Sidler, H., Economic Incentives & Safety Regulations: An Analytical Framework, *American Economist*, Spring 1984.

Simonds, R. H., *The Development and Use of a Method for Estimating the Cost to Producers of Their Industrial Accidents*, Ph.D. dissertation, Northwestern University, 1949.

Simonds, R. H., and Grimaldi, J. V., *Safety Management*, 3rd Ed., Homewood, IL: Irwin, 1975.

Smith, R. S., *The Occupational Safety and Health Act*, Washington, D.C., American Enterprise Institute for Public Policy Research, 1976.

Smith, R. S., The Impact of OSHA Inspections on Manufacturing Injury Rates, *Journal of Human Resources*, Spring 1979, p. 145.

Sugden, R. and Williams, A., *The Principles of Practical Cost-Benefit Analysis*, Oxford, England, Oxford University Press, 1978.

Sykes, J. T., Find the Break-Even Point for Safety Investment, *Safety Maintenance*, **28,** February 1967.

Tarrants, W. E., *The Measurement of Safety Performance*, New York: Garland Press, 1980.

Texas Department of Health, *Occupational Illnesses and Injuries in Texas 1972 and 1973*, Austin, TX: Department of Health, Occupational Safety Division, 1974.

Thaler, R. and Rosen, S., The Value of Saving a Life: Evidence from the Labor Market in *Household Production and Consumption*, N. E. Terleckyj, ed., New York: Columbia University Press, 1976.

Thompson, M. S., *Benefit-Cost Analysis for Program Evaluation*, Beverly Hills, CA: Sage Publications, 1980.

U.S. Department of Labor and Department of Health, Education and Welfare, *The President's Report on Occupational Safety and Health*, 1972, Washington, D.C.: Government Printing Office, 1973.

U.S. National Commission on State Workmen's Compensation Laws, *The Report of the National Commission on State Workmen's Compensation Laws*, Washington, D.C., Government Printing Office, 1972.

U.S. National Commission on State Workmen's Compensation Laws, *Supplemental Studies for the National Commission on State Workmen's Compensation Laws, Volumes I and III*, Washington, D.C., Government Printing Office, 1973.

Viscusi, W. K., *Employment Hazards: An Investigation of Market Performance*, Ph.D. dissertation, Harvard University, 1976.

Viscusi, W. K., The Impact of Occupational Safety and Health Regulations, *Bell Journal of Economics*, Spring 1979, p. 117.

Walters, N. K., Safety Management Accountability Process, *Professional Safety*, August 1983.

Zeckhauser, R., Harberger, A. C., Haveman, R. H., Lynn, L. E. Jr., Niskanen, W. A., and Williams, A., *Benefit-Cost and Policy Analysis*, 1974, Chicago: Aldine, 1975.

Zeckhauser, R. and Nichols, A., The Occupational Safety and Health Administration: An Overview, *Study on Federal Regulation, Senate Document No. 96-14*, Washington, D.C., Government Printing Office, 1978.

CHAPTER 23

How to Develop Strategies to Reduce Health and Workers' Compensation Costs

BRADFORD L. BARICK
Wausau Insurance Companies
Wausau, Wisconsin

JOHN W. JONES
The St. Paul Insurance Companies
St. Paul, Minnesota

23.1 PERSPECTIVES ON HEALTH CARE COSTS

The cost for workers' compensation and health care benefits is one of the greatest management concerns. While the insured cost continues to escalate, the indirect cost soars many times higher. In an effort to conserve financial assets, employers are being forced to be more creative in their efforts to contain the costs for health care, since the conventional methods of insurance programs have not been fully effective.

The cost-containment issue has become a national crisis that involves labor, management, government, and the entire health care delivery industry. Historically, management action to control injuries, illnesses, and costs has generally been retrospective rather than prospective. This is contrary to good loss prevention principles, in that it is reactive—losses occur before there is any action. As the benefits under workers' compensation and health care have been expanded, the costs have increased dramatically. In fact, health care is recognized as being the most expensive employee benefit.

Over $300 billion was spent on health care in 1982 when the medical component of the Consumer Price Index increased by almost 12%, which was more than double the general rate of inflation for that year. This represents a three-fold increase over the $100 billion spent in 1967. The cost of health care in 1981 amounted to 9.8% of the gross national product and has since exceeded 10%. Employer's direct share of the total health care expense is 25% to 33%; the remaining expense is borne by taxpayers through the federal government (BNA, 1983). As changes continue in the tax credits and federal medical care programs, private sector employers face increased expenses for health care. If this trend continues, health care expenditures may reach $850 billion by 1990 (Figure 23.1). Double-digit increases in health care spending will continue at least through 1990 due to the *lack of* (PCC, 1984):

1. Comprehensive strategic plans for managing health care expenses.
2. Centralized responsibility for all health related programs.
3. Sufficient information regarding cost containment measures.
4. Adequate record keeping and data for management information systems.
5. Knowledge and understanding about and/or recognition of the integrated effect of health care expenses on the economy.

Before the upward trend of health care expenses can be controlled, the multiple issues and concerns must be identified and definitive control measures implemented. Traditional expenses for work related injuries and illnesses, as noted by the National Safety Council (NSC) and National Council on Compensation Insurance (NCCI), represent less than 5% of the total dollars spent for health care (Figure 23.1). Contrary to this salient factor, less than 5% of the efforts of safety and health professionals have been directed toward nonoccupational injuries and illnesses. When employees leave the safety of their workplace, they enter the high risk environment—streets, highways, homes, and recreational areas. National Safety Council statistics indicate that off-the-job injuries occur 18 times more frequently than on the job, and that there are 35 times the number of fatalities away from the workplace. As with on-the-job accidents, off-the-job injuries and illnesses also result in lost sales, downtime, and less production, in addition to the excessive expenditures for health care.

The high cost for health care is universally credited to health care providers, insurers, and government. Although numerous reasons for rising costs exist, each entity, in addition to employers and employees, has an active responsibility to contain the cost. The essence of rising cost lies in the following five factors:

1. The effects of the labor force and bargaining representative on employers for improved health care services.
2. The expanded medical benefit programs by employers attempting to remain competitive in the labor market.
3. The increased workers' compensation benefits mandated by states.
4. The response by health care providers to consumer (employees, dependents, and employers) demands for increased technology and more specialized care.
5. The response by insurers to increase premiums to cover insured costs.

Thus, for several years there has been a proliferation in the number of persons and services covered, which resulted in a higher frequency, severity, and unit cost of claims. Industry was in need of programs to contain these costs. Unfortunately, only after rampant inflation in the mid-1970s drove health care costs sharply upward was any thought given to such "new" concepts as "stress management," "wellness," and "fitness," as viable cost-containment methods.

Ironically, Dr. W. Irving Clark, a recognized pioneer in industrial medicine and the organizer of the first occupational health department for Norton Company in 1911, observed that, "The modern factory is the epitome of human effort. So much depends on the physical condition of the men employed that it

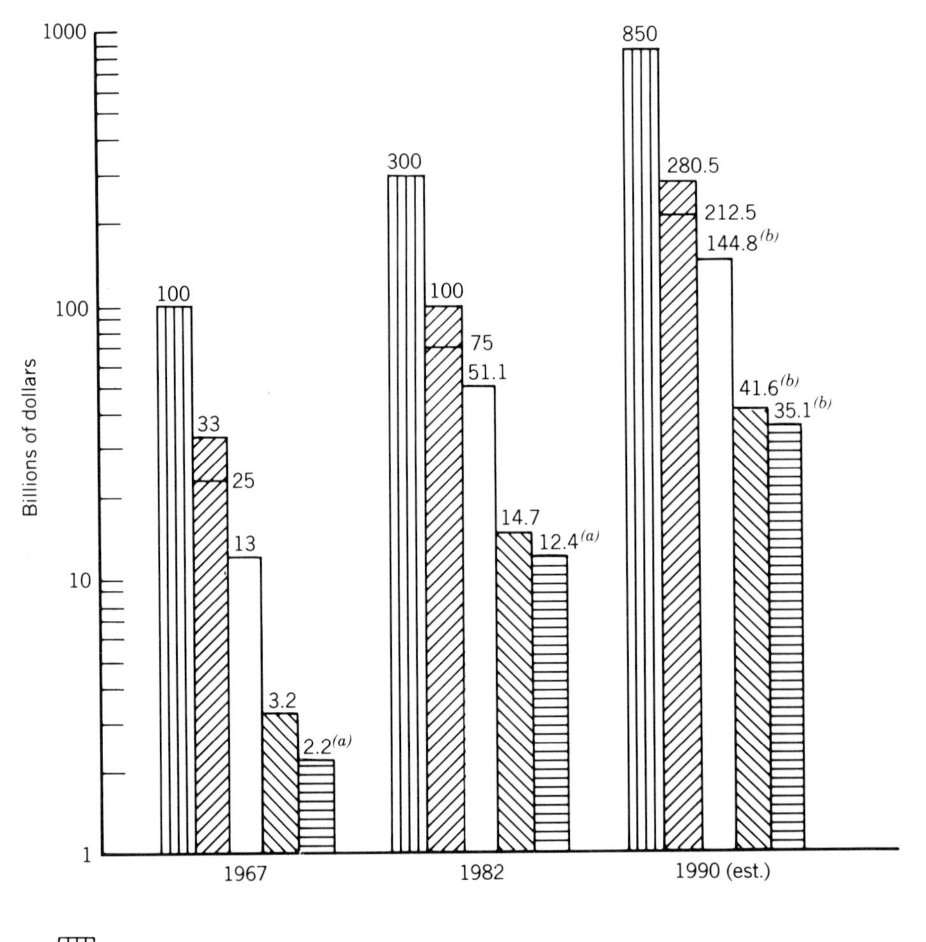

FIGURE 23.1. Comparison of health care expense.

is curious that only recently has the employee received any attention." He said, furthermore, "If we can keep the health of our men in perfect condition, we have done much for them and we naturally create a more productive and efficient force of operators."

Herein lies the answer to health care cost containment for workers' compensation and medical benefits: the achievement and maintenance of an optimal level of health and fitness for each employee, dependent, and retiree will effect a significant reduction in the frequency and severity of injuries and illnesses both on and off the job. To realize this goal of containing health care costs, three integrated objectives must be achieved:

1. Improved health status of employees and dependents through stress management, wellness, and fitness programs.
2. Prevention of injuries and illnesses, e.g. immunizations, driver improvement, physical examinations, and so on.
3. Control of medical and health care costs for those injured or ill, for example, utilization management, coalitions, financial planning, and so on.

23.2 ASSESSING THE NEED FOR HEALTH CARE COST-CONTAINMENT MEASURES

Health care cost containment is a risk management challenge that should be approached through the application of various risk management strategies. A first step in attempting to develop and implement cost containment measures is the identification and prioritization of specific risk areas for cost control and programs.

An assessment to identify the scope and magnitude of health care need and utilization would include:

1. A statistical analysis of injuries and illnesses (workers' compensation and medical benefits) to

   Identify:

   - Total number of incidents (frequency).
   - Total number of lost time claims (severity).
   - Total number of hospitalizations.
   - Total dollars paid.
   - Total dollars incurred.

   Evaluate:

   - Reporting effectiveness.
   - Claims handling procedures.
   - Health care provider utilization.
   - Medical monitoring procedures.

   Determine:

   - Cost effectiveness of treatment.
   - Rehabilitation effectiveness.
   - Causal factors.

2. An appraisal of the cost and effectiveness of prevention and control measures currently in effect relating to hiring and placement, on- and off-the-job health and safety functions and programs for employees and dependents, and post-illness/-injury procedures.
3. An evaluation of environmental, organization, and personal stresses on employees and dependents to identify high-risk persons. Higher levels of occupational and personal stress have consistently been linked to higher rates of accidents, injuries, illnesses, and corporate losses in general.
4. An inventory of employee interests in health promotion programs and their willingness to participate.

Results of the initial assessment should:

Provide information to determine what specific cost-containment measures and preventive programs are needed.

Serve as a base-line measurement to statistically validate the impact of new and/or enhanced programs on losses.

This prospective risk management approach is designed to assess the problem and then develop a plan of action to prevent, or at least minimize, those inherent risks. Goals and objectives can be established to the extent that an employer can project cost savings if its loss control goal is achieved, which will also reflect the effectiveness of the programs implemented. Conversely, employers may simply react to the cost problem and seek a quick fix. Rather than developing an effective plan to reduce the frequency, severity, and cost, the cost is merely shifted to employees or controlled through other reactive measures.

Through the assessment process, results from several components can be analyzed and correlated. The statistical loss analysis provides claims history; the appraisal of current programs in effect provide an overview of management's philosophy, policies, and organization; the evaluation of stressful conditions provides feedback from employees about their employer, job, and personal lifestyles. Thus, the basis is formed for health care cost-containment plans.

Specific cost-containment measures are discussed in the following sections of this chapter. However, when defining cost-containment measures to be implemented, three salient factors must be considered:

1. Each enterprise will be different and therefore the design strategies utilized must be tailored to fit the respective needs.
2. In establishing the cost-containment program, each component must be integrated and coordinated within the program as well as with internal and external functions and resources.
3. The development and implementation of each component must provide for significant employee involvement, communication, and education.

23.3 STATISTICAL DATA MANAGEMENT TO CONTROL COSTS

Better data management systems are required to monitor, evaluate, and control costs for health care benefits and workers' compensation. Improvement is needed from initial input to the final summary. Chronic and serious clerical errors such as improper coding of claims account for a significant portion of the often inaccurate data. "Dirty" health care data, information full of inaccuracies and omissions, is perhaps the greatest obstacle to developing reliable statistics that can be used as a basis for health care cost control.

The validity and reliability of data needs to be improved at all levels. Raw data acquired for analysis begins with initial claims reports. As the file on the claim develops, it should include detailed information on:

1. *Costs by providers:* These are payments for such services as medical procedures, medications, types of services (room and board, physician, auxiliary), and hospital costs (by average cost per day, average cost per stay).
2. *Utilization:* This should focus on the frequency of services utilized within the system. The information should include a comparison of the services and rates, those available and those used.

The lack of information to quantify and analyze the costs has been a primary obstacle to financial analysis and the development of effective management strategies. Management information systems need to be refined further to facilitate more timely and functional data for program records and monitoring, analysis and evaluation, and availability of information to management. Input for the development of such a system would involve the corporate users, health insurance plan administrators and companies, and the health care provider systems.

The resulting system for data analysis and evaluation should in effect convert accounting records into health care records. Cost containment will be effected by incorporating greater efficiencies into the system, as well as through the identification of health problems, which leads to preventive programs.

Most currently available statistical data can be meaningful if it is accurate and properly manipulated. To do so, begin the statistical analysis by determining what data and comparative factors are significant. Whether comparing workers' compensation or health care benefits data, meaningful comparative factors and a common denominator for comparison must be established.

Statistical data is the "score," and in effect, reflects the win or lose status. To analyze multivariable data such as workers compensation and/or health care benefit data:

1. *Develop and sort raw data by category:* For example, number of claims, lost time claims, paid costs, total incurred costs, causation factors, body parts affected, nature of injury/illness, utilization of information, provider costs.
2. *Determine comparative factors:* For example, premium, payroll, revenues, employees, hours, customers, units of work or items manufactured, events, National Center for Health Statistics, National Safety Council statistics.
3. *Figure raw comparative rates:* For example, number of claims/$1 million revenue, paid costs/$1 million revenue, incurred cost/$1 million revenue.

4. *Statistical analysis:* Statistically analyze each comparative variable by developing a Statistical *t*-value for each one.
5. *Statistical comparison:* Total the statistical *t*-values for each category of each unit for total comparison and ranking purposes (see Table 23.1).

Merely comparing loss ratios, number of claims, type of claims, paid cost, incurred cost, and so on, will not provide valid, reliable data because there are numerous variables. However, once a common denominator has been identified, an unlimited number of comparative factors can be viewed independently as well as collectively. This information is useful for:

1. Identification of key areas through objective ranking so that control efforts can be targeted.
2. Predicting future events.
3. Charging or crediting the share of insurance costs, loss costs, and so on to the appropriate unit.

23.4 HEALTH AND PROMOTIONAL PROGRAMS

All evidence points to personal lifestyles as the primary deciding factor concerning an individuals quality of health. Health promotion programs, which identify "wellness" as the ultimate goal, provide a long-term permanent solution for rising health care cost as well as approaches to healthier lifestyles. A very persuasive aspect of health promotion is the communication to employees and dependents of clear directions on how to improve their own health (Drury, 1983).

Wellness as the goal of health promotion programs is "a state of positive well-being for the body, mind, and spirit" (Drury, 1983). It is a way of life that is absent of injury, illness, and infirmity. Wellness is an exercise of personal responsibility for one's own health. A healthful environment leads to greater productivity and higher morale, with an increase in the quality and quantity of life.

Properly administered efficient health promotion programs will yield effective results; *efficient* means productive and cost-effective, and *effective* means that those involved achieve and maintain well-

TABLE 23.1. Summary of Statistical t-Values and Ranking for 8 Comparative Variables

Unit No.	1	2	3	4	5	6	7	8	Total	Rank
1	49.3	49.1	38.4	48.3	47.7	45.5	45.5	45.9	379.7	8
2	46.1	44.7	40.9	43.4	45.9	41.4	41.0	44.0	347.4	2
3	75.0	62.5	73.1	56.8	43.4	74.3	64.1	49.6	503.9	19
4	36.5	40.2	41.5	48.6	51.2	41.6	44.5	48.0	352.0	3
5	47.7	49.1	45.5	47.6	47.2	50.0	53.0	48.8	388.8	9
6	44.4	44.7	50.0	50.8	50.0	45.2	46.1	47.0	378.0	7
7	48.3	62.5	41.0	45.1	45.7	42.3	44.2	44.4	373.4	6
8	41.4	49.1	40.5	44.4	46.4	45.2	50.0	48.5	365.3	5
9	49.1	49.1	42.4	44.0	46.1	42.1	41.5	44.2	358.4	4
10	45.5	40.2	60.1	50.4	53.0	50.1	46.3	50.0	395.5	14
11	60.0	80.2	46.6	51.0	46.3	44.1	46.7	44.3	419.2	16
12	46.7	35.8	47.8	44.0	48.2	74.2	55.4	64.1	416.1	15
13	52.7	58.0	48.5	50.1	47.2	44.6	45.6	45.0	391.6	12
14	50.0	53.6	51.9	51.6	48.5	45.8	46.4	46.0	393.6	13
15	55.1	53.6	46.3	46.2	46.5	48.0	47.1	46.1	388.8	10
16	37.0	40.2	71.5	90.9	92.9	56.7	79.0	86.9	555.1	20
17	58.7	49.1	51.5	46.6	47.2	47.9	44.6	45.6	391.6	11
18	44.4	44.7	46.1	47.6	48.2	65.5	71.7	62.1	430.1	17
19	39.5	40.2	38.9	42.1	45.6	40.6	40.3	43.9	331.1	1
20	72.6	53.6	67.7	50.8	48.1	54.9	47.3	46.1	441.0	18

1 = Claims/200,000 work hours
2 = Claims/$100,000 revenue
3 = Paid cost/200,000 work hours
4 = Paid cost/$100,000 revenue
5 = Paid cost/claim
6 = Incurred cost/200,000 work hours
7 = Incurred cost/$100,000 revenue
8 = Incurred cost/claim

ness. Health promotion programs and terminology addressed in this chapter represent most progressive thinking, although some concepts may be contrary to traditional opinion. Health promotion and the goal of wellness is to be efficient and effective 24 hours a day. The following definitions are provided to classify and identify the scope of employer sponsored health programs more traditionally known as occupational health programs.

Occupational health: The prevention and control of disease, injury, or any condition that impairs the health and fitness of employees or their dependents to cause them to lose time from work or to work at less than full capability.

Health program: Usually referred to as an "occupational health program, this is an employer sponsored program of health promotion services and activities to address positively the health of employees and their dependents in their work and personal lives.

The logic in expanding health programs to focus on off-the-job as well as on-the-job wellness is not only humanitarian, it is cost containment. As pointed out earlier in this chapter, the cost of health care for off-the-job illness and injuries is much greater than for on-the-job incidents. Not only is the employer usually paying an excessive amount of money for insurance, incurring downtime and reduced productivity for on-the-job accidents, these losses are also created by off-the-job injuries and illnesses of employees and their dependents.

The scope of health promotion programs will vary by firm. The programs most frequently implemented include:

- Accident prevention
- Alcohol and drug abuse
- Back care
- Cardiopulmonary resuscitation and choke saver training
- First aid
- Hypertension screening
- Nutrition
- Physical fitness
- Smoking cessation
- Stress management
- Weight control

Health promotion programs have been considered adjunct functions to traditional occupational health, which usually addressed only work-related issues. The work-related health programs as well as health promotion programs have been implemented in a piecemeal fashion over time and operated as segregated entities. In perspective, wellness is a goal; employee and dependent wellness can be achieved effectively through employer sponsored health programs. The programs, whether traditionally on the job or off the job, have overlapping purposes and multiple benefits for employees and their dependents. Therefore, the management of health programs for cost containment must be integrated and the focus must be on preventive measures for the cost–benefit ratio to be positive. Basic objectives of an employer sponsored health program are to:

1. Provide guidance, motivation, encouragement, and a means for employees and dependents to achieve and maintain their optimal level of wellness.
2. Protect employees against health hazards in their work environment and personal lives.
3. Facilitate job placement and insure the suitability of individuals according to their physical capacities, mental abilities, and emotional makeup to work they can perform with an acceptable degree of efficiency and without endangering their health and safety or that of their fellow workers.
4. Provide for health screening, testing and monitoring, medical care, and rehabilitation of the injured and ill.
5. Maintain record-keeping and reporting systems to measure and evaluate the "wellness" of employees and dependents as well as to comply with legal, insurance, and company requirements.

By the definitions of wellness, occupational health, health programs, and the objective stated above, it is clear that a comprehensive approach to health programs encompasses numerous subjects, functions, and activities. A partial listing is contained in Table 23.2.

TABLE 23.2. Health Program Functions, Subjects, and Activities

Administrative
- Program objectives
- Policies and procedures
- Resources and guidelines
- Staffing
- Job responsibilities
- Facilities
- Equipment
- Supplies
- Costs
- Records
- Program audit and evaluation

Promotion
- Accident Prevention
- Aging
- Alcohol and alcoholism
- Allergies
- Arthritis
- Back injury prevention and control
- Cardiopulmonary resuscitation
- Cumulative trauma disorders
- Communicable disease
- Dental health
- Dermatitis
- Diabetes
- Digestive diseases
- Drugs and drug abuse
- Environmental evaluation and monitoring
- First aid
- Health counseling and education
- Hearing conservation
- Hypertension screening
- Immunization
- Medical emergency preparedness
- Medical examinations
- Medical monitoring and intervention
- Mental health
- Nutrition
- Physical fitness and exercise
- Prenatal health care
- Rehabilitation and return to work
- Sight conservation
- Smoking cessation
- Stress management
- Weight control

23.4.1 Health Promotion Programs in Operation

Many firms are taking the most progressive step to contain health care costs: they are setting up programs to screen for various health problems and programs to maintain as well as to improve health. In addition to health care cost-containment considerations, there are other very motivating factors for this action:

1. It is very expensive to replace corporate executives who are prematurely disabled or die.
2. Health promotion programs are effective in reducing employee absenteeism and increasing morale and productivity among many other things.
3. Prevention has proven to be the most powerful and cost-effective medicine for keeping healthy as well as containing cost for health care.

Health promotion programs will effect wellness for employees and dependent participants. As noted in Table 23.2, health programs cover a range of functions, subjects, and activities. Firms that have established some form of health promotion program with noted success include:

Baxter Travenol Laboratories, Inc.: Provides a convenient exercise facility with equipment and professional services to promote employee physical activity. Employees receive a fitness assessment, individual counseling, a program manual, and health education (India, 1984).

Burlington Industries: Focused upon a "healthy back" program designed to relieve pain and prevent further injury and time away from work.

Channing L. Bete: Uses the Parcourse Fitness Circuit to help employees and community neighbors stay in shape. It features individual exercise stations with signs posted to illustrate the correct exercise technique for each apparatus along a 1 1/3 mile trail. The four areas of fitness, with a built-in heart check system to allow participants to use their pulse as a guide to safe exercise, are (1) warm-up and stretching, (2) strengthening and toning, (3) cardiovascular conditioning, and (4) cool-down activities (National Safety Council, November 1984).

CIBA-Geigy: Developed a wellness program based on seven comprehensive communication objectives to provide employees with clear directives and incentives to improve their own health (Drury, 1983).

City of Eugene, Oregon: Eugene began a program for employees on an annual budget of $11,000. After a survey was conducted to determine what kind of fitness program would help to avoid injuries, a small fitness center was developed. The needs of workers were assessed and each was given a program to follow on a voluntary basis (Shapiro, 1984).

Continental Bank: Offers special programs every month, each aimed at a particular subject, in addition to sponsoring counseling for health functions.

Copps Corporation: Has had a health awareness fitness program in conjunction with the local YMCA. The voluntary program is open to all salaried employees. It includes a fitness test and regular physical fitness activities, stop smoking clinics, healthy back and stress management courses, a nutrition seminar and cardiopulmonary resuscitation training. Incentive awards are given and the company sponsors a monthly newsletter.

Ford: Initiated a cardiovascular risk intervention program through which extensive research has been conducted. Thousands of employees have participated in various model programs, and cafeterias have changed to offer a "heart healthful" menu.

Goodyear Tire & Rubber: Has maintained a sports fitness center for over 60 years. The philosophy is that fit employees are healthier and work better. Employees who become "fit" are most likely not to smoke, to eat properly, and to maintain weight control. Program includes newsletters, health history questionnaire, stress tests, and exercise principles (Kendall, 1985).

Fred. S. James & Co., Inc.: Sponsors the *Total Fitness* newsletter for employees and counseling services on health functions. Additionally, profit center offices offer incentives for employees to maintain health and fitness through community exercise programs and health clubs.

Johnson & Johnson: Provided health promotion facilities. Their "Live for Life" program includes resources to help individual employees determine how to improve their lifestyle.

Johnson Wax: Employed a specialist with a master's degree in exercise physiology to develop the Johnson Mutual Benefit Association (JMBA) Fitness Program. It includes scheduled classes on aerobic activities and related health functions. The medical department administers stress tests. Records are maintained on results: weight loss, pulse rate, diet, blood pressure, mileage, and so on (National Safety Council, April 1984).

Kimberly-Clark: Invested $2.5 million in facilities, including an olympic size swimming pool, a running track, various exercise areas and equipment rooms. The three-stage preventive program includes a physical exam and health risk profile, a conference to establish personal health goals, and an orientation to the facilities.

Revco: Implemented an employee exercise program designed to reduce workers' compensation costs. Warehouse workers take time out to stretch, bend, twist, and turn twice each day (Gordon, 1984).

Scherer Brothers Lumber: With only 135 employees, Scherer developed a very simple program that works. All cigarette machines and most ash trays were removed; the local Health Care Coalition was utilized as a resource to identify other basic but cost-effective programs to promote health programs at the work site.

Sentry Insurance Company: Has established a comprehensive approach to wellness for employees and family members. Flex-time allows ample time to use fitness facilities, which include the fitness center, pool, gym, racquetball courts, indoor driving range, cross-country jogging paths, and ski trails. Supervised programs and classes are provided on related health and fitness functions.

Tenneco: Developed an extensive health and fitness center including a large variety of supervised exercise and cardiovascular fitness programs.

Wausau Insurance Companies: Developed a series of twenty stretch exercises for its corporate clients called "Just Minutes a Day." It was created by Jennifer Locke, a former member of the U.S. Ski Team and a registered nurse. The philosophy is that to perform properly, workers need stretching and muscle conditioning just as athletes do.

Xerox Corporation: Provides, in addition to its exercise labs, a health maintenance program aimed at employee fitness.

23.4.2 Implementing a Health Promotion Program

Health promotion programs do not have to be elaborate and expensive, affordable only by Fortune 500 companies, to achieve effective results. There are as many alternatives as there are organizations developing health promotion programs. Programs can be conducted:

- Entirely by the organization and staffed by full-time medical, allied medical, and health and fitness staff.
- Through consultants in selected specialty areas full time or part time.
- Utilizing community resources such as the local "Y" or health and fitness clubs.
- "Co-op" with other firms in office complexes, sharing facilities and expenses.

Although the voluntary programs are most successful and, ideally, offered to all employees, programs may be restricted to certain groups by age or high risk groups. For obvious reasons, high risk groups should be identified and strongly encouraged to take an active part. Generally, employees with high risk factors are underinvolved in health promotion programs. Special efforts to bring those persons into programs will increase the potential effectiveness.

Planning objectives should be based upon data gathered as a result of the health care cost-containment assessment discussed in Section 23.2. However, as with any other program, management must establish objectives. These objectives will strongly influence the types and scope of health promotion programs offered. Pilot programs can be effective in testing program components and in estimating costs prior to full-scale implementation.

Organizations with most successful programs have several common characteristics (Brennan, 1985):

- Top management involvement and role-modeling.
- Effective planning and strong leadership.
- Concern for employees' welfare.
- High level of confidentiality along with voluntary participation.
- Services that are accessible and available.
- Programs that reflect cost-containment needs and employee interests.
- A support system to recognize and motivate employee progress in challenging ongoing programs.
- Keeping records of individual activity and progress.
- Periodic program evaluation to assess the success of the program.

Current success rate data for healthy promotion programs shows significant variations. The variations in success appear to correlate with the participants' perception of program relevance and accessibility. Gaining the acceptance and participation of employees and dependents in health promotion programs are primary considerations for success (Feldman, 1985). Programs that are easily accessible, that is, those that have been work site programs, have resulted in better participation and have been less costly than similar community-based programs (Logan, 1981). Exceptions would be programs of a highly confidential nature such as alcohol and drug treatment. However, utilization of efficient schedules, reduced waiting time, and continuity of care, facilitate program success (Feldman, 1985). Confidentiality of health and medical records for all participants, regardless of the type of program, is another important aspect. All participants must feel that confidentiality of their records is maintained and that participation in programs is their prerogative.

23.4.3 Economics of Health Promotion Programs

Health promotion programs serve as a basis to modify positively the health behavior of participants and thus improve the odds of staying healthy and productive. A multitude of health programs exist in which results show very positive trends for containing health care costs (see Table 23.3). In the cases identified, it is apparent that not only is there significant savings in health care, there are other positive results as well, for instance greater employee morale, use of program and facilities for recruiting new employees,

TABLE 23.3. Economics of Health Promotion Programs

Company or Research Sponsor	Discussion	Results
Hypertension Screening and Control:		
National Heart, Lung and Blood Institute Westinghouse Electric Company University of Michigan University of Maryland	Three demonstration projects sponsored by the Institute to evaluate effectiveness of workplace high blood pressure control programs.	Control rates ranged from 68% to 98%. By comparison, community high blood control rates are about 31.4%.
Campbell Soup Co.	Hypertension screening and treatment program. Annual cost of $300 per person for private care and $100 for on-site treatment.	Estimate annual saving for 1000-employee company would be: $12,500 from stroke mortality; $50,000 from stroke morbidity; and $90,000 from heart attacks.
Massachusetts Mutual Insurance Company	Three-year hypertension program for screening, referral, follow-up, and treatment cost $120 per hypertensive, or $92,000 for the program.	Savings generated: $56,000 in reduced absenteeism; $80,000 in direct medical care costs; reduction of 8 strokes and heart attacks per prediction based on national data.
Columbia University College of Physicians and Surgeons	710 state employees were treated for high blood pressure at a cost of approximately $2 per patient.	80% control rate was achieved (diastolic BP below 90 mmHg).
New York State[a]		
New York City	The work site treatment program to check blood pressure and monitor workers with high blood pressure.	Saved an estimated $1 million in health care costs. Only three patients out of 100 had stroke or heart problems compared to 5.4 of the general population.
Fitness Program:		
Prudential Insurance Company[b]	Five-year prospective fitness program involving 184 participants in Houston to decrease personal sick leave. The program included a medical examination and prescribed, tailored exercise. The average operational cost was $120.60.	Participants recorded a 45% reduction in major medical costs. The study indicated a 20.1% reduction in disability days and a 37.1% decrease in direct disability costs. Average combined savings per participant was $353.38.
City of Eugene, Oregon[c]	Public Works Department employees experienced most of the 46 on-the-job injuries that cost the city $40,578 in 1979. With $11,000, a fitness center was built and a jogging trail laid.	During a subsequent nine-month period of review, there were only eight claims, costing $3000.
Batelle, Inc.[d]	Offers employees a fitness program designed to improve cardiovascular efficiency, strength and flexibility.	Participating employees are absent 2.8 days per year less than nonparticipants. Battelle saved $150,000 in employee absenteeism costs.

TABLE 23.3. (*continued*)

Company or Research Sponsor	Discussion	Results
Control Data	Fitness is a part of the Stay Well Program.	Claim costs for males under 40 who exercised regularly were 39% less than for males of the same age who did not exercise.
Tenneco[c]	Fitness program includes exercise and classes on stress management, weight control, smoking cessation, back care, and nutrition.	1984 medical costs for male and female non-exercisers were $442 and $896 greater than those of male and female exercisers respectively. Absenteeism among non-exercisers increases with age, while it decreases with age for exercisers.
Comprehensive Health Promotion:		
New York Telephone Company	The program included blood pressure screening, alcohol abuse, lung disease, cancer, stress, and low back disability.	An analysis revealed a saving of over $5.5 million per year in the cost of illness and a net gain of $2.7 million.
Cancer:		
Campbell Soup Company	Instituted a colon-rectal cancer prevention program.	An estimated saving is $200,000 per 1000 employees each ten years.
Weight Control:		
Control Data	Stay Well was initiated in 1979 and very extensive evaluations of health care claims have been developed.	Claim cost for males over 40 who were 30% or more over their ideal weight, was 39% greater than for males of the same age who were within 20% of their ideal weight.
Back Care:		
Burlington	The back care program included counseling, back exercises, and job instruction to prevent stress on the back. About one-third of the 300 employees had experienced back problems resulting in over 400 days of absenteeism.	After one year, back problems were reduced 95% and the same employees had only 19 days of absenteeism.
Employee Assistance Programs (EAP):		
Kennicott Copper	Implemented an EAP for employees.	Estimated a six-to-one benefit/cost ratio per year.
Phillips Petroleum Company[a]	Implemented an EAP for employees.	Reported savings of $8 million a year in reduced accidents, less sick leave, and higher productivity.

TABLE 23.3. (*continued*)

Company or Research Sponsor	Discussion	Results
Kimberly-Clark[a]	Sampled employees who participated in a corporate drug and alcohol abuse program for one year.	Experienced a 43% reduction in absenteeism and 70% fewer accidents.
Stop Smoking Programs:		
Control Data Corporation[a]	Over a life time, a one-pack-plus per day smoker may cost the employer $335 to $600 per year in extra expenses than a nonsmoker.	Smokers experience 25% more claims and 114% longer hospital stays than non-smokers.

[a]Polakoff, 1985.
[b]Boune, 1984.
[c]Shapiro, 1984.
[d]Brennan, 1985.

reduced absenteeism, increased productivity, improved self-image, increased self-esteem and self-confidence. As people become more involved in the health programs, they start to take charge of their own lives and begin to feel they are in control.

23.5 CORPORATE STRESS MANAGEMENT FOR COST CONTAINMENT

Stress is a costly business expense, affecting both employee health and company profits. However, companies can reduce stress and its effects through comprehensive "human factors" risk management programs.

Consider these stress facts gleaned from various safety and insurance industry research:

- In 1982, the total cost (direct and "hidden") of work-related accidents in the United States alone was $32 billion.
- The cause of about 75–80% of all industrial accidents is an inability to cope with emotional stress and conflict.
- Psychological or psychosomatic problems contribute to over 60% of long-term employee disability cases.
- About 11% of all workers' compensation claims relate to stress-caused emotional illness, and the number is rapidly escalating.

But what is stress? By definition, stress is the adverse emotional and physical reactions to any source of pressure in one's environment. Stress reactions negatively affect personal health and organizational effectiveness and create losses (see Table 23.4).

Employees continually confront various pressures or "stressors." They experience stress if they are unable to cope effectively with such stressors as poor management, lack of job security, work overload, excessive deadline pressure, unrealistic expectations, insufficient pay, and uncertainty about job duties and responsibilities. Organizations must learn to identify and control employee stress in order to control costs.

23.5.1 Stress and Accidents

Injuries and related illness are a major cost source for employers. It follows that stress is a leading cause of employee injuries and illnesses. The personal and organizational effects of occupational stress are illustrated in Table 23.4. For example, stressed employees may have one or more of several symptoms predisposing them to accidents: fatigue, poor judgment, impaired physical coordination, inattentiveness, distorted perception, indecision, and intoxication from alcohol or drug abuse. These stress reactions can contribute to increased risk-taking by employees.

Stressed employees are also more susceptible to overexertion injury. Dr. Hans Kraus studied the relationship of emotional stress to back pain and injury. His research revealed that no organic disease,

TABLE 23.4. Personal and Organizational Effects of Occupational Stress

Personal	
Alcoholism	Anxiety
Drug abuse	Psychosomatic diseases
Emotional instability	Eating disorders
Lack of self-control	Boredom
Fatigue	Mental illness
Marital problems	Suicide
Depression	Health breakdowns (cardiovascular, etc.)
Insomnia	Irresponsibility
Insecurity	Violence
Frustration	
Organizational	
Accidents	Inflated health care costs
Thefts	Unpreparedness
Reduced productivity	Lack of creativity
High turnover	Increased sick leave
Increased errors	Premature retirement
Absenteeism	Organizational breakdown
Disability payments	Disloyalty
Sabotage	Job dissatisfaction
Damage and waste	Poor decisions
Replacement costs	Antagonistic group action

such as a ruptured disc, could be found in 80% of the back patients studied. Yet, many patients were experiencing increased muscle tension caused by emotional stress.

Tense muscles lose their suppleness, become tighter, and constrict, thus becoming more susceptible to injury. Dr. Kraus's patients suffered back pain because they did not know how to relieve their muscle tension effectively through stress management and exercise. The case study presented below illustrates the stress-accident connection.

CASE STUDY

The nurse was frustrated. She had too much work and too little time. Her station had been understaffed for two weeks, but this day seemed particularly hectic. She had just one more patient to attend before her break. Not wanting to wait for help from an aide, she impatiently decided to take a shortcut and move the last patient, an older woman, by herself.

The patient was fairly mobile, and everything was going fine as the nurse helped the patient move from the bed to the wheelchair. But, as she lifted the woman, she felt a sharp pain in her back. Her work day ended right there, and her back injury kept her from work for weeks. The medical treatment and time off for the injured nurse added to the hospital's workers' compensation statistics.

23.5.3 Breaking the Distress Cycle

The major goal of a stress management program is to help employers interrupt what is called the *distress cycle.* Such action would provide a method to *actively* contain health care and insurance costs.

Figure 23.2 illustrates how this damaging cycle evolves. Research by St. Paul Fire and Marine Insurance Co. has shown that there are two basic approaches to breaking the distress cycle. One is to identify and to modify the stressors, the other is to increase an employee's ability to cope with stress. The methods can be used individually or in combination.

For example, organizational stressors can be identified and corrected. Consider one production unit with a very high stress level, a high number of accidents, and lost productivity. A series of personnel department interviews showed that poorly defined job responsibilities caused stress in the unit members. After each person's job was analyzed and defined, production increased and accidents stopped. The interviews also revealed other stressors that needed controlling, including poor communications, undefined pay raise systems, and employee drug abuse.

The second way to break the distress cycle—increasing the ability to cope—consists of the more commonly known stress-control techniques such as physical fitness programs, relaxation techniques, biofeedback, weight loss, drug and/or alcohol rehabilitation, and periodic physical examinations.

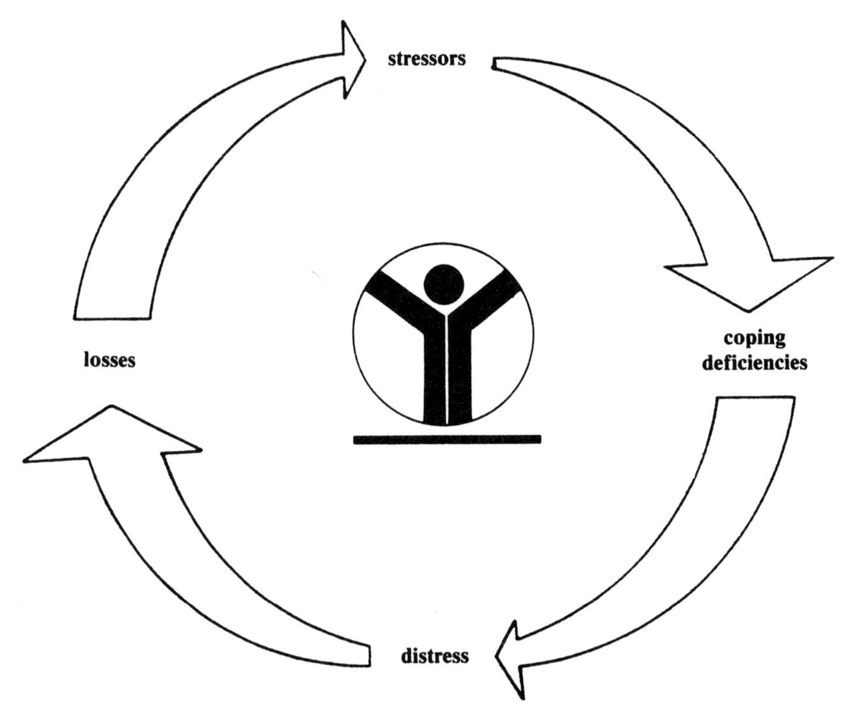

FIGURE 23.2. The distress cycle.

These techniques are not intended to alter stressors, but to increase an individual's ability to cope with stressors in their environment.

23.5.3 Assessing Occupational Stress

To better control stress-related losses in industry, employers must periodically use stress surveys that assess: (1) the organizational stressors that cause stressful, pressured work environments and (2) the organizational—and employee—coping skills that can serve as "stress buffers."

The St. Paul Insurance companies have developed a new approach in making these assessments. The assessment comes from a comprehensive organizational survey, the Human Factors Inventory (HFI). The HFI examines employee attitudes in seven risk areas:

- Job stress
- Job dissatisfaction
- Organizational stress
- Personal stress
- Accident risks
- Health and life risks
- Technostress

The 30-minute HFI is given to all company employees. Participation is both anonymous and voluntary. The survey results are computer scored and compared to a national norm of thousands of workers. An organizational "stress quotient" is then computed for each employer. This comparison allows the organization to determine if their employees are above or below average in terms of their stress reactions and coping skills. The inventory also indicates in which jobs or departments employees are experiencing the most stress.

The second step is to assess the range and effectiveness of the company's health and stress management services. In St. Paul's approach, this is done by administering the Human Factors Audit (HFA) one to three key management executives. The HFA is a 200-item checklist of organization health and stress management programs that examines programs, their effectiveness, and the program utilization rate.

These two surveys reveal: (1) the type, level, and location of stressors in the organization; and (2) the coping skills and resources currently in place to manage them. By using this survey information with corporate loss data and taking available resources into consideration, truly effective decisions can be made. Opportunities for improvement can be prioritized and programs can be selected based on a thorough assessment of the subject organization's true human resource needs.

To summarize, the St. Paul approach is based upon the following steps:

1. *Employee assessment:* Conduct an organizational survey of employees' emotional, attitudinal, and physical reactions to all types of personal, job, and organizational stressors. Their scores are compared to national norms in order to yield a "stress quotient."
2. *Program review:* Compile a list of all employer health and stress-management services currently available within the organization, the perceived effectiveness of each program, and the level of current employee participation.
3. *Analyze data:* Using the information above, assess both the type and level of organizational stressors and employee coping skills. Computer analyze this information by job type, by department, and by company location.
4. *Prepare report:* From the analyses, prioritize the organization's improvement options, target the job levels and departments that are at highest stress levels, and select the appropriate programs to address those needs.

This approach has been field-tested by a large number of companies ranging in size from 50 to over 6000 employees. A summary of the ways in which organizational stress profiles may be used to control health and workers' compensation costs is provided below.

1. *Focus efforts:* Employee groups at greatest risk to have stress-related accidents, injuries, or illnesses are identified. Some possible solutions to their situation are provided. Employers then direct their training and development dollars to where the need is greatest.
2. *Pinpoint strengths and weaknesses:* Employers get a clear picture of how well the employees and managers are coping with stress compared to a national norm group. Employers determine whether certain jobs or departments experience more or less stress than others. They can see if important human factors, such as job stress and employee wellness, cause their employees to be more susceptible to accidents, illness, poor productivity, and premature death.
3. *Program review:* Employers receive an evaluation of their organization health and stress-management services so they can build on their strengths and work on any areas in need of improvement.
4. *Create awareness:* Just by administering the HFI, management conveys to employees their interest in improving the quality of their work life. In turn, employees become more motivated to manage stress and seek wellness in their lives.
5. *Employee involvement:* The HFI opens up an invaluable communication channel between all levels of employees and management. Such employee involvement leads to improved morale, especially when employees see that their input helped to facilitate new human factors programs.
6. *Evaluate progress:* Results presented in one year's human factors profile can be compared with the results of future employee profiles to develop a clear measurement of progress. Study after study indicates that a reduction in employee and organizational stress, followed by an increase in both job satisfaction and employee wellness, should lead to a decrease in medical claims and accidents, illness, turnover and absenteeism, theft, sabotage, and poor productivity. Such decreases should be reflected in improved employee morale, better organizational efficiency, and higher profitability.
7. *Prevention:* Finally, the HFI can be used to identify potential human factor loss areas before they cause any significant level of loss.

23.5.4 Cost-Containment Programs

Accurately assessing stressors is the first step in controlling stress-related losses. The critical step, however, is the implementation of human factors loss control programs to control or actually prevent stress-related losses.

What should a good human factors loss control program look like? Sample program components are illustrated in Figure 23.3. This model program includes five general areas: (1) management and administration, (2) preemployment screening, (3) current employee risk management, (4) injured employee management, and (5) program evaluation.

First, management must be committed to address those human factors that cause losses. Personnel must be assigned to oversee human factors program activities and the program must have a reasonable budget to meet its objectives.

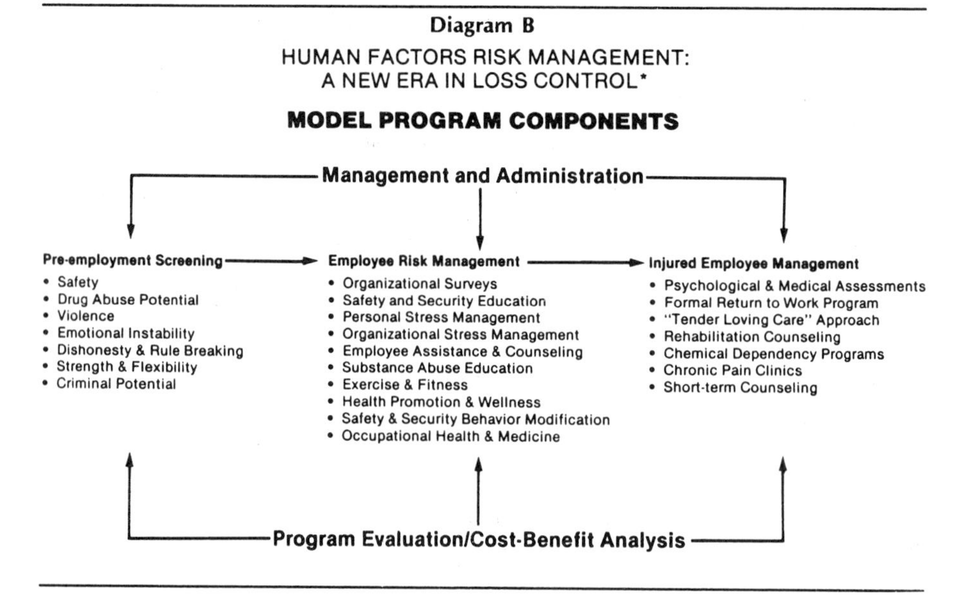

FIGURE 23.3. Human factors risk management: a new era in loss control. Model program components. Adapted from an invited address made by Dr. John W. Jones at the 72nd Annual National Safety Congress, Conrad Hilton, Chicago, October 15–18, 1984, "New Directions in Safety and Health."

Preemployment screening is the next part of the model. Research shows that employers tend to have fewer losses if they screen job applicants for tendencies toward theft, violence, drug abuse, and accident proneness. Preemployment screening reduces organizational stress by facilitating the hiring of honest, safe, and emotionally stable employees.

Human factors risk management programs for current employees range from organizational stress management to employee assistance and counseling to occupational health and medicine. For example, managers must learn how to identify and control work situations that cause employees excessive stress. Companies can begin to implement employee assistance and counseling programs that are designed to help troubled employees (e.g., alcoholics, drug addicts, emotionally unstable workers, etc.) and their families to receive professional counseling for their problems. Furthermore, occupational health and medical programs can save lives by offering such services as blood pressure screening, first aid, and periodic health checkups.

All the programs listed below can save money in health care and insurance costs. For example, companies can use Table 23.5 to estimate how much alcoholism and problem drinking costs their company. Yet research has shown that Employee Assistance Programs (EAPs) can help to control this cost source and others related to chronic employee distress. Empirical research on EAPs shows that they can reduce tardiness, absenteeism, visits to in-plant medical facilities, sick leave, accidents on the job, and use of workers' compensation and disability insurance. One such study is summarized in Table 23.6.

23.5.5 Management of Injured or Ill Employees

Employers can control the cost of worker disabilities better by improving their management of injured employees. Injured employees are exposed to tremendous amounts of stress, both personal and work-related. The major goal of injured-employee-management programs is a safe, prompt, and permanent return to work for the injured employee. A return to work plan should consist of the following activities:

- Complete an on-site job analysis of the position to which the injured employee will return, listing the job's physical demands.
- Obtain the treating physician's evaluation of the injured employee, outlining the employee's current physical abilities.
- Compare the employee's job analysis with his or her functional capacities. Job modification may be necessary to ensure that the employee's physical abilities meet the job's physical demands. Any change in a job's hours, duties, or expectations is considered job modification.

TABLE 23.5. Cost Analysis for Employee Alcohol Misuse

1. ______ Total # of Employees × 5%–10% = ______ Probable # of employees with alcohol problem

An estimated 5–10% of the nation's workforce suffers from alcoholism.

2. $ ______ Average Yearly Income × 24.8% Cost Efficiency Reduction = $ ______ Loss per alcoholic employee per year

The highest executive is as likely to develop a drinking problem as the lowest-paid laborer. Therefore, be sure to average the yearly income of all employees. The cost efficiency reduction figure is based on research that indicates an employee's alcohol problem will cost the employer an average of 24.8% of his or her yearly income. That loss results from bad decisions, absenteeism, on-the-job accidents, mistakes, wasted material, cost of grievance procedures, and decreased job performance in general.

3. $ ______ Loss per alcoholic employee year × ______ Probable # of alcoholic emp. = ______ Total estimated loss per year

4. 70% Recovery rate average × ______ Total estimated loss per year = ______ Estimated cost reduction per year

Research indicates that 60–80% of the alcoholic employees identified under an early identification program will recover and bring job performance back to normal.

Source: National Institute on Alcohol Abuse and Alcoholism, New Hampshire Municipal Association.

TABLE 23.6. Comparison of Employee Job Performance Indicators Before and After Contacting the EAP
(*N* = 60 Distressed Employees)

Number of:	1 Year Before EAP	1 Year After EAP	Amount of Change
Times tardy for work	231	148	−83 times
Times left work early	494	255	−239 times
Days absent from work	1,327.5	684	−643.5 times
Visits to in-plant medical facility	101	77	−24 times
On the job injuries			
Lost time	8	5	−3 injuries
No lost time	52	51	−1 injury
Times used disability insurance	29	20	−9 times
Grievances	6	15	+9 grievances

- Some job modifications will be permanent, others temporary, depending on the functional capacities of the employee. A modification that is temporary with a gradual resumption of hours, duties, or expectations is called "work hardening." An example of work hardening is allowing an employee to return to work four hours a day during the first week, five hours a day the second week, and gradually increase to a full work week. This process reduces stress and maximizes the potential for a successful return to work, especially for employees who have been off work for more than two months or who have sustained a severe injury.
- Consider internal transfer if job modification or work hardening is not feasible in the employee's former department. Transfers demonstrate management's total commitment to loss control and encourage supervisors to accommodate the return of recuperating employees.

- Consider light duty workstations as a last resort to help the employee remain in the work environment. This is important physically, because it keeps the worker active, and psychologically, because it reinforces a "wellness" role.
- Use a "tender loving care" approach. Encourage coworkers and supervisors to contact the injured employee during the recuperative period. Express sincere concern with the employee's welfare and earliest possible recovery.

Rehabilitation specialists can also improve injured-employee management. A rehabilitation specialist may be needed to work with an employer to help alleviate the employee's fears or to confront an employee who has become "comfortable" on workers' compensation benefits.

A rehabilitation specialist is a trained professional with a degree in a discipline such as nursing or vocational rehabilitation. The specialist works as a "teammate" with the employer, the employee, and the insurance company to help return the injured worker to a productive role. The specialist offers expertise in:

- Communicating with medical professionals to obtain treatment goals and return-to-work dates.
- Supervising home therapy programs to insure follow-through by the insured employee.
- Initiating vocational testing if a transfer is necessary to accommodate an injured employee.
- Coordinating short-term training, if needed, to help the injured employee acquire new skills.
- Referring injured employees to chemical dependency programs, chronic pain clinics, or short-term counseling, if needed.

A rehabilitation specialist has strong counseling skills and can provide both support and an occasional "push" to keep the injured person active in the rehabilitation process. At times, an employee may want to *be* rehabilitated, rather than actively participate in his or her own recovery. Because the injured person must be an enthusiastic participant in the rehabilitation process, a coach is sometimes needed to motivate the injured workers. The rehabilitation specialist can serve as that coach.

23.6 HEALTH CARE BENEFIT PLANS TO CONTAIN COSTS

Everyone would agree that something must be done about the dramatic rise in the cost of health care. One way to decrease health care cost is to reduce the amount of service rendered without curbing access to necessary care. This can be accomplished through wellness programs. However, the most popular strategy to contain health care costs has been to reduce costs at the point of delivery through competition, volume purchasing, or unique benefit configurations. This latter approach is basically a restructuring of health care benefits, which tends merely to shift the costs rather than reduce them. Historically, once the savings potential had been exhausted, the costs began to spiral upward again. To implement a permanent solution for rising health care costs, the frequency and severity of injuries and illnesses must be reduced and this effort integrated with the most cost-effective health care benefit plans.

A wide variety of health care benefit design changes for cost containment have been implemented since 1983. There are also numerous combinations of the plans, which are usually tailored to the financial requirements of the specific organization. In establishing a health care benefit plan, two major categories must be considered:

1. *Plan design:* Most of the plan dollars (85–90%) are distributed as health care benefits. It is vitally important that the program design is such that it accomplishes the intent of the organization while at the same time controls costs. The proper design will educate and encourage participants to use prudent judgment and low cost alternatives when choices exist.
2. *Plan funding:* To achieve optimum cash flow, it is necessary that the plan provide required payments only when they must be made. However, there are situations that demand funding flexibility. To establish a funding plan, funding objectives must be reviewed. Once the plan is operational, there should be periodic review to assure it is operating efficiently and as intended. Annual audits and reports reviewing the financial operation of the plan are used for verification purposes.

In the past few years a major shift in attitude in the design of health care benefits has occurred, in addition to the drastic change toward self-funding. Many employers have found that the typical prepackaged insurance product can be retailored to suit the needs of their own employees better. The movement towards "tailor-making" one's own benefit plan has been supported by evidence proving that benefit design affects utilization patterns and unnecessary medical and dental services. This, in addition to escalating costs, has been the incentive for employers to seek more effective plan design.

In developing the most effective plan design, there are numerous difficult questions to answer including:

- What is the most effective plan design?
- What is the most cost-effective way to administer health care benefits?
- Will dollar savings in a program compensate for possible employee animosity?
- Will there be an impact on the quality of health care?
- Will the plan be affordable in its present configuration five or ten years from now?

Various techniques and elements of plan design are included in the following discussion. All the techniques and elements of plan design have some value or potential for cost containment. However, the potential for cost containment is dependent upon the way variables that go into developing a comprehensive plan and integrating the plan design into the organization.

23.6.1 Cost Containment Through Plan Design

Outpatient surgery: Provide coverage for free-standing surgical facilities and consider limiting reimbursement for waiving deductibles for surgery performed on an inpatient basis that could be performed without a hospital stay.

It is possible to identify the surgical procedures that doctors feel can be performed on outpatient basis, resulting in savings of $300 to $500 on each surgery.

Second-opinion surgery: This can be provided for all surgery, elective surgery only, or to the 15 or 20 surgical procedures most frequently overutilized. In effect, the second-opinion (or third opinion) is provided at no out-of-pocket expense to the employee to determine if an operation is required. In some plans, failure to obtain a second opinion results in a lower surgical fee reimbursement.

"Focused" deductible: Deductibles apply only to the conditions or procedures that are most often abused or overutilized.

Outpatient lab, X-ray and other outpatient services: Try to encourage use of outpatient services rather than inpatient services through deductibles and/or coinsurance.

Provide for preventive care: Provide "limited" benefits for physical exams, immunizations, and so on, attempting to reduce or eliminate treatment costs.

Flexible benefit or "cafeteria" plans: Develop plans and offer options that will allow changes in plan design that are cost effective and structured to reduce utilization.

Preadmission testing: A program designed to eliminate unnecessary hospital admissions as well as admissions testing only one day in advance of surgery.

"Stay-well" plans: Plans designed to control rising health costs through the employee by discouraging unnecessary health care. Some programs are based upon educating the employee, disease prevention, risk assessment, and promoting good health.

Generic drugs: A program designed to provide incentive to purchase generic drugs rather than "brand name" drugs in order to reduce cost.

Cost shifting: Complete redesign usually leads to shifting cost away from the Plan to the employees. This is generally more palatable than increasing an employee's contribution, avoids selection problems, and reduces utilization and abuse.

Other programs are designed to create a direct incentive—cash—to control utilization. Emerging evidence indicates that incentive plans may lead to underuse of needed care, which can offset the savings from overutilization and unnecessary health care. Consideration of such appropriate preventive health measures *i.e.*, as immunizations, Pap smear, and so on, are usually recommended. This would be compatible with the concept behind stay-well plans.

It is obviously difficult to quantify actual dollar savings, particularly in areas of increased well-being, improved work attitudes, and better family relations. New York Telephone estimated that nine wellness programs saved them almost $3 million in employee absences and treatment costs alone. The biggest savings program was in smoking cessation, which brought about reductions in coronary diseases. Disease prevention and health promotion programs offer real long-term savings and are not merely cost-shifting arrangements.

23.6.2 Cost Containment Through Administration

The following plans and programs involve administrative changes with only limited impact on plan design:

Concurrent Hospital Review Program

Contract with an agency providing in-hospital concurrent review of the necessity of admission and length of stay. Programs utilizing trained personnel *not* employed by the hospital are showing savings in reduced length of stay four to ten times the program cost. These programs also reveal unnecessary admissions for minor surgery that should be performed outpatient.

Health Counseling or Employee Assistance Plans

Establish a program in the Personnel department to deal primarily with substance abuse problems or any other sickness that affects on-the-job behavior. Substantial reductions in absenteeism, turnover, and wasted time can be achieved. Increased and improved morale can also be achieved. Such plans replace ineffective punitive disciplinary measures with more effective problem-solving approaches and demonstrate a general concern for the welfare of an employee and his or her family.

Professional Standards Review Organizations

These organizations, which were created to review programs for federal patients, are becoming available to private companies. The services provided are:

Concurrent review.

Medical care evaluation studies.

Retrospective profiling.

Preadmission certification.

Concurrent review is the most common program offered by PSROs. Medical experts review patterns while in the hospital aimed toward suggesting earlier discharges when medically appropriate.

Medical care evaluation studies are generally specific studies to compare the quality of care provided and to compare differences between hospitals and providers.

Retrospective profiling is usually a study of discharges after the care has been rendered and the bills have been paid.

Preadmission certification requires that, for certain elective procedures, the hospitalization is approved in advance. This holds much promise for saving money because it can prevent unnecessary admissions. Less than one fourth of PSROs provide such a program.

Return-to-Work Evaluations

In some areas, arrangements can be made for setting up a pool of doctors to settle disputes between carrier or employer doctors, claim investigation programs, and the treating doctor regarding return-to-work dates. The purpose is to avoid disputes, while encouraging earlier return to work.

Data Collection System For Cost Analysis

Few administrators can provide such basic data as average hospital stays for the common diagnoses and procedures. If the suggested cost-containment measures are to be cost justified, an adequate database should be considered. Within the past few years computer programs have been designed to provide information about such things as DRGs (diagnostic related groupings), allowing one to measure how much care is being provided to employees for what kind of services, and how this profile of services compares to national and regional standards. This kind of data enables one to pinpoint areas of overutilization by medical specialty, and number of hospital days that exceed the norm.

Health Maintenance Organizations (HMOs)

HMOs are usually viewed as the traditional way to cut health care costs. Comprehensive health care is provided for a fixed, prepaid fee. Through the structure, an HMO can provide the delivery of high-quality, appropriate, and efficient medical care. The success of an HMO is predicated on the members' understanding of how the delivery system works as well as the convenience and availability to the services. Members can be educated about health maintenance as the plan is utilized. Cost-containment features of HMOs were initially most progressive, with an emphasis on wellness or prevention of illness. HMO programs include such options as second opinions, coverage for extended care facilities, and home health care services to control costs to employers and subscribers.

Preferred Provider Organizations (PPOs)

Recently, several new arrangements have been attempted. Employers have sought and obtained set-price and even discount arrangements from providers. Panels of physicians and/or hospitals have been established, where the employee can receive a financial incentive for using panel doctors. Such organizations of preferred providers are emerging in several varieties and under various auspices. It is likely that employers will develop even greater varieties of panels in the next few years, including options where companies provide direct care in some situations.

Coalitions

Another area for employer action is the rapidly growing coalition movement, which has come into national prominence within the last three years. There are currently over 60 coalitions, with estimates that the number will double during the next two years.

These groups offer the opportunity to create a critical mass of employers, large enough so that health care providers will have to deal with them. In any given area, few employers account for more than 1–2%

of hospital admissions. In such a situation, it is not likely that a hospital will meet with or negotiate with employers individually. However, if 15 or 20 employers account for 20% or more of hospital admissions, it is likely to get the attention of the hospitals, the medical society, and the individual providers.

In many ways, coalitions can do more expeditiously many of the things that individual employers attempt to do by themselves. There are advantages in combining expertise of employers. Coalition members can look into different problem areas and explore different opportunities for new program developments.

Combined Comp With Group Health Plans

This is a hybrid program through which workers' compensation and group health coverages are packaged in an integrated program, which allows more efficient and effective administration for claims, coordination of benefits, and injury and illness monitoring for on- and off-the-job claims. It is an effective cost-containment measure and will save employers money through lower personnel administrative cost, reduction of duplicate claims, and streamlined claims processing.

Although the concept of combined comp with group health is attractive, the programs in effect are too new to track actual cost savings. Likewise, how well the programs actually work has not been determined. Claims management that relies on sophisticated data collection and analysis is the primary purpose of this new plan.

Prorated Group Health Plan

This is a plan that is not known to be in use at the present time. The concept of the plan is that the employee's share of plan cost would be prorated, based on individual health risk factors or level of health, by employee and dependent. The same or similar guides used for underwriting individual health plans can be modified to determine the "fair share" for each employee and dependent. This would effect cost shifting, in which the heaviest users and those with the highest risk factors would pay more into the plan than those with low risk factors and those who used the plan least. Under the current system there is usually a standard cost for single and family coverages.

The system should encourage and support those high risk participants to do whatever is practical to reduce the inherent risk factors, for example, reduce blood pressure, reduce weight, and/or stop smoking. Such a system would automatically reward those with low risk factors and those who lowered their risk factors. The parameters for evaluation are currently available. Health history questionnaires, preplacement physical examinations, and other elements of health screening programs could easily be implemented to acquire the information needed to rate each individual accordingly.

23.6.3 Communicating About the Plan

The participants' understanding of the health care benefits is fundamental to the success of the plan. This is dramatically important when considering cost containment. An effective communication plan is an essential element for any cost-containment program to be successfully implemented without adverse employee reactions. Employers who have done an effective job in communicating changes in the health care program have encountered less resistance to the changes from employees. Increased deductibles and second surgical opinion programs were the most frequently instituted cost-containment changes. However, the changes that employers felt were most effective in controlling health care costs were the ones that employees found least acceptable, since the changes shifted costs to employees and resulted in immediate savings for the employer (Kittrell, 1985).

The communication program must accomplish at least two things. First, employees must understand that any plan changes are not reductions in coverages, or "take-aways," but rather, the cost containment plan changes are meant to encourage the delivery of health care benefits more efficiently and economically—and that the changes are in their best long-term interest.

Second, a health education program for employees and their dependents on health care financing, health care delivery, and other health care postures must be implemented. This aspect of the communication program should be combined with the "stay-well" concept.

Effective communication is necessary to educate plan participants properly to enable reduction or prevention of excessive costs. Through proper communications, employees can be introduced to a specific action plan to utilize benefits in a cost-effective way. For example, plan features such as second opinions for elective surgery, same-day surgery, and home care may be utilized to provide more convenient and thorough care, while at the same time bringing a lower cost to the plan. Employee meetings, payroll stuffers, bulletin-board posters, descriptive brochures, and benefit statements are all ways of communicating health care benefits and procedures.

23.6.4 How to Proceed

Two major changes are taking place in the financing of health care. The first is that companies are increasingly self-insuring. This has been the major thrust in the last decade. However, no matter how innovative or effective the funding techniques are, they will only impact on 5–10% of the total cost of your benefit plan. Controlling the amount of benefit payments through an effective cost containment plan provides the greatest source of savings, since 85–90% of your health plan cost is generated by claims.

Continuous consideration of a successful program for cost containment is salient. In order to teach new concepts, instill new attitudes, new lifestyles, and new responsibilities, a hit-or-miss approach will not work. A single benefit statement, booklet, meeting, and so on, is not enough. The employer and service organization must:

- Determine what resources can be committed, both in money and people.
- Identify the internal and community resources to draw upon.
- Establish a communications plan with a theme based upon employees' needs, perceptions, and values.
- Have management endorsement and commitment.
- Encourage employee involvement in both the planning and implementation phases.
- Encourage feedback. One-way communication is not as effective as two-way communication.

The range of cost containment is extensive. Health care costs can be managed effectively without impacting on the quality of health care delivered to employees and their families.

23.6.5 Claims Management

Administering a benefit program involves direct, continuous contact with employees. The appreciation of employees and the credibility of the plan depend upon timely and equitable consideration of the needs of employees and their dependents. At the same time, the financial commitment involved in these plans may be substantial enough to warrant the most careful possible review of how efficiently the money involved is spent. The emphasis on cost containment and claims control is vital to the success of the program.

The objective of providing a good benefit program for employees, while maintaining strict financial controls, presents a challenge to an employer whether insured or self-funded. Unless the employer is very large and has its own risk management/insurance department, it is common practice to purchase various services from the insurance carrier, third party administrator, or consultant. Services provided may include:

- Analysis of existing health care benefit program and advice on how to contain cost with respect to adequacy of the program, volume and patterns of utilization, and effective application of any coordination of benefits provisions.
- Evaluation of claims and claim administrative procedures to identify current and potential problem areas and to establish most effective administration procedures for sick leave programs, short- and long-term disability income plans, and medical reimbursement plans.
- Review of health care benefit delivery system and the development of administrative procedures and standards relative to the system for direct claim submissions by employees, employer certification and submissions, self-administered operations, or centralized control of multilocations.

Health care benefit plan design and administration procedures is only part of the answer to containing health care costs. Whereas health care benefits offer an avenue to reduce costs, it is likely that the reduction may be only temporary and there is no universal consensus on the most effective method, technique, or plan to reduce health care costs.

23.7 CONCLUSIONS

Cost containment requires a concerted effort by all involved parties, including health care providers, business and industry, insurance carriers, and government, as well as employees and their dependents. However, viewing the cost-containment problem pragmatically, employers are in the best position to influence all other parties, because the employer is the one buying and paying for health care related insurance and services. Therefore, organizations need to evaluate the effectiveness of all their programs and services.

The employer should exercise the options and evaluate the numerous funding alternatives in terms of design and administration for health care as well as workers' compensation. The short- and long-term financial impact of health care on the organization needs to be considered carefully. Whereas the per unit cost for health care may continue to increase due to medical and technological advances, cost containment will be achieved through a reduction in frequency and severity of injuries and illnesses both on and off the job.

For years loss control programs and health promotion programs have been "invisible profit centers," that is, they have saved organizations money. Yet these cost savings have not been attributed to the respective programs. Cost-benefit analysis indicates that positive measured benefits far outweigh the cost of prevention programs. Furthermore, the time has come to reinforce the proven capability of individuals to be responsible for their own health. For example, there have been significant gains in the reduction of cigarette smoking, people have recognized the link between diet and heart disease and thus consume significantly less fat, and fitness and weight control are "in." The paradox is that most Americans are born healthy; it is through misconduct and environmental conditions that premature death and disability occur.

Compromising factors are likely to impact cost-containment efforts. First, the point of diminishing returns is very close in terms of costs for workers' compensation and health care benefit plans. Second, the fact remains that injuries and illness will not be eliminated. Third, the population is aging, with more people living longer and requiring the use of high-cost technological equipment and facilities.

Lifestyle and behavior change is believed by many professionals to be the most important strategy to improve health and to contain health care costs. Lifestyle and behavior changes can be influenced by an organization supporting effective health promotion programs. The savings generated from plan design and funding changes must be reinvested in health promotion and risk reduction programs to negate future escalation of health care costs. Therefore, creative funding alternatives integrated with comprehensive health programs and health care provider price controls will collectively yield the most effective cost-containment results.

REFERENCES

Benedict, K. T., A Case History, Industrial Health Protection—From 1911 to OSHAct, *National Safety News*, June 1975, p. 115.

Bowne, D. W., Reduced Disability and Health Care Cost in an Industrial Fitness Program, *Journal of Occupational Medicine*, November, 1984, pp. 809–816.

Brennan, A. J. J., Health and Fitness Boom Moves Into Corporate America, *Occupational Health and Safety*, July 1985, pp. 38–45.

Bureau of National Affairs, *Controlling Health Care Costs: Crisis in Employee Benefits, A BNA Special Report*, Washington, D.C., 1983.

Drury, S. J., Wellness Can be the Solution to Rising Health Care Costs, *Business Insurance*, November 21, 1983.

Editor, Blood Pressure Check Saves Money According to NYC Statistics, *Occupational Health and Safety*, July 1985.

Feldman, R. H. L., Successful Health Promotions Key on Worker Acceptance, Use, *Occupational Health and Safety*, April 1985, pp. 46–52.

Gordon, D., Revco Hopes Workouts Will Reduce Comp Costs, *Business Insurance*, February 27, 1984.

India, D. M., Fitness Center's Club Theme Builds, Strengthens Employee Morale, *National Safety News*, February 1984.

Kendall, R., Business Joins the Nation's Fitness Boom, *Occupational Hazards*, May 1985, pp. 97–100.

Kittrell, A., Taking Their Medicine, *Business Insurance*, July 8, 1985, p. 3.

Logan, A. G., Cost-Effectiveness of a Worksite Hypertension Treatment Program, *Hypertension*, **3**, pp. 211–218, 1981.

National Council on Compensation Insurance, Compilation Insurance Expense Exhibit, 1967, 1982.

National Safety Council, *Accident Facts*, Chicago, Ill., 1968, 1983.

National Safety Council, Employees Wax Enthusiastically About Their Fitness Program, *National Safety News*, April 1984.

National Safety Council, Fitness: It's Not Only for Big Companies, *National Safety News*, November 1984, pp. 51–52.

PCC Newsletter, PCC/Drug Data Systems, Inc., November/December, 1984.

Polakoff, P. L., Worksite Economical Setting for Preventive Health Care Programs, *Occupational Health and Safety*, June 1985, pp. 75–76.

Shapiro, S., Prove to Superiors that Fitness Programs Work, *Business Insurance*, April 9, 1984, p. 26.

PART IX

SAFETY AND THE LAW

CHAPTER 24

How to Survive Workplace Litigation

BARBARA PFEFFER BILLAUER, ESQ.
Stroock & Stroock & Lavan
New York, New York

A recent article in *The National Law Journal* suggested over 60 theories of liability that might relate to workplace injuries.[1] The list was not exhaustive. The number continues to grow in the hands of creative plaintiffs' lawyers as they propose even more theories regarding workplace-related injuries and diseases. Target defendants for these lawsuits include the corporate employer, individual corporate officers, directors, plant managers, health and safety professionals, and other fellow employees of the injured party.

Unfortunately, in recent years judges and juries alike seem less inclined to require plaintiffs to conform to hitherto inviolate rules of law when it would deprive an injured party of recompense. Recovery in these lawsuits has therefore become easier than it once was; and the amount awarded often bears little relationship to the actual injury suffered by the plaintiff. This tendency to favor even the questionable claim is due partly to the belief that corporate America (as opposed to the government, for example) may be in a better position to provide financially for the injured worker. In fact, instead of requiring the plaintiff to affirmatively prove his case as is legally mandated, juries now expect the defendant to *dis*prove the plaintiff's case; and in some cases judges have even changed the law and shifted the "burden of proof" entirely to the defendant.

One situation that must be recognized, at the outset, is the changing legal status of the employer. The employer was once comfortably protected from civil suits, other than those for workers' compensation—or, in the very rare instance, suits for intentional acts. Today the employer may be subject to direct lawsuits by employees for fraud, recklessness, negligence, and even punitive damages if its actions are egregious enough. Even criminal convictions for murder have been returned against companies and their officers and directors.[2]

The company must also recognize that it functions in various roles—employer, supplier of raw materials, manufacturer, consultant, vendee, and sponsor of employee recreation, entertainment, and health programs—and as such may risk incurring various forms of liability other than workers' compensation from its customers, suppliers, the inhabitants of the environs of the plant, and the public at large.

Safety professionals and corporate management are often in a good position to prevent workplace-related lawsuits. At the very least, with planning, imagination, and foresight they can help to minimize or circumscribe recovery and possibly avoid the results of the recent proliferation of these lawsuits: bankruptcies, inopportune precedents, and astronomical costs to the employer.[3,4]

This chapter will explore ways to prevent workers' compensation claims from exploding into more serious civil lawsuits, and ways to prevent claimants from seeking unfettered damages for pain, suffering, and punitive damages, which are not components of recovery under the workers' compensation system.

This chapter will also address some of the legal theories commonly raised against an employer functioning in that and the other capacities, and some of the defenses and strategies available against such lawsuits. Proper handling of these lawsuits will minimize unjust recovery and also help to assure that the truly injured worker gets fair and immediate compensation.

While punitive damages and criminal sanctions are the more unusual legal repercussions of workplace injuries, they can do the most visible damage to the company's image and even threaten its continued viability. Moreover, they have a devastating impact on the individuals named as defendants. Even mere allegations, without proof, can engender negative press. If successful, such cases can result in fines or possible prison sentences for the individual coemployees, managers, officers, and directors. Besides the cost of millions of dollars in damages, once punitive damages are awarded or criminal sanctions levied, the companies must bear in addition the resultant legal and moral stigma and economic loss.

Fortunately, lawsuits for punitive damages and criminal sanctions may be the easiest to prevent, and, if alleged, the simplest to defeat. Consequently, this chapter will devote much attention to what constitutes punitive damages and criminal sanctions, and how to prevent them.

The mass-injury workplace lawsuits we have seen thus far have generated some helpful—if bitter—lessons for the health and safety professional. At least we now have a rough gauge of the effects of corporate and legal actions in the litigation process and their consequences. By way of example, the asbestos litigation has been studied in depth; the government solution to the black lung problem has been discussed and rejected as inappropriate for other industries. The settlement of the Agent Orange suit is still being evaluated. Related cases, such as the cigarette and drug litigation, also have yielded valuable information regarding the array of strategies available to the corporate and individual defendant to avoid or, if necessary, to defend workplace-related lawsuits.

24.1 PRELITIGATION STRATEGIES

24.1.1 A Pence for Prevention Is Worth a Pound for Cure

The first and best strategy for preventing a lawsuit is preventing the underlying claim. Company safety personnel should be actively involved in developing and reevaluating comprehensive programs to identify potential hazards and in designing safety protocols to prevent injury and disease.

Unfortunately, even if it were possible to eliminate culpability, this would not confer litigation-proof status upon prospective defendants. Any bad result—a serious injury or disabling disease suffered by an employee, regardless of who or what caused it—is an impetus for the injured party to seek someone else to hold responsible. The company, which is presumed to have a rich corporate bank account or a large insurance policy (commonly known as "Deep Pockets"), is a likely target as defendant in a lawsuit.

Additional actions taken prior to lawsuit or workplace injury can minimize the chances of both. These include preemployment physicals and screening, working with the unions, and hiring safety consultants[5] or inviting safety inspections by governmental agencies offering this service (such as NIOSH*

*The Hazard Evaluations and Technical Assistance Branch of NIOSH conducts field investigations of possible health hazards in the workplace. These investigations are conducted under the authority of Section 20(a)(6) of the Occupational Safety and Health Act of 1970, 29 U.S.C. § 699(a)(6), which authorizes the Secretary of Health and Human Services, following a written request from any employer or

and OSHA) and various other service agencies, such as risk control units of the company's insurance company.

Implementation of no-smoking policies has been found to cut absenteeism, decrease worker sick days,[6] and cut lung cancer incidence, and it may help limit the effects of toxic workplace chemicals, which seem to be increased by cigarette smoke.[7]

24.1.2 Knowledge Is Important: The Bad, The Good, and The Ugly

The second strategy is for the company to keep itself advised of all laws, statutes, ordinances, and formal and informal industry standards that may affect the worker and/or the workplace. New regulations appearing in government publications should be studied and relevant suggestions swiftly implemented. Proposed regulations should be evaluated and contested if inappropriate. New standards discussed in health and safety periodicals should be considered, and, if implemented, adopted industry-wide.

The right-to-know laws should be carefully reviewed and provisions made to assure compliance. The same goes for OSHA and EPA monitoring and record-keeping requirements.

While compliance with laws and regulations is no guarantee against a lawsuit, noncompliance is often grounds for a presumptive verdict in favor of the plaintiff. The more cavalier a company is in disregarding legal requirements, the higher the financial retribution may be.

The Bad: Plaintiff's Lawsuit Theories

To help prevent lawsuits it is also important to become aware of the more common types of lawsuits, the theories upon which they are based, and what is necessary for a plaintiff to prove before he can win (see Appendix), and to determine what steps might be taken to avoid these lawsuits.

The most common types of lawsuits likely to be brought as a consequence of a work-related injury are:

Workers' Compensation. Workers' compensation claims may be brought against either the employer or a coemployee. Under the workers' compensation system, employees are compensated for injury arising out of or in the course of the employment of the worker, hence work-related, without regard to fault of the worker or the employer. If the worker deliberately harms himself or does so while intoxicated, for example, workers' compensation may be denied. Intentional actions on the part of the employer or a coemployee may not be covered by workers' compensation. The award is limited to lost wages, medical expenses, and a circumscribed value of the injury based on a specified schedule.

Negligence. Lawsuits based on claims of negligence may be brought against the supplier of raw materials, equipment, or services, or against the manufacturer of equipment. Proving negligence requires proof that a duty existed on the part of the defendant towards the particular plaintiff, that the defendant acted carelessly, and that such actions directly caused injury and damages to the plaintiff. When the injured party contributes to the injury by his own actions, this may be raised by the employer as a partial defense. Recovery includes all factors allowed for workers' compensation, plus unfettered amounts for pain and suffering and loss of services by the worker's spouse. If the acts of the defendant were reckless, willful, or wanton, claims for punitive damages may be sought. Punitive damages are designed to punish the wrongdoer and need not bear any relationship to the severity to the injury sustained by the worker.

Strict Liability in Tort. Lawsuits based on strict liability may be brought against the supplier of equipment or raw materials or against the manufacturer of equipment. However, most states do not allow strict liability suits to be raised against the supplier of a service. These lawsuits focus on the product that allegedly caused the harm, rather than on the actions of the defendants. Defects in product design or manufacture that cause the plaintiff's injury will cast the defendant in liability. Defenses are more limited than with negligence, but include proving a lack of causal relationship between the product's defect or faulty manufacture and the plaintiff's injuries. (See Appendix.) That the benefits derived from the use of the product outweigh the risk of injury is usually a good defense—either if the danger is obvious or if there is an adequate warning affixed to the product. Damages allowed are the same as in the negligence case and may include punitive damages.

---

authorized representative of employees, to determine whether any substance normally found in the place of employment has potentially toxic effects in such concentrations as used or found.

The Hazard Evaluations and Technical Assistance Branch also provides, upon request, medical, nursing, and industrial hygiene technical and consultative assistance (TA) to federal, state, and local agencies; labor; industry; and other groups or individuals to control occupational health hazards and to prevent related trauma and disease.

Claims Based on OSHA/EPA Inspections. Claims based on OSHA/EPA inspections may be brought against the company or against the inspector for failure to correct a safety citation. These claims are relatively new theories, which have not met with success in all jurisdictions. Usually they are used in conjunction with a negligence case.

Industry on Trial. Sometimes it appears that industry itself is on trial.[8] New theories such as enterprise liability, market-share liability, concerted action, and joint liability[9] have implicated an entire industry even if only one manufacturer may have been at fault.

Though not a lawsuit based on workplace injury, the case of *Bichler v. Eli Lilly & Co.*[10] is noteworthy. In this case, the plaintiff was allowed to maintain an action against Eli Lilly even though she could not prove specifically that it was Lilly that manufactured the particular drug* that her mother ingested, a hitherto required element of proving a negligence case.

The action was allowed to be maintained against Lilly alone on the theory that Lilly was jointly liable with all other major companies that manufactured the drug, based on a variation of the "concerted action" theory. The New York Court of Appeals affirmed the jury's decision, holding that there was concerted action among Lilly and all other major manufacturers either by agreement or by substantial assistance in manufacturing and testing.

The evidence submitted was the *parallel* conduct of all the companies that manufactured the drug in question without first testing it on pregnant mice. The substantial assistance consisted of Lilly's having aided and encouraged the other manufacturers in this practice.

Reliance on industry-wide knowledge and standards, if inappropriate or anachronistic, may not save a company from lawsuit today, though at one time this may have been a viable defense.

In *Hall v. E.I. Du Pont De Nemours & Co., Inc.*,[11] the court shifted the burden of proof to the defendants (the plaintiffs normally carry the burden of proving their case by a fair preponderance of the credible evidence) because the product (dynamite caps) that injured plaintiff's children was obliterated by the explosion causing the injury. In a hybrid theory of "concerted action" and "joint liability," the entire industry was implicated. The court formulated a theory based on the premise that industry-wide standards adhered to by all the manufacturers were deficient in that they failed to include adequate warnings, and the defendants were responsible because they knew of the dangers.

The doctrine of *res ipsa loquitur* has also been applied to multiple-defendant cases.[12] Under this doctrine, when a plaintiff cannot establish specifically which of the defendants is at fault, and when all the factors that could have caused the injury were under the exclusive control of defendants, an inference of negligence against all the defendants is permitted, which then requires the defendants to come forward with an explanation.

Newer Theories of Liability. It is also prudent to be kept informed of newer theories being advocated and potential defenses that might be utilized. Often these new theories are hybrids of the more conventional ones, or they utilize an old type of lawsuit in a new way. Thus, negligence suits that previously were barred in suits against employers and coemployees by workers' compensation now are meeting with varying degrees of success when a particular quantum of intent or recklessness can be shown. Negligence lawsuits against OSHA inspectors and insurance company risk assessment teams in performing workplace inspections, violations of the Consumer Products Safety Code, and "nuisance" claims are also being raised. "Fraud" by an employer in withholding data about deleterious workplace conditions is another example of new theories currently being used in litigation.

Ordinarily, conventional negligence theory requires the wrongdoing to be causally tied into actual damages suffered by an employee either *now* or in the past. Creative attorneys, however, have managed to bypass this step by alleging that future injuries will occur, that is that the plaintiff faces an increased risk of injury in the future as compared to the general population, or that the worker presently suffers a *fear* of future injury so as to justify an award against a defendant even though there is no present physical injury. This has not garnered legal approval from all jurists in all states, but it has met with some success in some jurisdictions, and may signify a trend.

The Good: The Defenses

The factual and legal defenses available to the defendants must be well researched so that they can be adequately implemented, if need be. A more illustrative, though not exhaustive, listing of legal defenses is contained in the Appendix. Creative use of available theories and positions should be thoroughly considered.[13]

The favorable aspects of the company's safety procedures should be researched and documented. Witnesses who can testify to favorable safety practices and precautions utilized by the company should be identified and their addresses kept current. It is important to have a good handle on the company's marketing practices and product safety literature so that individuals do not give information that is

*DES, a drug designed to prevent spontaneous abortion, but alleged to cause cancer in the offspring.

inconsistent with general company policy. If profits for a particular product were small, it may be worthwhile to document this fact in order to help avoid being assessed punitive damages.

Scientific information that may explain company decisions should be accumulated. Often decisions made years ago were based on valid scientific theory of the day, but this may have changed drastically in the ensuing years. In many states the "state of the art" at the time of the defendant's actions is the standard against which they are judged. Compiling proof of what the "state of the art" was 40–50 years ago can be difficult and time consuming. Contemporaneous accumulation of medical and scientific articles to sustain safety decision-making practices often will save time, aggravation, and expense later on.

Safety manuals should contain citations to contemporaneous medical and scientific literature that substantiates their recommendations. They should be dated and saved. This is true for warning labels as well. Photographs of optimum working conditions should be taken and retained.

In sum, it should be accepted that workplace accidents and hence workplace litigation are inevitable. Preparation for those accidents that happen despite precautions involves gathering and preserving as much data as possible to demonstrate safety compliance. Concern should become a company routine.

The Ugly: Egregious Acts

The third strategy for preventing workplace-related lawsuits is to become aware of what factors increase the chances of a lawsuit and what factors maximize the dollars awarded the plaintiff (such as the reputation of the plaintiff's attorney). It is important to realize that these factors may have little to do with the severity or even the existence of the plaintiff's injury.

Real or perceived "evildoing" on the part of the company is the most infamous factor, which invariably increases the chances of a lawsuit. Sometimes the "evildoing" seems repugnant enough to the judge or jury to justify a high-value punishment award (punitive damages), even though it may have no relation to the dollar value of the injury claimed. "Evildoing" usually consists of reckless, wanton, and willful disregard of the safety of the workers by failure to correct known safety hazards after they have already caused injury, especially if safeguards are not in place because they would cost money and eat up company profits.

The error of "evildoing" is often compounded by hiding or ignoring past errors. Where the company attempts to hide past wrongdoing, perhaps by concealing the existence of these hazards or disregarding presently known deleterious effects of its products or manufacturing processes,* the retribution it faces often is far greater than the results any "owning up" would have caused.

Any combination of three factors—horrible injury or death; wrongdoing on the part of a defendant; or concealment of wrongdoing and disregard for untoward deleterious workplace conditions—multiplies the chances of lawsuit and the level of recompense awarded to the plaintiff employee.

Avoid Dirty Laundry

> My answer to the problem is: If you have enjoyed a good life while working with asbestos products, why not die from it? There has to be some cause.
>
> E. A. Martin, Director of Purchases, Bendix Company, to Noel Hendry of Johns-Manville, Sept. 12, 1966 (*Mealey's Litigation Reporter: Asbestos,* July 27, 1984, p. 982)

Irresponsible statements such as this should be avoided. Any unfortunate statements contained in company files must be uncovered early on in the litigation picture and considered, for—if they do exist—one can be virtually certain the plaintiff's attorneys will discover and use them. Playing ostrich is not good strategy, as Johns-Manville found after Vandiver Brown, whom Manville had claimed to be deceased, was discovered alive and well.† The casualty list of companies or firms that hid, disregarded, or otherwise disclaimed the existence of harmful or sensitive documents is well known, and the consequent results disastrous.

While not a workplace-related case, the decision in *Berkey Photo Inc. v. Eastman Kodak*[14] is noteworthy. In that case an expert witness, and, later, a lawyer representing Kodak, attempted to suppress the discovery of a document that might have hindered the defense. In the final days of trial, the existence of the previously unrevealed document came to light, and the credibility of the expert witness was destroyed on cross-examination. While the document was in some ways compromising to the witness's

*On the other hand, continuous and vigorous contest of a substance's toxicity may be another tactic available to a defendant.

†Vandiver Brown was the general counsel of Johns-Manville in the 1930s. Johns-Manville had successfully fought the introduction of damaging correspondence between Brown and Sumner Simpson, President of Raybestos Manhattan, as trial evidence by arguing that Brown was dead and therefore his signature could not be authenticated. However, that tactic proved futile after plaintiff's lawyers found Brown alive and well in Scotland. Use of this correspondence by plaintiff began in 1977.

testimony, its proper production, and the proper preparation of the witness by a lawyer in handling questions relating to it, might have been far more acceptable to both judge and jury than the discovery that the witness had committed perjury regarding its existence. In fact, an award was rendered against Kodak for $113 million (which later was reversed).[15]

The defendant should be aware of any such incriminating evidence before deciding about its posture regarding settlement, trial, or appeal.

The embarrassment to a defendant of being confronted with callous, reckless, and irresponsible statements from its own files is of minor significance, however, when compared with the increased jury verdicts and punitive damages that they necessarily engender. These statements may serve as the basis for criminal sanctions. They may also serve as the basis for fraudulent concealment, which may allow an employee to bypass the workers' compensation system and sue his employer for compensatory damages, loss of consortium, and punitive damages, based on a showing of reckless, willful, and wanton disregard of the welfare of employees or product users by an employer or manufacturer. Further, in over half of the 50 states, punitive damages are not covered by employer's or manufacturer's liability insurance.

Avoid Falsifying or Withholding Data. The MER 29 controversy is an excellent example of the effects of witholding data.[16] MER 29 was an anti-cholesterol drug manufactured by Richardson-Merrell. Though animal and human experimental data indicated that the drug could cause serious side effects, including cataracts, the FDA application did not contain this data. The FDA, however, ultimately did discover the withheld information. (A confidential memorandum to doctors from Dr. W. H. Kessenich, Medical Director of the FDA, dated May 7, 1962, states in part: "As you may know we have recently obtained evidence that the Wm. S. Merrell Co. falsified data submitted as part of the New Drug Application for MER 29.")[17]

While thousands of lawsuits were filed and ultimately settled, what was novel about the MER 29 litigation was the repeated quest for punitive damages. Though bona fide arguments were raised against allowing such claims to be filed,* the argument that multiple punitive damages were certain, and could destroy the company, received the most attention—yet ultimately went unheeded. The drug company argued that it would be bankrupted because each jury, acting without knowledge of punitive damage awards by other juries, would award punitive damages undiminished by any previous awards, although the company had done only one wrong, and therefore should be punished only once.

In fact, after a New York jury awarded the plaintiff $2 million, which was then the highest personal injury verdict in New York history, one of the persons was quoted as saying "the jury felt the company 'had to be punished not only for what [it] had done, but also as a warning to all drug companies, that they could not do things like this' " (Rheingold, 1968, quoting Morton Mintz, *The Washington Post,* Dec. 12, 1966, p. 12, col. 1).

Three other juries also awarded punitive damages, which were later reduced or rejected by the trial court or an appellate court on review for one of several reasons: the jury did not know and could not have known about other cases which were pending; the amount was excessive; or multiple and repeated awards of punitive damages were certain, and could destroy the company.

The ultimate judicial determination of whether punitive damages would lie in the MER 29 cases was not based on the actual falsification of the data, but on the determination of whether management was involved in such falsification. It would seem that such machinations could not have occurred without management's active or tacit approval.† The dichotomy of court decisions on this point serves only to underline the lack of clear-cut judicial policy. In *Roginsky v. Richardson-Merrell, Inc.*,[18] the appellate court reversed the trial court's punitive damage award on the ground that there was no evidence in the record that management had participated in the admittedly wrongful conduct. On the same set of facts, however, a California court in *Toole v. Richardson-Merrell, Inc.* came to the opposite conclusion about management's knowledge and expressly rejected the New York decision, stating, "We respectfully differ from that holding."[19]

In the California case, the court concluded, "In our case there is evidence from which the jury could conclude that appellant [drug company] brought its drug to market, and maintained it on the market, in reckless disregard of the possibility that it would visit serious injury upon persons using it." By comparison, the New York court that rejected the punitive damage claim stated: "Moreover, New York demands, as it might have to before punishing a defendant if fines [are] similar to those imposed on a criminal charge, that the quality of the conduct necessary to justify punitive damage must be 'clearly established.' "

*Two arguments raised were: (1) punitive damages should never be claimed against drug companies, which have the special role of making new drugs for sick people; and (2) punitive damages are so similar to criminal fines that the higher standard of proof required in criminal cases should be employed to determine punitive damages.

†Under the legal theory of *respondeat superior* an employer is liable for the acts of his employee, a principal for the acts of his agent, and a master for the acts of his servant.

In sum, it is therefore far better to acknowledge the existence of the sensitive documents or set of facts and claim a legal privilege,* if possible; get to work developing a defense; or advise the witness how to handle the existence of the document or situation on cross-examination, rather than to hide its existence.

Legal Consequences of Egregious Acts

Punitive Damages

> Punitive damages by definition are not intended to compensate the injured party, but rather to punish the tortfeasor whose wrongful action was intentional or malicious and to deter him and others from similar extreme conduct.
>
> *City of Newport v. Fact Concerts Inc.*, 1981[20]

Certain elements must be proven before punitive damages are assessed. "The question is, whether a defendant's conduct 'contains elements of intentional wrongdoing or conscious disregard' for plaintiff's rights." Thus the court in *Knippen v. Ford Motor Co.* rejected a claim for punitive damages, explaining that such damages are awarded to punish and deter outrageous conduct. "Actual malice is not required but [the] defendant must [be found to have] act[ed] with such conscious and deliberate disregard of the consequences of his actions to others that his conduct is wanton. W. Prosser, Law of Torts § 2, at 9–10 (1971)."[21] In *In re Marine Sulphur Queen*,[22] the court ruled that, before punitive damages can be awarded, a defendant must be guilty of gross negligence, actual malice, or criminal indifference (which is usually the equivalent of reckless and wanton misconduct).

Bases for Punitive Damage Claims. The type of gross misconduct required for imposition of punitive damages can be inferred from (1) callous remarks in company correspondence, (2) intentionally falsifying or withholding data (*The New York Times*, Feb. 26, 1985, p. A24), (3) corporate indifference in the face of knowledge of serious hazards in the workplace,[23] and (4) concealment of such hazards from employees.[24] Clearly, avoidance of such actions is critical.

Defenses to Multiple Punitive Damage Awards. It is interesting to note that in the recent asbestos litigation, the defendants' claims of the financial impact† and inherent unfairness of multiple punitive damage verdicts have not met with much receptivity. Arguing that because repeated awards of punitive damages unconstitutionally subjected it to double jeopardy, Johns-Manville sought to have the punitive damage claims dismissed. In *Hansen v. Johns-Manville Products Corp.* the court ruled that a defendant who was consciously, knowingly indifferent to the plaintiff's welfare and safety could indeed be subjected to multiple awards for punitive damages under Texas law, and that punitive damages are guaranteed to claimants by the Texas constitution.[25]

Again by way of contrast, in a similar suit filed in Mississippi, the court disallowed repeated punitive damage claims because the policy behind strict products liability was " . . . 'not merely to assure compensation for the individual plaintiff, but to achieve the broader societal objective of equitable loss distribution,' "[26] which could only be achieved by way of viable enterprise.

> Because of the number of similar suits already filed, asbestos defendants face repeated assessments of punitive damages that threaten to exhaust their resources. . . . [A]t the point at which awards of punitive damages destroy the viability of the enterprise, the remedy of punitive damages becomes antithetical to the policy of strict liability and thus, that punitive damages must not be allowed. Furthermore, . . . the basic policy objectives of punitive damages—punishment and deterrence—are satisfied by multiple exposure to compensatory damages.[27]

Thus, a divergence of opinion exists as to whether multiple punitive awards, which could easily undermine its financial integrity in one short time period, will be allowed to lie against a company.‡

*Trade-secret, attorney-client work-product.

†In 1981 and the first half of 1982, Johns-Manville had ten punitive damage verdicts against it, totalling $6.16 million. In *Janssens v. Johns-Manville*, a Florida appeals court upheld a $1.8 million award against Johns-Manville, marking the fourth time punitive damages were upheld against Manville on appeal. The court held that evidence indicated Manville persisted in "suppressing information" that its asbestos would cause asbestosis (*The Wall Street Journal*, Sept. 27, 1984, p. 15). Settlements and jury verdicts against Johns-Manville went from $300,000 in 1975 to $35 million in 1981 (E. Chen, The Law: Asbestos Litigation Is a Growth Industry, *The Atlantic*, **254**, No. 1 (1984):24).

‡In 1979, the Alaska Supreme Court stated that *punitive* damages would be allowed to injured parties with *minimal* actual damages, despite the otherwise prohibitive cost of litigation (*Litigation News*, **19**, No. 4, 1984).

Criminal Sanctions. A second consequence of a company's negligence, indifference, and misconduct is the possibility of criminal charges. While this usually does not have the same economic impact on a company as verdicts allowing punitive damages or multiple mass tort claims, the negative press it receives can be devastating to a company's image. Further, a criminal conviction can be used later to sustain a civil claim for punitive damages.

Five top officers of a Chicago electrical wire manufacturer, Chicago Magnet Wire Corp., were indicted for injuring employees through repeated exposure to toxic chemicals. They were charged with 400 counts of aggravated battery, conspiracy, and reckless conduct. Richard Daly, the State's Attorney, said, "We believe that the company and its officers lied to workers who raised questions about the safety of the chemicals they were handling."[28] The indictments charge that the officers and directors knew of these dangerous conditions and failed to correct them.

In March, 1985, five officials of the now-defunct Film Recovery Systems were charged with murder in the cyanide poisoning death of a Polish immigrant who had worked for the firm.[29]

In July of 1984, indictments were issued against officers of Illinois Brick Co. charging the silicosis deaths of two former brick company workers as homicide.[30] The basis of these criminal indictments was fraudulent concealment of known safety hazards, or implied knowledge of safety deficits that had caused previous deaths and had never been corrected.

These cases are still pending and hence their impact is as yet unclear.

Unusual Fraudulent Concealment. While the criminal approach is somewhat unusual, the same "fraudulent concealment" theories also have been used to bypass the "fellow servant" protection of the workers' compensation "exclusivity" doctrine and to allow direct negligence actions against an employer. Normally only intentional acts are the basis for allowing direct suits against an employer. The claim that the employer fraudulently concealed known hazards has been used with some success to establish the requisite level of "intent" and "willfulness" necessary to bypass the workers' compensation system.

Several civil cases have been attempted under the theory of fraudulent concealment of known hazards and have met with varying degrees of success.[31] What is of great interest is that the very allegations that serve as the basis for the civil negligence suit for fraudulent concealment are virtually identical to the criminal allegations. Moreover, while the criminal allegations brought to date have been leveled either at the top officers of the corporation or at the corporation itself, or both, the civil suits have been targeted against company physicians and health and safety personnel as well.[32]

Determination of "Intent." The question of whether the requisite level of "intent" has been met will ultimately be resolved by a jury in both the civil and the criminal actions. It is clear, however, that civil intent can be presumed or implied from a criminal conviction. In addition, knowledge of similar instances of hazards causing injury will probably suffice to impute recklessness or willfulness to the defendant.[33]

Proof of intent can come from other unusual sources. In a recent deposition, a former health consultant to a major asbestos manufacturer testified that management personnel did not have an adequate degree of concern for employees' health and safety. He stated that he reached this conclusion not by talking to them (or by reviewing company documents) but from the evidence of apathy, lack of diligence in responding to safety recommendations, and the observed day-to-day supervision, which did not appear to be concerned with matters of health.[34]

Similar testimony by former medical consultants or employees also has been used against companies to sustain punitive damage awards. Gerrett Schepers, former director of Saranac Laboratory, a private research facility funded by seven asbestos companies, testified that he had warned the industry about the dangers of asbestos and even sought industry funding for additional studies in the mid-50s. Schepers contended that a 1948 Saranac study apparently advised the defendant in question that its product was "dangerous."[35] (The study was actually published in 1955.)

In *Hansen v. Johns-Manville Products Corp.*, a former medical officer of Johns-Manville "testified that he informed Johns-Manville of the hazards asbestos presented to asbestos insulation workers and recommended placing warnings on insulation products as early as the 1940's, but that Johns-Manville did not act on his recommendations."[36] The court ruled that if the jury believed Dr. Smith, the former medical officer of Johns-Manville, "they could reasonably find Johns-Manville had been grossly negligent in failing to warn the plaintiff of the dangers related to its products" and could justifiably award punitive damages based on that finding.

24.1.3 The Company Informing the Worker

RIGHT-TO-KNOW LAWS

Emphasis on disseminating information to the employer has received much attention. Right-to-know laws require informing employees about toxic substances to which they may be exposed. Failing to comply with these requirements is not only illegal, but also may preclude proper emergency care geared to the specific chemicals to which the worker had been exposed.[37,38]

Warnings and Safety Precautions

The manufacturer and supplier of products is obligated to warn users of certain dangers incident to their use and to provide warnings, safety guards, and other safety mechanisms where appropriate. Failure to furnish adequate warnings could subject a manufacturer to negligence claims or claims in strict products liability.

The role of the employer in conveying warnings on these products is not yet resolved. In one recent case, however, the manufacturer was absolved of his duty to warn because it was found that he reasonably relied on the employer to carry out this function.[39] The air of bravado often common to the workplace (the "it can't get me" syndrome), as well as the employer's insistence on productivity, may motivate the employee to bypass recommended or routine safety procedures and lead to injury and, later, lawsuit. Employees often refuse to wear safety gear because it is uncomfortable, or it interferes with smoking (for example, respirators), and generally "it is a pain." Egregious acts of employers or fellow employees in removing safety guards to speed up productivity have been reported. In addition, employers often have failed to provide personal protective equipment out of ignorance.

Warning labels belong on all hazardous equipment and material, and should be affixed in such a way that they are not easily removable. Adequacy of warnings should be reviewed by an attorney familiar both with the product and with personal injury litigation. The critical issue, however, is not where and how the warning should be affixed, or what it should say, but: When is it necessary to warn the user?

If current legal decisions are any guide for preventing future claims, it is recommended that a general warning be placed when there is a general acceptance within the scientific community of the possibility of a relationship between the product and disease.* Epidemiologists of repute usually say that this occurs one year after two or three reputable epidemiological studies of sufficient size are reported.[40] Clearly the emergence of a single case report is not sufficient to generate the requirement of a warning.[41]

Plaintiffs' attorneys often seek to base the requisite time for issuance of a warning on the emergence of a few case reports. This is easy to do in retrospect, after a causal condition has been clearly established, but should be contested vigorously if scientifically unacceptable. The existence of epidemiological studies demonstrating an absence of causation is favorable evidence for the defendants to introduce in cases where warnings were not placed.

Just as the geography rule has been abrogated in medical malpractice (i.e., a physician can no longer avail himself of the standards of practice in his community if they are at odds with the standards in the worldwide medical community at large), so it seems that manufacturers should keep informed of epidemiological data gathered worldwide. While often various considerations exist that warrant disregard of these studies in deciding whether to place a warning, as a whole these studies should be considered, if they reasonably measure comparable conditions and products.

The excuse "I didn't know" of the existence of major epidemiological studies will not be given much credence by a court. The requirements for proving negligence are "knew" or "*should have* known." If the industry knows, then all companies in the industry would be presumed to know. Size of the company or remote geography probably no longer will be recognized as an adequate defense.

If a manufacturer has any doubts about the propriety of a warning, in today's legal climate he would be well advised to decide in favor of placing the warning on the product. Lawyers are currently acting under the constraints of the *Beshada v. Johns-Manville Products Corp.* decision,[42] which requires the defendant to warn of dangers both known and "unknowable." While this is a New Jersey case, which has been rejected by some other jurisdictions, its holding is so broad and far-reaching that it is not yet clear how much credence it will be given. It is wise to warn.

What constitutes a proper warning is usually a question for the jury to decide.[43] However, the acid test is that the user is actually warned. If the user already knows of the danger, or never reads warnings, or cannot read, then the manufacturer probably will not be liable. A noted legal scholar has written that "if the basis for recovery under strict liability is inadequacy of warnings or instructions about dangers, then plaintiff would be required to show that an adequate warning or instruction would have prevented the harm."[44] If proof shows that plaintiff would not have read the warnings had they been present, then presumably the manufacturer would not be held liable. Alternatively, in *Bolm v. Triumph Corp.*,[45] where the danger was open and obvious, no requirement to warn was found. Finally, if the plaintiff is a *sophisticated user* and is knowledgeable of the risks, there may be no duty to warn.[46,47]

Involve the Union

In order to assure that the union is cooperating in advising its members of safety precautions and to monitor union knowledge, it may be wise to have a company liaison attend union meetings, if possible. Union handouts should be kept and logged. If the union is advising its members of the hazards of a particular aspect of the workplace, then the employer should evaluate and determine whether action is warranted, especially if it means advising workers of the use of proper safety precautions, or of notifying the union of the falsity of their data or information.

*Unfortunately, this may occur even before the manufacturer is aware of such a relationship, as these cases are tried in retrospect. Wise employers should therefore make it their business to keep up with current scientific and medical developments that affect their products and workplace.

In the event of lawsuit by a worker alleging failure to warn of hazards or failure to advise of proper use of personal protective equipment, consideration might be given to evaluating the union's responsibility. As the union may have a concurrent duty to advise its members of safety precautions in the workplace, management might even consider assigning responsibility for safety to the union during collective bargaining seasons. While this would in all probability not absolve management, it might divide the responsibility, and perhaps stem what has been regarded as unions' "stirring up of frivolous suits."

Unions should assist in safety matters as much as possible. Advice and discretion in the use of personal protective equipment may be received better by workers when a union representative recommends their use.

24.1.4 Pay Attention to Corporate Records

Once litigation has been instituted, the plaintiff's attorneys often have obtained court permission for an unfettered search of company records, most of which are admissible at trial. Hiding or destroying sensitive documents, besides being illegal and inviting contempt or perjury charges against the individual directing or carrying out such a maneuver, usually is discovered and only incites the ire of judge and jury and often invokes greater economic sanctions.

Marking papers "confidential" does nothing to insure the confidentiality of their contents. In fact, this is a red flag to the plaintiff's attorney that the document contains data that may be used *against* the defendant.

Valid legal objections do exist that may protect certain documents from discovery during the course of litigation, among them that the documents may contain trade secret information, or fall under certain other "exceptions," or may be privileged communications such as those between attorney and client. In planning prevention programs it is not wise to rely upon these, but these valid objections to discovery should be raised the first moment plaintiff seeks access to these documents, as a number of courts have held that an inadvertent production of privileged matter is a waiver of relying on the privilege later on.

Sometimes attorneys are misled by misplaced reliance on the "deposition-stage exception," which does *not* require all objections to be raised at the time of deposition and preserves the right to make most objections at the time of trial. For *document* production, this does necessarily follow. To prevent waiver, vehement, strenuous, written objection and motions for protective orders should be raised the moment the plaintiff's attorney demands production of privileged documents.

In the event the protective order is not granted, the decision should be appealed if the documents are sensitive in nature *and* if a valid legal ground exists. *In camera* inspection—that is, allowing the judge to inspect the documents in closed chambers, to determine if they are nondiscoverable—is another alternative to appeal. In the event discovery is ordered, consideration should be given to requesting a type of "gag order," directing the plaintiff's attorney not to share the data with the bevy of other plaintiffs' lawyers in similar litigation who are eager to share the information.

How To Avoid Production of Sensitive Documents

Use the Attorney–Client Privilege. The purpose of the attorney–client privilege is to promote freedom of consultation between a lawyer and his client and to encourage full discussion of facts without fear that disclosure can later be compelled from the lawyer without the client's consent.[48] A noted legal scholar (Dean Wigmore) articulated this widely recognized definition of attorney–client privilege:

> Where legal advice of any kind is sought from a professional legal advisor in his capacity as such, the communications relating to that purpose, made in confidence by the client are at his insistence permanently protected from disclosure by himself or by the legal advisor, except if the protection is waived.[49]

When (1) corporate employees seek legal advice under the order of a superior for the purposes of the corporation, (2) the advice sought concerns matters within the scope of the employees' corporate duties, (3) the employees are aware that the communications with counsel were ordered so that the corporation could obtain legal advice, and (4) the legal communications were ordered to be confidential and *they remain confidential,* then the communications will remain privileged.

The party seeking to assert the privilege has the burden of establishing it. A common error is that the communication with counsel is divulged outside the company and thus does not remain confidential, and the privilege is therefore lost.

To help prevent the loss of the privilege, the corporate employee should specifically request legal advice in writing; insure that the communication remains privileged by not divulging it to outsiders; and meet with the lawyer acting in his legal capacity, not as a business advisor or policy maker. The request should specifically authorize a legal analysis, rather than a mere factual investigation (Miller, 1981, citing *Securities & Exchange Commission v. Gulf & Western Industries, Inc.*, 518 F. Supp. 681 (D.D.C. 1981)). General memoranda and business decisions are not privileged even if directed to company lawyers. If senior management needs a written report offering factual and legal analysis, the report should be written by a lawyer working on the matter. If appropriate, the request should include a specific reference to the likelihood of litigation. Legal titles should be used in all correspondence, and documents containing legal advice should relate only to legal subjects.

Communications among in-house counsel may not qualify for the attorney-client privilege (Miller, 1981, citing *In re D.H. Overmyer Telecasting Co.*, 470 F. Supp. 1250, 1255 (S.D.N.Y. 1979)). Therefore the company may wish to retain outside counsel at the earliest hint of a potential lawsuit to direct investigation and coordinate legal posture.

Loss of Attorney-Client Privilege

Inadvertent Production of Privileged Material. Disclosure of one privileged document may forever waive the privilege regarding other documents on the same subject. Also, confidential materials used to refresh the recollection of a witness before trial or deposition may cause waiver of the privilege as well (Miller, 1981, citing *James Julian, Inc. v. Raytheon Co.*, 93 F.R.D. 138 (D. Del. 1982). *But see contra Jos. Schlitz Brewing Co. v. Muller Phipps (Hawaii) Ltd.*, 85 F.R.D. 118 (W.D. Mo. 1980), where an attorney's reviewing his correspondence file before deposition did not waive privilege).

In complex cases with large volumes of papers available for discovery, inadvertent production of privileged information can occur, causing waiver of the attorney-client privilege.

This inadvertent production waives the privilege unless good cause exists.[50] Courts have recognized that in complex litigation inadvertent production of documents does occur in large-scale discovery. If reasonable precautions have been taken and the party acts promptly to recover the privileged documents, the courts may conclude that good cause is shown and the privilege should not be lost.

Waiver by Computer. In our computer era the possibility of waiving the attorney-client privilege is even greater than before. Outsider access to computer information is, unfortunately, something that must be acknowledged. Often, routine precautions are not sufficient to prevent computer code cracking. (In one recent publicized case a computer leak occurred at a well-known New York law firm.)[51] Restricted access to the data bank probably is required before a court will uphold the confidentiality of and prevent waiver of the attorney-client privilege.

Communications with Governmental Agencies. It is recommended that, in turning over privileged documents to governmental agencies, a stipulation of confidentiality, or a protective order reserving the right to assert the privilege later on, be negotiated. This would bolster a claim for confidentiality in future litigation. In *Teachers Insurance & Annuity Association of America v. Shamrock Broadcasting Co.*,[52] the court held the disclosure to the SEC to be a complete waiver, unless it could be established that plaintiff's right to assert privilege in subsequent proceedings had been specifically reserved at the time disclosure was made.

It must be recalled that the privilege belongs to the company, rather than a particular employee or counsel seeking to assert it. Thus, if an employee raises information that could be used personally against him in the future, he should be cautioned not to divulge the information without the advice of personal counsel. (The "privileged documents" also can be used against counsel, who also would have no standing to raise the privilege.)

Maintain Confidentiality. In World War II, posters urging discretion bore the motto "Loose lips sink ships." This rule applies also to corporate ships, many of which have been sunk through the "loose lips" (or typewriters) of their employees or management.

Circulation of memos or inclusion in conferences of persons who are unnecessary to the decision-making process may waive the attorney-client privilege, especially if the memos or conferences include persons outside the company.

Use the Work-Product Privilege. Where the attorney directs a particular function or information gathering "in anticipation of litigation," the material may be protected as attorney work-product. Experts' reports may or may not fall under a corollary of attorney-client privilege called "work-product."* Experts therefore should be retained by legal counsel. The same holds true for hiring legal assistants and investigators.

The experts' report should be divided into two sections: data compilation and opinions. Occasionally, experts' reports may be discoverable where the data or subject of investigation is no longer in existence. The *opinion* of the experts will remain privileged and disclosure-free even if the factual text is discoverable, unless the expert will testify at trial. Even then, in some courts, any reports made may be protected from disclosure if the expert has been hired by an attorney.

*Unless the expert will be testifying. In New York, however, even the opinion of the testifying expert is privileged in state court.

Avoiding Problems. Nonlegal company personnel should refrain from predictions that involve legal assessments, such as that the company will lose a case if it is forced to go to trial, or that the company is ethically or morally culpable. Such predictions may be made upon a misunderstanding of the law, or with an incomplete compilation of the facts. Imprecise language or misuse of legal terminology also can come back to haunt the company.*

Any document by an employee or ex-employee criticizing the company's actions would have a good chance at admissibility at a "trial." Comments by someone outside the company that are deleterious to both the company and the author also are often admissible as "declarations against interest."

On the other hand, helpful documents or correspondence in company files may be inadmissible as evidence on behalf of the company because they could be considered as self-serving documents. (Clearly the plaintiff will try not to admit these documents as they would be harmful to his case. Sometimes, if there is a codefendant, an agreement may be made where otherwise inadmissible self-serving, but helpful, memos of each company will be admitted into evidence by the other codefendant.)

Written concurrent reports are helpful in refreshing the recollection of the parties as to why a particular course of conduct was undertaken, or a particular decision made. During the course of a trial, or in preparation, a witness may refer to these writings to refresh his recollection regarding whether and how a particular action occurred, even if the actor has no present recollection of that particular act. He may testify regarding the act if a document records its occurrence.† However, once he refers to the document, any privilege attached to that document is waived.

Further Suggestions for Avoiding Problems. In summary, to avoid unnecessary problems caused by damaging documents or those carelessly written, the following steps might be taken.

1. Consult with counsel regularly before preparing negative or sensitive reports or evaluating a transaction that might have legal repercussions. Letters to injured persons should be reviewed by company counsel, as they undoubtedly will find their way to the claimant's attorney should a lawsuit be instituted.
2. Don't write unnecessary "regular reports," if they have legal implications, without legal advice.
3. Be careful in writing correspondence to outsiders when it might be construed against the company. If it has legal ramifications, consult with the company attorney.
4. Make sure that accident and incident reports usually compiled by safety or health personnel never contain speculation or assumptions. They should state only the facts known to the report's author. Thus, if the injured employee comes in complaining he fell on a banana peel, the report should state "patient *states* he fell on banana peel," and not "patient fell on banana peel," unless the nurse actually saw this happen. Of course all medical treatment rendered should be reported in the employee's chart.

   The company nurse then should inform the company's attorney that a "potential claim may have occurred" and request instruction from the attorney as to further follow-up. Copies of the incident reports then should be sent to the attorney's office, since it should be part of the attorney's responsibility to investigate all potential claims.
5. Employees' medical records should be accurate and complete. All medications given to employees should be logged with notations as to specific warnings, if any, given by the company health care professional to the patient. Better yet, the employee should be required to sign an "informed consent sheet" specifying the warnings he was given. The sheet should be kept in the medical file.
6. Answer written complaints with self-serving letters. Letters not rebutted by the recipient may serve as the predicate for an argument that they were not rebutted because they contained the truth. By the same token, unrebutted letters to outsiders by employees give the company's version of the facts, may be admissible into evidence and may be able to bolster the company's position.

*By way of example, the following exchange during a radio interview of a high-level Union Carbide executive, Mr. Browning, was reported by *The New York Times* (Dec. 14, 1984, p. D2, col. 1):

"I think you've said the company was not liable to the Bhopal victims," Mr. Kilmer, the radio reporter, said.

"I didn't say that," Mr. Browning replied.

"Does that mean you are liable?" Mr. Kilmer asked.

"I didn't say that either," Mr. Browning responded.

"Then what did you say?" the reporter asked.

"Ask me another question," the Carbide spokesman said.

†It is important to note that any document used by a witness to refresh his recollection during preparation for testimony by the company counsel is discoverable by the opposing party.

7. Letters alleging lack of safety precautions in the workplace or poor health conditions are often quite sensitive, and should be referred to an attorney for handling. The attorney should respond only after he or she has conferred with company health and safety personnel and fully understands the health and safety considerations involved, and the state-of-the-art knowledge in that area as well.
8. Keep all safety and health reports together in a separate, single location, and advise the legal department periodically of the number of reports received and the types of injury or disease involved.

   It is very important to remember that while the first claim of injury due to health and safety defects may be insignificant in and of itself, the accumulation of a few of these reports may well serve as the basis for later allegations that the employer or manufacturer had knowledge of deleterious effects of its products, machinery, manufacturing process, or workplace hazards, and either fraudulently concealed the health effects from the workers or knowingly subjected them to "poisonous" chemicals or hazardous conditions. These claims and reports can serve to undermine what is commonly called the "state of the art" defense—that is, that, based on the state of knowledge in the field at the time of the incident, the employer (manufacturer) did not know, nor should he have been on notice od, the particular hazards of this situation.

   "Knowledge" that may be attributed to the company may arise by way of a collection of workers' compensation claims, incident reports, notations by company health personnel, absenteeism, OSHA violations, and correspondence with the company from employees and their attorneys. Unfortunately, records of these claims and reports may be kept in different places within the company. While one report here and one report there may not cause concern, and thus may be not be brought to management's attention, if they are then collected in one place, the company may find six or seven different reports of injury or disease from the same substance. While management may therefore not have "actual" knowledge of these incidents, their existence may attribute "constructive" knowledge to the company. A separate clearinghouse may help bring these problems to light sooner so that adequate steps may be taken to deal with them.
9. *Avoid cavalier reports of health problems.* As mentioned earlier, cavalier reports of health problems do not belong in writing. Nor do they belong in open speeches made by company personnel.
10. Reliance on invoking confidentiality rules is not a good litigation tactic.

24.2 DOUBLE-EDGED SWORDS: DARNED IF I DO, DARNED IF I DON'T

24.2.1 Workplace Inspections

Safety inspections by a company, if done improperly, can also sustain lawsuits. Reliance on outside safety professionals for safety inspections may not protect the company if the plaintiffs can prove that the selection of the safety contracts was poor, or if the recommendations were not implemented immediately. Moreover, in the case of an injury that was the result of an alleged negligent inspector, the safety inspector becomes a target for suit, both from the plaintiff and from the employer who relied on this inspection. Thus, it is important to select competent safety services and to document in writing the basis for the selection.

24.2.2 Responsibility of the "Inspected"

If safety inspections contain information as to possible or potential hazards, the company must react immediately, either by implementing recommended safety steps or by disputing the report or citation.

1. If management finds the recommendations are warranted, not only should they be followed immediately, but compliance should be documented in written form. Training manuals and seminars, checklists, and the like should be instituted and preserved.
2. If not all the recommendations can feasibly be implemented immediately, a follow-up report by the consultant should list recommendations in order of priority, giving time parameters and reasons.
3. If management takes issue with the validity of the report—either in whole or in part—this should be communicated to the consultant in writing by someone within the company with appropriate knowledge and the consultant's comments sought. If the matter cannot be resolved, management should document in writing why it found the results of the study unsound. Another independent study might be considered.

   While the existence of a pending contested OSHA citation ordinarily would not be admissible in a court of law, there are several conceivable methods whereby a creative plaintiff's attorney could attempt to have this information brought before a jury.*

4. Disagreements over the economic propriety of implementing safety recommendations will not protect an employer unless it can be shown that it is technologically and economically unfeasible, especially when an alleged OSHA violation exists.[53]† Further "riskutility" is the standard used in products liability cases. Under this standard, there is really no justification for not placing a warning on a hazardous product.[54]

If a claim for an injury or disease occurs after the company is put on notice of a hazard, a greater onus is placed on the employer (e.g., if a second accident occurs after one has already taken place). Thus, the company now may be open to claims for punitive damages as well as criminal sanctions.

While the company has little practical choice in allowing OSHA inspections (though it legally could refuse an inspection without a warrant), there is no absolute requirement to bring in outside safety services. Serious consideration should be given before doing so. An "all-clear" is, of course, helpful to the company should a claim be brought later—*if* the inspection was done properly. Carefully conducted inspections may yield helpful recommendations, which if implemented may prevent workplace injury and lawsuits. If it can be proved, however, that the inspection was negligently done, or that the selection of the inspector was "negligently made," the results of the inspection will be of little use to the company,‡ and in fact could conceivably serve as another theory in the plaintiff's lawsuit.

24.2.3 Lessons To Be Learned from Past Cases

In *Rosenhack v. State of New York,*[55] an injured worker brought a suit against his employer's workers' compensation carrier, alleging negligent inspection of the workplace. The suit was *disallowed.* The court discussed the first of the five elements the plaintiff must prove to sustain a negligence claim (see Appendix): whether the alleged tortfeasor has a *duty* to act and, if so, whether that duty extends directly to the claimant.[56] It is well settled that in the absence of any duty there cannot be any liability.[57] The duty must extend to a person or class of persons seeking recovery.

The court found that while the workers' compensation policy allowed the carrier the *right* to inspect the workplaces and equipment, it also contained the following exculpatory clause:

> Neither the right to make inspections nor the making thereof nor any reports thereon shall constitute an undertaking on behalf of or for the benefit of the employer or others, to determine or warrant that such work places, operations, plants, machinery, appliances and equipment are safe.

Citing the case of *Nieto v. Investors Insurance Co. of America,*[58] the court rejected the claimant's argument that the making of the *inspection* itself affirmatively created a duty on the carrier to *make the premises safe.* The court concluded, "[I]f in fact the carrier undertook these inspections it did [so] solely for its own benefit. . . . It is well settled that one may not be liable for such conduct if it's for one's own protection and not for the purpose of aiding another."[59] (New York does not recognize the existence of third-party beneficiaries to a contract, and, hence, the court held the plaintiff could not pursue her claim, on the basis that the contract of insurance was for her benefit.[60] Where other state law is contrary to this position, the results also may be different.)

Insurance carriers do not necessarily enjoy the same immunity from suit that workers' compensation usually affords the employer. Thus, if it is found they do have an independent duty to the employee,[61] they may be sued separately for negligence. Recent case law demonstrates a trend toward holding the insurance company liable for negligent inspections under certain circumstances.

A duty also can be created where none legally exists if the actor gratuitously proceeds to act without due care and is acting for the benefit of another—that is, a duty can be gratuitously assumed even if none legally existed.

In *Nallan v. Helmsley-Spear, Inc.,*[62] the landowner had no duty to provide a safe place to live. When a guard was provided, however, the court held that the landowner had assumed this duty and was held liable for negligence in the nonperformance of this service.

Mountain Pride Farms, Inc. v. Dow Chemical Co.[63] held that an insurance company that agreed to render its client loss-control and fire-protection services was under a duty to exercise reasonable care in fulfilling the agreement, and if it failed it could be held liable. In this case, Dow Chemical Company was

*For example, in certain situations he conceivably could subpoena the OSHA inspector to testify as an expert witness on behalf of plaintiff.

†While cost-benefit analysis is still the legal standard for negligence cases, it is often acknowledged more in its breach than in its observance.

‡It may, however, be possible to bring a suit against the independent safety contractor.

sued by the purchaser of its styrofoam insulation for failure to disclose the flammability of the material.* Dow was allowed to "implead" (bring a suit against) its insurer, claiming that the insurer was negligent in rendering loss-control services, that Dow relied on the insured's services, and that the insurer thus was jointly liable for plaintiff's fire loss on the same theory.†

The court ruled that an insurance company that was alleged to have undertaken loss control and fire prevention solely as a service to its client, and *not* for its own protection "to reduce risks that might give rise to liability on the policy," could be held liable.

Thus, while an employer has a duty to provide a safe place to work,[64] he has no duty to be a guarantor of safety.[65] Extraordinary actions are not necessary, but if the employer assumes them—and carries them out negligently—he may be subject to liability.

Should a company decide to seek outside expertise, two routes are available that later may help protect the employer from liability either for negligent selection of the expert services or for failure to follow the recommendations of the safety contractor:

1. Retain the independent contractor expert expressly for the employer's company's benefit, incidental to procurement of insurance and lowering of premiums.
2. Retain the expert through an attorney (company or outside) incidental to investigation and defense of a possible claim, perhaps in response to a case report of incidents in other similar industries.

Understand the difference between the use of the words *survey* and *inspection*. Insurance companies wishing to protect themselves from lawsuits should insist upon the "disclaimer" mentioned earlier. Also, they should use the phrase "safety survey," which in the industry implies providing general advice, as opposed to "inspecting" for specific hazards. Employers seeking to rely on the insurance "inspections" both factually and legally should take note.

24.2.4 Preemployment Physicals and Screenings

Other common and frequently necessary precautions taken by employers, including preemployment physicals and screenings, could subject the employer to suit. For example, NIOSH recommends preemployment screenings for workers who may be assigned tasks involving exposure to fluoride welding fumes.

Negligent performance of a preemployment physical and failure to detect an underlying condition—even one that may be unrelated to the purpose of the screening—may subject the company doctor and the employer to lawsuit for malpractice.[66] Thus, the employer should consider advising the prospective employee that the company doctor and nurse are performing a preemployment screening designed only to determine if the employee is fit to carry out a particular job.[67] If any other condition is found, the employee may be advised, but must then be told to seek his own medical care. The employee should be told that in no event is he to rely on the screening as a substitute for an examination by his own personal physician, and that if no adverse condition is reported to the employee he should not assume that he is otherwise healthy. He should be advised that the screening is being performed solely for the benefit of the employer, so that the employee is placed in a position where he can best perform company duties.[68] Ideally, the disclaimer should be in writing, written by a lawyer familiar with the company, in conjunction with the company doctor. It should be signed by the employee and a copy kept in his personnel file.

Preemployment screenings also have been criticized for their possible inappropriate use. Genetic and gender-connected screening[69] has come under attack as being a method of discrimination. Proponents of such screening claim, however, that it can provide vital protection for workers by enabling their removal or exclusion from jobs involving risks, such as exposure to hazardous substances, to which they are particularly susceptible.

Several companies that instituted policies of excluding fertile women from certain jobs because of the possible teratogenic effects of some hazardous substances have been criticized for using these guides as invidious methods of sex discrimination, since male workers are permitted to work in these environments. What will occur if there is found to be a teratogenic effect causing sperm defects has yet to be determined, both legally and medically. Will the company be at increased legal risk from suit from those offspring whose fathers were exposed to teratogenic chemicals, where the company evidenced knowledge of potential hazards of these chemicals by excluding women from such exposure?

Most experts conclude that since genetic screening is in its infancy, it is inappropriate for use in a routine occupational setting. Misuse can subject the employer to claims of sex and race discrimination and violation of employee privacy, and to outrage from unions. On the whole, most employers and industries, even if they once had genetic screening programs, have discontinued their use.

*Failure to warn.

†This requirement of liability on the same theory is not necessary in an impleader action.

24.3 WHEN IT'S TOO LATE FOR PREVENTION

In mass tort cases, such as Agent Orange, asbestos, and Love Canal, the number of claims filed and the frequency of their filing can be virtually overwhelming. Case-handling strategy must be evaluated at the outset in light of the possible repercussions the handling of one case may have on the thousands of others pending or yet to be filed.

It frequently may be tempting to settle a case at the outset, thereby avoiding most of the cost. However, settlement often begets settlement, and an initial policy of using settlement to foreclose further costs of trial and appeal sometimes defeats that very purpose by inviting further lawsuits. This seems to be especially so in the case of mass chemical or toxic substance exposures.

24.3.1 Selection of Counsel

The first step once the summons and complaint are received is to notify the insurance carrier, broker, and legal department. Selection of counsel is the next step. Some insurance policies allow the policyholder to select its own counsel. Some insurance companies will invite the policyholder's input even if not required. Even if the insurance company insists upon the selection of counsel, the company's legal department should monitor all cases that may have broad impact or engender negative publicity. The same goes for cases concerning punitive damage claims or any other claim that may not be covered by insurance. Where a potential conflict of interest exists, the insured may be able to insist on its choice of selection of counsel to be paid by the insurer.[70]

In terms of complexity and cost, products causing workplace injuries have become the major products liability issue of the 1980s. In addition to traditional damages for pain and suffering, where a company causes environmental hazards, such as what might result from dumping toxic chemicals, damages may be obtainable for impairment of quality of life, for emotional injury caused by fear of cancer, and for continued medical surveillance and future medical services.[71*]

In order to handle these cases adequately, especially the complex question of what caused the alleged harm, a team of lawyers with integrated specialties is advised. The proposed team would have experts in Products Liability Law, Environmental Law, Occupational Health Law, and Insurance Law.

Where mass tort claims are involved, or claims in any way based on epidemiological data (signifying that additional claimants will soon appear), consideration should be given to appointing national counsel to monitor and coordinate all defenses (i.e., assure a consistent legal defense posture), and to insure that counsel in each jurisdiction does not inadvertently make an admission on behalf of the company, or does not waive any defenses or privileges, which might have far-reaching impact in binding other lawyers throughout the country. If national counsel is not appointed, a unit of the in-house lawyer's office should be appointed for this purpose.

Planning Legal Strategy

Before any specific legal strategy is implemented (including whether to settle or to try), consideration must be given to:

1. Existence of "dirty laundry" in company records, which might motivate early settlement.
2. A change of forum (location of the suit) because of the following factors.
    a. Adverse publicity.
    b. Adverse case law (precedent).
    c. Adverse jury findings (i.e., a history of high jury verdicts).
    d. Judicial bias.
3. Gathering of all possible defenses, old, new, and "creative."
4. Timing and availability of legal motion practice.

Other Considerations

How to handle even one claim may be a sensitive matter that has far-reaching ramifications. Balanced in a wise decision are: (a) the legal defenses available; (b) the factual defenses available; (c) a knowledge of the seriousness of plaintiff's claim, his likelihood of rehabilitation, his past medical and litigation history, and the reputation of the plaintiff's attorney; (d) considerations of ultimate company aims and policy; (e) the cost of settling versus the cost of trial; (f) if a decision is made to settle, the timing of the settlement; and (g) the extent of insurance coverage available, and the relationships between policyholder and insurance company.

*The *Ayers* case also demonstrates that courts may continue to bend the legal rules in favor of plaintiffs.

INVESTIGATION OF PLAINTIFF'S CLAIM

In order to defend or assess the plaintiff's case, it is necessary to gather as much information as possible through investigation, as quickly as possible. Examples of such information, which may later be used as trial evidence, are witness deposition testimony; documents; company records; reports; hospital, medical, and pharmaceutical records, including x-rays and spirographs relating to the plaintiff's injury or disease; workers' compensation records; company personnel records, including attendance records; governmental studies; and scientific data and articles.

Where the seriousness of plaintiff's claim is disputed, hidden movies have been known to be taken demonstrating plaintiff's "miraculous" recovery.

Other factors that may account for the plaintiff's illness or disease should be investigated, including other occupational or environmental exposures, past medical or familial conditions, and geographic prevalence of a particular disease (such as a higher evidence of cancer in New Jersey). Appropriate investigation often includes interviews of witnesses and photographs.

Some people think that an assessment of how the accident happened with emphasis on prevention and implementation of additional safety measures is not a good idea, fearing that this might be construed as an admission of negligence.

On the other hand, reckless disregard of the condition causing the incident and failure to correct obvious safety hazards would normally be much more burdensome to the company. It is not a good idea to delay immediate implementation of necessary safety measures or changes in procedure to incorporate additional safeguards just to prevent plaintiff's attorney from trying to use this "conduct" of implementing new safety features as an admission of prior poor safety practice, although this may indeed happen. Thus, on balance, it is probably better to investigate rather than put one's head in the sand.

One suggestion is that the past-accident-assessment meetings be conducted with a company attorney, with an emphasis on addressing potential litigation and preventing future litigation. This may be sufficient to invoke the attorney-client privilege to prevent its being used at trial.

24.3.2 To Settle or To Try

Different industries have had very different levels of success in litigation. To date, the tobacco industry has very successfully countered claims against it. By contrast, the asbestos industry is being fiscally bled by its claimants. What is interesting is that the injury claimed by plaintiffs in both types of lawsuits is the same disease: lung cancer.

SETTLEMENT

Stand Up and Say "Ready for Trial." If your company has decided upon settlement rather than trial, one very efficient method for precipitating settlement is for you to be ready, willing, and able to try the case—preferably before the plaintiff is.

One veteran litigator has voiced an overall litigation philosophy that boils down to one simple directive: "Stand up and say 'ready for trial!' " This eminent trial lawyer noted that "insecure" clients as well as perfectionist litigators have to "bite the bullet sooner." In condemning discovery abuse (i.e., taking years for mutual discovery processes), he stated that "clients have to play a part in solving abuse. They must be willing to [take] risks. The [discovery] practice in the last ten years has been contrary to the interests of clients and the bar."[72]

It must be noted that most good trial lawyers have the capacity to assimilate vast quantities of information in a short period of time. This phenomenon of data compression is accelerated under the pressures of trial. Many defense lawyers take advantage of this by forcing the plaintiff to try a case before they themselves are prepared, knowing that they will likely have an additional two to three weeks to continue getting ready while the plaintiff's attorney is trying his case.

Sometimes it may be wiser to surprise the plaintiff and force him to trial unexpectedly, especially if settlement is the ultimate objective. The unprepared and surprised plaintiff is often quite eager to settle.

THE INDUSTRY SETTLES OR FIGHTS BACK

One factor that must be considered in evaluating the cost of a settlement is its long-range value. While a mass "settlement" may be considered reasonable—or even cheap—at face value, if a company has a valid legal or medico-legal argument that will arise in other cases, or in cases involving other products manufactured by the company, it should consider the feasibility of litigating the legal point (usually done on motion papers), and appealing this point, if necessary, as far as possible. The precedential value of a favorable decision that may apply to other cases (and other products) may far outweigh the benefits of settling one case or one group of cases.

It must be noted, however, that some venerable lawyers believe a bad settlement is better than a good lawsuit.

JOINT SETTLEMENTS

It has been estimated that 25,000 Americans have filed lawsuits against asbestos companies. In the wake of the asbestos controversy, four companies have filed for bankruptcy. "Highly publicized awards have encouraged some personal injury lawyers to recruit as clients people with little or no discernible asbestos illness. As a result, defense attorneys like to point out, numerous meritless claims have been filed against the industry."[73]

Moreover, in an effort to settle these very questionable cases, there have been reports that some plaintiffs' attorneys engage in "joint settlement requests." In these joint settlements a demand is made to settle a group of cases—some obviously frivolous, some questionable, and some clear-cut cases with high potential for jury sympathy. While a defendant would seriously consider settling the case with a severely injured plaintiff, settling the frivolous case is unfair both to the defendant and to the rest of the group of plaintiffs, who must share their award with someone who has no legally valid claim. This practice not only is questionable but also has the effect of forcing higher and higher settlements, and may ultimately deprive a seriously injured worker of any compensation if the "pot of money" is used up.

If the group consists of similarly injured and exposed plaintiffs, then a group settlement can be a time-saver for all concerned. If it consists of some egregiously injured claimants and some with no injury at all, the former can be used as a coercive tactic to force defendant's lawyers to overpay the value of all cases. Because of the volatile nature of one or two of the cases in the group, which defendant's lawyers may be fearful of trying, plaintiff may be able to squeeze more money out of the company for the entire group.

24.3.3 Trial

COMMITMENT TO APPEAL

While it may seem strange to address the question of "appeal" before discussing the actual trial, it is recommended that that is precisely what should be done. If it is decided that a case should be tried to a conclusion (and the first few cases that are to be tried on a particular issue should be very carefully selected), commitment to appeal should be made at the outset in the case of an untoward verdict, if reversible error is found to exist. This commitment to appeal should be made at the onset. If the decision is made to try to verdict, the ultimate decision sought is a favorable one for the defendant. If this is not reached at the trial court, then all efforts should be made to obtain this goal at an appellate level. Abandoning the litigation at the trial level is tantamount to abandoning ship and may be more detrimental to the company than never having tried the case at all.

While *juries* may not be strict in requiring that plaintiff prove each element of his lawsuit, more often *appellate courts* are. Thus, while a manufacturer or employer may be disappointed at losing a lawsuit at the trial court or hearing level, it may find a more receptive tribunal on appeal.

The appeal, however, must often reach the highest level to be resolved with "blind justice"—that is, on the law, without regard to any emotional considerations.[74] Reliance on numerous appeals, however, not only is costly but also can generate negative publicity in the process, and thus the risks of appeal must be balanced against the benefits (e.g., costs of appeal, the possibility of a decision in favor of the plaintiff, or a judicial directive permitting the plaintiff to try the case again) before an informed decision can be made on whether trial or settlement is the appropriate route.

EARLY MOTION PRACTICE

A motion to change the forum must be made at the earliest possible moment. To avoid waiving this option, it may be wise to have the defense machinery in place *before* a lawsuit is brought. And since all defenses and some dispositive legal motions usually must be made early in the proceedings, those on the defense team should be familiar with the industry, the company, the diseases or issues claimed, and the legal defenses possible. A creative approach also pays off.[75]

CHOOSING EXPERTS

Sometimes it may be helpful to have two expert teams—one to work "behind the scenes" and one to appear at trial. There are several reasons for this. The "behind the scenes" experts, if retained by an attorney for the purpose of helping to prepare for trial, may be able to prepare reports that are not discoverable. These experts can pinpoint trouble spots and weak points in the case, which then can be addressed by the legal defense team.

The trial witness can be asked to address only specific issues. While, clearly, the trial expert witness must be prepared to deal with the problem areas at deposition and trial, at least he has not committed any negative comments to writing. Also, by dealing with two experts, the company has the benefit of getting two approaches to deal with any problem and to develop defenses.

Often, some scientists are reluctant to appear at trial for such reasons as a bad experience, or fear of being regarded as partisan or biased by the scientific community if they appear on one "side" or the other. However, these same scientists sometimes will agree to work behind the scenes, and their insights can be invaluable.

Choosing a Trial Expert

While the more minds one has working on a case the better it is, it is often best to limit the number of trial experts to the bare minimum. The more witnesses who testify, the greater the likelihood they will contradict each other, say something adverse to your case, or simply bore the jury or the judge. Also, parading in a battery of experts before a jury often inflates the value of a case in their eyes.

Only *one* witness should testify on a particular subject to avoid contradictory or redundant testimony. If there is more than one defendant and no conflict of positions, the defendants should consider "sharing the expert," perhaps having a "backup" in case the initial expert "bombs."

Legal Certainty vs. Scientific Certainty. For an expert to testify in a civil case (negligence, malpractice, strict liability in tort), the expert must be convinced to a reasonable degree of scientific or medical certainty. This means that, based on his own experimentation, research, and reading customarily done in the field, "more probably than not" his opinion is valid. He cannot express an opinion of some other scientist unless he has relied upon it in formulating his own opinion. He may not base his opinion purely on what another scientist told him. The quantum of certainty necessary if expressed in mathematical terms would be 51% or "more probably than not."

This legal requirement invites two problems. The first is the scientist, used to a 95% margin of confidence requirement, who has no opinion on anything, because, after all, it is virtually impossible to be 95% certain about any new issue in science.

The second is the scientist who, when asked if he has "an opinion" regarding a certain matter, has an opinion about everything, not realizing that what is meant is that he must have "an opinion with a reasonable degree of medical certainty." He does not realize he must have the expertise to back up his opinions. The opposition lawyer often can embarrass a witness who is such an "expert" that he has an opinion (with a reasonable degree of medical certainty) on every subject from ambulance care to x-ray therapy.

It is thus wise to explain carefully the differences between legal, scientific, and lay parlance to prospective experts.

The proper format to elicit any conclusion on the part of an expert is to ask, "Do you have an opinion with a reasonable degree of scientific certainty on ______?" (whatever the issue). Savvy lawyers routinely object unless the question is phrased in this form.

The words "with a fair or reasonable degree of scientific certainty" used in a trial or deposition often remind the overeager or overcautious expert of the requisite quantum of expertise and certainty necessary in order for the witness to render an opinion.

Causation

At present, *causation* remains the common element the plaintiff must prove in workers' compensation, negligence, and strict liability in tort (products liability) claims—that is, that the defendant's acts, its product, or its workplace actually caused the plaintiff harm. In toxic chemical cases, the most common harm alleged is cancer.

In current theories of carcinogenesis, scientists have proposed that multiple causative factors are required before irreversible cancer cells occur. It is often impossible to determine which agents are the substantial causative factors, and even if that is established scientifically, it is almost more difficult to convince a jury. Long latency periods between job exposure and disease manifestation further cloud the issue. Most doctors are not sufficiently attuned to the myriad of occupations that have been associated with an increased risk of cancer. Rather than taking an in-depth occupational history, they often are satisfied if they can identify one well-known "popular" culprit—such as asbestos or dioxin—even if the individual's exposure to such a product was minimal or short-term.[76]

Because of hysteria often accompanying alleged exposure to toxic substances, many claimants attribute imagined or ubiquitous diseases (such as lung cancer) with many other causes to the workplace exposure.

The plaintiff usually introduces into evidence epidemiological studies demonstrating an increased incidence of disease associated with exposure to the workplace substance over that which would be expected in the general population. While this data in no way demonstrates causation in a particular individual (rather, it demonstrates population susceptibility), this commonly has been accepted as sufficient evidence for a jury to find the defendant's product *caused* the particular plaintiff's disease.

In defending cases based on epidemiological evidence, it is necessary to investigate plaintiff's medical history or employment history to determine whether he may have been exposed to other causative elements. Such an investigation requires thorough questioning of the plaintiff at deposition, as well as obtaining and carefully reviewing the plaintiff's medical records, hospital charts, workers' compensation records, and other records.

Before a realistic assessment of the plaintiff's assertions can be made, a thorough search of all relevant medical literature should be completed. This thorough literature search also needs to be done to evaluate a plaintiff's "state of the art" claim—that the defendant knew or should have known of the hazards incident to its particular product and failed to place adequate and appropriate warnings. Usually the initial articles are isolated case reports,[77] which usually are not sufficient to attribute knowledge to a manufacturer.

Because the acceptance of scientific knowledge by the medical, legal, and regulatory communities is a slow, tortuous, and circuitous path, the arguments on "state of the art" are vigorously pursued by plaintiffs' attorneys. Experts frequently disagree on exactly when an industry or company should have known of the existence of a particular hazard. Other than the general epidemiological rule that adequate dissemination and acceptance of potentiality of causation occur after two or three epidemiological surveys are published, there are no precise guidelines. Even this rule is punctuated by *caveats* such as the size of the study, reproducibility, method error, existence of studies with contrary conclusions, the reputation of the investigator, and the like.

Publicity, Communication, Juror Surveys

Hiring a communications expert to work in conjunction with the legal staff is something that should be considered. Remember that ultimately the decision makers (judge or jury) may well be influenced before they ever reach the courtroom. It must be noted the publicity should not be employed to influence sitting jurors. All publicity should be employed before the jurors reach the courtroom.

While extensive *voir dire* (pretrial questioning) of prospective jurors is permitted in state courts, this is not the case in federal court. Nor is *voir dire* allowed of judges. One never knows if the judge suffered personally (or via a loved one) the same disease as the claimant. Even the state court *voir dire* may not elicit all existing juror prejudices. Further, even nonbiased jurors may be influenced by pretrial publicity—the usual "blood and gore" stories by investigative reporters. To counter this "subtle" means of influencing prospective jurors before they reach the courtroom, some members of the tobacco industry* maintain an active (and effective) publicity campaign (lately *very* active) coupled with a "no settlement" policy on lawsuits.

Jury surveys have been gaining receptivity. One possible use for jury survey teams is to investigate possible biases and to design publicity to counter misconceptions of the public *before* any cases go to trial and to address jury concerns adequately *during* trial, as well as to recruit expert witnesses who will be received favorably by the jury. The jury survey team should thus be involved with any publicity either before or during trial. Publicity during trial can be very dangerous from a legal perspective—it can be misconstrued as an attempt to influence the jury. Publicity should be limited to general portrayal of the company in a popular light and countering popular misconceptions of the company and its products.

Another innovative idea is to have a "shadow jury" while the case is in progress. This is a "mock jury" of like demography, sex, and attitudes as the real one. The attorneys can question the shadow jury each day at the conclusion of testimony to determine what was clear, what points made the greatest impact, and what questions the jurors have, and then attempt to address these items on the following days of trial.

Postscript: The Relationship Between Smoking and Workplace Chemicals

It has been said that cigarette smoking is responsible for 85% of lung cancers in the United States.[78] Epidemiological studies claim that a combination of cigarette smoking and workplace chemicals causes significant amounts of lung cancer and increases the carcinogenic effects of the workplace chemical acting alone. Whether this places a burden on the employer to ban smoking in the workplace or for unions to educate their members to the dangers of the increased potency of cigarette smoke in conjunction with workplace chemicals remains to be seen.[79]

The dangers of passive smoking are being debated. Nonsmokers are beginning to sue employers, seeking to ban smoking in the workplace. They argue that an employer has a duty to provide employees with a reasonably safe place to work.[80] In *Shimp v. New Jersey Bell Telephone Co.*,[81] the court took judicial notice of the toxic nature of tobacco smoke. The court stated, "[I]t is reasonable to expect an employer to foresee health consequences and to impose upon him a duty to abate the hazard." In *Smith v. Western Electric Co.*,[82] the court held the plaintiff employee was entitled to an injunction preventing his employer from subjecting him to tobacco smoke. *But see contra Gordon v. Raven Systems & Research Inc.*,[83] where the court held that "the employer's general duty to provide a reasonably safe place to work, does not mean the employer . . . owes a duty to adapt his workplace to the particular sensitivities of an individual employee."

While these cases revolve around the allergic or hypersensitive nonsmoker, it is not unforeseeable that a nonallergic nonsmoker could claim workplace disease was caused by being exposed to a combination of noxious chemicals and cigarette smoke at work. Management, therefore, should seriously consider banning smoking in all areas where toxic chemicals are used.

*It has countered successfully every civil lawsuit, notwithstanding very vocal scientific and medical positions that smoking is responsible for 85% of all lung cancer.

24.4 CONCLUSIONS

The recent proliferation of lawsuits relating to workplace injuries will surely have an impact on safety personnel, management, and in-house legal teams, who may have not had previous experience in these matters. Bankruptcies, inopportune precedents, and a national wave of hysteria are just a few of the results of a lack of overall direction in handling workplace injuries. Acknowledging this fact may be the first step in handling this vexing problem.

Although steps will have been taken to minimize suits, the likelihood of some lawsuits is almost inevitable. The real goal is to prevent claims by maximizing safety precautions, implementing employee education, and encouraging use of personal protective devices where appropriate and necessary, and to mount an aggressive and knowledgeable defense.

REFERENCES

1. Atkinson, L. M., The Search for Liability: Checklist for Trial Lawyers, *The National Law Journal,* March 1, 1982, p. 30.
2. Lewin, T., Criminal Onus on Executives, *The New York Times,* March 5, 1985, p. D2.
3. *Borel v. Fibreboard Paper Products Corp.,* 493 F.2d 1076 (5th Cir. 1973).
4. Chen, E., The Law: Asbestos Litigation Is a Growth Industry, *The Atlantic,* **254,** No. 1 (1984):24.
5. Chenoweth, D., Health Management Consultants Use Diverse Programs in Achieving Goals, *Occupational Health & Safety,* **54,** No. 2 (1985):53-56.
6. Gruson, L., Employers Get Tough on Smoking at Work, *The New York Times,* March 14, 1985, p. B1.
7. Blackwood, M. J., Health Risks of Smoking Increased by Exposure to Workplace Chemicals, *Occupational Health & Safety,* **54,** No. 2 (1985):23-27, 81.
8. Lewin, T., Pharmaceutical Companies Are Hit the Hardest, *The New York Times,* March 10, 1985, p. F1.
9. *Sindell v. Abbott,* 26 Cal. 3d 588, 607 P.2d 924, *cert. denied sub nom. Squibb & Sons, Inc. v. Sindell,* 449 U.S. 912 (1980), enterprise liability (*Hall v. E.I. Du Pont De Nemours & Co., Inc.,* 345 F. Supp. 353 (E.D.N.Y. 1972)), the alternative liability (*Summers v. Tice,* 33 Cal. 2d 80, 5 A.L.R.2d 91 (1948).
10. *Bichler v. Eli Lilly & Co.,* 79 A.D.2d 317, 436 N.Y.S.2d 625 (1st Dep't 1981), *aff'd,* 55 N.Y.2d 571 (1982).
11. *Hall v. E.I. Du Pont De Nemours & Co., Inc.,* 345 F. Supp. 353 (E.D.N.Y. 1972).
12. *Loch v. Confair,* 63 A.2d 24, 361 Pa. 158 (1949).
13. *Keene Corp. v. INA,* 667 F.2d 1034 (D.C. Cir. 1981), *cert. denied,* 455 U.S. 1007 (1982).
14. *Berkey Photo Inc. v. Eastman Kodak,* 457 F. Supp. 404 (S.D.N.Y. 1978), *aff'd in part and rev'd in part,* 603 F.2d 263 (2d Cir. 1979), *cert. denied,* 444 U.S. 1043 (1980).
15. Kiecheck, W. III, The Strange Case of Kodak's Lawyers, *Fortune,* May 8, 1978, p. 188.
16. SmithKline Is Fined for Failing to Report Side Effects of Drug, *The New York Times,* Feb. 26, 1985, p. A24.
17. Rheingold, P. D., The MER/29 Story—An Instance of Successful Mass Disaster Litigation, 56 Cal. L. Rev. 116, 120 n.15 (1968) (quoting *Hearings Before the Subcomm. on Reorganization and International Organizations,* 88th Cong., 1st Sess., pt. 3, at 908 (1963)).
18. *Roginsky v. Richardson-Merrell, Inc.,* 378 F.2d 832 (2d Cir. 1967).
19. *Toole v. Richardson-Merrell, Inc.,* 251 A.C.A. 785, 810-811 & n.3, 60 Cal. Rptr. 398, 416-417 & n.3 (1967), *reh'g denied* (Sept. 7, 1967).
20. *City of Newport v. Fact Concerts Inc.,* 453 U.S. 247, 267 (1981). *See also* Restatement (Second) of Torts § 908(1).
21. *Knippen v. Ford Motor Co.,* 546 F.2d 993, 1002 (D.C. Cir. 1976).
22. *In re Marine Sulphur Queen,* 460 F.2d 89, 105 (2d Cir.), *cert. denied sub nom. United States Fire Ins. v. Marine Sulphur Transport Corp.,* 409 U.S. 982 (1972).
23. Letz, G. A., et al., Two Fatalities After Acute Occupational Exposure to Ethylene Dibromide, *J. A.M.A.,* **252,** No. 17 (1984):2428-31.
24. *Blankenship v. Cincinnati Milacron Chemicals, Inc.,* 69 Ohio St. 2d 608, 433 N.E.2d 572, *cert. denied,* 459 U.S. 857 (1982).
25. *Hansen v. Johns-Manville Products Corp.,* 734 F.2d 1036 (5th Cir. 1984), *cert. denied,* 470 U.S. 1051 (1985).

26. *Id.* at 1041 (quoting *Jackson v. Johns-Manville Sales Corp.*, 727 F.2d 506 (5th Cir. 1984), *cert. denied*, 106 S. Ct. 3339 (1986)).
27. *Id.*
28. Silicosis Death Called Homocide, *Business Insurance*, **18,** No. 29, 1984:2.
29. Lewin, T., Criminal Onus on Executives, *The New York Times*, March 5, 1985, p. D2.
30. Goerth, C. R., Employers Indicted for Murder in Connection with Worker Death, *Occupational Health & Safety*, **53,** No. 9. (1984):59-62.
31. *Mandolidis v. Elkins Industries, Inc.*, 161 W. Va. 695, 246 S.E.2d 907 (1978); *Blankenship v. Cincinnati Milacron Chemicals, Inc.*, 69 Ohio St. 2d 608, 433 N.E.2d 572, *cert. denied*, 459 U.S. 857 (1982). *But see contra Kofron v. Amoco Chemicals Corp.*, 441 A.2d 226 (Del. 1982).
32. *Kofron v. Amoco Chemicals Corp.*, 441 A.2d 226 (Del. 1982).
33. *Nallan v. Helmsley-Spear, Inc.*, 50 N.Y.2d 507, 407 N.E.2d 451, 429 N.Y.S.2d 606 (1980).
34. Dr. Corn's Deposition: Testimony Continues on Unibestos Plant, Management's Attitude Towards Health, *Asbestos Litigation Reporter*, June 15, 1984, pp. 8496, 8530-40 (Pittsburgh Corning).
35. Chen, E., Justice Takes Slow Path in Asbestos Cases, *The Los Angeles Times*, Feb. 13, 1984, p. 1.
36. *Hansen v. Johns-Manville Products Corp.*, 734 F.2d 1036, 1043 (5th Cir. 1984), *cert. denied*, 470 U.S. 1051 (1985).
37. Melton, C. M., Jr., Emergency Preparedness Assures Minimal Damage to Personnel, Plant, *Occupational Health & Safety*, **53,** No. 7 (1984):50-52, 61.
38. UAW's Right-to-Know Bpeakthrough, *Chemical Week*, Jan. 23, 1985, pp. 19-20.
39. *Adams v. Union Carbide Corp.*, 737 F.2d 1453 (6th Cir. 1984).
40. Lave, L. B., *Quantitative Risk Assessment in Regulation* (Washington, D.C.: The Brookings Institution, 1982).
41. MacMahon, B., et al., Coffee and Cancer of the Pancreas, *New England Journal of Medicine*, **304,** No. 11 (1981):630-33.
42. *Beshada v. Johns-Manville Products Corp.*, 90 N.J. 191, 447 A.2d 539 (1982).
43. *Cover v. Cohen*, 61 N.Y.2d 261, 276-277, 461 N.E.2d 864, 473 N.Y.S.2d 378 (1984); *Radziul v. Hooper, Inc., Cryovac Division of W.R. Grace & Co.*, 125 Misc. 2d 362, 479 N.Y.S.2d 324 (Sup. Ct. Monroe County 1984).
44. Keeton, W. P., Products Liability—Inadequacy of Information, 48 Tex. L. Rev. 414 (1970). But see *W.T. Jacobs v. Technical Chemical Co.* (Tex. Civ. App. Aug. 4, 1971).
45. *Bolm v. Triumph Corp.*, 33 N.Y.2d 151, 305 N.E.2d 769, 350 N.Y.S.2d 644 (1973).
46. *Johnston v. Upjohn Co.*, 442 S.W.2d 93 (Mo. Ct. App. 1969), which holds that justifiable ignorance by the manufacturer imposes no duty to warn.
47. *Rosenbrock v. General Electric Co.*, 236 N.Y. 227, 140 N.E. 571 (1923); *Littlehale v. E.I. Du Pont De Nemours & Co.*, 268 F. Supp. 791 (S.D.N.Y. 1966); *Martinez v. Dixie Carriers, Inc.*, 529 F.2d 457 (5th Cir. 1976).
48. Miller, S. R., Preserving the Privilege, *Litigation*, **10,** No. 4 (1984):20-23, citing *Upjohn Co. v. United States*, 449 U.S. 383, 390 (1981).
49. 8 Wigmore, Evidence § 2292 (McNaughton rev. 1961).
50. *See Control Data Corp. v. International Business Machines Corp.*, 16 Fed. R. Serv. 2d (Callaghan) 1233 (D. Minn. 1972); *IBM v. United States*, 471 F.2d 507 (2d Cir. 1972), *rev'd on jurisdictional grounds*, 480 F.2d 293 (2d Cir. 1973) (en banc), *cert. denied*, 416 U.S. 979 (1974).
51. Quade, V., Sealing the Computer Leaks, **70,** No. 25 *A.B.A. J.* (1984):25-27.
52. *Teachers Ins. & Annuity Ass'n of America v. Shamrock Broadcasting Co.*, 521 F. Supp. 638, 644-646 (S.D.N.Y. 1981).
53. *American Textile Mfrs. Inst. v. Donovan*, 452 U.S. 490 (1981).
54. *Beshada v. Johns-Manville Products Corp.*, 90 N.J. 191, 447 A.2d 539 (1982).
55. *Rosenhack v. State of New York*, 112 Misc. 2d 967, 447 N.Y.S.2d 856 (Ct. Cl. 1982).
56. *Bivas v. State of New York*, 97 Misc. 2d 524, 411 N.Y.S.2d 854 (Ct. Cl. 1978).
57. *Pulka v. Edelman*, 40 N.Y.2d 781, 358 N.E.2d 1019, 390 N.Y.S.2d 393 (1976).
58. *Nieto v. Investors Ins. Co. of America*, N.Y.L.J., Dec. 14, 1981, at 18, col. 5 (Sup. Ct. Kings County 1981).
59. *Home Mut. Ins. Co. v. Broadway Bank & Trust Co.*, 53 N.Y.2d 568, 428 N.E.2d 842, 444 N.Y.S.2d 436 (1981).
60. *Ramos v. Shumavon*, 21 A.D.2d 4, 247 N.Y.S.2d 699 (1st Dep't), *aff'd*, 15 N.Y.2d 610, 203 N.E.2d 912, 255 N.Y.S.2d 658 (1964).

61. *Cline v. Avery Abrasives, Inc.*, 96 Misc. 2d 258, 409 N.Y.S.2d 91 (Sup. Ct. Monroe County 1978).
62. *Nallan v. Helmsley-Spear, Inc.*, 50 N.Y.2d 507, 407 N.E.2d 451, 429 N.Y.S.2d 606 (1980).
63. *Mountain Pride Farms, Inc. v. Dow Chemical Co.*, 95 F.R.D. 400 (S.D.N.Y. 1982).
64. N.Y. Labor Law §§ 240, 241 (McKinney 1987).
65. *American Textile Mfrs. Inst. v. Donovan*, 452 U.S. 490 (1981).
66. *See Rastaetter v. Charles S. Wilson Memorial Hospital*, 80 A.D.2d 608, 436 N.Y.S.2d 47 (2d Dep't 1981); *Matter of Allen v. American Airlines*, 78 A.D.2d 917, 433 N.Y.S.2d 512 (3d Dep't 1980); *Botwinick v. Ogden*, 87 A.D.2d 293, 451 N.Y.S.2d 141 (1st Dep't 1982), *rev'd*, 59 N.Y.2d 909, 466 N.Y.S.2d 291 (1983).
67. Billauer, B. P., Will Workers' Compensation Protect the Company Doctor? N.Y.L.J., Feb. 1, 1985, at 1.
68. Billauer, B. P., The Legal Liability of the Occupational Health Professional, *Journal of Occupational Medicine*, **27**, No. 3 (1985):185–88.
69. Schechter, D., Genetic Screening in the Workplace, *Occupational Health & Safety*, **52**, No. 4 (1983):8–12.
70. *Rimar v. Continental Casualty Co.*, 50 A.2d 169, 376 N.Y.S.2d 309 (4th Dep't 1975).
71. *Ayers v. Township of Jackson*, 189 N.J. Super. 561, 461 A.2d 184 (Law Div. 1983). *See also* Ashman, A. and L. R. Reskin, Defendants Must Pay for Plaintiffs' Pretrial Medical Exams,________ A.B.A. J. **71**, No. 134 (1985):134–135.
72. *Legal Times of New York*, Dec. 26, 1984, p. 26.
73. Chen, E., The Law: Asbestos Litigation Is a Growth Industry, *The Atlantic*, **254**, No. 1 (1984):24.
74. *See Botwinick v. Ogden*, 87 A.D.2d 293, 451 N.Y.S.2d 141 (1st Dep't 1982), *rev'd*, 59 N.Y.2d 909, 466 N.Y.S.2d 291 (1983).
75. *See Keene Corp. v. INA*, 667 F.2d 1034 (D.C. Cir. 1981), *cert. denied*, 455 U.S. 1007 (1982).
76. Hadler, N. M., Occupational Illness: The Issue of Causality, *Journal of Occupational Medicine*, **26**, No. 8 (1984):587–93.
77. MacMahon, B., et al., Coffee and Cancer of the Pancreas, *New England Journal of Medicine*, **304**, No. 11 (1981):630–33.
78. Doll, Sir R., and Peto, R., *The Causes of Cancer* (London: Oxford University Press, 1981)
79. Blackwood, M. J., Health Risks of Smoking Increased by Exposure to Workplace Chemicals, *Occupational Health & Safety*, **54**, No. 2 (1985):23–27, 81.
80. Gruson, L., Employers Get Tough on Smoking at Work, *The New York Times*, March 14, 1985 p. B1.
81. *Shimp v. New Jersey Bell Tel. Co.*, 145 N.J. Super. 516, 368 A.2d 408 (1976).
82. *Smith v. Western Electric Co.*, 643 S.W. 2d 10 (Mo. Ct. App. 1982).
83. *Gordon v. Raven Systems & Research Inc.*, 462 A.2d 10 (D.C. 1983).

APPENDIX. Theories of Recovery

I. Negligence
  A. Elements
    1. Duty
      Examples:
      a. Of employer:
        i. To maintain a reasonably safe place to work.
        ii. To act as reasonably as any prudent employer under the same circumstances.
        iii. To advise employees and key personnel of risks and hazards incident to employment (to comply with right-to-know laws).
      b. Of manufacturer:
        i. To act as a reasonably prudent manufacturer under the same circumstances.
        ii. To comply with statutes, laws, and regulations incident to manufacture, warnings, testing of product, reporting of defects, and recall letters.
      c. Of insurer:
        i. To provide insurance and services contracted for.
        ii. To familiarize itself with risks incident of the business of insured for the purpose of underwriting.
        iii. To defend insured in case of suit.
        iv. If specifically requested and agreed, to provide adequate safety inspections and/or surveys of the workplace.
        v. To indemnify insured where provided.

d. Of owner of premises/general contractor:
   i. To maintain a safe place to work (i.e., N.Y. Labor Law §§ 240, 241).
   ii. To act as a reasonably prudent landlord, etc., under the circumstances.
e. Of coemployee (professional):
   i. To follow the company protocol and not to exceed its parameters.
   ii. To act as a reasonably prudent professional under the circumstances.
f. Of independent safety contractor, OSHA inspectors:
   i. To act as a reasonably prudent inspector under the circumstances.

B. Standard of Care
To act as a reasonably prudent: employer, owner/general contractor/manufacturer/coemployee/independent contractor under the circumstances.

C. Causation
The careless act or omission caused the injury or disease alleged.

D. Proximate Cause
The negligent act or omission directly caused injury or disease without any intervening or superseding cause, and the injury or disease was foreseeable.

E. Damages
The claimant suffered actual (not projected), provable damages; usually this occurs in the form of lost wages, medical expenses, pain and suffering, loss of services of the spouse, and property damage. Psychological damages are allowed under certain conditions if they are accompanied by physical injury.

F. Defenses

1. Workers' Compensation
When the Workers' Compensation Act applies, employers are strictly liable to employees for work-related injuries, without regard to the negligence of either party. However, the employee is limited to the remedy provided by the Act, and is barred from suing his employer.
Intentional torts are never covered by workers' compensation. Exceptions may also exist in cases where the employer is guilty of fraudulent concealment or gross negligence.

2. Contributory Negligence
When a plaintiff behaves below the standard of a reasonable person, and his act or omission is a contributing cause to harm he suffers, he might be barred completely from recovering for negligence by the defendant. Today, most jurisdictions are reluctant to allow a negligent defendant to escape liability on the basis of the plaintiff's contributory negligence, and consequently most cases will be resolved on the theory of comparative negligence, pursuant to which the relative fault of both parties is taken into consideration.

3. Comparative Negligence
Defendants may relieve themselves of at least some of the costs of a judgment against them when the focus of attention is shifted from liability to damages. Under the theory of "pure" comparative negligence, the damages awarded a plaintiff are reduced in proportion to the plaintiff's degree of fault.
Some courts completely bar plaintiff's recovery if the plaintiff's contributory negligence is equal to or greater than that of the defendant, and diminish recovery if the plaintiff's contributory negligence is less than that of the defendant.
An example of a plaintiff's negligence can include the fact that the plaintiff failed to use the product(s) in the manner and for the purpose for which it was intended.

4. Assumption of Risk
Under the theory of assumption of risk, a plaintiff may not recover for an injury caused by a danger he accepted voluntarily.
The requirements for this defense include:
a. The plaintiff knew of the danger;
b. He appreciated the nature and extent of the danger; and
c. He voluntarily exposed himself to the danger.

5. Failure To Mitigate Damages (The Rule of Avoidable Consequences)
After a legal wrong on the part of the defendant has occurred, a plaintiff is expected to behave as a reasonable person would to eliminate or reduce any subsequent injury. Many courts will bar recovery for damages that could have been avoided by reasonable conduct by the plaintiff.

6. Government Immunity (Sovereign Immunity)
Traditionally federal, state, and local governments cannot be sued in tort, except where they have consented to suit by statute.
The federal government has waived its immunity in certain cases by the Federal Tort Claims Act. Most states have enacted similar statutes on both the state and the city level.
Further, in 1983, the Supreme Court ruled that local governments could "henceforth"

be sued directly as employers. *See Monell v. Dep't of Social Services of New York*, 429 U.S. 1071 (1983).

7. Sophisticated User
   A plaintiff with expert knowledge may be expected to use that knowledge. Thus, a defendant sued for injury caused by a product or instrument could be spared liability if he proved that the user did not exercise care commensurate with his knowledge of the danger.
   A manufacturer may claim that he cannot be liable for failure to warn of any risk of danger in the use of his product if the employer or the contracting agent (e.g., the government) had equal or greater knowledge about the nature of any risk or danger involved and was in a better position to warn the plaintiff.
8. Government Contracts
   A defendant may try to escape liability by proving that its acts or omissions were mandated by a government contract. In such a case, the defendant merely followed official government specifications, the drafters of which should have known of any danger involved.
   This defense could be applied in other cases where purchaser specifications ("cites") were involved.
9. Defect Occurred After Product Left Control of Manufacturer
   If the defendant could prove that the defect in question was not present at the time the product left his custody, control, and possession, he might bar plaintiff's recovery.
10. State-of-the-Art
    A defendant cannot be held liable for failing to warn of a risk of danger if, at the time of production and distribution of the product, the industry neither knew nor should have known of the potential danger. *But see Beshada v. Johns-Manville Products & Corp.*, 90 N.J. 191, 447 A.2d 539 (1982).
11. Lack of Causation
    A defendant cannot be liable for an injury if the plaintiff fails to prove that his allegedly defectively dangerous product was the cause of the damage.
12. Procedural Problems
    A defendant may avoid liability if the plaintiff's suit is faulty on other legal rules. For example, the plaintiff might not be suing in a court that has jurisdiction over the defendant, or the plaintiff might be suing too late and be barred by a time limitation.

II. Workers' Compensation

A worker who was disabled (or died) during the course of employment as the result of either an accidental injury or an occupational disease can file a workers' compensation claim if he can prove that:

1. The injury arose out of and in the course of employment.
2. The occupational disease was the result of a distinctive feature of the kind of work performed by the claimant and others similarly employed.
3. If the claimant (injured worker) had a preexisting injury, aggravation of the injury was caused by an accident that arose out of and in the course of employment.

An employer can use the defenses listed below to mitigate or dismiss the employee's workers' compensation claim.

A. Statutory Defenses

1. Willful failure by employee to use safety device.
2. Violation of statute or commission of crime.
3. Statute of limitations has expired.
4. Willful misconduct defense (which is an active defense in only a minority of the states and even where enacted it is hardly ever enforced).

B. Defenses Apart from Statutory Defenses

1. Deviation from the course of employment.
2. The employee is/was expressly prohibited from doing the work of another.
3. Intoxication (can also be statutory defense).
4. Suicide or intentional self-injury (compensable only if the injury produces mental derangement and the mental derangement produces suicide).

III. Strict Liability in Tort

In many states, strict liability in tort is one of three alternative theories available to a claimant seeking damages from a merchant, the other two alternatives being negligence and breach of warranty, express or implied. Under the theory of strict liability in tort, a manufacturer is strictly liable for defective or hazardous products that unduly threaten a consumer's personal safety. The plaintiff in such a case does not have to prove that the manufacturer was negligent in any way, and any claim by the manufacturer that he was not negligent will be of no avail.

The justification for this stringent treatment of products liability cases is threefold. First, the costs of injuries from dangerously defective products can best by borne by the manufacturers, who can offset this liability by increasing the cost of potentially dangerous products. Second, some believe that the adoption of strict liability promotes accident prevention, although this view is debatable. Third, proving negligence is too difficult, costly, and unnecessary when a product defect is established.

The most challenging aspect of strict liability in tort is defining when a product is defective in such a way that renders it unreasonably dangerous. It can be unreasonably dangerous for any of the following reasons:

1. A product defect is present at the time of sale.
2. Failure of the manufacturer to warn of a risk related to the nature of the product.
3. A defective design.

Each of these possible defects warrants separate explanation.

A. Product Defect

A defect or flaw in a product is defined as an abnormality that was unintended and makes the product more dangerous than it would be without the flaw. Any danger that is greater than the intended risk of the product is considered, by law, an unreasonable danger.

A strict liability case based on a product defect insinuates that the manufacturing process suffered from a lack of quality control.

B. Failure to Warn

A product is considered unreasonably dangerous if the manufacturer fails to warn the public adequately of a hazard related to the way the product is designed. In other words, it is necessary that the manufacturer knew or should have known of the hazard, and failed to take precautions as a reasonable person would, in adequately warning the public.

C. Design Defect

Defects in the design of a product may be analyzed by two different approaches:

1. *The consumer-contemplation test.* Defines a design as defective when the product is more dangerous than the ordinary consumer would contemplate. This test is often considered ambiguous and difficult to apply.
2. *The danger-utility test.* Defines a design as defective when the magnitude of the danger outweighs the utility of the product. This test considers the feasibility of alternative designs in lieu of current state-of-the art knowledge.

IV. Defenses to Strict Liability in Tort

1. Defendants in strict liability cases would be wise to consider whether any contributory negligence, misuse, or intervening misconduct on the part of the claimant was a proximate cause of the resulting injury. Any such superseding cause might bar or diminish recovery, although such defenses accepted by courts are limited and rare.

   There are four categories of conduct that might affect a claimant's recovery:

   a. Use or maintenance misconduct by the purchaser/user.
   b. Misconduct by an intermediate seller or supplier.
   c. Alteration of a product.
   d. Misconduct by a third party.
2. At the time the defendant manufactured the product, medical, scientific, and state-of-the-art knowledge was ignorant of any significant risk of harm to the plaintiff.
3. As a matter of public policy, the plaintiff's claim is barred because the social utility and public benefit of the product outweigh the alleged risk.

V. Breach of Warranty

A warranty is a statement or representation made by a seller of goods referring to the quality or title of the goods.

A warranty can be express or implied. An action in contract may lie for a breach of either an express or an implied contract.

A. Express Warranty

An affirmation, description, sample, or written statement by the seller promising the product will conform to a certain standard of quality.

B. Implied Warranty

By law, a product sold is assumed to be merchantable and fit for the purpose for which the seller knows it is required.

1. Implied Warranty of Fitness

   When a retailer, distributor, or manufacturer has reason to know the purpose for which a product is required, and that the buyer is relying on the skill of the seller to provide suitable goods, it is assumed that the goods are fit for that purpose.
2. Implied Warranty of Merchantability

   A product sold is assumed to meet each of the following requirements:

   a. It meets the contract description.
   b. It is fit for the ordinary purposes for which the goods are used.
   c. It is adequately packaged and labeled.
   d. It conforms to the promises made on the container.

VI. Intentional Torts

A. Assault

One commits an assault when he causes another to fear that injury will be inflicted. The necessary elements include the intent of the actor to threaten or to injure, the apparent present ability to do so, and an intentional display of force. No contact is necessary for an assault to occur.

B. Battery

The actual use of the force threatened by an assault. A battery has not occurred without an unlawful contact.

Consent of the victim, under some circumstances, may constitute a dedense.

C. Nuisance

An annoyance that interferes with another's enjoyment of life or property. There are public, private, and mixed nuisances.

D. Trespass

An unlawful interference with another's person, property, or rights. Trespass usually involves intentional, unauthorized use of another's real property.

E. Defenses to Intentional Interference with Person or Property

1. Privilege

A defendant may escape liability for an intentional tort if the "social" importance of his action outweighs the damage suffered by the plaintiff.

2. Mistake

If one commits a battery, assault, or trespass or cpeates a nuisance under an erroneous belief that circumstances existed that would justify his conduct, he may defend a claim against him on the grounds that it was an unavoidable mistake.

3. Consent

A defendant may escape liability for an intentional interference with another's person or property if the plaintiff exhibited a willingness for the defendant to engage in the conduct. The plaintiff must be of sufficient mental capacity to give consent.

4. Self-defense

A person may take reasonable steps to protect himself from harm when there is no time to resort to the law. A defendant accused of an intentional tort may escape liability if he can prove that it was reasonable fear of present danger, and that he responded with only reasonable force.

5. Defense of Others

If a defendant inflicts harm on a plaintiff to prevent an immediate battery or imprisonment of a third person, his conduct may be justified.

6. Defense of Property

A person in rightful possession of property may use reasonable force to prevent its being unlawfully taken.

7. Recapture of Chattels

Some courts permit someone wrongfully deprived of his property to use reasonable force to get it back. However, it is limited by the restrictions that a verbal demand was fruitless and that the action was immediate and necessary.

8. Necessity

One may be spared liability for an intentional tort if one acted reasonably to prevent a public or private disaster.

9. Legal Process

The defense of legal process protects public officers from charges due to their intentional interference with a person or his property as dictated by law.

10. Discipline

A parent or teacher may use reasonable force to restrain a child without liability for an intentional tort.

11. Criminal Liability

A harmful or injurious act may be both a crime against the state and a tort against an individual. When such an act occurs the state will be able to bring criminal charges for the act committed, while the injured party may be compensated for his loss via the civil courts. The hearing(s) of the criminal and civil actions may be conducted successively or at the same time. A decision reached in the criminal action is not conclusive of the civil action and vice versa. However, if an issue tried in a criminal case is also decisive of some aspect of a tort action, a conviction of the defendant in the criminal action is ordinarily preclusive in favor of the injured party (plaintiff) in the civil action.

The state can never sue in tort in its political or governmental capacity unless some injuries have been committed to its properties or it needs to recover the properties themselves.

VII. Defenses to Breach of Warranty

1. The plaintiff was not a person who could reasonably be expected to use, consume, or be affected by the product.
2. The defendant does not meet the definition of a seller.
3. For certain products, breach of warranty claims may be barred for lack of privity (a contractual relationship between the parties).

Index